AQUATIC POLLUTION
Edward A. Laws

MODELING WASTEWATER RENOVATION: Land Treatment
I. K. Iskandar

INTRODUCTION TO INSECT PEST MANAGEMENT
Robert L. Metcalf and William H. Luckman, Editors

THE MEASUREMENT OF AIRBORNE PARTICLES
Richard D. Cadle

CHEMICAL CONTROL OF INSECT BEHAVIOR: THEORY AND APPLICATION
H. H. Shorey and John J. McKelvey, Jr., Editors

MERCURY CONTAMINATION: A HUMAN TRAGEDY
Patricia A. D'Itri and Frank M. D'Itri

POLLUTANTS AND HIGH RISK GROUPS
Edward J. Calabrese

METHODOLOGICAL APPROACHES TO DERIVING ENVIRONMENTAL AND OCCUPATIONAL HEALTH STANDARDS
Edward J. Calabrese

NUTRITION AND ENVIRONMENTAL HEALTH—Volume I: The Vitamins
Edward J. Calabrese

NUTRITION AND ENVIRONMENTAL HEALTH—Volume II: Minerals and Macronutrients
Edward J. Calabrese

SULFUR IN THE ENVIRONMENT, Parts I and II
Jerome O. Nriagu, Editor

COPPER IN THE ENVIRONMENT, Parts I and II
Jerome O. Nriagu, Editor

ZINC IN THE ENVIRONMENT, Parts I and II
Jerome O. Nriagu, Editor

ENVIRONMENTAL ENGINEERING AND SANITATION

ENVIRONMENTAL ENGINEERING AND SANITATION

THIRD EDITION

JOSEPH A. SALVATO, P.E.

Assistant Commissioner, Division of Sanitary Engineering (Retired)
New York State Department of Health, Albany, N.Y.
Sanitary and Public Health Engineer

A Wiley-Interscience Publication

JOHN WILEY & SONS
NEW YORK CHICHESTER BRISBANE TORONTO SINGAPORE

Library of Congress Cataloging in Publication Data:

Salvato, Joseph A.
Environmental engineering and sanitation.

(Environmental science and technology, ISSN 0194-0287)
"A Wiley-Interscience publication."
Includes bibliographies and index.
1. Environmental health. 2. Environmental engineering. 3. Sanitary engineering. 4. Sanitation. I. Title. II. Series.

RA565.S3 1982 628 81-11509
ISBN 0-471-04942-5 AACR2

Printed in the United States of America

10 9 8 7 6 5 4 3 2 1

To Joseph

SERIES PREFACE

Environmental Science and Technology

The Environmental Science and Technology Series of Monographs, Textbooks, and Advances is devoted to the study of the quality of the environment and to the technology of its conservation. Environmental science therefore relates to the chemical, physical, and biological changes in the environment through contamination or modification, to the physical nature and biological behavior of air, water, soil, food, and waste as they are affected by man's agricultural, industrial, and social activities, and to the application of science and technology to the control and improvement of environmental quality.

The deterioration of environmental quality, which began when man first collected into villages and utilized fire, has existed as a serious problem under the ever-increasing impacts of exponentially increasing population and of industrializing society. Environmental contamination of air, water, soil, and food has become a threat to the continued existence of many plant and animal communities of the ecosystem and may ultimately threaten the very survival of the human race.

It seems clear that if we are to preserve for future generations some semblance of the biological order of the world of the past and hope to improve on the deteriorating standards of urban, suburban, and rural public health, environmental science and technology must quickly come to play a dominant role in designing our social and industrial structure for tomorrow. Scientifically rigorous criteria of environmental quality must be developed. Based in part on these criteria, realistic standards must be established and our technological progress must be tailored to meet them. It is obvious that civilization will continue to require increasing amounts of fuel, transportation, industrial chemicals, fertilizers, pesticides, and countless other products; and that it will continue to produce waste products of all descriptions. What is urgently needed is a total systems approach to modern civilization through which the pooled talents of scientists and engineers, in cooperation with social scientists and the medical profession, can be focused on the development of order and equilibrium in the presently disparate segments of the human environment. Most of the skills and tools that are needed are already in existence. We surely have a right to hope a technology that has created such manifold environmental problems is also capable of solving them. It is our hope that this Series in Environmental Sciences and Technology will not only serve to make this challenge more explicit to the established professionals, but that it also will help to stimulate the student toward the career opportunities in this vital area.

Robert L. Metcalf
Werner Stumm

PREFACE

Workers in environmental health who have had experience with environmental sanitation and engineering problems have noted the need for a book that is comprehensive in its scope and more directly applicable to conditions actually encountered in practice.

Many standard texts adequately cover the specialized aspects of environmental engineering and sanitation; but little detailed information is available in one volume dealing with the urban, suburban, and rural community.

In this text emphasis is placed on the practical application of sanitary science and engineering theory and principles to environmental control. This is necessary if available knowledge is to be of benefit for man now and for future generations. In addition, and in deliberate contrast to complement other texts, empirical formulas, rule of thumb, and good practice are identified and applied when possible to illustrate the "best" possible solution under the particular circumstances. It is recognized that this may be hazardous in some instances when blindly applied by the practitioner. It is sincerely hoped, however, that individual ingenuity and investigation will not be stifled by such practicality, but will be challenged and stimulated to consider alternatives.

A special effort was made to include general design, construction, maintenance, and operation details as they relate to plants and structures. Examples and drawings are used freely to help in the understanding and use of the subject matter. The reader is referred to the references and bibliography in each chapter for more information on complex designs and problems that are beyond the scope of this text.

Since the field is a very broad one, the following subjects are specifically covered in this new revised edition.

1. Control of Communicable and Certain Noninfectious Diseases.
2. Environmental Engineering Planning and Impact Analysis.
3. Water Supply.
4. Wastewater Treatment and Disposal.
5. Solid Waste Management.
6. Air Pollution and Noise Control.
7. Radiation Uses and Protection.
8. Food Protection.
9. Recreation Areas and Temporary Residences.

10. Vector and Weed Control and Pesticide Use.
11. The Residential and Institutional Environment.
12. Administration.

The control of diseases is discussed to emphasize the importance of the proper application of sanitary and epidemiological principles in disease prevention. This is particularly important for areas of the world where communicable and related diseases have not yet been brought under control; what can happen in the more advanced countries when basic sanitary safeguards are relaxed is also discussed. Without discussing this important subject, this text would be as incomplete as one dealing with preventive medicine or planning that failed to discuss water supply, sewage disposal, solid waste management, air pollution control, and other phases of environmental engineering and sanitation. Also, in view of the progress made in most developed areas of the world in the control of the communicable diseases, attention is given to the environmental factors associated with the noninfectious diseases. Although complete knowledge concerning the specific cause and the prevention of many noninfectious diseases is lacking, it is considered desirable to identify and apply such knowledge that does exist so that it can be expanded and refined with time.

Teachers and students of environmental health and the environmental sciences, as well as of civil, chemical, mechanical, environmental, sanitary, municipal, and public health engineering, will find much of direct value in this text. Others too will find the contents especially useful. These would include the health officer, professional sanitarian, social scientist, ecologist, biologist, conservationist, public health nurse, health educator, environmental health technician, and sanitary inspectors of towns, villages, cities, counties, states, and federal governments both in the United States and abroad. City and county engineers and managers, consulting engineers, architects, planners, equipment manufacturers and installers, contractors, farm extension personnel, and institution, resort, and camp directors can all benefit from the contents. The many environmentalists who are interpreting and applying the principles of sanitary science and environmental engineering, sanitation, and hygiene to both the advanced and the developing areas of the world will find the material in this text particularly helpful in accomplishing their objectives.

In the preparation of the first edition (1958) the following people were especially helpful: Col. William A. Hardenbergh, Professor William C. Gibson, Dr. Wendell R. Ames, Arthur Handley, Professor John E. Kiker, Gerald A. Fleet, Professor Nicholas A. Milone, and Richard M. McLaughlin. The assistance they gave is again acknowledged.

Four new chapters were added to the second edition (1972), thus expanding *Environmental Sanitation* to *Environmental Engineering and Sanitation.* Most of the material in the original chapters was reorganized, consolidated, updated, and rewritten. The new chapters covered environmental engineering planning, solid waste management, air pollution control, and radiation uses and protection. For chapter review and suggestions on air pollution control I am grateful to Alexander Rihm, Director of Air Pollution Control, Harry Hovey, Associate

Director, and staff members of the New York State Department of Environmental Conservation; and for the chapter on radiation uses and protection to Sherwood Davies, Director, Bureau of Radiological Health, New York State Department of Health. In addition, I am appreciative of the general professional encouragement and stimulation received from Meredith H. Thompson, Assistant Commissioner, Division of Sanitary Engineering, New York State Department of Health.

This third edition updates much of the material and adds new material to reflect the state of the art. In particular, Chapter 1 gives greater emphasis to the environmental factors associated with the chronic and noninfectious diseases. Chapter 2 expands on environmental impact analysis. Chapter 3 has added materials on organic and inorganic compounds in water, and their significance, treatment, and removal. Chapter 4 covers in greater detail alternative onsite sewage disposal systems, water reclamation and reuse, land disposal, and precautions to be taken to protect the public health and the groundwater resources. Chapter 5 has added material on hazardous wastes, resource recovery, and energy conservation. Chapter 6 updates information including reference to acid rain and the pollutant standards index. Noise control has also been included. Chapter 7 emphasizes further the elimination of exposure to unnecessary ionizing radiation and expands on nonionizing radiation. Chapter 8 reflects the revised *Food Service Sanitation Manual* printed in 1978, information on chemical and microbiological guidelines or action levels, and the *Grade A Pasteurized Milk Ordinance-1978*. Chapter 9 has been generally updated with additions on campgrounds, highway rest areas, marinas, and marine sanitation. Chapter 10 places greater emphasis on integrated pest management and source reduction and elimination or reduction of pesticide use. Chapter 11 updates references to the APHA-PHS *Recommended Housing Maintenance and Occupancy Ordinance*, and adds environmental health information on mobile home parks, hospitals and nursing homes, schools, colleges, universities, and correctional institutions. Chapter 12 recognizes the changing scope of environmental concerns and suggests preventive environmental program activities, refines information on program supervision and enforcement, expands on emergency sanitation, and adds information on travel and carrier sanitation.

JOSEPH A. SALVATO

Troy, New York

ACKNOWLEDGMENTS

As in previous editions, I am indebted to many individuals for their technical reviews and suggestions for this third edition. However, the author must of course bear the full responsibility for the content and errors, if any.

Special thanks are due Daniel W. Stone, Consulting Engineer, Milton Chazen Associates, for his review of the chapter on water supply.

I am grateful to the following in the New York State Department of Environmental Conservation: David Mafrici, Director, Bureau of Waste Disposal and Charles N. Goddard, Director, Bureau of Hazardous Waste, both of whom are in the Division of Solid Waste Management, for their comments on certain sections of the chapter on solid waste management; Salvatore Pagano, Director, Bureau of Water Quality, and Warren Schlickenrieder, Chief, Performance Evaluation Section, for their review of the chapter on wastewater disposal and treatment; and Harry Hovey, Director, Division of Air, S. Marlow, Director, Bureau of Source Control, and Donald E. Gower, Director of Air Quality Surveillance for their suggestions and review of the chapter on air pollution; and for the noise control section to Fred G. Haag, former Director, Bureau of Noise Control.

Richard Hogan, Director, Division of Environmental Health, Rensselaer County Department, was also helpful in the review of the chapter on air pollution control as was Lovell Camnitz for his review of the section on noise control.

I am appreciative of the reviews made by Louis Lanzillo, Principal Sanitarian, Rensselaer County Health Department, and Lawrence B. Czech, Principal Radiological Health Specialist, New York State Department of Health of the chapter on radiation uses and protection.

For review of the chapter on vector and weed control and pesticide use I thank Alan M. Bowerman, Ph.D., Assistant Director, Bureau of Community Sanitation and Safety, and Stephen C. Frantz, Ph.D., Director, Rodent Control Evaluation Laboratory, New York State Department of Health.

As in the first and second editions, I am especially grateful to my wife Hazel for the typing, manuscript and galley reading, and extreme patience. Without her assistance this third edition would not have been possible.

J.A.S.

CONTENTS

ENVIRONMENTAL ENGINEERING AND SANITATION

INTRODUCTION

In many parts of the world, simple survival or prevention of disease and poisoning are still serious concerns. In other areas, maintenance of an environment that is suited to man's efficient performance and to the preservation of comfort and enjoyment of living are the goals for the future. These levels of life and progress can be the basis for action programs in environmental health.[1]

As urbanization increases, man's impact on the environment, and the impact of the environment on man, must be controlled to protect the human and natural resources essential to life and at the same time enhance man's well-being. However the simultaneous movement of people out of cities to suburban areas and especially to rural areas, often to uncontrolled environments, makes it important that the environmental sanitation lessons of the past are not lost and that the environmental impact of human activity on man and on our natural resources is controlled.

The environment encompasses "the sum of all external influences and conditions affecting life and development of an organism (including man)." Included is the air, water, and land and the interrelationship which exists among and between air, water, and land and all living things. The achievement of an environment to enhance man's well-being requires the application of environmental science and engineering principles. This means that our goal should be "the control of all those factors in man's physical environment which exercise or may exercise a deleterious effect on his physical development, health, and survival,"[2] and "the application of engineering principles to the control, modification or adaptation of the physical, chemical, and biological factors of the environment in the interest of man's health, comfort, and social well-being."[3] This in turn requires application of the principles of physical, biological, and social sciences to the improvement, control, and management of man's environment,[4] with consideration of the impact of the control measures applied.

[1]Frank M. Stead, "Levels in Environmental Health," *Am. J. Public Health*, March 1960, p. 312; *WHO Tech. Rep. Ser.*, **77**, 9 (1954).

[2]WHO Expert Committee on Environmental Sanitation, *WHO Tech. Rep. Ser.* **10**, 5 (1950).

[3]"The Education of Engineers in Environmental Health," *WHO Tech. Rep. Ser.* **376**, 6 (1967).

[4]*Health Resources Statistics—Health Manpower and Facilities, 1969*, U.S. Dept. of Health, Education, and Welfare (DHEW), Rockville, Md. 20852, May 1970.

The chapters that follow attempt to point out major areas of concern to be attacked by a host of disciplines working in close harmony. Effective control and management of these conditions will help achieve the highest possible quality of environment and living, keeping in mind the physical, social,* and economic factors involved and their interdependence.

*This includes political, cultural, educational, biological, medical, and public health. Social health is the "ability to live in harmony with other people of other kinds, with other traditions, with other religions, and with other social systems throughout the world." [World Health Organization (WHO)].

1

CONTROL OF COMMUNICABLE AND CERTAIN NONINFECTIOUS DISEASES

GENERAL

Definitions

Certain terms with which one should become familiar are frequently used in the discussion of communicable and noninfectious or noncommunicable diseases. Some common definitions are given here.

Disease In its broadest sense the communicable and noninfectious diseases. Disease may be considered the antithesis of health, defined as "a state of physical, mental and social well-being and ability to function, and not merely the absence of illness or infirmity."[1]

Communicable Disease An illness due to a specific infectious agent or its toxic product which arises through transmission of that agent or its products from a reservoir to a susceptible host, either directly, as from an infected person or animal, or indirectly, through the agency of an intermediate plant or animal host, vector, or the inanimate environment.[2] Illness may be caused by bacteria, bacterial toxins, viruses, protozoa, spirochetes, parasitic worms (helminths), poisonous plants and animals, chemical poisons, fungi, rickettsias, certain yeasts, and molds. In this text the communicable diseases are grouped into respiratory diseases, water- and foodborne diseases, insect- and rodentborne diseases and zoonoses, and miscellaneous diseases.

Noninfectious or **Noncommunicable Disease** The chronic and insidious diseases which develop usually over an extended period and whose cause may not be

[1]The WHO definition as modified by M. Terris, "Approaches to an Epidemiology of Health," *Am. J. Pub. Health*, October 1975, pp. 1037–1045.
[2]*Control of Communicable Diseases in Man*, 12th ed., Am. Pub. Health Assoc., Washington, D.C., 1975.

entirely clear. Cancer, alcoholism, blindness, mental illnesses, tooth decay, ulcers, and lead poisoning are regarded as diseases (dis-ease).[3] Noninfectious diseases would include cardiovascular diseases, nutritional deficiency diseases, carbon monoxide poisoning, and illnesses associated with toxic organic and inorganic chemicals in air, water, and food. For the purposes of this text, discussion of noninfectious diseases emphasizes the environmental media or factors serving as the vehicle for the transmission primarily of organic and inorganic chemicals associated with or contributing to disease. The usual media are air, food, water, and land (soil, flora, fauna); contact may also be involved. It is not intended to imply that the chemicals identified are necessarily the sole factor contributing to disease.

In contrast to the communicable diseases, *chronic diseases* may be caused by a variety or combination of factors which are difficult to identify, treat and control. The resulting illness may cause protracted or intermittent pain and disability with lengthy hospitalization.

Carrier An infected person (or animal) that harbors a specific infectious agent in the absence of discernible clinical disease and serves as a potential source of infection for man. The carrier state may occur in an individual with an infection that is inapparent throughout its course (commonly known as a *healthy* or *asymptomatic carrier*), or during the incubation period, convalescence, and postconvalescence of an individual with a clinically recognizable disease (commonly known as an *incubatory carrier* or *convalescent carrier*). Under either circumstance the carrier state may be of short or long duration (*temporary* or *transient carrier* or *chronic carrier*).[2]

Contact A person or animal that has been in such association with an infected person or animal or a contaminated environment as to have had opportunity to acquire the infection.[2]

Contamination The presence of an infectious agent on a body surface; also on or in clothes, bedding, toys, surgical instruments or dressings, or other inanimate articles or substances including water, milk, and food. Pollution is distinct from contamination and implies the presence of offensive, but not necessarily infectious matter, in the environment. Contamination of a body surface does not imply a carrier state.[2]

Disinfection Killing of infectious agents outside the body by chemical or physical means, directly applied.[2]

Concurrent disinfection is the application of disinfective measures as soon as possible after the discharge of infectious material from the body of an infected person, or after the soiling of articles with such infectious discharges, all personal contact with such discharges or articles being minimized prior to such disinfection.

Terminal disinfection is the application of disinfective measures after the patient has been removed by death or to a hospital, or has ceased to be a source

[3]From *Prospectus—Manual on Prevention of Disease*, Task Force on Prevention, Am. Pub. Health Assoc., Washington, D.C., December 1976, p. 11.

of infection, or after hospital isolation or other practices have been discontinued. Terminal disinfection is rarely practiced; terminal cleaning generally suffices along with airing and sunning of rooms, furniture, and bedding. It is necessary only for diseases spread by indirect contact; steam sterilization or incineration of bedding and other items is desirable after smallpox (now rare).

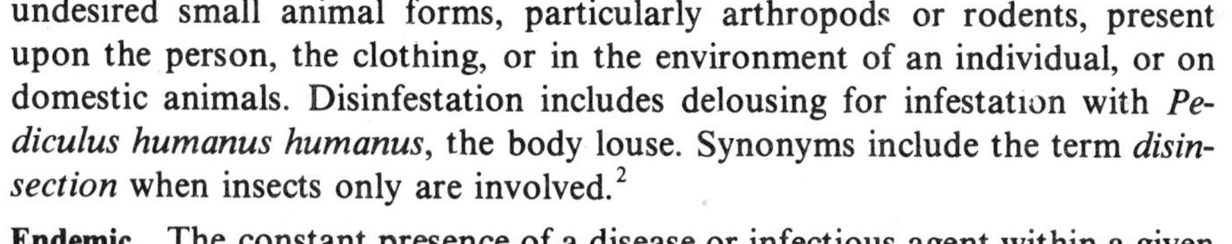

Disinfestation Any physical or chemical process serving to destroy or remove undesired small animal forms, particularly arthropods or rodents, present upon the person, the clothing, or in the environment of an individual, or on domestic animals. Disinfestation includes delousing for infestation with *Pediculus humanus humanus*, the body louse. Synonyms include the term *disinsection* when insects only are involved.[2]

Endemic The constant presence of a disease or infectious agent within a given geographic area; may also refer to the usual prevalence of a given disease within such area. *Hyperendemic* expresses a persistent intense transmission, e.g., malaria.[2]

Epidemic The occurrence in a community or region of cases of an illness (or an outbreak) clearly in excess of normal expectancy and derived from a common or a propagated source. The number of cases indicating the presence of an epidemic will vary according to the infectious agent, size and type of population exposed, previous experience or lack of exposure to the disease, and time and place of occurrence; epidemicity is thus relative to the usual frequency of the disease in the same area, among the specified population, at the same season of the year. A single case of a communicable disease long absent from a population (as smallpox in a traveler through New York City in 1962) or the first invasion by a disease not previously recognized in that area (as American trypanosomiasis in Arizona) is to be considered sufficient evidence of a potential epidemic to require immediate reporting and full field investigation.[2]

Transmission of Infectious Agents Any mechanism by which a susceptible human host is exposed to an infectious agent. These mechanisms are[2]

1. *Direct Transmission*: Direct and essentially immediate transfer of infectious agents (other than from an arthropod in which the organism has undergone essential multiplication or development) to a receptive portal of entry by which infection of man may take place. This may be by direct contact by touching, as in kissing or sexual intercourse, or by the direct projection (droplet spread) of droplet spray onto the conjunctiva or onto the mucous membranes of the nose or mouth during sneezing, coughing, spitting, singing, or talking (usually limited to a distance of about 1 m or less). It may also be by direct exposure of susceptible tissue to an agent in soil, compost, or decaying vegetable matter in which it normally leads a saprophytic existence, for example, the systemic mycoses; or by the bite of a rabid animal.
2. *Indirect Transmission*
   a. VEHICLEBORNE. Contaminated materials or objects such as toys, handkerchiefs, soiled clothing, bedding, surgical instruments, or dressings

(indirect contact); water, food, milk, biological products including serum and plasma, or any substance serving as an intermediate means by which an infectious agent is transported and introduced into a susceptible host through a suitable portal of entry. The agent may or may not have multiplied or developed in or on the vehicle before being introduced into man.

b. VECTORBORNE.

i. *Mechanical*: Includes simple mechanical carriage by a crawling or flying insect through soiling of its feet or proboscis, or by passage of organisms through its gastrointestinal tract. This does not require multiplication or development of the organism.

ii. *Biological*: Propagation (multiplication) cyclic development, or a combination of these (cyclopropagation) is required before the arthropod can transmit the infective form of the agent to man. An incubation period (extrinsic) is required following infection before the arthropod becomes *infective*. Transmission may be by saliva during biting, or by regurgitation or deposition on the skin of agents capable of penetrating subsequently through the bite wound or through an area of trauma following scratching or rubbing. This is transmission by an infected nonvertebrate host and must be differentiated for epidemiological purposes from simple mechanical carriage by a vector in the role of a vehicle. An arthropod in either role is termed a *vector*.

c. AIRBORNE. The dissemination of microbial aerosols with carriage to a suitable portal of entry, usually the respiratory tract. Microbial aerosols are suspensions in air of particles consisting partially or wholly of microorganisms. Particles in the 1 to 5 micron range are quite easily drawn into the lungs and retained there. They may remain suspended in the air for long periods of time, some retaining and others losing infectivity or virulence. Not considered as airborne are droplets and other large particles, which promptly settle out (see 1. *Direct transmission*, above); the following are airborne, their mode of transmission indirect:

i. *Droplet nuclei*: Usually the small residues which result from the evaporation of droplets emitted by an infected host. Droplet nuclei also may be created purposely by a variety of atomizing devices, or accidentally, in microbiology laboratories or in abattoirs, rendering plants, autopsy rooms, and so on. They usually remain suspended in the air for long periods of time.

ii. *Dust*: The small particles of widely varying size which may arise from contaminated floors, clothes, bedding, other articles; or from soil (usually fungus spores separated from dry soil by wind or mechanical stirring).

Host A man or other living animal, including birds and arthropods, affording under natural conditions subsistence or lodgment to an infectious agent. Some protozoa and helminths pass successive stages in alternate hosts of different species. Hosts in which the parasite attains maturity or passes its sexual stage are *primary* or *definite hosts*; those in which the parasite is in a larval or asexual state are *secondary* or *intermediate hosts*. A transport host is a carrier in which the organism remains alive but does not undergo development.[2]

Incubation Period The time interval between exposure to an infectious agent and appearance of the first sign or symptom of the disease in question.[2]

Reservoir of Infectious Agents Any human beings, animals, arthropods, plants, soil, or inanimate matter in which an infectious agent normally lives and multiplies and on which it depends primarily for survival, reproducing itself in such manner that it can be transmitted to a susceptible host.[2]

Source of Infection The person, animal, object, or substance from which an infectious agent passes immediately to a host. The source of infection should be clearly distinguished from the source of contamination, such as the overflow of a septic tank contaminating a water supply, or an infected cook contaminating a salad.[2]

Susceptible A person or animal presumably not possessing sufficient resistance against a particular pathogenic agent and for that reason liable to contract a disease if or when exposed to the disease agent.[2]

Carcinogen Any factor or combination of factors which increases the risk of cancer in humans. Cancer is many diseases in which derangement of body cells is involved. The effects of carcinogens on human tissue, if exposure is sufficient, are irreversible. Carcinogens that produce cancer in experimental animals are found in low concentrations in food, in some air and water pollutants, in certain pesticides, and in food additives. The carcinogenic potential of many carcinogenic substances acting singly or in combination with other carcinogens and chemicals is not known.[4]

Mutagen A chemical which is capable of producing a heritable change in genetic material. Many chemicals which pollute the environment in large doses are mutagenic, but their hazard is not known for the levels found in the environment.[4]

Teratogen An agent which acts during pregnancy to produce a physical or functional defect in the developing offspring. Substances that have caused defects are methylmercury and thalidomide. Some environmental pollutants may be both carcinogenic and teratogenic.[4]

Toxicity The intrinsic quality of a chemical to produce an adverse effect. The term includes the capacity to induce teratogenic, mutagenic, and carcinogenic effects.

Pollution The undesirable change in the physical, chemical, or biological characteristics of air, land, and water that may or will harmfully affect human life or that of other desirable species, our industrial processes, living condi-

Legal Def.

[4]*Pollution and Your Health*, U.S. Environmental Protection Agency (USEPA), Office of Public Affairs (A-107), Washington, D.C., May 1976, p. 17.

tions, and cultural assets; or that may or will waste or deteriorate our raw material resources.[5] Examples of pollution are inadequately treated municipal and industrial waste discharges, leachate and runoff from a landfill or other disposal site, pesticides and agricultural runoff, toxic discharges from incinerators and industrial processes, discharge of a liquid which causes an increase in water temperature which does not support sport fish, algae and decaying aquatic weeds in bathing areas, and drainage from abandoned mines. See **Contamination.**

Epidemiology The study of the occurrence, frequency, and distribution of disease in human populations, leading to the discovery of the cause and an informed basis for preventive action—social, biological, chemical, or physical.

Prevention, Primary (First Degree) Prevention of an etiologic agent from causing disease or injury in man; intervention; regulation of exposure to environmental hazards, which cause disease or injury, to decrease morbidity and mortality.

Prevention, Secondary (Second Degree) Early detection and treatment to cure or control disease. Surveillance and monitoring the environment. Also measures to protect the public, that is, treatment of public water supplies; fluoridation for dental carries control.

Prevention, Tertiary (Third Degree) Amelioration of a disease to reduce disability or dependence resulting from it. Voluntary action by the individual.

Primary Health Care "Essential care made universally accessible to individuals and families in the community by means acceptable to them through their full participation, and at a cost that the community and country can afford."[6]

Life Expectancy and Mortality

The life expectancy at birth has varied with time, geography, and the extent to which available knowledge concerning disease prevention and control could be applied. Table 1-1*a* shows the trend in life expectancy with time. The gains in life expectancy between 1900 and 1974 shown in Table 1-1*b* have occurred mostly in the early years, 22.7 years at birth and 13.3 years at age 5, reducing to 6.1 years at age 45 and 3.1 years at age 70. The life expectancy gains are due primarily to better sanitation and nutrition and to the conquest of the major epidemic and infectious diseases, including immunization and chemotherapy.

The vital statistics in Table 1-2*a* are of interest in that they show the changes in major causes of death in 1900 related to 1970 and the net reduction in the total death rate. Table 1-2*b* shows the leading causes of death as of 1975. The leveling off in life expectancy that is apparent in the United States is due in

[5] *Waste and Management Control*, National Academy of Sciences, National Research Council, publication 1400, Washington, D.C., 1966, p. 3.

[6] "Primary Health Care," A joint report by the Director-General of the WHO and the Executive Director of the United Nations Children's Fund, New York, 1978.

Table 1-1a Life Expectancy at Birth

Period or Year	Life Expectancy
Neanderthal (50,000 B.C.–35,000 B.C.)	29.4[a]
Upper Paleolithic (600,000 B.C.–15,000 B.C.)	32.4[a]
Mesolithic	31.5[a]
Neolithic anatolia (12,000 B.C.–10,000 B.C.)	38.2[a]
Bronze age—Austria	38[a]
Greek Classical (700 B.C.–460 B.C.)	35[a]
Roman Classical (700 B.C.–A.D. 200)	32[a]
Roman empire (27 B.C.–A.D. 395)	24
1000	32
England (1276)	48[a]
England (1376–1400)	38[a]
1690	33.5
1800	35
1850	40
1870	40
1880	45
1900	48
1910	50
1920	54
1930	59
1940	63
1950	66
1960	68
1970	71
1980	73

Source: "Environmental Health," Joseph A. Salvato, Jr., *Encyclopedia of Environment Science and Engineering*, Edward N. Ziegler and James R. Pfafflin, Eds. Gordon Braech Science Publishers, Inc., 42 William IV St., London WC2, 1976, p. 286.

Note: The 1975 life expectancy reported by the United Nations for Sweden was 72.1 for males and 77.6 for females and for the United States 68.5 for males and 76.4 for females.

Life expectancy figures from 1690 to 1980 are for the United States. The average life expectancy for the world in 1975 was 55 years and for Africa 45 years. The world population was reported by the UN as four billion in 1975 and projected to six billion in 2000.

[a]E. S. Deevy, Jr., "The Human Population," *Sci. Amer.*, **203**(3). 200 (September 1960).

part to our inability thus far to identify the causes and to control the chronic, noninfectious diseases such as heart disease and cancer.

The prevention of deaths from a particular disease does not increase the life expectancy in direct proportion to its decreased mortality.[7] Keyfitz gives an

[7]Nathan Keyfitz, "Improving Life Expectancy: An Uphill Road Ahead," *Am. J. Publ. Health*, October 1978, pp. 954–956.

Table 1-1*b* Increase in Life Expectancy between 1900 and 1974 at Selected Ages, United States Total Population

Age	1900[a]	1974[b]	Gain During 1900–1974
0	49.2	71.9	22.7
1	55.2	72.1	16.9
5	55.0	68.3	13.3
15	46.8	58.6	11.8
25	39.1	49.2	10.1
35	31.9	39.9	8.0
45	24.8	30.9	6.1
55	17.9	22.6	4.7
60	14.8	18.9	4.1
65	11.9	15.5	3.6
70	9.3	12.4	3.1
75	7.1	9.8	2.7

[a]Department of Commerce, U.S. Bureau of the Census, *United States Life Tables 1890, 1901, 1910, and 1901–1910*, James W. Glover (Washington, GPO, 1921), pp. 52–53.
[b]U.S. Department of Health, Education, and Welfare, *Vital Statistics of the United States, 1974*, Vol. II–Section 5, "Life Tables" National Center for Health Statistics, Rockville, Md. 1976, p. 8.

Table 1-2*a* Leading Causes of Death, 1900, 1960, and 1970 in the United States

Rank	Cause of Death	Deaths per 100,000 Population	Percent of All Deaths
	1900		
	(All causes)	(1,719)	
1	Pneumonia and influenza	202.2	11.8
2	Tuberculosis (all forms)	194.4	11.3
3	Gastritis, etc.	142.7	8.3
4	Diseases of the heart	137.4	8.0
5	Vascular lesions affecting the central nervous system	106.9	6.2
6	Chronic nephritis	81.0	4.7
7	All accidents[a]	72.3	4.2
8	Malignant neoplasma (cancer)	64.0	3.7
9	Certain diseases of early infancy	62.5	3.6
10	Diphtheria	40.3	2.3
11	All other and ill-defined causes	615.3	36

Table 1-2*a* (*Continued*)

Rank	Cause of Death	Death per 100,000 Population	Percent of All Deaths
	1960		
	(All causes)	(955)	
1	Diseases of the heart	366.4	38.7
2	Malignant neoplasms (cancer)	147.4	15.6
3	Vascular lesions affecting the central nervous system	107.3	11.3
4	All accidents[b]	51.9	5.5
5	Certain diseases of early infancy	37.0	3.9
6	Pneumonia and influenza	36.0	3.5
7	General arteriosclerosis	20.3	2.1
8	Diabetes mellitus	17.1	1.8
9	Congenital malformations	12.0	1.3
10	Cirrhosis of the liver	11.2	1.2
11	All other and ill-defined causes	148.4	15
	1970		
	(All causes)	(945.3)	
1	Diseases of the heart	362.0	38.3
2	Malignant neoplasms (cancer)	162.8	17.2
3	Cerebrovascular diseases (stroke)	101.9	10.8
4	Accidents	56.4	6.0
5	Influenza and pneumonia	30.9	3.3
6	Certain causes of mortality in early infancy[c]	21.3	2.2
7	Diabetes mellitus	18.9	2.0
8	Arteriosclerosis	15.6	1.6
9	Cirrhosis of the liver	15.5	1.6
10	Bronchitis, emphysema, and asthma	15.2	1.6
11	All other and ill-defined causes	144.8	15

[a]Violence would add 1.4 percent; horse, vehicle, and railroad accidents provide 0.8 percent.
[b]Violence would add 1.5 percent; motor vehicle accidents provide 2.3 percent; railroad accidents provide less than 0.1 percent.
[c]Birth injuries, asphyxia, infections of newborn, ill-defined diseases, immaturity, etc.
Source: President's Science Advisory Committee Panel on Chemicals, Chemicals and Health (Washington, D.C., GPO, 1973), p. 152; DHEW, PHS, "Facts of Life and Death," DHEW Pub. No. (HRA) 74-1222 (Washington, D.C., GPO, 1974), p. 31.

example showing that if a general cure for cancer were found there would be nearly 350,000 fewer deaths per year (cancer deaths in 1970). It would seem then that the mortality would be lowered by one-sixth, since cancer deaths were one-sixth of all deaths, and the life expectancy increased by one-sixth. But this would hold true only for a homogeneous population. "Only in such a population would the reduction of the deaths and of the death rate by one-sixth

Table 1-2*b* Causes of Death in the United States, 1975

Cause	Total Number of Deaths	Deaths per 100,000 Population
Diseases of the heart	716,215	336.2
Malignant neoplasms	365,693	171.7
Cerebrovascular diseases	194,038	91.1
Accidents	103,030	48.4
Motor vehicle accidents	(45,853)	(21.5)
All other accidents	(57,177)	(26.8)
Influenza and pneumonia	55,664	26.1
Certain diseases of early infancy	26,616	12.5
Arteriosclerosis	28,887	13.6
Diabetes mellitus	35,230	16.5
Other diseases of the circulatory system	25,607	12.0
Other bronchopulmonic diseases	25,468	12.0
Cirrhosis of liver	31,623	14.8
Suicide	27,063	12.7
Congenital malformations	13,245	6.2
Homicide	21,310	10.0
Other hypertensive diseases	6,300	3.0
Other and ill-defined	216,608	101.7
Total—All causes	1,892,879	888.5

Source: From Vital Statistics of New York State, 1975 published by New York State Department of Health, as well as in-house data available in the Office of Biostatistics. Courtesy New York State Department of Health, Office of Biostatistics.

extend the expectation of life by one-sixth. Only then could each of us expect to live 12 more years (assuming a life expectancy of 72 years) as a result of the discovery of a cure for cancer." But because the population is not homogeneous and the risk factors for cancer, and other diseases, vary with age [such as for a 20-year-old man (1:10,000) compared to a 70-year-old man (1:100)], the "universal elimination of cancer would increase life expectancy by only about two years—not the 12 years that would apply if the population were homogeneous." Keyfitz goes on to say, "But even the gain so calculated (two years if cancer is eliminated) is almost certainly an overestimate of the benefit. For within any given age group, the people subject to any one ailment tend to have higher than average risks from other ailments." To extend average life expectancy beyond 80 years, which seems to be the present limit, Keyfitz feels it is necessary to focus on prevention of "deterioration and senescence of the cells of the human body." It would seem then that a general improvement in the "quality of life"*

*In addition to elimination, insofar as possible, of the communicable and noninfectious diseases, a desirable quality of life implies a decent home, medical care, adequate and safe water and food supply, proper waste disposal, clean air and water, absence of poverty, a suitable level of education and cultural opportunity, balanced diet, nondestructive life style, and safe and adequate recreation and transportation facilities. See Chapter 2, Statement of Goals and Objectives.

to slow down premature aging, together with prevention and control of the noninfectious as well as communicable diseases, will accomplish a greater increase in life expectancy than concentrating *solely* on elimination of the major causes of death. This appears to be a sound approach since it is known that "mortality levels are determined by the complicated interplay of a variety of sociocultural, personal, biological, and medical factors."[8] On the other hand, if the causes of a disease are also contributing factors to other diseases, then elimination of the cause of one disease may, at the same time, eliminate or reduce morbidity and mortality from other diseases thereby resulting in an additional overall increase in life expectancy.

There seems to be a concensus that further increase in life expectancy is dependent on the extent to which personal behavior will be changed—obesity, poor nutrition, lack of exercise, smoking, stress—and the environmental pollutants are controlled—industrial and auto emissions, chemical discharges into our waters, use of pesticides and fertilizers, interaction of harmless substances forming hazardous compounds[9]—together with reduction of accidental and violent deaths and improvement in living and work conditions.

It must also be recognized that although life expectancy is a measure of health progress, it does not measure the morbidity levels and the quality of life.

Disease Control

The communicable diseases and malnutrition are considered the core health problems of developing countries. However, illnesses associated with contaminated drinking water and food are not uncommon in the so-called developed countries. The major diseases of the developed countries are diseases of the heart, malignant neoplasms (cancer), cerebrovascular diseases (stroke), accidents, influenza and pneumonia, diabetes mellitus, and others, primarily the noninfectious diseases. In 1975 diseases of the heart, cancers, and cerebrovascular diseases caused two-thirds of all deaths in the United States.

Sound factual information upon which to base programs for the prevention and control of morbidity and mortality associated with chronic diseases, aging, mental stress, destructive life styles, environmental hazards, and injury in many cases is not adequate or available. Multiple causes of disease and delayed effects compound the uncertainties. Nevertheless it is prudent to apply, and update, known preventive environmental, physiologic-medical, and health education-motivational measures with the full knowledge of their limitations, and without raising unreasonable expectations of the public. The environmental preventive measures are elaborated on here, but the importance of physiologic-medical and health education-motivational measures is not to be minimized.

[8]Shan Pou Tsai, et al., "The Effect of a Reduction in Leading Causes of Death: Potential Gains in Life Expectancy," *Am. J. Pub. Health*, October 1978, pp. 966–971.

[9]Charles Warren, "Current Knowledge on the Environment," *National Conference on the Environment and Health Care Costs*, August 15, 1978, House of Representatives Caucus Room, Washington, D.C.

The goal of environmental health programs is not only the prevention of disease, disability, and premature death, but also the maintenance of an environment that is suited to man's efficient performance, and to the preservation of comfort and enjoyment of living today and in the future. The goal is not only the prevention of communicable diseases, but also prevention of the noncommunicable diseases, the chronic and acute illnesses, and the hazards to life and health. This requires better identification and control of the contributing environmental factors in the air, water, food, at the home, and at the place of work and recreation, as well as changes in personal behavior and reduced individual assumption of risk. Lacking complete information, the best possible standards based on the available knowledge must be applied for the public good. Standards adoption and regulatory effort should be based on the risk that society or the individual is willing to assume and pay for, taking into consideration other risk factors and needs.

Communicable and certain noninfectious diseases may be controlled or prevented by taking steps to regulate the "source," the "mode of transmission," or the "susceptibility" of persons as appropriate and feasible based on the knowledge available. This is shown in Figure 1-1 and for communicable diseases is sometimes pictured as a three-link chain. Although the diseases can be brought under control by eliminating one of the links, it is far better to direct one's attack simultaneously toward all three links and erect "barriers" or "dams" where possible. Phelps called this the "principle of multiple barriers." It recognizes as axiomatic the fact that "all human efforts, no matter how well conceived or conscientiously applied, are imperfect and fallible."[10] Sometimes it is only practical to partially break one link in the chain. Therefore the number and type of barriers or interventions should be determined by the practical-

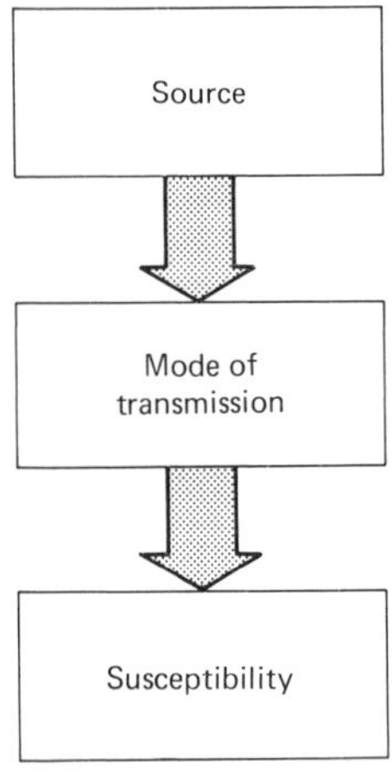

Figure 1-1 Spread of communicable and noninfectious diseases. Animals include humans and arthropods. Arthropods include insects, arachnids, crustaceans, and myriapods (invertebrate animals with jointed legs and segmented body). Physical agents may be heat, cold, causes of accidents. Biologic agents include arthropods, helminths, protozoa, fungi, bacteria, rickettsiae, and viruses. Environmental pollutants may be transmitted by air, water, food, and contact. Personal behavior may involve cigarette smoking, poor nutrition, stress, lack of exercise, cultural habits, and obesity. *Source* (agent factors—physical, chemical, biologic): infected or infested animals; poisonous plants and animals; parasites; toxic solid, liquid and gaseous wastes and deposits; genetic and inherited substances. *Mode of transmission* or contributing factor (environmental factors): environmental pollutants, contact, animals; personal behavior; work, recreation, travel, home. *Susceptibility* (host factors): all animals or susceptibles, resulting in acute, chronic, or delayed effects.

[10]Earle B. Phelps, *Public Health Engineering*, John Wiley & Sons, New York, 1948, p. 347.

ity and cost of providing the protections, the benefits to be derived, and the probable cost if the barriers are not provided. Cost is used in the sense not only of dollars but also in terms of human misery, loss of productiveness, ability to enjoy life, and loss of life. Here is a real opportunity for applying professional judgment to the problems at hand to obtain the maximum return for the effort expended.

Communicable and certain noninfectious diseases can usually be regulated or brought under control. A health department having a complete and competent staff to prevent or control diseases that affect individuals and animals is usually established for this purpose. The preventive and control measures conducted by a health department might include supervision of water supply, wastewater, and solid wastes; housing and the residential environment; milk and food; stream pollution; recreation areas including camps, swimming pools, and beaches; occupational health and accident prevention; insects and rodents; rural and resort sanitation; air pollution; noise; radiological hazards; hospital, nursing homes, jails, schools, and other institutions; medical clinics, maternal and child health services, school health, dental clinics, nutrition, and medical rehabilitation; medical care; disease control, including immunizations, cancer, heart disease, tuberculosis, and venereal diseases; vital statistics; health education; epidemiology; and nursing services. Personnel, fiscal, and public relations support functions would be carried out by the office of business management. In some states, certain environmental and medical activities are combined with the activities of other agencies and vice versa, making achievement of a comprehensive and coordinated preventive services program difficult. See also Chapter 12.

Control of Source (Agent Factors)

General sources of disease agents are noted in Figure 1-1. Elimination or control of the source and environmental exposure to disease agents or vectors is a primary step, carried out to the extent feasible. Individuals frequently are not aware that they are being exposed to a potential source of disease, particularly when it is a minute, insidious and cumulative substance, such as certain chemicals in the air, water, and food. This calls for regulatory action as noted in the discussion of noninfectious diseases under Prevention and Control.

In many instances control at the source is not only possible but practical. Some measures that might be taken are:

1. Change the raw material or industrial process to eliminate or adequately minimize the offending substance. For example, use low sulfur fuel or substitute gas; terminate production of a chemical such as PCB; or remove waste products, such as by means of air pollution control devices or wastewater treatment plants, to reduce toxic discharges into the en-

vironment to acceptable levels. The USEPA "zero discharge" *goal* is a step in this direction.

2. Select the cleanest available source of drinking water, as free as possible from microbiological and toxic organic and inorganic chemicals.
3. Make available water with optimum mineral content, such as through fluoridation and water hardness control.
4. Prohibit taking of fish and shellfish from contaminated (pathogens, methylmercury, PCB) waters.
5. Regulate food production and processing to assure freedom from toxic substances and pathogens and to assure food of good nutritional content.
6. Provide decent housing in a suitable living environment.
7. Provide a safe and healthful work and recreation environment.
8. Promote recycling, reuse, and "zero" discharge of hazardous wastes.
9. Eliminate disease vectors (arthropods and other animals, including rodents) at the source. Practice integrated pest management.
10. Isolate infected persons and animals from others during their period of communicability and treat to eliminate disease reservoir.
11. Educate polluters, legislators and the public to the need for regulation and funding where indicated.
12. Adopt and enforce sound standards.
13. Support comprehensive environmental health, engineering and sanitation surveillance and regulation programs at the state and local levels.

See also Disease Control and Control of Susceptibles, this chapter, and Future Preventive Environmental Program Activities in Chapter 12.

Control of Mode of Transmission or Contributing Factor (Environmental Factors)

The means whereby specific agents or factors may become the vehicle or vector for the transmission of disease are numerous. Prevention of disease requires the continual application of control measures such as those listed below and elimination of the human element to the extent feasible.

1. Prevent the travel of disease vectors and control disease carriers.
2. Assure that all drinking water is at all times safe to drink and adequate for drinking, culinary, laundry, and bathing purposes.
3. Provide adequate spatial separation between sources of disease (and pollution) and receptors.
4. Assure that food processing, distribution, preparation, and service does not cause disease.
5. Control air and water pollution, hazardous wastes, accidents, carcinogens, and toxics.
6. Prevent access to disease sources—polluted bathing waters and disease vector infested areas.
7. Adopt and enforce noise standards.
8. Educate polluters, legislators, and the public to the need for regulation and funding where indicated.

9. Support comprehensive environmental health, engineering, and sanitation surveillance and regulation programs at the state and local levels.
10. Adjust personal behavior to counteract cigarette smoking, poor nutrition, stress, overeating, and lack of exercise.

Control of Susceptibles (Host Factors)

The more susceptible individuals are the very young, the elderly, those with cardiovascular and respiratory disease, those occupationally exposed to air pollutants, those who smoke heavily, the obese, and those who overexercise. There are many diseases to which all persons are considered to be generally susceptible. Among these are measles, streptococcal diseases caused by group A streptococci, the common cold, ascariasis, chickenpox, amebic dysentery, bacillary dysentery, cholera, malaria, trichinosis, and typhoid fever. There are other diseases such as influenza, meningococcus meningitis, pneumonia, human brucellosis (undulant fever), and certain water- and foodborne illnesses to which some people apparently have an immunity or resistance. To these should be added the noninfectious diseases such as diseases of the heart, malignant neoplasms, and cerebrovascular diseases.

In order to reduce the number of persons who may be susceptible to a disease at any one time, certain fundamental disease prevention principles are followed to improve the general health of the public. This may be accomplished by instructions in personal hygiene and immunization; avoidance of smoking; maintenance of proper weight; minimal liquor consumption; and conserving or improving the general resistance of individuals to disease by a balanced diet and nutritious food, fresh air, moderate exercise, sufficient sleep, rest, and the avoidance of stress, fatigue and exposure. In addition, all individuals should be educated and motivated to protect themselves to the extent feasible from biological, physical, chemical, and radiation hazards and environmental pollutants.

Immunization can be carried out by the injection of vaccines, toxoids, or other immunizing substances for the prevention or lessening of the severity of specific diseases. Smallpox, typhoid and paratyphoid fevers, poliomyelitis, and tetanus are some of the diseases against which all in the armed forces are routinely immunized. Children are generally immunized against diphtheria, tetanus, pertussis (whooping cough), poliomyelitis, rubeola (measles), mumps, rubella (German measles), and smallpox. It is now possible to discontinue mass smallpox vaccination as a routine measure in view of the global eradication of smallpox.[11]

Typhoid immunization is about 70 to 90 percent effective, depending on de-

[11]WHO Director-General Dr. Halfdan Mahler, World Health Assembly, May 5–23, 1980, Geneva: *PAHO Reports*, July–August 1980.

gree of exposure.[12,13] Routine typhoid vaccination is indicated only where a person is in intimate contact with a known carrier, or travels in areas where there is a recognized risk of exposure, but precautions should still be taken with water and food. There is no reason to use typhoid vaccine for persons in areas of natural disaster such as floods, or for persons attending rural summer camps.[13] Immunization requires a series of two injections several weeks apart before it becomes effective. The vaccination is then effective only against a small infective dose and requires a booster injection usually once every three years. Cholera vaccine provides 50 percent effectiveness in reducing clinical illness for three to six months. Yellow fever vaccine offers about 95 percent protection for about 10 years.[14]

A WHO Expert Committee[15] points out that

> observations and cost-effect analyses have shown that good housing and sanitation are far more effective measures for the control of cholera, typhoid, and similar diseases than is immunization.

Good housing and sanitation (water, sewerage, solid wastes, and vermin control) protect against many diseases whereas an immunization protects only against a specific disease. Individual and community performance, and environmental, hygiene, and economic levels are also improved.[16]

Typical Epidemic Control

Outbreaks of illnesses such as influenza, measles, dysentery, poliomyelitis, and other diseases can still occur. At such times the people become apprehensive and look to the health department for guidance, assurance, and information to calm their fears.

An example of the form health department assistance can take is illustrated in the precautions released June 1, 1951, in the Illinois Health Messenger for the control of poliomyelitis. These recommendations are quoted here, even though the disease now can be controlled, for the principles are generally applicable to other outbreaks of disease.

[12]B. Cyjetanovic and K. Uemura, "The Present Status of Field and Laboratory Studies of Typhoid and Paratyphoid Vaccine," *WHO Bull.*, **32,** 29–36 (1965).

[13]Recommendations of the Public Health Service Advisory Committee on Immunization Practices *Morbidity and Mortality Weekly Report* (*MMWR*), Center for Disease Control (CDC), Atlanta, Ga., July 7, 1978, and Dec. 15, 1978.

[14]*Health Information for International Travel 1980*, *MMWR*, Dept. of Health and Human Resources (HHS), Public Health Service (PHS), CDC, Atlanta, Ga. 30333, pp. 74 and 76.

[15]"Uses of Epidemiology in Housing Programmes and in Planning Human Settlements," *WHO Tech. Rep. Ser.*544, 13 (1974).

[16]Dennis Warner and Jarir S. Dajani, *Water and Sewer Development in Rural America*, D. C. Heath and Company, Lexington, Mass., 1975.

General Precautions during Outbreaks

1. The Illinois Department of Public Health will inform physicians and the general public as to the prevalence or increase in incidence of the disease.
2. *Early diagnosis* is extremely important. Common early signs of polio are headache, nausea, vomiting, muscle soreness or stiffness, stiff neck, fever, nasal voice, and difficulty in swallowing, with regurgitation of liquids through the nose. Some of these symptoms may be present in several other diseases, but in the polio season they must be regarded with suspicion.
3. *All children with any of these symptoms should be isolated in bed, pending diagnosis.* Early medical care is extremely important.
4. Avoid undue fatigue and exertion during the polio season.
5. *Avoid unnecessary travel and visiting in areas where polio is known to be prevalent.*
6. Pay special attention to practice of good personal hygiene and sanitation:
   a. Wash hands before eating.
   b. Keep flies and other insects from food.
   c. Cover mouth and nose when sneezing or coughing.

Surgical Procedures

Nose, throat, or dental operations, unless required as an emergency, should not be done in the presence of an increased incidence of poliomyelitis in the community.

***General Sanitation**, Including Fly Control*

1. Although there has been no positive evidence presented for spread of poliomyelitis by water, sewage, food, or insects, certain facts derived from research indicate that they might be involved in the spread.
   a. *Water.* Drinking water supplies can become contaminated by sewage containing poliomyelitis virus. Although no outbreaks have been conclusively traced to drinking water supplies, only water from an assuredly safe source should be used to prevent any possible hazards that might exist.
   b. *Sewage.* Poliomyelitis virus can be found for considerable periods of time in bowel discharges of infected persons and carriers and in sewage containing such bowel discharges. Proper collection and disposal facilities for human wastes are essential to eliminate the potential hazard of transmission through this means.
   c. *Food.* The infection of experimental animals by their eating of foods deliberately contaminated with poliomyelitis virus has been demonstrated in the laboratory, but no satisfactory evidence has ever been presented to incriminate food or milk in human outbreaks. Proper handling and preparation of food and pasteurization of milk supplies should reduce the potential hazard from this source.
   d. *Insects.* Of all the insects studied, only blowflies and houseflies have shown the presence of the poliomyelitis virus. This indicates that these flies might transmit poliomyelitis. It does not show how frequently this might happen; it does not exclude other means of transmission; nor does it indicate how important fly transmission might be in comparison with other means of transmission.
2. Fly eradication is an extremely important activity in maintaining proper sanitation in every community.
3. Attempts to eradicate flies by spraying of effective insecticides have not shown any special effect on the incidence of polio in areas where it has been tried. Airplane spraying is not considered a practical and effective means in reducing the number of flies in a city. The best way to control flies and thus prevent them from spreading any disease is to eliminate fly breeding places. Eradicate flies by:
   a. Proper spreading or spraying of manure to destroy fly breeding places.

b. Proper storage, collection, and disposal of garbage and other organic waste.
c. Construction of all privies with fly- and rodent-proof pits.

Proper sanitation should be supplemented by using effective insecticide around garbage cans, manure piles, privies, etc. Use effective insecticide spray around houses or porches or paint on screen to kill adult flies.

Swimming Pools

1. Unsatisfactorily constructed or operated swimming pools should be closed whether or not there is poliomyelitis in the community.
2. On the basis of available scientific information, the State Department of Public Health has no reason to expect that closure of properly equipped and operated swimming pools will have any effect on the occurrence of occasional cases of poliomyelitis in communities.
3. In communities where a case of poliomyelitis has been associated with the use of a swimming pool, that pool and its recirculation equipment should be drained and thoroughly cleaned. (The State Department of Public Health should be consulted about specific cleansing procedures.) After the cleaning job is accomplished, the pool is ready for reopening.
4. Excessive exertion and fatigue should be avoided in the use of the pool.
5. Swimming in creeks, ponds, and other natural waters should be prohibited if there is any possibility of contamination by sewage or too many bathers.

Summer Camps

Summer camps present a special problem. The continued operation of such camps is contingent on adequate sanitation, the extent of crowding in quarters, the prevalence of the disease in the community, and the availability of medical supervision. Full information is available from the Illinois Department of Public Health to camp operators and should be requested by the latter.

1. Children should not be admitted from areas where outbreaks of the disease are occurring.
2. Children who are direct contacts to cases of polio should not be admitted.
3. The retention of children in camps where poliomyelitis exists has not been shown to increase the risk of illness with polio. Furthermore, return of infected children to their homes may introduce the infection to that community if it is not already infected. Similarly, there will be no introduction of new contacts to the camp and supervised curtailment of activity will be carried out, a situation unduplicated in the home. This retention is predicated upon adequate medical supervision.
4. If poliomyelitis occurs in a camp it is advisable that children and staff remain there (with the exception of the patient, who may be removed with consent of the proper health authorities). If they do remain:
   a. Provide daily medical inspection for all children for two weeks from occurrence of last case.
   b. Curtail activity on a supervised basis to prevent overexertion.
   c. Isolate all children with fever or any suspicious signs or symptoms.
   d. Do not admit new children.

Schools

1. Public and private schools should not be closed during an outbreak of poliomyelitis, nor their opening delayed except under extenuating circumstances and then only upon recommendation of the Illinois Department of Health.
2. Children in school are restricted in activity and subject to scrutiny for any signs of illness. Such children would immediately be excluded and parents urged to seek medical attention.

3. Closing of schools leads to unorganized, unrestricted, and excessive neighborhood play. Symptoms of illness under such circumstances frequently remain unobserved until greater spread of the infection has occurred.
4. If poliomyelitis occurs or is suspected in a school:
   a. Any child affected should immediately be sent home with advice to the parents to seek medical aid, and the health authority notified.
   b. Classroom contacts should be inspected daily for any signs or symptoms of illness and excluded if these are found.

Hospitals

1. There is no reason for exclusion of poliomyelitis cases from general hospitals if isolation is exercised—rather, such admissions are necessary because of the need for adequate medical care of the patient.
2. Patients should be isolated individually, or with other cases of poliomyelitis in wards.
3. Suspect cases should be segregated from known cases until the diagnosis is established.
4. The importance of cases to hospitals in a community where poliomyelitis is not prevalent has not been demonstrated to affect the incidence of the disease in the hospital community.

Recreational Facilities

1. Properly operated facilities for recreation should not be closed during outbreaks of poliomyelitis.
2. Supervised play is usually more conducive to restriction of physical activities in the face of an outbreak.
3. Playground supervisors should regulate activities so that overexertion and fatigue are avoided.

RESPIRATORY DISEASES

Definition

The respiratory diseases are a large group of diseases spread by discharges from the mouth, nose, throat, or lungs of an infected individual. The disease-producing organisms are spread by coughing, sneezing, talking, spitting, by dust, and by direct contact as in kissing, eating contaminated food, using contaminated eating and drinking utensils or common towels, drinking glasses, and toys. Included are the insidious diseases associated with the contaminants in polluted air.

Group

A list of respiratory diseases and their incubation periods is shown in Tables 1-3 and 1-4. Many are transmitted in ways other than through the respiratory tract. Scarlet fever, streptococcal sore throat, and diphtheria, for example, may also be spread by contaminated milk, particularly raw milk. Infectious hepatitis may be transmitted from person to person and may be carried by sewage-contaminated water, uncooked clams and oysters, milk, sliced meats, salads, and bakery products. Smallpox, chickenpox, mumps, infectious mono-

Table 1-3 Respiratory Diseases

Disease	Communicability (days)[a]	Incubation Period (days)	Disease	Communicability (days)[a]	Incubation Period (days)
Chickenpox (v)	5 to +6	14 to 21	Pertussis (b)	−7 to +21	7 to 10
Common cold (v)	−2 to +5	½ to 3	(whooping cough)		
Diphtheria (b)	14	2 to 5	Plague,		
German (Rubella)			pneumonic (b)	In illness	2 to 4
measles (v)	−7 to +4	14 to 21	Plague, bubonic (b)	—	2 to 6
Influenza (v)	3	1 to 3	Pneumonia (v)	[b]	1 to 3
Measles (v)	3 to +4	8 to 13	Smallpox (v)[c]	7 to 21	7 to 17
(Rubeola)			Scarlet fever and		
Meningococcal			streptococcal		
meningitis (b)	[b]	2 to 10	sore throat (b)	10 to 21	1 to 3
Mumps (v)	−6 to +9	12 to 26			

Note: (b) bacteria; (v) virus.
[a]Period from onset of symptoms.
[b]Meningococci usually disappear within 24 hr after appropriate chemotherapeutic treatment. Pneumococcus eliminated within 3 days after penicillin treatment. Streptococcus transmission eliminated within 24 hr after penicillin treatment.
[c]Declared by WHO officially "eradicated" in 1978, if no new cases discovered.

Table 1-4 Respiratory Diseases, Other

Disease	Communicability (days)[a]	Incubation Period (days)	Disease	Communicability (days)[a]	Incubation Period (days)
Coccidioido-	No direct		Poliomyelitis (v)	−7 to 42	3 to 21
mycosis (f)	transmission	7 to 28	Psittacosis (v)	In illness	4 to 15
Histoplas-	No direct		Q Fever (r)	No direct trans.	14 to 21
mosis (f)	transmission	5 to 18	Vincent's		
Infectious			infection (b) (s)	—	Unknown
mononucleosis	?	14 to 42	Tuberculosis (b)	Extended	28 to 84

Note: (f) fungus; (v) virus; (?) unknown; (s) spirochete; (b) bacteria; (r) rickettsias, airborne. For greater details see *The Control of Communicable Diseases in Man*, Amer. Pub. Health Assoc., 12th ed., 1975.
[a]Period from onset of symptoms.

nucleosis, meningococcal meningitis, and others may also be transmitted by contact with infected persons. Eye irritations and emphysema are associated with air pollution.

Control

When the source of a respiratory disease is an infected individual, control would logically start with that person. The individual should be taught the importance of personal hygiene and cleanliness, particularly when ill, to prevent the spread of disease. Such things as avoiding spitting, covering up a cough or sneeze with paper tissue, and staying away from people while ill are some of the simple yet important precautions that are not always followed. Every effort should be made to detect and treat the carriers and promptly hospitalize the seriously ill. Identification of the reservoir or agent of disease and its control or elimination should be the goal. When this is impractical or not completely effective, attention should be given to control of the mode of transmission and to the susceptible persons, animals, or arthropods.

A major procedure for the prevention of respiratory diseases is immunization of susceptible persons, with care to exclude those to whom administration of a particular vaccine is contraindicated. Schedules for the use of specific vaccines are available from the Public Health Service, Center for Disease Control, and state and local health departments. Unfortunately, not all students, for example, have been adequately immunized; hence continuous programs to obtain comprehensive coverage are necessary to build up group resistance to infection. The same holds true for health care workers and persons handling infectious materials. Travelers to areas where certain diseases are epidemic or endemic are also advised to check their immunization record with their physician.

WATER- AND FOODBORNE DISEASES

General

The disease agents spread by water and food not only incapacitate large groups of people, but sometimes result in serious disability and death. The diseases are caused by the disregard of known fundamental sanitary principles and hence are in most cases preventable. In some instances, as among the very young, the very old, and those who are critically ill with some other illness, the added strain of a water- or foodborne illness might be disastrous.

The water- and foodborne diseases are sometimes referred to as the intestinal or filth diseases because they are frequently transmitted by food or water contaminated with feces. Included as foodborne diseases are those caused by poisonous plants and animals used for food, toxins produced by bacteria, and foods accidentally contaminated with chemical poisons. They are usually, but not always, characterized by diarrhea, vomiting, nausea, or fever. Symptoms may appear in susceptible persons within a few minutes, several hours, several days, or longer periods, depending on the type and quantity of deleterious material swallowed and the resistance of the individual.

Water may be polluted at its source by excreta or sewage, which is almost certain to contain pathogenic microorganisms and cause illness by draining into an improperly protected and treated surface or groundwater supply. Food may also be contaminated by unclean food handlers who can inoculate the food with infected excreta, pus, respiratory drippings, or other infectious discharges by careless or dirty personal habits. In addition, food can be contaminated in processing, in preparation, by unclean equipment and practices, and by flies carrying the causative organisms of such diseases as salmonellosis, dysentery, or gastroenteritis from an open privy or overflowing cesspool to the kitchen. The role that the fly plays, and the roach may play, in disease transmission is treated separately in Chapter 10. Briefly then, the intestinal diseases can be transmitted by feces, fingers, flies, food, equipment, and water.

Survival of Pathogens

The survival of pathogens in soil is affected by the type of organism, the presence of other organisms, the soil temperature, moisture, nutrients, pH and sunlight. The amount of clay and organic matter in the soil affect the movement of pathogens; but porous soils, cracks, fissures, and channels in rocks permit pollution to travel long distances. See Chapter 3, Travel of Sewage Pollution through the Ground.

Some organisms are more resistant than others. Soil moisture of about 10 to 20 percent of saturation appears to be best for survival of pathogens; drier conditions increase die-off. Nutrients increase survival. The pH is not a major factor. Exposure to sunlight increases the death rate. Low temperatures favor

Table 1-5 Survival of Certain Pathogens in Soil and on Plants

Organism	Media	Survival Time (days)
Coliforms	Soil surface	38
	Vegetables	35
	Grass and clover	6 to 34
Salmonella	Soil	1 to 120
	Vegetables and fruits	less than 1 to 68
Shigella	On grass (raw wastewater)	42
	Vegetables	2 to 10
	In water containing humus	160
Tubercle bacilli	Soil	greater than 180
	Grass	10 to 49
Entamoeba histolytica cysts	Soil	6 to 8
	Vegetables	less than 1 to 3
	Water	8 to 40
Enteroviruses	Soil	8
	Vegetables	4 to 6
Ascaris ova	Soil	up to 7 yrs
	Vegetables and fruits	27 to 35

Source: D. Parsons et al., "Health Aspects of Sewage Effluent Irrigation," Pollution Control Branch, British Columbia Water Resources Services, Victoria, 1975, cited by Eliot Epstein, Rufus L. Chancy, "Land disposal of toxic substances and water-related problems," *J. Water Pollut. Control Fed.*, August 1978, pp. 2037–2042. (The survival of pathogens can be quite variable.)

Table 1-6 Survival of Certain Pathogens in Water

Organism	Survival Time[a]
Vibrio cholerae	5 to 16 days (greater than 3 weeks if frozen)[b]
Salmonella typhi	1 day to 2 months
Leptospira ichterohemorrhagiae	3 to 9 days
Shigella	1 to 24 months
Pasteurella tularensis	1 to 6 months
Entamoeba histolytica	1 month

Source: Arthur P. Miller, *Water and Man's Health*, U.S. Administration for Internal Development, 1961, reprinted 1967.
Note: See also references 16–20.
[a]The survival of pathogenic organisms is dependent on the environment in which they are found, and can be quite variable. The survival data is therefore approximate and should only be used as a guide. Viruses may survive long periods.
[b]In fish and shellfish.

survival.[17,18] The survival of pathogens in soil, on foods, and following various wastewater unit treatment processes as reported by various investigators is summarized by Bryan[19] and others.[20] Tables 1-5 and 1-6 list the survival of certain pathogens in soil, on plants, and in water. Most enteroviruses pass through sewage treatment plants, survive in surface waters, and may pass through water treatment plants providing conventional treatment. But water treatment plants maintaining a free residual chlorine and low turbidity (less than 1 NTU) in the finished water, as noted under Chlorine Treatment for Operation and Microbiological Control in Chapter 3, or using ozone treatment, can accomplish satisfactory virus destruction. Viruses may live up to 200 days in soil.

Substance Dose to Cause Illness

The development of illness is dependent on the toxicity or virulence of a substance, the amount of the substance or microorganisms ingested (at one time or intermittently), and the resistance or susceptibility of the individual. The result may be an acute or long-term illness. Sometimes two or more substances may be involved which may produce a synergistic, additive, or antagonistic effect.

[17]William Rudolfs, Lloyd L. Falk, and R. A. Ragotzkie, "Literature Review on the Occurrence and Survival of Enteric, Pathogenic, and Relative Organisms in Soil, Water, Sewage, and Sludges, and on Vegetation," *Sewage Ind. Wastes*, **22**, 1261–1281 (October 1955).
[18]Robert A. Phillips and Cynthia Sartor Lynch, *Human Waste Disposal on Beaches of the Colorado River in Grand Canyon*, Tech. Rep. 11, U.S. Dept. of the Interior, National Park Service.
[19]Frank L. Bryan, "Diseases Transmitted by Foods Contaminated by Wastewater," *J. Food Protection*, January 1977, pp. 45–56.
[20]*Health Aspects of Excreta and Wastewater Management*, The International Bank for Reconstruction and Development/The World Bank, Washington, D.C., October 1978, pp. 25, 122–123, 128–129, 175–235.

Table 1-7 Substance Dose to Cause Illness

Microorganism	Approximate Number of Organisms (Dose) Required to Cause Disease
[a]*Staphylococcus aureus*	10^6 to 10^7 viable enterotoxin-producing cells per gram of food or milliliter of milk
[a,b]Shigella	10^1 to 10^2
[a,b]*Salmonella typhi*	10^5
[b]*Salmonella typhimurium*	10^3 to 10^4
[a]*Escherichia coli*	10^8
[a]*Vibrio cholerae*	10^8
[b]*Vibrio cholerae*	10^9
[a]*Coxiella burneti*	10^7
[c]*Giardia lamblia*	10^2 (10^1)
Virus, pathogenic	1 pfu or more
Entamoeba histolytica	20 cysts

Sources: [a]Herbert L. Dupont and Richard B. Hornick, "Infectious Disease from Food," *Environmental Problems in Medicine*, William C. McKee, Ed., Charles C. Thomas, Springfield, Ill., 1974.
[b]Eugene J. Gangarosa, "The Epidemiologic Basis of Cholera Control," *Bull. Pan Am. Health Organ.*, Vol. VIII, No. 3, 1974.
[c]R. C. Rendtorff, "Experimental Transmission of *Giardia lamblia*," *Am. J. Hyg*, 59:209 (1954).

Table 1-7 lists various microorganisms and the approximate number (dose) of organisms required to cause disease. Bryan[19] has summarized the work of numerous investigators giving the clinical response of adult humans to varying challenge doses of enteric pathogens. For example, a dose of 10^9 *Streptococcus faecalis* was required to cause illness in 1 to 25 percent of the volunteers, 10^8 *Clostridium perfringens* type A (heat resistant) to cause illness in 26 to 50 percent of the volunteers, and 10^9 *Clostridium perfringens* type A (heat sensitive) to cause illness in 76 to 100 percent of the volunteers.

It is believed that ingestion of one virus particle can cause infection in man. In that case it would appear that viral infections should be readily spread through drinking water, shellfish, and water contact recreational activities. Fortunately, the tremendous dilution that wastewater containing viruses usually receives on discharge to a watercourse and the treatment given drinking water greatly reduce the probability of an individual receiving an infective dose. However some viruses do survive and present a hazard to the exposed population. Not all viruses are pathogenic.

Information concerning the *acute* effect of ingestion of toxic substances is available in toxicology tests.[21] Substances causing or associated with long-term illnesses and deaths such as diseases of the heart, malignant neoplasms (cancer), and cerebrovascular diseases (stroke) are not clear but are being slowly identified.

[21]N. Irving Sax, *Dangerous Properties of Industrial Materials*, 5th ed., Van Nostrand Reinhold, N.Y., 1979.

An indication of the difficulty involved is given by Kennedy.[22] "A typical chronic toxicology test on compound X, done to meet a regulatory requirement with an adequate number of animals and an appropriate test protocol, costs $250,000 to 300,000," and requires two to three more years to complete.

Summary of Characteristics and Control of Water- and Foodborne Diseases

In view of the fact that the water- and foodborne diseases result in discomfort, disability, and even death, a better understanding of their source, method of transmission, control, and prevention is desirable. A concise grouping and summary of the characteristics and control of a number of these diseases is given in the summary sheets, Figure 1-2, for easy reference. Although extensive, it should not be accepted as being complete or final, but should be used as a starting point for further study.

Gastroenteritis is a vague disease that has also been listed. The term is often used to designate a water- or foodborne disease for which the causative agent has not been determined. Much remains to be learned about this broad catch-all classification. As the term implies, it is an inflammation of the stomach and intestines, with resultant diarrhea, nausea, vomiting, low grade fever, and extreme discomfort. The occurrence of a large number of diarrheal cases indicates that there has been a breakdown in the sanitary control of water or food and may be followed by cases of salmonellosis, typhoid fever, infectious hepatitis, dysentery, or other illness. There are undoubtedly many bacterial toxins, bacteria, viruses, protozoa, chemicals, and others, that are not suspected or that are not examined for or discovered by available laboratory methods.

The diseases listed in Figure 1-2 under bacterial toxins are also known as bacterial intoxications or food poisoning, to distinguish them from food infections.

The headings in Figure 1-2 are defined here briefly. Bacteria are single-celled plantlike microscopic organisms. Viruses are organisms that pass through filters that retain bacteria. They can be seen only with an electron microscope, grow only inside living cells, and possess other characteristics to distinguish them from microorganisms such as bacteria and protozoa. Rickettsias resemble bacteria in shape, are Gram-negative, nonmotile, and difficult to stain with ordinary dyes. They are considered to be related and intermediate between bacteria and viruses. Spirochetes are a family of spiral organisms without a nucleus that multiply by transverse division. Protozoa are single-cell animals that reproduce by fission and have other special characteristics. Helminths include intestinal worms or wormlike parasites: the roundworms (nematodes), tapeworms (cestodes), and flatworms or flukes (trematodes). Poisonous plants contain toxic substances that, when consumed by man or other animals, may

[22]Donald Kennedy, "Future Directions and Trends," *National Conference on the Environment and Health Care Costs*, August 15, 1978, House of Representatives Caucus Room, Washington, D.C.

Figure 1-2 Characteristics and control of waterborne and foodborne diseases.

	Disease	Specific Agent	Reservoir	Common Vehicle	Symptoms in Brief	Incubation Period	Prevention and Control
Bacterial Toxins	Botulism food poisoning	*Clostridium botulinum* and *C. parabotulinum* that produce toxin.	Soil, dust, fruits, vegetables, foods, mud.	Improperly processed canned and bottled foods containing the toxin.	Gastrointestinal pain, diarrhea or constipation, prostration, difficulty in swallowing, double vision, difficulty in respiration.	2 hr to 8 days, usually 12 to 36 hr.	Boil home canned nonacid food 5 min; thoroughly cook meats, fish, dried foods held over. Do not taste suspected food!
	Staphylococcus food poisoning	Staphylococci that produce enterotoxin, *Staphylococcus albus*, *S. aureus* (toxin is stable at boiling temperature).	Skin, mucous membranes, pus, dust, air, sputum, and throat.	Contaminated custard pastries, cooked or processed meats, poultry, dairy products, hollandaise sauce, salads, milk.	Acute nausea, vomiting, and prostration; diarrhea, abdominal cramps. Usually explosive in nature, followed by rapid recovery of those afflicted.	1 to 6 hr or longer, average 2 to 4 hr.	Refrigerate prepared food in shallow containers at a temperature below 45°F immediately upon cooling. Reuse leftover food within 4 hr. Avoid handling food. Educate food handlers in personal hygiene and sanitation.
	Clostridium perfringens food poisoning	*Clostridium perfringens* (*C. welchi*), a spore-former (certain strains are heat resistant).	Soil, gastrointestinal tract of man and animals, cattle, poultry, pigs, vermin, and wastes.	Contaminated food, inadequately heated meats including roasts, stews, beef, poultry.	Sudden abdominal pain, then diarrhea and nausea.	8 to 22 hr, usually 10 to 12 hr.	Cook foods thoroughly, cool and refrigerate promptly foods not consumed. Store foods in shallow containers, cut up large pieces. Reheat thoroughly to 165°F before reserving. Educate cooks.
Bacteria	Salmonellosis (Salmonella infection)	*Salmonella typhimurium*, *S. newport*, *S. enteritidis*, *S. montevideo*, others.	Hogs, cattle, and other livestock, poultry, pets, eggs, carriers, powdered eggs, turtles.	Contaminated sliced cooked meat, salads, uncooked meats, equipment, warmed-over foods, milk, milk products. Contaminated water.	Abdominal pain, diarrhea, chills, fever; vomiting and nausea. Diarrhea usually persists several days.	6 to 48 hr, usually 12 to 24 hr.	Protect storage of food. Thoroughly cook food. Eliminate rodents, pets, and carriers. Similar measures as in Staphylococcus. Poultry, water and meat sanitation. Do not eat raw ground beef.
	Typhoid fever	Typhoid bacillus, *Salmonella typhosa.*	Feces and urine of typhoid carrier or patient.	Contaminated water, milk and milk products, shellfish, and foods. Flies.	General infection characterized by continued fever, usually rose spots on the trunk, diarrheal disturbances.	Average 14 days, usually 7 to 21 days.	Protect and purify water supply; pasteurize milk and milk products; sanitary sewage disposal; educate food handlers; food, fly, shellfish control; supervise carriers; immunize. Personal hygiene.
	Paratyphoid fever	*Salmonella paratyphi* A, *S. schott-*	Feces and urine of carrier or patient.	Contaminated water, milk and milk prod-	General infection characterized by continued fever, diarrheal	1 to 10 days for gastro-	Similar preventive and control measures as in typhoid fever

		mulleri B, *S. hirschfeldii* C.		ucts, shellfish, and foods. Flies.	disturbances, sometimes rose spots on trunk, other symptoms.	enteritis; 1 to 3 weeks for enteric fever.	and salmonellosis.
Bacteria	Streptococcal food poisoning	Alpha-*strepto-cocci*, Beta-hemolytic S., *S. fecalis*, *viridans*.	Human mouth, nose, throat, respiratory tract.	Contaminated meats, milk, croquettes, cheese, dressings.	Nausea, sometimes vomiting, colicky pains, and diarrhea.	2 to 18 hr, average 12 hr.	Similar preventive and control measures as in Staphylococcus. Pasteurize milk and milk products.
	Shigellosis (Bacillary dysentery)	Genus, *Shigella*, i.e., *flexneri*, *sonnei*, *boydii*, *dysenteriae*.	Bowel discharges of carriers and infected persons.	Contaminated water or foods, milk and milk products. Flies.	Acute onset with diarrhea, fever, tenesmus, and frequent stools containing blood and mucus.	1 to 7 days, usually less than 4 days.	Food, water, sewage sanitation as in typhoid. Pasteurize milk (boil for infants); control flies; supervise carriers.
	Cholera[a]	*Vibrio cholera*, *Vibrio comma*.	Bowel discharges, vomitus; carriers.	Contaminated water, raw foods. Flies.	Diarrhea, rice-water stools, vomiting, thirst, pain, coma.	A few hours to 5 days, usually 3 days.	Similar to typhoid. Immunize, quarantine; isolate patients.
	Melioidosis[a]	*Pseudomonas pseudomallei*.	Rats, guinea pigs, cats, rabbits, dogs, horses.	Contact with or ingestion of contaminated excreta, soil, or water.	Acute diarrhea, vomiting, high fever, delerium, mania.	Less than 2 days or longer.	Destroy rats; protect food; thoroughly cook food; control biting insects; personal hygiene.
	Brucellosis (Undulant fever)	*Brucella melitensis*-goat, *Br. abortus*-cow, *Br. suis*-pig.	The tissues, blood, milk, urine, infected animals.	Raw milk from infected cows or goats, also contact with infected animals.	Insidious onset, irregular fever, sweating, chills, pains in joints and muscles.	5 to 21 days or longer.	Pasteurize all milk; eliminate infected animals. Handle infected carcasses with care.
	Streptococcal sore throat	Hemolytic *Streptococci*.	Nose, throat, mouth secretions.	Contaminated milk or milk products.	Sore throat and fever, sudden in onset, vomiting.	1 to 3 days.	Pasteurize all milk, Inspect contacts. Exclude carriers.
	Diphtheria	*Corynebacterium diphtheriae*.	Respiratory tract, patient, carrier.	Contact and milk or milk products.	Acute febrile infection of tonsils, throat, and nose.	2 to 5 days or longer.	Pasteurize milk, disinfect utensils. Inspect contacts, immunize.
	Tuberculosis	*Mycobacterium tuberculosis* (hominis and bovis).	Respiratory tract of man. Rarely cattle.	Contact, also eating and drinking utensils, food, and milk.	Cough, fever, fatigue, pleurisy.	4 to 6 weeks.	Pasteurize milk, eradicate TB from cattle. X-ray. Control contacts and infected persons. Selective use of BCG.
	Tularemia	*Pasteurella tularensis* (*Bacterium tularense*).	Rodents, rabbits, horseflies, wood ticks, dogs, foxes, hogs.	Meat of infected rabbit, contaminated water, handling wild animals.	Sudden onset, with pains and fever, prostration.	1 to 10 days, average of 3.	Thorough cooking of meat of wild rabbits. Purify drinking water. Use rubber gloves (care in dressing wild rodents).
	Gastroenteritis (Diarrhea)	Microorganisms unknown.	Probably man and animals.	Water, food, including milk, air.	Diarrhea, nausea, vomiting, cramps, possibly fever.	Variable, 8 to 12 hr average.	Environmental sanitation, education, personal hygiene.
	Bacillus cereus gastroenteritis	*Bacillus cereus*.	Dust and soil, also milk.	Cereal products, rice, meat loaf, custards.	Nausea, abdominal cramps, vomiting, possibly diarrhea later.	8 to 16 hr or less.	Proper refrigeration. Spores survive boiling; serve rice and other cooked foods promptly.

Figure 1-2 (*Continued*)

	Disease	Specific Agent	Reservoir	Common Vehicle	Symptoms in Brief	Incubation Period	Prevention and Control
Bacteria	Campylobacter enteritis	*Campylobacter fetus.*	Chickens, swine, dogs, raw milk, contaminated water.	Undercooked chicken, also pork.	Watery diarrhea, abdominal pain, fever, chills.	1 to 4 days	Thorough cooking of chicken and pork.
	Vibrio parahaemolyticus gastroenteritis	*Vibrio parahaemolyticus.*	Marine fish, shellfish, mud, sediment, salt water.	Raw seafoods or seafood products; inadequately cooked seafoods, and cross-contamination between raw and cooked products and sea water.	Nausea, headache, chills, fever, vomiting, severe abdominal cramps, watery diarrhea, sometimes with blood.	2 to 48 hr, usually 12 to 24 hr.	Properly cook all seafood (shrimp 7 to 10 min). Avoid cross-contamination or contact with sea water or preparation surfaces used for uncooked foods; refrigerate prepared seafoods promptly if not immediately served.
	Diarrhea enteropathogenic *Escherichia coli*	Enteropathogenic *Escherichia coli* invasive and enterotoxic strains.	Infected persons.	Food, water, and fomites contaminated with feces.	Fever, mucoid, occasionally bloody diarrhea; or watery diarrhea, cramps, acidosis, dehydration.	12 to 72 hr.	See Typhoid. Scrupulous hygiene and formula sanitation in hospital nursery.
Rickettsias	Q Fever	*Coxiella burneti.*	Dairy cattle, sheep, goats.	Slaughterhouse, dairy employees, handling infected cattle; raw cow and goat milk.	Heavy perspiration and chills, headache, malaise.	2 to 3 weeks.	Pasteurization of milk and dairy products, elimination of infected animal reservoir, cleanliness in slaughterhouse and dairies. Pasteurize at 145°F for 30 min or 161°F for 15 sec.
Viruses	Choriomeningitis, lymphocytic	*L. choriomeningitis* virus.	House mice urine, feces, secretions.	Contaminated food.	Fever, grippe. Severe headache, stiff neck, vomiting, somnolence.	8 to 13 days, 15 to 21 days.	Eliminate or reduce mice. General cleanliness, sanitation.
	Infectious hepatitis (hepatitis A)	Viruses unknown.	Discharges of infected persons.	Water, food, milk, oyster, clams, contacts.	Fever, nausea, loss of appetite; possibly vomiting, fatigue, headache, jaundice.	10 to 50 days, average 30 to 35 days.	Sanitary sewage disposal, food sanitation, personal hygiene; coagulate and filter water supply, and plus 0.6 mg/l free Cl_2. Obtain shellfish from certified dealers. Steam clams 4 to 6 minutes.
	Gastroenteritis, viral	Probably parvovirus, or reovirus-	Man.	Water, food including milk, possibly fecal-	Nausea, vomiting, diarrhea, abdominal pain, low fever; pri-	24 to 48 hr.	Same as Salmonellosis.

	Disease	Causative agent	Source of infection	Mode of transmission	Symptoms	Incubation period	Control
		like agent.		oral or fecal-respiratory route.	marily children.		
Protozoa	Amebiasis (Amebic dysentery)	*Entamoeba histolytica.*	Bowel discharges of carrier, and infected person, possibly also rats.	Cysts, contaminated water, foods, raw vegetables and fruits. Flies, cockroaches.	Insidious and undetermined onset, diarrhea or constipation, or neither; loss of appetite, abdominal discomfort; blood, mucus in stool.	5 days or longer, average 3 to 4 weeks.	Same as Shigellosis. Boil water or coagulate, set, filter through diatomite 5 gpm/ft^2, Cl_2. Usual Cl_2 and high rate filtration not 100% effective. Slow sand filtration plus Cl_2, or conventional RSF OK. Pres. sand filt. ineffective. Also sanitation and hygiene.
	Giardiasis	*Giardia lamblia*	Bowel discharges of carrier and infected persons; dog, beaver.	Cysts, contaminated water, food, raw fruits; also hand-to-mouth route.	Prolonged diarrhea, abdominal cramps, severe weight loss, fatigue, nausea, gas. Fever is unusual.	6 to 22 days, avg. 9 days.	Same as Amebiasis.
Spirochetes	Leptospirosis (Weil's disease) (Spirochetosis icterohemorrhagic)	*Leptospira icterohaemorrhagiae, L. hebdomadis, L. canicola, L. pomona,* others.	Urine and feces of rats, swine, dogs, cats, mice, foxes, sheep.	Food, water, soil contaminated with excreta or urine of infected animal. Contact.	Fever, rigors, headaches, nausea, muscular pains, vomiting, thirst. Prostration, jaundice.	4 to 19 days, average 9 to 10 days.	Destroy rats; protect food; avoid polluted water; treat abrasion of hands and arms. Disinfect utensils, treat infected dogs.
Helminths	Trichinosis (Trichiniasis)	*Trichinella spiralis.*	Pigs, bears, wild boars, rats, foxes, wolves.	Infected pork and pork products, bear and wild boar meat.	Nausea, vomiting, diarrhea, muscle pain, swelling of face and eyelids, difficulty in swallowing.	2 to 28 days, usually 9 days.	Thoroughly cook pork (150°F), pork products, bear and wild boar meat; destroy rats. Feed hogs boiled garbage or discontinue feeding. Store meat 20 days at 5°F or 36 hr at −27°F.
	Schistosomiasis (Bilharziasis)[b] (blood flukes)	*Schistosoma haematobium, S. mansoni, S. japonicum, S. intercalatum.*	Venous circulation of man; urine, feces, dogs, cats, pigs, cattle, horses, field mice, wild rats, water buffalo.	Cercariae-infested drinking and bathing water (lakes and coastal sea waters).	Dysenteric or urinary symptoms. Rigors. Itching on skin, dermatitis. Carrier state 1 to 2 years and up to 25 years. Swimmer's itch schistosomes do not mature in man.	4 to 6 weeks or longer.	Avoid infested water for drinking or bathing; coagulation, sedimentation, and filtration plus Cl_2 1 mg/1; boil water; 10 mg/1 $CuSO_4$, and impound water 48 hr, Cl_2. Slow sand filtration plus Cl_2. Sanitation. 1 mg/1 $CuSO_4$ to kill cercariae and 20 mg/1 to kill snails.

Figure 1-2 *(Continued)*

	Disease	Specific Agent	Reservoir	Common Vehicle	Symptoms in Brief	Incubation Period	Prevention and Control
Helminths	Ascariasis (intestinal roundworm)	*Ascaris lumbricoides.*	Small intestine of man, gorilla, ape.	Contaminated food, water; sewage.	Worm in stool, abdominal pain, skin rash, protuberant abdomen, nausea, large appetite.	About 2 months.	Personal hygiene, sanitation.[a] Boil drinking water in endemic areas. Sanitary excreta disposal.
	Echinococcosis (Hydatidosis)	*Echinococcus granulosus*, dog tapeworm.	Dogs, sheep, wolves, dingoes, swine, horses, monkeys.	Contaminated food and drink; hand to mouth; contact with infected dogs.	Cysts in tissues—liver, lung, kidney, pelvis. May give no symptoms. May cause death.	Variable, months to several years.	Keep dogs out of abattoir and do not feed raw meat. Mass treatment of dogs. Educate children and adults in the dangers of close association with dogs.
	Taeniasis (pork tapeworm) (beef tapeworm)	*Taenia solium* (pork tapeworm), *T. saginata* (beef tapeworm).	Man, cattle, pigs, buffalo, possibly rats, mice.	Infected meats eaten raw. Food contaminated with feces of man, rats, or mice.	Abdominal pain, diarrhea, convulsions, insomnia, excessive appetite.	8 to 10 weeks.	Thoroughly cook meat, control flies, properly dispose of excreta; food-handler hygiene. Use only inspected meat. Store meat 10 days at 15°F or lower.
	Fish Tapeworm (broad tapeworm)	*Diphyllobothrium latum*, other.	Man, frogs, dogs, cats, bears.	Infected freshwater fish eaten raw.	Abdominal pain, loss of weight, weakness, anemia.	3 to 6 weeks.	Thoroughly cook fish, roe, (caviar). Proper excreta disposal.
	Dracontiasis (Guinea worm disease)	*Dracunculus medinensis*, a nematode worm.	Man.	Water contaminated with copepods-*Cyclops*.	Blistering of feet, legs, and burning and itching of skin; fever, nausea, vomiting, diarrhea.	About 12 mo.	Use only filtered or boiled water for drinking, or a safe well-water supply.
	Paragonimiasis (lung flukes)	*Paragonimus ringeri, P. westermani, P. kellicotti.*	Respiratory and intestinal tract of man, cats, dogs, pigs, rats, wolves.	Contamin. water, freshwater crabs or crayfish.	Chronic cough, clubbed fingers, dull pains, and diarrhea.	Variable.	Boil drinking water in endemic areas; thoroughly cook freshwater crabs and crayfish.[a]
	Clonorchiasis[b] (liver flukes)	*C. sinensis, Opisthorchis felineus.*	Liver of man, cats, dogs, pigs.	Contamin. freshwater fish.	Chronic diarrhea, night blindness.	Variable.	Boil drinking water in endemic areas; thoroughly cook fish.[a]
	Fascioliasis (sheep liver flukes)	*Fasciola hepatica.*	Liver of sheep.	Sheep liver eaten raw.	Irregular fever, pain, diarrhea.	Several months.	Thoroughly cook sheep liver.[a]
	Trichuriasis (whipworm)	*Trichuris trichiura.*	Large intestine of man.	Contaminated food, soil.	No special symptoms, possibly stomach pain.	Long and indefinite.	Sanitation. Boil water, cook well, properly dispose feces.[a]

Helminths	Oxyuriasis (pinworm, threadworm, or enterobiasis)	*Oxyuris vermicularis*, or *Enterobius vermicularis*	Large intestine of man, particularly children.	Fingers. Ova-laden dust. Contaminated food, water; sewage. Clothing, bedding.	Nasal itching, anal itching, diarrhea.	3 to 6 weeks; months.	Wash hands after defecation, keep fingernails short, sleep in cotton underwear. Sanitation.
	Fasciolopsias[b] (intestinal flukes)	*Fasciolopsis buski.*	Small intestine of man, dogs, pigs.	Raw freshwater plants; water, food.	Stomach pain, diarrhea, greenish stools, constipation. Edema.	6 to 8 weeks.	Cook or dip in boiling water roots of lotus, bamboo, water chestnut, caltrop.
	Dwarf tapeworm (rat tapeworm)	*Hymenolepis nana* (*diminuta*).	Man and rodents.	Food contaminated with ova; direct contact.	Diarrhea or stomach pain, irritation of intestine.	1 month.	Sanitary excreta disposal, personal hygiene, food sanitation, rodent control; treat cases.
Poisonous plants and animals	Ergotism[c]	*Ergot*, a parasitic fungus (*Claviceps purpurea*).	Fungus of rye and occasionally other grains.	Ergot-fungus contaminated meal or bread.	Gangrene involving extremities, fingers, and toes; or weakness and drowsiness, headache, giddiness, painful cramps in limbs.	Gradual, after prolonged use of diseased rye in food.	Do not use discolored or spoiled grain (fungus grows in the grain). Meal is grayish, possibly with violet-colored specks.
	Rhubarb poisoning	Probably oxalic acid.	Rhubarb.	Rhubarb leaves.	Intermittent cramplike pains, vomiting, convulsions, coma.	2 to 12 hr.	Do not use rhubarb leaves for food.
	Mushroom poisoning	Phalloidine and other alkaloids; also other poisons in mushroom.	Mushrooms—*Amanita phalloides* and other Aminita.	Poisonous mushrooms (Amanita phalloides, Amanita muscaria, others).	Severe abdominal pain, intense thirst, retching, vomiting, and profuse watery evacuations.	6 to 15 hr or 15 min to 6 hr with muscaria	Do not eat wild mushrooms; warn others. Amanita are very poisonous, both when raw or cooked.
	Favism[b]	Poison from *Vicia faba* bean, pollen.	*Vicia faba.* Plant and bean.	The bean when eaten raw, also pollen.	Acute febrile anemia with jaundice, passage of blood in urine.	1 to 24 hr.	Avoid eating bean, particularly when green, or inhalation of pollen.
	Fish poisoning	Poison in fish, ovaries and testes, roe (heat-stable).	Fish, pike, carp, sturgeon roe in breeding season.	Fish—Tedrodon, Meletta, Clupea, Pickerel eggs, mukimuki.	Painful cramps, dyspnea, cold sweats, dilated pupils, difficulty in swallowing and breathing.	30 min to 2 hr or longer.	Avoid eating roe during breeding season. Heed local warnings concerning edible fish.
	Ciguatera poisoning	Toxin concentrated in certain fish flesh possibly from dinoflagellate.	Warm-water fish, possibly barracuda, snapper, grouper, amberjack.	Warm-water fish caught near shore from Pacific and Caribbean.	Progressive numbness, tetanus-like spasms, heavy tongue, facial stiffness; also nausea, vomiting, metallic taste, dryness of the mouth, abdominal cramps.	1 to 8 hr, usually 1 to 6 hr.	Avoid warm-water fish caught near shore in Pacific and Caribbean. The toxin ciguatera is not destroyed by cooking; toxin is not poisonous to fish.
	Shellfish poisoning (Paralytic)	Neurotoxin produced by (*Gonyaulax catenella*) *Gymnodinium breve*	Clams and mussels feeding on specific dinoflagellates.	Mussels and clams, associated with so-called "red tides."	Respiratory paralysis. In milder form, trembling about lips to loss of control of the extremities and neck.	5 to 30 min and longer.	Obtain shellfish from certified dealers and from approved areas. Toxicity reduced 70 to 90% by steaming or frying. Toxin not destroyed by routine cooking.

Figure 1-2 *(Continued)*

	Disease	Specific Agent	Reservoir	Common Vehicle	Symptoms in Brief	Incubation Period	Prevention and Control
Poisonous Plants	Snakeroot poisoning	Trematol in snakeroot (*Eupatorium urticaefolium*.	White snakeroot jimmy weed.	Milk from cows pastured on snakeroot.	Weakness or prostration, vomiting, severe constipation and pain, thirst; temperature normal.	Variable. Repeated with use of the milk.	Prevent cows from pasturing in wooded areas where snakeroot exists.
	Potato poisoning	*Solanum tuberosum*; other S.	Sprouted green potatoes.	Possibly green sprouted potatoes.	Vomiting, diarrhea, headache, abdominal pains, prostration.	Few hours.	Do not use sprouts or peel of sprouted green potatoes.
	Water-hemlock poisoning	Cicutoxin or resin from hemlock (*Cicuta maculata*).	Water hemlock.	Leaves and roots of water hemlock.	Nausea, vomiting, convulsions, pain in stomach, diarrhea.	1 to 2 hr.	Do not eat roots, leaves, or flowers of water hemlock.
Chemical Poisons	Antimony poisoning	Antimony.	Gray-enameled cooking utensils.	Foods cooked in cheap enameled pans.	Vomiting, paralysis of arms.	Five minutes to an hour.	Avoid purchase and use of poor quality gray-enameled, chipped enamel utensils.
	Arsenic poisoning	Arsenic.	Arsenic compounds.	Arsenic-contaminated food or water.	Vomiting, diarrhea, painful tenesmus. (A cumulative poison.)	10 min and longer.	Keep arsenic sprays, etc., locked; wash fruits, vegetables. Avoid substances with concentrations greater than 0.05 mg/l.
	Cadmium poisoning	Cadmium.	Cadmium-plated utensils.	Acid food prepared in cadmium utensils.	Nausea, vomiting, cramps, and diarrhea.	15 to 30 min.	Watch for cadmium-plated utensils, racks, and destroy. Inform manufacturer.
	Cyanide poisoning	Cyanide, sodium.	Cyanide silver polish.	Cyanide-polished silver.	Dizziness, giddiness, dyspnea, palpitation, and unconsciousness.	Rapid.	Select silver polish of known composition. Prohibit sale of poisonous polish.
	Fluoride or sodium fluoride poisoning	Fluoride or sodium fluoride.	Roach powder.	Sodium fluoride taken for baking powder, soda, flour.	Acute poisoning, vomiting, abdominal pain, convulsions; paresis of eye, face, finger muscles, and lower extremities; diarrhea.	Few minutes to 2 hr.	Keep roach powder under lock and key; mark "Poison"; color the powder, apply with care, if use is permitted.
	Lead poisoning	Lead.	Lead pipe, sprays, oxides, and utensils. Lead-base paints.	Lead-contaminated food or acid drinks; toys, fumes, paints.	Abdominal pain, vomiting, and diarrhea. (A cumulative poison.)	30 minutes and longer.	Do not use lead pipe; $Pb < 0.05$ mg/l. Wash fruits. Label plants. Avoid using unglazed pottery. Screen child; remove lead paint.
	Mercury poisoning	Mercury—Methyl mercury and other alkylmercury compounds.	Contaminated silt, water, aquatic life.	Mercury-contaminated food, fish.	Fatigue, mouth numbness, loss of vision, poor coordination and gait, tremors of hands, blindness, paralysis.	2 to 30 min. or longer.	Keep mercuric compound under lock and key; do not consume: fish with concentrations of mercury more than 0.5 ppm, water

	Disease	Specific Agent	Source/Reservoir	Common Vehicle	Symptoms	Incubation Period	Prevention and Control
							with more than 0.002 ppm, food with more than 0.05 ppm. Eliminate discharges to the environment.
Chemical poisons	Methyl chloride poisoning	Methyl chloride.	Refrigerant, methyl chloride.	Food stored in refrigerator having leaking unit.	Progressive drowsiness, stupor, weakness, nausea, vomiting, pain in abdomen, convulsions.	Variable.	Use nontoxic refrigerant, such as Freon, water, brine, dry ice.
	Selenium poisoning	Selenium.	Selenium-bearing vegetation.	Wheat from soil containing selenium.	Gastrointestinal, nervous, and mental disorders; dermatitis in sunlight.	Variable.	Avoid semiarid selenium-bearing soil for growing of wheat, or water with more than 0.05 mg/l Se.
	Zinc poisoning	Zinc.	Galvanized iron.	Acid food made in galvanized iron pots and utensils.	Pain in mouth, throat, and abdomen followed by diarrhea.	Variable.	Do not use galvanized utensils in preparation of foods or drink, or water with more than 15.0 mg/l zinc.
	Methemoglobinemia	Nitrate nitrogen, plus nitrite.	Ground water; shallow dug wells, also drilled wells.	Drinking water from private wells.	Vomiting, diarrhea, and cyanosis in infants.	2 to 3 days.	Use water with less than 45 mg/l NO_3 for drinking water and in infant formula. Properly develop and locate wells.
	Sodium nitrite poisoning	Sodium nitrite.	Impure sodium nitrate and nitrite.	Sodium nitrate taken for salt.	Dizziness, weakness, stomach cramps, diarrhea, vomiting, blue skin.	5 to 30 min.	Use USP sodium nitrate in curing meat. Nitrite is poisonous, keep locked.
	Copper poisoning	Copper.	Copper pipes and utensils.	Carbonated beverages and acid foods in prolonged contact with copper.	Vomiting, weakness.	1 hr or less.	Do not prepare or store acid foods or liquids or carbonated beverages in copper containers. Cu should not exceed 0.3 mg/l. Prevent CO_2 backflow into copper lines.

Source: This figure represents a summary of information selected from: 1. G. M. Dack, *Food Poisoning*, 251 pp., University of Chicago Press, 1956. 2. C. E. Dolman, "Bacterial Food Poisoning," 46 pp., *Canad. Pub. Health J. Assoc.*, 1943. 3. V. A. Getting, "Epidemiologic Aspects of Food-Borne Disease," 75 pp., *New Eng. J. Med.*, 1943. 4. F. A. Korff, "Food Establishment Sanitation in a Municipality," *Am. J. Pub. Health* **32,** 740 (1952). 5. P. Manson-Bahr, *Synopsis of Tropical Medicine*, 224 pp., Williams & Wilkins Co., Baltimore, 1943. 6. New York State Department of Health, *Health News*. 7. Miscellaneous military and civilian texts and reports. 8. R. P. Strong, *Stitt's Diagnosis, Prevention and Treatment of Tropical Diseases*, 2 vols., Blakiston Co., Philadelphia, 1942. 9. *The Control of Communicable Diseases in Man*, Am. Pub. Health Assoc., Washington, D.C., 1981. (Sept. 1944, Revised May 1945, 1946, 1952, 1971, 1980. Copyright 1946, Joseph A. Salvato, Jr. MCE.) More complete characteristics, preventive and control measures, and modes of transmission, other than food and water, have been omitted for brevity as has been the statement "epidemiological study" and "education of the public" opposite each disease under the heading "Prevention and Control." Milk and milk products are considered foods. Under "Specific Agent" and "Common Vehicle" above, only the more common agents are listed.

[a]Take same precautions with drinking, culinary, and bathing water as in Schistosomiasis. [b]Does not originate in the U.S. [c]Many other fungi that produce toxin are associated with food and feedstuffs. The mycotoxins cause illness in humans and animals; see text. For more information see F. L. Bryan, *Diseases Transmitted by Foods*. DHEW, PHS, Atlanta, 1971, 58 p., *Procedure for the Investigation of Foodborne Disease Outbreaks*, and *Procedures to Investigate Waterborne Illness*, International Association of Milk, Food, and Environmental Sanitarians, P.O. Box 701, Ames, Iowa 50010, 1976 and 1979. Also *Control of Communicable Diseases in Man*, Am. Pub. Health Assoc.

cause illness or even death. Poisonous animals include fish whose meat is poisonous when eaten in a fresh and sometimes cooked state. Poisonous meat is not to be confused with decomposed food. The toxic substance in some poisonous fish meat appears to be heat-stable. Chemical poisons are certain elements or compounds, usually metallic, that may cause illness or death when consumed by man or animals in food and water. Fungi are small plantlike organisms that are found in s il, rotting vegetation, and bird excreta. Systemic fungal diseases are transmitted through the soil, vegetation, or excreta by contact or ingestion. Illnesses associated with the consumption of poisonous plants and animals, chemical poisons, and poisonous fungi are strictly not communicable diseases but more properly noninfectious, or noncommunicable diseases. They are listed and discussed here for convenience.

Reservoir or Source of Disease Agents

Intestinal diseases occur more frequently where there are low standards of hygiene and sanitation and where there is poverty and ignorance. Contamination of food and drink either directly with human or domestic animal feces or indirectly by flies that have had contact with infected waste is usually necessary for disease transmission. A high incidence of these diseases in developed areas of the world is an indication of lack of or faulty sanitation and shows a disregard of fundamental hygienic practices.

The feces of infected persons may be the reservoir of a broad group of intestinal diseases. (Urine is usually sterile, except for urinary schistosomiasis, typhoid, and leptospirosis carriers,[23] which are not common in developed countries.) A certain percentage of the population is always infected with one or more of the intestinal diseases. For example, the incidence of amebic dysentery in the native populations of tropical countries varies between 10 and 25 percent and may be as high as 60 percent; shigellosis may be higher. Stoll has ventured to hazard a guess of the incidence of helminthic infections in the world.[24] He estimates that at least 500 million persons harbor ascarids, 400 million other worms.* Actually, if a person is ill with a helminthic disease, he probably is infected with more than one parasite, since the conditions conducive to one infection would allow additional species to be present.

The WHO estimates (1978) that 200 million humans are afflicted by schistosomiasis (bilharziasis). In some endemic areas more than half the population is affected.[25] In addition, it is estimated that almost a quarter of the world's

*WHO estimates that there are 10 to 48 million cases of dracunculiasis (guinea worm) in rural regions of Southeast Asia, Africa, and the Eastern Mediterranean. ("International Water Supply and Sanitation Decade," *MMWR, CDC*, Atlanta, Ga., May 1, 1981, pp. 194–195.)

[23] *Health Aspects of Excreta and Wastewater Management*, The International Bank for Reconstruction and Development/The World Bank, Washington, D.C., October 1978, pp. 16–17.

[24] Norman R. Stoll, "Changed Viewpoints on Helminthic Disease: World War I vs. World War II," *Ann. N.Y. Acad. Sci.*, **44**, 207–209 (1943).

[25] Abram S. Benenson, Editor, *Control of Communicable Diseases in Man*, Am. Pub. Health Assoc., Washington, D.C., 1975, p. 282.

population suffers from one of four water-related diseases: gastroenteritis, malaria, river blindness (onchocerciasis), or schistosomiasis.[26]

A survey of U.S., state, and territorial public health laboratories by the PHS Center For Disease Control (CDC) in 1976 for frequency of diagnosis of intestinal parasitic infections in 414,820 stool specimens showed 15.6 percent contained one or more pathogenic or nonpathogenic intestinal parasites, 3.8 percent were positive for *Giardia lamblia*, 2.7 percent for *Trichuris trichiura*, 2.3 percent for *Ascaris lumbricoides*, 1.7 percent for *Enterobius vermicularis*, and 0.6 percent for *Entamoeba histolytica*.[27]

A study made at a missionary college in east central China showed that 49 percent of the students harbored parasitic worms, and a survey made in an elementary school in New Jersey showed that 23 percent of the children were infected. In 1970 Lease[28] reported on the study of the day care and elementary school programs in four counties in South Carolina involving 884 children. He found that 22.5 percent of the Negro children harbored Ascaris intestinal roundworms and, of the 52 white children in the group, 13.5 percent had worms. Central sewerage and water supply was lacking. The rural infection rate was higher and the infected rural child had twice the number of worms as the infected city child. Since parasitic infection plus poor diet or other causes may result in serious debility and perhaps death, preventive measures, including better sanitation, are essential.

The mouth, nose, throat, or respiratory tract of man is a reservoir of organisms that directly or indirectly causes a large group of illnesses. Staphylococci that produce enterotoxin are also found on the skin, on the mucous membranes, in pus, feces, dust, air, and in insanitary food processing plants. They are the principal causes of boils, pimples, and other skin infections and are particularly abundant in the nose and throat of a person with a cold. It is no surprise, therefore, that staphylococcus food poisoning is one of the most common foodborne diseases. Scrupulous cleanliness in food processing plants, in the kitchen, and among food handlers is essential if contamination of food with salmonellas, staphylococci, and other organisms is to be reduced or prevented.

Man is in intimate contact with the sources of the water- and foodborne diseases at all times. Man, rodents, livestock, dogs and cats, soil, certain plants, fish, insecticides, and metal are sources, or reservoirs, of the deleterious substances causing illness. By knowing more about these sources of infection or poisoning, it is possible to take precautions to prevent the harmful substances harbored by these sources from gaining access to food and drink. This is not easily done. Practically speaking, it is not yet possible to remove man, for example, from the kitchen where food is prepared. The entrance into the kitchen

[26] *The Nation's Health*, Am. Pub. Health Assoc., Washington, D.C., March 1979.

[27] "Intestinal Parasite Surveillance—United States, 1976," *MMWR*, CDC, Atlanta, Ga., May 19, 1978, pp. 167–168.

[28] E. John Lease, "Study Finds Carolina Children Afflicted by Intestinal Parasites," *New York Times*, March 28, 1970, p. 30.

of cats and dogs, unauthorized persons, mice and rats, and persons who are ill can, however, be prevented or controlled so as not to be hazardous.

The U.S. Department of Agriculture has indicated that between one and two percent of all the swine examined have living trichinae. In 1948 the incidence of trichinosis in grain-fed hogs was 0.95 percent and in garbage-fed hogs 5.7 percent.[29] A 1967 report shows the rates reduced to 0.12 percent and 2.2 percent, and a 1977 report shows the rates reduced to 0.125 percent and 0.5 percent, respectively.[30,31] Wild animals including bears, martens, wolverines, bobcats, and coyotes are also carriers.

Salami, cervelat, mettwurst, and Italian-style ham may be considered acceptable when stamped "U.S. inspected for wholesomeness," but this is no guarantee that the product is absolutely safe. This is particularly true of raw meat and poultry, which frequently contain salmonellas and other pathogenic organisms, even though stamped "inspected." Such raw products require hygienic handling and adequate cooking. Uncooked summer sausage (ground fresh pork, beef, and seasoning plus light smoking) and raw or partially cooked pork product should be avoided.[32] The incidence of adult *Trichinella* infection in the United States is estimated at four percent or less.[33] Swine are a reservoir of the organisms causing:

1. Fascioliasis (intestinal fluke) and fasciolopsiasis.
2. Paragonimiasis (lung fluke).
3. Salmonella infection (salmonellosis).
4. Taeniasis (pork or beef tapeworm) and cysticercosis.
5. Trichinosis (trichiniasis).
6. Trichuriasis (whipworm).
7. Tularemia.
8. Brucellosis (undulant fever).

The spread of all these diseases can be prevented by thoroughly cooking all foods. Using only inspected meats, prohibiting the feeding of uncooked garbage or offal to hogs, and practicing good sanitation will also help. Storage of pork 20 days at 5°F or 36 hours at −27°F is believed adequate to kill trichina larvae. Swine and other livestock are also reservoirs of infection for clonorchiasis, echinococcosis, leptospirosis, lymphocytic choriomeningitis, schistosomiasis, and toxoplasmosis.

The excreta or urine of rats or mice contain the causative organisms of

1. Choriomeningitis, lymphocytic
2. Leptospirosis (Weil's disease)
3. Salmonella infection (salmonellosis)

[29] *Pub. Health Rep.*, April 9, 1948, pp. 478–488.

[30] W. J. Zimmerman, "The Incidence of *Trichinella spiralis* in Humans of Iowa," *Pub. Health Rep.*, 1967, pp. 127–130.

[31] "Outbreak of Trichinosis—Louisiana," *MMWR*, CDC, Atlanta, Ga., August 3, 1979, p. 357.

[32] "Trichinosis—Louisiana," *MMWR*, CDC, Atlanta, Ga., July 4, 1980, pp. 309–310.

[33] Harry Most, "Trichinellosis in the United States," *J. Am. Med. Assoc.*, **193**, 11, 871–878, (September 13, 1965).

Because of the practical difficulty of permanently eliminating all mice and rats, the threat of contaminating food with the organisms causing the foregoing diseases is ever-present. This emphasizes the necessity of also keeping all food covered or protected. In addition, rats are frequently infested with fleas that, if infected, can spread murine typhus. Sodoku and Haverhill fevers transmitted by the bite of infected rats are also to be guarded against. Sodoku is not common in the United States.

Dust, eggs, poultry, pigs, sheep, cattle and other livestock, rabbits, rats, cats, and dogs may harbor salmonella and other causative organisms. Shelled eggs and egg powder may also contain salmonella. Salmonella food infection is common. Poisonous plants, fish, and chemicals should also be guarded against.

Food Spoilage

When fresh foods are allowed to stand at room temperature, they begin to deteriorate. Enzymes develop in decomposing food or are introduced by bacteria into the food. The changes in the food, brought about by the action of enzymes, cause the food to change in its composition. Such favorable outside factors as oxygen, sunlight, warmth, and moisture cause an acceleration of the rate of decomposition. This shows up in the unpleasing appearance of the food, by the loss of freshness, by the color, and by the odor. Food that has been permitted to decompose loses much of its nutritive value. Oxidation causes reduction of the vitamin content and breaking down of the fats, then the proteins, to form hydrogen sulfide, ammonia, and other products of decay. Ptomaine poisoning is a misnomer for food poisoning or food infection. Ptomaines are basic products of decay formed as the result of the action of bacteria on nitrogenous matter such as meat. Ptomaines in and of themselves are not considered dangerous unless toxic amines are formed due to the advanced decomposition of food. Contamination, which almost always accompanies putrefaction, is dangerous. In certain instances, the bacterial activity will produce a toxin (*staphylococcus aureus* strain) which even ordinary cooking will not destroy.

Mycotoxins

Mycotoxins are poisonous chemicals produced by molds (minute fungi). Mycotoxins are hazardous to man and animals. Ingestion of contaminated feed by farm animals may permit carry-over of toxins into meat and milk. There are about 15 types of dangerous mycotoxins. One common type, aflatoxins, is produced by the molds *Aspergillus flavus* and *Aspergillus parasiticus*. The mold has been detected in peanuts and peanut butter, corn, figs, cereals, cottonseed products, milk and milk products, and other foods that are not properly dried and stored thereby favoring fungus contamination and growth on the food. Fortunately, the mere presence of a toxic mold does not automatically mean the presence of mycotoxins. Contamination may result also before harvest. Most fungal toxins, including aflatoxins, are not destroyed by boiling and autoclaving.[34] Improperly stored leftover foods may also be a source of aflatoxins. Aflatoxins fluoresce under long-wave ultraviolet light.

Aflatoxin is suspected to be a cause of liver cancer. Mycotoxins can also damage the liver, brain, bones, and nerves with resultant internal bleeding. Aflatoxins cause cancer in rats. Toxins that attack nerves are *neurotoxins*. Toxins that attack the intestinal tract are *enterotoxins*.

A concise summary of mycotoxins and some mycotoxicoses of man and animals is found in a Report of a WHO Expert Committee with the participation of Food and Agriculture Organization of the United Nations (FAO),[35] and a paper by Bullerman.[36]

The Food and Drug Administration (FDA) reports that it is not safe to scrape off mold and eat the remaining food; that the toxins produced are not always destroyed by cooking; that freezing prevents mold growth; and that mold grows at refrigerator temperature although at a slower rate. The inside of refrigerators should be washed and dried regularly to keep down mold growths and musty odors; commercial deodorants are not a substitute for cleanliness. Some cheeses, such as Roquefort and Blue, are processed with special species of molds, similar to those from which penicillin is made, and have been consumed with safety for hundreds of years. Foods (vegetables, meats, fruits, and cheeses) with abnormal mold should be discarded.[37]

The Vehicle or Means by which Water- and Foodborne Diseases are Spread

The means by which the water- and foodborne disease agents are transmitted to individuals include water, milk, milk products, and other foods. These vehicles are not, however, the only methods by which diseases are spread. Some, as previously discussed, are also spread through the air, some by contact with persons who are ill, some by insects, and others by hands or articles soiled with infectious discharges. This discussion will cover the role of water, milk, milk products, and food as bearers or carriers of disease-producing organisms and poisonous substances. The lack of water for washing also contributes to poor personal hygiene and sanitation and to disease spread.

The reporting of water- and foodborne illnesses has, with rare exceptions, been very incomplete. Various estimates have been made in the past indicating that the number reported represented only 10 to 20 percent of the actual number.

[34]Benjamin J. Wilson, "Hazards of Mycotoxins to Public Health," *J. Food Prot.*, May 1978, pp. 375–384.

[35]"Microbiological Aspects of Food Hygiene," *WHO Tech. Rep. Ser.*, **598** (1976).

[36]L. B. Bullerman, "Significance of Mycotoxins to Food Safety and Human Health," *J. Food Prot.*, January 1979, pp. 65–86.

[37]Jane Heeman, "Please Don't Eat the Mold," *FDA Consumer*, Washington, D.C., November 1974.

Hauschild and Bryan, in an attempt to establish a better basis for estimating the number of people affected, compared the number of cases initially reported with the number of cases identified by thorough epidemiological investigations or to the number estimated. They found that for 51 outbreaks of bacterial, viral, and parasitic disease (excluding milk), the median ratio of estimated cases to cases initially reported to the local health authority, or known at the time an investigating team arrived on the scene, was 25 to 1. On this basis and other data, the annual food- and waterborne disease cases for 1974 to 1975 were estimated to be 1,400,000 to 3,400,000 in the United States and 150,000 to 300,000 in Canada. The annual estimate for the United States for 1967 to 1976 was 1,100,000 to 2,600,000.[38] The authors acknowledge that the method used to arrive at the estimates is open to criticism. However, it is believed that the estimates come closer to reality than the present Center for Disease Control reporting would indicate, particularly to the nonprofessional. The estimates would also serve as a truer basis for justifying regulatory and industry program expenditures for water- and foodborne illness prevention, including research and quality control.

Waterborne Disease Outbreaks

The number of waterborne disease outbreaks has been summarized periodically.[39–42] In the United States they averaged 45 per year for the years 1938–1940, 38 for 1941–1945, 23 for 1946–1950, 10 for 1951–1955, 12 for 1956–1960, 11 for 1961–1965, 14 for 1966–1970, 26 for 1971–1975, 35 for 1976, 34 for 1977, and 32 for 1978. The apparent increase in the number of outbreaks is probably due to better reporting rather than an actual relative increase in number.

Weibel et al. studied the incidence of waterborne disease in the United States for 1946–1960.[42] They reported 22 outbreaks (10%) with 826 cases due to use of untreated surface water; 95 outbreaks (42%) with 8811 cases due to untreated groundwater; 3 outbreaks (1%) with 189 cases due to contamination of reservoir or cistern; 35 outbreaks (15%) with 10,770 cases due to inadequate control of treatment; 38 outbreaks (17%) with 3344 cases due to contamination of distribution system; 7 outbreaks (3%) with 1194 cases due to contamination of collection or conduit system; and 28 outbreaks (12%) with 850 cases due to miscellaneous causes, for a total of 228 outbreaks with 25,984 cases.

[38,39] A. H. W. Hauschild and Frank L. Bryan, "Estimate of Cases of Food- and Waterbone Illness in Canada and the United States," *J. Food Prot.*, June 1980, pp. 435–440.

[40] *Water-Related Diseases Surveillance Annual Summary 1978*, HHS Publ. No. (CDC) 80-8385, HHR, PHS, CDC, Atlanta, Ga. 30333, p. 17.

[41] A. Wolman and A. E. Gorman, "Waterborne Typhoid Fever Still a Menace," *Am. J. Pub. Health*, **21,** 2, 115–129 (February 1931); *Water-Borne Outbreaks in the United States and Canada 1930–36 and Their Significance*, Eighth Annual yearbook, Am. Pub. Health Assoc., 1937–1938, p. 142; Gainey and Lord, *Microbiology of Water and Sewage*, Prentice-Hall, Englewood Cliffs, N.J., 1952.

[42] S. R. Weibel, F. R. Dixon, R. B. Weidner, and L. J. McCabe, "Waterborne Disease Outbreaks, 1946–60," *J. Am. Water Works Assoc.*, **56,** 947–958 (August 1964) (as revised).

Wolman and Gorman showed that the greatest number of waterborne diseases occurred among population groups of 1000 and under and among groups from 1000 to 5000, that is, predominantly in the rural communities.[41] Weibel reported the greatest number of outbreaks and cases in communities of 10,000 population or less. The need for emphasis on water control and sewage disposal at small existing and new communities, as well as at institutions, resorts, and rural places, is apparent and was again confirmed in the 1970 PHS study,[43] and in a 1978 summary.[40] Between 1971 and 1978, 58 percent of the outbreaks occurred at small noncommunity water systems. These figures show what can happen when there is a breakdown in the water distribution system, treatment plant, or water source protection. See also Chapter 3.

Drinking water contaminated with sewage is the principal cause of waterborne diseases. The diseases that usually come to mind in this connection are typhoid fever, dysentery, gastroenteritis, infectious hepatitis (hepatitis A), and giardiasis. However, because of the supervision given public water supplies and control over a lessening number of typhoid carriers, the incidence of typhoid fever has been reduced to a low residual level. Occasional outbreaks, due mostly to carriers, remind us that the disease is still a potential danger.

In 1940 some 35,000 cases of gastroenteritis and six cases of typhoid fever resulted when about five million gallons of untreated, grossly polluted Genesee River water was accidentally pumped into the Rochester, New York, public water supply distribution system. A valved cross-connection between the public water supply and the polluted Genessee River fire-fighting supply had been unintentionally opened. In order to maintain the proper high pressure in the fire supply, the fire pumps were placed in operation and hence river water entered the potable public water supply system. The check valve was also inoperative.

At Manteno State Hospital in Illinois, 453 cases of typhoid fever were reported resulting in 60 deaths in 1939.[44] It was demonstrated by dye and salt tests that sewage from the leaking vitrified clay tile hospital sewer line passing within a few feet of the drilled well-water supply seeped into the well. The hospital water supply consisted of four wells drilled in creviced limestone. The state sanitary engineer had previously called the hospital administrator's attention to the dangerously close well location to the sewer and made several very strong recommendations over a period of eight years; but his warning went unheeded until after the outbreak. Indictment was brought against three officials, but only the director of the Department of Public Welfare was brought to trial. Although the county court found the director guilty of omission of duty, the Illinois Supreme Court later reversed the decision.

An explosive epidemic of infectious hepatitis in Delhi, India, started during the first week of December 1955 and lasted about six weeks. A sample survey

[43]Leland J. McCabe, James M. Symons, Roger D. Lee, and Gordon G. Robeck, "Survey of Community Water Systems," *J. Am. Water Works Assoc.*, November 1970, pp. 670–678.

[44]Illinois Department of Public Health, "A Report on a Typhoid Fever Epidemic at Manteno State Hospital in 1939," *J. Am. Water Works Assoc.*, **38**, 1315–1316 (November 1946).

showed about 29,300 cases of jaundice in a total population of 1,700,000. (The authorities estimated the total number of infections at 1,000,000.) No undue incidence of typhoid or dysentery occurred. Water was treated in a conventional rapid sand filtration plant; but raw water may have contained as much as 50 percent sewage. Inadequate chlorination (combined chlorine), apathetic operation control, and poor administration apparently contributed to the cause of the outbreak, although the treated water was reported to be well clarified and bacteriologically satisfactory.[45]

An outbreak of gastroenteritis in Riverside, California, affected an estimated 18,000 persons in a population of 130,000. Epidemiological investigation showed that all patients were carriers of *Salmonella typhimurium*, serological type B and phage type II. The water supply was implicated. There was no evidence of coliform bacteria in the distribution system, although 5 of 75 water samples were found positive for *S. typhimurium*, type B, phage II. The cause was not found in spite of an extensive investigation.[46]

In 1974–75 a waterborne outbreak of giardiasis occurred in Rome, N.Y. About 5357 persons out of a population of 46,000 were affected. The source of water was an upland surface supply receiving only chlorine-ammonia treatment. The coliform history was generally satisfactory. Giardiasis outbreaks have also been reported in Grand County and near Estes Park, Colorado; in Camas, Washington in 1976[47,48]; in Portland, Oregon; in the Unita Mountains of Utah[49]; in Berlin, New Hampshire in 1976[50]; in California, Pennsylvania,[51] and Baffin Island, Canada. Between 1969 and 1976 a total of 18 outbreaks with 6198 cases were reported. An additional five outbreaks with approximately 1000 cases are estimated for 1977. Acceptable turbidity and coliform tests do not assure absence of giardia or entamoeba.

The reporting of outbreaks of waterborne giardiasis is becoming more common in the United States. The source of *Giardia lamblia* cyst is man, and possibly the beaver, muskrat, and domestic animals probably infected from man's waste. The cyst is resistant to normal chlorination, similar to the cyst of *Entamoeba histolytica*. Conventional rapid sand filtration of surface water including coagulation, sedimentation, sand filtration, and disinfection is considered effective in removing the giardia cyst. Pressure sand filtration is not

[45]Joseph M. Dennis, "1955–56 Infectious Hepatitis Epidemic in Delhi, India," *J. Am. Water Works Assoc.*, **51,** 1288–1298 (October 1959).

[46]Everett C. Ross and Howard L. Creason, "The Riverside Epidemic," *Water Sewage Works*, **113,** 128–132 (April 1966).

[47]J. C. Kirner, J. D. Littler, and L. A. Angelo, "A Waterborne Outbreak of Giardiasis in Camas, Wash.," *J. Am. Water Works Assoc.*, January 1978, pp. 35–40.

[48]"Waterborne Giardiasis Outbreaks—Washington and New Hampshire," *MMWR*, CDC, Atlanta, Ga., May 27, 1977, p. 169.

[49]Gunther F. Craun, "Waterborne Outbreaks," *J. Water Pollut. Control Fed.*, June 1977, pp. 1268–1279.

[50]Edwin C. Lippy, "Tracing a Giardiasis Outbreak at Berlin, New Hampshire," *J. Am. Water Works Assoc.*, September 1978, pp. 512–520.

[51]"Waterborne Giardiasis—California, Colorado, Oregon, Pennsylvania," *MMWR*, CDC, Atlanta, Ga., March 21, 1980, pp. 121–123.

reliable and should not be used, as the cyst penetrates the filter. Superchlorination, as for amebic cysts, is believed adequate to kill giardia cysts, as is bringing drinking water to a boil. See also Table 3-16.

Legionnaires' disease, caused by *Legionella pneumophila*, affected a large number of convention-goers in Philadelphia. The organism was isolated from water taken at the site from cooling towers and evaporative condensers. The air-conditioning units were implicated in the dissemination of the organism. A quaternary ammonium compound and calcium hypochlorite appeared to be effective in inactivating *L. pneumophila*, but their efficacy in inhibiting growth in cooling tower and evaporative condenser water and preventing transmission of the organism remains to be demonstrated.[52] Other incidents of the disease have been reported since the organism was first discovered. The organism is believed to be ubiquitous.

Foodborne Disease Outbreaks

Raw or improperly pasteurized milk, poor milk-handling practices, and carriers, especially contaminated food and improper refrigeration, have been the principal causes of the milkborne diseases.

Fuchs states that from 1923 to 1937, inclusive, 639 milkborne disease outbreaks were reported, involving 25,863 cases and 709 deaths.[53] In other words, there was an average of 43 outbreaks involving 1724 cases and 47 deaths reported each year. Between 1938 and 1956 an average of 24 milkborne disease outbreaks per year, with 980 cases and 5 deaths, were reported to the PHS. Between 1957 and 1960 the outbreaks averaged 9, and the cases, 151 per year. There were no deaths reported between 1949 and 1960. The 1978 PHS/FDA reported that milk and fluid milk products were associated with less than one percent of all foodborne disease outbreaks.[54] The dramatic decrease in the number of outbreaks, cases, and deaths is due to better equipment and more effective control over the pasteurization of milk and milk products. However illnesses can recur if controls are relaxed, particularly if the sale of raw or improperly pasteurized milk is permitted.

Raw milk was incriminated in an outbreak of gastroenteritis involving over 500 persons in Australia in February 1976.[55] *Salmonella typhimurium* was isolated from 78 of the 273 persons investigated. Two of the cows and one of the employees were found to be excreting the same phage type *S. typhimurium*. Pasteurized milk was suspected as the vehicle of transmission in the United States in 1976; one was caused by *Shigella flexneri* and the other was caused

[52]"Preliminary Studies on Environmental Decontamination of *Legionella pneumophila*," *MMWR*, CDC, Atlanta, Ga., June 22, 1979, pp. 286–287.

[53]A. W. Fuchs, address before Philadelphia College of Physicians, February 5, 1940.

[54]Grade A Pasteurized Milk Ordinance, 1978 Recommendation, Public Health Service/FDA, DHEW, Washington, D.C., p. 5.

[55]"Outbreak of Milkborne Salmonella Gastroenteritis—South Australia," *MMWR*, CDC, Atlanta, Ga., April, 1977, p. 127.

by *S. newport*. Several salmonella outbreaks have been traced to certified raw milk,[56] and in 1976, an outbreak of yersiniosis was caused by milk to which chocolate syrup had been added after pasteurization. Three salmonella episodes of milkborne infection caused by the consumption of raw milk were reported in England during 1974 and 1975.[57] *Salmonella typhimurium* was associated with post contamination of pasteurized milk in Arizona in 1978.[58] Salmonellosis has also been associated with the consumption of nonfat powdered milk.[59]* More than half (57 percent) of the reported cases of brucellosis from 1967 to 1978 were in abattoir workers.[60]

Whereas milkborne diseases have been brought under control, foodborne illnesses remain unnecessarily high. Between 1938 and 1956 there were reported 4647 outbreaks, 179,773 cases, and 439 deaths. In 1967, 273 outbreaks were reported, with 22,171 cases and 15 deaths. The bacteria associated with foodborne illnesses are commonly *Staphylococcus aureus*, Salmonellae, and *Clostridium perfringens*. *Vibrio parahaemolyticus* was the most significant agent in Japan. Banquets accounted for over half of the illnesses reported; schools and restaurants made up most of the rest. The largest number of outbreaks occurred in the home.[61]

Another analysis of foodborne illnesses based on 1969 and 1970 CDC, DHEW information reported on 737 outbreaks with 52,011 cases. It was found that 33.0 percent of the outbreaks occurred at restaurants, cafeterias and delicatessens; 39.1 percent occurred at homes; 8.7 percent occurred at schools; 5.2 percent at camps, churches, and picnics; and 14 percent at other places. However, 48 percent of the *cases* were at schools, and 28 percent at restaurants, cafeterias and delicatessens.[62]

Bryan,[63] in a summary of foodborne diseases in the United Staes from 1969 to 1973, reported 1665 outbreaks with 92,465 cases. During this same period it was found that food service establishments accounted for 35.2 percent of the outbreaks; homes accounted for 16.5 percent of the outbreaks; food processing establishments accounted for 6.0 percent of the outbreaks, and 42.1 percent of the outbreaks occurred at unknown places.

*Present technology cannot produce raw milk (including that listed as certified) that can be assured to be free of pathogens; only with pasteurization is there this assurance. (MMWR, CDC, Atlanta, Ga., March 6, 1981, p. 97.)

[56] "CDC Salmonella Surveillance," December Report, CDC, Atlanta, Ga., 1964.

[57] "Milkborne Salmonella Infection—United Kingdom," *MMWR*, CDC, Atlanta, Ga., July 2, 1976, p. 203.

[58] "Salmonella Gastroenteritis Associated with Milk—Arizona," *MMWR*, CDC, Atlanta, Ga., March 16, 1978, p. 117.

[59] "Salmonellosis Associated with Consumption of Nonfat Powdered Milk—Oregon," *MMWR*, CDC, Atlanta, Ga., March 23, 1979, p. 129.

[60] "Brucellosis—United States, 1978," *MMWR*, CDC, Atlanta, Ga., September 21, 1979, p. 438.

[61] William E. Woodward, et al., "Foodborne Disease Surveillance in the United States, 1966 and 1967," *Am. J. Pub. Health*, **60**, 130–137 (January 1970).

[62] Madean Horner, "Safe Handling of Foods in the Home," *FDA Papers*, June 1972, DHEW Publication (FDA) 73-2002.

[63] Frank L. Bryan, "Status of Foodborne Diseases in the United Staes," *J. Environ. Health*, September/October 1975, pp. 74–83.

Todd,[64] in an analysis of foodborne diseases, reported 1440 outbreaks with 14,573 cases in Canada (1973–1975), 1199 outbreaks with 49,214 cases in the U.S.A. (1973–1975), 1957 outbreaks with 14,246 cases in England and Wales (1973–1975), 6109 outbreaks with 182,900 cases in Japan (1968–1972), and 48 outbreaks with 2500 cases in Australia (1967–1971). The numbers are considered more a reflection of the efficiency of reporting rather than an indication of the problem in each country.[38] Foodborne disease is a worldwide problem.

Between 1963 and 1975 there were 651 reported outbreaks of salmonella with 38,811 cases. Poultry, meat (beef, pork), and eggs were the three most common vehicles. Eggs were not incriminated in 1974 and 1975, probably due to hygienic processing, pasteurization,* and quality control.[65] Gangarosa reported that 23,300 cases of salmonella food poisoning (infection) were reported to the Communicable Disease Center in 1976 but that the actual number was about 2.5 million Americans.[66] Hauschild and Bryan found that for a total of 26 outbreaks of salmonellosis, the median ratio of estimated cases to initial human isolations of salmonella was 29.5. On this basis the actual number of cases of human salmonellosis for the period from 1969 to 1978 was estimated to be 740,000 in the United States and 150,000 in Canada annually.[38] Although estimates differ they do show the seriousness of the problem and the need for more effective control methods. The overall national salmonellosis morbidity has remained relatively constant.[67] The average number of isolates has actually increased.[68]

The Control and Prevention of Water- and Foodborne Diseases

Design, operation, and control measures dealing with food protection are covered in Chapter 8. Water system control and details are covered in Chapter 3.

Many health departments, particularly on a local level, are placing greater emphasis on water quality and food protection at food-processing establishments, catering places, schools, restaurants, institutions, and the home, and on the training of food management and staff personnel. An educated and observant public, a systematic inspection program with established management responsibility, coupled with a selective water and food quality laboratory surveillance system, and program evaluation, can help greatly in making health department food protection programs more effective.

*Pasteurization of liquid whole egg at 140°F (60.0°C) for 3.5 min. and salted egg products at 146°F (63.3°C) for 3.5 min. is required by USDA.

[64] Ewen C. D. Todd, "Foodborne Disease in Six Countries—A Comparison," *J. Food Prot.*, July 1978, pp. 559–565.

[65] Mitchell L. Cohen and Paul A. Blake, "Trends in Foodborne Salmonellosis Outbreaks: 1963–1975," *J. Food Prot.*, November 1977, pp. 798–800.

[66] "Salmonella Still Problem in the United States," Washington (AP), Jan. 12, 1978.

[67] John H. Silliker, "Status of salmonella—Ten Years Later," *J. Food Prot.*, April 1980, pp. 307–313.

[68] *MMWR, Annual Summary 1977*, CDC, Atlanta, Ga.

Prevention of Foodborne Diseases

The application of known and well established sanitary principles has been effective in keeping foodborne diseases under control, but it is apparent that more effective measures are needed. Refrigeration, hygienic practices, food preparation planning, hot or cold holding of potentially hazardous food, proper cooking, and general sanitation are most important. These precautions apply as well to prepared frozen dinners, reconstituted foods and drinks. See Figure 1-3.

The approximate optimal temperature for growth of the principal organisms associated with foodborne illnesses are: salmonella 99°F (maximal 114°F),

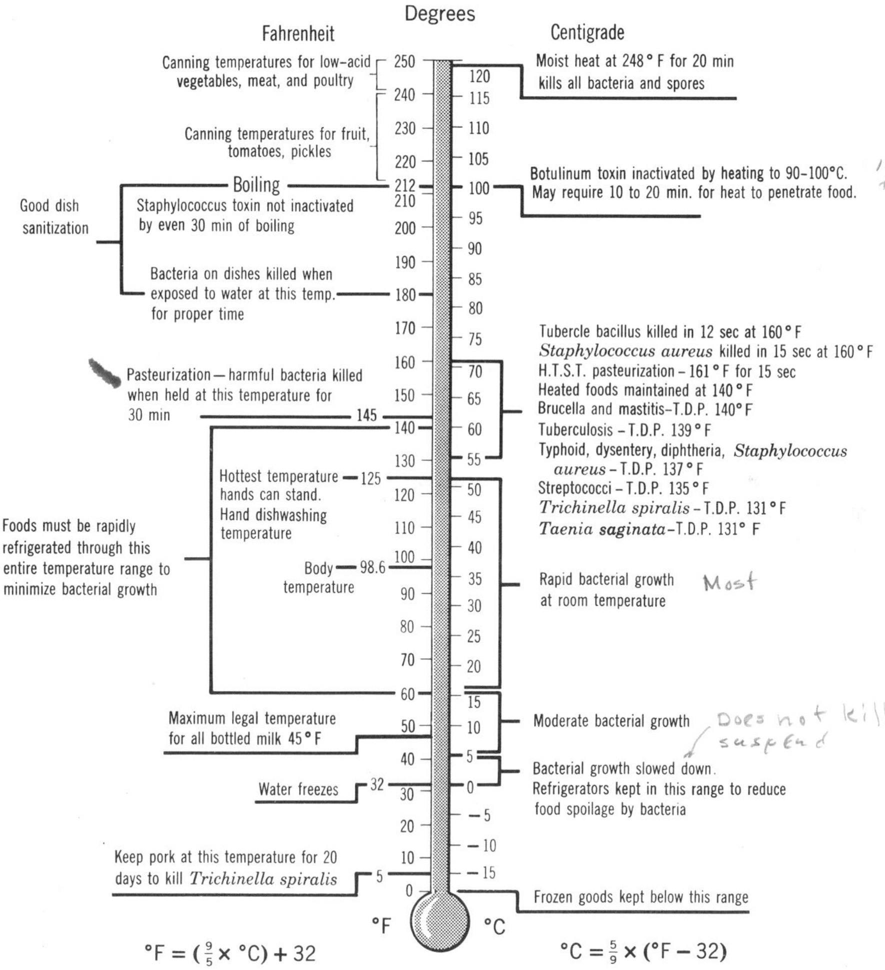

$$°F = (\tfrac{9}{5} \times °C) + 32 \qquad °C = \tfrac{5}{9} \times (°F - 32)$$

Note: T.D.P. = thermal death point temperature that will completely destroy a test culture in the presence of moisture in 10 min.
H.T.S.T. = high temperature, short time.

Figure 1-3 Food sanitation temperature chart.

Staphylococcus aureus 99°F (maximal 114°F), *Clostridium perfringens* 115°F (maximal 112°F), and enterococci (maximal 126°F).[69]

Salmonellae are widely distributed in nature and found in many raw food products, especially poultry and beef. Tables and surfaces used in preparing raw poultry and other meats can serve as vehicles for the spread of salmonellae unless they are thoroughly cleaned and sanitized between each use. *Clostridium perfringens* and *Staphylococcus aureus* are also frequently found in samples of raw beef, and on workers' hands, knives, and cutting boards.

All cooked and precooked beef and beef roasts must be heated to a minimum internal temperature of 145°F (62.7°C) to comply with U.S. Department of Agriculture (USDA) regulations to assure destruction of all salmonellae. At this temperature it would not be possible to make available "rare" roast beef. However, USDA permits other time-temperature relations for processing of water or steam cooked and dry roasted beef.[70] Study shows that salmonella-free "rare" roast beef can be produced for example at internal temperature-times ranging from 130°F for 121 minutes to 136°F for 32 minutes. The elimination of salmonella from the surface of dry oven-roasted beef (at least 10 pounds uncooked in size) requires a minimum internal temperature of 130°F (54.4°C) in an oven set at 250°F (121.0°C) or above.[71]

Cooked beef roasts and turkeys, because of their size, are rarely rapidly cooled to 45°F or less. If not consumed or sold immediately, they should be reheated as noted before use. Cooked roasts that have been rolled or punctured should be reheated to 160°F (71.1°C). Cooked roasts that have been cut up into small pieces should be reheated to 165°F (73.9°C) because the handling introduces greater possibility of contamination. Cooked roasts that are solid muscle should be reheated to assure pasteurization of the surface of the roast.[72]

There is a danger of cooking large masses of raw meat on the outside but leaving the interior of the food underdone, thereby permitting survival of salmonellae,[73] spores introduced in handling, or those intrinsically present that can germinate and cause *Clostridium perfringens* food poisoning.[74] However, if the meat is cooked as noted above and eaten immediately after cooking, there is usually no risk of bacterial foodborne illness.[72]

Food poisoning may also result from the presence of a bacterial toxin in the food. Certain specific strains of staphylococcus (*Staphylococcus aureus*) commonly found in skin infections and in discharges from the nose and throat are

[69]Frank L. Bryan, "Impact of Foodborne Disease and Methods of Evaluating Control Programs," *J. Environ. Health*, May/June 1978, pp. 315–323.

[70]*Fed. Reg.*, 43:30793, July 8, 1978.

[71]S. J. Goodfellow and W. L. Brown, *J. Food Prot.*, August 1978, pp. 598–605.

[72]Frank L. Bryan and Thomas W. McKinley, "Hazard Analysis and Control of Roast Beef Preparation in Foodservice Establishments," *J. Food Prot*, January 1979, pp. 4–18.

[73]"Salmonellae in Precooked Roasts of Beef—New York," *MMWR*, CDC, Atlanta, Ga., August 25, 1978, p. 315; also *MMWR*, August 26, 1977, p. 277.

[74]M. Ingram, "Meat Preservation—past, present and future," *J. R. Soc. Health*, June 1972, pp. 121–130.

frequently associated with food poisoning. Such staphylococci multiply under favorable conditions and produce a toxin that is stable at boiling temperature. The consumption of food containing sufficient toxin, therefore, even after heating, may cause food poisoning. The botulinus bacillus in improperly canned or bottled low-acid food will also produce a toxin, but this poison is destroyed by boiling (three minutes) and cooking. The botulinus bacillus is rarely found in commercially canned foods but may be of some significance in home canned foods.

Incomplete cooking of stews, contaminated meat, and large cuts of meat that have been rolled or penetrated with skewers, and failure to provide prompt and thorough refrigeration can lead to infection with *Clostridium perfringens*. *C. perfringens* spores are not completely destroyed by normal cooking. Therefore, foods contaminated with spores which are cooked and not promptly refrigerated can permit the germination of spores and the multiplication of vegetative cells with the danger of food poisoning on consumption. Heating *C. perfringens* enterotoxin at 60°C in cooked turkey showed a gradual decrease in serological activity with no detectable toxin being present after 80 minutes.[75]

The essential elements of health protection in food establishments are:[76]

1. Adequate refrigeration equipment and prompt and proper refrigeration at 45°F (7°C) or less of perishable and prepared foods.
2. Cooking to proper (minimum) internal temperature of pork (150°F, 66°C), beef roasts (145°F, 63°C),* poultry and all stuffed meats (165°F, 74°C), holding of hot foods at 140°F (60°C) and thorough reheating to 165°F of precooked (refrigerated) potentially hazardous foods,† and holding at or above 140°F or refrigerating at 45°F until served, heating of custard and pastry filling to 165°F and cold holding at 45°F. Stock should be brought to a boil and kept at 140°F or above. Leftover food that has been served should not be reused.
3. Planning of food preparation to coincide as closely as possible with serving time.
4. Use of wholesome food and food ingredients; purchase and use of shellfish from approved safe sources.
5. Cleanliness and good habits of personal hygiene in employees, who should be free from communicable disease or infection transmissible through food or food service; minimum handling of food.
6. Clean dishware, utensils, equipment, and surfaces for food preparation; adequate properly constructed equipment that is easily cleaned and sanitized, and is kept clean. Avoid cross-contamination.
7. An adequate supply of potable water (hot and cold), detergents, and equipment for cleaning and sanitization of dishes and utensils; elimination of cross-connections or conditions that

*See text for other time-temperature ranges.

†Potentially hazardous foods are foods (including dressings) consisting in whole or in part of milk or milk products, eggs, meat, poultry, fish, shellfish or edible crustacea, or other ingredients, including synthetic ingredients in a form capable of supporting rapid and progressive growth of infectious or toxigenic microorganisms. The term does not include clean, whole, uncracked, odor-free eggs or foods which have a pH level of 4.6 or below or a water activity (a_w) value of 0.85 or less. (*Food Service Sanitation Manual*, 1976, PHS/FDA, Washington, D.C., p. 21.)

[75]H. S. Naik and C. L. Duncan, "Thermal Inactivation of *Clostridium perfringens* Enterotoxin," *J. Food Prot.*, February 1978, pp. 100–103.

[76]J. A. Salvato, *Guide to Sanitation In Tourist Establishments*, 1976, p. 75.

may permit backflow or backsiphonage of polluted or suspect water into the water supply piping or equipment.

8. Proper storage and disposal of all liquid and solid wastes.
9. Control of rodents, flies, cockroaches, and other vermin and proper use and storage of pesticides, sanitizers, detergents, solvents, and other toxic chemicals.
10. Protection of dry food stores from flooding, sewage backup, drippage, and rodent and insect depredations.
11. Structurally sound, clean facilities in good repair, and adequately lighted and ventilated premises that can be properly cleaned.

Open self-service food counters or buffets require a physical barrier such as a canopy or guard which will effectively prevent contamination by persons assisting themselves to the displayed food. In any case, the potentially hazardous food should be held either at or above 140°F or at or below 45°F at all times. Such foods remaining should not be reused. See Chapter 8.

Sandwiches containing potentially hazardous food which remain unrefrigerated for more than two or three hours at room temperature can cause nausea, vomiting, and diarrhea. Prior refrigeration, or freezing where appropriate, and consumption within four hours will minimize the hazard. Cheese, peanut butter, and jelly sandwiches, and hard-boiled eggs will keep better. Commercial mayonnaise (pH below 4.1 to 4.6) will inhibit the surface growth of salmonellae and staphylococci on food, but the pH of all the ingredients or mass of the food, such as egg, meat, or potato salad, must be reduced to inhibit bacterial growth. Vinegar can accomplish the same objective, provided the food ingredients do not neutralize the acidity of the mixture. Hygienic food preparation practices, proper cooking, and *prompt refrigeration*, if the food is not immediately consumed, should be the guiding principles.

If cooked foods are to be kept for later use, they must be refrigerated immediately without preliminary cooling at room temperature. And if the food, such as a roast or other food in a container, is greater than 4 in. in diameter, thickness, or depth, it should be reduced to 4 in. or less to permit proper cooling and to prevent the rapid growth of microorganisms and the production of toxins.

Prevention of Waterborne Diseases

A primary requisite for the prevention of waterborne disease is the ready availability of an adequate supply of water that is of satisfactory sanitary quality—microbiological, chemical, physical, and radiological. It is important that the water be attractive and palatable to induce its use, for otherwise, water of doubtful quality from some nearby unprotected stream, well, or spring may be used. Where a municipal supply is available it should be used, as such supplies are usually of satisfactory quality, ample in quantity, and under competent supervision. However, this is not always the case. Because of the excellent water service generally available in the United States, and in many developed areas of the world, the people and public officials have tended to become complacent and take for granted their water supply. As a result, in some instances, funds have been diverted to other more popular causes rather than to the maintenance, operation, and upgrading of the water supply system. Safeguards to protect and maintain the integrity of a public water supply system are necessary in such instances.

Adequate drinking water statutes and regulations and surveillance of public water supply systems are necessary for their regulatory control. This is usually a state responsibility which may be shared with local health or environmental regulatory agencies. USEPA recommendations for a minimum state program include the following.[77,78]

1. A drinking water statute should define the scope of state authority and responsibility with specific statutory regulations and compliance requirements. Regulations should be adopted for drinking water quality standards; water supply facility design and construction criteria; submission, review, and approval of preliminary engineering studies and detailed plans and specifications; approval of a water supply source and treatment requirements; establishment of a well construction and pump installation code; operator certification; provision for state laboratory services; and cross-connection and plumbing control regulations.
2. The surveillance of public water supply systems should involve water quality sampling—bacteriological, chemical, and radiological also turbidity and residual chlorine; supervision of operation, maintenance, and use of approved state, utility, and private laboratory services; cross-connection control; bottled and bulk water safety.
3. Surveillance and disease prevention is recommended with periodic, onsite fact finding as part of a comprehensive sanitary survey of each public water supply system, from the source to the consumer's tap, made by a qualified person to evaluate the ability of the water supply system to *continuously* produce an adequate supply of water of satisfactory sanitary quality. The qualified person may be a professionally trained public health, sanitary, or environmental engineer, or a sanitarian, to make sanitary surveys of the less complex water systems such as well-water supplies. The USEPA suggests that the sanitary survey, as a minimum, cover quality and quantity of the source; protection of the source (including the watershed drainage area); adequacy of the treatment facilities; adequacy of operation and operator certification; distribution storage; distribution system pressure; chlorine residual in the distribution system; water quality control tests and records; cross-connection control; and plans to supply water in an emergency. The WHO has similar suggestions.[79]

Details concerning water supply quality and quantity, source protection, design, and treatment are given in Chapter 3.

Schistosomiasis

The global prevalence of schistosomiasis, estimated at 200 million cases, is expected to increase as new impoundments and irrigation canal systems are built, if known preventive precautions are not taken. Cooperation in the planning through the construction phases in endemic areas, or potentially endemic areas, between the health and water resources agencies can reverse the trend.

[77] *A Manual For the Evaluation of a State Drinking Water Supply Program*, USEPA, Water Supply Division, EPA-430/9-74-009, 1974.
[78] *Manual For Evaluating Public Water Supplies*, USEPA, Office of Water Programs, PHS Pub. 1820, Reprinted 1971, GPO, Washington, D.C., 20402.
[79] "Surveillance of Drinking Water Quality," *WHO Monogr. Ser.*, **63,** 1976.

Long-term schistosomiasis control would involve an appropriate combination of chemotherapy; molluscicding; basic sanitation including biological intervention and the supply of potable water at the village level; and socioeconomic development.[80] See also Chapter 9.

INSECTBORNE DISEASES AND ZOONOSES

General

The diseases transmitted by arthropods, commonly known as insectborne disseases, are those diseases that are usually transmitted by biting insects from man to man or from animal to man. The ordinary housefly which acts as a mechanical carrier of many disease agents, is discussed separately in Chapter 10. Zoonoses are defined as "those diseases and infections which are naturally transmitted between vertebrate animals and man."[81]

Insectborne Diseases

A list of insectborne diseases together with their important reservoirs is given in Tables 1-8, 1-9, and 1-10. The list is not complete but includes some of the common as well as less known diseases. Tick- or fleaborne diseases may be spread directly by the bite of the tick or flea and indirectly by crushing the insect into the wound made by the bite. Usually mosquitoes, lice, ticks, and other blood-sucking insects spread disease from man to man, or animal to man, by biting a person or animal carrying the disease-causing organisms. By taking blood containing the disease-producing organisms, the insect is in a position to transmit the disease organism when biting another person or animal. Insect and other arthropod control is discussed in Chapter 10.

Since many diseases are known by more than one name, other nomenclature is given to avoid confusion. As time goes on and more information is assembled, there will undoubtedly be greater standardization of terminology. In some cases there is a distinction implied in the different names that are given to very similar diseases. The names by which the same or similar diseases are referred to are presented below.

> Bartonellosis includes oroya fever, Carrion's disease, and verruga peruana. Dengue is also called dandy fever, breakbone fever, bouquet, solar, or sellar fever. Endemic typhus, fleaborne typhus, and murine typhus are synonymous. Epidemic typhus is louseborne typhus, also known as classical, European, and Old-World typhus. Brill's disease is probably epidemic typhus. Plague, black death, bubonic plague, and flea-

[80] "Epidemiology and control of schistosomiasis," *WHO Tech. Rep. Ser.*, **643,** 1980.

[81] *WHO Tech. Rep. Ser.*, **169,** 6 (1959). *WHO Tech. Rep. Ser.*, **378,** 6 (1967) considers the definition too wide but recommends no change.

Table 1-8 Some Exotic Insectborne Diseases (Not Normally Found in the United States)

Disease	Incubation Period	Reservoir	Vector
Bartonellosis	16 to 22 days	Man	Sandflies (*Phlebotomus*)
Leishmaniasis, cutaneous	Days to months	Animals, dogs	Sandflies (*Phlebotomus*)
Leishmaniasis, visceral	2 to 4 months	Man, dogs, cats, wild rodents	Sandflies (*Phlebotomus*)
Loiasis (Loa loa)	Years	Man	Chrysops-blood-sucking flies
Onchocerciasis	1 year or more	Man, gorillas	Blackflies (*Simulium*)
Sandfly fever (Phlebotomus fever)	3 to 4 days	Man, sandfly	Sandfly (*Phlebotomus*)
Relapsing fever	5 to 15 days	Man, ticks, rodents	Lice, crushed in wound; ticks
Trench fever	7 to 30 days	Man	Lice, crushed in wound (*Pediculus humanus*)

Reference: *Control of Communicable Diseases in Man*, op. cit.

borne pneumonic plague are the same. Filariasis or mumu, an infestation of *Wuchereria bancrofti*, may after obstruction of the lymph channels cause elephantiasis. Loa loa and loiasis are the same filarial infection. Cutaneous leishmaniasis, espundia, uta, bubas and forest yaws, aleppo, Baghdad or Delhi boil, chiclero ulcer, and oriental sore are synonyms. Visceral leishmaniasis is also known as kala azar. Malaria, marsh miasma, remittent fever, intermittent fever, ague, and jungle fever are synonymous. Blackwater fever is believed to be associated with malaria. Onchocerciasis is also known as blinding filarial disease. Sandfly fever is the same as phlebotomus fever, three-day fever, and pappataci fever. Q fever is also known as nine-mile fever. Febris recurrens, spirochaetosis, spirillum fever, famine fever, and tick fever are terms used to designate relapsing fever. Rocky Mountain spotted fever, tick fever of the Rocky Mountains, tick typhus, black fever, and blue disease are the same. Tsutsugamushi disease, Japanese river fever, scrub typhus, and miteborne typhus are used synonymously. Trench fever is also known as five-day fever, Meuse fever, Wolhynian fever, and skin fever. Plague-like diseases of rodents, deer-fly fever, and rabbit fever are some of the other terms used when referring to tularemia. Other forms of arthropodborne infectious encephalitis in the United States are the St. Louis type, the Eastern equine type, and the Western equine type; still other types are known. Nasal myiasis, aural myiasis, ocular myiasis or myiases, cutaneous myiases, and intestinal myiases are different forms of the same disease. Sleeping sickness, South American sleeping sickness, African sleeping sickness, Chagas' disease, Negro lethargy, and trypanosomiasis are similar diseases caused by different species of trypanosomes. Tick-bite fever is also known as Boutonneuse fever, Tobia fever, and Marseilles fever; Kenya typhus and South African tick fever are related. Scabies, "the itch," and the "seven-year itch" are the same disease.

The complete elimination of rodents and arthropods associated with disease has been a practical impossibility. Man, arthropods, and rodents, therefore, offer ready foci for the spread of infection unless controlled. Rodent and arthropod control is discussed in Chapter 10.

Table 1-9 Characteristics of Some Insectborne Diseases

Disease	Etiologic Agent	Reservoir	Transmission	Incubation Period	Control[a]
Endemic typhus (Murine) (Fleaborne)[b]	*Rickettsia, typhi* (*R. mouseri*) also possibly *Ctenocephalides felis*	Infected rodents, *Rattus rattus* and *Rattus norvegicus*, also fleas, possibly opossums	Bite or feces of rat flea *Xenopsylla cheopis*. Also possibly ingestion or inhalation of dust contaminated with flea feces or urine.	7 to 14 days, usually 12	First, elimination of rat flea by insecticide applied to rat runs, burrows, and harborages, then rat control. Spray kennels, beds, floor cracks.
Epidemic typhus (Louseborne)	*Rickettsia prowazeki*	Infected persons and infected lice	Crushing infected body lice *Pediculus humanus* or feces into bite, abrasions, or eyes. Possibly louse feces in dust.	7 to 14 days, usually 12	Insecticidal treatment of clothing and bedding; personal hygiene, bathing, elimination of overcrowding. Immunization. Delousing of individuals in outbreaks.
Bubonic plague	*Pasteurella pestis*, plague bacillus (*yersinia pestis*)	Wild rodents and infected fleas	Bite of infective flea *Xenopsylla cheopis*, occasionally bedbug and human flea. Pneumonic plague spread person-to-person.	2 to 6 days	Immunization. Surveys in endemic areas. Chemical destruction of flea. Community sanitation; rat control. (Plague in wild rodents called sylvatic plague.)
Q fever	*Coxiella burneti* (*Rickettsia burneti*)	Infected wild animals (bandicoots); cattle, sheep, goats, ticks, carcasses of infected animals	Airborne rickettsias in or near premises contaminated by placental tissues. Raw milk from infected cows, direct contact with infected animals or meats.	2 to 3 weeks	Immunization of persons in close contact with rickettsias or possibly infected animals. Pasteurization of all milk at 145°F for 30 min or 161°F for 15 sec.
Rocky Mountain spotted fever	*Rickettsia rickettsii*	Infected ticks, dog ticks, wood ticks, Lone Star ticks	Bite of infected tick or crushed tick blood or feces in scratch or wound.	3 to 10 days	Avoid tick-infested areas and crushing tick in removal; clear harborages; insecticides.
Colorado tick fever	Colorado tick fever virus	Infected ticks and small animals	Bite of infected tick, *Dermacentor andersoni*.	4 to 5 days	See Rocky Mountain spotted fever.

Tularemia	*Francisella tularensis* (*Pasteurella tularensis*)	Wild animals, rabbits, muskrats; also wood ticks	Bite of infected flies or ticks, handling infected animals. Ingestion of contaminated water or insufficiently cooked rabbit meat.	1 to 10 days, usually 3 days	Avoid bites of ticks, flies. Use rubber gloves in dressing wild animals, avoid contaminated water; thoroughly cook rabbit meat.
Rickettsial-pox	*Rickettsia akari*	Infected house mice; possibly mites	Bite of infective rodent mites.	10 to 24 days	Mouse and mite control. Application of miticides to infested areas; incinerators.
Scabies	*Sarcoptes scabiei*, a mite	Persons harboring itch mite; also found in dogs, horses, swine (called mange), do not reproduce in skin of man	Contact with persons harboring mite and use of infested garments or bedding. Also during sexual contact.	Several days or weeks	Personal hygiene, bathing, chemical treatment, clean laundry; machine laundering. Exclude children from school until treated. Prevent crowded living.
Trypanosomiasis, American	*Trypanosoma cruzi*	Infected persons, dogs, cats, wood rats, opossums	Fecal material of infected insect vectors, conenosed bugs in eye, nose, wounds in skin.	5 to 14 days	Screening and rat-proofing of dwellings; destruction of vectors by insecticides and on infected domestic animals.
Scrub typhus	*Rickettsia tsutsugamushi*	Infected larval mites, wild rodents	Bite of infected larval mites.	10 to 12 days	Eliminate rodents and mites; use repellents; clear brush.
Trypanosomiasis, African (sleeping sickness)	*Trypanosoma gambiense*	Man, wild game, and cattle	Bite of infected tsetse fly.	2 to 3 weeks	Fly control; treatment of population; clear brush; education in prevention.

Reference: *Control of Communicable Diseases in Man*, op. cit.

[a]Investigation and survey usually precede preventive and control measures. See also Chapter 10.

[b]"The association of seropositive opossums with human cases of murine (endemic) typhus in southern California and the heavy infestation of the animals with *Ctenocephalides felis* which readily bite man, suggest that opossums and their ectoparasites are responsible for some of the sporadic cases of typhus in man." William H. Adams, Richard W. Emmons, and Joe E. Brooks, "The Changing Ecology of Murine (Endemic) Typhus in Southern California," *Am. J. Trop. Med. Hyg.*, March 1970, pp. 311–318.

Table 1-10 Mosquitoborne Diseases

Disease	Etiologic Agent	Reservoir	Transmission	Incubation Period	Control
Dengue or Breakbone fever[a]	Viruses of dengue fever	Infected vector mosquitoes, man, and possibly animals, including the monkey	Bite of infected *Aëdes aegypti*, *Aëdes albopictus*, *Aëdes scutellaris* complex.	3 to 15 days, commonly 5, to 6 days.	Eliminate aëdes vectors and breeding places; screen rooms; use mosquito repellents.
Encephalitis, anthropodborne viral	Virus of Eastern Equine, Western, St. Louis, Venezuelan Equine, Japanese B, Murray Valley, West Nile, and others	Possibly wild and domestic birds and infected mosquitoes, Ring-necked pheasants, rodents, bats, reptiles	Bite of infected *Culex tarsalis*, also *Culiseta melanura* suspected. *Culex tritaeniorhynchus*, *Culex pipiens*, *Culex pipiens-quinquefasciatus*. Possibly aëdes, psorophora and mansonia. Also *Culex nigripalpus*.	Usually 5 to 15 days	Destruction of larvae and breeding places of culex vectors. Space spraying, screening of rooms; use mosquito bed-nets where disease present. Avoid exposure during biting hours or use repellents. Public education on control of disease. Vaccination of equines.
Filariasis[a] (Elephantiasis after prolonged exposure)	Nematode worms, *Wuchereria bancrofti*, and *W. malayi*	Blood of infected person bearing microfiliariae, mosquito vector	Bite of infected mosquito: *Culex fatigans*, *C. pipiens*; *Aëdes polynesiensis* and several species of anopheles.	3 months; microfilariae do not appear in blood until at least 9 months	Antimosquito measures. Determine insect vectors, locate breeding places and eliminate. Spray buildings. Educate public in spread and control of disease.
Malaria[a]	*Plasmodium vivax*, *P. malariae*,	Man and infected mosquitoes, found between 45°N. and	Bite of certain species of infected anopheles and injection or transfusion of blood of infected person	Average of 12 days for falciparum, 14 for	Residual insecticide on inside walls and places where anopheles rests. Community spraying.

	P. falciparum, *P. ovale*	45°S. latitude and where average summer temperature is above 70°F or where the average winter temperature is above 48°F		vivax, 30 for malariae; sometimes delayed for 8 to 10 months	Screen rooms and use bed-nets in endemic areas. Apply repellents to skin and clothing. *Eliminate breeding places by drainage and filling*; use larvicides: oil and Paris green. Suppressive drugs, treatment, health education. *Gambusia affinis* fish for larvae control.
Rift Valley fever[a]	Virus of Rift Valley fever	Sheep, cattle, goats, monkeys, rodents	Probably through bite of infected mosquito or other blood-sucking arthropod. Laboratory infections and butchering.	Usually 5 to 6 days	Precautions in handling of infected animals. Protection against mosquitoes in endemic areas. Care in laboratory.
Yellow fever[a]	Virus of yellow fever	Infected mosquitoes, persons, monkeys, marmosets, and probably marsupials.	Bite of infected *Aëdes aegypti.* In S. Africa, forest mosquitoes, *H. Spegazzinii* and others. In E. Africa, *Aëdes simpsoni*, *A. africanus*, and others. In forests of S. America, by bite of several species of the genus *Naemagogus* and *A. leucocelaenus*.	3 to 6 days	Control of aëdes breeding places in endemic areas. Intensive vaccination in S. and E. Africa. Immunization of all persons exposed because of residence or occupation. In epidemic area spray interior of all homes, apply larvicide to water containers; mass vaccination, evaluation surveys.

Reference: Various sources and *Control of Communicable Diseases in Man*, op. cit.
[a]Normally not found in United States.

Zoonoses and Their Spread

The Pan American Health Organization lists as the major zoonoses in the Americas encephalitis (arthropodborne), psittacosis, rabies, jungle yellow fever, Q fever, spotted fever (Rocky Mountain, Brazilian, Colombian), typhus fever (murine), leishmaniasis, trypanosomiasis (Chagas' disease), anthrax, brucellosis, leptospirosis, plague, salmonellosis, tuberculosis (bovine), tularemia, hydatidosis, taeniasis (cysticercosis), and trichinosis.[82] Others are ringworm, crytococcosis, toxoplasmosis, yersiniosis, cat scratch fever, tetanus; also tapeworm, hookworm, and roundworm infections.[83] It will be recognized that many of these diseases are also classified with water-, food-, or insect-borne diseases. A very comprehensive summary of zoonoses was prepared by Steele.[83]

The rodentborne diseases include rat-bite fever, Haverhill fever, leptospirosis, choriomeningitis, salmonellosis, tularemia, possibly amebiasis or amebic dysentery, rabies, trichinosis (indirectly), and tapeworm. Epidemic typhus, endemic or murine typhus, Rocky Mountain spotted fever, tsutsugamushi disease, and others are sometimes included in this group. Although rodents are reservoirs of these diseases (typhus, spotted fever, etc.), the diseases themselves are actually spread by the bite of an infected flea, tick, or mite or by the blood or feces of an infected flea or tick on broken skin, as previously discussed. Immunization with plague vaccine reduces the incidence and severity of disease.[84] Rats are also carriers of *Staphylococcus aureus*, *Escherichia coli*, *Yersinia enterocolitica*, *Yersinia pseudotuberculosis*,[85] as well as leptospirae and other pathogens.

Sodoku and Haverhill fever are two types of rat-bite fever. The incubation period for both is 3 to 10 days. Contaminated milk has also been involved as the cause of Haverhill fever. The importance of controlling and destroying rats, particularly around dwellings and barns, is again emphasized.

The causative organism of leptospirosis, also known as Weil's disease, spirochetosis icterohemorrhagic, leptospiral jaundice, spirochaetal jaundice, hemorrhagic jaundice, canicola fever, mud fever, and swineherd's disease, is transmitted by the urine of infected rodents, cattle, dogs, swine, and wild animals. Direct contact or the consumption of contaminated food or water, or direct contact with waters containing the leptospira, may cause the infection after 4 to 19 days.

Dogs are carriers of many microorganisms and parasites which are discharged in the feces and urine and which may be transmitted to man, particularly children. These include *Toxocara canis*, an ascarid roundworm (the larval stage in humans is called visceral larva migrans), also found in cats (*Toxocara cati*); *Ancylostoma caninum*, a canine hookworm which may affect

[82] Miscellaneous Publication No. 74, PAHO, Washington, D.C., 1963.

[83] *CRC Handbook Series in Zoonoses*, Editor-in-Chief, James H. Steele, CRC Press, Inc., 2255 Palm Beach Lakes Blvd., West Palm Beach, Florida, 33409, 1958.

[84] *MMWR*, CDC, Atlanta, Ga., July 21, 1978, p. 255.

[85] R. S. Nakashima, T. F. Wetzler, and J. P. Nordin, "Some Microbial Threats Posed by Seattle Rats to the Community's Health," *J. Environ. Health*, March/April 1978, pp. 264–267.

humans, also found in cats (*Ancylostoma brazillense*); *Dipylidium caninum* and *Taenia pisiformis*, two common canine tapeworms; *Toxoplasma gondii*, a protozoa causing toxoplasmosis, (also carried by cats, goats, pigs, rats, pigeons, and humans).[86] Salmonella can also be transmitted to man and from man to dogs. Dogs may be the reservoir of tapeworms, hookworms, and of many other diseases.[87] Stray and pet dogs are carriers of *Brucella canis;* stray dogs have a higher rate of infection, 9 percent as compared to 1 percent for pet dogs.[88] A significant number of dogs and cats excrete the toxocara ova and the hazard to human health is reported to be considerable.[89] General control measures include proper disposal of dog feces, avoidance of contact with the feces such as in children's play areas, deworming of dogs, and regulation of dogs in urban areas.

The virus causing lymphocytic choriomeningitis (LCM) is found in the mouth and nasal secretions, urine, and feces of infected house mice, which in turn can infect guinea pigs and hamsters. The virus is probably spread by contact, bedding, or consumption of food contaminated by the discharges of an infected mouse. The disease occurs after an incubation period of 8 to 21 days. Precautions include destruction of infected mice, hamsters, and guinea pigs and burning their bodies and bedding. This is followed by cleansing of all cages with water and detergent, disinfection, rodent-proofing pet stores and animal rooms in laboratories and hospitals, and waiting one week before restocking with LCM-free animals.[90]

During 1968–1977, wild and domestic pigeons were associated with 13 percent (88 of 657) of the psittacosis cases in humans reported in the United States.[91]

Rodents, poultry, eggs, and livestock are sources of salmonella infection in addition to human carriers, as previously noted. It has been demonstrated that salmonellosis can be transmitted from man to cattle and then from cattle to man. An intensive investigation in Yorkshire, England showed that a human carrier of *Salmonella paratyphi B* discharged his wastes via septic tank effluent to a stream which flowed through a pasture. Cows grazing in the pasture became infected and subsequently 7 of 13 persons living or working on the farm became infected with the same organism.[92]

Melioidosis is an uncommon disease in man. It is a disease primarily of rodents and small animals. The rat, cat, dog, horse, rabbit, and guinea pig are

[86]J. L. Burt, "The Epidemiology of Selected Animal Parasites, *J. Environ. Health*, November/December 1976, pp. 199–200.

[87]Abram S. Benenson, *Control of Communicable Diseases in Man*, Am. Pub. Health Assoc., 1015 Fifteenth Street NW, Washington, D.C. 20036, 1975.

[88]John Brown, et al.,"*Brucella canis* Infectivity Rates in Stray and Pet Dog Populations," *Am. J. Pub. Health*, September 1976, pp. 889–891.

[89]A. J. Blowers, "Toxocariais in Perspective," *J. R. Soc. Health*, Dec. 1977, pp. 279–280.

[90]John P. Woodall, "LCM—Lymphocytic Choriomeningitis," *Proceedings of The Seminar on Environmental Pests and Disease Vector Control*, Karl Westphal, Ed., New York State Dept. of Health, Albany, N.Y., January 20–24, 1975, pp. 1–16.

[91]*MMWR*, CDC, Atlanta, Ga., February 9, 1979, p. 49.

[92]J. F. Harbourne, "Intestinal Infections of Animals and Man," *J. R. Soc. Health*, June 1977, pp. 106–114.

reservoir hosts of the disease-producing organisms. The incubation period is less than 10 days.

Tularemia may be transmitted by handling infected rodents with bruised or cut hands, particularly rabbits and muskrats, by the bite of infected deer-flies, ticks, and other animals, and by drinking contaminated water. Freezing may not destroy the organism.

Trichinosis is ordinarily spread by the consumption of undercooked infected pork, pork products, and, less frequently, by bear or wild boar meat.

Taeniasis includes beef tapeworm (beef measles) and pork tapeworm (pork measles). The infective larva is found in beef and pork; when consumed in the raw or partially cooked state, it develops into the beef tapeworm and pork tapeworm, respectively, in 2 to 3 months. Cysticercosis is caused by eating or drinking the pork tapeworm egg. Larval forms of the pork tapeworm develop into cysts that may locate in any organ of the human body and cause serious disability. Eating raw beef or pork is dangerous. Cattle can contract the disease if permitted to pasture in fields upon which human feces containing tapeworm segments or eggs have been spread. Prevention and control measures include protection of animal feed, pasture, and drinking water from human feces and from wastewater (sewage) effluent. Freezing of beef carcasses at −10°C (15°F) for 10 days and cooking beef to above 56°C (133°F) [USDA says to at least 60°C (140°F)] inactivates the *Taenia saginata* (beef tapeworm) cysticerci larvae.[93] Freezing of pork carcasses 6 days at 15°F or lower is effective in inactivating *Taenia solium* (pork tapeworm) cysticerci as is cooking to at least 150°F.

Dwarf tapeworm (*Hymenolepis nana*) is found in man, rats, and mice and has worldwide distribution. Consumption of food and drink contaminated with their feces is the major cause of infection. It has been estimated that about 1 to 2 percent of the population, especially children, are infected in southern United States.[94] Higher incidences ranging between 18 and 10 percent are reported from India, Portugal, Spain, and Sicily.

In addition to the human diseases mentioned above, rats may transmit hog cholera, swine erysipelas, fowl tuberculosis, and probably hoof-and-mouth disease to livestock or domestic animals.

Anthrax, also known as woolsorter's disease, malignant pustule, and charbon, is an infectious disease principally of cattle, swine, sheep, and horses that is transmissible to man. Many other animals may be infected. (See Table 1-11.)

Animal Rabies Control

Rabies is a disease of many domestic and wild animals and biting mammals, including bats. In 1977 3182 laboratory-confirmed cases were reported in the

[93] "Tapeworms, Meat and Man: A Brief Review and Update of Cysticercosis Caused by *Taenia Saginata* and *Taenia Solium*," *J. Food Prot.*, January 1979, pp. 58–64.

[94] Richard P. Strong, *Stitt's Diagnosis, Prevention and Treatment of Tropical Diseases*, 6th ed., Blakiston Co., Philadelphia, 1942, p. 1470.

United States and areas under U.S. jurisdiction. Ninety-seven percent of the reported cases occurred in 7 kinds of animals: skunks, 51%; bats, 20%; raccoons, 9%; cattle, 6%; foxes, 4%; dogs, 4%; and cats, 3%.[95] Every sick-looking dog or other animal that becomes unusually friendly, or ill-tempered and quarrelsome, should be looked on with suspicion. One should not place one's hand in the mouth of a dog, cat, or cow that appears to be choking. The animal may be rabid. A rabid animal may be furious or it may be listless, depending on the form of the disease. A person bitten or scratched by a rabid animal, or an animal suspected of being rabid, should immediately wash and flush the wound and surrounding area thoroughly with soap and warm water, a mild detergent and water, or plain water if soap or detergent is not available, and seek immediate medical attention. The physician will notify the health officer or health department of the existence of the suspected rabid animal and take the required action.

The animal should be caged or tied up with a strong chain and isolated for 10 days to see if the symptoms appear. A dog or cat bitten by or exposed to a rabid animal should be confined for 4 months or destroyed, if it has not been vaccinated against rabies or if the vaccination is not current.[96] A wild animal, if suspected, should be killed without unnecessary damage to the head. Gloves should be worn when handling the carcass of a suspected rabid animal since rabies can be introduced through a cut or scratch in the hands. The dead animal should be wrapped in newspaper or other covering and taken to a veterinarian or local health department. The head should be immediately delivered or packed in ice and shipped to the nearest equipped health department laboratory where the brain is examined for evidence of rabies.

Several immunization products, vaccines and globulins, are available and used for postexposure prophylaxis. If treatment has been initiated, and subsequent testing of the animal shows it to be negative, treatment can be discontinued. Preexposure prophylaxis is also practical for persons in high-risk groups. These include veterinarians, animal handlers, certain laboratory workers, and persons, especially children, living or visiting countries where rabies is a constant threat. Persons whose vocational or avocational pursuit bring them in contact with potentially rabid dogs, cats, foxes, skunks, or other species at risk of having rabies should also be considered for preexposure prophylaxis.[97] The CDC, HHS, PHS, provides detailed recommendations for rabies prevention.

Vaccination of dogs and cats in affected areas, stray animal control, and public information are important for a good control program. In areas where rabies exists, mass immunization of at least 70 percent of the dog and cat population in the county or similar unit within a 2- or 3-week period is indicated. The "chick embryo vaccine" was used to vaccinate 250,000 dogs between 1952 and 1954; only two animals later developed rabies, as compared to 76 known

[95] *MMWR*, CDC, Atlanta, Ga., December 15, 1978, pp. 499–500.

[96] *Control of Bats*, New York State Dept. of Health, Albany, N.Y., May 1979.

[97] "Rabies Prevention," Recommendation of the Immunization Practices Advisory Committee, *MMWR*, CDC, Atlanta, Ga., June 13, 1980, pp. 265–280.

cases in nonvaccinated dogs, and only one canine death was attributed to the vaccine.[98] Where rabies exists, or where it might be introduced, a good program should include vaccination of all dogs at 5 months of age and older. Vaccines are available for dogs, cats, and cattle; special vaccines can be used in certain animals. See also Table 1-11.

Bat Rabies

The vampire bat (*Desmodus rotundus*) is a rabies carrier spreading death and diseases among cattle and other livestock and endangering humans in Latin America from Mexico to northern Argentina. The anticoagulant diphenadione is effective against the vampire bat species. The chemical may be injected directly in cattle and then taken by the bat when it gets its blood meal or spread as a petroleum jelly mixture on captured bats which when released spread the chemical by contact throughout a bat colony. In either case the diphenadione enters the bloodstream and the bat bleeds to death.[99] Annual livestock production losses is estimated at $250 million.

DDT 50 percent wettable powder may be used for rabid bat control. The powder is applied directly to the specific indoor roosting area. USEPA and CDC guidelines state that before use of DDT will be authorized it must be shown that rabies is known to exist in the bat colony to be treated and that the bats exist in places which constitute an imminent hazard to health. The most effective means of control of bats is to build them out insofar as possible. Fiberglass insulation will keep bats out of spaces so insulated. See Chapter 10 for control measures.

Any person bitten by a bat should receive antirabies therapy without delay and until the bat is found negative by laboratory test. Any person who has handled a bat, dead or alive, may also have to undergo antirabies therapy as the bat saliva, containing the rabies virus, may enter a patient's body through open cuts in the skin or through mucous membranes.[100]

Control of Zoonoses and Insectborne Diseases

To eliminate or reduce the incidence of zoonoses and insectborne diseases it is necessary to control the environment and reservoirs and the vehicles or vectors. This would include control of water and food, carriers of disease agents, and the protection of persons and domestic animals from the disease (immunization). Where possible and practical the reservoirs and vectors of disease should be destroyed and the environment made unfavorable for their propagation.

[98]Donald J. Dean, *The New York State Conservationist*, April–May 1955, pp. 8–12.

[99]Val Montanari, "Latin American Countries Take Over Bat Control Programs," *War on Hunger*, February 1974, p. 5.

[100]"Rabies and Animal Bites," *Health Facilities Memorandum*, Series 78-45, May 19, 1978, New York State Department of Health, Albany.

Theoretically, the destruction of one link in the chain of infection should be sufficient. Actually, efforts should be exerted simultaneously toward elimination and control of all the links, since complete elimination of one link is rarely possible and protection against many diseases is difficult, if not impossible, even under ideal conditions. Personnel, funds, and equipment available will frequently determine the action taken to secure the maximum results or return on the investment made. The control of insects and rodents is discussed in some detail in Chapter 10.

MISCELLANEOUS DISEASES AND ILLNESSES

These are a divergent group, are not all communicable in the usual sense, and are discussed separately below.

Ringworm

This disease includes a group of fungus infections that develop on the surface of the skin and may involve the nails, scalp, body, and feet. See Table 1-11. Practically all persons have or have had one or more of the fungus infections. Thickening and scaling of the skin, raw inflamed areas, cracked skin, and blistering accompanied by severe itching are usual symptoms of dermatophytosis. (Bacteria predominate at this point.) The microscopic fungi grow best under conditions of warmth, darkness, and moisture. Damp floors and cracks in floors, mats, benches and chairs at swimming pools, shower rooms, bathrooms, and athletic clubs present ideal conditions for the incubation and spread of the fungi that come into contact with the bare skin if not controlled.

The provision of smooth, nonslip, impervious floors constructed to remain well drained and facilitate quick drying is recommended. The direct entrance of sunlight that has not passed through window glass to take advantage of the sterilizing properties of solar ultraviolet rays, and the provision of ample window area with a southern exposure for adequate ventilation, should be given serious consideration in swimming pool, shower, and bathhouse design. This will encourage prompt drying of the floor and destruction of fungi and their spores. Indoor floors and benches can be disinfected with a solution of sodium hypochlorite (500 mg/l available chlorine) or cresol, but the strong odor may discourage their use unless followed by rinsing with clear water. Cresol should not be used where it can be tracked into a pool. Daily scrubbing with a strong detergent in hot water followed by a rinse containing a quaternary ammonium compound (0.1 percent) would seem to be the most practical fungicide treatment where material might be attacked by chlorine or cresol. A strong hot alkaline solution is also effective. Hexachlorophene solution should not be used where skin contact and absorption is possible.

Simple laundering of stockings or towels is not effective in destroying the pathogenic fungi responsible for athlete's foot; sterilization is necessary. This

Table 1-11 Some Characteristics of Miscellaneous Diseases

Disease	Etiologic Agent	Reservoir	Transmission	Incubation Period	Control
Ringworm of scalp (Tinea capitis)	Microsporum and Trichophyton	Infected dogs, cats, cattle, children	Contact with contaminated barber clippers, toilet articles or clothing, dogs, cats, cattle, backs of seats in theaters, planes, and railroads.	10 to 14 days	Survey of children with Wood lamp; education about contact with dogs, cats, infected children; reporting to school and health authorities, treatment of infected children, pets, and farm animals; investigation of source.
Ringworm of body (Tinea corporis)	*Epidermophyton floccosum*, Microsporum, Trichophyton	Skin lesions of infected man or animal	Direct contact with infected person or contaminated floors, shower stalls, benches, towels, etc.	10 to 14 days	Sterilization of towels; fungicidal treatment of floors, benches, mats, shower stalls with creosol or equal. Exclusion of infected persons from pools and gyms. Treatment of infected persons, pets, and animals. Cleanliness, sunlight, dryness.
Ringworm of foot (Tinea pedis) (Athlete's foot)	Same as Ringworm of body	Skin lesions of infected man or animal	Contact with infected persons, contaminated floors, shower stalls, benches, mats, towels.	10 to 14 days	In addition to above, drying feet and between toes with individual paper towels; use of individual shower sandals, foot powder, and clean sterilized socks. Well-drained floors in bathhouses, pools, etc.
Ringworm of nails (Tinea unguium)	Epidermophyton and Trichophyton	Skin or nails of infected persons	Probably from infected feet, contaminated floors.	Unknown	Same as above.
Ancylostomiasis (Hookworm disease)	*Necator americanus* and *Ancylostoma duodenale*	Feces of infected persons, soil containing infective larvae	Larvae hatching from eggs in contaminated soil penetrate foot. Larvae also swallowed. (Cat and dog larvae cause a dermatitis.)	About 6 weeks	Prevention of soil pollution; sanitary privies or sewage disposal systems; wearing of shoes; education in method of spread; treatment of cases; sanitary water supply.

Rabies (Hydrophobia)	Virus of rabies	Infected dogs, foxes, cats, squirrels, cattle, horses, swine, goats, wolves, vampires, and fruit-eating bats	Bite of rabid animal, or its saliva, on scratch or wound.	2 to 6 weeks, or up to 6 months	Detention and observation for 10 days of animal suspected of rabies. Immediate destruction or 6 months detention of animal bitten by a rabid animal. Vaccination of dogs, dogs on leashes, dogs at large confined. Education of public. Avoid killing animal; if necessary, save head intact. Reduce wild life reservoir in cooperation with conservation agencies. If bitten, wash wound immediately and obtain medical attention.
Tetanus	*Clostridium tetani*, tetanus bacillus	Soil, street dust, animal feces containing bacillus	Entrance of tetanus bacillus in a wound.	4 days to 3 weeks	Immunization with tetanus toxoid plus reinforcing dose and booster within 5 years. Allergic persons should carry record of sensitivity. Thorough cleansing of wounds. Safety program.
Anthrax	*Bacillus anthracis*	Cattle, sheep, goats, horses, swine	Contaminated hair, wool, hides, shaving brushes, ingestion or contact with infected meats. Inhalation of spores. Flies possibly; laboratory accidents. (Shaving brushes are under USPHS regulations. Bristles soaked 4 hr in a 10% formalin at a temperature of 110°+F destroys anthrax spores.)	7 days, usually less than 4 days.	Isolation and treatment of suspected animals. Postmortem examination by veterinarian of animals suspected of anthrax and deep burial of carcass, blood, and contaminated soil at a depth of at least 6 ft and above ground water, or incineration. Spores survive a long time. Vaccination of workers handling animals, hair, hides, or meats; personnel hygiene; prompt treatment of abrasions. Treatment of trade wastes.

Table 1-11 (*Continued*)

Disease	Etiologic Agent	Reservoir	Transmission	Incubation Period	Control
Trachoma	Virus of trachoma	Tears, secretions, discharges of nasal mucous membranes of infected persons	Direct contact with infected persons and towels, fingers, handkerchiefs, clothing soiled with infective discharges.	5 to 12 days	Routine inspections and examinations of schoolchildren. Elimination of common towels and toilet articles and use of sanitary paper towels. Education in personal hygiene and keeping hands out of eyes.
Psittacosis (Ornithosis)	Viruses of psittacosis	Infected parrots, parakeets, love birds, canaries, pigeons, poultry, other birds	Contact with infected birds, their wastes and surroundings. Virus is airborne.	6 to 15 days	Importation of birds from psittacosis-free areas. Quarantine of pet shops having infected birds until thoroughly cleaned. Education of public to dangers of parrot illnesses.
Impetigo contagiosa	Probably streptococci, also staphylococci	Skin lesions of infected person, possibly nose and throat discharges	Direct contact with infected skin lesions or articles such as towels and pencils soiled by discharges.	Within 5 days, often 2 days	Personal hygiene, avoidance of common use of toilet articles; prompt recognition and treatment of illness. Inspection of children at camps, nurseries, institutions, schools.
Chancroid (Soft chancre)	*Hemophilus ducreyi*, *Ducrey bacillus*	Discharges from open lesions and pus from buboes from infected persons	Prostitution, indiscriminate sexual promiscuity and uncleanliness	3 to 5 days or longer	Character guidance, health and sex education, premarital and prenatal examinations. Improvement of social and economic conditions; elimination of slums, housing rehabilitation and conservation, new housing, neighborhood renewal. Suppression of commercialized prostitution, personal prophylaxis, facilities for early
Syphilis (Pox, lues)	*Treponema pallidum*	Exudates from lesions of skin, mucous membrane, body fluids, and secretions of infected persons	By direct contact in sexual intercourse, kissing, fondling of children.	10 days to 10 weeks, usually 3 weeks	

Gonorrhea (Clap, dose)	*Neisseria gonorrhoeae*, gonococcus	Exudate from mucous membranes of infected persons	Sexual intercourse, infection during birth, careless use of rectal thermometer.	3 to 9 days, sometimes 14 days	diagnosis and treatment, public education concerning symptoms, modes of spread, and prevention.
Lymphogranuloma venereum (Tropical bubo)	*Lymphogranuloma venereum*	Lesions of rectum, urethra, and sinuses and ulcerations of infected persons	Sexual intercourse. Contact with contaminated articles.	5 to 21 days, usually 7 to 12 days	Case-finding, patient interview, contact-tracing and serologic examination of special groups known to have a high incidence of venereal disease, with follow-up. Report to local health authority.
Granuloma inguinale (Tropical sore)	*Donovania granulomatis*	Probably active lesions of infected persons	Presumably by sexual intercourse.	Unknown, probably 8 days to 12 weeks	Same as above.

Reference: *Control of Communicable Diseases in Man*, op. cit.

Note: Genital herpes infections, sexually transmitted diseases, are very common. No effective treatment is currently available; periodic recurrences are the rule.

can be accomplished by boiling cotton socks 15 min. Air drying in sunlight may also be effective. Shoes frequently become a reservoir of infection, since leather is difficult to sterilize without ruining it. Investigators found that leather shoes placed in a closed container with ethylene oxide maintained at a concentration of 150 mg/l of gas destroyed the fungi in leather shoes.[101] The gas presents a serious fire and explosion hazard. Placing the shoes in a box containing formaldehyde vapor for several hours followed by airing to prevent skin irritation is also suggested.[102] Emphasis should be placed on keeping the feet dry and clean, especially between the toes. The use of cotton or wool socks, a foot powder, and leather shoes will help keep the feet dry.

Hookworm Disease or Ancylostomiasis

Hookworm disease is most common in those areas where the winters are mild and where moist sandy soil is encountered. The parasite develops best at a temperature of around 81°F, although it will live at temperatures between 57 and 100°F. A hard frost will kill the eggs and larvae.

Strongyloidiasis is an infection of the nematode *Strongyloides stercoralis*. It is similar to hookworm disease. Hookworm is the greatest cause of anemia.

Tetanus or Lockjaw

This is a disease caused by contamination of a wound or burn with soil, street dust, or animal excreta containing the tetanus bacteria. The bacillus lives in the intestines of domestic animals. Gardens that are fertilized with manure, barnyards, farm equipment, and pastures are particular sources of danger. The tetanus germ produces a toxin that affects the nervous system, causing spasms, convulsions, and frequently death. It is a spore-former that survives in soil and is resistant to ordinary boiling. Deep puncture-type wounds are most dangerous, regardless of whether they are made by clean or rusty objects. See Table 1-11.

NONINFECTIOUS AND NONCOMMUNICABLE DISEASES ASSOCIATED WITH THE ENVIRONMENT INCLUDING AIR, WATER, AND FOOD

Background

Treatment of the environment is supplementing treatment of the individual, but more effort and knowledge is needed. The total environment is "the most

[101] John D. Fulton and Roland B. Mitchell, "Sterilization of Footwear," *U.S. Armed Forces Med. J.*, **3**, No. 3, 425–439 (March 1952).

[102] *Control of Communicable Diseases in Man*, op. cit., p. 94.

important determinant of health." A review of more than 10 years of research conducted in Buffalo, New York, showed that the overall death rate for people living in heavily polluted areas was twice as high, and the death rates for tuberculosis and stomach cancer three times as high, as the rates in less polluted areas.[103] Dr. Rene Dubos points out that "many of man's medical problems have their origin in the biological and mental adaptive responses that allowed him earlier in life to cope with environmental threats. All too often the wisdom of the body is a shortsighted wisdom."[104] In reference to air pollution, he adds that "while the inflammatory response is protective (adaptive) at the time it occurs, it may, if continuously called into play over long periods of time, result in chronic pathological states, such as emphysema, fibrosis, and otherwise aging phenomena."

Human adjustment to environmental pollutants and emotional stresses due to crowding and other factors can result in later disease and misery with reduced potential for longevity and a productive life.[105]

USEPA Administrator Barbara Blum, in an address to the Sierra Club stated:[106]

> Inner-city people—white, yellow, brown and black—suffer to an alarming degree from what are euphemistically known as diseases of adaptation. These are not healthy adaptations, but diseases and chronic conditions resulting from living with bad air, polluted water, excessive noise, and continual stress. Hypertension, heart disease, chronic bronchitis, emphysema, sight and hearing impairment, cancer, and congenital anomalies are all roughly fifty percent higher (for inner-city people) than the level for suburbanites. Behavioral, neurological and mental disorders are about double. . . .

Whereas microbiological causes of most communicable diseases are known and are under control or being brought under control in many parts of the world (with the possible exception of malaria and schistosomiasis), physiological and toxicological effects on human health of the presence, or absence, of certain chemicals in air, water, and food in trace amounts has not yet been clearly demonstrated. The cumulative body burden of all deleterious substances, especially organic and inorganic chemicals, gaining access to the body must be examined both individually and in combination. The synergistic, additive, and neutralizing effects must be learned in order that the most effective preventive measures may be applied. Some elements, such as fluorine for the control of tooth decay, iodine to control goiter, and iron to control iron-deficiency anemia, have been recognized as being beneficial in proper amounts. But the action of trace amounts ingested individually and in combination, of lead, cadmium, hexavalent chromium, nickel, mercury, manganese, and or-

[103]Warren Winkelstein, Address at the 164th Annual Convention of the Medical Society of the State of New York, 1969.

[104]*The Fitness of Man's Environment*, Smithsonian Institution Press, Washington, D.C., 1968.

[105]Rene Dubos, *Man Adapting*, Yale University Press, New Haven, Conn., 1965, p. 279.

[106]*Environmental Health Letter*, May 15, 1978, Gershon W. Fishbein, Publisher, 1097 National Press Building, Washington, D.C. 20045.

ganic chemicals are often insidious. Their carcinogenic, mutagenic, and teratogenic effects are extended in time, perhaps for 10, 20, or 30 years, to the point where direct relationships with morbidity and mortality are difficult to conclusively prove in view of the many possible intervening and confusing factors.

An interesting analysis was made by Dever[107] for use in policy analysis of health program needs. He selected 13 causes of mortality and allocated a percentage of the deaths, in terms of an epidemiological model, to four primary divisions, namely, System of Health Care Organization; Life-Style; (self-created risks); Environment; and Human Biology. He envisioned the environment as comprised of a physical, social, and psychological component. Environmental factors were considered to be associated with 9 percent of the mortality due to diseases of the heart, with the rest due to causes associated with health care, life-style, or human biology. Similarly, environmental factors were considered the cause of 24 percent of the cancer deaths, 22 percent of the cerebrovascular deaths, and 24 percent of the respiratory system deaths.

Of added interest is Dever's analysis showing that environmental factors were considered to be the cause of 49 percent of all deaths due to accidents, 20 percent of the influenza and pneumonia deaths, 41 percent of the homicides, 15 percent of the deaths due to birth injuries and other diseases peculiar to early infancy, 6 percent of the deaths due to congenital anomalies, and 35 percent of the deaths due to suicides.

There are an estimated 2 million recognized chemical compounds with more than 60,000 chemical substances in past or present commercial uses. Approximately 600 to 700 new chemicals are introduced each year; but only about 15,000 have been animal-tested with published reports. Limited trained personnel and laboratory facilities for carcinogenesis testing in the United States by government and industry will permit testing of no more than 500 chemicals per year. Each animal experiment requires 3 to 6 years and a cost of at least $300,000.[108] Another estimate is $500,000 just to establish the carcinogenicity of one compound with the National Cancer institute test protocol requiring at least two species of rodents and three years' time.* A full toxicological test, including those for carcinogenicity, can take five years and cost in excess of $1.25 million for each compound. The chemicals are viewed by Harmison[109] as falling into four groups.

1. Halogenated hydrocarbons and other organics, polychlorinated biphenyls (PCBs); chlorinated organic pesticides such as DDT, Kepone, Mirex, and endrin; polybrominated biphenyls (PBBs); fluorocarbons; chloroform, and vinyl chloride. These chemicals are persistent, often bioaccumulate in food organisms, and may in small quantities cause cancer, nervous disorders, and toxic reactions.

*Stephen Nesnow quoted by Julian Josephson in "Is predictive toxicology coming?" *Environ. Sci. Technol.*, April 1981, pp. 379–381.

[107]G. E. A. Dever, *An Epidemiological Model for Health Policy Analysis*, Atlanta, Ga., Department of Human Resources, 1976, pp. 453–466.

[108]HEW Agencies Fight Environmental Threat," *FDA Consumer*, Dec. 1978–Jan. 1979, p. 26.

[109]Lowell T. Harmison, "Toxic Substances and Health," *Pub. Health Rep.*, January–February 1978, pp. 3–10.

2. Heavy metals: lead, mercury, cadmium, barium, nickel, and vanadium, selenium, beryllium. These metals do not degrade; they are very toxic, and may build up in exposed vegetation, animals, fish, and shellfish.
3. Nonmetallic inorganics: arsenic and asbestos for example are carcinogens.
4. Biological contaminants such as aflatoxins and pathogenic microorganisms; animal and human drugs such as diethylstilbestrol (DES) and other synthetic hormones; and food additives such as Red dye No. 2.

Evaluation of the toxicity of existing and new chemicals on the workers, users, and the environment, and their release for use represents a monumental task, as noted above. Although short-term testing of chemicals, such as the microbial Ames test, is very valuable to screen inexpensively for carcinogens and mutagens, the problem of further testing of the chemicals found positive still remains.

Selected chemical pollutants known or suspected of causing or aggravating certain noninfectious diseases are shown in Figure 1-4. Environmental pollutants subject to Federal regulation are given in Table 1-12. Agents, pollu-

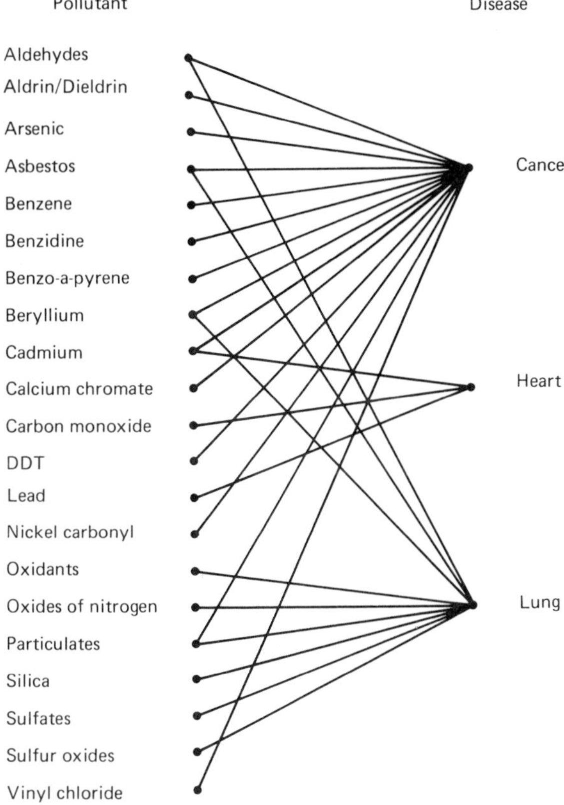

Figure 1-4 Known or suspected links between selected pollutants and disease. (Source: First Annual Report by The Task Force on Environmental Cancer and Heart and Lung Disease, Printing Management Office, USEPA, Washington, D.C., 20460, August 7, 1978, p. 9.)

Table 1-12 Federal Regulation of Levels of Environmental Pollutants

	Ambient air quality standards	
Particulates	Lead (proposed)	Hydrocarbons
Sulfur dioxide	Nitrogen dioxide	
Carbon monoxide	Photochemical oxidants	
	Air emission standards	
Acid mist	Nitrogen dioxide	Sulfur dioxide
Carbon monoxide	Particulates	Total reduced sulfur
Fluorides	Beryllium	Mercury
Hydrocarbons	Vinyl chloride	Asbestos
	Toxic substances control	
Several substances have been recommended to the USEPA by the TSCA Interagency Testing Committee for further testing.		
	Occupational standards	
Permissible exposure limits for approximately 400 toxic and hazardous substances. Occupational Safety and Health Standards for 20 designated carcinogens.		
	Drinking water standards	
Arsenic	Cyanide	Nitrate
Barium	Fluoride	Selected pesticides
Cadmium	Lead	Selenium
Chromium	Mercury	Silver
	Others—regulatory actions to limit environmental damage from	
Effects of food additives	Pesticides	Solid waste
Radioactive materials	Noise	Effects of smoking

Source: First Annual Report by The Task Force on Environmental and Heart and Lung Disease, Printing Management Office, USEPA, Washington, D.C. 20460, August 7, 1978, p. 9.

tants, or sources having definite and possible health effects are summarized in Table 1-13 and suggest possible environmental program activities.

Prevention and Control

The prevention and control of chemical pollutants generally involves the following.

1. Elimination or control of the pollutant at the source: minimize or prevent production and sale; substitute nontoxic or less toxic chemical; materials and process control and changes; recover and reuse; waste treatment, separation, concentration, incineration, detoxification, neutralization. See Chapters 4 and 5.
2. Interception of the travel or transmission of the pollutant: air and water pollution control and prevention of leachate travel. See Chapters 4, 5, and 6.
3. Protection of man to eliminate or minimize the effects of the pollutant: water treatment, air conditioning, land use planning, occupational protection. See Chapters 2, 3, and 6.

At the same time the air, sources of drinking water, food, aquatic plants, fish and other wildlife, surface runoff, leachates, precipitation, surface waters, and man himself should be monitored. This should be done for potentially toxic and deleterious chemicals as indicated by specific situations.

Control Legislation

Legislation establishing national standards and controls for the discharge of pollutants to the environment and for consumer and worker protection are listed here.

1. *The Clean Air Act.* To improve the quality of the nation's air; the USEPA is to establisn national air quality standards to protect the public health and welfare from harmful effects of air pollution, and to ensure that existing clean air is protected from significant deterioration by controlling and preventing harmful substances from entering the ambient air.
2. *Occupational Safety and Health Act.* P.L. 91-956. To prevent occupational disease and accidents and to establish workplace standards as well as national standards for significant new pollution sources and for all facilities emitting hazardous substances.
3. *Federal Food, Drug, and Cosmetic Act.* 1938 (as amended). The FDA is responsible for the safety and effectiveness of foods, drugs, medical devices, and cosmetics; it is also responsible for radiological health and toxicologic research and establishes allowable limits for pesticides on food and feed crop.
4. *Consumer Product Safety Act.* P.L. 92-573. The Consumer Product Safety Commission (CPSC) is responsible for reducing injuries associated with consumer products, including the development of safety standards and investigation of product-related morbidity and mortality.
5. *Noise Control Act of 1972.* Makes the federal government responsible for the regulation of noise emissions from a broad range of sources.
6. *Federal Water Pollution Control Acts.* P.L. 92-500 and P.L. 95-217 (Clean Water Act of 1977). Controls water pollutants and other related factors to make surface waters fishable and swimmable. Intended to restore and maintain the chemical, physical, and biological integrity of the nation's waters.
7. *Marine Protection, Research, and Sanctuaries Act.* 1972. P.L. 92-532. Controls the dumping of materials, including toxic substances, into the oceans.
8. *Safe Drinking Water Act of 1974.* P.L. 93-523. Establishes regulations for drinking water in public water systems, including microbiological, radiological, organic, and inorganic standards, and turbidity levels in drinking water, to protect the public's health.
9. *Resource Conservation and Recovery Act of 1976.* P.L. 94–580. Requires a regulatory system for the treatment, storage, and disposal of hazardous wastes, that is, hazardous to human health or to the environment. Conserves natural resources directly and through resource recovery from wastes. The EPA defines hazardous wastes "as any solid waste or combination of solid wastes, which because of its quantity, concentration, or physical, chemical, or infectious characteristics poses a substantial present or potential hazard to human health or to the environment as associated with the method of disposal."
10. *Hazardous Materials Transportation Act of 1974.* P.L. 93-633. Regulates the transportation in commerce of hazardous materials by all means of transportation.
11. *Ports and Waterways Safety Act of 1972.* P.L. 92-340. Regulates, through the Coast Guard, the bulk of shipment of oil and hazardous materials by waters, also under the authority of the Tanker Act and the Dangerous Cargo Act.

Table 1-13 Definite and Possible Health Effects of Environmental Pollutants and Exposures

Agent, Pollutant, or Source	Definite Effect	Possible Effect
	Community air pollution	
Sulfur dioxide (effects of sulfur oxides may be due to sulfur, sulfur trioxide, sulfuric acid, or sulfate salts) Sulfur oxides and particulate matter from combustion sources	1. Aggravation of asthma and chronic bronchitis 2. Impairment of pulmonary function 3. Sensory irritation 4. Short-term increase in mortality 5. Short-term increase in morbidity 6. Aggravation of bronchitis and cardiovascular disease 7. Contributory role in etiology of chronic bronchitis and emphysema 8. Contributory role to respiratory disease in children	9. Contributory role in etiology of lung cancer
Particulate matter (not otherwise specified)		10. Increase in chronic respiratory disease
Oxidants	11. Aggravation of emphysema, asthma, and bronchitis 12. Impairment of lung function in patients with bronchitis-emphysema 13. Eye and respiratory irritation and impairment in performance of student athletes	14. Increased probability of motor-vehicle accidents
Ozone	15. Impairment of lung function	16. Acceleration of aging, possibly due to lipid peroxidation and related processes
Carbon monoxide	17. Impairment of exercise tolerance in patients with cardiovascular disease	18. Increased general mortality and company mortality rates 19. Impairment of central nervous system function 20. Causal factor in atherosclerosis
Nitrogen dioxide		21. Factor in pulmonary emphysema 22. Impairment of lung defenses such as mast cells and macrophages or altered lung function
Lead	23. Increased storage in body	24. Impairment of hemoglobin and porphyrin synthesis
Hydrogen sulfide	25. Increased mortality from acute exposure 26. Sensory irritation	

Mercaptans		27. Headache, nausea, and sinus affections.
Asbestos	28. Pleural calcification 29. Malignant mesothelioma, asbestosis	30. Contributory role in chronic pulmonary disease (asbestos and lung cancer)
Organophosphorus pesticides	31. Acute fetal poisoning 32. Acute illness 33. Impaired cholinesterase activity	
Other odorus compounds		34. Headache and sinus affections
Beryllium	35. Berylliosis with pulmonary impairment	
Airborne microorganisms	36. Airborne infections	
	Food and water contaminants	
Bacteria	1. Epidemic and endemic gastrointestinal infections (typhoid, cholera, shigellosis, salmonellosis, leptospirosis, etc.)	2. Secondary interaction with malnutrition and with nitrates in water (cf., No. 13)
Viruses	3. Epidemic hepatitis and other viral infections	4. Eye and skin inflammation from swimming
Protozoa and metazoa	5. Amoebiasis, schistosomiasis, hydatidosis, and other parasitic infections	
Metals	6. Lead poisoning 7. Mercury poisoning (through food chains) 8. Cadmium poisoning (through food chains) 9. Arsenic poisoning 10. Chromium poisoning	11. Epidemic nephropathy 12. "Blackfoot" disease
Nitrates	13. Methemoglobinemia (with bacterial interactions)	
"Softness" factor		14. Increase in cardiovascular disease
Sulfates and/or phosphates	15. Gastrointestinal hypermotility	
Fluorides	16. Fluorosis of teeth when in excess	
	Land pollution	
Human excreta	1. Schistosomiasis, taeniasis, hookworm, and other infections	

Table 1-13 (*Continued*)

Agent, Pollutant, or Source	Definite Effect	Possible Effect
	Land pollution	
Sewage		2. Infectious diseases
Industrial and radioactive waste	3. Storage and effects from toxic metals and other substances through food chains	
Pesticides: lead arsenate	4. Increased storage of heavy metals in smokers of tobacco grown on treated areas	
	Thermal exposures	
Cold damp	1. Excess mortality from respiratory disease and fatal exposure	
		2. Contribution to excess mortality and morbidity from other causes
	3. Excess morbidity from respiratory and related diseases and morbidity from exposure	
		4. Rheumatism
Cold dry	5. Mortality from frostbite and exposure	
		6. Impaired lung function
	7. Morbidity from frostbite and respiratory disease	
Hot dry	8. Heatstroke mortality	
	9. Excess mortality attributed to other causes	
	10. Morbidity from heatstroke and from other causes	
	11. Impaired function; aggravation of renal and circulatory diseases	
Hot damp	12. Increase in skin affections	
		13. Increase in prevalence of infectious agents and vectors
	14. Heat-exhaustion mortality	
	15. Excess mortality from other causes	
	16. Heat-related morbidity	
	17. Impaired vigor and circulatory function	
	18. Aggravation of renal and circulatory disease	

	Radiation and microwaves	
Natural sunlight	1. Fatalities from acute exposure 2. Morbidity due to "burn" 3. Skin cancer 4. Interaction with drugs in susceptible individuals	5. Increase in malignant melanoma
Diagnostic X-ray	6. Skin cancer and other skin changes	7. Contributing factors to leukemia 8. Alteration in fecundity
Therapeutic radiation	9. Skin cancer 10. Increase in leukemia	11. Increase in other cancers 12. Acceleration of aging 13. Mutagenesis
Industrial uses of radiation and mining of radio-active ores	14. Acute accidental deaths 15. Radiation morbidity 16. Uranium nephritis 17. Lung cancer in cigarette-smoking miners	
Nuclear power and reprocessing plants		17. Increase in adjacent community morbidity or mortality 19. Increase in cancer incidence 20. Community disaster 21. Alteration in human genetic material 22. Tissue damage
Microwave		
	Noise and vibrations	
Traffic		1. Progressive hearing loss
Aircraft (including sonic boom)	2. Permanent hearing loss	
Vibrations		3. Aggravation or cause of mental illness 4. Articular and muscular disease 5. Adverse effects on nervous system
	Housing and household agents	
Heating, cooking, and refrigeration	1. Acute fatalities from carbon monoxide, fires and explosions, and discarded refrigerators	2. Increase in diseases of the respiratory tract in infants
Fumes and dust	3. Acute illness from fumes	

Table 1-13 *(Continued)*

Agent, Pollutant, or Source	Definite Effect	Possible Effect
	Housing and household agents	
	4. Aggravation of asthma	
		5. Increase in chronic respiratory disease
Crowding	6. Spread of acute and contribution to chronic disease morbidity and mortality	
Structural factors (including electrical wiring, stoves, and thin walls)	7. Accidental fatality	
	8. Accidental injury	
	9. Morbidity and mortality from lack of protection from heat or cold	
	10. Morbidity and mortality due to fire or explosion	
Paints and solvents	11. Childhood lead-poisoning fatalities, associated mental impairment, and anemia	
	12. Renal and hepatic toxicity	
	13. Fatalities	
Household equipment and supplies (including pesticides)	14. Fatalities from fire and injury	
	15. Morbidity from fire and injury	
	16. Fatalities from poisoning	
	17. Morbidity from poisoning	
Toys, beads, and painted objects	18. Mortality and morbidity	
Urban design	19. Increased accident risks	
		20. Contribution to mental illness

Source: *Vital and Health Statistics—Series 4—No. 20*, DHEW Publication No. (HRA) 77-1457, National Center for Health Statistics, Hyattsville, Md., July 1977, pp. 27–29.
Note: Items in parentheses refer to effects other that those directly affecting human health status.

12. *Toxic Substances Control Act of 1976.* P.L. 94-469. Grants to the USEPA authority: to control chemical substances that present an unreasonable risk of injury to health or the environment, except for pesticides, foods, drugs, cosmetics, tobacco, liquor, and several additional categories of chemicals regulated under other federal laws; to develop adequate data and knowledge on the effects of chemical substances and mixtures on health and the environment; to establish an inventory and selectively act on those that appear to pose potential hazard; and to devise a system to examine new chemicals before they reach the marketplace.
13. *Atomic Energy Act of 1954.* P.L. 83-703. Regulates the discharge of radioactive waste into the environment.
14. *National Environmental Policy Act of 1969.* P.L. 91-190. To encourage productive and enjoyable harmony between man and his environment; to promote efforts which will prevent or eliminate danger to the environment and biosphere and stimulate the health and welfare of man; to enrich the understanding of the ecological systems and natural resources important to the Nation; and to establish a Council on Environmental Quality.
15. *Federal Insecticide, Fungicide, and Rodenticide Act of 1974* (as amended). Requires that pesticides be used strictly in accordance with label instructions; gives government (USEPA) authority to prohibit or restrict a pesticide to special uses for application only by a person trained in an approved program; extends control to intrastate products, to container storage and disposal methods, and to direct or indirect discharge of pesticides to surface waters; also requires proof that pesticides will cause no harm to people, wildlife, crops, or livestock when used as directed.

Lead Poisoning

Lead is a cumulative poison ending up in the bones and tissue. It may cause mental retardation, blindness, chronic kidney diseases, fatigue, anemia, gastroenteritis, muscular paralysis, and other impairments. Lead poisoning is commonly associated with children living in old substandard housing who eat lead-based paint that peels or flakes from walls (both inside and outside), ceilings, and woodwork. It is the most common high-dose source of lead for children with lead toxicity. However, there are other environmental sources of lead.

Lead in gasoline discharged with auto exhaust in urban areas, lead fumes and ashes produced in burning lead battery casings, soft corrosive water standing and flowing in lead pipe, natural or added lead in food and drink (a major source), lead in soil and dust,[110] and lead in some household products all contribute to the body burden. The phasing out of tetraethyl lead from gasoline has introduced a potential unknown problem associated with manganese compounds used as a replacement for lead, which are emitted at low levels in various forms including the toxic manganese tetroxide.[111] On the other hand, a Department of Housing and Urban Development study between 1970 and 1976 in New York City showed a drop in blood lead levels in children from

[110]Anthony J. Yankel, Ian H. von Linden, and Stephen D. Walter, "The Silver Valley Blood Levels and Environmental Exposure," *J. Air Pollut. Control Assoc.*, August 1977, pp. 763–767.

[111]Morris M. Joselow, et al., "Manganese Pollution in the City Environment and its Relationship to Traffic Density," *Am. J. Pub. Health*, June 1978, pp. 557–560.

30 to 21 μg/100 ml of blood. The drop paralleled a recorded decrease of lead in the ambient air, suggesting a significant relationship.[112]

Serious illnesses and at least one death have been attributed to the use of earthenware pottery with improper glaze. Such glaze dissolves in fruit juice, coffee, wine, soda pop, and other acid drinks.[113] Most of the glaze applied to earthenware pottery contains lead. When the pottery is not fired long enough at the correct temperature, the glaze will not seal completely and its lead component can be leached or released.

Control of lead poisoning is approached through mass screening of ghetto children between the ages of one and six for excessive lead in the blood—30 μg/100 ml with an erythrocyte protoporphyrin $\geqslant$ 50 μg/100 ml*—and treatment of those affected, with priority to the 1 to 3-year-olds. The "normal" human blood is 20 to 30 μg/100 ml. In 1979 seven percent of the children screened were identified as having lead toxicity. Three percent of the dwellings inspected had lead paint that was related to children having lead toxicity.[114]

Additional controls include identification through selective systematic inspection of housing having lead-based paint on the walls, woodwork or ceilings; removal or covering of paint containing more than 0.5 percent lead by weight; prohibition of sale of toys or baby furniture containing lead paint; education of parents, social workers, sanitarians, health guides, and owners of old buildings to the hazard and its control.[115] The X-ray fluorescence lead paint analyzer has improved hazard identification. The use of lead paint with more than 0.06 percent lead is prohibited for most consumer uses and products by the CPSC.[116] Children two- to three-years-old absorb 40 to 50 percent of their lead from food and water as compared to five to ten percent for adults.[117]

The maximum limit for lead in drinking water is 50 μg/l according to the National Interim Primary Drinking Water Regulation. The national ambient air quality standard for airborne lead is 1.5 $\mu g/m^3$ of air averaged over a three-month period. The OSHA permissible exposure level averaged over an eight-hour day is 50 $\mu g/m^3$. The daily ingestion from *all* sources over any period of years should not exceed 600 μg per day.[118] However, a safe level of daily intake

*35 μg/100 ml maximum in European Economic Community. There is evidence that newborn babies have accumulated lead poisoning from their mothers, due to ingestion of lead from very soft water supplies in old lead pipes. ("Lead threat to new-born babies," *World Water*, January 1981, p. 9).

[112]"Lead Count in the Blood of New York's Children Down 30% in Six Years," *New York Times*, February 5, 1978.

[113]"Family Health," *Fam. Health Mag.*, February 1971, p. 12.

[114]"Surveillance of Childhood Lead Poisoning—United States," *MMWR*, CDC, Atlanta, Ga., April 18, 1980, pp. 170–171.

[115]For discussion of the problem and of planning and implementing a control program, see *Control of Lead Poisoning in Children*, Bureau of Community Environmental Management, U.S. Public Health Service, Dept. of HEW (December 1970).

[116]42 *Fed. Reg.* 44193 (1977).

[117]"Drinking Water and Health, Recommendations of the National Academy of Sciences," *Fed. Reg.* 42:132:35764, July 11, 1977.

[118]Maxcy-Rosenau, *Preventive Medicine and Public Health*, Edited by Philip E. Sartwell, Appleton-Century-Crofts, New York, 1965, p. 796.

according to the WHO is 5 μg lead per kg per person. Based on this, and the intake from air and food,* the National Academy of Sciences suggests that the maximum contaminant level in drinking water be reduced to 25 μg/l to provide an increased factor of safety.[117] See also Chemical Examination (Lead), Chapter 3.

Carbon Monoxide Poisoning

Carbon monoxide poisoning is sometimes confused with food poisoning; nausea and vomiting are common to both. In carbon monoxide poisoning additional symptoms include headache, drowsiness, dizziness, flushed complexion, and general weakness; carbon monoxide is found in the blood. Excessive exposure results in reduced oxygen availability to the heart, brain, and muscles leading to weakness, loss of consciousness, and possibly death. Persons with cardiovascular diseases are very sensitive to carbon monoxide in low concentrations.

Carbon monoxide combines readily with blood hemoglobin to form carboxyhemoglobin, thereby reducing the amount of hemoglobin available to carry oxygen to other parts of the body. Hemoglobin has a greater affinity for carbon monoxide than for oxygen—about 210 to 1. Fortunately the formation of carboxyhemoglobin is a reversible process.

The gas is odorless, tasteless, and colorless. It is a product of incomplete combustion of carbonaceous fuel. Poisoning is caused by leaks in an automobile exhaust system; running a gasoline or diesel engine indoors, or while parked; unvented gas, oil, coal or wood-burning space or water heater, gas range-oven, or gas-fired floor furnace; use of charcoal grill indoors; clogged chimney or vent; *inadequate fresh air for complete combustion*; improperly operating gas refrigerator; incomplete combustion of liquefied petroleum gas in recreational and camping units. The work environment may also be a hazardous source of carbon monoxide.

Motor vehicles are the principal source of carbon monoxide air pollution. Room space heaters are a major potential hazard indoors. See Venting of Heating Units, Chapter 11. Cigarette smoke is a significant source of carbon monoxide to the smoker.

Education of the public and physicians by the utilities and the health department, standards for appliances, and housing-code enforcement can reduce exposure and deaths from this poisoning.[119]

Concentrations of 70 to 100 ppm carbon monoxide are not unusual in city traffic. The federal ambient air quality standard maximum 8-hr concentration is 10 mg/m^3 (9 ppm); the maximum 1-hr concentration is 40 mg/m^3 (35 ppm). These levels can reduce mental efficiency. The standard recommended by the

*Estimated to average 200 to 300 μg/day; air may average 15 to 20 μg/day. Dust and soil may contribute an additional 10 μg/day.

[119]Eugene L. Lehr, "Carbon Monoxide Poisoning: A Preventable Environmental Hazard," *Am. J. Pub. Health*, February 1970, pp. 289–293.

National Institute for Occupational Safety and Health is 40 mg/m^3 (35 ppm) time-weighted average for a 40-hour workweek with 8 hours' exposure per day and with a ceiling value of 220 mg/m^3 (200 ppm). This latter value is to limit carboxyhemoglobin formation to 5 percent in a nonsmoker engaged in sedentary activity at normal altitude.[120]

Mercury Poisoning

Mercury poisoning in man has been associated with the consumption of methylmercury-contaminated fish, shellfish, bread, and pork and, in wildlife, through the consumption of contaminated seed. Fish and shellfish poisoning occurred in Japan in the Minamata River and Bay region and at Niigata between 1953 and 1964. Bread poisoning occured as a result of the use of wheat seed, treated with a mercury fungicide, to make bread in West Pakistan in 1961, in Central Iraq in 1960 and 1965, and in Panorama, Guatemala, in 1963 and 1964. Pork poisoning took place in Alamagordo, New Mexico, when methylmercury-treated seed was fed to hogs that were eaten by a family. In Sweden, the use of methylmercury as a seed fungicide was banned in 1966 in view of the drastic reduction in the wild bird population attributed to treated seed. In Yakima, Washington, early recognition of the hazard prevented illness when 16 members of an extended family were exposed to organic mercury poisoning in 1976 by the consumption of eggs from chickens fed mercury-treated seed grain. The grain contained 15,000 ppb total mercury, an egg 596 and 1902 ppb respectively organic and inorganic mercury. Blood levels in the family ranged from 0.9 to 20.2 ppb in a man who ate 8 eggs per day. A whole blood level above 20 ppb may pose a mercury poisoning hazard.[121]

It is also reported that crops grown from seed dressed with minimal amounts of methylmercury contain enough mercury to contribute to an accumulation in the food chain reaching man. The discovery of moderate amounts of mercury in tuna and most freshwater fish, and relatively large amounts in swordfish, by many investigators in 1969 and 1970, tended to further dramatize the problem.[122]

The organic methylmercury and other alkylmercury compounds are highly toxic. Depending on the concentration and intake, they can cause unusual weakness, fatigue, and apathy followed by neurological disorders. Numbness around the mouth, loss of side vision, poor coordination in speech and gait, tremors of hands, irritability, and depression are additional symptoms leading possibly to blindness, paralysis, and death. The methylmercury also attacks

[120]Irving R. Tabershaw, H. M. D. Utidjian, and Barbara L. Kawahara, *Occupational Diseases*, DHEW (NIOSH) Publication No. 77-181, HEW, GPO, Washington, D.C. 20402, June 1977, p. 418.

[121]"Organic Mercury Exposure—Washington," *MMWR*, CDC, Atlanta, Ga., May 7, 1976.

[122]Goran Löfroth, "Methylmercury," *Redaktionstajänsten*, Natural Science Research Council, Stockholm, Sweden, March 20, 1969.

vital organs such as the liver and kidney. It concentrates in the fetus and can cause birth defects.

Methylmercury has an estimated biological half-life of 70 to 74 days in man, depending on such factors as age, size, and metabolism, and is excreted mostly in the feces at the rate of about one percent per day. Mercury persists in large fish such as pike from one to two years.

Elemental mercury is generally stored under water; it vaporizes on exposure to air. Certain compounds of mercury may be absorbed through the skin, gastrointestinal tract, and respiratory system (up to 98%), although elemental mercury and inorganic mercury compounds are not absorbed to any significant extent through the digestive tract because they do not remain in the body.

Mercury is ubiquitous in the environment. The sources are both natural and man-made. Natural sources are leachings, erosion, and volitalization from mercury-containing geological formations. Carbonaceous shales average 400 to 500 ppb Hg; up to 0.8 ppm in soil. Man-made sources are waste discharges from chlor-alkali and pulp manufacturing plants, mining, chemical manufacture and formation, the manufacture of mercury seals and controls, treated seeds, combustion of fossil fuels, fallout, and surface runoff. The mercury ends up in lakes, streams, tidal water, and in the bottom mud and sludge deposits.

Microorganisms and macroorganisms in water and bottom deposits can transform metallic mercury, inorganic divalent mercury, phenylmercury and alkoxialkylmercury into methylmercury. The methylmercury thus formed, and perhaps other types, in addition to that discharged in wastewaters, are assimilated and accumulated by aquatic and marine life such as plankton, small fish, and large fish. Alkaline waters tend to favor production of the more volatile dimethylmercury, but acid waters are believed to favor retention of the dimethyl form in the bottom deposits. Under anaerobic conditions, the inorganic mercury ions are precipitated to insoluble mercury sulfide in the presence of hydrogen sulfide. The process of methylation will continue as long as organisms are present and have access to mercury. It is a very slow process, but exposure of bottom sediment such as at low tide permits aerobic action causing methylation of the inorganic mercury.[123]

The form of mercury in fish has been found to be practically all methylmercury, and there are indications that a significant part of the mercury found in eggs and meat is in the form of methylmercury.

The concentration of mercury in fish and other aquatic animals and in wildlife is not unusual. Examination of preserved fish collected in 1927 and in 1939 from Lake Ontario and Lake Champlain in New York State have shown concentrations up to 1.3 ppm mercury (wet basis). Fish from remote ponds, lakes, and reservoirs have shown 0.05 to 0.7 ppm or more mercury, with the larger and older fish showing the higher concentration.

In 1970 the amount of mercury in canned tuna fish averaged 0.32 ppm; in fresh swordfish, 0.93 ppm; in freshwater fish, 0.42 ppm (up to 2.0 and 3.0 ppm

[123]Tadao Ishikawa and Yoshikazu Ikegahi, "Control of mercury pollution in Japan and the Minamata Bay cleanup," *J. Water Pollut. Control Fed.*, May 1980, pp. 1013–1018.

in a few large fish such as walleyed pike); and as high as 8 to 23 ppm in fish taken from heavily contaminated waters. The mercury in urban air is generally in the range of 0.02 to 0.2 $\mu g/m^3$; in drinking water, less than 0.001 ppm; in rainwater, about 0.2 to 0.5 ppb ($\mu g/l$); in ocean water, 0.12 ppb; in Lake Superior water, 0.12 ppb; in Lake Erie, water 0.39 ppb; in soil, 0.04 ppm.[124] Reports from Sweden and Denmark (1967–1969) indicate a mercury concentration of 3 to 8 ppb (ng/g) in pork chops, 9 to 21 ppb in pig's liver, 2 to 5 ppb in beef filet, 9 to 14 ppb in hen's eggs, and 0.40 to 8.4 ppm in pike.

In view of the potential hazards involved, steps have been taken to provide standards or guidelines for mercury. The maximum allowable concentration for 8-hr occupational exposure has been set at 0.05 mg metallic vapor and inorganic compounds of mercury per cubic meter of air. For organic mercury the threshold limit is 0.01 mg/m^3 of air. The suggested limit for fish is 0.5 ppm and, for shellfish, 0.2 ppm. The interim standard for drinking water is 0.002 mg/l (2 ppb) as total mercury. A standard of 0.05 ppm has been suggested for food.

A maximum allowable steady intake (ADI) of 0.03 mg for a 70-kg (154-lb) man would provide a safety factor of ten. If fish containing 0.5 ppm mercury were eaten daily, the limit of 0.03 mg would be reached by the daily consumption of 60 g (about 2 oz) of fish.[125] The safe level for whole blood would be 2 $\mu g/100$ ml and, for hair, 6 ppm.[126]

There is no evidence to show that the mercury in the current daily dietary intake has caused any harm, although this does not rule out possible nondetectable effects on brain cells or other tissues. Nevertheless, from a conservative health standpoint, it has been recommended that pregnant women not eat any freshwater fish until more is known about the effects of methylmercury on fetal brain tissue.[127] The general population should probably not eat more than one freshwater-fish meal per week.

Since mercury comes from man-made and natural sources, every effort must be made to eliminate mercury discharges to the environment. The general preventive and control measures applicable to chemical pollutants were summarized previously under Background, but the goal should be "zero discharge."

Habashi[128] has summarized techniques for the removal of mercury at metallurgical plants in the United States, Europe, and Japan. The author reports that "the removal and recovery of traces of mercury from SO_2 gases or from sulfuric acid has been proved to be technically and economically feasible." In-

[124]"Sources of Mercury to the Environment," Symposium, Mercury in Man's Environment, Royal Society of Canada, February 15, 1971.

[125]*Hazards in Mercury*, a Special Report to the Secretary's Pesticide Advisory Committee, EPA, DHEW, Washington, D.C., November 1970.

[126]Statement of Roger C. Herdman, Deputy Commissioner for Research and Development, New York State Dept. of Health, before the Subcommittee on the Environment of the Committee on Commerce of the U.S. Senate, May 20, 1971.

[127]Statement by Roger C. Herdman, op. cit.

[128]Fathi Habashi, "Metallurgical plants: How mercury pollution is abated," *Environ. Sci. Technol.*, December 1978, pp. 1372–1376.

sofar as water supply is concerned, approximately 98 percent inorganic mercury may be removed by coagulation and settling at a pH of 9.5 followed by filtration through granular activated carbon filter.

Illnesses Associated with Air Pollution—Lung Diseases

The particulate and gaseous irritants in polluted air are reported to increase susceptibility to upper respiratory diseases, such as the common cold, and to aggravate existing illnesses. Diseases mentioned as being *also* associated with air pollution include bronchial asthma (restriction of the smaller airways or bronchioles and increase in mucous secretions), chronic bronchitis (excessive mucous and frequent cough), pulmonary emphysema (shortness of breath), lung cancer, heart diseases, and conjunctivitis (inflammation of the lids and covers of the eyeballs).

A direct single cause-and-effect relationship is often difficult to prove because of the many other causative factors and variables usually involved. Nevertheless, the higher morbidity and mortality associated with higher levels of air pollution and reported episodes (Table 6-2) is believed to show a positive relationship.

Certain air contaminants, depending on the body burden, may produce systemic effects. These include arsenic, asbestos, cadmium, beryllium compounds, mercury, manganese compounds, carbon monoxide, fluorides, hydrocarbons, mercaptans, inorganic particulates, lead, radioactive isotopes, carcinogens, and insecticides. They require attention and are being given consideration in the development of air-quality criteria. See Chapter 6.

Bronchial asthma affects susceptible sensitive individuals exposed to irritant air contaminants and aeroallergens. The aeroallergens include pollens, spores, rusts, and smuts. There also appears to be a good correlation between asthmatic attacks in children and adults and air pollution levels.

Chronic bronchitis has many contributing factors including a low socioeconomic status, occupational exposure, and population density; smoking is a major factor. Air pollution resulting in smoke, particulates, and sulfur dioxide is an additional factor.

Emphysema mortality rates in U.S. urban areas are approximately twice the rural rates, indicating an association with air pollution levels (sulfur oxides). Asthma and bronchitis often precede emphysema.

Lung cancer rates are reported to be higher among the urban populations than rural. The dominant factor in lung cancer is smoking. Air pollution plays a small but continuous role.

Some generalized effects of common air pollutants and their possible relationship to the above are of interest. Sulfur dioxide and sulfuric acid in low concentrations irritate the lungs, nose, and throat. This can cause the membrane lining of the bronchial tubes to become swollen and eroded, with resultant clotting in the small arteries and veins. Children are more susceptible to bronchitis and croup. Carbon monoxide can affect the cardiovascular system; in high concentrations the heart, brain, and physical activity can be impaired. It can reach dangerous levels where there is heavy auto traffic and little wind. Smokers are at greater risk. Acute carbon monoxide poisoning causes a lowered concentration of oxygen in the blood and body tissues. (See separate discussion.) Ozone and other organic oxidants, known as photochemical oxidants, are produced by the reaction of hydrocarbons and nitrogen oxides in sunlight. Ozone is believed to be responsible for a large portion of the health problems associated with photochemical oxidants.[129] Ozone irritates the eyes and air passages, causing chest pain, cough, shortness of breath, and nausea. Ozone can cause severe damage to the lungs. Nitrogen dioxide in high concentrations can result in acute obstruction of the air passages and inflammation of the smaller bronchi. Nitrogen dioxide at low levels causes eye and bronchial irritation. In the presence of strong sunlight, nitrogen dioxide breaks down into nitric oxide and atomic oxygen, and this latter combines with molecular oxygen in air to form ozone. Benzopyrene and related compounds are known to cause some types of cancer under laboratory conditions and are incriminated as carcinogens. Olefins have an injurious effect on certain body cells and are apt to cause eye irritation.[130] USEPA National Emission Standards for Hazardous Air Pollutants identify two organic chemicals, vinyl chloride and benzene, as hazardous. Considerable evidence has been assembled linking air pollution with adverse health effects.[131]

Asbestos Diseases

Asbestosis is caused by fine silicate fibers retained in the lungs. There are six grades of asbestos: crocidolite, amosite, and chrysotile being the most common, coming from serpentine, and actinolite, tremolite, and anthrophyllite. The crocidolite fibers are straight and stiff, the amosite are less so, and the chrysotile are curly. Fibers that are stiff and elongated lodge across the bronchi and eventually pass in to the lung tissue and the pleural cavity. Hence more of the crocidolite is retained in the lungs and may be the cause of most asbestosis.[132]

The most common diseases that might result from asbestos (usually prolonged) exposure are:

1. Asbestosis: a diffuse interstitial nonmalignant, scarring of the lungs;

[129] Basil Dimitriades, "EPA's view of the oxidant problem in Houston," *Environ. Sci. Technol.*, June 1978, pp. 642–43.
[130] *Act Now for Clean Air*, a program to reduce air pollution in New York State, New York State Dept. of Health, Albany, 1968.
[131] *Health Effects of Air Pollution*, American Thoracic Society, American Lung Association, New York, 1978.
[132] P. C. Elmes, "Health Risks From Inhaled Dusts and Fibres," *J. P. Soc. Health*, June 1977, pp. 102–105; also Kelly, Elmes, and Wagg, *J. R. Soc. Health*, December 1976, pp. 246–252.

2. Bronchogenic carcinoma: a malignancy of the interior of the lungs;
3. Mesothelioma: a diffuse malignancy of the lining of the chest cavity (pleural mesothelioma), or of the lining of the abdomen (peritoneal mesothelioma);
4. Cancer of the stomach, colon, and rectum.[133]

A potential health risk exists when asbestos fibers become airborne. The deterioration and exposure of asbestos in old acoustic plaster ceilings, in decorative and textured-spray finishes or paints, and in fire retardant coatings on steel beams, and the demolition of old buildings are examples. Spackling and other patching compounds may contain asbestos which would be released to the ambient air in mixing and sanding to prepare the surface for painting. Fireplaces that simulate live embers and ash usually contain asbestos in a form that can be inhaled. Other sources include furnace patching compounds, old steam pipe covers, floor materials, brake linings, paints, and certain domestic appliances. However, occupational exposure is the major risk.

Airborne asbestos is potentially hazardous where asbestos-containing materials are loosely bound or deteriorating, including areas subject to vibration or abrasion, permitting fibers to be released. Control measures in buildings include removal, coating or sealing (with butyl rubber in inaccessible locations) the surface, enclosure to prevent escape of fibers, or surveillance and taking action when the asbestos material begins to lose its integrity. The coating or sealer must be flame resistant, must not release toxic gases or smoke when burned, or contain asbestos. Removal poses added risks to workers and occupants; it may also cause air pollution and dangers in handling (respirators, disposable garments; showering needed) and disposal. Special precautions must be taken.[134]

Barrett[135] reports that in the occupational field in England, the standard for chrysotile asbestos is 2 fibers/ml of air and for crocidolite 0.2 fibers/ml. The acceptable level for ambient air would become 0.05 fibers/ml for chrysotile and 0.005 fibers/ml for crocidolite. This allows a factor of safety of four to convert from occupational to whole-time exposure and a factor of ten to allow for susceptible members of the population, for a total allowable of one-fortieth of the occupational standard. The USEPA regulates asbestos which is released into the air and into water.[136] The FDA regulates asbestos in food and drugs.

The Occupational Safety and Health Administration of the Department of Labor regulates exposure of workers to asbestos, except agricultural:[136] the 8-hour time-weighted average airborne concentration of asbestos shall not exceed two fibers, longer than 5 μm, per cm^3; at no time during the 8-hour period shall the airborne concentrations exceed 10 fibers per cm^3.

[133] *Fed. Reg.*, Vol. 42, No. 146, Friday, July 29, 1977.

[134] *Asbestos-Containing Materials in School Buildings: A Guidance Document*, Part 1 and 2, USEPA, Washington, D.C., March 1979.

[135] R. S. Barrett, "Ambient Asbestos Levels in Perspective," *J.* R. *Soc. Health*, February 1978, p. 25.

[136] USEPA Regulations for Asbestos, Code of Federal Regulations Title 40, Part 61, Subparts A and B.

Malignant Neoplasms (Cancer)

Cancer is any malignant growth in the body. It is an uncontrolled multiplication of abnormal body cells. The cause of the various types of cancer is unknown, circumstantial, or unclear except for cigarette smoking and exposure to ionizing radiation. There does not appear to be a dosage or level of exposure to cigarette smoking or ionizing radiation below which there is *no* risk. Viruses, genetic background, poor health, and exposure to various agents in our air, water, food, drugs, and cosmetics are believed to contribute to the disease. According to Hamburg,[137] some environmental substances become carcinogenic only after metabolism within the body. "Individual differences in metabolism of these carcinogens may be influenced both by genetic factors and by interaction with other environmental influences."

Figure 1-4 lists known or suspected links between selected pollutants and cancer, as well as heart and lung diseases. Some of the cancers and their associated agents, in addition to cigarette smoking and exposure to ionizing radiation, are tars in smoked fish (cancer of the stomach), asbestos and beryllium aspiration (cancer of the lungs), polyvinyl chloride (cancer of the liver), and aflatoxin.[138]

Mortality data for the United States shows that deaths due to malignant neoplasms in the year 1900 were 64 per 100,000 population, increasing to 147 in 1960, 163 in 1970, and 172 in 1975. This apparent increase may be due to better diagnosis and reporting. Handler[139] points out that

> bronchiogenic carcinoma due to cigarette smoking has risen sharply and the incidence of primary gastric carcinoma has declined dramatically for entirely unknown reasons. These two have more or less offset each other and the age-corrected incidence rate for the total of all forms of cancer has remained approximately constant for a half century.

There are, however, geographic differences for different types of cancer, probably due to cultural patterns and unknown environmental factors. In any case, exposure to known or suspected man-made and natural carcinogenic chemicals should be eliminated or minimized to reduce the incidence of cancer.

Cardiovascular Diseases

The following are the major cardiovascular diseases.

Ischemic Heart Disease (*Coronary Heart Disease*). A deficiency of the blood supply; the principal disease of the heart.

[137]David A. Hamburg, "Disease Prevention: The Challenge of the Future, Sixth Annual Matthew B. Rosenhaus Lecture," *Am. J. Pub. Health*, October 1979, pp. 1026–1033.

[138]*Health Hazards*, Medical Datamation, Belleviue, Ohio, 1976.

[139]From a Dedication Address: "Tracking the cancer epidemic," Northwestern University Cancer Center, May 18, 1979, by Dr. Philip Handler, President National Academy of Sciences, as reported in *Environ. News Dig.*, September/October 1979.

Cerebrovascular Disease (*Stroke*). An occlusion or rupture of an artery to the brain.

Arteriosclerosis. A thickening or hardening of the walls of the arteries, as in old age. Atherosclerosis is the most common form; fatty substances (containing cholesterol) deposited on inner lining restrict the flow of blood in the arteries, causing coronary thrombosis (an occlusion of arteries supplying heart muscle).

Hypertension (*High Blood Pressure*) *and Hypertensive Heart Disease.*

Rheumatic Fever and Rheumatic Heart Disease.

The risk factors associated with cardiovascular diseases include cigarette smoking, poor nutrition, socioeconomic status, age, sedentary way of life, family history, severe stress, personality type, and high blood pressure. Cardiovascular diseases have also been linked to high amounts of saturated fats and cholesterol in the diet. Persons with cardiovascular diseases are more sensitive to carbon monoxide in low concentrations.

The Council on Environmental Quality in its 1977 Annual Environmental Quality Report[140] confirmed reports showing the death rates from cardiovascular diseases tend to decrease as the hardness of drinking water increases; but the factor is not considered to be hardness per se. The direct relationship between cardiovascular death rates and the degree of softness or acidity of water, according to Schroeder, points to cadmium as the suspect.[141] Large concentrations of cadmium may also be related to hypertension in addition to kidney damage, chronic bronchitis, and emphysema. Cadmium builds up in the human body. See also Cadmium, Chapter 3. The indications are that the effects of soft water on cardiovascular diseases may be relatively small. Nevertheless the water association deserves close attention since cardiovascular disease deaths account for about half of all deaths in the United States. Additional substances such as high lithium content in hard water are being correlated with low cardiovascular mortality.[142] See also Hardness, Chapter 3.

There is also evidence associating the ingestion of sodium with heart disease as well as with kidney disease and cirrhosis of the liver. Soft waters and reused waters generally contain higher concentrations of sodium than hard waters. Incidentally, diet drinks generally contain more sodium than regular soft drinks as do sodium-containing dried milk preparations and cream substitutes. High sodium contributes also to disease in infants.[143,144] Home drinking water supplies softened by the ion exchange process (most home softeners) contain too much sodium for persons on sodium restricted diets. This can be avoided by having the cold water line bypass the softener and using only the cold water for drinking and cooking. Other sources of sodium in drinking water are road salt contamination of surface and ground water supplies, the sodium hydroxide, sodium carbonate, and sodium hypochlorite used in water treatment, and

[140]*Environmental Quality*, The eighth annual report of the Council on Environmental Quality, December 1977, Supt. of Documents, GPO, Washington, D.C. 20402, pp. 256–267.

[141]Henry A. Schroeder, *New York State Environment*, N.Y. State Dept. of Environmental Conservation, Albany, N.Y., April 1976.

[142]*J. Am. Water Works Assoc.*, November 1978, p. 8a.

[143]"Treatment Agents and Processes in Drinking Water," *J. R. Soc. Health*, June 1978, p. 137.

[144]"Sodium and heart disease: Are they related?" *Water Wastes Eng.*, November 1978, p. 97.

of course natural minerals in sources of drinking water. The total body burden including that from food and drink must be considered. See also Chemical Examination (Sodium) in Chapter 3.

Methemoglobinemia

The presence of more than 45 mg/l nitrates (10 mg/l as N), the standard for drinking water, appears to be the cause of methemoglobinemia or "blue babies." The disease is largely confined to infants less than three months old, but may affect children up to age six. It is caused by the bacterial conversion of the nitrate ion ingested in water, formula, and other food to nitrite.

> Nitrite then converts hemoglobin, the blood pigment that carries oxygen from the lungs to the tissues, to methemoglobin. Because the altered pigment no longer can transport oxygen, the physiologic effect is oxygen deprivation, or suffocation.[145]

Methemoglobinemia is not a problem in adults as the stomach pH is normally less than four, whereas the pH is generally higher in infants allowing nitrate-reducing bacteria to survive.

The boiling of water containing nitrates would cause the concentration of nitrates to be increased. Parsons[146] presents evidence showing the standard is too low. Also, certain respiratory illness may in themselves cause an increase in methemoglobin levels in infants. A better epidemiological basis for the standard is apparently needed. The inclusion of nitrite ion and nitrates ingested through food and air, in addition to those ingested through water, would give a more complete basis for evaluating dietary intake. Spinach, for example, is a high source of nitrate nitrogen.

Dental Caries

Fluoride deficiency is associated with dental caries and osteoporosis. Water containing 0.8 to 1.7 mg/l natural or artificially added fluoride is beneficial to children during the period they are developing permanent teeth. The incidence of dental cavities or tooth decay is reduced by about 60 percent. The maximum fluoride concentrations based on the annual average of the maximum daily air temperature for the location in which the community water system is situated are established in the National Interim Drinking Water Regulations (Table 3-4.) See also Fluoridation, Chapter 3. An alternate to community water fluoridation is a one-minute mouthrinse by children once a week; it is reported to reduce tooth decay by about one-third or more. Other alternatives include fluoridation of school water supplies if there is an onsite water supply, use of fluoride toothpaste, drops and tablets, and topical application.

[145] E. F. Winton, R. G. Tardiff, and L. J. McCabe, "Nitrates in Drinking Water," *J. Am. Water Works Assoc.*, February 1970, p. 95.

[146] M. L. Parsons, "Is the Nitrate Drinking Water Standard Too Low," *J. Am. Water Works Assoc.*, April 1978, p. 413.

A federal study involving almost one million people in 46 American cities showed virtually no difference in death rates, including from cancer, between 24 cities using fluoridated water and 22 without fluoridated water.[147]

The long-term consumption of water high in fluoride (8–20 mg/l) is reported to cause bone changes. An intake of 20 mg fluoride per day for 20 or more years may cause crippling fluorosis, and death can come from a single dose of 2250 to 4500 mg.[148] On the other hand, optimal concentrations of fluoride in drinking water and food appear to be beneficial in preventing osteoporosis.

Hypothermia

The maintenance of a normal body temperature at or near 98.4°F (37°C) is necessary for proper body function. When the body core temperature drops to 95°F (35°C) or below, the vital organs (brain, heart, lungs, kidneys) are affected causing what is known as hypothermia, a relatively rare condition. Rectal temperature measurement is necessary to get a correct reading. Predisposing conditions for hypothermia include old age, poor housing, inadequate clothing, poverty, lack of fuel, illness, cold weather, alcohol, and drugs.[149]

Proper body temperature requires a balance between body heat generated and heat loss. Bald people lose a great deal of heat; fat people are better insulated and lose less heat on a weight-body surface basis. Disease and drugs affect heat loss. Wind and dampness increase coldness. The maintenance of warmth and comfort is related to the prevailing temperature, building design and construction, clothing, heating and cooling facilities, and food consumed and also to air movement, radiant heat, relative humidity, the tasks performed, and the age and health status of individuals. At greater risk are babies and the elderly, particularly those already suffering from an acute or chronic illness. In view of the above, a minimum temperature of 70°F (21°C) is recommended, with provision for heating and cooling above and below that temperature.[150]

Signs of hypothermia are bloated face; pale and waxy skin, or pinkish color; drowsiness; low blood pressure; irregular and slow heart beat; shallow very slow breathing; and trembling of leg, arm, or side of body, but no shivering.

High Environmental Temperatures

Heat waves have been associated with marked increases in morbidity and mortality in the U.S. but these are preventable. The measures that have been shown to be effective to reduce heat stress include:

[147] "Fluoridated water found safe," *Water Wastes Eng.*, July 1978, p. 12 (As reported in the *New Eng. J. Med.*).

[148] Thomas J. Sorg, "Treatment Technology to Meet the Interim Primary Drinking Water Regulations for Inorganics," *J. Am. Water Works Assoc.*, February 1978, pp. 105–112.

[149] Michael Green, "Home Heating," *J. R. Soc. Health*, September 1977, p. 177.

[150] *Housing: Basic Health Principles & Recommended Ordinance*, American Public Health Association, 1015 Fifteenth Street, N.W., Washington, D.C., 20005, 1971, pp. 1–5.

1. Keep as cool as possible.
   a. Avoid direct sunlight.
   b. Stay in the coolest available location (it will usually be indoors).
   c. Use air conditioning, if available.
   d. Use electric fans to promote cooling.
   e. Place wet towels or ice bags on the body or dampen clothing.
   f. Take cool baths or showers.
2. Wear lightweight, loose-fitting clothing.
3. Avoid strenuous physical activity, particularly in the sun and during the hottest part of the day.
4. Increase intake of fluids, such as water and fruit or vegetable juices. Thirst is not always a good indicator of adequacy of fluid intake. Persons for whom salt or fluid is restricted should consult their physicians for instructions on appropriate fluid and salt intake.
5. Do not take salt tablets unless so instructed by a physician.
6. Avoid alcoholic beverages (beer, wine, and liquor).
7. Stay in at least daily contact with other people.

Special precautions should be taken for certain higher-risk groups. These safeguards may include increased efforts to keep cool or closer observation by others for early signs of heat illness. The high-risk groups are infants less than a year of age; persons over 65 years of age; persons who are less able to care for themselves because of chronic mental illness or dementia of any cause; persons with chronic diseases, especially cardiovascular or kidney disease; and those taking any of the three classes of medication that reduce the ability to sweat: diuretics ("water pills"), tranquilizers, and drugs used for the treatment of gastrointestinal disorders.[151] Add building insulation and ventilation.

Nutritional Deficiency and Related Diseases

Severe examples of diseases caused by deficiencies in the diet are not common in the United States. This cannot be said of many less developed countries of the world. There are, however, many people whose diet is slightly deficient in one or more nutrients but who show no clinically detectable symptoms for many years. Most malnutrition takes the form of protein deficiency.* Social, cultural, and emotional patterns, and ignorance contribute to the problem, regardless of the economic status of individuals. Diarrheal diseases and consequent malabsorption compound nutritional deficiency, hence basic environmental sanitation including safe water- and food-handling, refrigeration, and hygiene, are an essential part of a good nutrition program.

The vague and insidious nature of these diseases is good reason to apply known preventive and control measures continuously. Recommended daily

*M. B. Stoch and P. M. Smythe reported that children who were grossly undernourished from birth had smaller heads, lower intelligence quotients, and poorer body coordination than did a control group from the same socioeconomic environment. "The Effect of Undernutrition During Infancy on Subsequent Brain Growth and Intellectual Development," *South African Med. J.* (October 28, 1967).

[151]Extracted from "Heat Wave-Related Morbidity and Mortality—Missouri," *MMWR*, CDC, Atlanta, Ga., August 15, 1980, pp. 390–392.

dietary allowances for the maintenance of good nutrition, to be consumed in a variety of foods to provide other less defined required nutrients, are shown in Table 1-14. Of the more than 60 mineral elements found in living things, nine are considered essential to human life.[152] These are iron, iodine, fluoride, copper, manganese, zinc, selenium, chromium, and cobalt. The role of other minerals is not well established. Several of the nutritional diseases are mentioned below.

Scurvy

This is a disease caused by a deficiency of vitamin C or ascorbic acid, which is found in citrus fruit, fresh strawberries, tomatoes, raw peppers, broccoli, kale, potatoes, and raw cabbage. Weakness, anemia, spongy and swollen gums that bleed easily, and tender joints are some of the common symptoms. Vitamin C also strengthens body cells and blood vessels and aids in absorption of iron and in healing wounds and broken bones.

Pellagra

This disease is due to a prolonged deficiency of niacin (nicotinic acid) or tryptophan (amino acid). Niacin is found in eggs, lean meats, liver, whole-grain cereals, milk, leafy green vegetables, fruits, and dried yeast. Recurring redness of the tongue or ulcerations in the mouth are primary symptoms, sometimes followed by digestive disturbances, headache and mental depressions.

Rickets

This is a childhood disease (under two years) caused by the absence of vitamin D, which is associated with proper utilization of calcium and phosphorous. Vitamin D is found in liver, fortified milk, butter, eggs, and fish of high body oil content such as sardines, salmon, and tuna. An inadequate supply of vitamin D in the diet will probably show in knock-knees or bowed legs, crooked arms, soft teeth, potbelly, and faulty bone growth. Sunshine is a good source of Vitamin D, as are vitamin D fortified foods. Vitamin D helps build strong bones and teeth.

Beriberi

A prolonged deficiency of thiamin or vitamin B_1, found in whole-grain cereals, dried beans, peas, peanuts, pork, fish, poultry, and liver, may cause changes in the nervous system, muscle weakness, loss of appetite, and interference with digestion. Change from unpolished to polished rice in the diet can cause the disease in some countries where the diet is not varied.

Ariboflavinosis

This disease is due to a deficiency of riboflavin, known also as vitamin B_2 or G. Riboflavin is found in liver, milk, eggs, dried yeast, enriched white flour, and leafy green vegetables. An inadequate amount of this vitamin may cause

[152] Jean Mayer, "Those Mysterious Minerals," *Family Health*, September 1980, pp. 36–37.

Table 1-14 Food and Nutrition Board, National Academy of Sciences—National Research Council Recommended Daily Dietary Allowances,[a] Revised 1980

							Fat-Soluble Vitamins			Water-Soluble Vitamins							Minerals					
	Age (years)	Weight (kg)	Weight (lb)	Height (cm)	Height (in)	Protein (g)	Vitamin A (μg RE)[b]	Vitamin D (μg)[c]	Vitamin E (mg α-TE)[d]	Vitamin C (mg)	Thiamin (mg)	Riboflavin (mg)	Niacin (mg NE)[e]	Vitamin B-6 (mg)	Folacin[f] (μg)	Vitamin B-12 (μg)	Calcium (mg)	Phosphorus (mg)	Magnesium (mg)	Iron (mg)	Zinc (mg)	Iodine (μg)
Infants	0.0–0.5	6	13	60	24	kg × 2.2	420	10	3	35	0.3	0.4	6	0.3	30	0.5[g]	360	240	50	10	3	40
	0.5–1.0	9	20	71	28	kg × 2.0	400	10	4	35	0.5	0.6	8	0.3	45	1.5	540	360	70	15	5	50
Children	1–3	13	29	90	35	23	400	10	5	45	0.7	0.8	9	0.9	100	2.0	800	800	150	15	10	70
	4–6	20	44	112	44	30	500	10	6	45	0.9	1.0	11	1.3	200	2.5	800	800	200	10	10	90
	7–10	28	62	132	52	34	700	10	7	45	1.2	1.4	16	1.6	300	3.0	800	800	250	10	10	120
Males	11–14	45	99	157	62	45	1000	10	8	50	1.4	1.6	18	1.8	400	3.0	1200	1200	350	18	15	150
	15–18	66	145	176	69	56	1000	10	10	60	1.4	1.7	18	2.0	400	3.0	1200	1200	400	18	15	150
	19–22	70	154	177	70	56	1000	7.5	10	60	1.5	1.7	19	2.2	400	3.0	800	800	350	10	15	150
	23–50	70	154	178	70	56	1000	5	10	60	1.4	1.6	18	2.2	400	3.0	800	800	350	10	15	150
	51+	70	154	178	70	56	1000	5	10	60	1.2	1.4	16	2.2	400	3.0	800	800	350	10	15	150
Females	11–14	46	101	157	62	46	800	10	8	50	1.1	1.3	15	1.8	400	3.0	1200	1200	300	18	15	150
	15–18	55	120	163	64	46	800	10	8	60	1.1	1.3	14	2.0	400	3.0	1200	1200	300	18	15	150
	19–22	55	120	163	64	44	800	7.5	8	60	1.1	1.3	14	2.0	400	3.0	800	800	300	18	15	150
	23–50	55	120	163	64	44	800	5	8	60	1.0	1.2	13	2.0	400	3.0	800	800	300	18	15	150
	51+	55	120	163	64	44	800	5	8	60	1.0	1.2	13	2.0	400	3.0	800	800	300	10	15	150
Pregnant						+30	+200	+5	+2	+20	+0.4	+0.3	+2	+0.6	+400	+1.0	+400	+400	+150	[h]	−5	+25
Lactating						+20	+400	+5	+3	+40	+0.5	+0.5	+5	+0.5	+100	+1.0	+400	+400	+150	[h]	+10	+50

Source: *Recommended Dietary Allowances*, 9th ed., 1980, with the permission of the National Academy of Sciences, Washington, D.C.

Note: "Designed for the maintenance of good nutrition of practically all healthy people in the U.S.A." See source publication for discussion of dietary allowances, nutrients, suggested energy intakes, etc.

[a]The allowances are intended to provide for individual variations among most normal persons as they live in the United States under usual environmental stresses. Diets should be based on a variety of common foods in order to provide other nutrients for which human requirements have been less well defined.

[b]Retinol equivalents. 1 retinol equivalent = 1 μg retinol or 6 μg β carotene.

[c]As cholecalciferol. 10 μg cholecalciferol = 400 IU of vitamin D.

[d]α-tocopherol equivalents. 1 mg d-α tocopherol = 1 α-TE.

[e]1 NE (niacin equivalent) is equal to 1 mg of niacin or 60 mg of dietary tryptophan.

[f]The folacin allowances refer to dietary sources as determined by *Lactobacillus casei* assay after treatment with enzymes (conjugases) to make polyglutamyl forms of the vitamin available to the test organism.

[g]The recommended dietary allowance for vitamin B-12 in infants is based on average concentration of the vitamin in human milk. The allowances after weaning are based on energy intake (as recommended by the American Acadaemy of Pediatrics) and consideration of other factors, such as internal absorption.

[h]The increased requirement during pregnancy cannot be met by the iron content of habitual American diets nor by the existing iron stores of many women; therefore the use of 30–60 mg of supplemental iron is recommended. Iron needs during lactation are not substantially different from those of nonpregnant women, but continued supplementation of the mother for 2–3 months after parturition is advisable in order to replenish stores depleted by pregnancy.

greasy scales on the ear, forehead, and other parts of the body, drying of the skin, cracks in the corners of the mouth, and sometimes partial blindness. Riboflavin is essential for many enzyme systems.

Vitamin A Deficiency

This disease causes night blindness, skin and mucous membrane changes, and increases susceptibility to colds. Vitamin A is also needed for bone growth. The diet should be adjusted to include foods rich in vitamin A or carotene, such as dry whole milk and cheese, butter, margarine, eggs, liver, carrots, dandelions, kale, and sweet potatoes.

Liver Cirrhosis

This disease is caused by malnutrition and probably aggravated by alcohol. A high protein diet plus choline and dried brewer's yeast is indicated.

Iron-Deficiency Anemia

Lack of vitamin B_{12} or folic acid, repeated loss of blood, and increased iron need during pregnancy cause weakness, irritability, brittle fingernails, cuts and sores on the face at the mouth, and other debility. Prevention of blood loss and treatment with iron salts is suggested. Iron combines with protein to make the hemoglobin of the red blood cells which distribute oxygen from the lungs to body tissues. Consumption of liver, lean meats, poultry, shellfish, eggs, oysters, dried fruits, dark green leafy vegetables, flour, and cereal foods will contribute iron to the diet.

Goiter

The WHO reports that an estimated 200 million people throughout the world are unnecessarily affected by goiter. This is a thyroid disorder usually caused by deficient iodine content in food and water and inadequate iodine absorption. Universal use of iodized or iodated salt has practically eliminated goiter. Seafoods are good sources of iodine.

Kwashiorkor

Kwashiorkor is a protein deficiency nutritional disease common among children under about six years of age living in underdeveloped areas of the world. Changes in the color and texture of the hair, diarrhea, and scaling sores are some of the clinical signs. A diet rich in animal proteins, including dry skim milk, meat, eggs, fish, and cheese, and vegetables, including soybeans and Incaparina, also WSDM (sweet whey: 41.5 percent, full fat soy flour: 36.5 percent; soy-bean oil: 12.2 percent; corn syrup: 9 percent; and vitamins and minerals) and CSM (corn soya-milk) can control the disease. But the people affected must be taught to accept the high protein foods, to use a safe water supply in reconstituting manufactured foods, and the importance of refrigeration and environmental hygiene to prevent debility due to diarrhea and parasitism and thus also reduce susceptibility to kwashiorkor and other diseases.

Osteoporosis

In osteoporosis, the bone is decalcified and becomes porous and brittle, particularly in women after menopause, possibly due to dietary factors and a decrease in female hormones. Maintenance of an adequate level of calcium and vitamin D may offset the disease. Fluorides in proper amounts in drinking water and food also appear to be helpful in preventing osteoporosis. Calcium helps build teeth and bones, aids in bloodclotting, and helps maintain muscles and nerves. Major sources of calcium are milk, cheese and other milk products, egg yolk, and dark green leafy vegetables.

Diseases of the Bowel

Diets low in fiber have been associated with diseases of the bowel, including diverticulosis and cancer, but more confirmation information is needed. The consumption of more cereal products, potatoes, raw fruits, and vegetables has been suggested.

Obesity

Obesity is primarily the result of consuming large amounts of food containing high concentrations of sugar and fat, coupled with sedentary living. It contributes to heart disease, diabetes and other degenerative diseases.

INVESTIGATION OF WATER- AND FOODBORNE DISEASE OUTBREAKS

General

The promptness with which an investigation is started will largely determine its success. In this way the cause of the disease may be determined and precautions for the prevention of future incidents learned. One water- or foodborne outbreak at an institution, camp, hotel, or eating place, or associated with a food processing plant, will cause severe repercussions and may result in great financial loss.

Upon learning of the existence of an unusual incidence of a disease, it is the duty of the owner or manager and physician called, or the water superintendent as the case may be, to immediately notify the health department. They are required to cooperate with the investigating authority. The health department, in turn, should report to the owner or operator of the establishment and to the PHS upon completion of the investigation. Such notification and reporting, however, have been sporadic. The PHS Center for Disease Control compiles statistical reports for the entire country from such information and provides special investigatory assistance on request.

A general knowledge of the cause, transmission, characteristics, and control of diseases is necessary in order to conduct an epidemiological investigation. This is usually made under the direction of an epidemiologist and a sanitary engineer or sanitarian if available. The background material was discussed earlier in this chapter. Reference to Figure 1-2 will give, in easily accessible form, the characteristics of various water- and foodborne diseases and perhaps a clue as to what might be responsible for a disease outbreak. It is not a simple matter to quickly determine the cause of illness due to water or food, but a preliminary study of the symptoms, incubation periods, food and water consumed, housing, bathing area, and sanitary conditions may give a lead and form a basis for the immediate control action to be instituted. Publications summarizing disease outbreak investigation procedures are very helpful.[153,154]

A common method of determining the probable offending food is a tabulation as shown in Figure 1-5, which is made from the illness questionnaire, Figure 1-6. Comparison of the attack rates for each food will usually implicate or absolve a particular food.

A summation can be made in the field to take the place of individual questionnaires when assistance is available. The tabulation horizontal headings would include:

1. Names of persons served.
2. Age.
3. Ill—yes or no.
4. Day and time ill.
5. Incubation period in hours (time between consumption of food and first signs of illness).
6. Foods served at suspected meals—previous 12 to 72 hours (check foods eaten).
7. Symptoms—nausea, vomiting, diarrhea, blood in stool, fever, thirst, constipation, stomach ache, sweating, sore throat, headache, dizziness, cough, chills, pain in chest, weakness, cramps, other.

A simple bar graph, with hour and days as the horizontal axis and number ill each hour or other suitable interval plotted on the vertical axis, can be made from the data. The time between exposure and illness or between peaks represents the incubation period. The average incubation period is the sum of the incubation periods of those ill (time elapsing between the initial infection and the clinical onset of a disease) divided by the number of ill persons studied.

Sanitary Survey

The sanitary survey should include a study of all environmental factors that might be the cause, or may be contributing to the cause, of the disease outbreak. These should include water supply, food, housing, sewage disposal, bathing, and any other relevant factors. Each should be considered responsible for the illness until definitely ruled out. Figure 1-2 and Chapters 1, 3, 4, 8, 9,

[153] *Procedures To Investigate Foodborne Illnesses*, International Association of Milk, Food, and Environmental Sanitarians, P.O. Box 701, Ames, Iowa 50010, 1976.

[154] Frank L. Bryan, *Guide for Investigating Disease Outbreaks and Analyzing Surveillance Data*, DHEW, PHS, CDC, Atlanta, Ga. 30333, 1973.

FORM APPROVED
OMB NO. 68-R557

INVESTIGATION OF A FOODBORNE OUTBREAK

1. Where did the outbreak occur?

State__________(1,2) City or Town __________ County __________

2. Date of outbreak: (Date of onset 1st case)

__________(3-8)

3. Indicate actual (a) or estimated (e) numbers:

Persons exposed__________(9-11)
Persons ill __________ (12-14)
Hospitalized__________(15-16)
Fatal cases __________(17)

4. History of Exposed Persons:

No. histories obtained __________ (18-20)
No. persons with symptoms __________(21-23)
Nausea______(24-26) Diarrhea______(33-35)
Vomiting______(27-29) Fever______(36-38)
Cramps______(30-32) Other, specify______
__________(39)

5. Incubation period (hours):

Shortest ______(40-42) Longest ______(43-45)
Approx. for majority__________(46-48)

6. Duration of Illness (hours):

Shortest ______(49-51) Longest ______(52-54)
Approx. for majority __________(55-57)

7. **Food-specific attack rates: (58)**

Food Items Served	Number of persons who ATE specified food				Number who did NOT eat specified food			
	Ill	Not Ill	Total	Percent Ill	Ill	Not Ill	Total	Percent Ill

8. Vehicle responsible (food item incriminated by epidemiological evidence): (59,60)__________

9. Manner in which incriminated food was marketed: (Check all applicable)

(a) Food Industry (61)
Raw . . . . ☐ 1
Processed . . . . . . ☐ 2
Home Produced
Raw . . . . ☐ 3
Processed . . . . . . ☐ 4

(b) Vending Machine . . . ☐ 1 (62)

(c) Not wrapped . . . . ☐ 1 (63)
Ordinary Wrapping . . . . . . ☐ 2
Canned . . . . ☐ 3
Canned--Vacuum Sealed . . ☐ 4
Other (specify) . . . . ☐ 5
____

(d) Room Temperature . . . . . ☐ 1 (64)
Refrigerated . . . . ☐ 2
Frozen . . . . ☐ 3
Heated . . . . ☐ 4

If a commercial product, indicate brand name and lot number
____

10. Place of Preparation of Contaminated Item: (65)

Restaurant . . . . ☐ 1
Delicatessen . . . . ☐ 2
Cafeteria . . . . ☐ 3
Private Home . . . . ☐ 4
Caterer . . . . ☐ 5
Institution:
School . . . . ☐ 6
Church . . . . ☐ 7
Camp . . . . ☐ 8
Other, specify . . . . ☐ 9
____

11. Place where eaten: (66)

Restaurant . . . . . . ☐ 1
Delicatessen . . . . . ☐ 2
Cafeteria . . . . ☐ 3
Private Home . . . . ☐ 4
Picnic . . . . ☐ 5
Institution:
School . . . . ☐ 6
Church . . . . ☐ 7
Camp . . . . ☐ 8
Other, specify . . . . ☐ 9
____

DEPARTMENT OF HEALTH, EDUCATION, AND WELFARE*
PUBLIC HEALTH SERVICE
CENTER FOR DISEASE CONTROL
BUREAU OF EPIDEMIOLOGY
ATLANTA, GEORGIA 30333

*Renamed Department of Health and Human Services

CDC 4.245
1-74

(Over)

Figure 1-5 Investigation of a foodborne outbreak.

LABORATORY FINDINGS (Include Negative Results)

12. Food specimens examined: (67)
Specify by "X" whether food examined was original (eaten at time of outbreak) or check-up (prepared in similar manner but not involved in outbreak)

Item	Orig.	Check up	Findings Qualitative	Quantitative
Example: beef	X		C. perfringens, Hobbs type 10	$2X10^6$/gm

13. Environmental specimens examined: (68)

Item	Findings
Example: meat grinder	C. perfringens, Hobbs Type 10

14. Specimens from patients examined (stool, vomitus, etc.): (69)

Item	No. Persons	Findings
Example: stool	11	C. perfringens, Hobbs Type 10

15. Specimens from food handlers (stool, lesions, etc.): (70)

Item	Findings
Example: lesion	C. perfringens, Hobbs type 10

16. Factors contributing to outbreak (check all applicable):

	Yes	No
1. Improper storage or holding temperature	☐ 1	☐ 2 (71)
2. Inadequate cooking	☐ 1	☐ 2 (72)
3. Contaminated equipment or working surfaces	☐ 1	☐ 2 (73)
4. Food obtained from unsafe source	☐ 1	☐ 2 (74)
5. Poor personal hygiene of food handler	☐ 1	☐ 2 (75)
6. Other, specify	☐ 1	☐ 2 (76)

17. Etiology: (77, 78)
Pathogen ______ Suspected ☐ 1 (79)
Chemical ______ Confirmed ☐ 2
Other ______ Unknown ☐ 3

18. Remarks: Briefly describe aspects of the investigation not covered above, such as unusual age or sex distribution; unusual circumstances leading to contamination of food, water; epidemic curve; etc. (Attach additional page if necessary)

Name of reporting agency: (80)

Investigating official: | Date of investigation:

NOTE: Epidemic and Laboratory Assistance for the investigation of a foodborne outbreak is available upon request by the State Health Department to the Center for Disease Control, Atlanta, Georgia 30333.

To improve national surveillance, please send a copy of this report to:
Center for Disease Control
Attn: Enteric Diseases Section, Bacterial Diseases Branch
Bureau of Epidemiology
Atlanta, Georgia 30333

Submitted copies should include as much information as possible, but the completion of every item is not required.

CDC 4.245 (BACK)
1-74

Figure 1-5 (*Continued*)

Please answer the questions asked below to the best of your ability. This information is desired by the health department to determine the cause of the recent sickness and to prevent its recurrence. Leave this sheet, after you have completed it, at the desk on your way out. (If mailed, enclose self-addressed and stamped envelope and request return of completed questionnaire as soon as possible.)

1. Check any of the following conditions that you have had:

Nausea	Fever	Sore throat	Cough	Chills
Vomiting	Constipation	Headache	Pain in chest	Weakness
Diarrhea	Stomach ache	Dizziness	Laryngitis	Cramps
Thirst	Sweating	Paralysis	Bloody stool	Other

2. Were you ill .... Yes .... No.
3. If ill, first became sick on: Date.............. Hour.............. A.M./P.M.
4. How long did the sickness last?..............................
5. Check below (✓) the food eaten at each meal and (×) the food not eaten. Answer even though you may not have been ill.

Meal	Tuesday	Wednesday	Thursday
Breakfast	Apple juice, Corn flakes, Oatmeal, Fried eggs, Bread, Coffee, Milk, Water	Orange, Pancakes, Wheaties, Syrup, Coffee, Milk, Water	Grapefruit, Wheatina, Shredded wheat, Boiled egg, Coffee, Milk, Water
Lunch	Baked salmon, Creamed potatoes, Corn, Apple pie, Lemonade, Water	Roast pork, Baked potatoes, Peas, Rice pudding, Milk, Water, Chef salad	Swiss steak, Home fried potatoes, Turnips, Spinach, Chocolate pudding, Orange drink, Milk, Water
Dinner	Gravy, Hamburger steak, Mashed potatoes, Salmon salad, Cookies, Pears, Cocoa, Water	Roast veal, Rice, Beets, Peas, Jello, Coffee, Water	Fruit cup, Meat balls, Spaghetti, String beans, Pickled beets, Sliced pineapple, Tea, Coffee, Milk

6. Did you eat food or drink water outside?.......... If so, where and when?
....
7. Name.............................. Tel. .... Age .... Sex ....
8. Remarks (Physician's name, hospital)..............................
.... Investigator ....

Figure 1-6 Questionnaire for illness from food, milk, or water.

10, and 11 should be referred to for guidance and possible specific contributing causes to an outbreak and their correction. Figure 1-7 will be helpful. Water system, food service, housing, and swimming pool sanitary survey report forms are usually available from the state or local health department. A WHO publication also has a water system reporting form[155] and the USEPA an evaluation manual.[156] See Figure 1-8.

Medical Survey

The medical survey should develop a clinical picture to enable identification of the disease. Typical symptoms, date of onset of the first case, date of onset of last case, range of incubation periods, number of cases, number hospitalized, number of deaths, and number exposed are usually determined by the epidemiologist. To assemble this information and analyze it carefully, a questionnaire should be completed, by trained personnel if possible, on each person available or on a sufficient number to give reliable information. A typical simple, short question sheet is shown in Figure 1-6.

A very important part of the medical survey is examining all the food handlers and assembling a medical history on each one. Frequently a carrier or a careless infected food handler can be found at the bottom of a foodborne outbreak. The importance of animal reservoirs of infection should not be overlooked. Figure 1-2 summary sheets give, in condensed form, symptoms and incubation period of many diseases that, when compared to a typical clinical picture, may suggest the causative organism and the disease.

Samples

The collection of samples for laboratory examinations is a necessary part of the investigation of water- and foodborne disease outbreaks. Isolating the incriminating organism from the persons made ill, the allegedly responsible food and drink, and producing the characteristic symptoms in laboratory animals or human volunteers, and then isolating the same organisms from human volunteers or laboratory animals will confirm the field diagnosis and definitely implicate the responsible vehicle. In the early stages of the field investigation it is very difficult to determine just what samples to collect; yet if the samples of food, for example, are not collected at that moment, the chances are the samples will be gone when they are wanted. It is customary, therefore, to routinely collect samples of water from representative places and available samples of all leftover milk and food that had been consumed and place them under seal and refrigeration. Sterile spatulas or spoons boiled for 5 min can

[155] "Surveillance of Drinking Water Quality," *WHO Monog. Ser.*, **63** (1976), pp. 108–115.

[156] *Manual For Evaluating Public Drinking Water Supplies*, USEPA Office of Water Programs, Washington, D.C., 1971.

be used to collect samples. Sterile wide-mouth water bottles and petri dishes make suitable containers. In all cases, a sterile technique must be used. Since examination of all the food may be unnecessary it is advisable, after studying the questionnaires and accumulated data, to select the suspicious foods for laboratory examination and set aside the remaining food in protected sterile containers under refrigeration at a temperature of less than 40°F for possible future use.

Samples of water should be collected directly from the source, storage tanks, high and low points of the distribution system at times of high and low pressure, kitchens, and taps near drinking fountains for chemical and bacterial examinations. Samples of milk should be collected from unopened and opened bulk milk cartons or containers, and leftover milk in pitchers or other containers for phosphatase tests, coliform tests, direct microscopic counts, standard plate counts, presence of streptococcus, and any other specific tests that may be indicated. Samples of food possibly incriminated should be collected from the refrigerator, storeroom, or wherever available. As a last resort, samples have been collected from garbage pails or even dumps; however, results on such samples must be interpreted with extreme caution. Figure 1-2 will be of assistance in suggesting the common vehicle, and hence the foods to suspect, by using the incubation periods and symptoms of those ill as a guide. It should be remembered that the time elapsing before symptoms appear is variable and depends on the causative agent and size of dose, the resistance of individuals, the amount and kind of food or drink consumed. For example, an explosive outbreak with a very short incubation period of a few minutes to less than an hour would suggest a chemical poisoning. Antimony, arsenic, cadmium, cyanide, mercury, sodium fluoride, sodium nitrate, or perhaps shellfish poisoning, favism, fish poisoning, and zinc poisoning are possibilities. An explosive outbreak with an incubation period of several hours would suggest botulism or fish, mushroom, potato, rhubarb-leaf, shellfish, chemical, or staphylococcus food poisoning. An incubation period of 6 to 24 hours would suggest botulism, mushroom poisoning, rhubarb poisoning, salmonella infection, or streptococcus food poisoning. An incubation period of 1 to 5 days would suggest ascariasis, botulism, diphtheria, amebic dysentery, bacillary dysentery, leptospirosis, paratyphoid fever, salmonella infection, scarlet fever, streptococcal sore throat, or trichinosis. For other diseases with more extended incubation periods, refer to Figure 1-2. The laboratory examinations might be biologic, toxicologic, microscopic, or chemical, depending on the symptoms and incubation period.

Samples collected from food handlers may include stool, vomitus, urine, skin, nose, throat, and other membrane specimens, or swabs for biologic or microscopic examinations. Medical and laboratory study to determine the origin of the food contaminant may be carried over an extended period of time, since the absence of disease-producing organisms in three stools is generally required before a person is considered negative, although in screening, one stool specimen will reveal a high percentage of carriers.

Some of those ill with the severest symptoms, say about ten, should be selected for laboratory study to determine, if possible, the organism causing the illness. The samples collected and the examinations made would be indicated by the symptoms and dates of onset displayed by the sufferers as explained above. Stool, urine (chemical poisoning), rectal swab, serum, vomitus, skin, nose, and throat swabs may be indicated.[157] Samples from some not ill should also be taken.

When chemical poisoning is suspected, samples of flour, sugar, and other foods that might have been contaminated in transportation, handling, and preparation should be considered. Galvanized or zinc-, cadmium-, and antimony-plated utensils and liquids prepared in such containers should be collected for chemical and toxicologic tests. Also include samples of insecticides, rodenticides, ant powders or sprays, and silver polish.

The sanitary and medical surveys may involve the swimming pool or bathing beach. In that case, samples should be collected at the peak and toward the end of the bathing period for examinations.

It is customary to notify the laboratory in advance that an outbreak has occurred and that specimens will be delivered as soon as possible. All samples should be carefully identified, dated, sealed, and refrigerated. A preliminary report with the samples, including the probable cause, number ill, age spread, symptoms, incubation period, and so on, will greatly assist the laboratory in its work.

Laboratory analyses for water samples should include standard plate count, since large bacterial populations may suppress the growth of coliform organisms, and as appropriate campylobacter, salmonella, shigella, vibrio, *Yersinia enterocolitica*, *Escherichia coli*, and possibly leptospira, or naegleria isolates. Where large volumes of water are needed, use 2 to 5 gal sterile containers and store at 5°C. Sampling for recovery of viruses and giardia or entamoeba cysts require special onsite filters and equipment.[158]

Preliminary Recommendations

Before leaving the scene of an investigation, a preliminary study should be made of the data accumulated and temporary control measures instituted based on the findings summarized in Figures 1-6 and 1-7. Instructions should be left by the health officer and sanitary engineer or sanitarian with the owner or manager regarding those who are ill, the housing, water supply, food and milk supply, food handlers, sewage disposal, swimming and bathing place, and any other items that may be indicated.

[157]Frank L. Bryan, *Guide for Investigating Disease Outbreaks and Analyzing Surveillance Data*, DHEW, PHS, CDC, Atlanta, Ga. 30333, 1973, p. 21.

[158]Gunther F. Craun and Robert A. Gunn, "Outbreaks of Waterborne Disease in the United States: 1975–1976," *J. Am. Water Works Assoc.*, August 1979, pp. 442–428.

Date........................ Investigator........................
Name of place........................ Owner........................
Population........................ Manager........................
Onsets—day and hours........................ Incubation period........................
Number afflicted.............. Number hospitalized.............. Number deaths..............
Outbreak: explosive.............. gradual.............. undetermined..............
Samples collected........................

Underline symptoms most commonly reported:

Diarrhea, constipation, abdominal pains, stomach cramps, muscular cramps, prostration, high temperature, painful straining at stool or in urination, sore throat, chills, thirst, sweating, vomiting, nausea, swelling of face and eyelids, laryngitis, cough, pain in chest, enlarged tonsils or adenoids, pains in joints, eye movement difficult, swallowing difficult, headache, dizziness, other........................

Water

1. Water sources and treatment..............
2. Method of serving water..............
3. Interconnections: toilet.............. washbasin.............. bath tubs.............. tubs.............. other..............
4. Recent repairs..............
5. Cross-connections, with other supplies..............
6. Changes in water taste.............. color.............. odor..............

Milk and food

7. Source of milk (pasteurized)..............
8. Method of handling milk..............
9. Use of leftover foods..............
10. Source of fowl, meats, ice cream, shellfish, pastries..............
11. Food refrigeration and storage..............
12. Food handling and preparation..............
13. Ice source and handling..............
14. Thawing foods protected..............
15. Dressings, sauces, etc...............

Food handlers

16. Recent illness in food handlers..............
17. Hand-washing facilities..............
18. No. pyogenic skin infections..............
19. Personal hygiene..............

Kitchen and dining hall

20. Storage and use of insecticides.............. rat poison.............. roach powder.............. water paint.............. silver polish..............
21. Garbage storage and disposal..............
22. Prevalence of rodents and insects..............
23. Fly breeding controlled..............
24. Dish cleansing and disinfection..............
25. Premises and equipment clean..............
26. Food service well organized..............

Other

27. Housing overcrowding..............
28. Bathing beach or swimming pool operation, water source..............
29. Medical and nursing care..............
30. Other..............

Remarks (Comment on unsatisfactory items and probable cause, general impressions, etc.):

Figure 1-7 Outbreak investigation field summary.

Epidemiologic Report

The report of an investigation of a water- or foodborne disease outbreak should include the cause, laboratory findings, transmission, incidence, cases by dates of onset, average incubation period and range, typical symptoms, length of illness, age and sex distribution, deaths, secondary attack rate, and recommendations for the prevention and control of the disease. This can often

DEPARTMENT OF
HEALTH, EDUCATION, AND WELFARE
PUBLIC HEALTH SERVICE
CENTER FOR DISEASE CONTROL
BUREAU OF EPIDEMIOLOGY
ATLANTA, GEORGIA 30333

INVESTIGATION OF A WATERBORNE OUTBREAK

Form Approved
OMB No. 68-R0557

1. Where did the outbreak occur?

________ (1-2) City or Town ________ County ________

2. Date of outbreak: (Date of onset of 1st case)

________ (3-8)

3. Indicate actual (a) or estimated (e) numbers:

Persons exposed ________ (9-11)
Persons ill ________ (12-14)
Hospitalized ________ (15-16)
Fatal cases ________ (17)

4. History of exposed persons:

No. histories obtained ________ (18-20)
No. persons with symptoms ________ (21-23)
Nausea ________ (24-26) Diarrhea ________ (33-35)
Vomiting ________ (27-29) Fever ________ (36-38)
Cramps ________ (30-32)
Other, specify (39) ________

5. Incubation period (hours):

Shortest ________ (40-42) Longest ________ (43-4
Median ________ (46-48)

6. Duration of illness (hours):

Shortest ________ (49-51) Longest ________ (52-54)
Median ________ (55-57)

7. Epidemiologic data (e.g., attack rates [number ill/number exposed] for persons who did or did not eat or drink specific food items or water, attack rate by quantity of water consumed, anecdotal information) * (58)

ITEMS SERVED	NUMBER OF PERSONS WHO **ATE OR DRANK** SPECIFIED FOOD OR WATER				NUMBER WHO **DID NOT EAT OR DRINK** SPECIFIED FOOD OR WATER			
	ILL	NOT ILL	TOTAL	PERCENT ILL	ILL	NOT ILL	TOTAL	PERCENT ILL

8. Vehicle responsible (item incriminated by epidemiologic evidence): (59-60) ________

9. Water supply characteristics

(A) Type of water supply** (61)

- ☐ Municipal or community supply (Name ________)
- ☐ Individual household supply
- ☐ Semi-public water supply
  - ☐ Institution, school, church
  - ☐ Camp, recreational area
  - ☐ Other, ________
- ☐ Bottled water

(B) Water source *(check all applicable)*:

(C) Treatment provided *(circle treatment of each source checked in B)*:

(B)					(C)
☐ Well	a	b	c	d	a. no treatment
☐ Spring	a	b	c	d	b. disinfection only
☐ Lake, pond	a	b	c	d	c. purification plant – coagulation, settling, filtration, disinfection *(circle those applicable)*
☐ River, stream	a	b	c	d	d. other ________

10. Point where contamination occurred: (66)

☐ Raw water source ☐ Treatment plant ☐ Distribution system

*See CDC 4.245 Investigation of a Foodborne Outbreak, Item 7.
****Municipal** or community water supplies are public or investor owned utilities. **Individual** water supplies are wells or springs used by single residences. **Semipublic** water systems are individual-type water supplies serving a group of residences or locations where the general public is likely to have access to drinking water. These locations include schools, camps, parks, resorts, hotels, industries, institutions, subdivisions, trailer parks, etc., that do not obtain water from a municipal water system but have developed and maintain their own water supply.

CDC 4.461
11-78

This report is authorized by law (Public Health Service Act, 42 USC 241).
While your response is voluntary, your cooperation is necessary for the understanding and control of the disease.

*Renamed Department of Health and Human Services

Figure 1-8 Investigation of a waterborne outbreak.

be accomplished by careful study of the results of the sanitary and medical surveys and results of laboratory examinations, including the questionnaires and summaries given in Figures 1-5 to 1-8. Reports should be sent to the state health department and thence to the PHS.

11. Water specimens examined: (67)
(Specify by "X" whether water examined was original (drunk at time of outbreak) or check-up (collected before or after outbreak occurred)

ITEM		ORIGINAL	CHECK UP	DATE	FINDINGS		BACTERIOLOGIC TECHNIQUE (e.g., fermentation tube, membrane filter)
					Quantitative	Qualitative	
Examples:	Tap water	X		6/12/74	10 fecal coliforms per 100 ml.		
	Raw water		X	6/2/74	23 total coliforms per 100 ml.		

12. Treatment records: *(Indicate method used to determine chlorine residual):*
Example: **Chlorine residual – One sample from treatment plant effluent on 6/11/74 – trace of free chlorine**
Three samples from distribution system on 6/12/74 – no residual found

13. Specimens from patients examined (stool, vomitus, etc.) (68)

SPECIMEN	NO. PERSONS	FINDINGS
Example: Stool	11	8 *Salmonella typhi*
		3 negative

14. Unusual occurrence of events:
Example: Repair of water main 6/11/74; pit contaminated with sewage, no main disinfection. Turbid water reported by consumers 6/12/74.

15. Factors contributing to outbreak *(check all applicable):*

- ☐ Overflow of sewage
- ☐ Seepage of sewage
- ☐ Flooding, heavy rains
- ☐ Use of untreated water
- ☐ Use of supplementary source
- ☐ Water inadequately treated
- ☐ Interruption of disinfection
- ☐ Inadequate disinfection
- ☐ Deficiencies in other treatment processes
- ☐ Cross-connection
- ☐ Back-siphonage
- ☐ Contamination of mains during construction or repair
- ☐ Improper construction, location of well/spring
- ☐ Use of water not intended for drinking
- ☐ Contamination of storage facility
- ☐ Contamination through creviced limestone or fissured rock
- ☐ Other (specify) ____________

16. Etiology: (69-70)

Pathogen ____________
Chemical ____________
Other ____________

(71)
Suspected 1
Confirmed . . . . 2 *(Circle one)*
Unknown 3

17. Remarks: *Briefly describe aspects of the investigation not covered above, such as unusual age or sex distribution; unusual circumstances leading to contamination of water; epidemic curve; control measures implemented; etc. (Attack additional page if necessary)*

Name of reporting agency: (72)

Investigating Official: | **Date of investigation:**

Note: Epidemic and Laboratory assistance for the investigation of a waterborne outbreak is available upon request by the State Health Department to the Center for Disease Control, Atlanta, Georgia 30333.

To improve national surveillance, please send a copy of this report to: Center for Disease Control
Attn: Enteric Diseases Branch, Bacterial Diseases Division
Bureau of Epidemiology
Atlanta, Georgia 30333

Submitted copies should include as much information as possible, but the completion of every item is not required.

CDC 4.461 (BACK)
11-78

Figure 1-8 *(Continued)*

Reliable data from an accredited laboratory are essential to support administrative judgment. However, this is not the sole or necessarily the major factor in making an administrative decision. Equally important are proper container and specimen preparation, sample transportation and storage, proper collection of a representative sample, *and a comprehensive sanitary survey* of the water system and other environmental facilities and services involved. The sanitary survey is as important to the interpretation of an environmental sample and determining a sound course of action as is a medical history to the interpretation of results of an examination of a diagnostic specimen in determining proper treatment of a patient. Laboratories must refrain from attempting, or succumbing to the temptation, to make a judgment based solely on the laboratory results.

Other

Although only water- and foodborne disease investigations have been discussed, other agents might be suspected. Respiratory disease as related to overcrowding; ringworm infections as related to shower rooms and bathhouses; lousiness, impetigo, and scabies as related to personal hygiene and close contact; vectorborne diseases as related to arthropod prevalence; and carbon monoxide poisoning as related to unvented heaters or poor ventilation are other examples of possible investigations.

BIBLIOGRAPHY

An Evaluation of the Salmonella Problem, National Academy of Sciences, Washington, D.C., 1969, 207 pp.

Bowmer, Ernest J., "Salmonellae in Food—A Review," *J. Milk Food Tech.*, March 1965, pp. 74–86.

Bryan, Frank L., *Diseases Transmitted by Foods*, DHEW, PHS, CDC, Atlanta, Ga. 30333, 1971, 58 pp.

Bryan, Frank L., *Guide for Investigating Foodborne Disease Outbreaks and Analyzing Surveillance Data*, DHEW, PHS, CDC, Atlanta, Ga. 30333, 1975, 98 pp.

Bryan, Frank L. and Thomas W. McKinley, "Prevention of Foodborne Illness by Time-Temperature Control of Thawing, Cooking, Chilling, and Reheating Turkeys in School Lunch Kitchens," *J. Milk Food Tech.*, August 1974, pp. 420–429.

Bullerman, L. B., "Significance of Mycotoxins to Food Safety and Human Health," *J. Food Prot.*, January 1979, pp. 65–86.

Control of Communicable Diseases in Man, 13th ed., Am. Pub. Health Assoc., Washington, D.C., 1980.

Dack, G. M., *Food Poisoning*, University of Chicago Press, 1956, 251 pp.

Dolman, C. E., *Bacterial Food Poisoning*, Canadian Public Health Association, Toronto, 1943, 46 pp.

Getting, V. A., "Epidemiologic Aspects of Foodborne Diseases," *New Eng. J. Med.*, June 1943, 75 pp.

Hall, Herbert E., "Current Developments in Detection of Microorganisms in Foods—*Clostridium Perfringens*," *J. Milk Food Tech.*, November 1969, pp. 426–430.

Hanlon, John J., and George E. Pickett, *Public Health Administration and Practice*, 7th ed., C. V. Mosby Company, St. Louis, 1979, 787 pp.

Hilleboe, Herman E. and Granville W. Larimore, *Preventive Medicine*, W. B. Saunders Co., Philadelphia, 1965, 523 pp.

"Joint FAO/WHO Expert Committee on Zoonoses," *WHO Tech. Rep. Ser.*, **378,** 1967.

Lewis, K. H. and K. Cassel, Jr., *Botulism*, Proc. of Symposium, PHS Pub. 999-FP-1, Cincinnati, Ohio, December 1964, 327 pp.

Manson-Bahr, P., *Synopsis of Tropical Medicine*, Williams & Wilkins Co., Baltimore, 1943, 224 pp.

Maxcy-Rosenau Public Health and Preventive Medicine, John M. Last, Ed., Appleton-Century, Crofts, New York, 1980, 1926 pp.

"Microbiological Aspects of Food Hygiene," *WHO Tech. Rep. Ser.*, **399,** 1968.

Procedures to Investigate Foodborne Illnesses, International Assoc., of Milk, Food, and Environmental Sanitarians, Inc., P.O. Box 701, Ames, Iowa 50010, 1976, 52 pp.

Procedures to Investigate Waterborne Illnesses, International Association of Milk, Food, and Environmental Sanitarians, Inc., P.O. Box 701, Ames, Iowa, 50010, 1979, 68 pp.

Rieman, Hans, *Foodborne Infections and Intoxications*, Academic Press, New York, 1969, 698 pp.

Silliker, John H., "Status of Salmonella—Ten Years Later," *J. Food Prot.*, April 1980, pp. 307–313.

Strong, R. P., *Stitt's Diagnosis, Prevention and Treatment of Tropical Diseases*, 2 vols., Blakiston Co., Philadelphia, 1942, 1747 pp.

Taylor, Joan, *Bacterial Food Poisoning*, R. Soc. Health, London, England, 1969, 175 pp.

2

ENVIRONMENTAL ENGINEERING PLANNING AND IMPACT ANALYSIS

INTRODUCTION

Definitions

Comprehensive planning takes into account the physical, social, economic, ecologic, and related factors of an area and attempts to blend them into a single compatible whole that will support a healthful and efficient society. A *comprehensive plan* has been defined as

> (a) A model of an intended future situation with respect to: (i) specific economic, social, political and administrative activities; (ii) their location within a geographic area; (iii) the resources required; and (iv) the structures, installations and landscape which are to provide the physical expression, and physical environment for, these activities; and (b) A programme of action and predetermined co-ordination of legislative, fiscal, administrative and political measures, formulated with a review to achieving the situation represented by the model.[1]

The term *planning* is defined as the

> systematic process by which goals (policies) are established, facts are gathered and analyzed, alternative proposals and programs are considered and compared, resources are measured, priorities are established and recommendations are made for the deployment of resources designed to achieve the established goals.[2]

There is great need for emphasis on the area-wide, metropolitan, and regional approaches to planning and on the presentation of a total integrated and balanced appraisal of the essential elements for healthful living. The latter is difficult to achieve, as it requires a delicate synthesis of competing economic,

[1] *Report of Stage Three: The Seminar-Conference*, Toronto, August 6–16, 1967, p. 10; this definition modifies that of Ernest Weissman, in *Planning and Urban Design*, Bureau of Municipal Research, Toronto, 1967, p. 40.

[2] *Local and Regional Planning and Development*, Governor's Task Force Report, West Virginia, Regional Review, National Service to Regional Councils, Washington, D.C., January 1970, p. 2.

social, and physical environmental goals. However, a reasonable balance must be sought and these goals can only be approached if realistic objectives are set and *all* the facts are analyzed and presented with equal force. Regional planning should recognize national and local planning. In many instances, no matter how logical and sound the regional planning, the most that can be achieved is local implementation, hopefully within the context of the regional plan.

Environmental Engineering Considerations

Some of the essential environmental health, engineering, and sanitation objectives, frequently overlooked in planning, will be identified to show how they can be blended into the usual planning process to achieve more comprehensive community plans. People are demanding, as a fundamental right, not only clean air and pure water and food, but also decent housing, an unpolluted land, freedom from excessive noise, adequate recreation facilities, and housing in communities that provide comfort, privacy, and essential services. The usual single-purpose planning for a highway, a housing development, or a sanitary landfill must be broadened to consider the environmental, economic, and social purposes to be served *and the effects* (*beneficial and adverse*) *produced by the proposed project.* The planner must ensure that potential resultant problems are avoided in the planning stages and that projects are so designed as to be pleasing and to enhance the community well-being. This is discussed further under Environmental Factors to be Evaluated in Site Selection and in Environmental Impact Analysis, this chapter.

As an example, major highway planning must take into consideration the need for service and rest areas with picnic spaces, toilet facilities, drinking water, insect and weed control, wastewater disposal, and solid waste storage and disposal facilities; it must also consider traffic counts, landscaping, buffer zones, and the effects of routes, entrances, and exits on adjoining communities. Environmental effects during and after construction must also be taken into consideration: erosion control during construction, travel of sediment, dust, and possible flooding; pollutional effects on water supply intakes, streams, lakes, fish, and wildlife; and the disposal of trees, stumps, and demolition material without causing air pollution. Problems to be resolved in the planning stages are control of noise, vibration, and air pollution; dust and weed suppression; and control of deicing chemicals, herbicides, and other pollutants because they might affect nearby water wells, reservoirs, watersheds, and recreational areas, as well as wildlife, vegetation, streams, and lakes.

Another example is the design and construction of an incinerator. Many engineering and environmental considerations and interrelations, if recognized in the planning stage, can greatly minimize the design, social, political, and ecological problems that may delay or prevent project implementation. Figure 2-1 shows the process involved in the treatment and disposal of solid

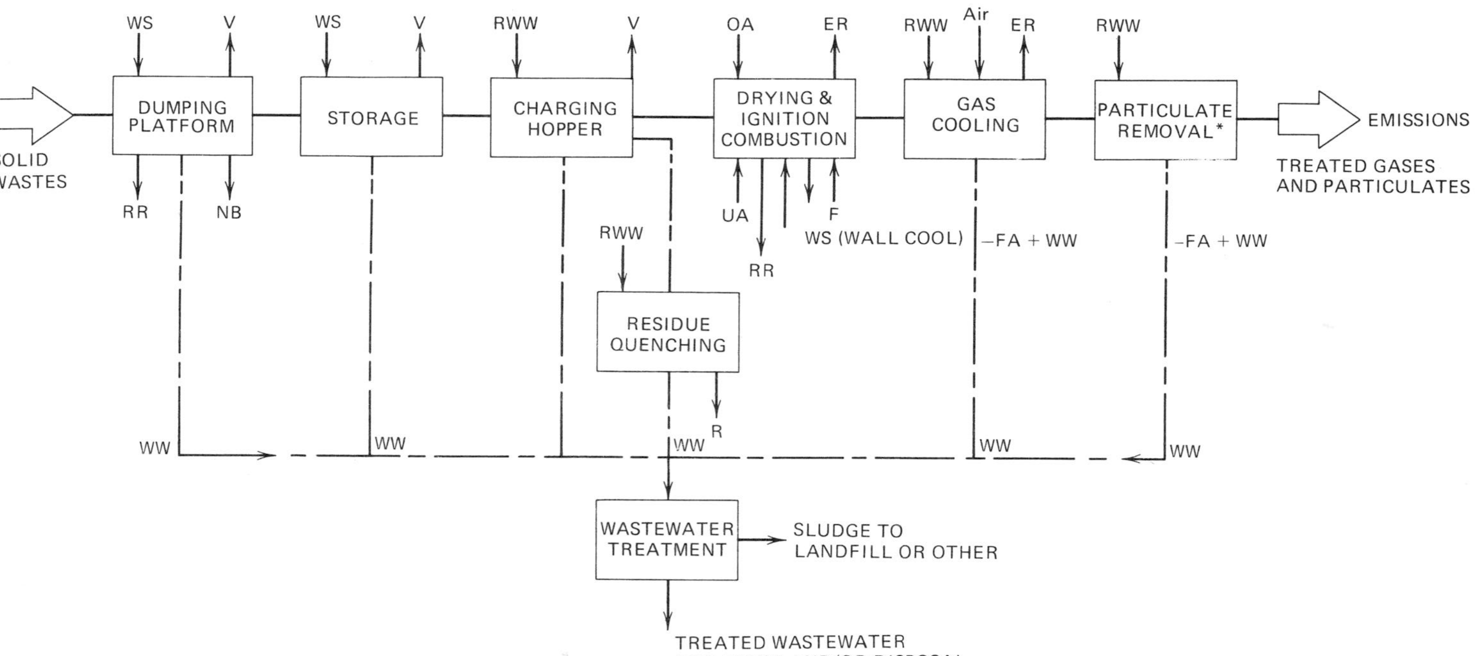

Figure 2-1 Incinerator Processes and Environmental Engineering Controls. *Legend*: WS—water supply; WW—wastewater; RWW—reclaimed wastewater; OA—overfire air; UA—underfire air; F—fuel, if needed: R—residue—landfill, metal salvage; FA—fly ash; V—ventilation, odor, and dust control; NB—nonburnables; ER—energy recovery; and RR—resource recovery. *Settling chamber, wetted baffles, fabric filters, cyclones, wet scrubbers, electrostatic precipitators. (Adapted from De Marco, et al., *Incinerator Guidelines-1969*, DHEW, PHS, Pub. No. 2021, p. 47.)

wastes by incineration, the potential liquid, solid, and gaseous waste products, energy and resource recovery, and the supporting needs. Listed below are factors that require study and resolution to help assure proper final construction, operation, and minimal deleterious effects.

Environmental, Engineering, Planning and Site Selection Factors

1. *Geology.* Suitable foundation soils, safe from earthquake, active faults and slides; avoid channeled and creviced formations, that is, limestone, dolomite or gypsum, and shale.
2. *Drainage and flooding.* Site above 100-year flood level; access roads well drained and above flood level.
3. *Transportation.* Access roads proper design, safe entrance and exit lanes; noise, dust, and other air pollution on transportation routes minimal and compatible with adjacent land uses.
4. *Aesthetics.* Site is screened and landscaped.
5. *Noise control.* At dumping platform, charging hopper, truck delivery and leaving; buffers.
6. *Water supply.* Drinking and sanitary purposes, cooling water, make-up water; cross-connection control. Also water resources; recreation, wildlife. Compliance with Safe Drinking Water Act; if source is on site, also state and local regulations.
7. *Wastewater disposal.* From dumping platform, storage and charging areas, residue quenching, gas cooling, particulate removal; toilet, shower, and dressing rooms; receiving stream classification, treatment and permit required, wastewater reuse, cross-connection control. Compliance with Clean Water Act as amended; also state and local regulations.
8. *Solid waste disposal.* Incinerator residue; fly ash from particulate removal; nonincinerable solid wastes; sludge from wastewater treatment; resource recovery and reuse of residue; sanitary landfill; air, surface and groundwater pollution control. Compliance with Solid Waste Disposal Act and Resource Conservation and Recovery Act; also state and local regulations.
9. *Air pollution control.* At dumping, storage, and charging areas; odors; air pollution control devices on stack; effect on surrounding areas; permits required. Compliance with Clean Air Act.
10. *Occupational health.* Safety, toilet, shower, and locker facilities, lunch room, clean air, safe water, accident and explosion hazards, pathogens and dust in work areas, ventilaton, general sanitation. Compliance with Occupational Safety and Health Act.
11. *Vermin control.* Flies, roaches, rats, mice, inside and outside.
12. *Environmental impact on surrounding area.* Compliance with Endangered Species Act, National Environmental Policy Act, and other applicable laws.

The same principles would apply to other construction such as a landfill site, wastewater treatment plant, factory, shopping center, or real estate development.

In addition, communication should be maintained in the planning, design, construction, and operation stages with official agencies and the people affected to assure compliance with special regulations such as the National Environmental Policy Act (NEPA—discussed later), and to obtain financial and other assistance that might be available. Federal agencies may include the USEPA, the Department of Labor, and the Department of Transportation. State agencies may include the environmental protection department, the health department, labor department, department of transportation. Local agencies may include planning and zoning, highway, building, and fire departments and the local county, city, village, or town government.

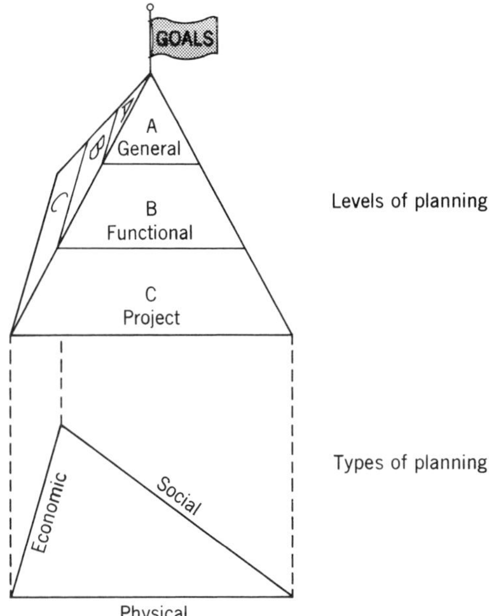

Figure 2-2 Types and levels of planning (applicable to national, state, regional, and local planning).

A. *General, Overall Policy Planning*—Identification of goals; aspirations and realistic objectives. Establishment of functional priorities.
B. *Functional Planning*—Such as for transportation, water supply, wastewater, recreation, air pollution, solid wastes, or medical care facilities in which *alternative functional solutions are presented*, including the economic, social, and ecological factors, advantages and disadvantages.
C. *Project Planning*—Detailed engineering and architectural specific project plans, specifications, drawings, and contracts for bidding. Implementation follows. Construction, Operation and Maintenance—Plan adjustment as constructed: updating and planning for alterations and new construction.

TYPES OF PLANNING

There are many kinds and levels of planning for the future. They may range from family planning to national planning for survival. The discussion here will deal with "comprehensive* community planning," also referred to as "general planning" from which policy decisions can be made; "comprehensive functional planning," also called "preliminary or feasibility studies" dealing with a single facility or service; and "definitive planning" for a specific project, which includes final engineering and architectural reports, plans, contract drawings, and specifications leading to construction. See Figure 2-2. The levels

*The term "comprehensive" referred to here and in the discussion that follows is used with considerable reservation because truly comprehensive planning is an ideal that will rarely if ever be achieved.

of planning become more detailed and specific as the objective or scope of the final construction is approached.

Comprehensive Community Planning

In this phase, an attempt is made to take an overall look at the total region. If only a part of the total region is involved, the area, or community, is studied within the context of the larger region of which it is a part. This is described later as "the process of comprehensive community planning." Report recommendations should lead to policy decisions or statements, including maps, that can be used to establish priorities to implement specific projects, with an estimate of their cost. At this stage cost is a secondary consideration, although cost may become a major factor in the specific planning and implementation phase. A combination of talents is needed to make a proper study. These may include the architect, artist, attorney, biologist, ecologist, civil engineer, economist, environmental engineer, geographer, geologist, hydrologist, landscape architect, mathematician, operations research scientist, planner, political scientist, sociologist, soils scientist, systems engineer, and others.

Comprehensive Functional Regional Planning

If the priority project is a new or improved municipal water system (it could be a sewerage system and treatment plant, park and recreation facilities, a solid waste collection and disposal system), then the next step is a functional comprehensive engineering planning study to consider in some detail the *several ways* in which the regional or area-wide service or facility can be provided, together with approximate costs.[3] Such a study is also referred to as a preliminary or feasibility report; it is discussed in greater detail under the heading Regional Planning for Environmental Engineering Controls. For example, the water system study alternatives for the region might include purchase of water to serve possible service areas and combinations; a well-water supply with water softening and iron removal; an upland lake or reservoir with multipurpose uses requiring land acquisition, water rights, and a conventional water treatment plant; or a nearby stream requiring preliminary settling and an elaborate water treatment plant to adequately handle the known pollution in the stream.

At this stage no detailed engineering or architectural construction plans are prepared. However, the engineering, political, legal, economic, and social

[3]For a guide to the selection of a consulting engineer see *Consulting Engineer—A Guide for the Engagement of Engineering Services*, No. 45, American Society of Civil Engineers (ASCE), New York City (1972). The same general principles would apply to selection of an architect or other planning consultant.

feasibility or acceptance of each alternative is presented together with the advantages, disadvantages, environmental impact, recommendations, cost estimates, and methods of financing each alternative. The study report should be sufficiently complete and presented so that the officials can make a decision (political) and select one alternative for definitive planning. It should also meet the purposes of the NEPA and public information needs.

Definitive, or Project Planning

The next step for the example given (a new water system) would be the establishment of a legal entity to administer the project as provided for by state or local law, followed by the acquisition of necessary right-of-way and water rights, resolution of any legal constraints, establishment of service districts, approval of bond issues, rate-setting and financing of operation, maintenance, and debt retirement. This step is followed by selecting a consulting engineer (it could be the same engineer who made the preliminary study), preparing plans, specifications, and contract drawings, advertising for bids, and awarding the contract to a contractor. Construction should be under the supervision of the consulting engineer, and a resident engineer responsible to the consultant or municipality should be employed.

The consulting engineer should ensure that the municipality is provided with revised drawings showing the works and location of facilities as constructed and in place. This person would also normally be expected to provide operation manuals and guides, to take responsibility for placing the plant in operation, and, during the first year, to train personnel to take over full operation.

THE PROCESS OF COMPREHENSIVE COMMUNITY PLANNING*

For comprehensive community planning, the process includes a (1) statement of goals and objectives, (2) basic studies, mapping, and data analysis, (3) plan preparation, (4) plan implementation, (5) public information and community action, and (6) re-evaluation and continual planning. The process is shown in Figure 2-3 and is outlined below.†[4]

*Tailored to fit the particular characteristics and needs of the area under study with consideration of social, environmental, and economic factors.

†Sometimes the formulation of goals and objectives is placed second. Goals are the final purpose or aim, the ends to which a design tends. Objectives are the realistically attainable ends. In any case, the goals and objectives should be continually adjusted as the planning study progresses and as the quality of life desired and demanded changes.

[4]Joseph A. Salvato, Jr., "Environmental Health and Community Planning," *J. Urban Plann. Dev. Div., ASCE,* **94,** No. UP 1, Proc. Paper 6084, 22-30 (August 1968). See also W. E. Bullard "Water-Related Land-Use Planning Guidelines," Interstate Sanitation Commission on the Potomac River, Bethesda, Md. 20014, March 1974.

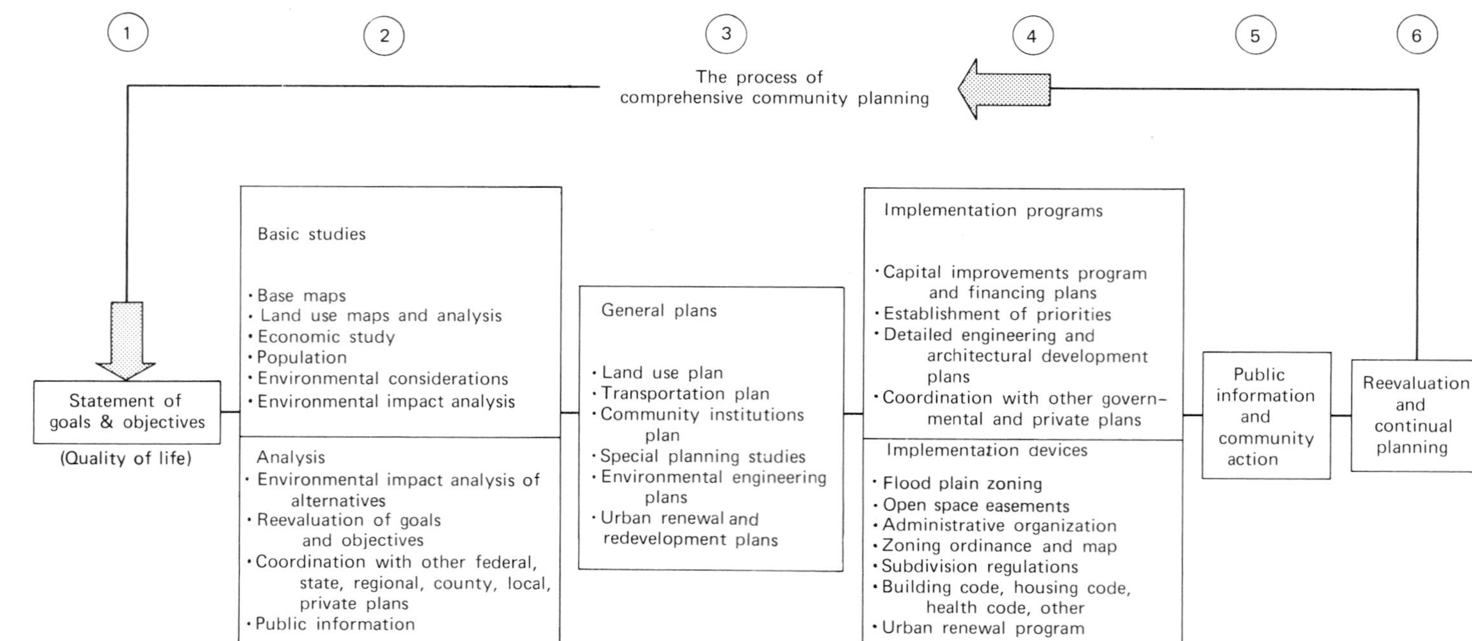

Figure 2-3 An example of the planning process. Source: J. A. Salvato, Jr., "Environmental Health and Community Planning," *J. Urban Plan. Dev. Div., ASCE*, **94**, No. UP 1, 22–30 (August 1968).

Statement of Goals and Objectives (Step 1)

Community Aspirations and Environmental Quality. A first step is the preparation of a tentative statement, to guide the planning staff, which reflects the goals and objectives the community expects will be achieved through the planning process. Public input is required. The statement should recognize the economic, social, and physical community aspirations, including the environmental health quality goals and objectives. Depending on the need, as confirmed or modified by Step 2, Basic Studies, these may be a water supply of satisfactory quality adequate for domestic, industrial, recreational, and fire-fighting purposes; clean air; proper sewage and other wastewater collection and disposal; water pollution abatement; proper solid waste collection and disposal; adequate and safe parks and recreation facilities, including swimming pools or bathing beaches; noise abatement and control; a convenient and acceptable transportation system; elimination of accident hazards; preservation of good housing and residential areas, rehabilitation of sound substandard housing, and construction of new sound housing in a healthful and pleasing environment; elimination of sources and causes of mosquitoes, ticks, blackflies, termites, rats, and other vermin; adequate schools; adequate hospitals, nursing homes, and other medical care facilities; adequate gas and electricity; and adequate cultural facilities.

Numerous attempts have been made to identify and measure the factors associated with a desirable quality of life (QOL). This is difficult because it must consider compliance with administrative or legal standards (objective measure) and the value of weight (subjective measure) which the individual assigns to a factor as perceived by the individual. From this comes the need for sound, defensible standards and public participation before quality of life objectives and community aspirations are established. Hornback, et al.[5] investigated several proposals to establish a QOL index for different purposes. He selected one method and grouped certain QOL factors and associated components as follows:

1. *Economic Environment.* Work satisfaction, income, income distribution, economic security.
2. *Political Environment.* Informed constituency, civil liberties, electoral participation, non-electoral participation, government responsibilities.
3. *Physical Environment.* Housing, transportation, material quality, public services, aesthetics.
4. *Social Environment.* Community, social stability, culture, physical security, family.
5. *Health.* Physical, mental, nourishment.
6. *Natural Environment.* Air quality, water quality; radiation, solid waste, toxicity, and noise protection.

The interaction of various factors in the environment on man is suggested in Figure 2-4. Control of the forces around us and the extent to which they are

[5]Kenneth E. Hornback, Joel Guttman, Harold L. Himmelstein, Ann Rappaport, Roy Reyna, *Studies in Environment, Vol. II, Quality of Life*, Office of Research and Development, USEPA, Washington, D.C., November 1974.

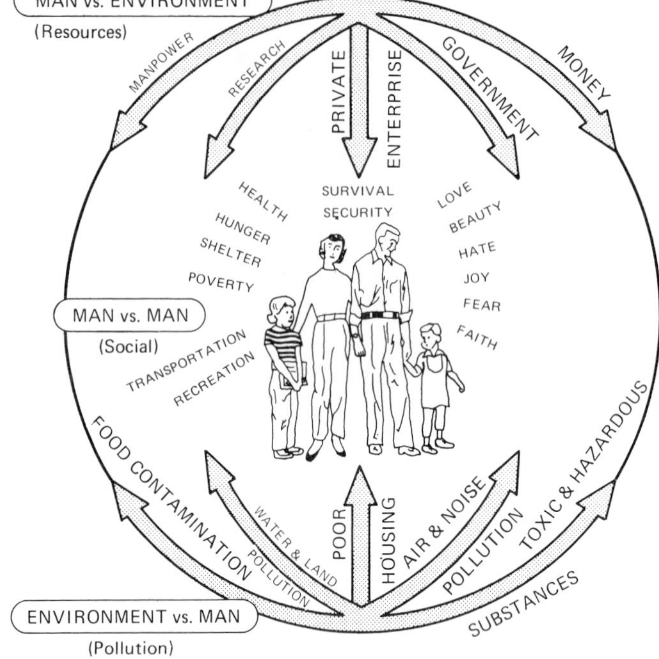

Figure 2-4 Forces to be balanced to achieve a favorable quality of life.

properly balanced will determine the quality of life achieved at any point in time.

A QOL index can be useful in determining the success achieved by policy decisions as related to the public's perception of the results. But before an action is taken on a policy decision, objective indicators that are measurable must be established so that progress or lack of progress to achieve stated objectives in a definite period of time can be noted in an evaluation. The reader is referred to Chapter 12 for a discussion of the evaluation process and the precautions necessary to obtain valid and reliable data.

The Overseas Development Council has created a Physical Quality of Life Index by combining literacy rates, life expectancy, and infant mortality figures and ranking them on a scale from 0 to 100. The ranking does not necessarily correlate directly with average per capita GNP levels.[6]

The National Environmental Policy Act, discussed later in this chapter, addresses the federal effort to avoid adverse consequences of proposed actions and thus obtain a better quality of life for present and future generations. Many states have a similar act.

[6] *Environmental Quality*, The ninth annual report of the Council on Environmental Quality, December 1978, GPO, Washington, D.C. 20402, p. 453.

Basic Studies, Mapping, and Data Analysis (Step 2)

Research and Problem Identification. Having tentatively agreed on a statement of goals and objectives, the next step is evaluation of the community. This is done by and includes the following:

1. Mapping. Preparation of a base map, which will be the basis of other detail maps referred to below, and of land use maps which can serve as policy references. Map scales found convenient for base maps are: State or region 1:500,000; metropolitan area 1:250,000; county 1:100,000; detail and work maps 1:100,000 and 1:20,000 or 1:10,000.
2. Land-Use Analysis. Collection, plotting, and analysis of data on residential, commercial, industrial, public, agricultural, and recreational land uses; blighted and deteriorated structures, inefficient and conflicting land uses, and desirable land uses; areas available for future population growth and industrial development; and availability and adequacy of supporting services.
3. Population and Demographic Studies. Present and future trends, locations, and amounts of populations; social composition and characteristics of the population; age distribution and changes.
4. Economic Studies and Proposals. Existing sources of income, future economic base, labor force, markets, industrial opportunities, retail facilities, stability of economy, family income.
5. Transportation Systems. Existing systems, their location and adequacy; effects of air and water pollution; aesthetic, zoning, noise, and vibration controls; population growth; industrial and recreation development of the systems; and modifications or protective features needed in existing and proposed systems.
6. Community Institutions. Description, location, and adequacy of educational, recreational, and cultural facilities; medical, public health, and environmental protection facilities; religious and other institutions; and public buildings such as post office, fire and police stations, auditoriums and civic centers, public markets, and government offices.
7. Environmental Health and Engineering Considerations. Define and show on base maps and charts both favorable and unfavorable natural and man-made environmental conditions and factors such as meteorology, including wind and solar radiation studies; topography, hydrology, flooding, tidal effects, seismology, and geography including soil drainage, soil characteristics such as percolation, and bearing characteristics; natural pollution of air, land, and water; background radiation and the flora and fauna of the area; man-made pollution of air, water, and land; noise and vibrations, ionizing radiations, and unsightly condition of housing and community facilities and utilities. See Comprehensive Environmental Engineering Planning, later in this chapter.
8. Environmental Impact Assessment and Cost-Benefit. Analyze the effects of the proposed action on the factors identified in item 7; also the alternative to eliminate or minimize deleterious effects and enhance beneficial effects. Environmental impact analyses are an integral part of project planning and preliminary design processes.[7] See Environmental Impact Analysis of Environmental Engineering Considerations.
9. Coordinate with Other Planning. Federal, state, regional, county, and local agencies and municipalities usually have some plans already completed and certain plans under way or proposed. Private enterprises such as individuals, business establishments, industrial plants, and others, are simultaneously making plans for renovation, expansion, or relocation and in many instances are in a position to support and implement the official planning. Hence it is extremely important, insofar as is possible, to coordinate all the planning and obtain the participation essential to the realization of the planning goals and objectives.

[7]Recommendations of the Conference on Environmental Impact Statement sponsored by the Interprofessional Council on Environmental Design, *Civil Eng.*, *ASCE*, February 1973, pp. 64–65.

10. Public Information. Acquaint the public, and specifically, individual representatives of organizations that may be participating in the implementation of the comprehensive community plans, with the goals, objectives, basic studies, mapping and data analysis, and results. Encourage and solicit feedback and carefully consider suggestions. Ask for supporting information to clarify suggestions made.
11. Reevaluate Goals and Objectives. Confirm or adjust as indicated by the basic studies, analyses, and feedback.

Plan Preparation (Step 3)

General plans, including area-wide and regional plans, that present alternatives and rough costs to help people understand what is involved and help officials make policy decisions should be prepared. The environmental, economic, and social effects of the actions proposed must also be considered. Detailed engineering and architectural drawings and specifications come later, when actual construction is scheduled as determined by implementation (Step 4). The plans should include the following:

1. *Land-Use Plan.* This type of plan shows the existing uses to be retained and future patterns and areas for residential, commercial, industrial, agricultural, recreation, open space or buffer, and public purposes. Such plans are based on the findings in Steps 1 and 2 and are integrated with the plans that follow.

2. *Transportation or Circulation Plan.* This shows major and minor highways and streets, transit systems, waterways, ports, marinas, airports, service areas, terminals, and parking facilities. The plan should clearly show facilities to be retained, those to be improved or altered, and new facilities proposed.

3. *Community Institutions Plan.* This shows location of existing or proposed new, expanded, or remodeled educational and cultural facilities; health, welfare, religious, and other institutions; public buildings and facilities. The plans for these institutions are considered in the light of their adequacy to meet present and future needs based on Steps 1 and 2.

4. *Special Planning Studies.* These include neighborhood analyses and plans for urban renewal, clearance, rebuilding, rehabilitation, code enforcement, housing conservation, preservation of sites and structures, central business, parking, parks and recreation areas, as well as private enterprise plans and their integration with community plans.

5. *Environmental Engineering Plans.* These plans are concerned with the adequacy and needs for water supply, sewerage, solid waste disposal, air and water pollution control, housing and a healthful residential environment, realty subdivision and construction controls, gas, electricity, and nuisance control. They are also concerned with the environmental, economic, and social effects of industrial, agricultural, residential, commercial, highway, airport, recreational, and power development; storm water, drainage, and flood control; forest, open space, soil, estuary, and wildlife conservation; and planning for more aesthetic structures and public buildings. See Regional Planning for Environmental Engineering Controls.

The plans should take into consideration the environmental factors that need to be improved and those that can be developed to eliminate certain hazardous or annoying conditions while at the same time making a needed or desired improvement. Incompatible or nonconforming structures or uses should be eliminated or adjusted to harmonize with the most desirable and obtainable environment. Additional details on environmental factors and controls are given later in this chapter under Comprehensive Environmental Engineering and Health Planning and Environmental Factors to be Evaluated in Site Selection and Planning.

6. Urban Renewal and Redevelopment Plans. Federal grants and assistance may be available to a municipality that formulates an acceptable "workable program" for community improvement.

A workable program is one that "will include an official plan of action . . . for effectively dealing with the problem of urban slums and blight within the community and for the establishment and preservation of a well-planned community with well-organized residential neighborhoods of decent homes and suitable living environment for family life."[8] A workable program includes the following seven elements. They are basic to the elimination of slums and blight and to the prevention of their spread in any city.

a. Up-to-date codes and ordinances including building, electrical, plumbing, housing, health codes and the like and zoning and subdivision regulations that provide sound standards governing land and building use and occupancy.
b. A comprehensive community plan, often referred to as a general or master plan and made up of policy statements and plans to guide community growth and development (as outlined in this chapter). It includes three essential features: a land-use plan, a circulation plan, and a community facilities plan.
c. Neighborhood analysis. An examination of the physical resources, a pinpointing of deficiencies, a notation of environmental problems of the area under study, and recommendations concerning the steps to be taken to eliminate physical and environmental shortcomings.
d. Administrative organization. To insure that codes and ordinances are enforced, the community must demonstrate that it has an effective organizational structure and adequate personnel to carry out these functions.
e. Financing. The community must demonstrate that it has adequate financing and sound budgetary policies to assure that public improvement projects related to urban renewal are carried out.
f. Housing for displaced persons. Most projects result in temporary or permanent displacement, and relocation assistance must be available to residents requiring it. There is also need to assist businesses that have a relocation problem. Both of these problems must be faced by the community to prevent hardship to those who are displaced.
g. Citizen participation. Broad citizen involvement, representative of all segments of community life, is required as part of the democratic process of program formulation.

Plan Implementation

Construction; *Problem Correction, Prevention and Control.* Plans, to yield a return, must be implemented. Implementation involves political and governmental decisions and public acceptance. This means:

1. *Capital Improvement Program and Financing Plans.* Project priorities are established. Approximate costs, sources of revenue, and financing for five years or longer are determined. Existing and contemplated public and private planning and construction (including industrial, commercial, urban renewal, and slum clearance), local, state, and federal assistance programs, the feasibility of area-wide or regional solutions, and the total tax burden are all recognized and coordinated. The program as implemented will largely determine the future quality of the environment and the economic health of the community. Government action usually follows adoption of a capital improvement program. The program is reviewed and updated annually.
2. *Detailed Engineering, Architectural, and Development Plans.* These are plans for specific projects as determined in item 1, and may be preceded by feasibility studies. In contrast to general plans, specific drawings, detail plans, and specifications are prepared from which accurate estimates or bids are obtained and construction contracts are let. Included are federal, state, urban renewal, and private construction projects such as a new city hall, school, shopping center, water

[8]U.S. Dept. of Housing and Urban Development (HUD), *Urban Renewal Fact Sheet*, GPO, Washington, D.C. 1966 as revised.

system, sewerage and treatment plant, marina, sanitary landfill, and housing and park and recreation area development.

3. *Regulation, Laws, Codes, and Ordinances.* Zoning controls and subdivision regulations are commonly provided. Also considered are flooding and drainage, flood plain management, special districts, environmental protection, and provision for long-term land leases and tax deferral for open space preservation. There is a need for standards and regulations for water supply, sewage disposal, and house connections; air pollution abatement and emission standards; control of noise and vibration nuisances; control of mining and excavations; regulations for solid waste storage, collection, and disposal. Also needed are a modern building code* (including plumbing, fire and safety, electrical, heating, and ventilation regulations), sanitary code,† and a housing occupancy and maintenance code.‡

4. *Administrative Organization.* The entire process of comprehensive community planning can at best be only of limited value unless provision is made for competent direction, personnel staffing, and administration. Adoption of an official map, land-use plans, an improvement program, codes, ordinances, rules, and regulations has little meaning unless implemented by competent personnel who are adequately compensated.

Public Information and Community Action (Step 5)

For a community plan to be effective, it is necessary that public participation and information be involved at all stages, without public interference with the technical planning operations. In keeping the public informed, it is also necessary to stimulate and provide channels for individuals to respond with information and ideas. Public participation can result in additional types of physical improvement and programs that help achieve the public objectives.

Involvement of the public, including community organizations and influential citizens, during the planning process makes possible better reflection and appreciation of the community goals and objectives and the problems to be overcome. In this way information presented by the planner has a better chance of reaching more people and in being discussed by individuals, groups, and news media. All this helps to further understanding and gain support for implementing definitive plans and bond issues.

*A building code deals primarily with, and contains standards for, the construction or alteration of buildings, structural and fire safety, light and ventilation, materials, electrical, and air conditioning, and prevention of related hazards. The code authorizes plan and specification review functions, inspection for approval of the construction or alteration, and the issuance of permits.

†A sanitary code is concerned with environmental sanitation and safety and control and prevention of communicable diseases for protection of the public health, safety, and general welfare. It may set standards and regulations for such matters as disease control; qualifications of personnel including those responsible for the operation of certain facilities; water supply; wastewater disposal; air pollution prevention; solid wastes; food sanitation, including milk; radiation; noise; vectors; accident prevention; housing, hospitals, nursing homes, and institutions; recreation areas and facilities; schools, camps, and resorts; trailer and mobile home parks; bathing beaches and swimming pools; occupational health; emergency sanitation; and other preventive measures that may be required to ensure that the public health is protected.

‡A housing code is concerned primarily with, and contains minimum standards for, the provision of safe, sanitary, decent dwellings for human habitation. It sets standards for the supplied utilities and facilities, occupancy, and maintenance. The code requires inspection to determine compliance and usually is applicable to all dwellings, regardless of when constructed. It is also a tool to obtain housing and community rehabilitation, conservation, and maintenance.

Reevaluation and Continual Planning (Step 6)

Not to be forgotten is the continual need for public information, coordination, and administration of planning activities locally on a day-to-day basis, and liaison with state and federal agencies. Studies and analyses in Step 2, the plans developed in Step 3, and the improvement program and the controls in Step 4 must all be kept reasonably current. This is necessary to prevent obsolescence of the comprehensive community plan as well as the community.

Community objectives and goals can be expected to change with time. New and revised land-use concepts, means of transportation, housing needs and designs, public desires and aspirations, and economic, technological, and sociological developments will require periodic reevaluation of the general plan and revision as indicated. It should be kept in mind that human wants are insatiable. As soon as a goal or objective is approached, another will be sought. This is as it should be, but must be kept in balance and within realistic bounds, which comprehensive planning should help accomplish.

Conclusion

There is an urgent need for more engineering in community planning and more comprehensive planning in engineering, with emphasis on area-wide, metropolitan, and regional approaches. The environmental health, sanitation, and engineering factors that are essential for community survival and growth, and the environmental, economic, and social effects of any proposed projects must not be overlooked in planning and engineering.

Public awareness demands a quality of environment that provides such fundamental needs as pure water, clean air, unpolluted land, pure foods, decent housing, privacy, safe recreation facilities, and open space. A balanced appraisal must be made, and the planning process must blend these goals in its objectives, analyses, plans, and in the capital budgeting.

State and local health and environmental protection departments have vital planning, plan approval, and regulatory responsibilities to assure that the public health, environment, and welfare is protected. These responsibilities usually deal with water pollution abatement; wastewater treatment and disposal; safe and adequate water supply; air pollution control; solid hazardous and nonhazardous waste disposal; X-ray and nuclear facility operations; housing and realty subdivision development, including temporary residences such as motels, trailer parks, and resorts; vector control; milk and food protection; medical care facilities; and recreational facilities including bathing beaches and swimming pools. Responsibilities of the health and environmental protection departments extend to the issuance of permits and the continual monitoring of operational results to protect the public and enhance the home, work, and recreation environment. Hence, they have an important stake in all planning (preventive environmental sanitation and engineering) to assure that the public does not inherit situations that are impossible or costly to correct.

Comprehensive community planning that gives proper attention to the environmental health considerations, followed by phased detail planning and capital budgeting, is one of the most important functions a community can engage in for the immediate and long-term economy and benefit of its people.

The achievement of the WHO health goal requires the control of all those factors in man's physical environment that exercise or may exercise a deleterious effect on his physical, mental, or social well-being. It is necessary, therefore, that single-purpose and general economic, social, and physical planning take into full consideration the environmental factors affected and the facilities and services required for healthful living.

REGIONAL PLANNING FOR ENVIRONMENTAL, HEALTH, AND ENGINEERING CONTROLS

General

Air, water, and land pollution, inefficient transportation facilities, urban and rural blight, and disease do not respect political boundaries. Adequate highways, land-use controls, park and recreation facilities, water, sewers, solid waste disposal, and other services necessary for proper community functions are usually best designed within the context of a regional plan. Such planning, however, must recognize federal and state agency planning and, to the extent feasible, local planning (including planning by private enterprise). The provision of services, however, can be hampered if legal impediments prevent regional solutions. As long as each city, town, or village can obstruct or curtail intermunicipal cooperation, planning for the future cannot be completely effective. Coordination of planning among smaller communities within the context of the applicable county, metropolitan, and regional plan is essential. The smallest practical planning unit for the development and administration of environmental engineering controls appears to be the county.

Local governments normally do not have jurisdiction and hence cannot plan and budget for an entire metropolitan region. The key to the solution of metropolitan problems is the establishment of an area-wide organization that can investigate, plan, and act on an area-wide basis. Unless this is done regional plans that are developed will probably remain on the shelf for a long time.

Although planning the solution of regional problems on a regional basis is generally accepted as being basically sound, it does not necessarily follow that a special organization or authority must be established to carry out the actual construction, operation, and maintenance of the utility or facility. Local experiences and sentiment may dictate that the only way a regional project can be carried out, in whole or in part, is on an individual community basis regardless of the additional costs involved. Most important is that the particular project or facility be constructed in general conformance with the regional plan and that it be put into service. If this is done, it will be possible at a later date, when the "climate" is propitious, to realize the more efficient and economical consolidated arrangement.

Content of a Regional Planning Report

At the very least it would appear logical to combine into one comprehensive environmental engineering study the evaluation of the regional water supply, wastewater, solid wastes, air quality, and related items. These all have in common the components listed below and are closely interrelated.

The ideal comprehensive regional plan would be one that combines all the project physical planning studies into one tied in with a comprehensive economic and social development plan. In the absence of such a triad of planning, the project or functional regional plans should take into consideration the economic, environmental, and social factors affecting the planning. In addition the effects of comprehensive or overall regional planning should be recognized and steps taken in the project planning to prevent or alleviate potential deleterious side effects.

An outline of a regional or area-wide planning study and report showing the elements that are common to most functional studies follows.

1. Letter of transmittal to the contracting agency.
2. Acknowledgments.
3. Table of contents.
   a. List of tables.
   b. List of figures.
4. Findings, conclusions, and recommendations.
5. Purpose and scope.
6. Background data and analysis, as applicable, including base maps, reports, and special studies.
   a. Geography, hydrology, meteorology, geology, groundwater levels.
   b. Population density and characteristics—past, present, future.
   c. Soil characteristics; flora and fauna.
   d. Transportation and mobility; adequacy and effects produced—present and future.
   e. Residential, industrial, commercial, recreational, agricultural, and institutional development and redevelopment.
   f. Land use—present and future; spread of blight and obsolescence; inefficient and desirable land uses.
   g. Drainage and flood control management.
   h. Water resources, multiuse planning, and development with priority to water supply; environmental impact.
   i. Air and water pollution, sewerage, solid waste management.
   j. Public utilities (electricity, gas, oil, heat) and their adequacy.
   k. Educational and cultural facilities, size, location, effects, adequacy.
   l. Economic studies: present sources of income, future economic base and balance, labor force, markets, industrial opportunities, retail facilities, stability.
   m. Sociological factors: characteristics, knowledge, attitudes, behavior of the people and their expectations.
   n. Local government, political organizations, and laws, codes, ordinances.
   o. Special problems, previous studies and findings, background data, including tax structure and departmental budgets.

7. Project study. This would be a regional or area-wide in-depth study of one or more projects or functions such as solid waste management, water supply, recreation, vector control, wastewater, or environmental health. Several examples of comprehensive project studies are outlined below under Project Study.
8. The comprehensive regional plan.
   a. Alternative solutions and plans.
   b. Economic, social, and environmental analysis and evaluation of alternatives, including adverse and beneficial environmental consequences.
   c. The recommended regional plan.
   d. Site development and reuse plans.
9. Administration and financing.
   a. Public information.
   b. Administrative and institutional arrangements, management, and costs.
   c. Capital improvement program and financing methods: general obligation bonds, revenue bonds, special assessment bonds; also grants, incentives, federal and state aid.
   d. Cost distribution, service charges and rates; capital costs: property, equipment, structures, engineering, legal services. Annual costs to repay capital costs, principal and interest, taxes. Regular and special service charges and rates.
   e. Legislation, standards, inspection, enforcement.
   f. Evaluation, research, replanning.
10. Appendices.
   a. Applicable laws.
   b. Special data.
   c. Charts, tables, illustrations.
11. Glossary.
12. References.

Project Study

Some examples of specific regional or area-wide comprehensive single-purpose, functional, or project studies are outlined below. These expand on item 7, "Project study." A complete project study and report would cover items 1 to 12, listed above.

Comprehensive Solid Waste Study (Expands on item 7 above)

1. Additional background information and data analysis, including residential, commercial, industrial, and agricultural, solid wastes.
   a. Field surveys and investigations.
   b. Existing methods and adequacy of collection, treatment, and disposal and their costs.
   c. Characteristics of the solid wastes.
   d. Quantities, summary tables, and projections.
   e. Waste reduction at source; salvage and reuse.
2. Solid waste collection, including transportation.
   a. Present collection routes, restrictions, practices, and costs.
   b. Equipment and methods used.
   c. Handling of special wastes.
   d. Recommended collection systems.
3. Preliminary analyses for solid waste treatment and disposal.
   a. Resource recovery: salvaging, recycling, refuse derived fuel and energy conservation, economic viability.

b. Available treatment and disposal methods (advantages and disadvantages)—compaction, shredding, sanitary landfill, incinerator, high-temperature incinerator, pyrolysis, fluidized bed oxidation, bulky waste incinerator, waste heat recovery, composting, garbage grinders.
c. Pretreatment devices: shredders, hammermills, hoggers, compaction, and their applicability.
d. Disposal of special wastes—automobile, water and wastewater treatment plant sludges, scavenger wastes, commercial and industrial sludges and slurries, waste oils, *toxic and hazardous wastes*, rubber tires, agricultural wastes, pesticides, forestry wastes.
e. Treatment and dispoal of commercial and industrial wastes. This would normally require independent study by the industry when quantities are large or when hazardous wastes and special treatment problems are involved.
f. Transfer stations, facilities, and equipment.
g. Rail haul; barge haul; other.
h. Alternative solutions, costs, advantages, and disadvantages.

4. Review of possible solutions.
   a. Social, political, and economic factors.
   b. Beneficial and adverse environmental consequences of proposed actions.
   c. Existing and potential best land use within 1500 ft of treatment and disposal site, and aesthetic considerations.
   d. Site development and reuse plans.
   e. Special inducements needed.
   f. Preliminary public information and education.

Comprehensive Wastewater Study (Expands on item 7 above)

1. Additional background information and data analysis.
   a. Field surveys and investigations including physical, chemical, biological, and hydrological characteristics of receiving waters.
   b. Existing methods of municipal and industrial wastewater collection, treatment, and disposal.
   c. Characteristics of municipal wastes and wastewater volumes, strengths, flow rates.
   d. Characteristics of industrial wastes, quantities, and amenability to treatment with municipal wastes. Identify each.
   e. Water pollution control requirements; federal, state, and interstate receiving water classifications and effluent standards.
   f. Wastewater reduction, reclamation, reuse.
   g. Extent of interim and private, on-lot sewage disposal (adequacy—present and future), including soils suitability.
2. Wastewater collection.
   a. Existing collection systems (condition and adequacy), including infiltration, surface, and storm-water flows.
   b. Existing collection systems with treatment (condition and adequacy), including infiltration, surface, and storm-water flows.
   c. Areas or districts needing collection systems, and construction timetables.
   d. Soils, rock, groundwater conditions.
   e. Routing and right-of-way.
   f. Storm-water separation feasibility, holding tanks, special considerations, local ordinances, and enforcement.
   g. Need for storm-water drainage and collection systems.
3. Preliminary analyses for wastewater treatment and disposal.
   a. Treatment plant sites, pollution load, degree of treatment required, land requirements including buffer zone, foundation conditions, outfall sewer, hydrologic and oceanographic considerations. Environmental impact analysis.

b. Areas served.
c. Trunk lines and pumping stations.
d. Property and easement acquisition problems.
e. Design criteria.
f. Industrial waste flows and pretreatment required, if any, at each industry.
g. Effect of storm-water flows on receiving waters and need for holding tanks or treatment.
h. Treatment plant and outfall sewer design considerations.
i. Grit, screening, and sludge disposal.
j. Alternative solutions, total costs, and annual charges.

Comprehensive Water Supply Study (Expand on item 7 above)

1. Additional background information and data analysis.
   a. Field surveys and investigations of distribution system for cross-connections, water pressure, breaks, water usage, storage adequacy, leaks, fire requirements.
   b. Occurrence of waterborne diseases and complaints.
   c. Leak survey of existing system, flow tests, and condition of mains, and water use.
   d. Existing and future land uses; service areas, domestic and industrial water demands.
   e. Areas and number of people served by individual well-water systems, sanitary quality and quantity of water, chemical and physical quality, cost of individual treatment.
   f. Recommendations of the Insurance Services Offices (formerly National Board of Fire Underwriters) and others.
   g. Existing fire rates and reductions possible.
2. Alternative sources of water.
   a. Chemical, bacteriological, and physical quality ranges.
   b. Average, minimum, and safe yields; source development.
   c. Storage needed at source and on distribution system.
   d. Flow requirements for fires.
   e. Service areas, hydraulic analysis, and transmission system needs.
   f. Preliminary designs of system and treatment required for taste, odor, turbidity, color, organic and inorganic chemicals control.
   g. Right-of-way and water rights needed.
   h. Preliminary study of total construction and operation costs of each alternate; advantages and disadvantages of each; annual cost to user and how apportioned.
   i. Improvements needed in existing system: source, storage, transmission, treatment, distribution system, operation and maintenance, costs, to meet state and federal requirements of the Safe Drinking Water Act.

Comprehensive Environmental Engineering and Health Planning (Expands on item 7 above)

1. Epidemiological and demographic survey including mortality, morbidity, births and deaths, age and sex distribution, communicable, noninfectious and chronic diseases, and incidences of specific diseases by age groups; population distribution of diseases and people most at risk; social, economic, and environmental relationships; respiratory, water-, insect-, and foodborne diseases; domestic and wild animal and animal-related diseases; airborne and air-related diseases and illnesses; indices of disease vectors, pesticide and other chemical poisonings; congenital malformations; mental disorders; health services and their availability; adequacy of data and programs.
2. Public water supply, treatment and distribution including population served, adequacy, operation, quality control, cross-connection control, storage and distribution protection,

operator qualifications. For individual systems: population served, special problems, treatment and costs, adequacy, control of well construction. Extension of public water supply based on a comprehensive regional plan, including fire protection, to replace inadequate and unsatisfactory small community water systems and individual well-water supplies in built-up areas. See Chapter 3 and Comprehensive Water Supply Study above.

3. Wastewater collection, treatment, and disposal; adequacy of treatment and collection system, population served, operator qualifications, sewer connection control. For individual systems: population served, special problems, control of installations. Water pollution control. Provision of sewerage meeting surface water and groundwater classifications based on a drainage area or regional plan to eliminate pollution by existing discharges, including inadequate sewage and industrial waste treatment plants and septic tank systems. See Chapter 4 and Comprehensive Wastewater Study above.
4. Solid waste management: storage, collection, transportation, processing and disposal, adequacy. Resource recovery, salvaging and recycling, including municipal refuse, industrial and agricultural wastes; handling of hazardous wastes, their environmental impact, prevention of contact, and air, water, and land pollution. Use of solid wastes to accelerate construction of open-space buffer zones and recreation areas. See Chapter 5 and Comprehensive Solid Waste Study above.
5. Air resources management and air pollution control including sources, air quality and emission standards, topographical and meteorological factors; problems and effects on man, livestock, vegetation, and property; regulation and control program. See Chapter 6.
6. Housing and the residential environment: control of new construction, modern building code including plumbing, electrical, heating; housing conservation and rehabilitation, and enforcement of housing occupancy and maintenance code; effectiveness of zoning controls, urban renewal, and redevelopment. Quality of housing, installed facilities, occupancy and overcrowding. Realty subdivision and mobile home park development and control, also effect of a development on the regional surroundings and effect of the region on the development, including the environmental impact of the development. See Chapter 11 and Fringe and Rural Area Housing Developments in this chapter.
7. Recreation facilities and open-space planning, including suitability of water quality and adequacy of sewerage, solid waste disposal, water supply, food service, restrooms, safety, and other facilities. See Chapter 9 and Environmental Factors to be Evaluated in Site Selection and Planning in this chapter.
8. Food protection program: adequacy from source to point of consumption. See Chapter 8.
9. Nuclear energy development; radionuclide and radiation environmental control, including fallout, air, water, food, and land contamination; thermal energy utilization or dissipation and waste disposal; naturally occurring radioactive materials; air, water, plant, and animal surveillance; federal and state control programs; standards; site selection and environmental impact; plant design and operation control; emergency plans. See Chapters 7 and 9.
10. Planning for drainage, flood control, and land-use management. Surface water drainage to eliminate localized flooding and mosquito breeding. Development of recreation sites, including artificial lakes, parks, swimming pools, bathing beaches, and marinas.
11. Health care institutions and adequacy of medical care facilities such as hospitals; nursing homes; public health, mental health, and rehabilitation centers; clincs; service agencies; jails and prisons; day care centers. Staffing, budgets, work load.
12. Noise and vibration regulations, abatement, and control.
13. Noxious weed, insect, rodent, and other vermin control, including disease vectors, nuisance arthropods; regulation, control, and surveillance, including pesticide use for control of aquatic and terrestrial plants and vectors; federal, state, and local programs; effects of water, recreation, housing, and other land resource development. See Chapter 10.
14. Natural and man-made hazards, including slides, earthquakes, brush and forest fires, reservoirs, tides, sand storms, hurricanes, tornadoes, high rainfall, fog and dampness, high winds, gas and high-tension transmission lines, storage and disposal of explosive and flammable substances and other hazardous materials.
15. Aesthetic and environmental considerations; wooded and scenic areas, prevailing winds and sunshine. Solar energy utilization.

16. Laws, codes, ordinances, rules, and regulations including environmental health criteria and standards.
17. Environmental health and quality protection; adequacy of organization and administration. See Chapter 12.

Financing

Financing for a municipal capital improvement is generally done by revenue bonds or general obligation bonds. Sometimes, as for small projects, funds on hand are used or special assessment bonds are issued to be paid by a special tax levy on properties directly benefitted, or a user service charge is levied. Other arrangements include a combination of revenue and general obligation bonding; the issuance of mortgage bonds using the physical utility assets as collateral; the creation of a nonprofit corporation or authority with the power to sell bonds; and contract with a private investor to build a structure or facility for lease at an agreed-on cost, with ownership reverting to the municipality at the end of a selected time period.

Usually the constitution or other law of a state contains a specific limitation in regard to the amount of debt that a municipality may incur. The debt limit may be set at approximately 5 to 10 percent, and the operating expenses at about 2 percent, of the average full value of real estate. The debt margin established generally does not include bonded indebtedness for schools. In most cases, bond issues for capital improvements require approval of the state fiscal officer to determine if the proposed project is in the public interest and if the cost will be an undue burden on the taxpayer. The ability to pay or per capita financial resources can be expected to vary from one municipality to another. Sometimes indebtedness for an essential revenue-producing service, such as water supply, is excluded from the constitutional limit.

Revenue bonds are repaid from a specific source of revenue, such as water and sewer charges. The service made possible by the bond issue is therefore directly related to the monthly, quarterly, or annual billing for a specific service. The bonds are not backed by the full credit of the municipality. Interest rates can therefore be expected to be somewhat higher than for general obligation bonds.

General obligation bonds are issued by a governmental agency, and their payment is guaranteed by the municipality through its taxing powers. The money thus borrowed is repaid by all of the people in a community, usually as additions to the real property tax. The total amount of general obligation bonds that may be issued by a municipality is generally limited by law. General obligation bonds may also be used to pay off a revenue-producing capital improvement. Approval by the voters in a special referendum may be required.

A combination of revenue and general obligation bonds would generally use revenue to pay off principal and interest, but would also be paid from general tax funds to the extent needed if the revenue were inadequate. This can result in lower interest rates. Other arrangements and cost apportionment are possible.

Assessment of users to pay off bonds may be based on the property tax, known as an *ad valorem* tax; on the service provided, such as metered water; on the potential service provided, such as a trunk sewer line that may be connected to 10 or 20 years in the future; or some combination. The *ad valorem* tax is not a tax proportionate to the service received. For example, a home assessed at $100,000 would pay four times the tax paid by a home assessed at $25,000, but it would not normally receive four times the service.

Notes are short term borrowings, that is, promises to pay, for a year or less, to tide a community over gaps in cash flow. They are issued in anticipation of taxes to be collected, or revenues to be received from the state or federal government, or of the proceeds of a long term bond sale to be held at a later date. See Appendix, Finance or Cost Comparisons.

Other financing alternatives include use of bond banks, water banks, industrial development bonds and federal and state government loans and grants.[9]

ENVIRONMENTAL FACTORS TO BE EVALUATED IN SITE SELECTION AND PLANNING

Certain general basic information should be known before a suitable site can be selected for a particular purpose. One should know the present and future capacity and the total land area desired. The type of establishment or facility to be maintained, the use to which it will be put—whether a new land subdivision, shopping plaza, adult or children's camp, park and picnic area, resort hotel, dude ranch, marina, trailer park, campground, country club, factory, industrial and scientific complexes, water treatment plant, wastewater treatment plant, solid waste resource recovery-treatment-disposal facility, power plant, school, institution, or private home—and the functions, activities and programs to be carried on during each season, or all year, are to be decided on before any property is investigated. In addition one should know the radius in miles or time of travel within which the area must be located; the accessibility to roads, airports, waterways, or railroads; the availability of utilities, permanency of the project; the money and manpower available; whether lake, river, or stream frontage is necessary; and whether the site need be mountainous, hilly, flat, wooded, or open. Not to be forgotten is evaluation of the environmental, social, political, and economic impact of the proposed project, the numerous federal and state environmental laws controlling development and construction, and minimization of the undesirable effects to acceptable levels. Environmental impact assessment is discussed later in this chapter.

[9] "Report of the Joint AWWA-NAWC Committee on Financing Water Industry Projects," *Willing Water*, January 1980, pp. 9–12.

Desirable Features

It probably will not be possible to find a site that will meet all the conditions authorities recommend as being essential. Desirable features include the following.

1. It is best to have an adequate groundwater or surface water supply not subject to excessive pollution that can be developed into a satisfactory supply at an accessible and convenient location on the property, if an adequate public water supply is not available. The water supply source should meet federal and state standards.

2. A permeable soil that will readily absorb rainwater and permit the disposal of sewage and other wastewater by conventional subsurface means is most desirable, if not essential, for the smaller establishment where public sewerage is not available. Such soil should contain relatively large amounts of sand and gravel, perhaps in combination with some silt, clay, broken stones, or loam. The underground water should not be closer than 4 ft to the ground surface at any time and there should be a porous earth cover of not less than 4 to 5 ft over impervious subsoil or rock. A suitable receiving stream or land area is needed if a sewage treatment plant is required.

3. Land to be used for housing or other structures must have suitable soil-bearing characteristics, and be well above flood or high-water level. There should be no nearby swamps.

4. Elevated, well-drained, dry land open to the air and sunshine part of the day, on gently sloping, partly wooded hillsides or ridges, should be available for housing and other buildings. The cleared land should have a firm, grass-covered base to prevent erosion and dust. A slope having a southern or eastern exposure protected from strong winds on the north and west is generally desirable.

5. The area of the property should be large enough to provide privacy, avoid crowding, accommodate a well-rounded program of activities, and allow for future expansion. The property should be accessible by automobile and bus and convenient to airports, superhighways, railroads, and waterways, if needed, and recreation facilities.

An allowance of 1 acre per camper has been suggested as being adequate for children's camps. For elementary schools, 1 acre per 100 pupils, and, for high schools, 10 acres plus 1 acre per 100 pupils is recommended. The play area should provide 1000 ft^2 per child using the area at any one time. For a residential area, $\frac{3}{4}$ acre of public playgrounds, $1\frac{1}{4}$ acres of public playfields, and 1 acre of public park land per 1000 population are considered minimums. Other suggested standards for parks are given in Table 2-1. Standards (more correctly, criteria) should be considered points of departure, to be interpreted in the light of the economic and social structure of the people affected and should be adjusted accordingly. Consider also the recreation potential of abandoned railroad and barge canal rights-of-way for hiking, horseback trails, bicycle paths, snowmobile trails, and other recreation purposes.

6. A satisfactory area should be available for bathing and swimming and other water sports at recreational sites. This may be a clean lake, river, or stream or an artificial swimming pool. A river or stream should not have a strong current or remain muddy during its period of use. An artificial swimming pool equipped with filtration, recirculation, and chlorination equipment may be substituted to advantage.

7. Noxious plants, poisonous reptiles, harmful insects, excessive dust, steep cliffs, old mine shafts or wells, dangerous rapids, dampness, and fog should be absent. All this is not usually possible to attain; however, the seriousness of each should be evaluated.

8. A public water supply, sewerage system, and solid waste disposal system, if available and accessible, would be extremely desirable.

9. For residential and industrial development, electricity, gas, and telephone service; a sound zoning ordinance and a land use plan that provides for and protects compatible uses; fire protection; and modern building construction and housing codes vigorously enforced by competent people should all be assured.

10. Air pollution, noise, and traffic problems from the site and from adjoining areas should not

Table 2-1 Suggested Standards for Parks

Type of Park	Area of Facility	Service Radius	Population Served
Play or tot lot	2500 ft^2 minimum	$\frac{1}{8}$ mi	25 to 75 children
Neighborhood	8 to 15 acres	$\frac{1}{4}$ mi—high density $\frac{3}{8}$ mi—low density	2000 to 5000
District	15 to 40 acres	$\frac{1}{2}$ to 1 mi average	15,000 to 35,000
Community	100 to 500 acres	$1\frac{1}{2}$ to $2\frac{1}{2}$ mi	1000/$\frac{1}{4}$ acre
Special area (golf course, marina, stadium, parkway)	35 to 175 acres	1 to $1\frac{1}{2}$ mi	—
School			
Elementary	8 to 15 acres	$\frac{1}{4}$ to $\frac{3}{8}$ mi	—
Junior High	10 to 25 acres	$\frac{3}{8}$ to 1 mi	—
Senior High	25 to 50 acres	$\frac{1}{2}$ mi minimum	—

Sources: *Suggested Standards for Parks and Recreation*, National Recreation and Park Association; U.S. Dept. of the Interior, *Outdoor Recreation Space Standards*, GPO, Washington, D.C., 1970.

interfere with the proposed use. For example, airborne motor vehicle exhaust lead at the site should not exceed 1.5 $\mu g/m^3$ averaged over a 3-month period and carbon monoxide should not exceed 40 mg/m^3 for one hour or 10 mg/m^3 for eight hours (Federal Air Quality Standard).

Topography and Site Surveys

A boundary survey of the property with contours shown at 5-ft intervals, in addition to roads, watercourses, lakes, swamps, woodlands, structures, railroads, power lines, rock outcrops, and any other significant physical features indicated would be of very great value in studying a property. If such a map is not available, a U.S. Geological Survey sheet (scale: 1 in. = $\frac{1}{2}$ mi. or 1 mi, contour interval = 10 ft) that has been blown up or an aerial photograph (approximately 1 in. = 1660 ft) from the U.S. Department of Agriculture or other agency may be used instead for the preliminary study. The outline of the property should be marked on the map, using deed descriptions. Old plot plans of the property that are available and distinctive monuments or other markings that can be found by inspection of the site would add valuable information. A long-time resident or cooperative neighbor may also be of assistance.

With the map as a beginning, one should hike over as much of the area as possible and carefully investigate the property. Be sure to keep complete notes that refer to numbers placed on the topographic map showing beautiful views and other desirable features. Supplementary freehand sketches and rough maps of possible camp, recreation, or building sites, with distances paced off, will be valuable details.

The bathing area should be sounded and slope of the bottom plotted. The need for cleaning and removal of mud, rocks, and aquatic growths should be noted. The drainage area tributary to a lake or stream to be used for bathing should be determined and the probable minimum contribution or flow computed to ascertain if an ample quantity of water will be available during the dry months of the year.

In most cases the watershed area tributary to a beach on a stream or lake will extend beyond the boundary of the land under consideration. It is important therefore to know what habitation, agriculture, and industry is on the watershed. The probable land usage, pollution of tributary streams, and probable water use should then be determined because persons owning land have the right to reasonable use of their property, including streams flowing through it. This will bring out whether the stream or lake is receiving chemical, bacterial, or physical pollution that would make it dangerous or unsuitable to use for bathing or water supply purposes. In order to obtain this valuable information it is necessary to make a survey on foot of every stream and brook on the watershed. The local health, agriculture, or conservation department sanitary engineer, agent, conservation officer, and sanitarian may be able to give assistance. In addition to quantity and quality, the water should be relatively clear and slow moving. Study of the stream bottom and float or weir measurements to determine the velocity and quantity of water will give this information.

While a survey is being made, one should look for signs indicating the high water level of the lake or stream. The topography of the ground, presence of a flood plain, stranded tree trunks and debris, discolorations on rocks and trees, width, depth, and slope of the stream channel, coupled with probable maximum flow, evaluation of nearby railroad beds, and type of vegetation growing, may give good clues. Valuable information can be obtained by discussing this point with long-time residents.

Geology, Soil, and Drainage

The soil should be sampled at representative locations to determine its characteristics. Borings should be made to a depth of about 15 ft in order to record variations in the strata penetrated and the elevation of the groundwater level, if encountered, with respect to the ground surface. Borings, posthole, and earth auger tests will also indicate the depth to rock and the presence and thickness of clay or hardpan layers that might interfere with proper drainage or foundations for structures.

In addition to borings, soil studies and soil percolation tests should be made in areas that, as indicated by the topography, are probably suitable for subsurface sewage disposal, if needed. Explanation of the soil's charcteristics and percolation tests and application of the results are discussed in Chapter 4. Public sewerage should be used if possible.

If a proposed development or building is to be located on the side of a long hill or slope, the necessity and feasibility of providing an earth berm or dam, or deep surface water drainage ditches to divert surface water around the site, should be kept in mind. The possibility of earth slides, flooding, erosion, and washout should be given careful study. Slopes with a greater than eight percent incline require special engineering study and treatment, such as water infiltration and percolation control to prevent slides, erosion control, drainage, and vegetative cover. The need for and practicability of constructing special foundations, a surface and curtain drain, or a subsurface drainage system to lower the groundwater table, or the need for draining wet and swampy areas (or the feasibility of making an artificial lake or protecting the swamp or marsh) will be apparent from the data accumulated in the topography study.

Utilities

The existence of or need for a water supply system, a sewage disposal system, a solid waste collection and disposal system, roads, electricity or a generator of electricity, gas, oil, coal, wood, and telephones should be studied, for they will determine the type of establishment, services, and sanitary facilities that could be provided.

It is needless to say if an adequate, satisfactory, and safe water supply is not obtainable at a reasonable cost, with or without treatment, the site should be abandoned. This is particularly important to a factory or industry dependent on large volumes of water. See Chapter 3.

The probable cost of a wastewater collection and treatment system should be estimated before any commitments are made. Water classification and effluent standards for water pollution abatement will govern the degree of treatment required and hence the cost of construction and operation. For a large project, an elaborate wastewater treatment plant may be required. For small establishments a subsurface sewage disposal system may suffice if the soil conditions are satisfactory. An alternative at pioneer-type camps might be the use of privies. See Chapter 4.

Electricity for lighting or for the operation of water and wastewater pumps, kitchen equipment, refrigerator compressors, and other mechanical equipment is usually taken for granted. If the provision of electricity means the running of long lines, purchase of an electric-generating unit, or gasoline motor-driven equipment, then the first cost, cost of operation, maintenance, and replacement should be estimated.

Roads to the main buildings are needed for access and for bringing in supplies. The distance from the main roads to the property and the length and condition of secondary roads within the property should be determined, also the need for road culverts and bridges.

If a power plant or industrial process will cause air pollution, air pollution control requirements, prevailing winds, temperature, and related factors will have to be studied and the cost of treatment devices determined in evaluating the suitability of a proposed site. See Chapter 6. If a plant process will result in the production of large quantities of solid wastes, the treatment and disposal of the residue must also be considered. See Chapter 5.

Meteorology

Slopes having an eastern or southern exposure in the United States are to be preferred for building locations, to get the benefit of the morning sun. This possibility can be ascertained by inspection of available topographic maps and by field surveys. Information about the direction of prevailing winds is of value if a summer or winter place is proposed. An indication of the wind direction can be obtained by observing the weathering of objects and the lean of the trees. This information, plus average monthly temperature, humidity, and rainfall data, may be available at local universities, nearby government weather stations, airfields, and at some water or power company offices. Where rainfall data is not available it may be possible to utilize stream flow measurements to judge the general pattern of precipitation in the area. Consider also potential sources of air pollution, prevailing winds, and possible effects on the proposed land use.

Location

The relative location of the property can best be appreciated by marking its outline on a recent U.S. Geological Survey Sheet. In this way the distance to airports, railroad and bus stations, first-class roads, shopping centers, neighbors, resort areas, schools, hospitals, and doctors is almost immediately apparent. State and county highway department road plans should also be investigated and proposed roads marked on the topographic plan to see what their probable effect will be.

Resources

To determine the resources on a property (when it is a large tract), one should seek the assistance of a person who is intimately familiar with wildlife, forestry, geology, hydrology, and engineering. Since this is not always possible, the next best thing would be for several persons having a broad knowledge to make the survey on foot. The location, size, and type of woodland, pasture, rock, sand, and gravel should be carefully noted. The woodland may also serve as building material or firewood; the rock as roadbed or foundation material; the pasture as a recreation area or golf course; the sand and gravel for concrete and cement work or road surfacing with admixtures if needed. The availability of such material near the construction site will result in a considerable saving. Not to be forgotten as resources are surface and underground water, for domestic, recreational, and power purposes, and unspoiled scenery.

Animal and Plant Life

Here again, a knowledge of these is necessary before a worthwhile study can be made. The presence of poison ivy, for example, must be accepted as a potential source of skin irritations; ragweed and other noxious plants must be accepted as sources of hay fever. It is therefore important to mark the location of the infested areas on a topographic map so that attention is directed to their existence in the construction and planning program. The same would apply to the presence of mosquitoes, flies, ticks, chiggers, rodents, poisonous reptiles, and dangerous animals. On the other hand, equal emphasis can be placed on the presence of wild flowers, useful reptiles and wild animals, and native trees. A broad knowledge and extensive investigation is needed to properly evaluate the importance of animal and plant life and on their conservation.

Improvements Needed

Inasmuch as the ideal school, institution, industry, camp, or housing site is rarely if ever found, the work that needs to be done to make an ideal site should be determined. An undesirable feature may be a low swampy area. This may be filled in if the area is not too large and suitable fill is available on the property; the ground may be drained if the topography or subsoil strata make this possible; the area may be cleared, dammed up, and made into a lake; or the swamp may be retained as a wildlife preserve. Each possibility should be examined, and reviewed with the regulatory agency having jurisdiction, the cost of the work estimated, and the probable value of the improvement appraised. Needed clearing, seeding, or reforestation and the extent of poison ivy, ragweed, thistle, and other objectionable plants should also be considered. If a natural bathing area is not available on a lake or stream on the property, it is advisable to compare the cost of developing and maintaining one with the cost of an artificial swimming pool. If a lake is available, then thought must be given to clearing, the construction of a beach, and perhaps dredging and shore development to keep down heavy aquatic growths and plant life such as algae. The need for new roads inside the property connecting with town, county, or state roads is another consideration. In some cases this may involve blasting, fill, bridges, and special construction, all of which could mean high costs. Other needed physical improvements may be surface water diversion ditches, culverts, groundwater drainage tile, brush clearing, boat docks, parking areas, and the preparation of areas for recreational purposes. Also important is the environmental impact of the proposed land development and use.

Site Planning

After properties have been explored and studied and a site is selected, the next step is to have prepared, if not already available, a complete large-scale topographic map of the purchased property to a scale of 1 in. equal to 100 ft, with contours at least at 5-ft intervals, incorporating the details already discussed. After this, put down on paper what the ideal place is to consist of. Lay out on the map the approximate location of the proposed establishment, making maximum use of the natural advantages offered by the site. Camps, for example, occupied during the summer months should have an eastern exposure. Locate camps on the west shore of a lake or stream to receive the benefit of early morning sun and afternoon shade. Then, without losing sight of the purpose the property is to serve and the program to be followed, prepare cutouts made to scale to represent plans of present and future buildings, roads, parking spaces, recreation areas, campsites, water supply, sewage and refuse disposal areas, bathing beaches, and other facilities. The preparation of a scale model also has great promotional possibilities. It will help visualize the buildings and their relationship to each other and to roads, streams, lakes, hills, and neighboring communities. Changes can be easily made on paper. This method has been used successfully to help obtain funds for major improvements and new developments.

It is advisable to seek the advice of an engineer, architect, planner, or other consultant trained and experienced in planning work. One should confer with the state and local planning, health, and building departments to learn of the regulations for the protection of the life, health, and welfare of the people, and with environmental protection and conservation agencies for protection of the flora, fauna, and ecology, and the need for an environmental impact assessment. This is particularly important when a housing development, industrial plant, camp, hotel, shopping center, school, or other public place is contemplated. The agency staff may also be in a position to offer valuable assistance. The plan should show, in addition to the items mentioned, such other details that may be required by the departments having jurisdiction to permit their staff to review the plans and pass on them favorably. After approved plans are received, bids can be received, a construction schedule established, and the work carried on in accordance with a preconceived, carefully thought-out plan. There will then be no looking back at useless, inefficient, wasteful structures; instead, the eventual realization of the "ideal" place by adding to what has already been accomplished will be anticipated.

The importance of obtaining competent professional advice cannot be emphasized too strongly. All too often charitable camps and institutions seek and expect free engineering and architectural services. This is not only unfair to the individuals consulted, but is also unfair to the camp or institution. The consultants would unconsciously try to conclude the planning and the design as rapidly as possible; and the recipient of the service would proceed on the basis of possibly incompletely studied and conceived plans, detailed drawings, specifications, and contract documents. A proper place for free expert advice would be in the selection of the engineer, architect, contractor, and equipment, and in helping to evaluate the bids received for the particular job. The fee for a consultant's services, in comparison to the cost of a project, is relatively small. The reduction of waste caused by incorrect size of buildings, improper materials, poor planning of equipment and facilities, foundation and structural weaknesses, and bad location or exposure of buildings, as well as savings in the selection of proper materials and equipment for minimum operating costs and the assurance that plans and specifications are being complied with, far outshadows the professional fee.

ENVIRONMENTAL IMPACT ANALYSIS

Concurrent with site selection and planning is the necessity to consider the effects of all the proposed land uses, actions, and required services and facilities, on the environment or geography* of the area. It is extremely difficult to identify and evaluate in depth *all* possible factors that may affect and be affected by a particular project or action. Reference to the material that follows, the pertinent federal and state legislation, and source documents such as footnoted and listed in the bibliography at the end of this chapter should assist in selection of items to be explored in greater depth in particular situations. A good regional plan, including the environmental, engineering, and health factors previously noted will lead to the collection of pertinent data and greatly simplify preparation of an environmental impact assessment, and, if necessary, the environmental impact statement.

The National Environmental Policy Act

Planning, design, and construction or implementation of a project, without regard to its environmental effects, not to mention the social, political, economic, and other consequences, calls attention to the custodial responsibility and moral obligation of society to protect the environment for future generations and ensure their healthful survival. The wave of public concern over environmental pollution, aided by scientific and professional prodding and support, led to federal and state legislation mandating consideration and documentation of the beneficial and adverse effects of proposed actions in the project planning stage for official and public scrutiny and indicated adjustments. This concept was given national recognition by the National Environmental Policy Act (NEPA) of 1969,[10] as amended. It is quoted in part below. The reader should refer to the entire Act and applicable state law if involved in a project or action coming under the Act.

*"The descriptive science dealing with the surface of the earth, its division into continents and countries, and the climate, plants and animals, natural resources, inhabitants, and industries of the various divisions." Webster's New World Dictionary.

[10]Regulations For Implementing the Procedural Provisions of the National Environmental Policy Act, Reprint 43 FR 55978–56007, November 29, 1978, 40 CFR Parts 1500–1508, Superintendent of Documents, GPO, Washington, D.C., 20402.

THE NATIONAL ENVIRONMENTAL POLICY ACT OF 1969, AS AMENDED*

An Act to establish a national policy for the environment, to provide for the establishment of a Council on Environmental Quality, and for other purposes.

Be it enacted by the Senate and House of Representatives of the United States of America in Congress assembled, That this Act may be cited as the "National Environmental Policy Act of 1969."

Purpose

SEC. 2. The purposes of this Act are: To declare a national policy which will encourage productive and enjoyable harmony between man and his environment; to promote efforts which will prevent or eliminate damage to the environment and biosphere and stimulate the health and welfare of man; to enrich the understanding of the ecological systems and natural resources important to the Nation; and to establish a Council on Environmental Quality.

Title I

Declaration of National Environmental Policy

SEC. 101. (a) The Congress, recognizing the profound impact of man's activity on the interrelations of all components of the natural environment, particularly the profound influences of population growth, high-density urbanization, industrial expansion, resource exploitation, and new and expanding technological advances and recognizing further the critical importance of restoring and maintaining environmental quality to the overall welfare and development of man, declares that it is the continuing policy of the Federal Government, in cooperation with State and local governments, and other concerned public and private organizations, to use all practicable means and measures, including financial and technical assistance, in a manner calculated to foster and promote the general welfare, to create and maintain conditions under which man and nature can exist in productive harmony, and fulfill the social, economic, and other requirements of present and future generations of Americans.

(b) In order to carry out the policy set forth in this Act, it is the continuing responsibility of the Federal Government to use all practical means, consistent with other essential considerations of national policy, to improve and coordinate Federal plans, functions, programs, and resources to the end that the Nation may—

(1) fulfill the responsibilities of each generation as trustee of the environment for succeeding generations;

(2) assure for all Americans safe, healthful, productive, and aesthetically and culturally pleasing surroundings;

(3) attain the widest range of beneficial uses of the environment without degradation, risk to health or safety, or other undesirable and unintended consequences;

(4) preserve important historic, cultural, and natural aspects of our national heritage, and maintain, wherever possible, an environment which supports diversity, and variety of individual choice;

(5) achieve a balance between population and resource use which will permit high standards of living and a wide sharing of life's amenities; and

(6) enhance the quality of renewable resources and approach the maximum attainable recycling of depletable resources.

(c) The Congress recognizes that each person should enjoy a healthful environment and that each person has a responsibility to contribute to the preservation and enhancement of the environment.

SEC. 102. The Congress authorizes and directs that, to the fullest extent possible: (1) the policies, regulations, and public laws of the United States shall be interpreted and administered in accordance with the policies set forth in this Act, and (2) all agencies of the Federal Government shall—

*Pub. L. 91-190, 42 U.S.C. 4321–4347, January 1, 1970, as amended by Pub. L. 94-52, July 3, 1975, and Pub. L. 94-83, August 9, 1975. See also Regulations For Implementing the Procedural Provisions of the National Environmental Policy Act.

(A) Utilize a systematic, interdisciplinary approach which will insure the integrated use of the natural and social sciences and the environmental design arts in planning and in decisionmaking which may have an impact on man's environment;

(B) Identify and develop methods and procedures, in consultation with the Council on Environmental Quality established by Title II of this Act, which will insure that presently unquantified environmental amenities and values may be given appropriate consideration in decisionmaking along with economic and technical considerations;

(C) Include in every recommendation or report on proposals for legislation and other major Federal actions significantly affecting the quality of the human environment, a detailed statement by the responsible official on—

(i) The environmental impact of the proposed action.

(ii) Any adverse environmental effects which cannot be avoided should the proposal be implemented,

(iii) Alternatives to the proposed action,

(iv) The relationship between local short-term uses of man's environment and the maintenance and enhancement of long-term productivity, and

(v) Any irreversible and irretrievable commitments of resources which would be involved in the proposed action should it be implemented.

Prior to making any detailed statement, the responsible Federal official shall consult with and obtain the comments of any Federal agency which has jurisdiction by law or special expertise with respect to any environmental impact involved. Copies of such statement and the comments and views of the appropriate Federal, State, and local agencies, which are authorized to develop and enforce environmental standards, shall be made available to the President, the Council on Environmental Quality and to the public as provided by section 552 of title 5, United States Code, and shall accompany the proposal through the existing agency review processes;

(d) Any detailed statement required under subparagraph (c) after January 1, 1970, for any major Federal action funded under a program of grants to States shall not be deemed to be legally insufficient solely by reason of having been prepared by a State agency or official, if:

(i) the State agency or official has statewide jurisdiction and has the responsibility for such action,

(ii) the responsible Federal official furnishes guidance and participates in such preparation,

(iii) the responsible Federal official independently evaluates such statement prior to its approval and adoption, and

(iv) after January 1, 1976, the responsible Federal official provides early notification to, and solicits the views of, any other State or any Federal land management entity of any action or any alternative thereto which may have significant impacts upon such State or affected Federal land management entity and, if there is any disagreement on such impacts, prepares a written assessment of such impacts and views for incorporation into such detailed statement.

The procedures in this subparagraph shall not relieve the Federal official of his responsibilities for the scope, objectivity, and content of the entire statement or of any other responsibility under this Act; and further, this subparagraph does not affect the legal sufficiency of statements prepared by State agencies with less than statewide jurisdiction.

(e) Study, develop, and describe appropriate alternatives to recommended courses of action in any proposal which involves unresolved conflicts concerning alternative uses of available resources;

(f) Recognize the worldwide and long-range character of environmental problems and, where consistent with the foreign policy of the United States, lend appropriate support to initiatives, resolutions, and programs designed to maximize international cooperation in anticipating and preventing a decline in the quality of mankind's world environment;

(g) Make available to States, counties, municipalities, institutions, and individuals, advice and information useful in restoring, maintaining, and enhancing the quality of the environment;

(h) Initiate and utilize ecological information in the planning and development of resource-oriented projects; and

(i) Assist the council on Environmental Quality established by Title II of this Act.

SEC. 103. All agencies of the Federal Government shall review their present statutory authority, administrative regulations, and current policies and procedures for the purpose of determining whether there are any deficiencies or inconsistencies therein which prohibit full compliance with the purposes and provisions of this Act and shall propose to the President not later than July 1, 1971, such measures as may be necessary to bring their authority and policies into conformity with the intent, purposes, and procedures set forth in this Act.

SEC. 104. Nothing in section 102 or 103 shall in any way affect the specific statutory obligations of any Federal agency (1) to comply with criteria or standards of environmental quality, (2) to coordinate or consult with any other Federal or State agency, or (3) to act, or refrain from acting contingent upon the recommendations or certification of any other Federal or State agency.

SEC. 105. The policies and goals set forth in this Act are supplementary to those set forth in existing authorizations of Federal agencies.

Title II of NEPA establishes a Council on Environmental Quality; requires the President to submit an annual Environmental Quality Report to Congress; authorizes personnel; states the duties, powers, and functions of the Council; and authorizes certain annual appropriations.

The Environmental Quality Improvement Act of 1970 was enacted

(1) To assure that each Federal department and agency conducting or supporting public works activities which affect the environment shall implement the policies under existing law; and

(2) To authorize an Office of Environmental Quality, which, notwithstanding any other provision of law, shall provide the professional and administrative staff for the Council on Environmental Quality established by Public Law 91-190.

By Executive Order, Protection and Enhancement of Environmental Quality, March 5, 1970, as amended, the President directed federal agencies to provide leadership and meet the national environmental goals. The Council on Environmental Quality was made responsible for advising and assisting the President in leading this national effort.[10]

Other federal agencies and many states have also adopted rules or legislation and prepared procedures for environmental quality reviews of proposed actions and for preparation of environmental impact statements.*

Terminology

Certain terms to aid in the interpretation of NEPA are explained below.

"*Categorical exclusion* means a category of actions which do not individually or cumulatively have a significant effect on the human environment and which have been found to have no effect in procedures adopted by a Federal agency in implementation of these regulations and for which, therefore, neither an environmental assessment nor an environmental impact statement is required. (Sec. 1508.4)[10]

"*Cumulative impact* is the impact on the environment which results from the incremental impact of the action when added to other past, present, and reasonably foreseeable future actions

*HUD, U.S.G.S., California, Illinois, Michigan, New York, Vermont, Virginia, Washington, others.

regardless of what agency (Federal or non-Federal) or person undertakes such other actions. Cumulative impacts can result from individually minor but collectively significant actions taking place over a period of time. (Sec. 1508.7)

Effects include:

(a) Direct effects, which are caused by the action and occur at the same time and place.

(b) Indirect effects, which are caused by the action and are late in time or farther removed in distance, but are still reasonably foreseeable. Indirect effects may include growth-inducing effects and other effects related to induced changes in the pattern of land use, population density or growth rate, and related effects on air and water and other natural systems, including ecosystems.

"Effects" and "impacts" as used in these regulations are synonymous. Effects include ecological (such as the effects on natural resources and on the components, structures, and functioning of affected ecosystems), aesthetic, historic, cultural, economic, social, or health, whether direct, indirect, or cumulative. Effects may also include those resulting from actions which may have both beneficial and detrimental effects, even if on balance the agency believes that the effect will be beneficial. (Sec. 1508.8)

*Environmental Assessment**:

(a) Means a concise public document for which a Federal agency is responsible that serves to:

(1) Briefly provide sufficient evidence and analysis for determining whether to prepare an environmental impact statement or a finding of no significant impact.

(2) Aid an agency's compliance with the Act when no environmental impact statement is necessary.

(3) Facilitate preparation of a statement when one is necessary.

(b) Shall include brief discussion of the need for the proposal, of alternatives as required by Sec. 102(2)(E), of the environmental impacts of the proposed action and alternatives, and a listing of agencies and persons consulted. (Sec. 1508.9)

Environmental document includes the documents specified in Secs. 1508.9 (environmental assessment), 1508.11 (environmental impact statement), 1508.13 (finding of no significant impact), and 1508.22 (notice of intent). (Sec. 1508.10)

*Environmental Impact Statement** means a detailed written statement as required by Sec. 102(2)(C) of the Act. (Sec. 1508.11)

Federal agency means all agencies of the Federal Government. It does not mean the Congress, the Judiciary, or the President, including the performance of staff functions for the President in his Executive Office. It also includes for purposes of these regulations States and units of general local government and Indian tribes assuming NEPA responsibilities under Sec. 104(h) of the Housing and Community Development Act of 1974. (Sec. 1508.12)

Finding of No Significant Impact means a document by a Federal agency briefly presenting the reasons why an action, not otherwise excluded (1508.4), will not have a significant effect on the human environment and for which an environmental impact statement therefore will not be prepared. It shall include the environmental assessment or a summary of it and shall note any other environmental documents related to it. If the assessment is included, the finding need not repeat any of the discussion in the assessment but may incorporate it by reference. (Sec. 1508.13)

Human Environment shall be interpreted comprehensively to include the natural and physical environment and the relationship of people with that environment. [See the definition of "effects" (Sec. 1508.8).] This means that economic or social effects are not intended by themselves to require preparation of an environmental impact statement. When an environmental impact statement is prepared and economic or social and natural or physical environmental effects are interrelated, then the environmental impact statement will discuss all of these effects on the human environment. (Sec. 1508.14)

*An environmental impact assessment is similar to an environmental impact statement but not necessarily as comprehensive or in the same depth. It can determine if a negative declaration is indicated or if an environmental impact statement is required. An environmental assessment is not required if it has been decided to prepare an environmental impact statement. (Sec. 1501.3)

Summary Definitions for Proposed Interpretation of NEPA 102(C)*

Describe present conditions. Requires a description of present conditions of the proposed project area, including specifics on surrounding terrain and ecosystems, existing and proposed land use, and other existing environmental and cultural features. A description of the project objective should be provided, including local, State, or Federal plans, and social, economic, and natural environmental goals of the area in question. Information and data adequate to permit careful assessment of the project area by commenting agencies is necessary. Where relevant, maps and/or photographs should be provided.

Describe alternative actions. Require the responsible agency to study, develop, and describe appropriate alternatives relevant to the proposed objective. Consideration should be given not only to engineering, design, location, institutional, and operation alternatives, but also to maintaining the status quo. Information and data adequate to permit careful assessment of the characteristics of each alternative by commenting agencies is necessary. Where relevant, maps and/or photographs would be provided.

Describe probable impacts of each alternative. Requires a description of primary and secondary impacts, including beneficial and detrimental impact on aesthetic, socio-economic, and ecological systems. This section also requires a description of the environmental interrelationships in the direct project area and the total affected area.

In particular, long-range impacts are to be evaluated regarding the extent to which actions taken now are decreasing sustained yield or carrying capacity of environmental components. Actions which once made cannot be withdrawn or reversed must also be specifically highlighted.

Identify alternative chosen and indicate evaluation which led to choice. Requires a statement of the action chosen to be proposed, including a more detailed development of its characteristics. The choice made implies tradeoffs which must be considered both for their relative value implications and the relationship of these values to particular constituencies.

Describe probable impacts of proposed action in detail. Requires a more detailed description of the probable effects, both beneficial and adverse. In particular, those adverse effects which will ensue even from this best alternative, and are therefore unavoidable in this context, should be highlighted. Evidence of compliance with local, State, and Federal environmental control regulations should be provided.

Describe techniques for minimizing harm. Requires a description of actions taken to minimize harm, including techniques employed to curb air pollution, water pollution, noise, distribution of economic and social patterns, or visual pollution. This applies to both the construction and the operation of the facility.

Recommended Format for Environmental Impact Statement (NEPA)

§ 1502.10 Recommended format.[10]

Agencies shall use a format for environmental impact statements which will encourage good analysis and clear presentation of the alternatives, including the proposed action. The following standard format for environmental impact statements should be followed unless the agency determines that there is a compelling reason to do otherwise:

(a) Cover sheet.
(b) Summary.†
(c) Table of Contents.
(d) Purpose of and Need for Action.

*Source: *Environmental Impact Statements: A Handbook for Writers and Reviewers*, Illinois Institute for Environmental Quality, Lewis D. Hopkins, Project Director, Chicago, Ill., August 1973, pp. 15–16.

†See Section 1500.8 and Appendix 1 of NEPA.

(e) Alternatives Including Proposed Action (Secs. 102(2)(C)(iii) and 102(2)(E) of the Act).
(f) Affected Environment.
(g) Environmental Consequences [especially Secs. 102(2)(C)(i), (ii), (iv), and (v) of the Act].
(h) List of Preparers.
(i) List of Agencies, Organizations, and Persons to Whom Copies of the Statement Are Sent.
(j) Index.
(k) Appendices (if any).

An explanation of each section of the environmental impact statement is given in sections 1502.11–1502.18 of the Regulations for Implementing the Procedural Provisions of the National Environmental Policy Act.[10]

A suggested outline for the content of an environmental impact statement is shown in Figure 2.5). The content is explained in section 1500.8, quoted below.

1. PROJECT DESCRIPTION
   a. Purpose of action
   b. Description of action
      (1) Name
      (2) Summary of activities
   c. Environmental setting
      (1) Environment prior to proposed action
      (2) Other related federal activities.
2. LAND-USE RELATIONSHIPS
   a. Conformity or conflict with other land-use plans, policies and controls
      (1) Federal, state, and local
      (2) Clean Air Act and Federal Water Pollution Control Act Amendment of 1972
   b. Conflicts and/or inconsistent land-use plans
      (1) Extent of reconciliation
      (2) Reasons for proceeding with action
3. PROBABLE IMPACT OF THE PROPOSED ACTION ON THE ENVIRONMENT
   a. Positive and negative effects
      (1) National and international environment
      (2) Environmental factors
      (3) Impact of proposed action
   b. Direct and indirect consequences
      (1) Primary effects
      (2) Secondary effects
4. ALTERNATIVES TO THE PROPOSED ACTION
   a. Reasonable alternative actions
      (1) Those that might enhance environmental quality
      (2) Those that might avoid some or all adverse effects
   b. Analysis of alternatives
      (1) Benefits
      (2) Costs
      (3) Risks
5. PROBABLE ADVERSE ENVIRONMENTAL EFFECTS WHICH CANNOT BE AVOIDED
   a. Adverse and unavoidable impacts
   b. How avoidable adverse impacts will be mitigated

Figure 2-5 Outline for CEQ-prescribed EIS content. Source: *Handbook for Environmental Impact Analysis*, Headquarters, Dept. of the Army, April 1975, p. 20, Supt. of Documents, GPO, Washington, D.C. 20402.

6. RELATIONSHIP BETWEEN LOCAL SHORT-TERM USES OF MAN'S ENVIRONMENT AND THE MAINTENANCE AND ENHANCEMENT OF LONG-TERM PRODUCTIVITY
   a. Trade-off between short-term environmental gains at expense of long-term losses
   b. Trade-off between long-term environmental gains at expense of short-term losses
   c. Extent to which proposed action forecloses future options
7. IRREVERSIBLE AND IRRETRIEVABLE COMMITMENTS OF RESOURCES
   a. Unavoidable impacts irreversibly curtailing the range of potential uses of the environment
      (1) Labor
      (2) Materials
      (3) Natural
      (4) Cultural
8. OTHER INTERESTS AND CONSIDERATIONS OF FEDERAL POLICY THAT OFFSET THE ADVERSE ENVIRONMENTAL EFFECTS OF THE PROPOSED PLAN
   a. Countervailing benefits of proposed action
   b. Countervailing benefits of alternatives

Figure 2-5 (*Continued*)

The Environmental Impact Statement (EIS)

The environmental impact statement required under NEPA, Section 102(2)(C), is a detailed written statement by a responsible official on every proposal for legislation or other major federal action significantly affecting the quality of the human environment. It is to include the information required by Section 102(2)(C) quoted earlier.

Guidelines for the preparation of environmental impact statements published by the Council on Environmental Quality[11] are reported below in Section 1500.8. Many states have adopted similar guidelines. A draft environmental impact statement is generally first prepared and circulated for comment. It must fulfill and satisfy the requirements established in Section 102(2)(C). The comments received are evaluated and considered in the decision process. The EIS may be for a proposed concept or program, for a planning study, and/or a proposed construction project.

§1500.8 CONTENT OF ENVIRONMENTAL STATEMENTS[12]

(a) The following points are to be covered:

(1) A description of the proposed action, a statement of its purposes, and a description of the environment affected, including information, summary technical data, and maps and diagrams where relevant, adequate to permit an assessment of potential environmental impact by commenting agencies and the public. Highly technical and specialized analyses and data should be avoided in the body of the draft impact statement. Such materials should be attached as appendices or footnoted with adequate bibliographic references. The statement should also succinctly describe the environment of the area affected as it exists prior to a proposed action, including other Federal activities in the area affected by the proposed action which are related to the proposed action. The interrelationships and cumulative environmental impacts of the proposed action and other related Federal projects shall be presented in the statement. The amount of detail provided in such descrip-

[11]38 *Fed. Reg.* 20550–20562 (1973). Also Code of Federal Regulations in Title 40, Chapter V, at Part 1500.

[12]*Environmental Quality*, the eighth annual report of the Council on Environmental Quality, December 1978, Supt. of Documents, GPO, Washington, D.C., 20402, pp. 405–407.

tions should be commensurate with the extent and expected impact of the action, and with the amount of information required at the particular level of decisionmaking (planning, feasibility, design, etc.). In order to ensure accurate descriptions and environmental assessments, site visits should be made where feasible. Agencies should also take care to identify, as appropriate, population and growth characteristics of the affected area and any population and growth assumptions used to justify the project or program or to determine secondary population and growth impacts resulting from the proposed action and its alternatives [see paragraph (3) (ii), of this section]. In discussing these population aspects, agencies should give consideration to using the rates of growth in the region of the project contained in the projection compiled for the Water Resources Council by the Bureau of Economic Analysis of the Department of Commerce and the Economic Research Service of the Department of Agriculture (the "OBERS" projection). In any event it is essential that the sources of data used to identify, quantify, or evaluate any and all environmental consequences be expressly noted.

(2) The relationship of the proposed action to land use plans, policies, and controls for the affected area. This requires a discussion of how the proposed action may conform or conflict with the objectives and specific terms of approved or proposed Federal, State, and local land use plans, policies, and controls, if any, for the area affected, including those developed in response to the Clean Air Act or the Federal Water Pollution Control Act Amendments of 1972. Where a conflict or inconsistency exists, the statement should describe the extent to which the agency has reconciled its proposed action with the plan, policy, or control and the reasons why the agency has decided to proceed notwithstanding the absence of full reconciliation.

(3) The probable impact of the proposed action on the environment.

(i) This requires agencies to assess the positive and negative effects of the proposed action as it affects both the national and international environment. The attention given to different environmental factors will vary according to the nature, scale, and location of proposed actions. Among factors to consider should be the potential effect of the action on such aspects of the environment as those listed in Appendix II* of these guidelines. Primary attention should be given in the statement to discussing those factors most evidently impacted by the proposed action.

(ii) Secondary or indirect, as well as primary or direct, consequences for the environment should be included in the analysis. Many major Federal actions, in particular those that involve the construction or licensing of infrastructure investments (e.g., highways, airports, sewer systems, water resource projects, etc.), stimulate or induce secondary effects in the form of associated investments and changed patterns of social and economic activities. Such secondary effects, through their impacts on existing community facilities and activities, through inducing new facilities and activities, or through changes in natural conditions, may often be even more substantial than the primary effects of the original action itself. For example, the effects of the proposed action on population and growth may be among the more significant secondary effects. Such population and growth impacts should be estimated if expected to be significant [using data identified as indicated in § 1500.8(a) (1)] and an assessment made of the effect of any possible change in population patterns or growth upon the resource base, including land use, water, and public services, of the area in question.

*Air quality, water quality, marine pollution, commercial fishers conservation, and shellfish sanitation, water regulation and stream modification, fish and wildlife, solid waste, noise, radiation, toxic materials, food additives and contamination of foodstuffs, pesticides, transportation and handling of hazardous materials, energy supply and natural resources development, petroleum development, extraction, refining, transport, and use, natural gas development, production, transmission, and use, coal and minerals development, mining, conversion, processing, transport and use, renewable resource development, production, management, harvest, transport and use, energy and natural resource conservation, land use and management, public land management, protection of environmentally critical areas—flood plains, wetlands, beaches and dunes, unstable soils, steep slopes, aquifer recharge areas, etc., land use in coastal areas, redevelopment and construction in built-up areas, density and congestion mitigation, neighborhood character and continuity, impact on low income populations, historic, architectural, and archeological preservation, soil and plant conservation and hydrology, outdoor recreation.

(4) Alternatives to the proposed action, including, where relevant, those not within the existing authority of the responsible agency. [Section 102(2) (D) of the Act requires the responsible agency to "study, develop, and describe appropriate alternatives to recommended courses of action in any proposal which involves unresolved conflicts concerning alternative uses of available resources."] A rigorous exploration and objective evaluation of the environmental impacts of all reasonable alternative actions, particularly those that might enhance environmental quality to avoid some or all of the adverse environmental effects, is essential. Sufficient analysis of such alternatives and their environmental benefits, costs and risks should accompany the proposed action through the agency review process in order not to foreclose prematurely options which might enhance environmental quality or have less detrimental effects. Examples of such alternatives include: the alternative of taking no action or of postponing action pending further study; alternatives requiring actions of a significantly different nature which would provide similar benefits with different environmental impacts (e.g., nonstructural alternatives to flood control programs, or mass transit alternatives to highway construction); alternatives related to different designs or details of the proposed action which would present different environmental impacts (e.g., cooling ponds vs. cooling towers for a power plant or alternatives that will significantly conserve energy); alternative measures to provide for compensation of fish and wildlife losses, including the acquisition of land, waters, and interests therein. In each case, the analysis should be sufficiently detailed to reveal the agency's comparative evaluation of the environmental benefits, costs, and risks of the proposed action and each reasonable alternative. Where an existing impact statement already contains such an analysis, its treatment of alternatives may be incorporated provided that such treatment is current and relevant to the precise purpose of the proposed action.

(5) Any probable adverse environmental effects which cannot be avoided [such as water or air pollution, undesirable land use patterns, damage to life systems, urban congestion, threats to health or other consequences adverse to the environmental goals set out in section 101(b) of the Act]. This should be a brief section summarizing in one place those effects discussed in paragraph (a) (3) of this section that are adverse and unavoidable under the proposed action. Included for purposes of contrast should be a clear statement of how other avoidable adverse effects discussed in paragraph (a) (2) of this section will be mitigated.

(6) The relationship between local short-term uses of man's environment and the maintenance and enhancement of long-term productivity. This section should contain a brief discussion of the extent to which the proposed action involves tradeoffs between short-term environmental gains at the expense of long-term losses, or vice versa, and a discussion of the extent to which the proposed action forecloses future options. In this context short-term and long-term do not refer to any fixed time periods, but should be viewed in terms of the environmentally significant consequent consequences of the proposed action.

(7) Any irreversible and irretrievable commitments of resources that would be involved in the proposed action should it be implemented. This requires the agency to identify from its survey unavoidable impacts in paragraph (a) (5) of this section and the extent to which the action irreversibly curtails the range of potential uses of the environment. Agencies should avoid construing the term "resources" to mean only the labor and materials devoted to an action. "Resources" also means the natural and cultural resources committed to loss or destruction by the action.

(8) An indication of what other interests and considerations of Federal policy are thought to offset the adverse environmental effects of the proposed action identified pursuant to paragraphs (a) and (5) of this section. The statement should also indicate the extent to which these countervailing benefits could be realized by following reasonable alternatives to the proposed action (as identified in paragraph (a)(4) of this section) that would avoid some or all of the adverse environmental effects. In this connection, agencies that prepare cost-benefit analyses of proposed actions should attach such analysis, or summaries thereof, to the environmental impact statement, and should carefully indicate the extent to which environmental costs have not been reflected in such analyses.

(b) In developing the above points agencies should make every effort to convey the required information succinctly in a form easily understood, both by members of the public and by public decisionmakers, giving attention to the substance of the information conveyed rather than to the particular form, or length, or detail of the statement. Each of the above points, for example, need not always occupy a distinct section of the statement if it is otherwise adequately covered in discussing the impact of the proposed action and its alternatives—which items should normally be the focus of the statement. Draft statements should indicate at appropriate points in the text any underlying studies, reports, and other information obtained and considered by the agency in preparing the statement including any cost-benefit analyses prepared by the agency, and reports of consulting agencies under the Fish and Wildlife Coordination Act, 16 U.S.C. 661 et seq., and the National Historic Preservation Act of 1966, 16 U.S.C. 470 et seq., where such consultation has taken place. In the case of documents not likely to be easily accessible (such as internal studies or reports), the agency should indicate how such information may be obtained. If such information is attached to the statement, care should be taken to ensure that the statement remains an essentially self-contained instrument, capable of being understood by the reader without the need for undue cross reference.

(c) Each environmental statement should be prepared in accordance with the precept in section 102(2)(A) of the Act that all agencies of the Federal Government "utilize a systematic, interdisciplinary approach which will insure the integrated use of the natural and social sciences and the environmental design arts in planning and decisionmaking which may have a impact on man's environment." Agencies should attempt to have relevant disciplines represented on their own staff; where this is not feasible they should make appropriate use of relevant Federal, State, and local agencies or the professional services of universities and outside consultants. The interdisciplinary approach should not be limited to the preparation of the environmental impact statement, but should also be used in the early planning stages of the proposed action. Early application of such an approach should help assure a systematic evaluation of reasonable alternative courses of action and their potential social, economic, and environmental consequences.

(d) Appendix I prescribes the form of the summary sheet which should accompany each draft and final environmental statement.

APPENDIX I SUMMARY TO ACCOMPANY DRAFT AND FINAL STATEMENTS

(Check one) ( ) Draft. ( ) Final Environmental Statement.
Name of responsible Federal agency (with name of operating division where appropriate). Name, address, and telephone number of individual at the agency who can be contacted for additional information about the proposed action or the statement.

1. Name of action (Check one) ( ) Administrative Action. ( ) Legislative Action.
2. Brief description of action and its purpose. Indicate what States (and countries) particularly affected, and what other proposed Federal actions in the area, if any, are discussed in the statement.
3. Summary of environmental impacts and adverse environmental effects.
4. Summary of major alternatives considered.
5. (For draft statements) List all Federal, State, and local agencies and other parties from which comments have been requested. (For final statements) List all Federal, State, and local agencies and other parties from which written comments have been received.
6. Draft statement (and final environmental statement, if one has been issued) made available to the Council and the public.

The statement covering environmental impacts, effects, and alternatives could take the following form.

1. Environmental considerations
   a. Status of environmental conditions. See Tables 2-2, 2-3 and Figures 2-1, 2-4, 2-6, 2-7, 2-8.
2. Environmental analysis and assessment
   a. Matrix. See Figure 2-8.

   b. Checklist. See Figures 2-6, 2-7.
   c. Weighting. See Table 2-3.
   d. Other. See Figure 2-4.
3. Alternative solutions.
   a. Advantages
   b. Disadvantages
4. Recommended solution

Environmental Impact Statement for an Alternative Action. Example

An environmental impact statement (EIS) for an alternative action, such as landfill disposal of solid wastes,[13] would include analysis and documentation of the related beneficial and adverse environmental (also related social and economic) consequences of the proposed landfill as related to the following factors.

Scope Practices and considerations to protect the public health and the environment. Performance criteria. Types of waste; hazardous waste disposal.

Definitions Aquifer, attenuation, base flood, cell, contamination, contingency plan, cover material, disposal, facility structures, floodplain, groundwater, hazardous waste landfill, leachate, liner, monitoring well, open burning, open dump, periodic application of cover material, permafrost, plans, potential zone of influence, recharge zone, responsible agency, runoff, salvaging, sanitary landfill, sludge, sole source aquifers, solid waste, state, ten-year 24-hour precipitation event, vector, water table, wetlands.

Site Selection Consideration of ground and surface water conditions; geology; soils and topographic features; solid waste types and quantities; social, geographic, and economic factors; and aesthetic and environmental impacts. Locations in environmentally sensitive areas require special studies and approvals.

Design Evaluation and documentation of the landfill for acceptance of the expected solid wastes; types and quantities; effect on current and projected groundwater resources; seasonal variations of surface waters; floodplain; water balance study to determine leachate control and surface runoff system; gas control; analysis of trade-offs among environmental impacts, economic considerations, future use alternatives, and nature and quantities of the waste to be disposed of; plans for design, construction, operation, and maintenance of new sites or modifications to existing sites.

Leachate Control Surface runoff diversion; grading, diking; daily and final cover; synthetic liners; natural clay liners; leachate collection; leachate treatment; leachate recycling; leachate monitoring, revegetation.

Gas Control Surface runoff diversion; grading; diking; daily and final cover; synthetic liners, natural clay liners; impermeable barriers; permeable trenches; vertical risers; gas collection systems, gas monitoring.

Runoff Control Surface runoff diversion; grading; diking; ponding; daily and final cover; revegetation.

Operation Compaction; shredding; baling; daily and final cover; access control; safety; fire control; vector control; litter control; revegetation.

Monitoring Gas monitoring; leachate monitoring; facility structure monitoring.

Summary Conclusions, areas of controversy, issues to be resolved including the choice among alternatives.

[13]"Landfill Disposal of Solid Waste, Proposed Guidelines," EPA, *Fed. Reg.*, Monday, March 26, 1979, Part II.

Reference to Chapter 5, Solid Waste Management, and the Comprehensive Solid Waste Study outline in this chapter will be found helpful. Similarly, reference to other chapters will be found useful in the study, analysis, and evaluation of other alternatives and proposed courses of action.

Some Methods for Comprehensive Analysis

Various methods and techniques have been devised to assist in making a comprehensive impact analysis. Each method is adapted to the particular project or action under consideration to properly reflect the action and assess its impact. It is unlikely that any analysis, no matter how carefully made, will satisfy everyone reading the report; but this should not deter one from doing a thorough job.

The methods used for analysis have certain steps in common.

1. Identification of project actions which may or will have an impact or effect on the environmental categories or factors such as water, land, air, ecology, and human settlements and socioeconomic, and related sub-categories. Another category grouping might be physical/chemical, ecological, aesthetic, social, with sub-categories under each (as shown in Figure 2-7).
2. Selection of parameters within each sub-category which will measure the quality of the category.
3. Measurement and interpretation of the significance of the parameters and effect on each of the categories.

These are discussed further below.

The U.S. Department of Housing and Urban Development (HUD) developed guidelines to assist urban areas and cities assess the environmental impacts of housing and urban development actions. Their broad point of view recognizes the social as well as the physical components for good environmental quality, in addition to the sub-categories under each. It is shown in Figure 2-6.

Leopold, et al., prepared a circular[14] to help determine "the probable impact of the proposed action on the environment" which can also serve as a guide and basis for preparation of the environmental impact statement called for under Sec. 102(2)(C) of NEPA. A generalized matrix is included which calls for value judgments* (numerical 1 to 10) to evaluate the effects on the environment of the proposed action including their *magnitude* and *importance*. The authors recognize that the matrix may not be adequate to cover all situations, but is nevertheless quite extensive. Those effects that are believed to be significant (cause death or serious harm or have a substantial probability of causing death or serious harm) or have a large magnitude and importance are then identified for comprehensive discussion in the statement.

*Value judgments: evaluation of scientific knowledge by an expert and the value society places on a factor or action.

[14]Luna B. Leopold, Frank E. Clarke, Bruce B. Hanshaw, and James R. Balsley, "A Procedure for Evaluating Environmental Impact," *Geological Survey Circular 645*, Washington 1971.

Social

Services
- Education facilities
- Employment
- Commercial facilities
- Health care/social services
- Liquid waste disposal
- Solid waste disposal
- Water supply
- Storm-water drainage
- Police
- Fire
- Recreation
- Transportation
- Cultural facilities

Safety
- Structures
- Materials
- Site hazards
- Circulation conflicts
- Road safety and design

Physiological well-being
- Noise
- Vibration
- Odor
- Light
- Temperature
- Disease

Sense of community
- Structural organization
- Homogeneity and diversity
- Physical stock and facilities

Psychological well-being
- Physical threat
- Crowding
- Nuisance

Historic value
- Historic structures
- Historic sites and districts

Visual quality
- Visual content
- Formal coherence
- Apparent access

Physical

Geology
- Unique features
- Resource value
- Slope stability/rockfall
- Foundation stability
- Depth of impermeable layers
- Subsidence
- Weathering/chemical release
- Tectonic activity/vulcanism

Soils
- Slope stability
- Foundation support
- Shrink-swell
- Frost susceptibility
- Liquefaction
- Erodibility
- Permeability

Special features
- Sanitary landfill
- Wetlands
- Coastal zones/shorelines
- Mine dumps/spoil areas

Water
- Hydrologic balance
- Aquifer yield
- Groundwater recharge
- Groundwater flow direction
- Depth to water table
- Drainage/channel form
- Sedimentation
- Impoundment leakage and slope failure
- Flooding
- Water quality

Biota
- Plant and animal special lists
- Vegetative community types
- Diversity
- Productivity
- Nutrient cycling

Climate and air
- Macroclimate hazards
- Forest and range fires
- Heat balance
- Wind alteration
- Humidity and precipitation
- Generation and dispersion of contaminants
- Shadow effects

Energy
- Energy requirements
- Conservation measures
- Environmental significance

Figure 2-6 Principal Components of Environment. (Source: Salvatore J. Bellomo, "Environmental Assessment Guidelines for HUD: Interim Summary," *J. Urban Plann. Dev. Div., ASCE*, May 1978, pp. 21–36).

The matrix lists 100 proposed actions horizontally which may cause environmental impact and 88 environmental characteristics and conditions or categories vertically for a total of 8,800 possible interactions. The matrix would show the broad concerns considered and reduce discussion primarily to those effects determined to be significant and to those repeatedly affected.

Another example[15] to assure consideration of potential impacts starts with expertises needed and then a generalized listing to screen broad categories that might be affected, namely,

Land Geography, Geology, Soils, Geomorphology, Land Resources Economics.
Air Meteorology, Bioclimatology.
Water Hydrology, Limnology.
Plant Botany, Forestry, Microbiology.
Animal Zoology, Wildlife.
Man Anthropology, Sociology, Medicine, Economics, Geography.

Having selected the broad areas affected, the next step is an in-depth analysis of those areas.

Another procedure for the preparation of an environmental impact assessment categorizes environmental attributes (parameters or variables) under Physical/Chemical, Ecological, Aesthetic, and Social as shown in Figure 2-7.[16] The more significant or critical parameters or variables affected by the proposed actions or activities are selected for further study to identify significant activities and then measure the effects (primary and indirect) of man's activities on man and on the environment (and the environment on man). The effects of the actions or activities are evaluated as positive or negative (work sheets are found useful for this purpose) and summarized as in Figure 2-8; the results are weighed and the total effects noted as none, moderate, or significant. Tables 2-2 and 2-3 can be helpful in this regard. Evaluation and interpretation of the data assembled, including methods to mitigate adverse effects, can then assist in making the environmental impact assessment and in preparing the environmental impact statement, if found necessary.

An example showing air quality parameters used and their significance is given in Table 2-2. The quality ratings of high, moderate or desirable, and poor or undesirable are not precise and do not vary uniformly on a straight line from 1.0 to 0.0, that is, from high to poor. They do however serve to give an indication of the changing air quality. Table 2-3 summarizes, interprets, and weighs selected attributes or parameters used to measure water quality which may be affected by a particular project.

Since the goal is a single measure which also reflects the quantitative impact of the project being studied on the environmental attributes, a great deal of expert professional judgement is needed to identify, evaluate, and weigh the significance of the information assembled.

In projects such as a new highway, industry, or a housing development, the possible effects and their measurement must be considered first in the plan-

[15]L. S. Hopkins, R. Bruce Wood, and Debra Brochmann, *Environmental Impact Statements: A Handbook for Writers and Reviewers*, State of Illinois, Illinois Institute for Environmental Quality, August 1973, pp. 18–44.

[16]*Environmental Quality Handbook for Environmental Impact Analysis*, Headquarters, Dept. of the Army, April 1975, Supt. of Documents, GPO, Washington, D.C. 20402.

Physical/Chemical

Water
Biochemical oxygen demand
Groundwater flow
Dissolved oxygen
Fecal coliforms
Inorganic carbon
Inorganic nitrogen
Inorganic phosphate
Heavy metals
Pesticides
Petrochemicals
pH
Stream flow
Temperature
Total dissolved solids
Toxic substances
Turbidity

Land
Soil erosion
Floodplain usage
Buffer zones
Soil suitability for use
Compatibility of land uses
Solid waste disposal

Air
Carbon monoxide
Hydrocarbons
Nitrogen oxides
Particulate matter
Photochemical oxidants
Sulfur oxides
Methane
Hydrogen and organic sulfides
Other

Noise
Intensity
Duration
Frequency

Ecological

Species and populations
Game and nongame animals
Natural vegetation
Managed vegetation
Resident and migratory birds
Sports and commercial fisheries
Pest species

Habitats and Communities
Species diversity
Rare and endangered species
Food chain index

Ecosystems
Productivity
Biogeochemical cycling
Energy flow

Aesthetic

Land
Geologic surface material
Relief and topography

Air
Odor
Visual
Sounds

Water
Flow
Clarity
Interface land and water
Floating materials

Biota
Animals—wild and domestic
Vegetation type
Vegetation diversity

Man-made objects
Man-made objects
Consonance with environment

Composition
Composite effect
Unique composition
Mood atmosphere

Figure 2-7 Assessment parameters. Source: *Environmental Assessments of Effective Water Quality Management and Planning*, EPA, Washington, D.C., April, 1972, p. 21, as cited in Reference 15, pp. 22 and 23.)

Social

Individual environmental interests	*Individual well-being*
Educational/scientific	Physiological health
Cultural	Psychological health
Historical	Safety
Leisure/recreation	Hygienic
Social Interactions	*Community well-being*
Political	Community well-being
Socialization	
Religious	
Family	
Economic	

Figure 2-7 (*Continued*)

ning and design phase, and then during the construction phase, and during the operational phase. The same tabulations, check lists, and matrices should be made for possible alternatives. In all cases, a combination of appropriate talents is needed to assure adequate consideration of all factors that might be impacted, including mitigation of adverse effects and promotion of beneficial effects.

The reader will find it useful to refer to pertinent chapters in this text to help identify and determine the significance of a particular factor or standard. For example, see Chapter 1 for health effects, Chapter 3 for drinking water standards and source protection, Chapter 4 for wastewater treatment and surface and groundwater pollution prevention, Chapter 5 for solid waste management. Chapter 6 for air pollution and noise, and the other chapters as appropriate. The *Environmental Quality Handbook for Environmental Impact Analysis*, is also useful to help interpret the significance of information gathered in the environmental analysis.

FRINGE AND RURAL-AREA HOUSING DEVELOPMENTS

Growth of Suburbs and Rural Areas

The rate of population growth of many cities has been leveling off, but the populations of suburbs have been increasing. In 1966 the Bureau of the Census estimated that 60 million Americans lived in metropolitan-area central cities as compared to 58 million in 1960; 66 million lived in suburbs as compared to almost 59 million in 1960. The 1970 census shows a continuation of the 1966 trends. As of 1974, according to the Bureau of the Census, 151 million people, or 73 percent, live in metropolitan areas. Although metropolitan and nonmetropolitan areas are growing at the same rate, population in central cities in metropolitan areas of one million or more decreased. Between 1970 and 1974 cities lost about 4.6 million persons to the suburbs, with the suburbs

Category		ATTRIBUTE NUMBER	Attribute	*Net Positive Impact +	Net Negative Impacts X
Air		1	Diffusion factor		
		2	Particulates		
		3	Sulphur oxides		
		4	Hydrocarbons		
		5	Nitrogen oxide		
		6	Carbon monoxide		
		7	Photochemical oxidants		
		8	Hazardous toxicants		
		9	Odor		
Water		10	Aquifer safe yield		
		11	Flow variations		
		12	Oil		
		13	Radioactivity		
		14	Suspended solids		
		15	Thermal pollution		
		16	Acid and alkali		
		17	Biochemical oxygen demand		
		18	Dissolved oxygen (DO)		
		19	Dissolved solids		
		20	Nutrients		
		21	Toxic compounds		
		22	Aquatic life		
		23	Fecal coliform		
Land		24	Erosion		
		25	Natural hazard		
		26	Land use patterns		
Ecology		27	Large animals (wild and domestic)		
		28	Predatory birds		
		29	Small game		
		30	Fish, shell fish, and water fowl		
		31	Field crops		
		32	Threatened species		
		33	Natural land vegetation		
		34	Aquatic plants		
Sound		35	Physiological effects		
		36	Psychological effects		
		37	Communication effects		
		38	Performance effects		
		39	Social behavior effects		
Socioeconomic	Human	40	Life styles		
		41	Psychological needs		
		42	Physiological systems		
		43	Community needs		
	Economic	44	Regional economic stability		
		45	Public sector revenue		
		46	Per capita consumption		

☐ No Significant Impact

▨ Moderate Impact

▩ Significant Impact

Project Name ________________

Project Number ________________

Alternative ________________

***Positive impacts are shown above the attribute number and negative impacts below.**

Figure 2-8 Summary of Impacts. (Source: *Environmental Quality Handbook for Environmental Impact Analysis*, Headquarters, Dept. of the Army, April 1975, Sup. of Documents, G.P.O., Washington, D.C., 20402, p. 46.)

showing a continued increase in population and a migration of 7.7 million whites from the central cities to the suburbs and 3.4 million whites from the suburbs to the cities. Some cities (south and southwest) have shown an increase.

There is a natural urge within the family to have its own home in the suburbs or in rural areas. Land in the suburbs and rural areas is considered inexpensive and the taxes low. But the cost, if needed, of constructing a private well and sewage disposal system; the lack of adequate fire, police, street cleaning, and refuse collection service; and the future need for constructing new schools and providing additional sanitary facilities are at first overlooked, as are the increased taxes to provide these services and facilities. Also, whereas the construction cost of community sewerage is federally and sometimes state subsidized, the cost of individual onsite septic tank systems has not been subsidized, although certain financing arrangements are becoming possible for housing group or hamlet alternative systems.

Facilities and Services Needed

Every 1000 new people in a community will require; for example,[17]

1. An additional supply of 100,000 to 200,000 gal of water daily; or 35 to 70 million gal/yr, or 300 individual well-water systems, many equipped with water conditioners.
2. The collection and disposal of 3000 to 4000 lb of solid wastes daily; or 548 to 730 tons/yr.
3. Recreational facilities to serve more people with more leisure time.
4. Sewage treatment works to handle 100,000 to 150,000 gpd, or 35 to 53 million gal/yr containing 170 lb of organic matter (biochemical oxygen demand) per day or 62,000 lb/yr, and 70,000 lb of dry sewage solids per yr or 300 additional septic tanks and appurtenant subsurface disposal facilities.
5. Expenditure to control the sources of air pollution and to offset the physical damage caused by the lack of air pollution control.
6. 4.8 new elementary schoolrooms, 3.6 new high school rooms, and additional teachers.
7. 10.0 or more acres of land for schools, parks, and play areas.
8. 1.8 policemen and 1.5 firemen; also new public service employees in public works, welfare, recreation, health, and administration.
9. More than a mile of new streets.
10. More streets to clean, free of snow and ice, and drain.
11. Two to four additional hospital beds, three nursing home beds, and appurtenant facilities.
12. 1000 new library books.
13. More automobiles, retail stores, services, commercial and industrial areas, county or state parks, and other private enterprises.

To obtain the fiscal impact of new development it is necessary to determine the total cost of providing the expected or required services and the total revenues from the development. The types of services are listed above; their costs can be estimated based on local conditions. Sources of revenues include prop-

[17]Adapted and updated from *U.S. Munic. News*, **22,** 16 (August 15, 1955); and "Environmental Health in Community Growth," *Am. J. Pub. Health*, **53,** No. 5 (May 1963). This is intended to be illustrative and not necessarily typical.

Table 2-2 Air quality parameters

Pollutant	High	Moderate	Poor	Federal air quality standards	Notes
Particulates[a] ($\mu g/m^3$)	0 to 80	80 to 230	230 to 500+	75	Visibility affected as low as 25 $\mu g/m^3$; human health effects begin at about 200 $\mu g/m^3$; condensation nuclei less desirable in concentrations of less than 25 $\mu g/m^3$; all based on 24-hr average annual concentration.
Sulfur oxides[b] (ppm)	0 to 0.10	0.10 to 0.17	0.17 to 0.25+	0.03	The minimum SO_2 concentration for vegetation damage is 0.03 ppm; less than 0.03 ppm can denote a safe environment; increased mortality observed at 0.2 ppm SO_2.
Hydrocarbons[c] (ppm)	0 to 0.19	0.19 to 0.27	0.27 to 0.4+	0.24	Conditions for smog development approached at 0.15 to 0.25 ppm.
Nitrogen oxides[d] (ppm)	0 to 0.025	0.025 to 0.075	0.075 to 0.20+	0.05	Nitrogen dioxide is about four times more toxic than nitric oxide. Nitrogen dioxide below 0.05 ppm does not pose a health problem, but above that level begins to act as a toxic agent.
Carbon monoxide[e] (ppm)	—	—	—	9	Concentrations of 10 to 15 ppm for 8 hr or more can cause adverse health effects; 30 ppm can cause physiologic stress in patients with heart disease; 8 to 14 ppm correlated with increased fatality in hospitalized heart patients.

Photochemical oxidants[f] (ppm) (Ozone)	—	—	—	0.12	Data on animal and human effects inadequate. Leaf injury in sensitive species after 4-hr exposure to 0.005 ppm. Polymers and rubber adversely affected. Smog develops at concentrations of 0.15 to 0.25 ppm.
Asbestos	0.0	None visible	Visible	—	Longterm exposure to high concentrations of asbestos dust can cause asbestosis.
Beryllium (mg/m^3)	0.01	0.10	Greater than 0.10	—	Above 0.01 $\mu g/m^3$ produce disease; at 0.10 or above larger number develop disease; should not exceed 0.01 $\mu g/m^3$ over 30-day period.
Mercury (mg/m^3)	0.0	0.1	Greater than 1.0		Mercury should not exceed 1.0 $\mu g/m^3$ over 30-day averaging period.
Odor	No odor to odor threshold	Odor threshold to slight odor	Slight odor to strong odor	—	Odor threshold can be detected by 5 to 10 percent of panelists. Moderate odor can be detected by about 40 percent; strong odor by 100 percent.

Source: Information abstracted from *Environmental Quality Handbook for Environmental Impact Analysis*, Headquarters, Department of the Army, April 1975, pp. A-1 to A-21. Superintendent of Documents, GPO, Washington, D.C. 20402.

[a]Based on 24-hr annual geometric mean.

[b]Based on 24-hr annual arithmetic mean.

[c]Based on 3-hr average annual concentration 0600 to 0900.

[d]Based on average annual concentration.

[e]Based on maximum 8-hr concentration not to be exceeded more than once a year.

Note: Lead average over 3-month period not to exceed 1.5 $\mu g/m^3$. See also Federal Air Quality Standards, Chapter 6.

Table 2-3 Selected Attributes (Variables) and Environmental Impact Categories—Water Quality

Selected Attributes	Observed Condition	Environmental Impact Category[a] 1	2	3	4	5
Physical						
Aquifer safe yield[b]	Changes occurring in physical attributes of aquifer (porosity, permeability, transmissibility, storage coefficient, etc.)	No change	No change	Slight change	Significant change	Extensive change
Flow variation[c]	Flow variation attributed to activities: Q_{max}/Q_{min}	None	None	Slight	Significant	Extensive
Oil[d]	Visible silvery sheen on surface, oily taste and odor to water and/or to fish and edible invertebrates, coating of banks and bottom or tainting of attached associated biota	None	None	Slight	Significant	Extensive
Radioactivity[d]	Measured radiation limit 10^{-7} μci/ml	Equal to or less	Equal to or less	Exceed limit	Exceed limit	Exceed limit
Suspended solids[c]	1. Sample observed in a glass bottle	Clear	Clear	Fairly clear	Slightly turbid	Turbid
	2. Turbidity in Jackson Turbidity Units	3 or less	10	40	60	140
	3. Suspended solids mg/l	4 or less	10	15	20	35
Thermal discharge[c]	Magnitude of departure from natural condition	0	2	4	6	10
Chemical						
Acid and alkali[d]	Departure from natural condition, pH units	0	1	2	3	4
BOD[d]	mg/l	1	2	3	5	10
DO[c]	percent saturation	100	85	75	60	Low
Dissolved solids[d]	mg/l	500 or less	1000	2000	5000	High
Nutrients[c]	Total phosphorus, mg/l	0.02 or less	0.05	0.10	0.20	Large
Toxic compounds[d]	Concentration, mg/l	Not detected	Traces	Small	Large	Large

Biological						
Fecal coliforms[d]	Number per 100 ml	50 or below	5000	20,000	250,000	Large
Aquatic life[c]	Green algae	Scarce	Moderate quantities in shallows	Plentiful in shallows	Abundant	Abundant
	Gray algae	Scarce	Scarce	Scarce	Present	Plentiful
	Delicate fish; trout, grayling	May be plentiful	Plentiful	Probably absent	Scarce	Absent
	Coarse fish; chub, dace, carp, roach	May be present	Plentiful	Plentiful	Scarce	Absent
	Mayfly naiad, stonefly nymph	May be plentiful	Plentiful	Scarce	Absent	Absent
	Bloodworm, sludge worm, midge larvae, rat-tailed maggot, sewage fly larvae and pupa	May be absent	Scarce	May be present	Plentiful	Abundant

Source: *Environmental Quality Handbook for Environmental Impact Analysis*, Headquarters, Department of the Army, April 1975, Sup. of Documents, GPO, Washington, D.C. 20402.

[a] *Environmental Impact Category*. Category 1 indicates most desirable condition; Category 5 indicates an extensive adverse condition. Because all attributes are related to environmental quality between 0 and 1 it is possible to compare the different attributes and five categories on a common base. Each category is equivalent to approximately 20% of the overall environmental quality. In the physical sense, water quality for five categories will be very clean, clean, fairly clean, doubtful and bad. Environmental impact may be adverse or favorable. Adverse impact will deteriorate the environmental quality while favorable impact will improve the quality. Proper signs and weights must be used to achieve overall effects.

[b] Applies to groundwater systems only.

[c] Applies to surface water systems only.

[d] Applies to both the groundwater and surface water.

EPA stream water quality indicators are: fecal coliform (200 colonies per 100 ml), dissolved oxygen (5 mg/l), total phosphorus (0.1 mg/l), total mercury (2 μg/ml), total lead (50 μg/ml), and total cadmium (4 to 10 μg/ml).

erty taxes, school taxes, fire protection taxes, sales taxes, income taxes, property transfer and other legal fees, business taxes, transient occupancy taxes, interest earnings, permit or license fees, and user charges for recreation, health, water, sewerage, solid waste, and other property services. Include also federal and state allocations and grants, which can be quite substantial. The total revenue can therefore also be estimated. An analysis of cost of services vs. revenues, if carefully done, before development takes place, will provide important information on the fiscal impact of a new development on the community.[18] However, other factors must be taken into consideration, such as compatibility with the community comprehensive plan; with the environmental impact statement, including the favorable and unfavorable aspects and alternatives as discussed later in this chapter; and with social, governmental, and other factors of local importance.

It has been acknowledged that in general the average home does not yield enough in property taxes to pay for the municipal services provided and demanded by the people unless supplemented by user charges and other revenues. Encouragement of industry to locate in selected areas in a town can relieve the tax burden. Desirable types of industries are those that will not cause air, noise, or water pollution or make special demands and that will have a high assessed valuation.

Industry also contributes to the economic growth of an area through increased personal income, bank deposits, and retail sales by attracting professional, skilled, and semiskilled manpower. This means more households, people including schoolchildren, automobiles, retail stores, and hence jobs.

Many communities desire to maintain the character of the community and not overtax the facilities and services. Some have adopted a policy of controlled or limited growth. This is accomplished by:

1. Zoning regulation: large lot zoning and open space preservation.
2. Limiting growth to the capacity of the water supply.*
3. Limiting growth to the capacity of sewerage facilities.*
4. Limiting growth to the capacity of the educational system.*
5. Extra charges to developers for schools, parks, water supply, sewerage, streets.
6. Preservation of agricultural land.
7. Encouraging development of vacant land that is served by streets, water, sewer, and other public services. Use land immediately adjacent to existing developed areas.

Causes and Prevention of Haphazard Development

Fringe and rural-area sanitation is an aspect of healthful living that, like the housing problem in cities, requires the combined talents of many private as

*The California Supreme Court upheld a law that prohibits construction of new homes in the San Francisco Bay area community of Livermore until educational, sewage disposal, and water supply facilities are built. ("California Court Approves Laws Limiting Building," *New York Times*, December 16, 1976—San Francisco, December 18, Associated Press.)

[18]Robert W. Burchell and David Listokin, *The Fiscal Impact Guidebook, Estimating Local Costs and Revenues of Land Development*, GPO, Washington, D.C., March 1979, HUD-PDR-371.

well as official agencies and individuals to prevent insanitary conditions and promote good health and a pleasing environment. Those who can contribute to this goal are the developers and builders, consulting engineers and land planners, health department and environmental protection agency, sanitary engineers and sanitarians, local officials, private lending institutions, federal agencies, planning and zoning boards, legislators and attorneys, the press, and an informed home-buying public.

Where comprehensive planning and controls are lacking, subdivisions are in many cases poorly planned and without utilities, drainage, or streets worthy of the name. Some lots may be under water, on rock shelves, steep slopes or tight clay soil, or poorly drained. Overflowing septic-tank systems and polluted wells become commonplace. Many people "get stuck," and when they seek legal redress they are confronted with a maxim of English and American law, *caveat emptor*—"let the purchaser beware," that is, let him examine the article he is buying and act on his own judgment and at his own risk. An official agency operating in the public interest cannot adopt such a principle.

The premature development of subdivisions causes an accumulation of tax arrears on vacant lots, thereby shifting the financial burden for governmental expenditures on other properties.[19] The owners of other properties are therefore forced to make up the tax money lost by paying higher taxes or lose their home or place of business, even though they had no part in the land speculation and could not have benefited had the venture been successful.

If the cost and liabilities for subdivision are borne by those who engage in subdivision for profit, rather than by the general public, bona fide developers are aided and irresponsible unscrupulous developers are discouraged. It is therefore proper that reasonable controls be invoked to control land subdivision for the general health and welfare of the people.

The generally accepted methods of preventing subdivision problems and obtaining orderly community growth are education, comprehensive land-use planning, and effective regulation. But laws must be understood and be enforceable; they must be reasonable and fair to the developer and investor, to the purchasers of homes, and to the local community. It must be recognized that the owner of land for sale is primarily desirous of subdividing for profit. The banks and mortgage insurers are concerned that dwellings in a subdivision and in surrounding areas maintain a high value so as to guarantee a continuing satisfactory return on their investments. The homeowner wants to live and bring up his or her children in a pleasant, healthful environment. The community is interested in seeing that a subdivision does not turn out to be a liability.

Many problems can be prevented or made less difficult to solve if the conditions surrounding the problem are more fully understood. Giving publicity to existing laws, including explanation of their purposes and how to comply with them, is one approach. Attorneys, lending institutions, realtors, builders

[19]Philip H. Cornick, "Problems Created by Premature Subdivision of Urban Lands in Selected Metropolitan Districts," State of New York Dept. of Commerce, Albany, 1938.

and subdividers, local officials and service clubs, land planners and consulting engineers, the home-buying public, and the local press are some of the important groups that can make a subdivision law more effective. The local health department and planning agency, through consultation, advice, leaflets, and letters and by personal contact, can explain the intent of the law and expedite compliance with it. Some excellent bulletins (see the bibliography at the end of this chapter) are available to the consulting engineer, land planner, builder, and subdivider for their information and guidance.

When considering the many educational measures that can be applied to control or solve subdivision problems, one must not forget the opportunities available to the administrative and enforcing agencies to prevent the problems in the first place. Immediate notification of the subdivider at the first sign of activity, followed by strict surveillance, can prevent many problems from getting out of hand. Informal conferences with interested parties, explanation of the advantages of filing proper plans for public sewer and water supply, and the offer of advice to help solve problems will usually convince the reasonable individual he has little to lose and much to gain by complying with the law. In some instances a hearing before the proper board or commissioner will give the desired result; in rare instances legal action is necessary.

A city, village, or town may also adopt a local sewage and wastes ordinance as part of its sanitary code, building code, plumbing code, zoning, or planning ordinance, depending on which is available and effective. This is done in cooperation with the health department and other agencies having jurisdiction. The individual planning to put in an individual sewerage system is required to make application to a designated department for a permit to build, at which time he or she also makes application to the health department for approval of a private sewage disposal or treatment system based upon soil percolation tests, soils maps, and a recommended design. The health department, registered architect, or professional engineer makes design recommendations and then inspections during construction. A certificate of compliance is issued by the health department when the system is properly installed. No dwelling or structure is permitted to be occupied or any facility or appurtenance used until the issuance of the health department certification. Potential subdivisions are then brought to the attention of the health department and control measures instituted.

An educational and persuasive tool is reference to legal decisions and opinions to help convince the skeptic. Numerous decisions have been handed down affirming the right of the health department to control subdivision development, to require engineering plans of proposed water supply and sewerage, and in some instances to require installation of sewers rather than septic-tank systems.[20]

Health department realty subdivision laws may be enacted on a state level and enforced on both a state and local level, or the laws may be enacted and enforced on a local level, where a competent engineering division is provided in the

[20] J. A. Salvato, Jr., "Experiences with subdivision regulations," *Pub. Works*, p. 126 (April 1957).

local health department. The former procedure may be preferred when political and administrative problems are anticipated. Sample laws, regulations, and guidelines are available from many state and local health and planning agencies.

Cooperative Effort Needed

Developers and builders can simplify their sanitation problems by selecting vacant land that is within or contiguous to communities already having water and sewer services. This makes possible the economical extension of these utilities; smaller lots, if desired, than would be required with individual wells and septic-tank leaching systems and hence more lots; and a more desirable development. If the extension of water or sewer lines is not possible, then the developer or builder can in many cases construct a water system and or sewers and treatment plant, with the cooperation of local officials. In such cases maintenance and operation of the systems should be guaranteed. It is also possible to construct these facilities on a planned step-by-step construction basis so as to reduce or spread out the financial burden of the initial plant cost.

Consulting engineers are sometimes reluctant to take a small subdivision job involving the design of water supply and sewerage systems. The time and effort needed to work out satisfactory plans may appear to be too great for the fee that can be reasonably charged. Actually this is where the professional engineers can perform a real public service by wise advice and sound design. Here is an opportunity for the application of imagination and initiative by not only engineers, but also land planners, builders and contractors, investors, and equipment manufacturers in designing and producing less expensive and more efficient sanitary devices as well as a more desirable living environment for population groups involving 50 to 1000 or more persons.

Local officials can encourage and direct proper development by sympathetic cooperation, by assisting in the formation of water and sewer districts, and through the enforcement of planning and zoning ordinances. Enlightened public officials are no more desirous of encouraging insanitary and haphazard construction than the health department. The submergence of sectional jealousies, the formation of drainage-area sanitary districts crossing town, city, or village boundary lines, and annexation, if necesary, are additional aides. In this connection, county and regional planning offer a great deal of promise. Policy agreements regarding certain services can be made among contiguous cities, villages, and townships through a county planning commission or similar agency. Development of certain lands could be encouraged by mutual understanding to extend more complete fire and police protection, sewers and water mains, snow plowing service, and refuse collection only to areas designated for development. The county real estate division can cooperate by encouraging the purchase of vacant tax-delinquent land in these delineated areas. Developers and builders would be required to do their part by agreeing to certain deed restrictions, improvements, and zoning, and also to share in the cost of constructing water and sewer lines. Participating communities could properly require larger lots outside of areas without water and sewer lines and agree to the review and approval of all subdivision plots by the coordinating agency. Prior to making this determination, the local soil conservation service, health department, and planning board should give a preliminary opinion on the suitability of the soil and geology for onlot subsurface sewage disposal and well-water supply. A consultant can also be retained to make this determination.

Involved federal agencies and lending institutions are obligated to comply with state and local sanitary code regulations. They usually withhold payments where sanitary facilities do not meet existing legal requirements. Attorneys, realtors, and mortgage and title investigators can protect their clients' interests and help make sounder investments if the adequacy of existing or proposed water supply and sewage disposal facilities at a property are also investigated before investment is recommended.

County, city, and state health department environmental engineers and sanitarians have the responsibility to enforce the state and local public health laws and sanitary codes. They can give assistance and guidance to the developer, builder, and engineer and help interpret the intent and compliance with the laws. Consultation with the health and/or environmental protection department having jurisdiction concerning the supply of water and disposal of sewage, and with the highway department concerning the drainage of surface water, will expedite the submission and approval of satisfactory subdivision plans. Health department personnel are often called on to talk before service clubs, professional groups, and community associations. This offers opportunities to explain the protection provided by existing laws and the services rendered by the health department to promote a better way of life through a healthful environment.

Subdivision Planning

Haphazard development can be controlled by comprehensive local and regional planning if broad public powers and regulatory authority are provided. In the absence of such control, public and private acquisition of large potential sites and their preparation for multipurpose uses, including residential development, has merit. Raw land would be planned for residential, recreational, open-space, commercial, industrial, agricultural, or supporting uses. This would then be followed by development and improvement with roads, water supply, drainage, sewers, and other utilities including wastewater and solid waste disposal facilities. The land would then be sold to private enterprise for development as *planned.*

In the planning of a housing development careful consideration should be given to the site selection and planning factors discussed earlier in this chapter.

The subdivision is dependent on the region and its central community for employment and cultural needs, and the central community is dependent on the surrounding development for human resources and economic survival. Controlled environmental conditions in a newly developed subdivision can only be assured if the neighboring inhabitants live under equally desirable conditions. No matter how remote a new subdivision may be from the urban centers, it is still an extension of the existing environment and is affected by it. The subdivision, in turn, influences the surrounding area environment by the character of its growth and the standards of its facilities and services. This interdependence must always be kept in mind.

Planning and zoning regulations are based on the rights of government to exercise its police power to control the private uses of land for the promotion of the general health, safety, and welfare. A planning board can be empowered to prepare a comprehensive plan to guide the physical development of land and public services within a delineated area. In carrying out this function the planning board may prepare an official map; make investigations, sketches, and reports relating to planning and community development; review and make recommendations for approval of plans of proposed subdivisions; advise concerning capital budget and long-term capital improvement programs; define problems; and conduct research and analyses.

Enforcement tools available to the planning board are subdivision regulation, the zoning code, and guidance of improvement programs relating to parks, major highways, sewers and water supply, including treatment plants, and desirable type of land development and redevelopment.

Land planning is a specialized activity. Frequently the importance of comprehensive design is ignored by the developer or its significance is not realized, yet nothing could be more fundamental in the subdivision of land. A future community being created can be a good place in which to live and work and in which to invest in a home; or it can be an undesirable subdivision right from the start. The planning board should determine if the land is suitable for housing, business, or industry, or should be retained as farmland. The demand and need for the intended use should govern whether immediate or future development is indicated. Then the subdivision designer should fit the proposed project in with the community zoning, major streets and parkways, recreational areas, churches, schools, shopping centers, and so forth, and design the interior elements of the subdivision accordingly. Due regard must be given to topography, roads, lot sizes and shapes, water supply, sewage disposal, gas and electricity, drainage, recreational areas, schools, easements for utilities, trees, and landscaping, as well as the environmental impact of the proposed land use.

Figure 2-9 illustrates the different ways in which a plot of land may be subdivided. An increased number of lots and a reduced cost of public improvements per lot are common results of good design. This of course is advantageous to the developer, the homeowner, and the community.

The legal steps and authority for the formation of a planning board are usually given in state laws. Ordinances creating local planning boards are ob-

Figure 2-9 Types of subdivision development. (*a*) Standard subdivision. The area is completely subdivided into 46 lots. No open space and only a few lots have access to the creek. (*b*) Density zoning. Same number of conventional lots but with half the area in open space. (*c*) True cluster development. Again 46 lots on 23 acres, but lots are grouped. (*d*) 110 townhouses on same parcel. Over half the land is untouched, and at 4.7 units per acre, this would be rather crowded if these were single-family units. On clear, flat land, quality townhouse developments can run 10 to 14 units per acre with low cost projects even higher. Various possibilities exist for subdividing this net area of 23 acres. The one chosen will affect the amount of open space regardless of the number of units. [Source: "Trends in Residential Development" by George C. Bestor, *Civil Eng., ACSE,* (September 1969).]

tainable from state planning agencies where established. Planning-board suggested subdivision regulations should also be available from this source. Any regulations considered for adoption should be adapted to the local conditions.

Design of Water and Sewerage Service for Subdivisons

Water and sewerage service for a subdivision can be provided in many different ways. The preferred method is the extension of existing water lines and sewers. If the property is not located within the boundaries of an existing service district, the facilities may become available by annexation to the central municipality, through the formation of an improvement district, or through a utility corporation if permitted by the existing laws. Where a county or regional service district or authority exists, it would be the controlling agency. If a public water supply or sewerage system is not available, the developer can construct his own facilities, including treatment plants.

The central city fortunate enough to have an adequate water and sewage works has an obligation to serve its suburbs. But it also has the right to maintain sufficient control over the extensions to protect itself and the right to charge for the service what is necessary to obtain a reasonable return on its investment. Mathews has discussed the pros and cons of water service extension to the suburbs in some detail.[21] He points out that the obligation to serve suburban consumers on the part of the city is mixed with resentment and resistance, but the usual result is that the service is furnished. Sometimes annexation is made a prerequisite. As to the right to control the distribution system beyond city limits, it is realistically stated that the individual suburban consumer is inclined to blame the city water department for all failures in service, even though the cause may be inadequate distribution facilities in the suburban area. Since inhabitants of the area generally have business and other connections in the city, pressure is brought to bear on the water department to rectify an impossible condition.

The design of small water systems is given in Chapter 3. Experiences in new subdivisions show that peak water demands of six to ten times the average daily consumption rate are not unusual. Lawn-sprinkling demand has made necessary sprinkling controls, metering, or the installation of larger distribution and storage facilities and, in some instances, ground storage and booster stations. As previously stated, every effort should be made to serve a subdivision from an existing public water supply. Such supplies can afford to employ competent personnel and are in the business of supplying water, whereas a subdivider is basically in the business of developing land and does not wish to become involved in operating a public utility.

In general, when it is necessary to develop a central water system to serve the average subdivision, consideration should first be given to a drilled well-water supply. Infiltration galleries or special shallow wells may also be practical sources of water. Such water systems usually require a minimum of supervision and can be developed to produce a known quantity of water of a satisfactory sanitary quality. Simple chlorination treatment will normally provide the desired factor of safety. Test wells and sampling will indicate the most probable dependable yield and the chemical and bacterial quality of the water. Well logs should be kept.

Where a clean, clear lake supply or stream is available, chlorination and slow sand filtration can provide reliable treatment with daily supervision for the small development. The turbidity of the water to be treated should not exceed 30 ppm. Preliminary settling may be indicated in some cases.

[21]C. K. Mathews, "Water Service Policies for Suburban Areas," *J. Am. Water Works Assoc.*, **48,** 174–178 (February 1956).

Other more elaborate types of treatment plants, such as rapid sand filters, are not recommended for small water systems unless specially trained operating personnel can be assured. Pressure filters have limitations as explained in Chapter 3.

The design of small slow sand filter and well-water systems is explained and illustrated in Chapter 3.

Where a public sewerage system is not available or accessible, design of a central sewerage system including treatment should be given careful study. The cost of sewers per dwelling can usually be divided in half, and lot sizes may be made smaller than is necessary when a well and septic tank are installed on every lot, thereby making available more lots. Such property is usually more desirable, a better investment, and less likely to cause public health nuisances due to overflowing septic-tank systems or polluted wells which have been improperly designed, located, or constructed.

Various wastewater treatment methods have been used to serve subdivisions, depending on the number of persons served, the degree of treatment required, and the type of supervision provided. For the small development of up to 50 homes, a plant consisting of a septic tank, dosing tank, open or covered sand filters, and chlorine contact tank—or a package-type aeration plant, has been found practical. For a housing development of 50 to 300 homes, a plant consisting of an Imhoff tank or primary settling tank, standard filter with provision for recirculation, and a secondary settling tank, or a package-type plant, followed by chlorination has been preferred. For larger developments, methods that have been used are the standard-rate trickling filter plant; the high-rate trickling filter plant; the activated sludge plant; the aerobic digestion plant; and variations of these processes, followed in some instances by oxidation ponds or intermittent sand filters with chlorination. See Chapter 4 for design details.

There are many rural areas that are remote from population centers and public water supplies and sewers. In such cases individual well-water and septic-tank sewage disposal systems offer the only practical answer for the immediate future. However, to be acceptable individual sanitary facilities must be carefully designed, constructed, and maintained in accordance with good standards. Where the soil is unsuitable for the disposal of sewage by conventional subsurface means, every effort must be made to prevent the subdivision of land until public sewers, including treatment and drainage if needed, can be provided, unless a satisfactory alternate can be agreed to.

In some situations public water supply is available, leaving only the problem of sewage disposal. In rarer situations public sewers are available, making necessary construction of individual wells. Athough the sanitary problems are simplifed in such cases, there is still need for careful design, construction, and maintenance of the individual facility that needs to be provided.

When either individual well or septic-tank systems are required, or when both are required, the preparation of a proper subdivision plan makes possible the proper and orderly installation of these facilities on individual lots. When homes are built in accordance with the approved plan, the intent of a subdivision control law is accomplished in the interest of the future homeowners and community. Some health departments have issued practical instructions to designing engineers regarding the preparation and submission of plans for realty subdivisions.[22] Such information simplifies the submission of satisfactory plans and also has a salutary educational effect on the designing engineer who has not had extensive experience in subdivision planning and design.

Details included on a subdivision plan are a location map, topography, typical soil-boring profile, lot sizes, roads, drainage, and location of soil tests and results; typical lot layouts showing the location of the house, well, sewage disposal system, and critical materials and dimensions; sketches showing drilled well development and protection, pump connection and sanitary seal, depth to water and yield, or a community water system; and sewage disposal units, design, and sizes, including septic tank, distribution box, absorption field, or leaching pit. Construction details of drilled wells, including pump connections and sanitary seals, are given and illustrated in Chapter 3. Septic-tank, distribution box, absorption field, and leaching pit details are given and illustrated in Chapter 4.

There is a danger in the preparation of subdivision plans, as in other engineering and architectural plans, to copy details blindly. Such carelessness defeats the purpose of employing a consultant or designing engineer and is contrary to professional practice. Subdivision plans involving individual wells and sewage disposal systems must be adapted to the topography and geological formations existing at the particular property. It is well-known that no two properties are exactly alike; hence each requires careful study and adaptation of general principles and typical details so that a proper engineering plan results. For example, the soil percolation tests and soils information determine the type and size of the required sewage disposal system. The slope of the ground determines the relative locations of the wells and sewage disposal systems on each lot; in most cases these must be established on the plot plan to prevent further interference. The type of well, required minimum depth of casing, and the need for cement grouting of the annular space around the outside of the casing, and the sealing of the bottom of the casing in solid rock vary with each property. These are just a few considerations to bring out the need for the application of engineering training and experience so as to perform a proper professional service.

BIBLIOGRAPHY

Anderson, Richard T., *Comprehensive Planning for Environmental Health*, Division of Urban Studies, Cornell University, Ithaca, N.Y., 1964, 203, pp.

Camp Sites and Facilities, Boy Scouts of America, New York, 1950, 90 pp.

Chapin, Stewart F., *Urban Land Use Planning*, University of Illinois Press, Urbana, 1965, 498 pp.

The Comprehensive Plan: A Guide for Community Action, State of New York, Office of Planning Coordination, Albany, 1966, 22 pp.

[22] *Planning the Subdivision as Part of the Total Environment*, Division of General Engineering and Radiological Health, New York State Dept. of Health, Albany, 1970.

DeMarco, et al., *Incinerator Guidelines—1969*, DHEW, PHS Pub. 20112, Washington, D.C. 1969.

Dickerson, Bruce W., "Selection of Water Supplies for New Manufacturing Operations," *J. Am. Water Works Assoc.*, **62,** 10, 611–615 (October 1970).

Draft Environmental Impact Statement, Office of Solid Waste, USEPA, Washington, D.C. 20460, March 1979.

Environmental Health Aspects of Metropolitan Planning and Development, *WHO Tech. Rep. Ser.*, **297** (1965).

"Environmental Health in Community Growth," *Am. J. Pub. Health*, **53,** 5, 802–822 (May 1963).

Environmental Health Planning, PHS Pub. 2120, HEW, Washington, D.C., 1971, 139 pp.

Environmental Impact Assessment Methods: An Overview, NTIS, PB-226 276, U.S. Dept. of Commerce, Illinois Institute for Environmental Quality, August 1973.

Environmental Impact Statement Guidelines, Environmental Protection Agency, Region X, Seattle, Washington, April 1973.

Environmental Planning and Geology, HUD, U.S. Dept. of Interior, DCPD-32, December 1971.

Financing and Charges for Wastewater Systems, A Joint Committee Report, American Public Works Association, ASCE, Water Pollut. Control Fed., 1973.

Gallion, Arthur B. and Simon Eisner, *The Urban Pattern, City Planning and Design*, D. Van Nostrand Company, Inc., New York, 1963, 424 pp.

Garing, Taylor & Associates, *A Handbook Approach to the Environmental Impact Report For the Design Professional*, 141 South Elm St., Arroyo Grande, Calif. 93420.

Handbook For Environmental Impact Analysis, Department of the Army, April 1975, Superintendent of Documents, GPO, Washington, D.C. 20402.

Logan, John A., Paul Opperman, and Norman E. Tucker. *Environmental Engineering & Metropolitan Planning*, Northwestern University Press, Evanston, Ill., 1962, 265 pp.

McKeever, Ross J., *Community Builders Handbook*, Urban Land Institute Publication, Washington, D.C., 1968, 526 pp.

McLean, Mary, *Local Planning Administration*, The International City Managers' Association, Chicago, 1959, 467 pp.

Outdoor Recreation Space Standards, Bureau of Outdoor Recreation, Dept. of the Interior, Washington, D.C., April 1967, 67 pp.

Part III, Appraisal of Neighborhood Environment, Am. Pub. Health Assoc., New York, 1950, 132 pp.

Plants/People/and Environmental Quality, U.S. Department of the Interior, Washington, D.C. 1972.

Robinette, Gary O., *Plants/People/and Environmental Quality*, U.S. Dept. of the Interior, National Park Service, Washington, D.C., 1972.

Salomon, Julian Harris, *Camp Site Development*, Girl Scouts of America, New York, 1959, 160 pp.

Salvato, Joseph A., Jr., "Environmental Health and Community Planning," *J. Urban Plan. Dev. Div., ASCE*, **94,** UP 1, Proc. Paper 6084, 23–20 (August 1968).

Salvato, Joseph A., Jr., Peter J. Smith, and Morris M. Cohn, *Planning the Subdivision as Part of the Total Environment*, New York State, Dept. of Health, Albany, 1970, 77 pp.

Shomon, Joseph James, *Open Land for Urban America*, an Audubon book published in cooperation with the National Audubon Society by The Johns Hopkins Press, Baltimore, 1971, 171 pp.

Soil, Water, and Suburbia, a Report of the Proceedings of the Conference Sponsored by the United States Dept. of Agriculture and the United States Dept. of Housing and Urban Development, June 15 and 16, 1967, Washington, D.C., March 1968, 160 pp.

Urban Planning Guide, ASCE, Manuals and Reports on Engineering Practice, No. 49, American Society of Civil Engineers, New York, 1969, 299 pp.

"Water and Related Land Resources," Water Resources Council, *Fed. Reg.*, Monday, September 10, 1973, Washington, D.C.

Weinstein, Norman J. and Arthur T. Goding, "Liquid and Solid Effluents and their Control," *Pub. Works*, October 1974, pp. 76–79.

Where Not to Build, a Guide for Open Space Planning, Tech. Bull. 1, U.S. Dept. of the Interior, Washington, D.C., April 1968, 160 pp.

3

WATER SUPPLY

Introduction

A primary requisite for good health is an adequate supply of water that is of satisfactory sanitary quality. It is also important that the water be attractive and palatable to induce its use; otherwise, consumers may decide to use water of doubtful quality from a nearby unprotected stream, well, or spring. Where a municipal water supply passes near a property, the owner of the property should be urged to connect to it because such supplies are usually under competent supervision.

When a municipal water supply is not available, the burden of developing a safe water supply rests with the owner of the property. Frequently private supplies are so developed and operated that full protection against dangerous or objectionable pollution is not afforded. Failure to provide satisfactory water supplies in most instances must be charged either to negligence or to ignorance, because in the long run it generally costs no more to provide a satisfactory installation that will meet with good health department standards.

It is of historical interest to note, for later comparison, that in 1962 there were 19,272 public (community) water supplies in the United States serving approximately 150 million people.[1]* Seventy-five percent were groundwater supplies, 18 percent surface water supplies, and 7 percent were a combination. Seventy-five million people in communities under 100,000 population were served by 18,873 public water supplies; 399 supplies served larger communities and 16,350 supplies (85 percent) served communities of 5000 or less.

The PHS completed a study in 1970 covering 969 small to large public surface and groundwater supply systems serving 18.2 million persons (12 percent of the total U.S. population served by public water supplies) and 84 spe-

*Leaving about 30 million people dependent on individual water supplies. Cornelius W. Krusé estimates that in 1960 31.4 million people were served by home and farm water systems and 12.6 million people relied on nonpiped water supplies; *Advances in Environmental Sciences Technology*, Vol. 1, James N. Pitts, Jr. and Robert L. Metcalf, Eds., Wiley-Interscience, New York, 1970. A 1965 report shows that about 153 million people in the United States and Puerto Rico were served by public water supplies and 42 million people by individual water supplies; *Estimated Use of Water in the United States, 1965*, Circular 556, Geological Survey, U.S. Dept. of the Interior, 1968.

[1]*Statistical Summary of Municipal Water Facilities in the United States*, PHS Pub. 1039, DHEW, Washington, D.C., January 1, 1963.

cial systems serving trailer and mobile home parks, institutions, and tourist accommodations.[2] Although the drinking water supplies in the United States rank among the best in the world, the study showed the need for improvements. Based on the 1962 PHS Drinking Water Standards it was found that in 16 percent of the 969 communities surveyed the water quality exceeded one or more of the mandatory limits established for coliform organisms (120 systems), fluoride (24), or lead (14). It is of interest to note that of the 120 systems that exceeded the coliform standard, 108 served populations of 5000 or less and that 63 of these were located in a state where disinfection was not frequently practiced or was inadequate. An additional 25 percent of the systems exceeded the recommended limits for iron (96 systems), total dissolved solids (95), manganese (90), fluoride (52), sulfate (25), and nitrate (19). The study also showed that 56 percent of the systems were deficient in one or more of the following: source protection, disinfection or control of disinfection, clarification (removal of suspended matter) or control of clarification, and pressure in the distribution system. It was also reported that 90 percent of the systems did not have sufficient samples collected for bacteriological surveillance; 56 percent had not been surveyed by the state or local health department within the last 3 years. In 54 percent cross-connection prevention ordinances were lacking; in 89 percent reinspection of existing construction was lacking; in 61 percent the operators had not received any water treatment training; in 77 percent the operators were deficient in training for microbiological work and 46 percent of those who needed chemistry training did not have any. The smaller communities had more water quality problems and deficiencies than the larger ones, showing the advisability of consolidation and regionalization when feasible.

The definition of a public water system was changed when the Safe Drinking Water Act was adopted. A "public water system" under the Section 1412 of the Public Health Service Act, as amended by the Safe Drinking Water Act (P.L. 93-523) on December 16, 1974, means a system for the provision to the public of piped water for human consumption, if such system has at least fifteen service connections or regularly serves at least 25 individuals daily at least 60 days out of the year. This definition would include the usual "community water system," serving a residential population including subdivisions, company towns, mobile home parks, apartment complexes and resident institutions; as well as "noncommunity water systems" such as camps, hotels, motels, rest stops, restaurants, recreation areas, schools, gasoline stations, roadside eating places, places of public assemblage, and the like. There are, under the new definition, approximately 261,500 public water supply systems in the United States, of which 200,000 serve noncommunity transient-type places noted above, which often are not under close surveillance. About 25 percent of the community systems and 5 percent of the noncommunity systems use surface water sources. Of the 61,500 community water systems,

[2]Leland J. McCabe, James M. Symons, Roger D. Lee, and Gordon G. Robeck, "Survey of Community Water Supply Systems," *J. Am. Water Works Assoc.*, 670–687 (November 1970).

53,000 serve populations less than 25,000 and 39,000 serve populations between 25 and 500.[3]

A survey made between 1975 and 1977 showed that 13 to 18 million people in communities of 10,000 and under used individual wells that had high rates of contamination.[4] The effectiveness of state and local well construction standards and health department programs would have a direct bearing on the extent of contaminated home well-water supplies in specific areas.

A safe and adequate water supply for 2 billion people, about one-half of the world's population, is still a dream. For example, approximately 20 million people in nearly 100,000 villages, one-sixth of all the villages in India, do not have a pipe, pond, or well source of drinking water within a mile. One government's goal is to bring water to every village by 1983, at a cost of nearly $1 billion.[5] The availability of any reasonably clean water in the less developed areas of the world just to wash and bathe would go a long way toward the reduction of such scourges as scabies and other skin diseases, yaws and trachoma, and high infant mortality. The lack of safe water makes commonplace high incidences of shigellosis, amebiasis, schistosomiasis,* leptospirosis, infectious hepatitis, giardiasis, typhoid, and paratyphoid fever.[6] It has been estimated that there are 250 million new cases of waterborne disease per year and 25,000 people (mostly children) die daily from them throughout the world.[7] (See also Chapter 1.) It is believed that the provision of safe water supplies, accompanied by a program of proper excreta disposal, and birth control, could vastly improve the living conditions of millions of people in developing countries of the world.[8]

Groundwater Contamination Hazard

Table 3-1 shows a classification of sources and causes of groundwater pollution. The 16 million (1970 U.S. Census) residential cesspool and septic-tank soil absorption systems alone discharge about 800 billion gal of sewage per year into the ground, which in some instances may contribute to groundwater pollution. This is in addition to sewage from restaurants, motels, and other structures not on public sewers. The contribution from industrial and other sources shown in Table 3-1 is unknown and is being inventoried by USEPA,

*200 million cases in the world estimated in 1976; spread mostly through water contact.

[3]"Committee Report Defines Small Utility Needs," *Willing Water*, December 1980.

[4]Alan Levin, "The Rural Water Survey," *J. Am. Water Works Assoc.*, August 1978, pp. 446–452.

[5]William Borders, "India Is Trying to Supply Water To All Its Villages in Five Years," *New York Times*, Sunday, August 6, 1978.

[6]Abel Wolman, "Water Supply and Environmental Health," *J. Am. Water Works Assoc.*, December 1970, pp. 746–749.

[7]*Water: Life or Death*, report of the International Institute for Environmental Development in preparation for the Mar Del Plata Water Conference, March 1976. Argentina.

[8]G. E. Arnold, "Water Supply Projects in Developing Countries," *J. Am. Water Works Assoc.*, December 1970, pp. 750–753.

Table 3-1 Classification of Sources and Causes of Groundwater Pollution Used in Determining Level and Kind of Regulatory Control.

Wastes		Nonwastes	
Category I: Systems, facilities, or activities designed to discharge waste or wastewaters (residuals) to the land and groundwaters	Category II: Systems, facilities, or activities which may discharge wastes or wastewaters to the land and groundwaters	Category III: Systems, facilities, or activities which may discharge or cause a discharge of contaminants that are not wastes to the land and groundwaters	Category IV: Causes of groundwater pollution which are not discharges
Land application of wastewater. Spray irrigation, infiltration-percolation basins, overland flow	*Surface impoundments.* Waste holding ponds, lagoons, and pits	*Buried product storage tanks and pipelines*	*Saltwater intrusion.* Seawater encroachment, upward coning of saline ground water
Sub-surface soil absorption systems. (Septic systems)	*Landfills and other excavations.* Landfills for industrial wastes, sanitary landfills for municipal solid wastes, landfills for municipal water and wastewater treatment plant sludges, other excavations (e.g., mass burial of livestock)	*Stockpiles.* Highway de-icing stockpiles, ore stockpiles	*River infiltration*
Waste disposal wells and brine injection wells	*Animal feedlots*	*Application of highway de-icing salts*	*Improperly constructed or abandoned wells*
Drainage wells and sumps	*Leaky sanitary sewer lines*	*Product storage ponds*	*Farming practices.* (e.g., dry land farming)
Recharge wells	*Acid mine drainage*	*Agricultural activities.* Fertilizers and pesticides, irrigation return flows	
	Mine spoil pipes and tailings	*Accidental spills*	

Source: The Report to Congress, *Waste Disposal Practices and Their Effects on Ground Water*, Executive Summary, January 1977, USEPA, Washington, D.C. 20460, p. 39.

but is estimated at 900 billion gallons per year.[9] A groundwater protection strategy is being developed. Groundwater pollution problems have been found in many states across the country. With about 100 million people in the United States dependent on groundwater sources for drinking water, it is apparent that groundwater resources must be protected from contamination.

Whereas surface water travels at velocities of feet per second, groundwater moves at velocities that range from several feet per day to less than a fraction of a foot per day. Consequently, surface water contamination is of relatively short duration, if sediment is not a problem and the source is eliminated or brought under control. But groundwater organic, and inorganic chemical contamination in particular, may persist for decades or longer because of the generally slow rate of movement of groundwater, and may go undetected for many years. Factors that influence the movement of groundwater include the type of geological formation and its permeability, the rainfall and the infiltration, and the water table slope. The slow uniform rate of flow* provides little opportunity for mixing and dilution and the usual absence of air in the groundwater to decompose or break down the contaminants add to the long-lasting problem usually created. However, conditions of dilution, microbial activity, and soil adsorptive characteristics might exist which could have the opposite effect. Hence each problem should be considered individually and on the basis of the hydrogeological, chemical, and microbiological factors of the site affected and the characteristics of the contaminant involved. Research is underway on using natural and "manufactured" organisms to assimilate and break down some organics in groundwater.

TRAVEL OF POLLUTION THROUGH THE GROUND

Since the character of soil and rock, quantity of rain, depth of groundwater, rate of groundwater flow, amount of pollution, microbial growth medium, and other factors beyond control are variable, one cannot say with certainty through what thickness or distance sewage must pass to be purified. Microbiological pollution travels a short distance through sandy loam or clay; but it will travel indefinite distances through coarse gravel, fissured rock, dried-out cracked clay, or solution channels in limestone. Acidic conditions and lack of organics and certain elements such as iron, manganese, aluminum, and calcium in soil increase the potential of pollution travel.

The PHS conducted experiments at Fort Caswell, North Carolina, in a sandy soil with groundwater moving slowly through it. The sewage organisms (coliform bacteria) traveled 232 ft, and chemical pollution as indicated by uranin dye traveled 450 ft.[10] The chemical pollution moved in the direction of

*Usually in an elongated plume.

[9]The Report to Congress, *Wastes Disposal Practices and their Effects on Ground Water*, Executive Summary, Jan. 1977, USEPA, Washington, D.C. 20460.

[10]C. W. Stiles, H. R. Crohurst, and G. E. Thomson, *Experimental Bacterial and Chemical Pollution of Wells via Ground Water, and the Factors Involved*, PHS Bull. 147, DHEW, Washington, D.C., June 1927.

the groundwater flow largely in the upper portion of the groundwater and persisted for 2½ years. The pollution band did not fan out but became narrower as it moved away from the pollution source. It should be noted that in these tests there was a small draft on the experimental wells and that the soil was a sand of 0.14 mm effective size and 1.8 uniformity coefficient.

Studies of pollution travel were made by the University of California using twenty-three 6-in. observation wells and a 12-in. gravel-packed recharge well. Diluted primary sewage was pumped through the 12-in. recharge well into a confined aquifer having an average thickness of 4.4 ft approximately 95 ft below ground surface. The aquifer was described as pea gravel and sand having a permeability of 1900 gal/ft^2/day. Its average effective size was 0.56 mm and uniformity coefficient, 6.9. The medium effective size of the aquifer material from 18 wells was 0.36 mm. The maximum distance of pollution travel was 100 ft in the direction of groundwater flow and 63 ft in other directions. It was found that the travel of pollution was not affected by the groundwater velocity but by the organic mat that built up and filtered out organisms, thereby preventing them from entering the aquifer. The extent of the pollution then regressed as the organisms died away and as pollution was filtered out.[11]

Butler, Orlob, and McGauhey made a study of the literature and reported the results of field studies to obtain more information about the underground travel of harmful bacteria and toxic chemicals.[12] The work of other investigators indicated that pollution from dry pit privies did not extend more than 1 to 5 ft in dry or slightly moist fine soils. However, when pollution was introduced into the underground water, test organisms (*B. coli*) traveled to wells up to 232 ft away.[10] Chemical pollution was observed to travel 300 to 450 ft, although chromate was reported to have traveled 1000 ft in three years, and other chemical pollution 3 to 5 mi. Leachings from a garbage dump in groundwater reached wells 1476 ft away, and a 15-year-old dump continued to pollute wells 2000 ft away. Studies in the Dutch East Indies (Indonesia) report the survival of coliform organisms in soil two years after contamination and their extension to a depth of 9 to 13 ft, in decreasing numbers, but increasing again as groundwater was approached. The studies of Butler et al. tend to confirm previous reports and have led the authors to conclude

> that the removal of bacteria from liquid percolating through a given depth of soil is inversely proportional to the particle size of the soil.

Knowledge concerning viruses in groundwater is limited, but better methodology for the detection of viruses is improving this situation. Keswick and Gerba[13] reviewed the literature and found nine instances in which viruses were

[11] *Report on the Investigation of Travel of Pollution*, Pub. No. 11, State Water Pollution Control Board, Sacramento, Calif., 1954.

[12] R. G. Butler, G. T. Orlob, and P. H. McGauhey, "Underground Movement of Bacterial and Chemical Pollutants," *J. Am. Water Works Assoc.*, **46,** 2, 97–111 (February 1954).

[13] Bruce H. Keswick and Charles P. Gerba, "Viruses in Groundwater," *Environ. Sci. Technol.*, November 1980, pp. 1290-1297.

isolated from drinking water wells and 15 instances in which viruses were isolated from beneath land treatment sites. Sand and gravel did not prevent the travel of viruses long distances in groundwater. However, fine loamy sand over coarse sand and gravel effectively removed viruses. Soil composition is very important in virus removal as it is in bacteria removal. The movement of viruses through soil and in groundwater requires further study.

When pumping from a deep well, the direction of groundwater flow around the well will be toward it. Since the level of the water in the well will probably be 25 to 150 ft, more or less, below the ground surface, it will exert an attractive influence on groundwater perhaps as far as 400 to 1000 ft away from the well, regardless of the elevation of the top of the well. In other words, distances and elevations of sewage disposal systems must be considered relative to the elevation of the water level in the well while it is being pumped, and its circle of influence.

A WHO Report reminds us that, in nature, atmospheric oxygen breaks down accessible organic matter and that topsoil (loam) contains organisms that can effectively oxidize organic matter.[14] However, these benefits are lost if wastes are discharged directly into the groundwater by way of sink holes, pits, or wells, or if a subsurface absorption system is water-logged.

From the investigations made, it is apparent that the safe distance between a well and a sewage disposal system is dependent on many variables, including chemical, physical, and biological processes.[15] Factors to be considered in arriving at a satisfactory answer include the following.

1. The amount of sand, clay, organic (humus) matter, and loam in the soil, the soil structure and texture, the effective size and uniformity coefficient, and soil depth largely determine the ability of the soil to remove bacterial pollution deposited in the soil.
2. The volume, strength, type and dispersion of the polluting material, as well as the distance, elevation, and time for pollution to travel with relation to the groundwater level and flow and soil penetrated, are important. Also important is the volume of water pumped and well drawdown.
3. The well construction, tightness of the pump connection, depth of well casing, and sealing of the annular space have a very major bearing on whether a well will be polluted by sewage and surface water.

Considerable professional judgment is needed to select a proper location for a well. The limiting distances given in Table 3-2 should therefore be used as a guide. Experience has shown them to be reasonable and effective in most instances *when coupled with proper interpretation of available hydrologic and geologic data and good well construction, location, and protection.* See Figure 3-1 for groundwater terms.

[14]S. Buchan and A. Key, "Pollution of Ground Water in Europe," *WHO Bull.*, **14**, 5–6, 949–1006 (1956).

[15]A summary of the distances of travel of underground pollution is also given in the Task Group Report, "Underground Waste Disposal and Control," *J. Am. Water Works Assoc.*, **49**, 1334–1341 (October 1957).

Table 3-2 Separation Distances from Wastewater Sources

Wastewater Sources	To Well or Suction Line[a]	To Stream, Lake, or Water Course	To Property Line or Dwelling
House Sewer (watertight joints)	25′ if cast iron pipe or equal, 50′ otherwise	25′	—
Septic Tank	50′	50′	10′
Effluent Line to Distribution Box	50′	50′	10′
Distribution Box	100′	100′	20′
Absorption Field	100′[b]	100′	20′
Seepage Pit, or Cesspool	150′[b] (more in coarse gravel)	100′	20′
Dry Well (roof and footing)	50′	25′	20′
Fill or Built-up System	100′	100′	20′
Evapotranspiration-Absorption System	100′	50′	20′
Sanitary Privy Pit	100′	50′	20′
Privy, Watertight Vault	50′	50′	10′
Septic Privy or Aqua Privy[c]	50′	50′	10′

[a]Water service and sewer lines may be in the same trench if cast iron sewer with lead-caulked joints is laid at all points 12 in. below water service pipe; or sewer may be on dropped shelf at one side at least 12 in. below water service pipe, provided that sewer pipe is laid below frost with tight and root-proof joints and is not subject to settling, superimposed loads or vibration. Water service lines under pressure shall not pass closer than 10 ft of a septic tank, absorption tile field, leaching pit, privy, or any other part of a sewage disposal system.

[b]Sewage disposal systems located of necessity upgrade or in the general path of drainage to a well should be spaced 200 ft or more away, and not in the direct line of drainage. Wells require a minimum 20 ft of casing extended and sealed into an impervious stratum.

[c]Evapotranspiration system separation distances same as absorption field; with liner, same as septic tank.

Disease Transmission

Water, to act as a vehicle for the spread of a specific disease, must be contaminated with the associated disease organism or hazardous chemical. Disease organisms can survive for periods of days to years depending on their form (cyst, ova), environment (moisture, competitors, temperature, soil, and acidity), and the treatment given the wastewater. All sewage-contaminated waters must be presumed to be potentially dangerous. Other impurities such as inorganic and organic chemicals and heavy concentrations of decaying organic

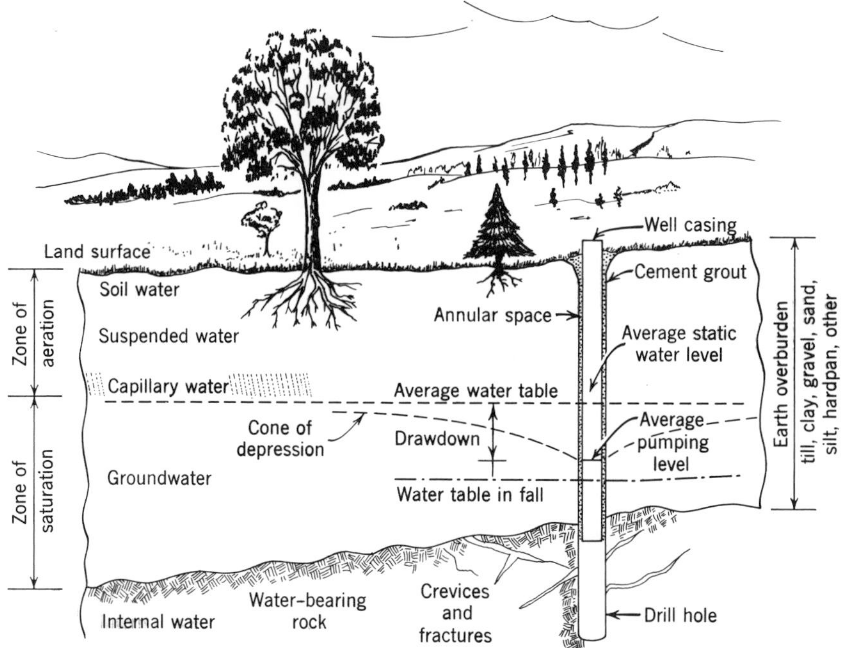

Figure 3-1 A geologic section showing groundwater terms. (From *Rural Water Supply*, New York State Dept. of Health, Albany, 1966.)

matter may also find their way into a water supply, making the water hazardous, unattractive, or otherwise unsuitable for domestic use unless adequately treated. The inorganic and organic chemicals which may cause illness include mercury, lead, chromium, nitrates, asbestos, PCB, PBB, mirex, Kepone, vinyl chloride, trichloroethylene, benzene, and others.

Communicable and noninfectious diseases that may be spread by water are discussed in Chapter 1 and are listed in the folded insert (Figure 1-2).

WATER QUANTITY AND QUALITY

Water Cycle and Geology

The movement of water can be best illustrated by the hydrologic or water cycle shown in Figure 3-2. Using the clouds and atmospheric vapors as a starting point, moisture condenses out under the proper conditions to form rain, snow, sleet, hail, frost, fog, or dew. Part of the precipitation is evaporated while falling; some of it reaches vegetation foliage, the ground, and other surfaces. Moisture intercepted by surfaces is evaporated back into the atmosphere. Part of the water reaching the ground surface runs off to streams, lakes, swamps, or oceans whence it evaporates; part infiltrates into the ground and percolates down to replenish the groundwater storage, which also supplies lakes, streams,

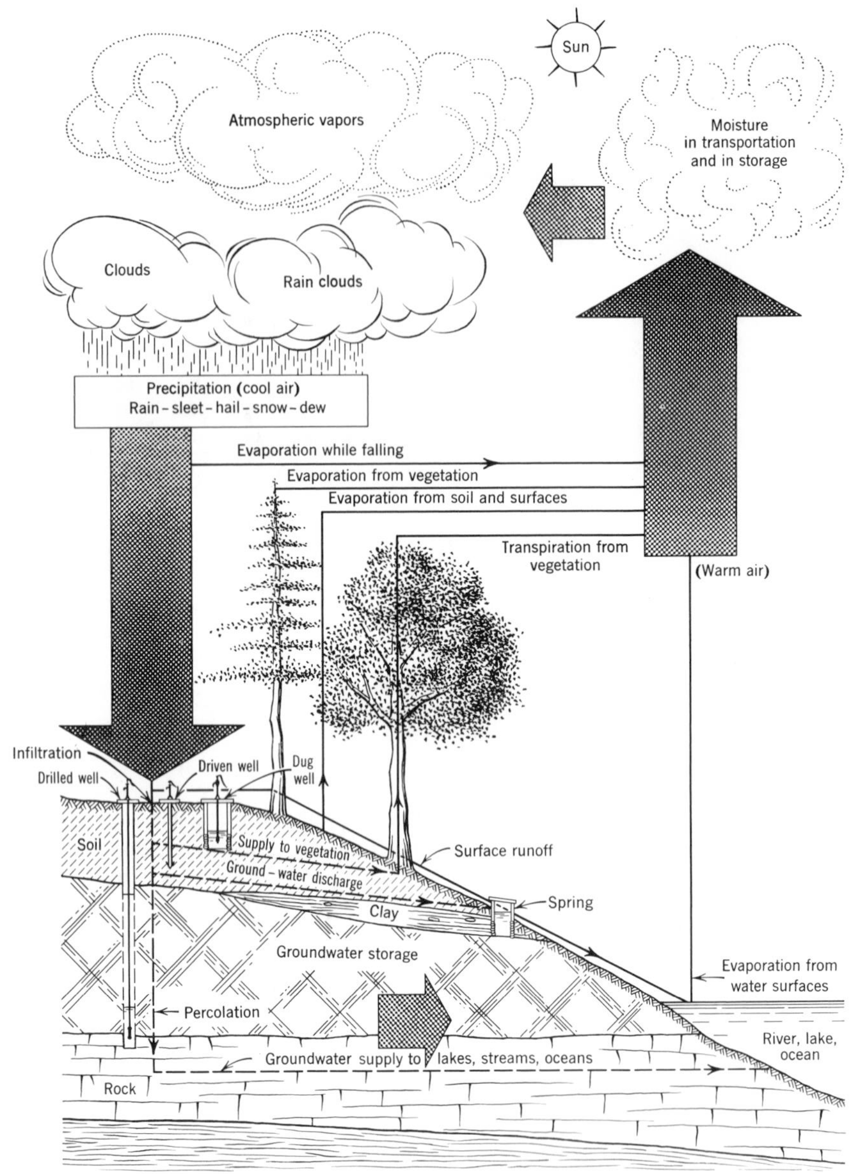

Figure 3-2 The hydrologic or water cycle. The oceans hold 317,000 cubic miles of water. 97 percent of the earth's water is salt water. 3 percent of the earth's fresh water is groundwater, snow and ice, fresh water on land, and atmospheric water vapor. 75 percent of the fresh water is in polar ice caps and glaciers. Total precipitation equals total evaporation plus transpiration. Precipitation on land equals 24,000 cubic miles per year. Evaporation from the oceans equals 80,000 cubic miles per year. Evaporation from lakes, streams, and soil, and transpiration from vegetation equals 15,000 cubic miles.

and oceans by underground flow. Groundwater in the soil helps to nourish vegetation through the root system. It travels up the plant and comes out as transpiration from the leaf structure and then evaporates into the atmosphere. In its cyclical movement, part of the water is temporarily retained by the earth, plants, and animals to sustain life. The average annual precipitation in the United States is about 30 in., of which 72 percent evaporates from water and land surfaces and transpires from plants, and 28 percent contributes to the groundwater recharge and stream flow.[16] See also Septic-Tank Evapotranspiration System, Chapter 4.

When speaking of water, we are generally concerned primarily with surface water and groundwater, although rainwater and saline water are also considered. In falling through the atmosphere rain picks up dust particles, plant seeds, bacteria, dissolved gases, ionizing radiation, and chemical substances such as sulfur, nitrogen, oxygen, carbon dioxide, and ammonia. Hence, rain water is not pure water as one might think; it is, however, very soft. Water in streams, lakes, reservoirs, and swamps is known as surface water. Water reaching the ground and flowing over the surface carries anything it can move or dissolve. This may include waste matter, bacteria, silt, soil, vegetation, and microscopic plants and animals. The water accumulates in streams or lakes. Sewage, industrial wastes, and surface and groundwater will cumulate, contribute to the flow, and be acted upon by natural agencies. Water flowing over the ground may also find its way to lakes or reservoirs where bacteria, suspended matter, and other impurities settle out. On the other hand, microscopic as well as macroscopic plant and animal life grow and die, thereby removing and contributing impurities in the cycle of life.

Part of the water reaching and flowing over the ground infiltrates and percolates down to form the groundwater, also called underground water. In percolating through the ground, water will dissolve materials to an extent dependent on the type and composition of the strata through which the water has passed and the quality (acidity) and quantity of water. Groundwater will therefore usually contain more dissolved minerals than surface water. The strata penetrated may be unconsolidated, such as sand, clay, and gravel, or consolidated, such as sandstone, granite, and limestone. A brief explanation of the classification and characteristics of formations is given below.

Igneous rocks are those formed by the cooling and hardening of molten rock masses. The rocks are crystalline and contain quartz, feldspar, mica, hornblende, pyroxene, and olivene. Igneous rocks are not usually good sources of water, although basalts are exceptions. Small quantities of water are available in cracks and fissures.

Sedimentary formations are those resulting from the deposition, accumulation, and then consolidation of materials weathered and eroded from older rocks by water, ice, or wind and the remains of plants, animals, or material precipitated out of solution. Sand and gravel, clay, silt, chalk, limestone, fos-

[16] *Hydrology Handbook*, Manual 28, ASCE, New York, 1949.

sils, gypsum, peat, shale, loess, and sandstone are examples of sedimentary formations. Deposits of sand and gravel generally yield large quantities of water. Sandstones, shales, and certain limestones may yield abundant groundwater, although results may be erratic depending on bedding planes and joints, density, porosity, and permeability of the rock.

Metamorphic rocks are produced by the alteration of igneous and sedimentary rocks, generally by means of heat and pressure. Gneisses and schists, quartzites, slates, marble, serpentines, and soapstones are metamorphic rocks. A small quantity of water is available in joints, crevices, and cleavage planes.

Porosity is a measure of the amount of water that can be held by a rock or soil in its pores or voids, expressed as a percentage of the total volume. The amount of water that will *drain* out of a saturated rock or soil by gravity is the *effective porosity* or *specific yield.* The amount of water retained is the *specific retention.* This is due to water held in the interstices or pores of the rock or soil by molecular attraction (cohesion) and by surface tension (adhesion). For example, plastic clay has a porosity of 45 to 55 percent but a specific yield of practically zero. In contrast, a uniform coarse sand and gravel mixture has a porosity of 30 to 40 percent with nearly all of the water capable of being drained out. On the average, it is estimated that 37 percent of the theoretical capacity of stratified rocks is taken up by water and 50 percent in igneous rocks.[17]

The *permeability* of a rock or soil, expressed as the standard coefficient of permeability, is the rate of flow of water at 60°F, in gallons per day, through a cross section of 1 ft^2 under a head of one foot per foot of water travel. There is no direct relationship between permeability, porosity, or specific yield.

The porosity of some materials are [17,18]

Soils	50 to 60 percent
Clay	45 to 55
Silt	40 to 50
Sand, medium to coarse (0.5 to 2.0 mm)	35 to 40
Sand, fine to medium (0.25 to 0.5 mm)	30 to 35
Sand, uniform	30 to 40
Gravel and sand (2.0 to 10.0 mm)	20 to 35
Gravel	30 to 40
Sandstone	10 to 20
Shale	1 to 10
Limestone	1 to 10
Igneous rock	1

[17]H. Ries and Thomas L. Watson, *Engineering Geology*, John Wiley & Sons, New York, 1931.
[18]D. K. Todd, *Ground Water Hydrology*, John Wiley & Sons, New York, 1967.

Water Quality

The cleanest available sources of groundwater and surface water should be preserved and used for potable water supply purposes. Numerous parameters are used to determine the suitability of a water and the health significance of contaminants that may be found in the untreated and treated water.

Bacterial, physical, chemical, and microscopic examinations are discussed and interpreted in this chapter under those respective headings. Water quality can be best assured by maintaining water clarity, a chlorine residual in the distribution system, confirmatory absence of indicator organisms, and low bacterial population in the distributed water.[19]

Tables 3-3, 3-4, and 3-5 show the regulations that drinking water coming out of a tap served by a public water system must meet. These are based on the interim primary regulations developed under the Safe Drinking Water Act of 1974. Secondary regulations, shown in Table 3-6, have also been adopted but these are designed to deal with taste, odor, and appearance of drinking water and are not mandatory unless adopted by a state.

Table3-7 shows the European Economic Community Council parameter limits for surface water *sources* for drinking water. Included is the degree of treatment required for different types of water.

Table 3-8 shows the criteria surface water should meet in the United States for use as a *source* for public water supply. The criteria are also applicable to groundwater sources.

Sampling and Quality of Laboratory Data

There is a tendency to collect more samples and laboratory data than is needed. The tremendous resources in money, manpower, and equipment committed to the proper preparation, collection, and shipment of the samples and to the analytical procedures involved is lost sight of or is not understood. Actually a few samples of good quality can usually serve the intended purpose.

The purpose or use to which laboratory data are to be put should determine the number of samples and quality (precision) of the laboratory work. The data should be correct but need not always be precise. Data of high quality is needed to support enforcement action or to support a health effects study while data of lesser quality may be acceptable for trend, screening, or monitoring purposes.

> The goal of the QA* program is to obtain scientifically valid, defensible data of known precision and accuracy to fulfill the Agency's responsibility to protect and enhance the nation's environment.[20]

*Quality Assurance.

[19]"Disinfection—Committee Report," *J. Am. Water Works Assoc.*, April 1978, pp. 219–222.

[20]Thomas R. Hauser, "Quality assurance update," *Environ. Sci. Technol.*, November 1979, pp. 1356–1366.

Table 3-3 Comparison of Interim Primary Drinking Water Regulations (IPDWR) and WHO International Water Regulations

Type of Contaminant	Name of Contaminant	Type of Water System	Maximum Contaminant Level	
			IPDWR[a]	WHO Tentative[b]
Inorganic Chemicals	Arsenic	Community	0.05 mg/l	0.05 mg/l
	Barium		1.	
	Cadmium		0.010	0.01
	Chromium		0.05	
	Lead		0.05	0.1
	Mercury		0.002[c]	0.001[c]
	Selenium		0.01	0.01
	Silver		0.05	
	Fluoride[d]			(Recommended)
	53.7°F & below		2.4	0.9 to 1.7
	53.8 to 58.3		2.2	0.8 to 1.5
	58.4 to 63.8		2.0	0.8 to 1.3
	63.9 to 70.6		1.8	0.7 to 1.2
	70.7 to 79.2		1.6	0.7 to 1.0
	79.3 to 90.5		1.4	0.6 to 0.8
	Nitrate (as N)	Community & Noncommunity	10.	45 (as NO_3)
	Cyanide			0.05
Organic Chemicals	Endrin	Community	0.0002 mg/l	
	Lindane		0.004	
	Methoxychlor		0.1	
	Toxaphene		0.005	
	2,4-D		0.1	
	2,4,5-TP, Silvex		0.01	
Turbidity[e]	Turbidity at representative entry point to distribution system.	Community & Noncommunity	1 TU monthly average and 5 TU average of two consecutive days (5 TU monthly average may apply at state option.)[f]	
Chloro-Organics	Trihalomethane	Greater than 10,000 population	100 ppb	

[a] Refer to text of regulations (USEPA) for full explanation, 40 CFR 141.

[b] *International Standards for Drinking-Water*, WHO, 1971, p. 32.

[c] Total as Hg.

[d] Determined by the annual average of the maximum daily air temperature for the location in which the community water system is situated. These levels are maximum, *not* optimum.

[e] Refers to surface water supplies at point of entry to distribution system, after treatment.

[f] Provided the turbidity does not interfere with disinfection, maintenance of chlorine residual throughout the distribution system, or microbiological determinations.

Table 3-4 Maximum Permissible Microbiological Contaminants (IPDWR)[a]

Coliform Method	Per Month	Less than 20 Samples per Month	20 or More Samples per Month
	Number of coliform bacteria shall not exceed		
Membrane filter (100-ml portions)	1/100 ml average density	4/100 ml in one sample	4/100 ml in 5 percent of samples
	Coliform bacteria shall not be present in more than		
Multiple tube fermentation (10-ml portions)	10 percent of portions	3 portions in one sample	3 portions in 5 percent of samples

Coliform Method	Per Month	Less than 5 Samples per Month	5 or More Samples per Month
	Coliform bacteria shall not be present in more than		
Multiple tube fermentation (100-ml portions)	60 percent of portions	5 portions in more than one sample	5 portions in more than 20 percent of the samples

[a] *International Standards for Drinking-Water*, WHO, 1971, page 17 states that "water circulating in the distribution system, whether treated or not, should not contain any organisms that may be of faecal origin."

Table 3-5 Maximum Permissible Radioactivity (IPDWR)

Contaminant	Maximum Contaminant Level Picocurie per liter (pCi/l)[a]
Natural	
Combined Radium-226 and Radium-228	5
Gross alpha particle activity, including Radium-226 but excluding Radon and Uranium	15[b]
Man made[c]	
Tritium (total body)	20,000
Strontium −90 (bone marrow)	8
Gross beta particle activity (applicable to surface water sources)	50[d]

[a] Average of four samples obtained at quarterly intervals.
[b] If gross alpha activity does not exceed 5 pCi/l, measure of combined radium may be omitted.
[c] The average annual concentration of beta particle and photon radioactivity from man-made radionuclides in drinking water shall not produce an annual dose equivalent to the total body or an internal organ greater than four millirems per year. Refer to text of USEPA regulations (40CFR 141) for full explanation, i.e., Part 141, National Interim Primary Drinking Water Regulations.
[d] Provided tritium and strontium average annual levels are not exceeded.

Table 3-6 Secondary Maximum Contaminant Levels—Advisory, Federal Safe Drinking Water Act (SDWA) and WHO International Standards

	Level[a]	
Contaminant	WHO[b]	SDWA
Chloride	600	250
Color	50 units[c]	15 Color Units
Copper	1.5	1
Corrosivity[d]		Noncorrosive
Foaming Agents		0.5
Hydrogen Sulfide		0.05
Iron[e]	1.0	0.3
Manganese[e]	0.5	0.05
Odor	Unobjectionable	3 Threshold Odor Number
pH	6.5 to 9.2	6.5 to 8.5
Sulfate	400	250
TDS (Total Dissolved Solids)	1500	500
Zinc	15	5

[a]In mg/l unless otherwise stated.
[b]Maximum permissible level. *International Standards for Drinking-Water*, WHO, 1971, pp. 38–40.
[c]On the platinum-cobalt scale.
[d]Corrosivity shall be determined by calcium carbonate saturation or equal method.
[e]If iron and manganese are both present, the total concentration of both substances should not exceed 0.5 mg/l.

The laboratory is an essential part of the environmental program effectiveness. However, the laboratory must resist the tendency to become involved in program operation and regulation activities since its function does not involve program enforcement, responsibility, regulation continuity, and effectiveness. In addition, its limited resources would be misdirected and diluted to the detriment of its primary function. This does not mean that the laboratory should not be involved in the solution of difficult water plant operational problems.

Sanitary Survey and Water Sampling[21,22]

A sanitary survey is necessary to determine the reliability of a water system to continuously supply safe and adequate water to the consumer. It is also necessary to properly interpret the results of water analyses and to evaluate the effects of actual and potential sources of pollution on water quality. The value of the survey is dependent on the training and experience of the person making

[21]See also, "Surveillance of Drinking Water Quality," *WHO Monog. Ser.* **63,** 1976.
[22]See also, *Manual For Evaluating Public Drinking Water Supplies*, DHEW, Bureau of Water Hygiene, Cincinnati, Ohio 45202, 1971.

Table 3-7 EEC Mandatory Limits for Subsurface Water Intended for the Abstraction of Drinking Water

Parameters		Treatment Categories		
		A1[a]	A2[b]	A3[c]
Coloration (after simple filtration)	mg/l Pt scale	20[d]	100[d]	200[d]
Temperature	°C	25[d]	25[d]	25[d]
Nitrates	mg/l No_3	50[d]	50[d]	50[d]
Fluorides	mg/l F	1.5	—	—
Dissolved Iron	mg/l Fe	0.3	2	—
Copper	mg/l Cu	0.05[d]	—	—
Zinc	mg/l Zn	3	5	5
Arsenic	mg/l As	0.05	0.05	0.1
Cadmium	mg/l Cd	0.005	0.005	0.005
Total chromium	mg/l Cr	0.05	0.05	0.05
Lead	mg/l Pb	0.05	0.05	0.05
Selenium	mg/l Se	0.01	0.01	0.01
Mercury	mg/l Hg	0.001	0.001	0.001
Barium	mg/l Ba	0.1	1	1
Cyanide	mg/l Cn	0.05	0.05	0.05
Sulphates	mg/l SO_4	250	250[d]	250[d]
Phenols	mg/l C_6H_5OH	0.001	0.005	0.1
Dissolved or emulsified hydrocarbons	mg/l	0.05	0.2	1
Polycyclic aromatic hydrocarbons	mg/l	0.0002	0.0002	0.001
Total pesticides	mg/l	0.001	0.0025	0.005
Ammonia	mg/l NH_4	—	1.5	4[d]

Source: "Britania Waives the Rules," *World Water*, London, June 1978, pp. 24–27. Adopted by the European Economic Community Council, 16 June 1975.

[a]Simple physical treatment and disinfection, e.g., rapid filtration and disinfection.

[b]Normal physical treatment, chemical treatment, and disinfection, e.g., prechlorination, coagulation, flocculation, decantation, filtration, disinfection.

[c]Intensive physical and chemical treatment, extended treatment, and disinfection, e.g., chlorination, to break-point, coagulation, flocculation, decantation, filtration, adsorption (activated carbon), disinfection (ozone, final chlorination).

[d]May be waived under exceptional climatic or geographical conditions.

the investigation. When available, one should seek the advice of the regulatory agency sanitary engineer or sanitarian.

If the source of water is a lake, attention would be directed to the entire drainage basin and location of sewage and other solid and liquid waste disposal or treatment systems, bathing areas, storm-water drains, sewer outfalls, swamps, cultivated areas, pastures, and wooded areas in reference to the pump intake, for each would contribute distinctive characteristics to the water. When water is obtained from a stream or creek, all land and habitation above the water supply intake should be investigated. This means inspection of the entire watershed

Table 3-8 Surface-Water Criteria for Public Water Supplies—Sources

Constituent or Characteristic	Permissible Criteria	Desirable Criteria
Physical		
Color (color units)	75	<10
Odor	[b]	Virtually absent
Temperature[a]	Do	[b]
Turbidity	Do	Virtually absent
Microbiological		
Coliform organisms	20,000/100 ml[c]	<100/100 ml[c]
Fecal coliforms	2,000/100 ml[c]	<20/100 ml[c]
Inorganic chemicals	(mg/l)	(mg/l)
Alkalinity	[b]	[b]
Ammonia	0.5 (as N)	<0.01
Arsenic[a]	0.05[d]	Absent
Barium[a]	1.0[d]	Do
Boron[a]	1.0	Do
Cadmium[a]	0.01[d]	Do
Chloride[a]	250	<25
Chromium,[a] hexavalent	0.05[d]	Absent
Copper[a]	1.0	Virtually absent
Dissolved oxygen	⩾4 (monthly mean) ⩾3 (individual sample)	Near saturation
Fluoride[a]	[b]	[b]
Hardness[a]	Do	Do
Iron (filterable)	0.3	Virtually absent
Lead[a]	0.05[d]	Absent
Manganese[a] (filterable)	0.05	Do
Nitrates plus nitrites[a]	10 (as N)[d]	Virtually absent
pH (range)	6.0–8.5	[b]
Phosphorus[a]	[b]	Do
Selenium[a]	0.01[d]	Absent
Silver[a]	0.05[d]	Do
Sulfate[a]	250	<50
Total dissolved solids[a] (filterable residue)	500	<200
Uranyl ion[a]	5	Absent
Zinc[a]	5	Virtually absent
Organic chemicals		
Carbon chloroform extract[a] (CCE)	0.15	<0.04
Cyanide[a]	0.20	Absent
Methylene blue active substances[a]	0.5	Virtually absent
Oil and grease[a]	Virtually absent	Absent
Pesticides		
Aldrin[a]	0.017	Do
Chlordane[a]	0.003	Do
DDT[a]	0.042	Do
Dieldrin[a]	0.017	Do

Table 3-8 (*Continued*)

Constituent or Characteristic	Permissible Criteria	Desirable Criteria
Pesticides (*Continued*)		
Endrin[a]	0.001 (0.0002)[d]	Do
Heptachlor[a]	0.018	Do
Heptachlor epoxide[a]	0.018	Do
Lindane[a]	0.056 (0.004)[d]	
Methoxychlor[a]	0.035 (0.01)[d]	Do
Organic phosphates plus carbamates[a]	0.1[e]	Do
Toxaphene[a]	0.005	Do
Herbicides		
2,4-D plus 2,4,5-T, plus 2,4,5-TP[a]	0.1[d]	Do
Phenols[a]	0.001	Do
Radioactivity	(pc/l)	(pc/l)
Gross beta[a] (Beta and photon)	1,000 (15, $\geqslant$3 mr/yr)[d]	$<$100
Radium-226[a] (and radium-228)	3 (5)[d]	$<$1
Strontium-90[a]	10	$<$2

Source: *Water Quality Criteria*, Report of the National Technical Advisory Committee to the Secretary of the Interior, Washington, D.C., April 1968, p. 20.

[a]The defined treatment process has little effect on this constituent. (Conventional coagulation, sedimentation, rapid sand filtration, and chlorination.)

[b]No consensus on a single numerical value that is applicable throughout the country. See Report and text.

[c]Microbiological limits are monthly arithmetic averages based on an adequate number of samples. Total coliform limit may be relaxed if fecal coliform concentration does not exceed the specified limit.

[d]National Interim Primary Drinking Water. Standards maximum for public water supplies at consumer's tap. The *total* mercury shall not exceed 0.002 mg/l.

[e]As parathion in cholinesterase inhibition. It may be necessary to resort to even lower concentrations for some compounds or mixtures. (Permissible levels are based on the recommendations of the Public Health Service Advisory Committee on Use of the PHS Drinking Water Standards.)

drainage area so that actual and potential sources of pollution can be determined and properly evaluated. All surface water supplies must be considered of doubtful sanitary quality unless given adequate treatment, depending on the type and degree of pollution received.

Groundwater supplies, such as wells or springs, should be investigated with a view toward finding ways whereby the source might be polluted. A complete sanitary survey should include inspection of the aquifer drainage area, land use and habitation, local geology and vegetation, nature of soil and rock strata, well logs, evidence of blasting, slope of water table, sources of pollution, and development of the source.

The sanitary survey would include, in addition to the source as noted above, the reservoir, intake, pumping station, treatment plant and adequacy of each unit process; operation records; distribution system carrying capacity, head losses, and pressures; storage facilities; emergency source of water and plans to supply water in emergency; integrity of laboratory services; connections with other water supplies; and actual or possible cross-connections with plumbing fixtures, tanks, structures, or devices that might permit back-siphonage or backflow. Where water treatment is provided, the integrity and competence of the person in charge of the plant is an important factor.

Water samples are collected as an adjunct to the sanitary survey as an aid in measuring the quality of the raw water and effectiveness of treatment given the water. Bacteriological examinations, chemical and physical analyses, and microscopic examinations may be made depending on the sources of water, climate, geology, hydrology, waste disposal practices on the watershed, problems likely to be encountered, and the purpose to be served. In any case, all tests should be made by an approved laboratory in accordance with the procedure given in the latest edition of *Standard Methods for the Examination of Water and Wastewater*,[23] or as approved by the USEPA.

A sanitary technique and a glass or plastic sterile bottle supplied and prepared by the laboratory for the purpose should be used when collecting a water sample for bacteriological examination. The hands or faucet must not touch the edge of the lip of the bottle or the plug part of the stopper. The sample should be taken from a clean faucet that does not have an aerator or screen and that is not leaking or causing condensation on the outside. Let the water run for about 10 min to get a representative sample. If a sample from a lake or stream is desired, the bottle should be dipped below the surface with a forward sweeping motion so that water coming in contact with the hands will not enter the bottle. When collecting samples of chlorinated water, the sample bottle should contain sodium thiosulfate to dechlorinate the water. It is recommended that all samples be examined promptly after collection and within 6 to 12 hr if possible.

The chemical and physical analyses may be for industrial or sanitary purposes, and the determinations made will be either partial or complete, depending on the information desired. Water samples for inorganic chemical analyses are collected in one-liter polyethylene containers, new or acid washed if previously used. Samples for organic chemical analyses are collected in one-gallon glass bottles with Teflon-lined closure.[24] Containers should be completely filled. A 500 gal total sample is needed for viral detection in water. A sample for *Giardia lamblia* detection requires collection of residue on a filter resulting from the filtration of up to 500 gal of water for microscopic examination. A membrane filter can be used when water is heavily contaminated with giardia and low in particulate matter. A special preservative is added for certain tests and delivery time to the laboratory sometimes is specified. Samples are also collected for selected tests to control routine operation of a water plant, to determine the treatment required and its effectiveness.

Samples for microscopic examination should be collected in clean widemouth bottles having a volume of 1 or 2 liters from depths that will yield representative organisms. Some organisms are found relatively close to the surface,

[23]Published by the A. P. Health Assoc., 1015 Fifteenth Street, N.W., Washington, D.C. 20005, 1980.

[24]Gunther F. Craun and Robert A. Gunn, "Outbreaks of Waterborne Disease in the United States, 1975–76," USEPA, Cincinnati, Ohio.

whereas others are found at mid-depth or near the bottom, depending on the food, type of organisms, clarity, and temperature of the water. Microscopic examinations can determine the changing types, concentrations and locations of microscopic organisms, the control measures or treatment indicated, and the time to start treatment. A proper program can prevent tastes and odors by eliminating the responsible organisms that secrete certain oils before they can cause the problem. In addition, objectionable appearances in a reservoir or lake are prevented and sedimentation and filter runs are improved. Attention should also be given to elimination of the conditions favoring the growth of the organisms. See also Microscopic Examination and Control of Microorganisms, this chapter.

Sampling Frequency

The frequency with which source and distribution system water samples are collected and used for bacteriologic, chemical, radiologic, microscopic, and physical analyses is usually determined by the regulatory agency and by special problems. Operators of public water systems, and operators of industrial and commercial water systems, will want to collect more frequent but carefully selected samples and make more analyses to detect changes in raw water quality to better control treatment and plant operation.

The number of distribution system samples is usually determined by the population served, and special problems. Table 3-9 shows the minimum required sampling frequency at community water systems in the United States for coliform density. At noncommunity water supplies a sample shall be collected in each quarter during which the system provides water to the (traveling) public. The sampling frequency proposed by the international drinking water standards are shown in Table 3-10.

Interpretation of Water Analyses

As indicated previously, the interpretation of water analyses is based primarily on a sanitary survey of the water supply. A water supply that is coagulated and filtered would be expected to be practically clear, colorless, and free of iron, whereas the presence of some turbidity, color, and iron in an untreated surface water supply may be accepted as normal. A summary is given below of the constituents and concentrations considered significant in water examinations. Other compounds and elements not mentioned are also found in water.

Bacterial Examination—The Standard Plate Count

The standard plate count is the total colonies of bacteria developing from measured portions (two 1 ml and two $\frac{1}{10}$ ml) of the water being tested, which

Table 3-9 Sampling Frequency for Coliform Density

Population Served:	Minimum Number of Samples per Month	Population Served:	Minimum Number of Samples per Month
Up to 1,000	1	90,001 to 96,000	95
1,001 to 2,500	2	96,001 to 111,000	100
2,501 to 3,300	3	111,001 to 130,000	110
3,301 to 4,100	4	130,001 to 160,000	120
4,101 to 4,900	5	160,001 to 190,000	130
4,901 to 5,800	6	190,001 to 220,000	140
5,801 to 6,700	7	220,001 to 250,000	150
6,701 to 7,600	8	250,001 to 290,000	160
7,601 to 8,500	9	290,001 to 320,000	170
8,501 to 9,400	10	320,001 to 360,000	180
9,401 to 10,300	11	360,001 to 410,000	190
10,301 to 11,100	12	410,001 to 450,000	200
11,101 to 12,000	13	450,001 to 500,000	210
12,001 to 12,900	14	500,001 to 550,000	220
12,901 to 13,700	15	550,001 to 600,000	230
13,701 to 14,600	16	600,001 to 660,000	240
14,601 to 15,500	17	660,001 to 720,000	250
15,501 to 16,300	18	720,001 to 780,000	260
16,301 to 17,200	19	780,001 to 840,000	270
17,201 to 18,100	20	840,001 to 910,000	280
18,101 to 18,900	21	910,001 to 970,000	290
18,901 to 19,800	22	970,001 to 1,050,000	300
19,801 to 20,700	23	1,050,001 to 1,140,000	310
20,701 to 21,500	24	1,140,001 to 1,230,000	320
21,501 to 22,300	25	1,230,001 to 1,320,000	330
22,301 to 23,200	26	1,320,001 to 1,420,000	340
23,201 to 24,000	27	1,420,001 to 1,520,000	350
24,001 to 24,900	28	1,520,001 to 1,630,000	360
24,901 to 25,000	29	1,630,001 to 1,730,000	370
25,001 to 28,000	30	1,730,001 to 1,850,000	380
28,001 to 33,000	35	1,850,001 to 1,970,000	390
33,001 to 37,000	40	1,970,001 to 2,060,000	400
37,001 to 41,000	45	2,060,001 to 2,270,000	410
41,001 to 46,000	50	2,270,001 to 2,510,000	420
46,001 to 50,000	55	2,510,001 to 2,750,000	430
50,001 to 54,000	60	2,750,001 to 3,020,000	440
54,001 to 59,000	65	3,020,001 to 3,320,000	450
59,001 to 64,000	70	3,320,001 to 3,620,000	460
64,001 to 70,000	75	3,620,001 to 3,960,000	470
70,001 to 76,000	80	3,960,001 to 4,310,000	480
76,001 to 83,000	85	4,310,001 to 4,690,000	490
83,001 to 90,000	90	4,690,001 or more	500

Source: National Interim Primary Drinking Water Regulations, Section 141.21, *Fed. Reg.* November 24, 1975, p. 59571.

Table 3-10 Maximum Interval Between Successive Samples and Minimum Number of Samples to be Taken[a]

Population Served	Maximum Interval Between Successive Samples	Minimum Number of Samples to be Taken From Whole Distribution System Each Month
Less than 20,000	1 month	1 sample per 5000 population per month
20,001 to 50,000	2 weeks	
50,001 to 100,000	4 days	
More than 100,000	1 day	1 sample per 10,000 population per month

Source: *International Standards for Drinking-Water*, WHO, 1971, p. 47

[a]Samples to be collected from distribution system, whether the water has been subjected to disinfection or not.

have been planted in petri dishes with a suitable culture media (agar), and incubated for 48 hr at 35°C. For bottled water incubate at 35°C for 72 hr.[25] Drinking water will normally contain some nonpathogenic bacteria; it is almost never sterile.

The test is of significance when used for comparative purposes under known or controlled conditions to show changes from the norm. It can monitor changes in the quality of the water in the distribution system and storage reservoirs; it can be used to detect the presence of *Pseudomonas flavobacterium* and other secondary invaders that could pose a health risk in the hospital environment; it can call attention to limitations of the coliform test when the average of standard plate counts in a month exceeds 100 to 500 per ml; it can show the effectiveness of distribution system residual chlorine and possible filter breakthrough; and it can show distribution system deterioration, main growth, and sediment accumulation. Large total bacterial populations may support or suppress growth of coliform organisms. Taste, odor, or color complaints may also be associated with bacterial or other growths in mains or surface water sources.[26]

Tests Indicating Contamination

The bacterial examination of drinking water should always include a quantitative estimation of total organisms of the coliform groups, which are *indicative* of fecal contamination or sewage pollution. The total coliform group of organisms include escherichia, klebsiella, and citrobacter-enterobacter bacteria. Coliform bacteria are not normally considered disease organisms. However, pathogenic strains of *Escherichia coli* have caused outbreaks in nurseries, institutions, and communities, associated with food, water, or fomites.

[25]*Standard Methods For the Examination of Water and Wastewater*, 15th ed., 1981, Am. Pub. Health Assoc., 1015 Fifteenth Street, NW, Washington, D.C. 20005.

[26]Edwin E. Geldreich, "Is the Total Count Necessary," *Proc. Am. Water Works Assoc. Water Quality Tech. Conf.*, Am. Water Works Assoc., Denver, 1974.

The test for *E. coli* (35° C) is recommended as being a more specific indicator of fecal contamination in Denmark, Belgium, England, and France.[27] More detailed laboratory procedures are needed to identify *E. coli* and the enteropathogenic *E. coli.*

The coliform group of organisms includes all of the aerobic and facultative anaerobic, gram-negative, non-spore-forming, rod-shaped bacteria that ferment lactose with acid and gas formation within 48 hr at 35° C. This is the presumptive test that can be confirmed and completed by carrying the test further as outlined in *Standard Methods.*[25] The results in the multiple tube method are reported as the most probable number (MPN) of coliform bacteria, a statistical number most likely to produce the test results observed, per 100 ml of sample. The MPN should be less than 2.2 using the multiple tube method (not greater than 1 per 100 ml using the membrane filter technique).

The coliform group is sometimes also referred to as the *B. coli* group and as the coli-aerogenes group. The count includes *Escherichia coli, Enterobacter aerogenes*, and other coliforms as noted above. The principal species are the *E. coli* which are common in the intestinal tract of man and other animals, and the *Aerobacter aerogenes* which are also common in soil and vegetation. If the membrane filter technique is used, the bacteria produce a dark colony with a metallic sheen within 24 hr on an Endo-type medium containing lactose, instead of fermenting lactose with gas formation in the multiple tube test. The MPN index per 100 ml of sample and the membrane filter count (a 100 ml sample is recommended) are not the same and cannot be directly compared.

When five 10-ml portions are used per sample and:

all are negative	MPN is <2.2 per 100 ml
one is positive	MPN is 2.2 per 100 ml
two are positive	MPN is 5.1 per 100 ml
three are positive	MPN is 9.2 per 100 ml
four are positive	MPN is 16.0 per 100 ml
five are positive	MPN is >16.0 per 100 ml.

When five 10-ml portions, one 1-ml portion, and one 0.1-ml portion are used per sample and:

all are negative	MPN is <2.2 per 100 ml
one 10-ml positive	MPN is 2.2 per 100 ml
two 10-ml positive	MPN is 5.0 per 100 ml
three 10-ml positive	MPN is 8.8 per 100 ml
four 10-ml positive	MPN is 15.0 per 100 ml
five 10-ml positive	MPN is 38.0 per 100 ml
five 10-ml and one 1-ml positive	MPN is 240 per 100 ml
five 10-ml, one 1-ml, and one 0.1-ml positive*	MPN is 2400 per 100 ml.

The isolation of *Salmonella typhosa* and other organisms by cultivation on a selected medium is also possible. Suspended matter, algae, and bacteria in

*MPN is 2400 per 100 ml or greater.

[27]A. P. Dufour, "*Escherichia coli*: The Fecal Coliform," *Bacterial Indicators/Health Hazards Associated with Water*, ASTM STP635, A. W. Hoadley and B. J. Dutka, Eds., American Society for Testing Materials, Philadelphia, Pa., 1977, pp. 48–58.

large amounts interfere with the membrane filter procedure. Bacterial overgrowth on the filter would indicate an excessive bacterial population which should be investigated as to cause.

The *fecal coliform test* involves incubation at 44.5°C for 24 hr and measures mostly *Escherichia coli* in a freshly passed stool of humans or other warm-blooded animals. A loop of broth from each positive presumptive tube incubated at 35°C in the total coliform test is transferred to EC (*E. coli*) broth and incubated at 44.5°C in a waterbath; formation of gas within 24 hr indicates the presence of fecal coliform and hence also possibly dangerous contamination. Maintenance of 44.5 ± 0.2°C is critical. Nonfecal organisms generally do not produce gas at 44.5°C. The test has greatest application in the study of stream pollution, raw water sources, sea waters, wastewaters, and the quality of bathing waters.

The analysis for total coliforms would indicate the presence of feces of human and warm-blooded animals, and coliform associated with vegetation, soils, air, joint and valve packing materials, slimes, swimming pool ropes, pump leathers, sewage, storm-water drainage, surface-water runoff, surface waters, and others. The tests for fecal coliforms, *E. coli*, and fecal streptococci are helpful in interpreting the significance of surface water tests and their possible hazard to public health.

The *fecal streptococci* test (enterococci) uses special agar media incubated at 35°C for 48 hr. Dark red to pink colonies are counted as fecal streptococci. They are also normally found in the intestinal tract of warm-blooded animals, including man. Most of the human fecal streptococci are *Streptococcus fecalis*; *S. bovis* are associated with cows, and *S. equinus* with horses. These organisms may be more resistant to chlorine than coliform and survive longer in some waters, but usually die off quickly outside the host. If found it would indicate recent pollution.

The test for *Clostridium perfringens* (*Cl. welchii*), which is found in the intestines of man and animals, may be of value in the examination of polluted waters containing industrial wastes. The clostridium sporulates under unfavorable conditions and can survive indefinitely in the environment; it is more resistant than the escherichia and streptococci. Its presence therefore indicates past or possibly intermittent pollution.

In domestic sewage, the fecal coliform concentration is usually at least four times that of the fecal streptococci and may constitute 30 to 40 percent of the total coliforms. In storm water and in wastes from livestock, poultry, animal pets, and rodents the fecal coliform concentration is usually less than 0.4 of the fecal streptococci. In streams receiving sewage, fecal coliforms may average 15 to 20 percent of the total coliforms in the stream. The presence of fecal coliform generally indicates fresh and possibly dangerous pollution. The presence of intermediate-aerogenes-cloacae (IAC) subgroups of coliform organisms suggests past pollution or, in a municipal water supply, defects in treatment or in the distribution system.[28]

[28]E. E. Geldreich, *Sanitary Significance of Fecal Coliforms in the Environment*, U.S. Fed. Water Pollut. Control. Adm. Pub. No. WP-20-3, Dept. of the Interior, Washington, D.C., November 1966.

The presence of any of the coliform organisms in drinking water is a danger sign; it must be carefully interpreted in the light of a sanitary survey and promptly eliminated. There may be some justification for permitting a low coliform density in developing areas of the world where the probability of other causes of intestinal diseases greatly exceeds those caused by water, as determined by epidemiological information.

It must be understood and emphasized that the absence of coliform organisms, or other indicators of contamination, does not in and of itself assure that the water is *always* safe to drink unless it is supported by a comprehensive sanitary survey of the drainage area, treatment unit processes, storage, and distribution system (backflow prevention). Nor does the absence of coliform assure the absence of viruses (infectious hepatitis virus), protozoa (*entamoeba* and giardia), or helminths (schistosomes and worms) unless the water is coagulated, flocculated, settled, gravity filtered, and chlorinated to yield a *free* residual chlorine of at least 0.5 mg/l, preferably for one hour before it is available for consumption. A free ozone of 0.4 mg/l for four minutes has been found to inactivate virus "but somewhat more rigorous treatment would be desirable because the resistance of hepatitis virus to ozone is unknown."[29]

A properly developed, protected, and chlorinated well-water supply showing the absence of coliform organisms can usually be assumed to be free of viruses, protozoa, and helminths if supported by a satisfactory sanitary survey. Chemical examinations however, are needed to assure absence of toxic organic and inorganic chemicals.

A final point—the results of a bacterial or chemical examination reflect the quality of the water only at the time of sampling and must be interpreted in the light of the sanitary survey. However, chemical examination results from well-water supplies are not likely to change significantly from day to day or week to week. Some bacterial and chemical analyses are shown in Table 3-11.

Virus Examination

The examination of water for viruses has not yet been simplified to the point where the test can be made routinely as for coliform. A large volume of water must be sampled and an effective system used to concentrate the viruses.

Physical Examinations

Odor

Odor should be absent or very faint for water to be acceptable, not greater than 3 Threshold Odor Number. Water for food processing, beverages, and pharmaceutical manufacture should be essentially free of taste and odor. The test is very subjective, being dependent on the individual senses of smell and taste. The cause may be decaying organic matter, wastewaters including indus-

[29] *International Standards for Drinking Water*, WHO, 1971, p. 27.

Table 3-11 Some Bacterial and Chemical Analyses

Source of sample	Dug well	Lake	Reservoir	Deep well	Deep well
Time of year	—	April	October	—	—
Treatment	None	Chlor.	None	None	None
Bacteria per ml Agar 35°C 24 hrs	—	3	—	1	>5000
Coliform MPN per 100 ml	—	<2.2	—	<2.2	2400 or >
Color, units	0	15	30	0	0
Turbidity, units	Trace	Trace	Trace	Trace	5.0
Odor, cold	2 veget.	2 aromatic	1 veget.	1 aromatic	3 disagreeable
Odor, hot	2 veget.	2 aromatic	1 veget.	1 aromatic	3 disagreeable
Iron, mg/l	0.15	0.40	0.40	0.08	0.2
Fluorides, mg/l	<0.05	0.005	—	—	—
Nitrogen as ammonia, free, mg/l	0.002	0.006	0.002	0.022	0.042
Nitrogen as ammonia, albuminoid, mg/l	0.026	0.128	0.138	0.001	0.224
Nitrogen as nitrites, mg/l	0.001	0.001	0.001	0.012	0.030
Nitrogen as nitrates, mg/l	0.44	0.08	0.02	0.02	0.16
Oxygen consumed, mg/l	1.1	2.4	7.6	0.5	16.0
Chlorides, mg/l	17.0	5.4	2.2	9.8	6.6
Hardness (as $CaCO_3$), total, mg/l	132.0	34.0	84.0	168.0	148.0
Alkalinity (as $CaCO_3$), mg/l	94.0	29.0	78.0	150.0	114.0
pH value	7.3	7.6	7.3	7.3	7.5

trial wastes, dissolved gases, and chlorine in combination with certain organic compounds such as phenols. Odors are sometimes confused with tastes. Carbon adsorption or aeration will usually remove odors.

Taste

The taste of water should not be objectionable for otherwise the consumer will resort to other sources of water which might not be of satisfactory sanitary quality. Algae, decomposing organic matter, dissolved gases, high concentrations of sulfates, chlorides, and iron, or industrial wastes may cause tastes and odors. Bone and fish oil and petroleum products such as kerosene and gasoline are particularly objectionable. Phenols in concentrations of 0.2 ppb in combination with chlorine will impart a phenolic or medicinal taste to drinking water. The taste test, like the odor test, is very subjective and may be dangerous to laboratory personnel.

Turbidity

The National Interim Primary Drinking Water Regulations require that the maximum contaminant level for turbidity not exceed one turbidity unit as determined by a monthly average. A two-day turbidity average may not exceed five turbidity units. A monthly average of five turbidity units may be permitted by the regulatory agency if it can be demonstrated that the turbidity does not interfere with disinfection, the maintenance of a chlorine residual throughout the distribution system, or the coliform determination. See Table 3-3. Turbidity measurements are made in terms of Nephelometric Turbidity Units (NTU), Formazin Turbidity Units (FTU), and Jackson Turbidity Units (JTU). The results are interchangeable if calibration has been based on the Formazin scale. NTU is the standard measure, requiring use of a nephelometer, which measures the amount of light scattered, usually at 90° from the light direction, by suspended particles in the water test sample.

The public demands a sparkling clear water. This implies a turbidity of less than 1 unit; a level of less than 0.1 unit, which is obtainable when water is coagulated, settled, and filtered, is practical. Turbidity is a good measure of sedimentation, filtration, and storage efficiency, particularly if supplemented by the total microscopic count. Increased chlorine residual and bacteriological sampling of the distribution system is indicated when the maximum contaminant level for turbidity is exceeded in the distribution system, until the cause is determined and eliminated. Turbidity will interfere with proper disinfection of water, harbor viruses, and cause tastes and odors.

Color

Color should be less than 15 color units, although persons accustomed to clear water may notice a color of only 5 units. Water for industrial uses should generally have a color of 5 to 10 or less. Color is caused by substances in solution, known as true color, and by substances in suspension, mostly organics causing organic color.

Water that has drained through peat bogs, swamps, forests, or decomposing organic matter may contain a brownish or reddish stain due to tannates and organic acids dissolved from leaves, bark, and plants. Excessive growths of algae or microorganisms may also cause color. Coagulation, settling, and rapid sand filtration should reduce color-causing substances in solution to less than 5 units. Slow sand filters should remove about 40 percent of the total color. True color is costly to remove. Oxidation or carbon adsorption may be needed.

Color resulting from the presence of organics in water may also cause taste, interfere with chlorination, induce bacterial growth, make water unusable by certain industries without further treatment, foul anion-exchange resins, interfere with colorimetric analyses, limit aquatic productivity by absorbing photosynthetic light, render lead in pipes soluble, hold iron and manganese in solution causing color and staining of laundry and plumbing fixtures, and interfere with chemical coagulation. Chlorination of natural waters containing organic water color (and humic acid) results in the formation of trihalomethanes including chloroform. This is discussed later.

Color can be controlled at the source by watershed management. Involved is identifying waters from sources contributing natural organic color and excluding them, controlling beaver populations, increasing water flow gradients, using settling basins at inlets to reservoirs, and blending of water.[30]

[30]Bill O. Wilen, "Options for Controlling Natural Organics," *Drinking Water Quality Enhancement Through Source Protection*, Robert B. Pojasek, Ed., Ann Arbor Science, Ann Arbor, Mich. 48106. See also "Research Committee on Color Problems Report for 1966," *J. Am. Water Works Assoc.*, August 1967, pp. 1023–1035.

Temperature

The water temperature should preferably be less than 60° F. Ground waters and surface waters from mountainous areas are generally in the temperature range of 50 to 60° F. Design and construction of water systems should provide for burying or covering of transmission mains to keep drinking water cool and to also prevent freezing in cold climates or leaks due to vehicular traffic.

Microscopic Examination

Microscopic organisms that may be found in drinking water sources include bacteria, algae, actinomycetes, protozoa, rotifiers, yeasts, molds, and small crustacea, worms, and mites. Most algae contain chlorophyl and require sunlight for their growth. The small worms are usually insect larvae. Larvae, crustacea, worms, molds or fungi, large numbers of algae, or filamentous growths in the drinking water would make the water aesthetically unacceptable. Immediate investigation to eliminate the cause would be indicated.

The term "plankton" includes algae and small animals such as cyclops and daphnia. Plankton are microscopic plants and animals suspended and floating in fresh and salt water and are a major source of food for fish. "Algae" include diatoms, cyanaphyceae or blue-green algae (bacteria), and chlorophyceae or green algae; they are also referred to as phytoplankton. Protozoan and other small animals are referred to as zooplankton. The microbial flora in bottom sediments are called the benthos.[31]

Algal growths increase the organic load in water, produce tastes and odors, clog sand filters, clog intake screens, produce slimes, interfere with recreational use of water, may cause fish kills when in "bloom" and in large surface "mats" by preventing replenishment of oxygen in the water, become attached to reservoir walls, form slimes in open reservoirs and recirculating systems, and contribute to corrosion in open steel tanks.[32]

Microscopic examination involves collection of water samples from specified locations and depths. The sample is preserved by the addition of formaldehyde if not taken immediately to the laboratory. At the laboratory the plankton in the sample is concentrated by means of a centrifuge or a Sedgwick-Rafter sand filter. A 1-ml sample of the concentrate is then placed in a Sedgwick-Rafter counting cell for enumeration using a compound microscope fitted with a Whipple ocular micrometer. The Lackey Drop Microtransect Counting Method is also used, particularly with samples containing dense plankton populations.[33]

Examinations of surface water sources, or water main and well-water supplies which are sources of difficulty, should be made weekly to observe trends

[31]Don F. Kincannon, "Microbiology in Surface Water Sources," *OpFlow*, AWWA, Denver, Co., December 1978, p. 3.

[32]C. Mervin Palmer, *Algae in Water Supplies*, DHEW, PHS Pub. 657, Cincinnati, Ohio, 1959.

[33]*Standard Methods For the Examination of Water and Wastewater*, 15th ed., A. E. Greenberg, Ed., Am. Pub. Health Assoc., 1015 Fifteenth Street N.W., Washington, D.C., 20005, 1981.

and to determine the need for treatment or other controls and their effectiveness. The measure of concentration of microorganisms present is the "areal standard unit." It represents an area 20 microns (μm) square or 400 μm^2. One micron equals 0.001 mm. Microorganisms are reported as the number of areal standard units per ml. Protozoa, rotifers, and other animal life are individually counted. Material that cannot be identified is reported as areal standard units of amorphous matter (detritus). The apparatus, procedure, and calculation of results and conversion to "Cubic Standard Units" is explained in *Standard Methods.*[33] Figure 3-3 is an example of a microscopic examination, showing organisms per ml, areal standard units per ml, and mm^3 per liter.

When more than 300 areal standard units, or organisms, per milliliter are reported, treatment with $CuSO_4$ is indicated to prevent possible trouble with tastes and odors or short filter runs. When more than 500 areal standard units per milliliter are reported, complaints can be expected and the need for immediate action is indicated. A thousand units or more of amorphous matter indicates probable heavy growth of organisms that have died and disintegrated or organic debris from decaying leaves and similar vegetable matter.

The presence of asterionella, tabellaria, synedra, beggiatoa, crenothrix, *Sphaerotilis natans*, mallomonas, anabaena, aphanizomenon, volvox, ceratium, dinobryon, synura, uroglenopsis, and other, some even in small concentrations, may cause tastes and odors that are aggravated where marginal chlorine treatment is used. Free residual chlorination will usually reduce the tastes and odors. More than 25 areal standard units per milliliter of synura, dinobryon, or uroglena, or 300 to 700 units of asterionella, dictyosphaerium, aphanizomenon, volvox, or ceratium in chlorinated water will usually cause taste and odor complaints. The appearance of even one areal standard unit of a microorganism may be an indication to start immediate copper sulfate treatment if past experience indicates that trouble can be expected.

The blue-green algae, anabaena, microcystis (polycystis), nodularia, gloeotrichia, coelosphaerium, *Nostoc rivulare*, and aphanizomenon in large con-

FORM LR 44G, (REV. 12/73)

NEW YORK STATE DEPARTMENT OF HEALTH
DIVISION OF LABORATORIES AND RESEARCH-ENVIRONMENTAL HEALTH CENTER
ALBANY, N.Y. 12201

MICROSCOPIC EXAMINATION

LAB ACCESS NO. 77 02 237 SAMPLE REC'D 9 7 ____
YEAR LAB ACC. NO. MONTH DAY HOUR

PROGRAM CODE 100 NAME Public Water Supply
STATION NO. 10319000 LOCATION Coxsackie V COUNTY Greene
COMMON NAME Coxsackie (V) PWS

Figure 3-3 Microscopic examination of a reservoir water.

EXACT DESCRIPTION OF SITE Raw Water Dip—Medway Res. (Upper Res.)
SAMPLING TIME 09 07 12 SAMPLE TYPE 00 DESCRIPTION Raw Water
MONTH DAY HOUR

REASONS FOR SUBMISSION Routine Surveil., Special Study
REPORT TO: CO (1) RO (2) LPHE (1) LHO (1) FED ( )
SUBMITTED BY: R. Lupe Sr. San. Eng. DATE REPORTED 9 15
TITLE MONTH DAY

CODE	PARTICLE DESCRIPTION	NUMBER PER ML.	AREAL STD UNITS PER ML.	VOLUME CU. MM PER LITER
90002	Detritus, amorphous 10 μ		1600	6.4000
90002	Detritus, amorphous 2 μ		2400	1.9000
10041	CYAN Coelosphaerium Naeg. col 56.0 μ	56	340	5.1000
70000	CILIOPHORA ciliates cell 63.0 μ	5	46	.7700
11010	unidentified flagellate cell 3.7 μ	4200	120	.1200
10080	CYAN Anabaena sp. fil 9.0 μ	6	69	.2000
20111	CHLOR Ankistrodesmus falcat. cell 4.0 μ	160	5	.1600
70000	CILIOPHORA ciliates cell 23.3 μ	20	21	.1300
40031	PYRRH Cryptomonas erosa cell 5.6 μ	240	28	.0420
10091	CYAN Chroococcus limneticus col 32.0 μ	420	840	7.2000
	ORGANISMS	5100	1500	14.0000
	DETRITUS		4000	8.3000
	TOTALS:		5500	22.0000

QUALITATIVE EXAMINATION (scarce particles):

10080	CYAN Anabaena sp. 4 μ	40032	PYRRH Cryptomonas ovata
10061	CYAN Microcystis aeruginosa	50010	EUGLEN Trachelomonas sp.
20121	CHLOR Eudorina elegans	20371	CHLOR Nephrocytium Agardh.
20221	CHLOR Pandorina morum	20045	CHLOR Scenedesmus brasilien.
20011	CHLOR Sphaerocystis Schro.		
10031	CYAN Anacystis incert.		
30130	CHRYS Navicula spp.		
20231	CHLOR Dictyosphaerium pulch.		
30110	CHRYS Cymbella sp. cell		
20290	CHLOR Quadrigula sp. col		
20020	CHLOR Oocystis sp.		

Algae are defined as to phylum, genus, species (if possible), form, and size. Abbreviations are: CYAN = CYANOPHYTA, CHLOR = CHLOROPHYTA, CHRYS = CHRYSOPHYTA, PYRRH = PYRRHOPHYTA, EUGLEN = EUGLENOPHYTA. Species names are often abbreviated. sp. in the place of the species name indicates identification to genus only, col = colony, fil = filament. Size measurement, when given, is in microns (μ) and refers to particle width, depth, or diameter. Dead organisms are considered as detritus (formerly: amorphous matter). Results are rounded to two figures with an approx. 95% confidence interval of ± 25%.

Figure 3-3 (*Continued*)

centrations have been responsible for killing fish and causing illness in horses, sheep, dogs, ducks, chickens, mice, and cattle.[34] Illness in man from this cause has been suspected. Gorham estimated that the oral minimum lethal dose of decomposing toxic microcystis bloom for a 150-lb man is 1 to 2 qt of thick, paintlike suspension, and concluded that toxic waterblooms of blue-green algae in public water supplies are not a significant health hazard. Red tides caused by the dinoflagellates *Gonyaulax monilata* and *Gymnodinium brevis* have been correlated with mass mortality of fish.[35]

Investigation of conditions contributing to or favorable to the growth of plankton in a reservoir, and their control, should reduce dependence on copper sulfate treatment.

Chemical Examination*

The significance of chemical elements and compounds in drinking water is discussed below. Their removal is reviewed later in this chapter.

Hardness

Hardness is due primarily to calcium and magnesium carbonates and bicarbonates (carbonate hardness, which can be removed by heating) and calcium sulfate, calcium chloride, magnesium sulfate, and magnesium chloride (noncarbonate hardness, which cannot be removed by heating). In general, water softer than 50 mg/l, as $CaCO_3$, is corrosive, whereas waters harder than about 80 mg/l lead to use of more soap. Lead, cadmium, zinc, and copper in solution are associated with soft water. Desirable hardness values, therefore, should be 50 to 80 mg/l, with 80 to 150 mg/l as passable, and over 150 mg/l as undesirable. Waters high in sulfates (above 600 to 800 mg/l calcium sulfate, 300 mg/l sodium sulfate, or 390 mg/l magnesium sulfate) are laxative to those not accustomed to the water. In addition to being objectionable for laundry and other washing purposes by causing curdling of soap, excessive hardness contributes to the deterioration of fabrics. Hard water is not suitable for the production of ice, soft drinks, felts, or textiles. Satisfactory cleansing of laundry, dishes, and utensils is made difficult or impractical. In boiler and hot-water tanks, the scale resulting from hardness reduces the thermal efficiency and eventually causes restriction of the flows or plugging of the pipes. Calcium chloride, when heated, becomes acidic and pits boiler tubes.

There seem to be higher mortality rates from cardiovascular diseases in people provided with soft water than in those provided with hard water. The

*Results are reported as milligrams per liter (mg/l) which for all practical purposes can be taken to be the same as parts per million (ppm) except when the concentrations of substances in solution approach or exceed 7000 mg/l, when a density correction should be made.

[34]William Marcus Ingram and C. G. Prescott, "Toxic Fresh-Water Algae," *Am. Midland Naturalist*, **52,** 1, 75–87 (July 1954).

[35]Paul R. Gorham, "Toxic Algae as a Public Health Hazard," *J. Am. Water Works Assoc.*, **56,** 1487 (November 1964).

low concentration of magnesium has been implicated; but low concentrations of chromium and high concentrations of copper have also been suggested as being responsible. See Cardiovascular Diseases in Chapter 1.

Alkalinity

The alkalinity of water passing through iron distribution systems should be in the range of 30 to 100 mg/l, as $CaCO_3$, to prevent serious corrosion; up to 500 mg/l is acceptable, although this factor must be appraised from the standpoint of pH, hardness, carbon dioxide, and dissolved-oxygen content. Corrosion of iron pipe is prevented by the maintenance of calcium-carbonate stability. The goal, according to the American Water Works Association, "is a measure of alkalinity decrease or increase in the distribution system, and also after 12 hr at 130°F in a closed plastic bottle, followed by filtration." Potassium carbonate, potassium bicarbonate, sodium carbonate, sodium bicarbonate, phosphates, and hydroxides cause alkalinity in natural water. Calcium carbonate, calcium bicarbonate, magnesium carbonate, and magnesium bicarbonate cause hardness as well as alkalinity. Sufficient alkalinity is needed in water to react with added alum to form a floc in water coagulation. Insufficient alkalinity will cause alum to remain in solution. Bathing or washing in water of excessive alkalinity can change the pH of the lacrimal fluid around the eye, causing eye irritation.

*pH**

The pH values of natural water range from about 5.0 to 8.5 and are acceptable except when viewed from the standpoint of corrosion. The pH is a measure of acidity or alkalinity, using a scale of 0.0 to 14.0, with 7.0 being the neutral point. The bactericidal, virucidal, and cysticidal efficiency of chlorine as a disinfectant increases with a decrease in pH. The pH determination in water having an alkalinity of less than 20 mg/l by using color indicators is inaccurate; use the electrometric method. The ranges of pH color indicator solutions are: methyl orange, 3.0 to 4.4; bromcresol green, 3.8 to 5.4; methyl red, 4.4 to 6.2; bromcresol purple, 5.2 to 6.8; bromthymol blue, 6.0 to 7.6; phenol red, 6.8 to 8.4; cresol red, 7.2 to 8.8; thymol blue, 8.0 to 9.6; and phenolphthalein, 8.2 to 10.0. Waters containing more than 1.0 mg/l chlorine in any form must be dechlorinated with 1 or 2 drops of $\frac{1}{4}$ percent sodium thiosulfate before adding the pH indicator solution. This is necessary to prevent the indicator solution from being bleached or decolorized by the chlorine and giving an erroneous reading. The germicidal activity is greatly reduced at a pH level above 8.5 and corrosion is associated with pH levels below 6.5. The formation of trihalomethanes is significantly increased at pH above 8.5.

Corrosivity

Corrosivity of water is related to its pH, alkalinity, hardness, dissolved oxygen, total dissolved solids, and other factors. Since a simple, rapid test for cor-

*pH is defined as the logarithm of the reciprocal of the hydrogen ion concentration.

rosivity is not available, test pipe sections or metal coupons (90-day test) are used supplemented where possible by water analyses such as calcium carbonate saturation and tests for alkalinity and pH, and dissolved solids and gases. The corrosion of copper tubing increases particularly when carrying water above 140°F. Schroeder[36] reports that pewter, britannia metal, water pipes, and cisterns may contain antimony, lead, cadmium, and tin which leach out in the presence of soft water or acid fluids. Soft water flowing over galvanized iron roofs, through galvanized iron pipes, or stored in galvanized tanks contains cadmium and zinc. Ceramic vessels contain antimony, beryllium, barium, nickel, and zirconium; pottery glazes contain lead, all of which may be leached out if firing and glazing is not proper. Corrosivity is controlled by pH, alkalinity, and calcium carbonate adjustment, use of chemicals, and other means. See Corrosion Cause and Control, this chapter.

Carbon Dioxide

The only limitation on carbon dioxide is that pertaining to corrosion. It should be less than 10 mg/l, but when the alkalinity is less than 100 mg/l, the CO_2 concentration should not exceed 5.0 mg/l.

Dissolved Oxygen

Water devoid of dissolved oxygen frequently has a "flat" taste, although many attractive well waters are devoid of oxygen. In general it is preferable for the dissolved oxygen content to exceed 2.5 to 3.0 mg/l to avoid secondary tastes and odors from developing and to support fish life. Game fish require a dissolved oxygen of at least 5.0 mg/l to reproduce, and either die off or migrate when the dissolved oxygen falls below 3.0 mg/l.

Lead

The acceptable concentration of lead in potable water is 0.05 mg/l. It is a cumulative poison. Concentrations exceeding this value occur when acid waters of low mineral content are piped through lead pipe; zinc galvanized iron pipe, copper pipe joints, and brass pipe may also contribute lead. The use of lead pipe to conduct drinking water should be prohibited. Lead, as well as cadmium, zinc, and copper, are dissolved by carbonated beverages which are highly charged with carbon dioxide. Limestone, galena, and food are natural sources of lead. Man-made sources are motor vehicle exhaust lead, certain industrial wastes, mines and smelters, lead paints, glazes, car battery salvage operations, cosmetics, and agricultural sprays. Fallout from airborne pollutants may also contribute significant concentrations of lead to water supply reservoirs and drainage basins. Only 10 percent of the lead ingested in water is absorbed; nearly half of the amount absorbed by urban dwellers comes from air. See Chapter 1 for health effects.

[36]Henry A. Schroeder, "Environmental Metals: The Nature of the Problem," *Environmental Problems in Medicine*, William D. McKee, Ed., Charles C. Thomas, Springfield, Ill., 1974.

Copper

The copper content should be less than 1.0 mg/l. Concentrations of this magnitude are not present in natural waters, but may be due to the corrosion of copper or brass piping; 0.5 mg/l in soft water stains porcelain blue-green. A concentration in excess of 0.2 to 0.3 mg/l will cause an "off" flavor in coffee and tea; 1 to 1.5 mg/l results in a bitter metallic taste; 1 mg/l may affect film and reacts with soap to produce a green color in water; 0.25 to 1.0 mg/l is toxic to fish. Copper appears to be essential for all forms of life; but excessive amounts, are toxic to fish. The estimated adult daily requirement is 2.0 mg, coming mostly from food. Copper deficiency is associated with anemia. Copper salts are commonly used to control algal growths in reservoirs and slime growths in water systems. Copper can be removed by ion exchange and, when caused by corrosion of copper pipes, by proper pH control.

Zinc

The concentration of zinc in drinking water should not exceed 1.0 mg/l. Zinc is dissolved by surface water. A greasy film forms in surface water containing 0.5 mg/l or more zinc. More than 5.0 mg/l causes a metallic bitter taste and 25 to 40 mg/l may cause nausea and vomiting. Zinc may contribute to the corrosiveness of water. Common sources of zinc in drinking water are brass and galvanized iron pipe. Zinc from zinc oxide in automobile tires is a significant pollutant in urban runoff.[37] The ratio of zinc to cadmium may also be of public health importance. Zinc deficiency is associated with dwarfism and hypogonicidism.[38]

Chlorides of Mineral Origin

The permissible chloride content of water depends on the sensitivity of the consumer. Many people notice a brackish taste imparted by 100 mg/l of chlorides, whereas others are satisfied with concentrations as high as 250 mg/l. Irrigation waters should contain less than 200 mg/l. When the chloride is in the form of sodium chloride, use of the water for drinking may be inadvisable for persons who are under medical care for certain forms of heart disease. See Chapter 1. Hard water softened by the ion exchange or lime-soda process (with Na_2CO_3) will increase the concentrations of sodium in the water. Salt used for highway deicing may contaminate groundwater and surface water supplies. Its use should be curtailed and storage depots covered. Chlorides can be removed from water by distillation, reverse osmosis, or electrodialysis. See *Sodium* (next page).

[37]Erick R. Christensen and Vincent P. Guinn, "Zinc from Automobile Tires in Urban Runoff," *J. Environ. Eng. Div., ASCE,* February 1979, pp. 165–168.

[38]Henry A. Schroeder, "Environmental Metals: The Nature of the Problem," *Environmental Problems in Medicine,* William D. McKee, Ed., Charles C. Thomas, Springfield, Illinois, 1974.

Chlorides of Intestinal Origin

Natural waters remote from the influence of ocean or salt deposits and not influenced by local sources of pollution have a low chloride content—usually less than 4.0 mg/l. Due to the extensive salt deposits in certain parts of the country, it is impractical to assign chloride concentrations that, when exceeded, indicate the presence of sewage, agricultural, or industrial pollution, unless a chloride record over an extended period of time is kept on each water supply. In view of the fact that chlorides are soluble, they will pass through pervious soil and rock for great distances without diminution in concentration, and thus the chloride content must be interpreted with considerable discretion in connection with other constituents in the water. The concentration of chlorides in urine is about 5000 mg/l; in septic-tank effluent, about 80 mg/l; and in sewage from a residential community 50 mg/l, depending on the water source.

Iron

Water should have a soluble iron content of less than 0.1 mg/l to prevent reddish-brown staining of laundry, fountains, and plumbing fixtures. Some staining of plumbing fixtures may occur at 0.05 mg/l. Precipitated ferric hydroxide may cause a slight turbidity in water that can be objectionable and cause clogging of filters and softener resin beds. In combination with manganese, concentrations in excess of 0.3 mg/l cause complaints. Iron in excess of 1.0 mg/l will cause an unpleasant taste. A concentration of about 1 mg/l is noticeable in the taste of coffee or tea. Chlorine will precipitate soluble iron. Iron is an essential element for human health.

Manganese

Manganese concentrations should be less than 0.05 mg/l to avoid the black-brown staining of plumbing and clothes, although soluble manganese bound to organic matter may be present in considerably higher concentrations without producing difficulties. Concentrations greater than 0.5 to 1.0 mg/l may give a metallic taste to water. Concentrations above 0.05 mg/l can sometimes build up coatings in piping which can slough off causing staining of laundry and brown-black precipitate. Sodium hexametaphosphate should not be used to prevent precipitation of manganese when the manganese exceeds 1 mg/l. Excess polyphosphate may prevent absorption of essential trace elements from the diet[39]; it is also a source of sodium.

Sodium

The American Heart Association's 500-mg- and 1000-mg-sodium-per-day diet recommends that distilled water be used if the water supply contains more than 20 mg/l of sodium. The consumption of 2.5 liters of water per day is as-

[39]Richard J. Bull and Gunther F. Craun, "Health Effects Associated With Manganese in Drinking Water," *J. Am. Water Works Assoc.*, December 1977, pp. 662–663.

sumed. Water containing more than 270 mg/l sodium should not be used for drinking by those on a moderately restricted sodium diet. Many groundwater supplies and most home-softened (using ion exchange) well waters contain too much sodium for persons on sodium-restricted diets. If the well water is low in sodium (less than 20 mg/l sodium) and the water is softened by the ion-exchange process because of excessive hardness, the house cold-water system can be supplied by a line from the well that bypasses the softener and low-sodium water can be made available at cold-water taps. A laboratory analysis is necessary to determine the exact amount of sodium in water. Persons suffering from heart, kidney, or circulatory illnesses should be guided by their physicians' advice. A maximum drinking water standard of 100 mg/l has been proposed for the general population. Common sources of sodium, in addition to food, are certain well waters, ion-exchange water softening units, water treatment chemicals, and road salt.[40]

Sulfates

Sulfates should not exceed 250 mg/l, although this would have to be modified with zeolite softening where calcium sulfate or gypsum is replaced by an equal concentration of sodium sulfate. Sodium sulfate (or Glauber salts) in excess of 200 mg/l, magnesium sulfate (or Epsom salts) in excess of 390 mg/l, and calcium sulfate in excess of 600 to 800 mg/l are laxative to those not accustomed to the water. Magnesium sulfate causes hardness; sodium sulfate causes foaming in steam boilers. Concentrations of 300 to 400 mg/l cause a taste. Sulfate can be removed by distillation, reverse osmosis, or electrodialysis. Sulfates are found in surface waters receiving industrial wastes such as those from sulfate pulp mills, tanneries, and textile plants. Sulfates also occur in many waters as a result of leaching from gypsum-bearing rock.

Total Dissolved Solids (TDS)

The total solid content should be less than 500 mg/l; however, this is based on the industrial uses of public water supplies and not on public health factors. Higher concentrations may cause physiological effects and taste. Dissolved solids can be removed by distillation, reverse osmosis, electrodialysis, or ion exchange. Water with more than 1000 mg/l of dissolved solids is classed as "saline" irrespective of the nature of the minerals present.[41]

Fluorides

Fluorides are found in many groundwaters as a natural constituent, ranging from a trace to 5 or more mg/l. Fluorides in concentrations greater than 3 mg/l can cause the teeth of children to become mottled and discolored, de-

[40]Edward J. Calabrese and Robert W. Tuthill, "Sources of Elevated Sodium Levels in Drinking Water . . . and Recommendations for Reduction," *J. Environ. Health*, November/December 1978, pp. 151–155.

[41]C. Richard Murray and E. Bodette Reeves, "Estimated Use of Water in the United States in 1975," *Geological Survey Ciruclar 765*, U.S. Dept. of the Interior.

pending on the concentration and amount of water consumed. Drinking water containing 0.8 to 1.7 mg/l natural or added fluoride is beneficial to children during the period they are developing permanent teeth. An optimum level is 1.0 mg/l in temperate climates. The incidence of dental cavities or tooth decay is reduced by about 60 percent. See also Dental Caries, Chapter 1. The maximum permissible concentration in drinking water has been established in the National Interim Primary Drinking Water Regulations at 1.4 to 2.4 mg/l depending on the annual average of the maximum daily air temperatures for the location in which the community water system is located. See Table 3-3. Fluoride removal methods include ion exchange, reverse osmosis, lime softening, and activated alumina and tricalcium phosphate adsorption. It is not possible to reduce the fluoride level to 1 mg/l using only lime.[42]

Methylene Blue Active Substances (MBAS)

The test for MBAS also shows the presence of alky benzene sulfonate (ABS), linear alkylate sulfonate (LAS), and related materials that react with methylene blue. It is a measure of the apparent detergent present. The composition of detergents varies. Household wash water in which ABS is the active agent in the detergent may contain 200 to 1000 mg/l. ABS has been largely replaced by LAS, which can be degraded under aerobic conditions; if not degraded it too will foam. Both ABS and LAS detergents contain phosphates that may fertilize plant life in lakes and streams. The decay of plants will use oxygen, leaving less for fish life and wastewater oxidation. Because of these effects, the use of detergents containing phosphates have been banned in some areas. In any case, the presence of MBAS in a well-water supply is an indication of contamination, the source of which should be identified and removed, even though it has not been found to be of health significance in the concentrations found in drinking water. Carbon adsorption can be used to remove MBAS from drinking water.

Oxygen-Consumed Value

This represents organic matter that is oxidized by potassium permanganate under the test conditions. Pollution significant from a bacteriological examination standpoint is accompanied by so little organic matter as not to significantly raise the oxygen-consumed value. For example, natural waters containing swamp drainage have much higher oxygen-consumed values than water of low original-organic content that are subject to bacterial pollution. This test is of limited significance.

Free Ammonia

Free ammonia represents the first product of the decomposition of organic matter; thus appreciable concentrations of free ammonia usually indicate "fresh pollution" of sanitary significance. The exception is when ammonium

[42]Won-Wook Choi and Kenneth Y. Chen, "The Removal of Fluoride From Waters by Adsorption," *J. Am. Water Works Assoc.*, October 1979, pp. 562–570.

sulfate of mineral origin is involved. The following values may be of general significance in appraising free ammonia content: low—0.015 to 0.03 mg/l; moderate—0.03 to 0.10 mg/l; high—0.10 mg/l or greater. Special care must be exercised to allow for ammonia added if the "chlorine-ammonia" treatment of water is used or if crenothrix organisms are present. Ammonia in the range of 0.2 to 2.0 mg/l is toxic to many fish; 0.2 mg/l for rainbow trout.

Albuminoid Ammonia

Albuminoid ammonia represents "complex" organic matter and thus would be present in relatively high concentrations in water supporting algae growth, receiving forest drainage, or containing other organic matter. Concentrations of albuminoid ammonia higher than about 0.15 mg/l, therefore, should be appraised in the light of origin of the water and the results of microscopic examination. In general, the following concentrations serve as a guide: Low—less than 0.06 mg/l; moderate—0.06 to 0.15 mg/l; high—0.15 mg/l or greater.

Nitrites

Nitrites represent the first product of the oxidation of free ammonia by biochemical activity. Unpolluted natural waters contain practically no nitrites, so concentrations exceeding the very low value of 0.001 mg/l are of sanitary significance, indicating water subject to pollution that is in the process of change associated with natural purification. The nitrite concentration present is due to the organic matter in the soil through which the water passes. Nitrites in concentrations greater than 1 mg/l in drinking water are hazardous to infants and should not be used for infant feeding.

Nitrates

Nitrates represent the final product of the biochemical oxidation of ammonia. Its presence is probably due to the presence of nitrogenous organic matter of animal and, to some extent, vegetable origin, for only small quantities are naturally present in water. Septic tank systems may contribute nitrates to the groundwater. Manure and fertilizer contain large concentrations of nitrates; thus the existence of fertilized fields or cattle feedlots near sources of supply must be carefully considered in appraising the significance of nitrate content. Furthermore, a cesspool may be relatively close to a well and be contributing pollution without a resulting high-nitrate content, because the anaerobic conditions in the cesspool would prevent biochemical oxidation of ammonia to nitrates. In fact, nitrates may be reduced to nitrites under such conditions. In general, however, nitrates disclose the evidence of "previous" pollution of water that has been modified by self-purification processes to a final mineral form. Allowing for these important controlling factors the following ranges in concentration may be used as a guide: low—less than 0.1 mg/l; moderate—0.1 to 1.0 mg/l; high—greater than 1.0 mg/l.

The presence of more than 10 mg/l of nitrate expressed as nitrogen, the maximum contaminant level in drinking water, appears to be the cause of methemoglobinemia or "blue babies." The standard was formerly expressed as 45 mg/l as nitrate. Methemoglobinemia is largely a disease confined to infants less than three months old, but may affect children up to age six. The boiling of water containing nitrates would increase the concentration of nitrates in the water. See also Methemoglobinemia in Chapter 1.

Nitrates may stimulate the growth of water plants, particularly algae if other nutrients such as phosphorous and carbon are present. Gould[43] and others point out that

> a more objective review of literature would perhaps indicate that without any sewage additions most of our waterways would contain enough nitrogen and phosphorous to support massive algal blooms and that the removal of these particular elements would have little effect on existing conditions.

See also Control of Microorganisms in this chapter and Eutrophication in Chapter 4.

The feasible methods for the removal of nitrates are ion exchange, reverse osmosis, and electrodialysis. Ion exchange is the only method in use (Nassau County, N.Y., since 1974).

Hydrogen Sulfide

Hydrogen sulfide is most frequently found in groundwaters as a natural constituent and is easily identified by a rotten-egg odor. It is caused by microbial action on organic matter or the reduction of sulfate ions to sulfide. A concentration of 70 mg/l is an irritant; but 700 mg/l is highly poisonous. In high concentration it paralyzes the sense of smell, thereby making it more dangerous. Black stains on laundered clothes and black deposits in piping and on plumbing fixtures are caused by hydrogen sulfide in the presence of soluble iron. Hydrogen sulfide in drinking water should not exceed 0.05 mg/l.

Phosphorus

High phosphorus concentrations together with nitrates and organic carbon are sometimes associated with heavy aquatic plant growth, although other substances in water also have an effect. Uncontaminated waters contain 10 to 30 μg/l total phosphorus, although higher concentrations of phosphorus are also found in "clean" waters. Concentrations associated with nuisances in lakes would not normally cause problems in flowing streams. About 100 μg/l complex phosphates interfere with coagulation.[44] Phosphorus from septic-tank subsurface absorption system effluents is not readily transmitted through sandy soil and the groundwater.[45] Most waterways naturally contain sufficient

[43]Richard H. Gould, "Growing data disputes algal treatment standards," *Water Wastes Eng.*, May 1978, pp. 78–83.

[44]National Technical Advisory Committee Report "Raw-Water Quality Criteria for Public Water Supplies," *J. Am. Water Works Assoc.*, **61,** 133 (March 1969).

[45]Rebecca A. Jones and G. Fred Lee, "Septic-Tank Disposal Systems as Phosphorus Sources for Surface Waters," Richardson Inst. for Environ. Sci., Texas Univ. at Dallas, Nov. 1977.

nitrogen and phosphorus to support massive algal blooms. See *Nitrates*, this section.

Mercury

Episodes associated with the consumption of methylmercury-contaminated fish, bread, pork, and seed have called attention to the possible contamination of drinking water. Mercury is found in nature in the elemental and organic form. Concentrations in unpolluted waters are normally less than 1.0 $\mu g/l$. The organic methylmercury and other alkylmercury compounds are highly toxic, affecting the central nervous system and kidneys. The maximum permissible contaminant level in drinking water is 2.0 $\mu g/l$ (0.002 mg/l) as total mercury. See also Mercury Poisoning, Chapter 1.

Arsenic

Arsenic is sometimes found in drinking water. Sources of arsenic are natural rock formations (phosphate rock), industrial wastes, arsenic pesticides, fertilizers, and detergent "presoaks," and possibly other detergents. It is also found in foods, including shellfish, tobacco, and in the air in some locations. There appears to be a relationship between skin cancer and high levels of arsenic in drinking water. Arsenic in elemental form is not considered particularly toxic although continual ingestion of 0.3 mg/l increased the incidence of skin cancer.[46] Arsenites are more toxic than arsenates. Arsenic may be converted to dimethylarsine by anaerobic organisms and accumulate in fish, similar to methylmercury.[47] OSHA has set a standard of 10 $\mu g/m^3$ for occupational exposure to inorganic arsenic in air over an 8-hr day. The concentration in drinking water should not exceed 0.05 mg/l. Arsenic removal is discussed later in this chapter.

Barium

Barium may be found naturally in groundwater and in surface water receiving industrial wastes; it is also found in air. It is a muscle stimulant and in large quantities may be harmful to the nervous system and heart. The fatal dose is 550 to 600 mg. The level should not exceed 1 mg/l in drinking water.

Selenium

Selenium is associated with industrial pollution (copper smelting) and vegetation grown in soil containing selenium. It is found in meat and other foods. Selenium causes cancers and sarcomas in rats fed heavy doses.[48] Chronic exposure to excess selenium results in gastroenteritis, dermatitis, and central nervous system disturbance.[49] Selenium is considered an essential nutrient

[46]*Quality Criteria for Water*, USEPA, Washington, D.C. 20460, July 26, 1976, p. 27; (Chen and Wu, 1962; Tseng, et al., 1968; Yeh, et al., 1968; and Trelles, et al., 1970.)

[47]M. M. Varma, Steven G. Serdahely, and Herbert M. Katz, "Physiological Effects of Trace Elements and Chemicals in Water," *J. Environ. Health*, September/October 1976, pp. 90–100.

[48]Henry A. Schroeder, "Environmental Metals: The Nature of the Problem," *Environmental Problems in Medicine*, William D. McKee, Ed., Charles C. Thomas, Springfield, Illinois, 1974.

[49]"Drinking Water and Health, Recommendations of the National Academy of Sciences," *Fed. Reg.* 42:132:35764, July 11, 1977.

and may provide some protection against certain types of cancer. Selenium in drinking water should not exceed 0.01 mg/l.

Cadmium

The federal drinking water maximum contaminant level for cadmium is 0.01 mg/l. Common sources of cadmium are water mains and galvanized iron pipes, tanks, metal roofs where cistern water is collected, industrial wastes (electroplating), tailings, pesticides, nickel plating, solder, incandescent light filaments, photography wastes, and cadmium in paints, plastics, inks; also nickel-cadmium batteries, and cadmium-plated utensils. It is also found in zinc and lead ores. Cadmium vaporizes when burned and salts of cadmium readily dissolve in water and can therefore be found in air pollutants, wastewater, wastewater sludge, land runoff, some food crops, tobacco, and drinking water. Beef liver and shellfish are very high in cadmium. Large concentrations may be related to kidney damage, hypertension (high blood pressure), chronic bronchitis, and emphysema. Cadmium builds up in the human body. The direct relationship between cardiovascular death rates in the United States, Great Britain, Sweden, Canada, and Japan and the degree of softness or acidity of water points to cadmium as the suspect.[50] The Joint WHO Food and Agriculture Organization Expert Committee on Food Additives set in 1972 a provisional tolerable weekly intake of 400 to 500 μg. Cadmium removal from water is discussed later in this chapter.

Chromium

Chromium should not exceed 0.05 mg/l in drinking water. Chromium is found in cigarettes, in some foods, in the air, and in industrial wastes. Chromium deficiency is associated with atherosclerosis. Hexavalent chromium dust can cause cancer of the lungs.[48]

Aluminum

Aluminum is not found naturally in the elemental form although it is one of the most abundant metals on the earth's surface. It is found in all soils, plants, and animal tissues. Aluminum-containing wastes concentrate in and can harm shellfish and bottom life.[51] Alum as aluminum sulfate is commonly used as a coagulant in water treatment. Precipitation may take place in the distribution system or on standing when the water contains more than 0.5 mg/l. Its presence in filter plant effluent is used as a measure of filtration efficiency.

Cyanide

Cyanide is found naturally and in industrial wastes. Cyanide concentrations as low as 10 μg/l have been reported to cause adverse effects on fish. Long-

[50] Henry A. Schroeder, *New York State Environment*, N.Y. State Dept. of Environmental Conservation, Albany, N.Y., April 1976.

[51] *Water Quality Criteria 1972*, National Academy of Sciences—National Academy of Engineering, Washington, D.C., 1972, p. 87.

term consumption of up to nearly 5 mg/day has shown no injurious effects.[52] The cyanide concentration in drinking water should not exceed 0.05 mg/l. Cyanide is readily destroyed by conventional treatment processes.[53]

Asbestos

Most diseases associated with asbestos have been related to breathing air containing asbestos fibers over a period of 20 years or more, such as working or living in the immediate vicinity of crocidolite mines, asbestos textile factories, and shipyards. See Chapter 1. A study (1935 to 1973) on the incidence of gastrointestinal cancer and use of drinking water distributed through asbestos cement (A/C) pipe reached the preliminary conclusion that "no association was noted between these asbestos risk sources and gastrointestinal tumor incidence.[54] A subsequent study concluded "the lack of coherent evidence for cancer risk from the use of A/C pipe is reassuring.[55]

Asbestos-cement pipe was found to behave much like other piping materials, except PVC, that are commonly used for the distribution of drinking water. It has been concluded that where "aggressive water conditions exist the pipe will corrode and deteriorate; if aggressive water conditions do not exist, the pipe will not corrode and deteriorate."[56] The American Water Works Association (AWWA) Standard C400-77 establishes criteria for type of pipe to use for nonaggressive water ($\geqslant 12.0$), moderately aggressive water (10.9–11.9), and highly aggressive water ($\leqslant 10.0$) based on the sum of the pH plus the log of the alkalinity times the calcium hardness, as calcium carbonate. It is believed prudent, based on available information, not to use asbestos-cement pipe to carry aggressive water.

Proper coagulation and filtration will remove asbestos fibers from a raw water containing asbestos.

Silver

The maximum allowable concentration of silver in drinking water is 0.05 mg/l. Silver is sometimes used to disinfect small quantities of water and in home faucet "purifiers." Colloidal silver may cause permanent discoloration of the skin, eyes, and mucous membranes, but the precise concentration needed to cause such effects is not known.

Specific Conductance

Specific conductance is a measure of the ability of a water to conduct an electrical current and is expressed in micromhos per centimeter at 25°C. Be-

[52] *Quality Criteria for Water*, USEPA, Washington, D.C., July 26, 1976, pp. 128–136.

[53] *International Standards for Drinking Water*, WHO, 1971, p. 33.

[54] J. Malcolm Harrington, et al., "An Investigation of the Use of Asbestos Cement Pipe for Public Water Supply and the Incidence of Gastrointestinal Cancer in Connecticut, 1935–1973," *Am. J. Epidemiology*, Vol. 107, No. 2, 1978, pp. 96–103.

[55] J. Wister Meigs, et al., "Asbestos Cement Pipe and Cancer In Connecticut 1955–1974," *J. Environ. Health*, January/February, 1980, pp. 187–191.

[56] Ralph W. Buelow, et al., "The Behavior of Asbestos-Cement Pipe Under Various Water Quality Conditions: A Progress Report." *J. Am. Water Works Assoc.*, February 1980, pp. 91–102.

cause the specific conductance is related to the number and specific chemical types of ions in solution, it can be used for approximating the dissolved-solids content in the water, particularly the mineral salts in solution if present. Commonly, the amount of dissolved solids (in milligrams per liter) is about 65 percent of the specific conductance. This relation is not constant from stream to stream or from well to well, and it may even vary in the same source with changes in the composition of the water. Specific conductance is used for the classification of irrigation waters. In general, waters of less than 200 micromhos/cm are considered acceptable, and conductance in excess of 300 micromhos/cm unsuitable. Good fresh waters for fish in the United States are reportedly under 1100 micromhos/cm.[57] Wastewater with a conductivity up to 1200 to 4000 micromhos/cm may be acceptable for desert reclamation.

Radioactivity

The maximum contaminant levels for radioactivity in drinking water are given in Table 3-5. The exposure to radioactivity from drinking water is not likely to result in a total intake greater than recommended by the Federal Radiation Council. Naturally occurring radionuclides include thorium-232, uranium-235, and uranium-238, and their decay series. They may be found in well waters, especially those near uranium deposits. (Radium is sometimes found in certain spring supplies). Since these radionuclides emit alpha and beta radiation (as well as gamma), their ingestion or inhalation may introduce a serious health hazard, if found in well-water supplies.[58] Possible man-made sources of radionuclides include fallout (in soluble form and with particulate matter) from nuclear explosions, in precipitation and runoff, releases from nuclear reactors and from manufacturers, and radionuclides. Further discussion can be found in Chapter 7.

Uranyl Ion

This ion may cause damage to the kidneys. Objectionable taste and color occur at about 10 mg/l. It does not occur naturally in most waters above a few micrograms per liter. The taste, color, and gross alpha maximum contaminant level will restrict uranium concentrations to below toxic levels, hence no specific limit is proposed.[59]

Phenols

These should not exceed 0.001 mg/l. In combination with chlorine a medicinal taste is produced in drinking water. The presence of phenols can cause serious problems in the food and beverage industries.

[57]Howard B .Brown, *The Meaning, Significance, and Expression of Commonly Measured Water Quality Criteria and Potential Pollutants*, Louisiana State University and Agricultural and Mechanical College, Baton Rouge, La., 1957, pp. 35–36.

[58]Louis J. Kosarek, "Radionuclides removal from water," *Environ. Sci. Technol.*, May 1979, pp. 522–525.

[59]*Water Quality Criteria 1972*, National Academy of Sciences—National Academy of Engineering, EPA·R3·73·033, March 1973, Washington, D.C., p. 91.

Carbon-Chloroform Extract (CCE) and Carbon-Alcohol Extract (CAE)

CCE may include chlorinated hydrocarbon pesticides, nitrates, nitrobenzenes, aromatic ethers, and many others. Water from uninhabited and nonindustrial watersheds usually show CCE concentrations of less than 0.04 mg/l. The taste and odor of drinking water can be expected to be poor when the concentration of CCE reaches 0.2 mg/l. Carbon-alcohol extract measures gross organic chemicals including synthetics. A standard of 1.0 to 3.0 mg/l has been proposed.

Pesticides

Pesticides include insecticides, herbicides, fungicides, rodenticides, regulators of plant growth, defoliants, or desiccants. Sources of pesticides in drinking water are industrial wastes, spills and dumping of pesticides, and runoff from fields, inhabited areas, farms, or orchards treated with pesticides. Surface and groundwater may be contaminated. Conventional water treatment does not adequately remove pesticides. Maximum permissible contaminant levels of certain pesticides in drinking water are given in Table 3-3.

Polynuclear Aromatic Hydrocarbons

Polynuclear aromatic hydrocarbons such as fluoranthene, 3,4-benzfluoranthene, 11,12-benzfluoranthene, 3,4-benzpyrene, 1,12-benzperyline, and indeno [1,2,3-cd] pyrene are known carcinogens and are potentially hazardous to man. The WHO has set a limit of 0.2 μg/l of these chemicals in drinking water pending further investigation of their significance.[60]

Polychlorinated Biphenyls (PCBs)

PCBs give an indication of the presence of industrial wastes containing mixtures of chlorinated byphenyl compounds having various percentages of chlorine. Organochlorine pesticides have similar chemical structure. PCBs cause skin disorders in humans and cancer in rats. They are stable, fire resistant, and have good electrical insulation capabilities. They are used in transformers, brake linings, canvas waterproofing, and other products. PCBs are not soluble in water but they do cumulate in bottom sediment and in fish and other animals on a steady diet of food contaminated with the chemical. Concentrations up to several hundred and several thousand mg/l have been found in fish, snapping turtles, and other aquatic life. PBB, a derivative of PCB, is more toxic than PCB. The FDA action levels are 1.5 mg/l in fat of milk and dairy products; 0.3 mg/l in eggs; 2 mg/l in fish and shellfish. An interim guideline for drinking water is 1 ppb (1 μg/l) with 0.00 μg/l as the USEPA goal. A maximum concentration of 0.002 μg/l is suggested to protect aquatic life.[61] PCBs are destroyed at 2000°F and 3 percent excess oxygen for 2 sec contact time.

[60] *International Standards for Drinking Water*, WHO, 1971, p. 37.

[61] *Water Quality Criteria 1972*, National Academy of Sciences—National Academy of Engineering, EPA·R3·73·033, March 1973, Washington, D.C., p. 177.

Halogenated Chloro-organic Compounds (Trihalomethanes)

Trihalomethanes (THMs) are believed to be formed by the interaction of chlorine with humic and fulvic substances and other precursors produced either by normal organic decomposition or by metabolism of aquatic biota. THMs include chloroform (trichloromethane), bromoform (tribromomethane), dibromochloromethane, bromodichloromethane, and iodoform (dichloroiodomethane). Toxicity, mutagenicity, and carcinogenicity have been suspected as being associated with the ingestion of trihalomethanes. The USEPA has stated that

> epidemiological evidence relating THM concentrations or other drinking water quality factors and cancer morbidity-mortality is not conclusive but suggestive. Positive statistical correlations have been found in several studies,* but causal relationships cannot be established on the basis of epidemiological studies. The correlation is stronger between cancer and the brominated THMs than for chloroform.[62,63]

Chloroform is reported to be carcinogenic to rats and mice in high doses. The Epidemiology Subcommittee of the National Research Council says that cancer and THM should not be linked.[64] However, the National Drinking Water Advisory Council, based on studies in the review and evaluation by the National Academy of Sciences, the work done by the National Cancer Institute, and other research institutions within the USEPA, has accepted the regulation of trihalomethanes on "the belief that chloroform in water does impose a health threat to the consumer."[65] A standard of 100 μg/l for total THM has been established by the EPA for public water supplies serving 10,000 or more people. For further discussion of causes of THM formation, its control, and removal see Prevention and Removal of Organic Chemicals, and Synthetic Organic Chemicals—Halogenated Chloro-organic Compounds (Products of Chlorination) this chapter.

Other Synthetic Organic Chemicals

Synthetic organic chemicals (SOCs) are found in dry cleaning compounds, metal degreasers, and other solvents including household drain cleaners, septic tank cleaners, paint removers, and in certain industrial wastes. They can percolate into the groundwater from spills, discharges onto the ground surface, septic tank and cesspool systems, sanitary landfills, and industrial waste dump leachates, and then appear in drinking water. Many of the compounds are animal carcinogens and suspected human carcinogens and hence are considered hazardous to human health. In the absence of specific standards, the New York State Department of Health established guidelines for maximum

*The reliability and accuracy of studies such as these are often subject to question.

[62]USEPA Statement, "Chlorinated and Brominated Compounds are not equal," *J. Am. Water Works Assoc.*, October 1977, p. 12.

[63]Joseph A Cotruvo and Chieh Wu, *J. Am. Water Works Assoc.*, November 1978, pp. 590–594.

[64]"Update," *J. Am. Water Works Assoc.*, December 1978, p. 9.

[65]Margaret Gaskie and Charles C. Johnson Jr., "Face to Face," *J. Am. Water Works Assoc.*, August 1979, p. 16.

permissible concentrations of SOCs in drinking water. These include 50 ppb of any one SOC, 100 ppb total of any mix of SOCs, 5 ppb benzene, 5 ppb vinyl chloride, 1 ppb PCB, with other compounds considered on a case by case basis.

Quantity

The quantity of water used for domestic purposes will generally vary directly with the availability of the water, habits of the people, cost of water, number and type of plumbing fixtures provided, water pressure, air temperature, newness of a community, types of establishments, metering, and other factors. Wherever possible, the actual water consumption under existing or similar circumstances and the number of persons served should be the basis for the design of a water and sewerage system. Special adjustment must be made for industrial use. The average per capita water use has increased from 150 gpd in 1960 to 168 gpd in 1975. Included is water lost in the distribution system and water supplied for fire fighting, street washing, municipal parks, and swimming pools. The per capita use for rural domestic use is about 66 gpd.[66]

Table 3-12 gives estimates of water consumption at different types of places. Additions should be made for car washing, lawn sprinkling, and miscellaneous uses. If provision is made for fire-fighting requirements, then the quantity of water provided for this purpose to meet fire underwriters' standards will be in addition to that required for normal domestic needs in small communities.

Water Conservation

Water conservation can effect considerable saving of water with resultant reduction in water treatment and pumping costs, and wastewater treatment. With water conservation, development of new sources of water and treatment facilities can be postponed or made unnecessary, and low distribution system water pressure situations are less likely.

Water conservation can be accomplished, where needed, by a continuing program of leak detection in the community distribution system and in buildings; use of low water-use valves and plumbing fixtures; water pressure and flow control in the distribution system, and in building services (orifices); universal metering and price adjustment; conservation practices by the consumer; and a rate structure that encourages conservation.

Leak detection activities would include metering water use and water production balance studies; routine leak detection surveys of the distribution system; investigation of water ponding or seepage reports and complaints; and reporting and prompt follow-up on leaking faucets, running flushometer valves and water closet ball floats, and other valves. Universal metering will

[66]C. Richard Murray and E. Bodette Reeves, "Estimated Use of Water in the United States in 1975," *Geological Survey Circular 765*, U.S. Dept. of Interior.

Table 3-12 Guides for Water Use in Design

Type of Establishment	gpd[a]
Residential	
Dwellings and apartments (per bedroom)	150
Rural	48
Suburban	80
Urban	170
Temporary quarters	
Boarding houses	65
Additional (or nonresident boarders)	10
Campsites (per site), recreation vehicle with individual connection	100
Campsites, recreation vehicle with individual connection	40 to 50
Camps without WCs, baths, or showers	5
Camps with WCs but without baths or showers	25
Camps with WCs and bathhouses	35 to 50
Cottages, seasonal with private bath	50
Day camps	15 to 20
Hotels	65 to 75
Mobile home parks (per unit)	125 to 150
Motels	50 to 75
Public establishments	
Restaurants (toilets and kitchens)	7 to 10
Without public toilet facilities	2½ to 3
With bar or cocktail lounge, additional	2
Schools, boarding	75 to 100
Day with cafeteria, gymnasium, and showers	25
Day with cafeteria, but without gymnasium and shower	15
Hospitals (per bed)	250 to 500
Institutions other than hospitals (per bed)	75 to 125
Places of public assembly	3 to 10
Turnpike rest areas	5
Turnpike service areas (per 10 percent of cars passing)	15 to 20
Amusement and commercial	
Airports (per passenger)	3 to 5
Country clubs, excluding residents	25
Day workers (per shift)	15 to 35
Drive-in theaters (per car space)	5
Gas station (per vehicle serviced)	10
Milk plant, pasteurization (per 100 lb of milk)	11 to 25
Movie theaters (per seat)	3
Picnic parks with flush toilets	5 to 10
Self-service laundries (per machine) (or 50 gal per customer)	400
Shopping center (per 1,000 ft^2 floor area)	250
Stores (per toilet room)	400
Swimming pools and beaches with bathhouses	10
Fairgrounds (based on daily attendance), also sports arenas	1 to 2

Table 3-12 (*Continued*)

Type of Establishment	gpd[a]
Farming (per animal)	
Cattle or Steer	12
Milking cow	35
Goat or Sheep	2
Hog	4
Horse or Mule	12
Cleaning milk equipment and tank	2
Cow washer, milking center	5 to 10
Liquid manure handling, cow	1 to 3
Poultry (per 100)	
Chickens	5 to 10
Turkeys	10 to 18
Cleaning and sanitizing equipment	4

Miscellaneous Water Use Estimates	Water Use	
Home	Gallons	
Water closet, tank	4 to 6 per use	(3½ per use)[b]
Water closet, flush valve 25 psi (pounds per square inch)	30 to 40/min	(3½ per use)[b]
Washbasin	1½/use	(3 gpm)
Bathtub	30/use	
Shower	10 to 30/use	(3 gpm)[b]
Dishwashing machine, domestic	9½ to 15½/load	
Garbage grinder	1 to 2/day	
Automatic laundry machine, domestic	34 to 57/load—Top load	
	22 to 33/load—Front load	
Garden hose		
⅝ in., 25-ft head	200/hr	
¾ in., ¼ in. nozzle, 25-ft head	300/hr	
Lawn sprinkler	120/hr	
3000 square ft lawn, 1 in. per week	1850/wk	
Air conditioner, water-cooled, 3-ton, 8 hr per day	2880/day	

Household Water Use:	Percent
Toilet flushing	40
Bathing	30
Drinking and cooking	5
Dishwashing	6
Clothes washing	15
Cleaning and miscel.	4

Water Demand per Dwelling Unit	Water Use (gpd)
Average day	400
Maximum day	800
Maximum hourly rate	2000
Maximum hourly rate with appreciable lawn watering	2800

Table 3-12 (*Continued*)

Miscellaneous Water Use Estimates	Water Use			
	Bedrooms			
Home Water System (Minimums)	2	3	4	5
Pump capacity, gal/hr	250	300	360	450
Pressure tank, gal minimum	42	82	82	120
Service line from pump, diameter in in.[c]	$\frac{3}{4}$	$\frac{3}{4}$	1	$1\frac{1}{4}$

Other Water Use	Gallons
Fire hose, $1\frac{1}{2}$ in., $\frac{1}{2}$ in. nozzle, 70-ft head	2400/hr
Drinking fountain, continuous flowing	75/hr
Dishwashing machine, commercial	
Stationary rack type, 15 psi	6 to 9/min
Conveyor type, 15 psi	4 to 6/min
Fire hose, home, 10 gpm at 60 psi for 2 hours, $\frac{3}{4}$ in.	600/hr
Restaurant, average	35/seat
Restaurant, 24-hr	50/seat
Restaurant, tavern	20/seat
Gas station	500/set of pumps

Developing areas of the world

One well or tap/200 persons; controlled tap or hydrant—Fordilla or Robovalve type
Average consumption, 5 gal/capita/day at well or tap
Water system design, 30 gal/capita/day (10 gal/capita is common)
Pipe size, 2 in. and preferably larger (1 and $1\frac{1}{2}$ in. common)
Drilled well, cased, 6 to 8 in. diameter
Water system pressure, 20 lb/sq in.
(Keep mechanical equipment to a minimum)

	Per capita per day	
Water Consumption in Rural Areas	(Liters)	(Gallons)
Africa	15 to 35	4 to 9
Southeast Asia	30 to 70	8 to 19
Western Pacific	30 to 90	8 to 24
Eastern Mediterranean	40 to 85	11 to 23
Europe (Algeria, Morocco, Turkey)	20 to 65	5 to 17
Latin America and Caribbean	70 to 190	19 to 51
World Average	35 to 90	9 to 24

(Assumes hydrant or handpump available within 200 meters; 70 Lpcd or more could mean house or central courtyard outlet.)

[a]Per person unless otherwise stated.

[b]Water conservation fixtures at 60 psi, maximum flow for test purposes. Lower pressure would reduce usage further.

[c]Service lines less than 50 ft long, brass or copper. Use next larger size if iron pipe is used. Use minimum $1\frac{1}{4}$ in. service with flush valves. Minimum well yield, 5 gal/min.

make possible water balance studies to help detect lost water and also to provide a basis for charging for water use. Reduction in water use, however, may be temporary in some instances; the affluent will probably not be affected.

Low water-use plumbing fixtures and accessories would include the low-flush ($3\frac{1}{2}$ gal) water closets;* water-saving shower head flow controls, spray taps and faucet aerators; water-saving clothes washers and dishwashers. In a dormitory study at a state university, the use of flow control devices (pressure level) on shower heads effected a 40 to 60 percent reduction in water use as result of reducing the shower head flow rates from 5.5 gpm to 2.0 to 2.5 gpm.[67] Plumbing codes should require water-saving fixtures and pressure control in new structures and in rehabilitation projects. For example, only water-efficient plumbing fixtures meeting the following standards are permitted to be sold or installed in New York State.[68]

Sink and lavatory faucets, maximum flow 3 gpm at 60 psi.

Shower heads, 3 gpm at 60 psi.

Urinals and associated flush-valve, if any, not greater than $1\frac{1}{2}$ gal of water per flush.

Toilets and associated flush-valve, if any, not greater than $3\frac{1}{2}$ gal of water per flush.

Special fixtures such as safety showers and aspirator faucets are exempt, and the commissioner may permit use of fixtures not meeting standards if necessary for proper operation of the existing plumbing or sewer system. One might also add to the list of water conservation possibilities, where appropriate, use of the compost toilets, recirculating toilets, chemical toilets, incinerator toilets, and the various privies.

Pressure-reducing valves in the distribution system (pressure zones of 20 to 60 psi) to maintain a minimum water pressure of 20 to 40 psi at fixtures will reduce water use. Adjust the pressure switch on home water systems to operate the pump in the 20 to 40 psi range. The potential water saving through pressure control is apparent from the basic hydraulic formulas† $Q = VA$, $Q = (2gp/w)^{1/2} \times A$, and $Q = (2gh)^{1/2} \times A$, which show that the quantity of water flowing through a pipe varies with the velocity or the square root of the pressure head.

The success of water-use conservation depends also largely on the extent to which the consumer is motivated. He can be encouraged to have leaking faucets and running toilets immediately repaired; to not waste water; to understand that a leak causing a $\frac{1}{8}$ in. diameter stream adds up to 400 gallons in 24

*Very low flush-volume toilets using 3 to 6 liters (0.8 to 1.6 gal) per flush are also available.

†Q = cfs, V = fps, A = ft^2, g = ft per sec per sec, p = lb/ft^3, w = lb/ft^2 (62.4), h = ft of water.

[67]William E. Sharpe, "Water and Energy Conservation With Bathing Shower Flow Controls," *J. Am. Water Works Assoc.*, February 1978, pp. 93–97.

[68]Environmental Conservation Law, Section 15-0314, Albany, New York, 1979. The Washington Suburban Sanitary District plumbing code has similar requirements. (Robert S. McGarry and Jim M. Brusnighan, "Increasing Water and Sewer Rate Schedules: A Tool For Conservation," *J. Am. Water Works Assoc.*, September 1979, pp. 474–479.)

hours which is about the amount of water used by a family of five or six in one day; to purchase a water-saving clothes washer and dishwasher; to add two quart bottles or a "dam" to the flush tank to see if the closet still flushes properly; to install water-saving shower heads and to not use the tub; to install mixing faucets with single lever control; install aerators on faucets. Consumer education and motivation must be a continual activity. In some instances reuse of shower, sink, and laundry wastewater for gardens is feasible.[69,70]

Water Reuse

Discussion of water reuse should clearly distinguish between direct reuse and indirect reuse. In *direct reuse*, the additional wastewater treatment (such as storage, coagulation, flocculation, sedimentation, sand or anthracite filtration or granular activated carbon filtration, and disinfection) is usually determined by the specific reuse. (See Figure 4-33.) The wastewater is reclaimed for *nonpotable* purposes such as industrial process or cooling water, agricultural irrigation, groundwater recharge, desert reclamation, fish farming, lawn and park watering, landscape and golf course watering, and toilet flushing. See Chapter 4, Wastewater reuse. The treated wastewater must *not* be used for drinking, culinary, bathing, or laundry purposes. In *indirect reuse*, wastewater receiving various degrees of treatment is discharged to a surface water or a groundwater aquifer where it is diluted and after varying detention periods may become a source of water for potable purposes, after suitable treatment.

Direct municipal wastewater reuse, where permitted, would require a dual water system; one carrying potable water and the other reclaimed wastewater for toilet flushing, lawn watering and other nonpotable purposes. The reclaimed water is usually bacteriologically safe but questionable insofar as other biological, or organic and inorganic chemical content is concerned. Okun emphasizes that the reclaimed or nonpotable water should

> equal the quality of the potable systems that many communities now provide—the health hazard that results from the continuous ingestion of low levels of toxic substances over a period of years would not be present.[71]

Advanced wastewater treatment, monitoring, and surveillance cannot yet in practice guarantee removal of all harmful substances from wastewater at all times. More knowledge is needed concerning acute and long-term effects on

[69]Peter W. Fletcher and William E. Sharpe, "Water-Conservation Methods to Meet Pennsylvania's Water Needs," *J. Am. Water Works Assoc.*, April 1978, pp. 200–203.
[70]*Rural Wastewater Management*, State of California Water Resources Board, Sacramento, 1979, pp. 11–14.
[71]Daniel A. Okun, "The Use of Polluted Sources for Water Supply," *APWA Reporter*, September 1976, pp. 23–25.

human health of wastewater reuse.[72–74] The significance of trace organic and inorganic chemicals in drinking water is discussed in the previous pages. In Windhoek, South West Africa, reclaimed sewage blended with water from conventional sources has been used for drinking without any apparent problems. The sewage is given very elaborate treatment involving some 18 unit processes.[75]

More emphasis is needed on the removal of hazardous substances at the source, and on adequate wastewater treatment prior to its discharge to surface and underground waters. This will at least reduce the concentrations of contaminants discharged from urban and industrial areas and, it is hoped, the associated risks.

In any case, it is axiomatic that in general the cleanest surface and underground water source available should be used as a source of drinking water, and water conservation practiced, before a polluted raw water source is even considered, with cost being secondary.

SOURCE AND PROTECTION OF WATER SUPPLY

General

The sources of water supply are divided into two major classifications: groundwater and surface water. To these should be added rainwater and demineralized water. The groundwater supplies include dug, bored, driven and drilled wells, rock and sand or earth springs, and infiltration galleries. The surface-water supplies include lake, reservoir, stream, pond, river, and creek supplies.

The location of groundwater supplies should take into consideration the tributary area,[76] the probable travel of pollution through the ground, the well construction practices and standards actually followed, and the type of sanitary seal provided at the point where the pump lines pass out of the casing. These factors are explained and illustrated later.

It is sometimes suggested that the top of a well casing terminate below the ground level or in a pit. This is not considered good practice except when the pit can be drained above flood level to the surface by gravity or to a drained

[72] "Use of Reclaimed Waste Water as a Public Water Supply Source," *1978–79 Officers and Committee Directory including Policy Statements and Official Documents*, Am. Water Works Assoc., Denver, Colorado, 1979, p. 78.

[73] "Water Should Be Segregated By Use," *Water Sewage Works*, March 1979, pp. 52–54.

[74] Daniel A. Okun, "Wastewater Reuse Dilemma," *ESE Notes*, The University of North Carolina at Chapel Hill, November, 1975.

[75] A. J. Clayton and P. J. Pybus, "Windhoek reclaiming sewage for drinking water," *Civ. Eng., ASCE*, September 1972, pp. 103–106.

[76] For more information see *Ground Water Basin Management*, ASCE Manual No. 40, New York, 1961; and *Large-Scale Ground-Water Development*, United Nations Water Resources Centre, New York, 1960.

basement. Frost-proof sanitary seals with pump lines passing out horizontally from the well casing are generally available. Some are illustrated in Figures 3-7 through 3-10.

In order that the basic data on a new well may be recorded, a form such as the "Well Driller's Log and Report" shown in Figure 3-4 should be completed by the well driller and kept on file by the owner for future reference.

Surface-water supplies are all subject to continuous or intermittent pollution and must be treated to make them safe to drink. One never knows when the organisms causing typhoid fever, gastroenteritis, giardiasis, infectious hepatitis, or dysentery, in addition to organic and inorganic pollutants, may be discharged or washed into the water source. The extent of the treatment required will depend on the results of a sanitary survey made by a competent sanitary engineer, including chemical and bacterial analyses. The minimum required treatment may be coagulation, flocculation, sedimentation, filtration, and chlorination. If elaborate treatment is needed it would be best to abandon the idea of using a surface-water supply and resort to a protected groundwater supply if possible and practical. Where a surface supply must be used, a reservoir or a lake that does not receive sewage or industrial pollution and that can be controlled would be preferred to a stream or creek, the pollution of which cannot from a practical standpoint be controlled. There are many situations where there is no practical alternative to the use of polluted streams for water supply. In such cases carefully designed water treatment plants must be provided.

Groundwater

It is estimated that at any one time, there is 20 to 30 times more water stored in the ground than in all its surface streams and lakes. Development of groundwater sources can significantly help meet the increasing water needs. Exploration techniques include use of data from USGS and state agencies, previous studies, existing well logs, gains or losses in stream flow, surface resistivity surveys, and exploratory test wells.

A relatively new technique for water well location is called "fracture-trace mapping." It is reported to be a highly effective method for increasing the ratio of successful to unsuccessful well-water drilling operations and to greatly improve water yields (up to 50 times). Aerial photographs give the skilled hydrogeologist clues of the presence of a zone of fractures underneath the earth's surface. Clues are

> abrupt changes in the alignment of valleys, the presence of taller or more lush vegetation, the alignment of sink holes or other depressions in the surface, or the existence of shallow longitudinal depressions in the surface overtop of the fracture zone.

The soil over fracture zones is often wetter and hence shows up darker in recently plowed fields. The aerial photograph survey is then followed by a field

Well at .... In .... County of ....
Name of place City, village or town

Owner .... P.O. Address ....

Depth of well .... Diameter .... Yield .... Was well disinfected? ....
ft in. gpm yes or no

Amt. of casing above ground .... Below ground .... Well seal ....
in. ft cement grout

Draw a well diagram in the space provided below and show the depth and type of casing, the well seal, kind and thickness of formations penetrated, water bearing formations, diameter of drill holes with dotted lines and casing(s) with solid lines.

Well Diagram			Formations Penetrated
Diameter, in.		Depth in ft Grade	Kind, thickness, and if water bearing
		25	
		50	
		75	
		100	
		150	
		200	
		250	

Remarks

Type of well ....
Drilling method ....
Was well dynamited? ....

Pumping Tests

Details	#1	#2	#3
Static water level, in feet below grade			
Pumping rate in gpm			
Pumping level in feet below grade			
Duration of test, in hours			

Water at end of test:
Clear ..... Cloudy ..... Turbid .....
Recommended depth of pump in well, ft below grade ....

Wells in sand & gravel:
Sand Eff. size .... mm
Unif. Coef. ....
Length of screen .... ft
Diam. of screen .... in.
Type of screen ....
Screen openings .... x ....

Comments:

Show cross-section of well & formations penetrated above. Draw a sketch of the property on the back of this sheet locating the well and sewage disposal systems within 200 ft.

Drilling started .... Completed ....
Well Driller ....
Signature

Figure 3-4 Well driller's log and report. "Well yield" is the volume of water per unit of time, such as gal per min, discharged from a well either by pumping to a stabilized drawdown or by free flow. The specific capacity of a well is the yield at a stabilized drawdown and given pumping rate, expressed as gallons per minute per foot of drawdown. Use chalked tape, electric probe, or known length of air line with pressure gauge. Test run is usually 4 to 8 hr for small wells; 24 to 72 hr for wells serving the public, or for 6 hr at a stabilized drawdown when pumping at 1.5 times the design pumping rate.

investigation and actual ground location of the fractures and potential well drilling sites.[77]

Dug Well

A dug well is one usually excavated by hand, although it may be dug by mechanical equipment. It may be 3 to 6 ft in diameter and 15 to 35 ft deep, depending on where the water-bearing formation or groundwater table is encountered. Wider and deeper wells are less common. Hand pumps over wells and pumplines entering wells should form watertight connections, as shown in Figures 3-5 and 3-6. Since dug wells have a relatively large diameter, they have large storage capacity. The level of the water in dug wells will lower at times of drought and the well may go dry. Dug wells are not usually dependable sources of water supply, particularly where modern plumbing is provided. In some areas, properly developed dug wells provide an adequate and satisfactory water supply.

Bored Well

A bored well is constructed with a hand- or machine-driven auger. Bored wells vary in diameter from a few to 36 in. and in depth from 25 to 60 ft. A casing of concrete pipe, vitrified clay pipe, metal pipe, or plastic pipe is usually necessary to prevent the relatively soft formation penetrated from caving into the well. Bored wells have characteristics similar to dug wells in that they have small yields, are easily polluted, and are affected by droughts.

Driven and Jetted Well

These types of wells consist of a well point with a sc een attached, or a screen with the bottom open, which is driven or jetted into a water-bearing formation found at comparatively shallow depth. A series of pipe lengths are attached to the point or screen as it is forced into position. The driven well is constructed by driving the well point, preferably through at least 10 to 20 ft of casing, with the aid of a maul or sledge, pneumatic tamper, sheet pile driver, drive monkey, hand-operated driver or similar equipment. In many instances, the casing is omitted; but then less protection is afforded the driven well, which also serves as the pump suction line. The jetted well is constructed by directing a stream of water at the bottom of the open screen, thereby loosening and flushing the soil up the casing to the surface as the screen is lowered.Driven and jetted wells

[77] *Water Well Location by Fracture Trace Mapping*, Technology Transfer, Office of Water Research and Technology, Dept. of Interior, GPO, 1978.

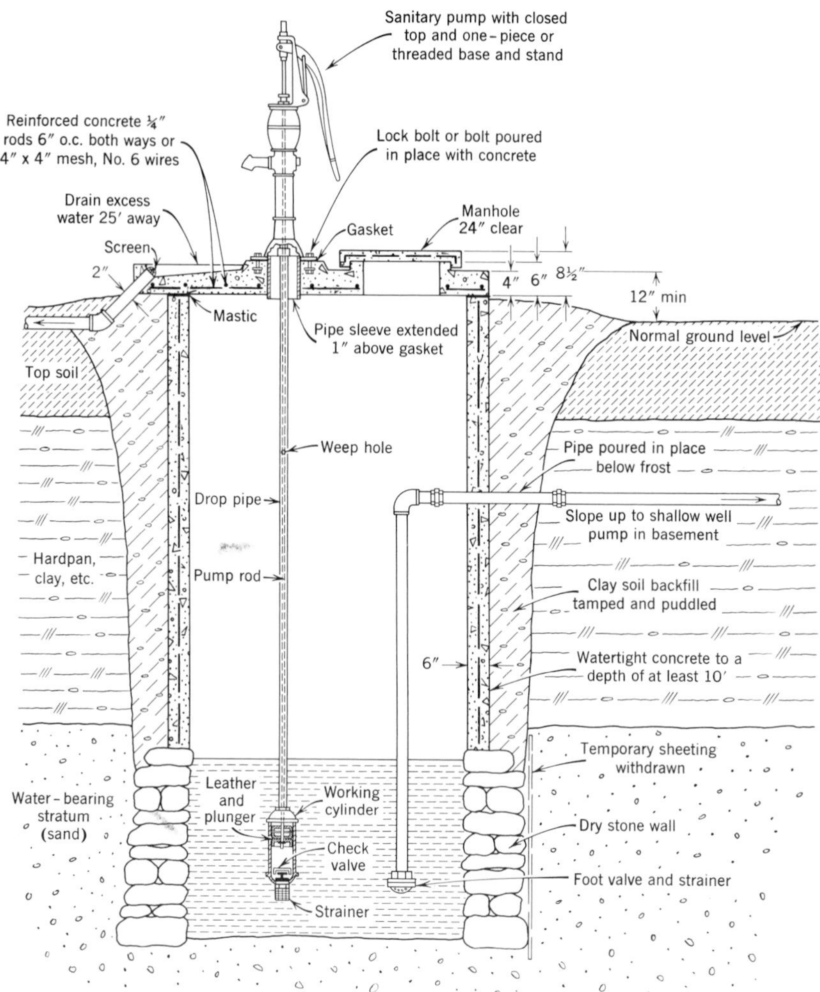

Figure 3-5 A properly developed dug well.

are commonly between 1¼ and 2 in. in diameter and less than 25 ft in depth, although larger and deeper wells can be constructed. In the small-diameter wells, a shallow well hand or mechanical suction pump is connected directly to the well. Large-diameter driven wells make possible installation of the pump cylinder close to or below the water surface in the well at greater depth, in which case the hand pump must be located directly over the well. In all cases, however, care must be taken to see that the top of the well is tightly capped, that the concrete pump platform extends 2 or 3 ft around the well pipe or casing, and that the annular space between the well casing and drop pipe, or pipes, is tightly sealed. This is necessary to prevent the entrance of unpurified water or other pollution from close to the surface.

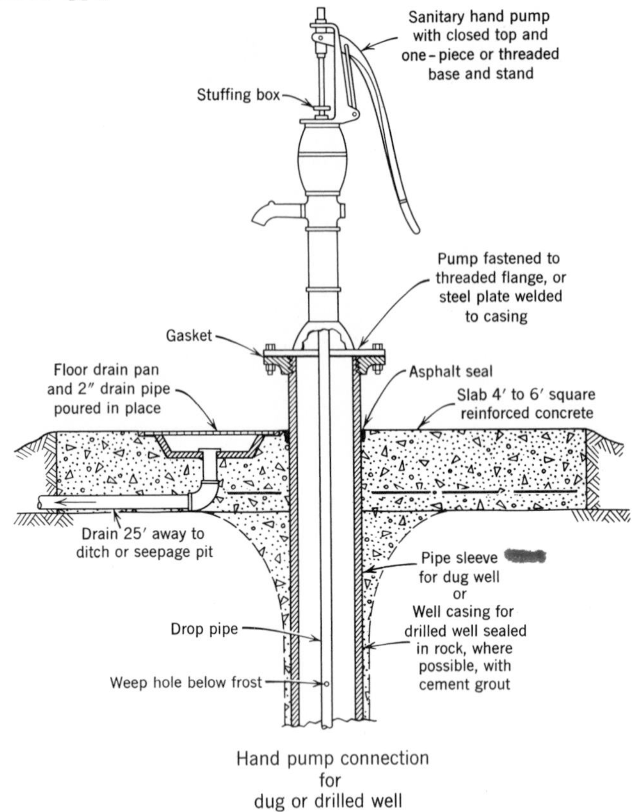

Figure 3-6 Sanitary hand pump and well attachment.

Drilled Well

Studies have shown that, in general, drilled wells are superior to dug, bored, or driven wells, and springs. There are some exceptions. Drilled wells are less likely to become contaminated and are usually more dependable sources of water. When a well is drilled, a hole is made in the ground with a percussion or rotary drilling machine. Drilled wells are usually 4 to 12 in. in diameter or larger. A steel or wrought-iron casing is lowered as the well is drilled to prevent the hole from caving in and to seal off water of a doubtful quality. Special plastic pipe is also used. Lengths of casing should be threaded and coupled or properly welded. The drill hole must, of course, be larger than the casing, thereby leaving an irregular space around the outside length of the casing. Unless this space or channel is closed by cement grout or naturally by formations that conform to the casing almost as soon as it is placed, pollution from the surface or from crevices close to the surface, or from polluted formations penetrated, will flow down the side of the casing and into the water source. Water can also move up and down this annular space in an artesian well and as the groundwater and pumping water level changes.

When the source of water is water-bearing sand and gravel, a gravel wall or gravel-packed well may be constructed. Such a well will usually yield more water than the ordinary drilled well with a screen of the same diameter and with the same drawdown. A slotted or perforated casing in a water-bearing sand will yield only a fraction of the water obtainable by the use of a proper screen. On completion, the well should be overpumped, surged and tested. Proper surging is an important part of well development. Surging or the forcing of water in and out through the screen, or around the bore hole to remove adhering mud, is done by pumping, use of compressed air, dry ice, or a surge block attached to a cable-tool drill rig. A high velocity jet is also used. To develop well yield and hydraulic data, a 24- to 72-hr pump test is desirable. But 4 to 8 hr tests are more common. See Figure 3-4.

Only water well casing of clean steel or wrought iron should be used. Plastic pipe may be permitted. Used pipe is not satisfactory. Standards for well casing are given in the American Water Works Association's publication *AWWA Standard for Deep Wells*, Am. Water Works Assoc. A100-66. Doubling the diameter of a casing increases the yield up to only 10 to 12 percent. The required well diameter is usually determined by the size of the discharge piping, fittings, pump, and motor that may be placed inside the well casing.

Extending the casing at least 5 ft below the pumping water level in the well—or if the well is less than 30 ft deep, 10 ft below the pumping level—will afford an additional measure of protection. In this way the water is drawn from a depth that is less likely to be contaminated. In some sand and gravel areas, extending the casing 5 to 10 ft below the pumping level may shut off the water-bearing sand or gravel. A lesser casing depth would then be indicated, but in no instance should the casing be less than 10 ft, provided sources of pollution are remote and provision is made for chlorination. The recommended depth of casing, cement grouting, and the need for double casing construction or the equivalent is given in Table 3-13.

A vent is necessary on a well because if not vented the fluctuation in the water level will cause a change in air pressure above and below atmospheric pressure in a well, resulting in the drawing in of contaminated water from around the pump base over the well or from around the casing if not properly sealed. Reduced pressure in the well will also increase lift or total head and reduce volume of water pumped.

It must be remembered that well construction is a very specialized field. Most well drillers are desirous of doing a proper job for they know that a good well is their best advertisement. However, in the absence of a state law dealing with well construction, the enforcement of standards, and the licensing of well drillers, price alone frequently determines the type of well constructed. Individuals proposing to have wells drilled should therefore carefully analyze bids received. Such matters as water quality, well diameter, type and length of casing, minimum well yield, type of pump and sanitary seal where the pump line or lines pass through the casing, provision of a satisfactory well log, method used to seal off undesirable formations and cement grouting of the well, plans to

Table 3-13 Standards for the Construction of Wells[a]

Water-Bearing Formation	Overburden	Oversize Drill Hole for Grout: Diameter	Oversize Drill Hole for Grout: Depth[b]	Cased Portion
1. Sand or gravel	Unconsolidated caving material; sand or sand and gravel	None required	None	2″ minimum, 5″ or more preferred
2. Sand or gravel	Clay, hardpan, silt, or similar material to depth of more than 20′	Casing size plus 4″	Minimum 20′	2″ minimum, 5″ or more preferred
3. Sand or gravel	Clay, hardpan, silt, or similar material containing layers of sand or gravel within 15′ of ground surface	Casing size plus 4″	Minimum 20′	2″ minimum, 5″ or more preferred
4. Sand or gravel	Creviced or fractured rock, such as limestone, granite, quartzite	Casing size plus 4″	Through rock formation	4″ minimum
5. Creviced, shattered or otherwise fractured limestone, granite, quartzite or similar rock types	Unconsolidated caving material, chiefly sand or sand and gravel to a depth of 40′ or more and extending at least 2000 in all directions from the well site	None required	None required	6″ minimum
6. Creviced, shattered or otherwise fractured limestone, granite, quartzite or similar rock types	Clay, hardpan, shale, or similar material to a depth of 40′ or more and extending at least 2000 in all directions from well site	Casing size plus 4″	Minimum 20′	6″ minimum

Source: *Recommended State Legislation and Regulations*, PHS, DHEW, Washington, D.C., July 1965.

Note: For wells in creviced, shattered, or otherwise fractured limestone, granite, quartzite, or similar rock in which the overburden is less than 40 ft and extends less than 2000 ft in all directions, and no other practical acceptable water supply is available, the well construction described in Line 7 of this table is applicable.

[a]Requirements for the proper construction of wells vary with the character of subsurface formations, and provisions applicable under all circumstances cannot be fixed. The construction details of this table may be adjusted, as conditions warrant, under the procedure provided for by the Health Department and in the Note above.

Table 3-13 (*Continued*)

Well Diameter		Minimum Casing Length or Depth[b]	Liner Diameter (If Required)	Construction Conditions[b]	Miscellaneous Requirements
Uncased Portion	Well Screen Diameter[c]				
Does not apply	2′ minimum	20′ minimum; but 5′ below pumping level[d]	2″ minimum		
Does not apply	2″ minimum	5′ below pumping level[d]	2″ minimum	Upper drill hole shall be kept at least one-third filled with clay slurry while driving permanent casing; after casing is in the permanent position annular space shall be filled with clay slurry or cement grout.	An adequate well screen shall be provided where necessary to permit pumping sand-free water from the well.
Does not apply	2″ minimum	5′ below pumping level[d]	2″ minimum	Annular space around casing shall be filled with cement grout.	
Does not apply	2″ minimum	5′ below overburden of rock	2″ minimum	Annular space around casing shall be filled with cement grout.	
6″ preferred	Does not apply	Through caving overburden	4″ minimum	Casing shall be firmly seated in the rock.	
6″ preferred	Does not apply	Through overburden.	4″ minimum	Annular space around casing shall be grouted. Casing shall be firmly seated in rock.	

[b] In the case of a flowing artesian well, the annular space between the soil and rock and the well casing shall be tightly sealed with cement grout from within 5 ft of the top of the aquifier to the ground surface in accordance with good construction practice.

[c] These diameters shall be applicable in circumstances where the use of perforated casing is deemed practicable. Well points commonly designated in the trade as $1\frac{1}{4}$″ pipe shall be considered as being 2″ nominal diameter well screens for purposes of these regulations.

[d] As used herein, the term "pumping level" shall refer to the lowest elevatoin of the surface of the water in a well during pumping, determined to the best knowledge of the water well contractor, taking into consideration usual seasonal fluctuations in the static water level and drawdown level.

Table 3-13 (*Continued*)

Water-Bearing Formation	Overburden	Oversize Drill Hole for Grout		Cased Portion
		Diameter	Depth[b]	
7. Creviced, shattered or otherwise fractured limestone, granite, quartzite or similar rock	Unconsolidated materials to a depth of less than 40′ and extending at least 2000 in all directions	Casing size plus 4″	Minimum 40′	6″ minimum
8. Sandstone	Any material except creviced rock to a depth of 25′ or more	Casing size plus 4″	15′ into firm sandstone or to 30′ depth, whichever is greater	4″ minimum
9. Sandstone	Mixed deposits mainly sand and gravel, to a depth of 25′ or more	None required	None required	4″ minimum
10. Sandstone	Clay, hardpan, or shale to a depth of 25′ or more	Casing size plus 4″	Minimum 20′	4″ minimum
11. Sandstone	Creviced rock at variable depth	Casing size plus 4″	15′ or more into firm sandstone	6″ minimum

Table 3-13 (*Continued*)

Well Diameter		Minimum Casing Length or Depth[b]	Liner Diameter (If Required)	Construction Conditions[b]	Miscellaneous Requirements
Uncased Portion	Well Screen Diameter[c]				
6″ preferred	Does not apply	40′ minimum	4″ minimum	Casing shall be firmly seated in rock. Annular space around casing shall be grouted.	If grout is placed through casing pipe and forced into annular space from the bottom of the casing, the oversize drill hole may be only 2″ larger than the casing pipe. Pipe 2″ smaller than the drill hole, and liner pipe 2″ smaller than casing shall be assembled without couplings.
4″ preferred		Same as oversize drill hole or greater	2″ minimum	Annular space around casing shall be grouted. Casing shall be firmly seated in sandstone.	
4″ preferred		Through overburden into firm sandstone	2″ minimum	Casing shall be effectively seated into firm sandstone.	
4″ preferred	2″ minimum, if well screen required to permit pumping sand-free water from partially cemented sandstone	Through overburden into sandstone	2″ minimum	Casing shall be effectively seated into firm sandstone. Oversized drill hole shall be kept at least one-third filled with clay slurry while driving permanent casing; after the casing is in the permanent position, annular space shall be filled with clay slurry or cement grout.	Pipe 2″ smaller than the oversize drill hole and liner pipe 2″ smaller than casing shall be assembled without couplings.
6″ preferred		15′ into firm sandstone	4″ minimum	Annular space around casing shall be filled with cement grout.	If grout is placed through casing pipe and forced into annular space from the bottom of the casing the oversize drill hole may be only 2″ larger than the casing pipe. Pipe 2″ smaller than the drill hole, and liner pipe 2″ smaller than casing shall be assembled without couplings.

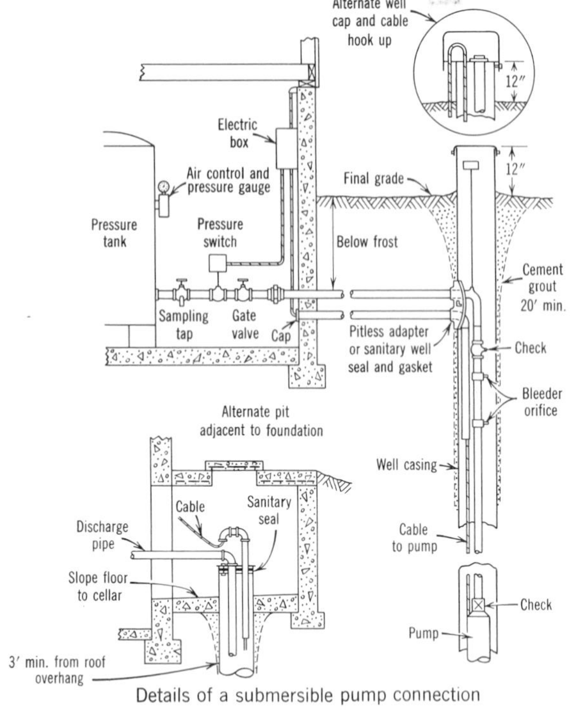

Figure 3-7 Sanitary well caps and seals and submersible pump connection.

pump the well until clear, and disinfection following construction should all be taken into consideration. See Figures 3-6 through 3-12.

Recommended water well construction practices and standards are given in this text. More detailed information, including contracts and specifications, is available in federal, state, and industrial publications.[78–80] Remember, the cheapest well is not necessarily the best buy.

Grouting

One of the most common reasons for contamination of wells drilled through rock, clay, or hardpan is failure to seal the well casing properly.

A contaminated well supply causes the homeowner or municipality considerable inconvenience and extra expense, for it is difficult to seal off contamina-

[78] *Manual of Water Well Construction Practices*, USEPA, Office of Water Supply, EPA—570/9-75-101, GPO, 1976, Washington, D.C.

[79] *Manual of Individual Water Supply Systems*, USEPA, WSD, EPA-430/9-74-007, GPO, Washington, D.C.

[80] *Recommended Standards For Water Works*, A Report of the Committee of the Great Lakes-Upper Mississippi River Board of State Sanitary Engineers, 1976 Edition, Health Education Service, P.O. Box 7283, Albany, N.Y. 12224.

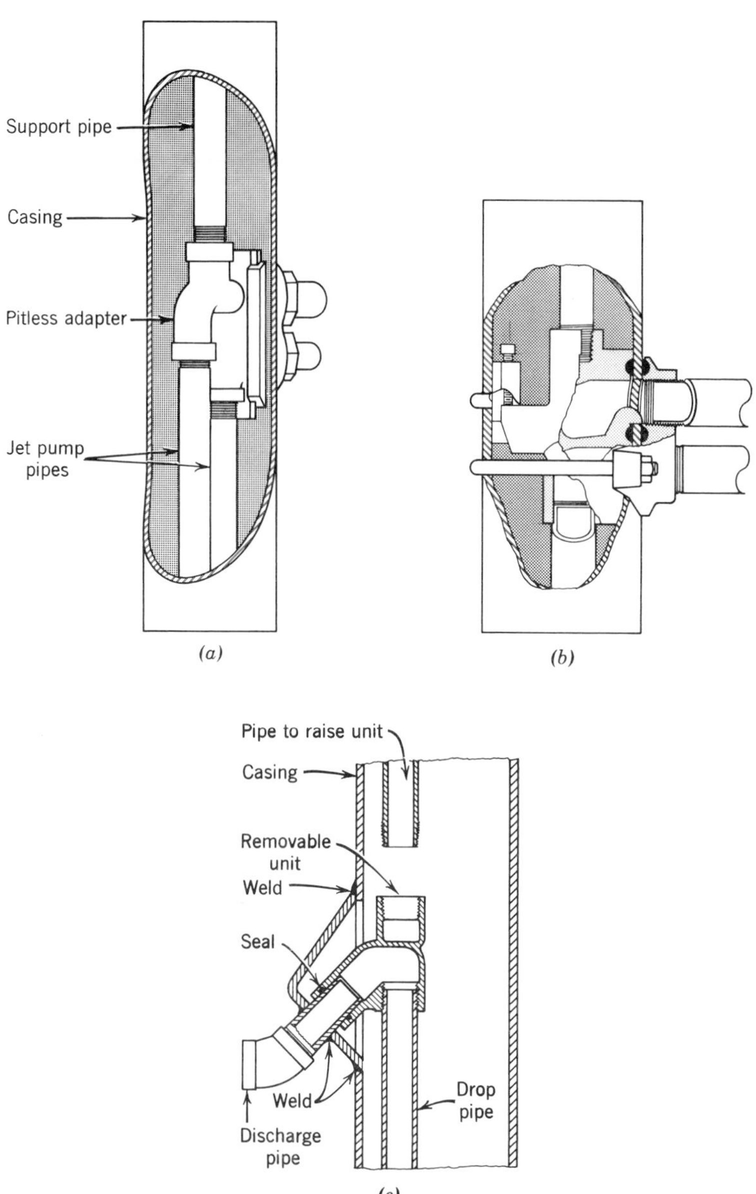

Figure 3-8 Pitless adapters. (*a*) Courtesy Martin Manufacturing Co., P.O. Box 256, Ramsey, N.J. (*b*) Courtesy Williams Products Co., Joliet, Ill. (*c*) Courtesy Herb Maass Service, 10940 West Congress Street, Milwaukee, Wis.

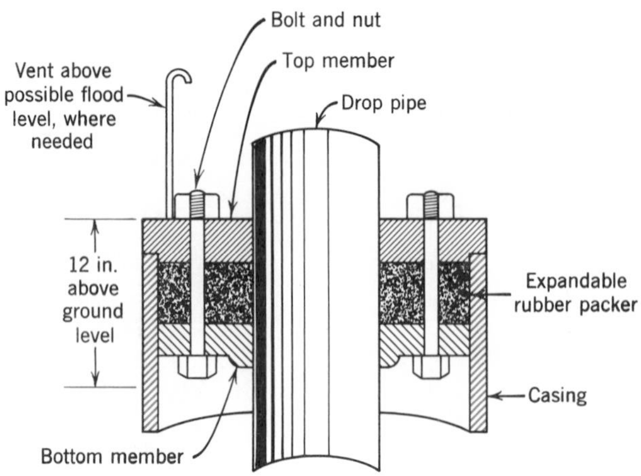

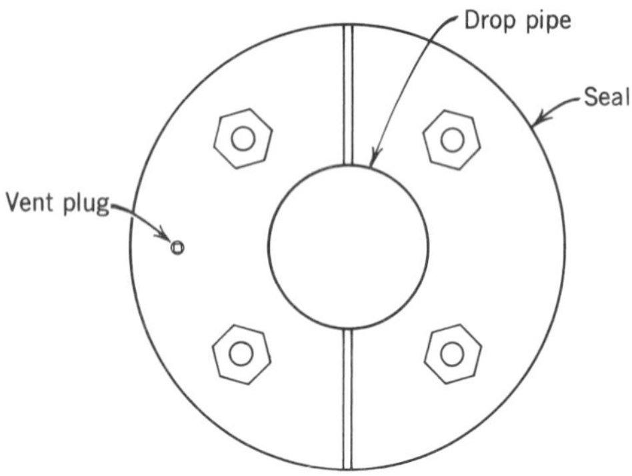

Figure 3-9 Sanitary expansion well cap.

tion after the well is drilled. In some cases the only practical answer is to build a new well.

Proper *cement grouting* of the space between the drill hole and well casing where the overburden over the water-bearing formation is clay, hardpan, or rock can prevent this common cause of contamination. (Table 3-13).

There are many ways to seal well casings. The best material is neat cement grout.* But to be effective the grout must be properly prepared (a proper mixture is $5\frac{1}{2}$ gal clean water to a bag of cement), pumped as one continuous mass, and placed upward from the bottom of the space to be grouted.

*Concrete grout-cement and sand 1:1, with not more than 6 gal water per sack of cement, or puddled clay if approved, are also used. The addition of up to 10 percent hydrated lime by volume will keep the grout more fluid while handling.

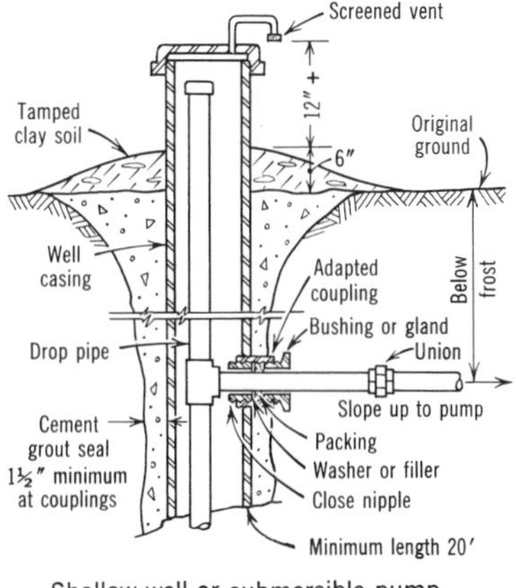

Figure 3-10 Improvised well seal.

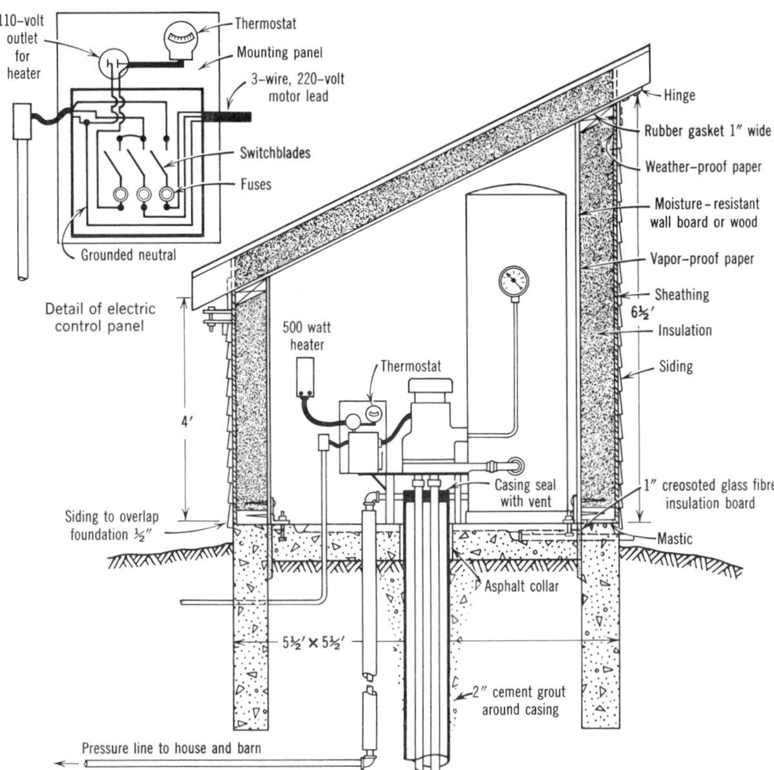

Figure 3-11 Insulated pumphouse. (From *Sewage Disposal and Water Systems on the farm*, University of Minnesota, Extension Bulletin 247, Revised 1956.)

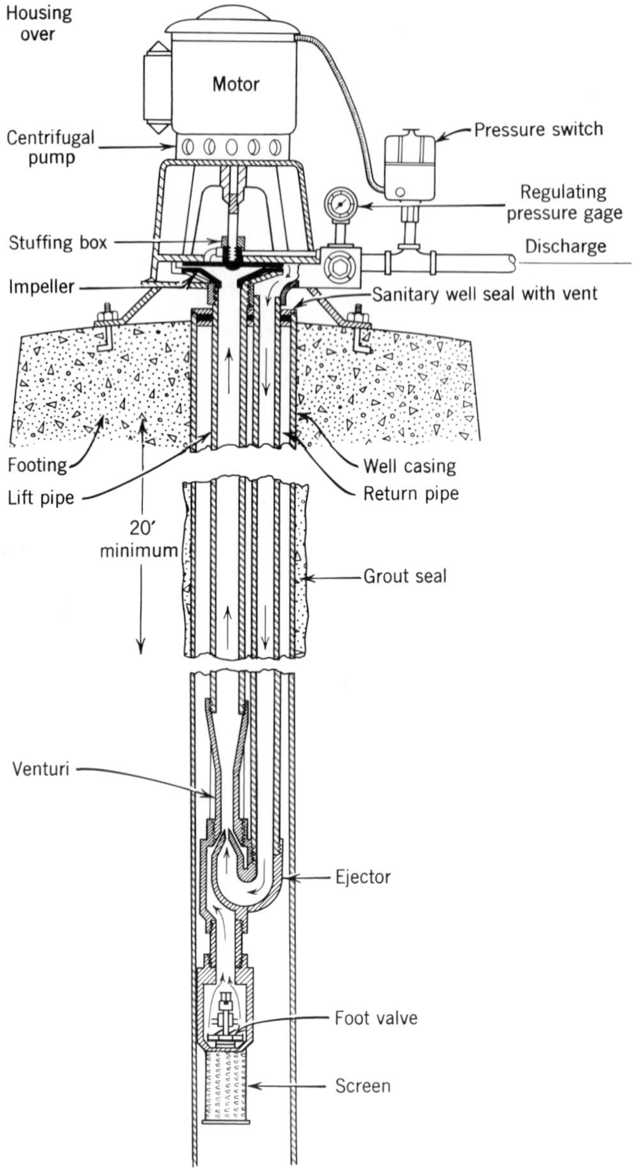

Figure 3-12 Sanitary well seal and jet pump.

The clear annular space around the outside of the casing and the drill hole must be at least $1\frac{1}{2}$ in. on all sides to prevent bridging of the grout. Guides must be welded to the casing.

Cement grouting of a well casing along its entire length of 50 to 100 ft or more is good practice but expensive for the average farm or rural dwelling. An alternative is grouting to at least 20 ft below ground level. This provides adequate protection for most installations, except in limestone and fractured formations. It also protects the casing from corrosion.

For a 6-in. diameter well a 10-in. hole is drilled, if 6-in. welded pipe is used, to at least 20 ft or to solid rock if the rock is deeper than 20 ft. If 6-in. coupled pipe is used, a 12-in. hole will be required. From this depth the 6-in. hole is drilled deeper until it reaches a satisfactory water supply. A temporary outer casing, carried down to rock, prevents cave-in until the cement grout is placed.

Upon completion of the well the annular space between the 6-in. casing and temporary casing or drill hole is filled from the bottom up to the grade with cement grout. The temporary pipe is withdrawn as the cement grout is placed—it is not practical to pull the casing after all the grout is in position.

The extra cost of the temporary casing and larger drill hole is small compared to the protection obtained. The casing can be reused as often as needed. In view of this, well drillers who are not equipped should consider adding larger casing and equipment to their apparatus

A temporary casing or larger drill hole and cement grouting are not required where the entire earth overburden is 40 ft or more of silt or sand and gravel, which immediately close in on the total length of casing to form a seal around the casing; however, this condition is not common.

Drilled wells serving public places are usually constructed and cement-grouted as explained in Table 3-13.

In some areas, limestone and shale beneath a shallow overburden represent the only source of water. Acceptance of a well in shale or limestone might be conditioned on an extended observation period to determine the sanitary quality of the water. Continuous chlorination should be required on satisfactory supplies serving the public and should be recommended to private individuals. However, chlorination should not be relied on to make a heavily contaminated well-water supply satisfactory. Such supplies should be abandoned and filled in with concrete or puddled clay, unless the source of contamination can be eliminated.

Well drillers may have other sealing methods suitable for particular local conditions, but the methods described above utilizing a neat cement grout will give reasonably dependable assurance that an effective seal is provided, whereas this cannot be said of some of the other methods used. Driving the casing, a lead packer, drive shoe, rubber sleeves, and similar devices do not provide reliable annular space seals for the length of the casing.

Well Contamination—Cause and Removal

Well-water supplies are all too often improperly constructed, protected, or located, with the result that bacteriological examinations show the water to be contaminated. Under such conditions all water used for drinking or culinary purposes should first be boiled or adequately treated. Boiling will not remove chemical contaminants; treating may remove some. Abandonment of the well and connection to a public water supply would be the best solution if this is practical. A second alternative would be investigation to find and remove the cause of pollution; however, if the aquifer is badly polluted, this may take considerable time. A third choice would be a new, properly constructed and located drilled well in a clean aquifer.

When a well shows the presence of bacterial contamination it is usually due to one or more of four probable causes: lack of or improper disinfection of a well following repair or construction; failure to seal the annular space between the drill hole and the outside of the casing; failure to provide a tight sanitary seal at the place where the pump line or lines pass through the casing; sewage pollution of the well through polluted strata or a fissured or channeled formation. On some occasions the casing is found to be only a few feet in length and completely inadequate.

If a new well is constructed or if repairs are made to the well, pump, or piping, contamination from the work is probable. Disinfect the well, pump, storage tank, and piping as explained in this chapter.

If a sewage disposal system is suspected of contamination, a dye such as water-soluble sodium or potassium fluorescein or ordinary salt can be used as a tracer. A solution flushed into the disposal system or suspected source may appear in the well water within 12 to 24 hr. It can be detected by sight or taste if a connection exists. Samples should be collected every few hours and set aside for comparision. If the connection is indirect, fluoroscopic or chemical examination for the dye or for chlorides is more sensitive. One part of fluorescein in 10 to 40 million parts of water is visible to the naked eye, and in 10 billion parts if viewed in a long glass tube, or if concentrated in the laboratory. The chlorides in the well before adding salt should, of course, be known. Where chloride determinations are routinely made on water samples, sewage pollution may be apparent without making the salt test. Dye is not decolorized by passage through sand, gravel, or manure; it is slightly decomposed by calcareous soils and entirely decolorized by peaty formations and by free acids, except carbonic acid.[81,82]

If the cause of pollution is suspected to be an underground seal where the pump line or lines pass through the side of the casing, a dye or salt solution or even plain water can be poured around the casing. Samples of the water can be collected for visual or taste test or chemical examination. The seal might also be excavated for inspection. Where the upper part of the casing can be inspected, a mirror or strong light can be used to direct a light beam inside the casing to see if water is entering the well from close to the surface. Sometimes it is possible to hear the water dripping into the well. Inspection of the top of

[81]R. B. Dole, "Use of Fluorescein in the Study of Underground Waters," excerpt form *U.S. Geol. Survey, Water-Supply Paper* 160, 73-85 (1906).

[82]S. Reznek, William Hayden, and Maria Lee, "Analytical Note—Fluorescein Tracer Technique for Detection of Groundwater Contamination," *J. Am. Water Works Assoc.*, October 1979, pp. 586–587.

the well will also show if the top of the casing is provided with a sanitary seal and whether the well is subject to flooding. See Figure 3-7.

The path of pollution entry can also be holes in the side of the casing, channels along the length of the casing leading to the well source, crevices or channels connecting surface pollution with the water-bearing stratum, or the annular space around the casing. A solution of dye, salt, or plain water can be used to trace the pollution, as previously explained.

The steps taken to provide a satisfactory water supply would depend on the results of the investigation. If a sanitary seal is needed at the top or side of the casing where the pump lines pass through, then the solution is relatively simple. On the other hand, an unsealed annular space is more difficult to correct. A competent well driller could be engaged to investigate the possibility of grouting the annular space and installing an inner casing or a new casing carefully sealed in solid rock. If the casing is found tight, it would be assumed that pollution is finding its way into the water-bearing stratum through sewage-saturated soil or creviced or channeled rock at a greater depth. It is sometimes possible, but costly, to seal off the polluted stratum and if necessary drill deeper.

Once a stratum is contaminated, it is very difficult to prevent future pollution of the well unless all water from such a stratum is effectively sealed off. Moving the offending sewage-disposal system to a safe distance or replacing a leaking oil or gasoline tank is possible, but evidence of the pollution may persist for some time. The same general principles apply to dug and driven wells.

Unless all the sources of pollution can be found and removed, it is recommended that the well be abandoned and filled with puddled clay or concrete to prevent the pollution from traveling to other wells. In some special cases, and under controlled conditions, use of a slightly contaminated water supply may be permitted provided approved treatment facilities are installed. Such equipment is expensive and requires constant attention. If a public water supply is not available and a new well is drilled, it should be located and constructed as previously explained.

Chemical contamination of a well, and the groundwater aquifer, can result from spills, leaking gasoline and oil tanks, or improper disposal of chemical wastes such as by dumping on the ground, in landfills, lagooning, or similar methods. Gasoline and oil tanks typically have a useful life of about 20 years, depending on the type of soils and tank coatings. Since many tanks have been in the ground 20 to 30 years or longer, their integrity must be uncertain and they are probably leaking to a greater or lesser degree. If not already being done, oil, gasoline, and other buried tanks containing chemicals should be tested periodically and of course promptly replaced at the first sign of leakage. The number of tanks, surreptitious dumpings, discharges to leaching pits, and other improper disposals make control a formidable task. This subject is discussed further in this chapter under the headings Groundwater Contamination Hazard, Travel of Sewage Pollution Through the Ground, Domestic

Well-Water Supplies—Special Problems (Gasoline or Fuel Oil in Water), Methods to Remove or Reduce Objectionable Tastes and Odors (Removal of Gasoline by Purging and Bioreclamation); also under Hazardous Wastes in Chapter 5.

Oil, gasoline, and hazardous chemical storage tanks should have suitable tight outer shells that can be monitored for leaks. Other hazardous wastes should be stored in tight lined pits or ponds that can be monitored until adequately treated or otherwise disposed of.

Spring

Springs are broadly classified as either rock springs or earth springs, depending on the source of water. It is necessary to *find the source*, properly develop it, eliminate surface water, and prevent animals from gaining access to the spring area, to obtain a satisfactory water.

Protection and development of a source of water is shown in Figure 3-13. A combination of methods may also be possible under certain ground conditions and would yield a greater supply of water than either alone.

In all cases, the spring should be protected from surface-water pollution by constructing a deep diverting ditch or the equivalent above and around the spring. The spring and collecting basin should have a watertight top, preferably concrete, and water obtained by gravity flow or by means of a properly installed sanitary hand or mechanical pump. Access or inspection manholes, when provided, should be tightly fitted (as shown) and kept locked. Water from limestone or similar type channeled or fissured rock springs is not purified to any appreciable extent when traveling through the formation and hence may carry pollution from nearby or distant places. Under these circumstances, it is advisable to have periodic bacteriological examinations made and to chlorinate the water.

Infiltration Gallery

An infiltration gallery consists of a system of porous, perforated, or open-joint pipe or other conduit draining to a receiving well. The pipe is surrounded by gravel and located in a porous formation such as sand and gravel below the water table. The collecting system should be located 20 ft or more from a lake or stream, or under the bed of a stream or lake if installed under expert supervision. It is sometimes found desirable, where possible, to intercept the flow of groundwater to the stream or lake. In such cases a coffer dam, cut-off wall, or puddled clay dam is carefully placed between the collecting conduit and the lake or stream to form an impervious wall. It is not advisable to construct an infiltration gallery unless the water table is relatively stable and the water intercepted is free of pollution. The water-bearing strata should not contain cementing material or yield a very hard water as it may clog the strata or cause

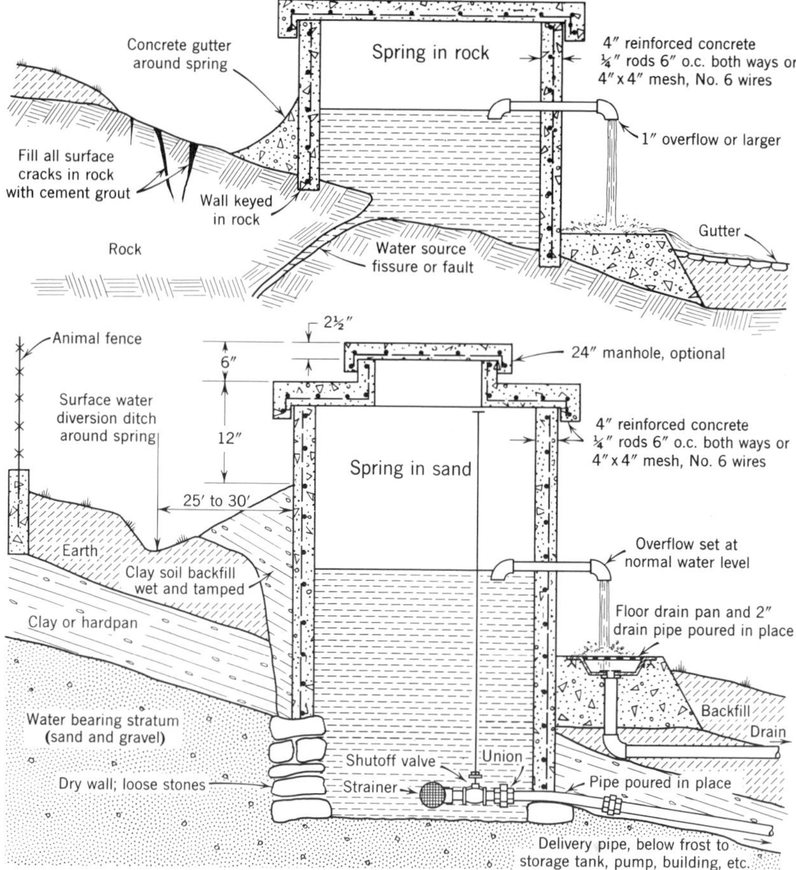

Figure 3-13 Properly constructed springs.

incrustation of the pipe, thereby reducing the flow. An infiltration gallery is constructed similar to that shown in Figure 3-14. The depth of the collecting tile should be about 10 ft below the normal ground level, and below the lowest known water table, to assure a greater and more constant yield. An infiltration system consisting of horizontal perforated or porous radial collectors draining to a collecting well can also be designated and constructed, where hydrogeological conditions are suitable (usually under a stream bed). The infiltration area should be controlled and protected from pollution by sewage and other wastewater and animals. Water derived from infiltration galleries should at the minimum, be given chlorination treatment.

Cistern

A cistern is a watertight tank in which rainwater collected from roof runoff is stored. When the quantity of groundwater or surface water is inadequate

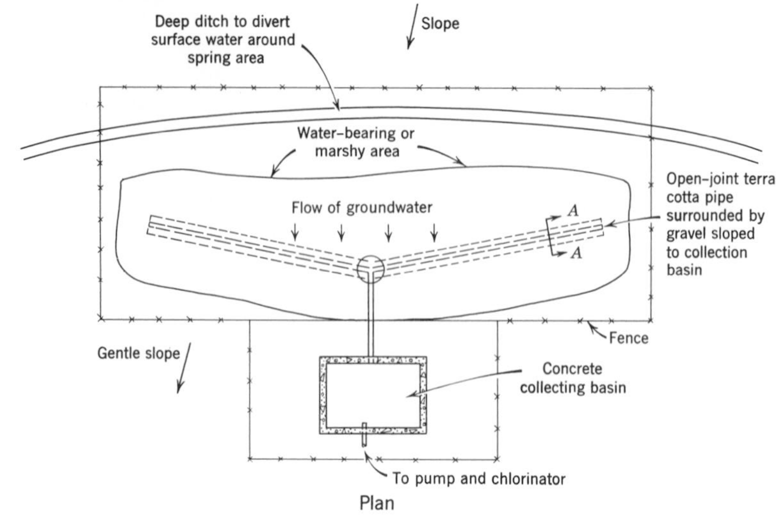

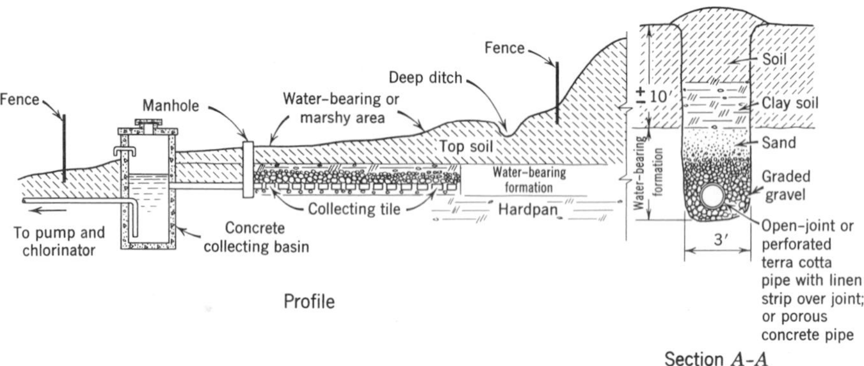

Figure 3-14 Development of a spring in a shallow water-bearing area.

or the quality objectionable, and where an adequate municipal water supply is not available, a cistern supply may be acceptable as a limited source of water. Because rainwater is soft, little soap is needed when used for laundry purposes. On the other hand rain will wash air pollutants, dust, dirt, bird and animal droppings, leaves, paint, and other material on the roof into the cistern unless special provision is made to bypass the first rainwater and to filter the water. The bypass may consist of a simple manually or float-operated damper or switch placed in the leader drain. When in one position, all water will be diverted to a float-control tank or to waste away from the building foundation and cistern; when in the other position water will be run into the cistern. The filter will not remove chemical pollutants.

The capacity of the cistern is determined by the size of the roof catchment area, the probable water consumption, the maximum 24-hr rainfall, the average annual rainfall, and maximum length of dry periods. Suggested rainwater cistern sizes are shown in Figure 3-15. The cistern storage capacity given al-

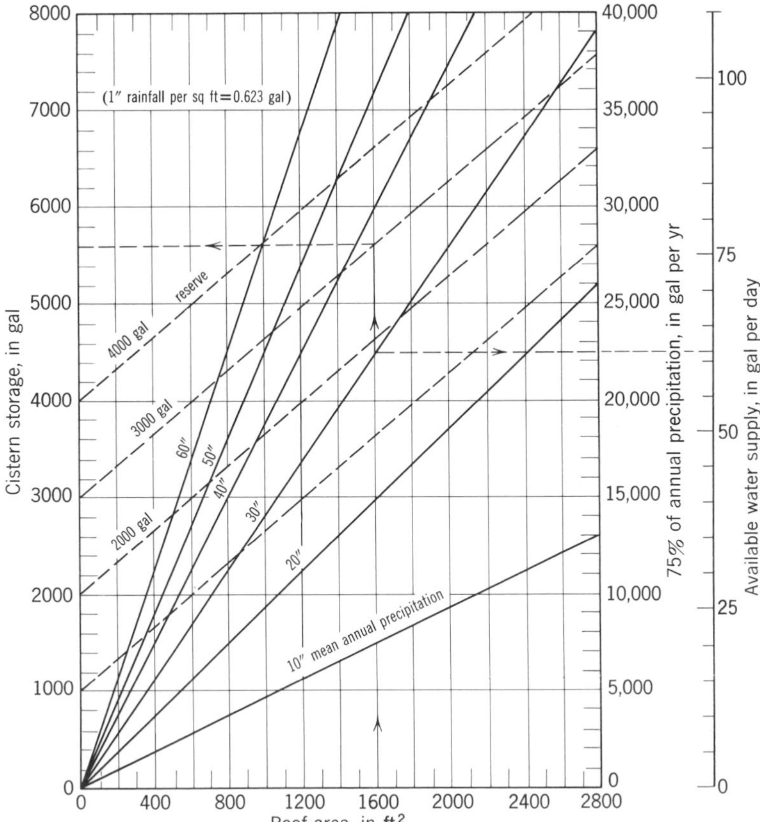

Figure 3-15 Suggested cistern storage capacity and available supply.

lows for a reserve supply, plus a possible heavy rainfall of 3½ in. in 24 hr. The calculations assume that 25 percent of the precipitation is lost. Weather bureaus, the *World Almanac*, airports, water department, and other agencies give rainfall figures for different parts of the country. Adjustment should therefore be made in the required cistern capacity to fit local conditions. The cistern capacity will be determined largely by the volume of water one wishes to have available for some designated period of time, the total volume of which must be within the limits of the volume of water that the roof catchment area and annual rainfall can safely yield. Monthly average rainfall data can be expected to depart from the true values by 50 percent or more on occasion. The drawing of a mass diagram is a more accurate method of estimating the storage capacity, since it is based on past actual rainfall in a given area.

It is recommended that the cistern water be treated after every rain with a chlorine compound to give a dosage of at least 5 mg/l chlorine. This may be accomplished by adding five times the quantities of chlorine shown in Table 3-14, mixed in 5 gal of water to each 1000 gal of water in the cistern. In areas affected by air pollution, fallout on the roof or catchment area will contribute

Table 3-14 Quantity and Type of Chlorine to Treat 1,000 Gallons of Clean Water at Rate of 1 mg/l

Chlorine Compound	Quantity
High test, 70 percent chlorine	$\frac{1}{5}$ oz or $\frac{1}{4}$ heaping tablespoon
Chlorinated lime, 25 percent chlorine	$\frac{1}{2}$ oz or 1 heaping tablespoon
Sodium hypochlorite, 14 percent chlorine	1 oz
Sodium hypochlorite, 10 percent chlorine	$1\frac{1}{3}$ oz
Bleach, $5\frac{1}{4}$ percent chlorine	$2\frac{3}{5}$ oz

Note: A jigger glass = $1\frac{1}{2}$ liquid ounces.

chemical pollutants that may not be neutralized by chlorine treatment. Soft water flowing over galvanized iron roofs, through galvanized iron pipe, or stored in galvanized tanks contain cadmium and zinc.[38]

Example

With a roof area of 1600 ft^2, in a location where the mean annual precipitation is 30 in. and it is desired to have a reserve supply of 3000 gal, the cistern storage capacity should be about 5600 gal. This should yield an average annual supply of about 62 gal per day.

In some parts of the world large natural catchbasins are lined to collect rain water. The water is settled and chlorinated before distribution. The amount of water is of course limited and may supplement groundwater, individual home cisterns, and desalinated water.

Domestic Well-Water Supplies—Special Problems[83]

Domestic well-water supply problems are discussed below. The local health department and commercial water-conditioning companies may be of assistance to a home owner.

Hard Water

Hard water makes it difficult to produce suds or to rinse laundry, dishes or food equipment. Water hardness is caused by dissolved calcium and magnesium bicarbonates, sulfates, and chlorides in well water. Pipes clog, and after a time equipment and water heaters become coated with a hard mineral deposit, sometimes referred to as lime scale. A commercial zeolite or synthetic resin water softener is used to soften water. The media must be regenerated periodically and disinfected with chlorine to remove contamination after each regeneration. Softeners do not remove contamination in the water supply. A filter should be placed ahead of a softener if the water is turbid. See also Water Softening, this chapter.

The sodium content of the water passing through a home water softener will be increased. Individuals who are on a sodium-restricted diet should advise their

[83]Adapted from: Joseph A. Salvato and Arthur Handley, *Rural Water Supply*, New York State Dept. of Health, Albany, N.Y. 1966, reprinted 1972, pp. 47–50.

physician that they are using home-softened water since such water is a continual source of dietary sodium. A cold-water by-pass line can be installed around the softener to supply drinking water and water for toilet flushing.

Turbidity or Muddiness

This usually occurs in water from a pond, creek, or other surface source. This water is polluted and requires coagulation, filtration, and chlorination treatment. Wells sometimes become cloudy from cave-in or seepage from clay or silt strata but usually clear up with prolonged pumping.

Sand filters can strain out mud, dirt, leaves, and foreign matter, but not all bacteria or viruses. Nor are charcoal, zeolite, or carbon filters suitable for this purpose, and in addition they clog. Iron and iron growths which sometimes cause turbidity in well water are discussed below. See also Filtration, this chapter.

Iron in Well Water

This may cause turbidity, red water, a bitter taste in tea or coffee, and, when exposed to the air, stains on plumbing fixtures, equipment, and laundry. A commercial zeolite water softener removes 1.5 to 2.0 mg/l and an iron removal filter removes up to 10 mg/l iron from well water devoid of oxygen. The water softener is regenerated with salt; the iron removal filter with potassium permanganate. Controlled addition of a polyphosphate can keep 1.0 to 2.0 mg/l iron in solution, but sodium is also added to the water.

With higher concentration of iron, the water is chlorinated to oxidize the iron, but the water should then be filtered to remove the iron precipitate before it goes to the softener. Raise the pH of the water to above 7.0 if the water is acid; soda ash is usually used for this purpose added together with the chlorine solution.

Another approach is to discharge the water to the air chamber of a pressure tank or to a sprinkler over a cascade above a tank. It is necessary to flush out the iron which settles in the tank and to filter out the remainder. Air control is needed.

Injecting a chlorine solution into the water at its source, where possible, controls the growth of iron bacteria, if this is a problem. See also Iron and Manganese Occurrence and Removal, this chapter.

Corrosive Water

This dissolves metal. shortens the life of water tanks, discolors water, and clogs pipes. Water can be made noncorrosive by passing it through a filter containing broken limestone or marble chips. The controlled addition of a polyphosphate, silicate, or soda ash (commercial units are available) usually prevents metal from going into solution. The water remains clear and staining is prevented. Use of a sodium polyphosphate would add sodium to water which would be undesirable for individuals on a low-sodium diet. See also Corrosion Cause and Control, this chapter.

Taste and Odors

Activated carbon filters are normally used to remove undesirable tastes and odors from domestic water supplies. They do not remove contamination. Hydrogen sulfide in water can be eliminated by aeration and chlorination, followed by an activated carbon filter. The activated carbon will have to be replaced when its capacity has been exhausted. Filtration alone, through a pressure filter containing a special synthetic resin, also removes up to 5 mg/l hydrogen sulfide in most cases. Use the water in question to check the effectiveness of a process before you purchase any equipment. See further discussion in this chapter.

Detergents

Detergents in water can be detected visually or by laboratory examination. When their concentration exceeds 1 mg/l, foam appears in a glass of water drawn from a faucet. Detergents themselves have not been shown to be harmful, but their presence is evidence that wastewater from one's own sewage disposal system or from a neighbor's system is entering the water supply source. In such circumstances, the sewage disposal system may be moved, a well constructed in a new area or the well extended and sealed into a deeper water-bearing formation not subject to pollution. There is no guarantee that the new water-bearing formation will not be or become polluted later. The solution of this problem is connection to a public water supply and/or to a public sewer. A granular activated carbon (GAC) filter may be used to remove detergent, but its effectiveness and cost should first be demonstrated.

Salty Water

In some parts of the country salty water may be encountered. Since the salt water generally is overlain by fresh water, the lower part of the well in the salt water zone can be sealed off. But when this is done, the yield of the well is decreased.

Sometimes, waste salt water resulting from the back-washing of a home ion exchange water softener is discharged close to the well. Since salt water is not filtered out in seeping through the soil, it may find its way into the well. The best thing to do is to discharge the wastewater as far as possible and downgrade from the well. Salt water is corrosive. It will damage grass and plants. It is a soil sterilant.

Special desalting units (using distillation, deionization and reverse osmosis) are available for residential use, but they are relatively expensive. Complete information, including effectiveness with the water in question and annual cost, should be obtained before purchase. See Desalination and Sodium (this chapter) for additional information.

Gasoline or Fuel Oil in Water

Gasoline or fuel oil may accidently get into a well. Leaking storage tanks, overflow from tank air vents, or accidental spillage near the well may be the cause. Correction requires elimination of the cause, followed by lowering of the pump drop pipe if possible, but the pollution is likely to persist in the source for a long time. The gasoline or fuel oil will gradually collect on the water surface in the well and will have to be separately pumped out until all accumulation is removed. An activated carbon filter will remove small amounts of oil or gasoline. It may become expensive if large quantities of oil or gasoline must be removed and the activated carbon replaced frequently. See also page 296.

Polluted Water

Sometimes chlorination or ultraviolet "sterilization" units are suggested to make polluted water safe for drinking without regard to the type, amount, or cause of pollution. This may be hazardous. Instead, every effort should be made to obtain water from a public system meeting the standards given in this chapter. Chlorination or the ultraviolet process is acceptable only for the treatment of clean, clear water not subject to chemical pollution.

Ultraviolet ray lamps are not considered satisfactory for the purification of water supplies which may be subject to pollution. Examples are surface water supplies such as ponds, lakes, and streams which usually vary widely in physical, chemical, and biological quality, and wells or springs in which the water may contain turbidity, color, iron, or organic matter. Pretreatment, usually including coagulation, flocculation, filtration, and chlorination, would be required ahead of the ultraviolet unit to remove substances which interfere with the effectiveness of the ultraviolet rays. In addition, certain controls are needed to ensure that the efficiency of the unit is not impaired by changes in light intensity, rate of water flow, condition of the lamp, slime accumulation, turbidity of the water, temperature conditions, etc.[84]

Similar pretreatment would be required prior to the disinfection of water which is not of good physical character when using only chlorination treatment. Check with your health department if your are considering the purchase of a chlorinator or ultraviolet unit.

DESALINATION

Desalination or desalting is the conversion of seawater or brackish water to fresh water. The conversion of treated wastewater to potable water using modified desalination processes is also being considered, but health effects and cost questions must first be resolved.

About seven-tenths of our globe is covered by seawater. The world oceans have a surface area of 139,500,000 mi^2 and a volume of 317,000,000 mi^3.[85] The oceans contain about 97 percent of the world water, brackish inland sites and polar ice make up 2.5 percent, leaving less than 0.5 percent fresh water to be used and reused for municipal, industrial, agricultural, recreational, and energy-producing purposes.[86] In addition, more than half of the earth's surface is desert or semi-desert. Under circumstances where adequate and satisfactory

[84]See "Policy Statement on use of the Ultraviolet Process for Disinfection of Water," HEW, PHS, April 1, 1966.

[85]J. H. Feth, "Water Facts and Figures for Planners and Managers," *Geological Survey Circular 601-1*, National Center, Reston Va. 22092, 1973, p. 14.

[86]*Desalting Water Probably Will Not Solve The Nation's Water Problems, But Can Help*, Report To The Congress, General Accounting Offices, Washington, D.C., May 1, 1979, p. 1.

groundwater, surface water, or rainwater is not available and a higher quality water is required, but where seawater or brackish water is available, desalination may provide an answer to the water problem. Cost of construction and energy however could be major deciding factors.

Desalting plants are in use all over the world. The Office of Water Research and Technology reports 1036 plants with capacity of 525 mgd in operation or under construction as of January 1, 1975.[87]*

Seawater has a total dissolved solids (TDS) concentration of about 36,000 mg/l. About 78 percent is sodium chloride, 11 percent magnesium chloride, 6 percent magnesium sulfate, 4 percent calcium sulfate, with the remainder potassium sulfate, calcium carbonate, and magnesium bromide, in addition to suspended solids and microbiological organisms. Water with 1000 to 4000 and up to 15,000 mg/l TDS is usually considered mildly to moderately brackish; 15,000 to 36,000 is considered heavily brackish. The source of brackish water may be groundwater or surface water sources such as oceans, estuaries, saline rivers, and lakes. Its composition can be extremely variable, containing different concentrations of sodium, magnesium, sulfate, calcium, chloride, bicarbonate, also fluoride, potassium, and nitrate. Iron, manganese, carbon dioxide, and hydrogen sulfide might also contribute to the variability of brackish water quality. Water containing more dissolved salt then seawater, such as the Great Salt Lake or the Dead Sea, is considered brine.

Desalting will remove dissolved salts and minerals such as chlorides, sulfates, and sodium in addition to hardness. Nitrates, nitrites, phosphates, fluorides, ammonia, and heavy metals are also removed to some degree. Very hard brackish water will require prior softening to make reverse osmosis or electrodialysis very effective.[88]

Some known methods for desalting water are:[89]

Membrane

Reverse Osmosis — Transport Depletion
Electrodialysis — Piezodialysis

Distillation

Multistage Flash Distillation — Vertical Tube Distillation
Multieffect Multistage Distillation — Solar Humidification
Vapor Compression

Crystallization

Vacuum Freezing-Vapor Compression — Eutectic Freezing
Secondary Refrigerant Freezing — Hydrate Formation

Chemical

Ion Exchange

*The 1981 estimate is 2200 plants, greater than 25,000 gpd, with a total capacity slightly less than two billion gpd.

[87] *The A-B-C of Desalting*, Dept. of the Interior, Office of Water Research and Technology, Washington, D.C., 1977, p. 30.

[88] Harry A. Faber, Sidney A. Bresler, and Graham Walton, "Improving Community Water Supplies with Desalting Technology," *J. Am. Water Works Assoc.*, November 1972, pp. 705–710.

[89] *The A-B-C of Desalting*, Dept. of the Interior, Office of Water Research and Technology, Washington, D.C., 1977, p. 2.

Distillation, particularly multistage, has been the process of choice for desalting seawater, with the vertical tube process gaining more acceptance. Electrodialysis and reverse osmosis appear to be more favorable for brackish water; also ion exchange for specific purposes. Only the distillation, membrane desalting, and ion exchange processes will be discussed.

Distillation

In distillation, sea water is heated to the boiling point and then into steam, usually under pressure, at a starting temperature of 250° F. The steam is collected and condensed in a chamber by coming into contact with tubes (condenser-heat exchanger) containing cool sea water. The heated saline water is passed through a series of distillation chambers in which the pressure is incrementally reduced and the water boils (made to "flash"), again at reduced temperature, with the production of steam which is collected as fresh water. The remaining, more concentrated, sea water (brine) flows to waste. In each step, the temperature of the incoming sea water is increased by the condenser-heat exchangers as it flows to the final heater. The wastewater (brine) and distilled water are also used to preheat the incoming sea water. This process is referred to as multistage flash distillation (MSF). There may be as many as 15 to 25 stages. A major problem is the formation of scale (calcium carbonate, calcium sulfate, and magnesium hydroxide) on the heat-transfer surfaces of the pipe or vessel in which the sea water is permitted to boil. This occurs at a temperature of about 160° F; but scale can be greatly minimized by pretreating the sea water to remove either the calcium or carbon dioxide.

Vertical tube distillation, multieffect multistage distillation, vapor compression distillation and solar distillation are distillation variations. Solar humidification (distillation) depends upon water evaporation at a rate determined by the temperature of the water and the prevailing humidity. The unit is covered with a peaked glass or plastic roof from which the condensate is collected. Distilled water is tasteless and low in pH if not aerated and adjusted before distribution.

Reverse Osmosis

Normally if saltwater and fresh water are separated by a semipermeable membrane, the fresh water diffuses through to the salt water as if under pressure, actually osmotic pressure. The process is known as osmosis. In reverse osmosis, pressure (typically 600 to 800 but up to 1500 psi) is applied to the salt water on one side of a special flat or cylindrical supported membrane or hollow fibre. The life of the membrane decreases with increasing pressure. In the process fresh water is separated out from the salt water into a porous or hollow channel from which the fresh water is collected. The concentration of TDS in the salt water flowing through the unit must be kept below the point at which calcium sulfate precipitation takes place. Some of the dissoved solids, 5 to 10 percent, will pass through the membrane. Chlorinated methanes and ethanes, which are common solvents, are not removed by reverse osmosis; air stripping however is effective.[90] An increase in the TDS will result in a small increase of solids in the fresh water.

[90] *Municipal Wastewater Reuse News*, AWWA Research Foundation, Denver, Colorado, August 1979, p. 7.

In reverse osmosis the salt water to be treated must be relatively clear and free of excessive hardness, iron, manganese, and organic matter to prevent fouling of the system membranes. The pretreatment may consist of softening to remove hardness; coagulation and filtration (sand, anthracite, multi-media, or diatomaceous earth) to remove turbidity, suspended matter, iron, and manganese; and filtration through activated carbon columns to remove dissolved organic chemicals. Acid is used if necessary to lower the pH and prevent calcium carbonate and magnesium hydroxide scale. Chlorine might also be used to control biological growths on the membranes.[91]

Electrodialysis

In electrodialysis the dissolved solids in the brackish water (less than 10,000 mg/l TDS) are removed by passage through a cell in which a direct electric current is imposed. Dissolved solids in the water contain positively charged ions (cations) and negatively charged ions (anions). The cations migrate to and pass through a special membrane which allows passage of the positive ions. Another special membrane allows the negative ions to pass through. The concentration of dissolved solids determines the amount of current needed. The partially desalted-demineralized water is collected and the wastewater is discharged to waste.

The plant size is determined in part by the desired amount of salt removal. However, a change in the total dissolved solids in the brackish water will result in an equal change in the treated water.[92] As in reverse osmosis, pretreatment of the brackish water is necessary to prevent fouling of the membranes and to prevent scale formation. The cost of electricity limits use of electrodialysis.

Transport depletion is a variation of the electrodialysis process. Piezodialysis is the research stage; it uses a new membrane desalting process.

Ion Exchange

In the deionization process, salts are removed from brackish water (2000 to 3000 mg/l TDS). Raw water passes through beds of special synthetic resins which have the capacity to exchange ions held in the resins with those in the raw water.

In the two-step process, at the first bed (acidic resin) sodium ions and other cations in the water are exchanged for cations (cation exchange) in the resin bed. Hydrogen ions are released and, together with the chloride ions in the raw water, pass through to the second resin bed as a weak hydrochloric acid solution. In the second resin bed, the chloride ions and other anions are taken up (anion exchange) from the water, are exchanged for hydroxide ions in the resin bed which are released, combine with the hydrogen ions to form water, and pass through with the treated water. The ion exchange beds may be in a series or in the same shell.

When the resins lose their exchange capacity and become saturated, the treatment of water is interrupted and the beds are regenerated, with acids or bases. The resins may become coated or fouled if the raw water contains excessive tur-

[91] Melvin E. Mattson, "Membrane Desalting Gets Big Push," *Water Wastes Eng.*, April 1975, pp. 35–42.

[92] Rod Chambers, "Electrodialysis or RO—how do you choose?", *World Water*, March 1979, pp. 36–37.

bidity, microorganisms, sediment, color, organic matter including dissolved organics, hardness, iron, or manganese. In such cases pretreatment to remove the offending contaminant is necessary. Chlorine in water would attack the cation resin and must also be removed prior to deionization.

Waste Disposal

The design of a desalting plant must make provision for the disposal of waste sludge from pretreatment and also of the concentrated salts and minerals in solution removed in the desalting process. The amount or volume of waste is dependent on the concentration of salts and minerals in the raw water and the amount of water desalted.

The waste from mildly brackish water (1000 to 3000 mg/l TDS) will contain from 5000 to 10,000 mg/l (TDS). The waste from a sea water desalting plant can contain as much as 70,000 mg/l (TDS).[93]

The waste disposal method will usually be determined by the location of the plant and the site geography. Methods that would be considered include disposal to the ocean, inland saline lakes and rivers, existing sewer outfalls, injection wells or sink holes where suitable rock formations exist, solar evaporation ponds, lined or tight-bottom holding ponds, or artificially created lakes. In all cases, prior approval of federal (USEPA) and state regulatory (water pollution and water supply) agencies have jurisdiction must be obtained. Surface and underground sources of drinking water and irrigation water must not be endangered.

Costs

The Office of Water Research and Technology reported that the cost of desalted water from some 1036 desalting plants around the world is upwards from 85 cents per 1000 gallons, except where fuel is available at very low cost.[94] Costs in the United States are about \$4 for sea water and \$1 for brackish water per 1000 gal, compared to 40 cents for conventional sources.[95]

An analysis was made by Miller[96] of fifteen municipalities in western United States demineralizing brackish water by reverse osmosis, electrodialysis, or ion exchange, and by combinations. Flows varied form 0.13 to 7.18 mgd and TDSs from 941 to 3236 mg/l. The demineralization cost varied from \$.37 to \$1.56 per 1000 gal. Reverse osmosis was found to be the least costly process by most of the communities. Reverse osmosis plant construction and operation cost for sea water desalting was reported to be less usually than for distillation.[97] This may not be the case however where large volumes of sea water are to be distilled, and

[93]William E. Katz and Rolf Eliassen, "Saline Water Conversion," *Water Quality & Treatment*, AWWA, McGraw-Hill, New York, 1971, p. 610.

[94]*The A-B-C of Desalting*, Dept. of the Interior, Office of Water Research and Technology, Washington, D.C., 1977, p. 1.

[95]*Desalting Water Probably Will Not Solve The Nation's Water Problems, But Can Help*, Report To The Congress, General Accounting Office, Washington, D.C., May 1, 1979, pp. i and 10.

[96]E. F. Miller, "Demineralization of Brackish Municipal Water Supplies—Comparative Costs," *J. Am. Water Works Assoc.*, July 1977, pp. 348–351.

[97]Robert A. Keller, "Seawater RO desalting moving into big league," *World Water* (Liverpool 3, London). March 1979, pp. 44–45.

where a convenient source of heat energy is available[98] such as from a power plant or incinerator or where fuel costs are low. In another report the energy break-even point of the reverse osmosis and electrodialysis treatment of brackish water and wastewater was approximately 1200 mg/l. Electrodialysis was more energy efficient below 1200 and reverse osmosis above that level.[95]

Construction and operating cost comparisons must be made with care. They are greatly influenced by location, energy costs, the TDS concentration, and the amount of pollutants such as suspended and other dissolved solids in the water to be desalted. Waste disposal and water distribution are additional factors usually considered separately.

General

The use of desalted water usually implies a dual water distribution and plumbing system, one carrying the potable desalted water and the other carrying non-potable brackish water or sea water. Obviously special precaution must be taken to prevent interconnections between these two water systems. The brackish water or sea water may be used for fire fighting, street flushing, and possibly toilet flushing.

The finished desalted water requires pH adjustment for corrosion control (lime, sodium hydroxide) and disinfection prior to distribution. It must contain not more than 500 mg/l total dissolved solids to meet drinking water standards. Up to 1000 mg/l dissolved solids might be acceptable in certain circumstances. Other standards would apply if the desalted water is used for industrial purposes. The USEPA considers a groundwater containing less than 10,000 mg/l TDS as a potential source of drinking water.[99]

Indirect benefits of desalting brackish water may include the purchase of less bottled water, use of less soap and detergents, no need for home water softeners and water-conditioning agents, and fewer plumbing and fixture repairs and replacements due to corrosion and scale buildup.[100]

TREATMENT OF WATER—DESIGN AND OPERATION CONTROL

Surface Water

The quality of surface water depends on the watershed area drained, land use, location and sources of pollution, and the natural agencies of purification, such as sedimentation, sunlight, aeration, nitrification, filtration, and dilution. These are variable and hence cannot be depended on to continuously purify water effectively. In addition increasing urbanization, industrialization, and intensive

[98] J. D. Sinclair "More efficient MSF plants are there to be specified," *World Water*, March 1979, pp. 33–34.

[99] *Fed. Reg.*, Thursday, June 14, 1979, USEPA, CFR 122, p. 34269.

[100] S. Louis Scheffer, "History of Desalting Operation, Maintenance, and Cost Experience," *J. Am. Water Works Assoc.*, November 1972, pp. 726–734.

farming have caused heavy organic and inorganic chemical discharges to streams, which are not readily removed by the usual water treatment. Treatment consisting of coagulation, flocculation, sedimentation, rapid sand filtration, and chlorination has little effect on the contaminants noted in Table 3-8. Because of these factors, heavily polluted surface waters should be avoided as drinking water supplies, if possible, and upland protected water sources should be used and preserved consistent with multipurpose uses in the best public interest.

> The American Water Works Association supports the principle that water of the highest quality be used as a source of supply for public water systems. Since each water utility is responsible for its product, determination of type and extent of recreational use of impounding reserviors shall be vested in the water utility.[101]

The growing demand for use of reservoirs for recreational purposes requires that the public understand the need for strict controls to prevent waterborne diseases and watershed disturbance. Involved are added capital and maintenance and operating costs that may increase the charges for the water and use of the recreational facilities, if the multipurpose uses are permitted.

Treatment Required

As an aid in determining the treatment that should be given water to make it safe to drink, water has been classified into several groups. The treatment required by this classification is based on the most probable number (MPN) of coliform bacteria per 100 milliliters (ml) of sample and is summarized in Table 3-15. It needs to be supplemented by information provided by chemical, physical, and microscopic examinations. For water to be generally acceptable, other treatment may be required in addition to that necessary for the elimination of disease-producing organisms. People expect the water to be safe to drink, attractive to the senses, soft, nonstaining, and neither scale-forming nor corrosive to the water system. The various treatment processes used to accomplish these results are briefly discussed under the appropriate headings below. In all cases, the water supply must meet the federal and state drinking water standards.[102] The untrained individual should not attempt to design a water-treatment plant, for life and health will be jeopardized. This is a job for a competent sanitary engineer.

Disinfection

The more common chemicals used for the disinfection of drinking water are chlorine (gas and hypochlorite), chlorine-ammonia, chlorine dioxide, and ozone. Chlorine is discussed below; the others are discussed in relation to the removal or

[101]Policy Statement Adopted by the Board of Directors on January 13, 1971 and reaffirmed on January 28, 1978.

[102]Interim Primary Drinking Water Regulations, pursuant to the Safe Drinking Water Act of 1974 (PL93-523).

Table 3-15 A Classification of Waters by Concentration of Coliform Bacteria and Treatment Required to Render the Water of Safe Sanitary Quality[a]

Group No.	Maximum Permissible Average MPN Total Coliform Bacteria per Month[b]	Treatment Required
1	Not more than 10% of all 10-ml or 60% of 100-ml portions positive; not more than 1.0 coliform bacteria/100 ml	None for protected underground water, but, at the minimum, chlorination for surface water
2	Not more than 50/100 ml	Simple chlorination or equivalent
3	Not more than 5000/100 ml and this MPN exceeded in not more than 20% of samples	Rapid sand filtration (including coagulation) or its equivalent plus continuous chlorination
4	MPN greater than 5000/100 ml in more than 20% of samples and not exceeding 20,000/100 ml in more than 5% of the samples	Auxiliary treatment such as 30 to 90 days storage, presettling, prechlorination, or equivalent plus complete filtration and chlorination
5	MPN exceeds Group No. 4	Prolonged storage or equivalent to bring within Groups 1 to 4

[a]Physical, inorganic, and organic chemicals, and radioactivity concentrations in the raw water and ease of removal by the proposed treatment must also be taken into consideration. See Table 3-8 and *Manual for Evaluating Public Drinking Water Supplies*, Environmental Control Administration, PHS, Cincinnati, Ohio, 1969, pp. 4–11.
[b]Fecal coliforms not to exceed 20% of total coliform organisms. The monthly geometric mean of the MPN for Group 2 may be less than 100 and, for Groups 3 and 4, less than 20,000/100 ml with the indicated treatment. (The total coliform density of 20,000/100 ml may be exceeded if fecal coliform do not exceed 2000/100 ml monthly geometric mean.) Complete treatment for Group 2 water is recommended.

reduction of objectionable tastes and odors and trihalomethanes. Other disinfectants that may be used under certain circumstances include ultraviolet radiation, bromine, iodine, silver, and chlorinated lime.

The National Research Council-National Academy of Science, in a study of disinfectants, concluded that there had not been sufficient research under actual water treatment conditions for the reactions of disinfectants and their by-products to be adequately understood and that the chemical side effects of disinfectants "should be examined in detail."*[103]

Chlorination is the most common method of destroying the disease-producing

*Greenburg points out that "the health effects of their (chlorine dioxide and ozone) reaction products, particularly the chlorite ion from chlorine dioxide and oxidized organic compounds from ozone are uncertain," and adds that "if unequivocal safety information becomes available, changes from chlorine to chlorine dioxide or ozone may be indicated but only if the manipulation of chlorination methods proves incapable of minimizing carcinogen hazard." (Arnold E. Greenburg, "Public health aspects of alternative water disinfectants," *J. Am. Water Works Assn.*, January 1981, pp. 31–33.)
[103]"Update," *J. Am. Water Works Assoc.*, July 1979, p. 9.

organisms that might normally be found in water used for drinking in the United States. The water so treated should be relatively clear and clean with an average monthly MPN of coliform bacteria of not more than 50/100 ml. Clean lake and stream waters and well, spring, and infiltration-gallery supplies not subject to significant pollution can be made of safe sanitary quality by continuous and effective chlorination; but surface sources usually also require complete treatment to protect against viruses, protozoa, and helminths.

Operation of the chlorinator should be automatic, proportional to the flow of water, and adjusted to the temperature and chlorine demand of the water. A spare machine should be on the line. A complete set of spare parts for the equipment will make possible immediate repairs. The chlorinator should provide for the positive injection of chlorine and be selected with due regard to the pumping head and maximum and minimum flow of water to be treated. The point of chlorine application should be selected so as to provide adequate mixing and at least 5 min, preferably 30 min, chlorine contact with the water to be treated before it reaches the first consumer.

Hypochlorinators are generally used to feed relatively small quantities of chlorine as 1 to 5 percent sodium or calcium hypochlorite solution. Positive feed machines are fairly reliable and simple to operate. Hypochlorite is corrosive and may produce severe burns. It should be stored in the original container in a cool, well-ventilated, dry place. Gas machines feed larger quantities of chlorine and require certain precautions as noted below.

Gas Chlorinator

When a dry feed gas chlorinator or a solution feed gas chlorinator is used, the chlorinator and liquid chlorine cylinders should be located in a separate gas-tight room that is mechanically ventilated to provide two air changes per minute, with the exhaust openings at floor level opposite the air inlets. Exhaust ducts must be separate from any other ventilating system of ducts and extend to a height and location that will not endanger the public, personnel, or property and ensure adequate dilution. The door to the room should have a glass inspection panel, and a chlorine gas mask or self-contained breathing apparatus, approved by the U.S. Bureau of Mines, should be available outside of the chlorinator and chlorine cylinder room.* The chlorine canister type of mask is suitable for low concentrations of chlorine in air. The self-contained breathing apparatus (pressure demand) is recommended as it can be used for high concentrations of chlorine.

The temperature around the chlorine cylinders should be between 50 and 85° F and cooler than the temperature of the chlorinator room to prevent condensation of chlorine in the line conducting chlorine or in the chlorinator. Cylinders must

*A chlorine container holding 100 lb of chlorine developing a leak that cannot be repaired can have the chlorine absorbed by 125 lb of caustic soda in 40 gal of water, 300 lb of soda ash in 100 gal of water, or 125 lb of hydrated lime in 125 gal of water continuously agitated. (Chlorine Institute, Inc., New York, N.Y.) Call the nearest supplier or producer in case of emergency.

be stored at a temperature below 140° F.* A platform scale is needed for the weighing of chlorine cylinders in use to determine the pounds of chlorine used each day and to anticipate when a new cylinder will be needed. Cylinders should be connected to a manifold so that chlorine may be drawn from several cylinders at a time and so that cylinders can be replaced without interrupting chlorination. Do not draw more than 35 to 40 lb of chlorine per day at a continuous rate from a 100- or 150-lb cylinder to prevent clogging by chlorine ice. Liquid chlorine comes in 100- and 150-lb cylinders, in 1-ton containers, and in 16- to 90-ton tank cars. Smaller cylinders are available. The major factors affecting withdrawal rates are ambient air temperature and size and type cylinder. The normal operating temperature is 70° F.

A relatively clear source of water of adequate volume and pressure is necessary to prevent clogging of injectors and strainers and to assure proper chlorination at all times. The water pressure to operate a gas chlorinator should be at least 15 psi and about three times the back pressure (water pressure at point of application, plus friction loss in the chlorine solution hose, and difference in elevation between the point of application and chlorinator) against which the chlorine is injected. About 40 to 50 gpd of water is needed per pound of chlorine to be added.

Testing for Residual Chlorine

The recommended tests for measuring residual chlorine in water are the DPD colormetric and the stabilized neutral orthotolidine (SNORT) methods.[104] The DPD method is approved by the USEPA. In any case, all tests should be made in accordance with accepted procedures such as in *Standard Methods for the Examination of Water and Wastewater* (see Bibliography).

The leuco crystal violet method is also satisfactory. It determines free available chlorine with minimal interference from combined chlorine, iron, nitrates, and nitrites;[105] but DPD is considered a more accurate field test. However, a study made for the USEPA[106]

> indicated that the best accuracy and precision was obtained by leuco crystal violet and the stabilized neutral orthotolidine (SNORT) procedures, followed by DPD-titrimetric, amperometric titration, DPD-colorimetric, and methyl orange. By far the poorest was the orthotolidine-arsenite (OTA) procedure.

Guter, et al., reported Syringaldazine the most specific for free available chlorine and DPD more accurate and precise over temperature and pH variations. Leuco crystal violet and SNORT performed with satisfactory accuracy. The SNORT

*The fusible plugs are designed to soften or melt at a temperature between 158 and 165° F. The chlorinator should have automatic shutoff if water pressure is lost or if chlorine piping leaks or breaks.

[104]"Disinfection—Committee Report," *J. Am. Water Works Assoc.*, April 1978, pp. 219–222.

[105]A. P. Black and G. P. Whittle, "Determination of Halogen Residuals. Part II. Free and Total Chlorine," *J. Am. Water Works Assoc.*, May 1967, p. 607.

[106]R. J. Lishka and E. F. McFarren, *Water Chlorine (Residual) No. 2*, Report Number 40, Analytical Reference Service, USEPA, Office of Water Programs, Cincinnati, Ohio, 1971.

procedure showed false positive readings for free available chlorine in the presence of combined chlorine.[107] Combined chlorine can also cause interference with the DPD method if readings are not made within one minute.

Chlorine Treatment for Operation and Microbiological Control

To assure that only properly treated water is distributed, it is important to have a competent and trustworthy person in charge of the chlorination plant. He should keep daily records showing the gallons of water treated, the pounds of chlorine or quarts of chlorine solution used and its strength, the gross weight of chlorine cylinders if used, the setting of the chlorinator, the time residual chlorine tests made, the results of such tests, and any repairs or maintenance, power failures, modifications, or unusual occurrences dealing with the treatment plant or water system. Where large amounts of chlorine are needed, the use of ton containers can effect a saving in cost, as well as in labor, and possibly reduce chlorine gas leakage.

The required chlorine dosage should take into consideration the appearance as well as the quality of a water. Pollution of the source of water, the type of microorganisms likely to be present, the pH of the water, and the temperature and degree of treatment a water receives are all very important.

The chlorine residual that will give effective disinfection of a *clear* water has been studied by Butterfield.[108] The germicidal efficiency of chlorine is primarily dependent on the percent free chlorine that is in the form of hypochlorous acid (HOCl), which in turn is dependent on the pH and temperature of the water, as can be seen in Table 3-16.

In a review of the literature, Greenberg and Kupka concluded that a chlorine dose of at least 20 mg/l with a contact time of 2 hr is needed to adequately disinfect a biologically treated sewage effluent containing tubercle bacilli.[109]

Laboratory studies by Kelly and Sanderson indicated that

> depending on pH level and temperature, residual chlorine values of greater than 4 ppm, with 5 min contact, or contact periods of at least four hours with a residual chlorine value of 0.5 ppm, are necessary to inactivate viruses, and that the recommended standard for disinfection of sewage by chlorine (0.5 ppm residual after fifteen min contact) does not destroy viruses.[110]

Another study showed that inactivation of partially purified poliomyelitis virus in water required a free residual chlorine after 10 min of 0.05 mg/l at a pH of 6.85 to 7.4. A residual chloramine value of 0.50 to 0.75 mg/l usually inactivated the

[107]K. J. Guter, et al., "Evaluation of Existing Field Test Kits for Determining Free Chlorine Residuals in Aqueous Solutions," *J. Am. Water Works Assoc.*, January 1974, pp. 34–43.

[108]C. T. Butterfield, "Bactericidal Properties of Chloramines and Free Chlorine in Water," *Pub. Health Rep.*, **63,** 934–940 (1948).

[109]Arnold E. Greenberg and Edward Kupka, "Tuberculosis Transmission by Waste Water—A Review," *Sewage and Ind. Wastes*, **29,** No. 5, 524–537 (May 1957).

[110]Sally Kelly and Wallace W. Sanderson, "Viruses in Sewage," *Health News*, **36,** 14–17 (June 1959).

Table 3-16 Chlorine Residual for Effective Disinfection of Demand-Free Water

pH	Approximate Percent at 68 to 32° F[b]		Bactericidal Treatment[a]		Cysticidal Treatment Free Available Chlorine After 30 min		
	HOCl	OCl^-	Free Available Chlorine After 10 min at 32 to 78° F	Combined Available Chlorine After 60 min at 32 to 78° F	36 to 41° F[c]	60° F[c]	78° F[b]
5.0	—	—	—	—	—	2.3	—
6.0	98 to 97	2 to 3	0.2	1.0	7.2	—	1.9[d]
7.0	83 to 75	17 to 25	0.2	1.5	10.0	3.1	2.5[d]
7.2	74 to 62	26 to 38	—	—	—	—	2.6[d]
7.3	68 to 57	32 to 43	—	—	—	—	2.8[d]
7.4	64 to 52	36 to 48	—	—	—	—	3.0[d]
7.5	58 to 47	42 to 53	—	—	14.0[d]	4.7	3.2[d]
7.6	53 to 42	47 to 58	—	—	—	—	3.5[d]
7.7	46 to 37	53 to 64	—	—	16.0[d]	6.0	3.8[d]
7.8	40 to 32	60 to 68	—	—	—	—	4.2[d]
8.0	32 to 23	68 to 77	0.4	1.8	22.0	9.9	5.0[d]
9.0	5 to 3	95 to 97	0.8	Reduce pH of water	—	78.0	20.0[d]
10.0	0	100	0.8	to below 9.0	—	761	170[d]

Note: Free chlorine = HOCl. Free available chlorine = HOCl + OCl^-. Combined available chlorine = chlorine bound to nitrogenous matter as chloramine. Only free available chlorine or combined available chlorine is measured by present testing methods; therefore to determine actual free chlorine (HOCl), correct reading by percent shown above. "Chlorine residual," as the term is generally used, is the combined available chlorine and free available chlorine = total residual chlorine. When the chlorine to ammonia reaches 15 or 20:1, nitrogen trichloride is formed; it is acrid and highly explosive. Ventilate!

Viricidal treatment requires a free available chlorine of 0.53 mg/l at pH 7 and 5 mg/l at pH 8.5 in 32° F demand-free water. For water at a temperature of 77 to 82.4° F and pH 7 to 9 a free available chlorine of 0.3 mg/l is adequate. (*Manual For Evaluating Public Drinking Water Supplies*, Environmental Control Administration, PHS Pub. 1820, Cincinnati, Ohio, 1969.) At a pH 7 and temperature of 77° F at least 9 mg/l combined available chlorine is needed with 30-min contact time. Turbidity should be less than one Jackson Unit.

The above results are based on studies made under laboratory conditions using water free of suspended matter and chlorine demand.

In practice, unless otherwise indicated, at least 0.4 to 0.5 mg/l free residual chlorine for 30 min or 2 mg/l combined residual chlorine for 3 hr should be maintained in a clear water before delivery to the consumer. The state health department may require more, dependent on source of raw water and sanitary survey.

[a] Butterfield, op. cit.

[b] *Water Treatment Plant Design*, AWWA, New York, 1969, pp. 153 and 165; and Edward W. Moore, "Fundamentals of Chlorination of Sewage and Waste," *Water Sewage Works* (March 1951) 130–136.

[c] S. L. Chang, "Studies on *Endamoeba histolytica*," *War Medicine*, **5**, 46 (1944); see also W. Brewster Snow, "Recommended Chlorine Residuals for Military Water Supplies," *J. Am. Water Works Assoc.*, **48**, 1510 (December 1956). Giardia cysts probably react same as amoeba cysts.

[d] Approximations. All residual chlorine results reported as mg/l. One mg/l hypochlorous acid gives 1.35 mg/l free available chlorine as HOCl and OCl^- distributed as noted above. The HOCl component is the markedly superior disinfectant, about 40 to 80 times more effective than the hypochlorite ion (OCl^-).

virus in 2 hr.[111] Destruction of Coxsackie virus required 7 to 46 times as much free chlorine as for *E. Coli.*[112] Infectious hepatitis virus was not inactivated by 1.0 mg/l total chlorine after 30 min, nor by coagulation, settling, and filtration (diatomite); but coagulation, settling, filtration, and chlorination to 1.1 mg/l total and 0.4 mg/l free chlorine was effective.[113] Bush and Isherwood suggest

> The use of activated sludge with abnormally high sludge volume index followed by sand filtration may produce the kind of control necessary to stop virus. Chlorination with five-tenths parts per million chlorine residual for an eight hour contact period seems adequate to inactivate Coxsackie virus.[114]

Malina[115] summarized the effectiveness of water and wastewater treatment processes on the removal of viruses. The virus concentration in untreated municipal wastewater was found to range from about 200 plaque-forming units per liter (PFU/l) in cold weather to about 7000 in warm months in the United States, with 4000 to 7000 PFU/l common. In contrast, the virus concentration in South Africa was found to be greater than 100,000 PFU/l. Virus removal in wastewater is related in part to particulate removal. Possible virus removal by various wastewater treatment systems are:

Primary sedimentation	0–55%
Activated sludge	64–99%
Contact stabilization	74–95%
Trickling filters	19–94%
Stabilization ponds	92–100%
Chlorine*	99–100%
Iodine*	100%
Ozone*	100%

Chemical coagulation, with adequate concentrations of aluminum sulfate or ferric chloride, of surface water used as a source of drinking water, or of wastewater which has received biological treatment, can remove 99 percent of the viruses. A high pH of 10.8 to 11.5, such as softening with excess lime, can achieve better than 99 percent virus removal.

Filtration using sand and/or anthracite, following coagulation, can remove 99 percent of the viruses, but viruses penetrate the media with floc breakthrough and turbidity at low alum feed. Diatomaceous earth filtration can remove better than 98 percent of the viruses, particularly if the water is pretreated. Activated

*As final treatment

[111]Serge G. Lensen et al., "Inactivation of Partially Purified Poliomyelitis Virus in Water by Chlorination, II," *Am. J. Public Health*, **37,** No. 7, 869–874 (July 1947).

[112]N. A. Clarke and P. W. Kabler, "The Inactivation of Purified Coxsackie Virus in Water," *Am. J. Hygiene*, **59,** 119–127 (January 1954).

[113]Greenberg and Kupka, op. cit.

[114]Albert F. Bush and John D. Isherwood, "Virus Removal in Sewage Treatment," *J. Sanit. Eng. Div., ASCE*, **92,** No. SA 1, Proc. Paper 4653, 99–107 (February 1966).

[115]Joseph E. Malina, Jr., "The Effect of Unit Processes of Water and Wastewater Treatment on Virus Removal," in *Virus and Trace Contaminants in Water and Wastewater*, Borchardt, Cleland, Redman, and Oliver, Eds., Ann Arbor Science, 1977, pp. 33–52.

carbon adsorption is not suitable for virus removal. Reverse osmosis and ultrafiltration, when followed by disinfection, can produce a virus-free water.

A conventional municipal biological wastewater treatment plant can produce an effluent with less than 10 PFU/l. When followed by conventional water treatment incorporating filtration and chlorination, a virus-free water can be obtained.[115]

The removal of nematodes requires prechlorination to produce 0.4 to 0.5 mg/l residual after 6 hours' retention followed by settling. The pathogenic fungus *Histoplasma capsulatum* can be expected in surface-water supplies, in treated water stored in open reservoirs, and in improperly protected well-water supplies. Fungicidal action is obtained at a pH of 7.4 and at a water temperature of 26°C with 0.35 mg/l free chlorine after 4 hours' contact and with 1.8 mg/l free chlorine after 35 minutes' contact. Complete rapid sand filter treatment completely removed all viable spores even before chlorination.[116]

Coliform bacteria can be continually found in a chlorinated surface-water supply (turbidity 3.8 to 84 units, iron particles, and microscopic counts up to 2000 units) containing between 0.1 and 0.5 mg/l of free residual chlorine and between 0.7 and 1.0 mg/l total residual chlorine after more than 30 minutes' contact time.[117]

It is evident from available information that the coliform index may give a false sense of security when applied to waters subject to intermittent doses of pollution. The effectiveness of proper disinfection, including inactivation of viruses, other conditions being the same, is largely dependent on the freedom from suspended material and organic matter in the water being treated. Treated water having a turbidity of less than 5 nephelometric turbidity units (NTUs) (ideally less than 0.1), a pH less than 8, and an HOCl residual of 1 mg/l after 30 min contact provides an acceptable level of protection.[118]

Free residual chlorination is the addition of sufficient chlorine to yield a free chlorine residual in the water supply in an amount equal to more than 85 percent of the total chlorine present. When the ratio of chlorine to ammonia is 5:1 (by weight), the chlorine residual is all monochloramine; when the ratio reaches 10:1, dichloramine is also formed; when the ratio reaches 15 or 20:1, nitrogen trichloride is formed. Nitrogen trichloride as low as 0.05 mg/l causes an offensive and acrid odor that can be removed by carbon, aeration, exposure to sunlight, or forced ventilation indoors.[119] It is also highly explosive. The reaction of chlorine in water is shown in Figure 3-16.

[116]Dwight F. Metzler, Cassandra Ritter, and Russell L. Culp, "Combined Effect of Water Purification Processes on the Removal of *Histoplasma capsulatum* from Water," *Am. J. Public Health*, **46,** No. 12, 1571–1575 (December 1956).

[117]Author's personal experience reported in *1956 Annual Report, Division of Environmental Hygiene*, Rensselaer County Department of Health, Troy, N.Y., p. 11. Confirmed by Wallace W. Sanderson and Sally Kelly, New York State Dept. of Health, Albany, in *Advances in Water Pollution Research Proceedings*, Vol. 2, Int. Conf., London, September 1962, Pergamon Press, London, pp. 536–541.

[118]"Engineering Evaluation of Virus Hazard in Water," *J. Sanit. Eng. Div., ASCE*, **96,** No. SA 1, Proc. Paper 7112, 111–150 (February 1970). (By the Committee on Environmental Quality Management of the Sanitary Engineering Division.)

[119]George C. White, "Chlorination and Dechlorination: A Scientific and Practical Approach," *J. Am. Water Works Assoc.*, May 1968, pp. 540–561.

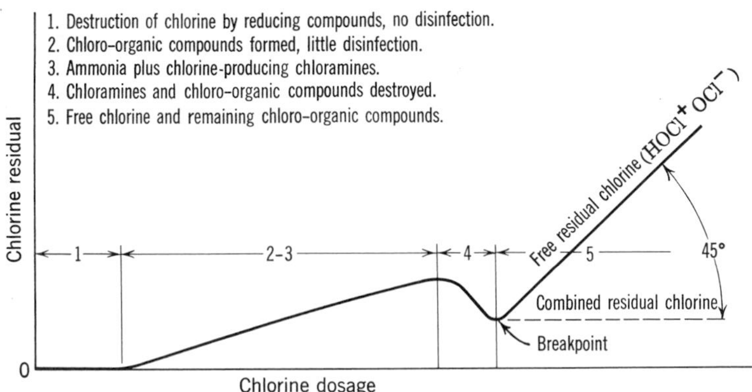

Figure 3-16 The reaction of chlorine in water. (Adapted from *Manual of Instruction for Water Plant Chlorinator Operators*, New York State Dept. of Health, Albany.)

In the presence of ammonia, organic matter, and other chlorine-consuming materials, the required chlorine dosage to produce a free residual will be high. The water is then said to have a high chlorine demand. With free residual chlorination, water is bleached, and iron, manganese, and organic matter are coagulated by chlorine and precipitated, particularly when the water is stored in a reservoir or basin for at least 2 hr. Most taste- and odor-producing compounds are destroyed; the reduction of sulfates to taste- and odor-producing sulfides is prevented; and objectionable growths and organisms in the mains are controlled or eliminated, provided a free chlorine residual is maintained in the water. An indication of accidental pollution of water in the mains is also obtained if the free chlorine residual is lost, provided chlorination is not interrupted.

The formation of trihalomethanes and other chloro-organics, their prevention, control, and removal is discussed later in this chapter.

Distribution System Contamination

Once a water supply distribution system is contaminated with untreated water, the presence of coliform organisms may persist for an extended period of time. A surface-water supply or an inadequately filtered water supply may admit into a distribution system organic matter, minerals, and sediment, including fungi, algae, macroscopic organisms, and microscopic organisms. These flow through or settle in the mains or become attached and grow inside the mains when chlorination is marginal or inadequate to destroy them. Suspended matter and iron deposits will intermingle with and harbor the growths. Hence the admission of contaminated water into a distribution system, even for a short time, will have the effect of inoculating the growth media existing inside the mains with coliform and other organisms. Elimination of the coliform organisms will therefore involve removal of the growth media and harborage material. Bacteriological control of the water supply is lost until the coliform organisms are removed, unless a free chlorine residual of at least 0.2 to 0.4 mg/l is maintained in active parts of a distribution system.

If a positive program of continuous heavy chlorination at the rate of 5 to 10 mg/l, coupled with routine flushing of the main, is maintained, it is possible to eliminate the coliform on the inside surface of the pipes and hence the effects of accidental contamination in 2 to 3 weeks or less. If a weak program of chlorination is followed, with chlorine dosage of less than 5 mg/l, the contamination may persist for 8 or 9 months. The rapidity with which a contaminated distribution system is cleared will depend on many factors: uninterruption of chlorination even momentarily; the chlorine dosage and residual maintained in the entire distribution system; the growths in the mains and degree of pipe incrustation; conscientiousness in flushing the distribution system; the social, economic, and political deterrents; and, mostly, the competency of the responsible individual.

Plain Sedimentation

Plain sedimentation is the quiescent settling or storage of water, such as would take place in a reservoir, lake, or basin, without the aid of chemicals, preferably for a month or longer. This natural treatment results in the settling out of suspended solids, reduction of hardness, ammonia, lead, cadmium, and other heavy metals, some synthetic organic chemicals, and fecal coliform, also removal of color (due to the action of sunlight), and death of bacteria principally because of the unfavorable temperature, lack of suitable food, and sterilizing effect of sunlight. Certain microscopic organisms, such as protozoa, consume bacteria, thereby aiding in purification of the water. Experiments conducted by Sir Alexander Houston showed that polluted water stored for periods of 5 weeks at 32° F, 4 weeks at 41° F, 3 weeks at 50° F, or 2 weeks at 64.4° F effected the elimination of practically all bacteria.[120] This treatment may, under certain conditions, be considered equivalent to filtration. Plain sedimentation, however, has some disadvantages that must be taken into consideration and controlled. The growth of microscopic organisms that cause unpleasant tastes and odors is encouraged, and pollution by surface wash, fertilizers, pesticides, recreational uses, birds, sewage, and industrial wastes may occur unless steps are taken to prevent or reduce these possibilities. Although subsidence permits bacteria to die off, it also permits bacteria to accumulate and grow in reservoir bottom mud under favorable conditions. In addition, iron and manganese may go into solution, carbon dioxide may increase, and hydrogen sulfide may be produced.

Presettling reservoirs are sometimes used to eliminate heavy turbidity or pollution and thus prepare the water for treatment by coagulation, settling, and filtration. Ordinarily, at least two basins are provided to permit one to be cleaned while the other is in use. A capacity sufficient to give a retention period of at least

[120] *Water Quality and Treatment*, 2nd ed. AWWA, New York, 1950, p. 94. Also P. K. Knoppert, G. Oskam, and E. G. H. Vreedenburgh, "An overview of European water treatment practice," *J. Am. Water Works Assoc.*, November 1980, pp. 592–599.

2 or 3 days is desirable. When heavily polluted water is to be conditioned, provision can be made for preliminary coagulation at the point of entrance of the water into the reservoirs followed by chlorination at the exit. However consideration must be given to the possible formation of trihalomethanes and their prevention.

Microstraining

Microstraining is a process designed to reduce the suspended solids, including plankton, in a water. The filtering media consist of very finely woven fabrics of stainless steel on a revolving drum. Applications to water supplies are primarily the clarification of relatively clean surface waters low in true color and colloidal turbidity, in which microstraining and disinfection constitute the pretreatment; and the clarification of waters ahead of slow or rapid sand filters and diatomite filters. Removals of the commoner types of algae have been as high as 95 percent. Wash-water consumption may run from 1 to 3 percent of the flow through the unit. Blinding of the fabric rarely occurs but may do so, from inadequate wash-water pressure or the presence of bacterial slimes. Cleansing is readily accomplished with commercial sodium hypochlorite.[121] Small head losses and low maintenance costs may make the microstrainer attractive for small installations.

Unit sizes start at about 2½ ft in diameter by 2 ft wide. These have a capacity varying between 50,000 and 250,000 gpd depending on the type and amount of solids in the water and the fabric used. Larger units have capacities in excess of 10 mgd.

Coagulation, Flocculation, and Settling

The addition to water of a coagulant such as alum (aluminum sulfate) permits particles to come together and results in the formation of a flocculent mass, or floc, which enmeshes microorganisms, suspended particles, and colloidal matter, removing and attracting these materials in settling out. The common coagulants used are alum, "black alum," activated alum, ammonium alum, sodium aluminate, copperas (ferrous sulfate), chlorinated copperas, ferric sulfate, ferric chloride, pulverized limestone, and clays.

To adjust the chemical reaction for improved coagulation, it is sometimes necessary to first add soda ash, hydrated lime, quicklime, or sulfuric acid. The mixing of the coagulant is usually done in two steps. The first step is rapid or flash mix and the second, slow mix, during which flocculation takes place. Rapid mix is a violent agitation for not more than 30 sec and may be accomplished by a mechanical agitator, pump impeller and pipe fittings, baffles, hydraulic jump, or

[121]George J. Turre, "Use of Micro-strainer Unit at Denver" and George R. Evans, "Discussion," *J. Am. Water Works Assoc.*, March 1959, pp. 354–362.

other means. Slow mix is accomplished by means of baffles or a mechanical mixer to promote formation of a floc and provide a detention of at least 30 min. The flocculated water then flows to the settling or sedimentation basin designed to provide a retention of 4 to 6 hr, an overflow rate of about 500 gpd per square foot of area, or 20,000 gpd per foot of weir length. Around 80 percent of the turbidity, color, and bacteria are removed by this treatment. It is always recommended that mixing tanks and settling basins be at least two in number to permit cleaning and repairs without interrupting completely the water treatment, even though mechanical cleaning equipment is installed.

For the control of coagulation, jar tests are made in the laboratory to determine the approximate dosage (normally between 10 and 50 mg/l) of chemicals that appear to produce the best results.[122] Then, with this as a guide, the chemical-dosing equipment, dry feed or solution feed, is adjusted to add the desired quantity of chemical proportional to the flow of water treated to give the best results. Standby chemical feed units and alarm devices are necessary to assure continuous treatment.

Zeta-potential is also used to control coagulation. It involves determination of the speed at which particles move through an electric field caused by a direct current passing through the raw water. Best flocculation takes place when the charge approaches zero, giving best precipitation when a coagulant such as aluminum sulfate, assisted by a polyelectrolyte if necessary, is added.

The use of alum, a polymer, and activated clay may assist coagulation and clarification of certain waters. A faster-settling and more filterable floc is reported which is less affected by temperature change or excessive flows. Less plugging of filters, longer filter runs, more consistent effluent turbidity, less backwash water, less sludge volume, and easier dewatering of sludge is claimed for polymer, clay-alum treatment.[123]

Another device for the coagulation and settling of water consists of a unit in which the water to be treated is introduced at the bottom and flows upward through a blanket of settled floc. The clarified water flows off at the top. These basins are referred to as upflow suspended-solids contact clarifiers. The detention period used in treating surface water is 4 hours, but may be as little as 1.5 to 2 hours depending on the quality of the raw water. The normal upflow rate is 1440 gpd per square foot of clarifier surface area and the overflow rate is 14,400 gpd per foot of weir length. A major advantage claimed, where applicable, is reduction of the detention period and hence savings in space. Disadvantages include possible loss of sludge blanket with changing water temperature and variable water quality.

Tube settlers are also coming into use. They are shallow tubes, usually inclined at an angle of approximately 60 degrees to the vertical. The tube cross section may be square, trapezoidal, triangular, or circular. Effective operation requires laminar flow, adequate retention, nonscouring velocities, and floc particle settling

[122]Charles R. Cox, *Water Supply Control*, Bull. 22, New York State Dept. of Health, Albany, 1952, pp. 38–53; See also *Standard Methods for the Examination of Water and Wastewater*, op. cit.

[123]Dale R. Lawson, "Polymer cuts cost of Rochester water," *Am. City County*, September 1977, pp. 97–98.

with allowance for sludge accumulation and desludging at maximum flow rates.[124] Pilot plant studies are advisable prior to actual design and construction.

Lamella Separators are similar to the tube settlers except that inclined plates are used instead of tubes.

Filtration

Filters are of the slow sand, rapid sand or other granular media (including multimedia), and pressure (or vacuum) type. Each has application under various conditions. The primary purpose of filters is to remove suspended materials. Of the filters mentioned, the slow sand filter is recommended for use at small communities and rural places, where adaptable. A rapid sand filter is not recommended, because of the rather complicated control required to obtain satisfactory results, unless competent supervision and operation can be assured. The pressure filter, including the diatomaceous earth type, is commonly used for the filtration of industrial water supplies and swimming pool water; it is not recommended for the treatment of drinking water, except where considered suitable under the conditions of the proposed use. Variations of the conventional rapid sand filter, which may have application where raw water characteristics permit, are direct filtration, deep-bed filtration (4 to 8 ft media depth and 1.0 to 2.0 mm size), declining flow rate filtration, and granular activated carbon filters. In all cases their feasibility and effectiveness should first be demonstrated by pilot plant studies at the site.

Filter units that are attached to faucets, porous stone filters, and unglazed porcelain (Pasteur filter) or Berkefield filters may develop hairline cracks. They are unreliable and should not be depended on to remove pathogenic bacteria. Ceramic filters may have limited application under certain conditions.[125] They should be cleaned and sterilized in boiling water once a week. Other faucet-type filters purported to "purify" water have at best a limited useful life in removing taste- and odor-causing organic compounds and turbidity. See Household Filters, this chapter.

Slow Sand Filter

A slow sand filter consists of a watertight basin, usually covered, built of concrete. The basin holds a special sand 24 to 48 in. deep, which is supported on a 12-in. layer of graded gravel placed over an underdrain system that may consist of open-joint, porous, or perforated pipe or conduits. The sand should have an effective size of 0.25 to 0.35 mm and a uniformity (nonuniformity) coefficient of 0.25 to 0.35. Operation of the filter is controlled so that filtration will take place at a rate

[124]Roderick M. Willis, "Tubular Settlers—A Technical Review," *J. Am. Water Works Assoc.*, June 1978, pp. 331–335.

[125]Robert Newton Clark, "The Purification of Water on a Small Scale," *Bull. Santé organization mondiale* (*WHO Bull.*), 14, 820–826 (1956).

of 1 to 4 million gal per acre per day, with $2\frac{1}{2}$ million gal as an average rate. This would correspond to a filter rate of 23 to 92 gal/ft^2 of sand area per day or an average rate of 57 gal. A rate of 10 million gal may be used if permitted by the approving authority.

From a practical standpoint, the water that is to be filtered should have low color, less than 30 units, low coliform concentration (less than 1000 per 100 ml), and be low in suspended matter with a turbidity of less than 50 units; otherwise the filter will clog quickly. A plain sedimentation basin, or other pretreatment, ahead of the filter can be used to reduce the suspended matter, turbidity and coliform concentration of the water if necessary. A loss-of-head gauge should be provided on the filter to show the resistance the sand bed offers to the flow of water through it and to show when the filter needs cleaning. This is done by draining the water out of the sand bed and scraping 1 to 2 in. of sand with adhering particles off the top of the bed. The sand is washed and replaced when the depth of sand is reduced to about 24 in. A scraper or flat shovel is practical for removing the top layer of clogged sand. The sand surface can also be washed in place by a special washer traveling over the sand bed. Slow sand filters should be constructed in pairs. These filters are easily controlled and produce a consistently satisfactory water, when followed by disinfection.

A well-operated plant will remove 98 to 99.5 percent of the bacteria in the raw water (after a film has formed on the surface of the sand, which will require slow filtration for several days to 2 weeks). Chlorination of the filtered water is necessary to destroy those bacteria that pass through the filter and to destroy bacteria that grow or enter the storage basin and water system. This type plant will also remove about 25 to 40 percent of the color in the untreated water. Chlorination of the sand filter itself is desirable either continuously or periodically to destroy bacteria that grows within the sand bed, supporting gravel, or underdrain system. Continuous prechlorination at a dosage to provide 0.3 to 0.5 mg/l in the water on top of the filter will not harm the filter film and will increase the length of the filter run.

A slow sand filter suitable for a small rural water supply is shown in Figure 3-17. Details relating to design are given in Table 3-17. The rate of filtration in this filter is controlled by selecting an orifice and filter area that will deliver not more than 50 gal/ft^2 of filter area per day and thus prevent excessive rates of filtration that would endanger the quality of the treated water. Where competent and trained personnel are available, the rate of flow can be controlled by manipulating a gate valve on the effluent line from each filter, provided a venturi, orifice, or other suitable meter, with indicating and preferably recording instruments, is installed to measure the rate of flow. The valve can then be adjusted to give the desired rate of filtration until the filter needs cleaning. Another practical method of controlling the rate of filtration is by installing a float valve on the filter effluent line as shown in Figure 3-18. The valve is actuated by the water level in a float chamber, which is constructed to maintain a reasonably constant head over an orifice in the float chamber. A hydraulically operated float can be connected to a control valve by tubing and hence be located at some distance from the valve.

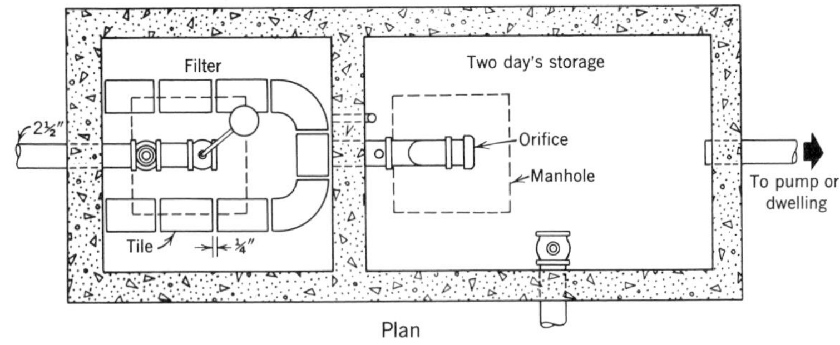

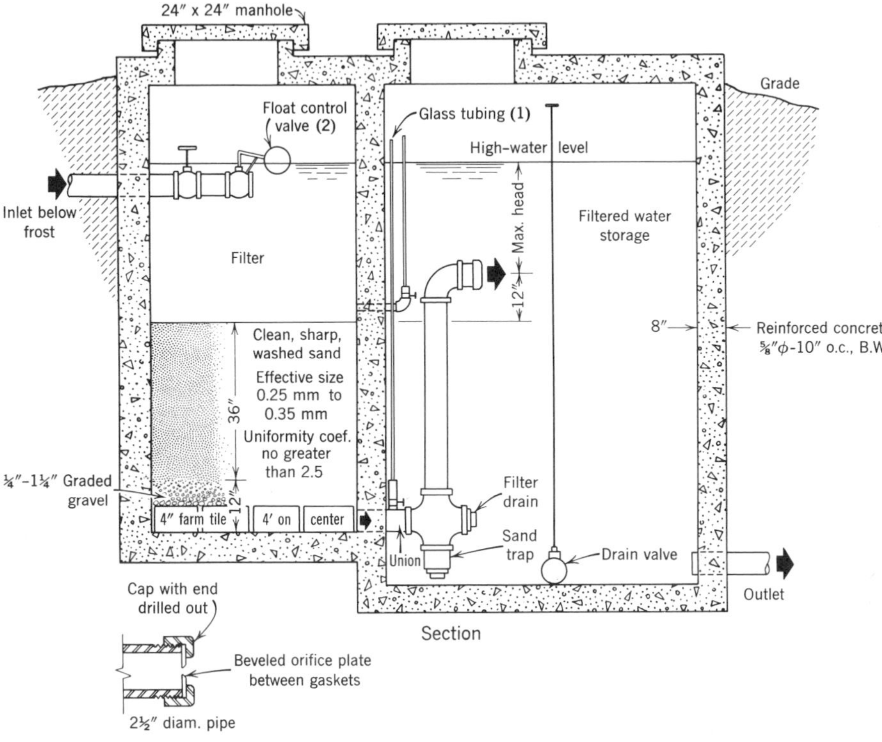

Figure 3-17 Slow sand filter for a small water supply. The difference in water level between the two glass tubes represents the frictional resistance to the flow of water through the filter. When this difference approaches the maximum head and the flow is inadequate, the filter needs cleaning. To clean, scrape the top 1 to 2 in. of sand bed off with a mason's trowel, wash in a pan or barrel, and replace clean sand on bed. Float control valve may be omitted where water on filter can be kept at a desirable level by gravity flow or by an overflow or float switch. Add a meter, venturi, or other flow-measuring device on the inlet to the filter. Rate of flow can also be controlled by maintaining a constant head with a weighted float valve over a triangular weir. See Figure 3-18. Filtered water should be disinfected before use. Allow sufficient head room for cleaning the filter.

Table 3-17 Flows from Orifices Under Various Heads of Water

Max. Head, (ft of water)	Diameter of Orifice (in.)														
	$\frac{1}{16}$	$\frac{3}{32}$	$\frac{1}{8}$	$\frac{3}{16}$	$\frac{1}{4}$	$\frac{5}{16}$	$\frac{3}{8}$	$\frac{7}{16}$	$\frac{1}{2}$	$\frac{9}{16}$	$\frac{5}{8}$	$\frac{11}{16}$	$\frac{3}{4}$	$\frac{7}{8}$	1
	Maximum Flow (gpd)[a]														
1	67	149	266	597	1,060	1,660	2,390	3,240	4,240	5,370	6,640	8,010	9,550	13,030	17,000
$1\frac{1}{2}$	82	183	326	732	1,305	2,040	2,930	3,990	5,220	6,580	8,130	9,850	11,700	15,950	20,800
2	96	213	380	852	1,520	2,370	3,410	4,650	6,060	7,680	9,480	11,480	14,300	18,600	24,200
$2\frac{1}{2}$	107	236	421	945	1,680	2,620	3,780	5,150	6,720	8,500	10,500	12,680	15,100	20,600	26,800
3	116	259	462	1,036	1,840	2,880	4,140	5,640	7,380	9,130	11,520	13,920	16,550	22,600	29,500
$3\frac{1}{2}$	126	279	498	1,120	1,990	3,100	4,470	6,060	7,950	10,050	12,420	14,900	17,900	24,300	30,700
4	135	300	534	1,200	2,125	3,330	4,790	6,530	8,520	10,800	13,300	16,100	19,150	26,100	34,000
$4\frac{1}{2}$	145	322	574	1,290	2,290	3,580	5,160	7,020	9,150	11,600	14,350	17,300	20,600	28,000	36,600
5	150	333	594	1,332	2,370	3,700	5,340	7,260	9,480	12,000	14,800	17,950	21,400	29,000	38,700
$5\frac{1}{2}$	157	350	624	1,400	2,490	3,890	5,600	7,620	9,960	12,600	15,550	18,850	22,400	30,500	39,300
6	164	366	650	1,460	2,600	4,070	5,850	7,980	10,400	13,150	16,250	19,900	23,400	31,800	41,500
$6\frac{1}{2}$	169	376	672	1,510	2,680	4,180	6,030	8,220	10,700	13,580	16,750	20,200	24,200	32,800	42,800

Note: The loss of head through a clean filter is about 3 in., hence add 3 in. to the "maximum head" in the table and sketch to obtain the indicated flow in practice. A minimum 2 to 3 ft of water over the sand is advised.

Example: To find the size of a filter that will deliver a maximum of 500 gpd: From Table 3-17, a filter with a $\frac{1}{8}$-inch orifice and a head of water of 3′9″ will meet the requirements. Filtering at the rate of 50 gpd/ft^2 of filter area, the required filter area $= \frac{500 \text{ gpd}}{50 \text{ gal}/(\text{ft}^2)(\text{day})} = 10 \text{ ft}^2$. Provide at least 2 days' storage capacity.

[a]No loss head through sand and gravel or pipe is assumed; flow is based on $Q = C_d VA$, where $V = \sqrt{2gh}$ and $C_d = 0.6$, with free discharge. (Design filter for twice the desired flow to assure an adequate delivery of water as the frictional resistance in the filter to the flow builds up. *Use two or more units in parallel.*)

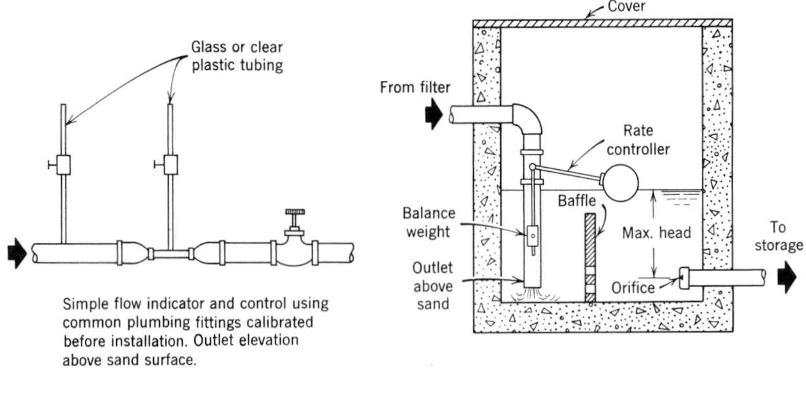

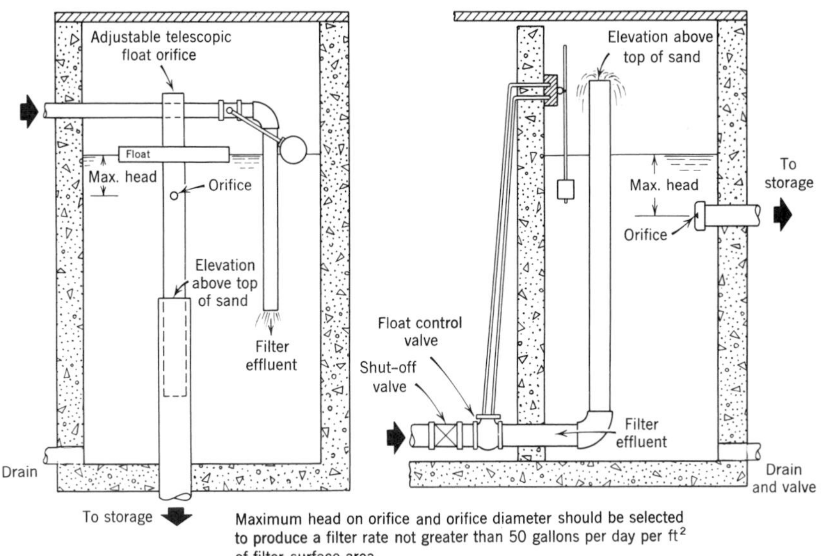

Figure 3-18 Typical devices for the control of the rate of flow or filtration. Plant capacity: 50 to 100% greater than average daily demand, with clear well.

A solenoid valve can accomplish the same type of control. A modulating float valve is more sensitive to water level control than the ordinary float valve. A remote float-controlled weighted butterfly valve, with spring-loaded packing glands and stainless steel shafts is described by Riddick.[126] A special rate-control valve can also be used if it is accurate within the limits of flow desired. The level of the orifice or filter outlet must be *above* the top of the sand to prevent the developing of a negative head. If a negative head is permitted to develop, the mat on the sur-

[126]Thomas M. Riddick, "An Improved Design for Rapid Sand Filter," *J. Am. Water Works Assoc.*, August 1952, pp. 733–744.

face of the sand may be broken and dissolved air in the water may be released in the sand bed, causing the bed to become air-bound. At least 6 in. of water over the sand will minimize possible disturbance of the sand when water from the influent line falls into the filter.

Rapid Sand (Granular Media) Filter

A rapid sand gravity filter, also referred to as a mechanical filter, is shown in Figure 3-19. Two important accessories to a rapid sand filter are the loss-of-head

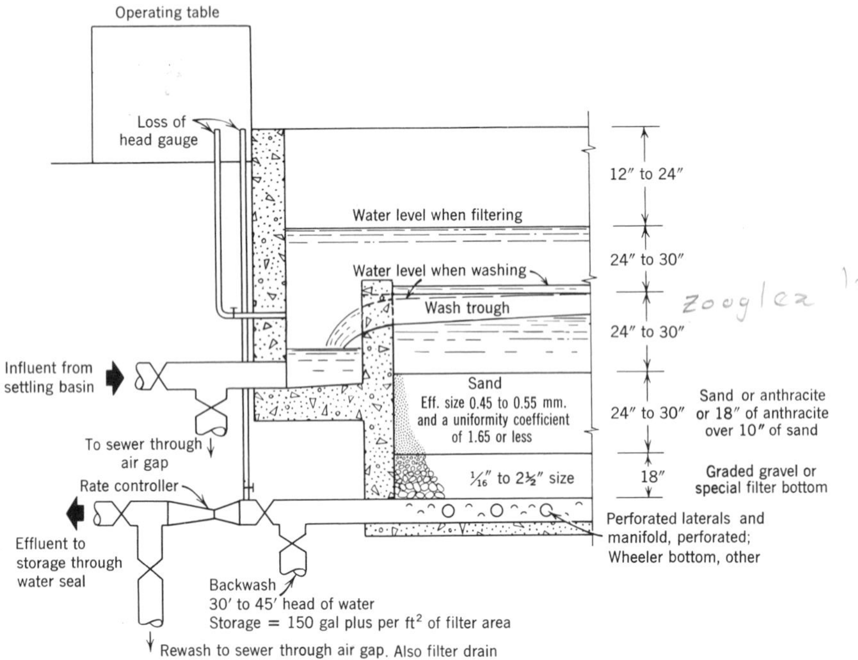

Figure 3-19 Essential parts of a rapid sand filter.

Rate of filtration $= \dfrac{7.48}{\text{minutes for water in filter to fall 1 ft}}$; fill filter with water, shut off influent, open drain.

Backwash time = 15 minutes minimum, until water entering trough is clear

Normal wash-water usage = 2 to 2.5% or less of water filtered

Sand expansion = 40 to 50% = 33.6 to 36 in. for 24-in. sand bed
= 25 to 35% for dual media, anthracite and sand

Rate of backwash $= \dfrac{7.48}{\text{minutes for water in filter to rise 1 ft}}$; lower water level to sand, open backwash valve, 15 to 20 gpm/ft^2 minimum

Orifice area = 0.25 to 0.30% of filter area

gauge and the rate controller. The loss-of-head gauge shows the frictional resistance to the flow of water through the sand, laterals, and orifices. When this reaches about 7 ft with sand and 5 ft with a dual media, it indicates that the filter needs to be backwashed. The rate controller is constructed to automatically maintain a uniform predetermined rate of filtration through the filter, usually about 3 gpm/ft^2, until the filter needs cleaning. Disturbance of filter rate or excessive head loss may cause breakthrough of suspended particles and filter floc. Filter design and operation should reduce the possible magnitude of filter fluctuations.[127] A filter rate of 3 to 4 gal or higher may be permitted with skilled operation, if pretreatment can assure water on the filter has a turbidity of less than about 10 and preferably 3 units and a coliform concentration of less than 2.2. Sand for the higher rate would have an effective size of 0.5 to 0.7 mm and a uniformity coefficient of 1.5 to 2.0.

In a combination anthracite over sand bed, use is made of the known specific gravity of crushed anthracite of about 1.5 and the specific gravity of sand of 2.5 to 2.65. The relative weight of sand in water is three times that of anthracite.* Fair and Geyer have shown that anthracite grains can be twice as large as sand grains and that after backwashing the sand will settle in place before the anthracite in two separate layers.[128] Combination sand-anthracite filters require careful operating attention and usually use of a filter conditioner to prevent floc passing through while at the same time obtaining a more uniform distribution of suspended solids throughout the media depth. Longer filter runs, such as 2 to 3 times the conventional filter, at rates of 4 to 6 gpm/ft^2 and up to 8 or 10 gpm/ft^2, and less wash-water are reported.

Treatment of the raw water by coagulation, flocculation, and settling to remove as much as possible of the pollution is usually a necessary and important preliminary step in the rapid sand filtration of water. The settled water, in passing to the filter, carries with it some flocculated suspended solids, color, and bacteria. This material forms a mat on top of the sand that aids greatly in the straining and removal of other suspended matter, color, and bacteria; but this also causes rapid clogging of the sand. Special arrangement is therefore made in the design for washing the filter by forcing water backward up through the filter at a rate that will provide a sand expansion of 40 to 50 percent based on the water temperature and sand effective size. For example, with a 0.4 mm effective size sand, a 40 percent sand expansion requires a wash-water rate rise of 21 in./min with 32° F water and a rise of 33½ in. with water at 70° F.[129] The dirty water is carried off to waste by troughs built in above the sand bed 5 to 6 ft apart. A system of water jets or rakes or a 1½- to 2-in. pressure line at 45 to 75 psi with hose connections should be provided to scour the surface of the sand to assist in loosening and re-

*Garnet sand has a specific gravity of 4.0 to 4.2.

[127]J. L. Cleasby, M. W. Williamson, and E. R. Baumann, "Effect of Filtration Rate Changes on Quality," *J. Am. Water Works Assoc.*, **55**, 869–880 (July 1963).

[128]G. M. Fair and J. C. Geyer, *Water Supply and Waste Water Disposal*, John Wiley & Sons, New York, 1954, p. 677.

[129]Robert W. Abbett, *American Civil Engineering Practice*, Vol. II, John Wiley & Sons, New York, 1956.

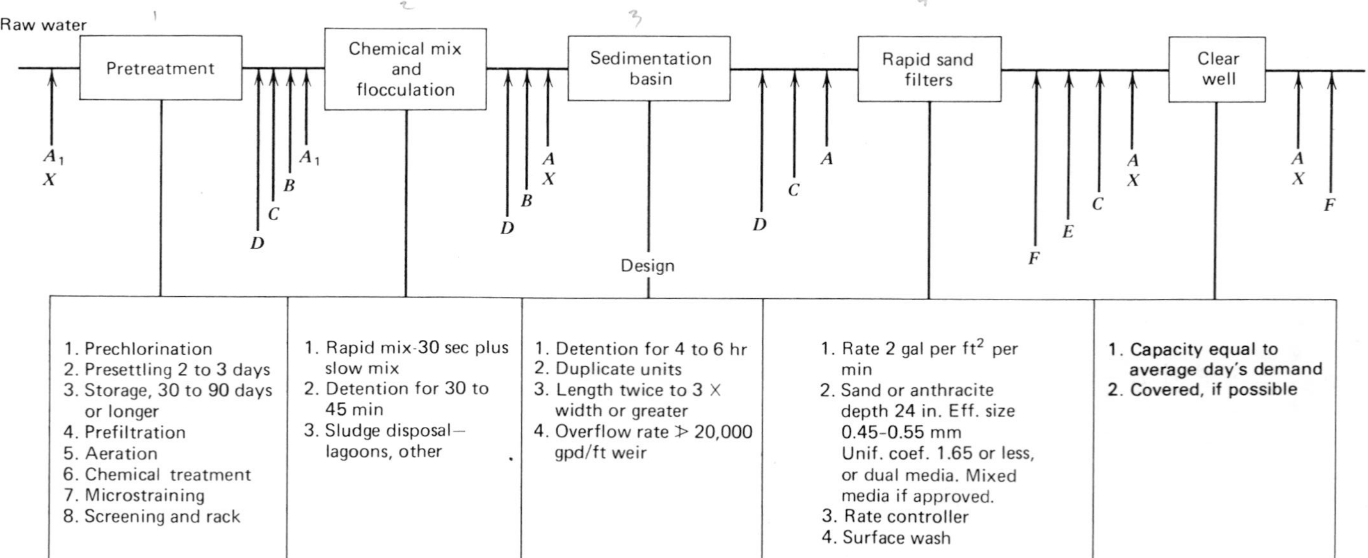

Figure 3-20 Conventional rapid sand filter plant flow diagram. Possible chemical combinations:

A Chlorine. A_1 Eliminate if THMs formed.
B Coagulant; aluminum sulfate (pH 5.5 to 8.0), ferric sulfate (pH 5.0 to 11.0), ferrous sulfate (pH 8.5 to 11.0), ferric chloride (pH 5.0 to 11.0), sodium aluminate, activated silica, organic chemicals (polyelectrolytes).
C Alkalinity adjustment; lime, soda ash, or polyphosphate.
D Activated carbon, potassium permanganate.
E Dechlorination; sulfur dioxide, sodium sulfite, sodium bisulfite, activated carbon.
F Fluoridation treatment.
X Chlorine dioxide, ozone, chlorine-ammonia.

Note that the chlorinator should be selected to prechlorinate surface water at 20 mg/l and post-chlorinate at 3 mg/l. Provide for a dose of 3 mg/l plus chlorine demand for groundwater. Additional treatment processes may include softening (ion exchange, lime-soda, excess lime and recarbonation), iron and manganese removal (ion exchange, chemical oxidation and filtration, ozone oxidation, sequestering), organics removal (activated carbon, superchlorination, ozone oxidation), and demineralization (distillation, electrodialysis, reverse osmosis, chemical oxidation and filtration, freezing).

moving the material on the sand. Effective washing of the sand is essential. A conventional rapid sand filter plant flow diagram and unit processes is shown in Figure 3-20.

When properly operated a filtration plant, including coagulation and settling, can be expected to remove about 98 percent of the bacteria, a great deal of the odor and color, and practically all the suspended solids. Nevertheless, chlorination must be used to assure that the water leaving the plant is safe to drink. Construction of a rapid sand filter should not be attempted unless it is designed and supervised by a competent sanitary engineer. The MPN of coliform organisms in the raw water to be treated should not exceed that listed in Table 3-15 unless the water is brought within the permissible limits by preliminary treatment. Adequate coagulation, flocculation, and settling in addition to granular media filtration and disinfection, is necessary to assure removal of protozoa (Giardia cyst) and viruses.

A flow diagram of a typical treatment plant is shown in Figure 3-21.

Direct Filtration

In recent years direct filtration of waters with low suspended matter and turbidity, color, coliform organisms, and plankton, and free of paper fiber, has been attractive because of the lower cost in producing a good quality water, if substantiated by prior pilot plant studies using the raw water available. In direct filtration, the sedimentation basin is omitted. The unit processes prior to filtration (dual or mixed media) may consist of only rapid mix, rapid mix and flocculation, or rapid mix and contact basin (1-hr detention) without sludge collector. Flocculation and a contact basin is recommended for better water quality control. A polymer is normally used in addition to a coagulant. Culp[130] considers direct filtration a good possibility if

> 1) the raw water turbidity and color are each less than 25 units; 2) the color is low and the maximum turbidity does not exceed 200 TU; or 3) the turbidity is low and the maximum color does not exceed 100 units. The presence of paper fiber or of diatoms in excess of 1000 areal standard units per milliliter (asu/ml) requires that settling (or microscreening) be included in the treatment process chain. Diatom levels in excess of 200 asu/ml may require the use of special coarse coal on top of the bed in order to extend filter runs.

Coliform MPNs should be low. Decreased chemical dosage and hence sludge production, but increased filter wash water, will usually result in reduced net cost as compared to conventional treatment.[131] Good operation control is essential.

[130] Russel L. Culp, "Direct Filtration," *J. Am. Water Works Assoc.*, July 1977, pp. 375–378.

[131] Garret P. Westerhoff, Alan F. Hess, and Michael J. Barnes, "Plant Scale Comparison of Direct Filtration Versus Conventional Treatment of a Lake Erie Water," *J. Am. Water Works Assoc.*, March 1980, pp. 148–155.

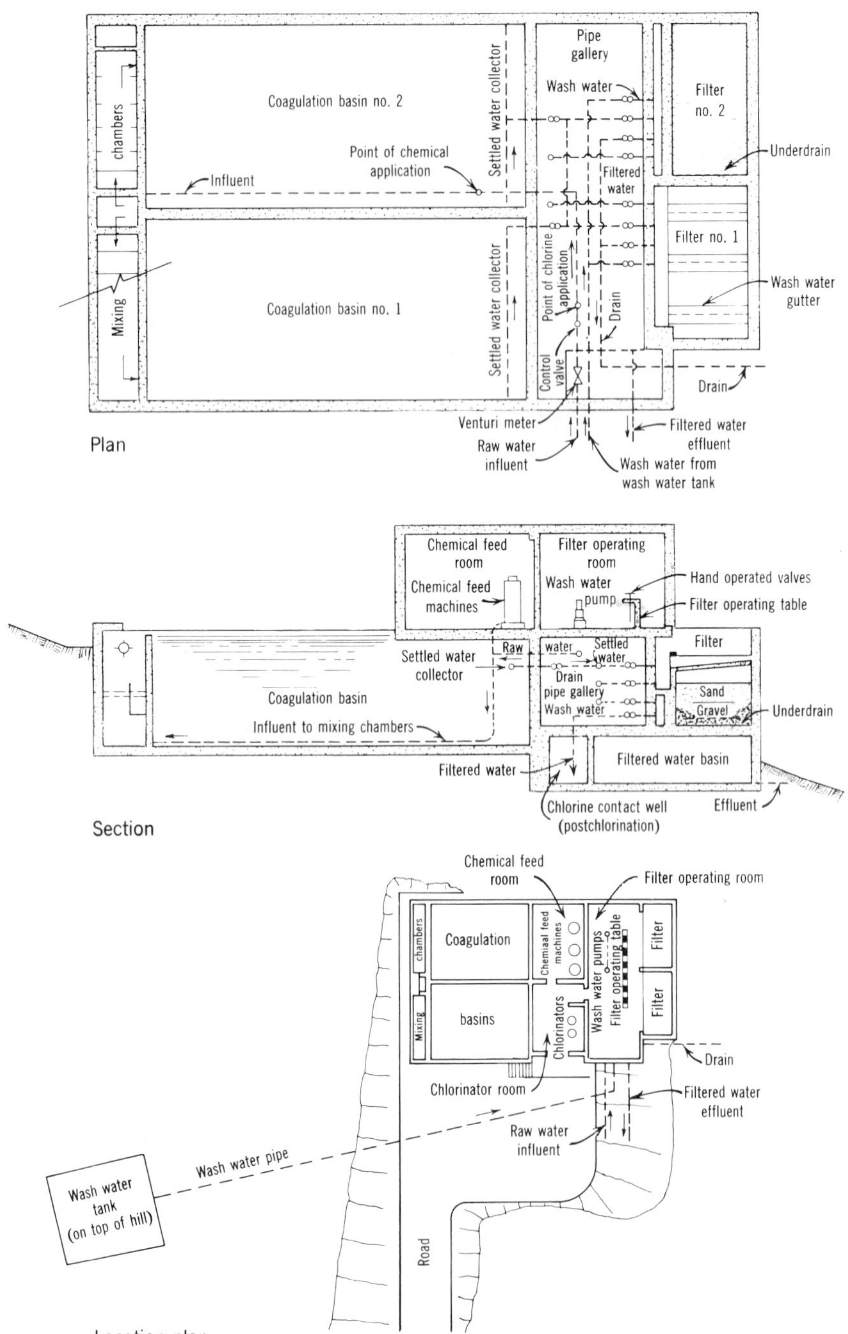

Figure 3-21 Flow diagram of typical treatment plant. This plant is compactly arranged and adaptable within a capacity range of 0.25 to 1.0 gpm. Operation is simple as the emphasis is on manual operation with only the essentials in mechanical equipment provided. Design data are described in the text. (Reprinted from the AWWA "Water Treatment Plant Design," by permission. Copyright 1969, the American Water Works Association.)

Pressure Sand Filter

A pressure filter is similar in principle to the rapid sand gravity filter except that it is completely enclosed in a vertical or horizontal cylindrical steel tank through which water under pressure is filtered. The normal filtration rate is 2 gpm/ft^2 of sand. Higher rates are used. Pressure filters are most frequently used in swimming pool and industrial plant installations. It is possible to use only one pump to take water from the source or out of the pool (and force it through the filter and directly into the plant water system or back into the pool), which is the main advantage of a pressure filter. This is offset by difficulty in introducing chemicals under pressure, inadequate coagulation facilities, and lack of adequate settling. The appearance of the water being filtered and the condition of the sand cannot be seen; the effectiveness of backwashing cannot be observed; the safe rate of filtration may be exceeded; and it is difficult to look inside the filter for the purpose of determining loss of sand or anthracite, need for cleaning, replacing of the filter media, and inspection of the wash-water pipes, influent, and effluent arrangements. Because of these disadvantages and weaknesses, a pressure filter is not considered dependable for the treatment of contaminated water to be used for drinking purposes. It may, however, have limited application for small, slightly contaminated water supplies and for turbidity removal. In such cases, the water should be coagulated and flocculated in an open basin before being pumped through a pressure filter. This will require double pumping.

Diatomaceous Earth Filter

The pressure filter type consists of a closed steel cylinder inside of which are suspended septa, the filter elements. In the vacuum type the septa are in an open tank under water that is recirculated with a vacuum inside the septa. Normal rates of filtration are 1 to $2\frac{1}{2}$ gpm/ft^2 of element surface. To prepare the filter for use a slurry of filter aid (precoat) of diatomaceous earth is introduced with the water to be treated at a rate of about $1\frac{1}{2}$ oz/ft^2 of filter septum area, which results in about $\frac{1}{8}$ in. depth of media being placed evenly on the septa, and the water is recirculated for at least 3 min before discharge. Then additional filter aid (body coat) is added with the water to maintain the permeability of the filter media. The rate of feed is roughly 2 to 3 mg/l per unit of turbidity in the water. Filter aid comes in different particle sizes. It forms a coating or mat around the outside of each filter element and is more efficient than sand because of smaller media pore size in removing from water suspended matter and such organisms as cysts, which cause amebiasis and giardiasis; cercariae, which cause schistosomiasis; flukes, which cause paragonimiasis and clonorchiasis; and worms, which cause ascariasis and trichuriasis. These organisms, except for giardia cysts, are not common in the United States.

Like the pressure sand filter, the diatomite filter has found greatest practical application in swimming pools and in industrial and military installations. It has a special advantage in the removal of oil from condensate water, since the diatomaeous earth is wasted. It should not be used to treat a public water supply unless pilot plant study results on the water to be treated meet the health department requirements.

A major weakness in the diatomite filter is that failure to add diatomaceous earth to build up the filtering mat, either through ignorance or negligence, will make the filter entirely ineffective and give a false sense of security. In addition, the septa will become clogged and require replacement or removal and chemical cleaning. During filtration, the head loss through the filter increases to 40 or 50 lb/in.^{2}, thereby requiring a pump and motor with a wide range in the head characteristics. The cost of pumping water against this higher head is therefore increased. Diatomite filters cannot be used where pump operation is intermittent, as with a pressure tank installation, for the filter cake will slough off unless sufficient continuous recirculation is provided by a separate pump. A reciprocating pump should not be used.

The filter is backwashed by reversing the flow of the filtered water back through the septum, thereby forcing all the diatomite to fall to the bottom of the filter shell, from which point it is flushed to waste. Only about 0.5 percent of the water filtered is used for backwash when the filter run length equals the theoretical or design length. The filter should not be used to treat raw water with greater than 2400 MPN per 100 ml, 30 turbidity units, or 3000 areal standard microscopic units per 100 ml. It does not remove taste- and odor-producing substances. In any case, chlorination is considered a necessary adjunct to filtration. The diatomite filter must be carefully operated by trained personnel in order to obtain dependable results.

Water Treatment Plant Wastewater and Sludge

Water treatment plant sludge from plain sedimentation and coagulation-flocculation settling basins and backwash wastewater from filters are required to be adequately treated by PL 92-500 prior to discharge to a surface water course. The wastes are characteristic of substances in the raw water and chemicals added in water treatment; they contain suspended and settleable solids, including organic and inorganic chemicals as well as trace metals, coagulants (usually aluminum hydroxide) and polymers, and clay, lime, powdered activated carbon, and other materials. The aluminum would interfere with fish survival and growth.

The common waste treatment and disposal processes include sand sludge drying beds where suitable, lagooning where land is available, natural or artificial freezing and thawing, chemical conditioning of sludge using inorganic chemicals and polymers to facilitate dewatering, and mechanical dewatering by centrifugation, vacuum filtration, and pressure filtration.[132]

Sludge dewatering increases sludge solids to about 20 percent. The use of a

[132] "Water Treatment Plant Sludges—An Update of the State of the Art: Part 2," Report of the AWWA Sludge Disposal Committee, *J. Am. Water Works Assoc.*, October 1978, pp. 548–554.

filter press involves a sludge thickener, polymer, sludge decant, lime, retention basin, addition of a precoat, and mechanical dewatering by pressure filtration. The filter cake solids concentration is increased to 40 percent (from the sludge 10 percent); it can be disposed of to a landfill if permitted. The use of a polymer with alum for coagulation would cut the amount of alum used to less than one-fifth, the cost of coagulant chemicals would be cut by one-third, and the sludge produced would be reduced by over 50 percent. Lime softening results in large amounts of sludge, increasing with water hardness. Recovery and recycling of lime may be economical at large plants. Sludge may be disposed of by lagooning, discharge to a wastewater treatment plant, or mechanical dewatering and landfilling, depending on feasibility and regulations.[133]

The ultimate disposal of sludge can be a problem in urban areas and in land disposal where the runoff or leachate might be hazardous to surface or underground waters.

Causes of Tastes and Odors

Tastes and odors in water supplies are caused by oils, minerals, gases, organic matter, and other compounds and elements in the water. Some of the common causes are oils and products of decomposition exuded by algae and some other microorganisms; wastes from gas plants, coke ovens, paper mills, chemical plants, canneries, tanneries, oil refineries, and dairies; high concentrations of iron, manganese, sulfates, and hydrogen sulfide in the water; decaying vegetation such as leaves, weeds, grasses, brush, and moss in the water; and chlorine compounds and high concentrations of chlorine. The control of taste- and odor-producing substances is best accomplished by eliminating or controlling the source when possible. When this is not possible or practical, study of the origin and type of the tastes and odors should form the basis for needed treatment.

Control of Microorganisms

Microorganisms that cause tastes and odors are for the most part harmless. They are visible under a microscope and include plankton, protozoa, fungi, and others as previously noted. Crenothrix, gallionella, and leptothrix, also known as iron bacteria, can also be included. *Thiobacillus thiooxidans,* the sulfur bacteria, have been implicated in the corrosion of iron. Algae contain chlorophyll, utilize carbon dioxide in water, produce oxygen, and serve as a basic food for fish life. All water is potentially a culture medium for one or more kinds of algae. Bacteria and algae are dormant in water below a temperature of about 48°F (10°C). Heavy algal growths cause a rise in pH and a decrease in water hardness during the day, and the opposite at night.

[133]Garret P. Westerhoff, "Treating waste streams: new challenge to the water industry," *Civ. Eng. ASCE,* August 1978, pp. 77–83.

Crenothrix and leptothrix are reddish-brown, gelatinous, stringy masses that grow in the dark inside distribution systems or wells carrying water devoid of oxygen but containing iron in solution. Control, therefore, may be effected by removal of iron from water before it enters the distribution system, maintenance of pH above 8.0, increase in the concentration of dissolved oxygen in the water above about 2 mg/l, or continuous addition of chlorine to provide a free residual chlorine concentration of about 0.3 mg/l, or 0.5 to 1.0 mg/l total chlorine. Chemical treatment of the water will destroy and dislodge growths in the mains, with resulting temporary intensification of objectionable tastes and odors, until all the organisms are flushed out of the water mains. Iron bacteria may grow in ditches draining to reservoirs. Copper sulfate dosage of 3 mg/l provides effective control.[134] The slime bacteria known as actinomycetes are also controlled by this treatment.

High water temperatures, optimum pH values and alkalinities, adequate food such as mineral matter (particularly nitrates, phosphorous, potassium, and carbon dioxide), low turbidities, large surface area, shallow depths, and sunlight favor the growth of plankton. Exceptions are diatoms, such as asterionella, which grow also in cold water at considerable depth without the aid of light. Fungi can also grow in the absence of sunlight. Extensive growths of anabaena, oscillaria, and microcystis resembling pea-green soup are encouraged by calcium and nitrogen. Protozoa such as synura are similar to algae, but they do not need carbon dioxide; they grow in the dark and in cold water. The blue-green algae do not require direct light for their growth; but green algae do. They are found in higher concentrations within about 5 ft of the water surface.

Sawyer has indicated that any lake having, at the time of the spring overturn, inorganic phosphorus greater than 0.01 mg/l and inorganic nitrogen greater than 0.3 mg/l can be expected to have major algal blooms.[135] Reduction of nutrients therefore should be a major objective, where possible. In a reservoir, this can be accomplished by removal of aquatic weeds before the fall die-off, minimizing the entrance of nutrients by watershed control, and draining the hypolimnion (the zone of stagnation) during periods of stratification since this water stratum has the highest concentration of dissolved minerals and nutrients. See also Aquatic Weed Control and Methods to Remove or Reduce Objectionable Tastes and Odors (Reservoir Intake Control).

Inasmuch as the products of decomposition and the oils given off by algae and protozoa cause disagreeable tastes and odors, preventing the growth of these microorganisms will remove the cause of difficulty. Where it is practical to cover storage reservoirs to exclude light, this is the easiest way to prevent the growth of those organisms that require light and cause difficulty. Where this is not possible, copper sulfate or chlorine should be applied to prevent the growth of the microorganisms. A combination of chlorine, ammonia, and copper sulfate has also been used with good results. However, in order that the proper chemical dosage required may be determined, it is advisable to make microscopic

[134]Kenneth M. Mackenthum, *Toward a Cleaner Aquatic Environment,* USEPA, Sup. of Documents, GPO, Washington, D.C., 20204, p. 231.

[135]C. N. Sawyer, "Phosphates and Lake Fertilization." *Sewage Ind. Wastes,* June 1952, p. 768.

examinations of samples collected at various depths and locations to determine the type, number, and distribution of organisms. This may be supplemented by laboratory tests using the water to be treated and the proposed chemical dose before actual treatment. In New England, diatoms usually appear in the spring, blue-green algae in the summer and fall; then diatoms reappear in fall and winter. The green algae appear between the diatoms and blue-green algae.

In general, the application of about 2½ lb of copper sulfate per million gal of water treated at intervals of 2 to 4 weeks between April and October, in the temperate zone, will prevent difficulties from most microorganisms. More exact dosages for specific microorganisms are given in Table 3-18. The required copper sulfate dose can be based on the volume of water in the upper 10 ft of a lake or reservoir, as most plankton are found within this depth. Bartsch[136] suggests an arbitrary dosage related to the alkalinity of the water being treated. A copper sulfate dosage of 2¾ lb per million gallons of water in the reservoir is recommended when the methyl orange alkalinity is less than 50 mg/l. When the alkalinity is greater than 40 mg/l a dosage of 5.4 lb per acre of reservoir surface area is recommended.[137] Higher doses are required for the more resistant organisms. The dose needed should be based on the type of algae making their appearance in the affected areas, as determined by periodic microscopic examinations. An inadequate dosage is of very little value and is wasteful. Higher dosages than necessary have caused wholesale fish destruction. For greater accuracy, the copper sulfate dose should be increased by 2½ percent for each degree of temperature above 59°F (15°C), and 2 percent for each 10 mg/l organic matter. Consideration must also be given to the dosage applied to prevent the killing of fish. If copper sulfate is evenly distributed, in the proper concentration, and in accordance with Table 3-19, there should be very little destruction of fish. Fish can withstand higher concentrations of copper sulfate in hard water. If a heavy algal crop has formed and then copper sulfate applied, the decay of algae killed may clog the gills of fish and reduce the supply of oxygen to the point that fish will die of asphyxiation, especially at times of high water temperatures. Tastes and odors are of course also intensified. Certain blue-green algae* may produce a toxin that is lethal to fish and animals. Other conditions may also be responsible for the destruction of fish. For example, a pH value below 4 to 5 or above 9 to 10; a free ammonia or equivalent of 1.2 to 3 mg/l; an unfavorable water temperature; a carbon dioxide concentration of 100 to 200 mg/l or even less; free chlorine of 0.15 to 0.3 mg/l, chloramine of 0.4 to 0.76 mg/l; 0.5 to 1.0 mg/l hydrogen sulfide and other sulfides; cyanogen; phosphine; sulfur dioxide; and other waste products are all toxic to fish.[138] Even a chlorine residual of

*Some belonging to the genera *Microcystis* and *Anabaena*. *Prymnesium parvum* is incriminated in fish mortality in brackish water. Marine dinoflagellates *Gymnodinium* and *Gonyaulax* toxins cause death of fish and other aquatic life.

[136]A. F. Bartsch, "Practical Methods for Control of Algae and Water Weeds," *Pub. Health Rep.*, August 1954, p. 749–757.

[137]Mackenthum, op. cit., p. 234.

[138]Peter Doudoroff and Max Katz, "Critical Review of Literature on the Toxicity of Industrial Wastes and their Components to Fish," *Sewage Ind. Wastes*, **22**, 1432–1458 (November 1950). Trout are usually more sensitive.

Table 3-18 Dosage of Copper Sulfate to Destroy Microorganisms, Pounds Per Million Gallons

Organism	Taste, Odor, Other	Dosage
Diatomaceae (Algae)	(Usually brown)	
Asterionella	Aromatic, geranium, fishy	1.0 to 1.7
Cyclotella	Faintly aromatic	Use chlorine
Diatoma	Faintly aromatic	—
Fragilaria	Geranium, musty	2.1
Meridon	Aromatic	—
Melosira	Geranium, musty	1.7 to 2.8
Navicula		0.6
Nitzchia		4.2
Stephanodiscus	Geranium, fishy	2.8
Synedra	Earthy, vegetable	3.0 to 4.2
Tabellaria	Aromatic, geranium, fishy	1.0 to 4.2
Chlorophyceae (Algae)	(Green algae)	
Cladophora	Septic	4.2
Closterium	Grassy	1.4
Coelastrum		0.4 to 2.8
Conferva		2.1
Desmidium		16.6
Dictyosphaerium	Grassy, nasturtium, fishy	Use chlorine
Draparnaldia		2.8
Entomophora		4.2
Eudorina	Faintly fishy	16.6 to 83.0
Gloeocystis	Offensive	—
Hydrodictyon	Very offensive	0.8
Miscrospora		3.3
Palmella		16.6
Pandorina	Faintly fishy	16.6 to 83.0
Protococcus		Use chlorine
Raphidium		8.3
Scenedesmus	Vegetable, aromatic	8.3
Spirogyra	Grassy	1.0
Staurastrum	Grassy	12.5
Tetrastrum		Use chlorine
Ulothrix	Grassy	1.7
Volvox	Fishy	2.1
Zygnema		4.2
Cyanophyceae (Algae)	(Blue-green algae)	
Anabaene	Moldy, grassy, vile	1.0
Aphanizomenon	Moldy, grassy, vile	1.0 to 4.2
Clathrocystis	Sweet, grassy, vile	1.0 to 2.1
Coelosphaerium	Sweet, grassy	1.7 to 2.8
Cylindrosphermum	Grassy	1.0
Gloeocopsa	(Red)	2.0
Microcystis	Grassy, septic	1.7

Table 3-18 (*Continued*)

Organism	Taste, Odor, Other	Dosage
Cyanophyceae (Algae) (*Continued*)		
Oscillaria	Grassy, musty	1.7 to 4.2
Rivularia	Moldy, grassy	—
Protozoa		
Bursaria	Irish moss, salt marsh, fishy	—
Ceratium	Fishy, vile (Red-brown)	2.8
Chlamydomonas		4.2 to 8.3
Cryptomonas	Candied violets	4.2
Dinobryon	Aromatic, violets, fishy	1.5
E. histolytica (cyst)		Use chlorine 5 to 25 mg/l
Euglena		4.2
Glenodinium	Fishy	4.2
Mallomonas	Aromatic, violets, fishy	4.2
Peridinium	Fishy, like clam-shells, bitter taste	4.2 to 16.6
Synura	Cucumber, musk-melon, fishy	0.25
Uroglena	Fishy, oily, cod-liver oil	0.4 to 1.6
Crustacea		
Cyclops		16.6
Daphnia		16.6
Schizomycetes		
Beggiatoa	Very offensive, decayed	41.5
Cladothrix		1.7
Crenothrix	Very offensive, decayed	2.8 to 4.2
Leptothrix	Medicinal with chlorine	—
Sphaerotilis natans	Very offensive, decayed	3.3
Thiothrix (Sulfur bacteria)		Use chlorine
Fungi		
Achlya		—
Leptomitus		3.3
Saprolegnia		1.5
Miscellaneous		
Blood worm		Use chlorine
Chara		0.8 to 4.2
Nitella flexilis	Objectionable	0.8 to 1.5
Phaetophyceae	(Brown algae)	—
Potamogeton		2.5 to 6.7
Rhodophyceae	(Red algae)	—
Xantophyceae	(Green algae)	—

Note: Chlorine residual 0.5 to 1.0 mg/l will also control most growths, except melosira, cysts of *Entamoeba histolytica,* Crustacea, and Synura (2.9 mg/l free).

Table 3-19 Dosage of Copper Sulfate and Residual Chlorine Which if Exceeded May Cause Fish Kill

Fish	Copper Sulfate lb/mil gal	Copper Sulfate mg/l	Free Chlorine (mg/l)	Chloramine (mg/l)
Trout	1.2	0.14	0.10 to 0.15	0.4
Carp	2.8	0.33	0.15 to 0.2	0.76 to 1.2
Suckers	2.8	0.33		
Catfish	3.5	0.40		
Pickerel	3.5	0.40		
Goldfish	4.2	0.50	0.25	
Perch	5.5	0.67		
Sunfish	11.1	1.36		0.4
Black bass	16.6	2.0		
Minnows			0.4	0.76 to 1.2
Bullheads				0.4
Trout fry				0.05 to 0.06
Gambusia				0.5 to 1.0

greater than 0.1 mg/l may be excessive.[139] Lack of food, overproduction, and species survival also result in mass "fish kills."

Copper sulfate may be applied in several ways. The method used usually depends on such things as the size of the reservoir, equipment available, proximity of the microorganisms to the surface, reservoir inlet and outlet arrangement, and time of year. One of the simplest methods of applying copper sulfate is the burlap-bag method. A weighed quantity of crystals (bluestone) is placed in a bag and towed at the desired depth behind a rowboat or, preferably, motor-driven boat. The copper sulfate is then drawn through the water in accordance with a planned pattern, first in one direction in parallel lanes about 25 ft apart and then at right angles to it so as to thoroughly treat the entire body of water. The rapidity with which the chemical goes into solution may be controlled by regulating the fabric of the bag used, varying the velocity of the boat, using crystals of large or small size, or by combinations of these variables. In another method, a long wedge shaped box (12 in. × 6 in.) is attached vertically to a boat. Two bottom sides have double 24 mesh copper screen openings one foot high; one has a sliding cover. Copper sulfate is added to a hopper at the top. The rate of solution of copper sulfate is controlled by raising or lowering the sliding cover over the screen, by the boat speed, and by the size copper sulfate crystals used. Where spraying equipment is available, copper sulfate may be dissolved in a barrel or tank carried in the boat and sprayed on the surface of the water as a $\frac{1}{2}$ or 1 percent solution. Pulverized copper sulfate may be distributed over large reservoirs or lakes by means of a mechanical blower carried on a motor-driven boat. Larger crystals are more effective against algae at lower depths. Where water flows into a reservoir, it is possible to add copper sulfate continuously and

[139] *J. Environ. Eng. Div., ASCE,* December 1973, pp. 761–772.

proportional to the flow, provided fish life is not important. This may be accomplished by means of a commercial chemical feeder, an improvised solution drip feeder, or a perforated box feeder wherein lumps of copper sulfate are placed in the box and the depth of submergence in the water is controlled to give the desired rate of solution. In the winter months when reservoirs are frozen over, copper sulfate may be applied if needed by cutting holes in the ice 20 to 50 ft apart and lowering and raising a bag of copper sulfate through the water several times. If an outboard motor is lowered and rotated for mixing, holes may be 1000 ft apart. Scattering crystals on the ice is also effective in providing a spring dosage when this is practical.

It is possible to control microorganisms in a small reservoir, where chlorine is used for disinfection and water is pumped to a reservoir, by maintaining a free residual chlorine concentration of about 0.3 mg/l in the water. However, chlorine will combine with organic matter and be used up or dissipated by the action of sunlight unless the reservoir is covered and there is a sufficiently rapid turnover of the reservoir water. Where a contact time of 2 hr or more can be provided between the water and disinfectant, the chlorine-ammonia process may be used to advantage. Chlorine may also be added as chloride of lime or in liquid form by methods similar to those used for the application of copper sulfate.

Mackenthum[137] cautions that the control of one nuisance may well stimulate the occurrence of another under suitable conditions and necessitate additional control actions. For example, the control of algae may lead to the growth of weeds. Removal of aquatic weeds may promote the growth of phytoplankton or bottom alga such as chara. The penetration of sunlight is thereby facilitated but nutrients are released by growth and then decay of chara.

Gnat flies sometimes lay their eggs in reservoirs. The eggs develop into larvae, causing consumer complaints of worms in the water. The best control measure is covering the reservoir or using fine screening to prevent the entrance of gnats.

Aquatic Weed Control

The growth of aquatic plants (and animals) is accelerated when the temperature of the surface water is about 59°F. Vegetation that grows and remains below the water surface does not generally cause difficulty. Decaying and emergent aquatic vegetation, as well as decaying leaves, brush, weeds, grasses, and debris in the water, can cause tastes and odors in water supplies. The discharge of organic wastes from wastewater treatment plants, storm sewers, and drainage from lawns, pastures, and fertilized fields contains nitrogen and phosphorus, which promote algal and weed growths. The contribution of phosphorus from sewage treatment plants can be relatively small compared to that from surface runoff. Unfortunately little can be done to permanently prevent the entrance of wastes and drainage or destroy growths of rooted plants, although certain chemical, mechanical, and biological methods can provide temporary control.

Reasonably good temporary control of rooted aquatic plants may be obtained by physically removing growths by dredging, with wire or chain drags, or rakes, and by cutting. Filling of marshy areas and deepening the edges of reservoirs, lakes, and ponds to a depth of 2 ft or more will prevent or reduce plant growths. Weeds that float to the surface should be removed before they decay.

Where it is possible to drain or lower the water level 6 ft to expose the affected areas of the reservoir for about one month, followed by drying the weeds and roots, and clearing and removal, it is of great value. Drying out of roots and burning and removal of the ash is effective for a number of years. Flooding 3 ft or more above normal is also effective where possible.

As a last resort, aquatic weeds may be controlled by chemical means. Tastes and odors may result if the water is used for drinking purposes; the chemical may kill fish and persist in the bottom mud, and it may be hazardous to the applicator. The treatment must be repeated annually or more often, and heavy algal blooms may be stimulated, particularly if the plant destroyed is allowed to remain in the water and return its nutrients to the water. Chemical use should be restricted and permitted only after careful review of the toxicity to humans and fish, the hazards involved, and the purpose to be served. Copper sulfate should not be used for the control of aquatic weeds, since the concentration required to destroy the vegetation will assuredly kill any fish present in the water. Diquat and endothal have been approved by the USEPA, if applied according to directions. Diquat use requires a 10-day waiting period. Endothal use requires a 14-day waiting period with the amine salt formulation and 7 days with the potassium or sodium salts formulation.[140] See also Control of Aquatic Weeds in Chapter 9.

Other Causes of Tastes and Odors

In new reservoirs, clearance and drainage reduce algal blooms by removing organic material beneficial to their growth. Organic material, which can cause anaerobic decomposition, odors, tastes, color, and acid conditions in the water, is also removed. If topsoil is valuable, its removal may be worthwhile.

Some materials in water cause unpleasant tastes and odors when present in excessive concentrations, although this is not a common source of difficulty. Iron and manganese, for example, may give water a bitter astringent taste. In some cases sufficient natural salt is present, or saltwater enters to cause a brackish taste in well water. It is not possible to remove the salt in the well water without going to great expense. Elimination of the cause by sealing off the source of the saltwater, groundwater recharge with freshwater, or controlling pump draw-down is sometimes possible.

Other causes of tastes and odors are sewage and industrial or trade wastes and spills. Sewage would have to be present in very large concentrations to be noticeable in a water supply. If this were the case, the dissolved oxygen in the water receiving the sewage would most probably be used up, with resultant nuisance conditions. On the other hand the billions of microorganisms introduced, many of which would cause illness or death if not removed or

[140] "Spraying Your Reservoir for Weed Control," *Opflow,* AWWA, Denver, Co., 80235, May 1978.

destroyed before consumption, are the greatest danger in sewage pollution. Trade or industrial wastes introduce in water suspended or colloidal matter, dissolved minerals and organic chemicals, vegetable and animal organic matters, harmful bacteria, and other materials that are toxic and produce tastes and odors. Of these, the wastes from steel mills, paper plants, and coal distillation (coke) plants have proved to be the most troublesome in drinking water, particularly in combination with chlorine. Tastes produced have been described as "medicinal," "phenolic," "iodine," "carbolic acid," and "creosote." Concentrations of one part phenol to 500 million parts of water will cause very disagreeable tastes even after the water has traveled 70 miles.[141] The control of these tastes and odors lies in the prevention and reduction of stream pollution through improved plant operation and waste treatment. Chlorine dioxide has been found effective in treating a water supply not too heavily polluted with phenols. The control of stream pollution is a function and responsibility of federal and state agencies, municipalities, and industry. Treatment of water supplies to eliminate or reduce objectionable tastes and odors is discussed separately below.

Sometimes high uncontrolled doses of chlorine produce chlorinous tastes and chlorine odors in water. This may be due to the use of constant feed equipment rather than a chlorinator, which will vary the chlorine dosage proportional to the quantity of water to be treated. In some installations chlorine is added at a point that is too close to the consumers, and in others the dosage of chlorine is marginal or too high, or chlorination treatment is used where coagulation, filtration, and chlorination should be used instead. Where superchlorination is used and high concentrations of chlorine remain in the water, dechlorination with sodium sulphite, sodium bisulfite, sodium thiosulphate, sulphur dioxide, or activated carbon is indicated. Sulphur dioxide is most commonly used in a manner similar to that used for liquid chlorine and with the same precautions; dosage must be carefully controlled to avoid lowering the pH and dissolved oxygen, as reaeration may then be necessary.

Methods to Remove or Reduce Objectionable Tastes and Odors

Some of the common methods used to remove or reduce objectionable tastes and odors in drinking water supplies, not in order of their effectiveness, are

1. Free residual chlorination or superchlorination.
2. Chlorine-ammonia treatment.
3. Aeration, or forced-draft degasifier,
4. Application of activated carbon.
5. Filtration through granular activated carbon, or charcoal filters.
6. Coagulation and filtration of water (also using an excess of coagulant).
7. Control of reservoir intake level.
8. Elimination or control of source of trouble.
9. Chlorine dioxide treatment.

[141] *Manual of Water Quality Control* AWWA, New York, 1940, p. 49.

10. Ozone treatment.
11. Purging and bioreclamation of the groundwater aquifer.

Free Residual Chlorination

Free residual chlorination will destroy by oxidation most taste- and odor-producing substances and inhibit growths inside water mains. Biochemical corrosion is also prevented in the interior of water mains by destroying the organisms associated with the production of organic acids. The reduction of sulfates to objectionable sulfides is also prevented. However chloro-organics, which are suspected of being carcinogenic, may be formed depending on the precursors in the water treated.

Nitrogen trichloride is formed in water high in organic nitrogen when a high free chlorine residual is maintained and the pH is less than 8.0. (See page 264.) It is an explosive, volatile, oily liquid that is removed by aeration or carbon. Nitrogen trichloramine exists below pH 4.5 and at higher pH in polluted waters.

Chlorine-Ammonia Treatment

Chlorine-ammonia treatment in practice is the addition of about three to four parts of chlorine to one part of ammonia. The ammonia is added a few feet ahead of the chlorine. Chloramines, which are weak bactericides, are formed. Because of this, chlorine-ammonia treatment is not recommended as the primary disinfectant. Chloramines prevent chlorinous tastes due to the reaction of chlorine with taste-producing substances in water. Chloroform is not formed as in free residual chlorination. Free residual chlorination followed by dechlorination and then chloramination of the water distributed is sometimes practiced with good bacteriological control.[142]

Aeration

Aeration is a natural or mechanical process of increasing the contact between water and air for the purpose of improving the chemical and physical characteristics of water. Some waters, such as water from deep lakes and reservoirs in the late summer and winter seasons, cistern water, water from deep wells, and distilled water, may have an unpleasant or flat taste due to a deficient dissolved-oxygen content. Aeration will add oxygen to such waters and improve their taste. In some instances the additional oxygen is enough to make the water corrosive. Aerators have other limitations.[143] Free carbon dioxide, hydrogen sulfide, volatile organic compounds, and odors due to volatile oils exuded by algae will also be removed or reduced. Aeration is advantageous in the treatment of water containing dissolved iron and manganese in that oxygen will change or oxidize the dissolved iron and manganese to insoluble ferric and manganic forms that can be removed by settling, contact, and filtration.

[142]Noel V. Brodtmann Jr., and Peter J. Russo, "The Use of Chloramine for Reduction of Trihalomethanes and Disinfection of Drinking Water," *J. Am. Water Works Assoc.*, January 1979, pp. 40–45.

[143]"Aeration of Water," revision of *Water Quality and Treatment*, Chapter 6 *J. Am. Water Works Assoc.*, **47**, No. 9, 873–885 (September 1955).

Aeration is accomplished by allowing the water to flow in thin sheets over a series of steps, weirs, splash plates, riffles, or waterfalls; by water sprays in fine droplets; by allowing water to drip out of trays, pipes, or pans that have been slotted or perforated with $\frac{1}{8}$- to $\frac{1}{4}$-in. holes; by causing the water to drop through a series of trays containing 6 to 9 in. of coke or broken stones; by means of spray nozzles; by using air-lift pumps; by introducing finely divided air in the water; by permitting water to trickle over 1-in. by 3-in. cypress wood slats with $\frac{1}{2}$- to $\frac{3}{4}$-in. separations in a tank through which air is blown up from the bottom; and by similar means. Coke will become coated and hence useless if the water is not clear. Slat trays are usually 8 to 12 in. apart. Many of these methods are adaptable to small rural water supplies; but care should be taken to protect the water from insects and accidental or willful contamination. Screening of the aerator is desirable to prevent the development of worms.

Activated Carbon—Powdered and Granular

The sources of raw material for activated carbon include bituminous coal, lignite, peat, wood, bone, petroleum-based residues, and nut shells. The carbon is activated in an atmosphere of oxidizing gases such as CO_2, CO, O_2, steam, and air at a temperature of between 300 and 1000°C, usually followed by sudden cooling in air or water. The micropores formed in the carbonized particles contribute greatly to the adsorption capacity of the activated carbon. Granular carbon can be reactivated in a multi-hearth furnace at a temperature of 820 to 930°C in a controlled atmosphere, where adsorbed impurities are volatilized and oxidized, followed by sudden cooling in water.

Granular activated carbon (GAC) filters (pressure type) are used for treating water for soft drinks and bottled drinking water. GAC filter beds are used at water treatment plants to remove taste- and odor-producing compounds as well as color and synthetic organic chemicals suspected of being carcinogenic. Colloids interfere with adsorption if not removed prior to filtration. The GAC filters or columns normally follow conventional rapid sand filters, but can be used alone if a clear clean water is being treated.

Granular activated carbon is of limited effectiveness to remove trihalomethane precursor compounds. It is effective for only a few weeks.[144] In contrast, granular activated carbon beds for taste and odor control need regeneration every three to six years.[145] When the GAC bed becomes saturated with the contaminant being removed, the contaminant appears in the effluent (an event known as breakthrough) if the GAC is not replaced or regenerated.

Activated carbon in the powdered form is used quite generally and removes by adsorption, if a sufficient amount is used, practically all tastes and odors found in water. The powdered carbon may be applied directly to a reservoir as a suspension with the aid of a barrel and boat (as described for copper sulfate), or released slowly from the bag in water near the propeller, but the reservoir should

[144] Alan A. Stevens, Clois J. Slocum, Dennis R. Seeger, and Gordon R. Robek, "Chlorination of Organics in Drinking Water," *J. Am. Water Works Assoc.*, November 1976, p. 615.

[145] Gene Dallaire, "Are cities doing enough to remove cancer-causing chemicals from drinking water?" *Civ. Eng. ASCE*, September 1977.

be taken out of service for one or two days, unless the area around the intake can be isolated. The application of copper sulfate within this time will improve settling of the carbon.

Doses vary from 1 to 60 lb or more of carbon to one million gal of water, with 25 lb as an average. In unusual circumstances as much as 1000 lb of carbon per million gal of water treated may be needed, but cost may make this impractical. Where a filtration plant is provided, carbon is fed by means of a standard chemical dry-feed machine, or as a suspension, to the raw water, coagulation basin, or filters. However, carbon can also be manually applied directly to each filter bed after each wash operation. A 10 to 15 min contact time between the carbon and water being treated and good mixing will permit efficient adsorption of the taste and color compounds. Activated carbon is also used in reservoirs and settling basins to exclude sunlight causing the growth of algae. This is referred to as "blackout" treatment. The dosage of carbon required can be determined by trial and error and tasting the water, or by a special test known as the "Threshold Odor Test," which is explained in *Standard Methods* (See Bibliography). If the water is pretreated with chlorine, after 15 to 20 minutes the activated carbon will remove up to about 10 percent of its own weight of chlorine, hence they should *not* be applied together if avoidable. Possible arrangements are activated carbon followed by chlorine, or prechlorination followed by application of carbon to the top of the filters. Careful operation control can make possible prompt detection of taste- and odor-producing compounds reaching the plant and the immediate application of corrective measures.

GAC filters are usually $2\frac{1}{2}$ to 3 ft deep and operate at rates of 2 to 5 gpm/ft^2. They are supported on a few inches of sand. Pressure filters containing sand and activated carbon are used on small water supplies. GAC columns are up to 10 ft deep. The water, if not clear, must be pretreated by conventional filtration including coagulation and clarification.

Charcoal Filters

Charcoal filters, either of the open-gravity or closed-pressure type, are also used to remove substances causing tastes and odors in water. The water so treated must be clear, and the filters must be cleaned, reactivated, or replaced when they are no longer effective in removing tastes and odors. Rates of filtration vary from 2 to 4 gpm/ft^2 of filter area, although rates as high as 10 gpm/ft^2 are sometimes used. Trays about 4 ft^2, containing 12 in. of coke, are also used. The trays are stacked about 8 in. apart, and the quantity is determined by the results desired.

Coagulation

Coagulation of turbidity, color, bacteria, organic matter, and other material in water, followed by flocculation, settling, and then filtration, will also result in the removal of taste- and odor-producing compounds, particularly when activated carbon is included. The use of an excess of coagulants will sometimes result in the production of a better tasting water. In any case, a surface water supply should be treated to produce a very clear water so as to remove the colloids which together with volatile odors account for the taste and odors of most finished waters.[146]

Reservoir Intake Control

The quality of reservoir and lake waters varies with the depth, season of the year or temperature, wave action, organisms and food present, condition of the bottom, clarity of the water, and other factors.

Temperature is important in temperate zones. At a temperature of 39.2°F (4°C) water is heaviest, with a specific gravity of 1.0. Therefore, in the fall of the year, the cool air will cause the surface temperature of the water to drop, and when it reaches 39.2°F this water, with the aid of wind action, will move to the bottom and set up convection currents, thereby forcing the bottom water up. Then in the winter the water may freeze, and conditions remain static until the spring when the ice melts and the water surface is warmed. A condition is reached when the entire body of water is at a temperature of about 39.2°F, but a slight variation from this temperature, aided by wind action, causes an imbalance, with the bottom colored, turbid water deficient in oxygen (usually also acidic and high in iron, manganese, and nutrient matter) rising and mixing with the upper water. The warm air will cause the temperature of the surface water to rise, and a temporary equilibrium is established, which is upset again with the coming of cold weather. This phenomenon is known as reservoir turnover.

In areas where the temperature does not fall below 39.2°F, and during warm months of the year, the water in a deep reservoir or lake will be stratified in three layers: the top mixed zone (epilimnion) which does not have a permanent temperature stratification; the middle transition zone (metalimnion or thermocline) in which the drop in temperature equals or exceeds 1°C (1.8°F) per meter (39.37 in.); and the bottom zone of stagnation (hypolimnion) which is generally removed from surface influence. The metalimnion is usually the source of best quality water. The euphotic zone, in the epilimnion, extends to the depth at which photosynthesis fails to occur because of inadequate light penetration. The reservoir or lake layer or region in which organic production from mineral substances takes place because of light penetration is called the trophogenic region. The layer where there is a light deficiency, where nutrients are released by dissimilation (the opposite of assimilation), is called the tropholytic region.

A better quality of water can usually be obtained by drawing from different depth levels. To take advantage of this, provision should be made in deep reservoirs for an intake tower with inlets at different elevations so that the water can be drawn from the most desirable level. Where an artificial reservoir is created by the construction of a dam, it is better to waste surplus water through a blowoff rather than over a spillway; then stagnant bottom water containing decaying organic matter, manganese, iron, and silt can be flushed out.

Chlorine Dioxide Treatment

Chlorine dioxide treatment was developed originally to destroy tastes produced by phenols, but it is also effective against other taste-producing

[146]Thomas M. Riddick in "Letters to the Editor," *Pub. Works,* May 1975, p. 144.

materials. Chlorine dioxide is manufactured at the water plant where it is to be used. Sodium chlorite solution and chlorine water are usually pumped into a glass cylinder where chlorine dioxide is formed and from which it is added to the water being treated, together with the chlorine water. A gas chlorinator is needed to form chlorine water, and for a complete reaction with full production of chlorine dioxide, the pH of the solution in the glass reaction cylinder must be less than 4.0. Where hypochlorinators are used, the chlorine dioxide can be manufactured by adding hypochlorite solution, a dilute solution of hydrochloric acid, and a solution of sodium chlorite in the glass reaction cylinder so as to maintain a pH of less than 4.0. Three solution feeders are then needed. Cox[147] gives the theoretical ratio of chlorine to sodium chlorite as 1.0:2.57 with chlorine water or hypochlorite solution, and sodium chlorite to chlorine dioxide produced as 1.0:0.74. Actually, more chlorine is usually needed to drop the pH of the reaction to 4 or less. A chlorine dioxide dosage of 0.2 to 0.3 mg/l will destroy most phenolic taste-producing compounds. Chlorine dioxide does not react with nitrogenous compounds or other materials having a chlorine demand and trihalomethane formation is reduced. It is an effective disinfectant over a wide pH range. However chlorine dioxide residual oxidants should be controlled to not exceed a total concentration of 0.5 mg/l as it leaves chlorate and chlorite residuals which can be very toxic. The use of chlorine dioxide is discouraged until such time as more information on the health effects of the reaction products, particularly the chlorite ion, is known.

Ozone Treatment

Ozone has been used for many years as a disinfectant and as an agent to remove color, taste, and odors from drinking water.[148] It is more effective in eliminating or controlling color, taste, and odor problems not amenable to other treatment methods. It also oxidizes and permits removal of iron and manganese and aids in turbidity removal. Like chlorine, ozone is a toxic gas.

Ozone is a powerful oxidizing agent over a wide pH and temperature range, in contrast to chlorine. It is an excellent virucide, is effective against amoebic cysts, and destroys bacteria and phenols. The potential for the formation of chlorinated organics such as THMs is reduced with preozonation; the removal of soluble organics in coagulation is also reported to be improved.[149] Ozone is reported to be 3100 times faster than chlorine in disinfection.[150] Ozonation provides no lasting residual in water treated; the ozone disappears in 7 to 8 min and is more expensive compared to chlorine and chlorine dioxide. The disadvantage of no lasting chlorine residual can be offset by adding chlorine to maintain a chlorine residual in the distribution system.

[147]Charles R. Cox, *Water Supply Control*, New York State Dep. of Health, Albany, N.Y., 1952, pp. 121–122.

[148]George E. Symons and Kenneth W. Henderson, "Disinfection—Where Are We?," *J. Am. Water Works Assoc.*, March 1977, pp. 148–154.

[149]Harvey M. Rosen, "Ozonation—Its time has come," *Water Wastes Eng.*, September 1978, pp. 106-108.

[150]J. Holluta, "Ozone in Water Chemistry," Gass-Wasserfach (Ger.), 104, (1963), *J. Water Pollut. Control Fed.*, December 1973, p. 2507.

New products can be formed during the ozonation of wastewaters; not all low molecular weight organic compounds are oxidized completely to CO_2 and H_2O.

> Careful consideration must be given to the possibility of the formation of compounds with mammalian toxicity during ozonation of drinking water.[151]

However, at least one study concludes that the probability of potentially toxic substances being formed is small.[150]

Ozone must be generated at the point of use; it cannot be stored as a compressed gas. Although ozone can be produced by electrolysis of perchloric acid and by ultraviolet lamps, the practical method for water treatment is by passage of dry clean air between two high-voltage electrodes. Pure oxygen can be added in a positive pressure injection system. The ozonized air is injected in a mixing and contact chamber with the water to be treated. The space above the chamber must be carefully vented after its concentration is reduced using an ozone destructive device, to avoid human exposure, as ozone is very corrosive and toxic. The vented ozone may contribute to air pollution. The same precautions must be taken in the storage, handling, piping, respiratory protection, and housing of ozone as with chlorine.

Hydrogen Sulfide, Sources and Removal

Hydrogen sulfide is undesirable in drinking water for aesthetic and economic reasons. Its characteristic "rotten egg" odor is well known; but the fact that it tends to make water corrosive to iron, steel, stainless steel, copper, and brass is often overlooked. The permissible 8-hr occupational exposure to hydrogen sulfide is 20 ppm, but only 10 min for 50 ppm exposure. Death is said to result at 300 ppm. See Safety, Chapter 4. As little as 0.2 mg/l in water causes bad taste and odor and staining of photographic film.

The sources of hydrogen sulfide are both chemical and biological. Water derived from wells near oil fields or from wells that penetrate shale or sandstone frequently contain hydrogen sulfide. Calcium sulfate, sulfites, and sulfur in water containing little or no oxygen will be reduced to sulfides by anaerobic sulfur bacteria or biochemical action, resulting in liberation of hydrogen sulfide. this is more likely to occur in water at a pH of 5.5 to 8.5, and particularly in water permitted to stand in mains or in water obtained from close to the bottom of deep reservoirs. Organic matter often contains sulfur that, when attacked by sulfur bacteria in the absence of oxygen, will release hydrogen sulfide. Another source of hydrogen sulfide is the decomposition of iron pyrites or iron sulfide.

The addition of 2.0 to 4 mg/l copper sulfate or the maintenance of at least 0.3 mg/l free residual chlorine in water containing sulfate will inhibit biochemical activity and also prevent the formation of sulfides. The removal of H_2S already formed is more difficult, for most complete removal is obtained at a pH of around 4.5. Aeration removes hydrogen sulfide, but this method is not entirely effective; carbon dioxide is also removed, thereby causing an increase in the pH of the water, which reduces the efficiency of removal. Therefore, aeration must

[151]Victor J. Elia, et al., "Ozonation in a wastewater reuse system: examination of products formed," *J. Water Pollut. Control Fed.*, July 1978, pp. 1727–1732.

be supplemented. Aeration followed by settling and filtration is an effective combination. Chlorination alone can be used without precipitation of sulfur; but large amounts, theoretically 8.4 mg/l chlorine to each milligram per liter of hydrogen sulfide, would be needed. The alkalinity (as $CaCO_3$) of the water is lowered by 1.22 parts for each part of chlorine added. Chlorine in limited amounts, theoretically 2.1 mg/l chlorine for each milligram per liter of hydrogen sulfide, will result in formation of flowers of sulfur, which is a fine colloidal precipitate requiring coagulation and filtration for removal. If the pH of the water is reduced by adding an acid to the water or by adding a sufficient amount of carbon dioxide as flue gas, for example, good hydrogen sulfide removal should be obtained. But pH adjustment to reduce the aggressiveness of the water would be necessary. Another removal combination is aeration, chlorination, and filtration through an activated carbon pressure filter.

Pressure tank aerators, that is, the addition of compressed air to hydropneumatic tanks, can reduce the entrained hydrogen sulfide in well water from 35 to 85 percent, depending on such factors as the operating pressures and dissolved oxygen in the hydropneumatic tank effluent.[152] The solubility of air in water increases in direct proportion to the absolute pressure. Carbon dioxide is not removed by this treatment. Air in the amount of 0.005 to 0.16 ft^3 per gal of water and about 15-min detention is recommended, with the higher amount preferred. The air may be introduced through perforated pipe or porous media in the tank bottom, or with the influent water. Unoxidized hydrogen sulfide and excess air in the tank must be bled off. Air relief valves or continuous air bleeders can be used for this purpose. It is believed that oxidation of the hydrogen sulfide through the sulfur stage to alkaline sulfates takes place, since observations show no precipitated sulfur in the tank. Objections to pressure tank aerators are milky water caused by dissolved air, and corrosion. The milky water would cause air binding or upset beds in filters if not removed.

A synthetic resin has been developed that has the property of removing hydrogen sulfide. It can be combined with a resin to remove hardness so that a low-hardness water can be softened and deodorized. The resin is manufactured by Rohm and Haas Company, Philadelphia, Pennsylvania 19106.

Removal of Gasoline by Purging and Bioreclamation

There are no simple inexpensive ways of removing gasoline or other chemical contamination from a well drawing water from a contaminated aquifer. The gasoline adheres to soil grains and, depending on rainfall and other hydrogeological conditions, will continue leaching out into the aquifer until dissipated through adsorption, dispersion, diffusion, or ion exchange.[153] Installation of a deep well, or well points, as indicated, at the source of the spill or pollution, or in the line of underground water flow, and pumping out as much of the gasoline or heavily gasoline-polluted water as possible, could be a first step,

[152] Sidney W. Wells, "Hydrogen Sulfide Problems of Small Water Systems," *J. Am. Water Works Assoc.*, **46**, 2, 160–170 (February 1954).

[153] Julian Josephson, "Groundwater Strategies," *Environ. Sci. Techno.* September 1980, pp. 1030–1035.

after the source and heavily contaminated soil has been identified and eliminated. Purging of the aquifer might require two years.

The pumping out might be followed by recirculating large volumes of a nontoxic biodegradable detergent water solution over the spill area to purge the aquifer of gasoline.

The gasoline polluted water, and the detergent-treated water, once pumped out would require treatment and proper disposal to prevent recontamination of the aquifer.

Another method that has been proposed, after pumping out as much of the gasoline-contaminated water as possible, is bioreclamation. In this process,

> nutrients such as oxygen, nitrogen, and phosphorous, are injected into the soil to induce naturally occuring bacteria to multiply. The bacteria are said to use the hydrocarbons for food and then die off as the gasoline is consumed.[154]

Aeration and activated carbon will remove gasoline and other volatile organics from water but pilot plant studies should first be made to determine the effectiveness and cost using the actual water in question. The use of manufactured organisms to assimilate and break down some organics in groundwater is also being considered.

Containment of a chemical contaminant in the aquifer may also be possible. This may be accomplished by the use of barriers such as a bentonite slurry trench, grout curtain, sheet piling, or freshwater barrier, and by the provision of an impermeable cap over the offending source if it cannot be removed. The method used would depend on the problem and the hydrogeological conditions. However there are many uncertainties in any method used and no barrier can be expected to be perfect or maintain its integrity forever.

Iron and Manganese Occurrence and Removal

Iron in excess of 0.3 to 0.5 mg/l will stain laundry and plumbing fixtures and cause water to appear rusty. When manganese is predominant the stains will be black. Neither iron nor manganese are harmful in the concentrations found in water. Iron may be present as soluble ferrous bicarbonate in alkaline well or spring waters; as soluble ferrous sulphate in acid drainage waters or waters containing sulfur; as soluble organic iron in colored swamp waters; as suspended insoluble ferric hydroxide formed from iron-bearing well waters, which are subsequently exposed to air; and as a product of pipe corrosion producing red water.

Most soils, including gravel, shale, and sandstone rock, contain iron and manganese in addition to other minerals. Decomposing organic matter in water removes the dissolved oxygen usually present in water; then the water dissolves mineral oxides, changing them to soluble compounds. Water containing carbon dioxide or carbonic acid will have the same effect. In the presence of air, however,

[154]"Massachusetts Town Is Seeking Funds to Save Its Water Supply," Provincetown, Mass., *New York Times*, Sunday, November 19, 1978.

soluble ferrous bicarbonate will change to ferric iron, which will settle out in the absence of interfering substances. Ferrous iron may be found in the lower levels of deep reservoirs, flooding soils, or rock containing iron or its compounds; hence it is best to draw water from a higher level, but below the upper portion, which supports microscopic growths like algae. This requires the construction and use of multiple-gate intakes, as previously mentioned.

The presence of as little as 0.1 mg/l iron in a water will encourage the growth of such bacteria as leptothrix and crenothrix. Carbon dioxide also favors their growth. These organisms grow in distribution systems and cause complaints. Mains, service lines, meters, and pumps may become plugged by the crenothrix growths. Gallionella bacteria can grow in wells and reduce capacity. Complaints reporting small gray or brownish flakes, masses of stringy or fluffy growths in water would indicate the presence of iron bacteria. The control of iron bacteria in well water is also discussed under Control of Microorganisms, Iron in Well Water, and Corrosion Cause and Control.

Corrosive waters that are relatively free of iron and manganese may attack iron-pipe and house plumbing, particularly hot-water systems, causing discoloration and other difficulties. Such corrosion will cause red water, the control of which is discussed separately.

Iron and manganese can be removed by aeration, sedimentation, the base-exchange process, filtration, the addition of chemicals, and combinations of these methods. The cation exchanger-sodium cycle, if water is clear and unaerated, removes up to 50 mg/l iron; use of a manganese filter removes up to 2 mg/l iron or manganese. A summary of the processes used to remove iron and manganese is given in Table 3-20. Use of a polyphosphate will prevent red water or iron deposit if iron concentration is less than 3 mg/l.

Most of the carbon dioxide in water is removed by aeration; then the iron is oxidized and the insoluble iron is removed by settling or filtration. If organic matter and manganese are also present, the addition of lime or chlorine will assist in changing the iron to an insoluble form and hence simplify its removal.

The open coke-tray aerator is the most common method used to remove iron and manganese. Two or more perforated wooden trays containing about 9 in. of coke are placed in tiers. A 20- to 40-min detention basin is provided beneath the stack of trays; there the heavy precipitate settles out. The lighter precipitate is pumped out with the water to a pressure filter, where it is removed. Carbon dioxide and hydrogen sulfide are liberated in the coke-tray aerator, and when high concentrations of carbon dioxide are present it may be necessary to supplement the treatment by the addition of soda ash, caustic soda, or lime to neutralize the excess carbon dioxide to prevent corrosion of pipe lines.

Open slat-tray aerators operate similar to the coke-tray type but are not as efficient; however, they are easier to clean than the coke tray, and there is no coke to replace. When the trays are enclosed and air under pressure is blown up through the downward falling spray, a compact unit is developed in which the amount of air can be proportioned to the amount of iron to be removed. Theoretically 0.14 mg/l oxygen is required to precipitate 1 mg/l iron. The unit may be placed indoors or outdoors.

Another method for iron removal utilizes a pressure tank with a perforated air distributor near the bottom. Raw water admitted at the bottom of the pressure tank mixes with the compressed air from the distributor and oxidizes the iron present. The water passes to the top of a pressure tank, at which point air is released and automatically bled off. The amount of air injected is proportioned to the iron content by a manually adjusted needle valve ahead of a solenoid valve on the air line.[155]

At a pH of 7.0, 0.6 parts of chlorine removes 1 part iron and 0.9 parts alkalinity. At a pH of 10.0, 1.3 parts of chlorine removes 1 part of manganese and 3.4 parts alkalinity.[156]

Corrosion Cause and Control

Corrosion in water supply usually means the dissolving of pipe in contact with soft water of low alkalinity containing oxygen. In serious cases water heaters are damaged, the flow of water is reduced, and the water is red or rusty where iron pipe is used. The inside surface of the pipe is dissolved, with consequent release of trace amounts of possibly harmful chemicals and weakening or pitting of the pipe and redepositing of iron with reduction of the pipe diameter and hence water flow. Stray electric currents, as from buried defective electric cable, the difference in electrical potential between the water and pipe or tank, dissimilar metals in contact, biochemical changes where iron bacteria such as crenothrix and leptothrix use iron in their growth, high water velocities, and high water temperatures all accelerate corrosion.

Although much remains to be learned concerning the mechanism of corrosion, a simple explanation may aid in its understanding. Water in contact with iron results in the formation of soluble ferrous oxide and hydrogen gas. The ferrous oxide combines with the water and part of the oxygen usually present in water to form ferric hydroxide, which redeposits in other sections of pipe or is carried through with the water. Gaseous hydrogen is attracted to the pipe and forms a protective film if allowed to remain. But gaseous hydrogen will combine with oxygen usually present in an "aggressive" water, thereby removing the protective hydrogen film and exposing the metal to corrosion. High water velocities also remove the hydrogen film. Another role is played by carbon dioxide. It has the effect of lowering the pH of the water, since more hydrogen ions are formed, which is favorable to corrosion.

The control of corrosion involves the removal of dissolved gases, treatment of the water to make it noncorrosive, building up of a protective coating inside pipe, use of resistant pipe materials or coating, cathodic protection, the insulation of dissimilar metals, prevention of electric grounding on water pipe, and control of growths in the mains. Therefore, if the conditions that are responsible for corrosion are recognized and eliminated or controlled, the severity of the

[155] H. R. Fosnot, "7 Methods of Iron Removal," *Pub. Works,* **86**, No. 11, 81–83 (November 1955).

[156] Edmund J. Laubusch, "Chlorination of Water," *Water Sewage Works*, **105**, No. 10, 411–417 (October 1958).

Table 3-20 Processes of Iron and Manganese Removal

Treatment Processes	Oxidation Required	Character of Water	Equipment Required	pH Range Required	Chemicals Required	Remarks
Aeration, Sedimentation, Sand filtration	Yes	Iron alone in absence of appreciable concentrations of organic matter	Aeration, settling basin, sand filter	Over 6.5	None	Easily operated. No chemical control required.
Aeration, contact oxidation, sedimentation, sand filtration	Yes	Iron and manganese loosely bound to organic matter, but no excessive carbon dioxide or organic acid content	Contact aerator of coke, gravel, or crushed pyrolusite, settling basin and sand filter	Over 6.5	None	Double pumping required. Easily controlled.
Aeration, contact filtration	Yes	Iron and manganese bound to organic matter, but no excessive organic acid content	Aerator and filter bed of manganese-coated sand, "Brim," crushed pyrolusite ore, or manganese zeolite	Over 6.5 ±	None	Double pumping required unless air compressor, or "snifler valve," is used to force air into water. Limited air supply adequate. Easily controlled.
Contact filtration	Yes, but not by aeration	Iron and manganese bound to organic matter, but no excessive carbon dioxide or organic acid content	Filter bed of manganese-coated sand, "Birm," crushed pyrolusite ore, or manganese zeolite	Over 6.5	Filter bed reactivated or oxidized at intervals with chlorine or sodium permanganate	Single pumping. Aeration not required.
Aeration, chlorination, sedimentation, sand filtration	Yes	Iron and manganese loosely bound to organic matter	Aerator and chlorinator or chlorinator alone, settling basin and sand filter	7.0 to 8.0	Chlorine	Required chlorine dose reduced by previous aeration but chlorination alone permits single pumping.

Aeration, lime treatment, sedimentation, sand filtration	Yes	Iron and manganese in combination with organic matter, and organic acids	Effective aerator, lime feeder mixing basin, settling basin, sand filter	8.5 to 9.6	Lime	pH control required.
Aeration, coagulation and lime treatment, sedimentation, sand filtration	Yes	Colored, turbid, surface water containing iron and manganese combined with organic matter	Conventional rapid sand filtration plant	8.5 to 9.6	Lime and ferric chloride or ferric sulfate, or chlorinated copperas, or lime and copperas	Complete laboratory control required.
Zeolite softening	No	Well water *devoid* of oxygen, and containing less than about 1.5 to 2 ppm iron and manganese	Conventional sodium zeolite unit, with manganese zeolite unit or equivalent for treatment of bypassed water[a]	Over 6.5 ±	None, added continuously, but bed is regenerated at intervals with salt solution	Only soluble ferrous and manganous compounds can be removed by base exchange, so aeration or double pumping is not required
Lime treatment, sedimentation, sand filtration	No	Soft well water *devoid* of oxygen containing iron as ferrous bicarbonate	Lime feeder, enclosed mixing and settling tanks and pressure filter	8.0 to 8.5	Lime	Precipitation of iron in absence of oxygen occurs at lower pH than otherwise. Absence of oxygen minimizes or prevents corrosion. Double pumping not required.

Source: Charles R. Cox, *Water Supply Control*, Bull. No. 22, New York State Dept. of Health, Albany, 1952, pp. 161–162.

[a]Manganese zeolite pressure filter (in series) removes iron and manganese in raw water; the optimum pH level is 7.5 to 8.5 which may require prior treatment of the raw water with caustic soda to raise pH. Potassium permanganate may be added continuously to maintain the zeolite filters regenerated.

problem will be greatly minimized. The particular cause or causes of corrosion should be determined by proper chemical analyses of the water, as well as field inspections and physical tests. The applicable control measures should then be employed.

Cement-asbestos, wood-stave, plastic, vitrified clay, and concrete pipe are corrosion-resistant. Polyvinyl chloride (PVC) pipe comes in diameters of 4 in. to 12 in., 20 ft lengths, and 150 and 200 psi working pressures. Fiberglass-reinforced plastic pipe is available in diameters up to 144 in. and lengths up to 60 ft. Fiberglass-epoxy pipe comes in 20 ft lengths, 2 in. to 12 in. diameters, and is easily installed. It combines light weight with high tensile and compressive strength. The pipe withstands pressures of 300 psi, electrolytic attack, as well as embrittlement associated with cold temperatures and aging. With soft waters, calcium carbonate tends to be removed from new concrete, cement-lined, and asbestos-cement pipe for the first few years. Wood-stave, vitrified clay, and concrete pipe have limited application. Iron and steel pipe are usually lined or coated with cement, tar, paint, or enamel, which resist corrosion. Polycyclic aromatic carbons, some of which are known to be carcinogenic, are picked up from bituminous lining of the water distribution system, but not from oil-derived tarry linings. On general principles, bituminous linings are being discontinued in England by the Department of the Environment.[157] Occasionally the coating spalls off or is imperfect, and isolated corrosion takes place. It should be remembered that even though the distribution system is corrosion-resistant, corrosive water should be treated to protect household plumbing systems.

The gases frequently found in water, and which encourage corrosion, are oxygen and carbon dioxide. Where practical, as in the treatment of boiler water or hot water for a building, the oxygen and carbon dioxide can be removed by heating and subjecting the water, in droplets, to a partial vacuum. Some of the oxygen is restored if the water is stored in an open reservoir or storage tank.

Dissolved oxygen can also be removed by passing the water through a tank containing iron chips or filings. Iron is dissolved under such conditions, but it can be removed by filtration. The small amount of oxygen remaining can be treated and removed with sodium sulfite. Ferrous sulfate is also used to remove dissolved oxygen.

All carbon dioxide except 3 to 5 mg/l can be removed by aeration, but aeration also increases the dissolved oxygen concentration, which in itself is detrimental. Sprays, cascades, coke trays, diffused air, and zeolite are used to remove most of the carbon dioxide. A filter rate of 25 gpm/ft^2 in coke trays 6 in. thick may reduce the carbon dioxide concentration from 100 to 10 mg/l and increase pH from about 6.0 to 7.0[158] The carbon dioxide remaining, however, is sufficient to cause serious corrosion in water having an alkalinity caused by calcium carbonate of less than about 100 mg/l. It can be removed where necessary by adding sodium carbonate (soda ash), lime, or sodium hydroxide (caustic soda). With soft waters having an alkalinity greater than 30 mg/l, it is easier to add soda ash or caustic soda in a small water system to eliminate the carbon dioxide and increase the pH

[157] A. W. Kenny, "Modern Problems in Water Supply," *J. R. Soc. Health*, June 1978, pp. 116–121.

and alkalinity of the water. The same effect can be accomplished by filtering the water through broken limestone or marble chips. Well water that has a high concentration of carbon dioxide but no dissolved oxygen can be made noncorrosive by adding an alkali such as sodium carbonate. Soft waters that also have a low carbon dioxide content (3 to 5 mg/l) and alkalinity (20 mg/l) may need a mixture of lime and soda ash to provide both calcium and carbonate for the deposition of a calcium carbonate film.*

Sodium hexametaphosphate, tetrasodium pyrophosphate, sodium septa-phosphate, sodium silicate (water glass), and lime are used to build up an artificial coating inside of pipe.

Sodium hexametaphosphate dissolves readily and can be added alone or in conjunction with sodium hypochlorite by means of a solution feeder. Concentrated solutions of metaphosphate are corrosive. A dosage of 5 to 10 mg/l is normally used for 4 to 8 wk until the entire distribution system is coated, after which the dosage is maintained at 1 to 2 mg/l. The initial dosage may cause precipitated iron to go into solution, with resultant temporary complaints; but flushing of the distribution system will minimize this problem. Calcium metaphosphate is a similar material, except that it dissolves slowly and can be used to advantage where this property is desirable. Inexpensive and simple pot-type feeders that are particularly suitable for small water supplies are available. Sodium pyrophosphate is similar to sodium hexametaphosphate. All these compounds are reported to coat the interior of the pipe with a film that protects the metal, prevents lime scale and red water trouble, and resists the corrosive action of water. However, heating of water above 140 to 150°F will nullify any beneficial effect.

Sodium silicate in solution is not corrosive to metals and can be easily added to a water supply with any type of chemical feeder to form calcium silicate, provided the water contains calcium. Doses vary between 25 and 240 lb/million gal, 70 lb being about average. The recommendations of the manufacturer should be followed in determining the treatment to be used for a particular water.

Adjustment of the pH and alkalinity of a water so that a thin coating is maintained on the inside of piping will prevent its corrosion. Any carbon dioxide in the water must be removed before this can be done, as previously explained. Lime† is added to water to increase the alkalinity and pH so as to come within the limits shown on Figure 3-22. The approximate dosage may be determined by the "marble test," but the Enslow stability indicator is a more accurate device.[159] Under these conditions, calcium carbonate is precipitated from the water and

*One grain per gallon (17.1 mg/l) of lime, caustic soda, and soda ash remove, respectively, 9.65 mg/l, 9.55 mg/l, and 7.20 mg/l free CO_2; the alkalinity of the treated water is increased by 23.1 mg/l, 21.4 mg/l, and 16.0 mg/l, respectively. One mg/l chlorine decreases alkalinity (as $CaCO_3$) 0.7 to 1.4 mg/l and one mg/l alum decreases natural alkalinity 0.5 mg/l.

†At a pH above 8.3 calcium carbonate is soluble to 13 to 15 mg/l.

[158]Ellsworth L. Filby, "The New Thomas H. Allen Pumping Station and Iron-Removal Plant of Memphis," *Water Sewage Works,* **99**, No. 4, 133–138 (April 1952).

[159]*Water Quality and Treatment*, AWWA, McGraw-Hill Book Company, New York, 1971, pp. 332–337. See also *Standard Methods.*

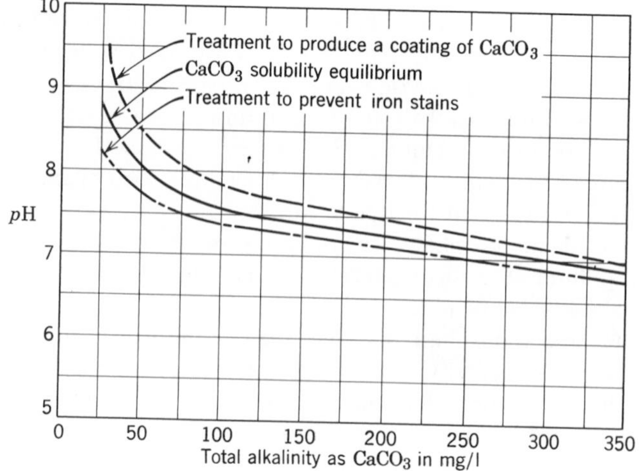

Figure 3-22 Solubility of $CaCO_3$ at 71°F (Baylis Curve).

deposited on the pipe to form a protective coating, provided a velocity of 1.5 to 3.0 fps is maintained to prevent heavy precipitation near the point of treatment and none at the ends of the distribution system. The addition of 0.5 to 1.5 mg/l metaphosphate will help obtain a more uniform calcite coating throughout the distribution system. The addition of lime must be carefully controlled so as not to exceed a pH of 9.2 to prevent caustic alkalinity being formed. Calcium carbonate is less soluble in hot water than in cold water. It should be remembered that the disinfecting capacity of chlorine (HOCl) decreases as the pH increases, hence the free available chlorine concentration maintained in the water should be increased with high pH. See Table 3-16. Also note that soft corrosive water with a high pH will increase corrosion of copper and zinc; old yellow brass plumbing can be dezincified and galvanizing can be removed from iron pipe.[160]

The Langelier index is also used to determine the point of calcium carbonate stability for corrosion control with waters having an alkalinity greater than 35 to 50 mg/l. The point of calcium carbonate stability is also indicated by the Ryznar index. The Caldwell-Lawrence diagrams[161] are useful for solving water-conditioning problems, but raw water concentration of calcium, magnesium, alkalinity, pH, and TDS values must be known.

The AWWA recognizes the coupon test to measure the effects of physical factors and substances in water on a small section of metal pipe inserted in a water line. Measurement of the weight loss due to corrosion or weight gained due to scale formation can thus be determined under the actual use conditions.

[160] William Wheeler, "Notes and Comments," *J. Am. Water Works Assoc.*, February 1979, p. 116.
[161] Douglas T. Merrill and Robert L. Sanks, "Corrosion Control by Deposition of $CaCO_3$ Films: Part 1, Part 2, and Part 3," *J. Am. Water Works Assoc.*, November 1977, pp. 592–599; December 1977, pp. 634-640; January 1978, pp. 12–18.

The danger of lead or zinc poisoning and off-flavors due to copper plumbing can be greatly reduced when corrosive water is conducted through these pipes by simply running the water to waste in the morning. This will flush out most of the metal that has had an opportunity to go into solution while standing during the night. Maintenance of a proper balance between pH, calcium carbonate level, and alkalinity as calcium carbonate is necessary to reduce and control lead corrosion by soft aggressive water. Formation and then *maintenance* of a carbonate film is necessary. See Figure 3-22. In a soft corrosive water, sodium hydroxide can be used for pH adjustment and sodium bicarbonate for carbonate addition. The use of zinc orthophosphate was found ineffective.[162] Lead pipe should not be used to conduct drinking water.

Biochemical actions, such as the decomposition of organic matter in the absence of oxygen in the dead end of mains, the reduction of sulfates, the biochemical action within tubercles, and the growth of crenothrix and leptothrix, all of which encourage corrosion in mains, can be controlled by the maintenance of at least 0.3 mg/l free residual chlorine in the distribution system.

Corrosion caused by electrolysis or stray electric currents can be prevented by making a survey of the piping and removing grounded electrical connections and defective electric cables. Moist soils will permit electric currents to travel long distances. A section of nonconducting pipe in dry soil may confine the current. In the vicinity of power plants this problem is very serious and requires the assistance of the power company involved. Corrosion of water storage tanks can be controlled by providing "cathodic protection," in which a direct current is imposed to make the metal more electronegative. But repainting the metal above the water line is necessary. Special equipment for this purpose is manufactured. Where dissimilar metals are to be joined, a plastic, hard rubber, or porcelain separating fitting can be used. It must be long enough to prevent the electric charge from jumping the gap. A polyethylene tube around cast-iron pipe will protect it from corrosive soils.

A common problem with wells is reduction in production capacity usually due to clogging or incrustation of the well screen openings and the formation immediately around the screen. This may be due to mineral scale precipitating in the formation around the screen, to bacteria which oxidize iron such as crenothrix and gallionella, and plugging when silt and clay build up in the formation around the well.

Treatment methods include the use of acids to dissolve mineral scale and bacterial iron precipitate. Chlorination (sodium hypochlorite) to disinfect a well will also remove the iron bacteria. Quaternary ammonium compounds might also be effective. Sodium polyphosphates have been found effective in unplugging wells caused by clay and silt particles.[163] Repeat treatment may be needed.

[162]James W. Patterson and Joseph E. O'Brien, "Control of Lead Corrosion," *J. Am. Water Works, Assoc.*, May 1979, pp. 264–271.

[163]David C. Schafer, "Use of Chemicals to Restore or Increase Well Yield," *Pub. Works*, April, 1975.

Water Softening

Water softening is the removal from water of minerals causing hardness. For comparative purposes, one grain per gal of hardness is equal to 17.1 mg/l. Hardness is caused primarily by the presence in water of calcium bicarbonate, magnesium bicarbonate, calcium sulfate (gypsum), magnesium sulfate (epsom salts), calcium chloride, and magnesium chloride in solution. In the concentrations usually present these constituents are not harmful in drinking water. The presence of hardness is demonstrated by the use of large quantities of soap in order to make a lather;* the presence of a gritty or hard curd in laundry or in a basin; the formation of hard chalk deposits on the bottom of pots and inside of piping causing reduction in the flow of water; and the lowered efficiency of heat transfer in boilers, caused by the formation of an insulating scale. Hard water is not suitable for use in boilers, laundries, textile plants, and certain other industrial operations where a zero hardness of water is used, in addition to the normal or partially softened water.

In softening water the lime-soda ash process, zeolite process, and organic resin process are normally used. The soluble bicarbonates and sulfates are removed by converting them to insoluble forms in the lime-soda ash method. The calcium and magnesium are replaced with sodium in the zeolite process, thereby forming sodium compounds in the water that do not cause hardness. With synthetic organic resins, dissolved salts can be almost completely removed. Table 3-21 gives ion exchange values.

Lime-soda ash softening requires the use of lime to convert the soluble bicarbonates of calcium and magnesium to insoluble calcium carbonate and magnesium hydroxide which are precipitated. The soluble calcium and magnesium sulfates or chlorides are converted to insoluble calcium carbonate and magnesium carbonate by the addition of soda ash and lime and precipitated. The sodium chloride and sodium sulfate formed remain in the water. Excess lime to achieve a pH of about 11 is needed to precipitate magnesium hydroxide; but pH adjustment is then needed to control calcium carbonate precipitation in the distribution system. A coagulant such as aluminum sulfate (filter alum), ferrous sulfate (copperas), ferric sulfate, or sodium aluminate is usually used to settle and coagulate the compounds formed and to remove turbidity and color. It is sometimes also necessary to add carbon dioxide to cause precipitation of the calcium hydroxide to calcium carbonate. Large volumes of sludge with high water content are produced. The lime-soda method is not suitable for the softening of small quantities of water because special equipment and technical control are necessary. The process is more economical for the softening of a moderately hard water. As water hardness increases, lime requirement increases making zeolite more attractive.

*With a water hardness of 45 mg/l the annual per capita soap consumption was estimated at 29.23 lb; with 70 mg/l hardness, soap consumption was 32.13 lb; with 298 mg/l hardness, soap consumption was 39.89 lb; and with 555 mg/l, soap consumption was 45.78 lb. (Merrill L. Riehl, *Hoover's Water Supply and Treatment,* National Lime Association, Washington, D.C., April 1957).

Table 3-21 Ion Exchange Materials and their Characteristics

Exchange Material	Exchange Capacity (grains/ft^3)	Effluent Contains	Regenerate with	Remarks
Natural zeolites	3,000 to 5,000	Sodium bicarbonate, chloride, sulfate	0.37 to 0.45 lb salt per 1000 grains hardness removed	Ferrous bicarbonate and manganous bicarbonate also removed from well water devoid of oxygen, pH of water
Artificial zeolites	9,000 to 12,000	Sodium bicarbonate, chloride,	0.37 to 0.45 lb salt per 1000 grains hardness removed	must be 6.0 to 8.5, moderate turbidity acceptable. Use 5 to 10% brine solution. Saturated brine is about 25%.
Carbonaceous zeolites	9,000 to 12,000	Carbon dioxide and acids, sodium chloride and sulfate	0.37 to 0.45 lb salt per 1000 grains hardness removed	Acid waters may be filtered. CO_2 in effluent removed by aeration, acid by neutralization with bypassed hard water or addition of caustic soda.
Synthetic organic resins	10,000 to 30,000	Carbon dioxide and acids	0.2 to 0.3 lb salt per 1000 grains hardness removed	Dissolved salts are removed by resins. To remove CO_2 and acids, add soda ash or caustic soda, or CO_2 by aeration, and acids by synthetic resin filtration.

Note: One gallon of saturated brine weighs 10 lb and contains 2.5 lb of salt. Hardness is caused by calcium bicarbonate, magnesium bicarbonate, calcium sulfate, magnesium sulfate, and calcium chloride. Natural zeolite is more resistant to waters of low pH than artificial zeolite.

The zeolite and synthetic resin softening methods are relatively simple and require little control. Only a portion of the hard water need be passed through a zeolite softener, since a water of zero hardness is produced by the zeolite filter. The softener effluent can be mixed with part of the untreated water to produce a water of about 50 to 80 mg/l hardness. The calcium and magnesium in water to be treated replace the sodium in the zeolite filter media, and the sodium passes through with the treated water. This continues until the sodium is used up, after which the zeolite is regenerated by bringing a solution of common salt in contact with the filter media. Units are available to treat the water supply of a private home or a community. Water having a turbidity of more than 10 mg/l will coat the zeolite grains and reduce the efficiency of a zeolite softener. Iron in the ferric form is also detrimental. Pretreatment to remove turbidity and iron would be indicated. The filters are 2 to 6 ft deep and downward flow filters operate at rates between 3 and 5 gpm/ft^2, upward flow filters operate at 4 to 8 gpm/ft^2.

Synthetic resins for the removal of salts by ion exchange are discussed under Desalination in this chapter.

Small quantities of water can be softened in batches for laundry purposes by the addition of borax, washing soda, ammonia, or trisodium phosphate. Frequently insufficient contact time is allowed for the chemical reaction to be completed, with resultant unsatisfactory softening.

Lime softening removes arsenic, barium, cadmium, chromium, fluoride, lead, mercury, selenium, radioactive contaminants, copper, iron, manganese, and zinc.

The extent to which drinking water is softened should be evaluated in the light of the relationship of soft water to cardiovascular diseases. In view of the accumulating evidence, the wisdom of constructing municipal softening plants is being questioned. There is also evidence associating ingestion of sodium with cardiovascular diseases, kidney disease, and cirrhosis of the liver. See discussions earlier in this chapter of *Hardness* and *Sodium*, also Cardiovascular Diseases in Chapter 1.

Fluoridation

The absence of dental caries in children with fluorosis led to discovery of a relationship with the concentration of fluorides in drinking water. Table 3-3 gives *maximum* permissible levels of fluorides in drinking water based on the annual average of the maximum daily air temperature for a particular location. See the discussion of Fluorides this chapter, also Dental Caries in Chapter 1.

Fluorides have been added to public water supplies in controlled amounts since about 1943 as an aid to the reduction of tooth decay. The compounds commonly used are sodium fluoride (NaF), sodium silicofluoride (Na_2SiF_6), and hydrofluosilicic acid (H_2SiF_6) also called fluosilicic acid. They are preferred because of cost, safety, and ease of handling. Ammonium silicofluoride may be used in conjunction with chlorine where it is desired to maintain a chloramine residual in the distribution system. Calcium fluoride (fluorspar) does not dissolve readily. Hydrofluoric acid is hazardous to handle and is not recommended.

Solution and gravimetric or volumetric dry feeders are used to add the fluoride, usually after filtration treatment and before entering the distribution system. Calcium hypochlorite and fluoride should not be added together as a calcium fluoride precipitate would be formed. Personnel handling fluorides should wear protective clothing and proper dust control measures should be included in the design where dry feeders are used.

The average annual per capita cost of fluoridation of a public water supply is small. It was estimated at 16 cents, with a range of 6 to 33 cents depending on population served.[164] Softened water should be used to prepare a sodium fluoride solution whenever the hardness of the water used to prepare the solution is greater than 75 mg/l, or even less, to prevent calcium and magnesium precipitation which clogs the feeder. Small quantities of water can be softened by ion exchange, or polyphosphates may be used.[165]

[164]William Berner, "Status of Fluoridation in New York State," *J. Am Water Works Assoc.* February 1969, pp. 68–72.

[165]Ervin Bellacks, *Fluoridation Engineering Manual,* USEPA, WSD, Washington, D.C., 20460, Reprint 1974.

Removal of Inorganic Chemicals

The sources, health effects, permissible concentrations, and control measures related to inorganic chemicals are discussed under the appropriate headings in Chapter 1 and earlier in this chapter. Fundamental to the control of inorganic chemicals in drinking water is a sanitary survey, identification of the sources, types, and amounts of pollutants, followed by their phased elimination as indicated, *starting at the source.*

Table 3-22 summarizes treatment methods for the removal of inorganic chemicals from drinking water. Several are discussed in some detail below.

Arsenic Removal

More than 90 percent removal of sodium arsenate can be achieved by coagulation with alum or ferric sulfate. Lime softening is very effective in removing sodium arsenate at pH 10.6 and above; but removal decreases with decreasing pH.

Sodium arsenite removal with ferric sulfate is 50 to 60 percent in the normal

Table 3-22 Most Effective Treatment Methods for Inorganic Contaminant Removal

Contaminant	Most Effective Methods	Contaminant	Most Effective Methods
Arsenic			
As^{3+}	Ferric sulfate coagulation, pH 6 to 8; Alum coagulation, pH 6 to 7; Excess lime softening; Oxidation before treatment required	Fluoride	Ion exchange with activated alumina or bone char media
		Lead	Ferric sulfate coagulation, pH 6 to 9; Alum coagulation, pH 6 to 9; Lime softening; Excess lime softening
As^{5+}	Ferric sulfate coagulation, pH 6 to 8; Alum coagulation, pH 6 to 7; Excess lime softening	Mercury	
		Inorganic	Ferric sulfate coagulation, pH 7 to 8
Barium	Lime softening, pH 10 to 11; Ion exchange	Organic	Granular activated carbon
		Nitrate	Ion Exchange
Cd^{3+}	Ferric sulfate coagulation, above pH 8; Lime softening; Excess lime softening	Selenium	
		Se^{4+}	Ferric sulfate coagulation, pH 6 to 7; Ion exchange; Reverse osmosis
Chromium		Se^{6+}	Ion exchange; Reverse osmosis
Cr^{3+}	Ferric sulfate coagulation, pH 6 to 9; Alum coagulation, pH 7 to 9; Excess lime softening	Silver	Ferric sulfate coagulation, pH 7 to 9; Alum coagulation, pH 6 to 8; Lime softening; Excess lime softening
Cr^{6+}	Ferrous sulfate coagulation, pH 7 to 9.5		

Source: Thomas J. Sorg, "Treatment Techniques for the Removal of Inorganic Contaminants from Drinking Water," *Manual of Treatment Techniques for Meeting the Interim Primary Drinking Water Regulations*, USEPA, Cincinnati, Ohio, May 1977, p. 3.

pH operating range. Lime softening will remove 70 to 80 percent above pH 11 dropping to 15 to 20 percent removal at pH 10.2. Alum coagulation will remove less than 20 percent of the sodium arsenite. Sodium arsenite can be oxidized to arsenate using chlorine and potassium permanganate, but use of chlorine may lead to the formation of trihalomethanes where the water contains organic material. Ozone should also be effective.

Activated alumina, ion exchange treatment using an anion resin, and bone char can remove sodium arsenite and sodium arsenate but pilot plants are suggested first. Reverse osmosis and electrodialysis should also be very effective.[166]

Cadmium Removal

Cadmium removal of greater than 90 percent can be achieved by iron coagulation at about pH 8 and above. Greater percentage removal is obtained in higher turbidity water. Lime and excess lime softening remove nearly 100 percent cadmium at pH 8.7 to 11.3. Ion exchange treatment with cation exchange resin should remove cadmium from drinking water. Powdered activated carbon is not efficient and granular activated carbon will remove 30 to 50 percent. Reverse osmosis may not be practical for cadmium removal.[167]

Lead Removal

Normal water coagulation and lime softening remove lead; 99 percent for coagulation at pH 6.5 to 8.5 and for lime softening at pH 9.5 to 11.3. Turbidity in surface water makes lead removal easier than in groundwater. Powdered activated carbon removes some lead; GAC effectiveness is unknown; reverse osmosis, electrodialysis, and ion exchange should be effective; but best removal by ion exchange is obtained near pH 5.[167]

Nitrate Removal

Treatment methods for removal of nitrates from drinking water include: chemical reduction, biological denitrification, ion exchange, reverse osmosis, and electrodialysis. Ion exchange is the most practical method based on experience at one community water system. The water has approximately 200 mg/l total dissolved solids; the nitrate-nitrogen levels are reduced from 20 to 30 mg/l to less than 2 mg/l.[168,169] Little plant-scale data is otherwise available.

[166]Thomas J. Sorg and Gary S. Logsdon, "Treatment Technology to Meet the Interim Primary Drinking Water Regulations for Inorganics: Part 2" *J. Am Water Works Assoc.*, July 1978, pp. 379–393.

[167]Thomas J. Sorg, Mihaly Casanady, and Gary S. Logsdon, "Treatment Technology to Meet the Interim Primary Drinking Water Regulations for Inorganics: Part 3," *J. Am. Water Works Assoc.*, December 1978, pp. 680–691. Also Thomas J. Sorg, "Compare nitrate removal methods," *Water & Wastes Engr.*, December 1980, pp. 26–31.

[168]Myra Sheinker and John P. Codoluto, "Making Water Supply Nitrate Removal Practicable," *Pub. Works*, June 1977, pp. 71–73.

[169]Thomas J. Sorg, "Treatment Technology to Meet the Interim Primary Drinking Water Regulations for Inorganics," *J. Am. Water Works Assoc.*, February 1978, pp. 105–109.

Reverse osmosis and electrodialysis are also effective (40 to 95 percent) but these methods are more costly than ion exchange.

Fluoride Removal

Treatment methods for the removal of fluorides from drinking water have been summarized by Sorg.[169] They include: high (250 to 300 mg/l) alum doses, activated carbon at pH 3.0 or less; lime softening if sufficient amounts of magnesium (79 mg/l to reduce fluoride from 4 to 1.5 mg/l) are present or added for coprecipitation with magnesium hydroxide; ion exchange using activated alumina, bone char, or granular tricalcium phosphate; and reverse osmosis. Of these methods, alum coagulation and lime softening are not considered practical. Reverse osmosis has not been demonstrated on a full-scale basis for this purpose, but ion exchange has been. Activated alumina and bone char have been successfully used, but the former seems to be the method of choice for the removal of fluoride from drinking water.[170]

Selenium Removal

Selenium is predominantly found in water as selenite and selenate. Selenite can be removed (40 to 80 percent) by coagulation with ferric sulfate dependent on the pH, coagulant dosage, and selenium concentration. Alum coagulation and lime softening are only partially effective, 15 to 20 percent and 35 to 45 percent respectively. Selenite and selenate are best removed by ion exchange, reverse osmosis, and electrodialysis but the effectiveness of these methods in removing selenium has not been demonstrated in practice.[166]

Radionuclide Removal

Coagulation and sedimentation are very effective in removing radioactivity associated with turbidity and are fairly effective in removing dissolved radioactive materials, with certain exceptions. The type of radioactivity, the pH of the treatment process, and the age of the fission products in the water being treated must be considered. For these reasons, jar-test studies are advised before plant-scale operation is initiated. A comprehensive summary of the effectiveness of different chemical treatment methods with various radionuclides is given by Straub.[171] The effectiveness of rapid and slow sand filtration, lime soda-ash softening, ion exchange, and other treatment processes is also discussed.

Studies for military purposes show that radioactive materials present in water as undissolved turbidity can be removed by coagulation, hypochlorination, and diatomite filtration. Soluble radioisotopes are then removed by ion exchange using a cation exchange column followed by an anion exchange column operated in series. Hydrochloric acid is used for regenerating the cation resin, and sodium carbonate the anion resin. The standard Army vapor compression distillation

[170] Frederick Rubel, Jr. and R. Dale Woolsey, "The Removal of Excess Fluoride From Drinking Water by Activated Alumina," *J. Am. Water Works Assoc.*, January 1979, pp. 45–49.

[171] Conrad P. Straub, "Radioactivity," *Water Quality and Treatment*, AWWA, McGraw-Hill Book Company, New York, 1971, pp. 443–462.

unit is also effective in removing radioactive material from water.[172] Ground-water sources of water can generally be assumed to be free of radioactive substances and should if possible be used in preference to a surface water source[172] in emergency situations.

Kosarek[173] reviewed the water treatment processes which have been used to reduce dissolved radium contamination to an acceptable level (5 pCi/l or less) in water for industrial and municipal purposes. Processes for industrial water uses are selective membrane mineral extraction, reverse osmosis, barium sulfate coprecipitation, ion exchange, lime-soda softening, and sand filtration. Processes for municipal water uses are reverse osmosis, ion exchange, lime-soda softening, aeration, greensand filtration, and sand filtration. Aeration, greensand filtration, and sand filtration have low radium removal efficiency. Lime-soda has a 50 to 85 percent efficiency; the other remaining processes have an efficiency of 90 to 95 percent or better. There are limits on the maximum allowable dissolved radium in the water being treated to achieve the 5 pCi/l standard.

Prevention and Removal of Organic Chemicals

Morris[174], in a statement concerning chlorinated organic compounds, cautions that

> one should differentiate clearly among four sources of such compounds in water supplies: 1) nonpoint sources, such as airborne pesticides or solvents, 2) industrial waste discharges, 3) products of municipal wastewater chlorination, and 4) chlorination of natural organic matter such as color, in municipal sources. Furthermore, the major areas of concern with regard to toxicity and carcinogenicity are sources 1) and 2); source 3) might well become a concern, but until now source 4) has not been demonstrated to be a problem, except by improper association with categories 1), 2), and 3).

He adds that

> chlorination of natural organic material in water occurs in specific limited ways, producing only a small number of substances like the relatively innocuous chloroform, and that we do not have to face an enormous spectrum of all sorts of chlorinated compounds with unknown toxic properties.

The significance of chloroform is dependent on its concentration and is debatable. See Halogenated Chloro-organic Compounds (Trihalomethanes) earlier in this chapter and the discussion that follows.

[172]Don C. Lindsten, "From Fission, Fusion, and confusion can come potable water," *Water Wastes Eng.*, July 1977, pp. 56–61.

[173]Louis J. Kosarek, "Radionuclide removal from water," *Environ. Sci. Technol.*, May 1979, pp. 522-525.

[174]George E. Symons and Kenneth W. Henderson, "Disinfection—Where Are We?" *J. Am Water Works, Assoc.*, March 1977, p. 149.

As noted for inorganic chemicals, the control of organic chemicals in drinking water should start with a sanitary survey to identify the sources, types, and amounts of pollutants, followed by their phased elimination as indicated by the associated hazard. Included would be watershed use regulation and protection; watershed management to minimize turbidity and organic and inorganic runoff; vigorous compliance with the natural and state water and air pollution elimination objectives; enforcement of established water and air classification standards; and complete effective drinking water treatment under competent supervision. It is obvious that selection of the cleanest available protected source of water supply, for the present and the future, would greatly minimize the problems associated not only with organic chemicals, but also with physical and microbiological pollution. In any case, water treatment plants must be upgraded where needed to consistently produce a water meeting the national drinking water standards.

Synthetic Organic Chemicals—Halogenated Chloro-organic Compounds (Products of Chlorination)

The halogenated, chloro-organic compounds include the trihalomethanes: trichloromethane (chloroform), bromodichloromethane, dibromochloromethane, and tribromomethane (bromoform). These chlorination by-products are formed by the reaction of *free* chlorine with certain organic compounds in water. The major cause of trihalomethane formation in drinking water that is chlorinated is probably humic and fulvic substances (natural organic matter in soil, peat, and runoff) and simple low molecular weight compounds, including algae, referred to as precursors. Other halo-organic compounds are also produced during chlorination. Chlorination of municipal wastewater also results in the formation of chloro-organics, but their concentration is very low when combined chlorine is formed,[175] which is usually the case. The reaction is dependent on pH, temperature, and contact time; the point of chlorination is critical in treating drinking water. Total trihalomethane concentration in treated drinking water has been found to be higher in the summer (and after reservoir turnover) and lowest in the winter. It correlates reasonably well with chlorine demand of untreated water, but not with organic carbon and chloroform extract.[176]

Prechlorination with long contact periods and sunlight increases formation of trihalomethanes as does the addition of chlorine prior to coagulation and the removal of precursors. GAC has been found to be of limited effectiveness to

[175] P. E. Gaffney, "Chlorobiphenyls and PCBs: formation during chlorination," *J. Water Pollut. Control Fed.*, March 1977, pp. 401–404.

[176] Robert C. Hoehm, Clifford W. Randall, Frank A. Bell, Jr., and Peter T. B. Shaffer, "Trihalomethane and Viruses In a Water Supply," *J. Environ. Eng. Div. ASCE,* October 1977, pp. 803–814.

remove precursor compounds; it is effective for only a few weeks.[177] In contrast, GAC beds for taste and odor control need regeneration every three to six years.[178] GAC is not suitable for the removal of trihalomethanes (THM) once formed. Treatment to remove as much suspended matter and colloidal and dissolved material as possible, usually by coagulation, flocculation, sedimentation, and filtration, should precede GAC, if used for taste and odor control, so as not to coat and thus reduce the adsorptive capacity of the carbon. Such treatment will also remove most THM precursors.

Morris[179] points out

> that methods other than carbon adsorption, including impoundment, coagulation, sedimentation, and even oxidation, are useful in removing organics.

European experience shows that a water treatment process for polluted waters consisting of screening, flocculation, sedimentation, rapid sand filtration, ozonation, GAC, and chlorine or chlorine dioxide will reduce organics concentration in drinking water to safe levels.[180] Aeration (air stripping) is also effective in removing trihalomethanes, as well as many other organics, at a reasonable cost. More than 90 percent removal can be achieved at air:water ratios equal to or greater than 10:1. Changing the point of chlorination from before to after the addition of chemicals, flocculation, and sedimentation reduced the chloroform concentration about 50 percent.[181,182] It is very desirable, if found effective, to chlorinate after coagulation, but prior to the sedimentation basin, to take advantage of the long contact period, particularly where the pH of the water is low or near 7.0 and the HOCl component of the free chlorine is highest, thereby providing the best bacteria, cyst, and virus kill. Care must be exercised to always assure that safety of the water is not endangered in efforts to reduce trihalomethane formation.*

A WHO Working Group of the Regional Office for Europe[183] advised consideration of the following action to reduce by-products of disinfection with chlorine:

(a) protection of raw water against pollution by precursors of such compounds;
(b) choice of the best available raw water;

*Boiling of water 3 to 5 min will release most of the trihalomethanes. This can be done in an emergency.

[177]Alan A. Stevens, Clois J. Slocum, Dennis R. Seeger, and Gordon R. Robek, "Chlorination of Organics in Drinking Water," *J. Am. Water Works, Assoc.*, November 1976, pp. 615–620.

[178]Gene Dallaire, "Are cities doing enough to remove cancer-causing chemicals from drinking water?" *Civ. Eng. ASCE*, September 1977.

[179]Charles Winkelhaus, "Chlorination: assessing its impact," *J. Water Pollut. Control Fed.*, December 1977, pp. 2354–2357.

[180]G. Wade Miller and Rip G. Rice, "European water treatment practices—the promise of biological activated carbon," *Civ. Eng. ASCE*, February 1978, pp. 81–83.

[181]John S. Young, Jr. and Philip C. Singer, "Chloroform Formation in Public Water Supplies: A Case Study," *J. Am Water Works Assoc.*, February 1979, pp. 87–95.

[182]David B. Babcock and Philip C. Singer, "Chlorination and Coagulation of Humic and Fulvic Acids," *J. Am. Water Works, Assoc.*, March 1979, pp. 149–152.

[183]"Treatment Agents and Processes for Drinking Water," *J. R. Soc. Health*, June 1978, p. 129.

(c) if necessary, provision of pretreatment for the raw water to at least reduce the precursors and, hence, a reduction in the quantities used;
(d) limitation of the quality of chlorine and other disinfectants to the maximum level to achieve the basic purpose of disinfection.

The maximum contaminant level for total trihalomethanes in drinking water in the United States is 100 μg/l. The goal is 10 to 25μg/l. The Canadian recommended maximum acceptable level is 350 μg/l; the objective level is 0.5 μg/l.[184]

Synthetic Organic Chemicals—Other (Man-made Pollution-Related)

The major sources of synthetic organic chemical pollution (also inorganic pollution in many places) are industrial wastewater discharges, air pollutants, municipal wastewater discharges, runoff from cultivated fields, spills, and waste storage sites, and leachate from sanitary landfills, industrial and commercial dump sites, ponds, pits, and lagoons. These affect both surface waters and groundwaters. It cannot be emphasized enough that *control of all pollutants must start at the source* including raw material selection, chemical formulation, and manufacturing process control. Separation of floating oils and collection of low solubility high density compounds in traps on building drains could reduce pollutant discharges and recover valuable products. Such actions would reduce the extent of needed plant upgrading, sophisticated water treatment and control, the burden on downstream water treatment plants, and hence the risks to the consumer associated with the ingestion of often unknown hazardous or toxic chemicals.

The more common water treatment methods considered to reduce the concentrations of synthetic organic chemicals in drinking water are aeration and filtration through granular activated carbon (GAC). Most experience with GAC has been for color, taste and odor control and has been found very effective. Bench scale tests using strongly basic anion exchange resins showed that most of the organics present in surface water can be removed.[185] However, because of the many variables involved, pilot plant studies with aeration, GAC, and resins are required to be carried out at the site to determine the effectiveness of a process and the basis for design before a treatment method is selected. This is also necessary to determine the GAC adsorption capacity before exhaustion and its reactivation cost. Organics in drinking water have different adsorptive characteristics on GAC.

Aeration will remove many volatile organic chemicals in water. The extent to which aeration is successful will depend on the concentration, temperature, solubility, and volatility of the compound in the water (Henry's Law). The rate of

[184]E. Somers, "Physical and Chemical Agents and Carcinogenic Risk," *Bull. Pan Am. Health Organ.*, 14(2), 1980, pp. 172–184.

[185]Claude T. Anderson and Walter T. Maier, "Trace Organics Removal by Anion Exchange Resins," *J. Am. Water Works Assoc.*, May 1979, pp. 278–283.

removal depends on the amount of air used, contact time, and temperature of the air and water. Very low efficiencies are obtained at freezing temperatures. Compounds reported to be removed by aeration include trichloroethylene, carbon tetrachloride, tetrachloroethylene, benzene, toluene, napthalene, biphenyl, methyl bromide, bromoform, chloroform, dibromochloromethane, bromodichloromethane, methylene chloride, vinyl chloride, sodium fluoroacetate, and others. Corrosion control is usually required after aeration, and airborne contamination, including worm growths in the aerator, must be avoided.

GAC is considered the best available broad spectrum adsorber of synthetic organic chemicals. The carbon is similar in size to filter sand. Adsorption is a complex process. It is influenced by the surface area of the carbon grains, the material being adsorbed or concentrated (adsorbate), the pH and temperature of the water being treated, the mixture of compounds present, and the nature of the adsorbent, that is, the carbon grain structure, surface area, and pores. The smaller the grain size, within the range of operational efficiency, the greater the rate of adsorption obtained.[186]

Treatment consisting of coagulation, filtration, and powdered activated carbon is reported[187,188] to remove 85 to 98 percent of the endrin and 90 to 98 percent 2-4-D, and 30 to 99 percent of the lindane at carbon dosages of 5 to 79 mg/l.

WATER SYSTEM DESIGN PRINCIPLES

Water Quantity

The quantity of water upon which to base the design of a water system should be determined in the preliminary planning stages. Future water demand is based on social, economic, and land-use factors, all of which can be expected to change with time. (See Chapter 2.) Population projections are a basic consideration. They are made using arithmetic, geometric, and demographic methods, and with graphical comparisons with the growth of other comparable cities or towns of greater population.[189,190] Adjustments should be made for hospital and other institution populations, industries, fire protection, military reservations, transients, and tourists.

[186]Walter J. Weber, Jr., *Physicochemical Processes For Water Quality Control*, Wiley-Interscience, New York, 1972, pp. 199–255.

[187]O. Thomas Love, Jr., "Let's drink to cleaner water," *Water Wastes Eng.*, September 1977, p. 136.

[188]O. Thomas Love, Jr., "Treatment Techniques for the Removal of Organic Contaminants from Drinking Water," *Manual of Treatment Techniques for Meeting the Interim Primary Drinking Water Regulations,* USEPA, Cincinnati, Ohio, May 1977, pp. 53–61.

[189]Meyer Zitter, "Population Projections for Local Areas," *Pub. Works,* June 1957.

[190]Gordon M. Fair, John C. Geyer, and Daniel A. Okun, *Water and Wastewater Engineering,* Vol. 1, John Wiley & Sons, New York, 1966.

Numerous studies have been made to determine the average per capita water use for water system design. Health departments and other agencies have design guides, and standard texts give additional information. In any case, the characteristics of the community must be carefully studied and appropriate provisions made. A study made in 1965 indicated that the average per capita water use in the United States was 155 gpd, with wide individual variations.[191] A more recent study of 795 utilities serving populations of 10,000 or more showed that the mean 1970 water production was 167 gpcd, with a low of 134 in the southeast central region and a high of 211 in the Pacific coast region of the United States.[192]

Design Period

The design period (the period of use for which a structure is designed) is usually determined by the future difficulties to acquire land or replace a structure or pipeline, the cost of money, and the rate of growth of the community or facility served. In general large dams and transmission mains are designed to function for 50 or more years; wells, filter plants, pumping stations and distribution systems for 25 yr; and water lines less than 12 in. in diameter for the full future life. When interest rates are high or when temporary or short-term use is anticipated, a lesser design period would be in order. Fair et al., suggest that the dividing line is in the vicinity of 3 percent per annum.[190]

Watershed Runoff and Reservoir Design

Certain basic information, in addition to future water demand, is needed upon which to base the design of water works structures. Long-term rainfall and stream-flow data, as well as groundwater information, are available from the Geological Survey, but these seldom apply to small watersheds. Rainfall data for specific areas are also available from local weather stations, airports, and water works. Unit hydrographs, maximum flows, minimum flows, mass diagrams, characteristics of the watershed, rainfall, evaporation losses, percolation, and transpiration losses should be considered for design purposes and storage determinations when these are applicable.

Watershed runoff can be estimated in different ways. The rational method for determining the maximum rate of runoff is given by the formula $Q = AIR$. Q is the runoff in ft^3/sec; A is the area of the watershed in acres; R is the rate of rainfall on the watershed in in./hr; and I is the imperviousness ratio, that is, the ratio of water that runs off the watershed to the amount precipitated on it. I will vary

[191] "Estimated Use of Water in the United States, 1965," Circular 556, compiled by the Geological Survey, Dept. of the Interior, 1968. Summarized in *Pub. Works,* April 1970, pp. 85–87.

[192] Harris F. Seidel, "A Statistical Analysis of Water Utility Operating Data for 1965 and 1970," *J. Am. Water Works Assoc.,* June 1978, pp. 315–323.

from 0.01 to 0.20 for wooded areas; from 0.05 to 0.25 for farms, parks, lawns, and meadows depending on the surface slope and character of the subsoil; from 0.25 to 0.50 for residential semirural areas; from 0.05 to 0.70 for suburban areas; and from 0.70 to 0.95 for urban areas having paved streets, drives, and walks.[193] $R = 360/t + 30$ for maximum storms and $R = 105/t + 15$ for ordinary storms in eastern United States; $R = 7/\sqrt{t}$ for San Francisco; $R = 56/(t + 5)^{85}$ for New Orleans; and $R = 19/\sqrt{t}$ for St. Louis, in which t = time (duration) of rainfall in min.[194]

Another formula for estimating the average annual runoff is by Vermuelé; it may be written as follows: $F = R - (11 + 0.29\ R)(0.035\ T - 0.65)$ in which F is the annual runoff in inches, R is the annual rainfall in inches, and T is the mean annual temperature in degrees Fahrenheit. This formula is reported to be particularly applicable to streams in northern New England and in rough mountainous districts along the Atlantic Coast.[195] For small water systems, it is suggested that design be based on the year of minimum rainfall, or on about 60 percent of the average.

In any reservoir storage study it is important to take into consideration the probable losses due to evaporation from water surfaces during the year. This becomes very significant in small systems when the water surfaces exceeds 6 to 10 percent of the drainage area.[196] In the north Atlantic states the annual evaporation from land surfaces averages about 40 percent, while that from water surfaces is about 60 percent of the annual rainfall.[197] A more general relationship of monthly transpiration to mean monthly air temperature is given in Figure 4-30 based on studies made by Langbein. The watershed water loss due to land evaporation, plant usage, and transpiration is significant and hence must be taken into consideration when determining rainfall minus losses. See Septic-Tank Evapotranspiration System in Chapter 4.

The minimum stream flow in New England has been estimated to yield 0.2 to 0.4 cfs/mi^2 of tributary drainage and an annual yield of 750,000 gpd/mi^2 with storage of 200 to 250 million gal/mi^2. New York City reservoirs located in upstate New York have a dependable yield of about 1 mgd/mi^2 of drainage area. For design purposes, long-term rainfall and stream flows should be used and a mass diagram constructed. See Figure 3-23*b* and Fair in Bibliography, p. 8-3.

Groundwater runoff at the 70 percent point (where flow is equaled or exceeded 70 percent of the time) for the United States land area averaged a yield of 0.23

[193] Frank W. MacDonald and Adam Mehn, Jr., "Determination of Run-off Coefficients," *Pub. Works,* 74–76 (November 1963).

[194] Leonard Church Urquhart, *Civil Engineering Handbook,* McGraw-Hill Book Co., New York, 1959.

[195] W. A. Hardenbergh, *Water Supply and Purification,* International Textbook Co., Scranton, Pa. 1953.

[196] Everett L. MacLeman, "Yield of Impounding Reservoirs," *Water Sewage Works,* April 1958, pp. 144–149.

[197] Merriman and Wiggin, *American Civil Engineering Handbook,* John Wiley & Sons, New York, 1946.

mgd/mi^2. In the Great Lakes Basin 25 to 75 percent of the annual flow of streams is derived from direct groundwater seepage.[198]

Intakes and Screens

Conditions to be taken into consideration in design of intakes include high- and low-water stages; navigation or allied hazards; floods and storms; floating ice and debris; water velocites, surface and subsurface currents, channel flows, and stratification; location of sanitary, industrial, and storm sewer outlets; and prevailing wind direction.

Small communities cannot afford elaborate intake structures. A submerged intake crib, or one with several branches and upright tee fittings anchored in rock cribs 4 to 10 ft above the bottom, is relatively inexpensive. The inlet fittings should have a coarse strainer or screen with about 1-in. mesh. The total area of the inlets should be at least twice the area of the intake pipe and should provide an inlet velocity less than 0.5 fps. Low-entrance velocities reduce ice troubles and are less likely to draw in fish or debris. Sheet ice over the intake structure also helps avoid anchor ice or frazil ice. If ice-clogging of intakes is anticipated, provision should be made for an emergency intake or for injecting steam, hot water, or compressed air at the intake. Back-flushing is another alternative that may be incorporated in the design. Fine screens at intakes will become clogged, hence they should not be used unless installed at accessible locations that will make regular cleaning simple. Duplicate stationary screens in the flow channel, with $\frac{1}{8}$- to $\frac{3}{8}$-in. corrosion-resistant mesh can be purchased.

Some engineers have used slotted well screens in place of a submerged crib intake for small supplies. The screen is attached to the end of the intake conduit and mounted on a foundation to keep it off the bottom, and, if desired, crushed rock or gravel can be dumped over the screen. For example, a 10-ft section of a 24-in. diameter screen with $\frac{1}{4}$-in. openings is said to be able to handle 12 mgd at an influent velocity of less than 0.5 fps. Attachment to the foundation should be made in such a way that removal for inspection is possible.

In large installations, intakes with multiple-level inlet ports are provided in deep reservoirs, lakes, or streams to make possible depth selection of the best water when the water quality varies with the season of the year and weather conditions.

For a river intake, the inlet is perpendicular to the flow. The intake structure is constructed with vertical slotted channels before and after the bar racks and traveling screens for the placement of stop planks if the structure needs to be dewatered. Bar racks, 1 by 6 in. vertical steel, spaced 2 to 6 in. apart, provided with a rake operated manually or mechanically, keep brush and large debris from entering. This may be followed by a continuous slow-moving screen traveling around two drums, one on the bottom of the intake and the other above the

[198] Roger M. Waller, "World's Greatest Source of Fresh Water," *J. Am. Water Works Assoc.*, April 1974, pp. 245–247.

operating floor level. The screen is usually a heavy wire mesh with square openings $\frac{3}{8}$ to 1 in.; and it is cleaned by means of water jets inside which spray water through the screen, washing off debris into a wastewater trough. In cold-weather areas, heating devices, such as steam jets, are needed to prevent icing and clogging of the racks and screens.

Pumping

When water must be pumped from the source or for transmission, electrically operated pumps should have gasoline or diesel standby units having at least 50 percent of the required capacity. If standby units provide power for pumps supplying chlorinators and similar units, the full 100 percent capacity must be provided where gravity flow of water will continue during the power failure.

The distribution of water usually involves the construction of a pumping station, unless one is fortunate enough to have a satisfactory source of water at an elevation to provide a sufficient flow and water pressure at the point of use by gravity. The size pump selected is based on whether hydropneumatic storage (steel pressure tank for a small system), ground level, or elevated storage is to be used; the available storage provided; the yield of the water source; the water usage; and the demand. Actual meter readings should be used, if available, with consideration being given to future plans, periods of low or no usage, and maximum and peak water demands. Metering can reduce water use by 25 percent or more. Average water consumption figures must be carefully interpreted and considered with required fire flows. If the water system is to also provide fire protection, then elevated storage is practically essential, unless ground-level storage with adequate pumps is available.

The capacity of the pump required for a domestic water system with elevated storage is determined by the daily water consumption and volume of the storage tank. Where the topography is suitable the storage tank can of course be located on high ground, although the hydraulic gradient necessary to meet the highest water demand may actually govern. The pump should be of such capacity as to deliver the average daily water demand to the storage tank in 6 to 12 hr. In very small installations the pump chosen may have a capacity to pump in 2 hr all the water used in one day. This may be desirable when the size of the centrifugal pump is increased to 60 gpm or more and the size electric motor to 5 to 10 hp or more, since the efficiencies of these units then approach a maximum. On the other hand, larger transmission lines, if not provided, would be required in most cases to accommodate the larger flow, which would involve increased cost. Due consideration must also be given to the increased electrical demand and the effects this has. A careful engineering analysis should be made.

Distribution Storage Requirements

Water storage requirements should take into consideration the peak daily water use and the maximum hourly demand, the capacity of the normal and standby pumping equipment, the availability and capacity of auxiliary power, the probable duration of power failure, and the promptness with which repairs can be made. Additional considerations are land use, topography, pressure needs, distribution system capacity, and demands.

Water storage is necessary to help meet peak demands, fire requirements, and industrial needs; to maintain relatively uniform water pressures; to eliminate the necessity for continuous pumping; to make possible pumping when the electric rate is low; and to use the most economical pipe sizes. Surges in water pressure due to water hammer are also dissipated. Other things being equal, a large-diameter shallow tank is preferable to a deep tank of the same capacity. It is less expensive to construct, and water pressure fluctuations on the distribution system are less. The cost of storage compared to the decreased cost of pumping, the increased fire protection and possibly lowered fire insurance rate, the greater reliability of water supply, and the decreased probability of negative pressures in the distribution system will be additional factors in making a decision.

In general, it is recommended that water storage equal not less than one-half the total daily consumption, with at least one-half the storage in elevated tanks. A preferred minimum storage capacity would be the maximum day usage plus fire requirements, less the daily capacity of the water plant and system for the fire-flow period. Another basis is

> to provide sufficient water storage capacity to supply the maximum daily rate for a 4-hr period without depleting storage by more than one-half. Additionally, the minimum amount of storage that usually should be reserved for fire protection and other emergencies is one-third of system storage.[199]

Hudson[200] suggests the provision of two tank outlets, one to withdraw the top third of tank water for general purposes and a second outlet at the bottom of the tank to withdraw the remaining two-thirds of tank water if needed to supply building sprinkling systems in developed areas with high-rise apartments, industries, shopping centers, office complexes, and the like. In small communities, real estate subdivisions, institutions, camps, and resorts, elevated storage should be equal to at least one full day's requirements during hot and dry months when lawn sprinkling is heavy. A 2- or 3-day storage is preferred. The amount of water required during peak hours of the day may equal 15 to 25 percent of the total maximum daily consumption. This amount in elevated storage will meet peak demands, but not fire requirements. Some engineers provide storage equal to 20 to 40 gal per capita or 25 to 50 percent of the total average daily water consumption. A more precise method for computing requirements for elevated storage is to construct a mass diagram. Two examples are shown in Figures 3-23*a* and 3-23*b*. Fire requirements should be taken into consideration.

It is good practice to locate elevated tanks near the area of greatest demand for water and on the side of town opposite from where the main enters. Thus peak demands are satisfied with the least pressure loss and smallest main sizes. All

[199]D. P. Proudfit, "Adequate Storage Capacity is Vital for Peak Demands," *OpFlow,* AWWA, Denver, Co., December 1977, p. 7.

[200]W. D. Hudson, "Elevated Storage vs Ground Storage," *OpFlow,* AWWA, Denver, Co., July 1978.

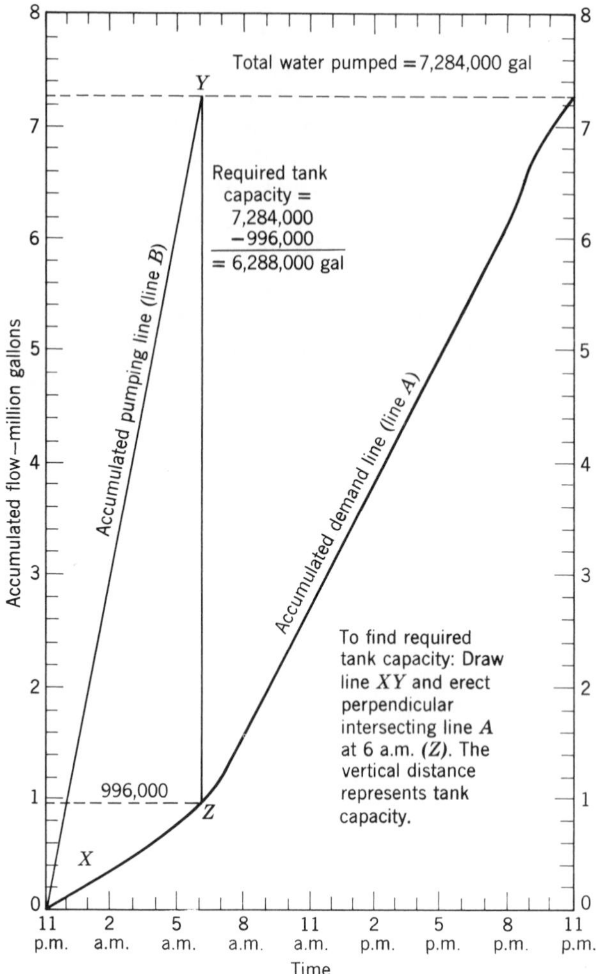

Figure 3-23*a* Mass diagram for determining capacity of tank when pumping 7 hours, from 11 p.m. to 6 a.m. [From John E. Kiker, Jr., "Design Criteria for Water Distribution Storage," *Pub. Works*, 102–104 (March 1964).]

distribution reservoirs should be covered; provided with an overflow that will not undermine the footing, foundation, or adjacent structures; and provided with a drain, water-level gauge, access manhole with overlapping cover, ladder, and screened air vent.

Water storage tanks are constructed of concrete, steel, or wood. Tanks may be constructed above or partly below ground, except that under all circumstances the manhole covers, vents, and overflows must be well above the normal ground level and the bottom of the tank above groundwater or flood water. Good drainage should be provided around the tank. Tanks located partly below ground must be at a higher level than any sewers or sewage disposal systems and

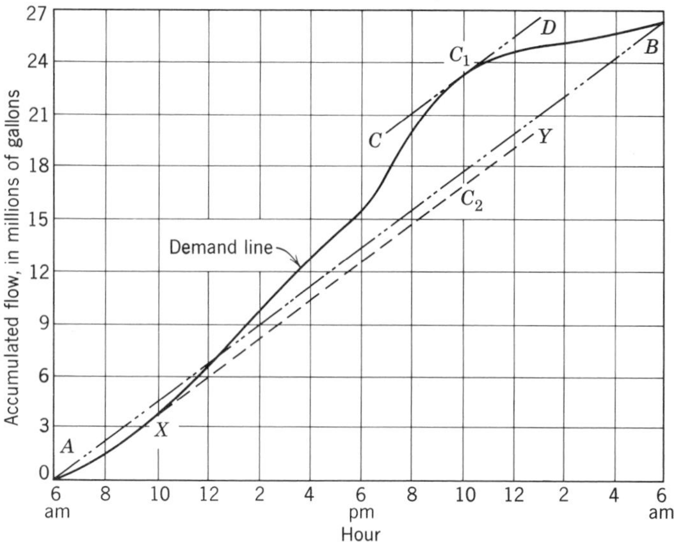

Figure 3-23*b* Mass diagram of storage requirements. The cumulative demand curve is plotted from records or estimates, and the average demand line, *AB*, drawn between its extremities. Line *CD* and *XY* are drawn parallel to line *AB* and tangent to the curve at points of greatest divergence from the average. At C_1—the point of maximum divergence—a line is extended down the coordinate to line *XY*. This line, C_1C_2, represents the required peak-hour storage: in this case, it scales to 6.44 mil gal. From George G. Schmid, "Peak Demand Storage," (Reprinted from *J. Am. Water Works Assoc.*, April 1956, by permission. Copyright 1956, American Water Works Association.)]

not closer than 50 ft. Vents and overflows should be screened and the tanks covered to keep out dust, rain, insects, small animals, and birds. A cover will also prevent the entrance of sunlight, which tends to warm the water and encourage the growth of algae. Manhole covers should be locked and overlap at least 2 in. over a 2- to 6-in. lip around the manhole. Partly below-ground storage is usually less costly and aesthetically more acceptable than elevated storage.

Properly constructed reinforced concrete tanks ordinarily do not require waterproofing. If tanks are built of brick or stone masonry, they should be carefully constructed by experienced craftsmen and only hard, dense material laid with full Portland cement mortar joints used. Two $\frac{1}{2}$-in. coats of 1:3 Portland cement mortar on the inside, with the second coat carefully troweled, should make such tanks watertight. A newly constructed concrete or masonry tank should be allowed to cure for about one month, during which time it should be wetted down frequently. The free lime in the cement can be neutralized by washing the interior with a weak acid, such as a 10 percent muriatic acid solution, or with a solution made up of 4 lb of zinc sulfate per gallon of water, and then flushed clean.

Wooden elevated storage tanks are constructed of cypress, fir, long-leaf yellow pine, or redwood. They are relatively inexpensive, easily assembled, and need not be painted or given special treatment; their normal life is 15 to 20 years. Wooden tanks are available with capacities up to 500,000 gal. The larger steel tanks start at 5000 to 25,000 gal; they require maintenance in order to prolong their life. Reinforced prestressed concrete tanks are also constructed. Underground fiberglass reinforced plastic tanks are also available up to 25,000 to 50,000 gal capacity.

Steel standpipes, reservoirs, and elevated tanks are made in a variety of sizes and shapes. As normally used, a standpipe is located at some high point to make available most of its contents by gravity flow and at adequate pressure; a reservoir provides mainly storage. The altitude of elevated tanks, standpipes, and reservoirs is usually determined, dependent on topography, to meet special needs and requirements. Elevated tanks rising more than 150 ft above the ground or located within 15,000 ft of a landing area, and in a 50-mi-wide path of civil airways, must meet the requirements of the Civil Aeronautics Administration.

Peak Demand Estimates

The maximum hourly or peak demand flow upon which to base the design of a water distribution system should be determined for each situation. A small residential community, for example, would have characteristics different from a new realty subdivision, central school, or children's camp. Therefore the design flow to determine distribution system capacity should reflect the pattern of living or operation, probable water usage, and demand of that particular type of establishment or community. At the same time consideration should be given to the location of existing and future institutions, industrial areas, suburban or fringe areas, highways, shopping centers, schools, subdivisions, and direction of growth. In this connection, reference to the city, town, or regional comprehensive or master plan, where available, can be very helpful. Larger cities generally have a higher per capita water consumption than smaller cities; but smaller communities have higher percentage peak demand flow than larger communities.

The maximum hourly domestic water consumption for cities above about 50,000 population will vary from about 200 to 700 percent of the average day annual hourly water consumption; the maximum hourly water demand in smaller cities will probably vary from 300 to 1000 percent of the average day annual hourly water consumption. The daily variation is reported to be 150 to 250 percent and the monthly variation 120 to 150 percent of the average annual daily demand in small cities.[201] A survey of 647 utilities serving populations of 10,000 or more in 1970 found the mean maximum daily demand to be 1.78 times the average day, with a range of 1.00 to 5.22[192] Studies in England showed that the peak flow is about ten times the average flow in cities of 5000 population.[202] It

[201]Charles P. Hoover, *Water Supply and Treatment,* Bull. 211, National Lime Association, Washington, D.C., 1934.

[202]Institution of Water Engineers, *Manual of British Water Supply Practice,* W. Heffer & Sons, Ltd., Cambridge, England, 1950.

can be said that the smaller and newer the community, the greater will be the probable variation in water consumption from the average.

Various bases have been used to estimate the probable peak demand at realty subdivisions, camps, apartment buildings, and other places. One assumption for small water plants serving residential communities is to say that, for all practical purposes, almost all water for domestic purposes is used in 12 hr[203]. The maximum hourly rate is taken as twice the maximum daily hourly rate, and the maximum daily hourly rate is $1\frac{1}{2}$ times the average maximum hourly rate. If the average maximum monthly flow is $1\frac{1}{2}$ times the average monthly annual flow, then the maximum hour's consumption rate is 9 times the average daily hourly flow rate.

Another basis used on Long Island is maximum daily flow rate = 4 times average daily flow rate; maximum 6-hr rate = 8 times average daily flow rate; and maximum 1-hr rate = $9\frac{1}{2}$ times average daily flow rate[204]

A study of small water supply systems in Illinois seems to indicate that the maximum hourly demand rate is six times the average daily hourly consumption.[205]

An analysis by Wolff and Loos showed that peak water demands varied from 500 to 600 percent over the average day for older suburban neighborhoods with small lots; to 900 percent for neighborhoods with $\frac{1}{4}$- to $\frac{1}{2}$-acre lots; to 1500 percent for new and old neighborhoods with $\frac{1}{3}$- to 3-acre lots.[206] Kuranz,[207] Taylor,[208] and many others[209] have also studied the variations in residential water use.

The results of a composite study of the probable maximum momentary demand are shown in Figure 3-24. It is cautioned, however, that for other than average conditions the required supply should be supplemented as might be appropriate for fire flows, industries, and other special demands.

Peak flows have also been studied at camps, schools, apartment buildings, highway rest areas, and other places.

The design of water requirements at toll road and superhighway service areas introduces special considerations that are typical for the installation. It is generally assumed that the sewage flow equals the water flow. In one study of national turnpike and highway restaurant experience, the extreme peak flow was

[203]Oscar G. Goldman, "Hydraulics of Hydropneumatic Installation," *J. Am. Water Works Assoc.*, **40**, 144 (February 1948).

[204]*Criteria for the Design of Adequate and Satisfactory Water Supply Facilities in Realty Subdivisions,* Suffolk County Dept. of Health, Riverhead, New York, 1956.

[205]C. W. Klassen, "Hydropneumatic Storage Facilities," *Tech. Release 10–8*, Div. Sanit. Eng., Illinois Dept. of Public Health (October 28, 1952).

[206]Jerome B. Wolff and John F. Loos, "Analysis of Peak Demands," *Pub. Works*, September, 1956.

[207]A. P. Kuranz, "Studies on Maximum Momentary Demand," *J. Am. Water Works Assoc.*, October 1942.

[208]D. R. Taylor, "Design of Main Extensions of Small Size," *Water Sewage Works,* July 1951.

[209]E. P. Linaweaver, Jr., *Residential Water Use, Report II on Phase Two of the Residential Use Research Project,* Dept. Sanit. Eng. Water Resources, Johns Hopkins University, Baltimore, Md., June 1965. *Maximum Design Standards for Community Water Supply Systems,* Federal Housing Administration, FHA No. 751, 52 (July 1965). E. P. Linaweaver, Jr., John C. Geyer, Jerome B. Wolff, *A Study of Residential Water Use,* HUD,Washington, D.C., 1967.

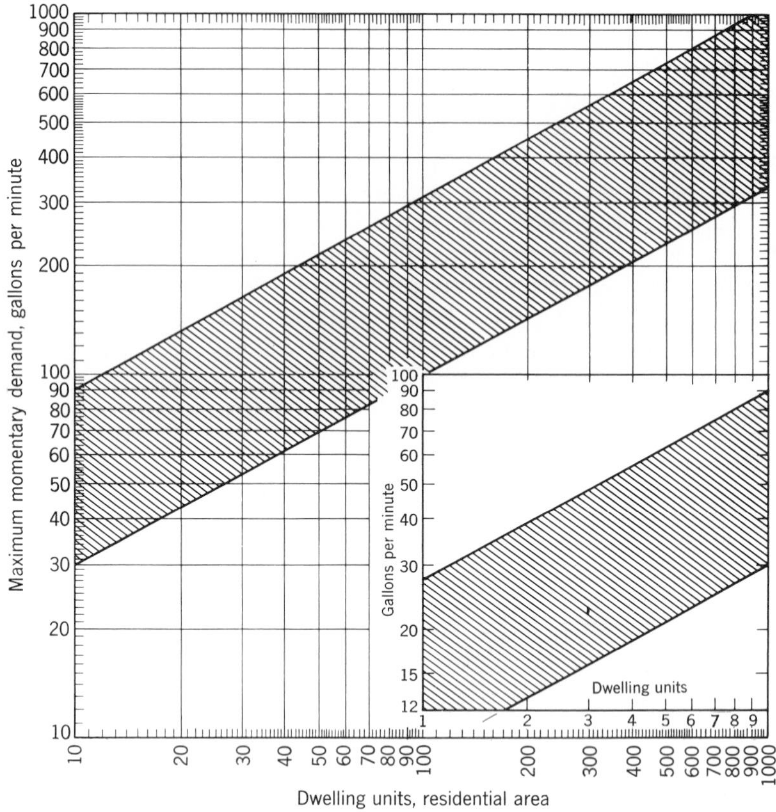

Figure 3-24 Probable maximum momentary water demand.

estimated at 1890 gpd per counter seat and 810 per table seat; but the peak day was taken as 630 gpd per counter seat and 270 gpd per table seat.[210] In another study of the same problem, the flow was estimated at 350 gpd per counter seat plus 150 gpd per table seat.[211] The flow was 200 percent of the daily average at noon and 160 percent of the daily average at 6 p.m. It was concluded that 10 percent of the cars passing a service area will enter and will require 15 to 20 gal per person. A performance study after one year of operation of the Kansas Turnpike service areas showed that 20 percent of cars passing service areas will enter; there will be $1\frac{1}{2}$ restaurant customers per car; average water usage will be 10 gal per restaurant customer, of which 10 percent is in connection with gasoline service; and plant flows may increase 4 to 5 times in a matter of seconds.[212]

Peak flows for apartment-type buildings can be estimated using the curves

[210]Gale G. Dixon and Herbert L. Kauffman, "Turnpike Sewage Treatment Plants," *Sewage Ind. Wastes* March 1956.

[211]Clifford Sharp, "Kansas Sewage Treatment Plants," *Pub. Works* August 1957.

[212]Mellville Gray, "Sewage from America on Wheels—Designing Turnpike Service Areas Plants to Serve Transient Flows," *Wastes Eng.* July 1959.

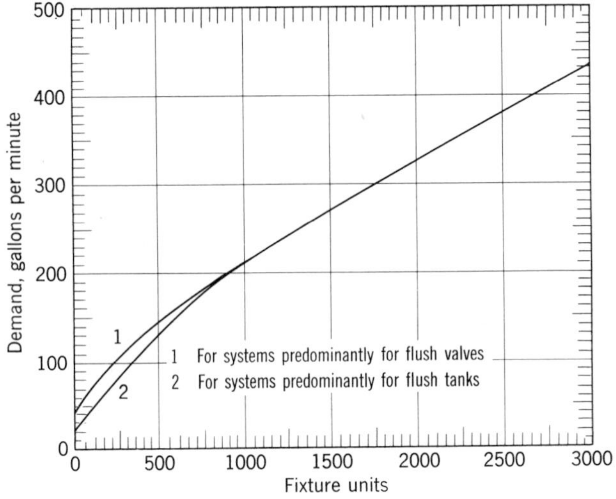

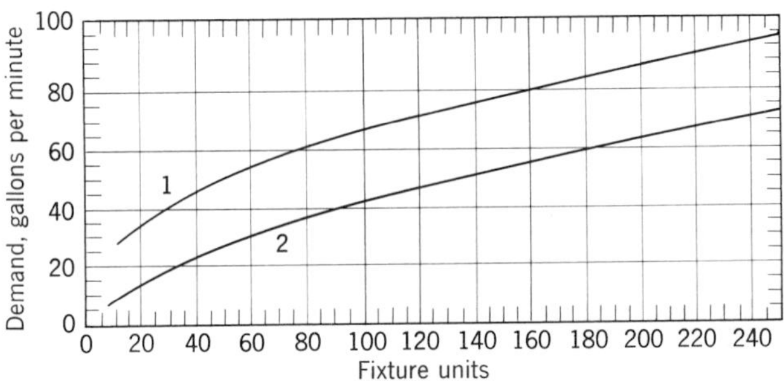

Figure 3-25 Estimate curves for demand load. (From R. B. Hunter, "Water-Distributing Systems for Buildings," National Bureau of Standards for Building Materials and Structures, Rep. BMS 79, November 1941.)

developed by Hunter.[213] Figure 3-25 and Tables 3-23 and 3-24 can be used in applying this method. Additions should be made for continuous flows. This method may be used for the design of small water systems, but the peak flows determined will be somewhat high.

At schools, peak flows would occur at recess and lunch periods, and after gym classes. At motels, peak flows would occur between 7 and 9 am and between 5 and 7 pm.

It must be emphasized that actual meter readings from a similar-type establishment or community should be used whenever possible in preference to an estimate. Time spent to obtain this information is a good investment, as each installation has different characteristics. Hence the estimates and procedures

[213]R. B. Hunter, *Water-Distributing Systems for Buildings*, National Bureau of Standards for Building Materials and Structures, Rep. BMS 79, November 1941.

Table 3-23 Demand Weight of Fixtures in Fixture Units[a]

Fixture or Group[b]	Occupancy	Type of Supply Control	Weight in Fixture Units[c]
Water closet	Public	Flush valve	10
Water closet	Public	Flush tank	5
Pedestal urinal	Public	Flush valve	10
Stall or wall urinal	Public	Flush valve	5
Stall or wall urinal	Public	Flush tank	3
Lavatory	Public	Faucet	2
Bathtub	Public	Faucet	4
Shower head	Public	Mixing valve	4
Service sink	Office, etc.	Faucet	3
Kitchen sink	Hotel or restaurant	Faucet	4
Water closet	Private	Flush valve	6
Water closet	Private	Flush tank	3
Lavatory	Private	Faucet	1
Bathtub	Private	Faucet	2
Shower head	Private	Mixing valve	2
Bathroom group	Private	Flush valve for closet	8
Bathroom group	Private	Flush tank for closet	6
Separate shower	Private	Mixing valve	2
Kitchen sink	Private	Faucet	2
Laundry trays (1 to 3)	Private	Faucet	3
Combination fixture	Private	Faucet	3

Source: R.B. Hunter, *Water-Distributing System for Buildings,* National Bureau of Standards Building Materials and Structures, Rep. BMS 79, November 1941.

[a]For supply outlets likely to impose continuous demands, estimate continuous supply separately and add to total for fixtures.

[b]For fixtures not listed, weights may be assumed by comparing the fixture to a listed one using water in similar quantities and at similar rates.

[c]The given weights are for total demand. For fixtures with both hot and cold water supplies, the weights for maximum separate demands may be taken as three-fourths the listed demand for supply.

mentioned here should be used as a guide to supplement specific studies and to aid in the application of informed engineering judgment.

Distribution System Design Standards

So far as possible distribution system design should follow usual good waterworks practice and provide for fire protection.[214,215] Mains should be designed on the basis of velocities of 4 to 6 fps with maximums of 10 to 20 fps, the

[214]*Recommended Standards For Water Works,* a Report of the Committee of the Great Lakes-Upper Mississippi River Board of State Sanitary Engineers, 1976 Edition, Health Education Service, Inc., P.O. Box 7126, Albany, N.Y. 12224.

[215]*Grading Schedule for Municipal Fire Protection* (1974) and *Guide for Determination of Required Fire-Flow* (1972), Insurance Services Office, 160 Water Street, New York, N.Y. 10038.

Table 3-24 Rate of Flow and Required Pressure During Flow for Different Fixtures

Fixture	Flow Pressure[a] (psi)	Flow Rate (gpm)
Ordinary basin faucet	8	3.0
Self-closing basin faucet	12	2.5
Sink faucet, $\frac{3}{8}$ in.	10	4.5
Sink faucet, $\frac{1}{2}$ in.	5	4.5
Bathtub faucet	5	6.0
Laundry-tub cock, $\frac{1}{2}$ in.	5	5.0
Shower	12	5.0
Ball cock for closet	15	3.0
Flush valve for closet	10 to 20	15 to 40[b]
Flush valve for urinal	15	15.0
Garden hose, 50 ft and sill cock	30	5.0
Dishwashing machine, commercial	15 to 30	6 to 9

Source: *Report of the Coordinating Committee for a National Plumbing Code,* U.S. Dept. of Commerce, Washington, D.C., 1951.

[a]Flow pressure is the pressure in the pipe at the entrance to the particular fixture considered.

[b]Wide range due to variation in design and type of flush-valve closets. (See Chapter 11 for fixture-supply pipe diameters.)

rates of water consumption (maximum daily demand), and fire demand, plus a residual pressure of not less than 35 psi nor more than 100 psi, using the Hazen and Williams coefficient C= 100, with a normal working pressure of about 60 psi.

Air-release valves or hydrants are provided where necessary where air can accumulate in the transmission lines, and blowoffs are provided at low drain points. These valves must not discharge to below-ground pits unless provided with a gravity drain to the surface above flood level. So far as possible dead ends should be eliminated or a blowoff provided, and mains tied together at least every 600 ft. Lines less than 6 in. in diameter should generally not be considered, except for the smallest system, unless they parallel secondary mains on other streets. In new construction 8-in. pipe should be used. In urban areas 12-in. or larger mains should be used on the principal streets and for all long lines that are not connected to other mains at intervals close enough for proper mutual support. Although the design should aim to provide a pressure of not less than 35 psi in the distribution system during peak flow periods, 20 psi minimum may be acceptable. A minimum pressure of 60 to 80 psi is desired in business districts, although 50 psi may be adequate in small villages with one- and two-story buildings. Thrust blocks and joint restraints must be provided on mains where indicated such as at tees, bends, plugs, and hydrants.

Valves are spaced not more than 500 ft apart in commercial districts and 800 ft in other districts, and at street intersections. A valve book, at least in triplicate, should show permanent ties for all valves, number of turns to open completely, left- or right-hand turn to open, manufacturer, and dates valves operated. A valve should be provided between each hydrant and street main.

Hydrants should be provided at each street intersection and spacing may range generally from 350 to 600 ft depending on the area served for fire protection and as recommended by the state Insurance Services Office. The connection to the street main should be not less than 6 in. in diameter. Operating nuts and direction of operation should be standard on all hydrants and conform with AWWA Standard. Hydrants should be set so that they are easily accessible to fire department pumpers; they should not be set in depressions, cutouts, or on embankments high above the street; pumper outlets should face directly toward the street; with respect to nearby trees, poles, and fences, there should be adequate clearance for connection of hose lines. Hydrants should be painted a distinguishing color so that they can be quickly spotted at night. Hydrant drains shall not be connected to or located within 10 ft of sanitary sewers or storm drains.

Main breaks occur longitudinally and transversely. Age is not a factor. Breaks are associated with sewer and other construction, usually starting with a leaking joint. The leak undermines the pipe making a pipe break due to beam action likely. Sometimes poor quality control in pipe manufacture contributes to the problem. Good pipe installation practice, including bedding and joint testing, followed by periodic leak surveys will minimize main breaks.

Water lines are laid below frost, separated from sewers a minimum horizontal distance of 10 ft and a vertical distance of 18 in. Water lines may be laid closer horizontally in a separate trench, or on an undisturbed shelf with the bottom at least 18 in. above the top of the sewer line, under conditions acceptable to the regulatory agency. It must be recognized that this type of construction is more expensive and requires careful supervision during construction. Mains buried 5 ft are normally protected against freezing and external loads.

The selection of pipe sizes is determined by the required flow of water that will not produce excessive friction loss. Transmission mains for small water systems more than 3 to 4 mi long should not be less than 10 to 12 in. in diameter. Design velocity is kept under 5 fps and head loss under 3 ft/1000 ft. If the water system for a small community is designed for fire flows, the required flow for domestic use will not cause significant head loss. On the other hand, where a water system is designed for domestic supply only, the distribution system pipe sizes selected should not cause excessive loss of head. Velocities may be $1\frac{1}{2}$ to $5\frac{1}{2}$ fps. In any case a special allowance is usually necessary to meet water demands for fire, industrial, and other special purposes.

Design velocities as high as 10 to 15 fps are not unusual, particularly in short runs of pipe. The design of water distribution systems can become very involved and is best handled by a competent sanitary engineer. When a water system is carefully laid out, without dead ends, so as to divide the flow through several pipes, the head loss is greatly reduced. The friction loss in a pipe connected at both ends is about one-quarter the friction loss in the same pipe with a dead end. The friction loss in a pipe from which water is being drawn off uniformly along its length is about one-third the total head loss. Also, for example, an 8-in. line will carry 2.1 times as much water as a 6-in. line for the same loss of head.*

*A 6-in. line carries 2.9 times as much as a 4-in. line; an 8-in. line carries 6.2 times as much as a 4-in. line, a 12-in. line carries 18 times as much as a 4-in. line, 6.2 times as much as a 6-in. line and 2.9 times

A water system that provides adequate fire protection is highly recommended where possible. This is discussed further below. The advantages of fire protection should at the very least be compared with the additional cost of increased pipe size, plant capacity, and water storage. If the cost of 8-in. pipe for example is only 20 percent more per ft than 6-in. pipe, the argument for the larger diameter pipe where needed is very persuasive since the cost of the trench would be the same. In any case only pipe and fittings that have a permanent-type lining or inner protective surface should be used.

Small Distribution Systems

In some communities, where no fire protection is provided, small diameter pipe may be used. In such cases, a 2-in. line should be no more than 300 ft long; a 3-in. line, no more than 600 ft; a 4-in. line, no more than 1200 ft; and a 6-in. line, no more than 2400 ft. If lines are connected at both ends 2- or 3-in. lines should be no longer than 600 ft; 4-in. lines, are not more than 2000 ft.

Transmission lines for rural areas have been designed for peak momentary demands of 2 to 3 gpm per dwelling unit and for as low as 0.5 gpm per dwelling unit with storage provided on the distribution system to meet peak demands. Adjustments are needed for constant or special demands and for population size. For example, Figure 3-24 shows a probable maximum demand of 3 to 9 gpm per dwelling unit for 10 dwelling units, 1 to 3.2 gpm per dwelling unit for 100 dwelling units and 0.33 to 1.1 gpm per dwelling unit for 1000 dwelling units.

Hudson suggests as a rule of thumb that a 6-in. main can be extended only 500 ft if the average amount of water of 1000 gpm is to be supplied for fire protection, or about 2000 ft if the minimum amount of 500 gpm is to be supplied.[216]

The minimum pipe sizes and rule-of-thumb guides mentioned above are not meant to substitute for distribution system hydraulic analysis but are intended for checking or rough approximation. Use of the equivalent pipe method, the Hardy Cross method, or one of its modifications should be adequate for the small distribution system. Computer analysis methods are used for large distribution system analysis.[217]

Fire Protection

Many factors enter into the classification of municipalities (cities, towns, villages, and other municipal entities) for fire insurance rate setting purposes.

The Insurance Services Office, their state representatives, and other authorized offices use the *Fire Suppression Rating Schedule* to classify municipalities

---

as much as a 8-in. line. The discharges vary as the 2.63 power of the pipe diameters being compared, based on the Hazen-Williams formula. See flow charts, nomograms, or Table 3-27.

[216]W. D. Hudson, "Design of Additions to Distribution Systems," *Water Sewage Works,* July 1959.

[217]Allen L. Davis and Roland W. Jeppson, "Developing a Computer Program for Distribution System Analysis," *J. Am. Water Works, Assoc.,* May 1979, pp. 236–241.

Table 3-25 Needed Duration for Fire Flow

Needed Fire Flow (gpm)	Needed Duration (hr)
2500 or less	2
3000	3
3500	3
4000 and greater	4

Source: *Fire Suppression Rating Schedule*, Insurance Services Office, 160 Water Street, New York, N.Y. 10038, 1980.

with reference to their fire defenses. This is one of several elements in the development of property fire insurance rates.

The municipal survey and grading work formerly performed by the National Board of Fire Underwriters, then by the American Insurance Association, as well as that formerly performed by authorized insurance-rating organizations, are continued under the Insurance Services Office. Credit is given for the facilities provided to satisfy the needed fire flows of the buildings in the municipality.[218] Since this discussion is intended only for familiarization purposes, the reader interested in the details of the grading system is referred to the references cited in this section for further information.

An adequate water system provides sufficient water to meet peak demands for domestic, commercial, and industrial purposes as well as for fire fighting. For fire suppression rating, the water supply has a weight of 40 percent; the fire department, 50 percent; and receiving and handling fire alarms, 10 percent. The water system considers the adequacy of the supply works, mains and hydrant spacing; size and type of hydrants; and the inspection and condition of hydrants.

To be recognized for fire protection a water system must be capable of delivering at least 250 gpm at 20 psi at a fire location for at least 2 hours with consumption at the maximum daily rate. The method of determining the needed fire flow for a building is given in the *Fire Suppression Rating Schedule.*[218] The needed fire flow will vary with the class of construction, its combustibility class, openings and distance between buildings, and other factors. Table 3-25 shows the needed duration for fire flow. There should be sufficient hydrants within 1000 feet of a building to supply its needed fire flow. Each hydrant with a pumper outlet and within 300 feet of a building is credited at 1000 gpm; 301–600 ft, 670 gpm; and 601–1000 ft, 250 gpm.

Where possible, water systems should be designed to also provide adequate fire protection and old systems should be upgraded to meet the requirements. This will also help assure the most favorable grading, classification, and fire insurance rates. Improvements in a water system resulting in a better fire protection grade and classification would generally be reflected in a reduced fire

[218] *Fire Suppression Rating Schedule*, Insurance Services Office, 160 Water Street, New York, N.Y. 10038, 1980.

insurance rate on specifically rated commercial properties, although other factors based on individual site evaluation may govern. However this is not always the case in "class-rated properties" such as dwellings, apartment houses, and motels. It generally is not possible to justify the cost to improve the fire protection class solely by the resulting savings in insurance premiums.[219] Nevertheless, the greater safety to life and property makes the value of improved fire protection more persuasive.

It is prudent for the design engineer to follow the state Insurance Services Office requirements.[218]

One must be alert to assure that fire protection programs do not include pumping from polluted or unapproved sources into a public or private water system main through hydrants or blowoff valves. Nor should bypasses be constructed around filter plants or provision made for "emergency" raw water connections to supply water in case of fire. In *extreme emergencies* the health department might permit a temporary connection under certain conditions, but in any case the water purveyor must immediately notify every consumer not to drink the water or use it in food or drink preparation unless first boiled or disinfected as noted at the end of this chapter.

Cross-Connection Control

There have been numerous instances of illnesses caused by cross-connections.[220,221] A discussion of water system design would not be complete without reference to cross-connection control and backflow prevention. The goal is to have no connection between a water of drinking water quality (potable) and an unsafe or questionable (nonpotable) water system, or between a potable system and any plumbing, fixture, or device whereby nonpotable water might flow into the potable water system.

A *cross-connection* is any physical connection between a potable water system and a nonpotable water supply; any waste pipe, soil pipe, sewer, drain; or any direct or indirect connection between a plumbing fixture or device whereby polluted water or contaminated fluids including gases or substances might enter and flow back into the potable water system. Backflow of nonpotable water and other fluids into the potable water system may occur by backpressure or backsiphonage. In *backpressure* situations the pressure in the nonpotable water system exceeds that in the potable water system. In *backsiphonage* the pressure in the potable water system becomes less than that in the nonpotable water system due to a vacuum or reduced pressure developing in the potable water system. Backflow prevention is also discussed in Chapter 11.

[219] Kenneth Carl, "Municipal Grading Classifications and Fire Insurance Premiums," *J. Am. Water Works Assoc.*, January 1978, pp. 19–22.

[220] *Cross-Connection Control Manual,* USEPA, WSD, 1973, pp. 3–8.

[221] "Final Report of the Committee on Cross-Connections," *J. New England Water Works Assoc.*, March 1979.

Negative or reduced pressure in a water distribution or plumbing system may occur when a system is shut off or drained for repairs, when heavy demands are made on certain portions of the system causing water to be drawn from the higher parts of the system, or when the pumping rate of pumps installed on the system (or of fire pumps or fire pumpers at hydrants) exceeds the capacity of the supply line to the pump. Backpressure may occur when the pressure in a nonpotable water system exceeds that in the potable water system, such as when a fire pumper at a dock or marina pumps nonpotable water into a hydrant or when a boiler chemical feed pump is directly connected to the potable water system.

The more common acceptable methods or devices to prevent backflow are air gap separation (as shown in Chapter 9, Figures 9-2, 9-5, and 9-6), backpressure units as shown in Figure 3-26 and 3-27, and vacuum breakers. The non-pressure-type vacuum breaker is always installed on the atmospheric side of a valve and is only intermittently under pressure, such as when a flushometer valve is activated (See Figure 11-13). The pressure-type vacuum breaker is installed on a pressurized system and will function only when a vacuum occurs. It is spring-loaded to overcome sticking and is used only where authorized. The vacuum breaker is not designed to provide protection against backflow resulting from backpressure and should not be installed where backpressure may occur.

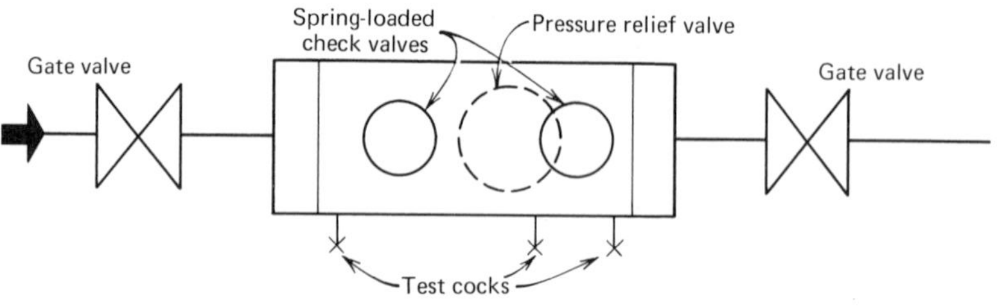

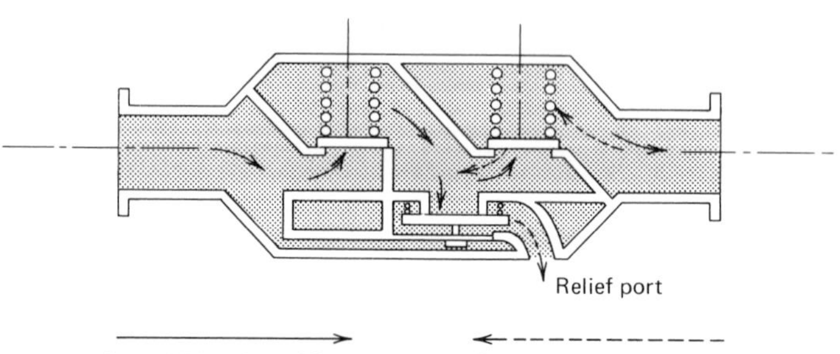

Figure 3-26 Reduced pressure zone backflow preventer—principle of operation. Malfunctioning of check or pressure relief valve is indicated by discharge of water from relief port. Preferred for hazardous facility containment. [Source: *Cross-Connection Control*, USEPA, WSD, p. 25 (EPA-430/9-73-002).]

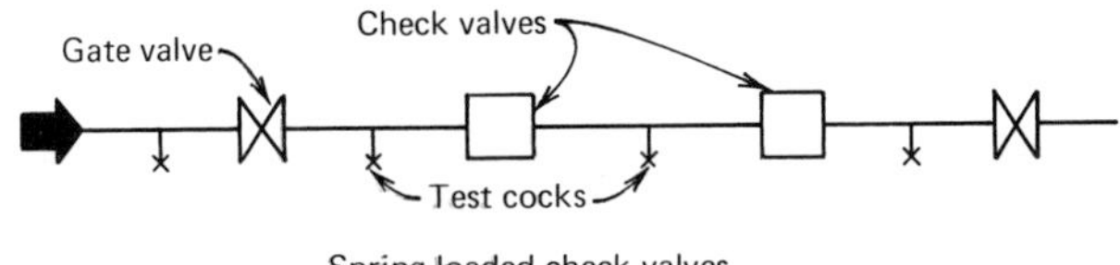

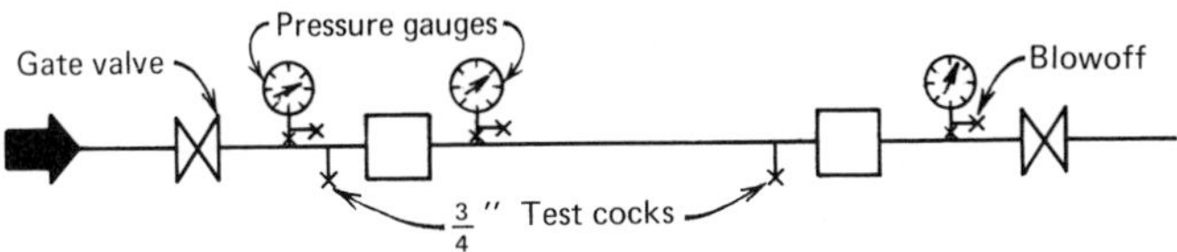

Figure 3-27 Double check valve—double gate valve assembly. For aesthetically objectionable facility containment.

The barometric or atmospheric loop which extends 34 to 35 ft above the highest outlet is not acceptable as a backflow preventer because a backpressure due to water, air, steam, hot water, or other fluid can negate its purpose. The swing joint, four-way plug valve, three-way two-port valve, removable pipe section, and similar devices are not reliable as nonpotable water can enter the potable water system at the time they are in use.[222,223]

An elevated or ground-level tank providing an air gap, the reduced pressure zone backflow preventer, and the double check valve assembly are generally used on public water system service connections to prevent backflow into the distribution system. The vacuum breaker is usually used on plumbing fixtures and equipment.

An approved backflow preventer or air break should be required on the water service line to every building or structure using or handling any hazardous substance that might conceivably enter the potable water system. In addition, building and plumbing codes should prohibit cross-connections within buildings and premises and require approved-type backflow preventers on all plumbing, fixtures, and devices which might cause or permit backflow. It is the responsibility of the designing engineer and architect, the building and plumbing inspector, the waterworks official, and the health department to prevent and prohibit possibilities of pollution of public and private water systems.

There are two major aspects to a cross-connection control program. One is

[222]"Use of Backflow-Preventers for Cross-Connection Control," Joint Committee Report, *J. Am. Water Works Assoc.*, December 1958, pp. 1589–1617.

[223]Gustave J. Angele Sr., *Cross-Connections and Backflow Prevention*, AWWA, Denver, Colorado, 1970.

protection of the water distribution system to prevent its pollution. The other is protection of the internal plumbing system used for drinking and culinary purposes to prevent its pollution.

The water purveyor has responsibility to provide its customers with water meeting drinking water standards. This requires control over unauthorized use of hydrants, blowoffs, and main connections or extensions. It also means requirement of a backflow prevention device at the service connection (containment) of all premises where the operations or functions on the premises involve toxic or objectionable chemical or biological liquid substances or use of a nonpotable water supply which may endanger the safety of the distribution system water supply through backflow. However, although these precautions may protect the water system, it is also necessary to protect the consumers on the premises using the water for drinking and culinary purposes. This reponsibility is usually shared by the water purveyor, the building and plumbing department, and the health department depending on state laws and local ordinances. The water purveyor has been held legally responsible for the delivery of safe water to the consumer and the Safe Drinking Water Act bases compliance with federal standards on the quality of water coming out of the consumer's tap. Under these circumstances, a cross-connection control program is needed in every community having a public water system to define and establish responsibility, and to assure proper installation and adequate inspection, maintenance, testing, and enforcement.

A comprehensive cross-connection control program according to Springer[224,225] must include the following:

1. An implementation ordinance that provides the legal basis for the development and complete operation of the program.
2. The adoption of a list of devices that are acceptable for specific types of cross-connection control.
3. The training and certification of qualified personnel to test and assure devices are maintained.
4. The establishment of a suitable set of records covering all devices.
5. Periodic seminars wherein supervisory, administrative, political, and operating personnel, as well as architects, consulting engineers, and building officials are briefed and brought up-to-date on the reason for the program as well as on new equipment in the field.

In some states the legal basis for the adoption of a local cross-connection ordinance is a state law or sanitary code, hence consultation with the state health department or other agency having jurisdiction is advised in the development of a local ordinance and program. Model ordinances and instruction manuals are available.[226] Enforcement is best accomplished at the local level.[227,228]

[224]E. Kent Springer, "Cross-Connection Control," *J. Am. Water Works Assoc.*, August 1976, pp. 405–406.

[225]V. W. Langworthy, "Persistant, these cross connections," *Am. City,* October 1974, p. 19.

[226]*Cross-Connection Control Manual,* USEPA, WSD, 1973, pp. 35–42.

[227]Fred Aldridge, "The Problems of Cross Connections," *J. Environ. Health*, July/August 1977, pp. 11–15.

[228]John A. Roller "Cross-Connection Control Practices in Washington State," *J. Am. Water Works Assoc.,* August 1976, pp. 407–409.

Implementation of a control program requires, in addition to the above five steps, that a priority system be established. Grouping structures and facilities served as "Hazardous," "Aesthetically Objectionable," and "Nonhazardous" can make inspection manageable and permit concentration of effort on the more serious conditions. Estimating the cost of installing backflow prevention devices is helpful in understanding what is involved and in obtaining corrections. Some devices are quite costly. An inspection program, with first priority to hazardous situations, is followed by review of findings with the local health department public health engineer or sanitarian; official notification of the customer; request for submission and approval of plans; establishment of a correction timetable; inspection and testing of the backflow device when installed; enforcement action if indicated; follow-up inspections; and testing of installed devices. The program progress should be reviewed and adjusted as needed every six months.[229]

Some practical applications of backflow prevention devices are illustrated and discussed in Chapters 9 and 11.

Hydropneumatic Systems

Hydropneumatic or pressure-tank water systems are suitable for small communities, housing developments, private homes and estates, camps, restaurants, hotels, resorts, country clubs, factories, and institutions, and as booster installations. In general only about 10 to 20 percent of the total volume of a pressure tank is actually available. Hydropneumatic tanks are usually made of $\frac{3}{16}$ in. or thicker steel and are available in capacities up to 10,000 or 20,000 gal. Tanks should meet American Society of Mechanical Engineers (ASME) code requirements. Small commercial size tanks are in 42, 82, 120, 144, 180, 220, 315, 525, and 1000 gal sizes.

The required size of a pressure tank is determined by peak demand and the capacity of the pump and source. The capacity of a well and pump should be at least 6 to 10 times the average daily water requirement.

A simple and direct method for determining the recommended volume of the pressure storage tank and size pump to provide is given by Figure 3-28. This figure is derived from Boyle's Law and is based on the formula

$$Q = \frac{Qm}{\left(1 - \frac{P_1}{P_2}\right)}$$

in which Q is equal to the pressure-tank volume in gal, Qm is equal to 15 minutes' storage at the maximum hourly demand rate, P_1 is the minimum absolute operating pressure (gauge pressure plus 14.7 lb/in.^{2}), and P_2 is the maximum absolute pressure.[230] The pump capacity given on the curve is equal to 125

[229] *Public Water Supply Guide, Cross-Connection Control*, Bureau of Public Water Supply, New York State Dept. of Health, Albany, N.Y., 1979.

[230] Joseph A. Salvato Jr., "The Design of Pressure Tanks for Small Water Systems," *J. Am. Water Works Assoc.*, June 1949, pp. 532–536.

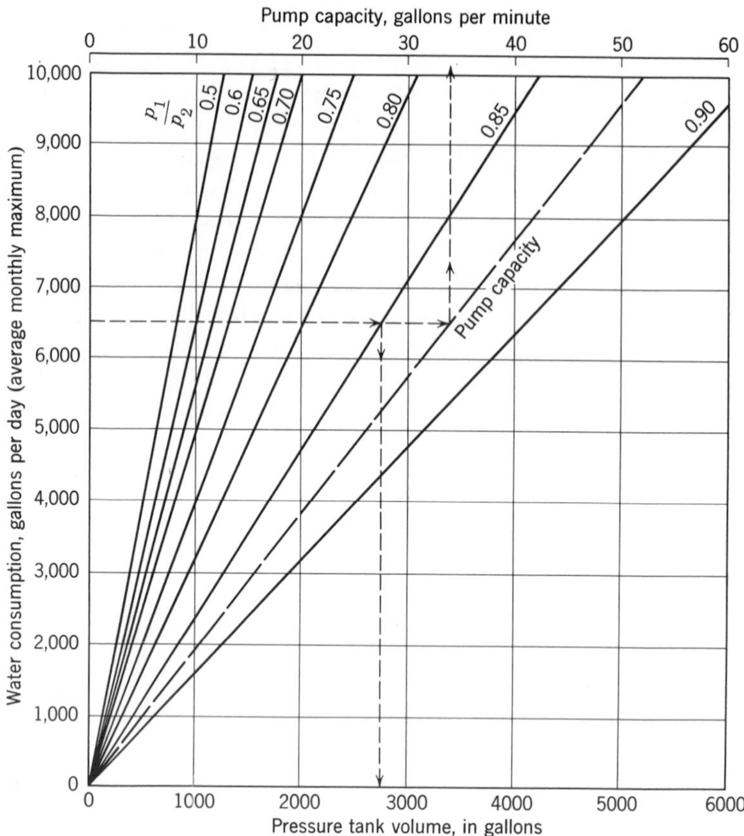

Figure 3-28 Chart for determining pressure storage tank volume and pump size. [From J. A. Salvato, Jr. "The Design of Pressure Tanks for Small Water Systems," *J. Am. Water Works Assoc.*, June 1949.]

percent of the maximum hourly demand rate. The maximum hourly demand is based on the following:

Average daily rate = Average water use per day ÷ 1440 min/day, in gpm; based on annual water use.

Average maximum monthly rate = 1.5 × average daily rate.

Maximum hourly demand rate = 6 × average maximum monthly rate or 9 × average daily rate.

Instantaneous rate (pump capacity = 1.25 × maximum hourly demand rate, or 11.25 × average daily rate.

The tank is assumed to be just empty when the pressure gauge reads zero. The required tank volume can be reduced proportionately if less than 15 minute storage is acceptable. For example, it can be reduced to one third if 5 minute storage is acceptable, with the available pump capacity, to meet momentary demands. An example is given on page 360.

The water available for distribution is equal to the difference between the dynamic head (friction plus static head) and the tank pressure. Because of the relatively small quantity of water actually available between the usual operating pressures, a higher initial (when the tank is empty) air pressure and range are sometimes maintained in a pressure tank to increase the water available under pressure. When this is done, the escape of air into the distribution system is more likely. Most home pressure tanks come equipped with an automatic air volume control (Figure 3-29), which is set to maintain a definite air-water volume in the

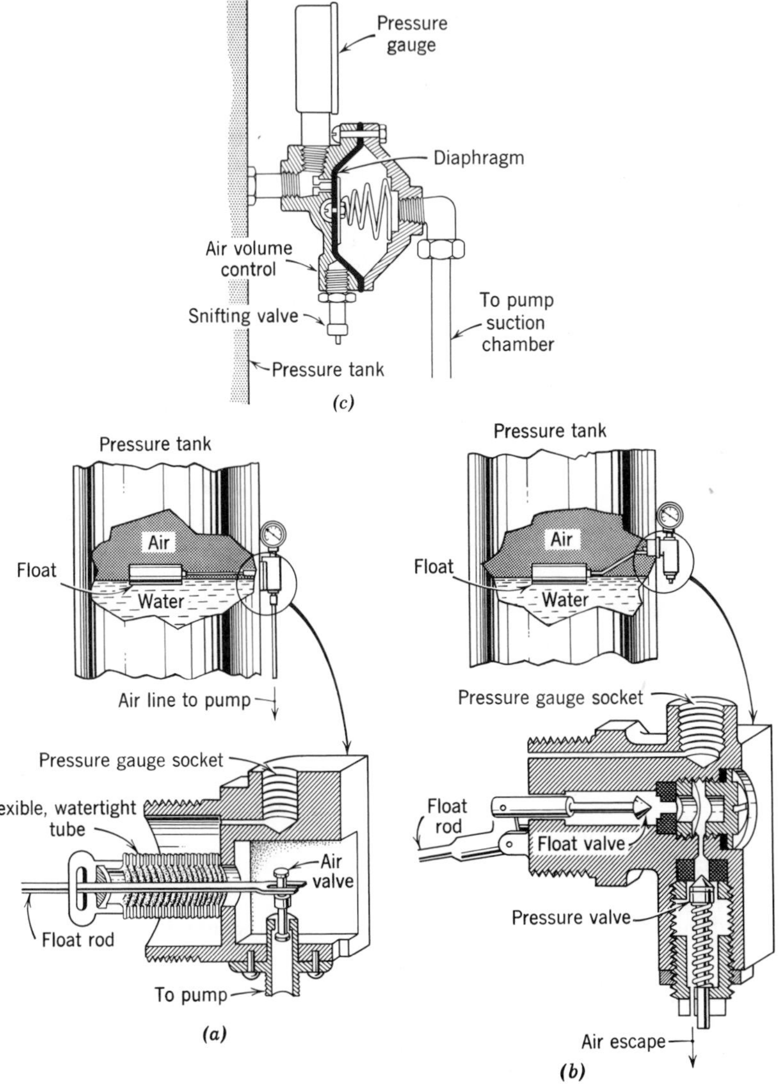

Figure 3-29 Pressure-tank air volume controls: (*a*) shallow-well-type for adding air; (*b*) deep-well-type for air release—used with submersible and piston pumps; (*c*) diaphragm-type in position when pump is not operating (used mostly with centrifugal pump). [From *Pumps and Plumbing for the Farmstead*, Agric. Eng. Dev. Div., TVA (November 1940).]

pressure tank at previously established water pressures, usually 20 to 40 psi. Air usually needs to be added to replace that absorbed by the water to prevent the tank from becoming waterlogged. Small pressure tanks are available with a diaphragm inside that separates air contact with the water, thereby minimizing this problem. Some manufacturers, or their representatives, increase the pressure tank storage slightly by precharging the tank with air. With deep-well displacement and submersible pumps, an excess of air is usually pumped with the water, causing the pressure tank to become airbound unless an air-release or needle valve is installed to permit excess air to escape.

In large installations an air compressor is needed, and an air-relief valve is installed at the top of the tank. A pressure relief valve should also be included on the tank. See Figure 3-30.

For small installations pump manufacturers recommend that the tank size equal 10 times the pump capacity (gpm) and that pump capacity equal the daily usage (gal) divided by 120 (min). However the *well yield* must be taken into consideration together with the operating pressure range and air-volume control.

Where a well yield (source) is inadequate to meet water demand with a pressure tank, then gravity or in-well storage, an additional source of water, or double pumping with intermediate storage may be considered. Intermediate ground-level storage can be provided between the well pump and the pressure-tank pump. The well pump will require a low-water cutoff and *its capacity must be related to the dependable well yield.* The intermediate storage tank (tightly covered) should have a pump stop and start device to control the well pump and a low-water sensor to signal depletion of water in the intermediate storage tank. A centrifugal pump would pump water from the intermediate tank to a pressure tank, with a pressure switch control, and thence to the distribution system.

Low-rate pumping to elevated storage, a deeper well to provide internal storage, or an over-size pressure tank may be possible alternatives to intermediate ground-level storage, depending on the extent of the problem and relative cost.

Pumps

The pump types commonly used to raise and distribute water are referred to as positive displacement, including reciprocating, diaphragm, and rotary; centrifugal, including turbine, submersible, and ejector jet; air lift; and hydraulic ram. Other types include the chain and bucket pump and hand pump.

Displacement Pump

In reciprocating displacement pumps, water is drawn into the pump chamber or cylinder on the suction stroke of the piston or plunger inside the pump chamber and then the water is pushed out on the discharge stroke. This is a simplex or

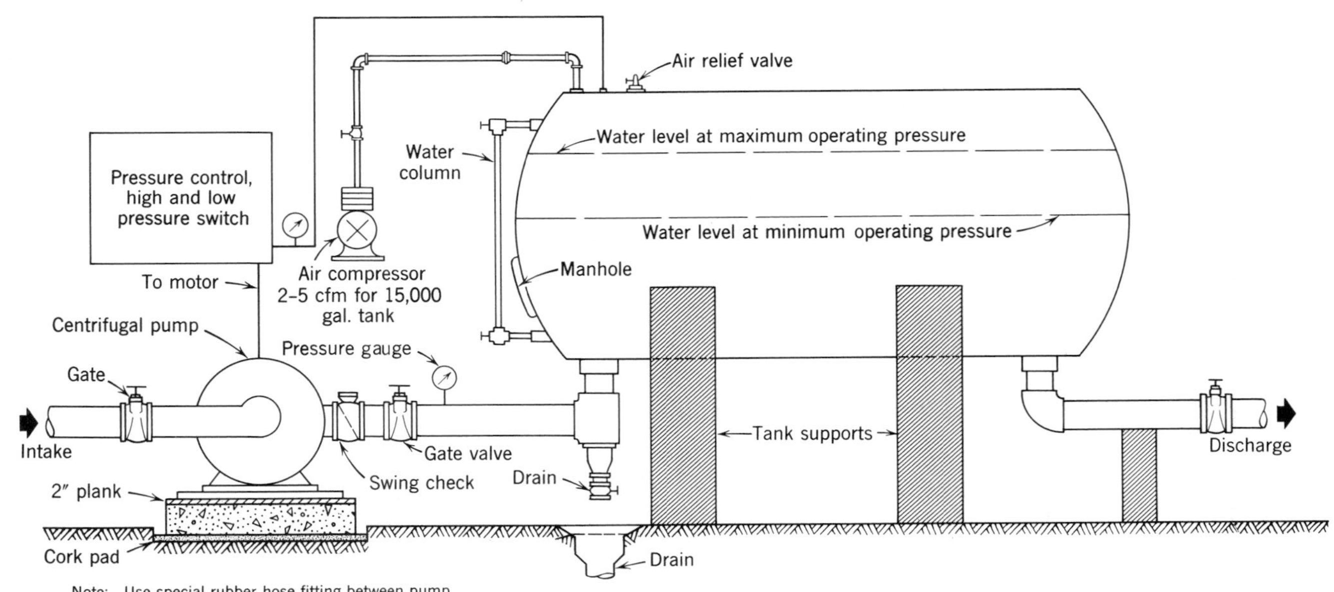

Figure 3-30 A typical large pressure-tank installation.

single-acting reciprocating pump. An air chamber (Figure 3-31) should be provided on the discharge side of the pump to prevent excessive water hammer caused by the quick-closing flap or ball valves. The air chamber will protect piping and equipment on the line and will tend to even out the intermittent flow of water. Reciprocating pumps are also of the duplex type wherein water is

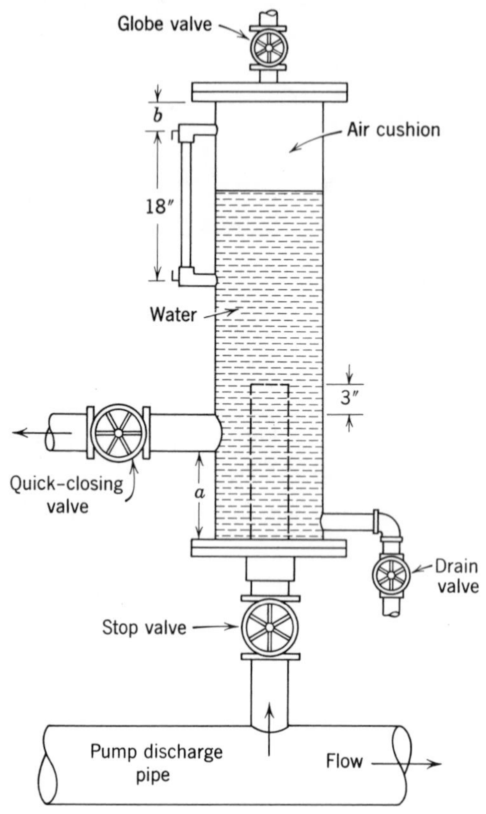

Figure 3-31 Air chamber dimensions for reciprocating pumps. (From *Water Supply and Water Purification*, T.M. 5-295, War Dept., Washington, D.C., 1942, 96 pp.)

Air Chamber Dimensions

Discharge Pipe	Inside Diameter of Air Chamber	Total Height	*a*	*b*
2″	8″	3′0″	4″	9″
2½″	8″	3′6″	4″	12″
3″	10″	4′0″	5″	15″
4″	10″	5′0″	6″	21″
5″	12″	6′0″	6″	27″
6″	16″	7′0″	6″	33″

pumped on both the forward and backward stroke, and of the triplex type in which three pistons pump water. The motive power may be manual, a steam, gas, gasoline, or oil engine, an electric motor, or a windmill. The typical hand pump and deep-well plunger or piston pumps over wells are displacement pumps.

A rotary pump is also a displacement pump, since the water is drawn in and forced out by the revolution of a cam, screw, gear, or vane. It is not used to any great extent to pump water.

Displacement pumps have certain advantages over centrifugal pumps. The quantity of water delivered does not vary with the head against which the pump is operating, but depends on the power of the driving engine or motor. A pressure-relief valve is necessary on the discharge side of the pump to prevent excessive pressure in the line and possible bursting of a pressure tank or water line. They are easily primed and operate smoothly under suction lifts as high as 22 ft. Practical suction lifts at different elevations are given in Table 3-26.

Displacement pumps are flexible and economical. The quantity of water pumped can be increased by increasing the speed of the pump, and the head can vary within wide limits without decreasing the efficiency of the pump. A displacement pump can deliver relatively small quantities of water as high as 800 to 1000 ft. Its maximum capacity is 300 gpm, although horizontal piston pumps are available in sizes of 500 to 3000 gpm. The overall efficiency of a plunger pump varies from 30 percent for the smaller sizes to 60 to 90 percent for the larger sizes with electric motor drive. It is particularly suited to the pumping of small quantities of water against high heads and can, if necessary, pump air with water. This type pump is no longer widely used.

Centrifugal Pump, also Submersible and Turbine

Centrifugal pumps are of several types depending on the design of the impeller. Water is admitted into the suction pipe or into the pump casing and rotated in the

Table 3-26 Atmospheric Pressure and Practical Suction Lift

Elevation Above Sea Level		Atmospheric Pressure		Design Suction Lift (ft)		
(ft)	(mi)	(lb/in.^{2})	(ft of water)	Displacement Pump	Centrifugal Pump	Turbine Pump
0		14.70	33.95	22	15	28
1,320	$\frac{1}{4}$	14.02	32.39	21	14	26
2,640	$\frac{1}{2}$	13.33	30.79	20	13	25
3,960	$\frac{3}{4}$	12.66	29.24	18	11	24
5,280	1	12.02	27.76	17	10	22
6,600	$1\frac{1}{4}$	11.42	26.38	16	9	20
7,920	$1\frac{1}{2}$	10.88	25.13	15	8	19
10,560	2	9.88	22.82	14	7	18

Note: The possible suction lift will decrease about 2 ft for every 10° F increase in water temperature above 60° F. One lb/in.^{2} = 2.31 ft head of water.

pump by an impeller inside the pump casing. The energy is converted from velocity head primarily into pressure head. In the submerged turbine-type pump used to pump water out of a well, the centrifugal pump is in the well casing below the water level in the well; the motor is above ground. In the submersible pump the pump and electric motor are in the well, requiring a minimum 3-in. (preferably 4-in.) diameter casing. It is a multi-stage centrifugal pump unit.

If the head against which a centrifugal pump operates is increased beyond that for which it is designed, and the speed remains the same, then the quantity of water delivered will decrease. On the other hand, if the head against which a centrifugal pump operates is less than that for which it is designed, then the quantity of water delivered will be increased. This may cause the load on the motor to be increased and hence overloading of the electric or other motor, unless the motor selected is large enough to take care of this contingency.

Sometimes two centrifugal pumps are connected in series so that the discharge of the first pump is the suction for the second. Under such an arrangement the capacity of the two pumps together is only equal to the capacity of the first pump, but the head will be the sum of the discharge heads of both pumps. At other times, two pumps may be arranged in parallel so that the suction of each is connected to the same pipe and the discharge of each pump is connected to the same discharge line. In this case the static head will be the same as that of the individual pumps, but the dynamic head, when the two pumps are in operation, will increase because of the greater friction and may exceed the head for which the pumps are designed. It may be possible to force only slightly more water through the same line when using two pumps as when using one pump, depending on the pipe size. Doubling the speed of a centrifugal pump impeller doubles the quantity of water pumped, produces a head four times as great, and requires eight times as much power to drive the pump. In other words the quantity of water pumped varies directly with the speed, the head varies as the square of the speed, and the horsepower as the cube of the speed. It is the usual practice to plot the pump curves for the conditions studied on a graph to anticipate operating results.

The centrifugal pump has no valves or pistons; there is no internal lubrication; it takes up less room and is relatively quiet. A single-stage centrifugal pump is generally used where the suction lift is less than 15 ft and the total head not over 125 to 200 ft. A single-stage centrifugal pump may be used for higher heads, but where this occurs a pump having two or more stages, that is, two or more impellers or pumps in series, should be used. The efficiency of centrifugal pumps varies from about 20 to 85 percent; the higher efficiency can be realized in the pumps with a capacity of 500 gpm or more. The peculiarities of the water system and effect they might produce on pumping cost should be studied from the pump curve characteristics. A typical curve is shown in Figure 3-32. All head and friction losses must be accurately determined in arriving at the total pumping head.

Centrifugal pumps that are above the pumping water level should have a foot valve on the pump suction line to retain the pump prime. However, foot valves

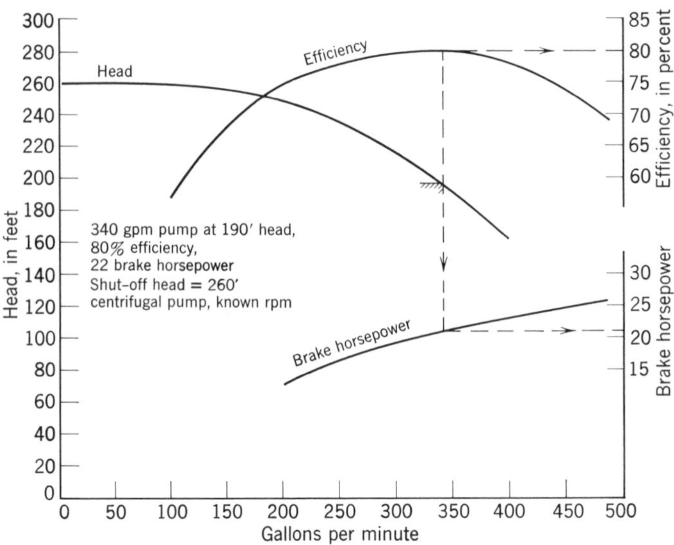

Figure 3-32 Typical centrifugal pump characteristic curves.

sometimes leak thereby requiring a water connection, or other priming device, or a new check valve on the suction side of the pump. The foot valve should have an area equal to at least twice the suction pipe. It may be omitted where an automatic priming device is provided. In the installation of a centrifugal pump, it is customary to install a gate valve on the suction line to the pump and a check valve followed by a gate valve on the pump discharge line near the pump. An air chamber, surge tank, or similar water-hammer suppression device should be installed just beyond the check valve, particularly on long pipe lines or when pumping against a high head. Make arrangements for priming a centrifugal pump, unless the suction and pump are under a head of water, and keep the suction line as short as possible. Slope the suction line up toward the pump to prevent air pockets.

Jet Pump

The jet pump is actually the combination of a centrifugal pump and a water ejector down in a well below or near the water level. The pump and motor can be located some distance away from the well, but the pipelines should slope up to the pump about $1\frac{1}{2}$ in. in 20 ft. In this type of pump, part of the water raised is diverted back down into the well through a separate pipe. This pipe has attached to it at the bottom an upturned ejector connected to a discharge riser pipe that is open at the bottom. The water forced down the well passes up through the ejector at high velocity, causing a pressure reduction in the venturi throat, and with it draws up water from the well through the riser or return pipe. A jet pump may be used to raise small quantities of water 90 to 120 ft, but its efficiency is lowered when the lift exceeds 50 ft. Efficiency ranges from 20 to 25 percent. The maximum capacity is 50 gpm. There are no moving parts in the well. Jet pumps are shallow well single-pipe-type (ejector at pump) and deep well single and multi-stage (ejector in well). Multi-stage pumps may have impellers horizontal or vertical.

The air-ejector pump is similar in operation to a water-ejector pump except that air is used instead of water to create a reduced pressure in the venturi throat to raise the water.

Air-Lift Pump

In an air-lift pump, compressed air is forced through a small air pipe extending below the water level in a well and discharged in a finely diffused state in a larger (eduction) pipe. The air-water mixture in the eduction pipe, being lighter than an equal volume of water, rises. The rise (weight of column of water) must at least equal the distance (weight of column of the same cross-sectional area) between the bottom of the eduction pipe and the water level in the well. For maximum efficiency, the distance from the bottom of the eduction pipe to the water level in the well should equal about twice the distance from the water surface to the point of discharge. The depth of submergence of the eduction pipe is therefore critical. The area in in.^2 of the eduction pipe is $A = Q/20$, in which Q is the volume of water discharged in gpm. Q depends on the V, the rate in cfm at which air is supplied. $V = Qh/125$, in which h is the distance in ft between the water surface and the point of discharge. Efficiencies vary from about 20 to 45 percent. The eduction pipe is about 1 in. smaller in diameter than the casing.[231, 232]

Hydraulic Ram

A hydraulic ram is a type of pump where the energy of water flowing in a pipe is used to elevate a smaller quantity of water to a higher elevation. An air chamber and weighted check valve are an integral part of a ram. Hydraulic rams are suitable where there is no electricity and the available water supply is adequate to furnish the energy necessary to raise the required quantity of water to the desired level. A battery of rams may be used to deliver larger quantities of water provided the supply of water is ample. Double-acting rams can make use of a nonpotable water to pump a potable water. The minimum flow of water required is 2 to 3 gpm with a fall of 3 ft or more. A ratio of lift to fall of 4 to 1 can give an efficiency

[231] W. A. Hardenbergh, *Water Supply and Purification,* International Textbook Company, Scranton, Pa., 1953, pp. 252–254.

[232] Harold E. Babbitt and James J. Doland, *Water Supply Engineering,* McGraw-Hill Book Company, Inc., New York, 1939, pp. 306–313.

of 72 percent, a ratio of 8 to 1 an efficiency of 52 percent, a ratio of 12 to 1 an efficiency of 37 percent, and a ratio of 24 to 1 an efficiency of 4 percent.[233,234] Rams are known to operate under supply heads up to 100 ft and a lift, or deliver heads, of 5 to 500 ft. In general, a ram will discharge from $\frac{1}{7}$ to $\frac{1}{10}$ of the water delivered to it. From a practical standpoint, it is found that the pipe conducting water from the source to the ram (known as the drive pipe) should be at least 30 to 40 ft long for the water in the pipe to have adequate momentum or energy to drive the ram. It should not, however, be on a slope greater than about 12 deg with the horizontal. If these conditions cannot be met naturally, it may be possible to do so by providing on the drive pipeline an open stand pipe, so that the pipe beyond it meets the conditions given. The diameter of the delivery pipe is usually about one-half the drive pipe diameter. The following formula may be used to determine the capacity of a ram:

$$Q = \frac{\text{supply to ram} \times \text{power head} \times 960}{\text{pumping head}}$$

where Q = gallons delivered per day.
supply to ram = gallons per minute of water delivered to and used by the ram.
power head = the available supply head of water, in feet or fall.
pumping head = the head pumping against, in feet, or delivery head.

NOTE: This information plus the length of the delivery pipe and the horizontal distance in which the fall occurs are needed by manufacturers to meet specific requirements.

Pump and Well Protection

A power pump located directly over a deep well should have a watertight well seal at the casing as illustrated in Figures 3-11 and 3-12. An air vent is used on a well that has an appreciable drawdown to compensate for the reduction in air pressure inside the casing, which is caused by a lowering of the water level when the well is pumped. The vent should be carried 18 in. above the floor and flood level and the end should be looped downward and protected with screening. A downward-opening sampling tap located at least 12 in. above the floor should be provided on the discharge side of the pump. In all instances the top of the casing, vent, and motor are located above possible flood level.

The top of the well casing or pump should not be in a pit that cannot be drained to the ground surface by gravity. In most parts of the country, it is best to locate pumps in some type of housing above the ground level and above any high water. Protection from freezing can be provided by installing a thermostatically

[233]R. T. Kent, *Mechanical Engineers' Handbook* Vol. II, John Wiley & Sons, New York, 1950.

[234]S. B. Watt, *A Manual on the Hydraulic Ram for Pumping Water,* Intermediate Technology Development Group, Water Development Unit, National College of Agricultural Engineering, July 1977, Silsoe, Bedford, MK45 4DT, U.K.

controlled electric heater in the pump house. Small, well-constructed and insulated pump housings are sometimes not heated but depend on heat from the electric motor and a light bulb to maintain a proper temperature. Some type of ventilation should be provided, however, to prevent the condensation of moisture and the destruction of the electric motor and switches. See Figure 3-11.

Use of a submersible pump in a well would eliminate the need for a pumphouse; but would still require that the discharge line be installed below frost. See Figure 3-7.

Pump Power and Drive

The power available will usually determine the type of motor or engine used. Electric power, in general, receives first preference, with other sources used for standby or emergency equipment.

Steam power should be considered if pumps are located near existing boilers. The direct-acting steam pump and single, duplex, or triplex displacement pump can be used to advantage under such circumstances. When exhaust steam is available, a steam turbine to drive a centrifugal pump can also be used.

Diesel-oil engines are good, economical pump-driving units when electricity is not dependable or available. They are high in first cost. Diesel engines are constant low-speed units.

Gasoline engines are satisfactory portable or standby pump power units. The first cost is low, but the operating cost is high. Variable speed control and direct connection to a centrifugal pump is common practice. Natural gas, methane, and butane can also be used where these fuels are available.

Electric motor pump drive is the usual practice when this is possible. Residences having low lighting loads are supplied with single-phase current, although this is becoming less common. When the power load may be three or more horsepower, three-phase current is needed. Alternating-current two- and three-phase motors are of three types: the squirrel-cage induction motor, the wound-rotor or slip-ring induction motor, and the synchronous motor. Single-phase motors are the repulsion-induction type having a commutator and brushes; the capacitator or condenser type, which does not have brushes and commutator; and the split-phase type. The repulsion-induction motor is, in general, best for centrifugal pumps requiring $\frac{3}{4}$ hp or larger. It has good starting torque. The all-purpose capacitor motor is suggested for sizes below $\frac{3}{4}$ hp. Make sure the electric motor is grounded to the pump. Check the electrical code.

The *squirrel-cage motor* is a constant-speed motor with low starting torque but heavy current demand, low power factor, and high efficiency. Therefore, this type motor is particularly suited where the starting load is large. Larger power lines and transformers are needed, however, with resultant greater power use and operating cost.

The *wound-rotor motor* is similar to the squirrel-cage motor. The starting torque can be varied from about one-third to three times that of normal, and the speed can be controlled. The cost of a wound-rotor motor is greater than a squirrel-cage motor; but where the pumping head varies, power saving over a long-range period will probably compensate for the greater first cost. Larger transformers and power lines are needed.

The *synchronous motor* runs at the same frequency as the generator furnishing the power. A synchronous motor is a constant-speed motor even under varying loads; but it needs an exciting generator to start the electric motor. Synchronous motors usually are greater in size than 75 to 100 hp.

An electric motor starting switch is either manually or magnetically operated. Manually operated starters for small motors (less than 1 hp) throw in the full voltage at one time. Overload protection is provided, but undervoltage protection is not. Full-voltage magnetic starters are used on most jobs. Overload and undervoltage control to stop the motor is generally included. Clean starter controls and proper switch heater strips are necessary. Sometimes a reduced voltage starter must be used when the power company cannot permit a full voltage starter or when the power line is too long. A voltage increase or decrease of more than 10 percent may cause heating of the equipment and winding, and fire.

Automatic Pump Control

One of the most common automatic methods of starting and stopping the operation of a pump on a hydropneumatic system is by the use of a pressure switch. This switch is particularly adaptable for pumps driven by electric motors, although it can also be used to break the ignition circuit on a gasoline-engine-driven pump. The switch consists of a diaphragm connected on one side with the pump discharge line and on the other side with a spring-loaded switch. This spring switch makes and breaks the electric contact, thereby operating the motor when the water pressure varies between previously established limits.

Water-level control in a storage tank can be accomplished by means of a simple float switch. Other devices are the float with adjustable contacts and the electronic or resistance probes control and altitude valve. Each has advantages for specific installations.[235]

When the amount of water to be pumped is constant, a time-cycle control can be used. The pumping is controlled by a time setting.

In some installations the pumps are located at some distance from the treatment plant or central control building. Remote supervision can be obtained through controls to start or stop a pump and to report pressure and flow data and faulty operation.

Another type of automatic pump switch is the pressure-flow control. This equipment can be used on ground-level or elevated water storage tanks.

[235]Lowell Wolfe, "Automatic Control of Level, Pressure, and Flow," *J. Am. Water Works Assoc.*, October 1973, pp. 654–662.

When pumps are located at a considerable distance from a storage tank, and pressure controls are used to operate the pumps, heavy draw-offs may cause large fluctuations in pressure along the line. This will cause sporadic pump starting and stopping. In such cases, and when there are two or more elevated tanks on a water system, altitude valves should be used at the storage tanks. An altitude valve on the supply line to an elevated tank or standpipe is set to close when the tank is full and is set to open when the pressure on the entrance side is less than the pressure on the tank side of the valve. In this way overflowing of the water tank is prevented, even if the float or pressure switch fails to function properly.

EXAMPLES

Design of Small Water Systems

Experiences in new subdivisions show that peak water demands of six to ten times the average daily consumption rate are not unusual. Lawn-sprinkling demand has made necessary sprinkling controls, metering, or the installation of larger distribution and storage facilities and, in some instances, ground storage and booster stations. As previously stated, every effort should be made to serve a subdivision from an existing public water supply. Such supplies can afford to employ competent personnel and are in the business of supplying water, whereas a subdivider is basically in the business of developing land and does not wish to become involved in operating a public utility.

In general, when it is necessary to develop a central water system to serve the average subdivision, consideration should first be given to a drilled well-water supply. Infiltration galleries or special shallow wells may also be practical sources of water. Such water systems usually require a minimum of supervision and can be developed to produce a known quantity of water of a satisfactory sanitary quality. Simple chlorination treatment will normally provide the desired factor of safety. Test wells and sampling will indicate the most probable dependable yield and the chemical and bacterial quality of the water. Well logs should be kept.

Where a clean, clear lake supply or stream is available, chlorination and slow sand filtration can provide reliable treatment with daily supervision for the small development. The turbidity of the water to be treated should not exceed 30 mg/l. Preliminary settling may be indicated in some cases.

Other more elaborate types of treatment plants, such as rapid sand filters, are not recommended for small water systems unless specially trained operating personnel can be assured. Pressure filters have limitations, as explained earlier in this chapter.

The design of small slow sand filter and well-water systems is explained and illustrated on pages 269 to 274, 362, and below.

An example (Figure 3-33) will serve to illustrate the design bases previously discussed. The problem is to determine the probable maximum hourly demand rate for a good summer development consisting of 60 two-bedroom dwellings.

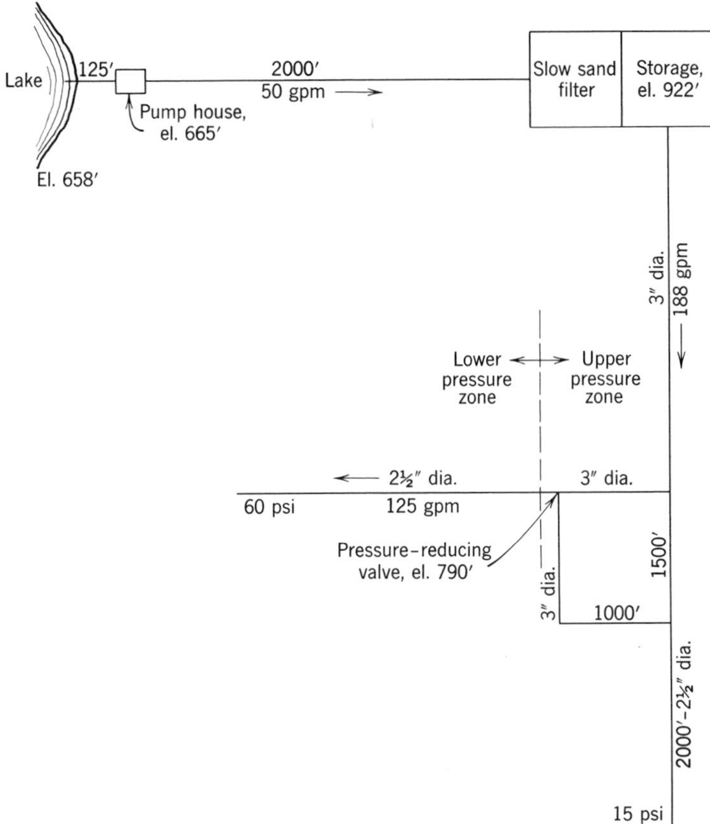

Figure 3-33 A water system flow diagram.

The population, at two persons per bedroom, is 240. From Figure 3-24, the demand can vary from 75 to 230 gpm. An average maximum demand would be 152 gpm. Adjustment should be made for local conditions.

Examples showing calculations to determine pumping head, pump capacity, and motor size follow.

In one instance, assume that water is pumped from a lake at an elevation of 658 ft to a slow sand filter and reservoir at an elevation of 922 ft. See Figure 3-32. The pump house is at an elevation of 665 ft and the intake is 125 ft long. The reservoir is 2000 ft from the pump. All water is automatically chlorinated as it is pumped. The water consumption is 30,000 gpd. With the reservoir at an elevation of 922 ft, a pressure of at least 15 lb/in.^{2} is to be provided at the highest fixture. It is required to find the size of the intake and discharge pipes, the total pumping head, the size pump, and motor. The longest known power failure is 14 hr and repairs can be made locally. Assume that the pump capacity is sufficient to pump 30,000 gal in 10 hr or 50 gpm. Provide one 50-gpm-pump and one 30-gpm-standby, both multistage centrifugal pumps, one to operate at any one time.

From the above, with a flow of 50 gpm, a $2\frac{1}{2}$ in. pipeline to the storage tank is indicated.

Table 3-27 Friction due to Water Flowing in Pipe

Capacity (gmp)	Friction Head Loss (ft/100 ft of pipe) Pipe Diameter (in.)														
	1	$\frac{3}{4}$	1	$1\frac{1}{4}$	$1\frac{1}{2}$	2	$2\frac{1}{2}$	3	4	5	6	8	10	12	14
1	2.1	—	—	—	—	—	—	—	—	—	—	—	—	—	—
2	7.4	1.9	—	—	—	—	—	—	—	—	—	—	—	—	—
3	15.8	4.1	1.3	—	—	—	—	—	—	—	—	—	—	—	—
4	27.0	7.0	2.1	0.57	—	—	—	—	—	—	—	—	—	—	—
5	41.0	10.5	3.2	0.84	0.40	—	—	—	—	—	—	—	—	—	—
8	98.0	25.0	7.8	2.0	0.95	—	—	—	—	—	—	—	—	—	—
10	—	38.0	11.7	3.0	1.4	0.50	—	—	—	—	—	—	—	—	—
15	—	80.0	25.0	6.5	3.1	1.1	—	—	—	—	—	—	—	—	—
20	—	136.0	42.0	11.1	5.2	1.8	0.61	—	—	—	—	—	—	—	—
25	—	—	64.0	16.6	7.8	2.7	0.95	0.40	—	—	—	—	—	—	—
30	—	—	89.0	23.5	11.0	3.8	1.3	0.54	—	—	—	—	—	—	—
35	—	—	119.0	31.2	14.7	5.1	1.7	0.75	—	—	—	—	—	—	—
40	—	—	152.0	40.0	18.8	6.6	2.2	0.91	—	—	—	—	—	—	—
50	—	—	—	60.0	28.4	9.9	3.3	1.4	—	—	—	—	—	—	—
60	—	—	—	85.0	39.6	13.9	4.6	1.9	0.47	—	—	—	—	—	—
70	—	—	—	113.0	53.0	18.4	6.2	2.6	0.63	—	—	—	—	—	—
80	—	—	—	—	68.0	23.7	7.9	3.3	0.81	—	—	—	—	—	—
90	—	—	—	—	84.0	29.4	9.8	4.1	1.0	—	—	—	—	—	—
100	—	—	—	—	102.0	35.8	12.0	5.0	1.2	0.41	—	—	—	—	—
125	—	—	—	—	—	54.0	18.2	7.6	1.9	0.64	—	—	—	—	—

150	—	—	—	—	—	76.0	26.0	10.5	2.6	0.87	—	—	—	—	—
175	—	—	—	—	—	102.0	33.8	14.0	3.4	1.2	—	—	—	—	—
200	—	—	—	—	—	129.0	43.1	17.8	4.4	1.5	0.62	—	—	—	—
225	—	—	—	—	—	—	54.0	22.0	5.3	1.8	0.72	—	—	—	—
250	—	—	—	—	—	—	65.0	27.1	6.7	2.2	0.92	—	—	—	—
300	—	—	—	—	—	—	92.0	38.0	9.3	3.1	1.3	—	—	—	—
400	—	—	—	—	—	—	—	65.0	16.0	5.4	2.2	—	—	—	—
500	—	—	—	—	—	—	—	98.0	24.0	8.1	3.3	0.83	—	—	—
600	—	—	—	—	—	—	—	—	33.8	11.7	4.7	1.2	—	—	—
700	—	—	—	—	—	—	—	—	45.0	15.2	6.2	1.5	0.52	—	—
800	—	—	—	—	—	—	—	—	57.6	19.4	8.0	2.0	0.67	—	—
900	—	—	—	—	—	—	—	—	71.6	24.2	10.0	2.5	0.83	—	—
1000	—	—	—	—	—	—	—	—	87.0	29.4	12.1	3.0	1.0	0.42	—
1500	—	—	—	—	—	—	—	—	—	62.2	25.6	6.3	2.1	0.88	0.42
2000	—	—	—	—	—	—	—	—	—	—	43.6	10.8	3.6	1.5	0.71
3000	—	—	—	—	—	—	—	—	—	—	—	22.8	7.7	3.2	1.5
4000	—	—	—	—	—	—	—	—	—	—	—	—	13.1	5.4	2.6
5000	—	—	—	—	—	—	—	—	—	—	—	—	19.8	8.2	3.8

Source: Hazen and Williams, $C = 100$, adapted from "Water Supply and Purification," WD FM 5—295 (1942).

Note: If the frictional resistance under certain known conditions is desired, multiply the friction head loss in feet by the following factors:

New cast iron (straight)	0.540	New riveted steel	0.840	Old wrought iron (over 15 yr)	1.51
Cement asbestos	0.540	Cast iron (C.I.) (5-yr-old)	0.715	Iron with very rough interior	2.58
New brass, copper	0.540	Steel (10-yr-old), C.I. (18-yr-old)	1.00	Smooth concrete or masonry	0.615
New cast iron (not straight)	0.615	Cast iron (30-yr-old)	1.51	Vitrified sewer	0.840
New wrought iron, wood (smooth)	0.715	Plastic pipe	0.54		

Table 3-28 Friction of Water in Fittings

Pipe Fitting	Friction Head Loss as Equivalent Number of Feet of Straight Pipe Size Pipe (in.) (nominal diameter)											
	$\frac{1}{2}$	$\frac{3}{4}$	1	$1\frac{1}{4}$	$1\frac{1}{2}$	2	$2\frac{1}{2}$	3	$3\frac{1}{2}$	4	5	6
Open gate valve	0.4	0.5	0.6	0.8	0.9	1.2	1.4	1.7	2.0	2.3	2.8	3.5
$\frac{3}{4}$ closed gate valve	40.0	60.0	70.0	100.0	120.0	150.0	170.0	210.0	250.0	280.0	350.0	420.0
Open globe valve	19.0	23.0	29.0	38.0	45.0	58.0	70.0	85.0	112.0	120.0	140.0	170.0
Open angle valve	8.4	12.0	14.0	18.0	22.0	28.0	35.0	42.0	50.0	58.0	70.0	85.0
Standard elbow or through reducing tee	1.7	2.2	2.7	3.5	4.3	5.3	6.3	8.0	9.3	11.0	13.0	16.0
Standard tee	3.4	4.5	5.8	7.8	9.2	12.0	14.0	17.0	19.0	22.0	27.0	33.0
Open swing check	4.3	5.3	6.8	8.9	10.4	13.4	15.9	19.8	24.0	26.0	33.0	39.0
Long elbow or through tee	1.1	1.4	1.7	2.3	2.7	3.5	4.2	5.1	6.0	7.0	8.5	11.0
Elbow 45°	0.75	1.0	1.3	1.6	2.0	2.5	3.0	3.8	4.4	5.0	6.1	7.5
Ordinary entrance	0.9	1.2	1.5	2.0	2.4	3.0	3.7	4.5	5.3	6.0	7.5	9.0

Note: The frictional resistance to flow offered by a meter will vary between that offered by an open angle valve and globe valve of the same size. See manufacturers for meter and check valve friction losses, also Table 3-29.

The head losses, using Table 3-27 and Table 3-28 are:

Intake, 125 ft of $2\frac{1}{2}$-in. pipe (3.3 × 1.25)	=	4.1 ft
For this calculation, assume the entrance loss and loss through pump are negligible. Say that there are four long elbows = 16.8 ft, two globe valves = 140 ft, one check valve = 16 ft, three standard elbows = 18.9 ft, and two 45° elbows = 6 ft. Total equivalent pipe = 198 ft of $2\frac{1}{2}$-in. pipe; head loss = 3.3 (1.98)	=	6.5 ft
Discharge pipe, 2000 ft of $2\frac{1}{2}$-in. = 3.3(20)	=	66.0 ft
Total friction head loss	=	76.6 ft
Suction lift = 7 ft to center of pump	=	7.0 ft
Static head, difference in elevation = (922 − 655)	=	257.0 ft
Pressure at point of discharge		NONE
Total head	=	340.6 ft

Add for head loss through meters if used (Table 3-29).

If a 3-in. intake and discharge line is used instead of a $2\frac{1}{2}$-in. intake, the total head can be reduced to about 300 ft. The saving thus effected in power consumption would have to be compared with the increased cost of 3-in. pipe over $2\frac{1}{2}$-in. pipe. The additional cost of power would be approximately \$76.65 per year, with the unit cost of power at 2 cents per kilowatt-hour. This calculation is shown below using approximate efficiencies.

Table 3-29 Head Loss Through Meters

Flow (gpm)	Head Loss through Meter (psi) Meter Size (in.)								
	$\frac{5}{8}$	$\frac{3}{4}$	1	$1\frac{1}{4}$	$1\frac{1}{2}$	2	3	4	6
4	1								
6	2								
8	4	1							
10	6	2	1						
15	14	5	2	2					
20	25	9	4	3	1				
30		20	8	7	2	1			
40			15	12	4	2			
50			23	18	6	3			
75					14	5	1		
100					25	10	3	1	
200							10	4	1
300							24	9	2
400								16	4
500								25	6

Source: Adapted from G. Roden, "Sizing and Installation of Service Pipes," Reprinted from *Am. Water Works Assoc.* **38**, 5 (May 1946), by permission. Copyright 1946, Am. Water Works Assoc.

340-foot head

$$\text{Horsepower to motor} = \frac{\text{gpm} \times \text{total head in ft}}{3960 \times \text{pump efficiency} \times \text{motor efficiency}}$$
$$= \frac{50 \times 340}{3960 \times 0.45 \times 0.83} = 11.5$$
$$11.5 \text{ hp} = 11.5 \times 0.746 \text{ kW} = 8.6 \text{ kW}$$

In 1 hr, at 50 gpm, 50 × 60, or 3000 gal, will be pumped and the power used will be 8.6 kW-hr. To pump 30,000 gal of water will require 8.6 × (30,000/3000) = 86 kW-hr.

If the cost of power is 2 cents per kW-hr, the cost of pumping 30,000 gal of water per day will be 86 × 0.02 = 1.72 or $1.72.

300-foot head

$$\text{Horsepower} = \frac{50 \times 300}{3960 \times 0.45 \times 0.83} = 10.1 \text{ hp} = 10.1 \times 0.746 \text{ kW} = 7.55 \text{ kW}.$$

In 1 hour, 3000 gal will be pumped as before, but the power used will be 7.55 kW-hr. To pump 30,000 gal will require 7.55 × 10 = 75.5 kW-hr.

If the cost of power is 2 cents per kW-hr, the cost of pumping 30,000 gal will be 75.5 × 0.02 = 1.51 or $1.51.

The additional power cost due to using $2\frac{1}{2}$-in. pipe is 1.72 − 1.51 = 0.21 per day or $76.65 per year.

At 4 percent interest, *i*, compounded annually for 25 years, *n*, $76.65, *d* set aside each year would equal about $3200, *S*, as shown below:

$$D = \frac{i}{(1+i)^n - 1} \times S; \qquad 76.65 = \frac{0.04}{(1+0.04)^{25} - 1} \times S$$
$$S = \frac{76.65}{0.024} = \$3{,}194, \text{ say } \$3{,}200$$

This assumes that the life of the pipe used is 25 years and the value of money 4 percent. If the extra cost of 3-in. pipe over $2\frac{1}{2}$-in. pipe, plus interest on the difference, minus the saving due to purchasing a smaller motor and lower head pump, is less than $1200 (present worth of $3200), then 3-in. pipe should be used.

The size of electric motor to provide for the 50-gpm pump against a total head of 340 ft is shown above to be 11.5 hp. Since this is a nonstandard size, the next larger size, a 15-hp motor, will be provided. If a smaller motor is used, it might be overloaded when pumping head is decreased.

$$\text{Horsepower to the motor drive} = \frac{\text{gpm} \times \text{total head in ft}}{3960 \times \text{pump efficiency} \times \text{motor efficiency}}$$

But the total head loss through $2\frac{1}{2}$-in. pipe when pumping 30 gpm would be:

Intake, 125 ft of $2\frac{1}{2}$-in. pipe	125	
Fittings, total equivalent pipe	198	
Discharge pipe, 2000 ft of $2\frac{1}{2}$-in.	2000	
Total friction head loss	$2323 \times \frac{1.3}{100}$	= 32 ft
Suction lift		= 9
Static head		= 255
Total head		= 296 ft

$$\text{Horsepower to the motor drive} = \frac{30 \times 296}{3960 \times 0.35 \times 0.85}$$
$$= 7.55. \text{ Use a } 7\tfrac{1}{2}\text{-hp electric motor.}$$

Because of the great difference in elevation, 658 ft to 922 ft, it is necessary to divide the distribution system into two zones, so that the maximum pressure in pipes and at fixtures will not be excessive. In this problem, all water is supplied the distribution system at elevation 922 ft. A suitable dividing point would be at elevation 790 ft. All dwellings above this point would have water pressure directly from the reservoir, and all below would be served through a pressure-reducing valve so as to provide not less than 15 lb/in.^{2} at the highest fixture nor more than 60 lb/in.^{2} at the lowest fixture. If one-third of the dwellings are in the upper zone and two-thirds are in the lower zone it can be assumed that the maximum hourly demand rate of flow will be similarly divided.

Assume the total maximum hourly demand rate of flow for an average daily water consumption of 30,000 gpd to be about 188 gpm. Therefore, 63 + 125 gpm can be taken to flow to the upper zone and 125 gpm to the lower zone. If a 3-in. pipe is used for the upper zone, and water is uniformly drawn off, the head loss at a flow of 188 gpm would be about $\frac{1}{3} \times 17$ ft per 100 ft of pipe. And if $2\frac{1}{2}$-in. pipe is used for the lower zone, and water is uniformly drawn off in its length, the head loss at a flow of 125 gpm would be about $\frac{1}{3} \times 18$ ft per 100 ft of pipe. If the pipe in either zone is connected to form a loop, thereby eliminating dead ends, the frictional head loss would be further reduced to $\frac{1}{4}$ of that with a dead end for the portion forming a loop.

In all of the above considerations, actual pump and motor efficiencies obtained from and guaranteed by the manufacturer should be used whenever possible. Their recommendations and installation detail drawings to meet definite requirements should be requested and followed if it is desired to fix performance responsibility.

In another instance, assume that all water is pumped from a deep well through a pressure tank to a distribution system. See Figure 3-34. The lowest water level in the well is at elevation 160, the pump and tank are at elevation 200, and the highest dwelling is at elevation 350. Find the size pump, motor and pressure

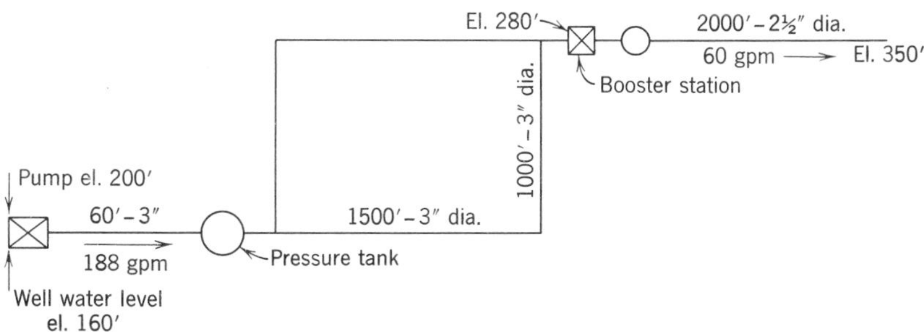

Figure 3-34 A water system flow diagram with booster station.

storage tank, the operating pressures, required well yield, and the size mains to supply a development consisting of 60 two-bedroom dwellings using an average of 30,000 gpd.

Use a deep-well turbine pump. The total pumping head will consist of the sum of the total lift, plus the friction loss in the well drop pipe and connection to the pressure tank, plus the friction loss through the pump and pipe fittings, plus the maximum pressure maintained in the pressure tank. If at the pump, the maximum pressure in the tank is equal to the friction loss in the distribution system plus the static head caused by the difference in elevation between the pump and the highest plumbing fixture, plus the friction loss in the house water system including meter if provided, plus the residual head required at the highest fixture.

With the average water consumption at 30,000 gpd, the maximum hourly demand was assumed to be 188 gpm (9 times the daily average hourly demand). The recommended pump capacity is taken as 125 percent of the maximum hourly rate, which would be 235 gpm. This assumes that the well can yield 235 gpm, which frequently is not the case. Under such circumstances, the volume of the storage tank can be increased two or three times, and the size of the pump correspondingly decreased to one-half or one-third the original size so as to come within the well yield. Another alternative would be to pump water out of a well, at a rate equal to the safe average yield of the well, into a large ground-level storage tank from which water can be pumped through a pressure tank at a higher rate to meet maximum water demands. This would involve double pumping and hence increased cost. Another arrangement, where possible, would be to pump out of the well directly into the distribution system, which is connected to an elevated storage tank. Although it may not be economical to use a pressure-tank water system, it would be of interest to see just what this would mean.

The total pumping head would be:

Lift from elevation 160 to 200 = 40 ft 40 ft

Figure 3-34 shows a distribution system that forms a rectangle 1000 × 1500 ft with a 2000-ft dead-end line serving one-third of the dwellings taking off at a point diagonally opposite the feed main. The head loss in a line connected at both ends is approximately one-fourth that in a dead-end line. The head loss in one-half the rectangular loop, from which water is uniformly drawn off, is one-third the loss in a line without drawoffs. Therefore, the total head loss in a 3-in. pipeline with a flow of about 188 gpm is equal to

$$\frac{1}{4} \times \frac{1}{3} \times \frac{17}{100} \times 2500 = 35 \text{ ft}$$

and the head loss through a 2000-ft dead-end line, with water being uniformly drawn off, assuming a flow of 60 gpm through $2\frac{1}{2}$-in. pipe is equal to

$$\frac{1}{3} \times \frac{4.6}{100} \times 2000 = 31 \text{ ft}$$

This would make a total of 35 + 31 or 66 ft. 66 ft

(For a more accurate computation of the head loss in a water distribution grid system by the equivalent pipe, Hardy Cross, or similar method, the reader is referred to standard hydraulic texts. However, the assumptions made here are believed sufficiently accurate for our purpose.)

The static head between pump and the curb of the highest dwelling plus the highest fixture is $(350 - 200) + 12 = 162$ ft.	162 ft
The friction head loss in the house plumbing system (without a meter) is equal to approximately 20 ft.	20 ft
The residual head at highest fixture is approximately 20 ft.	20 ft
The friction loss in the well drop pipe and connections to the pressure tank and distribution system with a flow of 235 gpm in a total equivalent length of 100 ft of 3-in. pipe is 25 ft.	25 ft
The head loss through the pump and fittings is assumed negligible.	—
Total pumping head	333 ft
	= 145 psi

Because of the high pumping head, and so as not to have excessive pressures in dwellings at low elevations, it will be necessary to divide the distribution system into two parts, with a booster pump and pressure storage tank serving the upper half.

If the booster pump and storage tank are placed at the beginning of the 2000 ft of $2\frac{1}{2}$-in. line, at elevation 280 ft, only one-third of the dwellings need be served from this point. The total pumping head here would be:

Friction loss in 2000 ft of $2\frac{1}{2}$-in. pipe with water withdrawn uniformly along its length and a flow of 60 gpm is

$$\frac{1}{3}\left(2000 \times \frac{4.6}{100}\right) = 31 \text{ ft} \qquad 31 \text{ ft}$$

The static head between the booster pump and the curb of the highest dwelling plus the highest fixture is

$$(350 - 280) + 12 = 82 \text{ ft} \qquad 82 \text{ ft}$$

The head loss in the house plumbing is 20 ft.	20 ft
The residual pressure at highest fixture is 20 ft.	20 ft
Booster pumping station total head	153 ft
	= 67 psi

The total pumping head at the main pumping station at the well would be:

Lift in well = 40 ft.	40 ft
Friction loss in distribution-system forming loop = 35 ft.	35 ft

The static head between the pump and the booster tank, which is also adequate to maintain a 20-ft head at the highest fixture is

$$(280 - 200) = 80 \text{ ft} \qquad 80 \text{ ft}$$

Friction loss in well drop pipe and connections to the distribution system = 25 ft.	25 ft
Main pumping station, total head	180 ft
	= 78 psi

With an average daily water consumption of 30,000 gal, the average daily maximum demand, on a monthly basis, would be $(30{,}000 \times 1.5) = 45{,}000$ gal. The ratio of the absolute maximum and minimum operating pressures at the main pumping station, using a 10-lb differential, would be

$$\frac{78 + 14.7}{88 + 14.7} = \frac{92.7}{102.7} = 0.905, \text{ say } 0.90$$

From Figure 3-28 the pressure tank volume should be (28,500 gal) about 30,000 gal if 15 min storage is to be provided at the maximum demand rate, or 10,000 gal if 5 min storage is acceptable. When the average monthly maximum water consumption exceeds that in Figure 3-28, multiply the vertical *and* horizontal axis by 5 or 10 or other suitable factor to bring the reading within the desired range. The pump capacity, as previously determined, should be 235 gpm. Use a 250-gpm pump. If a 20-lb pressure differential is used, $P_1/P_2 = 0.82$, and Figure 3-28 indicates a 16,000-gal pressure tank could be used to provide 15 min storage at the probable maximum hourly demand rate of flow.

The booster pumping station would serve one-third of the population, hence the average daily maximum demand on a monthly basis would be 1/3(45,000) or 15,000 gal. The ratio of the absolute maximum and minimum operating pressure at the booster pumping station, using a 10-lb differential, would be

$$\frac{67 + 14.7}{77 + 14.7} = \frac{81.7}{91.7} = 0.89, \text{ say } 0.90$$

From Figure 3-28 the pressure tank should have a volume of about 10,000 gal to provide 15-min storage at times of peak demand. The pump capacity should be 78 gpm. Use a 75-gpm pump. On the other hand, if the operating pressure differential is 20 lb and only 5-min storage at peak demand is desired, the required pressure tank volume would be 1,600 gal.

To determine the size motor required for the main pumping station and booster pumping stations, use the average of the maximum and minimum operating gauge pressures as the pumping head. The size motor for the main pumping station using manufacturer's pump and motor efficiencies:

$$\frac{250 \text{ gpm} \times (180 + 11\tfrac{1}{2}) \text{ ft avg. head}}{3960 \times 0.57 \times 0.85} = 24.7$$

Use a 25-hp motor. The size motor for the booster station is:

$$\frac{75 \text{ gpm} \times (153 + 11\tfrac{1}{2}) \text{ ft avg. head}}{3960 \times 0.50 \times 0.80} = 8$$

Use a 10-hp motor.

In the construction of a pumping hydropneumatic station, provision should be made for standby pump and motive power equipment.

The calculations are based on the use of a multistage centrifugal-type pump. Before a final decision is made, the comparison should include the relative merits and cost using a displacement-type pump. Remember that price and efficiency, although important when selecting a pump, are not the only factors to consider. The requirements of the water system and peculiarities should be anticipated and a pump with the desirable characteristics selected.

Design of a Camp Water System

A typical hydraulic analysis and design of a camp water system is shown in Figure 3-35.

Water System Cost Estimates

Because of the wide variations in types of water systems and conditions under which constructed, it is impractical to give reliable cost estimates. Some approximations are listed to give a feeling for some of the costs involved. Adjust costs using ENR or other appropriate construction cost index. (See Table 4-22).

1. The approximate costs (1980) of water pipe, valves, and hydrants, including labor and material, but not including engineering, legal, land, and administrative costs are:*

$\frac{3}{4}$-in. copper pipe, per ft	$6.00
1-in. copper pipe, per ft	8.00
$1\frac{1}{4}$-in. copper pipe, per ft	10.00
$1\frac{1}{2}$-in. copper pipe, per ft	12.50
$\frac{3}{4}$-in. corp. stop, per ft	50.00
6-in. ductile iron pipe, per ft	$10 to 13
8-in. ductile iron pipe, per ft	12 to 15
12-in. ductile iron pipe, per ft	16 to 19
16-in. ductile iron pipe, per ft	20 to 22
6-in. ABS or PVC pipe, per ft	$10 to 11
8-in. ABS or PVC pipe, per ft	12 to 13
10-in. ABS or PVC pipe, per ft	16 to 18
12-in. ABS or PVC pipe, per ft	21 to 26
6-in. double gate valve	$300 to 500
8-in. double gate valve	500 to 740
12-in. double gate valve	870 to 1100
6-in. hydrant assembly including valve, and tee on main	1600

2. Elevated storage, small capacity—20,000-gal capacity, $13,500 to $15,200; 50,000 gal, $20,000 to $35,800; 100,000 gal, $33,100 to $53,700. Ground-level storage—41,000 gal, $12,100; 50,000 gal, $14,300; 72,000 gal, $15,200; 92,000 gal, $16,900 (1969 cost).[236] For larger installations, standpipe costs may run $60,000 for 0.15 million gal capacity; $140,000 for 0.5 million gal; $230,000 for 1.0 million gal, and $500,000 for 3.0 million gal. For elevated tanks, cost may run $120,000 for 0.15 million gal capacity; $300,000 for 0.5 million gal; $540,000 for 1.00 million gal; and $1,300,000 for 3.0 million gal (1980 adjusted cost).
3. A complete conventional rapid sand filter plant including roads, landscaping, lagoons, laboratory, and low lift pumps may cost $300,000 for a 0.3-mgd plant; $440,000 for a 0.5-mgd

*The assistance of Kestner Engineers, P.C., Troy, N.Y. is gratefully acknowledged in arriving at the cost estimates.

[236]David H. Stoltenberg, "Construction Costs of Rural Water Systems," *Pub. Works,* **94** (August 1969).

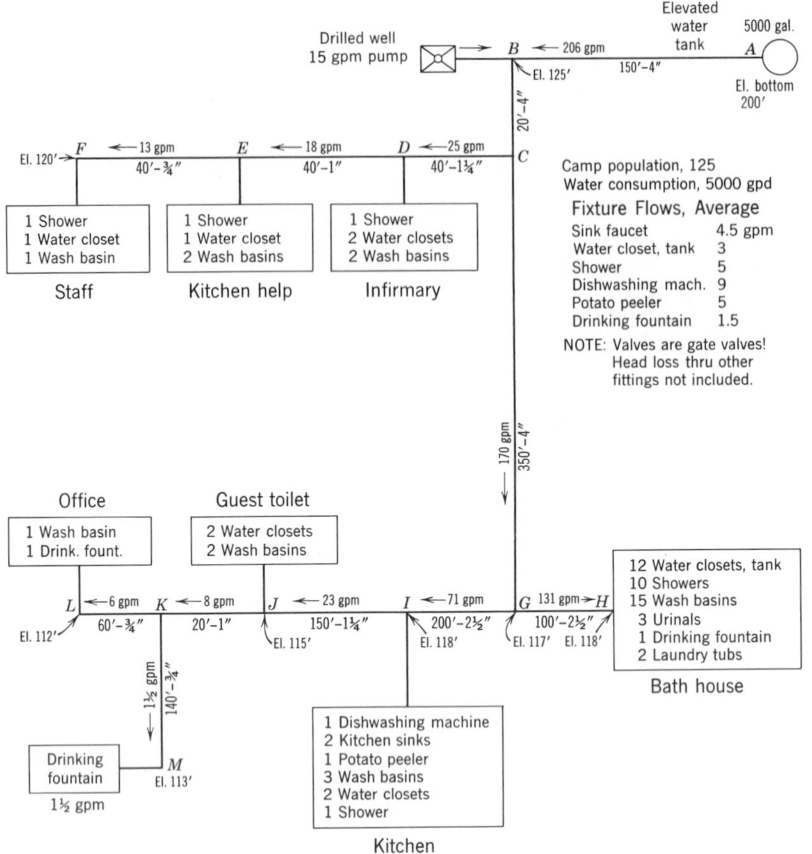

Figure 3-35 Typical hydraulic analysis of camp water system. (See Figure 4-35.)

Distance			Gpm Flow			Head Available (Ft)				Pipe	Head Loss		Head	
From	To	Ft	Max.	%	Probable	Initial	+ Fall	− Rise	Total	Size (in.)	Ft per 100 ft	Total	(Ft Remaining)	Facility Served
A	*C*	170	295	70	206	0	75	0	75	4	4.5	7.6	67.4	
C	*D*	40	50	50	25	67.4	1.6	0	69	1¼	16.6	6.6	62.4	Inf., kitch., staff
D	*E*	40	30	60	18	62.4	1.0	0	63.4	1	36	14.4	49.0	Kitch., staff
E	*F*	40	13	90	12	49.0	1.6	0	50.6	¾	52	21	29.6	Staff cabin
C	*G*	350	245	70	170	67.4	8.0	0	75.4	4	3.4	11.9	63.5	
G	*H*	100	174	75	131	63.5	0	1	62.5	2½	20	20	42.5	Bath house
G	*I*	200	71	100	71	63.5	0	1	62.5	2½	6	12	50.5	Kitch. guest, off.
I	*J*	150	23	100	23	50.5	3.0	0	53.5	1¼	15	22	31.5	
J	*K*	20	8	100	8	31.5	0	0	31.5	1	7.8	1.6	29.9	
K	*L*	60	6	100	6	29.9	3.0	0	32.9	¾	12	7.2	25.7	Office
K	*M*	140	1½	100	1½	25.7	2.0	0	27.7	¾	1.0	1.4	26.3	Drink. fount.

plant; $740,000 for a 1.0-mgd plant; $1,660,000 for a 3.0-mgd plant; $2,440,000 for a 5.0-mgd plant; $4,000,000 for a 10.0-mgd plant; and $6,800,000 for a 20.0-mgd plant (1980 adjusted cost).

4. The annual cost of water treatment plants (at 7 percent, 20 years) has been estimated at $41,630, $83,390, and $124,230 for a 70 gpm, 350 gpm, and 700 gpm complete treatment package plant; $396,610 for a 5 mgd plant; $161,060 and $482,000 for a 1 mgd and 10 mgd direct filtration

plant; and $248,890 and $1,064,870 for a 2 mgd and 20 mgd granular activated carbon plant (1979 cost).[237]

5. Iron and manganese removal plant, well supply, 3 mgd $1,150,000, including new well pumps and disinfection equipment, site work and treatment building (1980 adjusted cost).*
6. Well construction costs including engineering, legal, and site development have been estimated[238] as follows:

Year	1974	1974	1975	1976	1974
Yield, gpm	70	350	500	600	700
Type	Drilled	Gravel Pack	Gravel Pack	Drilled	Gravel Pack
Diameter, in.	10	16 to 12	18 to 12	16	16 to 12
Depth, ft	40	50	80	68	50
Pump	Submersible	Turbine	Turbine	Turbine	Turbine
Total Cost	$68,000 to $85,000	$108,000 to $135,000	$113,000 to $141,000	$164,000 to $205,000	2 wells
Average Cost	$78,000	$120,000	$128,000	$185,000	$235,000

7. The National Water Well Associated reported it costs $2000 to drill a private domestic well, $8000 to drill an irrigation well, and $30,000 to drill a municipal or industrial well.[239] The average cost of a 6 in. drilled well is estimated at $5 to $10 per ft plus $5 to $7 per ft for steel casing. A shallow well pump may cost $180 to $300 and a deep well pump $350 to $1250 (1980), plus installation.

CLEANING AND DISINFECTION

Special precautions must be taken before entering a well, spring basin, reservoir, storage tank, manhole, pump pit, or excavation to avoid accidents due to lack of oxygen (and excess carbon dioxide) or exposure to hazardous gases such as hydrogen sulfide or methane which are found in groundwater and underground formations. Hydrogen sulfide, for example, is explosive and very toxic. Methane is explosive. Open flames and sparks from equipment or electrical connections can cause the explosion and hence must be prevented. Make sure wells, tanks, etc., are well ventilated before entering. In any case, the person entering should use a safety rope and harness and two strong persons above the ground or above the tank should be ready to pull the worker out should he experience dizziness or other weakness. Self-contained breathing apparatus should be available.

Wells and Springs

Wells or springs that have been altered, repaired, newly constructed, flooded, or accidentally polluted should be thoroughly cleaned and disinfected after all the

*The assistance of Kestner Engineers, P.C., Troy, N.Y., is gratefully acknowledged in arriving at the cost estimates.

[237]"Cost Estimates for Water Treatment Plant Operation and Maintenance," *Pub. Works,* August 1980, pp. 81–83. Based on "Estimating Water Treatment Costs," USEPA Report 600/2-79-162a, August 1979, R. C. Gumerman, R. L. Culp, and S. P. Hansen.

[238]Daniel W. Stone, *Organic Chemical Contamination of Drinking Water,* New York State Health Dept. internal document, Albany, N.Y., June 1979.

[239]*Water Well Location by Fracture Trace Mapping,* Technology Transfer, Office of Water Research and Technology, Dept. of the Interior, GPO, 1978.

work is completed. The sidewalls of the pipe or basin, the interior and exterior surfaces of the new or replaced pump cylinder and drop pipe, and the walls and roof above the water line, where a basin is provided, should be scrubbed clean with a stiff-bristled broom or brush and detergent, insofar as possible, and then washed down or thoroughly sprayed with water followed by washing or thorough spraying with a strong chlorine solution. A satisfactory solution for this purpose may be prepared by dissolving 1 oz of 70 percent high-test calcium hypochlorite made into a paste, 3 oz of 25 percent chlorinated lime made into a paste, or 1 pt of $5\frac{1}{4}$ percent sodium hypochlorite, in 25 gal of water. The well or spring should be pumped until clear and then be disinfected as explained below.

To disinfect the average well or spring basin, mix 2 qt of $5\frac{1}{4}$ percent "bleach" in 10 gal water. Pour the solution into the well; start the pump and open all faucets. When the chlorine odor is noticeable at the faucets, close each faucet and stop the pump. It will be necessary to open the valve or plug in the top of the pressure tank, where provided, just before pumping is stopped in order to permit the strong chlorine solution to come into contact with the entire inside of the tank. Air must be readmitted and the tank opening closed when pumping is again started. Mix one more qt of bleach in 10 gal water and pour this chlorine solution into the well or spring. Allow the well to stand idle at least 12 to 24 hr; then pump it out to waste, away from grass and shrubbery, through the storage tank and distribution system, if possible, until the odor of chlorine disappears. *It is advisable to return the heavily chlorinated water back into the well, between the casing and drop pipe where applicable, during the first 30 min of pumping to wash down and disinfect the inside of the casing insofar as possible.* A day or two after the disinfection, *after all the chlorine has been removed*, a water sample may be collected for bacterial examination to determine whether all contamination has been removed. If the well is not pumped out, chlorine may persist for a week or longer.

A more precise procedure for the disinfection of a well or spring basin is to base the quantity of disinfectant needed on the volume of water in the well or spring. This computation is simplified by making reference to Table 3-30.

Although a flowing well or spring tends to cleanse itself after a period of time, it is advisable nevertheless to clean and disinfect all wells and springs that have had any work done on them before they are used. Scrub and wash down the basin and equipment. Place twice the amount of calcium hypochlorite or swimming pool chlorine tablets indicated by Table 3-30 in a weighted cloth sack or container fitted with a cover. Punch holes in the container and fasten a strong string or wire to the container to also secure the cover. Suspend the can near the bottom of the well or spring, moving it around in order to distribute the strong chlorine solution formed throughout the water entering and rising up through the well or spring. Addition of the chlorine solution into a garden hose or plastic tubing extended to the bottom of a well or spring would be more satisfactory.

It should be remembered that disinfection is no assurance that the water entering a well or spring will be free of pollution. The cause for the pollution, if present, should be ascertained and removed. Until this is done, all water used for drinking and culinary purposes should first be boiled.

Table 3-30 Quantity of Disinfectant Required to Give a Dose of 50 mg/l Chlorine

Diameter of Well, Spring, or Pipe (in.)	Gallons of Water per ft of Water Depth	Ounces of Disinfectant/10-ft Depth of Water		
		70% Calcium Hypochlorite[a]	25% Calcium Hypochlorite[b]	$5\frac{1}{4}$% Sodium Hypochlorite[c]
2	0.163	0.02	0.04	0.20
4	0.65	0.06	0.17	0.80
6	1.47	0.14	0.39	1.87
8	2.61	0.25	0.70	3.33
10	4.08	0.39	1.09	5.20
12	5.88	0.56	1.57	7.46
24	23.50	2.24	6.27	30.00
36	52.88	5.02	14.10	66.80
48	94.00	9.00	25.20	120.00
60	149.00	14.00	39.20	187.00
72	211.00	20.20	56.50	269.00
96	376.00	35.70	100.00	476.00

[a]$Ca(OCl)_2$, also known as high-test calcium hypochlorite. A heaping teaspoonful of calcium hypochlorite holds approximately $\frac{1}{2}$ oz. One liquid oz = 615 drops.
[b]$CaCl(OCl)$.
[c]$Na(OCl)$, also known as Bleach, Clorox, Dazzle, Purex, Javel Water, Regina, etc., can be purchased at most supermarkets, drug, and grocery stores.

Pipelines

The disinfection of new or repaired pipelines can be expedited and greatly simplified if special care is exercised in the handling and laying of the pipe during installation. Trenches should be kept dry and a tight fitting plug provided at the end of the line to keep out foreign matter. Lengths of pipe that have soiled interiors should be cleansed and disinfected before being connected up. Each continuous length of main should be separately disinfected with a heavy chlorine dose or other effective disinfecting agent. This can be done by using a portable chlorinator, a hand-operated pump, or a cheap mechanical pump throttled down to inject the chlorine solution at the beginning of the section to be disinfected through a hydrant, corporation cock, or other temporary valved connection. Hypochlorite tablets can also be used to disinfect small systems, but the lines cannot be flushed and water must be introduced very slowly to prevent the tablets being carried to the end of the line.

The first step in disinfecting a main is to flush out the line thoroughly by opening a hydrant or drain valve below the section to be treated until the water runs clear. A velocity of at least 3 fps should be obtained. (Use a hydrant flow gauge.) The valve is then partly closed so as to waste water at some known rate. The rate of flow can be estimated with a flow gauge, by running the water into a can, barrel, or other container of known capacity and measuring the time to fill it. With the rate of flow known, determine from Table 3-31 the strength of chlorine solution to be injected into the main at the established rate of 1 pt in 3 min. The rate of flow of water can, of course, be adjusted and should be kept low for small-diameter pipe. It is a simple matter to approximate the time, in minutes, it would take for the chlorine to reach the open hydrant or valve at the end of the line being treated by dividing the capacity of the main in gal (see Table 3-30) by the rate of flow in gpm. In any case, injection of the strong chlorine solution should be continued at the rate indicated until samples of the water at the end of the main show at least 50 mg/l residual chlorine. The hydrant should then be closed, chlorination treatment stopped, and the water system let stand at least 24 hr. At the end of this time the treated water should show the presence of 25 mg/l residual chlorine. If no residual chlorine is found, the operation should be repeated. Following disinfection, the water main should be thoroughly flushed out with the water to be used and samples collected for bacterial examination for a period of several days. If the laboratory reports the presence of coliform bacteria, the disinfection should be repeated until satisfactory results are received. Where poor installation practices have been followed, it may be necessary to repeat the main flushing and disinfection several times. The water should not be used until all evidence of contamination has been removed as demonstrated by the test for coliform bacteria.

If the pipeline being disinfected is known to have been used to carry polluted water, flush the line thoroughly and double the strength of the chlorine solution injected into the mains. Let the heavily chlorinated water stand in the mains at least 48 hr before flushing it out to waste and proceed as explained in the preceding paragraph. Cleansing of heavily contaminated pipe by the use of detergent, followed by flushing and then disinfection, may prove to be the quickest method.

Where pipe breaks are repaired, flush out the isolated section of pipe and dose the section with 200 mg/l chlorine and try to keep the line out of service at least 2 to 4 hr before flushing out the section and returning it to service.

Potassium permanganate can also be used as a main disinfectant. The presence and then the absence of the purple color can determine when the disinfectant is applied and then when it has been flushed out.[240]

Storage Reservoirs and Tanks

Before disinfecting a reservoir or storage tank it is essential to first remove from the walls (also bottom and top) all dirt, scale, and other loose material. The interior should then be flushed out (a fire hose is useful) and disinfected by one of the methods explained below.

If it is possible to enter the reservoir or tank, prepare a disinfecting solution by dissolving 1 oz of 70 percent calcium hypochlorite (HTH, Perchloron, Pitt-Chlor, etc.) made into a paste, 3 oz of 25 percent calcium hypochlorite

[240]John J. Hamilton, "Potassium Permanganate as a Main Disinfectant," *J. Am. Water Works Assoc.*, December 1974, pp. 734–735.

(chlorinated lime) made into a paste, or 1 pt of $5\frac{1}{4}$ percent sodium hypochlorite (Bleach, Clorox, Dazzle, etc.) in 25 gal of water. Apply this strong 250 mg/l chlorine solution to the bottom, walls, and top of the storage reservoir or tank using pressure-spray equipment. *Make sure the tank is adequately ventilated before entering.* Wear a chlorine gas mask and protective clothing during the work, including goggles. *Insist on all safety precautions.* The tank or reservoir may be filled an hour after the work is completed.

Another method, for small tanks, is to compute the capacity. Add to the empty tank 10 oz of 70 percent calcium hypochlorite, or 2 lb of 25 percent chlorinated lime completely dissolved, or 4 qt of $5\frac{1}{4}$ percent sodium hypochlorite for each 1000 gal capacity. Fill the tank with water and let it stand for 12 to 24 hr. This will give a 50 mg/l solution. Then drain the water through the distribution system to waste.

A third method for the small tank involves the use of a chlorinator or hand-operated force pump. Admit water to the storage tank at some known rate and add at the same time the chlorine solution indicated in Table 3-31 at a rate of 1 pt in 3 min. Let the tank stand full for 12 to 24 hr and then drain the chlorinated water through the distribution system to waste. Rinse the force pump immediately after use.

It should be remembered, when disinfecting pressure tanks, that it is necessary to open the air-relief or other valve at the highest point so that the air can be released and the tank completely filled with the heavily chlorinated water. Air should be readmitted before pumping is commenced. In all cases, a residual chlorine test should show a distinct residual in the water drained out of the tanks. If no residual can be demonstrated, the disinfection should be repeated.

Coliform bacteria, klebsiella and enterobacter, have been a problem in redwood water tanks. Klebsiella have been isolated from water samples extracted from redwood, which are apparently leached from the wood (especially

Table 3-31 Hypochlorite Solution to Give a Dose of 50 mg/l Chlorine for Main Sterilization

Rate of Water Flow in Pipeline, in gpm	Quarts of $5\frac{1}{4}$% Sodium Hypochlorite Made up to 10 gal with Water	Quarts of 14% Sodium Hypochlorite Made up to 10 gal with Water	Pounds of 25% Chlorinated Lime to 10 gal Water	Pounds of 70% Calcium Hypochlorite to 10 gal Water
5	4.6	1.7	2.0	0.7
10	9.1	3.4	4.0	1.4
15	13.7	5.1	6.0	2.1
20	18.3	6.8	8.0	2.9
25	22.8	8.5	10.0	3.5
40	36.6	13.7	16.0	5.7

Add hypochlorite solution at rate of 1 pt in 3 min. The 10-gal solution will last 3 hr if fed at rate of 1 pt in 3 min. Mix about 50 percent more solution than is theoretically indicated to allow for waste.

new tanks) when the tank is filled with water. Tanks are treated with soda ash to leach out wood tannins (seven days duration) and disinfected with 200 mg/l chlorine water prior to use. A free chlorine residual of 0.2 to 0.4 mg/l in the tank water when in use will keep bacterial counts under control.[241]

EMERGENCY WATER SUPPLY AND TREATMENT

Local or state health departments should be consulted when a water emergency arises. Their sanitary engineers and sanitarians are in a position to render valuable, expert advice based on their experience and specialized training. See also Emergency Sanitation, Chapter 12.

The treatment to be given a water used for drinking purposes depends primarily on the extent to which the water is polluted and the type of pollution present. This can be determined by making a sanitary survey of the water source to evaluate the significance of the pollution that is finding its way into the water supply. It must be borne in mind that all surface waters, such as from ponds, lakes, streams, and brooks, are almost invariably contaminated and hence must be treated. The degree of treatment required, based on the maximum permissible and average MPN coliform bacteria, is given in Table 3-15. However, under emergency conditions it is not practical to wait for the results of bacterial analyses and hence one should be guided by the results of sanitary surveys, diseases, epidemic, and such reliable local data as may be available. Using the best information on hand, select the cleanest and most attractive water available and give it the treatment necessary to render it safe. Water passing through inhabited areas is presumed to be polluted with sewage and industrial wastes and must be boiled or given complete treatment, including filtration and disinfection, to be considered safe to drink.

Boiling

In general, boiling water vigorously for one minute will kill most disease-causing bacteria and viruses. If sterile water is needed, water should be placed in a pressure cooker at 250°F (121°C) for 15 min.[242] Boil water 20 min where protozoal and helminthic diseases are endemic.

Chlorination

Chlorination treatment is a satisfactory method for disinfecting water that is not grossly polluted. It is particularly suitable for the treatment of a relatively clean

[241]Henry W. Talbot Jr., and Jan E. Morrow, and Tamon J. Seidler, "Control of Coliform Bacteria in Finished Drinking Water Stored in Redwood Tanks," *J. Am. Water Works Assoc.*, June 1979, pp. 349–353.

[242]Water Supply Guidance #51, "Emergency Disinfection of Drinking Water, Boiling," USEPA, Office of Drinking Water, Policy Statement, May 8, 1978.

lake, creek, or well water that is of unknown or questionable quality. Chlorine for use in hand chlorination is available in supermarkets, drugstores, grocery stores, and swimming pool supply places, and can be purchased as a powder, liquid, or tablet. Store solutions in the dark.

The powder is a calcium hypochlorite and the liquid a sodium hypochlorite. Both these materials deteriorate with age. The strength of the chlorine powder or liquid is on the container label and is given as a certain percent available chlorine. The quantity of each compound to prepare a stock solution, or the quantity of stock solution to disinfect one gallon of water, is given in Table 3-32. When using the powder, make a paste with a little water, then dissolve the paste in a quart of water. Allow the solution to settle and then use the clear liquid, without shaking. The stock solution loses strength and hence should be made up fresh once a week. It is important to allow the treated water to stand for 30 minutes after the chlorine is added before it is used. Double the chlorine dosage if the water is turbid or colored.

Chlorine-containing tablets suitable for use on camping, hunting, hiking, and fishing trips are available at most drugstores. The tablets contain 4.6 grains of chlorine; they deteriorate with age. Since chloramines are slow-acting disinfectants, the treated water should be allowed to stand at least 60 min before being used.

Table 3-32 Emergency Disinfection of Small Volumes of Water

Product	Available Chlorine, %	Stock Solution[a]	Quantity of Stock Solution to Treat 1 gal of Water[b]	Quantity of Stock Solution to Treat 1000 gal of Water[b]
Zonite	1	Use full strength	30 drops	2 qt
S.K., 101 Solution	$2\frac{1}{2}$	Use full strength	12 drops	1 qt
Clorox, White Sail, Dazzle, Rainbow, Rose-X	$5\frac{1}{4}$	Use full strength	6 drops	1 pt
Sodium Hypochlorite	10	Use full strength	3 drops	$\frac{1}{2}$ pt
Sodium Hypochlorite	15	Use full strength	2 drops	$\frac{1}{4}$ pt
Calcium Hypochlorite, "Bleaching Powder," or Chlorinated Lime	25	6 heaping tablespoonfuls (3 oz) to 1 qt of water	one teaspoonful or 75 drops	1 qt
Calcium Hypochlorite	33	4 heaping tablespoonfuls to 1 qt of water	one teaspoonful	1 qt
HTH, Perchloron, Pittchlor	70	2 heaping tablespoonfuls (1 oz) to 1 qt of water	one teaspoonful	1 qt

[a]One quart contains 135 ordinary teaspoonfuls of water.

[b]Let stand 30 min before using. To dechlorinate, use sodium thiosulfate in same proportion as chlorine. One jigger = $1\frac{1}{2}$ liquid oz. Chlorine dosage is approximately 5 to 6 mg/l. (See Chlorination.) (1 liquid oz = 615 drops.) Make sure chlorine solution or powder is fresh; check by making residual chlorine test. Double amount for turbid or colored water.

Homemade chlorinators may be constructed for continuous emergency treatment of a water supply where a relatively large volume of water is needed. Such units require constant observation and supervision as they are not dependable. Figures 3-36 and 3-37 show several arrangements for adding hypochlorite solution. In some parts of the country it may be possible to have a commercial hypochlorinator delivered and installed within a very short time.

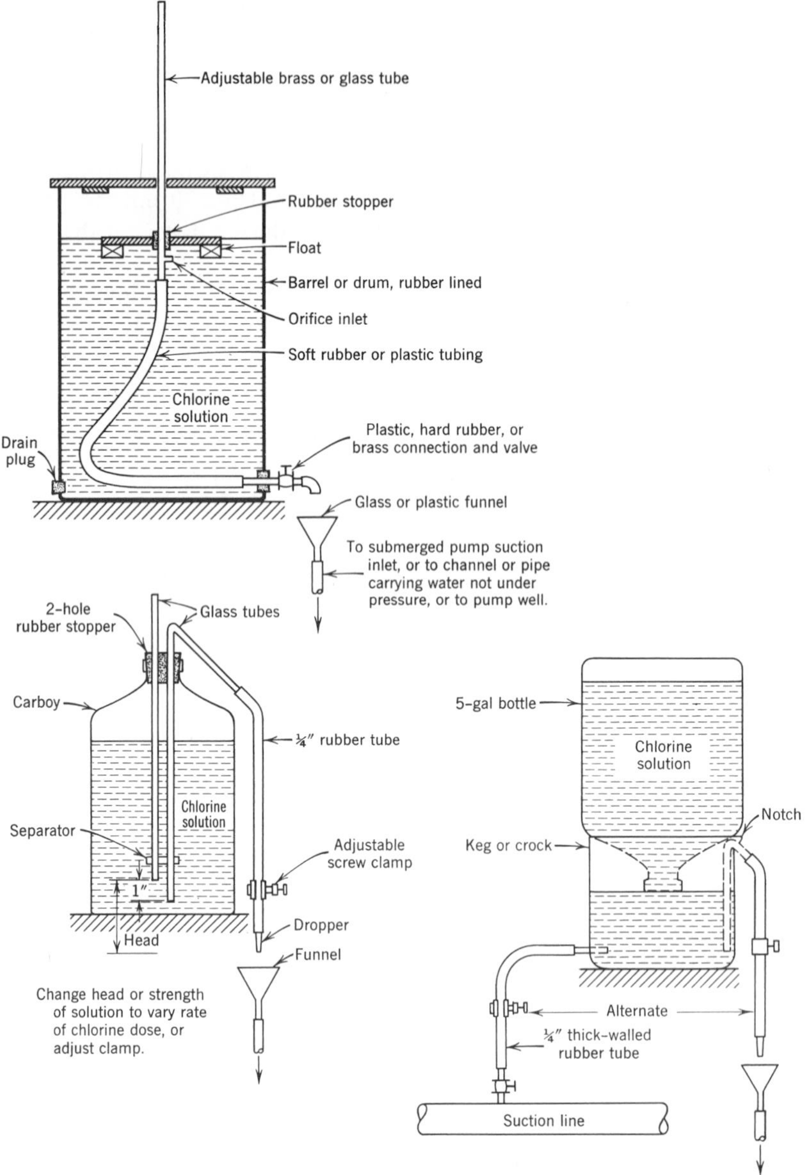

Figure 3-36 Homemade emergency hypochlorinators. To make chlorine solution, mix 4 pints of 5 percent hypochlorite to 5 gal of water.

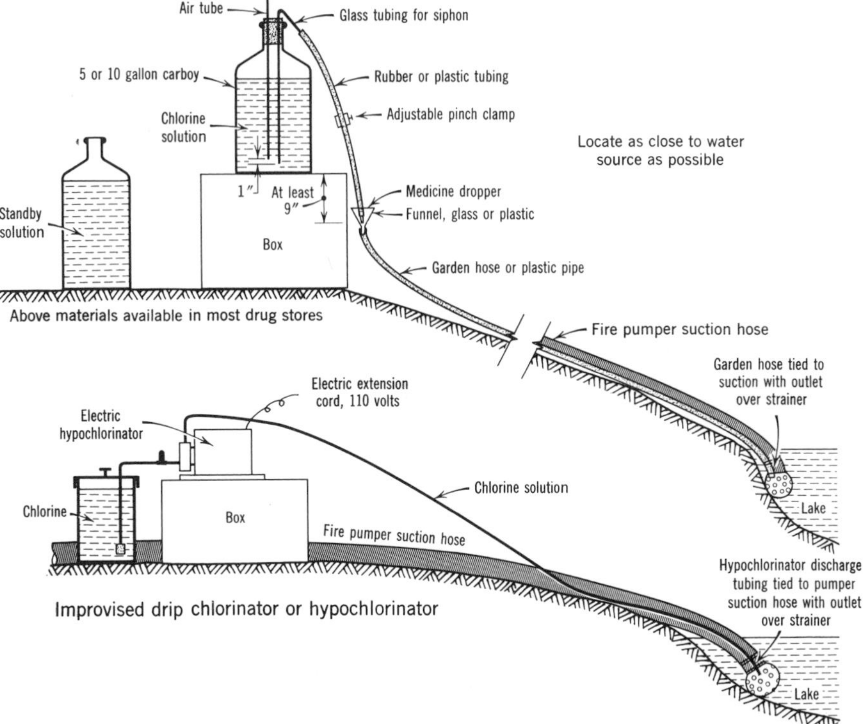

Figure 3-37 Emergency chlorination for fire supply, under health department supervision, for pumping into a hydrant on the distribution system, if necessary. Data for preparation and feed of chlorine solution: The asterisk below denotes that the paste should be made in a jar; add water and mix; let settle for a few minutes then pour into carboy or other container and make up to 5 gal. Discard white deposit; it has no value. Dosage is 5 mg/l chlorine. Double solution strength if necessary to provide residual of 4 to 5 mg/l.

Chlorine	Quantity to 5 gal Water	Rate of Feed 500 gpm Pumper
Perchloron or HTH, 70% available chlorine	4 lb*	4 oz or $\frac{1}{4}$ pt per min
Chlorinated lime, 24% available chlorine	12 lb*	4 oz or $\frac{1}{4}$ pt per min
Sodium hypochlorite, 14% available chlorine	20 pt	4 oz or $\frac{1}{4}$ pt per min

Some health departments have available a hypochlorinator for emergency use. Communicate with the local or state health department for assistance and advice relative to the manufacturers of approved hypochlorinators. Simple erosion type chlorinators can also be purchased or improvised for very small places. When large volumes of water are disinfected, a daily report should be kept showing the gallons of water treated, the amount of chlorine solution used, and the results of hourly residual chlorine tests.

Iodine

Eight drops of 2 percent tincture of iodine may be used to disinfect one qt of clear water (8 mg/l dose). Allow the water to stand at least 30 min before it is used. (Bromine can also be used to disinfect water, although its use has been restricted to the disinfection of swimming pool water.) Studies of the usefulness of elemental iodine show it to be a good disinfectant over a pH range of 3 to 8, even in the presence of contamination.[243] Combined amines are not formed to use up the iodine. A dosage of 5 to 10 mg/l, with an average of 8 mg/l for most waters, is effective against enteric bacteria, amoebic cysts, cercariae, leptospira, and viruses within 30 min. Tablets that can treat about 1 qt of water may be obtained from the National Supply Service, Boy Scouts of America, large camping supply centers, drug stores, and from the Army, in emergency. These tablets dissolve in less than 1 min and are stable for extended periods of time. They are known as iodine water purification tablets, of which Globaline, or tetraglycine hydroperiodide, is preferred. They contain 8.0 mg of active iodine per tablet. The treated water is palatable.

Filtration

Portable pressure filters are available for the treatment of polluted water. These units can produce an acceptable water provided they are carefully operated by trained personnel. Preparation of the untreated water by settling, prechlorination, coagulation, and sedimentation may be necessary, depending on the type and degree of pollution in the raw water. Pressure filters contain special sand, crushed anthracite coal, or diatomaceous earth. Diatomite filtration, or slow sand filtration, should be used where diseases such as amoebic dysentery, giardiasis, ascariasis, schistosomiasis, or paragonimiasis are prevalent, in addition of course to chlorination.

Slow sand filters (consisting of barrels or drums) may be improvised in an emergency. Their principles are given in Figure 3-17 and Table 3-17. It is most important to control the rate of filtration so as not to exceed 50 gpd/ft^2 of filter area and to chlorinate each batch of water filtered as shown in Table 3-32 in order to obtain reliable results.

Bottled, Packaged, and Tank-Truck Water

The bottled water industry has shown a large growth in many parts of the world because of public demand for a more palatable and "pure" water. It is not uncommon to find a wide selection of waters from various sources in the United

[243] Shih Lu Chang and J. Carrell Morris, "Elemental Iodine as a Disinfectant for Drinking Water," *Ind. Eng. Chem.*, **45**, 1009 (May 1953).

States and Europe in supermarkets and small grocery stores in almost all parts of the world. A major bottler in France was reported to have a capacity of 800 million bottles per year. The production in the United States was estimated at 388 million gallons per year.[244] There were an estimated 700 water bottling plants in the United States in 1972.[245] The demand for bottled water is of course minimized where a safe, attractive, and palatable public water supply is provided.

In an emergency it is sometimes possible to obtain bottled, packaged, or tank-truck water from an approved source, which is properly handled and distributed. Such water should meet the federal and state drinking water standards as to source, protection, and bacteriological, chemical, radiologic, and physical quality. Water that is transported in tank trucks from an approved source should be batch-chlorinated at the filling point as an added precaution. The tank truck should, of course, be thoroughly cleaned and disinfected before being placed in service. Detergents and steam are sometimes needed, particularly to remove gasoline or other gross pollution, followed by thorough rinsing with potable water and disinfection. Each tank of water should be dosed with chlorine at the rate of 1 to 2 mg/l and so as to yield a free chlorine residual of 0.5 mg/l. See Table 3-32.

Pasteurization plants and beverage bottling plants have much of the basic equipment needed to package water in paper, plastic, or glass containers. Contamination that can be introduced in processing (filtration through sand and carbon filters) and in packaging (pipelines, storage tanks, fillers) should be counteracted by germicidal treatment of the water just prior to bottling. Ozonation or ultraviolet treatment is sometimes used instead of chlorination. In any case, the source of water, equipment used, and operational practices must meet recognized standards.

Bottled water should meet the federal, USEPA, and state quality standards for drinking water and comply with the FDA regulations for the processing and bottling of drinking water.[246] Many states also have detailed regulations and standards.[247] The FDA microbiological quality standards are: 9 of 10 samples less than 2.2 coliforms per 100 ml with no sample showing greater than 9.2, with the arithmetic mean of all samples not greater than 1 per 100 ml by the membrane filter method.[248] The standard plate count of the bottled water at the retail outlet should be less than 100 per ml. FDA considers bottled water a "food," and regulates its purity. The regulation of drinking water additives is the responsibility of USEPA.[249]

[244]"Bottled water for gourmets with a thirst," *Changing Times*, December 1978, pp. 13–14.

[245]News of Environmental Research in Cincinnati, Water Supply Research Laboratory, USEPA, May 23, 1975.

[246]"FDA GMP Regulations for Processing and Bottling of Bottled Drinking Water," *Fed. Reg.* March 5, 1977.

[247]*Public Water Supply Guide for Owners and Operators of Bottled & Bulk Water Facilities*, New York State Dept. of Health, Albany, January 1971.

[248]*Fed. Reg.*, November 26, 1973.

[249]"EPA, FDA Agree on Water," *FDA Consumer*, October 1979, p.5.

Household Filters

Carbon, sand, and resin filters on a household water system cumulate organic matter which serves as a medium for the growth and then release of microorganisms in the drinking water if not supervised and regularly cleaned. Replacement, or washing and disinfection, of the filter media on a controlled effective basis is not normally feasible in a household filter system installed on a tap. A preliminary USEPA report on the efficiency of some 45 home filter water treatment devices showed trihalomethane (THM) removals of 6 to 93 percent and total organic carbon removals of 2 to 41 percent. In some cases higher bacteria counts were found in the water that had passed through the filter.[250] Another USEPA study of 31 home filters showed THM removals ranging from 4 to 98 percent (median of 41 percent) and nonpurgeable total organic carbons removal ranging from 2 to 87 percent (median of 13 percent) depending on the particular product.[251]

BIBLIOGRAPHY

Ameen, Joseph S., *Source Book of Community Water Systems,* Technical Proceedings, High Point, N.C., 1960, 214 pp.

Bennison, E. W., *Ground Water, Its Development, Uses and Conservation,* Edward E. Johnson, Inc., St. Paul, Minn., 1947, 509 pp.

Cairncross, Sandy and Richard Feachem, *Small Water Supplies,* The Ross Institute of Tropical Hygiene, Keppel Street, Grower Street, London WC1E7HT, January 1978, 78 pp.

Cox, Charles R., *Operation and Control of Water Treatment Processes,* WHO Mon. Ser., **49** 1964.

Cross-Connection Control Manual, USEPA, WSD, GPO, 20402, 1973, 57 pp.

"Engineering Evaluation of Virus Hazard in Water," *J. Sanit. Eng. Div. ASCE,* February 1970.

Fair, Gordon M., John C. Geyer, and Daniel A. Okun, *Water and Wastewater Engineering,* Vol. 1, John Wiley & Sons, New York, 1966, 505 pp.

Henderson, G. E. and Elmer E. Jones, *Planning for an Individual Water System,* Agricultural Research Service, U.S. Dept. of Agriculture, and USEPA, Water Programs Branch, American Association for Vocational Instructional Materials, Engineering Center, Athens, Ga. 30602, May 1973, 156 pp.

International Standards for Drinking Water, WHO, 1971, 70 pp.

Koch, Alwin G. and Kenneth J. Merry, *Design Standards for Public Water Supplies,* Dept. of Social and Health Services, Office of Environmental Programs, Olympia, Wash. 98504, May 1973, 42 pp.

Liguori, Frank R., *Manual Small Water Systems Serving the Public,* Conference of State Sanitary Engineers, June 1978, 320 pp.

Mackenthun, Kenneth M., and William M. Ingram, *Biological Associated Problems in Fresh Water Environments,* Federal Water Pollution Control Administration, U.S. Dept. of the Interior, Washington, D.C., 1967, 287 pp.

Manual for Evaluating Public Drinking Water Supplies, PHS Pub. 1820, DHEW, Bureau of Water Hygiene, Cincinnati, Ohio, 1971, 62 pp.

Manual of Individual Water Supply Systems, USEPA, Office of Water Programs, Washington, D.C., 1975, 118 pp.

Manual of Instruction for Water Treatment Plant Operators, New York State Dept. of Health, Albany, 1965, 308 pp.

[250] "Update," *J. Am. Water Works Assoc.,* July 1979, p. 8

[251] "Home filters to 'purify' water," *Changing Times,* February 1981, pp. 44–47.

Military Sanitation, Dept. of the Army Field Manual 21-10, May 1957, 304 pp.

Okun, Daniel A., "Alternatives in Water Supply," *J. Am. Water Works Assoc.*, **61**, No. 5, 215–221 (May 1969).

Quality Criteria For Water, USEPA, Washington, D.C. 20460, July 1976, 501 pp.

Recommended Standards for Water Works, Great Lakes Upper-Mississippi River Board of State Sanitary Engineers, Health Education Service, P.O. Box 7126, Albany, N.Y. 12224, 1976, 94 pp.

Standard Methods for the Examination of Water and Wastewater, 15th ed., A. E. Greenberg, Ed., American Public Health Association, Inc., 1015 Fifteenth St., N.W., Washington D.C., 20005, 1980.

State of the Art of Small Water Treatment Systems, USEPA, Office of Water Supply, Washington, D.C. 20460, August 1977, 154 pp.

"Surveillance of Drinking Water Quality," *WHO Monogr. Ser.* **63** 1976, 135 pp.

The Safe Drinking Water Act as amended, GPO, November 1977, 40 pp.

Van Dijk, J. L. and J.H.C.M. Oomen, *Slow Sand Filtration for Community Water Supply in Developing Countries*, A Design and Construction Manual, Technical Paper No. 11, WHO International Reference Centre for Community Water Supply, P.O. Box 140, 2260 AC Leidschendam, The Netherlands, December 1978, 175 pp.

Viruses in Waste, Renovated and Other Waters, Gerald Berg, Ed., Federal Water Quality Administration, U.S. Dept. of the Interior, Cincinnati, Ohio, 1969.

Wagner, E. G., and J. N. Lanoix, "Water Supply for Rural Areas and Small Communities," *WHO Monogr., Ser.,* **42**, 1959, 337 pp.

Water Quality Criteria, Fed. Water Pollut. Control Admin. U.S. Dept. of the Interior, Washington, D.C., April 1, 1968, 234 pp.

Water Quality Criteria 1972, National Academy of Sciences, National Academy of Engineering, Washington, D.C., 1972, 594 pp.

Water Quality and Treatment, prepared by AWWA, Inc., McGraw-Hill Book Company, New York, 1971, 654 pp.

Water Systems Handbook, Water Systems Council, 221 North LaSalle St., Chicago, Ill. 60601, 1977, 100 pp.

Well Drilling Operations, Technical Manual 5-297, Dept. of the Army and Air Force, September, 1965, 249 pp.

Wisconsin Well Construction Code, Division of Well Drilling, Wisconsin State Board of Health, 1951, 56 pp.

Wright, F. B., *Rural Water Supply and Sanitation*, Robert E. Krieger Publishing Co., Huntington, New York, 1977, 305 pp.

1980 Public Works Manual, Public Works Magazine, Ridgewood, N.J. 07451.

4

WASTEWATER TREATMENT AND DISPOSAL

DISEASE HAZARD

The improper disposal of human excreta and sewage is one of the major factors threatening the health and comfort of individuals in areas where satisfactory municipal, onsite, or individual facilities are not available. This is so because very large numbers of different disease-producing organisms can be found in the fecal discharges of ill and apparently healthy persons, as explained in Chapter 1. Surveys show that 5 to 10 percent of the population are carriers of *Entamoeba histolytica*, causing amebic dysentery (amebiasis),[1] and 25 percent of the population are carriers of ascarid, hookworm, or tapeworm.[2] The infection rate may be 50 percent in localized areas with primitive sanitation.[3] Handa, et al., reported a 74 to 82 percent parasitic infestation, admittedly very high, in a survey of Indian villages. The common parasites were ascaris, hookworm, *Entamoeba histolytica,* and *Entamoeba coli*[4] Studies in an American city showed that 9.1 percent of the local population harbored *Entamoeba histolytica* and that 23.1 percent harbored parasites.[5] The *Giardia lamblia* carrier rate in the United States may range between 1.5 and 20 percent depending on the community and group surveyed.[6] Giardiasis is a worldwide intestinal disease.

Knowing that numerous chemicals, and other organisms causing various types of diarrhea, bacillary dysentery (shigellosis), infectious hepatitis, salmonella infection, and many other illnesses are found in sewage and excreta, it becomes

[1]L. T. Coggeshall, "Current and Postwar Problems Associated with the Human Protozoan Diseases," *Ann. N. Y. Acad. Sci.,* **44,** 198 (1943).

[2]Norman R. Stoll, "Changed Viewpoint on Helminthic Disease: World War I vs. World War II," *Ann. N. Y. Acad. Sci.,* **44,** 207–209 (1943).

[3]Arthur P. Miller, *Water and Man's Health,* Agency for International Development, Washington, D.C., April 1962, p. 36.

[4]B. K. Handa, P.V.R.C. Panicker, S. W. Kulkarni, A. S. Godkari, and V. Raman, "The Impact of Sanitation in Ten Indian Villages," *Sanitation in Developing Countries*, Arnold Pacey, Ed., John Wiley & Sons, New York, 1978, p. 39.

[5]J. A. Kasper, E. J. Cope, M. Lyon, and M. White, "Report on the Results of Examinations for Intestinal Parasites," *Am. J. Pub. Health,* **40,** 1395–1397 (1950).

[6]Abram S. Benenson, *Control of Communicable Diseases in Man,* Am. Pub. Health Assoc. Washington, D.C., 1975, p. 129.

obvious that all sewage should be considered presumptively contaminated, beyond any reasonable doubt, with disease-producing organisms and toxic chemicals. In addition, it is known that some pathogenic organisms will survive from less than one day in peat to more than two years in freezing moist soil.[7] See Tables 1-5 and 1-6. Moist soil is favorable and dry soil is unfavorable for survival of many pathogens.

Numerous writers[8-10] have summarized the work of investigators who studied the survival of enteric pathogens in various wastewater treatment plant effluents. In general, available data shows that primary sedimentation removes up to 50 percent of coliform and pathogenic bacteria from sewage but is relatively ineffectual in removing viruses and protozoa. Activated sludge or trickling filter treatment removes about 90 percent of the coliform or pathogenic bacteria remaining after primary sedimentation. Viruses, although reduced, survive activated sludge and especially trickling filter treatment. Chemical coagulation, flocculation, sedimentation, and filtration will remove nearly all bacteria, viruses, protozoa, and helminths, particularly if supplemented by chlorination.

The waterborne microbiological agents of concern are the bacteria, viruses, helminths, protozoa, and spirochetes. The more important infectious bacterial agents are associated with shigellosis and salmonella infections. The viral agents are associated more commonly with infectious hepatitis and other enteric virus diseases. The helminths are associated with ascariasis and enterobiasis and other worm illnesses. The protozoa are generally associated with amebiasis and giardiasis. The spirochetes are associated with leptospirosis. To these should be added nonspecific diarrheas, secondary skin infections through scratches and open wounds, and infections of the eyes, ears, nose, and throat.

The transmission of microbiological agents of disease is dependent upon many factors; dose being most important. A sufficient number of organisms must be ingested to cause illness; data for some microbiological agents are available.[11-13] See Table 1-7.

[7]Willem Rudolfs, L. Lloyd Falk, and R. A. Ragotzkie, "Literature Review on the Occurrence and Survival of Enteric, Pathogenic, and Relative Organisms in Soil, Water, Sewage, and Sludges, and on Vegetation," *Sewage Ind. Wastes,* **22,** 1261 (1950).

[8]F. L. Bryan, "Diseases Transmitted by Foods Contaminated by Wastewater," *J. Food Prot.* January 1977, pp. 45–56. A historical review of medical and engineering literature covering 180 selected references.

[9]R. C. Cooper, "Health Considerations in Use of Tertiary Effluents," *J. Environ. Eng. Div., ASCE,* February 1977, pp. 37–47.

[10]S. A. Kollins, "The Presence of Human Enteric Viruses in Sewage and Their Removal by Conventional Sewage Treatment Methods," *Advances in Applied Microbiology,* Vol. 8, Academic Press, N.Y., 1966.

[11]M. D. Sobsey, "Methods for Detecting Enteric Viruses in Water and Wastewater," *Viruses in Water,* Am. Pub. Health Assoc., Washington, D.C., 1976, p. 122.

[12]J. C. N. Westwood and S. A. Sattar, "The Minimal Infective Dose," *Viruses in Water*, Am. Pub. Health Assoc., Washington, D.C., 1976, p. 68.

[13]O. J. Sproul, "Quality of Recycled Water: Fate of Infectious Agents," *Can. Inst. Food Sci. Technol.,* June 1973, pp. 91–95.

It has also been brought out that *Entamoeba histolytica* and giardia cysts, hepatitis viruses, and tapeworm eggs withstand the normal chlorination treatment given sewage. Since chlorination of wastewater treatment plant effluent may not protect against these diseases, but only minimize the probability, more advanced treatment may be needed in some circumstances such as where human contact or ingestion is probable. Other practical and more effective means of disinfection need to be investigated, in addition to better mixing and increased contact time, which will not result in the formation of chloro-organics or other substances potentially harmful to man and aquatic life. In the meantime, chlorination is acknowledged to far outweigh any risks it may have to human health.

Aside from the known disease outbreaks caused by drinking contaminated water and consuming contaminated shellfish, there have been incidents of disease transmission by swimming in contaminated waters, although not well-documented. For example, shigellosis was associated with swimming in the Mississippi River about five miles south of the Dubuque, Iowa sewage treatment plant.[14] In another example a Coxsackie virus B epidemic attributed to sewage contamination occurred at a boys' summer camp on Lake Champlain, N.Y.[15]

Infectious diseases vary in their clinical manifestations from severe to mild. Many do not come to the attention of practicing physicians or epidemiologists, or are not reported. Because of this it cannot be assumed that relationships between sewage discharges and diseases do not exist. On the contrary, knowing that they do exist, it is essential that unnecessary human suffering and illness be prevented by the application of existing knowledge.

Therefore the mere exposure of excreta, sewage, or other wastewater (including gray water) on the surface of the ground, or its improper treatment and disposal, immediately sets the stage for possible disease transmission by direct contact, by a vehicle or vector such as an individual or the housefly, by an inanimate object such as a child's toy, or by ingestion of excreta or sewage directly from soiled fingers or via contaminated water or food. This is especially true in developing countries where enteric diseases are prevalent and clean water and personal hygiene are wanting.

Sewage sludge accumulates the heavy metals in municipal wastewater. Many of the metals are very toxic; hence the use of such sewage sludge as a soil builder may result in higher levels of toxic metals in the vegetation treated and in the animals eating the vegetation. The use of pesticides containing lead, mercury, barium, and cadmium, as well as fallout from air pollution, may contribute additional toxic contaminants. Sewage sludge therefore should not be used as a soil builder or fertilizer supplement for crops for forage unless it is found to be free of significant amounts of toxic metals and parasite ova or other pathogens.

Awareness of these dangers, coupled with adequate treatment of sewage, provision of potable water, sanitation, and personal hygiene, are recognized as being primarily responsible for reducing intestinal and waterborne diseases to

[14] *MMWR*, CDC, Atlanta, Ga., November 16, 1974.
[15] *J. Am. Med. Assoc.*, 226(1):33–36, 10 Oct., 1973.

their present low level in many parts of the world. Maintenance of the disease barriers and vigorous application of the basic sanitary engineering and sanitation principles in less developed areas of the world are necessary for the enjoyment of a healthful environment and better quality of life. It may appear inconceivable, but there are still many urban areas, as well as suburban and rural areas, in the United States and abroad where the discharge of raw or inadequately treated sewage to roadside ditches and streams is commonplace. See Chapter 1 and Figure 1-2 for additional details on disease control.

Criteria for proper wastewater disposal. Proper disposal of sewage and other wastewater is necessary not only to protect the public's health and prevent contamination of groundwater and surface water resources, but also to preserve fish and wildlife populations and to avoid the creation of conditions that could detract from the attractiveness of a community, tourist establishment, resort, and recreation area. The following basic criteria should be satisfied in the design and operation of an excreta, sewage, or other wastewater disposal system.[16]

1. Prevention of pollution of water supplies and contamination of shellfish intended for human consumption.
2. Prevention of pollution of bathing and recreational areas.
3. Prevention of nuisance, unsightliness, and unpleasant odors.
4. Prevention of human wastes coming into contact with man, animals, and foods or being exposed on the ground surface accessible to children and pets.
5. Prevention of fly and mosquito breeding; exclusion of rodents and other animals.
6. Strict adherence to standards for groundwater and surface waters; compliance with local regulations governing wastewater disposal and water pollution control.

Failure to observe these basic principles can result in the development of health hazards and the degradation of living conditions, recreational areas, and natural resources that are essential for the well-being of the general public. Protection of land and water resources should be the national policy and every effort should be made to prevent their pollution by improper treatment and disposal of sewage and other wastewater.

DEFINITIONS

Before proceeding further it is desirable to define some commonly used terms.

Excreta The waste matter eliminated from the body; about 27 grams per capita per day dry basis (100 to 200 grams wet) with about 400 billion *E. coli.*[17] Mara[18] reports an average weight of 150 grams feces wet basis, 2000 million fecal coliform, and 450 million fecal streptococci per capita per day.

[16] J. A. Salvato, *Guide to Sanitation In Tourist Establishments,* WHO, 1976, pp. 36 and 37.
[17] F. W. Sunderman and F. Buerner, *Normal Values in Clinical Medicine,* W. B. Saunders and Co., 1949, p. 260, as cited in *Controlling the Effects of Industrial Wastes on Sewage Treatment,* the New England Interstate Water Pollution Control Commission, Wesleyan University, Conn., June 1970.
[18] D. D. Mara, *Bacteriology for Sanitary Engineers,* Churchill Livingstone, London, 1974, p. 102.

Domestic sewage The used water from a home or community. Includes toilet, bath, laundry, lavatory, and kitchen-sink wastes. (See Table 4-1.) Sewage from a community may include industrial and commercial wastes, groundwater, and surface water. Hence the more inclusive term *wastewater* is also in general usage. The terms are used interchangeably. Normal domestic sewage will average about 0.1 percent total solids in soft water regions. The strength of wastewater is commonly expressed in terms of 5-day biochemical oxygen demand (BOD), suspended solids, and chemical oxygen demand (COD).

Biochemical Oxygen Demand This characteristic of surface water, sewage, sewage effluents, polluted waters, industrial wastes, or other wastewaters is the amount of dissolved oxygen in milligrams per liter (mg/l) required during stabilization of the decomposable organic matter by aerobic bacterial action. Complete stabilization requires more than 100 days at 20°C. Incubation for 5 days (carbonaceous demand satisfied) or 20 days (carbonaceous plus nitrification demand satisfied) is not unusual but as used in this chapter BOD refers to the 5-day test unless otherwise specified. Generally for domestic sewage, one pound of 5-day BOD is roughly equivalent to $1\frac{1}{2}$ pounds of ultimate BOD. If one pound of 5-day BOD is completely aerated, requiring 1.3 pounds of oxygen, 0.14 pounds of inert residue will remain.

Suspended Solids Those solids that are visible and in suspension in water. They are the solids that are retained on the asbestos mat in a Gooch crucible.

Chemical Oxygen Demand This is usually measured in relation to certain industrial wastes. The COD is the amount of oxygen expressed in ppm or mg/l consumed under specific conditions in the oxidation of organic and oxidizable inorganic material in the water. The test is relatively rapid. It does not oxidize some organic pollutants (pyridine, benzene, toluene) but does oxidize some inorganic compounds that are not measured, that is, affected by the BOD analysis.

Facultative Bacteria These have the ability to live under both aerobic and anaerobic conditions.

Aerobic Bacteria These are bacteria that require elemental oxygen for their growth.

Anaerobic Bacteria These are bacteria that grow only in the absence of free elemental oxygen, and obtain oxygen from breaking down complex substances.

Total Organic Carbon (TOC) This test is a measure of the carbon as carbon dioxide; the inorganic carbon compounds present interfere with the test hence they must be removed before the analysis is made, or a correction applied.

Privy or one of its modifications, is the common device used when excreta is disposed of without the aid of water. When excreta is disposed of with water, a *Water-Carriage* sewage-disposal system is used; generally all other domestic liquid wastes are included.

When storm water and domestic sewage enter a sewer it is called a *Combined Sewer*. If domestic sewage and storm water are collected separately, in a *Sanitary Sewer* and in a *Storm Sewer*, the result is a *Separate Sewer System*. A *Sewer*

Table 4-1 Characteristics of Wastewater[a]

Constituents	Domestic Wastewater (Community)[b]	Household Wastewater[c]	Septic Tank Household Effluent[d]	Gray Water[e,f]	Black Water[e,f]	Septic Effluent Rest Area[g]
Color						
nonseptic	Gray					
septic	Blackish		3.5			
Odor						
nonseptic	Musty					
septic	H_2S		4.5			
Temperature (°F)	55° to 90°[h]		63			
Total Solids	800	968	820	528	621	
Total vol. Solids	425	514				
Suspended Solids	200	376	101	162	77	165
Vol. Sus. Solids	130					
Settleable Solids	5					
pH	7.5	8.1	7.4	6.8	7.8	8.2
Total Nitrogen	40	84	36	11.3	153	140
Organic Nitrogen	25					
Ammonia Nitrogen	0.5	64	12	1.7	138	
Nitrate Nitrogen	0.5		0.12	0.12	0.22	0.6
Total Phosphate	15	61	15	1.4	18.6	29
Total Bacteria, per 100 ml	30×10^8		76×10^6			
Total Coliform, MPN per 100 ml	30×10^6		110×10^6	24×10^6[g]	0.25×10^6	
Fecal Coliform, per 100 ml				1.4×10^6	0.04×10^6	
BOD, 5-day	200	435	140	149	90	165
COD		709	675	366	258	405
Total Organic Carbon				125	97	
Grease		65				

[a] Average, in mg/l unless otherwise noted.
[b] Peter F. Atkins, "Water Pollution By Domestic Wastes," *Selection and Operation of Small Wastewater Treatment Facilities—Training Manual*. Charles E. Sponagle, USEPA, Cincinnati, Ohio, April, 1973, p. 3-3.
[c] K. S. Watson, R. P. Farrell, and J. S. Anderson, "The Contribution from the Individual Home to the Sewer System," *J. Water Pollution Control Fed.*, December 1967, pp. 2039-2054.
[d] J. A. Salvato, Jr., "Experience with Subsurface Sand Filters," *Sewage Ind. Wastes*, **27**, 8, 909-916 (August 1955).
[e] Septic tank effluent. The higher concentration of coliform bacteria in the gray water effluent are attributed to the large amounts of undigested organic matter in kitchen wastewater.
[f] M. Brandes, *Characteristics of Effluents from Separate Septic Tanks Treating Grey Water and Black Water from the Same House*, Ministry of the Environment, Toronto, Canada, October 1977, pp. 9 and 27.
[g] Robert O. Sylvester and Robert W. Seabloom, *Rest Area Wastewater Disposal*, University of Washington, Seattle, Washington, January 1972, p. 30.
[h] For the Central States zone in United States.

System is a combination of sewers and appurtenances for the collection, pumping, and transportation of sewage, sometimes called *Sewerage*; when facilities for treatment and disposal of sewage, known as the *Sewage* or *Wastewater Treatment Plant*, are included the reference would be to a *Sewage Works*.

STREAM POLLUTION AND RECOVERY

The 5-day BOD is the best single strength measure of wastewater or polluted water containing degradable wastes. However, organic loading, aquatic organisms including the animal life (benthos) in the bottom sediments, the COD where indicated, the dissolved oxygen, and the sanitary survey taken all together with the BOD (carbonaceous and nitrogenous) are the best indicators of water pollution. Dissolved oxygen is the best indicator of a water body's ability to support desirable aquatic life. Other chemical, physical, and biological parameters provide additional information. Total and fecal coliform density, pH, nitrates, nitrites, Kjeldahl nitrogen, phosphates, chlorides, turbidity, suspended solids, total solids, specific conductivity, temperature, toxics, grease, fats, and oils are also significant in determining water quality for specific situations.

In a freshly polluted water, during the first stage, mostly carbonaceous matter is oxidized. This is demonstrated by an immediate increase in the stream BOD and in oxygen utilization in the area of pollution discharge, followed by the second, or nitrification, stage in which a lesser but uniform rate of oxygen utilization takes place for an extended period of time. This is accompanied by a related characteristic change in the stream biota, as illustrated in Figure 4-1 for an assumed condition and as discussed below. The amount of dissolved oxygen in a receiving water is also the single most important factor determining the waste assimilation capacity of a body of water.

Stream pollution (organic) is apparent along its length by a zone of degradation just below the source, a zone of active decomposition, and, if additional pollution is not added, a zone of recovery. In the zone of degradation the oxygen in the water is decreased, suspended solids may be increased, and the stream bottom accumulates sludge. The fish life changes from game and food species to coarse. Worms, snails, and other biota associated with pollution increase. In the zone of active decomposition, the dissolved oxygen is further reduced and may approach zero. The water becomes turbid and gives off foul odors. Fish disappear, anaerobes predominate in the bottom mud, and sludge worms become very numerous. In the zone of recovery the process is gradually reversed and the stream returns to normal. The zones mentioned should not be discernible or experienced where sewage has been given adequate treatment before discharge.

It is often difficult to show in true life and in easily understandable terms the improvement in a stream or lake water quality which can be attributed to wastewater treatment accomplished. A method originally proposed by Brown, et

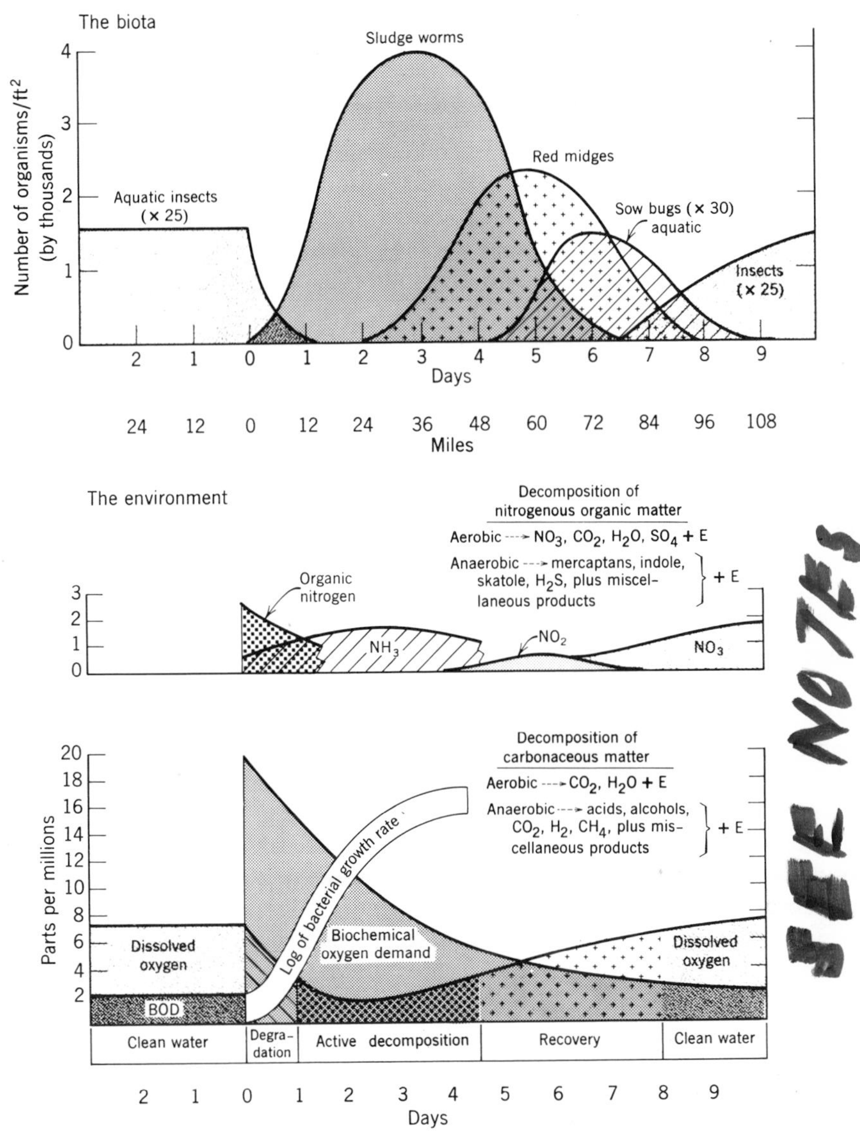

Figure 4-1 Stream degradation and recovery. The assumptions in the hypothetical pollution case under discussion are a stream flow of 100 cfs, a discharge of raw sewage from a community of 40,000, and a water temperature of 25°C, with typical variation of dissolved oxygen and BOD. The biota population curve is composed of a series of maxima for individual species, each multiplying and dying off as stream conditions vary. The equivalent curve shows that with a heavy influx of nitrogen and carbon compounds from sewage, the bacterial growth rate is accelerated and dissolved oxygen is utilized for oxidation of these compounds. As this proceeds, food is "used up" and the BOD declines. (From Alfred F. Bartsch and William Marcus Ingram, "Stream Life and the Pollution Environment," *Pub. Works*, July 1959.)

al.,[19,20] uses nine parameters to establish the "arithmetic water quality index" (WQI). The parameters are fecal coliform count, pH, biochemical oxygen demand, nitrate, phosphate, temperature (deviation from equilibrium), turbidity, total solids, and percent saturation of dissolved oxygen. The judgement of some 100 water experts was used in ranking the water quality parameters from 0 (worst) to 100 (best) which was then used to validate the WQI. The system can be applied to compare before and after treatment effects on a stream. Similar systems have been proposed by Prati, McDuffie, Harkins, and others.[21–25] The WQI parameters selected should be supplemented by indices of organic and inorganic chemicals such as mercury, PCB, mirex, trichloroethylene, vinyl chloride, and the like and also benthic and aquatic organisms.

EUTROPHICATION

Lakes go through a natural aging or maturing process. The rate of aging, or eutrophication, is dependent on the amount and type of nutrients received. The degree of eutrophication is indicated by the quantity of planktonic algae, reduced water transparency, and dissolved oxygen in the water near the surface.

A young lake is said to be *oligotrophic*. It is usually relatively clear, high in dissolved oxygen, deep, and receives few nutrients, thereby supporting little plant and animal life. As nutrients increase, together with siltation due to the acts of man and nature, plant and animal life increase. The lake then begins to mature and is referred to as a balanced *mesotrophic* lake. The continued siltation and accumulation of organic matter begin to fill up the lake making it shallower. This, together with proper nutrients, increases the growth of aquatic plants, particularly algae, and the lake becomes mature or *eutrophic* with low water transparency, large organic deposits colored brown or black, and often hydrogen sulfide odors. If there is an excess of nutrients the algal growths greatly increase ("bloom"), die, and decay. The decay process uses up more oxygen to the point of there not being enough for other forms of aquatic life. As the growth and decay progresses the lake fills with organic matter and silt to become a marsh and eventually, dry land.

[19] R. M. Brown, et al., "A Water Quality Index—Do We Dare?" *Water Sewage Works*, October 1970, p. 117.

[20] R. M. Brown, et al., "Validating the WQI," Paper presented at National meeting of ASCE on Water Resources Engineering, Washington, D.C., January 1973.

[21] L. Prati, et al., "Assessment of Surface Water Quality by a Single Index of Pollution," *Water Resources*, May 1971, p. 741.

[22] B. McDuffie and J. T. Haney, "A Proposed River Pollution Index," Paper presented to American Chemical Society, Division of Water, Air, and Waste Engineering, April 1973.

[23] R. D. Harkins, "An Objective Water Quality Index," *J. Water Pollut. Control Fed.*, March 1974, pp. 588–591.

[24] D. A. Dunnette, "A Geographically Variable Water Quality Index Used In Oregon," *J. Water Pollut. Control Fed.*, January 1979, pp. 53–61.

[25] Roy O. Ball, et al., "Water Quality Indexing," *J. Environ. Eng. Div. ASCE*, August 1980, pp. 757–771.

The aging process or eutrophication of a lake normally takes place over hundreds of years. With the addition of the proper combination of nutrients the aging process can be greatly accelerated.

The nutrients associated with eutrophication include phosphates, nitrates, and organic carbon, any one of which may be a limiting factor in algal growth. Phosphorus appears to be the most practical nutrient to control.[26] Their source may be compounds discharged to the atmosphere and precipitation, groundwater, lake sediment, tributary and stream drainage, agricultural runoff, forest runoff, urban drainage, and domestic wastewater. The nitrogen, phosphorus, and carbon from land drainage alone have been found to be adequate to support algal growths at the nuisance level. The upgrading of wastewater treatment alone will not, as a rule, reverse the aging process in a lake or pond water.[27] Nonpoint sources of nutrients must also be removed.

Lee, et al.[28] conclude that phosphorus is a key element in controlling growth of planktonic algae in fresh water and that nitrogen is generally the controlling element in marine waters. But for many water bodies phosphorus from nonpoint sources will also have to be controlled if the water quality is to be restored to desirable levels. Randall, et al.[29] found that eutrophication in a reservoir could not be controlled by reduction of point soures of phosphorus and biochemical oxygen demand alone. Over 85 percent of the nitrogen and phosphorus entering the reservoir came from storm-water runoff, mostly from urbanized areas. Sawyer* indicates that any lake having 0.01 mg/l inorganic phosphorus and 0.3 mg/l nitrogen can expect to have major algal blooms. Ahern and Weand[30] found that in general, higher phosphorus concentrations are associated with lower transparencies as measured with a Secchi disk. The desirable total phosphorus level in reservoir waters was 0.03 mg/l *total* phosphorus or less and, that based on transparency, there does not seem to be any reason to require phosphorus controls unless they can reduce phosphorus concentrations to 0.08 to 0.03 mg/l. See Control of Microorganisms and Chemical Examination (Phosphorus) and (Nitrates), Chapter 3.

SMALL WATERBORNE WASTEWATER DISPOSAL SYSTEMS

The provision of running water in a dwelling or structure immediately introduces the requirement for sanitary removal of the used water. Where sewers are

*C. N. Sawyer, "Phosphates and Lake Fertilization," *Sewage and Ind. Wastes*, June 1952, p. 768.

[26]Kuo-Chun Tsai and Ju-Chang Huang, "P, N & C head the critical list," *Water Wastes Eng.* April 1979, pp. 45–47.

[27]Thomas J. Grizzard and Ernest M. Jenelle, "Will Wastewater Treatment Stop Eutrophication?" *Diplomate*, December, 1973, pp. 8–10.

[28]G. Fred Lee, Walter Rast, and R. Anne Jones, "Eutrophication of water bodies: Insights for an age-old problem," *Environ. Sci. Technol.*, August 1978, pp. 900–908.

[29]C. A. Randall, T. J. Grizzard, R. C. Hoehn, "Effect of upstream control on a water supply reservoir," *J. Water Pollut. Control Fed.*, December 1978, pp. 2687–2702.

[30]John J. Ahern and Barron L. Weand, "Keep recreational waters algae-free," *Water Wastes Eng.*, October 1979, pp. 14–18.

available, connection to the sewer will solve a major sanitation problem. Where sewers are not provided or anticipated, as in predominantly rural areas and many suburban areas, consideration must be given to the proposed method of collection, removal, treatment, and disposal of wastewater. With a suitable soil, the disposal of wastewater can be simple, economical, and inoffensive; but careful maintenance, in addition to proper design and construction, is essential for continued satisfactory operation. Where rock or groundwater is close to the surface or the soil is a tight clay, it would be well to investigate some other property. Sometimes the groundwater level can be lowered by the proper design and construction of curtain drains.

The common system for wastewater treatment and disposal at a private home in a rural area consists of a proper septic tank for the settling and treatment of the wastewater and a subsurface absorption system for the disposal of the septic tank overflow, provided the soil is satisfactory. The soil percolation test and soil characteristics are commonly used as a means for determining soil permeability or the capacity of a soil to absorb settled wastewater. This and the quantity of wastewater from a dwelling are the bases upon which a subsurface sewage disposal system is designed. Sand filters, elevated systems in suitable fill, evapotranspiration-absorption systems, evapotranspiration beds, aeration systems, stabilization ponds or lagoons, recirculating toilets, and various types of toilets and privies are used under certain conditions and where the soil is unsuitable. These are discussed later in this chapter.

Wastewater Characteristics

The composition of wastewater can be expected to vary considerably depending on the community and water use, industries served, infiltration, and whether the sewers are combined sewers or sanitary sewers. However, the wastewater from a residential community is fairly uniform. The characteristics of average domestic wastewater, septic tank effluent, septic tank gray water effluent, and septic tank black water effluent are given in Table 4-1.

The wastewater from water closet and latrine or aqua privy flushing is referred to as *black water*. All other domestic wastewater is referred to as *gray water*.

For all practical purposes, and from a public health standpoint, gray water should be considered sewage or wastewater and should be treated as such. Gray water can be expected to contain pathogens from the bathroom shower and wash basin, and from clothes washing, including baby diapers, and clothing. Gray water is amenable for treatment in a septic tank and subsurface absorption system.

General Soil Characteristics

In its broadest sense, soil is made up of decayed or broken-down rock containing varying amounts of organic material such as animal and plant wastes.

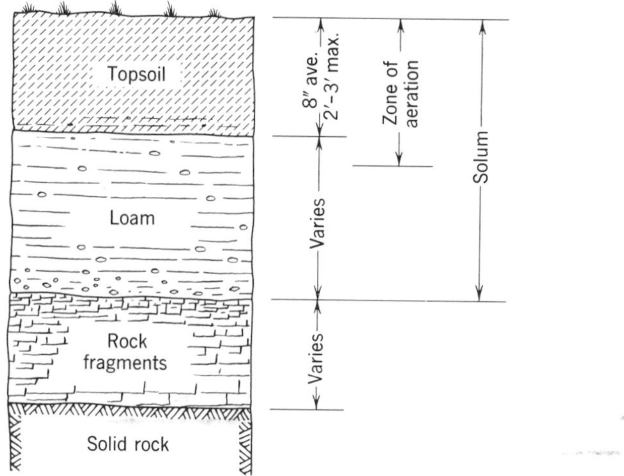

Figure 4-2 Typical earth formation.

Destruction of rock is accomplished by water, wind, glacial ice, chemical action, plant life, freezing and thawing, heat, and other forces to form soil. The soil may accumulate in place or may be transported by wind, water, or glacial ice. Soil that accumulates in place is representative of the rock from which it has been derived and the local flora and fauna. Soils may be divided, for reference purposes, into gravel, sand, silt, and clay, and, depending on which is predominant, into sandy loam, gravelly loam, silty loam, loam, clay loam, and clay, with and without large stones or boulders. Loam is a mixture of gravel, sand, silt, and clay containing decayed plant and animal matter or humus. Figure 4-2 illustrates a typical earth formation with the top layer, or topsoil, as much as 2 or 3 ft. deep, although on the average it will be about 6 to 8 in. or less. The topsoil is usually richer in humus. Clay loam and clay do not drain well and usually are considered unsuitable for the disposal of sewage and other wastewater by subsurface means. However, some of the clay chemicals in soil can be displaced by salts, acids and bases. This ion-exchange process, for example, accounts for soil acidity and alkalinity, the friability of some clays, the binding of potassium and ammonium in soil, and also the travel of nitrogen through soils.[31]

Soils Characteristics and Clues

The permeability of a soil, or the ability of the soil to absorb and allow water and air to pass through, is related to the chemical composition, texture, and granular structure of the soil. The soil texture refers to the proportion of clay, silt, and sand less than 2 mm in diameter. The soil structure refers to the agglomeration or clumping of particles of clay, silt, and sand and the intervening cleavage planes. The microbial population modifies the properties of the soil as noted below. A lump of soil with good structure will break apart, with little

[31]Witold Rybezynski, Chongrak Polprasert, and Michael McGarry, *Low-Cost Technology Options for Sanitation,* International Development Research Centre, Ottawa, Canada, 1978, p. 22.

pressure, along definite cleavage planes. If the color of the soil is yellow, brown, or red, it would indicate that air, and therefore water, passes through. However, if the soil is blue or grayish it would indicate a soil saturated for extended periods, and if mottled brown and red, it would indicate a fluctuating water table, or lack of aeration and therefore a tight soil that is probably unsuitable for subsurface absorption of wastewater. A grayish soil, however, may be suitable if drained. Magnesium and calcium tend to keep the soil loose, whereas sodium and potassium have the opposite effect. Sodium hydroxide, a common constituent of so-called septic tank cleaners, would cause a breakdown of soil structure with resultant smaller pore space and reduced soil permeability.

Role of Microorganisms and Macroorganisms

Aerobic oxidizing bacteria, that is, oxygen-loving bacteria, are found in the zone of aeration. This zone extends through the topsoil and into the upper zone of the subsoil, depending on the soil structure and texture, earthworm population, root penetrations, and other factors. The topsoil contains organic matter, minerals, air, water, supportive vegetative organisms such as bacteria, fungi, and molds, as well as protozoa, nematodes, actinomycetes, algae, rotifers, earthworms, insects, and larger animals. A gram of topsoil can be expected to contain millions of bacteria and other organisms. These organisms have the faculty of reducing complex organic matter to simpler forms through their life processes. Earthworms play an important role together with other soil flora and fauna in keeping soil aerated.

Septic tank effluent, for example, which contains material in solution, in colloidal state, and in suspension, when discharged into or close to the topsoil will be acted upon by these organisms and be reduced to "soil" as well as liquids and gases. This is accomplished provided the sewage is not discharged at too rapid a rate or in too great a volume and concentration into the earth in the zone of aeration. A waterlogged soil destroys the aerobic organisms, producing anaerobic conditions that tend to preserve the organic matter in septic tank effluent, thereby delaying its decomposition and increasing mechanical clogging of the liquid-soil interface with organic matter including slimes and sulfides. The nitrites in septic sewage are toxic to plants.

Maintenance of Soil Infiltrative Capacity

To assist in the maintenance of aerobic conditions, subsurface absorption fields are usually laid at depths of 24 to 30 in., although depths as great as 36 in. or more are sometimes used. The gravel around the open-joint tile or perforated pipe should extend up into the zone of aeration, usually within 12 to 18 in. of the ground surface.

McGauhey and Winneberger call attention to the recovery infiltrative capacity of trench sidewalls after resting a few hours; the need to have a permeable soil in the first instance; the insulating effect of impermeable lenses or strata to the downward percolation of water; the reduction of percolative capacity of soils containing colloids that swell; and the necessity for the groundwater table to be at a sufficient depth to permit the soil to drain during rest periods rather than water remaining suspended by surface tension and capillary phenomena.[32] The

> loss of infiltrative capacity is directly traceable to the organic fraction of sewage which leads to clogging of the soil surface by suspended solids, bacterial growth, and ferrous sulfide precipitation.

They report "conclusively that intermittent dosing and draining of the soil system is necessary to the maintenance of optimum infiltration rates." In a subsequent report, narrow trenches, 8 to 12 in. in width, and placement of the distributing line as high as possible are advised to provide a maximum effective sidewall surface area.[33] Full loading (siphon chamber or pump) on an intermittent basis will promote maintenance of aerobic conditions in the trench and soil. Prolonged resting periods (several months to a year) will permit restoration of infiltrative capacity; but two absorption systems and a diversion gate or valve are needed to permit alternate use of each system. Laak[34] believes that

> soil seepage beds will function forever if the system is properly designed, constructed, and maintained. There is no evidence to suggest that soil clogging is irreversible.

It has also been pointed out that most soils will reach a steady-state equilibrium percolation rate of about 0.1 ft/ft^2/day when the soil surface is completely inundated by untreated effluent. Soil adsorptive capacity is an important consideration in the design of a septic tank system for the protection of the groundwater. Robeck et al. advise that a soil must have a low permeability (effective size 0.1 to 0.3 mm) and some adsorptive capacity to allow organic material to be retained. A minimum soil organic content of 0.5 to 1 percent is suggested (found in practically all agricultural soil, together with some clay and silt, which add to the adsorptive capacity). Under such circumstances pathogenic and essential aerobic organisms, which are capable of degrading such food sources as ABS detergents and other organic matter, are retained.[35] A soil with low adsorption (coarse gravel) or a formation with solution channels, fractures, or fissures will permit pollution to travel long distances without purification. Careful consideration to these factors in the design of subsurface sewage-disposal systems is necessary.

Hydrogen peroxide has been used on an experimental basis to restore and maintain the infiltrative capacity of clogged absorption systems. Between 15 and

[32]P. H. McGauhey and John H. Winneberger, "Studies of the Failure of Septic-Tank Percolation Systems," *J. Water Pollut. Control Fed.* **36**, No. 5, 593–606 (May 1964).

[33]P. H. McGauhey and John H. Winneberger, "Final Report on a Study of Methods of Preventing Failure of Septic-Tank Systems," Sanitary Engineering Research Laboratory, College of Engineering and School of Public Health, University of California, Berkeley, October 31, 1965.

[34]Rein Laak, "Septic Tank and Leach Field Operation," *Wastewater Treatment Systems For Private Homes And Small Communities.* Paul S. Babiarz, Robert D. Hennigan, and Kevin J. Pilon, Eds., Central New York Regional Planning and Development Board, Syracuse, N.Y., 1978, p. 23.

[35]Gordon G. Robeck et al., "Factors Influencing the Design and Operation of Soil Systems for Waste Treatment," *J. Water Pollut. Control Fed.,* **36,** No. 8, 971 (August 1964).

40 gallons of commercial 50 percent hydrogen peroxide was used, diluted with about 300 to 600 gallons of water as the hydrogen peroxide was added to the absorption system.[36] The hydrogen peroxide must be applied directly to the clogged lines, after the ponded wastewater has been drained or pumped out. This can be done by adding the mixed solution through the distribution box in such a manner as to prevent it from draining back to the septic tank, or by augering holes down to the gravel and adding the peroxide through 3-in. PVC pipes inserted in the holes, or by adding 30 percent hydrogen peroxide to well points sunk into the absorption trench. The hydrogen peroxide should be handled with extreme care; protective clothing and eyeglasses must be worn. The septic tank should be pumped out to provide storage while the system is being restored. It should be noted that the peroxide treatment does not correct failures due to impermeable soils, high groundwater, or shallow bedrock.[37]

Site Investigation, Soil Profile, and Suitability

Prior to the construction of a sewage disposal system, subsurface explorations are necessary to determine the subsoil formation in a given area. An auger with an extension handle is often used for making the investigation but a backhoe is sometimes employed for large systems and deep test holes. The examination of road cuts, stream embankments, or building excavations provides useful information. Wells and well drillers' logs can also be used to obtain information on groundwater and subsurface conditions. In some areas subsoil strata vary widely within short distances and numerous borings are required in the selected site. Agricultural and highway soil maps, if available, can give an indication of soil characteristics, provided the depth of the soil cores or test holes used to prepare the soil maps is representative of the soil strata to be used for the subsurface disposal of wastewater. Aerial photograph maps are also very useful to experienced individuals. If the subsoil appears suitable (as judged by study of soil maps, the soil texture, and an investigation of the structure, color, and depth or thickness of permeable strata and their swelling characteristics), percolation tests should be made at points and elevations which are selected as typical of the area in which the disposal system is to be located to establish a settled sewage application rate for design purposes.

It is necessary to have at least 2 ft of suitable soil between the bottom of absorption trenches and leaching pits and the highest groundwater level, clay, rock, or other relatively impermeable layer. Some agencies require a minimum of 3 ft.

The design of leaching pits and cesspools is based on the ability of the soil found at the depth between 3 and 8 or 10 ft to absorb water. Sometimes pits are

[36]John M. Harkin, Michael D. Jawson, and Fred G. Baker, "Causes and Remedy of Failure of Septic-Tank Seepage Systems," *Second National Conference on Individual Onsite Wastewater Systems,* National Sanitation Foundation, Ann Arbor, Mich., Nov. 5–7, 1975, pp. 119–124.
[37]*Rural Wastewater Management in California,* The California Water Resources Control Board, Sacramento, California, 1979, p. 43.

made 20 to 25 ft or more in depth, using prefabricated sections, in order to reach permeable soil. It is not known to what extent bacteria, protozoa, and metazoa are active in leaching pits and cesspools. Since relatively large quantities of sewage would be discharged in a small area, designs incorporating cesspools or leaching pits must take into consideration the soil structure, direction and depth of groundwater flow, and the relative location of wells or springs with respect to their possible pollution. Cesspools and leaching pits should be prohibited in shale, course gravel, and limestone areas or where groundwater is high, and avoided when shallow wells or springs are in the vicinity unless adequate protecting distances and soils can be assured.

Where the soil is relatively nonpermeable at shallow and deep depths, an alternate treatment and disposal system as discussed later is needed in place of a conventional leaching system. Building should preferably be postponed in such situations until sewers are available.

It is extremely important while in the planning stage, before building construction is started, to consider:

1. The suitability of the soil to absorb settled sewage as determined by soil percolation tests, soil characteristics, and the type and size of the disposal system required. At least 4 to 5 ft of suitable soil should be available over clay, hardpan, rock, or groundwater for absorption trenches and 8 to 10 ft for leaching pits.
2. The area of land available for the sewage disposal system, and its adequacy for sewage disposal and water supply protection. This should include the location of existing and proposed onsite and offsite sewage disposal systems and underground sources of drinking water; type of well construction, underground strata, and depth of water-bearing source; slope of the groundwater table; and protecting distances between wells and sewage disposal systems.
3. The elevation of the sewage disposal units, the house sewer, the house drain, the first-floor level, and location of the lowest plumbing fixture. Sometimes the installation of a sewage pump, excavation at a greater depth, or the carting in of earth fill is made necessary because the slope of the ground surface, clay or rock level, and depth of the underground water are not considered in the planning stage.
4. The location of rock outcrops, hills, large trees, storm-water drains, watercourses, adjoining structures; also bathing beaches, shellfish-growing areas, and water supply intakes.
5. Surface and underground water drainage, including roof, cellar, and foundation drainage (this drainage must be excluded from the sewage disposal system to prevent the system becoming overloaded and waterlogged).
6. The average rainfall and temperature during the period of use.

Information on aquifer geology, well yields, well pump capacities, static and pumping water levels will help determine the circle of influence and travel of underground pollution. All these factors should be carefully evaluated by a trained person before a decision is made and construction is started. The state and county sanitary or public health engineers and sanitarians are trained to give sound advice.

Travel of Pollution from Septic-Tank Absorption Systems

Groundwater contamination potential from a septic tank leaching system is determined by the soil characteristics and the depth to groundwater. Studies show that unsaturated soil will remove a high percentage of total dissolved solids, 5-day BOD, soluble organic carbon, ammonia nitrogen, iron, coliforms, fecal coliforms, and fecal streptococci from septic tank effluent. Phosphate (total PO_4-P) removal may or may not be high (removal or retention is dependent upon the type of soil). Nitrates will increase in the vicinity of the leaching system as ammonia and nitrites convert to nitrates, but decrease downstream with distance and dilution.[38]

Soils containing loam will remove most of the phosphorus in sewage effluent. If the absorption trenches are kept shallow (top of gravel about 12 in. from the ground surface) as recommended, the vegetative (grass) cover root system over the absorption field penetrates and takes up much of the nitrogen during growing periods, which coincide with maximum system use at summer vacation homes, resorts, and recreation areas. Hence the danger of phosphates and nitrates passing any significant distance through the soil to the groundwater table and contributing significant amounts of nutrients that might reach a lake or other impoundment and accelerate its eutrophication can be greatly minimized. This is particularly so when considered in relation to the phosphorus and nitrogen contribution from surface runoff, storm water, and wastewater discharges to lakes and streams. There are however some ecologically critical waters where even minimal amounts of phosphorus and/or nitrogen, regardless of source, should be avoided. But this may be impossible, as natural sources of nitrogen from precipitation and/or phosporus from automobile exhaust, phosphorus and nitrogen from agricultural, uncultivated, and forest area, and storm-water overflows cannot in practice be eliminated. And even if minimized, there may still be sufficient amounts of nutrients reaching sensitive waters to cause algal growths and natural eutrophication, although it is hoped, at a slower rate.

Laboratory studies using nine types of soil indicate that passage through 40 to 50 cm of an agricultural-type soil is very effective in removing viruses from water, with soils at pH below 7.0 to 7.5 best.[39] Culp concluded that although not well established, soil with reasonable amounts of silt and clay removes viruses within the first 2 ft.[40] Wellings, et al. reported virus travel of 5 ft in sandy soil spray-irrigated with chlorinated activated sludge effluent and also travel to 6 and 20 ft-deep wells.[41] Slow sand filters dosed uniformly at standard loading rates removed 99 percent or more viruses from septic tank effluent over a two-year period under

[38]T. Viraraghavan and R. G. Warnock, *J. Am. Water Works Assoc.*, November 1976, pp. 611–614.

[39]William A. Drewry and Rolf Eliassen, "Virus Movement in Groundwater," *J. Water Pollut. Control Fed.*, **40,** No. 8, R271 (August 1968).

[40]Gordon Culp, *Summary of A "State-of-the-Art" Review of Health Aspects of Waste Water Reclamation for Ground Water Recharge,* Supplement to *Report of the Consulting Panel on Health Aspects of Wastewater Reclamation for Groundwater Recharge,* State of California, PB-268 540, U.S. Dept. of Commerce, National Technical Information Service, Springfield, Va. 22161, June 1976, p. 29.

[41]F. M. Wellings, A. L. Lewis, and C. W. Mountain, "Virus Survival Following Wastewater Spray Irrigation of Sandy Soils," cited in *Virus Survival in Water and Wastewater Systems,* J. F. Malina, Jr. and B. P. Sagik, (Eds.), Austin, Texas, Center for Research in Water Resources, 1974, pp. 253–260.

controlled experimental conditions.[42] Sorber and his co-workers found that enteric viruses can be recovered in soils at considerable distances from their point of application.[43] However organic and microbial pollution will not travel great distances (more than about 3 ft) in saturated soil, which is not coarse sand or gravel. It is apparent that the travel of pollution is not reduced to an exact science. Well Contamination—Cause and Removal is discussed in Chapter 3.

Soil Percolation Test

Background of the Soil Percolation Test

The suitability of soil for the subsurface disposal of sewage and other wastewater can be determined by a study of soil characteristics and the soil percolation test. The test is a measure of the relatively constant rate at which clear water maintained at a relatively constant depth (6 in.) will seep out of a standard size test hole (Ryon used a 12 in. square hole) that has been previously saturated; the bottom of the test hole is at the approximate depth of the proposed absorption system and greater than 2 ft above the seasonal high groundwater, rock, or tight soil. Henry Ryon first introduced this test in 1924 based on his investigation of subsurface disposal systems that had failed or were about to fail after 20 years of use in New York State[44] He plotted the results of his tests, covering a wide range of soils, and developed curves to interpret the results of those tests. The procedure developed by him has been used throughout the world and, except for some refinements,* remains as the only rational basis for the design of subsurface disposal systems. Ryon emphasized the importance of "taking care to wet the soil before pouring water in for the test if it appears dry" and that "the ground must be thoroughly wet before the test is made." *Saturation of the soil is essential* to obtain reproducible results, that is, a relatively constant rate of water drop in the test hole. The work done by Ryon was confirmed by the PHS in independent field tests and, in spite of its limitations, serves as the basis for present-day design of subsurface absorption systems, adjusted for the automatic home clothes washer and garbage grinder disposal unit.[45]

The reliability of the soil percolation test is increased by scientific evaluation of soils maps, as noted under the heading General Soil Characteristics, and in Site Investigation, Soil Profile, and Suitability. Special training and experience in soils is very helpful in making a proper interpretation of the soils data.

*Two in. of gravel in bottom of test pit and reemphasis on prior soaking of the hole.

[42]*Management of Small Waste Flows,* PB 286 560, University of Wisconsin, National Technical Information Service, Springfield, Va. 22161, September 1978, p. C-55.

[43]Paper given at Environmental Engineering Division Conference, ASCE, Seattle, Washington, July 1976.

[44]Henry Ryon, *Notes on the Design of Sewage Disposal Works With Special Reference to Small Installations;* also *Notes on Sanitary Engineering*, Albany, N.Y., 1924.

[45]*Studies on Household Sewage Disposal Units,* PHS Pub. No. 397, Part I, 1949; Part II, 1950; Part III, 1954; DHEW, Cincinnati, Ohio. *Manual of Septic-Tank Practice,* DHEW, PHS Pub. 526, Cincinnati, Ohio, 1967.

Other investigators have studied the test to determine the effect of the shape of the test hole and the saturation of the soil on the percolation test. It has been found that the shape of the hole has no effect; size or diameter does affect results. However, it is again emphasized that *saturation of the soil is essential* to obtain reproducible results, that is, a constant rate of water drop in the test hole. This was recognized by Ryon and needs to be continually emphasized.

One of the main significant differences of opinion in connection with the soil percolation test is its interpolation to determine the allowable rate of septic settled sewage application per square foot of leaching area. This rate is taken as a percentage of the actual volume of water a test hole accepts in 24 hr. Various investigators have stated that this rate should be 0.4 to 3.5 (Ryon), to 5.0 or 7.0 percent of the actual amount of water absorbed or accepted by the test hole.

The Percolation Test

Percolation tests help to determine the acceptability of the site and to establish the design size of the subsurface disposal system. The length of time required to carry out percolation tests will vary with different types of soil. The safest method is to make tests in holes that have been kept filled with water at least 4 hours, preferably overnight, except where the soil is clean sand or gravel. This is particularly desirable if the tests are to be made by an inexperienced person, and for some soils, such as those that swell on wetting, this precaution is necessary even if the person carrying out the test has had considerable experience. Percolation rates in such cases should preferably be calculated from test data obtained after soil has had an opportunity to swell overnight. Enough tests should be made to ensure a valid result.[46] Some agencies require that soil tests be made during the wet period of the year to minimize error.

The soil percolation test is performed as follows:

1. Dig a hole about 1 ft^2 (USEPA suggests a 6 in. diameter hole) and to the depth at which it is proposed to lay the drain tile or perforated pipe. Scrape the inside of the hole to remove all smooth or cemented patches. A good average depth is 24 to 30 in. About 2 in. of washed gravel should be placed in the bottom of the hole. See Figure 4-3.
2. Pour about 12 in. of water in the hole. If the soil is relatively tight, let the water soak for about 4 hr, adding additional water as necessary, and proceed as explained in Step 3. If the soil is very absorbent, allow the water to seep away. Add 8 in. more of water in the hole and proceed as in Step 3. The test can be expedited by routinely having all test holes filled with water the night before the tests are made to allow ample time for soil swelling and saturation.
3. Measure the rate at which the water surface drops. This can be done by placing a piece of 2- × 4-in. lumber across the hole, *being careful to anchor it in a firm position.* Then, using a point or line on the 2 × 4 as a reference (See Figure 4-3) for the remainder of the test, slide a pointed slat or similar measuring stick straight down until it just touches the water surface. Immediately read the exact time on your watch and draw a horizontal pencil line on the measuring stick using a point or line on the 2 × 4 as a guide. Repeat the test at one-min intervals, if the water level drops rapidly, or at five- or ten-min intervals if the water level drops slowly. Observe the space between the pencil markings. Keep the depth of water in the hole at about 6 in. above gravel. When at least three spaces become relatively equal,

[46] J. A. Salvato, *Guide To Sanitation In Tourist Establishments,* WHO, 1976, p. 128.

which may require as long as 3 or 4 hr of presoaking in clayey soils, the test is completed, since equilibrium conditions have been reached for all practical purposes.

4. With the aid of a ruler, measure the space between the *equal* pencil markings and reduce this to minutes for the water level to drop one inch. This can be approximated closely with a ruler or can be computed.

For example, if the interval between 5-min readings is $\frac{3}{8}$ in., the time for the water level to drop one inch (call this x) is calculated as follows:

$$\frac{\frac{3}{8}\text{ in.}}{5\text{ min}} = \frac{1\text{ in.}}{x\text{ min}}, \qquad \text{or} \quad \frac{3}{8}x = 5, \qquad \text{or} \quad x = 13\tfrac{1}{3}, \text{ say } x = 15\text{ min.}$$

Tests must be carefully made to minimize errors due to soil swelling, smearing, variations in soils characteristics, inaccurate readings, and water scour.

Make at least two or three tests for the average lot; six tests are preferable where soil is not relatively uniform. Consider the soil characteristics and excavate the test holes to a depth of 4 to 5 ft to ensure groundwater, rock, or tighter soil is not encountered. Interpret the percolation tests results in the light of soil characteristics and soil clues as previously described.

5. Use Table 4-2 to determine the allowable rate of settled sewage application in gal per day per square foot of bottom trench area (gpd/ft^2) in an absorption trench system.

An example will serve to illustrate how the soil test results are used.

Example

Number of bedrooms—3.
Required septic tank—1000 gal liquid volume.*
Average of soil tests for absorption field, from table = 0.9 gpd/ft^2 (1 in. in 30 min).
Estimated sewage flow at 150 gal per bedroom = 450 gpd.
Required leaching area = 450/0.9, plus 60% for garbage grinder and clothes washer = 800 ft^2.
The required area can be obtained by providing:

800/1.0 or 800 lineal ft of tile in trenches 12 in. wide, or
800/1.5 or 533 lineal ft of tile in trenches 18 in. wide, or
800/2.0 or 400 lineal ft of tile in trenches 24 in. wide.
} Recommended widths

NOTE No additional credit is given for bottom area for trenches wider than 2 ft; the trench may be wider. The required leaching area using the Great Lakes-Upper Mississippi River Board or USEPA recommendation would be: 450/0.6 = 750 ft^2.

A variation of the soil percolation test is to observe the time for the water level to drop from a depth of 6 to 5 in. as shown in Figure 4-3. Repeat the test, adding water as necessary, and if the time recorded in the second test is within 10 percent of the time for the first test, use this time to determine the allowable rate of settled sewage application per square foot per day using Table 4-2. If the "times" vary by more than 10 percent, repeat the test until the times for two successive tests do not vary by more than 10 percent. With a tight soil, this method will take somewhat longer to

*Assumes a home laundry machine and a garbage grinder are to be installed.

Table 4-2 Interpretation of Soil Percolation Test

Time for Water to Fall 1 in. (min)	Allowable Rate of Settled Sewage Application (gpd/ft^2)		
	USPHS[a]	USEPA[b]	GLUMR[c]
<1	5.0[d]	[b]	1.2
1	5.0[d]	1.2	1.2
2	3.5[d]	1.2	1.2
3	2.9[d]	1.2	1.2
4	2.5[d]	1.2	1.2
5	2.2[d]	1.2	1.2
6	2.0	0.8	0.9
7	1.9	0.8	0.9
8	1.8	0.8	0.9
9	1.7	0.8	0.9
10	1.6	0.8	0.9
11	1.5	0.8	0.6
12	1.4	0.8	0.6
15	1.3	0.8	0.6
16	1.2	0.6	0.6
20	1.1	0.6	0.6
25	1.0	0.6	0.6
30	0.9	0.6	0.6
31	0.8	0.45	0.5
35	0.8	0.45	0.5
40	0.8	0.45	0.5
45	0.7	0.45	0.5
46	0.7	0.45	0.45
50	0.7	0.45	0.45
60	0.6	0.45	0.45
61–120	[e]	0.2	[e]
>120	[e]	[e]	[e]

Note: Be guided by state and local regulations.

[a]USPHS, *Manual of Septic-Tank Practice*, PHS Pub. 526, HEW, Washington, D.C., 1967. Increase leaching area by 20 percent where a garbage grinder is installed and by additional 40 percent where a home laundry machine is installed. The required length of the absorption field may be reduced by 20 percent if 12 in. of gravel is placed under the distribution lateral, or by 40 percent if 24 in. of gravel is used, provided the bottom of the trench is at least 24 in. above the highest groundwater level.

[b]USEPA, *Design Manual Onsite Wastewater Treatment and Disposal Systems*, USEPA, Cincinnati, Ohio, October 1980. Soils with percolation rates <1 min/in. can be used if the soil is replaced with a suitably thick (>2 ft) layer of loamy sand or sand. Use 6 to 15 min/in. percolation rate.

Rates based on septic tank effluent from a *domestic* waste source, are used to determine the required *trench* bottom area, or *bed* bottom area only for sands and sandy loam. Additional area credit may be given for sidewall trench area above 6 in. (but below the distributor invert). The trench or bed bottom must be 2 to 4 ft above seasonally high water table or bedrock.

[c]GLUMR, *Recommended Standards for Individual Sewage Disposal Systems*, Great Lakes-Upper Mississippi River Board of State Sanitary Engineers, 1980 Edition. Absorption trench or bed shall not be constructed in soils having a percolation rate slower than 60 min/in., or where rapid percolation may result in contamination of water-bearing formation or surface water. The percolation rate is for *trench* bottom area. For absorption *bed* use 0.6 gpd/ft^2 for percolation rate of 0 to 5 min/in. and for 6 to 10 min/in. use 0.45 gpd/ft^2. Trench or bed bottom, or seepage pit bottom, not less than 3 ft above highest groundwater level. Maximum trench width credit shall be 24 in. for design purposes.

[d]Reduce rate of 2.0 gpd/ft^2 where a well or spring water supply is downgrade; increase protective distance, and place 6 to 8 in. sandy soil on trench bottom below gravel and between gravel and sidewalls.

[e]Soil not suitable.

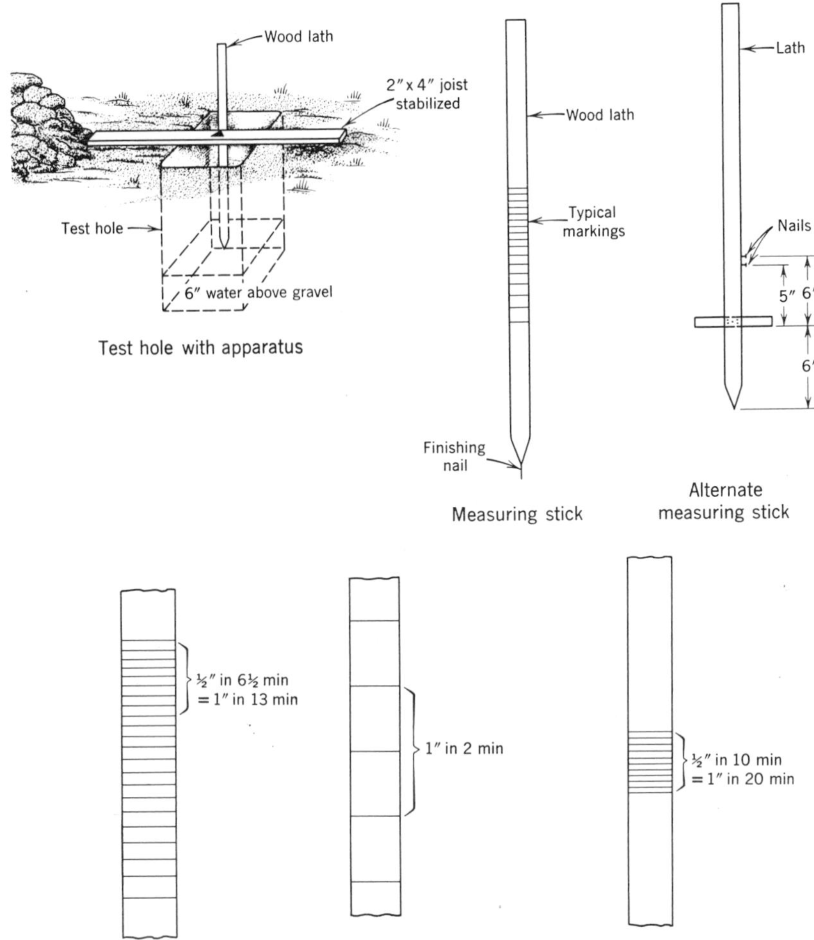

Figure 4-3 The soil percolation test. (Place 2 in. gravel in bottom of test hole.)

perform than the first method. Some sanitary engineers and sanitarians prefer to continue the test until the times for the water to fall 1 in. are within 3 to 5 percent.

Many variations and refinements of the soil percolation test, including the use of float gauges and permeability tests, have been proposed.[47,48] An ingenious method* described by Peterson[49] minimizes errors due to changes in water level in

*Simlar to the inverted carboy in a water cooler principle.

[47] *Manual of Septic-Tank Practice,* PHS Pub. 526, Washington, D.C., revised 1967, pp. 4–8, 75–76.

[48] John T. Winneberger, "Correlation of Three Techniques for Determining Soil Permeability," *J. Environ. Health,* September/October 1974, pp. 108–118. Also Rein Laak, *Wastewater Engineering Design for Unsewered Areas,* Ann Arbor Science Publishers, Inc., Ann Arbor, Mich., pp. 12–15.

[49] Michael E. Peterson, "Soil Percolation Tests," *J. Environ. Health,* January/February 1980, pp. 182–186.

the test hole and its measurement; does not require constant attention; and makes possible pre-soaking of the test hole. It is still necessary however to consider the differences in soils characteristics of a site, the need to investigate soil clues, and the depth to groundwater or an impermeable strata, and to remove the smear inside the test hole and place 2 in. of gravel on the bottom of the test hole. A perforated rigid plastic liner (5-in. diameter) surrounded by gravel is placed inside an 8-in. diameter test hole (D_h). A transparent cast acrylic pipe 3-ft high and 5-in. diameter inside (d_t) sealed at both ends, but with a rubber stopper in the fill hole at the top and a $\frac{1}{2}$-in. diameter valved tail pipe at the bottom extends to the water surface. The 5-in. diameter pipe is supported by a tripod over the test hole. The water level in the test hole is maintained at a constant depth of 8 in. by the water in the 3 ft pipe and the tail pipe. The time in minutes (t) for the water to drop in the cylinder a measured distance (h) in inches is recorded. The actual percolation time (T_a) in min/in. is determined from the equation

$$T_a = \frac{D_h^2 t,}{d_t^2 h} \qquad \text{or} \qquad T_a = 2.56\frac{t}{h}$$

In any case, a sufficient number of soil tests should be made that will give information representative of the soil. This will also make possible determination of an average percolation rate that can be used in design.

Where a small rural real estate subdivision is under consideration, approximately one hole per acre should be tested and soil borings made. More holes should be tested if the percolation results vary widely, say by more than 20 percent. If rock, clay, hardpan, or groundwater is encountered within 4 ft of the ground surface, the property should be considered unsuitable for the disposal of sewage by means of conventional subsurface absorption fields. This calls for the exercise of trained engineering judgment. A typical layout is shown in Figure 4-4.

In special cases, where tighter soil is encountered just below the bottom of the test hole, interpretation of the results might require adjustment of the allowable sewage application from an apparent, say 2.0 gpd/ft^2, to 0.5 gpd/ft^2 based on the tighter soil.

It should be pointed out that the ill repute of septic-tank absorption systems is due to improperly made and interpreted soil percolation tests, high groundwater, poor construction, or lack of maintenance, and to the use of septic tanks where they were never intended. Inadequate design and lack of inspection by regulatory agencies may contribute to the problem.

Absorption systems should not be laid in filled-in ground until it has been thoroughly settled or otherwise stabilized, and then trenches should be dug in the fill. Tests should be made in fill after at least a six-month settling period and after complete stabilization of the soil. Percolation tests cannot be made in frozen ground.

Where the ground is flat, provision should be made to drain surface water from off and around the absorption field to prevent the soil from becoming waterlogged. On steeply sloping ground, a surface water diversion ditch or berm

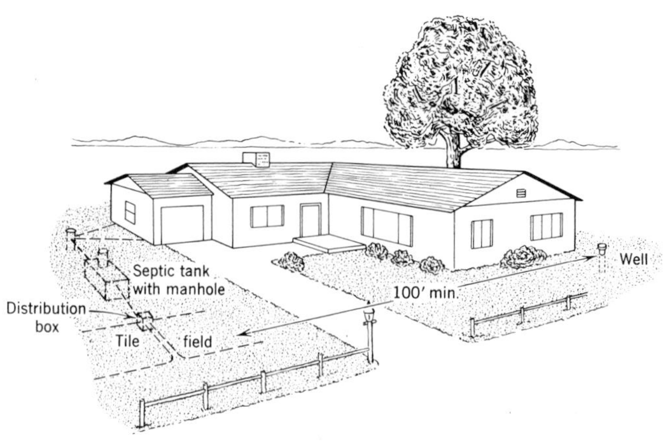

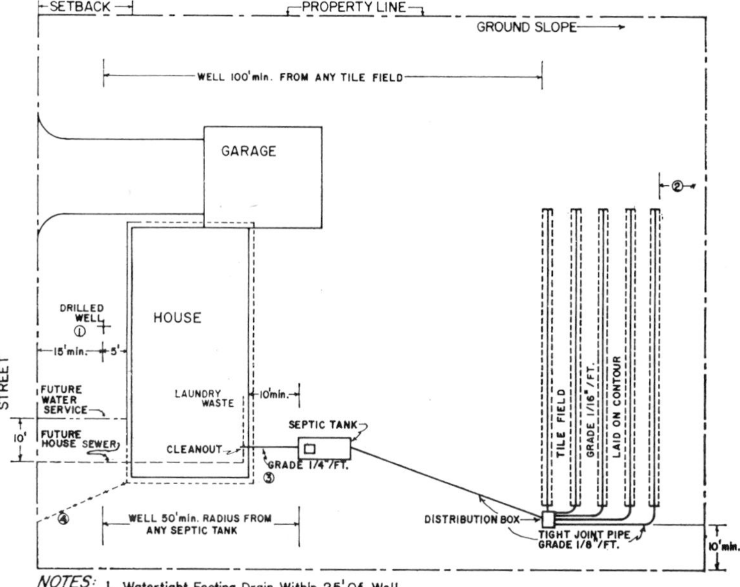

Figure 4-4 Typical private water supply and sewage disposal layouts.

Notes:
(1) Watertight footing drain within 25′ of well. (2) Tile field to be 50′ or more from any lake, swamp, ditch, or watercourse and 10′ or more from any water line under pressure. (3) Cast-iron pipe, lead caulked joints within 50′ of any well. (4) Discharge footing, roof, and cellar drainage away from sewerage system and well.

should be provided above and around the absorption field to minimize water infiltration and to prevent the absorption field from being washed out. This will also prevent silt and mud from washing into the trench during construction and coating the bottom with a relatively impervious film. It is also important to point out that the undesirable practice of walking in the bottom of trenches causes a compaction of the earth and a reduced percolation capacity. If this happens, rake the bottom to restore the original surface. Assure that the trench bottom is on a slope of $\frac{1}{16}$ in. per ft.

Where leaching pits are permitted, the soil test is made in a test hole about 1 ft^2 at the bottom and at a depth about one-half the proposed *effective depth* of the leaching pit. On completion of the test, the hole should be extended to the proposed full depth to assure that the soil at a greater depth is similar to that tested and that groundwater is not encountered. The test is made in the same manner as explained for an absorption system but at the greater depth. The results are interpreted in Table 4-2.

The design of septic tanks, subsurface absorption fields, and seepage pits is explained and illustrated in the discussion of these systems that follows.

Estimate of Sewage Flow

The sewage flow to be expected from a dwelling or other type of establishment is not constant each day. The day of the week, season of the year, habits of the people, water pressure, type and number of plumbing fixtures, and type of place or business maintained are some of the factors affecting the probable sewage flow. The design cannot be based on the minimum flow but must be based at least on the average maximum. Daily water-meter consumption figures from a similar type of establishment taken over an extended period of time, including weekends and maximum days, would be of value in arriving at a good average maximum daily flow estimate to be used in design. Caution must be used when interpreting quarterly, semiannual, or annual meter readings, as averages derived from these figures will be low unless corrected for vacation periods, weekends and holidays, seasons of the year, and so on. In the absence of actual figures, the per capita or unit estimated water flow given in Table 3-12 may be used as a guide.

Fixture bases of estimating sewage flow assume that all water used finds its way to the sewage disposal or treatment system. Adjustment should be made for lawn watering, car washing, and so forth. In one method, the total number of different types of fixtures is summarized. The sum of each type is multiplied by the usual flow from such a fixture per use or operation. The frequency of use per hour can be estimated for the type of establishment under study. Knowing the number of hours of operation daily, a rough estimate of the probable flow in gpd can be arrived at.

Another fixture basis of estimating sewage flow is given in Table 4-3. Although the fixtures refer to country clubs, public parks, and restaurants, they can be applied with modifications to similar types of establishments. The fixture bases of estimating sewage flow are useful in determining the required size drain or sewer

Table 4-3 A Fixture Basis of Estimating Sewage Flow

Type of Fixture	Gallons per Day per Fixture, Country Clubs[a]	Gallons per Hour per Fixture, Public Parks[b]	Gallons per Hour per Fixture (Average), Restaurants[c]
Shower	500	150	17
Bathtub	300	—	17
Washbasin	100	—	$8\frac{1}{2}$
Water closet	150	36	42 (flush valve) 21 (flush tank)
Urinal	100	10	21
Faucet	—	15	$8\frac{1}{2}$, 21 (hose bib)
Sink	50	—	17 (kitchen)

[a]John E. Kiker, Jr., "Subsurface Sewage Disposal," *Fla. Eng. Exp. Sta.*, Bull. No. 23 (December 1948).
[b]National Park Service.
[c]After M. C. Nottingham Companies, California.

line and also as a check on other methods. Fixture unit values in Table 4-4 can also be used with Figure 3-25 for the same purpose.

A third fixture unit basis of estimating sewage flow is that described in Chapter 3, Water Supply, using the probability curves developed by Hunter. These flows are somewhat high, being based on estimated peak discharge. An analysis made by Wyly[50] is based on the estimated average discharge. For two-bath houses with automatic dishwasher, clothes washer, and garbage grinder, for a total of 19 fixture units per home, the discharge may be 1.6 gpm for 1 home, 7.6 gpm for 5 homes, 15.2 gpm for 10 homes, 30.4 gpm for 20 homes, 76 gpm for 50 homes, and 152 gpm for 100 homes. See Peak Demand Estimates in Chapter 3.

House Sewer and Plumbing

The house or building sewer is that part of the building drainage system carrying sewage that extends from the septic tank or public sewer to a point 3 ft out from the foundation wall. That portion of the drainage system extending from the house sewer horizontally into the structure is the house or building drain. The recommended size of the house or building sewer and drain is given in Table 4-5, although the local plumbing or building code will govern, where one has been adopted. In general, the house drain and building sewer should be not less than 4-in. bell and spigot cast-iron pipe with lead-caulked or equal joints. Other plastic or composition pipe constructed of durable material and laid with tight joints is also used.

Increasing fittings with smooth joints should be used where the pipe size increases in diameter. Sewer lines should be laid in a straight line to the septic tank

[50]Robert S. Wyly, *Hydraulics of 6- and 8-inch Diameter Sewer Laterals,* Building Research Institute, BRAB, NAS-NRC, 2102 Constitution Ave., Washington, D.C., November 27, 1956.

Table 4-4 Sanitary Drainage Fixture Unit Values

Fixture or Group	Fixture Unit Value
Bathroom group consisting of a lavatory, bathtub or shower stall, and a water closet (direct flush valve)	8
Bathroom group consisting of a lavatory, bathtub or shower stall, and a water closet (flush tank)	6
Bathtub with $1\frac{1}{2}''$ trap	2
Bathtub with 2″ trap	3
Bidet with $1\frac{1}{2}''$ trap	3
Combination sink and wash tray with $1\frac{1}{2}''$ trap	3
Combination sink and wash tray with food waste grinder unit (separate $1\frac{1}{2}''$ trap for each unit)	4
Dental unit or cuspidor	1
Dental lavatory	1
Drinking fountain	$\frac{1}{2}$
Dishwasher, domestic type	2
Floor drain	1
Kitchen sink, domestic type	2
Kitchen sink, domestic type with food waste grinder unit	3
Lavatory with $1\frac{1}{2}''$ waste plug outlet	1
Lavatory with $1\frac{1}{4}''$ or $1\frac{3}{8}''$ waste plug outlet	2
Lavatory (barber shop, beauty parlor, or surgeon's)	2
Lavatory, multiple type (wash fountain or wash sink), per each equivalent lavatory unit	2
Laundry trap (1 or 2 compartments)	2
Shower stall	2
Showers (group) per head	3
Sink (surgeon's)	3
Sink (flushing rim type, direct flush valve)	8
Sink (service type with floor outlet trap standard)	3
Sink (service type with P trap)	2
Sink (pot, scullery, or similar type)	4
Urinal (pedestal, syphon jet, blowout type with direct flush valve)	8
Urinal (wall lip type, flush tank)	4
Urinal (stall, washout type, flush tank)	4
Water closet (direct flush valve)	8
Water closet (flush tank)	4
Swimming pools, per each 1000-gal capacity	1
Unlisted fixture, $1\frac{1}{4}''$ or less fixture drain or trap size	1
Unlisted fixture, $1\frac{1}{2}''$ fixture drain or trap size	2
Unlisted fixture, 2″ fixture drain or trap size	3
Unlisted fixture, $2\frac{1}{2}''$ fixture drain or trap size	4
Unlisted fixture, 3″ fixture drain or trap size	5
Unlisted fixture, 4″ fixture drain or trap size	6

Source: *Code Manual for the State Building Construction Code,* State of New York, Division of Housing and Community Renewal, New York, August 1, 1977, p. 5–7.

Note: Values given are for continuous flow. For a continuous or semicontinuous flow into a drainage system, such as from a pump, pump ejector, or similar device, two fixture units shall be allowed for each gpm/of flow. One fixture unit is equivalent to 7.5 gpm.

Table 4-5 Fixture-Unit Load to Building Drains and Sewers

Diameter of Pipe (in.)	Maximum Number of Fixture Units that may be Connected to Any Portion[a] of the Building Drain or the Building Sewer: Fall per ft			
	$\frac{1}{16}$-in.	$\frac{1}{8}$-in.	$\frac{1}{4}$-in.	$\frac{1}{2}$-in.
2	—	—	21	26
$2\frac{1}{2}$	—	—	24	31
3	—	20[b]	27[b]	36[b]
4	—	180	216	250
5	—	390	480	575
6	—	700	840	1,000
8	1,400	1,600	1,920	2,300

Source: Adapted from "Report of the Coordinating Committee for a National Plumbing Code," U.S. Dept. of Commerce, Washington, D.C., 1951.
[a]Includes branches of the building drain.
[b]Not over 2 water closets.

where possible. If bends are necessary, use one or two 45° ells, as may be needed, and provide a cleanout. A manhole is sometimes preferable. A cleanout should also be provided at the end of the building drain in the basement and ahead of the septic tank when the tank is located 30 or more ft from the cleanout on the building drain. A cleanout may be provided on a buried sewer line by installing a "T" fitting in the line with the vertical leg up and connecting to it a section of pipe extending to the ground surface. The fitting, sewer joints, and pipe extension should be encased in 6 in. of concrete. All cleanouts should have tight-fitting brass caps. See Figures 4-5 and 4-6.

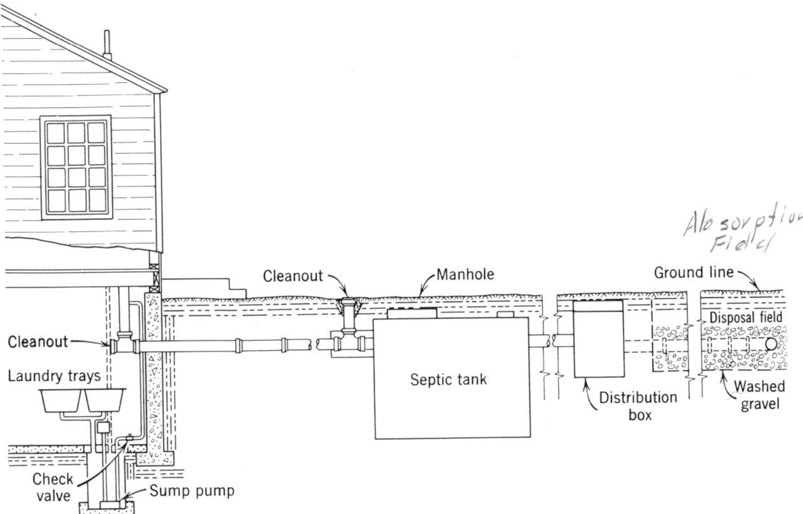

Figure 4-5 Selection showing sewage disposal system. (See also Figure 4-21.)

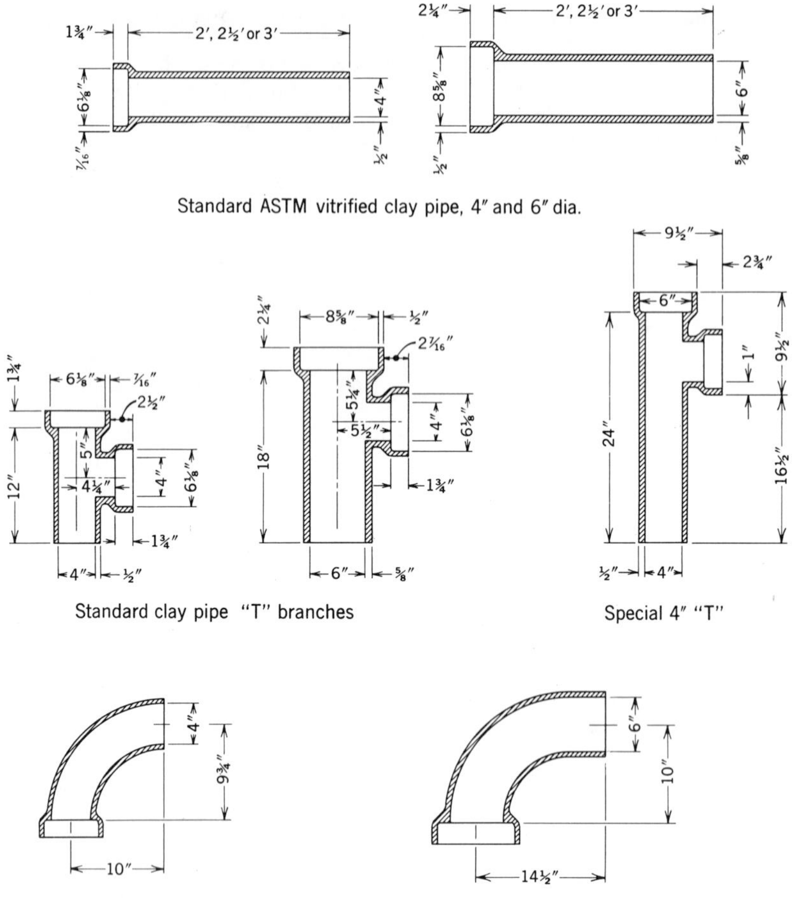

Standard clay pipe long radius elbows, 4″ and 6″ dia.

Figure 4-6 Details of some clay pipe fittings. (Cement mortar joints are unsatisfactory where sulfides are expected.)

Grease Trap

A grease trap, interceptor, or separator is a unit designed to remove grease and fat from kitchen wastes. Liquid wastes leaving properly designed and maintained units should not cause clogging of pipes nor have a harmful effect on the bacterial and settling action in a septic tank.

Grease traps are of the septic-tank and commercial types. The commercial types are of questionable value, particularly with the general use of detergents for dishwashing in restaurants. In the septic-tank type, use is made of the cooling separating, and congealing effect obtained when the warm or hot greasy liquid wastes from the kitchen mix with the cooler liquid standing in the tank. There is also a natural tendency, if mixing is not too rapid, for the warmer liquid to rise and the cooler liquid, from which grease has been separated, to settle and be carried out with the food particles. A grease trap is unnecessary in the average private home. The small quantity of fat and grease that does find its way into the kitchen drain is mixed with soaps and detergents and is difficult to separate. In any case, such small quantities would not be harmful when allowed to enter a proper size septic tank. Grease traps of the septic-tank type should be provided, however, in restaurants and similar establishments where the quantity of grease and fats in liquid wastes is likely to be large unless special provision is made in the treatment plant. All grease traps should discharge into the building sewer ahead of the septic tank. Grease traps should be located within 20 or 30 ft from the plumbing fixtures served to prevent congealing and clogging of waste lines.

The septic-tank type of grease trap's capacity is made equal to the maximum volume of water used in a kitchen during a mealtime period.* The type of meals served and kitchen equipment used should be taken into consideration. A figure of $2\frac{1}{2}$- or 3-gal capacity per meal served during a mealtime is frequently used. The tank should be located in an accessible place outside the building, using the same precautions as for the septic tank. It should have a tight-fitting cover and be light in weight. Heavy cast iron and steel make satisfactory covers. When covers do not fit tightly, a tar or rubber seal around the cover or a layer of clay soil over the cover may be necessary to eliminate odors.

The construction of several septic-tank-type grease traps is shown in Figure 4-7. Large tanks should have two compartments. Grease traps can be built on the job out of concrete or brick masonry, or can be prefabricated out of metal, concrete, asbestos cement, or terra cotta, with inlet and outlet arrangements as shown in the sketches. Because of the greater capacity of septic-tank-type grease traps they do not require cleaning as frequently as the commercial type. Nevertheless, they must be cleaned so as not to greatly reduce the liquid volume available for cooling of the greasy wastes entering and, of course, to prevent large quantities of grease being carried over to the septic tank. The frequency of cleaning should be determined at each establishment during operation, when the grease occupies one-half the liquid depth of the tank. Cleaning at monthly intervals may be sufficient; but experience should dictate the frequency. One person should be given this responsibility, and a supervisor should check to see that the job is done. Grease removed from a grease trap may be disposed of with the garbage, rendered and sold, or thoroughly buried as explained under Prives, Latrines, and Waterless Toilets, depending on local conditions.

Septic Tank

A septic tank is a watertight tank designed to slow down the movement of raw sewage and wastes passing through so that solids can separate or settle out and be broken down by liquefaction and anaerobic bacterial action. It does not purify

*Another design basis is given in *Design Manual Onsite Wastewater Treatment and Disposal Systems,* USEPA, October 1980, pp. 321–327.

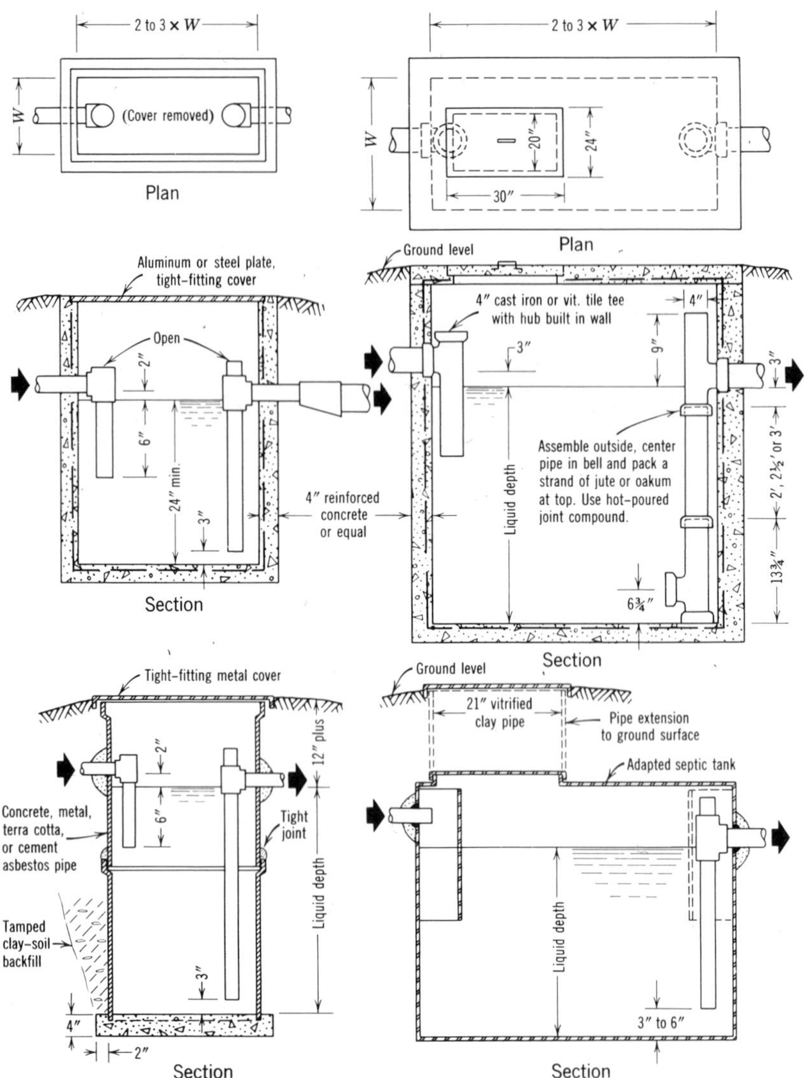

Figure 4-7 Grease trap details.

the sewage, eliminate odors, or destroy all solid matter. The septic tank simply conditions the sewage so that it can be disposed of normally to a subsurface absorption system without prematurely clogging the system. Suspended solids removal is 50 to 70 percent; 5-day BOD removal is about 60 percent.

Recommended septic tank sizes based on estimated daily flows are given in Tables 4-6 and 4-7. The septic tank should have a liquid volume of not less than 750 gal. When a tank is constructed on the job, its liquid volume can be increased at a nominal extra cost thereby providing capacity for possible future additional flow, garbage grinder, and sludge storage.

Table 4-6 Water Supply and Sewerage Schedule (Use Combination of Headings that Fit Local Conditions)

Population			Septic Tank-Minimum[b,c]				Tile Field Laterals[c,d]	
Bedrooms	Persons	Sewage Flow GPD[a]	Length (ft)	Width (ft)	Depth (ft)	Volume (gal)	No. Length (ft)	Trench Width (in.)
2 or less	4	300	$7\frac{1}{2}$	$3\frac{1}{2}$	4	750	(Determined by site and soil percolation test)	
3	6	450	9	4	4	1000		
4	8	600	11	4	4	1250		
5	10	750	$10\frac{1}{2}$	5	4	1500		

Population		Leaching Pit System[c,d]			Water Supply—Well, Drilled[f]		
Bedrooms	Persons	No. Pits	Size	Depth	Service[e]	Pump Size (gal/hr)	Pres. Tank (gal)
2 or less	4	(Determined by site and soil percolation test)			$\frac{3}{4}$ in.	250	42
3	6				$\frac{3}{4}$	300	82
4	8				1	360	82
5	10				$1\frac{1}{4}$	450	120

Population		Sand Filter System[c,g]			Chlorine Contact Tank[c]			Sump Pump Float Setting (gal)[c]
Bedrooms	Persons	Length (ft)	Width (ft)	Area (sq ft)	Length (ft)	Width (ft)	Depth (in.)	
2 or less	4	$21\frac{1}{2}$	12	260	3	2	8	20
3	6	21	18	390	3	2	12	30
3	6	$32\frac{1}{2}$	12	390	3	2	12	30
4	8	30	18	520	3	2	16	40
4	8	43	12	520	3	2	16	40
5	10	36	18	650	3	2	20	50
5	10	54	12	650	3	2	20	50

[a]The design basis is 75 gal per person and 150 gal per bedroom.

[b]Includes provision for home garbage grinder and laundry machine. Larger than minimum size septic tank is strongly recommended.

[c]See detail drawings for construction specifications.

[d]Based on the results of soil percolation tests. Discharge all kitchen, bath, and laundry wastes through the septic tank, but *exclude* roof and footing drainage, surface and groundwater, and softener wastes.

[e]Use next larger diameter house service line if water is corrosive or hard, if service line is 50 to 100 ft long, if two bathrooms are provided, or flush valve is used for water closet. These pipe sizes are based on the use of brass or copper pipe; use next larger size if iron pipe is proposed.

[f]The minimum dependable well yield should be 3 to 7 gpm.

[g]Sand filter normally should not be used. Reserve for compelling circumstances to relieve an impending or existing public health hazard.

Table 4-7 Suggested Tank Dimensions

Gallons	Width (ft)	Length (ft)	Depth (ft)
1,000	4	9	4
1,250	4	11	4
1,500	5	$10\frac{1}{2}$	4
1,750	5	12	4
2,000	5	14	4
2,250	6	13	4
2,500	6	$14\frac{1}{2}$	4
2,750	6	16	4
3,000	6	17	4
3,250	6	15	5
3,500	6	16	5
3,750	6	17	5
4,000	7	16	5
5,000	7	$19\frac{1}{2}$	5
6,000	8	$20\frac{1}{2}$	5
7,000	8	24	5
8,000	8	23	6
10,000	8	28	6

Concrete Details:

1. Concrete for top and bottom 4-in. thick for 2000-gal tank or smaller and 6 in. for 2,000- to 10,000-gal tank.
2. Concrete for sides and ends 6 in. thick for 6000-gal tank or smaller and 8 in. for 6,000- to 10,000-gal tank.
3. Reinforce with $\frac{3}{8}$-in. deformed rods 4 in. on center both ways for ordinary loading. Place rods 1 in. above bottom of top slab and 1 in. in from inside of tank for sides, ends, and bottom. Overlap $\frac{3}{8}$-in. rods 15 in. where needed. Adjust steel for local conditions.
4. Concrete mix: 1 bag cement to $2\frac{1}{4}$ ft^3 sand to 3 ft^3 gravel with 5 gal water for moist sand.

The detention time for large septic tanks* should not be less than 12 hr. A 24-to 72-hr detention time is recommended. Schools, camps, theaters, factories, and fairgrounds are examples of places where the total or a very large proportion of the daily flow takes place within a few hours. For example if the total daily flow takes place over a period of 6 hr ($\frac{1}{4}$ of 24 hr), the septic tank should have a liquid volume equal to four times the 6-hr flow to provide a detention of 24 hr over the period of actual use. The larger tank would minimize scouring of septic tank sludge and scum and carry-over of solids into the absorption system.

If the septic tank is to receive ground garbage, its capacity should be increased by at least 50 percent. Some authorities recommend a 30 percent increase.

Septic tanks can be constructed of concrete, concrete block, or brick or stone masonry on the job as explained in Figure 4-8 and 4-9 and Tables 4-6 and 4-7. Masonry tanks, other than concrete, require two $\frac{1}{2}$-in. cement plaster coats on the

*Inspect every 6 months and clean, if necessary.

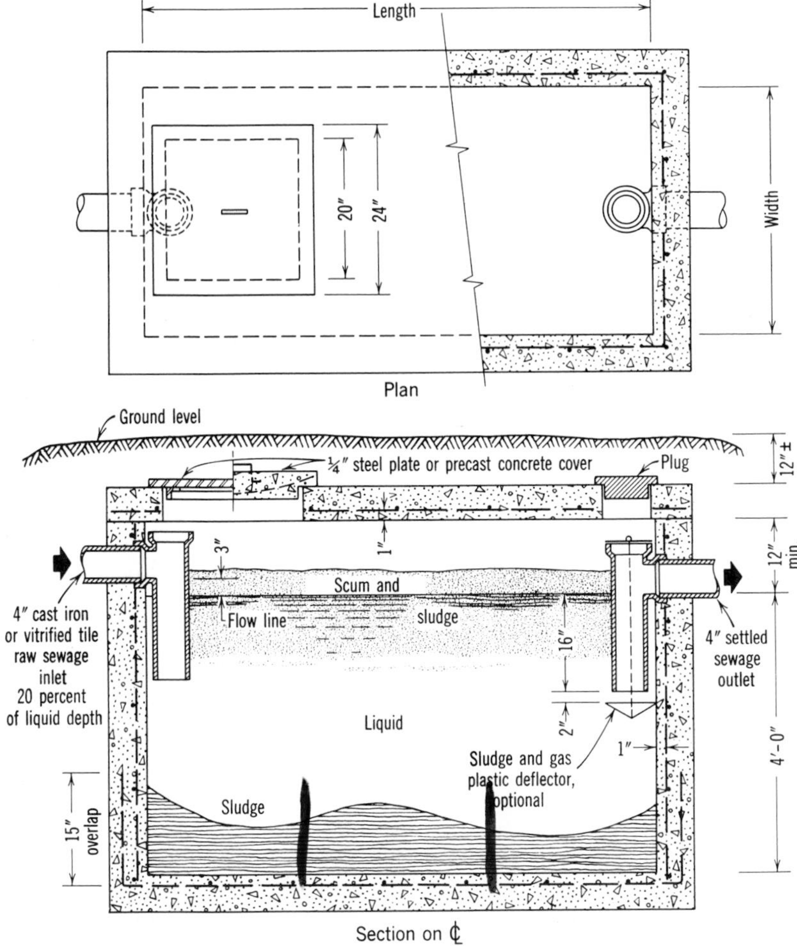

Figure 4-8 Details for small septic tanks. Recommended construction for small septic tanks:

Top—Reinforced concrete poured 3- to 4-in. thick with two $\frac{3}{8}$-in. steel rods per ft, or equivalent, and a 20 × 20-in. manhole over inlet, or precast reinforced concrete 1-ft slabs with sealed joints.

Bottom—Reinforced concrete 4-in. thick with reinforcing as in "top" or plain poured concrete 6-in. thick.

Walls—Reinforced concrete poured 4-in. thick with $\frac{3}{8}$-in. steel rods on 6-in. centers both ways, or equivalent; plain poured concrete 6-in. thick; 8-in. brick masonry with 1-in. cement plaster inside finish; or 8-in. stone concrete blocks with 1-in. cement plaster inside finish and cells filled with mortar.

Concrete Mix—One bag of cement to $2\frac{1}{4}$ ft^3 of sand to 3 ft^3 of gravel with 5 gal of water for moist sand. Use 1:3 cement mortar for masonry and 1:2 mortar for plaster finish. Gravel or crushed stone and sand shall be clean, hard material. Gravel shall be $\frac{1}{4}$ to $1\frac{1}{2}$ in. in size; sand from fine to $\frac{1}{4}$ in.

Bedding—At least 3 in. of sand or pea gravel, leveled.

Cover—Raise cover by means of collar if tank greater than 12 in. below ground.

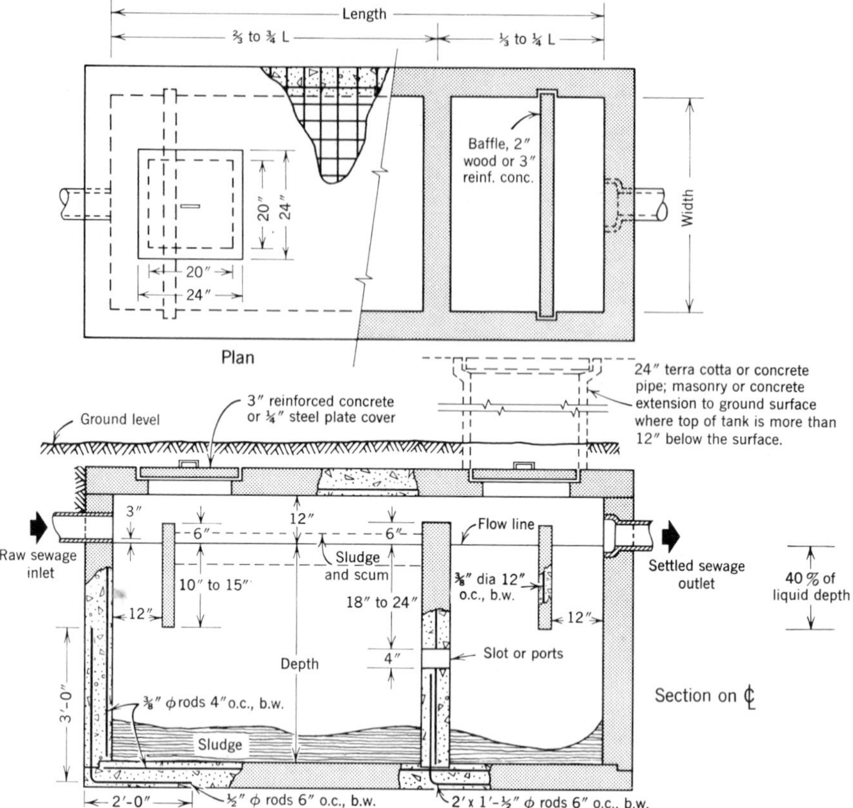

Figure 4-9 Details for large septic tanks. (See Table 4-7.)

inside to provide a smooth watertight finish. Precast reinforced concrete and commercial tanks of metal, fiberglass, and other composition materials are also available. Since some metal tanks have a limited life, it is advisable that their purchase be predicated on their meeting certain minimum specifications. These should include a guaranteed 20-yr life expectancy, not less than 12- or 14-gauge metal thickness, thorough covering both inside and outside with a heavy, continuous, protective coating resistant to acids, and a minimum liquid capacity of 750 gal. Metal septic tanks manufactured in accordance with the Department of Commerce Commercial Standard, provided with a minimum 16- to 18-in. manhole, represent a great improvement over the ordinary metal tank. Large metal tanks, greater than 1000 gal capacity, should be 10 to 8 gauge metal. In any case, it should be kept in mind that metal tanks, especially the baffles, have a limited life.

The depth of septic tanks and ratio of width to length recommended by most health departments are very similar. A liquid depth of 4 ft and a ratio of width to length of 1:2 or 1:3 is common. Depths as shallow as 30 in. and as deep as 6 ft have been found satisfactory. Compartmented tanks are somewhat more efficient. Open-tee inlets and outlets as shown in Figure 4-8 are generally used in small tanks, and high quality reinforced concrete baffled inlets and outlets as shown in Figure 4-9 are recommended for the larger tanks.[51] Precast open concrete tees or baffles have in some instances disintegrated or fallen off; vitrified clay, cast iron, PVC, ABS, or PE tees should be used. Cement mortar joints are unsatisfactory. A better distribution of flow and detention is obtained in the larger tank with the baffle arrangement. A minimum 16-in. manhole over the inlet of a small tank, and a 20-to 24-in. manhole over both the inlet and outlet of a larger tank, constructed with a top slab poured monolithically with the sides, is preferred to a sectional slab top. The sectional slab top can, however, be more easily purchased or constructed on a flat surface over a plastic sheet with a minimum of form lumber. Joints will require a seal to prevent the entrance of surface water into the tank.

An efficient septic tank design should provide for a detention period longer than 24 hr; an outlet configuration with a gas baffle (see Figure 4-8); maximized surface area/depth ratio for all chambers (ratio more than 2); and a multichamber tank with interconnections similar to the outlet design (open-tee inlet and outlet).[52]

The elevation of the septic tank and the inlet should be selected and established with regard to the landscaping; elevations of sewers that discharge to the septic tank; elevation of dosing tank outlet pipe inverts where used; location available and elevation of the area selected for the disposal or treatment system; and the high-water level of groundwater and nearby watercourses, to make pumping unnecessary, provided the topography makes this possible. On the other hand, if pumping is required it should be taken into consideration in the initial design and every advantage taken of this necessity to reduce excavations and shorten and straighten lines. Figure 4-10 gives design information in a form letter.

Care of Septic-Tank and Subsurface Absorption System

Proper maintenance of a properly designed and constructed septic-tank system is the best assurance of satisfactory operation of a subsurface sewage disposal or treatment system and prevention of sudden replacement expenses.

A septic tank for a private home will generally require cleaning every 3 to 5 years, but in any case it should be inspected about once a year. Septic tanks serving commercial operations should be inspected at least every 6 months. When the depth of settled sludge or floating scum approaches the depth given in Table 4-8, the tank needs cleaning.[52,53] Sludge accumulation in a normal home

[51] *Manual of Septic-Tank Practice*, PHS Pub. No. 526 (1967), states that "the outlet device should generally extend to a distance below the surface equal to 40 percent of the liquid depth. For horizontal, cylindrical tanks, this should be reduced to 35 percent." The inlet should penetrate at least 6 in. below the liquid level but not greater than the outlet. The distance between the liquid line and underside of the tank (air space) should be approximately 20 percent of the liquid depth. In horizontal, cylindrical tanks, the area should be 15 percent of the total circle. Sludge accumulation $S = 17 + 7.5t$, in which S = sludge in gal per capita and t = years after cleaning.

[52] Rein Laak, "Multichamber Septic Tanks," *J. Environ. Eng. Div., ASCE*, June 1980, pp. 539–546.

[53] S. R. Wiebel, C. P. Straub, and J. R. Thoman, *Studies on Household Sewage Disposal Systems, Part 1*, Report: Federal Security Agency, PHS, Environmental Health Center, Cincinnati, Ohio, 1949.

RENSSELAER COUNTY DEPARTMENT OF HEALTH
DIVISION OF ENVIRONMENTAL HYGIENE

Seventh Ave and State St. Troy, N. Y.

Re: Private Sewerage System

Dear Sir:

As you know, Mr. .... of this department made an investigation of your property, including soil percolation tests, on .... Our recommendations for an adequate system to serve your .... bedroom house are given on the attached sketches and are checked below.

☐ 1. The design is based upon soil tests indicating that .... gallons of settled sewage can be applied per square foot in a subsurface (tile field system) (leaching pit system) (sand filter system) (tile field system in fill).

☐ 2. The septic tank should have a liquid capacity of .... gallons, and be of an approved type with a manhole over the inlet. It should be followed by a small distribution box. Extension of the septic tank manhole and top of the distribution box close to the surface will make easy inspection and if necessary cleaning of the septic tank, required usually every 2 to 3 years.

☐ 3. The tile field following the septic tank and distribution box is to consist of .... laterals of open joint or perforated pipe laid in washed gravel or crushed stone at least .... feet on centers. Each lateral is to be .... feet long laid in trenches .... inches wide.

☐ 4. The leaching pit system following the septic tank and distribution box is to consist of .... pits. Each pit is to have a depth below the inlet of .... feet and an inside .... of .... feet. If standard 8 inch by 16 inch building blocks are used to construct a pit, then .... blocks will be needed below the inlet for each pit. Coarse gravel shall be placed around the outside wall of the pit.

☐ 5. Suitable gravelly loam fill containing topsoil and having an area of at least .... square feet may be provided in locations where the soil is unsuitable provided at least 12 to 18 inches of natural porous earth exists. Sufficient fill shall be provided to bring the bottom of trenches 2 feet above the clay, rock or ground water. After stabilization of the fill, a tile field is to be constructed with laterals not closer than 20 feet to the edge of the feathered fill. The tile field following the septic tank and distribution box is to consist of .... laterals of open joint or perforated pipe laid .... feet apart. Each lateral is to be .... feet long laid in trenches .... inches wide. Establish grades of all pipes to assure gravity flow of sewage from the house to the tile field without pumping.

☐ 6. The subsurface sand filter following the septic tank and distribution box is to be .... feet long and .... feet wide constructed as shown on the sketches, with a chlorine contact-inspection box on the outlet. The architect, builder and contractor shall establish all grades so as to assure gravity flow from the dwelling through the sewage treatment system to the outlet. Submit a 2-pound sand sample fifteen days before construction for Health Department approval.

The bottom of the tile field trenches and leaching pits must be at least 2 feet above ground water, clay, hardpan, rock, etc. and the bottom of the sand filter above ground water for the system to function properly. Additional important information is given on the back of this sheet and on the attached sketches. Please advise this office when the work is expected to start. We can then arrange to have our representative inspect the construction for compliance with the health department standards and thus help you obtain the best possible job.

HELP MAINTAIN A CLEAN, HEALTHFUL ENVIRONMENT

Figure 4-10 A typical form letter for design of a private sewerage system.

septic tank has been estimated at 69 to 80 liters (18 to 21 gal) per person per year.[53,54] The Ontario Ministry of the Environment set the permissible highest level of sludge at 0.45 m below the bottom of the outlet fitting. Gray water septic tank sludge was found to accumulate at the rate of 8.3 liters (2.2 gal) per person per year, and black wastewater sludge at the rate of 65.7 liters (17.4 gal) per person per year.[55] A long pole having a small board about 8 in.^{2} nailed to the bottom to make a plunger, with Turkish toweling wrapped around the lower 18

[54]M. Brandes, *Accumulation Rate and Characteristics of Septic Tank Sludge and Septage,* Ontario Ministry of the Environment, Research Report W63, 1977, Toronto, Canada. Also *J. Water Pollut. Control Fed.,* May 1978, pp. 936–943.

[55]M. Brandes, *Characteristics of Effluents from Separate Septic Tanks Treating Gray Water and Black Water from the Same House,* Ontario Ministry of the Environment, October 1977.

Table 4-8 Allowable Sludge Accumulation (in.)

Tank Capacity (gal)	Sludge Depth (in.) Tank Liquid Depth (in.) 30	36	48	60
250	4			
300	5	6		
400	7	9	10	
500	8	11	13	15
600	10	14	16	18
750	13	16	19	23
900	14	18	22	26
1000	14	18	23	28
1250		18	24	30

Source: Adapted from *Manual of Septic-Tank Practice,* PHS Pub. 526, DHEW, Cincinnati, Ohio, 1967.

This Table assumes the outlet baffle or tee depth below the flow line is 40 percent of the tank liquid depth. *Clean the tank when the bottom of the scum layer builds up to within 3 in. of the bottom of the baffle or tee outlet, or when the sludge depth approaches the depth given in this table.* For example, a tank 48 in. deep with a capacity of 750 gal will require cleaning when the sludge depth reaches 19 in.

in. of the pole, can be used to measure the sludge depth and floating-scum thickness. The appearance of particles or scum in the effluent from a septic tank going through a distribution box is also an indication of the need for cleaning. Routine inspection and cleaning will prevent solids from being carried over and clogging the treatment or leaching systems. The larger the septic tank above the minimum, the less frequent is the need for cleaning. A contractual arrangement for annual inspection and cleaning as needed and noted above is a good investment. Community cooperation in this regard should be encouraged. It is best to clean a septic tank during the dry months of the year. The groundwater level should be low (to prevent possible flotation of the tank) and bacterial adjustment will proceed faster in warm weather.

Septic tanks are generally cleaned by septic-tank cleaning firms. They mix and pump the entire contents, referred to as septage, out into a tank truck with special equipment. Care must be taken to prevent spillage and consequent pollution of the surrounding ground. The contents should be emptied into a sanitary sewer system or sludge digestion tank, if permissible, or in a shallow trench or pit at a point 200 ft or more and downgrade from water sources and covered with at least 24 in. of compacted earth, as approved by the regulatory agency. Sludge sticking to the inside of a tank that has just been cleaned would have a seeding effect and assist in renewing the bacterial activity in the septic tank. The septic tank should not be scrubbed clean.

Alternative methods[56] for the disposal of septage consist of ridge-and-furrow, spray irrigation, plow-furrow-cover, subsurface injection, and sanitary landfill under controlled conditions, with storage when necessary in colder climates when the ground is frozen to prevent runoff. Also possible are leaching lagoons in which the sludge is periodically removed, and disposal lagoons in which the sludge can be removed or allowed to dry for disposal in a sanitary landfill or covered over with 2 ft of earth. Septage treatment facilities include aerated lagoons, facultative lagoons, composting with dry organic matter for moisture control, chemical treatment using lime and ferric chloride, and other proprietary processes.

A precaution must be taken before a tank is emptied. If the groundwater level is high the tank might float to the surface causing considerable damage. Under such conditions the cleaning of the tank should be postponed until the groundwater level is lowered.

Safety and Other Precautions

Excavations such as for septic tanks, privies, leaching pits, pump wells, and cesspools can create a safety hazard to the worker and to passing people, especially children. No person should be permitted to work in a trench or pit 5 ft or more in depth which has sides or banks with slopes steeper than 45° unless the sides or banks are supported with sheeting or shoring. Any excavation in sand, silt, loam or clay 3 ft or more in depth needs side-wall protection to prevent cave-in; excavated material should be placed at least 24 in. back from the edges. Where the excavation is left unattended a fence or a barricade should be placed around the opening, or the opening covered with properly supported 2-in. planking or $\frac{3}{4}$-in. exterior-grade plywood. If sheeting is used and extended 42 in. above the adjacent ground, other barricades are not necessary.

An individual should not enter a septic tank, septic privy, pump well, or aqua privy tank that has been emptied. Cases of asphyxiation and death have been reported due to the lack of adequate oxygen in the emptied tank. If it should become necessary to inspect or make repairs, at least three strong individuals should be present. The tank should first be checked with a gas detector* and thoroughly ventilated using a blower. Then two persons should remain on top and the other making the inspection or repairs should have a safety harness or a strong rope tied around his waist with the other end connected to a pulley supported by a tripod so that he can be hauled out in case of trouble. The tank should not be left uncovered or unguarded as small children or pets would be attracted and possibly fall in. See Safety, later in this chapter.

Certain chemicals should not be added to septic-tank systems. The use of one gal of sulfuric acid to "unclog" a home septic tank system resulted in reaction with sulfides present and the release of toxic fumes (H_2S) which overcame three

*Hazardous gas monitoring equipment is available to detect the presence of and measure the concentration of methane, hydrogen sulfide, and lack of oxygen. Check also liquid manure tanks.

[56] Ivan A. Cooper and Joseph W. Razek, *Alternative for Small Wastewater Treatment Systems,* USEPA, EPA-625/4-77-011, October 1977, pp. 61–90.

and killed two persons.[57] The mixing of household cleaning compounds such as chlorine bleach and ammonia, caustic soda (lye), or similar cleaners is dangerous as toxic gases such as chlorine are released. This can cause injury to the throat and lungs and possibly permanent damage and death. Soap, drain solvents, disinfectants, and similar materials used individually for household purposes are not harmful to septic tank operation unless used in large quantities. Organic solvents and cleaners could contaminate the groundwater and well water supplies.

There may be occasions when the level of the contents in a septic privy drops below the overflow level, or where the water level in an aqua privy drops below the bottom of the squat plate funnel and pipe. In such circumstances the use of special chemical compounds to promote liquifcation and/or to control odors may be hazardous. The gases given off may be explosive and may be set off by a discarded match or lighted cigarette. It is safer to add water and mix the contents with a long pole.

Salt or brine from softening units in amounts as little as 1.2 percent retards temporarily the bacterial action in the septic tank and tends to build up in the sludge, but the salt is gradually flushed out as the sludge digests, rises, and falls. However, the salt (sodium) tends to cause soil clogging. Therefore, it is not prudent to discharge brine waste to the septic tank.

High weeds, brush, shrubbery, and trees, although consumers of groundwater, should not be permitted to grow over an absorption system or sand filter system. It is better to crown the bed, seed the area to grass, and build up a lawn. Sunlight and exposure to winds is beneficial as it encourages evapotranspiration.

If trees are near the sewage disposal system, difficulty with roots entering poorly joined sewer lines can be anticipated. Lead-caulked cast-iron pipe, a sulphur base or bituminous pipe joint compound, mechanical clay pipe joints, copper rings over joints, and lump copper sulfate in pipe trenches have been found effective in resisting the entrance of roots into pipe joints. Roots will penetrate first into the gravel in absorption field trenches rather than into the pipe. About 2 to 3 lb of copper sulfate crystals flushed down the toilet bowl once a year will destroy roots the solution comes into contact with, but will not prevent new roots from entering. The application of the chemical should be done at a time, such as late in the evening, when the maximum contact time can be obtained before dilution. Copper sulfate will corrode chrome, iron, and brass, hence it should not be allowed to come into contact with these metals. Cast iron is not affected to any appreciable extent. Some time must elapse before the roots are killed and broken off. Copper sulfate in the recommended dosage will not interfere with operation of the septic tank.[58] The cutting or mechanical removal of roots in sewers tends to increase root growth and size, leading to more problems and sewer repair.

[57] "Two dead, one critical from fumes inhalation," Endicott, N.Y., Associated Press, November 5, 1978.

[58] Don. E. Bloodgood, "Tree Roots, Copper Sulfate, and Septic Tanks," *Water and Sewage Works*, **99**. 190–193 (May 1952).

Flooding of sewer lines with scalding water (180° to 210°F) will kill roots subjected to a 30-min soak at 170°F. A portable steam generator is needed to reheat recirculated water to maintain the water temperature and compensate for heat loss. A temperature of 122°F will kill most plant tissue.[59]

Hydrogen peroxide has been used to oxidize the sludge and organic growths in clogged distribution lines and trenches. If used it should be handled with extreme care. Hydrogen peroxide is a strong oxidizing agent and is potentially explosive. See page 389.

Causes of Failure of Septic-Tank System and Corrective Measures

Common causes of septic tank system failures are seasonal high ground water; carry-over of solids into the absorption field due to use of septic tank cleaning compounds, lack of routine cleaning of the septic tank, or outlet baffle loss; leaking plumbing fixtures; excessive water use, connected roof and footing drains, and system overloading; uneven settlement of the septic tank, connecting pipe, or distribution box; and improper design and construction of the absorption system.

Corrective measures, once the cause is identified, might include water conservation measures such as reduced water usage, low-flush toilets, low-flow shower heads, reduced water pressure, faucet aerators, spray taps, and use of commercial laundromat; cleaning of septic tank and flushing out distribution lines (*do not empty tank if groundwater level is high—tank will float to surface*) and installation of additional leaching lines between existing lines; installation of a separate absorption system and division box or gate for alternate use with the existing system; lowering the water table with curtain drains; discontinuation of use of septic-tank cleaning compounds, replace corroded or disintegrated baffles with terra cotta, cast iron, ABS, or PVC tees, and clean septic tank every 3 years; and disconnect roof, footing, and area drains.

Use of Additives

Compounds that are supposed to make the cleaning of tanks unnecessary may actually cause solids to be carried over into the absorption or treatment system and the penetration of fine solids into the soil infiltrative surface with resultant clogging. A grab sample collected from a septic tank serving a 60-unit trailer park showed a total solids concentration of 15,058 mg/l one day after a septic-tank cleaner had been added. The effluent from the sand filter following the septic tank showed a total solids content of 1038 mg/l at the same time.

Some septic-tank cleaners (degreasing compounds) contain sodium or potassium hydroxide or sulfuric acid. Others contain methylene chloride, trichloroethane, or orthodichloro-benzene which are suspected of being carcinogenic. These toxic compounds may eventually reach the groundwater and endanger well-water supplies in the area. Their use is not advised.

Commercial compounds alleged to prevent septic-tank system clogging and

[59]James T. Conklin, "Thermal Kill of Roots in Sanitary Sewers," *Deeds and Data, Water Pollut. Control Fed.*, March 1977.

backup usually require regular application; weekly or monthly is not unusual. Temporary relief may be obtained, but the cost of the chemical on an annual basis could equal twice the cost of having a septic tank cleaned *annually*. The only acceptable treatment that has been found effective in unclogging septic tank leaching systems is hydrogen peroxide *when applied directly to the clogged soil.*[60] See page 389.

A starter, such as yeast, added to a septic tank does not speed up the digestion. The addition of 6 gal of digested sludge per capita to a new septic tank appears to have a beneficial effect. A new septic tank does not need a starter or other additive to function. The sewage it receives contains the organisms necessary to initiate and promote anaerobic digestion.

Division of Flow to Soil Absorption System

The overflow from a septic tank should be run to a distribution box to assure equal division of the settled sewage flow to all the leaching pits or laterals comprising the disposal or treatment system. The distribution box should have a removable cover extended to the surface to simplify inspection of the septic-tank effluent and flow distribution to the disposal or treatment system. *Outlets must leave the distribution box at exactly the same level.* A gravel fill or footing under the box, extending below frost, will help keep the box level. A baffle is usually necessary in front of the inlet to break the velocity of the incoming sewage and permit equal distribution to the outlets. Bricks or blocks are very useful for this purpose. Details of distribution boxes are given in Figures 4-11 and 4-13. If the outlets are constructed about 6 in. above the bottom of the distribution box, the liquid collected will have the effect of breaking the incoming velocity of the settled sewage. In any case, it is important to place all outlets at the same level and obtain equal distribution of flow to each of the laterals. A $\frac{1}{8}$-in. mesh plastic basket screen over each outlet would prevent large particles of septic-tank sludge from being carried into the absorption field. Backup would occur at the box and without ruining the absorption field would call attention to the need for cleaning the septic tank.

Serial distribution of sewage (Figure 4-12) by the use of tees and elbows is reported to have certain advantages over the use of distribution boxes.[61] It compensates for varying soils and absorptive capacity; it forces full use of a trench before overflow to the next and overcomes the hazard of overflow associated with the parallel system if one trench is overloaded due to uneven flow distribution from an improperly installed or disturbed distribution box. However greater expertise is needed in construction; this method of sewage distribution promotes anaerobic conditions and creeping failure of the absorption system.

[60] *Management of Small Waste Flows,* Municipal Research Laboratory, USEPA, Cincinnati, Ohio 45268, September 1978, pp. B-130 to B-141.

[61] *Manual of Septic-Tank Practice,* PHS, Pub. 526, DHEW, Washington, D.C., 1967.

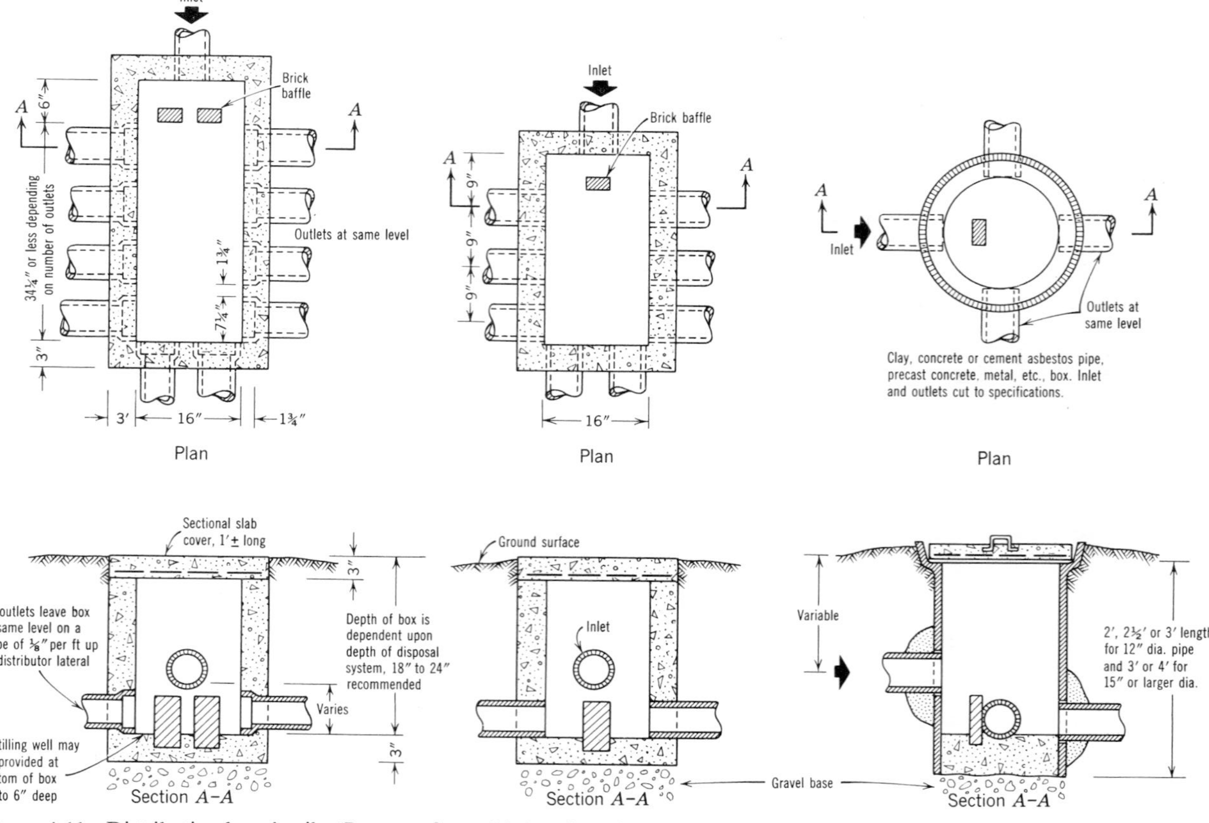

Figure 4-11 Distribution box details. (Bottom of gravel below frost; level box on 12 in. of well stabilized gravel.)

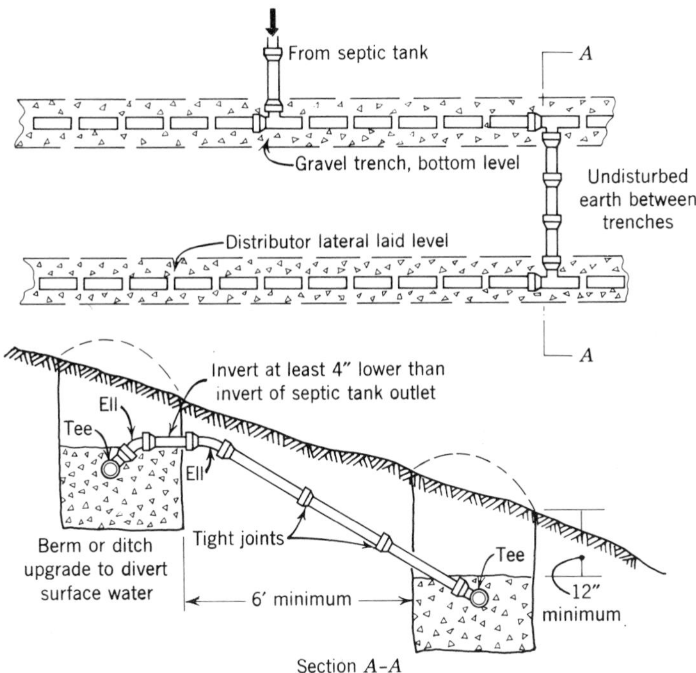

Figure 4-12 Serial distribution for sloping ground. (Adapted from U.S. PHS Pub. 526, DHEW, Washington, D.C., 1967.)

On steep grades special provision must be made for reducing the velocity of the sewage leaving the septic tank in order to get good distribution to the subsurface tile field or absorption system. Drop manholes are used for this purpose. The flow can be divided approximately in proportion to the length of the absorption system at each manhole. Drop-manhole details are shown in Figure 4-13.

A flow diversion box or two-port valve on the line leaving the septic tank to permit use of alternate absorption systems, say on a 6-month cycle, will permit maintenance of trench infiltrative capacity, aerobic conditions, and prolonged life if the septic tank is cleaned regularly, say every three years.

Subsurface Soil Absorption Systems

The conventional subsurface absorption system following the septic tank is the absorption field or leaching pit. The cesspool is still used for raw sewage, although generally prohibited, and the dry well is used for the disposal of rainwater, footing, roof, and basement floor drainage. Where the soil is not suitable for subsurface disposal, a sand filter, evapotranspiration system, modified tile field system, aeration system, system in fill, mound system, stabilization pond, or some combination may be used. These systems are discussed later.

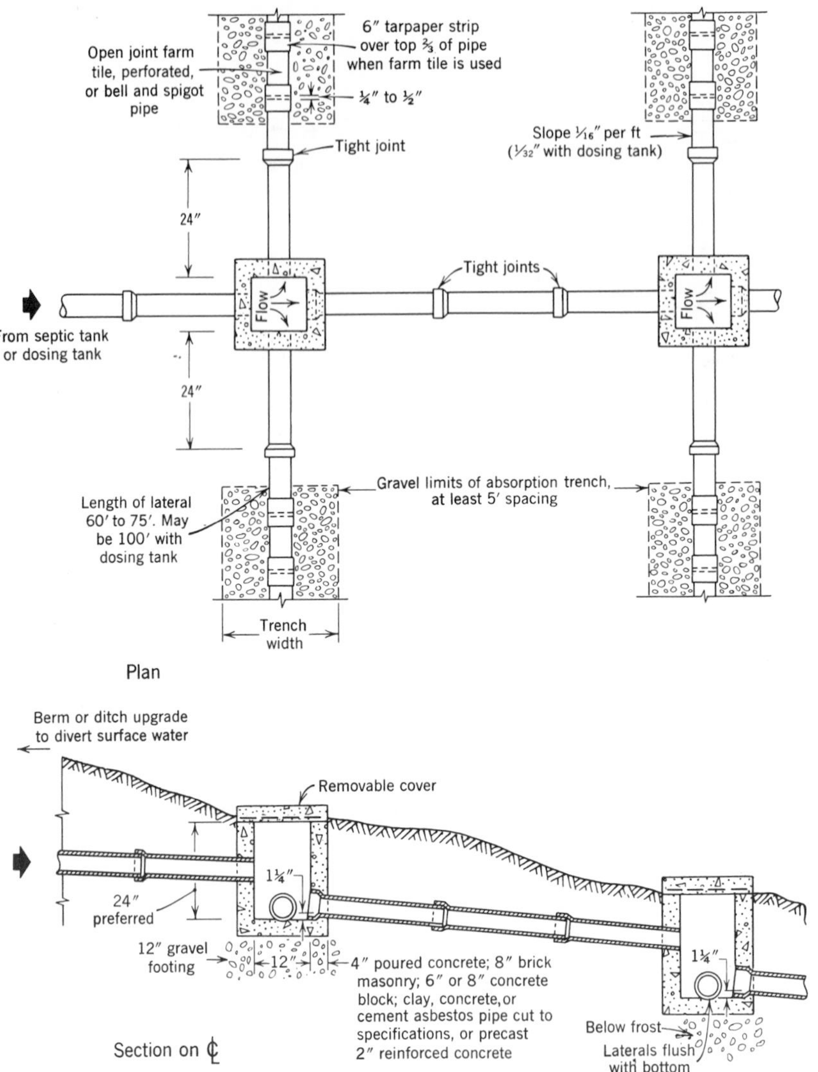

Figure 4-13 Drop-manhole details for tile field on sloping ground.

Absorption Field System

The soil percolation test basis for the design of absorption systems was given earlier in this chapter. Design standards and details for absorption systems are shown in Figures 4-4, 4-12, 4-13, 4-14*a*, and in Table 4-9.

The absorption field laterals should be laid in trenches preferably not more than 24 in. below the ground surface. Where laid at a greater depth, the gravel fill around the open joint or perforated lateral should extend at least to the topsoil and as shown in Figure 4-14*a*. The sunny and open side of a slope is the preferred

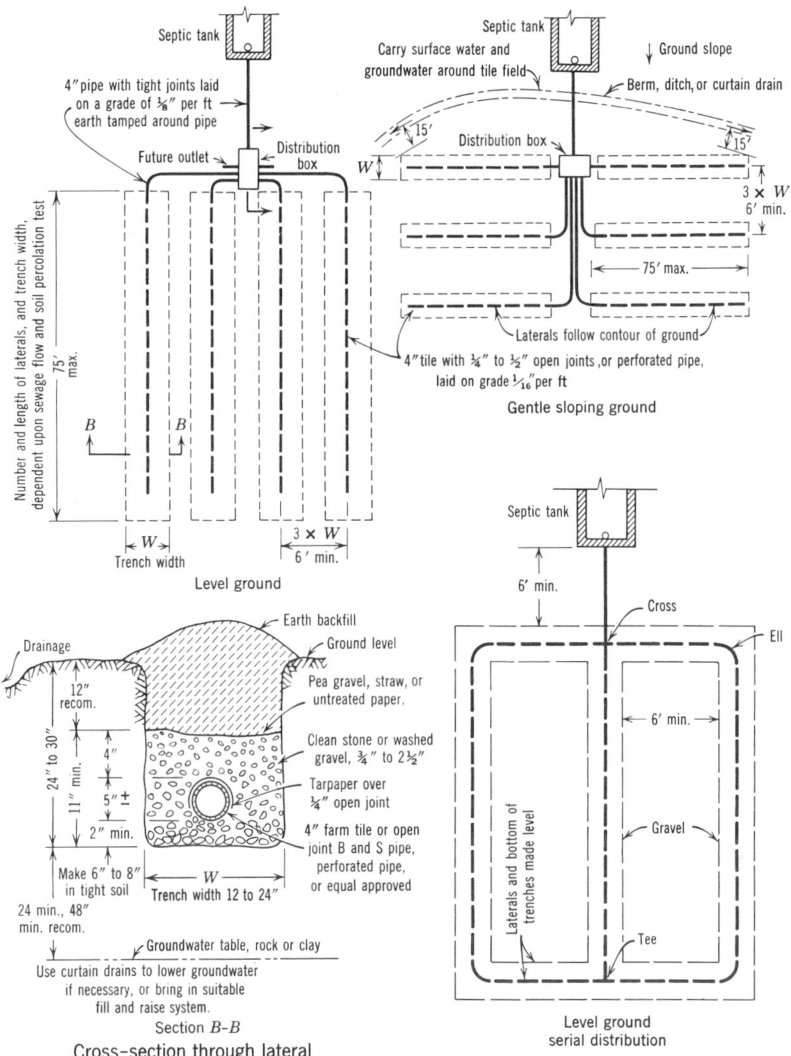

Figure 4-14a Arrangements and details for absorption field disposal systems. See local regulations.

location for an absorption field, if there is a choice. After settlement and grading, the absorption field area should be seeded to grass.

When the total length of the laterals to provide the required leaching area is 500 to 1000 linear ft, a siphon should be installed between the septic tank and absorption system to distribute the sewage to all the laterals. If the total required length of the laterals is 1000 to 3000 linear ft, the system should be divided into two or four sections with alternating siphons to feed each section, or each two sections when four are provided. Where the total length of laterals required is greater than 3000 linear ft it is advisable to investigate a secondary treatment

Table 4-9 Suggested Minimum Standards—Subsurface Sewage Disposal Systems

Item	Material	Size
Sewer to septic tank	Cast iron for 10′ from bldg. recommended.	4″ min. dia. recommended.
Septic tank	Concrete or other app'd matrl. Use a 1:$2\frac{1}{4}$:3 mix.	Min. 750 gal 4′ liquid depth, with min. 16″ M.H. over inlet.
Lines to distribution box and disposal system	Cast iron, vit. clay, concrete, or composition pipe.	Usually 4″ dia. on small jobs.
Distribution box	Concrete, clay tile, masonry, coated metal, etc.	Min. 12″ × 12″ inside carried to the surface. Baffled.
Absorption field[b]	Clay tile, vit. tile, concrete, composition pipe, laid in washed gravel or crushed stone, $\frac{3}{4}$″ to $2\frac{1}{2}$″ size, min. 12″ deep.	4″ dia., laid with open joint or perforated pipe. Depth of trench 24″ to 30″.
Sand filter[b]	Clean sand, all passing $\frac{1}{4}$″ sieve with effective size of 0.30 to 0.60 mm and uniformity coefficient less than 3.5. Flood bed to settle sand.	Send 2-lb sample to health dept. for analysis 15 days before construction.
Leaching or seepage pit[b]	Concrete block, clay tile, brick, fieldstone, precast.	Round, square, or rectangle.
Chlorine contact-inspection tank	Concrete, concrete block, brick, precast.	2′ × 4′ and 2′ liquid depth recommended.

Note: A slope of $\frac{1}{16}$″ per ft = 6.25′ per 100′ = 0.0052 ft per ft = 0.52 percent.

Note: All parts of disposal and treatment system shall be located above groundwater and *downgrade* from sources of water supply. The architect, builder, contractor, and subcontractor shall establish and verify all grades and check construction. Laundry and kitchen wastes shall discharge to the septic tank with other sewage. Increase the volume of the septic tank by 50 percent if it is proposed to also install a garbage grinder. No softening unit wastes, roof or footing drainage, surface water or groundwater shall enter the sewerage system. Where local regulations are more restrictive, they govern, if consistent with county and state regulations.

process, although larger absorption systems can operate satisfactorily. In some instances flat topography makes it impossible to install siphons and still obtain distribution of settled sewage by gravity to all the laterals. In such cases it would be necessary to install pumps or ejectors; but the design should permit gravity flow to the absorption system in case of pump failure, if possible. Dosing arrangements are discussed later.

Absorption field laterals must be laid on careful grades. *The bottom of trenches should be dug on the same grades as the laterals* to prevent the sewage running out at one end of a trench or on to the ground surface. Laterals for fields of less than 500 ft in total length, without siphons, should be laid on a slope of $\frac{1}{16}$ in./ft or 3 in./50 ft. When siphons are used, the laterals should be laid on a slope of 3 in./100 ft. Absorption fields for steep sloping ground are shown in Figures 4-12 and 4-13, and layouts for level and gently sloping ground are shown in Figures 4-14*a* and 4-14*b*.

Table 4-9 (*Continued*)

	Minimum Governing Distances		
Grade	To Building or Property Line	To Well or Suction Line	To Water Service Line
$\frac{1}{4}''$ per ft max., $\frac{1}{8}''$ per ft min.	5′ or more recommended.	25′ if cast-iron pipe, otherwise 50′.	10′ hor.[a]
Outlet 2″ below inlet.	10′	50′	10′
$\frac{1}{8}''$ per ft; but $\frac{1}{16}''$ per ft with pump or siphon.	10′	50′	10′
Outlets at same level.	10′	100′	10′
$\frac{1}{16}''$ per ft, but $\frac{1}{32}''$ per ft with pump or siphon.	10′ except when fill is used in which case 20′ is required.	100′	10′ (25′ from any stream; 50′ is recommended.)
Laterals laid on slope $\frac{1}{16}''$ per ft; but $\frac{1}{32}''$ per ft with pump or siphon.	10′	50′	10′ (25′ from any stream; 50′ is recommended.)
Line to pit $\frac{1}{8}''$ per ft.	20′	150′ plus in coarse gravel.	20′ (50′ from any stream).
Outlet 2″ below inlet.	10′	50′	10′

[a]Water service and sewer lines may be in same trench, if cast-iron sewer with lead-caulked joints is laid at all points 12″ below water service pipe; or sewer may be on dropped shelf at one side at least 12″ below water service pipe, provided sound sewer pipe is laid below frost with tight and root-proof joints which is not subject to settlement, superimposed loads, or vibration.

[b]*Manual of Septic-Tank Practice,* PHS Pub. 526 (1967), states that the leaching area should be increased by 20 percent where a garbage grinder is installed, and by 40 percent where a home laundry machine is also installed. It recommends that the gravel in the tile field extend at least 2″ above pipe and 6″ below the bottom of the pipe.

Leaching or Seepage Pit

Leaching pits, also referred to as seepage pits, are used for the disposal of settled sewage where the soil is suitable and a public water supply is used, or where private well-water supplies are preferably 150 to 200 ft away, at a higher elevation and not likely to be affected. The bottom of the pit should be at least 2 ft and preferably 4 ft above the highest groundwater level and channeled or creviced rock. If this cannot be assured, subsurface absorption fields should be used. In special instances, where suitable soil is found at greater depths, pits can be dug 20 to 25 or more ft deep, using precast perforated wall sections. The soil percolation test is made at mid-depth and at the bottom of the proposed leaching pit and interpreted for design purposes as explained earlier in this chapter and in Table 4-2. The effective leaching area provided by a pit is equal to the vertical wall area of

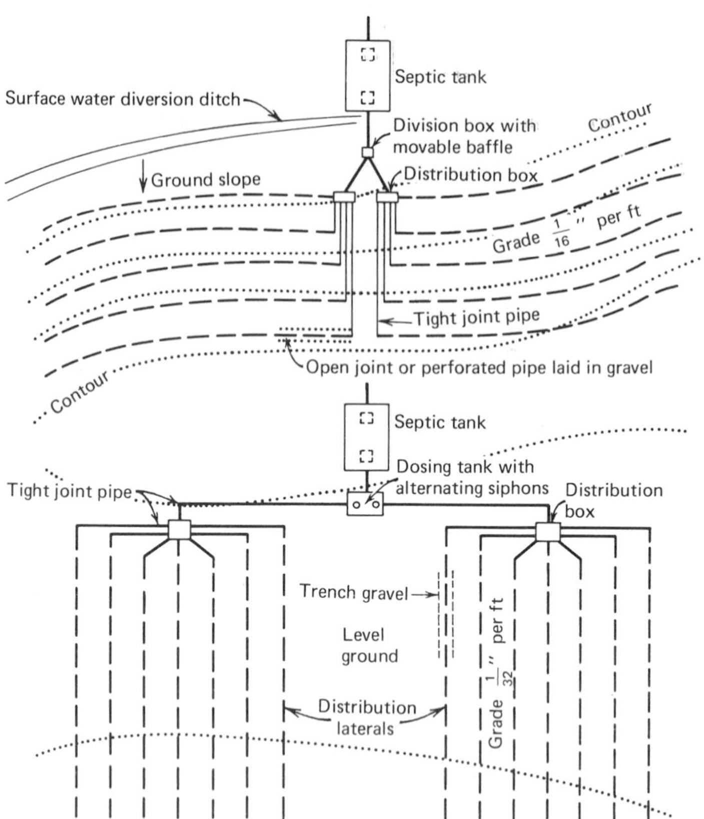

Figure 4-14*b* Absorption fields with division box, distribution box, and dosing tank.

the pit below the inlet. Credit is not usually given for the pit bottom. A leaching pit may be round, oval, square, or rectangular. The wall below the inlet is dry-wall construction; that is, laid with open joints, without mortar. Field stones, cinder or stone concrete blocks, precast perforated wall sections, or special cesspool blocks are used for the wall construction. Concrete blocks are usually placed with the cell holes horizontal. Crushed stone or coarse gravel should be filled in between the outside of the leaching pit wall and the earth hole. A nomogram to simplify determination of the sizes of circular leaching pits is shown in Figure 4-16. Sketches of leaching pits are given in Figures 4-15 and 4-17.

Cesspool

Cesspools are covered, open-joint or perforated walled pits that receive raw sewage. Their use is not recommended where the groundwater serves as a source of water supply. Many health departments prohibit the installation of cesspools

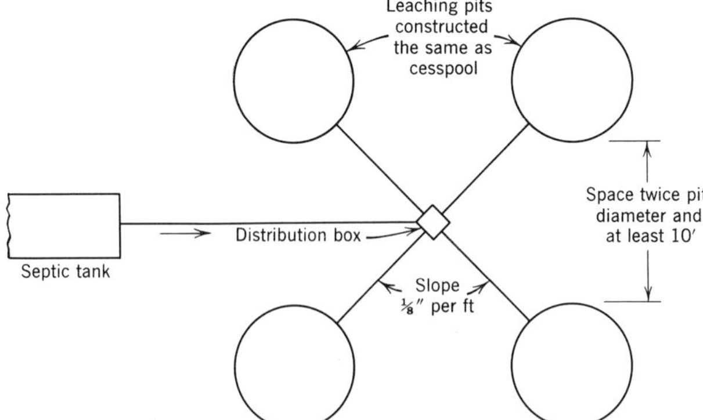

Septic tank – distribution box – leaching (seepage) pit plan

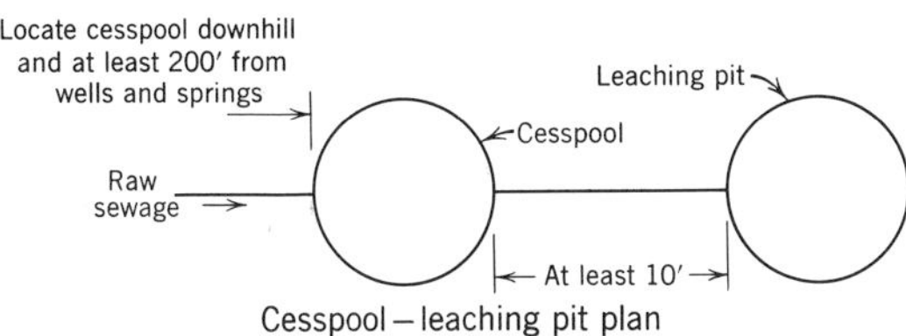

Cesspool – leaching pit plan

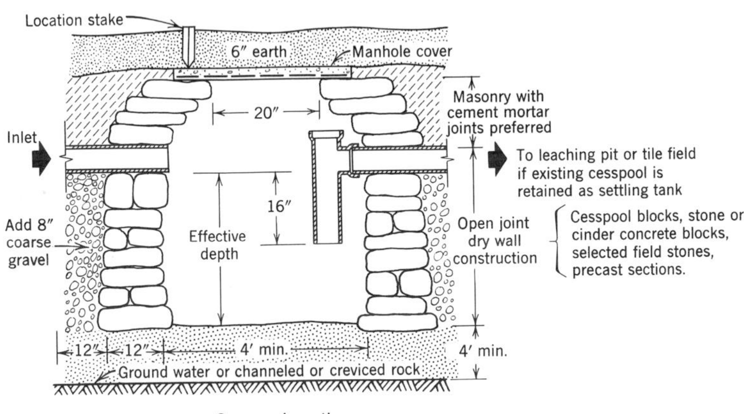

Cesspool section

Figure 4-15 Leaching pit and cesspool details.

where groundwater or creviced and channeled rock is close to the surface. Pollution could travel readily to wells or springs used for water supply. Where cesspools are permitted, they should be located downgrade from sources of water supply and 200 to 500 ft away. Even 500 ft may not be a safe distance in a coarse gravel unless the water-bearing stratum is below the gravel and separated by a thick clay or hardpan stratum. On the other hand, lesser distances may be

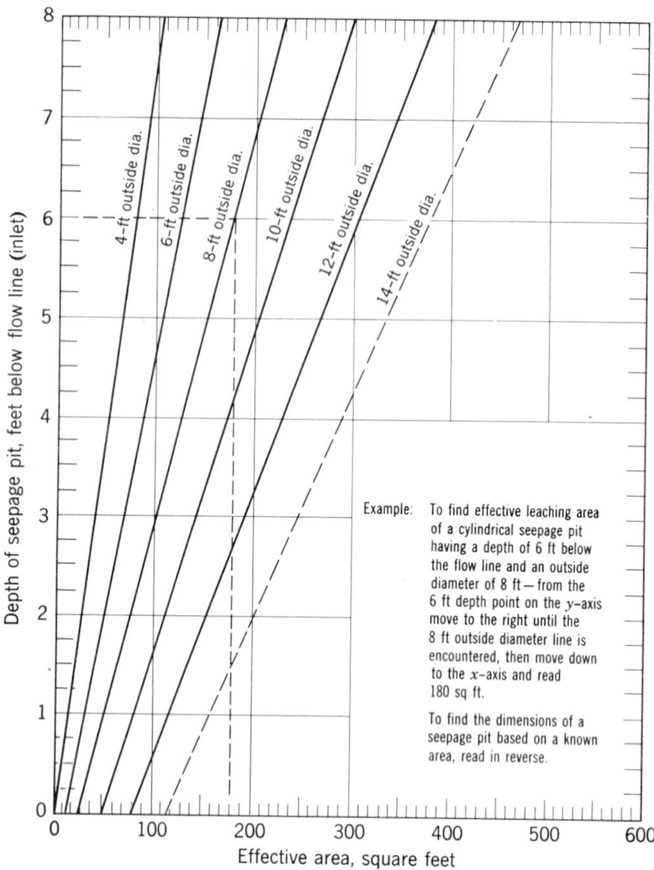

Figure 4-16 Effective areas of round seepage pits preceded by septic tanks or cesspools.

permitted where fine sand and no groundwater is involved. In all cases the bottom of the cesspool should be at least 4 ft above the highest groundwater level.

The construction of a cesspool is the same as a leaching pit, shown in Figure 4-15. In some areas, such as where sand and gravel deposits are found, cesspools have been in common use for many years before requiring cleaning. Cleaning the cesspool will not restore it to full use again since the space behind the wall cannot be effectively cleaned. Heavy chlorination may be of value. Special cesspool (or septic-tank, as previously noted) cleaning compounds may contain toxic or carcinogenic chemical compounds and should not be used. These compounds may persist for many years and contaminate the groundwater aquifer serving as the source of drinking water. The cesspool system can be made more efficient under such circumstances by providing a tee outlet, as shown in Figure 4-15, with the overflow discharging to an absorption field or leaching pit. Another alternative would be replacement of the cesspool with a septic tank followed by an absorption field or leaching pit. The required size of a cesspool can be

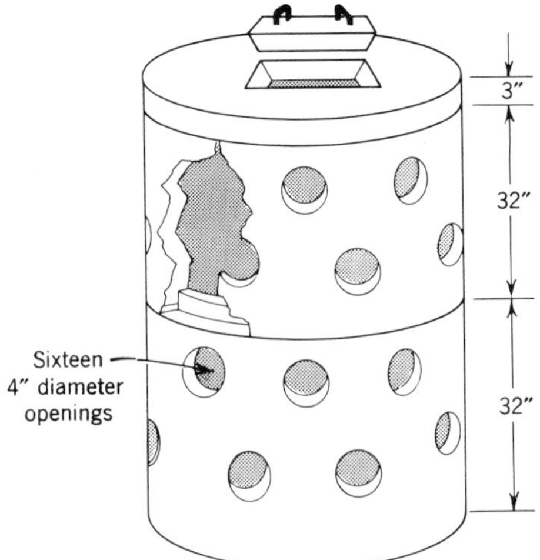

Figure 4-17 Precast leaching pit or dry well. (Courtesy of the Fort Miller Co., Inc., Fort Miller, N.Y.)

Each section: 32″ high, 4′ inside diameter (available in larger sizes); 3″ thick walls—place 8″ coarse gravel all around; 250-gal volume; 1,100-lb weight. Cover weight: Approximately 400 lb.

Manhole can be built up to grade using standard chimney blocks.

determined by making soil percolation tests as explained for a leaching pit. No credit is given for the bottom area of a cesspool and the required leaching area is arbitrarily doubled since raw rather than settled sewage is to be received.

Dry Well

A dry well is constructed similarly to a leaching or seepage pit and with the same care. A dry well is used where the subsoil is relatively porous, for the underground disposal of clear rainwater, surface water, or groundwater collected in footing, roof, and basement floor drains and similar places. Footing, roof, or basement floor drainage should never be discharged to a private sewage disposal or treatment system as the septic-tank and leaching system would be seriously overloaded and cause exposure of the sewage and premature failure of the leaching system. To design the sewerage system for this additional flow is uneconomical and unnecessary. If the soil at a depth of 6 to 10 ft or more is tight clay, gravel- or stone-filled trenches about 3 ft deep may be found more effective. Dry wells should not be used for the disposal of toilet, bath, laundry, or kitchen wastes. These wastes should be discharged through a septic tank. In some cases footing and roof drainage is discharged to a nearby watercourse, combined sewer, storm sewer, or roadside ditch, if permitted by local regulations, rather than to dry wells.

Dry wells should be located at least 50 ft from any water well, 20 ft from any leaching portion of a sewage disposal system, and 10 ft or more from building foundations or footings.

SMALL WASTEWATER DISPOSAL SYSTEMS FOR UNSUITABLE SOILS OR SITES

General

Waterborne systems in this category, also referred to as alternative systems, are usually more complex in design and costly than the conventional septic-tank subsurface absorption systems previously discussed and described. These systems include the modified absorption system, the absorption-evapotranspiration system, the sand filter system, the aerobic treatment unit, the mound system, the built-up system, the evapotranspiration system; also the oxidation pond system, the spray irrigation system, the overland flow system, the oxidation ditch system, and various combinations.

Alternative systems are considered when a conventional system cannot be expected to function satisfactorily because of high groundwater, or because rock, clay, or other relatively impervious formation is close to the surface, or where space is limited, or where a highly porous formation exists and protection of nearby well-water supplies is a major concern. The local sanitarian or sanitary inspector is advised to consult with a sanitary or public health engineer on his staff or with the state health department or environmental protection agency in such situations. In the case involving a public place such as a hotel, motel, campground, recreation area, commercial operation, or realty subdivision, the owner should be referred to a consulting engineer for advice on how to best solve his problem. Plans, specifications, and an engineer's report are normally required for review and approval *prior to construction,* before any decisions are made.

Modified Septic-Tank Soil Absorption System

The conventional subsurface soil absorption system is usually designed on the basis of soil percolation rates not exceeding 1 in. in 60 min. Some regulatory agencies arbitrarily establish 1 in. in 30 min as the maximum rate beyond which construction is prohibited, unless an acceptable alternative system is permitted.

Design and Construction Details

There is nothing sacred about the 60-min maximum soil percolation rate; most so-called "tight" soils are not entirely impermeable. This is simply common sense which Ryon recognized in his original notes.[62] He recommended the following application rates for 60-min or poorer soils.

[62] Henry Ryon, "Notes on Sanitary Engineering," New York State, Albany, N.Y., 1924, p. 33.

Time to fall one inch (hours)	Safe application rate (gpd/ft^2)
1	0.4
$1\frac{1}{2}$	0.3
2	0.2
3	0.14
5	0.07
10	0.03

Ryon further recommended that the required absorption area be doubled if an absorption bed is used instead of an absorption field.

The same precautions should be taken in the construction of a modified system as in the conventional system. Drainage of surface water, and attention to possible high groundwater and its control become particularly important in view of the tighter soil which would be involved. It is apparent that large absorption areas will be required, which for very tight soils may become impractical. The wastewater flow to be treated, the size of the absorption system, the problem to be resolved, the space available, and the cost will largely determine the practicality in individual situations.

Construction of the modified system would be the same as that for a conventional subsurface absorption field. Intermittent dosing (siphon, pump, tipping bucket), or a switch box (division box) to each of two absorption fields is good practice, particularly if the total length of distributors exceeds 500 ft. Alternating (twin) siphons are preferred where the topography permits. In any case, alternating dosage should be provided for lengths of 1000 to about 2500 ft. If a switch box (division box) is used instead of a dosing tank, care must be taken to assure that flows are in fact periodically switched, say every three or six months, and not forgotten.

An example showing a design for a "tight" soil site of fairly uniform composition to a depth of 4 ft or more follows.

The design for a relatively tight soil makes use of the conventional soil percolation test carried to the point of constant rate, beyond the 1 in./60 min test. The moisture loss due to evaporation and transpiration is not credited but is taken as a bonus.

Assume a soil with an actual percolation rate of $\frac{1}{4}$ in./hr. If the rate for a 60 min/in. soil is 0.40 gpd/ft^2, then for $\frac{1}{4}$in./hr soil the rate could be $\frac{1}{4} \times 0.40 = 0.10$ gpd/ft^2, as noted above by Ryon.

Example

Design a subsurface leaching system for a daily flow of 300 gal. The soil test shows $\frac{1}{4}$ in./hr and a permissible settled sewage application of 0.10 gpd/ft^2.

$$\text{Required leaching area} = \frac{300}{0.10} = 3000 \text{ ft}^2$$

If trenches 36-in. wide with 18-in. gravel underneath lateral distributors are provided, each linear ft of trench can be expected to provide 5 ft^2 of leaching area.

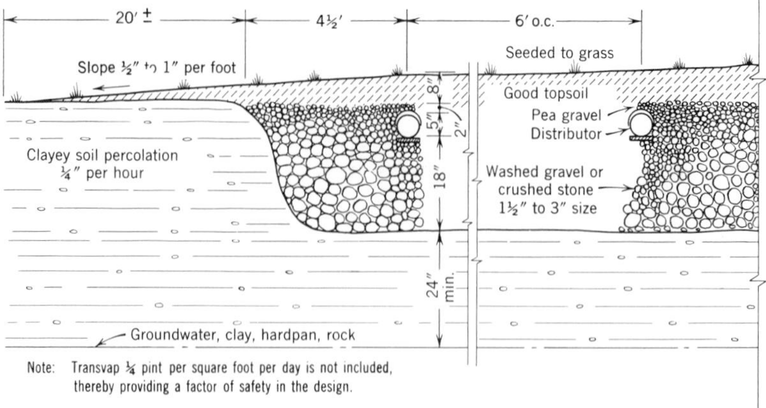

Figure 4-18 Absorption bed for a tight soil. Curtain drains may be needed to lower groundwater level. Crown bed to readily shed rain water.

The required trench =3000/5 = 600 linear ft, or 8 laterals each 75-ft long, spaced 9 ft on center. Provide a dosing arrangement.

The leaching area can also be provided by *two* gravel beds 50 ft × 60 ft to compensate for loss of sidewall trench infiltration area. See Figure 4-18. Use alternating dosing device. This occupies the same land area as the absorption field. Evapotranspiration can be enhanced by incorporating sand trenches or funnels in the gravel between the distributors. (See Figure 4-28.)

Built-Up Septic-Tank Absorption-Evapotranspiration System

If clay, hardpan, groundwater, or channeled, creviced, or solid rock is found within and below 4 ft of the ground surface, disposal of sewage by means of a conventional subsurface absorption system is not recommended. For practical purposes clay, hardpan, or solid rock cannot be expected to absorb sewage; and the disposal of sewage directly into the groundwater is to invite failure of the system and direct groundwater pollution or, if into channeled or creviced rock or coarse gravel, sewage pollution of nearby and distant wells. It is possible to artificially build up an earth area for sewage disposal *provided at least 12 to 18 in. of natural porous earth exists,* and thus spread the percolating sewage over a large absorbtive area, if approved by the regulatory agency. Sufficient suitable soil must be brought in so that the bottom of absorption trenches will be at least 2 ft above the highest groundwater level, rock, clay, or hardpan. The slope or dip of the rock, clay, or hardpan and the depth and limits of the fill must be such as to make improbable seepage of sewage to the ground surface or directly into creviced rock. The earth gravelly loam fill should be porous and contain topsoil and sand with not less than 20 percent or more than 40 percent clay. See General Soil Characteristics earlier in this chapter. Fill must be allowed to become stabilized before cutting the trenches and constructing the absorption system.

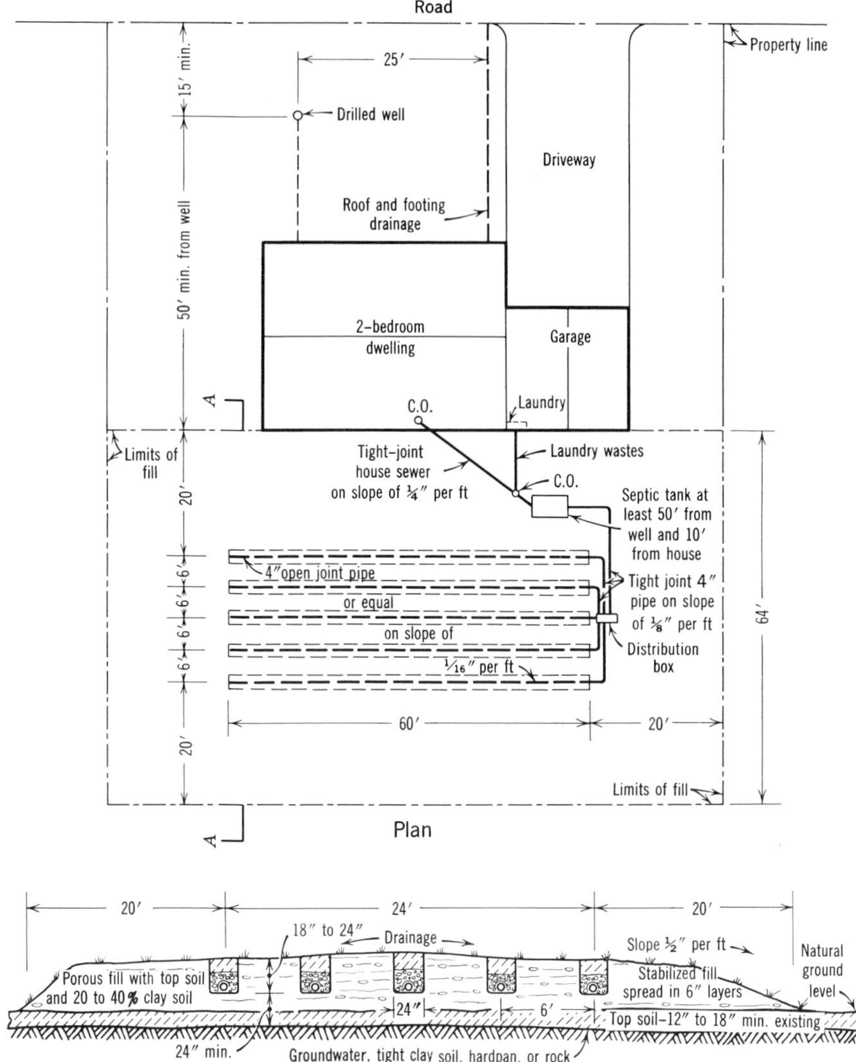

Figure 4-19 Built-up evapotranspiration-absorption sewage disposal system over clay soil or rock. First floor and house sewer elevations must be established to provide gravity flow to sewage disposal system, otherwise pumping will be required. Design basis: 300 gpd and a transvap-percolation rate of 0.5 gal for trench ft^2/day or 0.04 gal/ft^2 of gross area of fill (64′ × 100′), or 0.19 gal/ft^2 of bed area (26′ × 60′).

The limits of the fill should extend about 20 ft beyond the absorption field in all directions. On sloping ground the upgrade fill limit could be reduced but the downgrade fill limit might have to be extended more than 20 ft, particularly on steeper ground and where 12 to 18 in. of natural soil is not available, to prevent seepage out of the tee of the fill. A suggested sewage disposal system incorporating fill is illustrated in Figure 4-19. The design is based on 300 gpd and a transpiration-evaporation (transvap) rate plus percolation, or a transvap-percolation rate of 0.5 gpd/ft^2 of conventional bottom trench area, 0.2 gpd/ft^2 of bed area (26 ft × 60 ft), or 0.04 gpd/ft^2 of gross area of fill (64 ft × 100 ft). This principle of design may be used for small motels, private dwellings, and similar establishments *if necessary.*

As an alternative to the design method given above, the area of the absorption field and volume of fill can be estimated by means of preliminary soil percolation tests made first at selected borrow pits to obtain the best soil. The fill soil must be carefully selected and consist of a mixture of topsoil (the upper layer of soil, usually darker and richer than the subsoil) and loam containing sand, small stones, silt, and some clay as previously described. The tests (at least two) should be repeated in the graded fill, preferably during the wet season, after stabilization, which may require six to nine months. Since the original soil structure is radically changed in excavating, loading, dumping, and spreading, the goal is eventual restoration of a permeable soil structure with a healthy microbial population in the soil of the new fill area and a firm grass cover.

Light equipment should be used for soil spreading; the natural topsoil existing should not be removed but simply lightly plowed to provide a bond with the filled-in soil. The surface of the fill should be carefully graded to readily shed rain water and the border around the sides of the built-up absorption field for a distance of about 20 ft should be feathered to the natural soil so as to keep the hydraulic gradient of the wastewater entering the absorption trenches below the toe of the fill and ground surface. On sloping ground, the downgrade fill might have to be extended, as previously noted, as much as 40 ft or more, particularly where 12 to 18 in. of natural soil is not available over clay or rock, to prevent seepage to the ground surface.

A diversion ditch or berm should be provided upgrade to divert surface runoff around the absorption-evapotranspiration system. A curtain drain around the bed to intercept and lower the groundwater table, with discharge to the surface, may be needed in areas of high groundwater if the bottom of trenches cannot be kept at least 2 ft above groundwater. See Figure 4-20. It is important to retain as much of the natural topsoil as exists. *Do not remove.*

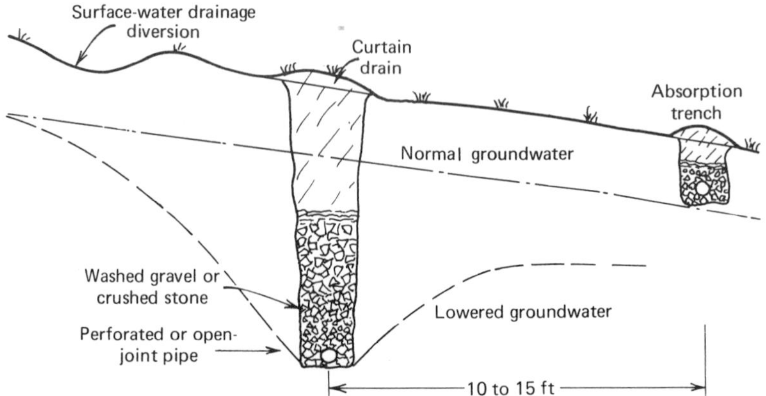

Figure 4-20 Curtain drain to lower groundwater level. The gravel or crushed stone *must* intercept the strata carrying the groundwater flow, and the collecting pipe *must* drain to the surface.

The system grade should be carefully established starting with the house sewer, through the septic tank, distribution box, the required invert of the distribution laterals and trenches, and still maintain a minimum distance of 2 ft between the bottom of trenches and the highest groundwater, rock, hardpan, or clay soil level. If the topography and location of the absorption system does not permit gravity flow then a pump, pump well, and, of course, electricity will be required. This will permit intermittent operation and full dosage of the distributor laterals, which is advantageous. However any time a pump is installed the chance for malfunction and inconvenience in the use of the household plumbing exists.

Septic-Tank Sand Filter System 

The sand filter following a septic tank may be earth-covered or open. The covered filter is generally used for small flows. Sand filters have particular application where conventional subsurface absorption systems could not be expected to function satisfactorily because of soil conditions or rock, or where space is very limited and discharge to a surface water or ditch is permissible.

Settled sewage is distributed over the top of a small sand filter bed by means of perforated or open-joint pipe. The sewage is filtered and oxidized in passing through 24 to 30 in. of carefully selected sand. Greater sand depths do not produce significant additional purification. A film containing aerobic and nitrifying organisms forms on the gravel and sand grains of the filter. Bacteria break down the organic matter in sewage. Protozoa and metazoa feed on bacteria thereby preventing clogging of the bed. The annelid worms appear to be most important in consuming sludges and slimes and are largely responsible for keeping sand filters open and active.[63]

A sand filter is an efficient treatment unit. Typical analyses of sanitary sewage applied to and leaving subsurface sand filters and efficiencies are shown in Table 4-10 and 4-11. A large reduction in bacteria, protozoa, helminths, turbidity, BOD, and suspended solids is obtained, in addition to a well-nitrified effluent containing dissolved oxygen. Such effluents would not cause a nuisance in undeveloped areas; but they should be chlorinated if discharged in locations accessible to children or pets because microorganisms associated with disease transmission, although greatly reduced, are still present.

The satisfactory operation of a sand filter is dependent on the rate and strength of sewage application and on the effective size and uniformity coefficient of the sand.* Some studies are reported here for guidance.

Filtration of an aerobic unit effluent at 3.5 gpd/ft^2 produced a good effluent with a 0.19 mm effective size and 3.31 uniformity coefficient sand which operated 9 months before clogging. Filtration of a septic-tank effluent at 5 gpd/ft^2 also

*Sands having uniformity coefficients between 1 and 5 will have practically the same hydraulic characteristics, provided the effective size of the sands is the same. (G. M. Fair and J. C. Geyer, *Water Supply and Wastewater Disposal,* John Wiley and Sons, New York, 1954.)

[63] Wilson T. Calaway, "Intermittent Sand Filters and Their Biology," *J. Water Pollut. Control Fed.*, January 1957, pp. 1–5.

Table 4-10 Typical Septic-Tank and Subsurface Sand Filter Effluent

Determination	Sewage Effluent[a]	
	Septic Tank	Subsurface Sand Filter
Bacteria per ml, Agar, 36°C, 24 hr	76,000,000	127,000
Coliform group MPN	110,000,000	150,000
Color (mg/l)	3.5	2
Turbidity (mg/l)	50	5
Odor[b]	4.5	1
Suspended matter[b]	3	1
pH	7.4	7.4
Temperature °C	17	14
BOD, 5-day (mg/l)	140	4
DO (mg/l)	0	5.2
DO saturation (%)	0	52
Nitrogen, total (mg/l)	36	21
Free ammonia	12	0.7
Organic	12	3.4
Nitrites	0.001	0.02
Nitrates	0.12	17
Oxygen consumed (mg/l)	80	20
Chlorides (mg/l)	80	65
Alkalinity (mg/l)	400	300
Total solids (mg/l)	820	810
Susp. solids (mg/l)	101	12

Source: J. A. Salvato, Jr., "Experience with Subsurface Sand Filters," *Sewage Ind. Wastes*, **27**, No. 8, 909–916 (August 1955).

Note: Sand effective size 0.30 to 0.60 mm and uniformity coefficient not greater than 3.5.

[a]Median results, using 51 samples from septic tanks and 56 from filters.

[b]1 = very slight, 2 = slight, 3 = distinct, 4 = decided, 5 = extreme. Normal municipal domestic sewage has an MPN of 50–100 million coliform bacteria per 100 ml.

Table 4-11 Typical Efficiencies of Subsurface Sand Filters[a]

Determination	Percent Reduction
Bacterial per ml, Agar, 36°C, 24 hr	99.5
Coliform group, MPN per 100 ml	99.6
BOD, 5-day (mg/l)	97
Susp. solids (mg/l)	88
Oxygen consumed (mg/l)	75
Total nitrogen (mg/l)	42
Free ammonia	94
Organic	72

Source: J. A. Salvato, Jr., "Experience with Subsurface Sand Filters," *Sewage Ind. Wastes,* **27**, No. 8, 909–916 (August 1955).

[a]Effluent will contain 5.2 mg/l dissolved oxygen and 17 mg/l nitrates.

produced a good effluent with a 0.45 mm effective size and 3.0 uniformity coefficient sand which operated for 3 to 5 months before clogging. Cleaning required removal of the top 2 to 5 in. of sand and replacement, which is not practical with an earth-covered filter. The filters matured in about two weeks. Complete nitrification was reported.[64]

Covered septic-tank sand filter systems with sand effective sizes of 0.15, 0.19, 0.24, 0.24, 0.30, 0.60, 1.0, and 2.5 mm and uniformity coefficients varying from 1.2 to 4.4 were studied for 30 to 42 months. The filters with 0.15 and 0.19 mm sand and 2.8 and 4.4 uniformity coefficient clogged after four months and were discarded. The other six filters, with uniformity coefficients of 1.2 to 3.9, operated at 1 and 1.5 gpd/ft^2, and produced good results: BOD of 10.0 mg/l or less and suspended solids of 10.5 mg/l or less in 85 percent of the samples. The 2.5 mm media filter showed a BOD increase from 10 mg/l to 13 mg/l when operated at 1.5 gpd/ft^2. Phosphorus removal (0.24 mm grain size) was in the range of 26.3 to 36.2 percent; but a filter (0.24 mm grain size) containing a "red mud" with oxides of calcium aluminum, and iron showed 73 to 90 percent phosphorus removal. The septic-tank effluent, which was fed to the filters on a trickle pattern, averaged 237 mg/l BOD and 139 mg/l suspended solids.[65]

A study of open intermittent sand filters loaded at 3 gpd/ft^2 using filter sands with effective grain sizes of 0.20 to 0.60 mm and uniformity coefficients of 3.2 to 6.3 showed a good quality effluent. The larger grain size filter (0.6 mm) produced lower removal of coliforms (78½%), COD (84%), and BOD (83%). The uniformity coefficient range studied did not seem to be critical to effluent quality.[66] A study in Ontario, Canada showed that 58 percent of the annual precipitation on an experimental sand filter left the filter through evapotranspiration; 15.4 percent of the inflow to the septic tank left the filter through evapotranspiration. The study confirmed the importance of designing filters also for maximum possible runoff to reduce precipitation infiltration, and for high sun exposure to maximize vegetative transpiration and soil evaporation. The degree of infiltration also affected the contaminant removal efficiency of the filter.[67]

Sand Filter Design

The recommended sizes of covered sand filters to serve private homes are shown in Figure 4-21 and 4-22. These systems are designed for a flow of 150 gpd/bedroom and a settled sewage application rate of 1.15 gal per ft^2 of sand

[64]David K. Sauer, William C. Boyle, and Richard J. Otis, "Intermittent Sand Filtration of Household Wastewater," *J. Environ. Eng. Div., ASCE,* August 1976, pp. 789–803.

[65]N. A. Chowdhry, *Domestic Sewage Treatment By Under-drained Filter Systems,* Ministry of the Environment, Toronto, Ontario, Canada, December 1974.

[66]Thomas M. Allen, *Effective Grain Size and Uniformity Coefficients' Role in Performance of Intermittent Sand Filters,* New York State Dept. of Environmental Conservation, Albany, N.Y., June 1971.

[67]Marek Brandes, "Effect of precipitation and evapotranspiration on a septic tank-sand filter disposal system," *J. Water Pollut. Control Fed.*, January 1980, pp. 59–75.

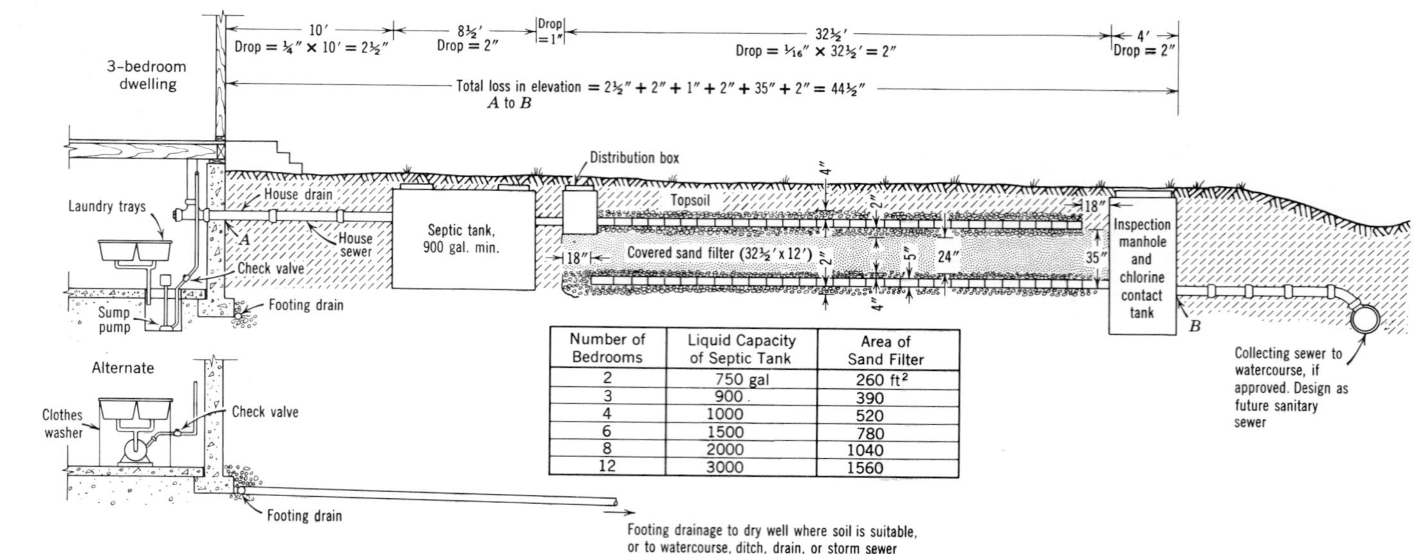

Number of Bedrooms	Liquid Capacity of Septic Tank	Area of Sand Filter
2	750 gal	260 ft^2
3	900	390
4	1000	520
6	1500	780
8	2000	1040
12	3000	1560

Figure 4-21 Section through covered sand filter system. Design basis is 150 gal per bedroom and filter rate 1.15 gpd/ft^2. Larger capacity septic tank (50 to 100 percent larger than minimum) is strongly recommended.

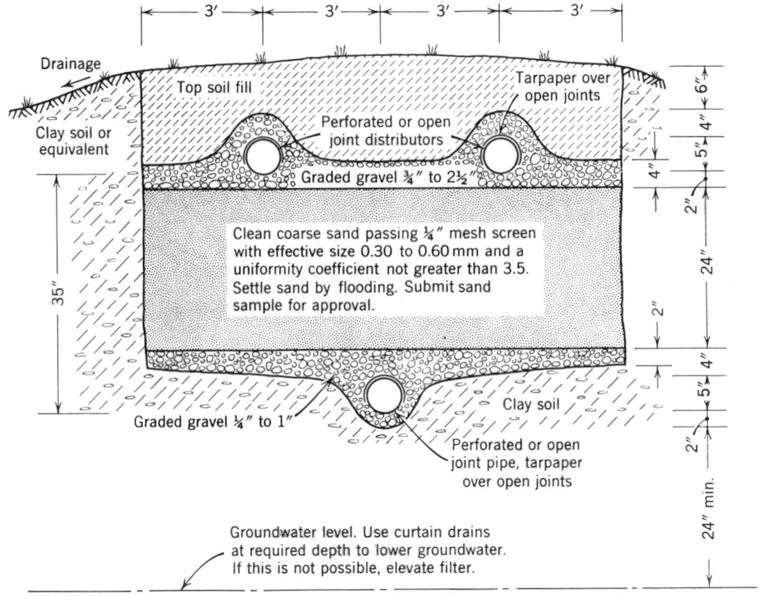

Number of Bedrooms	Capacity of Septic Tank	Size of Filter			Alternate Filter Size*		Sump Capacity Between Float Settings†
		Length	Width	Area	Length	Width	
2	750 gal	21½ ft	12 ft	260 ft²	43 ft	6 ft	20 gal
3	900	32½	12	390	65	6	30
4	1000	43	12	520	87	6	40
5	1250	54	12	650	—	—	50

* Use one distributor on top and one underdrain on bottom. † Where needed.

Required size of subsurface sand filter

Figure 4-22 Typical section of a subsurface sand filter. See Figure 4-21. The architect, builder, and contractors will determine invert elevations of the house sewer, septic tank outlet, distribution box, sand filter distributor and collector lines, chlorine contact tank, inspection manhole, and outlet sewer, drain, ditch, or watercourse so as to provide gravity flow through the system where possible. Exclude roof and footing drainage. Increase the volume of the septic tank by 50 percent if a garbage grinder is installed, and the area of the filter by 30 to 60 percent if both garbage grinder and home laundry machine are installed.

filter area per day. It is extremely important to *use a proper sand* meeting the specifications given in Figure 4-22. The sand grains should be somewhat uniform in size, that is, *not graded in size* from fine to coarse, as this will surely cause premature clogging of the filter. Assistance and approval of the regulatory agency should be sought before a sand is purchased and then again when delivered to the job. Some sand and gravel companies are equipped to make sieve analyses of sand and can assist individuals. The sand filter must be carefully constructed and settled by flooding, with distributor and collector lines laid at exact grade. The use of 1- × 4-in. boards under farm tile or perforated pipe distributors, laid on gravel, assist greatly in placing the distributor lines at proper grade. A topsoil cover preferably not exceeding 8 to 12 in. should be filled in over the gravel-covered distributor lines.

Design details for open and covered sand filters are summarized below for easy reference:

Filter rate: Earth or gravel covered—50,000 gpd/acre for settled domestic sewage; 100,000 gpd/acre for temporary summer use if a recommended sand size is used and the effluent does not enter a water supply source. Open filter—75,000 to 100,000 gpd/acre for settled sewage and 200,000 to 400,000 gpd/acre for secondary treated sewage. Loading should normally not exceed 2.5 pounds of either 5-day BOD or suspended solids per 1000 ft^2/day. Recommended filter rates related to climate and sand size are also given in the *Manual of Septic-Tank Practice.*[68] Recirculating open filter rate is 4 to 5 times daily design rate.

Dose: A dosing device is recommended when the total length of distributor laterals exceeds 300 linear ft or sand bed area exceeds 1800 ft^2. An alternating dosing device and separate beds are recommended when length of distributor laterals exceeds 800 linear ft or sand bed exceeds 4800 ft^2. Dose should be at least 90 gpm/1000 ft^2 of filter area at average head.

Volume of dose with distributor laterals = 60 to 75 percent of volume of distributors dosed. A 4-in. pipe holds 0.653 gal/ft. Volume on open filter = 2- to 4-in. flooding or 50,000 to 100,000 gal/acre. Design for 1 to 3 doses/day and a minimum 4-hr rest period between doses. Rotary distributors and spray nozzles have also been used to apply settled sewage to the sand beds.

Length of lateral distributor: Earth or gravel covered—75 ft or less; 100 ft acceptable with dosing device; for open filter, maximum distance of sewage travel from splash plate to edge of filter bed, 20 to 30 ft. See Figure 4-23.

Sand: Depth 24 to 30 in., effective size* recommended 0.35 to 0.50 mm, although 0.30 to 0.60 mm is usually acceptable. Uniformity coefficient† 3.5 or less recommended. Use clean silica sand passing $\frac{1}{4}$-in. sieve. A sand analysis is shown in Figure 4-24. Representative sand sampling is essential at the source and as delivered.

Grade of lateral distributor: $\frac{1}{16}$ in./ft, but $\frac{1}{32}$ in./ft with dosing device.

In freezing weather open filters will require greater operation control and maintenance. Scraping the sand before freezing weather into furrows about 8 in. deep with ridges 24 to 48 in. apart will help maintain continuous operation, as ice sheets will form between ridges and help insulate the relatively warm sewage in the furrows. Greenhouse covers are very desirable and will help assure continuous operation of the filters; however, they are expensive.

**Effective (grain) size* is a measure of the diameter of particles, when compared to a theoretical material having an equal transmission constant. It is the dimensions of that mesh which will permit 10 percent of the sample to pass and will retain 90 percent. The size of the grain in millimeters, such that 10 percent by weight are smaller.

†*Uniformity coefficient* is the ratio of the grain size that has 60 percent finer than itself to the size which has 10 percent finer than itself (effective size). See Figure 4-24.

[68] *Manual of Septic-Tank Practice*, PHS Pub. 526, DHEW, Cincinnati, Ohio, Revised 1967, p. 66.

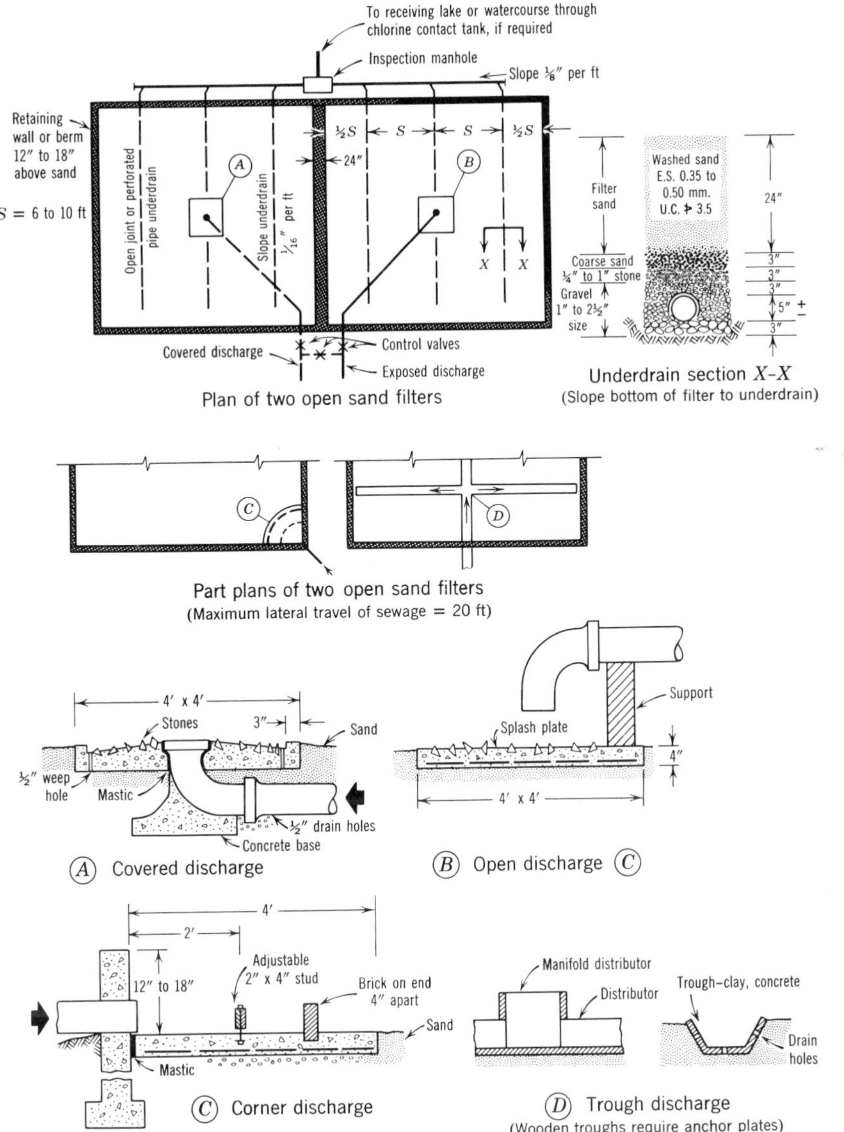

Figure 4-23 Open sand filter and distribution details.

Aerobic Sewage Treatment Unit

Another type treatment unit that can be used where subsurface absorption systems are not practical is the self-contained prefabricated aeration unit. The effluent from a properly operating unit is low in suspended solids and BOD and high in nitrates, but still requires further treatment and disposal. This may consist of sand filtration and/or chlorination prior to discharge to a stream, if

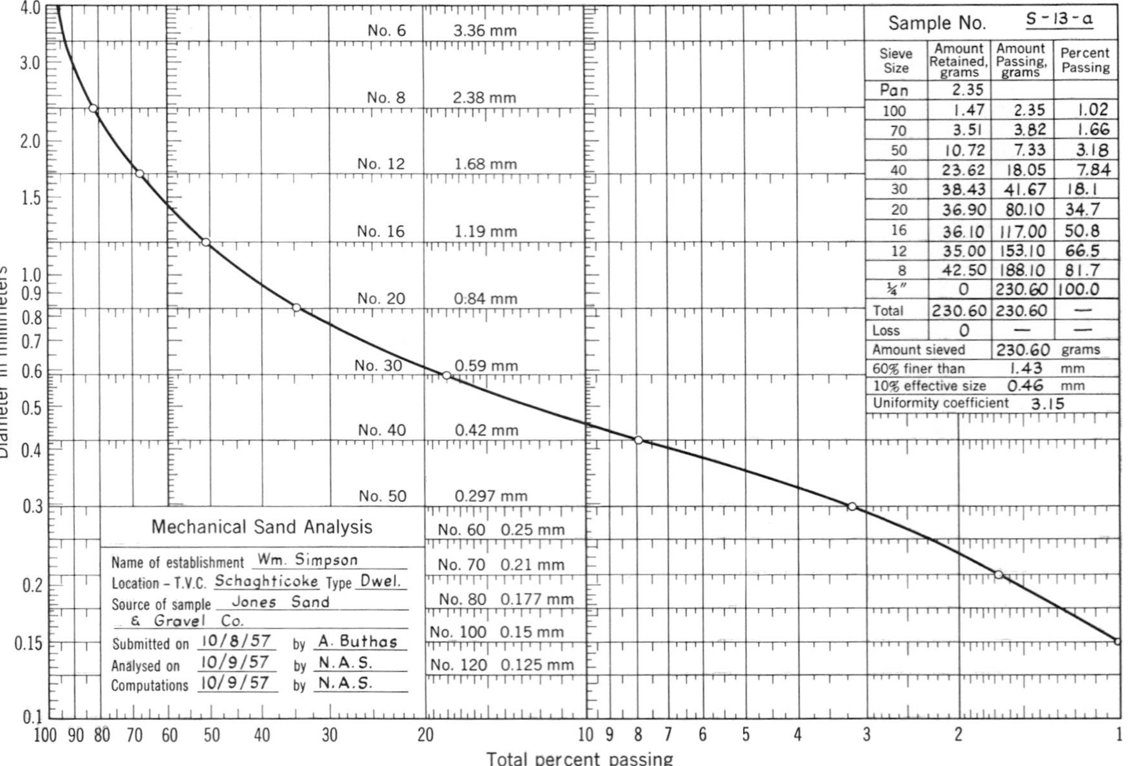

Figure 4-24 Mechanical sand analysis.

approved by the local regulatory authority, or discharge to a subsurface soil absorption system, oxidation pond, or irrigation system. If there is an electrical failure or mechanical malfunction, the effluent from the aeration unit is no better than that from a septic tank. Routine maintenance and operation of the unit must be assured by a maintenance contract or other means. Design details for extended aeration and activated sludge treatment plants are given on pages 466 and 500.

Waste Stabilization Pond and Land Disposal

In some locations, where tight soil exists and ample property is owned, the waste stabilization pond, irrigation, oxidation ditch, or overland flow system design principles may be adapted to small installations. Design information is given under Sewage Works Design—Small Treatment Plants.

Septic-Tank Mound System

The original mound system, called the NODAK system, was developed in North Dakota. Variations and refinements of the mound system have been described by Goldstein and Moberg,[69] Salvato,[70] and in studies reported by Bouma, et al.,[71] and Converse, et al.[72,73]

In the mound system the absorption area is raised above the natural soil to keep the bottom of trenches at least 2 ft above groundwater, creviced or porous rock, or relatively impermeable soil. In these respects it serves the same purpose as the fill or built-up soil absorption system previously described. It differs however in the type of fill material, size, and in the method used to apply septic-tank effluent to the mound system.

The Wisconsin mound system incorporates a 2 ft bed of clean medium sand* under the distribution trenches, a $1\frac{1}{2}$ in. to 1 in. perforated PVC pipe for pressure distribution of settled sewage, and at least 1 ft of natural topsoil. A basic

*Defined as 0.5 to 0.25 mm size, or Tyler standard sieve size No. 35 to 60 mesh. *Design Manual Onsite Wastewater Treatment and Disposal Systems,* USEPA, Washington, D.C., October 1980, p. 367.

[69]Steven N. Goldstein and Walter J. Moberg, Jr., *Wastewater Treatment Systems for Rural Communities,* Commission on Rural Water, Washington, D.C. 1973, pp. 57–64.

[70]Joseph A. Salvato, Jr., *Environmental Sanitation,* John Wiley & Sons, New York, 1958, pp. 221–225.

[71]J. Bouma, J. C. Converse, R. J. Otis, W. G. Walker, and W. A. Ziebell, "A Mound System for On-Site Disposal of Septic-Tank Effluent in Slowly Permeable Soils With Seasonally Perched Water Tables," *J. Environ. Qual.*, July–September 1975.

[72]J. C. Converse, R. J. Otis, and J. Bouma, *Design and Construction Procedures for Fill Systems in Permeable Soils With Shallow Creviced or Porous Bedrock,* Small Scale Waste Management Project, University of Wisconsin, Madison, April 1975.

[73]J. C. Converse, R. J. Otis, and J. Bouma, *Design and Construction Procedures for Fill Systems in Permeable Soils With High Water Tables,* Small Scale Waste Management Project, University of Wisconsin, Madison, April 1975.

objective is to assure that the mound system is constructed over at least 12 in. of permeable natural soil having an area large enough to permit lateral spreading of the percolating wastewater without surface seepage. The 2 ft of sand and 1 ft of natural topsoil are reported to remove pathogenic bacteria and viruses from the wastewater if properly applied.[74] The sand also promotes capillarity as explained in page 451.

Mound System Design Considerations[75]

1. To determine absorption area (bottom trench or bed area), with a medium sand fill material, use a design infiltration rate of 1.2 gpd/ft^2.

   Example

   For a daily flow of 300 gal, area required = 300/1.2 = 250 ft^2

2. Use a bed or parallel trenches:
   a. For a slowly permeable soil, use 2 or 3 narrow parallel trenches 2 to 4 ft wide. If groundwater may be a problem, elevate the bed or trenches.
   b. For permeable soil use a narrow, rectangular bed, not greater than 10 ft wide.
   c. For shallow soils over bedrock, use bed or trenches.

   Sufficient length of trenches and bed must be provided so all of the effluent infiltrates into the natural soil before it reaches the toe of the mound. The bottom of the bed or trenches must be level and at the same elevation.
3. Depth of mound and fill:
   a. A minimum depth below the distributor is 3 ft, consisting of medium sand fill and natural soil, for proper purification of the effluent.
   b. Provide a minimum of 1 ft of sand fill when the water table is greater than 2 ft beneath the soil surface.
   c. Provide a minimum 2 ft of sand fill for shallow permeable soils over creviced bedrock, where the natural soil depth over the bedrock is at the minimum of 24 in. Where the natural soil is greater than 24 in. the minimum depth of sand fill necessary is 1 ft.
4. The distribution trench or bed consists of 6 in. of gravel ($\frac{1}{2}$-2 in.) beneath the distribution pipe and 2 in. above the pipe for a total depth of about 9 in.
5. Cover the gravel with 4 to 5 in. of uncompacted straw or marsh hay, untreated building paper, or a synthetic fabric mesh.
6. Place soil over the covered gravel to a depth of 1 ft in the center and 6 in. at the outer edge of the trenches or bed. Then cover the entire area with at least 6 in. of good quality topsoil, graded so as to provide positive runoff, without erosion and with minimal infiltration.
7. Side slopes of the mound should be no steeper than 3:1.
8. The percolation rate of the natural soil (at 12 to 16 in. depth) determines the *basal* area of the system. On level ground, the basal area is the total area beneath the mound. For sloping ground, the basal area for design purposes is the area beneath and downslope of the bed or trenches.

[74]*Alternatives For Small Wastewater Treatment Systems,* USEPA Technology Transfer, Washington, D.C., October 1977, pp. 8, 47, 48.

[75]James C. Converse, *Design and Construction Manual for Wisconsin Mounds,* Small Scale Waste Management Project, University of Wisconsin, Madison, September 1978. See also *Recommended Standards for Individual Sewage Systems,* 1980 Edition, Great Lakes-Upper Mississippi River Board of State Sanitary Engineers, Health Education Service, P.O. Box 7126, Albany, N.Y. 12224, and *Design Manual Onsite Wastewater Treatment and Disposal Systems,* USEPA, Washington, D.C., pp. 239–259 and 278–296.

The design basal loading rates (infiltration rates) for certain soil percolation rates are:
1.2 gpd/ft^2 for 3 to 29 min/in. percolation
.74 gpd/ft^2 for 30 to 60 min/in. percolation
(.50 gpd/ft^2 for 46 to 60 min/in. percolation)
.24 gpd/ft^2 for 60 to 120 min/in. percolation

9. Distribution lateral diameter (1 to 3 in.), perforation spacing (2 to 10 ft o.c.), and perforation diameter ($\frac{1}{4}$ to $\frac{1}{2}$ in.) are selected so as to obtain uniform distribution of wastewater over the entire absorption area of the bed or trenches. Maximum lateral length from header recommended is 25 to 40 ft with ends capped. With longer laterals, increase pipe diameter and perforations.
10. Dosing frequency of four times daily is recommended. The volume of dose is therefore approximately $\frac{1}{4}$ the daily design flow. The dosing volume should be about 10 times the total lateral pipe volume. The pumping chamber size should be at least equal to the daily design flow. A pump, or siphon if the slope permits, is selected to provide sufficient capacity (volume of dose) and head (2 ft at end of lateral) to give good effluent distribution. USEPA and Converse[75] give details and charts to assist in pump selection for distributor pipe size, perforation diameter, and spacing.
11. Trench spacing is taken as the dose per trench in gpd divided by the natural soil loading rate (infiltration rate, see "8" above) in gpd/ft^2 times the trench length in ft.

Example

Given: Daily flow = 300 gal
Trench bottom infiltration rate for medium sand = 1.2 gal/ft^2/day
Trench width = 2 ft
Natural soil infiltration (60 to 120 min/in. soil) = .24 gpd/ft^2

Solution: Absorption trench area required = 300/1.2 = 250 ft^2
Total trench length = 250/2 = 125 ft
Use three trenches, each 42 ft; each trench to take 100 gpd
Trench spacing = 100/.24 × 42 = 9.9 ft, center to center
Basal area required = 300 gpd/.24 gpd/ft^2 = 1250 ft^2

See Figure 4-25 for construction details and specifications. *Recommended Standards for Individual Sewage Systems*, 1980 Edition, Great Lakes-Upper Mississippi River Board of State Sanitary Engineers, also gives design and specification details.

In view of the experiences reported with sand filters, sands with effective size less than 0.2 to 0.25 mm can be expected to clog with a dosage of 1 to 1.5 gpd/ft^2. Sand size can therefore be very critical.[64,65]

Septic-Tank Evapotranspiration System

In evaporation, surface water, soil water, and precipitation in falling and collecting on vegetation or other surfaces, are converted to atmospheric moisture. The rate of evaporation depends mostly on temperature, relative humidity, barometric pressure, and wind speed, also soil moisture, type of soil, and depth of moisture below the soil surface. In transpiration, water is taken in by plant roots, is used to build up plant tissue, moves up through the stem or trunk, and is released as vapor through the leaves. Transpiration is related to

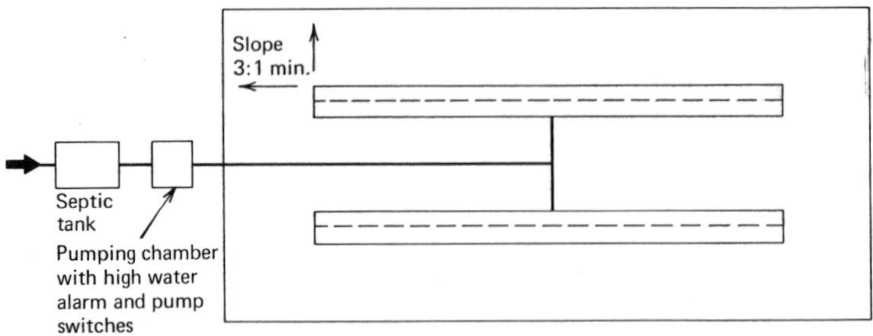

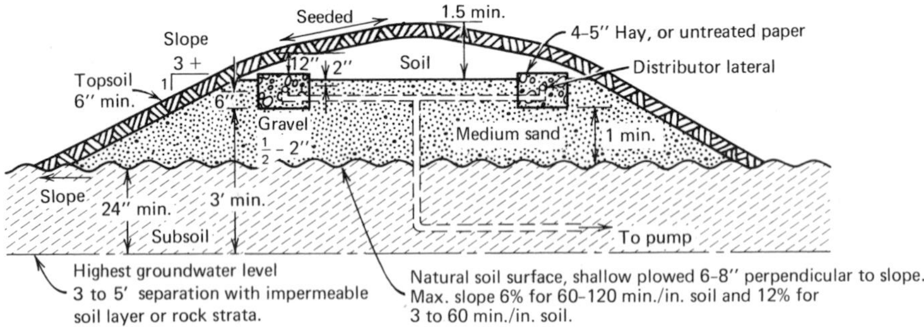

Figure 4-25 Details of a mound system using trenches. The asterisk denotes that the sand contains a minimum of 25 percent medium or coarser size grains. No grain size or stones larger than $\frac{1}{16}$ in. in more than 15 percent by volume of the sand material. It must be 2 ft minimum depth where permeable or creviced bedrock exists. Use perforated small diameter pipe for distributor laterals. Select pipe diameter and perforation spacing to give uniform distribution. See Converse for details, also text. (Based on James C. Converse, *Design and Construction Details for Wisconsin Mounds*, Small Scale Waste Management Project, University of Wisconsin, Madison, September 1978)

wind speed, soil moisture content, and type of vegetation. Empirical transpiration is shown in Figure 4-26. The quantity of water transpired and evaporated from a cropped area, or the normal loss of water from the soil by evaporation and plant transpiration, is known as the consumptive use or evapotranspiration (transvap for short). The largest proportion of consumptive use is usually transpiration. It is to be noted that evapotranspiration rates for crops and other vegetation in various areas of the world reflect the differences in meteorological conditions such as temperature, rainfall, humidity, and wind speed.

The evapotranspiration system can be used where the available soil has no absorptive capacity or where little or no topsoil exists over clay, hardpan, or rock. It can also be built where the groundwater level is high provided it is built with a watertight liner on the bottom and sides to exclude the groundwater from

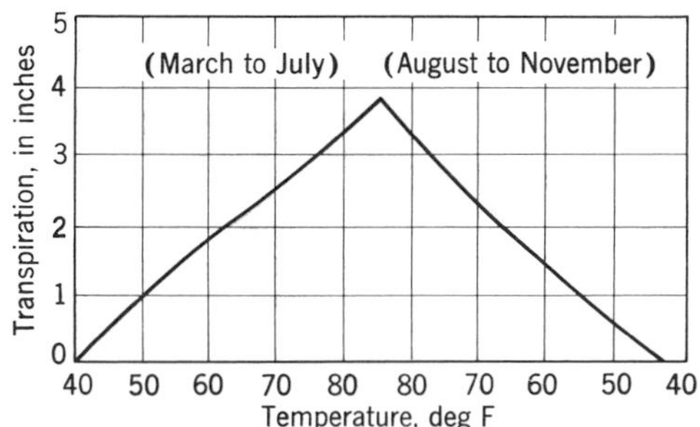

Figure 4-26 Empirical transpiration curve. The temperature scale refers to mean monthly air temperature and the transpiration scale refers to the corresponding transpiration in in. per month. [From "Computing Runoff from Rainfall and Other Physical Data," *ASCE Trans.*, **79,** 1094 (1915).]

the transvap bed. If an impermeable liner is not provided, elevation of the bed or curtain drains may be necessary if seasonal high water is a problem.

Evapotranspiration

The design of a transvap system is based on maintenance of a favorable input-output water balance. The precipitation less runoff (the infiltration), plus the wastewater flow, must be less than the evaporation plus transpiration. The year-round use of a transvap system where the average daily temperature is below freezing is probably impractical. A transvap bed design for summer use is shown in Figure 4-27. A more detailed design basis for year-round use follows. (See also Figures 4-28 and 4-29.)

Schwartz and Bendixen reported that vegetation on the surface of subsurface sewage disposal systems doubled the hydraulic longevity of conventional septic-tank absorption systems.[76]

Phelps states that evaporation from water surfaces varies from about 20 in. per year in the Northeastern United States to 90 in. per year in the Imperial Valley (with 100 to 120 in. in some southwestern areas), and that evaporation from land areas will be approximately one-third to one-half these values.[77] About two-thirds of the total annual evaporation takes place from April to September. Phelps also gives the following transpiration figures for several types of vegetation for the period in leaf:

Grain and grass crops	9 to 10 in.
Deciduous trees	8 to 12 in.
Small brush	6 to 8 in.
Coniferous trees	4 to 6 in.

[76]Warren A. Schwartz and Thomas W. Bendixen, "Soil Systems For Liquid Waste Treatment and Disposal: Environmental Factors," *J. Water Pollut. Control Fed.*, **42**, 624–630 (April 1970).
[77]Earle B. Phelps, *Public Health Engineering*, Vol. 1, John Wiley & Sons, New York, 1948, p. 264.

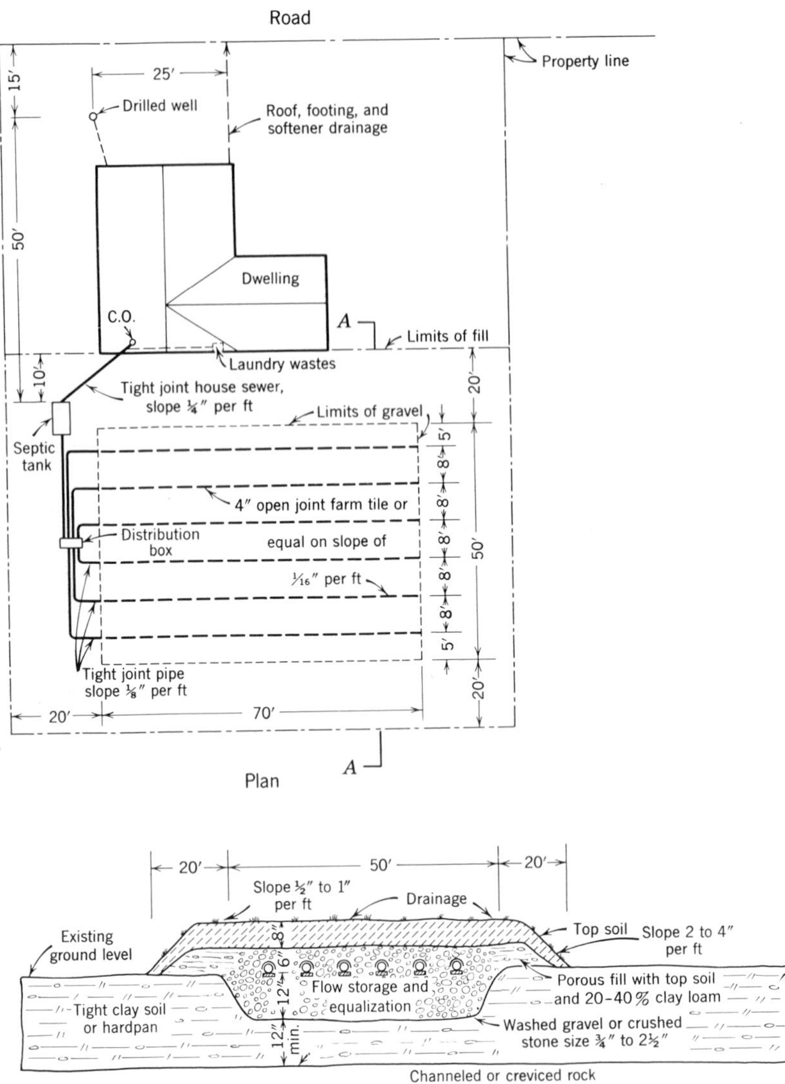

Figure 4-27 Transvap sewage disposal system on clay, for seasonal use only. Incorporation of clean sand (E.S. 0.10mm) wicks between laterals extending from bottom of bed to topsoil will enhance capillarity and hence evapotranspiration. See Figure 4-28. *Design basis:* Location—Northeastern U.S.; Season—Dry weather; Soil evaporation—8 in. per yr. in 5 to 6 months; Grass transpiration—9 in. per yr, in 5 to 6 month growing season; Soil percolation—Zero; Daily sewage flow—300 gpd; Design transvap: (Evaporation + Transpiration) × 0.623 gal/ft^2/in.

$= (8 + 9) \times 0.623$

$= 10.59$ gal/ft^2/yr

$= 0.03$ gal/ft^2/day;

Required transvap area for a 2-bedroom dwelling—(300 gpd) 300/0.03 = 10,000 ft^2 gross area = approx. 90′ × 110′; as illustrated.

Adjust design for rainfall and runoff, and occupancy period. Uniform gravel and sand have about 30 to 40% void space for storage. Bottom of bed at least 2 ft above groundwater. Curtain drain may be needed.

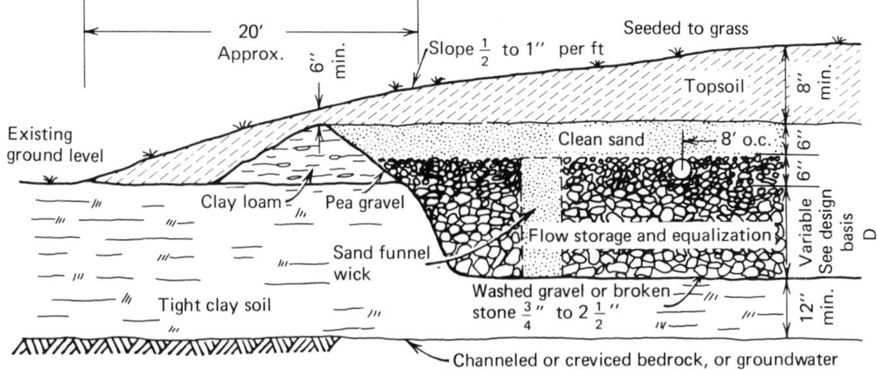

Figure 4-28 Transvap sewage disposal system in tight soil (Raise bed as necessary if groundwater or bedrock is a problem). Clean washed sand, 0.1mm effective size for up to 30 in. gravel depth and 0.2mm sand for up to 16 in. gravel depth. Sand wicks placed 8 ft on center; installed with aid of 12 in. stove pipe or other cylinder which is pulled out after the gravel is placed.

McGauhey gives the following average hydrologic water recycling distribution:[78]

Evaporation	30 percent
Evapotranspiration	40 percent (from soil mantle)
Surface runoff	20 percent
Groundwater storage	10 percent

Studies in England show that evaporation from a free water surface is 16½ in. of water per year and 14½ in. from bare soil.[79] Evaporation loss from turf or grassland is 0.6 to 0.8 times that from an open water surface. Lake water evaporation averages 27 in. per year in Portland, Me. and Toronto; 30.7 in. in Syracuse, N.Y.; 57 in. in Miami, Fla.; 74 in. in Phoenix; and 24 in. in Vancouver, B.C. See Weather Bureau for information on specific locations.

Blaney has reported on the studies of a number of investigators giving the consumptive use figures for some crops in California.[80] This information has been used in the development of Table 4-12. Table 4-13 gives some additional estimates of seasonal consumption of water by crops and vegetation. An example will show the possible use of this data for the design of a transvap system.

With a 150-day (5-month) growing season, using a cover of meadow or lucern grass, assume a very conservative 15- to 36-in. consumptive use and 18 in. of precipitation during the season, of which one-half (9 in.) or more runs off on a

[78]P.H. McGauhey, *Engineering Management of Water Quality*, McGraw-Hill Book Co., New York, 1968.
[79]*Manual of British Water Supply Practice*, Institute of Water Engineers, W. Heffer & Sons, Ltd., Cambridge, England, 1950.
[80]Harry F. Blaney, "Use of Water by Irrigation Crops in California," *J. Am. Water Works Assoc.*, **43,** No. 3, 189–200 (March 1951).

Table 4-12 Consumptive Use of Crops in California

Crop	Evaporation (soil) and Transpiration (plant)	Growing Season	Average Annual Transvap[a]
Alfalfa	5 in./month or 0.83 pt/ft²/day[b]	6 or 7 months	0.415 pt/ft²/day
Truck garden	4 in./month or 0.66 pt/ft²/day	5 months	0.275 pt/ft²/day
Cotton	4 in./month or 0.66 pt/ft²/day	7 months	0.38 pt/ft²/day
Citrus orchard	3 in./month or 0.50 pt/ft²/day	7 months	0.3 pt/ft²/day
Deciduous orchard	5 in./month or 0.83 pt/ft²/day	6 months	0.415 pt/ft²/day

[a]Soil evaporation for 7 to 5 months outside of growing season not included.
[b]pints per square foot per day = pt/ft²/day

Table 4-13 Approximate Evapotranspiration/Consumptive Uses

Growth[a]	Inches of Water[b]	Ref.
Alfalfa		
Brawley, Calif.—annual measured, Lysimeter	80	1
Kimberly, Idaho—April to October	55	4
Kimberly, Idaho—annual	53	1
Kimberly, Idaho—1 May to 30 September	36	1
Upham, N.D.—143 days observed	23	1
Reno, Nev.—124 days	40	1
Arvin, Calif.—annual	50	1
Mesa and Tempe, Ariz.—annual	74	1
Swift Current, Sask.—annual	25	1
Colorado—annual	26	2
Alberta—annual	22	3
Trees		
Coniferous trees	4 to 9	6
Deciduous trees	7 to 10	6
Oak—54 ft high, 25 in. dia., 34 gal summer, 6 gal winter[c,5]		
Pine—46 ft high, 15 in. dia., 40 gal summer, 10 gal winter[c,5]		
Apple—15 ft high, 6 in. dia., 18 gal summer, 3 gal winter[c,5]		
Rye grass		
Seabrook, N.J.—clipped grass, annual	38	1
Lompac, Calif.—hoed grass, annual	41	1
Davis, Calif.—hoed grass, annual	52	1
Copenhagen, Denmark—clipped clover and rye grass, annual	16	1
Aspendale, Australia,—clipped clover and rye grass, annual	51	1
Alberta—pasture grass, annual	18	3
Colorado—meadow grass, annual	23	3
Meadow grass—season	22 to 60	6

Table 4-13 *(Continued)*

Growth[a]	Inches of Water[b]	Ref.
Rye grass (Continued)		
Lucern grass—season	26 to 55	6
Canarygrass—season		
Hay		
Coshocton, Ohio—grass legume hay, annual	40	1
South Park, Colo.—native hay, annual	22	1
Clover		
Aspendale, Australia—annual	51	1
Prosser, Wash.—23 May to 28 Oct.	34	1

[1]*Consumptive Use of Water and Irrigation Water Requirements*, A report prepared by the Technical Committee on Irrigation Water Requirements, Marvin E. Jensen, ASCE, Ed., New York, September 1973.

[2]Harry F. Blaney, "Water and Our Crops," *The Yearbook of Agriculture 1955 Water*, USDA, Washington D.C., 1955.

[3]John R. Davis, *Evaporation and Evapo-Transpiration Research in the United States and Other Countries*, American Society of Agricultural Engineers, December, 1956. Based on report by W. L. Jacobson and L. G. Sonmor, Department of Agriculture Experiment Farms Service, Canada Department of Agriculture.

[4]James L. Wright and Marvin E. Jensen, "Peak Water Requirements of Crops in Southern, Idaho," *J. Irrig. Drain. Div., ASCE*, Proc. Paper 8940, June 1972.

[5]Alfred P. Bernhart, *Treatment and Disposal of Waste Water from Homes by Soil Infiltration and Evapo-transpiration*, University of Toronto Press, Canada, 1973, p. 146.

[6]Leonard C. Urquart, *Civil Engineering Handbook*, McGraw-Hill Book Co., New York, January 1950, p. 796.

[a]Well watered.

[b]Obtain more accurate data from local farm bureau or agricultural college. One in. of water = 0.623 gal/ft^2.

[c]Transpiration per day.

well crowned transvap bed. The water that can be disposed of during the growing season by evapotranspiration will range from:

$$\frac{(15-9)\times 0.623^{*}\times 8^{\dagger}}{5\times 30}=0.2\ \text{pt/ft}^2/\text{day, to}$$

$$\frac{(36-9)\times 0.623^{*}\times 8^{\dagger}}{5\times 30}=0.9\ \text{pt/ft}^2/\text{day}$$

This would correspond to an average of 0.08 to 0.37 pt/ft^2/day *on an annual basis*. A design on this basis, with a gravel bed 18 to 24 or more in. deep (including sand wicks, see Figure 4-28) to provide storage during the nongrowing season

*One in. of water/ft^2 in gal.

†Eight pt in one gal.

topped by a 6 to 12 in. bed of clean sand to promote capillarity, and then 8 in. of topsoil (well crowned) for growing grass, would have application where tight soil exists, groundwater is not too high (although the bed could be elevated), and there is no practical alternative. Credit, although small, can also be given for soil evaporation,* and possibly sublimation during the snow-covered nongrowing season. This design basis is much more conservative than that given by others. See examples in Figures 4-25, 4-27, and 4-28.

Beck,[81] based on studies made in San Antonio, Texas, recommends an evapotranspiration rate of 0.482 gpd/ft^2 in raised sand beds. Lomax found evapotranspiration systems satisfactory for the disposal of aerobic wastewater near Cambridge, Maryland during a year in which the annual precipitation was 55 in. The bed was lined, designed to dispose of 0.08 gpd/ft^2 of bed with 1.65 ft depth of sand, was crowned, and seeded to grass.[82]

Capillarity

For evaporation to take place, it is necessary to have upward movement of the water (capillarity) in the soil to the ground surface, and for transpiration to take place it is necessary for the capillary water (capillary fringe) to reach the surface vegetation root system.

Molecules within a liquid are attracted to one another (cohesion), and on a water surface create surface tension. Water molecules have a greater affinity for a solid they come into contact with than for other molecules (adhesion). The result is that water will climb up the surface of the solid, to an extent dependent on the diameter of the soil particle pore space or tube. This is referred to as capillary rise in soil and forms the capillary fringe above the underground water level or water table. Water rises higher in smaller pores. For example, water will rise by capillarity 28 cm in a soil having a cylindrical pore radius of 100 μm and 103 cm in a soil with a pore space of 30 μm. More specifically, a fill material consisting of uniform sand in the size range of 0.10 mm is capable of raising water about 3 ft by capillarity.[74] No capillary action occurs in a saturated soil.

Fair, et al.,[83] give an example for a capillary tube 0.1 mm in diameter and water at a temperature of 10° C, resulting in a rise of 31.6 cm based on

$$h = \frac{\sigma}{235d}$$

in which h is the capillary rise of the water in centimeters, d is the tube diameter in centimeters, and σ the surface tension in dynes/cm (74.9 at 5° C, 74.2 at 10° C,

*Approximately 0.07 to 0.10 pt/ft^2/day, where average lake evaporation is 30 in./yr and soil evaporation is $\frac{1}{3}$ to $\frac{1}{2}$ lake evaporation for 6 months of year.

[81]Arthur F. Beck, "Evapotranspiration Bed Design," *J. Environ. Eng. Div. ASCE*, April 1979, pp. 411–415.

[82]Kenneth M. Lomax, "Evapotranspiration Method Works for Wastewater Disposal along Chesapeake Bay," *J. Environ. Health*, May/June 1979, pp. 324–328.

[83]Gordon M. Fair, John C. Geyer, and Daniel A. Okun, *Water and Wastewater Engineering*, John Wiley & Sons, Inc., New York, 1966, pp. 9–10 and 9–11.

73.5 at 15° C, and 72.8 at 20° C). The equation assumes that the weight of air is insignificantly small and the weight of water close to unity.

Operation

The successful operation of a transvap system is largely dependent on runoff, surface vegetation, soil cover, capillarity, and evapotranspiration, in addition to controlled wastewater flow *to maintain a favorable water balance.* Plant roots can reach a depth of about 24 in. in well developed absorption beds and take up wastewater. Needless to say, the use of water-saving devices such as low-flush toilets, aerators on water outlets, water-saving shower heads, water pressure and flow controls would reduce water use and wastewater volume to be treated. Maintenance of a permeable soil structure and microbial population are essential, to minimize system clogging and failure as previously explained.

Design

The bed consists of washed gravel or crushed stone, topped with pea gravel, clean sand, and finally topsoil as shown in Figure 4-28. The area of the bed and depth of gravel are determined by a water balance study, that is, sewage flow, precipitation, infiltration and bed runoff, evaporation (temperature, humidity, winds) and transpiration (vegetation consumptive use) during the growing season, evaporation from the soil during the nongrowing season, and storage required. To promote capillary rise, sand should funnel down into the gravel to provide a wick for the water to rise into the topsoil and vegetation root system as shown in Figure 4-28. An example of a rational design for year round transvap bed is given in Figure 4-29.

A more precise water balance analysis to determine the storage required can be made by adapting the mass diagram or Rippl method. The weekly or monthly inflow consisting of the precipitation minus runoff or infiltration plus wastewater input flow would be balanced against the outflow evaporation and transpiration. When the cumulative difference between inflow and outflow is plotted against time it is possible to determine the storage required at any point in time. See Figures 3-23*a* and 3-23*b*, also standard civil-sanitary engineering texts showing the design of impounding reservoirs and water storage tanks.[84]

Dosing Arrangements

The size of a dosing tank is determined by the length of an absorption field or area of a sand filter to be dosed at any one time. The volume of the dose should equal 60 to 75 percent of the volume of lines dosed. In general, a siphon or other dosing device should be provided when the total length of the absorption field exceeds 300 to 500 ft, and when the sand filter distributor laterals exceed 300 linear ft or the area of the sand bed exceeds 1800 ft^2. Alternating siphons are

[84]Gordon M. Fair, John C. Geyer, and Daniel A. Okun, *Elements of Water Supply and Wastewater Disposal,* John Wiley & Sons, Inc., New York, 1971, pp. 74–82.

Assumptions—Based on each installation and location

1. Sewage flow = 300 gpd
2. Rainfall infiltration = 0.75 in. (25% of 3-in. average rainfall per month; 75% runoff and evaporation and 25% infiltration). *Crown bed to promote runoff.*
3. Land evaporation = 1.4 in./month (=0.6 of surface water evaporation of 22 to 35 in./yr or average of 1.1 to 1.7 in./month = 1.4 in.)
4. Gravel void space = 40%
5. Transvap bed = 10,000 ft^2, 7500 ft^2, or 5000 ft^2
6. Operation period = 12 months
7. Evapotranspiration, or consumptive use, occurs only over 5-month growing season.

Storage Required—During 7-month nongrowing season, the total volume of sewage and rainwater infiltration minus soil evaporation, to be stored with a 10,000 ft^2 bed = Y. (Actually some transpiration will also occur.)

Y = sewage flow + infiltration − soil evaporation

= 300 × 30 × 7 + [(0.75 − 1.4) × .623 × 7 × 10,000]

= 63,000 − 28,346

= 34,654 gal storage; 41,740 gal with 7500 ft^2 bed; 48,827 gal with 5000 ft^2 bed

During 5-month growing season, the total volume of sewage and water to be disposed of in a 10,000 ft^2 bed by transvap = Q = 5 month sewage flow + 5-month rainwater infiltration + Y (storage).

Q = (300 × 30 × 5) + (5 × 0.75 × .623 × 10,000) + 34,654

= 45,000 + 23,362 + 34,654

= 103,016 gal; 104,261 with 7,500 ft^2 bed; 105,508 gal with 5000 ft^2 bed.

Solutions

Find required vegetative consumptive use and depth of gravel for Transvap bed.

Consumptive use I for 5-month growing season over 10,000 ft^2 bed. 103, 016 gal = I × .623 × 10,000; I = 16.54 inches. I = 22.31 with 7500 ft^2 bed. I = 33.87 with 5000 ft^2 = bed.

Use alfalfa, oats, meadow grass, or lucern grass—see local soil conservation service for best vegetative cover for the area. See Table 4-13.

Bed depth D to provide 34,654 gal storage. 34,654 = 10,000 × D × 7.5 × .4 (void space.) D = 1.16 ft. D = 1.86 ft with 7500 ft^2 bed. D = 3.26 ft with 5000 ft^2 bed.

Note:

Can vary design sewage flow, bed area, bed depth, and vegetative cover (consumptive use) to meet particular need and obtain water balance.

Obtain monthly evapotranspiration (consumptive use), rainfall, water, and land evaporation rates from soil conservation service, agricultural college, and weather station for local conditions.

Design Considerations and Specifications

Topsoil, capillarity, texture, structure
Sewage distribution
Need for pumping
Gravel size, depth, area; also sand
Availability and cost of suitable gravel and sand
Bed surface slope, crown 1 in. per ft for good runoff

Vegetation selection
Bed edge feathering
Bed clay loam dam
Land area available and needed
Cover vegetation evapotranspiration rate
Land evaporation, nongrowing season
Runoff and actual infiltration
Rainfall; includes all precipitation

Figure 4-29 Rational Design of a Transvap Bed—Example.

recommended when the absorption field exceeds 1000 ft and when the sand filter distributors exceed 800 linear ft.

Dosing tanks are usually designed to operate automatically. Dosage is accomplished by a siphon, pneumatic ejector, tipping bucket,[85–87] or pump. Hand-operated gates, float valves, and motorized valves are also used. If the available grade does not permit the use of automatic siphons or a similar device, pumps or ejectors will have to be used. In such cases the system should be designed, if possible, to permit gravity flow through the dosing tank or pump well to the absorption field or filter, in case of pump or power failure, and while repairs are being made. If this cannot be done a standby gasoline-engine-driven pump should be provided, in addition to a high-water alarm to warn of pump failure. In larger systems, two siphons or two pumps with proper float setting are installed, each feeding a separate absorption field or sand filter section in alternation.

The size siphon or other dosing device that has a minimum discharge rate at least 125 percent greater and preferably twice the probable maximum rate at which settled sewage might enter the dosing tank is selected. This is necessary in order to exceed the rate of flow of the incoming sewage and vent the siphon, making possible its continued automatic operation. Where open sand beds are dosed, rapid discharge of the sewage gives better distribution over the sand bed. The head or fall available will also determine the size siphon that can be used. Special designs incorporating given drawdowns can be obtained from the manufacturer. The dosing device should be capable of applying the required volume on the filter in less than 10 min. Some design details of the Miller siphon are given in Figure 4-30. Prefabricated metal dosing tanks incorporating single or alternating siphons are also available.[88] The diameter of the carrier line, that is, the line between the dosing tank and the absorption field, can be developed with the aid of Figure 4-31, since the maximum discharge rate of the siphon is given in Figure 4-30.

Manufacturers' detail drawings of the specified siphon shown should be obtained and carefully followed during construction and critical elevations staked out. The siphon trap must be filled with water when the system is to be placed in operation. The small vent pipe on the bell must have airtight joints, with the bell perfectly level. Sometimes the floor under the siphon bell is depressed 4 or 6 in. below the dosing tank floor with satisfactory results. Another acceptable alternative is to move the siphon forward in the dosing tank and install a tee extension on the discharge or carrier line in the dosing tank extending about 2 in.

[85]Ehlers and Steel, *Municipal and Rural Sanitation*, McGraw-Hill Book Co., N.Y., 1927, p. 55.
[86]O.J.S. Macdonald, *Small Sewage Disposal Systems*, H.K. Lewis & Co. Ltd., Gower St., London, 1951, pp. 75 and 96.
[87]Leonard Metcalf and Harrison P. Eddy, *American Sewerage Practice, Vol. III, Disposal of Sewage*, McGraw-Hill Book Co., N.Y., 1935, p. 806.
[88]San-Equip, Inc., Syracuse, N.Y.; Kaustine Co., Inc., Perry, N.Y.; Pacific Flush-Tank Co., 4211 Ravenswood Ave., Chicago, Ill.; Fluid Dynamics Co., Box 4659, Boulder, Colo. 80302. See also *Sewage and Sewerage of Farm Homes*, U.S. Dept. of Agriculture Farmers' Bull. 1227, Washington, D.C., January 1922, revised October 1928.

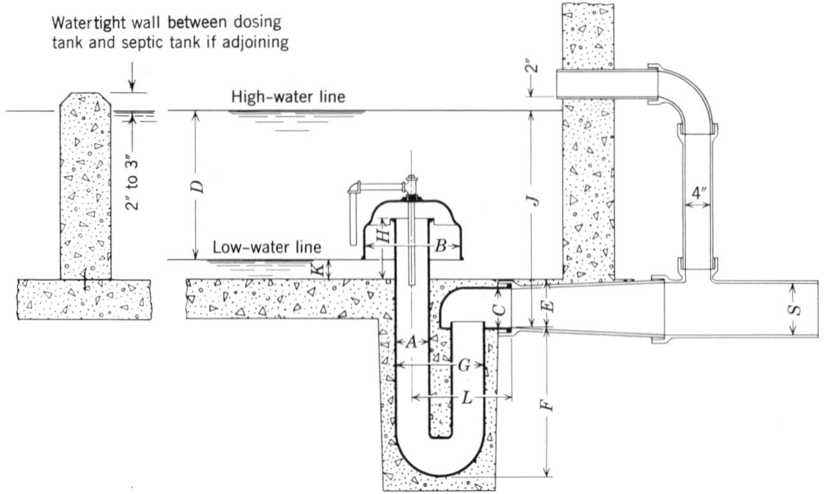

Figure 4-30 Design details of the Miller siphons. (See manufacturer's direction.) Two single siphons of this type set side by side in the same tank will alternate. The draft D will be 1 in. to 2 in. less in this case. One ft of 4-in. pipe holds 0.653 gal. Approximate dimensions (in.).

Diameter of siphon	A	3	4	5	6
Drawing depth	D	13	17	23	30
Diameter of discharge head	C	4	4	6	8
Diameter of bell	B	10	12	15	19
Invert below floor	E	$4\frac{1}{4}$	$5\frac{1}{2}$	$7\frac{1}{2}$	10
Depth of trap	F	13	$14\frac{1}{4}$	23	$30\frac{1}{4}$
Width of trap	G	10	12	14	16
Height above floor	H	$7\frac{1}{4}$	$11\frac{3}{4}$	$9\frac{1}{2}$	11
Invert to discharge = $D + E + K$	J	$20\frac{1}{4}$	$25\frac{1}{2}$	$33\frac{1}{2}$	44
Bottom of bell to floor	K	3	3	3	4
Center of trap to end of discharge ell	L	$8\frac{5}{8}$	$11\frac{3}{4}$	$15\frac{1}{2}$	$17\frac{1}{8}$
Diameter of carrier	S	4	4–6	6–8	8–10
Average discharge rate (gpm)	—	72	165	328	474
Maximum discharge rate (gpm)	—	96	227	422	604
Minimum discharge rate (gpm)	—	48	102	234	340
Shipping weight (lb)	—	60	150	210	300

above the high water line. This can take the place of the vent and overflow line passing through the dosing tank wall. Either the open tee or overflow are essential, however, to prevent siphoning out the water seal in the siphon trap and to assure continuous automatic operation of the siphon.

Lift Station

As in the case of the siphon, pump manufacturers' working drawings should be followed when a design calls for the installation of pumps. Typical sump pump details are shown in Figures 4-32 and 4-39. Horizontal pumps are also available and are sometimes preferred; a vacuum pump and air-relief valves permit

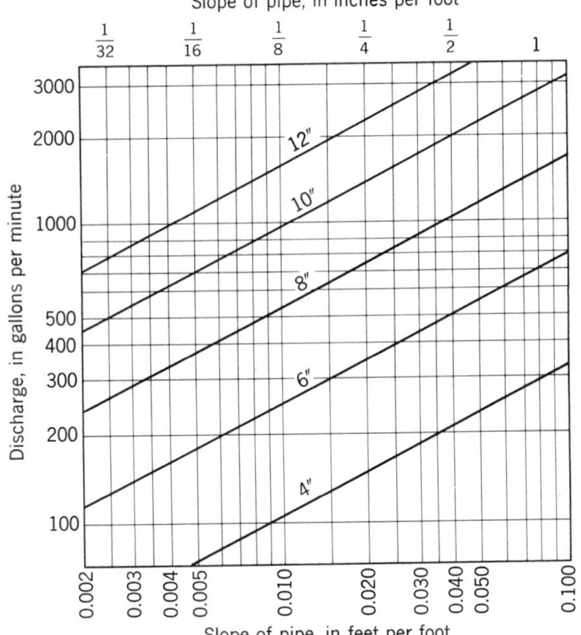

Figure 4-31 Discharge of clay pipe sewers. Discharge is based on Manning formula: $Q = A\ (1.486/n)\ R^{2/3}\ S^{1/2}$, with $n = 0.013$, sewer full.

automatic operation. Pumps should be selected so as to operate for at least 10-min intervals. Compressed air-operated sewage ejectors may be used to advantage. Obtain the manufacturers' recommendations for the most efficient pump and motor horsepower to meet job requirements. Submersible pumps are also available for pumping small and large quantities of settled sewage. Normally at least two pumps, each being capable of handling the maximum flow, should be provided.

Pressure, Vacuum, and Cluster Systems

Low-pressure, vacuum, and cluster systems are possible alternatives to serve areas where septic-tank systems are inappropriate or failing. Unsuitable soil, high groundwater, small lots, hilly terrain, and high-density recreational areas are situations where such systems may have application.

In the pressure sewer system the septic-tank effluent from one or more dwellings flows by gravity to a pumping station from which the sewage is pumped through small diameter pipe to an existing sewer or a new central treatment plant. The septic tanks require periodic cleaning. In some designs the individual septic tank is eliminated and a special collection tank-grinder pump and check valve assembly is used.

In the vacuum system a vacuum pump creates a vacuum in collector pipes. A valve opens when sewage from a dwelling presses against it. Sewage and a plug of

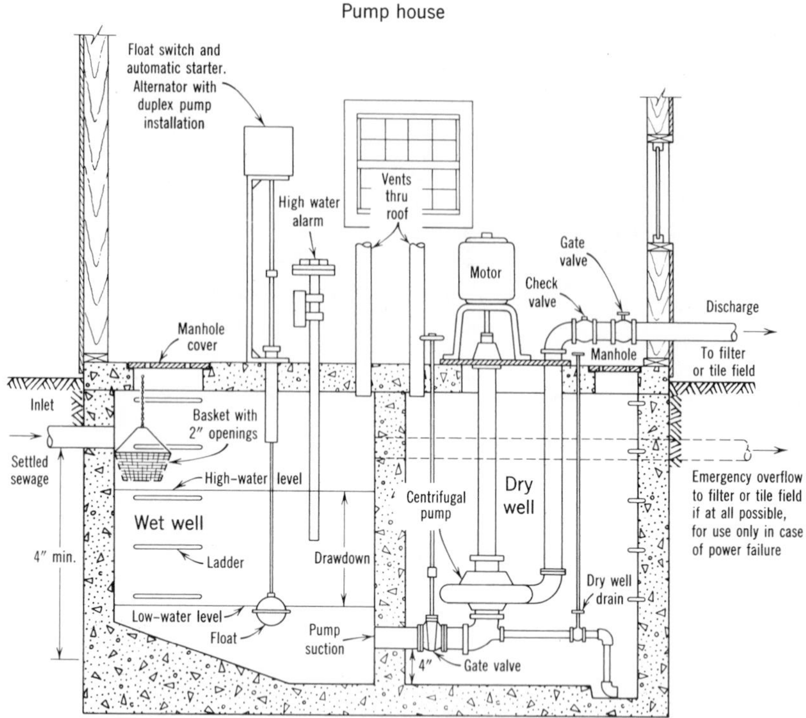

Figure 4-32 Sump pump for dosing sand filter or absorption field. When dry well is omitted, the pumps are submerged in the wet well. Sump may be attached to septic tank. (A comminutor may be substituted for the basket screen with raw sewage. Exhaust fans with lower inlets will provide more positive ventilation. Provide battery-powered alarm in case of electric power failure.)

air behind it enter the collection pipe. Air (vacuum) forces sewage to a collection tank. A sewage pump then pumps sewage from the tank to a treatment plant. Special vacuum valves and intermediate sumps (usually) are needed.

In the cluster system a group of dwellings is served by a common treatment and disposal system. Each dwelling makes connection to a common treatment system, or each house has its own septic tank or aerobic tank which connects to a common absorption field or other approved system. The septic tanks require periodic cleaning.

Both the pressure system and vacuum system require regular maintenance. All of the systems need to have a sewer district or equivalent to assure continued maintenance and operation in perpetuity.

SEWAGE WORKS DESIGN—SMALL TREATMENT PLANTS

Preparation of Plans and Reports

The design should be tailor-made to fit the local conditions and take into consideration probable future additions. Designs for small sewage treatment plants should be prepared by licensed professional engineers experienced in sanitary engineering work and be approved by the regulatory agency involved, before construction is started. Designs for typical small installations are sometimes made available through state or county health or environmental protection departments. For small jobs, pencil sketches drawn to scale with dimensions of all units and critical elevations can be prepared for checking by the approving agency. Changes during construction may be very costly. The line of demarcation between large and small sewage works is usually an arbitrary one and subject to local interpretation based on the value of the work, its difficulty, the danger to health and safety, and the reasonableness of requiring the services of a professional engineer or registered architect. A practical dividing point might be 25 persons or a sewage flow of 1500 gpd. In some states, when the value of the work exceeds $5000 or $10,000, or danger to health and safety is involved, it is required that plans be submitted by a licensed engineer or architect.

Most states require approval of plans and have rules, regulations, and standards to guide the designing engineer in the preparation of satisfactory plans. The review of preliminary drawings with the local sanitary engineer official will usually expedite approval of final plans. Federal permits (NPEDS) are needed. In some instances states are carrying out the federal permit program.

Plans submitted for approval should give all information necessary for review of the design and for construction of the disposal or treatment system as designed. The more complete the plans, the closer bids received for the job will be. Plans should include construction details, engineer's report, application or statement of information, specifications, and a topographic map showing the location of the disposal plant and outlet, if any. The first step, therefore, when a sewerage system is to be constucted, is to engage the services of an experienced licensed professional engineer to prepare preliminary plans and cost estimates and then complete plans for approval by the agency having jurisdiction and for construction of the system. The plans and engineer's report should include the following information.

1. A plot plan giving the boundaries of the property, drawn to a scale of 20 to 100 ft to the in., showing contour lines or critical elevations, streams, swamps, lakes, rock outcrops, wooded areas, structures, roads, play areas, and other significant features. Existing and proposed structures are also shown.
2. Location of existing and proposed sewer and water lines and other utilities, sources of water supply and storage facilities, grease traps, manholes, septic tanks, aeration tanks, and other disposal or treatment units, and outfall sewers if any.
3. Ground elevations and sizes, materials, slopes, and invert elevations of existing and new sewer lines leaving buildings and entering and leaving manholes, grease traps, septic tanks, siphons, trickling filters, settling tanks, aeration tanks, sludge tanks, drying beds, pump pits, distribution boxes, distributing laterals, sand filters, inspection manholes, chlorine contact tanks, and also the flood level of receiving streams. A profile through the sewerage system is very valuable.
4. Population served and present and future capacity. Where food or drink is served, the seating capacity, number of sales, and meals served should be taken into consideration.
5. Number of different plumbing fixtures including dishwashing machines, potato peelers, laundry machines, and continuously running and automatic flushing devices. Commercial and industrial devices should also be shown.

6. Location, depth, and results of soil percolation tests, with findings of tests made 2 or 3 ft deeper and description of the type of soil to a depth of at least 10 ft.
7. Construction details and sizes of all sewerage units, including structural details and specifications.
8. The highest groundwater levels, time of the year when tests were made, and how determined.
9. Location and yield or adequacy of the source, storage, and distribution of water. Actual or estimated daily water consumption and water pressures should also be given. Where the source of water is one or more wells or springs, the type of well or spring, strata penetrated, depth, and size, in addition to capacity and type of pump and motive power, should be included.
10. An environmental impact analysis, possible effects of treated sewage discharge on the aquifer if a subsurface absorption system is proposed, and possible effects downstream and on lakes and streams, and their uses, if a surface discharge is proposed. See also approving agency instructions and *Recommended Standards for Sewage Works* (see Bibliography).

Design Details

The provision of bar screens or comminutors and grit chambers ahead of pumping equipment or settling tanks is strongly recommended.

Pumping equipment is readily accessible for inspection and servicing when located in a dry pump pit. For lift stations, the pneumatic ejectors and "bladeless" sewage pumps seem to be the most suitable for small plants.

If secondary treatment will be needed, primary treatment units should be designed with the water level at a sufficient height to permit gravity flow to trickling filters, aeration units, oxidation ponds, or sand filters and to the receiving stream without additional pumping.

Trickling filter units for small installations may be built at ground level. The stone may be contained by reinforced concrete, heavy fencing, concrete block with reinforcing, or cypress or pressure-treated pine staves held with iron hoops or similar arrangements. Make provision for recirculation for treatment and filter fly control.

The secondary settling tank should be constructed with a sloping bottom. A bottom shaped like an inverted four-sided truncated regular pyramid will permit gravity collection and concentration of sludge for removal. Provide for a 2-ft hydrostatic head and proper diameter sludge drawoff pipe to give a flushing velocity when the control valve is opened and pumping started.

Chlorination of the final effluent can be accomplished in a chlorine contact tank or in the secondary settling tank. In small installations, manual-control continuous-feed chlorinators are sometimes used. Proportional-feed automatic chlorinators are found to be more economical in the larger installations. Other means for chlorination include dosing tank and pump-activated controls.

The location of the treatment plant should take into consideration the type of plant and available supervision, the location of the nearest dwelling, the receiving watercourse, the availability of submarginal low land not subject to flooding, prevailing winds and natural barriers, and the cost of land. A distance of 400 ft from the nearest dwelling is frequently recommended, although distances of 250 to 300 ft should prove adequate with good plant supervision. Some equipment manufacturers and designing engineers feel that covered sand filters, aeration-type plants, and high-rate trickling filters can be located closer to habitation without danger of odors or filter flies. Oxidation ponds and lagoons should be located at least $\frac{1}{4}$ to $\frac{1}{2}$ mi from habitation.

Some of the more common flow diagrams for small sewage treatment plants are illustrated in Figure 4-33 to suggest the different possibilities, dependent on local factors. Predesigned and prefabricated units are available.

Disinfection—Chlorination

Disinfection of sewage effluent is not always necessary. The need for disinfection should be predicated on the probability of disease transmission by ingestion of contaminated water or food, by contact, and by aerosols. The disease agent may be microbiological or chemical. See Disease Hazard, this chapter. It is known that wastewater treatment plants, including industrial ones, experience operational problems requiring the occasional release of untreated or inadequately treated wastewater. In addition, heavy rains may require a bypass of the treatment plant or sewer overflow to the receiving stream. Hence the design of sewers and treatment plants should recognize these realistic factors and provide adequate disinfection facilities, including retention basins, to protect downstream drinking water sources, fish and shellfish waters, recreation areas, and human habitation to the extent needed. Buxton and Ross reviewed the disinfection practices in Canada, the United States, and Europe. They and others, point out the deleterious effects of chlorination on the aquatic ecosystem and identify factors that should be considered in a wastewater disinfection policy.[89–91] Although chlorine as a disinfectant is discussed here, it does not preclude the use of sodium hypochlorite, ozone, or other disinfectant where indicated and permitted by the regulatory authority. Disinfection is not a substitute for adequate wastewater treatment; it is an added safeguard to reduce the risk of disease transmission where the probability exists. However, the wastewater must be adequately treated in the first place for the disinfectant (usually chlorine) to be effective.

The effectiveness of disinfection depends on the degree of treatment the sewage has received, the amount of chlorine used and residual chlorine maintained, the mixing and retention period, and the condition of the sewage, including the pH, temperature, nitrogen compounds, and organic and suspended matter. Sometimes chlorine is also added to sewage to control odors, undesirable growths, sewage flies, septicity, and chemical or bacterial reactions unfavorable to the treatment process. Chlorination will also reduce the BOD of the treated effluent, roughly in the ratio of one part chlorine to two parts BOD.

[89] G. Victor Buxton and S. A. Ross, "Wastewater disinfection—toward a rational policy," *J. Water Pollut. Control Fed.*, pp. 2023–2032.

[90] *Unnecessary And Harmful Levels of Domestic Sewage Chlorination Should Be Stopped*, Report to the Congress, The Comptroller General of the United States., Washington, D.C. 20548, August 30, 1977.

[91] *Disinfection of Wastewater*, USEPA, EPA-430/9-75-012, Washington, D.C. 20460, March 1976.

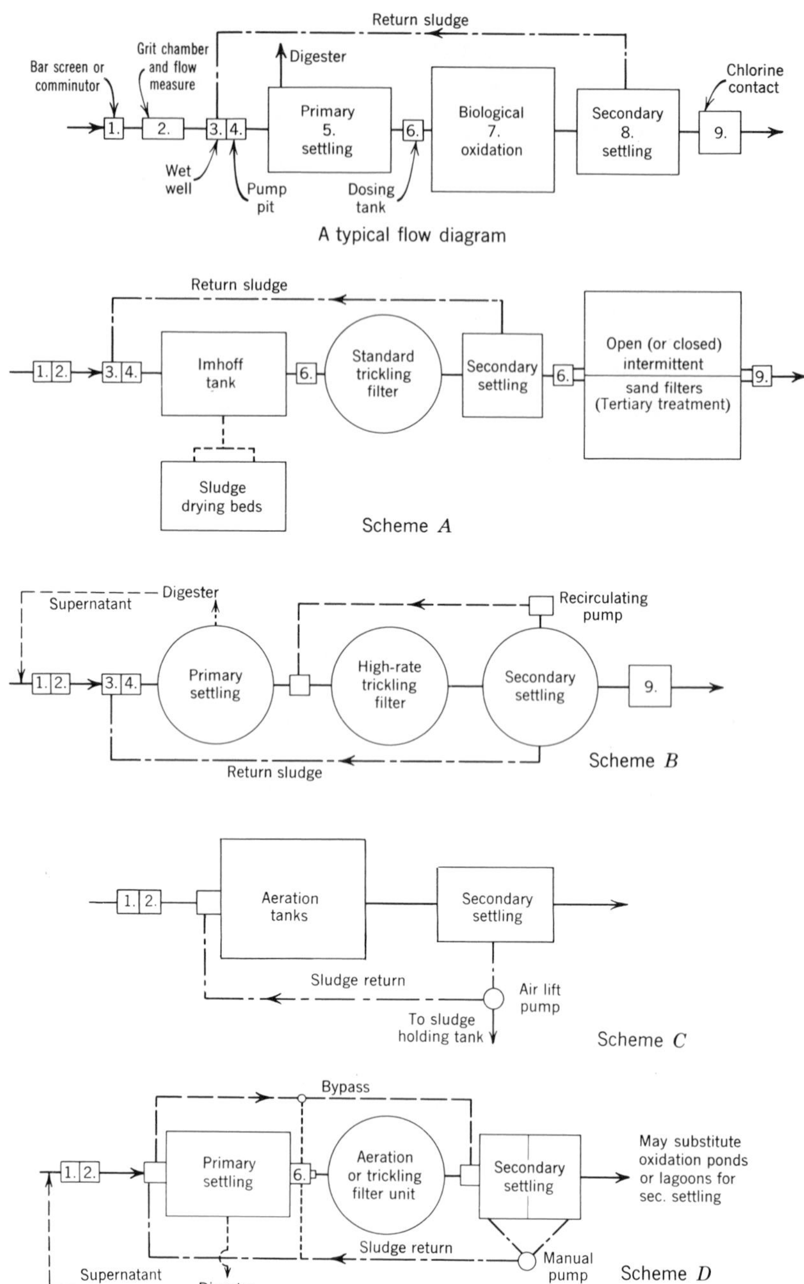

Figure 4-33 Typical flow diagrams.

Chlorination treatment of raw sewage is not reliable for the destruction of pathogenic organisms since solid penetration is limited. The required dosage of chlorine to produce a 0.5 mg/l residual after 15 min contact has been approximated in Table 4-14 for different kinds of sewage. Studies made using a domestic sewage show that less than 250 coliform organisms per 100 ml can be obtained in treated sewage 100 percent of the time if an orthotolidine chlorine residual of 2.1 to 4 mg/l is maintained in the effluent after 10 min contact.[92] Other experiments show that if the chlorine is first mixed with the sewage, and the treated sewage allowed to stand 10 min, an MPN of 300 coliform organisms per 100 ml or less can be obtained 100 percent of the time if a chlorine residual of 1.1 mg/l is maintained in the effluent.[93] These tests also show that with no mixing at least twice the chlorine residual must be maintained in the treated sewage for 10 min to give results approximately equal to the results obtained with mixing. Another study shows that an MPN around 1000 per 100 ml can usually be expected in the effluent if the product of the combined chlorine residual (as measured by the OT test) and detention time in minutes (based on average flow) is equal to or greater than 20.[94] Eliassen also reports a reduction in the MPN of aftergrowth values in a brackish water tidal basin, following discharge of a chlorinated combined sewage, of from 10 to 30 percent of those that would develop without chlorination.[95]

Combined chlorine, mostly monochloramine, is normally produced in the conventional chlorination of sewage effluent. This is to be expected since most secondary effluents contain substantially more than 1 mg/l of ammonia, which alone requires 8 to 10 mg/l of chlorine for neutralization, before there is a free residual chlorine. Nevertheless, although slow-acting, combined chlorine is effective in reducing fecal coliforms to 200 mg/l or less, with sufficient contact time. It is important to note, as found by Jolley[96] and reported by Smith,[97] that conventional chlorination of municipal wastewater to the combined chlorine residual level yields relatively small concentrations of chlorinated organic compounds suspected of being carcinogenic. This is in contrast to the chlorination of surface water supplies for drinking water to the free chlorine residual level with the formation of high concentrations (200 to 500 ppb) of chlorinated organic compounds such as trihalomethanes in some instances. Hence, con-

[92]H. Heukelekian and R. V. Day, "Disinfection of Sewage with Chlorine, III," *Sewage Ind. Wastes*, **23**, No. 2, 155–163 (February 1951).

[93]Rolf Eliassen, Austin N. Heller, and Herman L. Krieger, "A Statistical Approach to Sewage Chlorination," *Sewage Works J.*, **20**, 1008–1024 (November 1948).

[94]Nicholas W. Classen, "Chlorination of Wastewater Effluents," *Pub. Works*, **100**, 63–66 (January 1969).

[95]Rolf Eliassen, "Coliform Aftergrowths in Chlorinated Storm Overflows," *J. Sanit. Eng. Div., ASCE*, **94**, No. SA2, Proc. Paper 5913 (April 1968), pp. 371–380.

[96]Robert L. Jolley, "Chlorine-containing organic constituents in chlorinated effluents," *J. Water Pollut. Control Fed.*, March 1975, pp. 601–618.

[97]James W. Smith, "Wastewater disinfectants: Many called—few chosen," *Water Wastes Eng.*, June 1978, pp. 19–25.

Table 4-14 Probable Chlorine Dosages to Give a Residual of at Least 0.5 mg/l After 15 Min Retention in Average Sanitary Sewage or Sewage Effluent[a]

	Suggested Chlorine Dosages in Milligrams per Liter[b]					
Type of Sewage Effluent	N.Y. State[1]	Dunham[2]	ASCE-WPCF[3]	Griffin[4,5]	Imhoff and Fair[6]	GLUMRB[7]
Raw sewage				6–12 fresh to stale 12–25 septic	6–25 fresh to stale and strength	
Septic tank		10–25		12–24		
Imhoff tank or settled sewage	20–25	5–20	18–24	5–10 fresh to stale 12–40 septic	5–20	
Trickling filter	15	3–15	6–12	3–5 normal 5–10 poor	3–20	10
Activated sludge	8		6–12	2–4 normal 5–8 poor	2–20	8
Intermittent sand	6	2		1–3 normal 3–5 poor	1–10	6[c]
Chemical precipitation				3–6	3–20	

[1] *Manual of Instruction for Wastewater Treatment Operators,* Vol. 1, N.Y. State Dept. of Environmental Conservation, Albany, N.Y., May 1979, p. 6–9.
[2] *Military Preventive Medicine,* Military Publishing Co., Harrisburg, Pa., 1940.
[3] *Wastewater Treatment Plant Design,* Water Pollut. Control Fed. and ASCE, 1977, p. 382.
[4] *Public Works Magazine*, Ridgewood, N.J. (October 1949), p. 35.
[5] *Operation of Wastewater Treatment Plants*, Water Pollut. Control Fed., Washington, D.C., 1970, p. 144.
[6] *Sewage Treatment,* John Wiley & Sons, Inc., New York, 1956.
[7] *Recommended Standards for Sewage Works*, 1978 Edition, Great Lakes-Upper Mississippi River Board of State Sanitary Engineers.
[a] WHO suggests 0.5 mg/l free residual chlorine after one hour to inactivate viruses (after secondary treatment) with turbidity < 1.0 JTU. Combined chlorine, mostly monochloramine, is normally produced which is a slow-acting disinfectant. Eight to ten mg/l of chlorine is needed to neutralize each mg/l of ammonia before free chlorine is produced. Most secondary effluents contain more than 1.0 mg/l ammonia.
[b] 12 mg/l = 1 lb per 10,000 gal. Each mg/l chlorine in sewage effluent reduces the BOD about 2 mg/l. No appreciable industrial wastes.
[c] For tertiary filtration effluent and for nitrified effluent.

trolled chlorination of sewage, where this is indicated, to below the free residual chlorine level would seem to have public health and economic merit, although free chlorine is recognized as the more rapid, effective disinfectant.

Chlorine and chlorine products formed are toxic to freshwater, marine, and estuarine aquatic organisms in minute concentrations. (See Doudoroff and Katz, Control of Microorganisms, Chapter 3.) A maximum total residual chlorine limit of 0.002 mg/l in salmonoid fish areas and 0.01 mg/l for marine and other freshwater organisms has been recommended by USEPA. Dechlorination with sulfur dioxide will remove essentially all residual toxicity to aquatic life from chlorination. Alternative disinfectants are ozone and chlorine dioxide. The use of ultraviolet light is in the experimental stage. Ozone is nontoxic to aquatic organisms, a good viricide, and adds dissolved oxygen to treated wastewater effluents. Its cost is higher than chlorine disinfection. Chlorine dioxide added to wastewater does not result in formation of appreciable concentrations of trihalomethanes. Excess sulfur dioxide may cause a drop in pH and dissolved oxygen, requiring treatment adjustment. Hydrogen peroxide reacts readily with free available chlorine in the ratio of about $\frac{1}{2}$ lb hydrogen peroxide to 1 lb of free chlorine; but it is relatively ineffective against combined chlorine.

Chlorine is available as a relatively pure liquid in steel cylinders under pressure having a net weight of 100 or 150 lb. Larger cylinders such as ton containers and tank cars are also available. When the pressure is released, the liquid turns to a gas, in which form it is added (mixed with or without water) to sewage by means of a control and measuring device known as a chlorinator. Liquid chlorine is not ordinarily required or economical to use at very small sewage treatment plants. A separate gas-tight room above ground, a building with separate ventilation, an outside entrance, and special gas mask or self-contained breathing apparatus would be required. Calcium hypochlorite, which is a powder containing 70 percent or less available chlorine, or sodium hypochlorite, which is a solution containing 15 percent or less available chlorine, is more generally used. Both the powder and solution are mixed and diluted with water to make a 0.5 to 5.0 percent solution, in which form it is added to the sewage by means of a solution feeder known as a hypochlorinator. Positive feed hypochlorinators are preferred to other types because of their greater dependability.

Chlorine should be added in a manhole, approximately 20 ft ahead of the chlorine contact tank, or in a mixing box, in a manner to provide good mixing of the chlorine and sewage. The chlorine contact tank should be designed to provide 30 to 60-min retention at peak hourly flows. Detention time and the chlorine residual is usually specified by the regulatory agency. Chlorine contact tanks should be constructed with over-and-under or round-the-end baffles or equivalent obstructions to assure the required contact period and avoid short-circuiting. Serpentine flow with channel length-to-width ratio of 40 or 70 to 1 is reported to be most efficient. Provision should be made in the design for the collection of samples. Sometimes the required contact time can be obtained in a long outfall sewer or by adding chlorine in a final sedimentation tank when secondary treatment is provided.

Trickling Filter

A trickling filter may be used following a primary settling tank, a septic tank, or an Imhoff tank to provide secondary treatment of the sewage. Habitation should not be closer than 400 ft. Some odors and filter flies can be expected with a standard rate filter. A receiving stream providing adequate dilution and supervision over operation is required. Seeding of the filter stone is necessary before good results are produced. High BOD reduction is obtained within 7 days of starting a trickling filter, but as long as 3 months may be required to obtain equilibrium, including high nitrification.[98]

Small standard-rate trickling filters are usually 6-ft deep and designed for a dosage of 200,000 to 300,000 gpd/acre-ft, or not more than 1,800,000 gal for a 6-ft deep filter. Filter loading is also expressed, with greater accuracy, in terms of 5-day BOD in the sewage applied to the filter. It is usually assumed that 35 percent of the BOD in a raw sewage is removed by the primary settling unit. Standard-rate trickling filters are dosed at 200 to 600 lb of BOD/acre-ft/day. Average loadings are 400 lb in northern states and 600 lb in southern states. Since dosage must be controlled, dosing siphons or tipping trays may be used for very small jobs and dosing tanks with siphons or pumps with revolving distributors or stationary spray nozzles on the usual job. Continuous dosage at a higher rate, with recirculation of part of the effluent, may be suitable where good supervision is available and operation can be controlled to produce the intended results. Filter flies are reduced with recirculation.

A trickling filter should be followed by a secondary settling or humus tank; this unit will require the removal of sludge at least twice a day. The sludge is removed by pumping or by gravity flow if possible usually to the Imhoff tank or sludge digester, depending on the plant design. The discharge of the raw sludge to a sand drying bed is not advisable, as sludge drying will be slow and odors will result.

Chlorination of the final effluent, for odor control or for disinfection of the sewage for bacterial reduction, is additional treatment that is often required. Trickling-filter treatment can be supplemented by sand filtration, an oxidation pond, or chemical coagulation and settling where a higher quality effluent is necessary.

A typical design of an Imhoff tank standard-rate trickling filter plant is shown on pages 481 and 489. (See also Table 4-19.)

Rotating Biological Contactors

Rotating biological contactors, reactors, or surfaces, have closely spaced plastic disk drums which are rotated (around 1 to 4 rpm) partially submerged $\frac{1}{3}$ to $\frac{1}{2}$ in wastewater which has received primary treatment. In some instances prior trash

[98]G. R. Grantham and J. G. Seeger, Jr., "Progress of Purification During the Starting of a Trickling Filter," *Sewage Ind. Wastes*, **23**, 1486–1492 (December 1951).

and grit removal is considered necessary.[99] The biological growth (biomass) which forms on the wetted area of the disk surfaces, through contact with organic material in the wastewater, is maintained aerobic by the rotation and air contact which makes possible oxygen transfer to the wastewater as it trickles down the disks. The process is similar to that in a trickling filter. Some of the growth is stripped or sloughed off from the disk as it passes through the moving wastewater and is removed in the secondary settling tank. Design is based on organic and hydraulic loading using pilot plant or full scale data with the particular wastewater. Hydraulic loading rates are generally from 2 to 4 gpd/ft^2 of contactor surface; but organic loading, 1 to 4 lb BOD/day/1000 ft^2, should also determine design. Lower design loading (1 gpd/ft^2) is needed to produce a high quality (10 mg/l BOD and suspended solids in effluent). Multiple-stage design and operation of contactors with equal loading per square foot of disk and maintenance of aerobic conditions should be the objective.[99] Better effluent BOD quality and nitrification is possible with control of pH (8.4), dissolved oxygen, and raw wastewater alkalinity (7.1 times the influent ammonia concentration).[100] Plant efficiency may reach 95 percent at low design loading, but is reduced at lower temperatures. The rotating biological contactors are required to be covered or enclosed not only to protect them from low temperatures, but also from rainfall and heavy winds which would flush off growths, and from sunlight which would embrittle the plastic disks and promote the growth of algae. The process is reported to be reliable and to withstand shock loading; unusual variations may result in reduced efficiency, but appropriate flow equalization can minimize the effects. The units are also suitable to treat small residential sewage flows.

Physical-Chemical Treatment[101]

Physical-chemical treatment involves chemical coagulation, sedimentation, filtration, and activated carbon adsorption in addition to prior grit removal and comminution. The unit processes through filtration remove suspended matter; the activated carbon removes soluble organics. Phosphorus is normally removed by coagulation. Nitrogen can also be removed by adding to the process ion-exchange and breakpoint chlorination. Disposal of the large quantities of sludge resulting from coagulation can be a problem. However, the need for biological treatment may be eliminated.

Chemicals used for coagulation include ferric chloride or ferric sulfate, typically 45 to 90 mg/l ferric chloride for 85 to 90 percent phosphorus reduction; aluminum sulfate (alum) and sodium aluminate, typically 75 to 250 mg/l alum

[99] B. C. G. Steiner, "Take a new look at the RBS process," *Water Wastes Eng.*, May 1979, p. 41

[100] *Operation of Wastewater Treatment Plants*, MOP 11, Water Pollut. Control Fed., 2626 Pennsylvania Avenue, N.W., Washington, D.C. 20037, 1976, pp. 105–115.

[101] Gordon Culp, *Physical-Chemical Wastewater Treatment Plant Design*, USEPA, EPA 625/4-73-002a, August 1973.

for 55 to 90 percent phosphorus reduction; and lime in amounts dependent on wastewater alkalinity and hardness. At a pH of 9.5 the orthophosphate is converted to an insoluble form. Typically lime doses range from 200 to 400 mg/1 and pH at 10 to 11; polymers used alone are more costly, but are attractive as settling and filtration aids when used in conjunction with the coagulants previously mentioned. At average flow, rapid mix should be 2 min, flocculation 15 min, and sedimentation 900 gpd/ft^2 (1400 gpd/ft^2 at peak hourly flow). When alum or lime is used, pH control is necessary before filtration.

Filtration using a mixed media at a rate of 5 gpm/ft^2 is advised but up to 10 gpm/ft^2 may be used. A flow-equalization pond ahead of the filter would permit filtration at a relatively constant rate. Gravity and pressure filters are used.

Granular carbon adsorption is designed to provide about 30 min wastewater contact time in open tanks in either an upflow or downflow pattern, or in upflow countercurrent carbon columns in steel tanks. The countercurrent pattern is said to provide more efficient utilization of the carbon. The carbon is backwashed as needed and regenerated when its adsorption capacity is exhausted.

Weber (see Bibliography) has prepared a very comprehensive treatise on the principles and application of chemicals and physical-chemical methods of water and wastewater treatment.

Extended Aeration

Extended aeration plants, also referred to as aerobic digestion plants, have particular application for relatively small installations serving small subdivisions, schools, trailer parks, motels, shopping centers, and the like. Some basic design data are given below and in Table 4-20.

Average Sewage Flow. 400 gal/dwelling or 100 gpd/capita. See Table 3-12 for other unit flows.

Screening and comminutor. Recommended, bar screen minimum.

Aeration Tanks. At least two to treat flows greater than 40,000 gpd, 24 to 36-hr detention period at average daily flow, not including recirculation.

Air Requirements. 3 cfm/ft of length of aeration tank, or 200 ft^3/lb of BOD entering the tank daily, whichever is larger. Additional air is required if air is needed for air-lift pumping of return sludge from settling tank.

Settling Tanks. At least two to treat flows greater than 40,000 gpd, 4-hr detention period based on average daily sewage flow, not including recirculation. For tanks with hopper bottoms, upper third of depth of hopper may be considered as effective settling capacity.

Rate of Recirculation. At least 1:1 based on average daily flow.

Measurement of Sewage Flow. By V-notch weir or other appropriate device. Recording devices required for larger installations.

Sludge Holding Tanks. Provide 8 ft^3/capita. Sludge holding tanks should be required for all plants.

Daily operation control is essential. Air blowers must be operated continuously and sludge returned. Clogging of the air lift for return sludge is a common cause of difficulty. Grease that accumulates on the surface of settling tanks should be skimmed off and disposed of separately, not to the aeration tank. Aeration tubes or orifices require periodic cleaning. Dissolved oxygen level in the aeration tank and the mixed liquor suspended solids concentration must be watched. Odors should be minimal. A 90 to 97 percent BOD and suspended solids removal and good nitrification of ammonia nitrogen can be expected with proper control. (See also page 477 and Table 4-20.)

Waste Stabilization Pond 

In areas where ample space is available, 1000 ft or more from habitation, with consideration to the prevailing winds, a waste stabilization pond may be a relatively inexpensive and practical solution to a difficult problem. It is reported that small systems at resorts or motels designed with a septic tank ahead of the oxidation pond never produced an odor problem.[102] Where indicated, primary treatment with grit chamber, comminutor and rack, and duplicate ponds arranged for series or parallel operation are recommended. A BOD removal of 85 to 90 percent is not unusual. Ponds in open areas and in series give best results with additional detention time. Pond performance is affected by temperature, solar radiation, wind speed, loading, detention time, and other factors.[103] A summary of design criteria for waste stabilization ponds follows.

Pond loading	15 to 35 lb of BOD/acre/day, depending on climatic conditions; 3 ft^2 per gal settled sewage for small system.
Detention time	90 to 180 days, depending on climatic conditions; 180 days for controlled discharge pond; 45 days minimum for small systems.
Liquid depth	5 ft plus 2 ft freeboard. Minimum liquid depth 2 ft.
Embankment	top width 6 to 8 ft; inside and outside slope 3 horizontal to 1 vertical. Use dense impervious material; prepare bottom surface. Liner of clay soil, asphaltic coating, bentonite, plastic or rubber membrane, or other material required if seepage can be expected.
Pond bottom	level, impervious, no vegetation. Soil percolation should be less than $\frac{1}{4}$ in./hr after saturation.
Inlet	4-in. diameter minimum, at center of square or circular pond; at $\frac{1}{3}$ point if rectangular, with length not more than twice width. Submerged inlet 1 ft off bottom on a concrete pad or at least 1.5 ft above highest water level.
Outlet	4-in. minimum diameter; controlled liquid depth discharge using baffles, elbow, or tee fittings; drawoff about 6 in. below water surface and so as to avoid short-circuiting; permit drainage of pond and discharge to concrete or paved gutter.
General	round pond corners; provide fencing, warning signs, and means for flow measurement. Locate in isolated area. Seed top and sides of embankment to grass from above waterline. Keep grass cut and prevent growth of weeds, trees, and shrubs. Liners on inside slopes will prevent weed growth and permit good mixing of pond water by wind action. 4-ft fence to keep out livestock and discourage trespass. Post area.

[102] Harvey F. Ludwig, "Industry's Idea Clinic," *J. Water Pollut. Control Fed.*, **36**, 937 (August 1964).
[103] L. W. Canter and A. J. Englande, Jr., "States' Design Criteria for Waste Stabilization Ponds," *J. Water Pollut. Control Fed.*, **42**, 1840 (October 1970).

Table 4-15 Types of Lagoons

Type	Depth (Ft)	Loading, (lb/5-day/BOD/acre/day)	BOD removal or Conversion (percent)
High-rate aerobic pond	1 to 1.5	60 to 200	80 to 95
Facultative pond	3 to 8	15 to 80	70 to 95
Anaerobic pond	Variable[a]	200 to 1000	50 to 80
Maturation pond[b]	3 to 8	<15	Variable
Aerated lagoon	Variable[a]	Up to 400	70 to 95

Source: *Upgrading Lagoons,* USEPA, EPA-625/4-73-00 lb, (August 1973), p. 1.
[a]Usually 10- to 15-ft deep.
[b]Generally used for polishing effluents from conventional secondary treatment plants.

Stabilization ponds or lagoons, also called oxidation ponds, are operated as high-rate aerobic ponds, as aerobic-anaerobic (facultative) ponds (which are most common), as aerated ponds which are mechanically aerated, or as anaerobic ponds (lagoons). Normally the stabilization pond is aerobic at the surface and to some depth depending on wastewater clarity, sunlight penetration (algae), and mixing. In deeper ponds the wastewater at lower layers becomes facultative and then anaerobic at greater depths with anaerobic digestion of solids on the bottom and gas production. Anaerobic and aerated ponds are usually followed by aerobic ponds to reduce suspended solids and BOD to acceptable levels for surface discharge. In general, increased detention will increase BOD removal, and decreased BOD areal loading will increase BOD removal. Hence the required BOD and suspended solids removal and effluent quality will determine the detention time and areal loading. Pond efficiency can be improved by recirculation, inlet and outlet arrangements, supplemental aeration and mixing, and algae removal by various methods such as coagulation-clarification, filtration, and land treatment of the effluent.[104] Table 4-15 summarizes types of "lagoons," depths, loadings, and efficiencies. A design permitting parallel operation of ponds will simplify sludge removal.

Algae formed in ponds and lagoons and the seasonal blooms are the main cause of solids carry-over and increased oxygen demand in receiving streams due to the algae decomposition. Further treatment of pond effluent may therefore be required to remove the algae before discharge to a stream.

The state regulatory agency, Soil Conservation Service, local health department, or other regulatory agency may be of assistance in connection with needed soils studies, design, and effluent standards to be met.[105]

The practicability of using waste stabilization ponds, lagoons, or land treatment should be investigated in light of local conditions. The effect of

[104]D. H. Caldwell, et al., *Upgrading Lagoons*, USEPA, EPA-625/4-73-00lb, (August 1973).
[105]*Individual Sewage Systems*, Great Lakes-Upper Mississippi River Board of State Sanitary Engineers, 1979, Health Education Service, Inc., P.O. Box 7126, Albany, N.Y. 12224.

possible odor nuisance or health hazard should be evaluated before a treatment process is adopted. These processes should not be dismissed too quickly as they may sometimes provide an acceptable answer when no other treatment is practical and at a reasonable cost.

Wastewater Reuse—Hazards and Constraints

The treatment given wastewater which is to be reused should meet applicable regulatory requirements to protect the public health and natural resources. As noted in a discussion of Water Reuse in Chapter 3, treated (renovated or reclaimed) wastewater may be used directly for *nonpotable* purposes or indirectly when discharged to surface water or groundwater. The extent to which wastewater is reused will depend on the availability and cost of other suitable water, the wastewater treatment required for the proposed reuse and for its disposal, and the quality of the wastewater and its treatability. The hazards involved and treatment to minimize the risks to the public health are discussed below.

Public Health Hazard and Disease Transmission Aspects.

The public health hazards associated with the land disposal of wastewater effluent include:

1. Possible inhalation of aerosols containing pathogenic microorganisms, particularly bacteria and viruses from spray irrigation, activated sludge, and trickling filter treatment systems. Also contact with and ingestion of pathogens in non-disinfected wastewater by workers.
2. Consumption of raw or inadequately cooked vegetables from crops irrigated with wastewater, and the possible ingestion of heavy metals or other toxic materials taken up by crops during growth.
3. Contamination of groundwater through infiltration and percolation of wastewater microorganisms and chemicals into a groundwater aquifer serving as a source of drinking water.
4. Runoff, from land areas receiving wastewater effluent, to surface waters used as sources of drinking water, shellfish, bathing water, or other recreational purposes.
5. Possible cross-connection between potable and non-potable water systems.
6. Build-up of detrimental chemicals in the soil.

Although not strictly a public health hazard, odors, aesthetic, and nuisance factors should be considered for they would call attention to a land wastewater disposal facility and suggest psychosomatic illnesses. In addition, public accessibility to the disposal area could permit exposure and accidental ingestion of wastewater, particularly where wastewater has not been adequately treated and disinfected. Fencing to exclude livestock and prevent trespassing and appropriate warning signs around the site would be indicated. Fly and mosquito breeding could also become a problem. The role of wildlife, including migratory birds, in carrying infectious organisms great distances is not known.

As previously noted, pathogenic microorganisms in wastewater generally include certain viruses, bacteria, helminths, spirochetes, and protozoa. Experience shows that essentially all microbiological contaminants can be removed from wastewater as it infiltrates and percolates through an adequate distance of unsaturated loamy and sandy clay soil. However, microorganisms can travel several hundred feet in saturated soil and 1500 to 2000 feet or more in coarse gravel and creviced rock. Tight soils are obviously more effective than sandy and gravelly soils. It is difficult to predict the effectiveness of different soils for removal of bacteria and viruses. Heavy rains, for example, could cause viruses to be carried to the groundwater and thence to well-water supplies.

Acidic conditions and lack of organics and certain elements such as iron, manganese, aluminum, and calcium in soil is reported to increase water pollution potential.

Chemical contaminants in wastewater might include heavy metals (cadmium, copper, nitrates, lead, mercury, zinc, nickel, and chromium), pesticides (insecticides, fungicides, herbicides, rodenticides, nematicides, and other chemicals used to control animal and plant pests), and numerous commercial and industrial wastes. Mercury, PCB, kepone, mirex, and trichloroethylene are well-known examples. In addition, the nitrification of organic material in sewage will add nitrate-nitrogen to the ground water if not utilized immediately by plants, and endanger sources of drinking water used by infants (methemoglobinemia). Sodium accumulation may also be of concern. In a study of the effects of 20 years irrigation with secondary *domestic sewage* effluent (population 3800) containing no major industrial wastes, soil analyses showed no accumulation of nitrogen, lead, copper, zinc, nickel, chromium, or cadmium. There was a measurable increase in phosphorus; no harmful accumulation of heavy metals in alfalfa.[106]

The natural removal of organic chemicals from wastewater by land treatment is dependent on the physical, chemical, and biological factors in the environment favorable to the degradation, movement, and simplification of the complex substances in the wastewater. The natural removal of inorganic chemicals is dependent on the clay and organic content of the soil through which the liquid percolates—but the removal process is reversible.

And knowing that conventional wastewater and water treatment processes are ineffective, or only partially effective, in removing toxic organic and inorganic chemical contaminants, it is apparent that the land disposal of wastewater should not be permitted unless groundwater aquifers that may serve as sources of drinking water, and soils used for growing crops or pasturage, are positively protected and carefully monitored to meet existing health standards and guidelines. This requires peripheral monitoring wells around land disposal sites and that industrial waste discharges variations in quality and quantity be continuously monitored as a precondition to discharge, if approved. The burden for the removal of chemical contaminants from wastewater should remain at the plant source. Until this can be guaranteed, water treatment plants are in fact serving as the final wastewater treatment plant unit process.

In view of the many pathogens normally found in wastewater, it is recognized

[106]James H. Reynolds et al., "Long-term effects of irrigation with wastewater," *J. Water Pollut. Control Fed.*, April 1980, pp. 672–687.

that irrigation and spraying of crops with wastewater may be a source of infection to man and animals. Hence, if used, wastewater should be restricted to those foods that are not eaten raw. It has been suggested that wastewater receiving conventional secondary treatment, including chlorination, could be applied to crops eaten raw. This is not advisable as previously noted since many pathogenic microorganisms pass through activated sludge and trickling filter plants, although greatly reduced in number. Furthermore, it is again emphasized that chlorination as normally practiced does not destroy all pathogenic viruses, bacteria, protozoa, and helminths.

Many metals normally found in wastewater and applied to crops do not appear to be a problem in the food chain. However, heavy metals (cadmium, copper, molybdenum, nickel, and zinc) may accumulate in soil and become toxic unless good management practices are followed and the source is practically eliminated. Sewage sludges can be expected to contain higher concentrations of heavy metals. Crop tissue and grain analyses should therefore be required to monitor vegetation uptake. These include tobacco and food crops and pasture, forage, and feed grain for animals whose products are consumed by humans. Late crop irrigation with wastewater high in nitrate leads to a high nitrate concentration in both the soil and the crop. Excessive concentrations of nitrate (in forage) are injurious when fed to animals, resulting in cyanosis. Phosphates in wastewater should not be considered dangerous to crops.[107] Boron, on the other hand, a component of many commercial laundry detergents which is not normally removed by conventional wastewater treatment, is a well-known toxicant to citrus crops.[108]

There is always a risk of wastewater runoff from land treatment systems, dependent on topography, seasonal factors, heavy rains and thaws, wastewater loading, and soil characteristics. Such runoff can be expected to contain considerable amounts of pathogens. It therefore poses a hazard to downstream surface water supplies serving as sources of drinking water, to bathing beaches, to shellfish growing areas, and to recreational areas. Algae production may also be stimulated. Adequate dependable design and management safeguards are therefore required to prevent surface runoff from entering or leaving the land disposal site.

Aerosols Hazard

The potential hazard from aerosols is related primarily to wastewater treatment by the activated sludge, trickling filter, and spray irrigation processes. The presence of microbiological pathogens in sewage and its aeration products downwind for a short distance has been well and repeatedly documented (particularly for *E. coli*). The particle size is such that it would allow both upper and lower respiratory tract implantation. It is well-known that susceptible

[107]Dwight C. Baier and Wilton B. Fryer, "Undesirable Plant Responses With Sewage Irrigation," *J. Irrig. Drain. Div.*, *ASCE*, **99**, No. IR2, Proc. Paper 9783, June 1973, pp. 133–141.

[108]"Reuse of Effluents: Methods of Wastewater Treatment and Health Safeguards," *WHO Tech. Rep. Ser.*, **517**, 1973.

individuals may be successfully infected by very small numbers of organisms for several diseases. The major gap in knowledge is the lack of evidence that this type of exposure does, in fact, represent a human disease hazard. The only way to answer this question is through appropriately designed epidemiological studies. This is not an easy task, since the groups of human diseases likely to be involved have widely fluctuating seasonal patterns and characteristics, and subclinical infection may have immunized the exposed population (particularly wastewater treatment plant workers), making assessment difficult. The hazards associated with spray irrigation of wastewater have been studied. Although pathogens can be recovered in aerosols from the spray irrigation of treated wastewater, human disease from the aerosols has not been demonstrated.[109]* In view of the potential, adequate buffer zones (1000 feet or more) are advised as a precautionary measure.

Regulation

Many agencies and individuals have suggested standards and guidelines controlling the land disposal and use of wastewater.[108,110] See Table 4-16. Those established by the California Department of Public Health are among the most explicit.[111] Their standards cover

> wastewater constituents which will assure that the practice of directly using reclaimed wastewater† for the specified purposes does not impose undue risks to public health.

Wolman[112] has suggested seven constraints which can serve as a model to guide professional judgment in the review of proposals for wastewater disposal by land treatment.

1. Carefully, efficiently, and continuously managed.
2. Appropriate site—permeable and porous soil.
3. Hold-over storage for wet weather.
4. Crops not eaten raw.
5. Undue hazard to groundwater or drainage prevented.
6. Potential hygienic risks are detected and controlled.
7. The process is cost-effective.

*However, an increased risk among workers exposed to nondisinfected effluent application and/or spray irrigation has been demonstrated in India. (D. C. J. Zoeteman, "Reuse of Treated Sewage for Aquifer Recharge and irrigation," *Municipal Wastewater Reuse News*, AWWA Research Foundation, Denver, Colorado, November 1980, pp. 8–11.

†"Reclaimed Wastewaters" means waters, originating from sewage or other waste, which have been treated or otherwise purified so as to enable direct beneficial reuse or to allow reuse that would not otherwise occur. "Disinfected wastewater" means wastewater in which the pathogenic organisms have been destroyed by chemical, physical, or biological methods.

[109] *Process Design Manual for Land Treatment of Municipal Wastewater*, USEPA, U.S. Army Corps of Engineers, USDA, October, 1977, pp. D-1 to D-29.

[110] "The Utah State Division of Health Guidelines," *Municip. Wastewater Reuse News*, AWWA Research Foundation, Denver, Colorado, May 1978, pp. 3–5.

[111] *California Administrative Code*, Title 17—"Public Health, Statewide Standards For the Safe Direct Use of Reclaimed Waste Water for Irrigation and Recreational Impoundments," May 1968, Dept. of Public Health, Berkeley, California.

[112] Abel Wolman, "Public Health Aspects of Land Utilization of Wastewater Effluents and Sludges." *J. Water Pollut. Control Fed.*, November 1977, pp. 221–228.

Table 4-16 Existing Standards Governing the Use of Renovated Water in Agriculture

	California	Israel	South Africa	Federal Republic of Germany
Orchards and vineyards	Primary effluent; no spray irrigation; no use of dropped fruit	Secondary effluent	Tertiary effluent, heavily chlorinated where possible. No spray irrigation	No spray irrigation in the vicinity
Fodder, fibre crops, and seed crops	Primary effluent; surface or spray irrigation	Secondary effluent, but irrigation of seed crops for producing edible vegetables not permitted	Tertiary effluent	Pretreatment with screening and settling tanks. For spray irrigation, biological treatment and chlorination
Crops for human consumption that will be processed to kill pathogens	For surface irrigation, primary effluent. For spray irrigation, disinfected secondary effluent (no more than 23 coliform organisms per 100 ml)[a]	Vegetables for human consumption not to be irrigated with renovated wastewater unless it has been properly disinfected (<1000 coliform organisms per 100 ml in 80% of samples)	Tertiary effluent	Irrigation up to 4 weeks before harvesting only
Crops for human consumption in a raw state	For surface irrigation, no more than 2.2 coliform organisms per 100 ml. For spray irrigation, disinfected, filtered wastewater with turbidity of 10 units permitted, providing it has been treated by coagulation	Not to be irrigated with renovated wastewater unless they consist of fruits that are peeled before eating		Potatoes and cereals —irrigation through flowering stage only

Source: "Reuse of Effluents: Methods of Wastewater Treatment and Health Safeguards," *WHO Tech. Rep. Ser.*, **517**, p. 41, 1973.
[a]Also for milking animals' pasture and landscaping irrigation.

These brief comments emphasize the public health aspects of wastewater disposal to the land. The disease hazards associated with the ingestion of contaminated water and food or the inhalation of aerosols have been noted. Proper wastewater treatment, including secondary treatment, coagulation, clarification, filtration and effective disinfection, as indicated by the microbiological and chemical constituents in the wastewater, will minimize the aerosol, crop, groundwater, and hygienic health risks associated with the disposal of wastewater to the land. Each proposal must be carefully evaluated to avoid preventable health hazards and irreparable damage to our groundwater sources, particularly in regard to the fate of organic chemicals, heavy metals, and viruses in percolation through saturated and unsaturated soils.

Wastewater Disposal by Land Treatment and Reuse

Land treatment methods include spray or sprinkler irrigation; ridge-and-furrow and border strip (flooding) irrigation including subsurface and contour ditch irrigation; land over-land flow; subsurface disposal in a soil absorption system or subsurface sand filter; wetland treatment; and possibly a waste stabilization pond or lagoon. Careful evaluation of the site soils, hydrogeologic, and geographical conditions, and management practices is necessary to prevent heavy runoff and erosion, overloading, groundwater pollution, and nuisance conditions. Evapotranspiration will play an important role in reducing deep percolation of the wastewater, except for high rate irrigation. Planning and design details for various land treatment processes are briefly reviewed here. Health risk and water balance analyses are necessary.[113–115] See Wastewater Reuse—Hazards and Constraints.

Table 4-17 compares design features for land treatment processes and Table 4-18 shows expected quality of treated wastewater from land treatment processes. The degree of treatment required for the use of wastewater (renovated water) in agriculture is given in Table 4-16. Subsurface disposal, sand filters, and waste stabilization ponds have been previously discussed.

Irrigation

Spray irrigation is the most common method of applying wastewater to land. Ridge-and-furrow and border strip irrigation are also used. Application of wastewater in spray irrigation is generally limited to 8 hr followed by a 40-hr rest period to permit drainage of the soil, reaeration, plant nutrient uptake, and microbial readjustment. Other operating cycles are also used. Wastewater disposal is by evapotranspiration and percolation, except that for high-rate irriga-

[113]Herman Bouwer, W. J. Bauer, Franklin D. Dryden, "Land treatment of wastewater in today's society," *Civ. Eng. ASCE*, January 1978, pp. 78–81.

[114]*Process Design Manual for Land Treatment of Municipal Wastewater*, USEPA, U.S. Army Corps of Engineers, USDA, October 1977.

[115]George Tchobanoglous, *Wastewater Engineering: Treatment, Disposal, Reuse*, Metcalf & Eddy, Inc., McGraw-Hill Book Company, New York, 1979, pp. 760–828.

Table 4-17 Comparison of Design Features for Land Treatment Processes

Feature	Principal processes			Other processes	
	Slow rate	Rapid infiltration	Overland flow	Wetlands	Subsurface
Application techniques	Sprinkler or surface[a]	Usually surface	Sprinkler or surface	Sprinkler or surface	Subsurface piping
Annual application rate (ft)	2 to 20	20 to 560	10 to 70	4 to 100	8 to 87
Field area required (acres)[b]	56 to 560	2 to 56	16 to 110	11 to 280	13 to 140
Typical weekly application rate (in.)	0.5 to 4	4 to 120	2.5 to 6[c] 6 to 16[d]	1 to 25	2 to 20
Minimum preapplication treatment provided in United States	Primary sedimentation[e]	Primary sedimentation	Screening and grit removal	Primary sedimentation	Primary sedimentation
Disposition of applied wastewater	Evapotranspiration and percolation	Mainly percolation	Surface runoff and evapotranspiration with some percolation	Evapotranspiration, percolation, and runoff	Percolation with some evapotranspiration
Need for vegetation	Required	Optional	Required	Required	Optional

Source: *Process Design Manual for Land Treatment of Municipal Wastes*, USEPA, U.S. Army Corps of Engineers, USDA, October 1977, p. 2–2.

[a]Includes ridge-and-furrow and border strip.

[b]Field are in acres not including buffer areas, roads, or ditches for 1 million gpd (43.8 litre/s) flow.

[c]Range for application of screened wastewater.

[d]Range for application of lagoon and secondary effluent.

[e]Depends on the use of the effluent and the type of crop.

1 in. = 2.54 cm

1 ft = 0.305 m

1 acre = 0.405 ha

Table 4-18 Expected Quality of Treated Water from Land Treatment Processes (mg/l)

Constituent	Slow rate[a]		Rapid infiltration[b]		Overland flow[c]	
	Average	Maximum	Average	Maximum	Average	Maximum
BOD	<2	<5	2	<5	10	<15
Suspended solids	<1	<5	2	<5	10	<20
Ammonia nitrogen as N	<0.5	<2	0.5	<2	0.8	<2
Total nitrogen as N	3	<8	10	<20	3	<5
Total phosphorus as P	<0.1	<0.3	1	<5	4	<6

Source: *Process Design Manual for Land Treatment of Municipal Wastewater*, USEPA, U.S. Army Corps of Engineers, USDA, October 1977, p. 2-4.
[a]Percolation of primary or secondary effluent through 5 ft (1.5m) of soil.
[b]Percolation of primary or secondary effluent through 15 ft (4.5 m) of soil.
[c]Runoff of comminuted municipal wastewater over about 150 ft (45 m) of slope.

tion, percolation plays the major role. Phosphorus, cadmium, and other metals are absorbed by plants; some, cadmium in particular, may be hazardous if consumed. Dissolved solids and chlorides may cause a soil problem where the wastewater is high in these constituents.

Ridge-and-furrow ditches are on a grade from 100 to 1500 ft in length. Depth and spacing varies with type of crop and soil ability to transmit water laterally. Border strips are 30 to 60 ft long.[116]

The low-rate spray irrigation rate is 0.5 to 4 in. per week depending on soil permeability, climate, and wastewater strength. The ground slope should be less than 20 percent on cultivated land and less than 40 percent on noncultivated land. Soil permeability should be moderately slow to moderately rapid* and the depth to groundwater a minimum of 2 to 3 ft although 5 ft is preferred.

High-rate spray irrigation is 4 to 40 and up to 120 in. per week, depending on soil permeability, climate, and wastewater characteristics. The soil permeability should be rapid, and the permeable soil depth preferably 15 ft or more. Crops are not usually grown. Nitrogen and phosphorus removal is not complete. Treated water may be used for groundwater recharge and subsequent reuse, subject to regulatory control and approval, or to protect groundwater from salt water intrusion. The depth to groundwater should be at least 10 ft, although lesser depths are acceptable where underdrainage is provided to collect the water and thus keep the soil aerated.

Application rates for ridge-and-furrow (gpm/100 ft) and border strip irrigation are similar to spray irrigation and will vary with soil permeability, spacing, and slope of the furrow. The wastewater is discharged directly to the furrows or strips between elevated rows of crops.

*0.2 to 6.0 or more in./hr permeability corresponding roughly to a soil percolation rate of 1 in. in 45 min to less than 10 min.

[116]"Design of Effluent Irrigation Systems," *Municip. Wastewater Reuse News*, AWWA Research Foundation, August 1980, pp. 16–22.

Overland Flow

Wastewater is applied to sloped (2 to 8 percent), grassed, slightly permeable ground surface as sheet flow during which physical (grass filtration and sedimentation) and chemical-biological (oxidation) treatment is accomplished. Treated runoff is collected in ditches and discharged to a watercourse. Surface runoff may be 50 percent or more. Grasses, which have high nitrogen uptake capacity,* are usually chosen for cover vegetation. Viruses and bacteria are not removed.

Wetland Treatment

Wetlands include marshes, bogs, wet meadows, peatlands, and swamps; they are not lakes, but have enough water to prevent most agricultural and silviculture uses. Secondary wastewater effluent may be applied to existing wetlands, artificial wetlands, and peatlands, if approved by regulatory authorities having jurisdiction. An environmental impact analysis may be required. Hyacinths have been used in wetland lagoons to effect removal of BOD, suspended solids, and nutrients. Design data is being accumulated.

Advanced Wastewater Treatment

Advanced wastewater treatment (tertiary treatment) may be needed in some instances to protect the water quality of the receiving groundwaters and surface waters from added undesirable nutrients, toxic and hazardous chemicals, or pathogenic organisms not removed or inactivated by conventional biological secondary wastewater treatment. For example, nitrogen and phosphorus may promote the growth of plankton and nitrates may contaminate groundwater; toxic organic and inorganic chemicals may endanger fish and other aquatic life, contaminate edible fish and shellfish, and endanger the quality of sources of water for water supply, recreation, and shellfish growing; and pathogens such as the infectious hepatitis virus, giardia, entamoeba, ascaris, and certain worms, not removed or destroyed by the usual sewage treatment including chlorination, place an additional burden on water treatment plants and increase the probability of waterborne disease outbreaks.

Figure 4-34 shows wastewater treatment unit processes including advanced or tertiary treatment. See Physical-Chemical Treatment in this chapter for non-biological pretreatment.

Advanced wastewater treatment may include combinations of the following unit processes, following secondary treatment, depending on the water quality objectives to be met. These are examples and are not intended to be all-inclusive.

For Nitrogen Removal

Breakpoint chlorination—to reduce ammonia nitrogen level (nitrate and organic nitrogen are not affected)

*Bent grass, Bermuda grass, Reed Canary grass, Sorghum-Sudan, Vetch; also Alfalfa, Clover, Orchard grass, Broome grass, and Timothy.

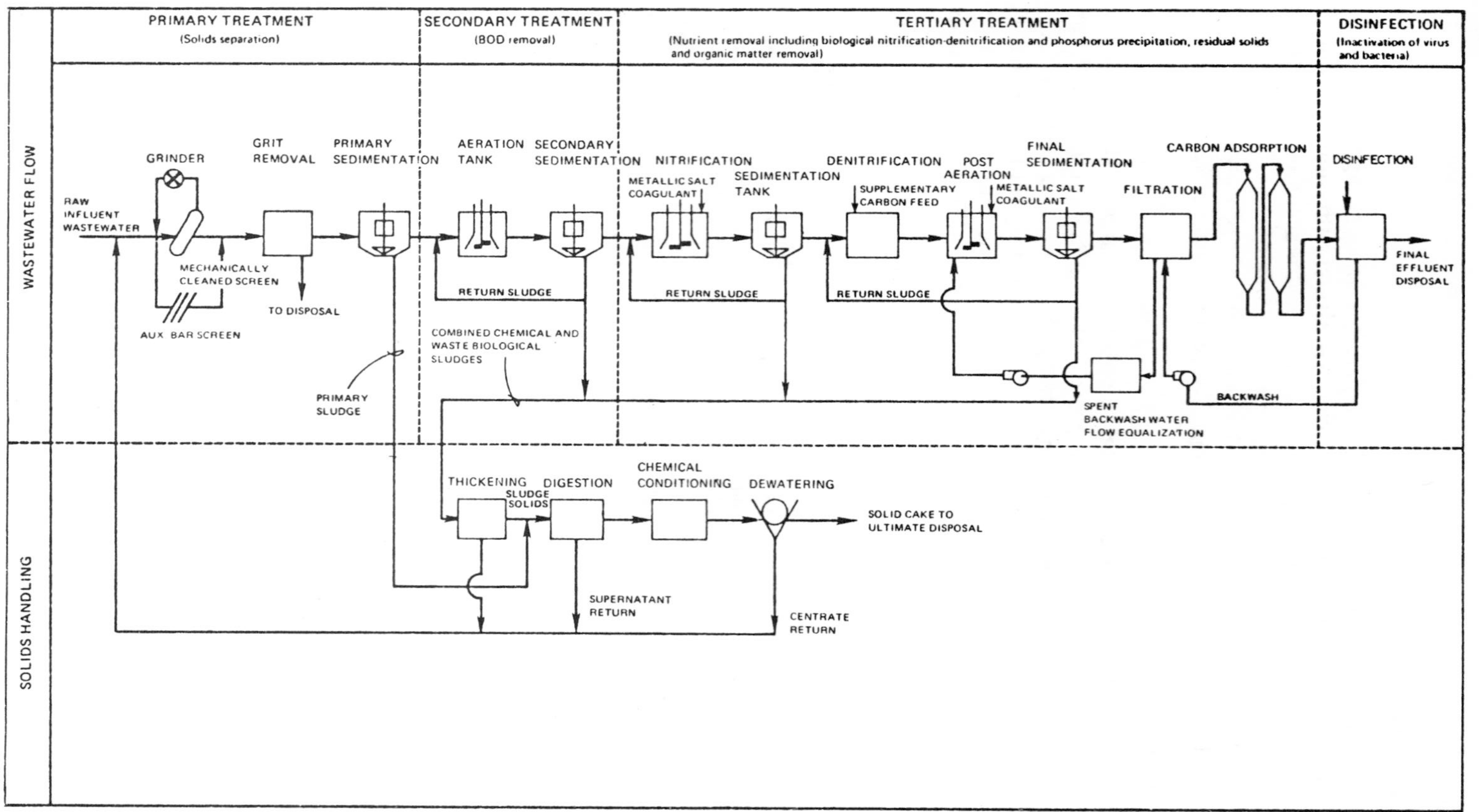

Figure 4-34 Typical schematic flow and process diagram. (Source: *Operation of Wastewater Treatment Plants*, MOP 11, Water Pollut. Control Fed., Washington, D.C., 1976, p. 17.)

Ion exchange, after filtration pretreatment—to reduce nitrate nitrogen and ammonium levels using selective resin for each; phosphate also reduced

Nitrification followed by denitrification, ammonia if present removed or converted to nitrate and then to nitrogen gas—ammonia stripping* (degasifying) to remove ammonia nitrogen, or biological oxidation of ammonia in the activated sludge process to nitrate; denitrification (organic nitrogen) achieved by filtration through sand or GAC, or by biological denitrification, usually under anaerobic conditions, following activated sludge treatment (nitrification and denitrification) (See design texts)

Methanol—to reduce nitrate level

Reverse osmosis, following pretreatment to prevent fouling of membranes—to reduce total nitrogen level; also dissolved solids

Electrodialysis, following pretreatment—to reduce ammonia, organic, and nitrate nitrogen levels; also dissolved solids

Oxidation pond—to reduce total nitrogen level

Land treatment, low-rate irrigation or overland flow—to reduce total nitrogen level; also phosphorous. Rapid infiltration also effective

For Phosphorus Removal

Coagulation (lime, alum, or ferric chloride, and polyelectrolyte) and sedimentation—to reduce phosphate level, TDS increased, additional nitrogen removal, also some heavy metals

Coagulation, sedimentation, and filtration (mixed media)—to further reduce phosphate level; also suspended solids, TDS increased, additional nitrogen also removed

Lime treatment, after biological treatment, followed by filtration—to reduce phosphorus (pH above 11), also suspended solids

Ion exchange, with selected specific resins—to reduce phosphate, also dissolved solids and nitrogen

For Dissolved Organics Removal

Activated carbon (granular or powdered) adsorption, following filtration—to reduce COD including dissolved organics; also chlorine

Reverse osmosis, following pretreatment—to reduce dissolved solids

Electrodialysis following pretreatment—to reduce dissolved solids

Distillation, following pretreatment—to reduce dissolved solids

Biological wastewater treatment—to reduce dissolved organics

For Heavy Metals Removal

Lime treatment—to reduce heavy metals level

Coagulation and sedimentation—to reduce heavy metals level

For Dissolved Inorganic Solids Removal (Demineralization)

Ion exchange, using anionic and cationic resins, following pretreatment—to reduce total dissolved solids

Coagulation and sedimentation—to reduce heavy metals

Reverse osmosis—to reduce TDS

Electrodialysis—to reduce TDS

For Suspended Solids Removal

Filtration (sand, anthracite, multimedia, diatomaceous earth, microstrainer)—to reduce suspended solids

Coagulation (alum, lime, or ferric chloride and possibly polyelectrolytes), sedimentation, filtration—to reduce suspended solids, also ammonia nitrogen, and phosphate if high alum or lime dosage used; adding ammonia stripping will reduce total nitrogen further; adsorption using activated carbon will reduce dissolved organics and total nitrogen

*Wastewater pH is raised to 10.0 to 10.5 or above, usually by the addition of lime, at which pH the nitrogen is mostly in the form of ammonia which can be readily removed by aeration, but pH adjustment of the effluent will be needed. Organic or nitrate nitrogen are not removed. Stripping equipment includes tray towers, cascade aerators, step aerators, and packed columns.

For Recarbonation

Carbon dioxide addition—to reduce pH where wastewater pH has been raised to 10 to 11; this is necessary to change dissolved ammonia gas which is toxic to fish to ammonium ion, and to reduce deposition of calcium carbonate in pipelines, equipment, or the receiving watercourse.[117]

For Heat Removal

Open reservoir or evaporative cooler—to lower temperature of wastewater prior to discharge.

Removal of toxic or hazardous substances must start at the source with in-house process change if possible, waste reclamation and reuse, waste control, and pretreatment before discharge to a municipal sewer or watercourse.

Inspection During Construction

The best of design is no better than the construction, and, one could add, operation. On installations for which engineering or architectural plans are prepared, it is advisable to make arrangements for retaining competent inspection and supervision during construction. This would include the checking of grades and elevations and pipe bedding, inspection of quality of material and construction, infiltration tests, and full compliance with the treatment plant, pumping station, and sewerage plans and specifications. On small installations the regulatory agency, if properly staffed, may be able to perform some of the inspection—but only in an advisory capacity for general compliance with design.

Operation Control

A common weakness in both small and large wastewater treatment plants is the failure to properly operate and maintain a well-designed and constructed plant. It is important for the design engineer to impress upon the owner, be it an individual, corporation, or a municipality, that wastewater treatment plants require daily attention by a qualified person or persons. Repairs should be made promptly and preventive maintenance should be the rule. If this is not done, the money spent is wasted; major damage requiring expensive repairs can result before the cause is detected, and the purpose of the treatment plant to prevent water pollution is nullified. Adequate salaries and good working conditions are essential to attract and keep competent people.

Wherever a treatment plant is in use, an operation report should be kept and entries made daily. The regulatory agency usually provides forms for this purpose. This will help assure daily inspection and continuous operation of the plant as it was designed to function. An equipment maintenance schedule or checklist is also found to be very worthwhile in reducing expensive repairs and equipment replacement. A complete set of spare parts for pumps, special dosing devices, and chlorinators is necessary if extended interruptions are to be kept at a minimum.

[117]Russell L. Culp, "The Operation of Wastewater Treatment Plants," *Pub. Works*, December 1970, pp. 51–52.

TYPICAL DESIGNS OF SMALL PLANTS

Design for a Small Community

Three design analyses are given below for a plant to serve 150 persons at 100 gal per capita per day (gpcd) = 15,000 gpd.

Standard-rate trickling filter plant with Imhoff tank

1. Flowing through channel provides 2.5-hr detention.

$$\frac{15{,}000}{24} \times 2.5 = 1560 \text{ gal} = 209 \text{ ft}^3$$

2. Sludge storage at 5 ft^3 per capita = 5 × 150 = 750 ft^3.
3. Sludge drying beds at 1.25 ft^2 per capita = 150 × 1.25 = 188 $ft.^2$
4. Trickling filter loading at 400 lb of BOD/acre-ft = 0.25 lb per yd^3. Loading based on 0.17 lb of BOD/capita with 35 percent removal in primary settling = 150 × 0.17 × 0.65 = 16.6 lb/day. Filter volume required $= \frac{400}{43{,}560} = \frac{16.6}{x}$; $x = 1800 \text{ ft}^3$. Hence the required filter diameter, assuming a 6-ft depth $= D = \sqrt{\frac{1800 \times 4}{\pi \times 6}} = 19.5$ ft, say 20 ft. The volumetric loading $= \frac{15{,}000}{\frac{\pi \times 20 \times 20}{4}} = \frac{x}{43{,}500}$; $x = 2{,}080{,}000$ gpd/acre on a 6-ft deep filter.
5. Final settling provides 2-hr detention. $\frac{15{,}000}{24} \times 2 = 1250$ gal $= 167 \text{ ft}^3$. With a surface settling rate $= 1000 \text{ gpd/ft}^2 = \frac{180 \times \text{tank depth}}{\text{2-hr detention}}$; tank depth = 11.1 ft.
6. If the BOD in the raw sewage is 200 mg/l, and the Imhoff tank removes 35 percent, the applied BOD = 0.65 × 200 = 130 mg/l. According to the National Research Council Sanitary Engineering Committee formulae* a filter loaded at 400 lb of BOD per acre-ft will produce an average settled effluent containing 14 percent of that applied, or 0.14 × 130 = 18 mg/l.

High-rate trickling filter plant with Imhoff tank

1. Flowing through channel same as with standard rate filter = 209 ft^3.
2. Sludge storage at 8 ft^3/capita = 8 × 150 = 1200 ft^3.
3. Sludge drying beds at 1.50 ft^2/capita = 150 × 1.50 = 225 ft^2.
4. Trickling filter loading at 3000 lb of BOD/acre ft = 1.86 lb/yd^3. Loading based on 0.17 lb of BOD per capita with 35 percent removal in primary settling = 150 × 0.17 × 0.65 = 16.6 lb/day. The BOD in the raw sewage is 150 × 0.17 = 25.5 lb. Filter volume required $= \frac{3000}{43{,}560} = \frac{25.5}{x}$; $x = 370 \text{ ft.}^3$ Hence the required filter diameter, assuming 3.25 ft depth × $D = \sqrt{\frac{370 \times 4}{\pi \times 3.25}} = 12.0$ ft. The volumetric surface loading on a 12.0-ft diameter filter with influent + recirculation ($I + R = 1 + 1 + = 2$) or 2(15,000) = 30,000 gal/day is

$$\frac{x}{43{,}560} = \frac{30{,}000}{\frac{\pi \times 12 \times 12}{4}}$$

$x = 11{,}500{,}000$ gpd/acre on a 3.25/ft deep filter.

$*E = \frac{100}{1 + 0.0085\sqrt{u}}$; E = percent BOD removed, standard filter and final clarifier.

u = filter loading in pounds BOD per acre-ft.

5. Final settling provides 2-hr detention at flow $I + R$.

$$\frac{30{,}000}{24} \times 2 = 2500 \text{ gal} = 334 \text{ ft}^3$$

6. Without recirculation, an applied BOD of 130 mg/l (0.65 × 200), at a rate of 3000 lb/acre-ft will be reduced to 0.32 × 130 = 42 mg/l in the settled effluent. With recirculation of $R/I = 1$ the efficiency of the high-rate filter and clarifier can be determined from the following formulas:[118]

$$F = \frac{1 + \frac{R}{I}}{\left(1 + 0.1\frac{R}{I}\right)^2}$$

where F = recirculation factor
R = volume of sewage recirculated = 1
I = volume of raw sewage = 1

$$F = \frac{1 + 1}{(1 + 0.1)^2} = 1.65$$

and from

$$u = \frac{w}{VF}$$

where u = unit loading on high-rate filter in lb of BOD/acre-ft
w = total BOD to filter in lb/day = 16.6
V = filter volume in acre-ft based on raw sewage strength = 0.0084
F = recirculation factor = 1.65

$$u = \frac{16.6}{0.0084 \times 1.65} = 1198 \text{ lb/acre-ft}$$

and from

$$E = \frac{100}{1 + 0.0085\sqrt{u}}$$

where E = percent BOD removed by a high-rate filter and clarifier.

$$E = \frac{100}{1 + 0.0085\sqrt{1198}} = 77 \text{ percent}$$

Hence, the BOD will be reduced to (1 − 0.77)130 = 30 mg/l.

Intermittent sand filter plant with Imhoff tank or septic tank

1. Flowing through channel of Imhoff tank provides 2.5-hr detention = 209 ft.^{3}
2. Sludge storage at 4 ft^3/capita = 4 × 150 = 600 ft^3
3. Sludge drying bed provides 188 ft^3. OR: Septic tank provides 24-hr detention = 15,000 gal = 2000 ft^3.
4. Sand filter, covered, designed for loading of 50,000 gpd/acre. Filter area is

$$\frac{50{,}000}{43{,}560} = \frac{15{,}000}{x}; \; x = 13{,}000 \text{ ft}^2$$

[118]W. A. Hardenbergh, *Sewerage and Sewage Treatment*, 3rd ed., International Textbook Co., Scranton, Pa., 1950, p. 328.

If filter is open, the required area = 6550 ft^2. Make filters in two sections. Provide dosing tank to dose each covered filter section at volume equal to 75 percent of the capacity of the distributor laterals or to dose each open filter section to depth of 2 to 4 in.

If the efficiency of BOD removal of a sand filter is 90 percent, the BOD of the effluent would be $130 \times 0.10 = 13$ mg/l.

Children's Camp Design

A typical sewerage layout to serve a camp dining room and central bathhouse is shown in Figure 4-35. This boys' camp is assumed to be located on the watershed of a public water supply. The soil is a clay loam. There are no unusual plumbing fixtures at the camp. A basis for the design is given below.

1. Camp capacity is 96 campers and 27 staff = 123 total — 123 total
2. Design flow based on 40 gal/capita,

$$123 \times 40 = 4920 \text{ gpd}$$

4920 gpd

3. Septic-tank capacity designed for a 12-hr detention period. Total flow takes place in 16 hr. Liquid volume = x.

$$\frac{4920 \text{ gal}}{16 \text{ hr}} = \frac{x \text{ gal}}{12 \text{ hr}}; \quad x = 3690 \text{ gal capacity}$$

3690 gal

Make septic tank $6' \times 17' \times 5'$ liquid depth. See Table 4-7 and Figure 4-9. A tank twice the calculated volume is recommended.

4. Grease trap designed for $2\frac{1}{2}$ gal/person served at mealtime.

$$\text{Capacity} = 2\tfrac{1}{2} \times 123 = 308 \text{ gal}$$

308 gal

Make grease trap $2\frac{1}{2}' \times 7' \times 2\frac{1}{2}'$ liquid depth. See Figure 4-7.

5. Sand filter designed for filtration rate of 1.15 gpd/ft^2

$$\text{The total required area} = \frac{4920}{1.15} = 4278 \text{ ft}^2$$

4278 ft^2

Make in two sections, each 2160 ft^2 = $2(30 \times 72)$. Each section will consist of 5 distributors 69-ft long, 6-ft apart. See Figures 4-22 and 4-35. See also open filter alternate, Figure 4-23.

6. Dosing tank to be provided with alternating siphons. Volume of dose is to be 60 percent of volume of 4-in. distributors in each section = $(5 \times 69) \times (0.653 \times 0.60) = 135$ gal — 135 gal
7. Size of siphon = twice maximum flow to the dosing tank. Probable maximum flow assumed as one-half daily flow in one hour. $\dfrac{4920 \times \frac{1}{2}}{60} = 41$ gpm. (Check with fixture unit basis.) A 4-in siphon has a minimum discharge of 102 gpm. Use 4-in. siphon; it has a drawdown of 17 in. If the dosing tank is made 6-ft wide, and the discharge liquid depth is 17 in. or 1.41 ft, the dosing tank length is

$$\frac{135}{7.48} = \frac{1}{6} = \frac{1}{1.41} = 2.14 \text{ ft}$$

2.14 ft

See Figure 4-30.

8. Provide a chlorine contact tank giving a minimum detention of 15 min. Say flow conditions given in paragraphs 6 and 7 are leveled off through filter and reach

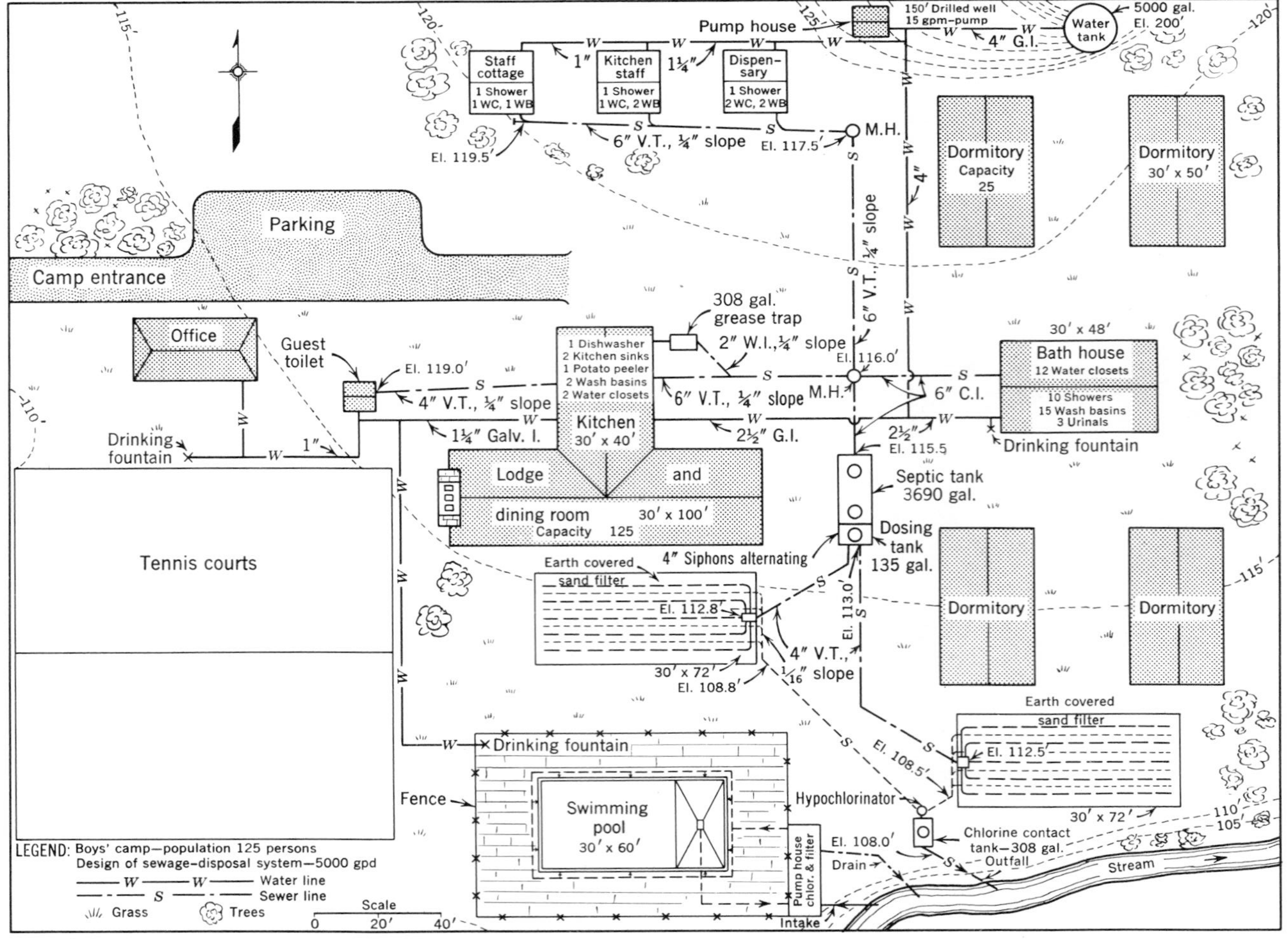

Figure 4-35 Typical camp compressed plan showing sanitary details. (Add recreation building, craft

contact tank at average rate of 20.5 gpm under peak conditions; make the contact tank equal to

$$20.5 \times 15 = 308 \text{ gal}$$ 308 gal

9. Chlorination treatment will be required to protect the receiving stream, which serves as a source of domestic water supply. Provide a hypochlorinator for the flow to be treated, having positive feed and operating continuously, with point of application in inspection manhole receiving filter drainage, 20 ft ahead of chlorine contact tank.

10. Increase the volume of the septic tank by 50 percent and the area of the sand bed by 20 percent if a kitchen garbage grinder is installed.

Referring to the above design, an alternate design based on soil percolation tests in clay loam is also investigated. Tests made in the proposed area for sewage disposal gave percolations of 1 in. in 40 min, 1 in. in 50 min, 1 in. in 10 min, and 1 in. in 60 min. Discard the 10-min test as not being representative. Use 1 in. in 50 min or 0.6 gpd/ft^2 in design. A design based on a subsurface absorption system for the camp mentioned above follows:

1. Camp capacity — 123 total
2. Design flow — 4920 gpd
3. Septic-tank capacity (Double this capacity is recommended) — 3690 gal
4. Grease trap capacity — 308 gal
5. Absorption area required:

$$\frac{4920}{0.6} = 8200 \text{ ft}^2$$ 8200 ft^2

Make in *two* sections *each* having an area of 4100 ft^2. This can be obtained if each section consists of 2050 linear ft of perforated pipe in trenches 2-ft wide. Space trenches 6-ft apart.

6. Dosing tank to be provided with alternating siphons. Volume of dose to be 60 percent of volume of 4-in. distributors in each section.

$$= 2050 \times 0.653 \times 0.6 = 803 \text{ gal}$$ 803 gal

(for 2-ft wide trench)

With a 6-ft wide dosing tank and a 4-in. siphon having a drawdown of 1.41 ft, for an 803-gal discharge, the tank length is

$$\frac{803}{7.48} \times \frac{1}{6} \times \frac{1}{1.41} = 12.7 \text{ ft}$$

7. Size of siphon should give twice the maximum flow to the dosing tank. Use a 4-in. siphon.

8. Increase the volume of the septic tank by 50 percent and the area of the tile field by 20 percent if a kitchen garbage grinder is installed.

Compare the cost of installation, operation, and maintenance of the sand filter system including chlorination and the subsurface absorption field system to determine which system to install. Other alternatives such as a trickling filter or aeration unit may also be studied and compared before a decision is made.

Specific construction details of grease traps, manholes, septic tanks, dosing tanks, sand filters, distribution boxes, and chlorine contact tanks are given elsewhere in this chapter.

Town Highway Building Design

1. Population: 70 persons
2. Fixtures: 5 showers
   5 water closets (flush valve)
   6 washbasins
   2 urinals (flush valve)
3. Soil tests: Percolation zero, tight clay soil
4. Design flow—Different bases:
   a. Fixture unit basis (see Table 4-4)
   5 showers @ 3 = 15
   5 water closets @ 8 = 40 (flush valve operated)
   6 washbasins @ 2 = 12
   2 urinals @ 4 = 8
   Total = 75 fixture units

For 8-hr operation, say one fixture unit is 35 gal, hence 75 × 35 = 2625 gpd.

b. Usage per capita basis
70 persons @ 30 gcd = 2100 gpd

c. Fixture hourly flow basis (from Table 4-3):
5 showers @ 150 gal/hr = 750
5 water closets @ 36 gal/hr = 180
6 washbasins @ 15 gal/hr = 90
2 urinals @ 10 gal/hr = 20
Total 1040 gal/hr

With an hourly peak in morning and afternoon, and usage during the day, say practically all flow takes place in $2\frac{1}{2}$ hr.

$$\text{Daily flow} = 1040 \times 2\tfrac{1}{2} = 2600 \text{ gpd}$$

d. The average estimated flow $= \dfrac{2625 + 2100 + 2600}{3} = 2442$ gpd average, say 2500 gpd.

5. Septic tank: Since all flow takes place in 8 hr, provide a minimum 2500-gal septic tank. (See Table 4-7 and Figure 4-9).

Make septic tank 6-ft wide, $14\frac{1}{2}$-ft long, 4-ft liquid depth. Dimensions are inside.

6. Sand filter:

$$\text{Required area} = \frac{2500}{1.15} = 2170 \text{ ft}^2$$

Make bed: 36′ × 60′. (See Figure 4-21 and 4-22.)

7. Dosing tank: Dose field to 70 percent of interior capacity of distributors, which extend to within 18 in. of end of bed, 6 ft o.c., 3 ft from edge of bed.

6 laterals × 57′ long × 0.653 (vol. 1′ of 4″ pipe) × 70 percent =
$$= 6 \times 57 \times 0.653 \times 0.7 = 156 \text{ gal}$$

8. Size of siphon: Siphon should preferably have minimum capacity $1\frac{1}{2}$ to 2 times peak flow. Using National Bureau of Standards probability curves in Figure 3-25, with 75 fixture units; demand = ±60 gpm. Use a 4-in. Miller siphon, which has a drawdown of 17 in. and a total loss head from high water to invert of discharge pipe of $25\frac{1}{2}$ in. (See Figure 4-30)

Make dosing tank 6-ft wide and 2.45 ft-long, attached to septic tank, with 17-in. liquid drawdown.

9. Size of chlorine contact tank, if required: Provide for at least 15 min detention. Make tank equal to dosing tank dose = 156 gal.

Small Restaurant and Hotel Designs

Figure 4-36 shows a similar design for a sand filter to serve a small restaurant and Figure 4-37 shows a leaching pit system.

Elementary School Design

1. Design basis: N. persons—600.

Soil tests—rock at 12 to 30 in.; soil gravelly loam with clay; shallow soil tests = 1 in. in 20 min. Suitable porous fill containing topsoil and gravelly loam is required to bring bottom of leaching system at least 2 ft above rock. Use assumed percolation rate of 0.5 gpd/ft^2. Pumping is required!

2. Design flow: 600 persons at 20 gcd = 12,000 gpd.

3. Septic tank: Minimum size to provide 12-hr detention period, assuming all flow takes place between 8:30 a.m. and 5:30 p.m., a period of 9 hr = x. Then $\frac{12{,}000}{9} = \frac{x}{12}$; x = 16,000 gal.

4. Absorption field:

$$\frac{12{,}000 \text{ gpd}}{0.5 \text{ gpd/ft}^2} = 24{,}000 \text{ sq ft} = 12{,}000 \text{ linear ft in 24-in. trench}$$

Divide into three sections, 4000 linear ft each

5. Absorption field dose:

4000 (linear ft) × 0.653 (gal/1-ft 4-in. perforated pipe) × 0.75 (percent dose)
= 1960 gal (make pump sump ample size)

6. Pump capacity: Probable flow on fixture basis using Table 4-4.

5 urinals @ 4	=	20
28 water closets @ 8	=	224
38 washbasins @ 2	=	76
27 service sinks @ 3	=	81
12 showers @ 3	=	36
4 drinking fountains @ ½	=	2
Total		439 fixture units

Probable maximum flow = 135 gpm. See Figure 3-25. Make pump capacity 150 gpm.

7. Controls: Provide three 150-gpm pumps, each discharging to a separate absorption field section, and one standby hooked up and valved for series operation to each field. Provide high-water alarm and timer in series with motor to shut off and throw in the next pump after 13 min of continuous operation.

NOTE: a. No receiving stream or sewer in vicinity.

b. If a kitchen garbage grinder is installed increase the volume of the septic tank by 50 percent and the area of the absorption field by 20 percent, and provide grease trap on kitchen waste line ahead of septic tank.

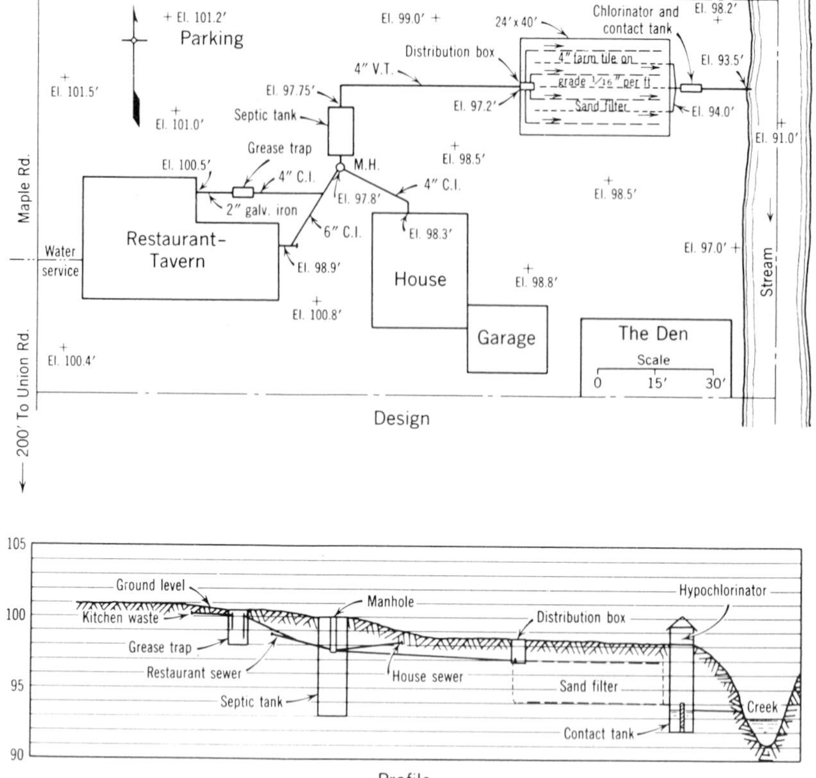

Figure 4-36 Typical plan and profile of restaurant treatment plant.

House, two-bedroom, at 150 gal = 300 gpd

Restaurant—fixtures:

2 flush toilets (tank)	=	10 fixture units
2 wash basins	=	4 fixture units
2 kitchen sinks	=	8 fixture units
Total		22 fixture units

One fixture unit = 35 gal per 8-hr operation

Flow from 22 fixture units = $22 \times 35 = 770$ gpd

Total flow = $770 + 300 = 1070$ gpd

Soil tests show no percolation.

(If period of operation is 16 hr double flow and septic tank and sand filter sizes.)

Septic tank capacity = 1100 gal (2500 gal recommended)

Grease trap = seat. cap. $\times$ 3 gal
$= 40 \times 3 = 120$ gal

$$\text{Sand filter} = \frac{1070 \text{ gal}}{1.15 \text{ gpd/ft}^2} = 930 \text{ ft}^2$$

Chlorine contact tank = 60 gal, 15-min detention at peak flow

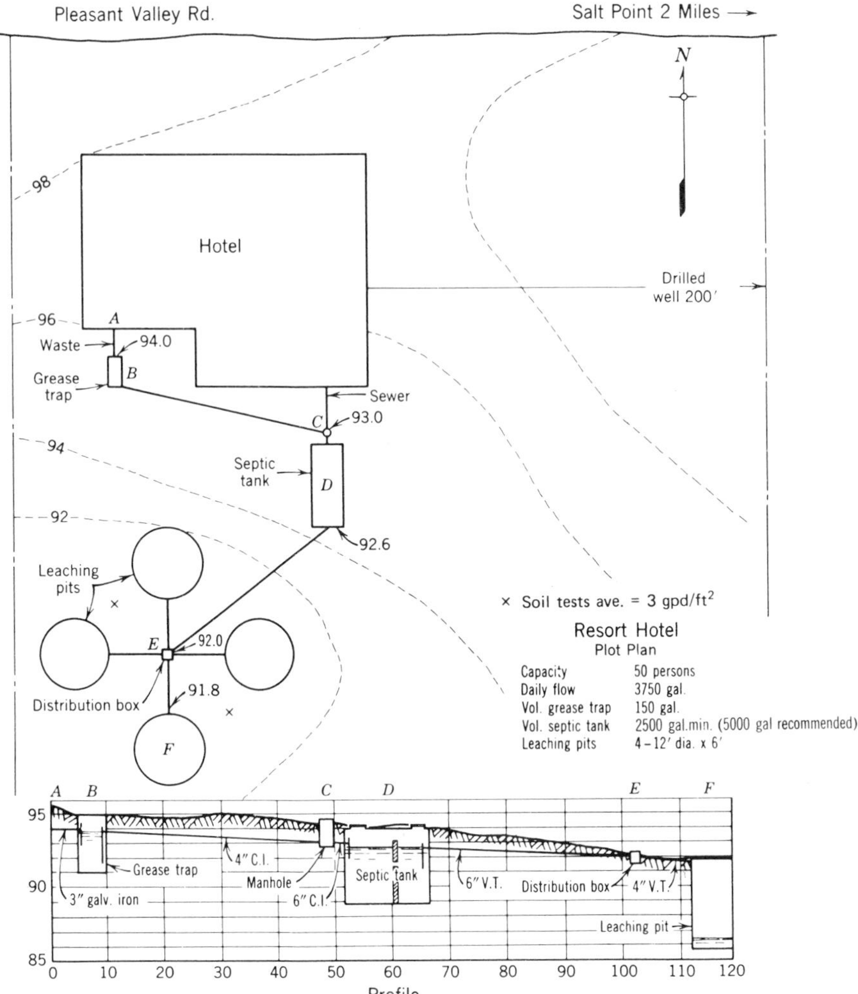

Figure 4-37 Plan and profile of hotel sewage disposal system.

Design for a Subdivision (See Figures 4-38, 4-39, and 4-40)

(Imhoff tank, standard-rate filter with secondary settling.)

1. Design population: 100 one-family homes = 350 persons.
2. Sewage flow: 100 gcd = 35,000 gpd.
3. Strength of sewage: 200 mg/l 5-day BOD or 0.17 lb 5-day BOD/capita.
4. Design flow: Assume total flow reaches plant in 16 hr.
5. Imhoff tank design: Required volume of settling compartment to provide $2\frac{1}{2}$-hr detention:

$$\frac{35{,}000}{16} \times 2.5 \times \frac{1}{7.48} = 732 \text{ ft}^3$$

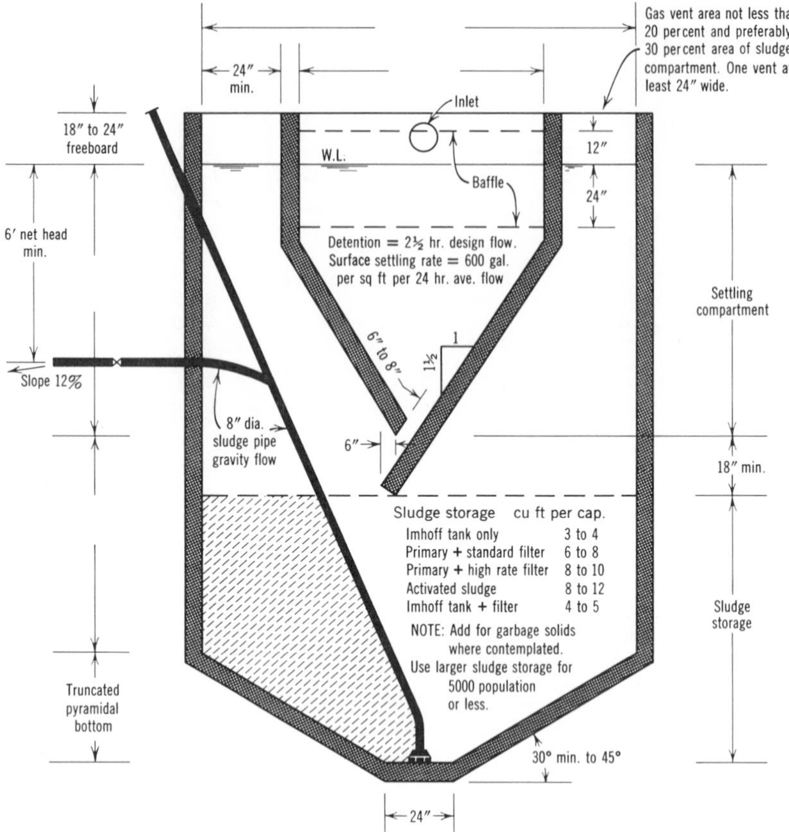

Figure 4-38 Section through Imhoff tank, with design details.

Required sludge storage to provide 8 ft^3 per capita:

$$350 \times 8 = 2800 \text{ ft}^3$$

6. Trickling filter design: Volume of filter stone required to provide a loading of 400 lb of BOD/acre-ft/day and not more than 300,000 gpd/acre-ft. Organic loading basis, assuming 35 percent BOD removal in the Imhoff tank = x.

$$\frac{0.65 \times 200}{1{,}000{,}000} = \frac{x}{35{,}000 \times 8.34}; \qquad x = 38.0 \text{ lb of BOD/day}$$

Volume of filter stone required based on organic load = Y_1:

$$\frac{400}{43{,}560} = \frac{38.0}{Y_1}; \qquad Y_1 = 4130 \text{ ft}^3$$

Volume of filter stone required based on volumetric load = Y_2:

$$\frac{300{,}000}{43{,}560} = \frac{35{,}000}{Y_2}; \qquad Y_2 = 5070 \text{ ft}^3$$

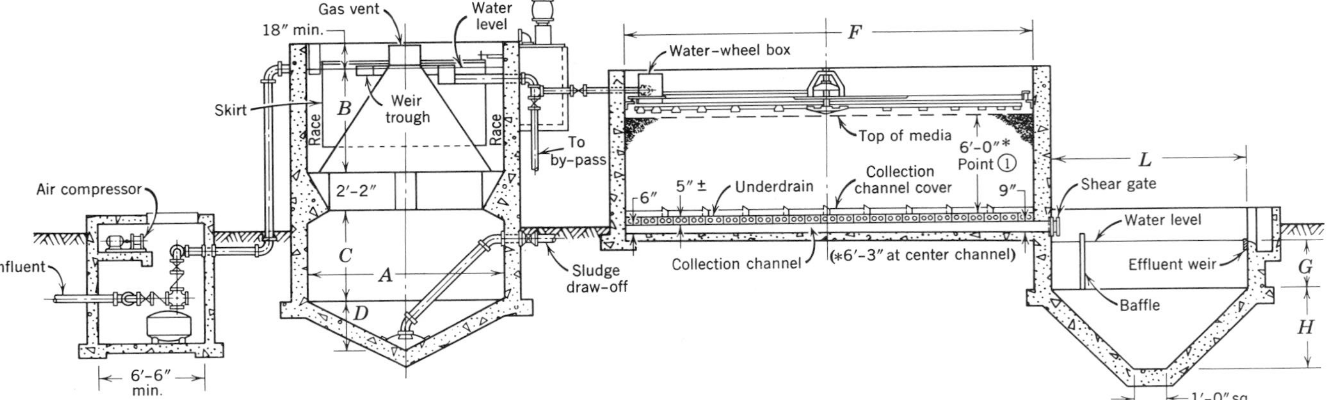

Figure 4-39 Typical arrangement of spiragester—"water-wheel" distributor system for complete treatment. (Courtesy Yeomans Brothers Co. Melrose Park, Ill.)

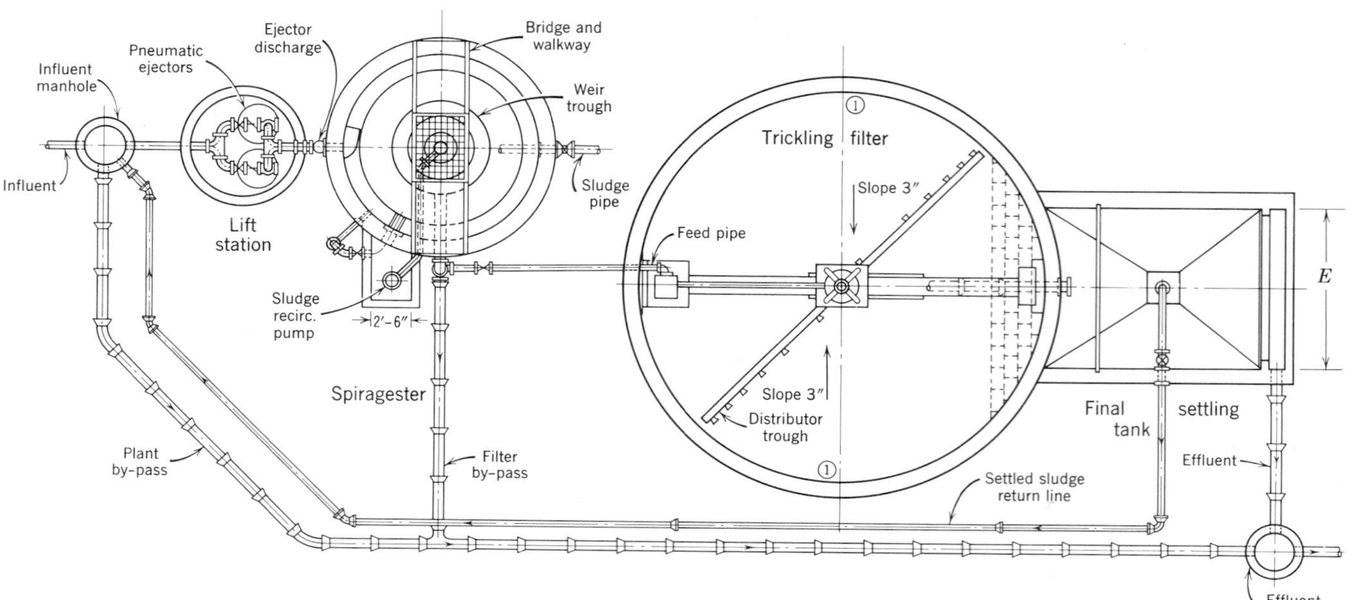

Figure 4-40 Plan view of spiragester—"water-wheel" distributor system.

Diameter of 6-ft deep filter to provide, say, 5000 $ft^3 = D$:

$$5000 = \frac{3.14 \times D^2}{4} \times 6; \qquad D = \sqrt{\frac{5000 \times 4}{3.14 \times 6}} = 32.6 \text{ ft}$$

If two filters proposed, each filter would have a diameter equal to

$$\sqrt{\frac{2500 \times 4}{3.14 \times 6}} = 23 \text{ ft}$$

7. Secondary settling: Required volume above sludge hopper to provide 2-hr detention:

$$\frac{35{,}000}{16} \times 2 \times \frac{1}{7.48} = 585 \text{ ft}^3$$

With a surface settling rate = 1,000 gpd/ft^2 = $\frac{180 \times \text{tank depth}}{\text{2-hr detention}}$; depth = 11.1 ft. Diameter = 8.2 ft. Make hopper bottom for temporary sludge storage and slope walls at 45°. Pump sludge to Imhoff tank inlet.

8. Sludge drying bed: Provide 1.25 ft^2 of open sand bed per capita. The required area = $350 \times 1.25 = 438$ ft^2. Provide two beds each 11′ × 20′.

Toll Road Service-Area Design

1. Wastes containing high concentrations of grease and detergents, typical of superhighway service areas, cannot be successfully treated on sand filters unless first given adequate preliminary treatment. High concentrations of detergents cause the emulsification of grease and an increase in the amount of material in colloidal suspension, thereby resulting in a carry-over of grease and solids with consequent clogging. On the other hand, if the design in a small installation provides for proper grease separation and primary septic tank treatment of 2 to 3 days, the carry-over of solids is reduced. Careful maintenance is of course essential. An Imhoff tank plant preceded by adequate grease separation, such as a septic-tank-type grease trap, has merit in the larger installation. Septic-tank subsurface absorption systems are used for small systems where soil is suitable and groundwater is not a problem.

2. Dixon and Kaufman report satisfactory results if the sewage is passed through an anaerobic digestion tank before being given high-rate trickling filter treatment with recirculation at a high ratio of final effluent to the primary settling tank following the digester.[119] A 24-hr displacement period based on average daily sewage flow is provided in the digester maintained at a temperature of 70°F. A 40 percent BOD reduction is reported in the digester.

3. Design of service areas in connection with toll roads was based on a BOD of 500 mg/l, suspended solids of 300 mg/l, grease of 130 mg/l, and commercial detergent concentration of 350 mg/l. The following design flows were used.

Flows	Per Counter Seat	Per Table Seat
Daily average	350	150
Peak day	630	270
Extreme peak	1890	810
Minimum flow	0	0

[119]Gale G. Dixon and Herbert L. Kaufman, "Turnpike Sewage Treatment Plants," *Sewage Ind. Wastes*, **28,** No. 3, 245-254 (March 1956).

4. Experience with superhighway service- and restaurant-area sewage treatment plants indicates that the design flows and sewage strengths given above are somewhat high. It also appears that because of the extreme variations in flows there is considerable advantage in the recirculation of primary and secondary settling tank effluents to the primary inlet. This would prevent the development of anaerobic conditions in these tanks and also add to the degree of treatment.

5. Kansas turnpike sewage treatment plant design used the following criteria:[120]

Flow = 350 gpd/counter seat plus 150 gpd/table seat. Ten percent of the cars passing a service area will enter and will contribute 15 to 20 gal/person or customer. The design flow is assumed to equal the water consumption. The flow is 200 percent of the daily average at noon and 160 percent of the daily average at 6 p.m.

Sewage characteristics: BOD = 600 mg/l, suspended solids = 300 to 450 mg/l with 90 percent volatile matter, pH = 9.5, grease = 100 mg/l, active detergent = 100 mg/l, temperature = 70 to 85°F.

Treatment: To consist of comminutor, Imhoff tank with overflow rate 1000 gpd/ft^2, trickling filter loaded at not greater than 650 lb BOD/acre-ft, final settling overflow rate not greater than 1000 gpd/ft^2 based on maximum rate of raw sewage flow, continuous sludge removal, and recirculation not less than 1:1. Sand filter dosage not greater than 80,000 gpd/acre when open, and 40,000 gpd/acre when covered. Sand ES 0.5 to 0.75 mm. Grease trap on kitchen waste line, 3 compartments and 500-gal capacity.

6. A study of New York State Thruway service areas showed the following:[121]

Average water use: 9.16 gal/vehicle stopping with a range of 7.38 to 11.58.

Average stay in parking area: 30 min/vehicle, 25 min for coffee shop, and 20 min for snack bar.

Peak hour factor: 8 percent of peak day traffic.

Peak capacity of women's facilities: 52 persons/hr for water closet.

Rest room usage: 53 percent women at peak periods; 42.5 percent at off-peak periods.

7. Another report gives the following information:[122]

Average flow from eight service areas: 100 gpm.

Recirculation flow: 100 gpm.

5-day BOD: 500 mg/l.

Suspended solids: 200 mg/l.

2-stage trickling filter: recirculation through primary settling tank and directly to secondary filter for continuous dosage; raw sewage enter digester for anaerobic digestion of grease and detergents.

Removal: BOD 95 to 99 percent, suspended solids 90 to 99 percent.

Final settling tank: detention 127 min, surface settling rate 841 gpd/ft^2.

Chlorine contact time: 21 min.

Sludge drying bed: 1.1 ft^2/capita.

Combined settling digestion tank: 4.7 ft^3/capita, surface settling rate 568 gpd/ft^2, detention 400 min.

Primary trickling filter: 159 gpd/ft^2, BOD 1.5 lb/yd^3.

Secondary trickling filter: 159 gpd/ft^2, BOD 0.39 lb/yd^3.

8. A study for the Federal Highway Administration[123] proposes the following design criteria:

Average daily traffic entering the rest area: 9 percent of vehicles passing by, based on six peak 3-day weekends or three peak months.

Average water use per vehicle: 6.7 gal.

Sixty-seven percent of average water use occurs between 8 a.m. and 4 p.m.

The peak hourly demand: 16 percent of average daily traffic entering the rest area × average water use per vehicle.

Design wastewater flow: 5.5 gal per vehicle.

Average BOD: 125 to 175 mg/l.

Average suspended solids: 125 to 200 mg/l.

[120]Clifford Sharp, "Kansas Turnpike Sewage Treatment Plants," *Pub. Works*, **88**, 142 (August 1957).

[121]Clayton H. Billings and Irwin P. Sander, "Determining the Future Needs of Highway Rest Areas," *Pub. Works*, **97**, 108–111 (October 1966).

[122]Henry W. Haunstein, "Sewage and Treatment for Turnpike Service Areas," *Water Sewage Works*, **107**, No. 3, 89–90 (March 1960).

[123]"Evaluating Wastewater Treatment at Highway Rest Areas," *Pub. Works*, April 1978, p. 74.

SEWAGE WORKS DESIGN—LARGE SYSTEMS

General

The need for sewerage studies that take into consideration the broad principles of comprehensive community planning and environmental impact analysis is discussed in Chapter 2. Also discussed is the importance of regional and area-wide sewerage planning (preliminary) that recognizes the extent of present and future service areas, the established water quality and effluent standards, and alternative solutions with their first costs and total annual costs. This information is needed to assist local officials in making a decision to proceed with the design and construction of a specific sewerage system, including treatment plant. These are essential first steps to ensure that the proposed construction will meet community, state, and national goals and objectives.

The degree and type of treatment to be provided is dependent on many factors, a major one being the water quality standards established for the receiving water. Also important, however, is the future as well as the existing upstream and downstream water usage, the minimum flows,* the types of sewers and wastewater characteristics, the assimilative capacity of the receiving waters, and the capability of the community to finance, operate, and maintain the facility as intended.

The design details of large sewage treatment plants and sewer systems is beyond the scope of this text. Some of the major design elements, however, are given here for general information. Federal and state regulatory agencies have recommended standards and guidelines.[124]

A preliminary basis of design, which has been used for many years where drinking water supplies are not directly involved and other local factors are not adversely affected, is given below. This dilution principle *by itself* is no longer acceptable in the United States.

Dilution water available[a]	Required degree of treatment	Dilution factor[b]
3.5 to 5.0 ft^3 sec or more	Effective sedimentation	22.6 to 32.3
2.0 to 3.0	Chemical precipitation	12.9 to 19.4
1.0 to 2.0	High-rate trickling filters	6.5 to 12.9
1.0 to 1.5	Conventional trickling filters	6.5 to 8.7
0.5 to 1.0	Activated sludge	3.2 to 6.5
0.1 to 0.5	Intermittent sand filter	0.65 to 3.2

[a]Per 1000 equivalent population.
[b]Based on 100 gpcd.

The dilution water available is in terms of ft^3/sec/1000 equivalent population, with the water 100 percent saturated with oxygen. Under special conditions a lesser volume of dilution water may be sufficient to prevent the development of

*Minimum average seven-consecutive-day flow of the receiving stream once in ten years.

[124]*Recommended Standards for Sewage Works*, Great Lakes-Upper Mississippi River Board of State Sanitary Engineers, Health Education Service, Inc., P.O. Box 7126, Albany, N.Y. 12224, 1978.

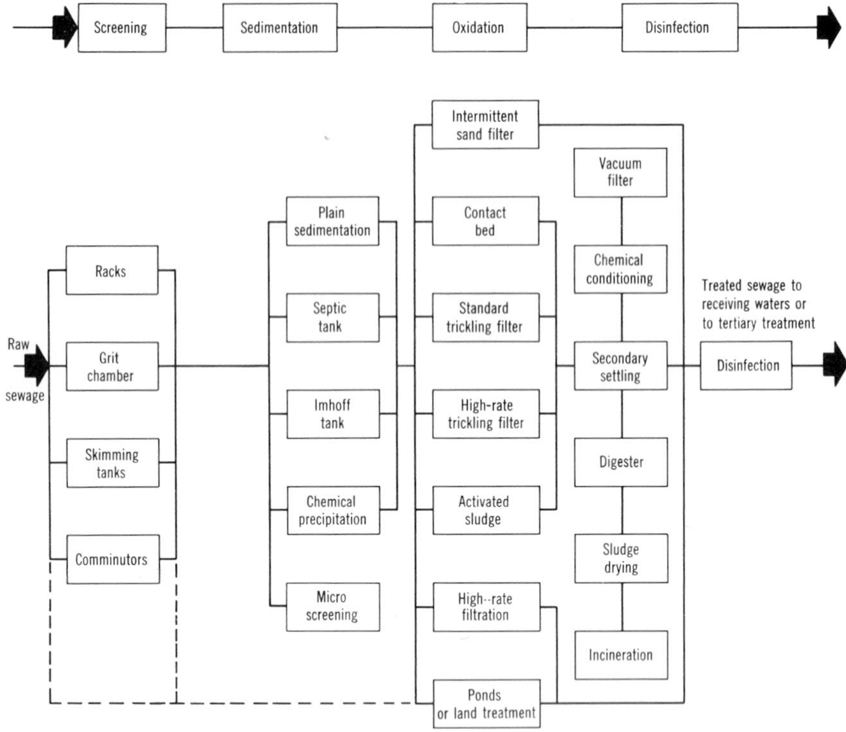

Figure 4-41 Conventional sewage treatment unit processes.

unsatisfactory conditions, such as when the stream has a turbulent flow or joins a larger watercourse after only a few hours flow. On the other hand five times the given dilution may be required if flow is through a densely populated area. In the final analysis the public health hazard, aquatic stream usage, and classification will determine the degree of treatment required. Present opinion in the United States is that all sewage should receive a minimum of secondary treatment.

In Britain, the Ministry of Housing and Local Government reaffirmed in 1966 the Royal Commission's "general standard" as a "norm" for sewage effluents: 5-day BOD 20 mg/l and suspended solids 30 mg/l with a dilution factor of 9 to 150 volumes in the receiving watercourse having not more than 4.0 mg/l BOD.[125] A higher effluent standard of 10 mg/l BOD and suspended solids may be required if indicated such as when the dilution is less than 8 or 9. For dilution of 150 to 500 a suspended solids not greater than 100 mg/l may be permitted; for dilutions greater than 500 very little treatment may be required. In any case sewage effluents should not contain any matter likely to render the receiving stream poisonous or injurious to fish.

The USEPA defined secondary treatment in July 1977 as one producing an effluent with a maximum monthly average of 30 mg/l and a maximum weekly

[125]W. R. Saunders, "Water and Wastes, The Public Health Viewpoint," *J. R. Soc. Health*, 247–250 (September/October 1968).

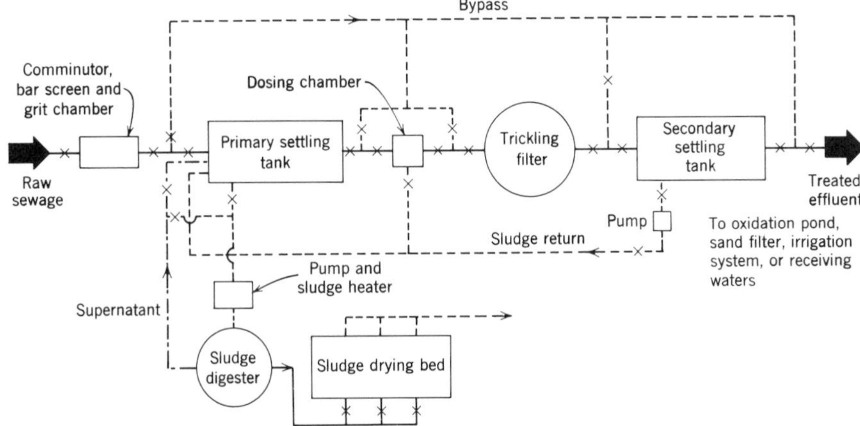

Figure 4-42 A secondary sewage treatment plant. (Units are usually in duplicate.)

average of 45 mg/l for BOD and suspended solids. For fecal coliform bacteria the goal is a monthly maximum average of 200/100 ml and the weekly average 400/100 ml which would probably require seasonal or year-round disinfection for certain water uses. The pH of the effluent shall be within the range of 6 to 9. Standards for phosphorus, oil, grease and COD are also proposed.

Sewage treatment processes and bases of design, are summarized in simplified form in Figure 4-41 and Tables 4-14, 4-19, and 4-20. They are not meant to be complete. Recommended standards for the design and preparation of plans and specifications for sewage works are given in government standards and various texts. Plant efficiencies are given in Table 4-21. A secondary sewage treatment process is shown in Figure 4-42. Typical flow diagrams are shown in Figures 4-33 and 4-34.

Plans and Report

Plans of the area to be sewered should be complete and should include specifications and an engineering report giving the problem, objectives, and design details. The plans must be prepared by a licensed professional engineer, drawn to a scale of 1 in. equal to not less than 100 ft or more than 300, with contours at 2- to 10-ft intervals. The discussion that immediately follows deals with sanitary sewers. Combined sewers should be redesigned as separate storm and sanitary sewers insofar as possible; extensions should be separate sewers.

Sewers

The location of the sewers, with surface elevations at street intersections and changes in grade, are indicated on the general plan. The size of sewers, outfalls, slope, length between manholes, and invert elevations of sewers and manholes to

Table 4-19 Conventional Sewage Treatment Plant Design Factors

Preliminary Treatment	Coagulation and Sedimentation Treatment
Racks	*Sedimentation*
Area: 200% plus sanitary sewer; 300% plus combined sewer. Bar space: 1″ to $1\frac{3}{4}$″, dual channels	Surface settling rates at peak flows: primary and intermediate set. tanks 1500 gpd/ft^2; final set. tanks 1200 gpd/ft^2 after trickling filters or rotating biological contactors, and for activated sludge for conventional, step aeration, contact stabilization, and the carbonaceous stage of separate-stage nitrification; following extended aeration 1000 gpd/ft^2; for physical-chemical treatment using lime: 1400 gpd/ft^2
Screens	Weir rates: 10,000 gpd per lin. ft for average flows to 1.0 mgd and up to 15,000 for larger flows
Net submerged area: 2 ft^2 per mgd for sanitary sewer; 3 ft^2 per mgd for combined sewer. Slot opening $\frac{1}{8}$″ min. Dual units, preceded by racks	Sludge hopper: 1 hor. to 1.7 vert
Grit chamber	Sludge pipe: 6 in. min
Sewage velocity: 1 fps mean, $\frac{1}{2}$ fps, min. Detention: 45 to 60 sec, floor 1 ft below outlet. Min. of 2 channels	*Chemical precipitation*
Skimming tank	Rapid mix, coagulation, sedimentation. Ferric chloride, ferrous sulfate, ferric sulfate, alum, lime, or a polymer
Air or mechanical agitation with or without chemicals. Detention: 20 min for grease removal, 5 to 15 min for aeration, 30 min for flocculation	*Imhoff tank*
Comminutors	Detention period: 2 to 2.5 hr. Gas vent: 20% total area of tank min. Bottom slope: $1\frac{1}{2}$ vert. to 1 hor. Sludge compartment: 3 to 4 ft^3 per capita 18 in. below slot; 6 to 10 ft^3 per capita secondary treatment. Bottom slope: 1 to 1 or 2. Slot & overlap: 8 in. Sludge pipe: 8 in. min. under 6 ft head. Velocity: 1 fpm. Surface settling rate: 600 gpd/ft^2
Duplicate or bypass. Downstream from grit chamber	
Flow basis	
100 gal per capita plus industrial wastes. Usual to assume total flow reaches small plants in 16 hr	
Flow equalization	
Based on 24-hr plot to smooth out hydraulic and organic loading	
Chemical treatment	
For odor control, oxidation, corrosion control, neutralization	

Note: Surface settling rate $= \text{gpd/ft}^2 = \dfrac{180 \times \text{tank depth in ft}}{\text{detention in hr}}$

[a]Sludge digestion will require approximately 120 days at 55°F, 68 days at 60°, 37 days at 71°, 33 days at 86°, 24 days at 95°, 14 days at 113°. Temperature of 140° causes caking on pipes.

[b]Gallons per acre per day = gpad.

[c]Million gallons per acre per day = mgad.

Biological Treatment	Sludge Treatment
Intermittent sand	*Digester*[a]
Filter rate: 50,000 to 100,000 (gpad)[b] with plain settling and 400,000 gpad with trickling filter or activated sludge. Sand: 24 in. all passing $\frac{1}{4}$-in. sieve, ES 0.35–0.6 mm. Unif. coef. less than 3.5	Capacity: with plain sedimentation 2 to 3 ft^3 per cap. heated, or 4 to 6 unheated. With standard trickling filter 3 to 4 ft^3 heated and 6 to 8 unheated; 4 to 5 ft^3 heated and 8 to 10 ft^3 unheated with a high-rate filter. With activated sludge 4 to 6 ft^3 per cap. heated and 8 to 12 ft^3 unheated. Bottom slope: 1 on 4, gravity
Contact bed	*Sludge drying bed*
Filter rate: 75,000 to 100,000 gpad per ft	Open: 1 ft^2 per capita with plain sedimentation, 1.5 ft^2 with trickling filter. 1.75 ft^2 with activated sludge, and 2 ft^2 with chemical coagulation. Glass covered: reduce area by 25%
Trickling filter	*Vacuum filtration*
Standard rate: 400 to 600 lb BOD per acre-ft per day; or 2 to 4 mgad[c], 6 ft depth. High rate: 3000 + lb BOD per acre-ft per day, or 30 mgad for 6-ft depth. Min. filter depth 5 ft, max. 10 ft	Lbs per ft^2 per hr dry solids. Prim. 6 to 10, trickling filter 1.5 to 2.0, activated sludge 1 to 2
Activated sludge	*Centrifuging*
See Table 4-20. Normally 2-hr retention in primary and final sed. and 6 to 8-hr aeration	Flow rate based on gpm per HP.
Rapid filtration	*Wet combustion*
1 to 2 gpm per ft^2, 1.2 mm media	Sludge thickener: Loading of 10 lb per day per ft^2
Land treatment	*Land disposal*
See Table 4-17	Stabilized sludge only. See text
Stabilization pond	*Incineration*
15 to 35 lb BOD per acre per day, 3- to 5-ft liquid depth, center inlet; variable withdrawal depth, 3-ft freeboard, detention 90 to 180 days; multiple units; winter flow retention. Use up to 50 lb BOD loading in mild climate and 15 to 20 in cold areas. See Table 4-15	Tons per hr depending on moisture and solids content. Temperature 1250° to 1400° F. Pyrolysis temperature higher
Rotating biological contactors	*Gas*
See text	A properly operated heated digester should produce about 1 ft^3 of gas per capita per day from a secondary treatment plant and about 0.8 ft^3 from a primary plant. The fuel value of the gas (methane) is about 640 Btu per ft^3
Disinfection	
Chlorine, ozone: See Table 4-14	

Table 4-20 Permissible Aeration Tank Capacities and Loading

Process	Aeration Tank Organic Loading (lb. BOD_5/day per 1000 ft^3)	F/M[a] Ratio (lb. BOD_5/day per lb. MLVSS[b])	MLSS[c]
Conventional Step aeration Complete mix	40	0.2 to 0.5	1000-3000
Contact stabilization	50[d]	0.2 to 0.6	1000-3000
Extended aeration Oxidation ditch	15	0.05 to 0.1	3000-5000

Source: *Recommended Standards for Sewage Works*, Great Lakes-Upper Mississippi River Board of State Sanitary Engineers, Health Education Service, Inc., P.O. Box 7126, Albany, N.Y. 12224, p. 80-6, 1978.
[a]Food to microorganism ratio (F/M)
[b]Mixed liquor volatile suspended solids (MLVSS)
[c]MLVSS values are dependent upon the surface area provided for sedimentation and the rate of sludge return as well as the aeration process.
[d]Total aeration capacity, includes both contact and reaeration capacities. Normally the contact zone equals 30 to 35 percent of the total aeration capacity.

the nearest 0.01 ft are also shown on the general plan. For all sewer laterals 12 in. or larger and all main, intercepting, and outfall sewers, profiles including manholes and siphons, stream crossings, and outlets must be included. The horizontal scale must be at least equal to that on the general plan and the vertical scale not smaller than 1 in. equal to 10 ft. Detail plans to a suitable scale are required of all appurtenances, manholes, flushing manholes, inspection chambers, inverted siphons, regulators, pumping stations, and any other devices, to permit thorough examination of the plans and their proper construction. The total drainage area and the area to be served by sewers should be shown on a topographic plan.

Diameter of Sewer (in.)	Grade (percent) $n = 0.013$*	$n = 0.012$	$n = 0.011$	Capacity Full = (mgd)	Population Served† at 250 gpd/capita	at 400 gpd/capita
4	0.65	0.625	0.60	0.12	—	—
6	0.60	0.51	0.42	0.27	—	—
8	0.40	0.32	0.25	0.46	1840	1150
10	0.28	0.23	0.18	0.72	2880	1800
12	0.22	0.18	0.14	1.00	4000	2500
15	0.15	0.125	0.10	1.60	6400	—
18	0.12	0.10	0.08	2.30	9200	—
21	0.10	0.08	0.063	3.20	12,800	—
24	0.08	0.066	0.053	4.00	16,800	—

*At minimum slope, $n = 0.013$, velocity 2 fps; n is the value in the Manning or Kutter formula.
†Courtesy Kestner Engineers, P.C., Consulting Engineers, Troy, N.Y.

Table 4-21 Sewage Treatment Plant Unit Combinations and Efficiencies

	Total Percent Reduction —Approximation	
Treatment plant	Suspended Solids	Biochemical Oxygen Demand
Sedimentation plus sand filter	90 to 98	85 to 95
Sedimentation plus standard trickling filter, 600 lb BOD per acre-foot maximum loading	75 to 90	80 to 95
Sedimentation plus single stage high-rate trickling filter	50 to 80	35 to 65[a]
Sedimentation plus two stage high-rate trickling filter	70 to 90	80 to 95[a]
Activated sludge	85 to 95	85 to 95
Chemical treatment	65 to 90	45 to 80
Preaeration (1 hr) plus sedimentation	60 to 80	40 to 60
Plain sedimentation	40 to 70	25 to 40
Fine screening	2 to 20	5 to 10
Stabilization (aerobic) pond	—	70 to 90
Anaerobic lagoon	70	40 to 70

[a]No recirculation. Efficiencies can be increased within limits by controlling organic loading, efficiencies of settling tanks, volume of recirculation, and the number of stages; however, effluent will be less nitrified than from standard rate filter, but will usually contain dissolved oxygen. Filter flies and odors are reduced. Study first cost plus operation and maintenance.

Sewers are usually designed for a future population 30 to 50 years hence and for a per capita flow of not less than 400 gpd for submains and laterals and 250 gpd for main, trunk, and outfall sewers, plus allowance for industrial wastes. Intercepting sewers are designed for not less than 350 percent of the average dry-weather flow. See the local regulatory agency. Sewer systems are designed for average daily flow of not less than 100 gpd/capita.

Street sewers should not be less than 8 in. in diameter and at a depth sufficient to drain cellars, usually 6 to 8 ft. Vitrified clay or concrete sewers are designed for a mean velocity of 2 fps when flowing full or half full, based on Kutter's formula, with $n = 0.013$. When ABS, PVC, cement asbestos, or cement- or enamel-lined cast-iron pipe is used, $n = 0.011$ or $n = 0.012$ may be used in design if permitted. In general, sewers should be laid on grades not less than those in the chart shown above.

In some special situations there may be justification for the use of 6-in. sanitary sewers with full knowledge of the limitations.[126] Compression-type joints for pipe 4 to 12 in. in diameter are highly recommended. Vitrified clay and cast-iron pipe have a life of more than 50 yr.

[126]*Small-Size Pipe for Sanitary Lateral Sewers*, Building Research Advisory Board, Division of Engineering and Industrial Research, National Academy of Science, National Research Council, Washington, D.C., February 28, 1957.

Manholes should be not more than 400-ft apart on 15-in. pipe or smaller and not more than 500-ft apart for 18- to 30-in. pipe. Manholes are installed at the end of each line, at intersections, and at changes in size, grade, or alignment.

"Y" connections should be installed for each existing and future service connection as the sewer is being laid. A tight connection must be assured.

Inverted Siphons

Where required, inverted siphons should not be less than 6 in. in diameter nor less than 2 in number. Sufficient head should be available to provide velocities of not less than 3 fps in sanitary sewers or 5 fps in combined sewers at average flows. Accessibility to each siphon and diversion of flow to either one would permit inspection and cleaning.

Pumping Stations

Pumping plants should contain at least three pumping units of such capacity to handle the maximum sewage flow with the largest unit out of service. The pumps should be selected so as to provide as uniform a flow as possible to the treatment plant. Two sources of motive power should be available. Small lift stations, under 1 mgd capacity, should have duplicate pumping equipment or pneumatic ejectors with auxiliary power. All stations should have an alarm system to signal power or pump failure. Every effort should be made to prevent or minimize overflow.

In all cases raw sewage pumps should be protected by screens or racks unless special devices are approved. Housing for electric motors above ground and in dry wells should provide protection against flooding and good ventilation, preferably forced air, and accessibility for repairs and replacements. All electrical equipment and wiring shall meet National Electrical Code requirements. Wet wells or sump pumps should have sloping bottoms and provide for convenient cleaning. See Figure 4-32. Select water-level pump controls with care because they are the most frequent cause of pump failure.[127] Submersible pump stations are also used.

Sewage Treatment

Sewage treatment works should be designed for a population at least 10 years in the future, although 15 to 25 years is preferred, and a per capita flow of not less than 100 gpd plus institutional wastes and *acceptable* industrial wastes. Actual flow studies should govern. Plants should be accessible from highways but as far

[127]George M. Ely, Jr., "Wastewater Pumping Station Design Criteria," *J. Water Pollut. Control Fed.*, **37**, 1547 (October 1965).

as practical from habitation and wells or sources of water supply and should be protected from damage at the 100-yr flood level. The required degree of treatment should be based on the water quality standards and objectives established for the receiving waters and other factors, as pointed out earlier.

The two major sewage treatment design parameters are the 5-day BOD and suspended solids of the wastewater to be treated, and the removal expected in the treatment process. The processes, efficiencies, and design factors are shown in Figures 4-33 to 4-42, and Tables 4-15 to 4-21.

The 5-day BOD is usually assumed to be 0.17 lb/capita/day and the suspended solids 0.20 lb with the average daily flow 100 gpd/capita for domestic wastes. Adjust the design basis and treatment for garbage grinders and industrial wastewater flows where indicated. Studies at 78 cities suggest 0.20 lb BOD, 0.23 lb suspended solids, and 135 gpd/capita as being more representative. Design should be based on actual wastewater strength, characteristics, and flow.[128]

Scale drawings of the units comprising the sewage treatment works, together with such other details as may be required to permit review of the design, examination of the plans, and construction of the system in accordance with the design, must be prepared and submitted to the regulatory agency having jurisdiction. Provision for measuring the flow and sampling the sewage and an equipped laboratory for examinations to control operation should be included in the original plans. The design must include adequate treatment and disinfection of the plant effluent if the receiving stream or body of water in the vicinity of the outfall is used for water supply, shellfish propagation, recreation, or other purposes that may be detrimentally affected by the sewage disposal.

Plants employing trickling filters, activated sludge, aeration, or spray irrigation, and those where dried sludge is handled, present a possible health hazard to plant workers and people downwind through inhalation of airborne microorganisms. See Wastewater Reuse—Hazards and Constraints, in this chapter.

Sludge Treatment and Disposal

Sludge treatment and disposal can be costly and present disposal problems. Sludge handling may involve the collection, thickening, stabilization, conditioning, dewatering, heat drying, and/or final disposal of the sludge.[129] Figure 4-33 shows some sludge treatment processes and Table 4-19 gives some treatment design parameters.

Thickening processes include gravity settling, flotation, and centrifugation. Sludge stablilization is accomplished by aerobic or anaerobic digestion, and by lime or chlorine stabilization. Sludge is conditioned, prior to thickening or dewatering, by the addition of chemicals. Heat treatment by means of a furnace or dryer reduces the moisture content of the sludge. Dewatering is accomplished by

[128]Raymond C. Loehr, "Variation of Wastewater Quality Parameters," *Pub. Works*, **99**, 81–83 (May 1968).

[129]Charles R. Scroggin, Don A. Lewis, and Paul B. Danheiser, "Developing a Cost-Effective Sludge Management Approach," *Pub. Works*, June 1980, p. 87–91.

means of drying beds, centrifuges, vacuum filters, continuous belt presses, or plate and frame presses.

Final disposal may be by composting, incineration, sanitary landfill, subsurface injection, or land application. Land application of sludge containing not more than 3 dry tons of nitrates or cadmium per acre per year should not cause a problem. Codisposal of sludge with solid wastes by means of composting, or by incineration with energy recovery, are also possibilities. All these processes require careful environmental control to prevent disease transmission, air pollution, surface and underground water pollution; nitrates, synthetic organics, and metals buildup in the soil, food crops, and groundwater. U.S. legislation requires the phase-out of "harmful" sewage sludge disposal by ocean dumping by December 1981; but the date has been extended pending further study.

Hydraulic Overloading of Sewers and Treatment Plant

The average dry-weather flow of sanitary sewers is approximately equal to the discharge rate of runoff from a rainfall having an intensity of 0.01 in./hr.[130] But 0.02 to 0.03 in. of rainfall is needed to wet the ground before there is a runoff. Camp concludes that the rate of flow in a combined sewer is approximately equal to 100 times the rainfall intensity in in./hr times the dry weather flow, up to the capacity of the sewer.[131] Therefore if the average dry-weather flow is 1 mgd and the rainfall intensity is 2 in./hr for a sufficient time to cause runoff from the area under consideration, the rate of flow would be 200 mgd. The economic futility of trying to design a treatment plant to treat combined sewer flows is obvious. On the other hand, combined sewers contribute significant amounts of sediment, oil, salts, and organic matter. Discharge, or overflow from combined sewers, immediately after a heavy rain may have a BOD of several hundred milligrams per liter and MPN of hundreds of thousands of coliform organisms. Even after the intital flushing the pollution discharged is still substantial. Hence complete stream pollution control cannot be realized unless combined sewer and storm-sewer overflows are temporarily retained and/or adequately treated before discharge.

The importance of sewer inspection before backfilling has not been given sufficient emphasis. Too many properly designed plants have created problems almost as serious as those they were intended to correct. Some plant flows equal or exceed the design flow before the system is completed because of improper sewer, manhole, and house connection construction; roof, footing, basement, and area drainage connections; street drainage and storm-sewer connections; cooling water discharges, and drainage from springs and swampy areas; and

[130] M. E. McKee, "Loss of Sanitary Sewage through Storm Water Overflows," *J. Boston Soc. Civ. Eng.*, **55** (April 1947); Clyde L. Palmer, "The Pollutional Effects of Storm Water Overflows from Combined Sewers," *Sewage Ind. Wastes*, **22,** 154 (February 1950); Frank C. Johnson, "Nation's Capital Enlarges its Sewerage System," *Civ. Eng.*, **56** (June 1958).

[131] Thomas R. Camp, "Chlorination of Mixed Sewage and Storm Water," Annual Convention, ASCE, Boston, October 11, 1960.

similar practices. Inspection during construction should assure tight service connections, manholes, and sewer construction by requiring full compliance with pipe, joint, bed, and backfill specifications. Precast concrete or poured-in-place manholes with liquid-tight joints and sewer connections are acceptable. Brick or concrete or cinder block manholes, lined or unlined, and without reinforcement are unacceptable.

Infiltration flow tests in wet ground (high groundwater or wet weather) and television inspection of each line before acceptance, or exfiltration flow tests in dry ground, will determine compliance with specifications. The exfiltration test is a more severe test; large volumes of water are needed and must be disposed of.[132] Under a head of 2 ft of water the exfiltration was found to be 4.8 percent greater than shown by the infiltration test; under a 4-ft head, 19.8 percent greater; and 27.3 percent greater under a 6-ft head. A low-pressure air test at 3.0 psi with air loss not greater than 0.0030 cfm/ft^2 of internal pipe surface has certain advantages and can also be used.[133]

The total leakage should not exceed 150 gpd/mi/in. of internal diameter of pipe over a 24-hr test period, in a well-laid line. *Recommended Standards for Sewage Works* states that leakage shall not exceed 200 gpd/mi/in. of pipe diameter. An infiltration rate up to 1500 gpd/mi/in. may be acceptable. A figure of 200 to 400 gal/acre of sewered area/day is also used. Rubber ring joints or the equivalent on vitrified clay tile, concrete, PVC, ABS, and asbestos cement pipe, and hot-poured joints in dry trenches can give good results.

A community, camp, institution, factory, school, or other establishment that constructs sanitary sewers and treatment plants should immediately set up rules and regulations prohibiting sewer connections by anyone other than responsible individuals. This must be supplemented by effective enforcement to guarantee exclusion of groundwater, surface water, and rainwater from the sanitary sewer system in order to protect the investment made and accomplish the objective of the treatment plant. Sanitary sewers that, in effect, become combined sewers almost always cause backing up in basements after storms and sometimes overflow through manholes onto the street and into stores, basements, and so on, with resultant damage to oil burners, electric motors, and personal property. Cellars remain damp and become contaminated with sewage. Treatment plants become overloaded, requiring the bypassing of diluted sewage to the receiving stream or body of water, resulting in danger to water supplies, recreational area, fish life, and property values. This, of course, nullifies in part the purpose of sewage treatment. Greater attention has been given to the proper construction of sanitary sewers and to the elimination of roof leader connections to the sanitary sewer, a major cause of hydraulic overloading.[134]

The effect of roof leader and submerged manhole flows is more fully appre-

[132]Sherwood Borland, "New Data on Sewer Infiltration-Exfiltration Ratios," *Pub. Works*, **87,** 7–8 (September 1956).

[133]Roy Edwin Ramseier and George Charles Riek, "Experience in Using the Low-Pressure Air Test for Sanitary Sewers," *J. Water Pollut. Control Fed.*, **38,** 1625–1633 (October 1966).

[134]Gerald L. Peters and A. Paul Troemper, "Reduction of Hydraulic Sewer Loading by Downspout Removal," *J. Water Pollut. Control Fed.*, **14,** 63–81 (January 1969).

ciated by some comparisons. A 4-in. rainfall in 24 hr can be expected in the United States roughly south of the Ohio River, once in 5 yr. If 180 dwellings (1000 ft^2 roof area) connected their roof leaders to the sanitary sewer, the resultant flow would equal the capacity of an 8-in. line; that is, 460,000 gpd with a slope of 0.4 ft/100 ft. Six manhole covers with six vent and pick holes, under 6 in. of water, will also admit enough water to fill an 8-in. sewer. Sixteen manholes under 1 in. of water will do the same.

Methods used to analyze sewer infiltration and inflow include dry weather and wet weather sewer and plant flows and rainfall measurements, physical inspection and smoke testing, groundwater level and dye testing, internal inspection, and closed circuit television.

Odor and Corrosion Control

Hydrogen sulfide is a common odor problem associated with septic sewage and sludge handling. Industrial wastes may also carry sulfur compounds including hydrogen sulfide and other odorous chemicals related to certain manufacturing processes. Hydrogen sulfide is formed when anaerobic bacteria convert inorganic sulfur compounds (sulfates) in wastewater, particularly sewer slimes and sludges, to sulfide (H_2S). The formation of hydrogen sulfide in flowing sewers, usually at less than 2 ft/sec, is increased by high temperatures, long sewers, low velocities, and low pH; also by strong sewage, high sulfate concentration, and anaerobic conditions in the submerged pipe wall slime.[135] Turbulence facilitates release of the hydrogen sulfide which combines with oxygen if present and with the moist air in the space above the flow line to form sulfur and water. The sulfur in turn is converted by thiobacillus bacteria, normally present, to sulfuric acid. The acid formed attacks metal pipe, concrete and asbestos-cement pipe, and the concrete and mortar in pipe joints and in brick masonry manholes, above the flow line.[136] Corrosion can be a serious problem in the arid climates. Where flow velocity, high temperatures, and septicity cannot be controlled, coal-tar epoxy, epoxy resins, and glass-reinforced linings, properly applied, have been found generally effective in protecting pipe.[137] Vitrified clay pipe jointed with acid-resistant material (not cement mortar), PVC, ABS, and PE pipe are all resistant to sulfuric acid attack. It should be noted that if suitable anaerobic conditions exist in the sewer, methane is also produced which, under certain conditions, may reach explosive concentrations. Hydrogen sulfide is toxic. See the discussion on Safety that follows.

Control of odors requires identification of potential odor sources and odor causes at a wastewater treatment plant and in the collecting sewers such as noted

[135]Elizabeth C. Price and Paul N. Cheremisinoff, "Sewage Treatment Plants Combat Odor Pollution Problems," *Water Sewage Works*, October 1978, pp. 64–69.

[136]Douglas C. Mathews, "Hydrogen Peroxide in Collection System Corrosion and Odor Control," *Deeds Data*, Water Pollut. Control Fed. April 1977, p. 2.

[137]D. G. M. Roberts, and A. P. Banks, "Designing For Arid Climates," *Water Sewage Works*, January 1980, p. 59.

above. The cause or source may also be related to deficiencies in plant design, operation, or maintenance; or to oxygen depletion below 1 mg/l in the mains, trunk, or outfall sewers.

Odor control at a treatment plant requires cleanliness and good housekeeping at all treatment units, pumping wells, and flow channels including prompt removal of grit and screenings, skimmings, slimes, and sludge deposits. The storage of septic sludge will result in hydrogen sulfide odors. Odor control in sewers usually depends on upstream chemical treatment, if the causes of anaerobic conditions cannot be eliminated. Chemicals used in gravity flow sewers include oxygen, chlorine, hydrogen peroxide, sodium nitrate (particularly at lagoons), zinc sulfate, potassium permanganate, activated carbon, calcium carbonate (lime), and sodium hydroxide. Activated carbon filters on indoor ventilating systems are effective to adsorb odors. Wet scrubbing and ozonation are also effective. Sometimes odor-masking or -modification compounds are used; however this is hazardous if toxic gases such as hydrogen sulfide are present and masked.

Oxygen as injected compressed air, in air lift, or as aspirated, and commercial oxygen are used in gravity sewers and in force mains to neutralize sulfide and prevent sulfide buildup. The USEPA and others have made exhaustive studies of sulfide and odor control in sewers and plants.[138,139]

Chlorination of sewage to remove hydrogen sulfide present and to inhibit bacterial action is commonly also used for odor control. Dosage ranges from 10 to 50 mg/l in sewer systems and from 10 to 20 mg/l for plant prechlorination.[135] Chlorine may be added as calcium or sodium hypochlorite or as liquid chlorine.

Hydrogen peroxide can oxidize a number of toxic and noxious substances in wastewater, and particularly hydrogen sulfide. It is an excellent source of dissolved oxygen; it attacks anaerobic organisms that produce sulfides; and hydrogen sulfide and mercaptans are readily oxidized. Industrial-strength hydrogen peroxide comes in 50 and 35 percent solutions (70 percent solution is hazardous).[136,140] Hydrogen peroxide has also been found effective in neutralizing hydrogen sulfide odors resulting from a pond gone septic. Hydrogen peroxide (50 percent) was injected in the pump discharge to recirculated pond water.[141] The addition of hydrogen peroxide upstream from a wastewater treatment plant, under controlled conditions, was found to be effective in eliminating hydrogen sulfide odors in wastewater.[142] Doses of 1 and 2 milligrams per milligram of sulfide present have been found adequate.[135] The theoretical addition is estimated at 4 to 8 parts hydrogen peroxide per part of sulfide.[138] A dosage

[138] *Process Design Manual for Sulfide Control in Sanitary Sewerage Systems*, USEPA, October 1974, pp. 6–11 to 6–17, and pp. 5–1 to 6–17

[139] *Odor Control for Wastewater Facilities*, Manual of Practice No. 22, Water Pollut. Control Fed., 2626 Pennsylvania Ave., N.W., Washington, D.C. 20037

[140] William K. Kibble, Jr., "Hydrogen Peroxide Helps Solve Industrial Wastewater Problems," *Ind. Wastes*, May/June 1978, pp. 27–29.

[141] ______ "Peroxide turns septic pond into community asset," *Am. City County*, June 1978, p. 93.

[142] ______ "Hydrogen Peroxide Wastewater Topics," *Water Sewage Works*, September 1978, p. 54.

of 40 mg/l was found adequate to control odors resulting from oxygen-deficient sludge from an activated sludge plant.[143]

Sodium nitrate at the rate of 10 lbs for each pound of sulfide in wastewater and in oxidation ponds has been successful for odor control.[135]

Little hydrogen sulfide is present at a pH above 8 to 9. The addition of calcium carbonate or sodium hydroxide to achieve that pH level will control hydrogen sulfide as well as other odors.

Safety

Gases in sewers, manholes, wet wells, lift stations, and other unventilated spaces associated with sewerage systems may be poisonous, asphyxiating, or explosive. The poisonous and asphyxiating gases include hydrogen sulfide, sulfur dioxide, chlorine, carbon dioxide, and carbon monoxide. The explosive types, which also burn, include methane, gasoline vapor, hydrogen sulfide, hydrogen, ammonia, carbon monoxide, and combinations of various sewer gases. Methane is produced in sewers under anaerobic conditions. Sewers may also carry very corrosive, highly volatile, hazardous liquids from industrial plants.

Hydrogen sulfide can be very toxic. The permissible 8-hour occupational exposure is 20 ppm, but 50 ppm for only a 10-minute exposure.[144] Above 200 ppm hydrogen sulfide deadens the sense of smell, and death is said to result at 300 ppm.[145] About half the deaths are would-be rescuers.[146] Methane is explosive in the range of 5 to 15 percent by volume. The oxygen concentration should be 20 percent by volume (air contains 20.94 percent).

Workers should not enter manholes, pump pits, or other enclosed spaces unless first assured that the space is thoroughly ventilated, that safety precautions are taken, and that calibrated detection equipment is used. Smoking in the vicinity of manholes or other enclosed spaces should, of course, be prohibited. A first step is the lowering of a calibrated gas detector (to measure the concentration of methane, hydrogen sulfide, and oxygen) with alarm into the work area to determine its safety; the detector should remain in place until all work is completed. The manhole or other space should be ventilated with a blower and kept in operation. The worker entering the manhole should carry an emergency air pack, for immediate use in case the detector alarm goes off, and wear a harness. The harness is connected to a pulley and tripod which is supervised by two other workers. In some instances an air pack with a breathing mask connected to an air hose and air cylinder is used if a sufficient supply of air is not available. A self-contained unit with a 45-min air supply is used for rescue

[143]George A. Brinsko and John A. Sheperd, "Sludge Treatment System Odor Controlled with Hydrogen Peroxide," *Deeds Data*, Water Pollut. Control Fed., April 1977, p. 1.

[144]Marcus M. Key, et al., *Occupational Diseases*, DHEW, PHS, June 1977, p. 423.

[145]*Process Design Manual for Sulfide Control in Sanitary Sewerage Systems*, USEPA, October 1974, p. 2-7.

[146]"Safety Tips for Hydrogen Sulfide in Manholes," *Pub. Works*, May 1980, p. 132.

operations.[147] Continuous hazardous gas monitoring systems for sewage treatment facilities can detect the presence of methane and hydrogen sulfide, and the lack of oxygen. The equipment includes work area and remote audio and visual indicators of unsafe conditions. See also Care of Septic-Tank and Subsurface Absorption System (Safety and Other Precautions) earlier in this chapter.

Cost of Sewerage and Treatment

Cost estimates can vary widely because of location, labor and material costs, volume of construction, season of the year, local characteristics, state of the economy, and degree of treatment required. In general, the cost can be divided into two parts: treatment and collection; operation and maintenance. Financing methods are discussed in Chapter 2. Cost estimates can be adjusted to present-day costs using the Engineering News Record or other appropriate construction cost index. (See Table 4-22.)

To arrive at the total *project* construction cost it is necessary to add to the construction cost 10 to 15 percent for contingencies; call this A. Then add 15 to 20 percent of A for engineering costs and 2 to 3 percent of A for legal and administrative costs; call this B. An additional cost is financing during construction, about 3 percent of B, making a total of C, plus interest during construction taken at 6 to 7 percent of C, depending on the cost of money. Therefore the total project cost can result in adding 36 to 48 percent to the construction cost. The added cost components can be expected to vary depending on the preliminary planning, cost of a project, and its complexity.

Suggested general design periods for sewers and treatment plants, and capacities, are given on pages 501 and 502, and also by regulatory agencies. The actual design period however should be related to the projected growth and land use of the tributary area, the cost of the additional sewer and treatment capacity, interest rate, and the local institutional arrangements.

Sewage Treatment Costs

Comparison of sewage treatment costs should consider the total annual cost, that is, the initial cost of construction and the cost of operation and maintenance (O and M). This means the annual principal and interest to pay off the bond issue, the cost of site development, engineering and legal fees, and the annual cost of operation and maintenance. The total cost of sewer service would be the cost of treatment plus the cost of all sewers, manholes, lift stations, and house connections.

Construction, annual, and unit costs of sewage treatment are given in Table 4-22. The economy of scale is very apparent. Economic comparison formulas are given in Appendix I.

Sometimes advanced wastewater treatment (Figure 4-33) is desired or re-

[147]Michael S. Macy, Roger Miller, and Stephen A. Kacmar, "Safety in Sanitary Sewers," *Deeds Data*, Water Pollut. Control Fed., May 1980, pp. 12–14.

Table 4-22 Estimated Total Annual and Unit Costs for Alternative Treatment Processes with a Design Flow of 1.0 Mgd

Process	Initial Capital Cost (Dollars)[a,b]	Annual Cost Dollars[b]			Unit Cost (Cents/1000 gal)[b]
		Capital[c]	O&M[d]	Total	
Imhoff tank	380,000	41,720	15,500	57,270	15.7
Rotating biological disks	800,000	87,832	57,680	145,512	39.9
Trickling filter processes	900,000	98,811	58,480	157,291	43.1
Activated sludge processes with					
external digestion	1,000,000	109,790	74,410	184,200	50.5
with internal digestion[e]	500,000	54,895	48,800	103,695	28.4
Stabilization pond processes[f]	250,000	27,447	23,680	51,127	14.0
Land disposal processes[g]					
basic system	340,000	37,328	41,540	78,869	21.6
with primary treatment	940,000	103,302	81,540	184,742	50.6
with secondary treatment	1,240,000	136,139	115,950	252,089	69.1
Land disposal processes					
basic system[h]	200,000	21,958	25,100	47,058	12.9
with primary treatment	800,000	87,832	65,100	152,932	41.9
with secondary treatment	1,000,000	109,790	99,510	209,300	57.3

Source: George Tchobanoglous, "Wastewater Treatment for Small Communities," *Water Pollution Control in Low Density Areas*, William J. Jewell and Rita Swan, Eds., University of Vermont, Burlington, 1975, p. 424. Published by the University Press of New England, Hanover, New Hampshire.

[a] Estimated average cost.

[b] Based on an Engineering News-Record Construction Cost index of 1900.

[c] Capital recovery factor = 0.10979 (15 years at 7 percent).

[d] Average values for variations in processes.

[e] Extended aeration, aerated lagoon, oxidation ditches.

[f] High-rate aerobic, facultative, and anaerobic.

[g] Irrigation and overland flow.

[h] Infiltration-percolation.

quired without fully realizing the large additional total cost to obtain a small incremental increase in plant efficiency. Advanced wastewater treatment (AWT) to remove an additional 3.8 to 10 percent BOD, 5.2 to 13 percent suspended solids, and approximately 61 to 68 percent phosphorus and ammonia-nitrogen was found to increase capital costs by 42 to 99 percent and operation and maintenance costs by 37 to 55 percent.[148] This suggests that the other more cost-effective alternatives should be explored where advanced wastewater treatment is requested, *if in fact it is actually needed.*

*Sewerage**

The approximate costs (1980) of sewers, manholes, pumping stations, septic tanks, absorption trenches, and related appurtenances are given below. This information is for guidance and comparative purposes only in view of the many

*The assistance of Kestner Engineers, P.C., Troy, N.Y. is gratefully acknowledged in arriving at the cost estimates.

[148] ______ "Advanced waste treatment: Has the wave crested?" *J. Water Pollut. Control Fed.*, July 1978, pp. 1706–1709.

variables encountered in practice. Table 4-23 will be found useful to make cost adjustments and comparisons.

Manhole, precast (4 ft dia.):	0 to 8 ft depth ea.	$750 to 1000
	8 to 10 ft	800 to 1100
	10 to 12 ft	920 to 1200
	12 to 14 ft	1050 to 1300
Manhole frame and cover		$145
Drop inside manhole		$50 to 100
8-in. sewer, ABS or PVC pipe (installed):*	0 to 8 ft cut, per ft	$15 to 20
	8 to 10 ft	17 to 21
	10 to 12 ft	19 to 22
	12 to 14 ft	21 to 24
	14 to 16 ft	23 to 26
"Y" branch		$25 to 40
6-in. house service pipe		12 to 14
8-in. sewer, ductile iron pipe:	0 to 8 ft cut, per ft	$19 to 22
	8 to 10 ft	21 to 24
	10 to 12 ft	23 to 26
	12 to 14 ft	25 to 28
	14 to 16 ft	27 to 30
	16 to 18 ft	29 to 34
Rock excavation, per yd^3:	blasting	$70
	soft rock	30
	air hammer	320
Restoring gravel and shoulder pavement, per yd^3		6
Restoring asphalt-concrete pavement, per ton		55
Top soil and seeding, per yd^3		8 to 12
Pumping station, below ground, including standby generator, dehumidifier, sump pump, and all necessary controls:	1.0 to 3.0 mgd	$115,500 to 173,250
	0.25 to 1.0	80,850 to 115,500
	0.05 to 0.25	69,300 to 80,850
Pumping station, above ground, including building and all necessary controls and equipment:		$288,750 to 346,500
Submersible pump station—small		$16,000 to 26,750

The above costs (1980) are based upon construction on unencumbered land. Where sewers are to be installed in an existing built-up area served by electricity, gas, streets, and perhaps water, the total cost of an 8-in. sewer including manholes and repairs may run to $40 to $60 per foot, using ABS or PVC pipe.

Septic-tank systems

The cost of an individual septic-tank absorption system will vary with the size of the dwelling, local conditions, and the type and size of the absorption or treatment system. Approximate cost estimates (1980), including labor and material, are given for rough comparisons.

*Polyvinyl chloride (PVC), acrylonitrile-butadiene-styrene (ABS), and polyethylene (PE) are resistant to sulfuric acid attack.

Table 4-23 Cost Indices (Average Per Year)

Year	Marshall & Stevens Installed Equipment Indices: 1926-100 (All Industry)	Engineering News-Record Construction Index: 1913-100	Handy-Whitman Index for Water Treatment Plants[a]: 1936-100 (Large Plant)	(Small Plant)	Engineering News-Record Building Cost Index: 1913-100	Chemical Engineering Plant Construction Index: 1957–1959 100	USEPA Sewage Treatment Plant Construction Index: 1957-1959-100
1950	168	510	210	213	375	74	
1951	180	543	225	229	401	80	
1952	181	569	235	235	416	81	
1953	183	600	246	246	431	85	
1954	185	628	251	251	446	86	
1955	191	660	258	257	469	88	
1956	209	692	275	276	491	94	
1957	225	724	288	289	509	99	
1958	229	759	296	296	525	100	102
1959	235	797	311	309	548	102	104
1960	238	824	317	317	559	102	105
1961	237	847	315	315	568	101	106
1962	239	872	324	322	580	102	107
1963	239	901	330	327	594	102	109
1964	242	936	340	336	612	103	110
1965	245	971	350	346	627	104	112
1966	252	1019	368	362	650	107	116
1967	263	1070	380[b]	374[b]	672	110	119
1968	273	1155	398	389	721	114	123
1969	285	1269	441	424	790	119	132
1970	303	1385	480	462	836	126	143
1971	321	1581			948	132	160
1972	332	1753			1048	137	172
1973	344	1895			1138	144	182
1974	398	2020			1204	165	217
1975	444	2212			1306	182	250
1976	472	2401			1425	192	262
1977	491[c]	2577			1545	199[d]	271[d]
1978		2776			1674		
1979		3003			1819		
1980		3200[e]			1932[e]		

Source: *Process Design, Wastewater Treatment Facilities For Sewered Small Communities*, USEPA, Environ. Research Inf. Center Techno. Trans., EPA-625/1-77-009, Cincinnati, Ohio 45628, October 1977, p. 17-9; also *Engineering News-Record*, March 20, 1980, p. 113.

[a]Based on July of year.

[b]Based on January of year.

[c]Based on first quarter of year.

[d]Based on March of year.

[e]Estimated annual average.

Septic tank: precast (delivered) 750 gal	$270
1000	290
1250	350
2000	540
Grease trap: cost same as septic tank plus (precast) cost of inlet and outlet fittings	
Excavation for septic tank or grease trap	$150
Leaching pit or dry well, with 8-in. washed stone, precast units, including cover; inside diameter:	
4 ft diameter and 5 ft deep	$250
6 ft diameter and 9 ft deep	350
8 ft diameter and 5 ft deep	550
Distribution box: 3 outlets	$25
6 outlets	40
8 outlets	60
10 to 12 outlets	100
Extension collar for septic tank or distribution box, 24-in. dia., per ft	$20
Sheeting and bracing, left in place, per ft^2	7
Absorption trench, including perforated pipe and 12-in. washed stone, per ft:	$8
24-in. trench	$3.75 to 6.00
30-in. trench	4.25 to 6.50

Type of system	O & M	Initial cost (1980 adjusted)
Septic tank - absorption system	$20/yr	$1,500– 3,500[a]
Septic tank - built-up absorption system	20/yr	6,000– 8,000[b]
Septic tank - subsurface sand filter, incl. chlor. and contact tank	30/yr	7,500–10,000[c]
Aerobic system — excl. absorption field: including service contract[149]	220/yr	2,700– 6,000
Chem. recir. toilet[150]	—	4,000– 6,000
Incinerator toilet	—	1,100– 1,275
Composting toilet — excl. gray water	5/yr	1,700– 3,400

[a] 3-bedroom home
[b] 9,000 to 10,000 ft^2
[c] 390 to 520 ft^2

LOW-COST SANITATION

The Problem

The population of developing countries has been estimated at 1721 million in 1970 and 2280 million in 1980. However only about 20 percent had "adequate"

[149] *Cleaning Up the Water, Private Sewage Disposal in Maine*, Maine Dept. of Environmental Conservation, Augusta, Maine, July 1974, pp. 16 and 17.

[150] Peter T. Silbermann, "Alternatives to Sewers," *Wastewater Treatment Systems For Private Homes and Small Communities*, Paul S. Babiarz, Robert D. Hennigan, and Kevin J. Pilon, Eds., Central New York Regional Planning and Development Board, Syracuse, 1978, pp. 127–188.

excreta disposal facilities such as a public sewer, privy, septic-tank system, or bucket latrine in 1970.[151] Cultural and institutional factors play an extremely important role in obtaining user acceptance of sanitary latrine facilities in developing countries.

In 1970 the U.S. Census of Housing[152] showed that 28.8 percent of the 67.7 million year-round housing units representing 58.5 million people in the United States were served by a septic tank, cesspool, privy, or other individual facility. Forty-nine million people, 24.5 percent, depended on septic-tank or cesspool systems and 8.7 million, 4.3 percent, depended on privies or other individual facilities. This does not include facilities at highway rest areas, parks, camps, and family vacation homes. The 1971 Census of Canada[153] showed that 17.0 percent of the 6.0 million dwellings were served by septic tanks and 9.6 percent by other onsite disposal facilities.

Public sewers and treatment plants are not necessarily the indicated or universal solution for all situations, particularly in developing areas of the world. Their construction, operation, maintenance, and repair costs can be prohibitive or impractical.[154] The unavailability of piped water, scattered housing, local customs, and slow economic and technological development may make low-cost alternative methods of excreta and wastewater disposal more feasible at a particular point in time. These include the various types of privies and latrines. See pages 514-520, also WHO publications.[155–158]

Privies, Latrines, and Waterless Toilets

Many types of privies, latrines, and waterless toilets have been in use all over the world for many years. If properly located, constructed, and maintained they can be acceptable, economical, and sanitary devices for the disposal of human excreta where waterborne sewage disposal systems are not provided or not practical, and particularly for temporary installations. The suitability, location, construction, and maintenance of various types of privies and latrines are summarized in Table 4-24. Figures 4-43, 4-44, and 4-45 illustrate essential design and construction features.

[151]C. S. Pineo and D. V. Subrahmanyam, *Community Water Supply and Excreta Disposal Situation in Developing Countries*, WHO, 1975, p. 39.

[152]*1970 Census of housing, detailed housing characteristics United States summary*, U.S. Dept. of Commerce, Bureau of the Census, Washington, D.C.

[153]*Housing: Source of Water Supply and Type of Sewage Disposal*, 1971 Census of Canada, Statistics Canada, Cat. 93-736, November 1973. (Mahdy, *J. Am. Water Works Assoc.*, Aug. 1979, p. 447.)

[154]Michael G. McGarry, "Sanitary sewers for undeveloped countries—necessity or luxury?" *Civ. Eng. ASCE*, August 1978, pp. 70–75.

[155]"Disposal of Community Wastewater," WHO Tech. Rep. Ser. **541,** 1974.

[156]D. A. Okun and G. Ponghis. *Community Wastewater Collection and Disposal*, WHO, 1975.

[157]S. Rajagopalan and M. A. Shiffman, *Guide To Simple Sanitary Measures For the Control of Enteric Diseases*, WHO, 1974.

[158]E. G. Wagner and J. N. Lanoix, *Excreta Disposal For Rural Areas and Small Communities*, WHO, 1958.

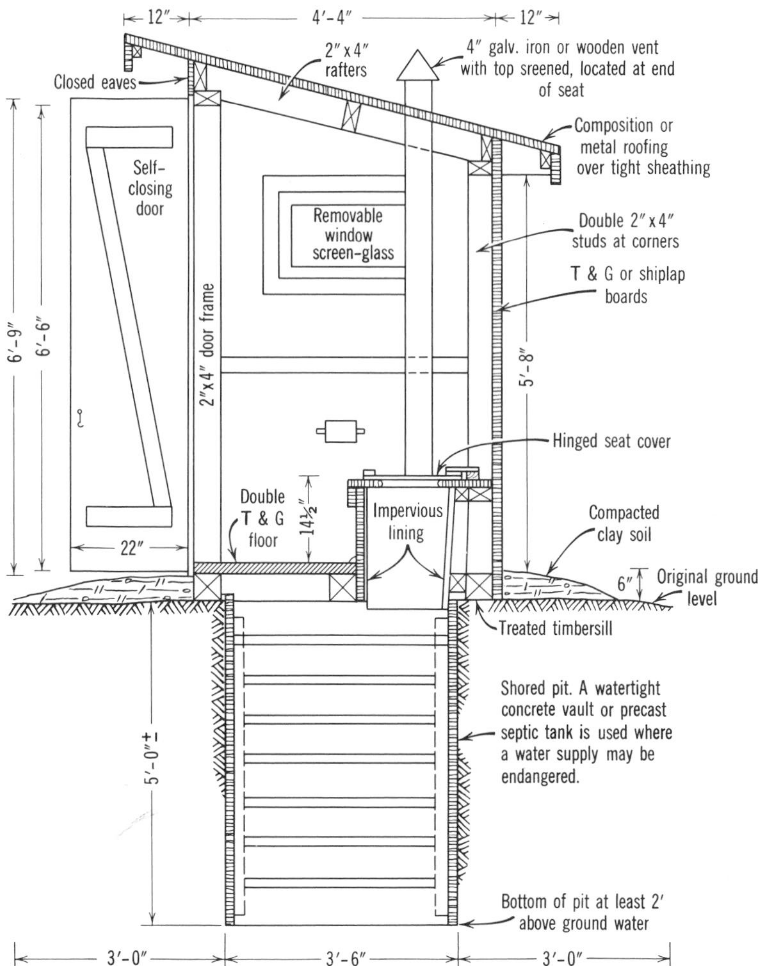

Figure 4-43 Construction details of a sanitary earth pit privy. Privy construction may also be concrete or cinder block, brick, stone, or other masonry, with reinforced concrete floor and riser. Make privy 3-ft wide for one seat or 6 ft wide for two seats. Locate privy within 150 ft of users, 100 ft or more from kitchens, 50 ft from any lake or watercourse, and not in direct line of drainage to or closer than 200 ft of any water supply.

In developing countries, where pit latrines are constructed, they should be planned and designed to promote their conversion to aqua privies, pour-flush latrines, septic privies, or septic tanks. It would be preferable, if this is possible, to construct a concrete block or poured concrete vault in the first instance with a "T"-outlet. A leaching pit, soakage pit, or absorption trench could then be easily added as water supply became more readily available, and as indoor plumbing is added, and people's social aspirations change. Eventually, the pit

Table 4-24 Sanitary Excreta Disposal Methods

Facility	Suitability	Location
Sanitary earth pit privy	Where soil available and groundwater not encountered. Earth mounded up if necessary to bring bottom of pit 2 ft above groundwater or rock.	Downgrade, 100 ft or more from sources of water supply; 100 ft from kitchens; within 50 to 150 ft of users; at least 2 ft above groundwater; 50 ft from lake, stream.
Masonry vault privy	To protect underground and surface water supplies.	Downgrade and 50 ft or more from sources of water supply; 100 ft from kitchens; within 50 to 150 ft of users.
Septic privy (Lumsden, Roberts, and Stiles: LRS privy)	Where cleaning of pit is a problem, odors unimportant, and water is limited.	Same as pit privy.
Excreta disposal pit	For disposal of pail privy and chemical toilet contents.	Downgrade and 200 ft from sources of water supply; 100 ft from kitchen.
Chemical toilet (cabinet and tank-type)	A temporary facility. To protect water supply, where other method impractical. Temporary camp, vehicle, boat.	May adjoin main dwelling. Tank type same as masonry vault privy. Cabinet - type usually within a facility.
Incinerator toilet	Where electricity or gas avail.	Within the facility.
Recirculating toilet	Airplanes, boats, fairgrounds, camps.	Within the facility.
Removable pail privy (bucket latrine) (portable toilet)	A temporary facility; to protect water supply, where pit privy impractical; for large gatherings.	Same as masonry vault privy.
Portable box, earth pit, latrine	At temporary camps.	Same as pit privy. Army recommends latrine 100 yd from kitchens.
Bored-hole latrine	In isolated place or when primitive, inexpensive, sanitary facility is needed.	Same as earth pit privy.
Straddle trench latrine	At temporary camp for less than one day.	Same as earth pit privy.
Cat hole	On hikes or in field.	Same as earth pit privy.
Squatting latrine	Where local conditions and customs permit.	Same as pit privy.
Recirculating oil[b] flush toilet	Where water not available or soil unsuitable.	Within facility.
Aqua privy	Where water is limited and local customs permit.	Same as pit privy.
Composting toilet	To conserve water; to convert excreta and garbage (food waste) to humus.	Storage chamber in basement or adjacent to house.

Criteria: Confines excreta; excludes insects, rodents, and animals; prevents contamination of water supply; provides convenience and privacy; clean and odor free.

Note: If privy seat is removable and an extra seat is provided, it is easier to scrub seats and set aside to dry. A commercial plastic or composition-type seat is recommended in place of improvised crudely made wooden seats. Deodorants that can be used if needed include chlorinated lime, cloroben, iron

Construction	Maintenance
Deep pit; insects, rodents, and animals excluded; surface water drained away; cleanable material; attractive; ventilated pit and building. Pit 3′×4′×6′ deep serves average family 3 to 5 yr.	Keep clean and flytight; supply toilet paper. Apply residual fly spray to structure. Natural decay and desiccation of feces reduce odors. Keep wastewater out. Scrub seat with hot water and detergent. Use commercial seat.
Watertight concrete vault; flytight building; cleanable material; ventilated vault and building. Capacity of 6 ft^3 per person adequate for 1 yr.	Keep clean, flytight, and attractive. Supply toilet paper. Apply residual spray. Clean pit when contents approach 18 in. of floor. Scavengers can be used.
Watertight vault with tee outlet to leaching pit, gravel trench, filter, vault, etc. Provide capacity of 250 gal plus 20 gal for each person over 8 yrs.	Add 1 to 2 gal water per seat per day. Keep clean and flytight. Supply toilet paper. Agitate after use. Clean vault when depth of sludge and scum is 12 to 18 in.
Shored pit with open-joint material. Tight top and access door.	Keep flytight and clean. Drain surface water away.
Same as masonry vault privy. Tank may be heavy gauge metal with protective coating. Provide capacity of 125 to 250 gal per seat. Cabinet-type seat and bucket usually prefabricated.	Use $\frac{1}{4}$ lb lye for each ft^3 of vault capacity made up to 6-in. liquid depth in vault, or 25 lb caustic per seat in 15 gal water. Keep clean. Clean vault when $\frac{2}{3}$ to $\frac{3}{4}$ full.[a] Agitate after each use. Empty bucket in pit or sewer.
Enclosed compartment, vent, and blower.	Keep clean and supply toilet paper.
Enclosed prefabricated unit, with filter and hand or mechanical pump.	Keep clean. Empty contents in approved facility and recharge with chemical.
Same as masonry vault privy. Provide easily cleaned pails. Available as prefabricated unit.	Provide collection service, excreta disposal facility or pit, and cleaning facilities, including hot water (backflow preventer), brushes, detergent, drained concrete floor.
Earth pit with portable prefabricated box.	Same as earth pit privy. Provide can cover to keep toilet tissue dry.
Bored hole 14- to 18-in. diameter and 15- to 25-ft deep with bracing if necessary. Seat structure may be oil drum, box, cement or clay tile riser with seat, or use squatting plate. Platform around hole.	Same as earth pit privy. Line upper 2 ft of hole; in a caving formation line hole to support earth walls.
Trench 1-ft wide, 2½-ft deep, and 4-ft long for 25 men.	Frequent inspection. Keep excreta covered. Provide toilet paper with waterproof cover.
Hole about 1 ft deep.	Carefully cover hole with earth.
Similar to privy or bored-hole latrine. See sketches.	Same as privies and latrines.
Special unit with tank, mineral oil, filter, chlorine, teflon-coated bowl.	Replace mineral oil and chlorine crystals every 6 mo; remove sludge as needed.
Squatting plate or hopper over tank, same as septic privy.	Keep clean. Maintain water seal in tank about 4 in. above bottom of plate or hopper drop pipe. See sketch.
Watertight tank with sloping bottom and chutes to kitchen and to toilet room.	Tank needs heat (70°F), moisture, and aeration to operate properly. Humus removed after 2 to 4 yr.

sulphate, copperas, activated carbon, and pine oil. Keep privy pits dry.

[a]Solutions for chemical toilets include lye (potassium hydroxide), caustic soda or potash (sodium hydroxide), chlorinated lime (1 lb in 2½ gal water), copper sulfate (1 lb in 2½ gal water), and a chlorinated benzene. Handle chemicals with care.

[b]May be wastewater receiving complete treatment.

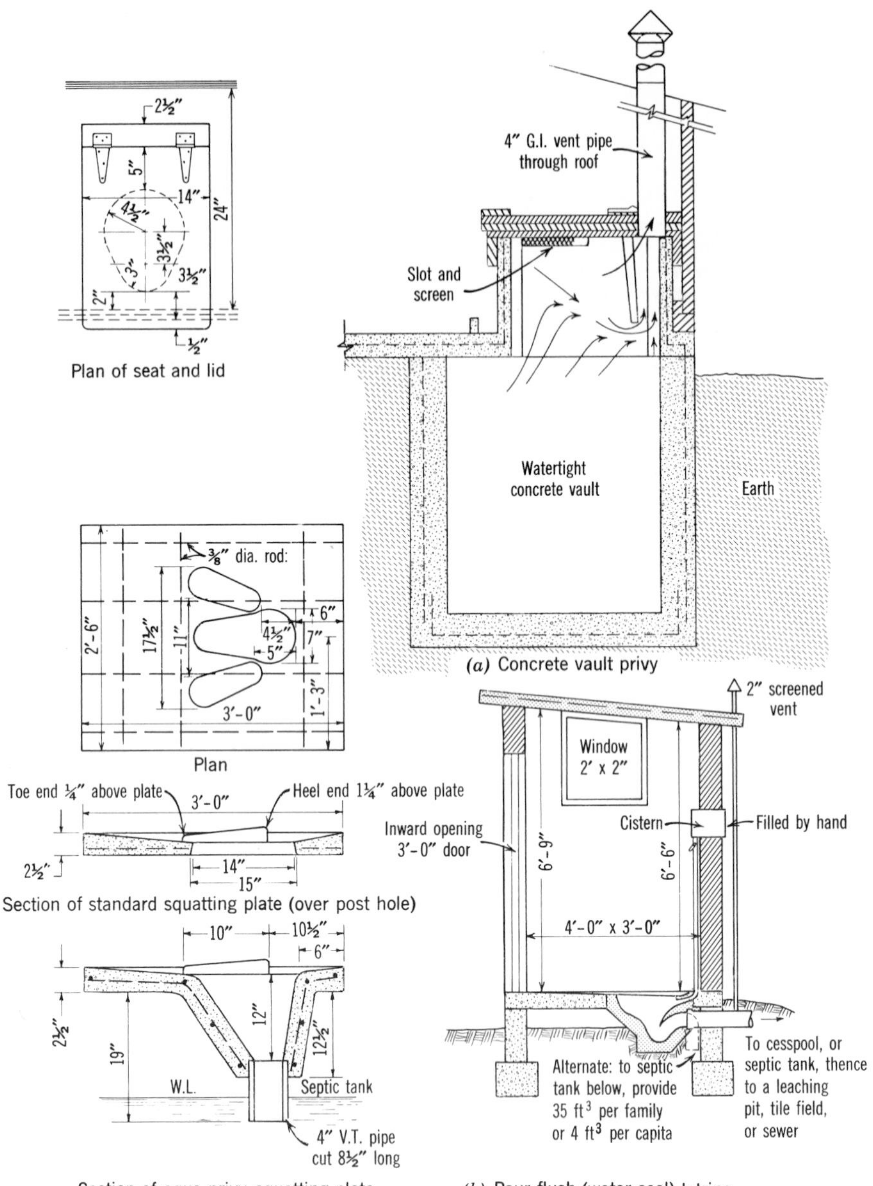

Figure 4-44 (*a*) Concrete vault and (*b*) squatting-type latrines. (Squatting latrine types adapted from O. J. S. Macdonald, *Small Sewage Disposal Systems*, H. K. Lewis & Co., Ltd., London, England, 1951.)

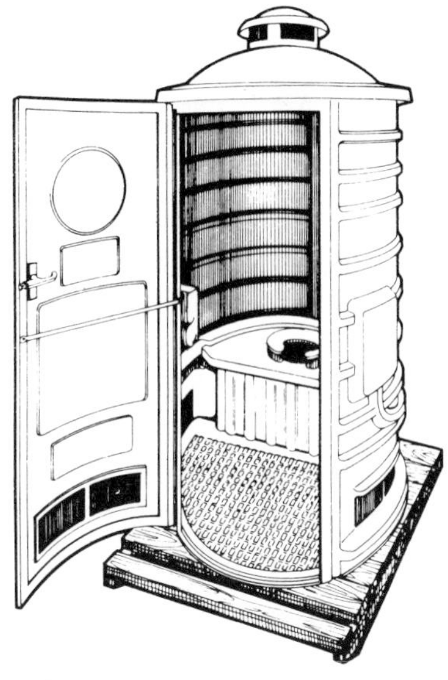

Figure 4-45 Portable toilet. (From *Environmental Health Practice in Recreational Areas*, DHEW, PHS, Washington, D.C., 1977, p. 26.)

or trench could be abandoned and connection made to a small diameter or conventional community sewer and treatment plant.[159]

Low-Cost Treatment

Studies show that waterborne sewage disposal systems including an oxidation pond can be provided at a cost that compares very favorably with a removal pail privy system.[160–162] Where aqua privies (with retention tanks) are acceptable, it is possible to collect the tank effluents and carry or conduct the wastewater to an oxidation pond for treatment. Operation is improved by connecting the laundry tub, the shower and washbasin drains to the aqua privy to maintain the water seal and thus prevent odors and fly and mosquito breeding. Since solids are removed in the aqua privy tank, 4-in. sewers designed for a flow velocity of 1 fps may be used. The flat grade makes possible reduced excavation cost but odors may be a problem. A 4-in. sewer designed for 15 gpd/capita and for peaks three times mean flow can serve a population of up to 1000. Oxidation ponds can also be used for the disposal of night soil. A loading of 145 lb of

[159]John M. Kalbermatten et al., *Volume I Appropriate Sanitation Alternatives: A Technical and Economic Appraisal*, The World Bank, Washington, D.C., October 1978, p. 124.

[160]G. J. Stander and P. G. Meiring, "Employing Oxidation Ponds For Low-Cost Sanitation," *J. Am. Water Pollut. Control Fed.*, **37**, No. 7, 1025–1033 (July 1965).

[161]L. J. Vincent, W. E. Algie, and G. van R. Marais, "A System of Sanitation for Low-Cost, High-Density Housing," Document submitted to WHO by the African Housing Board, Ridgeway, Lusaka, Northern Rhodesia, 1961, 29 pp., 18 figures.

[162]Stanley Sebastian and Ivan C. Buchanan, "Feasibility of Concrete Septic Privies for Sewage Disposal in Anguilla, B.W.I.," *Pub. Health Rep.*, December 1965, pp. 1113–1118.

BOD/acre/day appears reasonable. Conventional oxidation pond or lagoon design details are given on page 467. Other low-cost systems include land treatment systems such as irrigation systems and oxidation ditches.

See Table 3-12 for estimated water usage in developing areas of the world.

INDUSTRIAL WASTES

Industrial Wastewater Surveys

The design of new plants should incorporate separate systems for sewage, cooling water, industrial wastes, and storm- or surface-water drainage. In some cases combination of sewage and liquid industrial waste is possible, but this is dependent on the type and volume of waste. At an existing plant and at a new plant, a flow diagram should be made showing every step in a process, and every drain, sewer, and waste line. Dye or temperature tests will help confirm the location of flows. Radioactive isotope tracers might also be used under controlled conditions. Where pipelines are exposed, they can be painted definite colors to avoid confusion. In general, clean water from coolers, roof leaders, footing and area drains, and ice machines can be disposed of without treatment. Water conservation such as use of cooling towers and recirculation of the water, use of air-cooled exchangers, process modification, and industrial wastewater pretreatment, and reclamation and reuse will reduce the wastewater problem.

Industrial wastes that contain synthetic organic wastes or inorganic wastes that cannot be adequately treated, interfere with operation, or contaminate sewage sludge at a conventional municipal wastewater treatment plant must not be discharged to municipal sewers unless permitted and pretreated as required by the municipality and state regulatory agency. Included in the prohibition is the discharge of hazardous wastes or other deleterious substances to the sewers. Toxics of priority concern generally include mercury, cadmium, lead, chromium, copper, zinc, nickel, cyanide, phenol, and PCBs.[163] In addition, other metals and numerous halogenated organics are under study and may be regulated. In some instances the joint treatment of industrial wastes and municipal wastewaters may be mutually advantageous and should be considered on an individual basis.[164]

See Chapter 5, Hazardous Wastes, for additional discussion of this subject, including treatment and disposal.

A knowledge of the industrial process is fundamental in the study of a waste problem. The volume of flows from each step in a process, the strength, chemical characteristics, temperature, source, and variations in flow are some of the

[163]G. W. Foess and W. A. Ericson, "Toxic control–the trend of the future," *Water Wastes Eng.*, February 1980, pp. 21–27.

[164]*Joint Treatment of Industrial and Municipal Wastewater*, 1973 and *Pretreatment of Industrial Wastes*, MOP FD-3, 1981, Water Pollut. Control Fed., 2626 Pennsylvania Avenue, N.W., Washington, D.C. 20037.

details to be obtained. The existing or proposed methods of disposal, and opportunities for wastage, drippage, and spillage should be included. Sometimes revision of a chemical process can eliminate or reduce a waste problem. Possible waste-control measures are salvage, in-plant waste reduction, reclamation, concentration of wastes, flow equalization, and new methods. The recovery of materials such as copper, aluminum, iron, and silver, or liquids with specific gravity less than or greater than wastewater, from industrial wastes can help industry meet effluent standards and offset some of the costs of treatment. All of these possibilities should be evaluated.

A method of simplifying a waste problem is to spread its treatment and disposal over 24-hr rather than over a 4- or 6-hr period. This is accomplished by the use of a holding tank to equalize flows and strength of waste, accompanied by a constant uniform discharge over 24 hr. Where needed, aeration will prevent septic conditions and odors. If necessary, chemical mixing can be incorporated, followed by settling and uniform drawoff of supernatant by means of a flexible or swing-joint pipe. The pipe may be lowered uniformly by mechanical means such as a clock mechanism or motor, or a float with a submerged orifice may be used, to give the desired rate of discharge. The mass diagram approach may be used to determine the required storage to give a constant flow over a known length of time. Sludge is drawn from the hopper bottom of the settling tank to a drying bed or to a treatment device before disposal.

After investigating all possibilities of preventing or reducing the wastewater problem at the source, the next step would be provision of the required degree of treatment, based on existing laws and standards and competent engineering advice. Receiving waters' classification, minimum average seven-consecutive-day flow once in ten years of the receiving stream, and effluent standards will largely determine the type and degree of treatment required. Black proposed that the approving agency consider the following items in its review of engineering reports, plans, and specifications for industrial waste treatment systems.[165]

Engineering Report (Part I)

Project Delineation

1. Type of Industry
2. Waste-Pollution Load
3. Receiving Waters
4. Waste Treatment Requirements
5. Waste Abatement Plans
6. Map of Environment
7. Plot Plan and Hydraulic Profile

Plant and Process Description

8. General Description of Factory
9. Description of Principal Wet Processes
10. Process Flow Diagram Showing Sources of Liquid Waste
11. Sewer Map and Process Connections
12. Liquid Waste Control Measures
13. Experimental and New Processes
14. By-Product Recovery Systems

[165] Hayse H. Black, *Items Considered in Review of Engineering Reports, Plans, Specifications for Industrial Waste Treatment Systems*, New York State Dept. of Health, Albany, January 1969.

Factory Operations

15. Finished Products and Processes
16. Production in 24 Hours—Rated and Actual
17. Principal Raw Materials
18. Sources of Water Supply
19. Water Quantity and Quality Requirements
20. Process Water Reuse
21. Employees, Shifts, Days Per Week
22. Expansion—Planned and Potential

Development of Design Criteria

23. Comprehensive Industrial Waste Surveys
24. Parameters Specific for the Industry
25. Segregation of Cooling Water
26. Diversion of Storm Water
27. In-Plant Improvements and Waste Reduction
28. Waste Characterization and Treatability Studies
29. Pilot Plant Investigations
30. Avoidance of Nuisance to the Environment
31. Design Liberal and Flexible

Combined Treatment With Domestic Waste

32. Standards and Regulations for Controlling the Use of Municipal Sewers
33. Pretreatment at the Factory
34. Surcharge Agreements
35. Industrial Effluent Monitoring Systems
36. Influence on Sewage Treatment Efficiency

Engineering Report (Part II)

37. Solids Removal and Disposal
38. Chemical Precipitation
39. Chemical Treatment
40. Aerobic Biological Treatment
41. Anaerobic Biological Treatment
42. Role of Incineration
43. Alternate Proposals
44. Chronological Steps for Submission

The types of industrial wastes are of course numerous. Treatment processes vary from solids removal and disposal to involved chemical and biological processes including incineration, chemical neutralization and ion exchange. In view of the complexity of the problem, only those industrial wastes more commonly encountered in connection with small plant operations will be discussed. More detailed information concerning the treatment of specific wastes can be obtained from standard texts, periodicals, and other publications devoted to this subject.

A common method of expressing the strength of wastes is in milligrams per liter or parts per million of 5-day BOD, suspended solids, and chlorine demand, also pH. Another parameter is the COD. These are by no means the only measures of waste strength, as the particular waste and its characteristics must be considered. Specific organic and inorganic tests may also be indicated.

The BOD of an organic waste is frequently converted to its population equivalent. Since the waste from one person is said to equal 0.17 lb of 5-day BOD, the population equivalent X of, say, 1000 gal of a waste having a BOD of 600 mg/l is expressed by the proportion

$$\frac{600}{1{,}000{,}000} = \frac{X(0.17)}{1000 \times 8.34}, \qquad \text{or } X = 29 \text{ persons}$$

To determine the suspended-solids population equivalent, substitute 0.20 for 0.17.

The actual volume and strength of a waste should be individually determined. Many municipalities levy a charge for the handling and treatment of industrial wastes to help pay the cost of operating and maintaining the municipal treatment works. The basis for rental or assessment of cost should be volume and strength of the waste as determined by periodic analyses. Measures used are COD, BOD, chlorine demand, and other parameters. It is also common to require pretreatment in case the characteristics of waste exceed certain predetermined values, the waste as released is not amenable to treatment in a municipal treatment plant, or the waste would cause a hazardous condition in the sewers.

The Water Pollution Control Federation's Manual of Practice No. 3, entitled "Regulation of Sewer Use," suggests standards and recommendations for local sewer ordinances.[166] This manual provides sound regulations to exclude hazardous materials, protect sewers, and control the discharge of wastes to the sewer that may upset the municipal treatment plant operation. Unacceptable wastes include large volumes of uncontaminated wastes which may cause hydraulic overloading, storm waters, acid and alkali wastewaters, explosive and flammable substances, toxic substances, large volumes of organic wastes unless adequately pretreated, oil, and grease.[167]

In the United States, national guidelines for industry generally prohibit disposal solely by dilution.

Milk Wastes

Solution of a dairy waste problem starts with waste-saving and waste prevention. If judiciously carried out the waste can be drastically reduced, thereby making treatment of the waste remaining simpler and more practical of solution. Some of the saving and prevention methods found effective follow.

1. Reduce the volume and strength of floor wastes. Obtain management and employee cooperation to prevent overflow of cans and vats, spillage, leaks, and drips. Collect can, pipeline, and tank drips, and equipment-concentrated prerinse for salvage as animal feed or similar use. Make use of automatic shutoff nozzles on hoses.
2. Encourage bulk tank pickup rather than 10-gal cans, and in-place cleaning of dairy equipment and sanitary pipelines to reduce volume of waste.

[166] *Regulation of Sewer Use*, MOP 3, Water Pollut. Control Fed., 2626 Pennsylvania Avenue, N.W., Washington, D.C. 20037, 1975.
[167] *Operation of Wastewater Treatment Plants*, MOP 11, Water Pollut. Control Fed., Washington, D.C., 1976, pp. 445–461.

3. Eliminate steam-water mixing tees for making hot water. Use a water-heating system and distribute a tempered water for utility purposes.
4. Separate and dispose separately strong wastes, wash water, cooling water, and sanitary sewage. Discharge sulphuric acid used in butterfat testing into ash or cinder pit. Salvage caustic solution from bottle washer and rinse, after removing suspended matter.
5. Skim milk, buttermilk, cheese whey, whole milk, and milk products have very high BOD and hence are expensive to treat. If not used for by-product, investigate possibilities of disposal in a lagoon or oxidation pond located in an isolated area or, if permissible, in sewage treatment plant digester, provided the volume can be absorbed.
6. Use evaporator and drying equipment, or entrainment separator, where by-product production is possible.

All dairy plants should be equipped with a combination sand trap-grease trap and a screen with $\frac{3}{8}$-in. openings. It is good practice to incorporate a weir or other device for estimating flows. A trap $2\frac{1}{2}' \times 2\frac{1}{2}' \times 8'$ long is adequate for a 1000 gal/hr flow.[168]

A convenient method of dairy waste disposal is piping to a municipal sewer system if this is practical, as biological treatment plants can handle moderate amounts of dairy wastes. It is not advisable to dump large quantities of wastes into a sewer unless the sewage treatment plant operator is first notified and signifies it is prepared to accept the waste. Large quantities of whey are detrimental. In any case, by keeping the volume and strength of the waste low, it will be possible to realize a saving should a charge be made by the municipality. Dairy wastewater strength averages 2500 mg/l BOD and 700 mg/l suspended solids.

If it is not possible to discharge dairy wastes directly to a municipal sewer either partial or complete treatment may be required. A comprehensive study and state-of-the art for control and treatment of dairy wastes is given in a USEPA publication.[169] Some of the treatment and disposal methods available are outlined below. A coarse screen and grit chamber should generally precede the treatment and disposal process as previously noted.

Where a *dilution* of at least 50:1 is available it may be possible to discharge the waste to a stream through a holding tank with a controlled discharge, if approved by the regulatory agency. Aeration or chlorination and recirculation during the holding period will prevent souring of the waste. Aeration can be accomplished by means of splash plates, an aspirator, cascading, or other simple method. Disposal by dilution is not acceptable in the United States.

A *shallow oxidation pond*, 2 to 5 ft liquid depth, is an inexpensive and effective way of treating dairy wastes. A design based on a loading of 25 lb BOD per acre per day, preceded by a settling tank providing 72-hr detention or an anaerobic digestion chamber, should not cause a nuisance.

Ridge and furrow irrigation is another effective disposal method. Careful initial grading and maintenance of the furrows (1- to 3-ft deep, 1- to 3-ft wide, spaced 3- to 15-ft apart), with resting of sections is necessary. A loading of 3600 to 36,000 gpd/acre is used, depending on the prevailing temperature and soil

[168] H. A. Trebler and H. G. Harding, "Fundamentals of the Control and Treatment of Dairy Wastes," *Sewage & Ind. Wastes*, **27**, No. 12 (December 1955), pp. 1369–1382.

[169] W. James Harper, *Dairy Food Plant Wastes and Waste Treatment Practices*, Ohio State University Research Foundation, Columbus, Ohio, March 1971.

percolation. Odor nuisance is likely in warm weather. Oxidation ditches with brush or paddle rotors to provide mixing and aeration can also be used, but cost may be a limiting factor.

Spray irrigation with No. 80S Rainbird or ¾-in. Skinner sprinkler to pasture, field crops, or wooded areas at the rate of about 2000 to 6000 or more gpd/acre can also be used.[170] In favorable soil 10,000 to 15,000 gpd may be applied. The irrigation laterals and spray nozzles should be designed to give a hydraulic loading that can be safely absorbed. Sprinkler nozzle pressure may vary from 30 to 60 psi and discharge from 10 to 75 gpm. Intermittent dosing was extended rest periods and grading of the land to prevent ponding will prevent fly and odor nuisances. Odor can be a serious problem. Periodic cultivation of the land is usually necessary.

Conventional open sand filters can give satisfactory results for a time if preceded by a settling tank providing at least three days' detention. A loading of 50 to 100 lb of BOD/acre/day and 50,000 to 120,000 gal/acre/day are used in design. Odors can be expected.

High-rate trickling filters are one of the preferred treatment devices. Two complete stages in series are needed if a high-quality effluent is required. A low rate trickling filter (6-ft deep), loaded at the rate of one mgd/acre, will also give good results but will be accompanied by some odors and filter flies. Small rock or lath filter can be used for small flows. High-rate filters may be loaded at six times standard rate filters (3.0 lb BOD/yd^3).

For complete oxidation, *aeration* of the wastes for 24 hr at a temperature of 80 to 90° F accompanied by good agitation and 3- to 4-hr settling provides good treatment. For activated sludge treatment provide 3- to 6-hr detention. If batch treatment is proposed, provide ½ to 2 days' capacity, depending on the required degree of treatment, and air agitation to prevent septic action. Aerobic treatment causes a breakdown of milk wastes to CO_2 and H_2O with very little sludge production. Complete oxidation requires 1.25 lb of oxygen per pound of milk solids; initial assimilation requires 0.50 lb. An aerated lagoon is also suitable.

A *settling tank and shallow absorption field* can give satisfactory results at small plants if properly designed and maintained. Seeding of a new tank with lime and septic sludge, maintenance of the wastes at a temperature of 90° F, provision of *3 to 6 days' retention*, and design of an absorption field based on percolation tests are important considerations. If the settling tank is not cleaned the absorption field will soon fail.

Chemical precipitation is a suitable treatment method, but it is not advised for small plants unless intelligent operation can be assured; it is expensive. The Mallory and Guggenheim processes give satisfactory results. In the Mallory process, activated sludge plus chemical precipitation with lime and alum is used. In the Guggenheim process, storage with aeration, lime, ferric salt, return sludge, 4-hr aeration, and settling are used.

Combinations of processes are possible depending on the local conditions, volume and type of waste, personnel available, and other factors.

[170] Frank J. McKee, "Dairy Waste Disposal by Spray Irrigation," *Sewage Ind. Wastes*, **29**, No. 2, 157–164 (February 1957).

Poultry Wastes

As in all waste treatment operations, the most important step is good housekeeping. The major source of waste has been found to be the battery and feeding rooms. This is followed by the wastes from killing, scalding, and picking operations.

The waste reduction measures that are possible include:

1. Collect blood from the killing bench or trough, as blood has a very high BOD. An electric stunning plate is practical in reducing the spattering of blood. Running water to continuously carry away the blood and offal will greatly increase the volume of the waste and hence the cost of treatment. Squeegee the killing bench or trough and floor to concentrate the blood, feathers, and offal. Collect the waste in a container for separate disposal, then wash all surfaces with water.
2. Use securely fastened but removable screens or plates with $\frac{1}{8}$-in. opening at all floor drains. This will keep feathers, wax, and large particles from flowing into the sewer, which would add to the nuisance problem.
3. Study the scalding and chilling operations to reduce the volume of water caused by carelessness. Opportunities for improved or revised operations should not be overlooked.
4. Collect wastes from the evisceration process in separate cans. The entrails should not be permitted to enter the sewer.
5. Eliminate running hoses. Although desirable for sanitary reasons, the increase in waste volume will make the cost of treatment expensive. Substitute routine clean-up several times during the day.
6. Blood from poultry slaughtering may also be conveniently collected in floor catch-basins. But a routine procedure of skimming the coagulated blood for collection in containers and separate disposal must be in effect if this method is to be effective.

When arrangements cannot be made for discharge into a sewerage system, separate treatment must be considered. This should include holding tanks to equalize peak flows and screens. Holding tanks may require aeration or chlorination to prevent septic action. Grit chambers, if provided, should include aeration to separate out entrained organic matter. Pretreatment should include stationary or mechanical screens, depending on the size of the plant, and grease removal.

Treatment and disposal processes include trickling filters, activated sludge, extended aeration, chemical precipitation, spray irrigation, ridge and furrow, lagooning, and oxidation ponds (50 lb BOD/acre/day). The activated sludge process requires careful supervision and precautions against shock loads. The chemical precipitation process is expensive and also requires intelligent supervision.[171] Sludge digesters should be heated. Lagoons cause odors.

A typical spray irrigation plant may consist of a 75-gal screening tank equipped with a stationary or moving 10- to 20-mesh screen, followed by a storage tank providing 2 hours' detention and a pump to carry the wastes to the irrigation area or to a bypass to a lagoon.[172] The spray irrigation system would consist of a 125-gpm pump against a 90-ft head, and 3- and 4-in. aluminum laterals provided with revolving $\frac{7}{32}$-in. Rainbird spray heads spaced 60-ft apart.

[171]Ralph Porges and Edmund J. Struzeski, Jr., *Wastes From the Poultry Processing Industry*, PHS, DHEW, SEC TR W62-3, Cincinnati, Ohio, 1962.

[172]James W. Bell, "Spray Irrigation for Poultry and Canning Wastes," *Pub. Works*, **86**, 111–112 (September 1955).

Application rates vary from 0.125 in./day/acre with tight clay loam to 0.6 in./day/acre with a sandy loam having a soil percolation rate of 1 in. in 5 min.

Canning Wastes

Most canneries operate on a seasonal basis, usually during the dry months of the year when stream flows are lowest. The plants are commonly in rural areas near the supplying farms. Products packed may be limited to a few or to a full line of diversified products. The wastes consist largely of wash water plus soil, leaves, and parts of the products. The objectionable parts are the decomposing organic solids in suspension. Treatment should include screening and one of the following methods, depending on the stream and effluent requirements of the water pollution control agency.

Screening. A 40-mesh mechanically cleaned (screw- or bucket-type) screen, or perforated plate with $\frac{1}{16}$-in. diameter holes, and water jets at not less than 40 psi are commonly specified. The area of screen is based on the product; 10 ft^2/1000 cases of peas or 38 ft^2/1000 bushels of tomatoes per 12-hr operation. A 20-mesh screen is suggested for tomato wastes. The manufacturers of screening equipment should be consulted for their recommendations in connection with specific products. Provide for a conveyor and loading hopper. Larger screen openings may be considered, depending on the treatment and disposal proposed.

Dilution, not less than 1:100, if approved by the regulatory agency. Not acceptable in the United States.

Biological filtration. A trickling filter dosed at 0.5 mgd/acre removes 80 to 95 percent 5-day BOD. The effluent will not cause a nuisance with a dilution of 1:5, provided the stream is not seriously polluted. At a filtration rate of 2 mgd/acre 50 to 70 percent BOD can be removed. This effluent can be absorbed satisfactorily by a stream, providing a dilution of 1:20. Treatment with lime every 2 days at the rate of 100 lb per 100,000 gal of wastes will help prevent fungi clogging. Plants should be at least 200 ft from dwellings.

A *high-rate filter* with one settling tank providing 1-hr detention, based on the pump capacity, is also used. The loading could be about 1500 lb BOD per acre-ft with the pump capacity at least six times the maximum rate of waste flow at the plant. This has been recommended by the National Canners' Association. A removal of 70 percent for beet waste and up to 88 percent for tomato waste is reported.

Chemical precipitation. Chemical precipitation is particularly adaptable to canneries operating for short seasons. The fill-and-draw procedure is common. At least two hopper-bottom tanks are provided, each having a capacity to hold the maximum 4-hr waste flow. In this method, 30 min. is allowed for chemical feed, mix, and flocculation; 2 hrs are allowed for settling, and $1\frac{1}{2}$ hr are allowed for the drawing of sludge on a sand-drying bed and for the discharge of the treated waste. The chemical feed usually consists of 5 to 10 lb of lime in combination with ferric sulfate, ferric chloride, alum, zinc, chloride, or, in the case of tomatoes or carrots, lime alone. The expected BOD reduction is 33 to 75 percent, depending on the chemicals used. Control tests to determine the best chemical dosage should be made daily.

Spray irrigation. A holding tank providing about 2-hr storage is recommended following a stationary or moving 10- to 20-mesh screen. The waste is then pumped to the header and spray nozzles on lateral distributors at a rate that can be satisfactorily absorbed by the irrigation area. Portable 3- and 4-in. aluminum pipe and revolving Rainbird $\frac{7}{32}$-in. spray nozzles can be used. With a flow of 125 gpm and a head of 90 ft, a 60-ft diameter spray can be obtained. An application rate of 0.125 in./acre/day, 0.079 gal/ft^2, has been used where the soil is a tight clay to as much as 0.6 in./acre/day or 0.374 gal/ft^2 where the soil is a sandy loam having a soil percolation of 1 in. in 5 min. Design should include bypass to a lagoon.

Irrigation. Fields can also be used where the soil is porous. The field is plowed and trapezoidal channels, ditches, or furrows are formed having a depth of about 12 to 18 in. Careful grading and intermittent dosing will help assure satisfactory results. The dosage and rest periods should be based on the permeability of the soil. Ponding must be avoided.

Lagoons. Lagoons cause odors that carry $\frac{1}{2}$ to 1 mi. On the other hand, stream pollution can be completely eliminated. The most satisfactory treatment of wastes to be lagooned is the application of sodium nitrate. Agricultural sodium nitrate contains about 55 percent available oxygen, which helps maintain aerobic conditions during the decomposition of the wastes. Satisfaction of 20 percent of the BOD of the daily waste by the application of sodium nitrate should prevent a nuisance. The lagoon should have a capacity 25 percent greater than the total waste for the year, and the depth should be between 3 and 5 ft. About 25 percent of the waste should be left in the lagoon to provide seeding organisms for the next year.

The treatment methods mentioned can be expanded to include sand filtration, sedimentation, oxidation ponds, patented processes, and combinations of the methods given to produce the desired degree of treatment.

Packing-House Wastes

The treatment of slaughterhouse and meat-packing plant wastes can be a very difficult problem, particularly if good housekeeping practices are not followed. The reduction of losses requires proper equipment, a properly designed system of sewers, and constant vigilance. Some of the more common waste-prevention or -reduction procedures are outlined below.

By-product production. Edible fats, inedible fats, hides, skins, pelts, tankage, meat scraps, the condensate from the steam rendering of fats (called "stick"), dried blood, bone and bone products, hog hair and bristles, horns and hoofs, intestines, and glands are among the principal by-products. The salvaging of this material can reduce the amount of wastes but can also add wastes incidental to the production of the by-products.

Waste reduction. In addition to the salvaging of as much material as possible as explained above, other steps can be taken to reduce the volume and strength of the wastes. The methods used include the following:

Grease salvage. Screens, vacuators, baffled skimming basins, or traps should be used on all waste lines in rooms or spaces where greasy wastewater results. Grease basins or separators are designed to provide $\frac{3}{4}$- to 1-hr detention and be easily cleaned. Depths of 4 to 6 ft and a flow velocity of 1 to 2 ft/min are suggested. It is sometimes practical to install one large grease separator on a main drain. In any case, grease separators or traps, skimming basins, and screens require cleaning at least once a day. Other wastes that do not contain grease, including condenser water, paunch manure, boiler plant wastes, and wash water should be bypassed around grease separators or basins.

Wastes in yards and pens. Areas for the temporary holding of animals should be first cleaned by dry methods and then washed down.

Blood. All kill blood should be collected for separate disposal or use. Squeegees can be used to collect blood on floors and tables for transfer to the blood storage cans or tanks.

Paunch manure. A fine screen is recommended to collect paunch manure, sludge, hair, and trimmings. Special screens suitable for the purpose, either manually or mechanically cleaned, should be investigated. Fleshings and cuttings, grease, and solid particles are best collected by dry cleaning of floors.

Waste treatment. After all practical steps have been taken to reduce the volume and strength of the waste, consideration should be given to the degree and type of treatment required. This is usually dependent on the strength of the wastes, availability of municipal sewers and type of treatment, dilution available, stream usage and existent legal regulations. A special study by Mohlman showed that the water consumption from packing houses varied from 1900 to 8700 gal per ton of kill and averaged 4130.[173] The BOD averaged 28.9 lb per ton of kill, suspended solids 22.7 lb, nitrogen 3.49 lb, and grease 2.64 lb. The treatment and disposal methods possible are:

1. *Disposal to a municipal sewerage system*, if a satisfactory agreement can be made with the responsible officials. Screening, air flotation, and sedimentation may be required before discharge to the sewers. Pretreatment in an anaerobic pond is also feasible.[174]

2. *Coarse screens or fine screens* to take out hair, meat or fat particles, manure, and floating solids. Suspended solids removal of 80 percent is possible if followed by a grit chamber, flocculation and sedimentation.

3. Using an *Imhoff tank*, with 2 hours' detention, a suspended solids removal of approximately 65 percent and BOD removal of 35 percent can be expected.

4. *Trickling filters*,[175] at a rate of 0.5 to 1 million gal per acre per day, can give an overall removal of about 85 percent. The effluent can be expected to be well nitrified and the operation relatively trouble-free.

5. *Double filtration*,[175] two filters in series, with the first a high-rate 3 million gal per acre per day washable filter, and the second operated at 1.5 million gal per acre per day can yield a 95 percent BOD reduction. Such a plant would include fine screens, grit removal, 15-min grease flo-

[173] F. W. Mohlman, "Packing House Industry," *Ind. Eng. Chem.*, **39**, No. 5, 637–641 (May 1947).
[174] A. J. Steffen, "Waste disposal in the meat industry/2," *Water Wastes Eng./Ind.*, C-1-4 (May 1970).
[175] *An Industrial Waste Guide to the Meat Industry*, PHS Pub. No. 386, DHEW, Washington, D.C., Revised 1965.

tation, 40-min flocculation, 2-hr primary settling, 2-hr secondary settling, and final settling. Single filtration at the high rate would give between 65 and 75 percent overall removal.

6. *Using activated sludge processes*,[175] satisfactory results are obtained with 9 hours' aeration and 3.5 ft^3 of air per gal of waste. In cold months air is increased to 4 or 5 ft^3. Suspended solids and 10-day BOD removal averaged 95 percent.

7. Using *anaerobic and aerobic stabilization* ponds,[175] complete treatment can be obtained with ponds 8- to 17-ft deep, loaded at 0.011 to 0.015 lb BOD per day per ft^3 followed by conventional aerobic stabilization ponds loaded at 50 to 280 lb BOD per day per acre. Complete treatment can also be obtained in an aerobic stabilization pond loaded at 50 lb BOD per day per acre.

8. *Chemical precipitation*, using ferric chloride, zinc chloride, chlorine, ferric sulfate, ferric salts plus lime, and ferric chloride plus chlorine has been used for the chemical treatment of packing-house wastes. The cost of chemicals and operation are significant. Secondary treatment in trickling filters or in the activated sludge process produces a high-quality effluent. This method is used by small plants.[176]

9. Using *anaerobic digestion*, Schroepfer, et. al., report a 95 percent BOD and 90 percent suspended-solids removal at a temperature of 95° F, 60 turnovers per day, a loading of 0.20 lb of BOD per ft^3 of digester volume per day, and a 12-hr total detention in the digester.[177] The suspended solids of the mixed liquor was 15,000 mg/l. An 18-in. vacuum was maintained on the digester loading of 0.15 lb of BOD and supplementary treatment consisting of a high-rate trickling filter loaded at 2600 lb of BOD per acre-ft, and final settling providing a surface settling rate of 800 gal/ft^3 per day. Excess sludge is lagooned.

10. *Various combinations* of the treatment processes mentioned may be used, including anaerobic ponds, aerobic ponds, and anaerobic-aerobic systems. Final effluent can also be disposed of to an oxidation pond or by irrigation in isolated areas. Aeration and the addition of air to the waste under pressure can also be used.

Laundry Wastes

The treatment of laundry wastes where public sewers are not available is dependent on location, availability of a suitable stream, soil percolation characteristics, volume of the waste, and similar factors.

1. Small operations can use a plant consisting of a grease trap to remove floatable lint and grease, a dosing tank, trickling filter with recirculation, and settling. Dosing can be provided by means of a siphon, pump, or tipping bucket.

2. Where permeable soil is available and groundwater supplies are not endangered, a plant consisting of a large grease trap, trickling filter, settling tank, and leaching pits can give satisfactory results for a period of time. The grease trap will remove floating grease, and the settling tank will remove emulsified grease and other organic matter that has been agglomerated in passing through the trickling filter. The filter stone will have to be cleaned or replaced after extended use; the trap and settling tank will require frequent cleaning.

3. A modification of (2) is the substitution of chemical flocculation for the trickling filter.

4. Another treatment system consists of chemical precipitation using sulfuric acid to reduce the pH to about 6.0, alum to coagulate, followed by sedimentation, high-rate trickling filter, and secondary settling.

[176] A. J. Steffen, "Waste disposal in the meat industry/1," *Water Wastes Eng./Ind.*, B-20-2 (March 1970).

[177] George J. Schroepfer, W. J. Fullen, A. S. Johnson, N. R. Ziemke, and J. J. Anderson, "The Anaerobic Contact Process as Applied to Packinghouse Wastes," *Sewage Ind. Wastes*, **27**, 460–486 (April 1955).

BIBLIOGRAPHY

Alternatives for Small Wastewater Treatment Systems, USEPA, EPA-625/4-77-011, October 1977, 90 pp.

An Industrial Waste Guide to the Meat Industry, PHS Pub. 386, DHEW, Washington, D.C., 1965, 14 pp.

Barnes, D., and F. Wilson, *The Design and Operation of Small Sewage Works*, E. & F. N. Spon, Ltd., London, Halsted Press, John Wiley & Sons, New York, 1976, 180 pp.

Bernhart, Alfred P., *Treatment and Disposal of Waste Water from Homes by Soil Infiltration and Evapotranspiration*, University of Toronto Press, Canada, 1973, 173 pp.

Eckenfelder, W. Wesley, Jr., *Water Quality Engineering for Practicing Engineers*, Barnes & Noble, Inc., N.Y., 1970, 328 pp.

Design Manual Onsite Wastewater Treatment and Disposal Systems, USEPA, Office of Water Program Operations, Washington, D.C., October 1980, 391 pp.

Fair, Gordon M., and John C. Geyer, *Elements of Water Supply and Wastewater Disposal*, John Wiley & Sons, New York, 1966, 505 pp.

Fair, Gordon M., John C. Geyer, and Daniel A. Okun, *Water and Wastewater Engineering*, John Wiley & Sons, New York, 1966, 505 pp.

Flack, Ernest J., *Design of Water and Wastewater Systems*, Environmental Resource Center, Colorado State University, Fort Collins, Colo., 1976, 149 pp.

Goldstein, Steven N., and Walter J. Moberg, Jr., *Wastewater Treatment Systems for Rural Communities*, Commission on Rural Water, Washington, D.C., 1973, 340 pp.

Hardenbergh, William A., *Sewerage and Sewage Treatment*, 3rd ed., International Textbook Co., Scranton, Pa., 1950, 467 pp.

Harper, W. James, *Dairy Food Plant Wastes and Waste Treatment Practices*, USEPA, Washington, D.C., March 1971, 559 pp.

Imhoff, Karl, and Gordon Maskew Fair, *Sewage Treatment*, John Wiley & Sons, New York, 1956, 338 pp.

Individual Onsite Wastewater Systems, Vols. 1–6, Nina I. McClelland, Ed., Ann Arbor Science Publishers, Inc., P. O. Box 1425, Ann Arbor, Mich. 48106, 1974–1980.

Individual Sewage Systems, Great Lakes-Upper Mississippi River Board of State Sanitary Engineers, Health Education Services, Inc., P. O. Box 7126, Albany, N.Y. 12224, 1979, 51 pp.

Jensen, Marvin E. *Consumptive Use of Water and Irrigation Water Requirements*, ASCE, New York, September 1973, 215 pp.

Kreissl, James F., *Management of Small Waste Flows*, USEPA, Cincinnati, Ohio, 45268, September 1978, 800 pp.

Land Treatment of Municipal Wastewater, USEPA, U.S. Army Corps of Engineers, USDA, EPA-625/1-77-008, October 1977, 505 pp.

Manual of Septic-Tank Practice, PHS Pub. 526, DHEW, Washington, D.C., 1967, 92 pp.

Manuals of Practice: *Regulation of Sewer Use*—No. 3, 1975; *Sewer Maintenance*-No. 7, 1966; *Wastewater Treatment Plant Design*—No. 8, 1977; *Design and Construction of Sanitary and Storm Sewers*—No. 9, 1969; and *Operation of Wastewater Treatment Plants*—No. 11, 1976; Water Pollut. Control Fed., Washington, D.C.

Manual of Instruction for Wastewater Treatment Plant Operators, Vols. 1, 2, New York State Dept. of Environmental Conservation, Albany, N.Y., 1979.

Mara, Duncan, *Sewage Treatment in Hot Climates*, John Wiley & Sons, New York, 1976, 168 pp.

National Specialty Conference on Disinfection, ASCE, 345 East 47th Street, New York, N.Y., 1970, 705 pp.

Olson, Gerald W., *Application of Soil Survey to Problems of Health, Sanitation, and Engineering*, Cornell University, New York State College of Agriculture, Ithaca, N.Y., March 1964, 77 pp.

On-Site Wastewater Management, National Environmental Health Association, Denver, Colo., 1979, 108 pp.

Recommended Standards for Sewage Works, Great Lakes-Upper Mississippi River Board of State

Sanitary Engineers, Health Education Services, Inc., P. O. Box 7126, Albany, N.Y. 12224, 1978 ed., 104 pp.

Rural Wastewater Management, The California Resources Control Board, Sacramento, California, 1979, 52 pp.

Sanitation in Developing Countries, Arnold Pacey, Ed., John Wiley & Sons, New York, 1978, 238 pp.

Sawyer, George E., "New Trends in Wastewater Treatment and Recycle," *Chem. Eng.*, July 24, 1972, pp. 121–128.

Standard Methods for the Examination of Water and Wastewater, 15th ed., A. E. Greenberg, Ed., American Public Health Assoc., 1015 Fifteenth St., N.W. Washington, D.C. 20005, 1980.

Tchobanoglous, George, *Wastewater Engineering: Treatment, Disposal, Reuse*, Metcalf & Eddy, Inc., McGraw-Hill Book Company, N.Y. 1979, 920 pp.

Wagner, E. G., and J. N. Lanoix, "Excreta Disposal for Rural Areas and Small Communities," *WHO Mon. Ser.*, **39,** Geneva, (1958), 187 pp.

Wastewater Treatment Facilities For Sewered Small Communities, USEPA, EPA-625/1-77-009, October 1977, 465 pp.

Water Quality Criteria, Federal Water Pollution Control Administration, U.S. Dept. of the Interior, Washington, D.C., April 1, 1968, 234 pp.

Weber, Walter J., Jr., *Physicochemical Processes for Water Quality Control*, Wiley-Interscience, N.Y., 1972, 640 pp.

Winneberger, John H. Timothy, *Manual of Gray Water Treatment Practice*, Ann Arbor Science, P. O. Box 1425, Ann Arbor, Mich. 48106, 1974, 102 pp.

1980 Public Works Manual, Pub. Works Mag., Ridgewood, N.J. 07451.

5

SOLID WASTE MANAGEMENT

COMPOSITION, STORAGE AND COLLECTION

General

Aesthetic, land use, health, water pollution, air pollution, and economic considerations make proper solid waste storage, collection, and disposal municipal, corporate, and individual functions that must be taken seriously by all. Indiscriminate dumping of solid waste and failure of the collection system in a populated community for 2 or 3 weeks would soon cause many problems. Odors, flies, rats, roaches, crickets, wandering dogs and cats, and fires would dispel any remaining doubts of the importance of proper solid waste storage, collection, and disposal.

The complexities of solid waste management are not readily appreciated. Reference to Figure 5-1, however, will give a helpful perspective. As can be seen, there are numerous sources and types of solid wastes ranging from the home to the farm and from garbage to radioactive wastes, junked cars, and industrial wastes. Handling involves storage, collection, transfer, and transport. Processing includes incineration, densification, composting, separation, treatment, and energy conversion. Disposal methods show the environmental interrelation of air, land, and water, and the place of salvage and recycling. All these steps introduce constraints—social, political, economic, technological, ecological, legal, informational, and communicational—that must be considered in the analysis of the problem and in coming up with acceptable solutions. The discussion that follows will treat most of these factors in some depth. Chapter 2 contributes additional information on the planning and implementation aspects.

Definition

The term "solid waste" according to the USEPA includes any garbage, refuse, sludge from a waste treatment plant, water supply treatment plant, or air pollution control facility, and other discarded material, including solid, liquid, semisolid, or contained gaseous material resulting from industrial, commercial, mining, and agricultural operations and from community activities, but does not include solid or dissolved material in domestic sewage or solid or dissolved materials in irrigation return flows or industrial discharges, which are point sources subject to permit under section 402 of the Federal Water Pollution Act as amended, or source, special nuclear, or by-product material as defined by the Atomic Energy Act of 1954, as amended. Also excluded are agricultural wastes, including manures and crop residues, returned to the soil as fertilizers or soil conditioners, and mining or milling wastes intended for return to the mine.[1]

Composition, Weight, and Volume

Various estimates have been made of the quantity of solid waste generated and collected per person per day. The amount of municipal solid waste *generated* in 1968 was estimated to be 7 lb/capita/day and was expected to increase at about 4 percent/yr. The amount *collected* was found to average 5.32 lb (Table 5-1). The residential and commercial waste alone disposed of increased from 3.3 to 3.5 lb/capita/day between 1971 and 1973,[2] but decreased from 3.6 in 1974 to 3.4 in 1975. Averages are subject to adjustment dependent on many local factors, including time of the year; habits, education, and economic status of the people; whether urban or rural area; and location. The estimates should not be used for design purposes. Each community should be studied and actual weighings made to obtain representative information.

The volume occupied by solid waste under certain conditions determines the number and size or type of refuse containers, collection vehicles, and transfer

Table 5-1 Average Solid Waste Collected (lb per person per day)

Solid-Waste Type	Urban	Rural	National
Household	1.26	0.72	1.14
Commercial	0.46	0.11	0.38
Combined	2.63	2.60	2.63
Industrial	0.65	0.37	0.59
Demolition, construction	0.23	0.02	0.18
Street and alley	0.11	0.03	0.09
Miscellaneous	0.38	0.08	0.31
Totals	5.72	3.93	5.32

Source: Anton J. Muhich, "Sample Representativeness and Community Data," *Proc. Inst. Solid Wastes,* Am. Pub. Works Assoc. Chicago, Ill., 1968.
Post-consumer solid waste generation estimated at 136.1 million tons in 1975 and 144.1 million tons in 1974. (Fourth Report to Congress, *Resource Recovery and Waste Reduction,* USEPA, 1977.)

[1] "Solid Waste Disposal Facilities Proposed Classification Criteria," *Fed. Reg.*, February 6, 1978, pp. 4942–4943.

[2] *Third Report to Congress, Resource Recovery and Waste Reduction*, USEPA, Washington, D.C., 1975, p. 5.

Table 5-2 Solid Waste Disposal in New York State[a] (lb per capita per day)

Type of Waste	Range	Median
Residential	2.1 to 2.6	2.3
Commercial	0.4 to 2.4	0.9
Demolition and Construction	0.1 to 1.3	0.5
Industrial	0.2 to 1.4	0.3
Street Sweepings	0.1 to 0.8	0.4
Institutional	0.1 to 0.2	0.1
Trees and Brush	0.2 to 0.4	0.3
Other	0.3 to 0.9	0.6
Total Disposal	3.5 to 6.6	5.7

Source: *Sanitary Landfill—Planning, Design, Operation, Maintenance,* New York State Dept. of Health, Albany, 1969.
Cost of collection and disposal estimated at \$30/ton in 1977, once a week collection.
[a]Weighings from 8 counties, population 104,000 to 983,000.

stations. Transportation systems and land requirements for disposal are also affected. For example, Kaiser found that loose refuse weighed 108 to 135 lb/yd^3.[3] Under pressure equivalent to 30 ft of refuse, the weight was 349 lb when the moisture was 27 percent and 480 lb when the moisture was 42 percent. Settling caused a 7-percent increase in moisture. The composition of ordinary refuse is shown in Table 5-3. Solid waste characteristics and production rates from various sources are given in Tables 5-1 to 5-8.

Storage

Where refuse is temporarily stored on the premises, an adequate number of suitable containers should be provided to store the refuse accumulated between collections. Watertight rust-resistant containers with tight-fitting covers are needed. Containers for incinerator residue and ashes must, in addition, be fire-resistant. Cans for ordinary refuse and ash cans up to 20- or 32-gal capacity and rubbish containers up to 50-gal capacity are practical sizes. The 10-gal can for kitchen wastes alone is a good household size for the average family when there is twice-a-week collection. In any case the weight and shape of containers must be kept within the limits that can be easily and conveniently handled by the collection crew. The weight should preferably not exceed 70 lb.

Plastic liners for cans and wrapping of garbage in newsprint reduce the need for cleaning cans and bulk containers, keep down odors, rat and fly breeding, and minimize difficulty due to freezing. Can liners and sacks simplify and expedite collection.

Satisfactory storage and can-washing platforms for commercial and institutional operations are shown in Figures 5-2 and 5-3. Where outside platforms

[3]Elmer Kaiser, "Refuse is the sweetest fuel," *Am. City,* May 1967, pp. 116–118.

Table 5-3 Approximate Composition of Residential Solid Wastes

Component	Percent by Weight 1977[a]
Food waste	17.0
Paper products	33.5
Rubber, leather	2.6
Plastics	3.6
Rags	2.0
Metals	9.2[b]
Glass and ceramics	9.9
Wood	3.2
Yard waste	17.5
Rock, dirt, misc.	1.5

Source: *Environmental Quality-1979,* Sup. of Documents, GPO, Washington, D.C. 20402, pp. 256–309.

[a]Weights and percentages will vary with community, time of year, and geography. For design purposes, make actual weighings.

[b]About 87 percent ferrous and 10 percent aluminum.

1. The daily per capita solid wastes collected approximates:
   Municipal: 3.85 lb in 1978.
   Residential and commercial: 3.6 lb in 1974 and 3.4 lb in 1975.
2. The above figures do not include junked vehicles, water and sewage treatment plant sludges waste oil, pathological wastes, agricultural wastes, industrial wastes, mining or milling wastes.
3. The Council of Environmental Quality estimates that compliance with existing and proposed environmental quality standards will increase annual landfill disposal cost by about $4.50 from the national average of $4.39 per ton (1978). This is attributed mostly to new federal criteria for sanitary landfills.
4. Tipping fee at resource recovery facilities (1979) approximately $8.00 to $14.60 per ton.
5. Operation and maintenance cost (1978) of incinerator approximately $25.00 to $35.00 per ton. Energy recovery can reduce cost (1977) to $14.70 to $22.70 per ton.

Table 5-4 Approximate Solid Waste Production Rates

Source of Waste	lb per day
Municipal	5.5 per capita
Household	2.3 per capita, or
	5.0 per capita plus 1 lb per bedroom[a]
Apartment building[b]	4.0 per capita per sleeping room[a]
Seasonal home	2.5 per capita[c]
Resort	3.5 per capita[c]
Camp	1.5 per capita[c]
School, general	2.1 per capita
Grade school	0.25 per capita plus 10 lb per room[a]
High school	0.25 per capita plus 8 lb per room[a]
University	0.86 to 1.0 per student[d,e]
Institution, general	2.5 per bed
Hospital	8.0 per bed[a]
Hospital	9.5 per occup. bed and 3.7 if staff added[f]
Nurses' or interns' home	3.0 per bed[a]
Home for aged	3.0 per bed[a]
Rest home	3.0 per bed[a]

Table 5-4 (*Continued*)

Source of Waste	lb per day
Hotel, first class	3.0 per room[a]
Medium class	1.5 per room[a]
Motels	2.0 per room[a]
Day use facility, resort	0.5 per capita[c]
Trailer camp	6 to 10 per trailer[a]
Commercial building, office	1.0 per 100 ft^2 [a]
Office building	1.5 per worker[a]
Department store	40 per 100 ft^2 [a]
Shopping center	Survey required[a]
Supermarket	9.0 per 100 ft^2 [a]
Restaurant	2.0 per meal[a]
Drugstore	5.0 per 100 ft^2 [a]
Airport	0.5 per passenger[e]
Prison	4.5 per inmate[e]
Retail and service facility	13.0 per 1000 ft^2 [a]
Wholesale and retail facility	1.2 per 1000 ft^2 [a]
Industrial building, factory	Survey required
Warehouse	2.0 per 100 ft^2 [a]
National Forest recreation area[g]	
Campground	1.26 ± 0.08 per camper
Family picnic area	0.93 ± 0.16 per picnicker
Organized camps	1.81 ± 0.39 per occupant
Rented cabin, with kitchen	1.46 ± 0.31 per occupant
Lodge, without kitchen	0.59 ± 0.64 per occupant
Restaurant	0.71 ± 0.40 per meal served
Overnight lodge, winter sports area	1.87 ± 0.26 per visitor
Day lodge, winter sports area	2.92 ± 0.61 per visitor
Swimming beach	0.04 ± 0.01 per swimmer
Concession stand	0.14 per patron
Job Corps, Civilian Conservation	
Corps Camp, Kitchen Waste	2.44 ± 0.63 per corpsman
Administrative and Dormitory	0.70 ± 0.66 per corpsman

[a] *I.I.A. Incinerator Standards,* Incinerator Institute of America, New York, May 1966. Up to 20 to 30 lb per bed per day at teaching hospitals, total. Total solid waste production at general hospitals is estimated at 12 to 15 pounds per bed with an average patient occupancy of 80 percent.

[b] A study at a low-income high-rise multifamily housing project shows the average daily refuse disposal via trash chutes to be 1.48 lb per capita, or 6.23 lb per dwelling unit, and 5.6 lb/ft^3, not including yard or bulky wastes. R. Barry Ashby, "Experiment in New Haven," *Waste Age* (November-December 1970), p. 22.

[c] *Environmental Health Practice in Recreational Areas,* PHS Pub. No. 1195, DHEW, Washington, D.C., 1965.

[d] Yshu Chiu, Joan Eyster, George W. Gripe, "Solid Waste Generation Rates of a University Community," *J. Environ. Eng. Div., ASCE,* December 1976.

[e] Gary L. Mitchell and Charles W. Peterson, "Weighing Small-Scale Resource Recovery," *Waste Age,* March 1979, pp. 12–25.

[f] Damodar S. Airan, "Hospital Solid Wastes Management: A Case Study," *J. Environ. Eng. Div., ASCE,* August 1980, pp. 741–755.

[g] Charles S. Spooner, *Solid Waste Management in Recreational Forest Area,* USEPA, 1971. Average rate of waste generation 90 percent confidence interval.

Table 5-5 Miscellaneous Solid Waste Generation

Type	Amount
Tires	0.6 to 1.0 tires discarded per capita per year
Waste oil	2.5 gal per vehicle per year
Sewage sludge, raw	0.4 tons per day per 1000 people, dewatered to 25% solids, with no garbage grinders 0.8 tons per day per 1000 people, dewatered to 25% solids, with 100% garbage grinders
Sewage sludge, digested	0.25 tons per day per 1000 people, dewatered to 25% solids, or 3 lb per capita per day dry solids
Water supply sludge	200 pounds per million gal of raw water, on a dry-weight basis, with conventional rapid-sand filtration using alum; raw water with 10 JTU
Scavenger wastes	0.3 gal per capita per day*
Pathological wastes	0.6 to 0.8 lb per bed per day—hospital 0.4 to 0.6 lb per bed per day—nursing home
Junked vehicles	35 to 85 per 1000 population—1973–74

Source: *Comprehensive Solid Waste Planning Study—1969*, Herkimer-Oneida Counties, Malcom Pirnie Engineers, State of New York Dept. of Health, Albany.
*Reflects pumping out entire tank when cleaning a septic tank and not just the scum and sludge.

are impractical or undesirable, a refuse storage room may be used. Such a room should have a drained concrete floor, coved base, and waterproof wall at least to splash height. A hot-water hose bib outlet and ample natural or artificial light and ventilation are essential. The temperature of the storage room should be kept below 50° F for odor control.

Garbage stands should not be screened nor should cans be whitewashed or painted. Stands should be convenient to the kitchen, in an airy shaded location away from children and stray cats and dogs. If the cans, platform, and surrounding ground are kept clean and the cans are kept tightly covered, there should be no cause for complaint. Built-in garbage or trash boxes, bins or sheds are not recommended; the storage structure becomes soiled, is difficult to keep clean, and becomes an odor nuisance and a rat and fly feeding and breeding ground.

Bulk containers or refuse bins are recommended where large volumes of refuse are generated, such as at hotels, restaurants, motels, apartment houses, shopping centers, and the like. They can be combined to advantage with compactors in many instances. Containers should be placed on a level, hard, cleanable surface in a lighted, open area. The container and surrounding area must be kept clean, for the reasons previously stated. A concrete platform provided with a drain to an approved sewer with a water faucet at the site to facilitate cleaning should be satisfactory.

A nationwide survey by the Consumer Product Safety Commission found that about 40 percent of the half-million refuse bins appeared to be dangerously unstable. Bins or containers having a volume of one cubic yard or more which

Table 5-6 Agricultural Solid Wastes—Waste Production Rates

Category	Annual Waste Production Rate
Wet manures	
Chickens (fryers)	6.4 tons/1,000 birds[a]
Hens (layers)	67 tons/1,000 birds[a]
Hogs	3.2 tons/head[a]
Horses	12 tons/head[b]
Beef cattle (feedlot)	10.9 tons/head[a]
Dairy cattle	14.6 tons/head[a]
Fruit and nut crops	
Class 1 (grapes, peaches, nectarines)	2.4 tons/acre[c]
Class 2 (apples, pears)	2.25 tons/acre[c]
Class 4 (plums, prunes, misc.)	1.5 tons/acre[c]
Class 5 (walnuts, cherries)	1.0 tons/acre[c]
Field and row crops	
Class 1 (field and sweet corn)	4.5 tons/acre[c]
Class 2 (cauliflower, lettuce, broccoli)	4.0 tons/acre[c]
Class 3 (sorghum, tomatoes, beets, cabbage, squash, brussel sprouts)	3.0 tons/acre[c]
Class 4 beans, onions, cucumbers, carrots, peas, peppers, potatoes, garlic, celery, misc.)	2.0 tons/acre[c]
Class 5 (barley, oats, wheat, milo, asparagus)	1.5 tons/acre[c]

Source: Roy F. Weston, *A Statewide Comprehensive Solid Waste Management Study*, New York State Dept. of Health, Albany, February 1970.

[a]Raymond C. Loehr, "Animal Wastes—A National Problem," *J. Sanit. Eng. Div., ASCE,* April 1969.

[b]Estimated.

[c]"Status of Solid Waste Management in California, "California Dept. of Public Health, Sacramento, September 1968.

may be overturned under the weight of a child and possibly pin him or her underneath are considered illegal by the Commission.[4] The container must be able to withstand a hanging weight or force of 191 pounds and a horizontal force of 70 pounds without overturning, with the force applied where tipping of the bin is most likely. Bins less than 40-in. high and 2 yd^3 or less capacity are exempted.

Can Washing

If a properly drained can-washing platform is provided with a hot-water hose bib outlet, cleaning is simplified. The hose bib outlet should be provided with a

[4]"Many bulk trash bins 'dangerously unstable,'" Washington Associated Press, *Times Record,* Troy, N.Y., November 14, 1978, p. 6.

Table 5-7 Industrial Solid Waste Production Rates[a]

SIC Code[b]	Industry	Waste Production Rate (tons/employee/year)
201	Meat processing	6.2
2033	Cannery	55.6
2037	Frozen foods	18.3
Other 203	Preserved foods	12.9
Other 20	Food processing	5.8
22	Textile mill products	0.26
23	Apparel	0.31
2421	Sawmills and planing mills	162.0
Other 24	Wood products	10.3
25	Furniture	0.52
26	Paper and allied products	2.00
27	Printing and publishing	0.49
281	Basic chemicals	10.00
Other 28	Chemical and allied products	0.63
29	Petroleum	14.8
30	Rubber and plastic	2.6
31	Leather	0.17
32	Stone, clay	2.4
33	Primary metals	24
34	Fabricated metals	1.7
35	Nonelectrical machinery	2.6
36	Electrical machinery	1.7
37	Transportation equipment	1.3
38	Professional and scientific inst.	0.12
39	Miscellaneous manufacturing	0.14

Source: Roy F. Weston, *A Statewide Comprehensive Solid Waste Management Study*, New York State Dept. of Health, Albany, February 1970.
[a]Consultant's analysis.
[b]Standard Industrial Classification (SIC) Code.

Table 5-8 Relation of Domestic Solid Waste Production to Population Density

Persons/mi^2	lb per capita per day
10 to 200	2.4 to 3.1
200 to 2,000	2.8 to 3.8
2,000 to 7,000	3.2 to 4.5
7,000 to 10,000	5.0 to 5.5

Source: Adapted from Garret P. Westerhoff and Robert M. Gruninger, "Population Density vs. Per Capita Waste Production," *Pub. Works* February 1970, p. 87.
Note: Increase in lb per capita with population density is related to increase in commercial and industrial wastes and street sweepings.

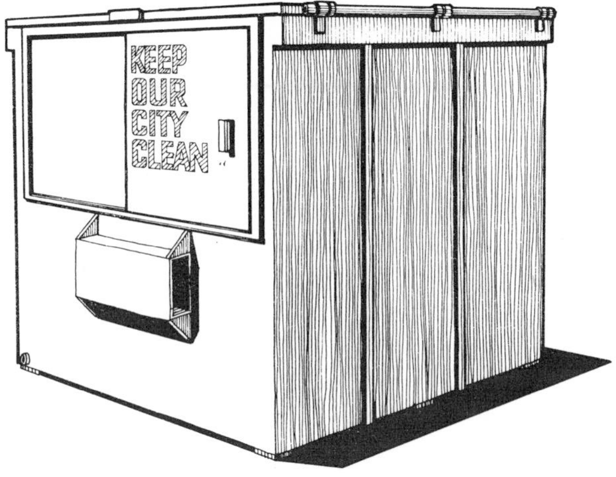

Figure 5-2 Good storage practices: bulk container (top) and can rack (bottom). (From *Refuse Collection and Disposal for the Small Community*, PHS and Am. Pub. Works Assoc., Washington, D.C., November 1953. Bulk container from *Environmental Health Practice in Recreational Areas*, DHEW, PHS, CDC, Atlanta, Ga., 30333; revised 1977, p. 38.)

backflow preventer. Hot water above a temperature of 120° F is necessary to dissolve grease. Grease and adhering food particles can be removed by scrubbing the can with a long-handled stiff-bristled brush and hot water. Then the container should be rinsed with a hot-water spray and allowed to dry. The platform drainage should pass through a grease trap and enter a sewerage system.

If a washing platform is not available, a detergent can be added to a small quantity of warm water in a pail to be used to scrub the inside and outside of the can. The wastewater should be collected in one garbage can and either disposed of with the garbage or discharged into a sewerage system ahead of a grease trap.

Wastewater and particles of garbage should not be allowed to run onto the surface of the ground, as an odor nuisance and fly-breeding problem will result.

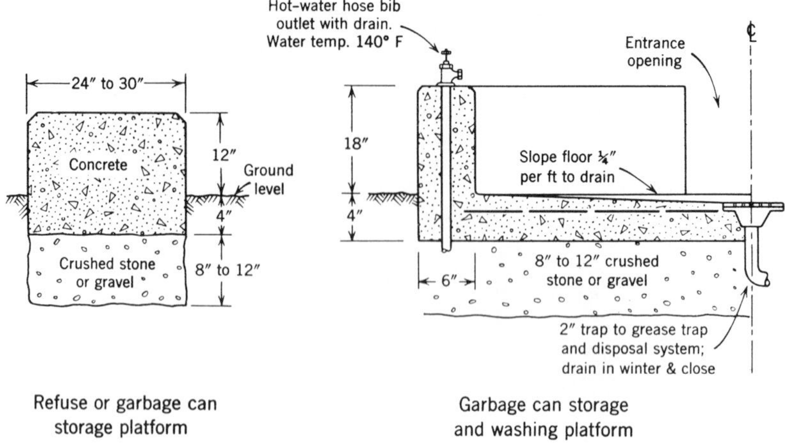

Figure 5-3 Sanitary refuse can storage platforms.

Where steam is available, emptied garbage cans can be inverted over steam jets to loosen adhering garbage. The steam jet can also be used for final "sterilization" somewhat similar to the handling of milk cans, after the cans have been cleaned.

As previously stated, a can liner will reduce the frequency for can cleaning, simplify collection, and reduce fly and rodent breeding and odor potential.

Apartment House, Commercial, and Institution Compactors, Macerators, and Pneumatic Tubes

The disposal and removal of solid wastes from apartment houses, and commercial places, and institutions have challenged the designer for some time. Compaction of wastes reduces the storage and handling volume. Maceration of wastes to a slurry form and transport to a central dewatering point and pneumatic transport make possible central collection for processing, incineration, or disposal. Each of these systems, and combinations, offer potential economies and conveniences as well as environmental sanitation improvements.

The limitations of on-site incinerators are well known and are discussed separately. Refuse compactors are in common use. Maceration of solid waste to a slurry at a central processing station followed by dewatering of the resulting pulp results in volume reduction, but the added moisture weight increases handling and haul costs. Odors can also be a problem. The pumping of crushed or shredded solid waste as a slurry in a pipeline requires individual hydraulic analysis and a feasibility study. The pneumatic transport of solid wastes in large pipelines (20 to 24 in.) from an apartment house complex to an incinerator also providing heat for space heating was first used in Sweden.[5]

[5]Iraj Zandi and John A. Hayden, "Collection of Municipal Solid Wastes in Pipelines," ASCE National Meeting on Transportation Engineers, San Diego, Calif., February 19–23, 1968.

Collection

Collection cost has been estimated to represent about 80 percent of the total cost of collection and disposal by sanitary landfill, and 60 percent when incineration is used.

The frequency of collection will depend on the quantity of solid waste, time of year, socioeconomic status of the area served, and municipal or contractor responsibility. In business districts refuse, including garbage from hotels and restaurants, should be collected daily except Sundays. In residential areas, twice-a-week refuse collection during warm months of the year and once-a-week at other times should be the maximum permissible interval. Slum and ghetto areas usually require at least twice-a-week collection. The receptacle should be either emptied directly into the garbage truck or carted away and replaced with a clean container. Refuse transferred from can to can will invariably cause spilling, with resultant pollution of the ground and attraction of flies. If other than curb pickup is provided, the cost of collection will be high. Some property owners are willing to pay for this extra service.

Collection equipment that simplifies the collection of refuse and practically eliminates cause for legitimate complaint is available. The tight-body open truck with a canvas or metal cover has been replaced in most instances by the automatic loading truck with packer to compact refuse dumped in the truck during collection, except for the collection of bulky items. Compaction-type bodies have twice the capacity of open trucks and a convenient loading height. Low-level closed-body trailers to eliminate the strain of lifting cans are also available. The weight of refuse under various conditions is shown in Table 5-9.

The number and size of the collection vehicles and the number of pickups in residential and business areas for communities of different population will vary with location, affluence, and other factors.

The people must understand that a good refuse-collection service also requires citizen cooperation in the provision and use of proper receptacles in order to keep the community clean and essentially free of rats, flies, and other vermin.

Table 5-9 Weight of Refuse Under Certain Conditions

Condition of Refuse	Weight (lb/yd^3)
Loose refuse at curb	125 to 240
As received from compactor truck at sanitary landfill	350
Normal compacted refuse in a sanitary landfill	750 to 850
After 5-yr settling in a sanitary landfill	1050
Well-compacted refuse in a sanitary landfill	1000 to 1250
In compactor truck	300 to 600
Shredded refuse, uncompacted	600
Shredded refuse, compacted	1600
Compacted and baled	1600 to 3200
Apartment-house compactor	700
In incinerator pit	300 to 550

Haul distance to the disposal facility must be taken into consideration in making a cost analysis. In some highly urbanized areas it is economical to reduce haul distance by providing large, specially designed trailers at transfer stations. In suburban and rural areas, container stations can be established at central locations. These stations usually include a stationary compactor for ordinary refuse and a bin for tires and bulk items.

The frequency and severity of injuries in the solid waste management industry are very high. The National Safety Council reported the solid waste collection workers have an injury frequency approximately 10 times the national average for all industry, higher than police work and mining underground. Workmen's compensation rates are 9 to 10 percent of payroll for all solid waste collectors.

Transfer Station

The urban areas around cities have been spreading, leaving fewer nearby acceptable solid waste disposal sites. This has generally made necessary the construction of incinerators, resource recovery facilities, or processing facilities in cities or in the outskirts, or the transportation of wastes longer distances to new sites. However, as the distance from the centers of solid waste generation increases, the cost of direct haul to a site increases. A "distance" is reached (in terms of time) when it becomes less expensive to construct an incinerator or a transfer station near the center of solid waste generation where wastes from collection vehicles can be transferred to large tractor-trailers for haul to more distant disposal sites. Ideally, the transfer station should be located at the centroid of the collection service area.

A comparison of direct haul versus use of a transfer station, for various distances, is shown in Table 5-10. This type of information is also useful in making an economic analysis of potential landfill sites, which should properly be considered with the physical development and social factors involved in site selection.

Table 5-10 Cost (per ton) Comparison of Direct Haul and Use of Transfer Station (1970)[a]

	Round-Trip Distance (mi)				
Method	10	20	30	40	60
Direct haul[b]	$1.80	3.60	5.40	7.20	10.80
Transfer station and haul[c]	$2.60	3.20	3.80	4.40	5.60

[a]Adjust using ENR or other cost index.

[b]Assumes a direct-haul cost of $.18 per ton-mile in a 20-yd^3 compactor truck (approximately 5-ton capacity) with a 2-man crew.

[c]Assumes a transfer station and operation cost of $2.00 per ton and a haul cost of $.06 per ton-mile in a 65-yd^3 tractor-trailer (approximately 17-ton capacity) with a 1-man operator. Refuse density in compactor truck and tractor-trailer is 450 to 500 lb/yd.^{3} Maximum allowable highway gross vehicle weight is 72,000 lb. A 75-yd^3 semi-trailer is also available for the transfer of solid waste.

A transfer station, resource recovery facility, or processing facility should be located and designed with the same care as described for an incinerator. Drainage of paved areas and adequate water hydrants for maintenance of cleanliness and for fire control are equally important. See Figure 5-4.

If the cost of disposal by sanitary landfill is added to the figures in Table 5-10, adjusted for inflation, one can compare the total relative cost of refuse transfer, transportation, and disposal by sanitary landfill with that of incineration. For example, if the total cost of modern incineration is $25.00/ton and sanitary landfill cost is $5.00/ton, the cost of incineration will become less expensive when the round-trip haul distance to a sanitary landfill exceeds 56 mi, under the conditions stated. This principle is also illustrated in the hypothetical example shown in Figure 5-5 for a population of 50,000. Figure 5-6 shows a similar comparison for the sanitation districts of Los Angeles County, in which distance is shown in terms of times of travel to the disposal site.

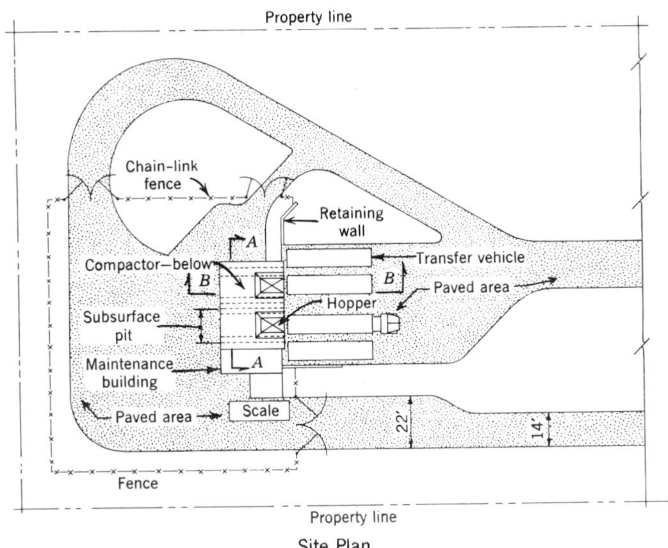

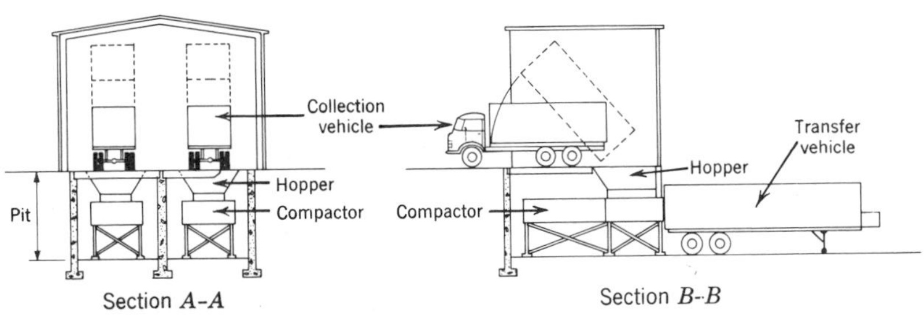

Figure 5-4 Transfer Station.
Herkimer-Oneida Counties comprehensive solid waste study. (Malcolm Pirnie Engrs., State of New York Dept. of Health, Albany, July 1969.)

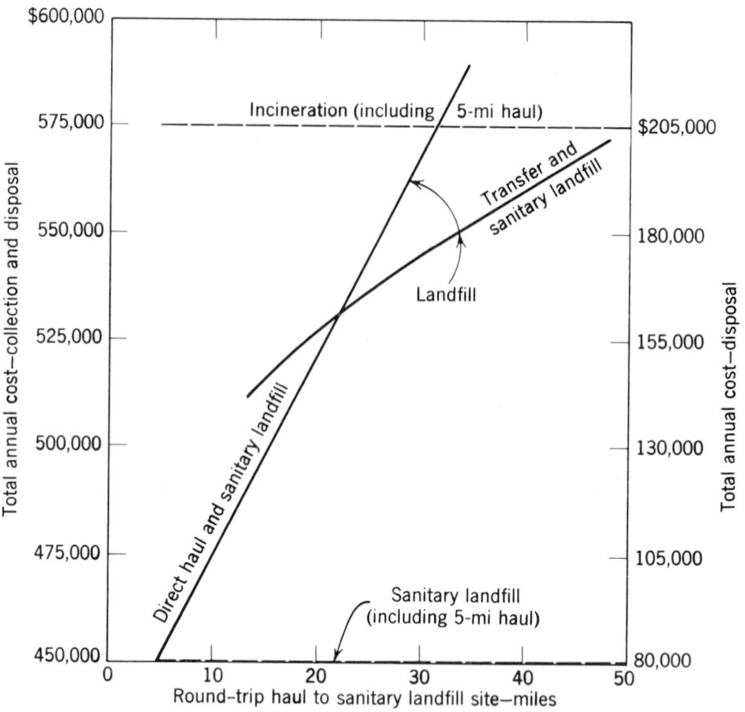

Figure 5-5 Effect of haul distances to site on cost of disposal by sanitary landfill compared to cost of disposal by incineration. 1964. Population 50,000–Density 12 people per acre. (From *Municipal Refuse Collection and Disposal,* Office for Local Government, Albany, N.Y., 1964.) Example: Annual Cost (Dollars) Collection and Disposal (Adjust using ENR or other cost index.)

Incineration	Sanitary Landfill			Cost Difference
(5-mile haul)	Haul	Direct	Transfer	(Additional Cost of Incineration)
	5	$450,000	NA	$125,000
	20	520,000	NA	55,000
$575,000	22	535,000	$535,000	40,000
	35	NA	550,000	25,000
	48	NA	575,000	equal

Cost Basis: Collection (5-mile Haul)—$370,000; Incineration—$205,000; Landfill—$80,000. Note: Each cost figure includes $370,000 for collection. N.A.: Not Applicable.

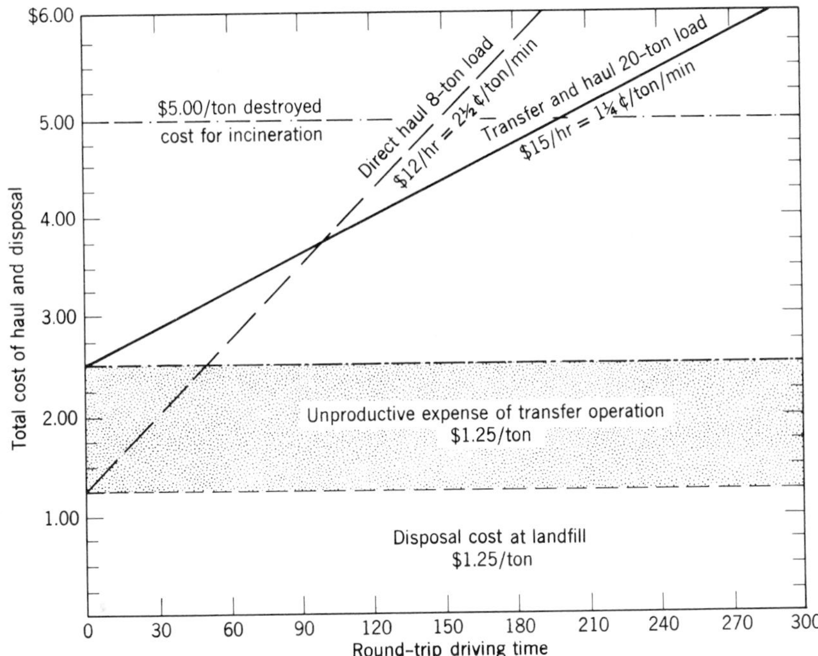

Figure 5-6 Cost comparison. Incineration vs. transfer and haul to landfill, 1968. Adjust cost using ENR or other cost index. (Courtesy Sanitation Districts of Los Angeles County, John D. Parkhurst, Chief Engineer and General Manager.)

Rail haul and barging to sea also involve the use of transfer stations. They may include one or a combination of grinding, compaction to various densities, and baling.

TREATMENT AND DISPOSAL OF SOLID WASTES

Solid waste disposal methods include the open dump, hog feeding, incineration, grinding and discharge to a sewer, milling, compaction, sanitary landfill, dumping and burial at sea, reduction, composting, pyrolization, wet oxidation, and anaerobic digestion. The common acceptable refuse disposal and treatment methods are incineration, sanitary landfill, and, in some parts of the world, composting. Resource conservation and recovery are receiving greater attention. Some relative solid waste handling and disposal costs are given in Table 5-11.

Open Dump

The open dump is all too common and needs no explanation. It is never satisfactory, as usually maintained. Refuse is generally spread over a large area,

Table 5-11 Solid Waste Handling and Disposal Costs (1968–1970)

Method	Cost ($ per ton)	Capital Cost ($ per ton)
Collection	10.00 to 13.00	—
Sanitary landfill[a]	0.75 to 2.50	8.00 to 14.00
Incineration[b]	4.00 to 8.00	8,000 to 12,500
Composting	5.00 to 10.00	8,000 to 12,000
Pyrolizer[c]	4.00 to 8.00	5,000 to 10,000
Transfer station	1.00 to 2.00	1,000 to 1,500
Haul by trailer and sanitary landfill	3.50 to 6.00	—
Compact, bale, rail haul to sanitary landfill	4.00 to 6.00	—
Rural container station[d]	7.00 to 10.00	—
Haul cost, 20-yd^3 compactor[e]	0.20 to 0.25 per mi.	—
Compaction, rail haul, disposal to sanitary landfill[f]	7.15 to 7.60	—
Compaction	1.00 to 1.35	9,000 to 12,000
Shredding	2.00 to 4.00	—
Shred, compact, bale	5.40 to 6.50	—
Barging to sea	4.00 to 6.00	—

Note: Costs are relative. Adjust using ENR or other cost index.

[a]Capital cost includes site, fencing, roadway, scale, hammermill or shredder, tractors, trucks, and engineering based on annual capacity. Can reduce by one-half if fencing, hammermill or shredder, and miscellaneous are omitted.

[b]The cost is estimated to vary from $9,000 to $15,000 for construction cost per ton per day rated capacity including precipitator, shredder, odor and dust control, and improved furnace and grate design. For a 1000-ton-per-day plant, the construction is estimated at $12,500 per ton per day capacity; the total annual cost including amortization at 4% interest, 40-year life of building and 15-year life of equipment is $7.80 per ton processed. [Casimir A. Rogus, *Am. Pub. Works Assoc. Reporter*, 3–5 (March 1969).]

[c]Experimental with solid wastes.

[d]Includes cost to build and operate.

[e]Will vary with compactor or trailer size, men per crew, and actual travel time to disposal site.

[f]Cost per ton estimated at $2.00 per ton for rail haul, $.25 for car leasing, $.75 for sanitary landfill disposal, and $4.00 for compaction and baling.

providing a source of food and harborage for rats, flies, and other vermin. It is unsightly, an odor and smoke nuisance, a fire hazard, and often a cause of water pollution. It should be eliminated or its operation changed to a sanitary landfill.

Hog Feeding

Where garbage is fed to hogs, careful supervision is necessary. The spread of trichinosis to man, hog cholera, the virus of foot-and-mouth disease, and vesicular exanthema in swine is encouraged when uncooked garbage is fed to hogs. In some instances tuberculosis, swine erysipelas, and stomatitis may also be spread by raw garbage. The boiling for 30 min of all garbage fed to hogs will prevent transmission of trichinosis and economic loss to the swine industry due to hog illness and death.

The federal interstate quarantine regulations require the heat treatment of garbage transported across state boundaries for hog feed. Methods for heating garbage are given in *Equipment for the Heat-Treatment of Garbage to be Used for Hog Feed,* issued jointly by the U.S. Dept. of Agriculture and the Federal Security Agency in 1952. Most states have the same requirements, but enforcement is weak and in some instances ineffective. For hog feeding to be satisfactory (in addition to the cooking of garbage) it is necessary to rat-proof concrete feeding platforms and structures; remove manure and leftover waste daily; dispose of the waste by sanitary landfill or incinerate or compost the waste; and clean the hog pens and flush the feeding platforms and troughs, frequently. Wastewater should discharge to a disposal system that will not pollute receiving waters or become a nuisance. More often than not, precautions are neglected. As a result fly and rat breeding is supported and bad odors are common.

Grinding

The grinding of garbage is an acceptable method of garbage "disposal." It is highly recommended from a convenience and public health standpoint, but the disposal of other refuse remains to be handled. The putrescible matter is promptly removed, thereby eliminating this as a source of odors and food for rats, flies, and other vermin. In one system, the home grinder is connected to the kitchen-sink drain. Garbage is shredded into small particles while being mixed with water and is discharged to the house sewer. In another system garbage is collected as before but dumped into large, centrally located garbage-grinding stations that discharge garbage to the municipal sewerage system. The strength of the sewage is increased and additional sludge digestion and drying facilities will be required when a large amount of garbage is handled. Greater energy recovery from the sludge may be possible.

Disposal at Sea

Where dumping at sea is permitted, all garbage and other refuse is dumped into large garbage scows or barges. The scows are towed by tugs and the garbage is taken out to sea and dumped a sufficient distance out to prevent the refuse being carried back to shore and causing a nuisance. Bad weather conditions hamper this operation and unless this method is kept under very careful surveillance, abuses and failures will result. Because of the cost of maintaining a small navy and difficulties in satisfactorily carrying out this operation, coastal cities have reverted to sanitary landfill and incineration. In recent years, consideration has been given to the compaction of refuse to a density greater than 66.5 lb/ft^3 prior to transport and then disposal by burial in the ocean at depths greater than 100 ft, based on oceanographic conditions, to ensure there will be no mixing with surface water. More research is needed to determine stability of the compacted refuse and effect on marine life. See High-Density Compaction. The dumping of garbage and other refuse at sea is prohibited in the United States.

Garbage Reduction

In the reduction method of garbage disposal, the garbage is cooked under pressure. Fats melt out and are separated from the remaining material. The fat is used in the manufacture of soaps or glycerines and the residue is dried, ground, and sold for fertilizer or cattle feed. Odor complaints are associated with this process and, where a solvent such as naphtha is used to increase the extraction of fat, a greater fire hazard exists. The use of synthetic detergents and chemical fertilizers and high operating costs have led to the abandonment of this process.

Composting

Composting is the controlled decay of organic matter in a warm, moist environment by the action of bacteria, fungi, molds, and other organisms. The organic matter may be municipal solid waste, sewage sludge, agricultural waste, or organic industrial waste. Moisture is maintained at 40 to 65 percent; 50 to 60 percent appears to be best. Composition of the refuse, disposal of refuse not composted, demand for compost and salvaged material, odor production and control, public acceptance, and total cost are factors to be carefully weighed. Compost is a good soil conditioner but a poor fertilizer. It may be used as an additive to fertilizer, fuel, or possibly in building materials. The process is very attractive, but because of the factors mentioned has met with only limited success in the United States.

The composting operation may involve a combination of steps. These may include weighing, separation of noncompostables and salvage by hand and by a magnetic separator; size reduction to 2 in. or less by means of a shredder, grinder, chipper, rasp mill, or hammermill; ballistic and magnetic separation; biological digestion by any one of a number of composting methods described below; screening and possible standardization of fertilizer value; and disposal by bagging for sale, spreading as a soil conditioner, or transporting to a sanitary landfill.

The Beccari method of garbage treatment has been used in Europe for many years. The garbage is placed in tightly sealed tanks and allowed to digest 10 days without air and then 10 to 20 days in the presence of air. Drainage from the garbage is collected at the bottom of the tank and is recirculated back over the garbage if necessary to keep it moist. The digested residue is relatively stable.

Naturizer composting uses sorting, grinding and mixing, primary and secondary composting including three grinding operations, aeration, and screening. Digested sewage sludge, raw sewage sludge, water, or segregated wet garbage is added at the first grinding for dust and moisture control. The total operation takes place in one building in about 6 days.[6]

The Dano composting (stabilizer) plant consists of sorting, crushing, biostabilization in a revolving drum to which air and moisture are added, grinding, air separation of nonorganics, and final composting in open windrows. Temperatures of 140° F are reached in the drum. Composting can be completed in 14 days by turning the windrows after the fourth, eighth, and twelfth days.[7] Longer periods are required if the windrows are not kept small, turned, and mixed frequently and if grinding is not thorough.

Windrow composting is a common method. The compost is piled in long rows in an open field and turned over every 4 or 5 days until the compost temperature drops to about 130° F or less. The compost should be dark brown to black in color and is referred to as humus.

The Bangalore process is used primarily in India. Layers of refuse and night soil are placed in piles which are turned frequently to maintain aerobic conditions and thus speed up the composting action.

The Fairfield-Hardy process handles garbage and trash and sewage sludge. The steps in the process are (1) sorting—manual and mechanical to separate salvageable materials, (2) coarse shredding, (3) pulping, (4) sewage sludge addition if desired, (5) dewatering to about 50 percent moisture, (6) digestion 3 to 5 days with mixing and forced air aeration, temperature ranges from 140 to 170° F, (7) air curing in covered windrows, and (8) pelletizing, drying, and bagging. Compost from the digester is reported to have a heat value of 4000 Btu per pound and when pelletized and dried 6450 Btu per pound.[8]

Studies reported at the University of California give time and temperature of composting separated municipal refuse alone and with raw and digested sludge.[9] The process involves sorting, grinding, stacking, aeration by turning or adding air, and regrinding. About 28 percent of the refuse is rejected and 72 percent is used in composting. A temperature of 158 to 178° F is reached after about 3 days. The temperature is maintained for about 3 to 5 days, which is believed to be adequate to kill pathogens, parasites, weeds, seeds, and fly ova *exposed to these temperatures*. After 10 or 12 to 21 days, when the temperature of the compost drops to 122 to 131° F, the process is essentially completed. Aerobic conditions prevent odors and accelerate the action. A curing period of about 3 weeks usually follows composting. This makes possible further breakdown of paper and wood fragments.

Experimental studies show that the degree of decomposition of compost,

[6]P. H. McGauhey, "Refuse Composting Plant at Norman, Oklahoma," *Compost Sci.* **1**, No. 3, 5–8 (Autumn 1960).

[7]Clarence G. Golueke, "Composting Refuse at Sacramento, California," *Compost Sci.* **1**, No. 3, 12–15 (Autumn 1960).

[8]*Solid Waste Management Technology Assessment*, General Electric Company, Schenectady, N.Y. 12345, 1975, pp. 179–186.

[9]C.G. Golueke and H. B. Gotaas, "Public Health Aspects of Waste Disposal and Composting," *Am. J. Pub. Health*, March 1954, pp. 339–348.

without special aeration, can be estimated by the maximum temperature reached by the refuse.[10] For example, a temperature of about 150° F can be reached in 2 to 7 days of operation; a temperature of 140 to 158° F in 7 to 21 days; a temperature of 112 to 140° F in 21 to 28 days; and a temperature of 86 to 113° F in 28 to 180 or more days. Refuse going through these temperatures can be classified in terms of decomposition as slight, moderate, medium, or good. As the paper content increases, time for decomposition increases.

Pathogenic organisms exposed to the higher temperatures for the times indicated will be destroyed. However, because of the nature of refuse and the range in temperature between the outside and inside of a mass of compost, this cannot be guaranteed.

The cost of composting usually does not include cost of disposing noncompostable wastes or value of compost salvage. The cost in 1969 was estimated to be $7 to $10/ton for a 100- to 200-ton/8 hr plant and $5 to $8 for a 200- to 500-ton/8 hr plant.[11]

Incineration

Incineration is a controlled combustion process for burning solid, liquid, or gaseous combustible waste to gases and a residue containing little or no combustible material *when properly carried out*. It is a volume reduction process suitable for about 70 percent of the municipal solid wastes. This subject is discussed later at greater length.

Sanitary Landfill

Sanitary landfilling is an engineered method of disposing of solid waste on land in a manner that protects the environment by spreading the waste in thin layers, compacting it to the smallest practical volume, and covering it with compacted soil by the end of each working day or at more frequent intervals as may be necessary.[12] This subject is discussed further later in this chapter.

Pyrolysis

Pyrolysis as applied to solid wastes (metal and glass removed) is an experimental thermochemical process for the conversion of complex organic solids, in the absence of added oxygen, to water, combustible gases, tarry liquids, and a stable

[10] G. Niese, "Experiments to Determine Degree of Decomposition of Refuse Compost by Its Self-Heating Capability," Agricultural Microbiology Institute, Justus-Liebig-Universitat, Giessen, *Inf. Bull.*, **17**, 18 (May-August 1963), PHS.

[11] Garret P. Westerhoff, "A Current Review of Composting," *Pub. Works*, **100**, 87–90 (November 1969).

[12] *Sanitary Landfill*, Manual No. 39, ASCE Solid Waste Management Committee of the Environ. Div. 345 East 47th St., New York, N.Y. 10017, 1976, p. 1–4.

residue. Intermediate products may be collected or may be used to contribute heat to support the process. The end products would be carbon, water, and carbon dioxide if carried to completion. If the raw material contains sulfur and nitrogen, these oxides will be formed with resultant air pollution unless provision is made for their removal. Temperatures of 900 to 1700° F have been used. In a variation, some oxygen and a temperature up to 2100° F is used. It is a process of destructive distillation similar to that used for making charcoal and for the recovery of organic by-products such as turpentine, acetic acid, and methanol from wood.[13] The Lantz converter, the Garrett Research and Development Co. unit, the Purox system by the Union Carbide Corp., and the Urban Research and Development Corporation unit are variations of the pyrolytic process. The Lantz system uses a temperature of 1200 to 1400° F with ground refuse. The Landgard system burns refuse in a rotary kiln using 40 percent of the required air supplemented by the burning of heating oil. Waste gases leave the kiln at a temperature of 1200° F. Pyrolysis can reduce the volume of municipal solid wastes by 95 percent or more. The normal residence time is 12 to 15 min.

High-Temperature Incineration

High-temperature incineration is carried out at 3000 to 3400° F (Melt-zit), 2500 to 2600° F (FLK Slagging Incinerator, Germany), 2600 to 3000° F (Torrax system), and 3000° F (American Design and Development Corporation). These units are in the developmental and pilot stage. Combustibles are destroyed and noncombustibles are reduced to slag or sand-like grit. A 97-percent volume reduction is reported. Special provision must be made for air pollution control.

Wet Oxidation

Wet oxidation and anaerobic digestion of refuse are in the experimental stage. In wet oxidation, the refuse can be processed under high pressures and temperatures or the refuse can be ground and aerated while in suspension in a liquid medium. In anaerobic digestion, decomposable material is separated from refuse, ground, and then digested at a controlled temperature in the absence of air. The resultant gases are mostly methane and carbon dioxide, and the residue can serve as a soil conditioner.

Size Reduction (Shredding, Grinding, Pulverizing)

Shredders, including hammermills, crushers, raspers, hoggers, chippers, and pulpers, are devices that reduce bulky refuse items and other solid wastes to a manageable size for disposal by landfill or incineration, or for processing preliminary to materials and energy recovery and composting.

[13]Donald A. Hoffman and Richard A. Fitz, "Batch Retort Pyrolysis of Solid Municipal Wastes," *Environ. Sci. Technol.*, November 1968, pp. 1023–1026.

At landfills, size reduction also has aesthetic benefits: odor, fly- and rodent-breeding problems are greatly minimized; waste is readily spread and compacted; and tire damage is reduced. Daily cover may not be required, but then the potential for leachate production due to precipitation and lack of drainage is increased. Shredding reduces the volume of waste to about 40 percent or less of the original bulk, and waste decomposition is accelerated. The cost of shredding equipment and its operation alone however may add $5 to $10 per ton.[14]

Volume reduction is also achieved by stationary compaction, pulping and dewatering, high-density compaction, and by landfill equipment.

Size reduction is a common initial processing step in solid waste resource recovery but explosions can be a problem. In 1975 The Factory Mutual Research Corporation survey showed 97 explosions with 69 causing significant damage resulting from municipal solid waste shredding, a frequency of one explosion per 85,000 tons processed. Dry shredding should be avoided. Explosion-venting systems over shredders and fine water spray in the shredders are minimum precautions, in addition to presorting to remove gasoline cans, ammunition and the like. Explosion detection and suppression devices are recommended.[15]

High-Density Compaction (Baling)

High-density compaction of solid wastes is accomplished by compression to a density of more than 66.5 lb/ft^3. The resulting bales may be enclosed in chicken wire, hot asphalt, vinyl plastic, plain or reinforced cement, or welded sheet metal, or tied with metal bands depending on the method of disposal or intended use of the bales or blocks. Bales need not be enclosed or tied if firmly compacted. Rogus, reporting on the early Tezuka-Kosan process in Tokyo, stated that the liquid release during compression ranged from 2 to 5 percent by weight and that it required treatment. The bales had a density of 70 to 109 lb/ft^3; sank in seawater; had good structural cohesiveness; resisted corrosion; were reasonably free of odors and insect and rodent hazard; and showed no evidence of aerobic or anaerobic decomposition.[16]

As with size reduction, there is an added cost to prepare the solid waste and to purchase and operate the compaction equipment. This may add $3 to $6 to the cost. On the other hand, landfill space requirements are reduced, the site is relatively free of debris, papers, and vermin, cover material is less, settlement is reduced simplifying maintenance and the solid waste bales are deposited easily and at less cost.

Disposal of Animal Wastes

Runoff from beef cattle feedlots and from land disposal of liquid cow manure, swine wastes, poultry manure, and wastes from other concentrated livestock

[14]James F. Mank, *Size Reduction of Solid Waste* (SW-117), USEPA, 1974.

[15]"Explosion Alleviation in Solid Waste Processing Equipment," based on a paper by A. R. Nollet and E. T. Sherwin, *Pub. Works*, September 1978, pp. 124–127.

[16]Casimir A. Rogus, "High Compression Baling of Solid Wastes," *Pub. Works*, June 1969, pp. 85–89.

production can contaminate surface water and groundwater supplies, kill fish, destroy aquatic biota, and in general degrade water quality. The odors associated with the handling and disposal of these wastes contribute to the problem and may determine the disposal method and its location. The primary control measure should be prevention of pollution at the source, followed by reclamation and reuse of wastes, composting and reuse, dehydration and sale as fertilizer, or treatment and disposal to prevent water pollution or other environmental degradation.

Systems for treatment and disposal include field spreading, plow-furrow cover, irrigation, aerobic digestion, anaerobic digestion, lagoon, aerated lagoon, oxidation ditch, lagoon and oxidation ditch, oxidation ditch and lagoon, drying, incineration, and wet oxidation. Each method has limitations and special application. The disposal of large quantities of animal wastes introduces special problems requiring individual study. Loehr suggests several alternatives.[17]

Where a small number of animals are kept, the manure should be collected each morning (not less than twice weekly) and spread on the fields, composted in a properly maintained compost pile, or stored in a flytight bin. Well-fed saddle horses pass 15 to 30 lb of manure and 4 to 7 qt of urine per day.[18] Cows will produce about twice as much. See Table 5-6.

Composting of manure takes advantage of the inclination that larvae have to move out of and away from moist manure in search of a dry place to pupate. Therefore, if manure is stacked on a slatted platform over about 12 in. of water on which kerosene has been sprayed, the crawling larvae will fall out of the compost heap into the water and drown. Waste oil may be used instead of water, and instead of a slatted platform a well-compacted earth or concrete base surrounded by a tight earth or concrete channel containing 12 in. or more of water may be used.

Manure is commonly spread on the field as a fertilizer in thin scattered layers to dry. If wet manure is plowed under, fly eggs almost always present will hatch and cause a nuisance. If weather conditions do not permit spreading of manure, it must be stored in an acceptable manner until it can be disposed of.

Dead animals are best disposed of at an incinerator, rendering plant, or in a separate area of a sanitary landfill. Large numbers should be buried in a special trench with due consideration to protection of groundwaters and surface waters. The earth cover should be at least 2 ft and then maintained.

RESOURCE RECOVERY

Background

Reclamation of materials and energy from solid wastes is not new. Scavengers have salvaged newsprint and cardboard, rags, copper, lead, and iron for years.

[17]Raymond C. Loehr, "Alternative for the Treatment and Disposal of Animal Wastes," *J. Water Pollut. Control Fed.*, **43**, 668 (April 1971).

[18]Seymour Barfield, "Practical Fly Control for Horse Keeping on Hillside Residential Lots," *J. Environ. Health*, **28**, No. 4 (January-February 1966).

These materials, together with aluminum, glass, and wood, are being reclaimed at many central collection and processing stations to a greater or lesser extent depending on the market available, tax policies, and public interest. See Table 5-3. In recent years energy recovery has been an important consideration in which raw refuse and shredded refuse, referred to as refuse-derived-fuel (RDF), is used to produce steam or electricity. Figure 5-7 shows a resource recovery system. The codisposal of wastewater and other sludges with RDF is also possible,[19] however liquids, dead animals, and toxic and hazardous wastes are excluded.

Resource recovery is not a municipal operation to be entered into just because it seems like the logical or proper thing to do. It is a complex economic and technical system with social and political implications all of which require competent analysis and evaluation before a commitment is made. Included are the capital cost, operating cost, market value of reclaimed materials and material quality, potential minimum reliable energy sales, assured quantity of solid wastes, continued need for a sanitary landfill for the disposal of excess and remaining unwanted materials and incinerator residue, and a site location close to the centroid of the generators of solid wastes.

Resource recovery is a partial waste disposal and reclamation process. Materials not recovered may amount to 70 to 80 percent of the original waste by weight, although a resource recovery system can separate up to 90 percent of the municipal waste stream into marketable components.[20] The USEPA believes that under the best conditions only 56 percent of the U.S.'s refuse will be recovered. In 1979, 7 percent was being recycled for materials or energy.[21] Resource recovery can be expected to achieve no more than 60 percent reduction in future landfill volume requirements.[20] It is inadvisable to plan on recovery of materials that do not pay for the cost of recovery.

Unit Processes or Operations

Unit processes used singly and in various combinations to recover and prepare wastes for reuse and/or disposal include:

1. Home separation of newspaper and cardboard (not magazines), glass and metals, and other refuse, with aluminum and bottles, and possibly newspapers delivered to a recycling center
2. Weighing
3. Unloading or tipping to remove large bulky items and hazardous containers (chemicals, gas tanks), and to even out the waste flow to other unit processes
4. Hand sorting
5. Coarse screening (trommel screen) to skim off bulky boxes, wood, metals, and miscellaneous items larger than about 1 ft in size; combustible items are separated and shredded for use as fuel as is, or are compacted or pelletized for use as fuel

[19]Allan Jacobs, Don Brailey, and Bryce Pickart, "Municipal Sludge Management For Recovering Energy—Part 2," *Water Sewage Works*, October 1979, pp. 48–52.
[20]Robert P. Stearns and John P. Woodyard, "Resource Recovery and the Need for Sanitary Landfills." *Pub. Works*, September 1977, pp. 106–109.
[21]"Tapping garbage and sludge for energy: U.S. prospects," *Civ. Eng. ASCE*, October 1979, pp. 73–75.

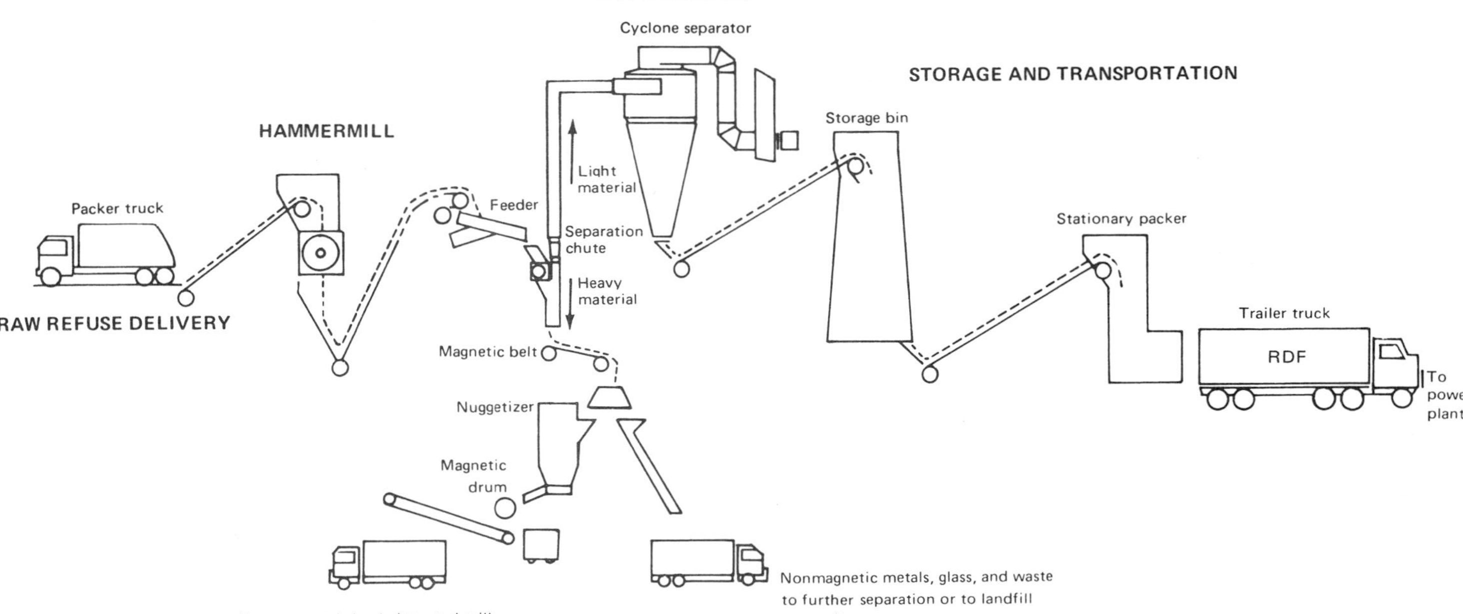

Figure 5-7 A resource recovery system. In the fuel-processing segment of the system (demonstrated by the city of St. Louis, the Union Electric Company, and the USEPA) fuel and ferrous metal are recovered from municipal solid waste that has been shredded and air-classified. (Source: *Third Report to Congress, Resource Recovery and Waste Reduction*, USEPA, Washington, D.C., 1975, p. 37.)

6. Finer screening (trommel screen) removes glass and other remaining wastes for disposal as fill or to a landfill
7. Air classifying to separate the light fractions, such as paper, plastics, leaves, textiles which are carried out with the air stream in a drum separator; the heavy fractions are collected at the bottom for further separation; other methods include cyclone separation
8. Magnetic separation to remove ferrous materials
9. Incineration of unprocessed solid waste using water-wall type to produce high pressure steam
10. Incineration of RDF for energy recovery
11. Wet pulping of solid waste and use of the pulp to generate steam and electricity, after first recovering valuable materials
12. Shredding, preceded by separation to avoid explosions due to dust or explosive gases and to remove highly flammable or explosive materials; the separation may be accomplished by manual sorting, rotary drums, coarse screening, and air classifying; primary shredders reduce solid waste to 4 to 6 in. in size; secondary shredders to $1\frac{1}{2}$ in. size
13. Transfer station may also serve as a presorting and preliminary processing location
14. Compaction and baling
15. Optical sorting to classify green, amber, and clear glass

Resource recovery must recognize what is worth recovering and the environmental benefits; not all concepts are viable. Plastics and glass have little value. Most aluminum is recovered at plants making aluminum products with some from consumer separation. Tires are a problem but shredded tires may find use as an asphalt additive to reduce pavement cracking and possible reuse in tire making.[22] Heat, paper, and ferrous metal recovery offer the greatest promise.

In a study of 31 operating or planned resource recovery systems, 22 incorporated shredding, 22 magnetic separation, 17 air classification, and 11 incineration.[23]

Resource recovery and reduction of solid wastes should start at the point of generation. Industrial material, process, and packaging changes can minimize the waste or substitute a less objectionable waste. The amount of waste can then be reduced and what is produced recycled at the source.

Energy Recovery

Refuse has been used for many years to produce steam for building heating and to generate electricity. The heat value of raw refuse averages about 4500 Btu/lb. The heat value of RDF from which ferrous and nonferrous materials, glass, rocks, dirt, and miscellaneous inorganics have been removed is about 5000 to 5600 Btu/lb. These figures compare with 8000 to 14,000 Btu/lb for coal, 150,000 Btu/lb for No. 6 heating oil, 1000 to 1100 Btu/ft^3 for natural gas, and 500 Btu/ft^3 for landfill methane. RDF made into pellets is reported to have a heat value of 6600 Btu/lb with 20 percent moisture. About 65 percent of municipal solid waste can end up as RDF. Presorting, shredding, possibly air separation, and magnetic separation of solid waste usually precede preparation of RDF.

[22]Richard W. Eldredge, "To Be Or Not To Be . . . ," *Waste Age,* March 1980, p. 14.
[23]*Prof. Eng.* January 1979, pp. 31–33. From data provided by the National Center for Resource Recovery, USEPA.

Energy can be recovered from refuse by burning it in a refractory lined incinerator or a water-wall incinerator. In the refractory lined incinerator the boiler tubes are installed after the combustion chamber. In the water-wall incinerator the boiler tubes are installed in the combustion chamber and take the place of the refractory lining. The heated water is used to produce steam. The steam may be used in a steam turbine to produce mechanical power or to operate an electric generator to produce electricity.

It is believed that any industrial plant requiring 1000 lb of steam per hour, and generating 1 ton of combustible waste per day should consider using waste-fuel firing as a supplementary source of energy. The cost of fuel, Btu value of the combustible waste, amount of waste generated per day, and proximity to point of use determine the feasibility in each situation. A controlled-air incinerator with a waste-heat boiler for energy recovery, using standard modular equipment to make up the system, can make the process economically attractive [24] not only for industrial plants but also for small communities and for institutions where there is a need or use for steam. See Controlled Air Combustion.

Methane Recovery

Methane is produced when anaerobic methane-producing bacteria are active. This condition may be reached in 6 months to 5 years depending on the landfill. Acidic conditions inhibit growth of methane-producing bacteria; alkaline conditions have the opposite effect. Methane production is quite variable depending on the amount and type of decomposable material in the landfill, moisture content, temperature, and resulting rate of microbial decomposition under anaerobic conditions. One estimate is a maximum gas production of 0.18 l/kg per day, or 5.4 l/kg dry weight per day; the effective life of a landfill for gas extraction at a rate of 20 ml/kg dry weight per day is approximately 17 years.[25] In another study, a rate of gas generation (CO_2 and CH_4) of 3.1 to 37 l/kg per year during the more active period of methane production, on the order of 10 years, was considered reasonable. This involved municipal refuse as received, with methane production at about 55 percent of the total gas.[26]

Methane is odorless, has a heat value of about 500 Btu/ft^3 compared to 1000 Btu for commercial gas, has a specific gravity less than air, and is nearly insoluble in water. The gases from landfills, after anaerobic conditions have been established, are quite variable, ranging from 50 to 60 percent methane (average 55 percent) and 40 to 50 percent carbon dioxide. Included are small amounts of nitrogen, oxygen, water, mercaptans (very odorous), and hydrocarbons. Hydrogen sulfide may also be released if large amounts of sulfates are in the

[24] W. H. Holmes, "Recovering Energy from Combustible Waste Materials," *Plant Eng.*, November 23, 1978, pp. 119–120.

[25] Floppe B. DeWalle, Edward S. K. Chian, and Edward Hammerberg, "Gas Production from Solid Waste in Landfills," *J. Environ. Eng. Div. ASCE*, June 1978, pp. 415–431, pp. 7–8.

[26] Robert K. Ham, "Predicting Gas Generation From Landfills," *Waste Age*, November 1979, pp. 50–57.

landfill. The presence of oxygen and nitrogen with methane gas would indicate the entrance of air into the landfill. This may be due to the rate at which methane is being pumped which, if not controlled to reduce or eliminate the entrance of oxygen and nitrogen, would slow down or stop methane production. In the early stages the landfill gases are primarily carbon dioxide with some methane. The carbon dioxide is heavier than air and can dissolve in water to form carbonic acid which is corrosive to minerals with which it comes into contact. Mercaptans, carbon dioxide, and water are usually taken out to upgrade the methane to "pipeline quality." Methane as it comes from a landfill is often very corrosive. Deep landfills, 30-ft or more deep, with a good cover are better methane producers. Gas can be extracted using plastic tube wells with perforations or well screens toward the bottom connected to a controlled vacuum pump.[27,28] The life of such a landfill is estimated at 12 to 20 years for California.[29] Actually, gas will be generated as long as biodegradable material remains.

It must be kept in mind that methane in the presence of air is explosive at concentrations between 5 and 15 percent. However a concentration above 15 percent is potentially hazardous as the gas may be quickly diluted by air movement to below 15 percent. Therefore gas monitoring and control must be included in the planning, design, and operation of a sanitary landfill. This is necessary to prevent the lateral migration of methane to buildings, tunnels, manholes, sewers, or other enclosed spaces, particularly if the surrounding soil is dry, well-drained, channeled or cracked and the landfill cover material is a tight soil which does not permit venting to the atmosphere. The control of methane migration is discussed under the Sanitary Landfill heading in this chapter.

Recovery of Used Oil

Large quantities of waste oil find their way into the environment. The means include spills, oiled roads, oil dumped in sewers and on the land, and oil deposited by motor vehicles. Used oils contain many toxic metals and additives which add to the pollution received by sources of drinking water, and aquatic life and terrestrial organisms. The lead content of oil is of particular concern. Much of the waste oil is used for dirt road dust control and stabilization. The collection and re-refining of waste oil for reuse as a lubricant has met with only limited success. Re-refined oil is so classified when it has had physical and chemical impurities removed and which by itself, or when blended with new oil or additives, is substantially equivalent or superior to new oil intended for the same purposes, as specified by the American Petroleum Institute.

Various methods are available for the treatment of oily wastewater and the removal of heavy metals and toxic substances. Methods for the disposal of oily wastes include land disposal in a diked area with periodic mixing of the wastes

[27] R. Patrick Caffrey and James Retzlaff, "Landfill Gas Collected by Negative Pressure System," *Pub. Works*, October 1979, pp. 64–65.

[28] Gregg Easterbrook, "Methane: Halting the Great Escape," *Waste Age*, October 1979, pp. 14–18.

[29] Gregg Easterbrook, "The Great Escape," *Waste Age*, September 1979, pp. 26–35.

with soil, depending on the bacterial action to break down the hydrocarbons. Incineration to recover the heat value, and incineration to destroy the oil and other organic matter is used in some cases. Scrubbers are needed to control air pollution.[30] Waste oil is also used as a fuel for heating, but air pollution controls are needed.

HAZARDOUS WASTES

Definition

The term "hazardous waste"[31] means a solid waste, or combination of solid wastes, which because of its quantity, concentration, or physical, chemical, or infectious characteristics may:

1. Cause, or significantly contribute to an increase in mortality or an increase in serious irreversible, or incapacitating reversible illness; or
2. Pose a substantial present or potential hazard to human health or the environment when improperly treated, stored, transported, or disposed of, or otherwise managed.

Hazardous wastes include chemical, biological, flammable, explosive, and radioactive substances. They may be in a solid, liquid, sludge, or gaseous state[32] and are further defined in various federal acts[33] designed to protect the public health and welfare including the land, air, and water resources.

> A waste is regarded as hazardous if it is lethal, nondegradable, persistent in the environment, can be biologically magnified (as in food chains), or otherwise causes or tends to cause detrimental cumulative effects.[34]

The USEPA lists four characteristics of hazardous wastes:[35]

1. *Ignitability*, which identifies wastes that pose a fire hazard during routine management. Fires not only present immediate dangers of heat and smoke but also can spread harmful particles (and gases) over wide areas.

[30] *Pollution Engineering Practice Handbook*, Paul N. Cheremisinoff and Richard A. Young, Eds, Ann Arbor Science, Ann Arbor, Mich., 1976, pp. 595–616.

[31] Public Law 94-580, October 21, 1976, The Solid Waste Disposal Act (Resource Conservation and Conservation Act of 1976), Sec. 1004 (4); also "Identification and Listing of Hazardous Wastes," *40 CFR 261*, May 19, 1980.

[32] Paul N. Cheremisinoff and William F. Holcomb, "Management of Hazardous and Toxic Wastes," *Pollut. Eng.*, April 1976, pp. 24–32.

[33] Water Pollution Control Act Amendments of 1972, The Hazardous Materials Transportation Act of 1974, Toxic Substances Control Act of 1976, as well as The Solid Waste Disposal Act as amended in 1976.

[34] *Cleaning Our Environment—A Chemical Perspective*, Am. Chem. Soc., Washington, D.C., 1978, p. 24.

[35] *Everybody's problem: hazardous waste*, USEPA, Office of Water & Waste Management, Washington, D.C. 20460, SW-826, 1980, pp. 12–13. See also "Identification and Listing of Hazardous Wastes," *40 CFR 261*, May 19, 1980, p. 33119.

2. *Corrosivity*, which identifies wastes requiring special containers because of their ability to corrode standard materials, or requiring segregation from other wastes because of their ability to dissolve toxic contaminants.
3. *Reactivity* (or explosiveness), which identifies wastes that, during routine management, tend to react spontaneously, to react vigorously with air or water, to be unstable to shock or heat, to generate toxic gases, or to explode.
4. *Toxicity*, which identifies wastes that, when improperly managed, may release toxicants in sufficient quantities to pose a substantial hazard to human health or the environment.

Legislation

The Resource Conservation and Recovery Act (RCRA) of 1976 expands the purposes of the Solid Waste Disposal Act. It promotes resource recovery and conservation and mandates government (federal and state) control of hazardous wastes from its point of generation to its point of ultimate disposal, including a manifest system. Legislation was prompted by the serious problems associated with the improper handling and disposal of hazardous wastes. The most common problem associated with the disposal of hazardous waste, in addition to public opposition, is groundwater pollution from lagoons, also leachate from landfills, dumps, sludge disposal, other land disposal systems, and spills and unauthorized dumping. This is followed by surface-water pollution resulting from unauthorized discharge or dumping into sewers and directly into surface waters. Air pollution, fires, explosions, volatilization, and direct contact poisoning are also reported as result of improper handling or disposal of hazardous wastes.*[36,37]

The Toxic Substances Control Act (TSCA) of 1976 regulates the production and use of chemical substances that may present an unreasonable risk of injury to health or environment. Manufacturers must give notice of plans to produce a new chemical or to market a significant new use for an old chemcial and may be required to provide and keep records and reports.

Priority Toxic Pollutants and Hazardous Wastes

Twenty-four toxic substances have been identified by the USEPA, the Consumer Products Safety Commission (CPSC), the FDA, and the Occupational Safety and Health Administration (OSHA) for *joint* attack. The National Institute for Occupational Safety and Health (NIOSH) is also concerned with the control of

*The Comprehensive Environmental Response, Compensation, and Liability Act of 1980, P.L.96-510, popularly known as the Superfund Act, provides funds for the cleanup of hazardous waste disposal sites and releases (spillage, discharges, injections, leachings, dumpings) and the recovery of money for releases that have damaged natural resources.

[36] *Environmental Quality*, The eighth annual report of the council on environmental quality, Sup. of Documents, GPO, Washington, D.C. 20402, 1977, p. 46.

[37] *Everybody's problem: hazardous waste*, USEPA, Office of Water & Waste Management, Washington, D.C., 20460, SW-826, 1980, pp. 1–8.

toxic substances. The substances include acrylonitrile, arsenic, asbestos, benzene, beryllium, cadmium, chlorinated solvents (trichloroethylene, perchloroethylene, methylchloroform, and chloroform), chlorofluorocarbons, chromates, coke oven emissions, diethylstilbestrol (DES), dibromochloropropane (DBCP), ethylene dibromide, ethylene oxide, lead, mercury and mercury compounds, nitrosamines, ozone, polybrominated biphenyls (PBBs), polychlorinated biphenyls (PCBs), radiation, sulfur dioxide, vinyl chloride and polyvinyl chloride, and toxic waste disposals that may enter the food chain.[38] The USEPA has listed 129 specific toxic pollutants for priority action. These have been divided into 10 categories as shown in Table 5-12. The list will no doubt be revised as new information is obtained.

The major generators of hazardous waste among 15 industries studied by the USEPA,[39] in million tons per year, are:

Primary metals	8.3	Textiles	1.8
Organic chemicals	6.7	Petroleum refining	1.8
Electroplating	5.3	Rubber and plastics	0.8
Inorganic chemicals	3.4	Miscel. (7 sectors)	0.7

Hazardous wastes that are characterized as ignitable, corrosive, explosive, or toxic should be removed from industrial wastes prior to discharge to a municipal sewer. Many toxic wastes upset biological wastewater treatment processes and are transferred to the sludge, adding to the disposal problem.

Treatment and Disposal

The *goal* for the management of hazardous wastes should be "zero discharge." Treatment and recovery should be given preference, with secure land burial or deep well disposal under carefully controlled conditions, where permitted, as a last resort. PCBs, PBBs, DBCP, DDT, and similar persistent long-lived chemicals are best destroyed in special incinerators at controlled temperatures, usually above 2000 to 2200°F dependent on residence time. Numerous techniques for the management of hazardous wastes have been in use and proposed, but all have their limitations.[32, 40–42] Because of the treatment and disposal complexities, and the social, economic, and environmental factors involved, regional processing centers could be advantageous to both the industry and public to help assure convenient and proper disposal of hazardous wastes.

[38]"4 U.S. Units Plan Fight On 24 Toxic Substances," *New York Times*, November 26, 1979, Washington, D.C., Nov. 25 (AP).

[39]*Solid Waste Facts*, (SW-694), USEPA, Washington, D.C., May 1978, p. 5.

[40]Gene Dallaire, "EPA's Hazardous waste program: will it save our groundwater?," *Civ. Eng. ASCE*, December 1978, pp. 39–45.

[41]Robert B. Pojasek, "Disposing of hazardous chemical wastes," *Environ. Sci. Technol.*, July 1979, pp. 810–814.

[42]*Technology for Managing Hazardous Wastes*, Rensselaer Polytechnic Institute, Troy, N.Y., September 1, 1979.

Table 5-12 The 129 Priority Toxic Pollutants

Pollutant	Characteristics	Sources	Remarks
Pesticides Generally chlorinated hydrocarbons	Readily assimilated by aquatic animals, fat soluble, concentrated through the food chain (biomagnified), persistent in soil and sediments	Direct application to farm- and forestlands, runoff from lawns and gardens, urban runoff, discharge in industrial wastewater	Several chlorinated hydrocarbon pesticides already restricted by USEPA; aldrin, dieldrin, DDT, DDD, endrin, heptachlor, lindane, and chlordane
Polychlorinated biphenyls (PCBs) Used in electrical capacitors and transformers, paints, plastics, insecticides, other industrial products	Readily assimilated by aquatic animals, fat soluble, subject to biomagnification, persistent, chemically similar to the chlorinated hydrocarbons	Municipal and industrial waste discharges disposed of in dumps and landfills	TSCA ban on production after 6/1/79 but will persist in sediments; restrictions on many freshwater fisheries as a result of PCB pollution (e.g., lower Hudson, upper Housatonic, parts of Lake Michigan)
Metals Antimony, arsenic, beryllium, cadmium, copper, lead, mercury, nickel, selenium, silver, thallium, and zinc	Not biodegradable, persistent in sediments, toxic in solution, subject to biomagnification	Industrial discharges, mining activity, urban runoff, erosion of metal-rich soil, certain agricultural uses (e.g., mercury as a fungicide)	
Asbestos	May cause cancer when inhaled, aquatic toxicity not well understood	Manufacture and use as a retardant, roofing material, brake lining, etc.; runoff from mining	
Cyanide	Variably persistent, inhibits oxygen metabolism	Wide variety of industrial uses	
Halogenated aliphatics Used in fire extinguishers, refrigerants, propellants, pesticides, solvents for oils and greases and in dry cleaning	Largest single class of "priority toxics," can cause damage to central nervous system and liver, not very persistent	Produced by chlorination of water, vaporization during use	Large volume industrial chemicals, widely dispersed, but less threat to the environment than persistent chemicals
Ethers Used mainly as solvents for polymer plastics	Potent carcinogen, aquatic toxicity and fate not well understood	Escape during production and use	Though some are volatile, ethers have been identified in some natural waters.

Phthalate esters Used chiefly in production of poly vinyl chloride and thermoplastics as plasticizers	Common aquatic pollutant, moderately toxic but teratogenic and mutagenic properties in low concentrations; aquatic invertebrates are particularly sensitive to toxic effects; persistent and can be biomagnified	Waste disposal vaporization during use (in nonplastics)
Monocyclic aromatics (excluding phenols, cresols and phthalates) Used in the manufacture of other chemicals, explosives, dyes, and pigments, and in solvents, fungicides, and herbicides	Central nervous system depressant; can damage liver and kidneys	Enters environment during production and by-product production states by direct volatilization; wastewater
Phenols Large volume industrial compounds used chiefly as chemical intermediates in the production of synthetic polymers, dyestuffs, pigments, pesticides, and herbicides	Toxicity increases with degree of chlorination of the phenolic molecule; very low concentrations can taint fish flesh and impart objectionable odor and taste to drinking water; difficult to remove from water by conventional treatment; carcinogenic in mice	Occur naturally in fossil fuels, wastewater from coking ovens, oil refineries, tar distillation plants, herbicide manufacturing, and plastic manufacturing; can all contain phenolic compounds
Polycyclic aromatic hydrocarbons Used as dyestuffs, chemical intermediates, pesticides, herbicides, motor fuels, and oils	Carcinogenic in animals and indirectly linked to cancer in humans; most work done on air pollution; more is needed on the aquatic toxicity of these compounds; not persistent and are biodegradable though bioaccumulation can occur	Fossil fuels (use, spills, and production), incomplete combustion of hydrocarbons
Nitrosamines Used in the production of organic chemicals and rubber; patents exist on processes using these compounds.	Tests on laboratory animals have shown the nitrosamines to be some of the most potent carcinogens.	Production and use can occur spontaneously in food cooking operations

Source: *Environmental Quality*, the ninth annual report of the Council on Environmental Quality, Sup. of Documents, GPO, Washington, D.C. December 1978, pp. 132–133.

As with most air, water, solid waste, and other pollution control activities, certain general and basic control principles can also be applied, as appropriate, to hazardous wastes. These include:

1. Elimination and reduction of waste at the source by prevention of leakage, by segregation of hazardous waste, and by process or materials change
2. Recovery, reuse, and recycling of wastes, including return to the manufacturer, energy recovery, and waste exchange among compatible industries
3. Concentration of waste by treatment—centrifugation, coagulation, sedimentation, filtration, flotation, surface impoundment, distillation, reverse osmosis, precipitation, solidification, encapsulation, evaporation, electrodialysis, absorption, or blending
4. Thermal decomposition—controlled incineration and proper disposal of residue; also ocean incineration
5. Chemical treatment—chemical oxidation, precipitation, reduction, neutralization, pyrolysis, chlorination, detoxification, ion exchange, or absorption
6. Burial in a secure landfill; storage or containment with proper monitoring and surveillance
7. Biological degradation; activated sludge, lagoon, or other biological treatment
8. Stabilization, solidification, or encapsulation
9. Deep well, mine, and ocean disposal under controlled conditions and if permitted; possibly composting and microwave decomposition; ocean disposal has been banned effective December 31, 1981, but this prohibition is being re-evaluated.

More details for some of the processes are given below as they apply to pesticides and other hazardous wastes. The methods available are not always completely effective and must be tailored to specific contaminants. Additional information as applied to industrial wastewaters is found in pages 520 and 578, and in Chapter 7 under waste disposal of radioactive wastes.

Thermal decomposition of pesticides requires temperatures of 1650 to 1830° F for varying periods of time. With very few exceptions this method appears adequate to degrade 99 percent or more of most commercial (organic) pesticidal formulations.[43] Volatile organic substances could be incinerated at lower temperatures. Care must be used to hold the pesticides at the required temperature long enough to decompose them, *as well as other toxic substances formed*, and to remove air pollutants before the gases are discharged to the atmosphere. A wet scrubber and filtration through a porous clay bed and carbon filter, with lagoon treatment of the wastewater, are suggested. Satisfactory methods are being developed. Mercury, arsenic, cadmium, chromium, lead, and similar toxic compounds should not be incinerated unless special residue handling, emission control, and disposal facilities are available. Arsenic in combination with lead or sodium cannot be adequately treated with present technology. Long-term secure storage is a temporary alternative.

PCBs are volatile, but only slightly soluble in water. They are extremely toxic and persistent in the environment, as are PBBs used in pesticides and plasticizers. PCBs are used in liquid electric insulators, transformers and capacitors,

[43]M. V. Kennedy, B. V. Stojanovic, and F. L. Shuman, Jr., "Chemical and Thermal Methods for Disposal of Pesticides," Mississippi Agricultural Experiment Station, Mississippi State University, State College, Miss., 1969.

hydraulic systems and other equipment, plastics, vinyl coated paper, waxes, and other uses. PCBs in landfills can volatilize and escape with methane. Some bacteria can apparently break down PCBs. PCBs can be destroyed by incineration at 2000°F and 3 percent excess oxygen for a 2-second contact time, or at 2700°F and 2 percent excess oxygen for a 1.5-second contact time. Incinerators require scrubbers to remove the hydrogen chloride formed.[44] This is a costly process.

The TSCA required that the manufacture of PCBs be phased out in 1979 but the problem of disposal of the existing pieces of equipment and other materials containing PCBs will remain for a long time.

Chemical processes must be tailored to each specific material. Neutralization is feasible with most of the organophosphate and carbamate insecticides, but not with the chlorinated hydrocarbons. Some pesticides are destroyed by nitric acid or sulfuric acid, whereas others are destroyed by sodium hydroxide or ammonium hydroxide. Still others are destroyed by chlorine compounds, peroxides, or other types of active chemicals. Calcium hypochlorite seems to have the broadest application. Strong acid or alkaline hydrolysis does not give complete treatment.[44] Ion exchange, oxidation, and precipitation are also suitable to remove heavy metals. Other processes include carbon or resin adsorption, distillation, reverse osmosis, ultrafiltration, flocculation, sedimentation, and centrifuging.

Burial introduces possible contamination of surface waters and groundwaters from leaching and runoff of the toxic waste. Careful shallow burial, with 18 in. of earth, of *small* quantities (less than 10 lb active ingredient dry material or 5 gal liquid) of pesticides in a clay soil may be acceptable but regulatory agencies prefer or require return to the manufacturer or collection and disposal by a licensed contractor. The location should be well above groundwater, downgrade, several hundred feet from any sources of water supply, and isolated to protect animals and children. The site should be fenced and have a permanent marker with the date, amount, and name of pesticide stated. *Do not* bury pesticides in sandy soils because the risk of leaching the pesticide into surface water or groundwater is too high. Larger quantities should be returned to the supplier or be disposed of in a high temperature incinerator approved for the purpose.

Before disposal to the soil, information and evaluation is needed of data on the soil permeability, natural neutralization, binding or detoxification, and bacteriological degradation of the wastes in the soil. Specially designed landfills—monitored and secured, with an impervious lining and leachate collection system,—may be used for certain classes of wastes, if approved by the regulatory authority. In general, hazardous wastes should be reused or rendered innocuous rather than stored in secure landfills where the waste remains for the foreseeable future as a problem and threat to man and the environment.

Biological or natural degradation of some shortlived materials is satisfactory,

[44]D. L. Russell, "PCBs: The Problem surrounding us and what must be done," *Pollut. Eng.*, August 1977, pp. 34–35.

while for other more persistent materials it is too slow. For example, the half-life of the metals class of pesticides such as lead, cadmium, and arsenic may be 10 to 30 years and herbicides such as 2, 4-D and benzoic acid herbicides 1 to 12 or 24 months. The organochlorine class of pesticides such as DDT may retain their chemical identity and biological activity in soil for more than a decade, whereas the organophosphorus class of insecticides such as methyl parathion may lose their identity and biological activity in a few days to a few weeks following application.[45] The persistence is therefore variable, and depends on the particular chemical's reactivity, water solubility, and susceptibility to biochemical degradation.

Biological processes to destroy organic compounds include activated sludge treatment, composting, trickling filters, and land treatment disposal.

Any organic compound that is synthesized by microorganisms is capable of being biodegraded in time by microorganisms. The formation of an organic compound (synthesis) and its simplification (biodegradation) is accelerated by the catalytic action of other organic substances in the living system, known as enzymes. Many carbon compounds, such as coal and crude oil, or manufactured chemicals, such as DDT and other hydrocarbons, require catalysis by suitable microbial enzymes or environmental adjustment to be readily degraded. Such adjustment includes exposure to sunlight, oxygen, suitable temperature, water or other solvent which weakens or makes more susceptible the carbon compound to microbial attack. Dioxin for example is reported to be destroyed in two days on leaves and soil in sunlight, but exposure to the chemical before degradation can take place; or in the absence of exposure to sunlight the chemical can be deadly.

Where possible and where facilities are available, surplus pesticides and other hazardous chemicals should be disposed of as described above. Small quantities of combustible containers or packaging may be incinerated or buried where permitted. Pesticide containers should first be rinsed at least three times, and the rinse returned to the spray tank for reuse, before disposal. Where these methods are impractical, large containers of pesticides (more than 10 lb dry active ingredient or 5 gal liquid) should be returned to the manufacturer or supplier. The alternative is to store these materials in a locked building or room until an acceptable method of disposal becomes available, with special precautions and procedures established in case of fires or explosions. The building should be clearly marked with a sign such as "DANGER," "POISON," "PESTICIDE STORAGE," or other appropriate notice.

Stabilization, according to Pojasek,[46] converts the identified toxic components

> to a chemical form that is more resistant to leaching in the ultimate disposal site. Stabilization also makes the waste compatible with the solidification step.

[45]Dale R. Bottrell, *Council on Environmental Quality Integrated Pest Management*, GPO, Washington, D.C. 20402, December 1979, p. 6.

[46]Robert B. Pojasek, "Stabilization, solidification of hazardous wastes," *Environ. Sci. Technol.*, Vol. 12, No. 4, April 1978, pp. 382–386.

The waste must have its pH adjusted for proper stabilization and solidification to maintain the integrity of the solid product. Cementation is a common stabilization technique. Codisposal of selected wastes can stabilize each other without the addition of other chemicals. Waste containing heavy metals codisposed with waste containing sulfur components is an example. Some hazardous wastes do not require stabilization, as determined by appropriate elutriation tests.

Solidification of hazardous wastes can be accomplished by chemical fixation and encapsulation. Materials used for solidification include Portland cement, urea-formaldehyde, asphalt, pozzolanic cements, polybutadiene, silicates, sulfur foams, soil-binding agents, and ion-exchange resins. In encapsulation the agent, such as asphalt, surrounds the waste particles; in chemical fixation a chemical reaction such as ion exchange is involved between the waste and the solidification agent. In all cases a durable product must be formed.

The adequacy of the stabilization and solidification treatment processes, and suitability of the product for land disposal can be determined by various methods including elutriate testing, leach testing, and resistance to environmental exposure. Better, standardized testing procedures are needed.[46] The U.S. Army Environmental Hygiene Agency selected a Portland cement-sodium silicate fixation method as the final treatment for small volumes of hazardous wastes usually containing toxic heavy metals. This method is not effective in binding some organic contaminants.[47] Only a small quantity of organics can destroy the solidification process.

Storage of Hazardous Wastes and the Use of Liners

Some hazardous wastes, both solid and liquid, may be stored in clay, asphalt, concrete, soil cement, or (sodium) bentonite-soil lined basins, or in polymeric membrane lined basins pending a decision on best methods for treatment and disposal. Membrane linings are made of special rubber, polyethylene, polyolefin, polychloroprene, and polyvinyl chloride.[48] All are usually suitable for sewage and biodegradable industrial wastes. However, solvents, strong caustics, and brines could damage clay or soil-based linings. Benzene and toluene, for example, are not contained by a clay liner; but when mixed with a small proportion of water and placed in a landfill, clay remains a good barrier. These chemicals, including pesticides, are best destroyed by controlled incineration. Petroleum-based organic wastes could damage some polymeric membranes and asphaltic materials. In view of the many limitations, *wastes and lining materials*

[47] H. Gladys Swope, "Hazardous Wastes Discussed At International Meeting of Chemical Societies," *Ind. Wastes*, September/October 1979, pp. 26–28. See also *Leachate Testing of Hazardous Chemicals From Stabilized Automotive Wastes*, National Sanitation Foundation, Ann Arbor, Mich. 48106, January 1979.

[48] *Draft Environmental Impact Statement on the Proposed Guidelines for the Landfill Disposal of Solid Wastes*, USEPA, Washington, D.C., 1979, pp. 175–181.

should be tested for compatibility before use. In any case, all liners should be carefully placed on well compacted subbases and, in addition, all basins storing hazardous wastes should incorporate a groundwater monitoring and surveillance system[49] including a leachate collection system and peripheral well monitoring. Two layers of linings with intermediate collection systems to collect possible leachate percolation may be required in some instances.[50]

INCINERATION

General

A properly designed and controlled incinerator is satisfactory for burning combustible refuse, provided air pollution standards can be met. Continuous operation six or seven days a week and a high controlled temperature are needed for efficiency, prevention of excessive air pollution, and odor control.

An operating design temperature range of 1500 to 1800°F is generally recommended. Excessively high temperatures and extreme variations cause cracking and spalling, with rapid deterioration of fire tile and brick linings (refractories). Batch feed or one-shift operation promote spalling and loosening of tile linings. Other types of lining and design may permit higher operating temperatures. Actually, temperature in the furnace may range from 2100 to 2500°F. When the gases leave the combustion chamber the temperature should be between 1400°F and 1800°F and the gas entering the stack 1000°F or less. The temperature will have to be lowered to 450 to 500°F before the gas is filtered or to 600°F or less if electrical precipitators are used. At a temperature of 1200 to 2000°F or higher, depending on temperature and residence time, oxides of nitrogen are formed which contribute to air pollution.

Refuse storage bins or pits providing at least 3 days' storage are necessary to provide sufficient refuse for a continuous period of operation. Incineration is not generally recommended for small towns, villages, apartment buildings, schools, institutions, camps, and hotels unless good design and supervision can be assured and cost is not a factor. Past experience indicates that incinerators generally are not economically feasible for communities of less than 50,000 to 100,000 population, but this can change with the use of modular units incorporating heat recovery.

Supplemental air pollution control equipment is required on all incinerators. Additional fuel is needed when 30 percent or less of the refuse is rubbish or when the refuse contains more than 50 percent moisture; however, this is not a problem with ordinary refuse in the more developed areas of the world.

[49] Chris Parks, "Getting A Line On Waste Pond Lining," *Water Sewage Works*, October 1979, p. 36.
[50] "Chemical Wastes Landfill Model," *Environ. Midwest*, January 1978, p. 12.

Nonincinerable Waste and Residue

Nonincinerable and bulky refuse may amount to 20 to 30 percent of the total weight of refuse collected (15 to 30 percent by volume). The residue after burning is 15 to 25 percent of the original weight (10 to 20 percent by volume). Thus 25 to 50 percent of the total waste by volume remains for disposal by landfill. Table 5-13 shows the physical and chemical characteristics of incinerator solid waste. Food for rats and flies all too frequently is still in the residue, although proper design and operation can greatly minimize this problem. In addition, there is always the problem of what to do with nonburnable refuse, such as street cleanings, construction and demolition wastes, abandoned automobiles, industrial wastes, junk, trees and trimmings, and so on. Burial of the incinerator residue and noncombustible refuse in a sanitary landfill can eliminate the objections mentioned. Where incinerators are carefully operated, the residue can be used as cover material. A land area (volume) of about one-quarter to one-third that required for sanitary landfill should be set aside for the burial of bulky solid wastes and incinerator residue, adequate for about 20 or more years in the future, based on population projections and industrial development. Salvage operations carried on at the incinerator are often economical and make possible elimination of the annoying on-site separation of refuse when this practice is still required.

The cost of incineration can vary greatly dependent on the factors included in

Table 5-13 Physical and Chemical Characteristics of Incinerator Solid Waste[a]

Constituents	Percent by Weight (as Received)
Proximate analysis	
Moisture	15 to 35
Volatile matter	50 to 65
Fixed carbon	3 to 9
Noncombustibles	15 to 25
Ultimate analysis	
Moisture	15 to 35
Carbon	15 to 30
Oxygen	12 to 24
Hydrogen	2 to 5
Nitrogen	0.2 to 1.0
Sulfur	0.02 to 0.1
Noncombustibles	15 to 25
Higher heating value	3,000 to 6,000 (Btu/lb as received)

Source: J. DeMarco, D. J. Keller, J. Leckman, and J. L. Newton, *Incinerator Guidelines 1969*, PHS Pub. No. 2012, DHEW, Washington, D.C., 1969, p. 6.

[a]Principally residential-commercial waste excluding bulky waste.

the cost and the method of reporting—the operation, maintenance, air pollution control requirements, and amortization. Some comparative costs are given in Table 5-11.

Site Selection, Plant Layout, and Building Design[51]

It is extremely important that a careful investigation be made of the social, physical, and economic factors involved when incineration is proposed. Some of the major factors are the following.

1. Public acceptance in relation to the surrounding land use and precautions to be taken in location and design to offset public objections should be considerations. A location near the sewage treatment plant, for example, may not meet with much objection. Heat utilization for sludge drying or burning and use of treated wastewater for cooling are possibilities.
2. Site suitability in reference to foundation requirements, prevailing winds, topography, surface water and groundwater, floods, adjacent land uses, and availability of utilities should be considered. A central location to the source of wastes for minimum haul distance, smooth movement of traffic in and out of the site, and location readily accessible from major highways without interrupting traffic are important considerations.
3. Plant layout should be arranged to facilitate tasks to be performed and provide for adequate space, one-way traffic, parking area, paving, drainage, and equipment maintenance and storage.
4. Building design should be attractive and provide adequate toilets, showers, locker room, and lunchroom. A control room, administrative offices, weighmaster office, maintenance and repair shops, and laboratory should be included. Adequate lighting contributes to attractiveness, cleanliness, and operating efficiency. Good landscaping will promote public acceptance.
5. Also to be evaluated are the availability and cost of providing electric power, water supply, sanitary sewers, and pretreatment required before plant wastewater can be discharged to the sewer, and availability of storm sewers, telephone, and fuels.
6. The proposed method and cost of handling bulky and nonincinerable wastes should be taken into consideration when incineration is proposed. Also to be determined are the location and size of the sanitary landfill and its ability to receive incinerator residue as well as the bulky and nonincinerable solid wastes.

The reader is referred to Chapter 2 for the broad aspects of community and facility planning and environmental impact analysis, and to Figure 2-3.

Incinerator Design

The basic components of an incinerator are shown in Figure 5-8. Incinerator design parameters are summarized in Table 5-14.

Design capacity. Incinerators are rated in terms of tons of burnable or incinerable waste per day. For example, an incinerator having a furnace capacity of 600 tons/day can theoretically handle 600 tons in 24 hr with three-shift

[51]Jack DeMarco, Daniel J. Keller, Jerold Leckman, and James L. Newton, *Incinerator Guidelines—1969*, PHS Pub. No. 2012, DHEW Washington, D.C., 1969, 98 pp.

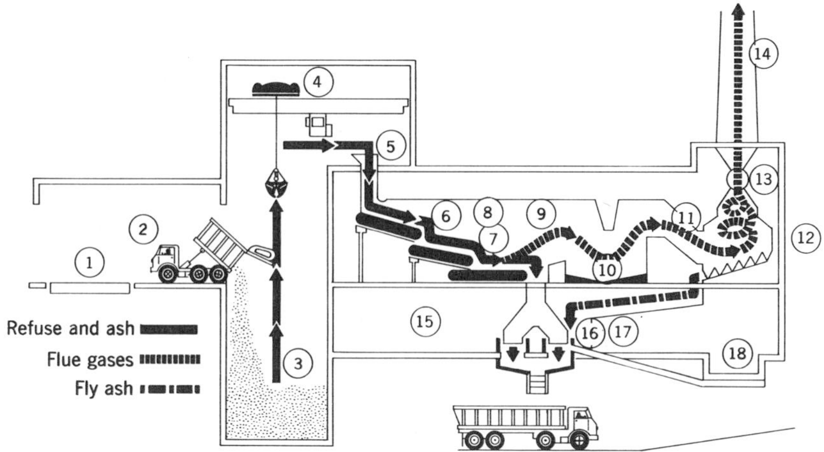

Figure 5-8 Basic incinerator design. (1) Scales; (2) Tipping floor; (3) Storage bin (Pit); (4) Bridge crane; (5) Charging hopper; (6) Drying grates; (7) Burning grates; (8) Primary combustion chamber; (9) Secondary combustion chamber; (10) Spray chamber; (11) Breeching; (12) Cyclone dust collector; (13) Induced draft fan; (14) Stack; (15) Garage—storage; (16) Ash conveyors; (17) Forced draft fan; (18) Fly ash settling chamber. (From *Solid Waste Management, 5, design and operation*, prepared by National Association of Counties Research Foundation for PHS, DHEW, Washington, D.C.)

operation; 400 tons in 16 hr with two-shift operation; and 200 tons in 8 hr with one-shift operation. Hence if 400 tons of incinerable wastes collected per day are to be incinerated in 8 hr, an incinerator with a rated capacity of 1200 tons/day will be required plus a 15-percent downtime allowance for repairs. Consideration must also be given in determining design capacity, to daily and seasonal variations, which will range from 85 to 115 percent of the median.[52]

The least expensive operation for a particular community would be determined by comparing the total annual cost, including operating costs and fixed charges on the capital outlay, for each method. It will generally be found that, for large cities, three-shift operation will be the least expensive. The two- or one-shift operation will be somewhat cheaper for the smaller community. The relative cost of maintenance, however, will be higher and the efficiency poor because of starting and shutting down of the furnace, with accompanying refractory brick spalling due to differential expansion and air pollution from fly ash.

Chimney. High chimneys, 150 to 200 ft above ground level, are usually constructed to provide natural draft and air supply for combustion. Stack heights of 300 to 600 ft are not uncommon. Discharge of gases at these heights also facilitates dilution and dispersal of the gases. In some designs short stacks are used for aesthetic reasons, and the equivalent effective stack height is obtained by induced draft. Meteorological conditions, topography, adjacent land use, air pollution standards, and effective stack height should govern. See Chapter 6.

[52]"Municipal Incineration of Refuse: Foreword and Introduction," Progress Report of the Committee on Municipal Refuse Practices, *J. Sanit. Eng. Div. ASCE*, 13–26 (June 1964).

Table 5-14 Incinerator Design Parameters

Item	Design Factor
Storage pit	Volume of 1 to $1\frac{1}{2}$ times the daily rated capacity; not less than 3 days' refuse collection recommended.
Grate loading	15 to 25 lb/ft^2/hr burning rate (commercial type); 50 to 70 lb/ft^2/hr burning rate, 55 lb average (other).
Grate area	Lb/hr solid waste to be burned ÷ lb/ft^2/hr solid waste the grates are capable of burning.
Combustion chamber, total	35 to 40 ft^3 (total)/ton rated capacity.
Air pollution control	Depending on local air pollution control regulations, 0.5 lb of dust/1000 lb of flue gas adjusted to 50% excess air or less. Dust emission can be reduced to less than 0.2 lb/1000 lb of flue gas.
Secondary, or combustion, chamber	Gas velocity of 10 to 40 fps.
Subsidence chamber	Gas velocity 5 to 10 fps.
Breeching	Gas velocity 20 to 40 fps.
Stack	Draft 2 to 4 in. water; gas velocity 25 to 50 fps.
Fuel value of refuse	9,000 to 10,000 Btu/lb of dry combustible solids or 3,000 to 6,000 Btu/lb of ordinary refuse (± 30% moisture). Determine Btu value for each community.
Noncombustible refuse	20 to 30% of refuse collected, by weight.
(Nonincinerable, bulky)	15 to 30% of refuse collected, by volume.
Residue	15 to 25% of original weight and 10 to 20% of original volume.
Operating temperature	1400 to 1800°F normal, 1700 to 1800°F optimum. Higher temperature recommended by some designers using special firewalls.
Supplemental fuel	When rubbish is less than 30%, or when refuse contains more than 50% mositure.
Design life	30 yr for plant; 15 yr for equipment.
Downtime for repairs	15%.
Process water	1,000 to 2,000 gal/ton solid waste processed. For residue quenching, ash conveying, wetted baffle dust collection. Can be reduced 50 to 80 percent with water treatment and recirculation. Residue is very abrasive and quench water is very corrosive.

For specific design details see *I.I.A. Incinerator Standards*, Incinerator Institute of America, New York, and standard texts.

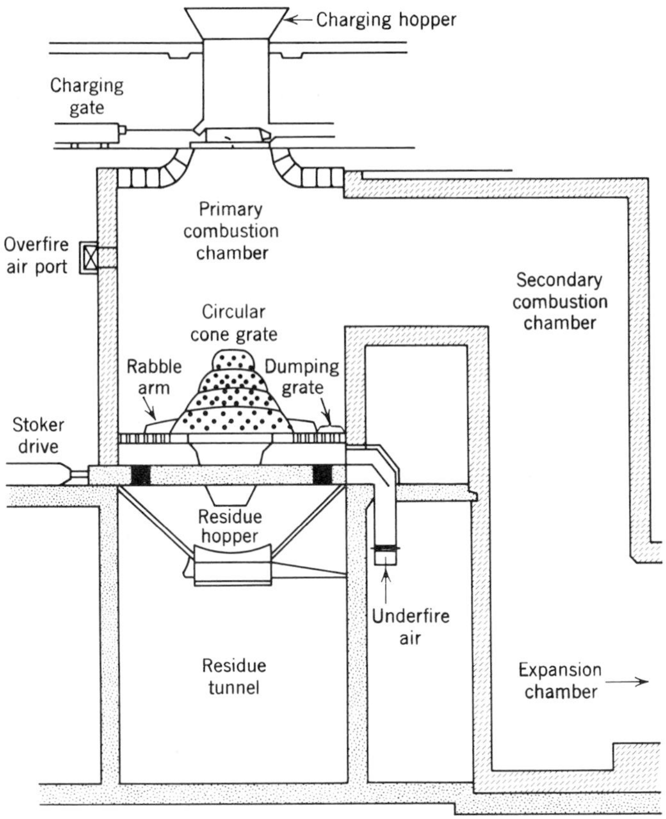

Figure 5-9 Vertical circular furnace. (From J. DeMarco, D. J. Keller, J. Leckman, and J. L. Newton, *Incinerator Guidelines 1969*, U.S. PHS Pub. 2012, DHEW, Washington, D.C., p. 27.)

Types of Furnaces

The common types of furnaces are the rectangular furnace (Figure 5-10), the vertical circular furnace (Figure 5-9), the rotary kiln furnace (Figure 5-11), and the rectangular furnace with water walls (Figure 5-12*a*). In the rectangular furnace two or more grates are arranged in tiers. The vertical circular furnace is fed from the top directly onto a circular grate. The rotary kiln furnace incorporates a drying grate ahead of a rotary drum or kiln where burning is completed. Water wall furnaces substitute water-cooled tubes for the exposed furnace walls and arches. Other types of furnaces are available. All furnaces should be designed for continuous feed. Reciprocating or moving grates are the most common.

Combustion essentials. The three essentials for combustion are time, temperature, and turbulence. There must be sufficient time to drive out the moisture,

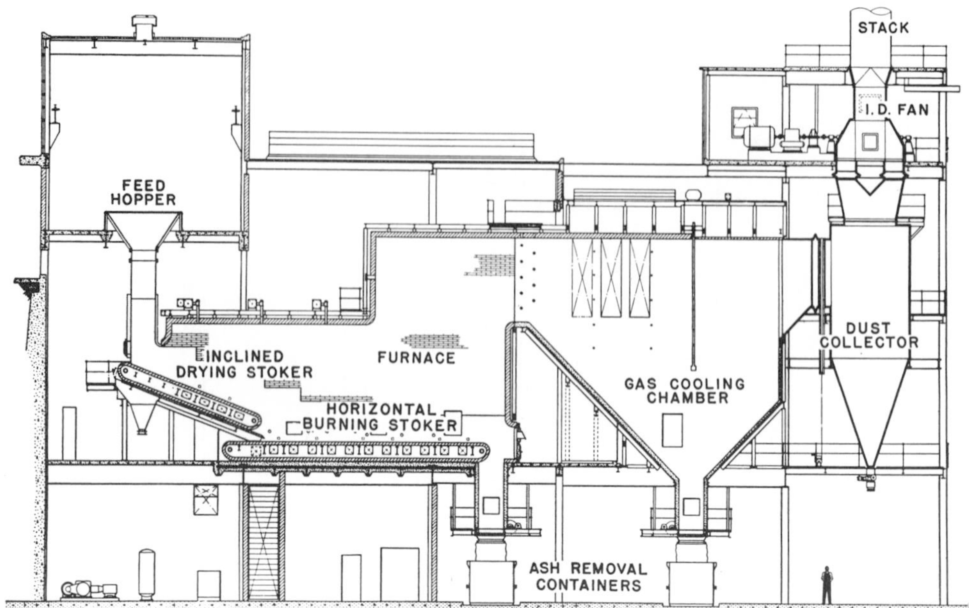

Figure 5-10 Modern continuous-feed, refractory-lined incinerator with traveling-grate stokers (rectangular type). (From Richard C. Corey, *Principles and Practices of Incineration*, Wiley-Interscience, New York, 1969, p. 183.)

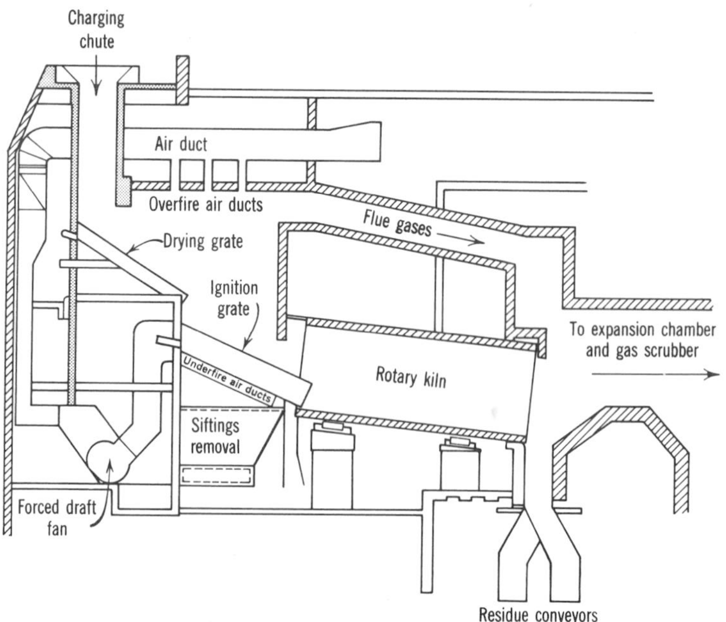

Figure 5-11 Rotary kiln furnace. (From J. DeMarco, D. J. Keller, J. Leckman, and J. L. Newton, *Incinerator Guidelines 1969*, PHS Pub. 2012, DHEW, Washington, D.C., p. 30.)

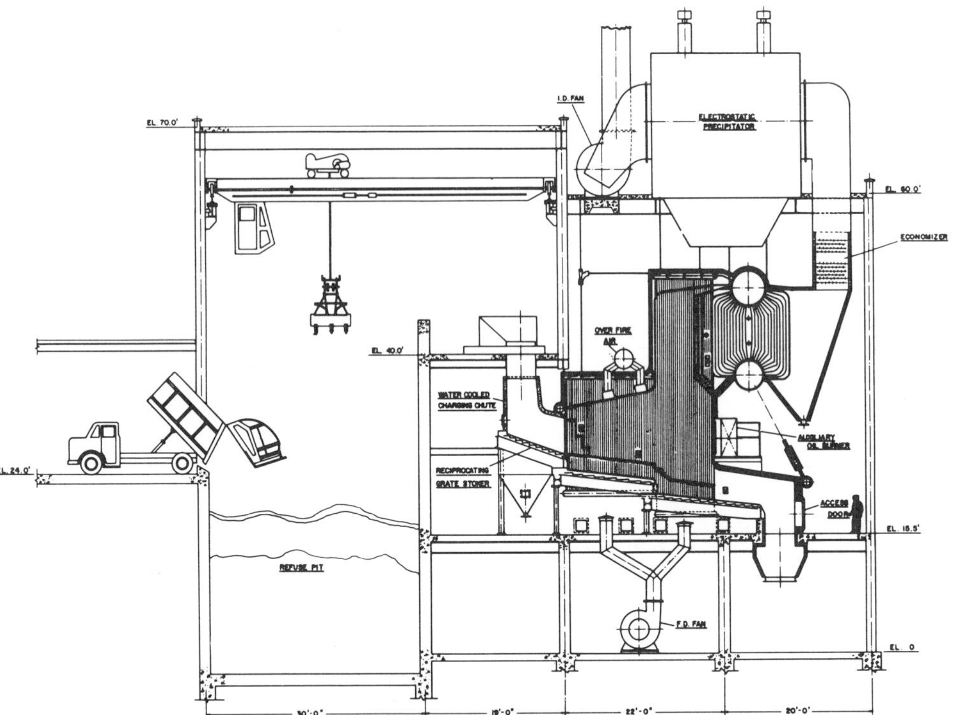

Figure 5-12*a* Continuous-feed incinerator with rocking-grate stoker and water walls. (From Richard C. Corey, *Principles and Practices of Incineration*, Wiley-Interscience, New York, 1969, p. 186.)

the temperature must be raised to the ignition point, and there must be sufficient turbulence to ensure mixing of the gases formed with enough air to completely burn the volatile combustible matter and suspended particulates. The combustion process involves first, drying, volatilization, and ignition of the solid waste; second, combustion of unburned furnace gases, elimination of odors, and combustion of carbon suspended in the gases. This second step requires a high temperature, at least 1400 to 1800°F, sufficient air, and mixing of the gas stream to give it turbulence until burning is completed.

Incinerators are of the batch- or intermittent-feed-type and the continuous-feed-type. Air pollution and odors are associated with the first type due to intermittent charging and changes in temperature and air supply. A continuous feed furnace permits addition of well mixed refuse at a uniform rate with controlled temperature and air supply.

Furnace walls. Modern furnace walls are usually lined with tile or have water walls. With tile refractories, repairs can be readily made without the need for expensive and time-consuming rebuilding of entire solid brick walls found in old plants. Special plastic or precast refractories can be used for major or minor repairs. Water walls in a furnace actually consist of water-cooled tubes that serve as heat exchangers, thereby reducing the outlet gas temperature and simplifying

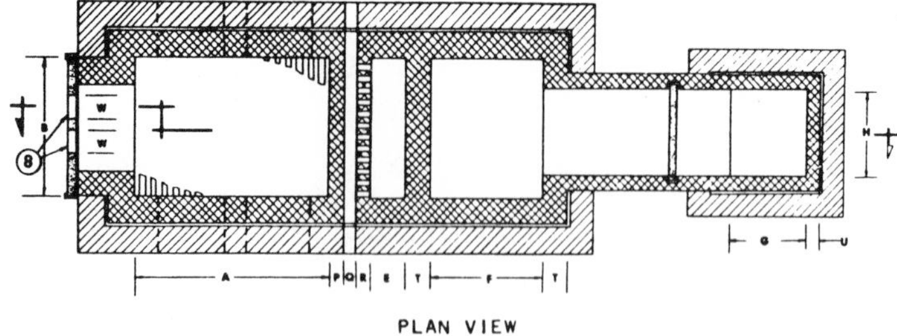

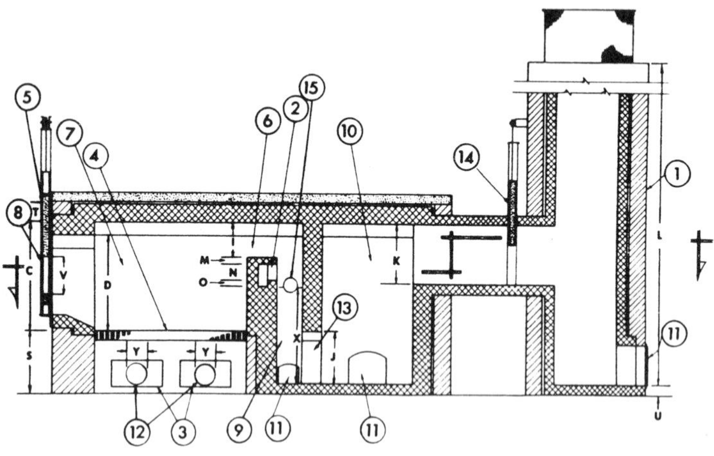

Figure 5-12*b* Design standards for multiple-chamber, in-line incinerators. (Source: *Air Pollution Engineering Manual*, Second Edition, Compiled and Edited by John A. Danielson, USEPA, Office of Air Quality Planning and Standards, Research Triangle Park, N.C. 27711, May 1973, p. 449.) (1) Stack; (2) Secondary air ports; (3) Ash pit cleanout doors; (4) Grates; (5) Charging door; (6) Flame port; (7) Ignition chamber; (8) Overfire air ports; (9) Mixing chamber; (10) Combustion chamber; (11) Cleanout doors; (12) Underfire air ports; (13) Curtain wall ports; (14) Damper: (15) Gas burners.

	Length (in.)																								
	A	B	C	D	E	F	G	H	I	J	K	L (ft)	M	N	O	P	Q	R	S	T	U	V	W	X	Y
Size of incinerator (lb/hr)																									
750	85½	49½	51½	45	15¾	54	27	27	9½	24	18	32	4½	5	7½	9	2½	2½	30	9	4½	5	11	51	7
1000	94½	54	54	47½	18	63	31½	31½	11	29	22½	35	4½	5	10	9	2½	2½	30	9	4½	7	12	52	8
1500	99	76½	65	55	18	72	36	36	12½	32	27	38	4½	5	7½	9	4½	4½	30	9	4½	8	14	61½	9
2000	108	90	69½	57½	22½	79½	40½	40½	15	36	31½	40	4½	5	10	9	4½	4½	30	9	4½	9	15	63½	10

dust collection. The tubes also cover and protect exposed furnace walls and arches. Less air is required: 100 to 200 percent excess air for refractory walls compared to less than 80 percent for water walls. External pitting of the water-cooled tubes may occur if the water temperature drops below 300°F due to condensation of the corrosive gases. Internal tube corrosion must also be prevented by recirculation of conditioned water.

Control of Incineration

The poor image that incineration has in the eyes of many people is due largely to the failure to control operation, with resultant destruction of the equipment and air pollution. A properly designed and operated incinerator requires control instrumentation for temperature, draft pressures, smoke emission, weights of solid wastes coming in and leaving the plant, and air pollution control equipment.

Temperature. It is necessary for control purposes to monitor the incoming air and gases leaving the combustion chamber at the settling chamber outlet, at the cooling chamber outlet, at the dust collector inlet and outlet, and the stack temperature. Furnace temperature can be controlled by adjusting the amount of overfire or underfire air. The temperature of the gases leaving the furnace is reduced by spraying with water (causes a white stack plume unless the flue gas is reheated before discharge), by dilution with cool air (high equipment cost to handle large volumes of diluted gases), or by passing through heat exchangers (ready market for heat, steam, or high-temperature water needed). Gas scrubbers using water sprays can be used to cool effluent gas so that an induced-draft fan can be used to reduce the chimney height; large particulates can also be removed.

Draft pressure is needed to control the induced-draft fan and the chimney draft. Measurements should be made at the underfire air duct, overfire air duct, stoker compartment, sidewall air duct, sidewall low furnace outlet, dust collector inlet and outlet, and induced fan inlet. Control of underfire air can provide more complete combustion with less fly-ash carryover up the stack.

Smoke density. The smoke emission can be controlled by continuous measurement of the particulate density in the exhaust gas. A photoelectric pickup of light across the gas duct is used, preferably located between the particulate collector and the induced fan duct.

Weigh station. Platform scales to weigh and record the incoming solid waste and outgoing incinerator residue, fly ash, siftings, and other materials are generally required.

Instrumentation should include devices to keep record of overfire and underfire air flow rates; temperatures and pressures in the furnace, along gas passages, in the particulate collectors, and in the stack; electical power and water use; and grate speed.

Odor control requires complete combustion of hydrocarbons, that is, excess air and a retention time of 0.5 seconds at 1500° F (above 1400° F at the exit of the furnace). Adequate dilution of gases leaving the stack by an effective stack height (actual stack height plus plume rise) is another possible method for odor control, but its effectiveness is related to meteorological conditions and persistence of the odors. Wet scrubbers can also be used to absorb odors while removing particulates.

Other gaseous emissions in addition to carbon dioxide, water vapor, and oxygen emissions include sulfur oxides, nitrogen oxides, carbon monoxide, and hydrogen chloride. Hydrogen chloride is released when plastic polyvinyl chloride is burned. The hydrogen chloride can cause corrosion of air pollution control equipment. Wet scrubbers are believed to be effective in removing the hydrogen chloride gas as well as fly ash.

Particulate emissions can be controlled by settling chambers, wetted baffle spray system, cyclones, wet scrubbers, electrostatic precipitators, and fabric filters. Their efficiencies and other details are discussed in Chapter 6. Apparently only wet scrubbers, electrostatic precipitators and bag filters can meet an air pollution code requirement of less than 0.5 lb of particulates per 1000 lb of flue gas at 50 percent excess air. Cyclones in combination with other devices might approach the standard.

On-Site Incineration[53]

When possible, a large municipal incinerator should be used in preference to a small on-site incinerator. Better operation at lower cost with less air pollution can usually be expected. However, on-site incinerators are used in homes, apartment houses, hospitals, schools, and commercial and industrial establishments. In recent years their use is being severely limited by air pollution control requirements. In Los Angeles, on-site incinerators are prohibited. In the City of New York, their use was required in buildings with more than four dwelling units, but strict air pollution standards make incinerators feasible only in large structures. Chutes and compaction units are finding application.

Residential incinerators need supplementary fuel to burn domestic garbage satisfactorily. After-burners are used to obtain smokeless and odorless operation. They are not entirely reliable in view of variable design, operation, and maintenance. Convenience, cleanliness, and reduced storage facilities make on-site units attractive.

Multiple-dwelling incinerators. In the single-flue type the refuse drops directly into the furnace. If a double flue is provided, the flue gases are separated from the charging duct. In a chute-fed incinerator, the refuse is dumped in a chute through hopper doors on each floor and is collected in a basement bin. The refuse is then fed into the incinerator mechanically or by hand. Auxiliary burners are needed in all types to maintain adequate temperature and attain complete combustion. Existing incinerators, unless designed to meet present-day standards, require major alterations.

Commercial and industrial incinerators are of the retort and in-line multiple-chamber types. See Figure 5-12*b*. The single-chamber incinerator is unsatisfactory. Auxiliary heat is usually needed to compensate for marginal or poor

[53]Richard C. Corey, *Principles and Practices of Incineration*, Wiley-Interscience, New York, 1969.

operation and for the burning of certain wastes. Some units need to be redesigned in order to meet modern air pollution control standards. The controlled-air incinerator with a waste-heat boiler for energy recovery can overcome most of the deficiencies.

The incineration of hazardous or toxic solid and liquid industrial wastes calls for special design and carefully controlled operation, with attention to permissible stack emissions, residue handling, and protection of the operator. See Hazardous Wastes, this chapter.

Controlled Air Combustion

Incorporation of an afterburner in a secondary burning chamber solves the black smoke, particulate, and gaseous pollution associated with the old conventional incinerator by maintaining combustion temperature at 1600 to 1800°F. Two controlled air combustion incinerator designs and operation methods are used to control combustion: *starved air* and *excess air*. In the starved air incinerator, just enough air, as calculated, is admitted in the primary chamber to provide (theoretically) the oxygen necessary for complete combustion of the particular waste being burned. This is known as the stoichiometric air requirement, which permits a controlled rate of burning. The gases then pass to the secondary chamber where a small amount of auxiliary fuel may be required to complete the combustion, although most of the fuel would come from the off gases from the primary chamber. In the excess air incinerator, more air than is theoretically needed for combustion in the primary chamber is admitted, but is controlled by temperature sensing. The primary chamber is larger than for the starved air incinerator and an afterburner is used to minimize emissions.

Modular combustion units are available for capacities of less than 700 lb/hr to 250 tpd with 70 to 200 tpd the most common. These units may be used for the incineration of municipal wastes, hospital wastes, commercial wastes, and industrial wastes. Volume reduction of 80 to 90 percent and energy recovery of about 55 percent is claimed. Life expectancy is not known; emission control (bag houses or scrubbers) may be needed and skilled operation required.[54]

SANITARY LANDFILL

Introduction

Sanitary landfill is a controlled method of refuse disposal in which refuse is dumped in accordance with a preconceived plan, compacted, and covered during and at the end of each day. *It is not an open dump.* The nuisance conditions associated with an open dump such as smoke, odor, unsightliness, and insect and

[54]"Modulars Made for the '80s," *Waste Age*, May 1980, p. 33.

rodent problems are not present in a properly designed and operated sanitary landfill. A sanitary landfill is as much an engineering project as is construction of a building. It must also provide for gas venting or recovery; prevent groundwater pollution; have a leachate collection system if leachate cannot be controlled; monitor adjacent groundwater for leachate travel; and be located above the 100-year flood level.

Municipalities have an obligation to provide or make available, by whatever means feasible, facilities and services to accomplish the above objectives. The purpose of this section is to set forth the proper methods and procedures necessary to ensure that a solid waste disposal project is a community asset rather than liability.

Sanitary Landfill Planning and Design[55]

Legal Requirements

State and local sanitary codes and environmental protection agency regulations or laws usually require that a new refuse disposal area not be established until the site and method of proposed operation have been approved in writing by the agency having jurisdiction. The agency should be authorized to approve a new refuse disposal area and require such plans, reports, specifications, and other data as are necessary to determine whether the site is suitable and the proposed method of operation feasible. Intermunicipal planning and operation on a county or regional basis should be given very serious consideration before a new refuse disposal site is acquired because larger operations result in more efficient and lower unit costs of operations.

Intermunicipal Cooperation—Advantages

County or regional area-wide planning and administration for solid waste collection, treatment, and disposal can help overcome some of the seemingly insurmountable obstacles to satisfactory solution of the problem. Some of the advantages of county or regional area-wide solid wastes management are the following:

1. It makes possible comprehensive study of the total area generating the solid wastes and consideration of an area-wide solution of common problems on short-term and long-term bases. This can also help overcome the mutual distrust that often hampers joint operations among adjoining municipalities.
2. There is usually no more objection to one large site operation than to a single town, village, or city operation. Coordinated effort can therefore be directed at overcoming the objections to one site and operation, rather than to each of several town, village, and city sites.
3. The unit cost for the disposal of a large volume of solid waste is less. Duplication of engineering, overhead, equipment, labor, and supervision is eliminated.
4. Better operation is possible in an area-wide service, as adequate funds for proper supervision, equipment, and maintenance can be more easily provided.

[55]Joseph A. Salvato and William G. Wilkie, "New York *Plans* for Solid Waste," *J. Environ. Health*, **32**, 2, 202–208 (September/October 1969).

5. More sites can be considered. Some municipalities would have to resort to the more costly method of incineration because suitable landfill sites may not be available within the municipality.

6. County or regional financing for solid waste disposal often costs less, as a lower interest rate can usually be obtained on bonds because of the broader tax base.

7. A county agency or a joint municipal survey committee, followed by a county or regional planning agency, and then an operating department, district, or private contractor, is a good overall approach because it makes possible careful study of the problem and helps overcome interjurisdictional resistance.

Social and Political Factors

An important aspect of refuse disposal site selection, in addition to engineering factors, is the evaluation of public reaction and education of the public so that understanding and acceptance are developed. A program of public information is also needed. Equally important are the climate for political cooperation, cost comparison of alternative solutions, available revenue, aesthetic expectations of the people, organized community support, and similar factors.

Films and slides that explain proper sanitary landfill operations are available from state and federal agencies and equipment manufacturers. Sites having good operations can be visited to obtain firsthand information, and to show beneficial uses to which a completed site can be put.

Experience shows that where open dumps have been operated, there will be opposition to almost any site proposed for sanitary landfill, or an incinerator for that matter. However, local officials will have to study all of the facts, resolve significant objections, and make a decision to fulfill their responsibility and exercise their authority for the public good in spite of some unresolved opposition. Usually the critical factor is convincing the public that a nuisance- and pollution-free operation will in fact be conducted.

Planning

Local officials can make their task easier by planning ahead together on a county or regional basis for 20 to 40 years in the future and by acquiring adequate sites at least 5 years prior to anticipated needs and use. The availability of federal and state funds for planning for collection, treatment, and disposal of refuse on an area-wide basis such as a county should be explored. The planning will require an engineering analysis of alternative sites including population projections, volume, and characteristics of all types of solid wastes to be handled; cost of land and site preparation; expected life of the site; haul distances from the sources of refuse to the site; cost of equipment; cost of operation; and possible use and value of the finished site. Consideration must also be given to the climate of the region, prevailing winds, zoning ordinances, geology, soils, hydrology, and topography. Location and drainage to prevent surface water on groundwater pollution, groundwater monitoring, access roads to major highways, and availability of suitable cover material are other considerations. The reader is referred to Chapter 2 for a discussion of the broad aspects of community and facility planning and environmental impact analysis.

Once a decision is made it should be made common knowledge and plans

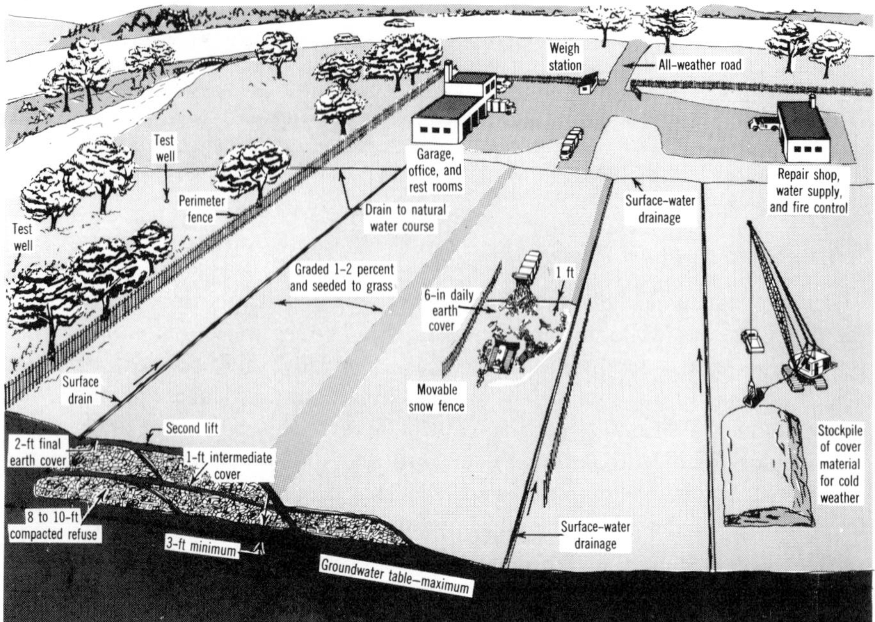

Figure 5-13 A sanitary landfill in a flat area with surface drainage control and monitoring wells. (Adapted from *Sanitary Landfill*, New York State Dept. of Health, Albany, 1969.)

developed to show how it is proposed to reclaim or improve and reuse the site upon completion. This should include talks, slides, news releases, question-and-answer presentations, and inspection of good operations. Artist's renderings are very helpful in explaining construction methods and final use of the land.

The general planning is followed by specific site selection and preparation. Site preparation requires that an engineering survey be made and a map drawn at a scale of not less than 200 ft to the in. with contours at 2-ft intervals, showing the boundaries of the property, location of structures within 1000 ft, adjoining ownership, topography, soil borings, groundwater levels, prevailing winds, and drainage. Proposed year-round access roads, direction of operation, borrow areas, finished grade and drainage with allowance for settlement, proposed windrows and fences, housing for men and equipment (trailer or building including locker room, showers, water supply, sewage disposal, or privies), monitoring wells, office, and telephone are also shown. See Figure 5-13.

It is essential that the refuse site be planned as an engineering operation and that it be under close engineering direction to assure that operation and maintenance follow the proposed plan and that the site does not degenerate into an open dump. The sanitary landfill supervisor should keep a daily record of the type and quantity of solid wastes received, sources of wastes, equipment use and repairs, personnel man-days, and other pertinent data.[56] He should be

[56] *Sanitary Landfill*, Manual No. 39, ASCE Solid Waste Management Committee of the Environ. Eng. Div. 345 East 47th St., New York, N.Y. 10017, 1976, pp. XI-1 to XI-19.

thoroughly familiar with the plan and be supplied the personnel and equipment to carry out the work properly. Consideration should be given to an attractive entrance and approach road.

Location

The site location directly affects the total refuse collection and disposal cost. If the site is remotely situated, the cost of hauling to the site may become high and the total cost uneconomical. It has been established that the normal economical hauling distance to a refuse disposal site is 10 to 15 mi, although this will vary depending on the volume of refuse and other factors. Actually, the hauling time is more important than the hauling distance. The disposal site may be as far as 30 to 40 mi away if a transfer station is used. Rail haul and barging introduce other possibilities. The cost of transferring the refuse per ton is used to compare refuse collection and disposal costs and to make an economic analysis. See pages 000 to 000. Holes left by surface mining operations may be considered for solid waste disposal as a means of reclaiming land and restoring it to productive use.

Accessibility

Another important consideration in site selection is accessibility. A disposal area should be located near major highways in order to facilitate use of existing arterial roads and lessen the hauling time to the site. Highway wheel load and bridge capacity and underpass and bridge clearances must also be investigated. It is not good practice to locate a landfill in an area where collection vehicles must constantly travel through residential streets in order to reach the site. The disposal area itself should normally be located at least 500 ft from habitation, although lesser distances have been successfully used to fill in low areas and improve land adjacent to residential areas for parks, playgrounds, or other desirable uses. Where possible, a temporary attractive screen should be erected to conceal the operation.

In order that vehicular traffic may utilize the site throughout the year, it is necessary to provide good access roads to the site so that trucks can move freely into and out of the site during all weather conditions and at all seasons of the year. Poorly constructed and maintained roads to a site can create conditions that cause traffic tie-ups and time loss for the collection vehicles.

Land Area (Volume) Required

The amount of refuse that will be produced by the communities served by the disposal area must be estimated in order to determine the amount of land that is needed. Land area for sanitary landfill should provide for a 20- to 40-yr period. Since the population in an area will not usually remain constant, it is essential that population projections and development be taken into account. These factors plus the probable solid waste contributions by industry and agriculture must be taken into account in planning for needed land. Disposal of industry wastes may double land requirements and require special handling and disposal controls. *Refuse production will vary, hence each community should make its own estimates.*

The space needed for refuse disposal is a function of population served, per capita refuse contribution, resource recovery, density and depth of the refuse in place, total amount of earth cover used, and time in use, adjusted for commercial and industrial wastes. This may be expressed as

$$Q = \frac{peck}{d}$$

in which

Q = space or volume needed in acre-ft per year

p = population served

e = ratio earth to compacted fill; use 1.25 if one part earth is used to four parts fill; use 1.20 if one part earth is used to five parts of fill; use 1.0 if no earth is used

c = pounds collected per capita per day

$k = 0.266 = \dfrac{365 \text{ days/yr} \times 27 \text{ ft}^3/\text{yd}^3}{43{,}560 \text{ ft}^3/\text{acre-ft}}$

d = density of compacted fill; a density of 800 to 1000 lb/yd^3 is readily achieved with proper operation; 600 or less is poor; 1200 or more is very good

Depth to Rock and Groundwater

In order to determine the depth at which a sanitary landfill can be operated, the location of bedrock, the groundwater table, and finished grade must be determined. This will require borings or test holes over the area under consideration. The location of bedrock and the highest groundwater table is of utmost importance in planning for a refuse disposal area. The bottom of sanitary landfills must be well above the high groundwater level and bedrock, and the intervening soil must assure protection of the groundwater.

Prevention and Control of Leachate Groundwater Pollution[57]

The best solution to the leachate problem is to prevent its development. Landfill leachate generation cannot in practice be entirely avoided except in some arid climates. Leachate control measures for groundwater and surface water quality protection should be incorporated in the site design. A water balance for the landfill disposal facility should be established to serve as a basis for the design of leachate control and surface runoff sytems.[58]

The precipitation less runoff and evaporation will determine the amount of infiltration. Infiltration and percolation less evapotranspiration, will, in the long term, after field capacity has been reached, determine the amount of leachate, if any, produced. This requires a water balance study. A major factor is a cover

[57]Joseph A. Salvato, William G. Wilkie, and Berton E. Mead, "Sanitary Landfill Leachate Prevention and Control," *J. Water Pollut. Control Fed.*, October 1971, pp. 2084–2100.

[58]"Landfill Disposal of Solid Waste," Proposed Guidelines, USEPA, *Fed. Reg.*, Monday, March 26, 1979, Part II. p. 18143(f).

material that is carefully graded, which ideally permits only limited infiltration and percolation to support vegetative cover and refuse decomposition, with optimal runoff but without erosion, to prevent significant leachate production. The soil cover should have a low permeability with low swell and shrink tendency upon wetting and drying. Runoff depends on rainfall intensity, and duration, permeability of the cover soil, surface slope (2 to 4 percent, not greater than 30 percent for side slopes), condition of the soil and its moisture content, and on the amount and type of vegetative cover. Evapotranspiration during the growing season for grasses and grains may be 20 to 50 in. See Chapter 4, Septic-Tank Evapotranspiration System (Evapotranspiration).

Experience indicates that the usual mixed *residential* refuse that is placed no deeper than 3 to 5 ft above the groundwater and bedrock in a clay loam soil will not present any serious hazard of groundwater pollution *provided surface water (and groundwater) and most of the precipitation is drained off the landfill and the landfill site.* However, if prevention of groundwater pollution cannot be assured, barrier protection may be required. Some authorities recommend a 10-ft separation, dependent on the type of soil and precipitation in the area.

If refuse is placed below the groundwater table or directly on bedrock, serious pollution of the groundwater can result. This pollution will mainly be the result of leachings from the refuse in contact with water, percolating water, and the transfer of gases such as carbon dioxide, methane, hydrogen sulfide, nitrogen, and ammonia (by diffusion and convection) which are produced during refuse decomposition. The carbon dioxide entering the groundwater accelerates the dissolving of calcium, magnesium, iron, and other substances that are undesirable at high concentrations in water.

If groundwater can come into contact with the refuse, it may be necessary to "line" the bottom and sides of the fill area and fill perimeter with a layer of coarse gravel or crushed stone to intercept the groundwater and drain it away before coming in contact with the refuse. An alternative is the construction of curtain drains around the landfill to intercept the water bearing formation and lower the groundwater table. It is also important to slope the finished grade of the fill to readily drain precipitation and to minimize ponding on the surface or seepage into the completed lifts as previously emphasized. Figures 5-13, 5-14, 5-15 suggest methods to control surface-water and groundwater flows. Monitoring wells as shown in Figure 5-13 are usually required.

The runoff from the drainage area tributary to the refuse disposal site must be determined to ensure that the surface-water drainage system, such as ditches, dikes, berms, or culverts, is properly designed and that flows are diverted to prevent flooding, erosion, infiltration, surface-water and groundwater pollution, both during operation and on completion. The design basis should be the maximum 10-year 24-hour precipitation. The topography and soil cover should be carefully examined to be sure that there will be no obstruction of natural drainage channels. Obstructions could create flooding conditions and excessive infiltration during heavy rains and snow melt. Flooding conditions can erode the cover material, expose the refuse, and cause the rapid travel of dissolved organic

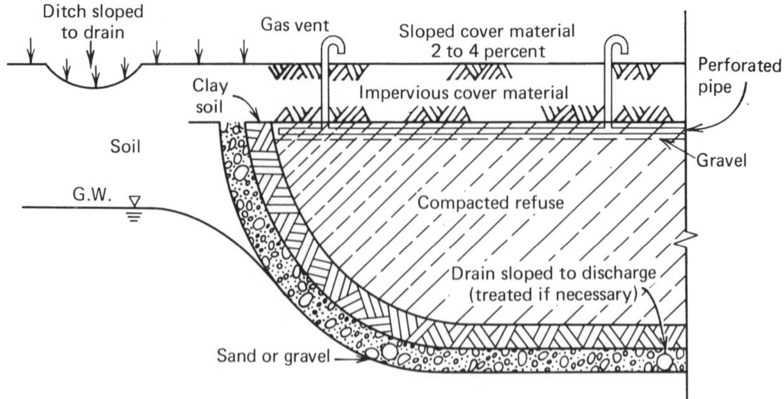

Figure 5-14 Landfill section showing groundwater control, leachate control, surface-water control, grading, cover material, ditching, and gas vents. (Adapted from Joseph A. Salvato, William G. Wilkie, and Burton E. Mead, "Sanitary Landfill Leachate Prevention and Control." *J. Water Pollut. Control Fed.*, October 1971, p. 2090.)

and chemical pollutants through the refuse to the groundwater table and to surface waters. The velocity of groundwater flow and its elevation will be increased as a result of the added flood water. In addition to water pollution problems, the erosion can create a condition in which insects and rodents can breed and find harborage.

Sites should be at least 200 ft from streams, lakes, or other bodies of water, and well above groundwater, or possible effects of floods up to the 100-year level. If a geological and hydrological study is made and special arrangements are made to contain and prevent the travel of pollution, a lesser distance may be used if acceptable to the agencies having jurisdiction. In any case special care must be exercised to provide for surface-water and groundwater drainage and natural runoff as shown in Figures 5-13, 5-14, and 5-15.

If leachate is, or may become, a problem in an existing landfill, and depending on the local situation and an engineering evaluation, the bottom and sides of the fill may be pressure-treated and sealed; surface-water drains may be provided upgrade and around the landfill area; curtain drains may be constructed to intercept and drain away the contributing groundwater flow; the leachate may be collected, and recycled or treated; or in special cases and if warranted, the material in the landfill can be excavated, treated and/or disposed of at a controlled site. This however may introduce other problems, particularly if hazardous wastes are involved, and hence must be carefully evaluated.

Liners are usually incorporated in the design of sanitary landfills where leachate production is considered inevitable and where groundwater quality may be impaired. Liners, one or more, are also used where hazardous or toxic chemicals or wastes are required to be deposited in a secure landfill or basin. Leachate collection and drainage systems are incorporated in the design to permit monitoring of the leachate and treatment if necessary. Monitoring wells

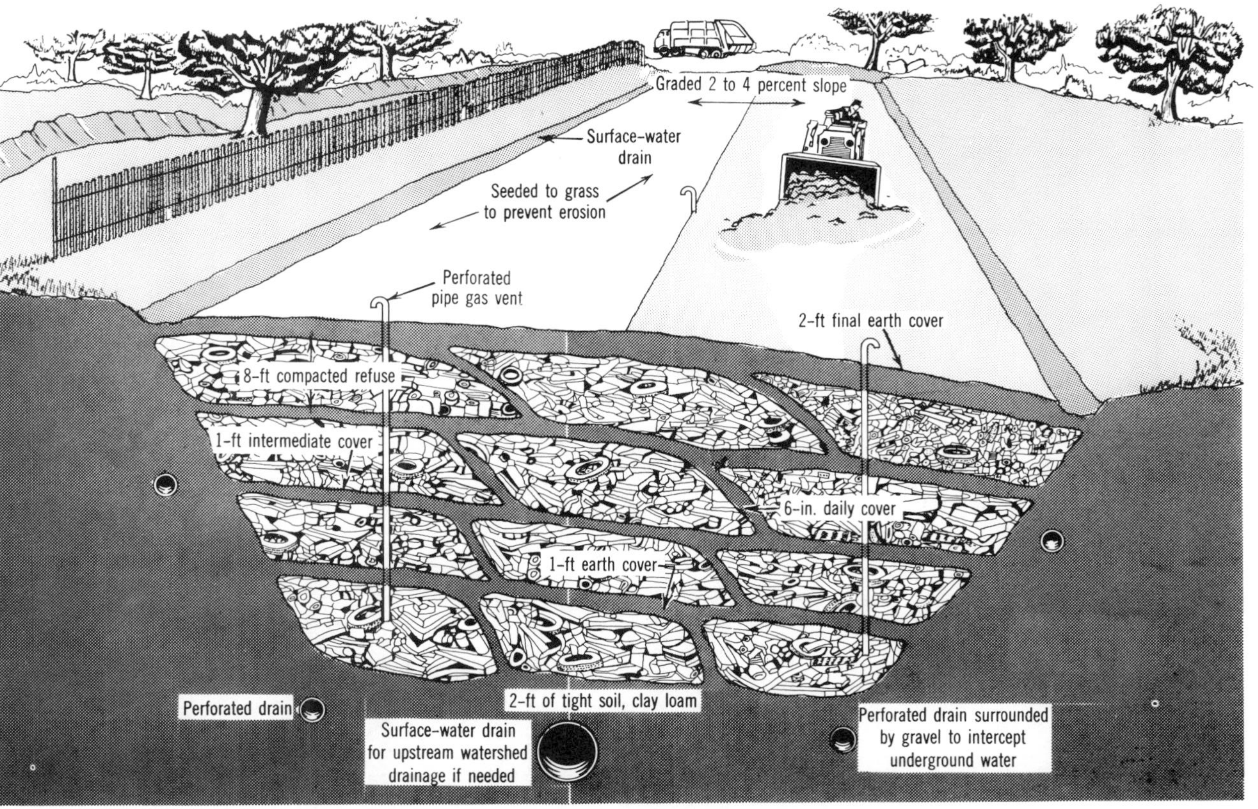

Figure 5-15 Sanitary landfill in ravine—fill starts at upper end of ravine bottom. Pipe gas-vent wells are perforated and gravel packed.

upgrade and downgrade are also provided to measure change in groundwater quality. Liners may be naturally occurring clays, soil cements, or artificial materials such as asphaltic materials and polymeric membranes. The characteristics of the wastes must be known to design a proper liner. Liners are discussed in Hazardous Wastes, this chapter.

After a landfill site is closed, it should be covered with at least 2 ft of compacted soil having a low permeability, graded to shed rainwater, melting snow, and surface water. The graded fill should be seeded to grass or other vegetation [see Cover Material, below, and Operation and Supervision (Maintenance) this chapter] and gas vents or barriers provided especially if there are any structures nearby that might be affected by methane migration or if gas collection is contemplated.

If not controlled or prevented, leachate might conceivably migrate considerable distances in the groundwater and persist for many years. The toxicity of landfill leachate is reduced as the fill ages, with high precipitation and infiltration, and with treatment using peat and physical-chemical treatment.[59]*

Leachates containing a significant fraction of biologically refractory high molecular weight organic compounds (i.e., those in excess of 50,000) are best treated by physical-chemical methods such as lime addition followed by settling. Leachates containing primarily low molecular weight organic compounds are best treated by biological methods such as activated sludge. Combinations of these methods may be required to achieve stream discharge standards.[60] In the final analysis, the treatment required will depend on the composition of the fill material and leachate and on the water pollution control standards to be met.

Control of Methane Gas Migration

Methane production and the hazards associated with it are discussed earlier under Methane Recovery.

The lateral migration of methane and other gases can be controlled by impermeable cutoff walls or barriers (Figure 5-16), or by the provision of a ventilation system such as gravel-filled trenches around the perimeter of the landfill (Figure 5-17). Gravel-packed perforated pipe wells or collectors (Figure 5-18) may also be used to collect and diffuse the gas to the atmosphere, if not recovered. To be effective, the system must be carefully designed, constructed, and maintained.[61] A relatively porous landfill earth cover will permit the escape of gases, except when the surface is frozen or saturated; but this will also permit greater infiltration and leachate production.

*Leachate recycling can greatly reduce residual contaminant concentration in the stabilized leachate after about one year. (Frederick G. Pohland, "Leachate Recycle as Landfill Management Option," *J. Envir. Engr. Div., ASCE*, December 1980, pp. 1057–1069.)

[59]Robert D. Cameron and Frederic A. Koch, "Toxicity of landfill leachates," *J. Water Pollut. Control Fed.*, April 1980, pp. 760–769.

[60]"Landfill Disposal of Solid Waste." Proposed Guidelines, USEPA, *Fed. Reg.*, March 26, 1979, pp. 18138–18148.

[61]Ralph Stone, "Preventing the underground movement of methane from sanitary landfills," *Civ. Eng. ASCE*, January 1978, pp. 51–53.

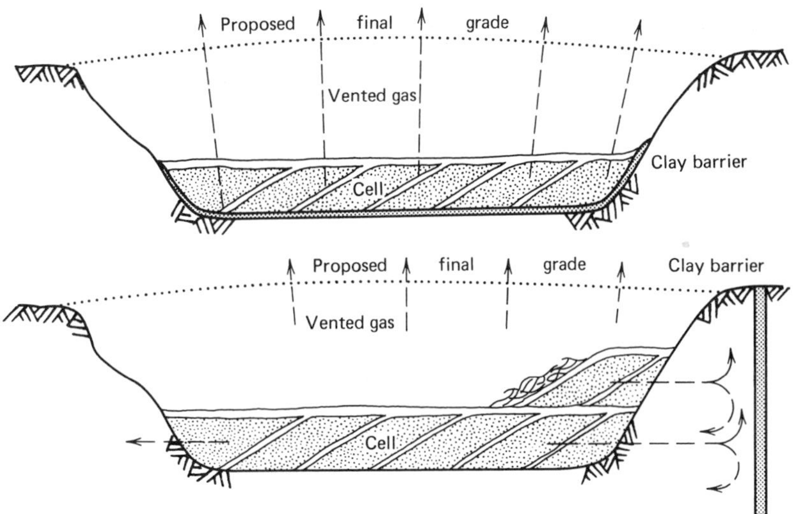

Figure 5-16 Clay placed as a liner in an excavation or installed as a curtain wall to block underground gas flow. (Source: *Sanitary Landfill Design and Operation*, USEPA, (SW-65ts), 1972, p. 26.)

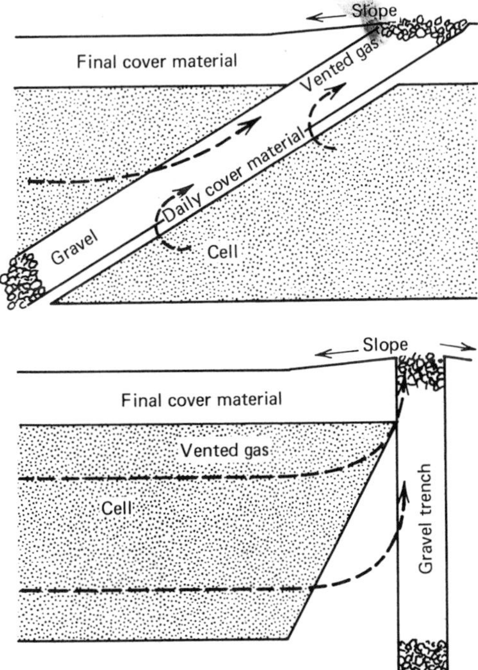

Figure 5-17 Gravel vents or gravel-filled trenches used to control lateral gas movement in a sanitary landfill. Gravel-filled trenches in combination with impermeable barrier provides good protection. Induced exhaust wells or trenches with perforated pipe and exhaust fan are reported to give good results. [Source: *Sanitary Landfill Design and Operation*, USEPA, (SW-65ts), 1972, p. 25.]

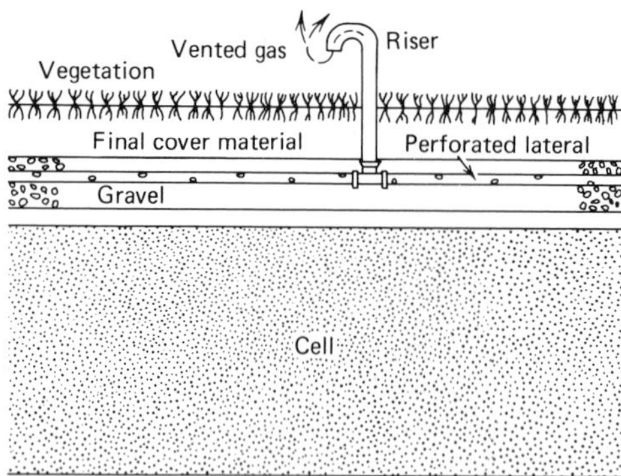

Figure 5-18 Gases vented out of sanitary landfill via pipes inserted through relatively impermeable top cover and connected to collecting laterals placed in shallow gravel trenches within or on top of the waste. [Source: *Sanitary Landfill Design and Operation*, USEPA, (SW-65ts), 1972, p. 26.]

Cutoff walls or barriers should extend from the ground surface down to a gas-impermeable layer such as clay, rock, or groundwater. Clay soils must be water-saturated to be effective. Perforated pipes have been shown to be of limited effectiveness and are not recommended for reduction of gas pressure, when used alone. Gravel-filled trenches may permit migration of gases across the trench, especially when covered by snow or ice; vertical perforated pipes reduce somewhat the effect of snow or ice. Gravel-filled trenches require removal of leachate or water from the trench bottom and are susceptible to plugging by biomass buildup. Gravel-filled trenches in combination with an impermeable barrier provide good protection against gas migration when keyed to a gas impermeable strata below the landfill. Induced exhaust wells or trenches with perforated pipes and pump or blower are reported to be very effective.[62] See Methane Recovery, this chapter.

Cover Material

The site should provide adequate and suitable cover material. A rough estimate for cover material requirements is 1 yd^3 per capita per year. Another estimate is that the volume of required earth cover is $\frac{1}{5}$ to $\frac{1}{4}$ the volume of compacted refuse. Actually, the amount of cover material required is a function of the depth of the compacted fill. From experience, the most suitable soil for cover material is one that is easily worked and yet minimizes infiltration; however, this is not always available. In any case the soil generally available can be made acceptable. It is good practice to stockpile topsoil for final cover and other soil for cold weather operation and access road maintenance.

Shredded (milled) solid waste in a landfill does not cause odors, rodent or insect breeding, or unsightliness and may not require daily earth cover. However precipitation will be readily absorbed and leachate produced unless the waste is covered with a low permeability soil which is well-graded to shed water.

[62]"Landfill Disposal of Solid Waste," USEPA, *Fed. Reg.*, March 26, 1979, Part II, pp. 18145–18147.

Table 5-15 Type and Depth of Earth Cover

Type of Cover	Depth	Period Left Open
Daily	6 in.	Day
Intermediate	12 in.	30 days
Final	24 in. or more	One year

The control of leachate and methane, and the role played by the final earth cover, including the importance of proper grading of the landfill final cover (2 to 4 percent slope) to minimize infiltration, promote runoff, and prevent erosion have been previously discussed. A final cover made up of 6 in. of clay soil and at least 18 in. of soil that will support vegetation and encourage evapotranspiration is recommended[62] See Table 5-15. The vegetation, such as seeded grass, will prevent wind and water erosion.

Landfill Vegetation

Four ft or more earth cover is needed if the area is to be landscaped, but the amount of cover depends on the plants to be grown. The carbon dioxide and methane gases generated in a landfill may interfere with vegetation root growth if not prevented or adequately diffused or collected and disposed of through specially provided sand, gravel, or porous pipe vents because oxygen penetration to the roots is necessary. Carbon dioxide as low or lower than 10 percent in the root zone can be toxic to roots, and methane-utilizing bacteria deplete the oxygen. Precautions to help maintain a healthy vegetation cover include selecting a tolerant species and seeking advice; avoiding areas of high gas concentrations; excluding gas from root zone—use mounds for planting, or line with membrane and vent trench and plant in suitable backfill soil; avoiding heavily compacted soil—loosen first if necessary and supplement soil fertility and improve its physical condition following good nursery practice; using smaller plant stock; and providing adequate irrigation.[63]

Fire Protection

The availability of fire protection facilities at a site should also be considered since fire may break out at the site without warning. Protective measures may be a fire hydrant near the site with portable pipe or fire hose, a watercourse from which water can be readily pumped, a tank truck, or an earth stockpile. The best way to control deep fires is to separate the burning refuse and dig a fire break around the burning refuse using a bulldozer. The refuse is then spread out so it can be thoroughly wetted down or smothered with earth. Limiting the refuse cells to about 200 tons, with a depth of 8 ft and 2 ft of compacted earth between cells (cells 15 × 112.5 ft or 20 × 85 ft assuming 1 yd^3 of compacted refuse weighs 800

[63] "Selecting Trees and Shrubs for Landfill Vegetation," *Pub. Works*, April 1980, p. 74. Based on report "Adapting Woody Species and Planting Techniques to Landfill Conditions—Field and Laboratory Conditions," Ida A. Leone, Franklin B. Flower, Edward F. Gilman, and John J. Arthur, Office of Research and Development, USEPA, Washington, D.C., August 1979.

lb), will minimize the spread of underground fires. The daily 6-in. cover will also minimize the start and spread of underground fires. Fires are a rare occurrence at a properly compacted and operated sanitary landfill.

Weigh Station

It is desirable to construct a weigh station at the entrance to the site. Vehicles can be weighed upon entering and, if necessary, billed for use of the site. Scales are required to determine tonnage received, unit operation costs, relation of weight of refuse to volume of in-place refuse, area work loads, personnel, collection rates, organization of collection crews, and need for redirection of collection practices. However, the cost involved in construction of a weigh station cannot be always justified for a small sanitary landfill handling less than 50 to 100 tons/day. Nevertheless, estimates of volume and/or weights received should be made and records kept on a daily or weekly basis to help evaluate collection schedules, site capacity, usage, and so forth. At the very least, an annual evaluation is essential.

Conversion of a Dump to a Sanitary Landfill

The typical open dump may be in a relatively flat low area or on a steep embankment. If the area (volume) available is limited, the dump should be graded and covered. A rat poisoning program should be instituted 2 weeks before the dump is covered and abandoned. If the site is adjacent to a stream, the refuse should be removed for an appropriate distance back from the high-water level to allow for the construction of an adequate and substantial, protected earth dike between the dump area and stream high-water level. The site is then finished off by covering with at least 2 ft of compacted earth.

Where adequate land is available and the site is suitable, steps should be taken to convert the dump into a properly operated and maintained sanitary landfill as described under Sanitary Landfill Methods. A rat-poisoning program should be carried on for 2 weeks before the conversion of the dump into a sanitary landfill. The conversion should be preceded by a plot layout drawn to scale showing the available land and its sequence of use. A plan of operation, supervision, and maintenance should accompany the layout. Conversion of a *suitable* existing site overcomes the problems associated with the selection of a new site and provides an opportunity to demonstrate dramatically the difference between an open dump and a sanitary landfill. The sequence for conversion of a dump to a sanitary landfill is shown in Figure 5-19.[64]

Sanitary Landfill Methods

General

There are many methods of operating a sanitary landfill. The most common are the trench, area, ramp, valley, and low-area fill methods. The trench method

[64]See also *Closing Open Dumps* and *Open Dump Closing—Alternative Procedures*, USEPA, Solid Waste Management Office, Cincinnati, Ohio, 1971.

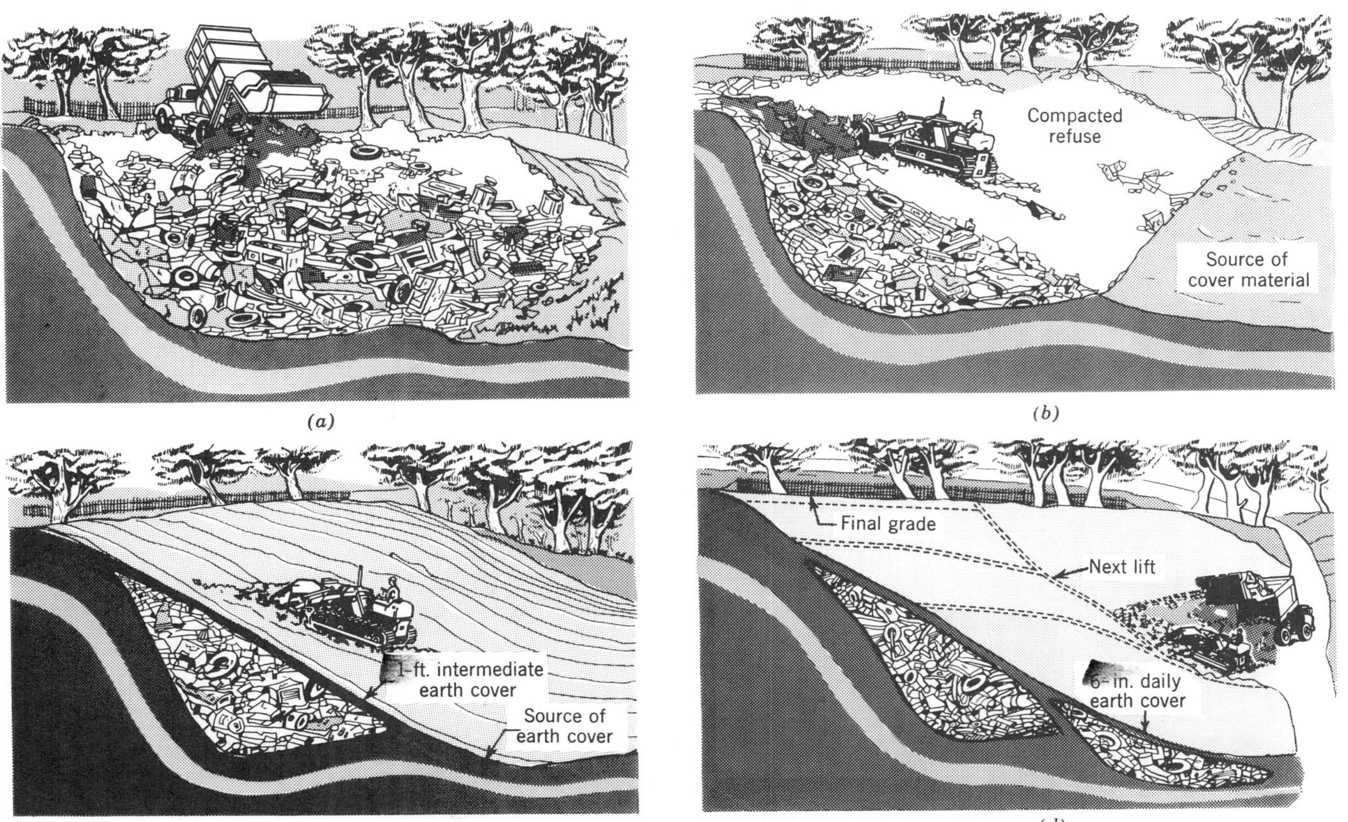

Figure 5-19 Conversion of an open dump to a sanitary landfill. (*a*) Existing open dump (*b*) Steep slope reduced to less than 2:1 to allow safe operation of equipment (*c*) Refuse compacted and covered (*d*) Refuse area operated as a sanitary landfill. (From *Sanitary Landfill*, New York State Dept. of Health, Albany, 1969.)

has the advantage of providing a more direct dumping control, which is not always possible with the area method. Since a definite place is designated for dumping, the scattering of refuse by wind is minimized and trucks can be more readily directed to the trench. The area method is more suitable for level ground. Here it is necessary to strip and stockpile sufficient cover material to meet the total need for earth cover; if this is not possible, earth must be hauled in. There are many variables. In all cases refuse should be spread and compacted in 12- to 18-in. layers as dumped and promptly covered, as explained previously under Prevention and Control of Leachate Groundwater Pollution (Cover Material). The spreading and compaction should be on a 30-percent slope, if possible.

Trench Method

The trench method is used primarily on level ground, although it is also suitable for moderately sloping ground. In this method, a trench is constructed by making a shallow excavation and using this excavated material to form a ramp above the original ground. Refuse is then methodically placed within the excavated area, compacted, and covered with suitable material at the end of the day's operation. Earth for cover material may be obtained from the area where the next day's refuse will be placed; hence a trench for the next day's refuse is completed as cover material is being excavated. Trenches are made 20- to 25-ft wide. The depth of fill is determined by the established finished grade and depth to groundwater or rock. If trenches can be made deeper, more efficient use is made of the available land area. Figure 5-20 shows the trench method.

Area or Ramp Method

On fairly flat and rolling terrain the area method can be utilized by using the existing natural slope of the land. The width and length of the fill slope are dependent on the nature of the terrain, the volume of refuse delivered daily to the site, and the approximate number of trucks that must be unloading at the site at one time. Side slopes are 30 percent; width of fill strips and surface grades are controlled during operation by means of line poles and grade stakes. The working face should be kept as small as practical to take advantage of truck compaction; restrict dumping to a limited area and avoid scattering of debris. In the ramp method earth cover is scraped from the base of the ramp. In the area method, cover material is hauled in from a nearby stockpile or from some other source. See Figure 5-13.

Low-Area Method

The sanitary landfill can also be used to improve marginal and hazardous areas, such as lowlands, depressions, and pits provided the spread of surface water and groundwater pollution is prevented and filling is not prohibited. The same basic method is used in this operation as in the area method except that cover material may have to be brought in. In addition, some means must be provided to prevent the scattering of refuse throughout the area. Depending on the condition of the area, this can be done by diking or "fencing."

Figure 5-20 Sanitary landfill, trench method. Crown fill area and provide drainage.

In wet areas, cover material may be obtained by the use of a dragline working ahead of the active face. The dragline can stockpile wet material in advance in order to let the material drain. This material will then be more stable and more easily handled by equipment for refuse cover. In this method the slope on the active face of the refuse must be gentle enough to allow a tractor to operate and compact refuse without becoming bogged down in mud.

If it is proposed to use a low or wet area, *and if its use is permitted*, arrangements must first be made to divert and drain surface water and to lower the groundwater table. As an alternative, clay soil fill can be placed in the low area to bring the bottom of solid waste well above the high-water level. Another method is to construct a watertight dike to isolate the fill area from the adjoining body of water and pump out the trapped water to facilitate filling with demolition and other inert material to above the surrounding water level. Park areas, marinas, and open spaces can be reclaimed along watercourses and coastal areas by this method, but a solid protected earth dam must be constructed along the shoreline to prevent leachate seepage and erosion. If the design engineer can justify that there will be no contravention of surface water, groundwater, or other environmental standards, a low-area fill may be approved. An environmental analysis and statement will probably be required.

Valley or Ravine Area Method

In valleys and ravines, the area method is usually the best method of operation. In those areas where the ravine is deep, the refuse should be placed in "lifts" from the bottom up with a depth of 8 to 10 ft. Greater depths are also used. Cover material is obtained from the sides of the ravine. It is not always desirable to extend the first lift the entire length of the ravine. It may be desirable to construct the first layer for a relatively short distance from the head of the ravine across its width. The length of this initial lift should be determined so that about one year's settlement can take place before the next lift is placed, although this is not essential if operation can be carefully controlled. Succeeding lifts are constructed by trucking refuse over the first lift to the head of the ravine. When the final grade has been reached (with allowance for settlement), the lower lift can be extended and the process repeated. See Figure 5-15.

Equipment for Disposal by Sanitary Landfill

General

In order to attain proper site development and ensure proper utilization of the land area, it is necessary to have sufficient proper equipment available at all times at the site. One piece of refuse-compaction and earth-moving equipment is needed for approximately each 80 loads per day received at the refuse site. The type of equipment should be suitable for the method of operation and the prevailing soil conditions. Additional standby equipment should be available for emergencies, breakdowns, and equipment maintenance. See Figure 5-21 and Table 5-16.

Tractors

Tractor types include the crawler, rubber-tired, and steel-wheeled types equipped with bulldozer blade, bullclam, or front-end loader. The crawler tractor with a front-end bucket attachment is an all-purpose piece of equipment. It may be used to excavate trenches, place and compact refuse, transport cover material, and level and compact the completed portion of the landfill. Some types can also be used to load cover material into trucks for transportation and deposition near the open face.

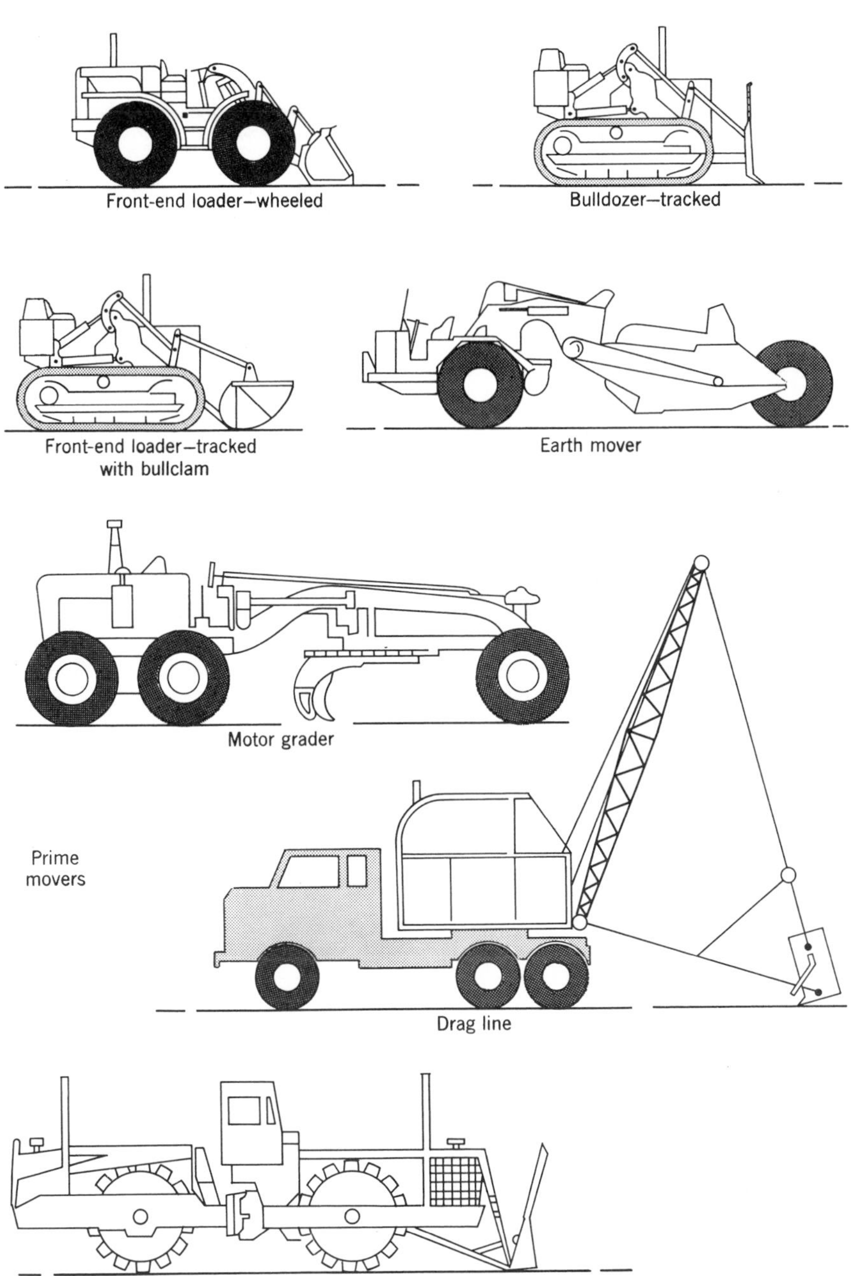

Figure 5-21 Sanitary landfill equipment.

Table 5-16 Average Equipment Requirements

Population	Daily Tonnage	No.	Type	Equipment Size (lb)	Accessory[a]
0 to 15,000	0 to 46	1	Tractor, crawler, or rubber-tired	10,000 to 30,000	Dozer blade, landfill blade, front-end loader (1- to 2-yd)
15,000 to 50,000	46 to 155	1	Tractor, crawler, or rubber-tired	30,000 to 60,000	Dozer blade, landfill blade, front-end loader (2- to 4-yd) multipurpose bucket
		[a]	Scraper, dragline, water truck		
50,000 to 100,000	155 to 310	1 to 2	Tractor, crawler, or rubber-tired	30,000 or more	Dozer blade, landfill blade, front-end loader. (2- to 5-yd), multipurpose bucket
		[a]	Scraper, dragline, water truck		
100,000 or more	310 or more	2 or more	Tractor, crawler, or rubber-tired	45,000 or more	Dozer blade, landfill blade, front-end loader, multipurpose bucket
		[a]	Scraper, dragline, steel-wheel compactor, road grader, water truck		

Source: *Sanitary Landfill Facts*, PHS Pub. No. 1792, DHEW, Washington, D.C., 1970, p. 21. See also *Sanitary Landfill Design & Operation*, USEPA, Washington, D.C. 1972, pp. 39–47.

[a]Optional. Dependent on individual need.

A bulldozer blade on a crawler tractor is good for landfills where hauling of cover material is not necessary. It is well suited for the area method landfill where cover material is taken from nearby hillsides. It can also be used for trench method operation where the trench has been dug with some other type of equipment. A bulldozer is normally used in conjunction with some other type of earth-moving equipment, such as a scraper, where earth is hauled in from a nearby source.

The life of a tractor is figured at about 10,000 hr. Contractors usually depreciate their equipment over a 5-year period. On a landfill, if it is assumed that the equipment would be used 1000 hr a year, the life of the equipment could be 10 yr. After 10 yr, operation and maintenance costs can be expected to approach or exceed the annual cost of new equipment. Lesser life is also reported. Equipment maintenance and operator competence will largely determine equipment life.

The size and type of machine needed at the sanitary landfill is dependent on the amount of solid waste to be handled, availability of cover material, compaction to be achieved, and other factors.[65] A rule that has been used is that a community with a population of less than 10,000 requires a $1\frac{1}{8}$-yd^3 bucket on a suitable tractor. Communities with a population between 10,000 and 30,000 should have a $2\frac{1}{4}$-yd^3 bucket, and populations of 30,000 to 50,000 should have at least a 3-yd^3 bucket. Larger populations will require a combination of earth-moving and compaction equipment depending on the site and method of operation. A heavy tractor (D-8) can handle up to 200 tons of refuse per day, although 100 to 200 tons per day per piece of equipment is a better average operating capacity. Tire fill foam and special tire chains minimize tire puncture and other damage on rubber-tired equipment.

Many small rural towns have earth-moving equipment that they use on highway maintenance and construction. For example, a rubber-tired loader with special tires can be used on a landfill that is open 2 days a week. On the other 3 days the landfill can be closed (with fencing and locked gate), and the earth-moving equipment can be used on regular road construction work and maintenance. The people and contract users of the site should be informed of the part-time nature of the operation so as to receive their cooperation. The public officials responsible for the operation should establish a definite schedule for the assignment of the equipment to the landfill site to make sure the operation is always under control and maintained as a sanitary landfill.

Other Equipment

The dragline is well adapted for work in low areas where soil is difficult to work. This piece of equipment is excellent for digging trenches, stockpiling cover material from swampy areas, and placing cover material over compacted refuse. An additional piece of equipment is necessary to spread and compact the refuse and cover material.

Although not commonly used, the backhoe is suitable for digging trenches on

[65]Dirk R. Brunner and Daniel J. Keller, *Sanitary Landfill Design and Operation*, (SW-65ts), USEPA, Washington, D.C., 1972, pp. 39–47.

fairly level ground and the power shovel is suitable for loading trucks with cover material.

In large operations earth movers can be used for the short haul of cover material to the site when adequate cover is not readily available nearby. Dump trucks may also be needed where cover material must be hauled in from some distance. Other useful equipment is a grader, sheepsfoot roller, and a water tank truck equipped with a sprinkler to keep down dust or a power sprayer to wet down the refuse to obtain better compaction.

Equipment Shelter

In cold areas it is necessary to construct an equipment shelter at the site. This will protect equipment from the weather and possible vandalism. The shelter can also be used to store fire-protection equipment and other needed materials. Operators of sanitary landfills have found a shelter to be of great value during the cold winter months since there is much less difficulty in starting motorized equipment.

Operation and Supervision

Operation Control

The direction of operation of a sanitary landfill should be with the prevailing wind. This will prevent the wind from blowing refuse back toward the collection vehicle and over the completed portion of the landfill. In order to prevent excessive wind scattering of refuse throughout the area, snow fencing or some other means of containing papers should be provided. The fencing can be utilized in the active area and then moved as the operation progresses. In some instances the entire area is fenced. Other sites have natural barriers around the landfill, such as is the case in heavily wooded areas. It is desirable to design the operation so that the work area is screened from the public line of sight. By taking this into consideration, scavenger wastes can be disposed of with the dry refuse without causing any particular problem.

The disposal of certain industrial and scavenger wastes at a sanitary landfill must take into consideration possible groundwater and surface-water pollution, insect and rodent breeding. Industrial wastes that introduce hazards, such as toxicity, explosion, and flammability *must be evaluated prior to disposal in a landfill.* The industry involved and the control agency should be contacted for advice before the wastes are accepted. Pretreatment or preparation of industrial wastes for disposal, if acceptable, may be required, and this is a proper responsibility of industry. Some industrial wastes including sludges and liquids should not be disposed of in an ordinary landfill. See Hazardous Wastes, this chapter.

Large items such as refrigerators, ranges, doors, and so forth, if not salvaged, can be placed directly into a sanitary landfill. However, in small operations this may be undesirable since these items are bulky and require considerable landfill volume and cover material. In these cases it is not objectionable to place such items in a separate area of the landfill and cover them periodically rather than daily. Compression or shredding of bulky objects will improve compaction of the fill, reduce land volume requirements, and allow more uniform settlement with less compaction. The cost of disposal, however, will be increased. Consideration should be given to resource recovery.

Drivers of small trucks and private vehicles carrying rubbish and other solid wastes interfere with the operation of a sanitary landfill. To accommodate these individuals on weekends and avoid traffic and unloading problems during the week, it is good practice to provide a special unloading area adjacent to the sanitary landfill entrance. Figure 5-22 shows a possible arrangement.

Personnel and Operation

Proper full-time supervision is necessary in order to control dumping, compaction, and covering. The supervisor should erect signs for direction of traffic to the proper area for disposal. It is essential that the supervisor be present at all hours of operation to ensure that the landfill is progressing according to plan. Days and hours of operation should be posted at the entrance to the landfill. A locked gate should be provided at the entrance to keep people out when closed. It is also advisable to inform the public of the days and hours of operation.

In supervising an operation, the length of the open face should be controlled since too large an open face will require considerably more cover material at the end of a day's operation. Too small an open face will not permit sufficient area for the unloading of the expected number of collection vehicles that will be present at one time. After vehicles have deposited the refuse at the top, or preferably, at the base of the ramp as directed, the refuse should be spread and compacted from the bottom up into a 12- to 18-in. layer with a tractor. *This should be done continually during the day's operation* to ensure good compaction, and vermin and fire control. If refuse is allowed to pile up without spreading and compaction for most of the day, proper compaction will not be achieved, resulting in uneven and excessive settling of the area and extra maintenance of the site after the fill is completed. At the end of each day, the

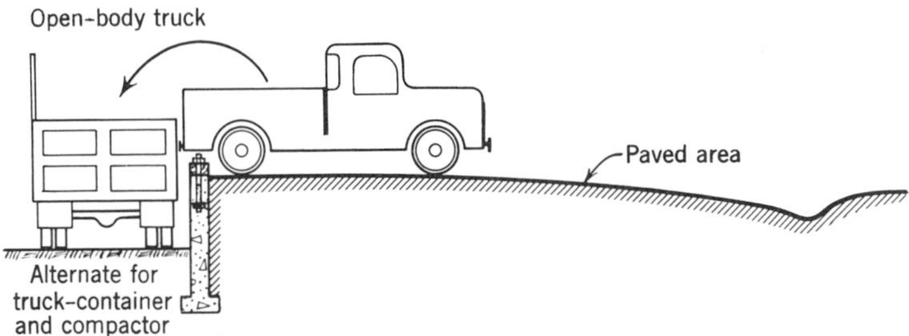

Figure 5-22 Small-load individual unloading area at the side of entrance area to sanitary landfill, or at a rural collection site. It may incorporate a compactor.

refuse should be covered with at least 6 in. of earth. For final cover of refuse, at least 2 ft of earth is required.

Adequate personnel are needed for proper operation. Depending on the size of the community, there should be a minimum of one man at a site and six men per 1000 yd^3 dumped per day that the site is open.

No Burning or Salvaging

Burning at a landfill site must not be permitted since air pollution problems due to smoke, fly ash, odors, and so on, would result. A nuisance condition would be created and public acceptance of the operation endangered. Air pollution standards and sanitary codes generally prohibit open burning. Fire regulations also prohibit open uncontrolled burning.

Limited controlled burning might be permitted in some emergency cases, but special permission would be required from the air pollution control agency, health department, and local fire chief. Arrangements for fire control, complete burning in one day, control over material to be burned (no rubber tires or the like), and restrictions for air pollution control would also be required.

Salvaging at sanitary landfills is not recommended since, as usually practiced, it interferes with the operation. It will slow down the entire operation and thus result in time loss. Salvaging can also result in fires and unsightly stockpiles of the salvage material in the area.

Area Policing

Since wind will blow papers and other refuse around the area as the trucks are unloading, it will be necessary to clean up the area and access road at the end of each day. One of the advantages of portable snow fencing is that it will usually confine the papers near the open face and thereby make the policing job easier and less time-consuming.

At many sanitary landfills, dust will be a problem during dry periods of the year. A truck-mounted water sprinkler can keep down the dust and can also be used to wet down dry refuse to improve compaction. The bulldozer operator should be protected by a dust mask, special cab, or similar device.

Insect and Rodent Control

An insect and rodent control program is not usually required at a properly designed and operated landfill. However, from time to time certain unforeseen conditions may develop that will make control necessary. For this reason, prior arrangement should be made to take care of such emergencies until the proper operating corrections can be made. Prompt covering of refuse is necessary. See Chapter 10.

Site Completion and Abandonment Precautions

If a disposal site is to be abandoned or closed, the users, including contractors, should be notified and an alternate site designated. A rat-poisoning program should be started 2 weeks before the proposed closing of a dump site that has not been operated as a sanitary landfill and continued until the dump has been completely closed. The site should be closed off so as to be inaccessible and covered with at least 2 ft of compacted earth on top and on exposed sides, graded to shed water, seeded to grass, and then posted to prohibit further dumping.

Maintenance

Once a sanitary landfill, or a lift of a landfill, is completed or partially completed it will be necessary to maintain the surface in order to take care of differential settlement. Settlement will vary, ranging up to approximately 20 to 30 percent, depending on the compaction, depth, and character of refuse. Maintenance of the cover is necessary to prevent ponding and excessive cracking, allowing insects and rodents to enter the fill and multiply.

It is also necessary to maintain proper surface-water drainage to prevent the seepage of contaminated leachate water through the fill to the groundwater table or to the surface. A 2- to 4-percent grade with culverts and lined ditches as needed is essential. The formation of water pockets is objectionable since vehicular traffic over these puddles will wash away the final cover from the refuse and cause trucks to bog down. The maintenance of access roads to the site is also necessary to prevent the formation of potholes, which will slow down vehicles using the site. Finally, the landfill surface should be prepared and planted with suitable tolerant vegetation as noted under Prevention and Control of Leachate Groundwater Pollution (Cover Material).

Use of Completed Sanitary Landfill

A sanitary landfill plan should provide for a specific use for the area after completion. It is expensive to excavate and regrade such an area. Final grades for a sanitary landfill should be established in advance to meet the needs of the proposed future use. For example, the use of the site as a golf course can tolerate rolling terrain while a park, playground, or storage lot would be best with a flat graded surface. In planning for the use of such an area, permanent buildings or habitable dwellings should not be constructed over the fill since gas production beneath the ground may escape into sewers and into the basements of such dwellings and reach explosive levels. Some structures that would not require excavation, such as grandstands, equipment shelters, and so on, can be built on a sanitary landfill with little resulting hazard. Buildings constructed on sanitary landfills can be expected to settle unevenly unless special foundation structures such as pilings are provided; but special provisions must be made to take care of gas production. When the final use is known beforehand, selected undisturbed ground islands or earth-fill building sites are usually provided to avoid these problems. Lateral movement of gases, particularly carbon dioxide and methane, must also be taken into consideration as previously noted.

Summary of Recommended Operating Practices

1. The sanitary landfill should be planned as an engineering project, operated and maintained by qualified personnel under technical direction, without causing air or water pollution, health hazards, or nuisance conditions.

2. The face of the working fill should be kept as narrow as is consistent with the proper operation of trucks and equipment in order that the area of exposed waste material be kept to a minimum.

3. All refuse should be spread as dumped and compacted into 12- to 18-in. thick layers as it is hauled in. Operate tractor up and down slope (3:1) of fill to get good compaction—four to five passes.

4. All exposed refuse should be covered with 6 in. of earth at the end of each day's operation.

5. The final earth covering for the surface and side slopes should minimize infiltration, be compacted, and be maintained at a depth of at least 24 in.

6. The final level of the fill should provide a 2- to 4-percent slope to allow for adequate drainage. Side slopes should be as gentle as possible to prevent erosion. The top of the fill and slopes should be promptly seeded. Drainage ditches and culverts are usually necessary to carry away surface water without causing erosion.

7. The depth of refuse should usually not exceed an average depth of 8 to 10 ft after compaction. In a landfill where successive lifts are placed on top of the preceding one, special attention should be given to obtain good compaction and proper surface-water drainage. A settlement period of preferably 1 yr should be allowed before the next lift is placed.

8. Control of dust, wind-blown paper, and access roads should be maintained. Portable fencing and prompt policing of the area each day after refuse is dumped are necessary. Design the operation if possible so that it is not visible from nearby highways or residential areas.

9. Salvaging, if permitted by the operator of the refuse disposal area, should be conducted in such a manner as not to create a nuisance or interfere with operation. Salvaging is not recommended at the site.

10. A separate area or trench may be desirable for the disposal of such objects as tree stumps, large limbs, refrigerators, water tanks, and so on.

11. Where necessary, provision should be made for the disposal, under controlled conditions, of small dead animals and septic tank wastes. These should be covered immediately. The disposal of sewage sludge, industrial or agricultural wastes, toxic, explosive, or flammable materials should not be permitted unless study and investigation show that the inclusion of these wastes will not cause a hazard or nuisance.

12. An annual or more frequent inspection maintenance program should be established for completed portions of the landfill to ensure prompt repair of cracks, erosion, and depressions.

13. Sufficient equipment and personnel should be provided for the digging, compacting, and covering of refuse. Daily records should be kept, including type and amount of solid wastes received. *At least annually*, an evaluation should be made of the weight of refuse received and volume of refuse in place as a check on compaction and rate at which the site is being used.

14. Sufficient standby equipment should be readily available in case there is a breakdown of the equipment in use.

15. The breeding of rats, flies, and other vermin; release of smoke and odors; pollution of surface waters and groundwaters; and causes of fire hazards are prevented by proper operation, thorough compaction of refuse in 12- to 18-in. layers, daily covering with earth, proper surface-water and groundwater drainage, and good supervision.

Modified Sanitary Landfill

A variation of the sanitary fill method of refuse disposal can be used at rural communities, camps, hotels, and other places where the quantity of refuse is small. This method consists of disposing of garbage and other refuse at one end of a trench. A bulldozer can be rented or borrowed from a contractor or the highway or street department to excavate the trench in advance. The trench may be 4- to 6-ft deep, 8- to 12-ft wide, and as long as needed, depending on the desired finished grade and quantity of refuse and equipment available. A trench 6-ft deep, 10-ft wide, and 20-ft long should serve a camp of 100 for at least one season.

The earth dug out of the trench should be piled alongside the trench so the refuse can be covered easily at the end of each day. The refuse is pushed to one end of the trench and compressed with a small bulldozer or tractor equipped with a blade or bucket. A camp truck or jeep provided with a homemade blade is also suitable. As the trench is filled, the refuse is completely covered with 24 in. of earth, which is packed down by driving over it.

Operation and maintenance of a modified landfill at a camp, hotel, or other resort is probably best avoided, if possible. It is better to contract with a private collector.

Where no collection service or community-operated sanitary refuse disposal area is provided, such as at a family summer camp, garbage, bottles, and tin cans from private dwellings can be satisfactorily disposed of on one's own property for a time. A trench should be dug about 4-ft deep and 2 ft-wide. The garbage is dumped at one end of the trench; cans, cartons, and bottles are smashed with a post and then covered with about 6 in. of earth. When the garbage approaches within about 12 in. of the ground surface, the pit should be filled in with earth tamped in layers to provide an earth cover of 24 in. over the garbage. A preferable procedure is to cart the refuse periodically to the town landfill, transfer station, or collection station.

BIBLIOGRAPHY

Brunner, Dirk R. and Daniel J. Keller, *Sanitary Landfill Design and Operation*, USEPA, (SW-65ts), Washington, D.C., 1972, 59 pp.

Composting of Municipal Solid Wastes in the United States, USEPA, 1971, GPO, Washington, D.C. 20402.

Corey, Richard C., *Principles and Practices of Incineration*, Wiley-Interscience, New York, 1969, 297 pp.

DeMarco, Jack, Daniel J. Keller, Jerold Leckman, and James L. Newton, *Incinerator Guidelines—1969*, PHS Pub. 2012, DHEW, Washington, D.C., 1969, 98 pp.

Ellis, H. M., W. E. Gilbertson, O. Jaag, D. A. Okun, H. I. Shuval, and J. Sumner, *Problems in Community Wastes Management*, WHO Health Paper 38, Geneva, 1969, 89 pp.

Glysson, Eugene A., James R. Packard, and Cyril H. Barnes, *The Problem of Solid-Waste Disposal*, University of Michigan, Ann Arbor, 1972, 153 pp.

Goleuke, C. G., and P. H. McGauhey, *Comprehensive Studies of Solid Wastes Management*, University of California, Berkeley, January 1969, 245 pp.

"Landfill Disposal of Solid Waste," Proposed Guidelines, USEPA *Fed. Reg.*, March 26, 1979, pp. 18138–18148.

Metry, Amir A., *The Handbook of Hazardous Waste Management*, Technomic Publishing Co., Inc., 1980, 265 Port Road West, Westport, CT 06880.

Municipal Refuse Collection and Disposal, Office for Local Government, Albany, N.Y., 1964, 69 pp.

Municipal Refuse Disposal, American Public Works Association, Public Administration Service, Chicago, 1970, 535 pp.

Pojasek, Robert B., *Toxic and Hazardous Waste Disposal*, Ann Arbor Science Publishers, Inc. Ann Arbor, Mich.
Volume 1: *Processes for Stabilization/Solidification*, 1979.
Volume 2: *Options for Stabilization/Solidifications*, 1979.
Volume 3: *Impact of Legislation and Implementation on Disposal Management Practices*, 1980.
Volume 4: *New and Promising Ultimate Disposal Options*, 1980.

Volume 5: *Perspectives on the Management of Hazardous Waste Disposal*, 1980.
Volume 6: *Development of State Hazardous Waste Management Programs*, 1980.

Proceedings: The Third National Conference on Air Pollution, DHEW, Washington, D.C., December 12–13, 1966, 667 pp.

Public Works Manual, Public Works Journal Corp., 200 South Broad St., Ridgewood, N.J. 07451, (annual publication).

Refuse Collection Practice, American Public Works Association, Public Administration Service, Chicago, 1966, 526 pp.

Sanitary Landfill, The ASCE Solid Waste Management Committee, of the Environmental Engineering Div., Manuals and Reports on Engineering Practice, No. 39, 345 East 47th St., New York, N.Y. 10017, 1976, 84 pp.

Solid Waste Management, Office of Science and Technology, Washington, D.C., May 1969, 111 pp.

Solid Waste Management, National Association of Counties Research Foundation, Washington, D.C., 1970, Ten Guides.

Solid Waste Management Technology Assessment, General Electric Company, Schenectady, N.Y., 1975, 348 pp.

Solid Wastes Management, Proceedings of the National Conference, University of California, Davis, April 4 and 5, 1966, 214 pp.

Tchobanoglous, George, Hilary Theisen, Rolf Eliassen, *Solid Wastes*, McGraw-Hill Book Co., New York, 1977, 621 pp.

Waste Management and Control, Committee on Pollution, National Academy of Science-National Research Council, Washington, D.C., 1969, 257 pp.

6

AIR POLLUTION AND NOISE CONTROL

THE AIR POLLUTION PROBLEM AND ITS EFFECTS

Air pollution is the presence of solids, liquids, or gases in the outdoor air in amounts that are injurious or detrimental to man, animal, plants, or property; or that unreasonably interfere with the comfortable enjoyment of life and property. Air pollution inside dwellings or places of assembly, although important, is not included in this discussion. The composition of clean air is shown in Table 6-1. The effects of air pollution are influenced by the type and quantity of pollutants and possible synergism;* also wind speed and direction, typography, sunlight, precipitation, vertical change in air temperature, photochemical reactions, height at which pollutant is released, and susceptibility of the individual and materials to specific contaminants—singularly and in combination. Air pollution is not a new or recent phenomena. It has been recognized as a source of discomfort for centuries as smoke, dust, and obnoxious odors.

Health Effects

Man is dependent on air. He breathes about 35 lb of air/day as compared to the consumption of 3 to 4 lb of water and 1½ lb (dry) of food. Pollution in the air may place an undue burden on the respiratory system, and contribute to increased morbidity and mortality, especially among susceptible individuals in the general population. Illnesses associated with air pollution are discussed in Chapter 1.

Some well-known air pollution episodes are given in Table 6-2. The illnesses were characterized by cough and sore throat, irritation of the eyes, nose, throat, and respiratory tract, plus stress on the heart. The weather conditions were typically fog, temperature inversion, and nondispersing wind. The precise levels at which specific pollutants become a health hazard are difficult to establish by existing surveillance systems but they probably are well in excess of levels currently found in the ambient air. Meteorological factors, sample site, fre-

*The combined effect is greater than the sum of the individual effects.

Table 6-1 Composition of Clean, Dry Air Near Sea Level

Component	% by Volume	Content (ppm)
Nitrogen	78.09%	780,900
Oxygen	20.94	209,400
Argon	0.93	9,300
Carbon dioxide	0.0318	318
Neon	0.0018	18
Helium	0.00052	5.2
Krypton	0.0001	1
Xenon	0.000008	0.08
Nitrous oxide	0.000025	0.25
Hydrogen	0.00005	0.5
Methane	0.00015	1.5
Nitrogen dioxide	0.0000001	0.001
Ozone	0.000002	0.02
Sulfur dioxide	0.00000002	0.0002
Carbon monoxide	0.00001	0.1
Ammonia	0.000001	0.01

Note: The concentrations of some of these gases may differ with time and place, and the data for some are open to question. Single values for concentrations, instead of ranges of concentrations, are given above to indicate order of magnitude, not specific and universally accepted concentrations.
From: *Cleaning Our Environment—The Chemical Basis for Action*, American Chemical Society, 1969, p. 4. Reprinted by permission of the copyright owner.
Sources: C. E. Junge, "Air Chemistry and Radioactivity" Academic Press, New York, 1963, p. 3; "Air Pollution," A. C. Stern, Ed., Vol. 1, 2nd ed., Academic Press, New York, 1968, p. 27; E. Robinson and R. C. Robbins, "Sources, Abundance, and Fate of Gaseous Atmospheric Pollutants," prepared for American Petroleum Institute by Stanford Research Institute, Menlo Park, Calif., 1968.

quency and measurement methods, including their accuracy and precision, all enter into data interpretation. Nevertheless, standards to protect the public health are necessary and have been established. (See Tables 6-6 and 6-7.)

It should be noted that whereas smoking is a major contributor to respiratory disease in the smoker, air pollution, climate, age, sex, and socioeconomic conditions affect the incidence of respiratory disease in the general population. Occupational exposure may also be a significant contributor in some instances.

Economic Effects

Pollutants in the air cause damage to property, equipment, and facilities, in addition to increased medical costs, lost wages, and crop damage. Sulfur pollution attacks copper roofs and zinc coatings; steel corrodes two to four times faster in urban and industrial areas; the usual electrical equipment contacts become unreliable unless serviced frequently; clothing fabric and leather are weakened; paint pigments are destroyed; and building surfaces, materials, and works of art are corroded. In addition particulates (including smoke) in polluted air cause erosion, accelerate corrosion, and soil clothes, buildings, cars, and

Table 6-2 Some Major Air Pollution Episodes

Location	Excess Deaths	Illnesses	Causative Agents
Meuse Valley, Belgium			
December 1930	63	6000	Probably SO_2 and oxidation products with particulates from industry—steel and zinc
Donora, Pa.			
October 1948	20	7000	Not proven; particulates and oxides of sulfur high; probably from industry—steel and zinc
Poza Rica, Mexico 1950	22	320	H_2S escape from a pipeline
London, England			
December 1952	4000	Increased	Not proven; particulates and oxides of sulfur high; probably from household coal-burning
January 1956	1000	—	—
December 1957	750	—	—
January 1959	200 to 250	—	—
December 1962	700	—	—
December 1967	800 to 1000	—	—
New York, N.Y.			
November 1953	165	—	Increased pollution
October 1957	130	—	Increased pollution
January-February 1963	200 to 400	—	SO_2 unusually high (1.5 ppm maximum)
November 1966	152-168	—	Increased pollution
New Orleans, La.			
October 1955	2	350	Unknown
1958	—	150	Believed related to smouldering city dump
Seveso and Meda, Italy[a]			
July 1976	Unknown, long—term	200 plus	Dioxin, an accidental contaminant formed in the manufacture of 2, 4, 5-T and hexachlorophene—a bactericide

[a]Reactor over-heated, safety valve opened, $4\frac{1}{2}$ pounds of dioxin discharged for 30 min to the atmosphere. About 50 persons hospitalized, 450 children had a skin disease, 200 families evacuated, 40,000 contaminated animals killed. (*Conserv. News*, Dec. 1, 1976, pp. 8, 9; and Associated Press, Seveso, Italy, July 10, 1977).

other property, making more frequent cleaning and use of air-filtering equipment necessary. Ozone reduces the useful life of rubber, discolors dyes, and damages textiles.

Air pollutants intercept part of the light from the sun, thereby increasing use of electricity in daytime. Unburned fuel coming out of a chimney or auto exhaust pipe as black smoke is wasted energy. A study prepared by the USEPA reported that air pollution was costing the residents of big cities as much as $6 billion a year in property and health damage. Crop and ornamental plant losses were estimated at $160 million.[1] Damage to buildings, clothing, and other property add $12.3 billion. Sickness alone from air pollution was estimated to cost Americans about $4.6 billion yearly in medical treatment, lost wages for sick workers, and lost work.[2] The American Lung Association estimates pollution-related losses in medical costs, lost wages, disability, and premature deaths to be more than $10 billion a year.[3] Carpenter, et al., estimated that exposure to air pollution (SO_2 and particulates) increased the average cost of hospitalization alone by $9.8 million in an area of 1.6 million people.[4] On the other hand, another study puts the benefits of air pollution control in 1978 at $21.4 billion. However, some feel that accurate estimates of cost are difficult to identify and compile and until more is known should not be used.[5] Hence the cost estimates should be considered only relative.

Effects on Plants

It has been suggested that plants be used as indicators of harmful contaminants because of their greater sensitivity to certain specific contaminants. Hydrogen fluoride, sulfur dioxide, smog, ozone, and ethylene are among the compounds that can harm plants. Assessment of damage shows that the loss can be significant, although other factors such as soil fertility, temperature, light, and humidity also affect production. Among the plants that have been affected are truck garden crops (New Jersey), orange trees (Florida), orchids (California), and various ornamental flowers, shade trees, evergreen forests, alfalfa, grains, tobacco, citrus, lettuce (Los Angeles), and many others. In Czechoslovakia more than 300 mi^2 of evergreen forests were reported severely damaged by sulfur dioxide fumes.[6] Sulfur dioxide, hydrogen fluoride, and the Los Angeles- and London-type smogs seem to be the pollutants causing the major effects on plants, as well as on man. Los Angeles smog is photochemical and oxidizing

[1]"Study Cites Damage Cost Of Cities Air Pollution," Associated Press, Kansas City, Mo., Sept. 3, 1977.

[2]Gerald Schneider, *Earth Trek—Explore Your Environment*, USEPA, Washington, D.C., 1977, p. 16.

[3]*Changing Times*, The Kiplinger Magazine, Washington, D.C., August 1978, p. 32.

[4]Ben H. Carpenter, et al., "Health Costs of Air Pollution: A Study of Hospitalization Costs," *Am. J. Pub. Health*, December 1979, pp. 1232–1241.

[5]Allen V. Kneese, "How Much Is Air Pollution Costing Us in the United States," *Proceedings: The Third National Conference on Air Pollution*, DHEW, Washington, D.C., December 12–14, 1966.

[6]Associated Press, Prague, Czechoslovakia, May 30, 1970.

(NO_x + hydrocarbons + 0). London smog is a mixture of smoke, fog, and sulfur oxides. Photochemical smog has also been reported in New York, Japan, Mexico City, Madrid, the United Kingdom, and in other congested areas with high motor vehicle traffic.

Injury to plants due to ozone shows up as flecks, stipple and bleaching, tip burns on conifers, and growth suppression. Peroxyacyl nitrates* (PAN) injury is apparent by glazing, silvering, or bronzing on the underside of the leaf. Sulfur dioxide injury shows up as bleached and necrotic areas between the veins, growth suppression, and reduction in yield. Hydrogen fluoride injury shows as plant leaf tip and margin burn, chlorosis, dwarfing, abrupt growth cessation, and lowered yield.[7]

Effects on Animals

Fluorides have caused crippling skeletal damage to cattle in areas where fluorides absorbed by the vegetation are ingested. Animal laboratory studies show deleterious effects from exposure to low levels of ozone, photochemical oxidants, and PAN. Lead and arsenic have also been implicated in the poisoning of sheep, horses, and cattle. All of the canaries and about 50 percent of the animals exposed to hydrogen sulfide in the Poza Rica, Mexico incident (see Table 6-2) were reported to have died. More controlled morbidity and mortality studies are needed to determine actual impacts of air pollutants on animals.

Aesthetic, Climatic, and Related Effects

Insofar as the general public is concerned smoke, dust, and haze, which are easily seen, cause the greatest concern. Reduced visibility not only obscures the view but is also an accident hazard to air, land, and water transportation. Soiling of statuary, clothing, buildings, and other property increases municipal and individual costs and aggravates the public to the point of demanding action on the part of public officials and industry. Correction of the air pollution usually results in increased product cost to the consumer, but failure to correct pollution is usually more costly.

Haze, dust, smoke, and soot reduce the amount of solar radiation (ultraviolet light) reaching the surface of the earth. Aerosol emissions from jet planes also intercept some of the sun's rays. The solar radiation as light energy that does reach the earth is absorbed and reradiated back to the atmosphere as heat energy. But the carbon dioxide in the lower atmosphere tends to trap the heat, causing reflection of heat and a warming of the atmosphere (the greenhouse effect) and a warming of the earth's surface. Hence the reduction in solar radiation or cooling effect of the haze, dust, smoke, and soot is offset by the warming effect

*Also cause eye irritation.

[7] *Air Pollution Control Orientation Course*, "The Effects of Air Pollution," Office of Air Programs, Air Pollution Training Institute, USEPA, Washington, D.C., 1972, p. 86.

of the carbon dioxide. In addition, the burning of fossil fuels contribute to the carbon dioxide in the atmosphere, but forests and other vegetation and humus remove carbon dioxide. Which effects will prevail is not known, although it appears that the carbon dioxide level is increasing.

Another factor is ozone which helps to shield the earth from harmful ultraviolet radiation. A deterioration in the ozone layer in the stratosphere can cause an increase in ultraviolet radiation reaching the earth. This could cause an increase in skin cancers and changes in our climate, animal, and plant life. Chlorofluorocarbons are believed to react with ozone in the upper atmosphere thereby reducing the total amount of ozone available to intercept ultraviolet radiation. Because of this, attempts are being made to phase out products containing fluorocarbons in the United States. The products include refrigerants, industrial solvents, and aerosol spray cans containing fluorocarbons. However existing refrigerating systems using chlorofluorocarbons remain in use. Unfortunately, worldwide use of fluorocarbons has been increasing, despite warnings of ozone damage, according to the National Academy of Sciences.[8] Nitrous oxide (from nitrogen fertilizer) also causes depletion of ozone.*

Air pollution, both natural and man-made, affects the climate in other ways. Dust and other particulate matter in the air provide nuclei around which condensation takes place, forming droplets and thereby playing a role in snowfall and rainfall patterns.

Certain malodorous gases interfere with the enjoyment of life and property. In some instances individuals are seriously affected. The gases involved include hydrogen sulfide, sulfur dioxide, aldehydes, phenols, polysulfides, and some olefins. Air pollution control equipment is available to reduce these gases to less objectionable compounds.

Releases of nitrogen and sulfur oxides, as well as other pollutants, are carried into the atmosphere and may be deposited as "acid rain" many miles from the source. Large regional emissions of SO_2 over a limited area exacerbate the acid rain problem such as in the northeast United States and Canada, although the Southeast, Midwest, West, and western Canada and some European countries are also being affected. In New York and the Northeast 60 to 70 percent of the acidity is reported to be due to sulfuric acid and 30 to 40 percent to nitric acid. The relative proportion of each is indicative of the probable preponderant pollutant sources.[9] Major sources of sulfur dioxide are coal- and oil-burning power plants, refineries, and metal smelters. Principal sources of nitrogen oxide emissions[10] are electric utility and industrial boilers (56 percent) and auto and truck engines. Nitrogen oxides from vehicle and high temperature combustion change in the atmosphere and return to the earth in acid form with rain. High stacks permit the discharge of pollutants into the upper air stream to be carried

*The National Aeronautics and Space Administration reported, based on measurements taken from weather satellites, that a change (depletion of the Earth's ozone shield by about 4 percent at 24 miles) appears to be taking place. (The Associated Press, New York, August 3, 1981.)

[8] "Report: Ozone depletion accelerating," Associated Press, Washington, November 9, 1979.

[9] Norman R. Glass, Gary E. Glass, and Peter J. Rennie, "Effects of acid precipitation," *Environ. Sci. Technol.*, November 1979, pp. 1350–1354.

[10] *Air Pollution and Your Health*, USEPA, Washington, D.C., March 1979, p. 10.

great distances with the prevailing winds. Natural sources such as active volcanoes, the oceans, and anaerobic emissions from decaying plants contribute to the problem. Acid rain contributes to deterioration of buildings, monuments and statues, trees, crops, timber, and other vegetation. It adversely affects lakes and streams (pH may be reduced to less than 5.0) with resultant reduced fish production and acidificaiton and demineralization of soils. The condition is more apparent in a lake when its buffering capacity, and that of surrounding soil, is reduced. This could also cause the release of toxic metals to water supply sources and their cumulation in fish as for example increased levels of mercury in walleyed pike, northern pike, and trout. Control measures should start with source reduction, such as at high sulfur oil- and coal-burning plants and in motor vehicles. Industry however states that its studies show the relationship between sulfur and nitrogen dioxide emissions and the amount of acid rain has not been proven, nor has acid rain been increasing.[11]

SOURCES AND TYPES OF AIR POLLUTION

The sources of air pollution may be man-made, such as the internal combustion engine, or natural, such as plants (pollens). The pollutants may be in the form of particulates, aerosols, and gases, or as microorganisms. Included are pesticides, odors, and radioactive particles carried in the air.

Particulates are from less than 0.01 to 1000 μm in size; generally smaller than 50 μm. Smoke is generally less than 0.1 μm size soot or carbon particles. Those below 10 μm can penetrate the respiratory tract; particles less than 3 μm reach deep parts of the lung. Particles over 10 μm are removed by the hairs at front of nose. (A micron (μm) is 1/1000 of a millimeter or 1/25,000 of an in. Particles of 10 μm and larger in size can be seen with naked eye.) Included are dust and inorganic, organic, fibrous and nonfibrous particles. Aerosols are usually particles 50 μm to less than .01 μm in size; generally less than 1 μm in diameter. Gases include organic gases such as hydrocarbons, aldehydes, ketones, and inorganic gases such as oxides of nitrogen and sulfur, carbon monoxide, hydrogen sulfide, ammonia, and chlorine.

Man-Made Sources

Air pollution in the United States is the result of industrialization and mechanization. The major sources and pollutants are shown in Tables 6-3 and 6-4. It can be seen that carbon monoxide is the principal pollutant by weight and that the motor vehicle is the major contributor, followed by industry and power plants. However, in terms of hazard it is not the tons of pollutant that is important but the toxicity or harm that can be done by the particular pollutant released.

Agricultural spraying of pesticides, orchard-heating devices, exhausts from various commercial processes, rubber from tires, mists from spray-type cooling

[11]Michael R. Deland, "Acid rain," *Environ. Sci. Technol.*, June 1980, p. 657.

Table 6-3 Sources and Emissions of Air Pollutants—1968 and 1975 (millions of tons per year)

	Total[a]		Carbon monoxide		Sulfur oxides		Nitrogen oxides		Hydro-carbons		Partic-ulates	
Source	1968	1975	1968	1975	1968	1975	1968	1975	1968	1975	1968	1975
Transportation	90.5 (42%)	101.9 (50%)	63.8	77.4	0.8	0.8	8.1	10.7	16.6	11.7	1.2	1.3
Fuel combustion in stationary sources	45.9 (21%)	47.9 (24%)	1.9	1.2	24.4	26.3	10.0	12.4	0.7	1.4	8.9	6.6
Industrial processes	29.3 (14%)	28.0 (14%)	9.7	9.4	7.3	5.7	0.2	0.7	4.6	3.5	7.5	8.7
Solid waste disposal	11.2 (5%)	5.1 (3%)	7.8	3.3	0.1	<0.1	0.6	0.2	1.6	0.9	1.1	0.6
Miscellaneous	37.3 (17%)	19.4 (10%)	16.9	4.9	0.6	0.1	1.7	0.2	8.5	13.4	9.6	0.8
Totals	214.2	202.3	100.1	96.2	33.2	32.9	20.6	24.2	32.0	30.9	28.3	18.0

Source: *Nationwide Inventory of Air Pollutant Emissions, 1968*, PHS, NAPCA Pub. AP-73, DHEW, Washington, D.C., August 1970. Totals are for the most part higher than 1966 estimates primarily because of the inclusion of sources not previously considered, that is, forest fires, burning of coal refuse banks, and an increased number of industrial process sources. Also *National Air Quality and Emissions Trends Report 1975*, EPA-450/1-76-002, USEPA, Research Triangle Park, N.C., 1976.

[a]Totals show relative weights and do not indicate hazard or severity of problem in a particular area.

Table 6-4 Estimated Nationwide Amounts of Air Pollutants (millions of tons)

Pollutant	1968[a]	1970[b]	1974[b]	1975[c]
Carbon monoxide	100.1	107.3	94.6	96.2
Particulates	28.3	27.5	19.5	18.0
Sulfur dioxide	33.2	34.3	31.4	32.9
Hydrocarbons	32.0	32.1	30.4	30.9
Nitrogen oxides	20.6	20.4	22.5	24.2
Total	214.2	221.6	198.4	202.2

[a] *Nationwide Inventory of Air Pollutant Emissions, 1968*, PHS, NAPCA Pub. AP-73, DHEW, Washington, D.C., August 1970.
[b] *Clean Air The Breath of Life*, USEPA, Washington, D.C., April 1976, p. 2.
[c] *National Air Quality and Emissions Trends Report, 1975*, EPA-450/1-76-002, USEPA, Research Triangle Park, N.C., 1976.

towers, and use of cleaning solvents and household chemicals add to the pollution load. Other potential sources are wastewater treatment plants, stack pollution from processing of nuclear fuels, nuclear reactors, and nuclear explosions.

Particulates that find their way into the air without venting through a stack are referred to as fugitive emissions. They include uncontrolled releases from industrial processes, street dust, and dust from construction and farm cultivation. These need to be controlled at the source on an individual basis.

Natural Sources

Discussions of air pollution frequently overlook the natural sources. These include dust; plant and tree pollens; arboreal emissions; bacteria and spores; gases and dusts from forest and grass fires; ocean sprays and fog; esters and terpenes from vegetation; ozone and nitrogen dioxide from lightning; ash and gases (SO_2, HCl, HF, H_2S) from volcanoes; natural radioactivity; and microorganisms such as bacteria, spores, molds, or fungi from plant decay. Most of these are beyond control or of limited significance.

Ozone is found in the stratosphere at an altitude beginning at 7 to 10 mi. The principal natural sources of ozone in the lower atmosphere are lightning discharges, intrusion of ozone from the stratosphere, and some small amount from reactions involving volatile organic compounds released by forests and other vegetation. Ozone is also formed by the action of sunlight on nitrogen oxides and hydrocarbons.

Types of Air Pollutants

The types of air pollutants are related to the original material used for combustion or processing, the impurities it contains, the actual emissions, and reactions in the atmosphere. See Tables 6-3 and 6-4. A *primary pollutant* is one that is found in the atmosphere in the same form as it exists when emitted from the stack; sulfur dioxide, nitrogen dioxide, and hydrocarbons are examples. A *secondary pollutant* is one that is formed in the atmosphere as a result of reactions such as hydrolysis, oxidation, and photo-chemistry; photochemical smog is an example.

Most combustible materials are composed of hydrocarbons. If the combustion of gasoline, oil, or coal, for example, is inefficient, unburned hydrocarbons, smoke, carbon monoxide, and to a lesser degree aldehydes and organic acids are released.

The use of automobile catalytic converters to control carbon monoxide and hydrocarbon emissions causes some increase in sulfates and sulfuric acid emissions but this is considered to be of minor significance. The elimination of lead from gasoline line has, in some cases, led to the substitution of manganese for antiknock purposes with consequent releases of manganese compounds which are also potentially toxic.

Impurities in combustible hydrocarbons (coal and oil), such as sulfur, combine with oxygen to produce SO_2 when burned. The SO_2 subsequently may form sulfuric acid and other sulfates in the atmosphere. Oxides of nitrogen, from high-temperature combustion in electric utility and industrial boilers and auto and truck engines (above 1200°F), are released mostly as NO_2 and NO. The source of nitrogen is principally the air used in combustion. Some fuels contain substantial amounts of nitrogen and these also react to form NO_2 and NO. Fluorides and other fuel impurities may be carried out with the hot stack gases.

Photochemical oxidants* are produced in the lower atmosphere (troposphere) as a result of the reaction of oxides of nitrogen and volatile organics in the presence of solar radiation as previously noted.

Of the sources noted above, automobile emissions are the principal source of volatile organics (hydrocarbons) with industrial sources such as petroleum refineries also large contributors. Automobiles and stationary fuel combustion plants are the major sources of nitrogen oxides. Ozone is the photochemical oxidant actually measured which is about 90 percent ozone. It is the principal component of modern smog.[12] Ozone and other photochemical products formed are usually found at some distance from the source of the precursor compounds.

Air pollutants that need more systematic investigation are the following:

Highest priority: Sulfur dioxide, carbon monoxide, carbon dioxide, fluoride, ozone, sulfuric acid droplets, oxides of nitrogen, carcinogens (various types), peroxyacyl nitrates, gasoline additives including lead, and asbestos particles.

High priority: Benzene and homologues, alkyl nitrites, aldehydes, ethylene, pesticides, auto exhaust (raw), amines, mercaptans, hydrogen sulfides, and beryllium particles.[13]

*Including ozone, PANs, formaldehydes, and peroxides. Nitrogen dioxide colors air reddish-brown.

[12] *Air Pollution and Your Health*, USEPA, Office of Public Awareness, Washington, D.C. 20460, March 1979, p. 9.

[13] *Restoring the Quality of Our Environment*, Report of Environmental Pollution Panel, President's Science Advisory Committee, Washington, D.C., November 1965.

In any case, sulfur oxides including SO_2 and sulfur-containing aerosols, carbon monoxide, photochemical oxidants particularly ozone, and nitrogen oxides such as NO_2, NO, and nitrite aerosols are of major concern. Progress has been made in the control of sulfur dioxide and particulates; greater emphasis is shifting to the control of photochemical oxidants, sulfates, nitrates, and carcinogens.

SAMPLING AND MEASUREMENT

Air-sampling devices are used to detect and measure smoke, particulates, and gases. The equipment selected and used, and its siting, is determined by the problem being studied and the purpose to be served. Representative samples free from external contamination must be collected and readings or analyses standardized to obtain valid data. Supporting meteorological and other environmental information is needed to properly interpret the data collected. Continuous sampling equipment should be selected with great care. The accuracy and precision of equipment needs to be demonstrated to ensure that it will perform the assigned task with a minimum of calibration and maintenance. Reliable instruments are available for the continuous monitoring of ambient air parameters such as listed in Table 6-5 and for carbon dioxide, carbon monoxide, opacity, and oxygen in stacks. Other instruments such as for smoke, hydrocarbons, and sulfur are also available.

Some continuous monitoring instruments for atmospheric measurement of pollution are quite elaborate and costly. The simplest readily available instrument should be selected that meets the required sensitivity and specificity. Power requirements, service, maintenance, the calibration frequency, and the time required to collect and transmit information are important considerations.

Table 6-5 Measurement Methods for Federal Ambient Air Quality Parameters

Pollutant	Measurement Methods
SO_2	Ultraviolet pulsed fluorescence, flame photometry, coulometric; dilution or permeation tube calibrators
CO	Nondispersive infrared tank gas and dilution calibration
O_3	Gas phase chemiluminescence ultraviolet (UV) spectrometry; ozone UV generators and UV spectrometer or gas phase titration (GPT) calibrators
NO_2	Chemiluminescence; permeation or GPT calibration
Lead	High-volume sampler and atomic absorption analysis
TSP[a]	High-volume sampler and weight determination
Sulfates, Nitrates	High-volume sampler and chemical analysis—deposit dissolved and analyzed colorimetrically
Hydrocarbons	Flame ionization and gas chromotograph;[b] calibration with methane tank gas

[a]Total suspended particulates.
[b]Not generally required to measure these if O_3 is measured.

A continuous air quality monitoring system for the measurement of selected gaseous air pollutants, particulates, and meteorological conditions over a large geographical area can make possible immediate intelligence and reaction when ambient air quality levels or emissions increase beyond established norms. In the system, each monitoring station sends data to a data reception center, say every 15 min, via telephone lines or other communication network. The collected data is processed by computer and visually displayed for indicated action. Field operators who can perform weekly maintenance and calibration checks, and a trained central technical staff which coordinates and scrutinizes the overall monitoring system operation and data validation every day are essential for the production of usable and valid "real time" data.[14]

The air monitoring data can be used to measure ambient air quality and its compliance with state and federal standards; detect major local source air quality violations; provide immediate information for a statewide air pollution episode alert warning system; provide long-term air quality data to meet public and private sector data needs such as for environmental planning and environmental impact analysis; determine long-term air pollution concentrations and trends in a state; and provide air quality information to the public.

A continuous air quality monitoring system requires use of continuously operating analyzers of a design which measure ambient concentrations of specified air pollutants in accordance with USEPA "reference methods" or "equivalent methods."[15] The USEPA designates air pollution analyzers after reviewing extensive test data submitted by the manufacturers for their instrumentation. Only anlayzers designated as reference or equivalent methods may be used in ambient air monitoring networks to define air quality. This is necessary to insure correct measurements and operation thereby promoting uniformity and comparability of data used to define ambient air quality across the country.

The USEPA has specified[16] a detailed ambient monitoring program for use by states, local government and industry. Included in the program are formal data quality assurance programs, monitoring network design, probe (air intake) siting, methodology, and data reporting requirements. The USEPA has specified a daily uniform air pollution index known as the Pollutant Standard Index (or PSI) for public use in comparing air quality. The PSI values are discussed later and summarized in Table 6-7. Through the quality assurance program the precision and accuracy of air quality data will be defined.

Types of anlayzers used to measure Federal ambient air quality parameters are summarized in Table 6-5. Continuous analyzers utilizing "gas phase" measurements with electronic designs, rather then "wet chemistry" measurements, are preferred as they are more accurate and reliable. However, inasmuch as not all regulatory agencies, particularly those at a local level, have the resources or need for sophisticated equipment, other devices are also mentioned below.

[14]Donald E. Gower, "Air Monitoring Methods and Instrumentation Performance Experiences in the New York State Monitoring System," paper presented to Mid-Atlantic States Section, Air Pollut. Control Assoc., Technical Conference, May 18, 1973.

[15]Environmental Monitoring Systems Laboratory, Department E (MD-77), USEPA, Research Triangle Park, N.C. 27711.

[16]*Fed. Reg.* 40 CRF Part 58, May 10, 1979.

Particulate Sampling—Ambient Air

Measurements needing much more development are in the area of particulates where inhalable particles sizing (less than 10 to 3 μm), identification, metals, sulfates, and nitrates are important. Particulates can also be collected and tested for their mutagenic properties. Of all the particulate ambient air sampling devices, the high-volume sampler is the one most commonly used in the United States. Other devices also have application for the collection of different sized particulates.

High-volume (*Hi-vol*) *samplers* pass a measured high rate of (40 to 60 cfm) through a special filter paper (or fiberglass), usually for a 24-hour period. The filter is weighed before and after exposure, and the change in weight is a measure of the suspended particulate matter in $\mu m/m^3$ of air filtered. The particulates can be analyzed for weight, particle size (usually between 100 to 0.1 μm), and composition (such as benzene solubles, nitrates, lead, and sulfates), and for radioactivity. Particle size selective inlets can be put on Hi-vol samplers, and samples can be separated into two parts using impactor principles, those in the particle size range above and below 2 to 3 μm.

Hi-vol is the USEPA reference method. Air flow measurement is very important. An orifice with a manometer is recommended for flow measurement.

Sedimentation and settling devices include fallout or dustfall jars, settling chambers or boxes, Petri dishes, coated metal sheets or trays, and gum-paper stands for the collection of particulates that settle out. Vertically mounted adhesive papers or cylinders coated with petroleum jelly can indicate the directional origin of contaminants. Dustfall is usually reported as mg/cm^2 per month. Particulates can also be measured for radioactivity.

The automatic (*tape*) *smoke sampler* collects suspended material on a filter tape that is automatically exposed for predetermined intervals over an extended period of time. The opacity of the deposits or spots on the tape to the transmission or reflectance of light from a standard source is a measure of the air pollution. This instrument provides a continuous electrical output which can be telemetered to give immediate data on particulates. Thus the data is available without the delay of waiting for laboratory analysis of the Hi-vol filter. The equipment is used primarily to indicate the dirtiness of the atmosphere and does not directly measure the particulate TSP ambient air quality standard.

Inertial or centrifugal collection equipment operates on the cyclone collection principle. Large particles above 1 μm in diameter are collected, although the equipment is most efficient for the collection of particles larger than 10 μm.

Impingers separate particles by causing the gas stream to make sudden changes in direction in passing through the equipment. The wet impinger is used for the collection of small particles, the dry impinger for the larger particles. In the dry impinger a special surface is provided on which the particles collide and adhere.

In the *cascade impactor* the velocities of the gas stream vary, making possible the sorting and collection of different sized particles on special microscopic slides. Particulates in the range of 0.7 to 50 μm are collected.

Electrostatic precipitator-type sampling devices operate on the ionization principle using a platinum electrode. Particles less than 1 μm in size collect on an electrode of opposite charge and then are removed for examination. Combustible gases, if present, can affect results.

Nuclei counters measure the number of condensation nuclei in the atmosphere. They are a useful reference for weather commentators. A sample of air is drawn through the instrument, raised to 100 percent relative humidity, and expanded adiabatically, with resultant condensation on the nuclei present. The droplets formed scatter light in proportion to the number of water droplets, which are counted by a photomultiplier tube. Concentrations of condensation nuclei may range from 10 to 10,000,000 particles per cubic centimeter.[17] Condensation nuclei are believed to result from a combination of natural and man-made causes, including air pollution. A particle count above 50,000 is said to be characteristic of an urban area.

Pollen samplers generally use petroleum jelly-coated slides placed on a covered stand in a suitable area. The slides are usually exposed for 24 hr, and the pollen grains are counted with the aid of a microscope. The counts are reported as grains per cm^2. See Chapter 10 for ragweed control and sampling.

Gas Sampling

Gas sampling requires separation of the gas or gases being sampled from other gases present. The temperature and pressure conditions under which a sample is collected must be accurately noted. The pressure of a gas mixture is the sum of the individual gas pressures, as each gas has its own pressure. The volumes of individual gases at the same pressure in a mixture are also additive. *Concentrations of gases when reported in terms of ppm are by volume rather than by weight.* Proper sampling and interpretation of results requires competency and experience, knowledge concerning the conditions under which the samples are collected, and an understanding of the limitations of the testing procedures. Automated and manual instruments and equipment for gas sampling and analysis include the following:

The Pulsed Fluorescent Analyzer[18] measures sulfur dioxide by means of absorption of UV light. Pulsating UV light is focused through a narrow-band pass filter which reduces the outgoing light to a narrow wavelength band of 230 to 190 nanometers and directs it into the fluorescent chamber. Ambient

[17]Alexander Rihm, "Air Pollution," in *Dangerous Properties of Industrial Materials*, N. Irving Sax, Ed., Van Nostrand Reinhold Co., New York, 1969.
[18]*New York State Air Quality Report*, Annual 1978 DAR-79-1, Division of Air, N.Y. State Dept. of Environmental Conservation, Albany, N.Y., September 1979, pp. 240–243.

air containing SO_2 flows continuously through this chamber where the UV light excites the SO_2 molecules which in turn emit their characteristic decay radiation. This radiation, specific for SO_2 passes through a second filter and onto a sensitive photomultiplier tube. This incoming light energy is transformed electronically into an output voltage that is directly proportional to the concentration of SO_2 in the sample air.

The Nitrogen Oxides Chemiluminescence Analyzer.[18] Nitric oxide (NO) is measured by the gas phase chemiluminescent reaction between nitric oxide and ozone. This technique is also used to determine nitrogen dioxide (NO_2) by catalytically reducing NO_2 in the sample air to a quantitative amount of NO. Sample air is drawn through a capillary into a chamber held at 25 in. Hg vacuum. Ozone produced by electrical discharge in oxygen is also introduced into the chamber.

The luminescence resulting from the reaction between NO and ozone is detected by a temperature-stabilized photomultiplier tube and wavelength filter. An automatic valving system periodically diverts the sample air through a heated activated carbon catalyst bed to convert NO_2 to NO before it enters the reaction chamber. The sample measured from the converter is called NO_x. Since it contains the original NO plus NO produced from the NO_2 conversion, the differences between the sequential NO_x and NO readings are reported as NO_x. Primary dynamic calibrations are performed with gas phase titration using ozone and nitric oxide standards and with NO_2 permeation tubes.

The Ozone Chemiluminescence Analyzer.[18] Ozone is measured by the gas phase chemiluminescence technique which utilizes the reaction between ethylene and ozone (O_3). Sample air is drawn into a mixing chamber at a flow rate of 1 l/min where it is mixed with ethylene gas, and introduced at a flow rate of 25 cc/min. The luminescence resulting from the reaction of the ethylene with ambient ozone in the air supply is detected by a temperature-stabilized photomultiplier tube. This signal is then amplified and monitored by telemetry and on-site recorders. These ozone instruments contain provision for weekly zero and span checks. Primary dynamic calibrations are periodically performed which require standardization against a known, artificially generated ozone atmosphere.

The Carbon Monoxide Infrared (IR) Analyzer[18] utilizes dual beam photometers with detection accomplished by means of parallel absorption chambers or cells which are separated by a movable diaphragm. IR energy passes into each chamber; one containing the sample with CO, the other containing the reference gas. The reference gas heats up more than the ambient air sample with CO since CO absorbs more of the IR energy. This results in higher temperature and hence the volume-pressure in the reference chamber which is transmitted to the separating diaphragm designed to provide an electrical output to measure the CO concentration. However, it is necessary to remove water vapor interference as the humidity in ambient air absorbed by IR can introduce a significant error in CO readings. In one instrument (the USEPA reference method) the interference due to water vapor is eliminated by first passing one portion of the ambient air sample through a catalytic converter where CO is converted to CO_2 prior to entry into the reference chamber. The other half of the air sample containing CO passes directly into the sample chamber. This procedure cancels out the effect of moisture since both gas streams are identical except for the presence of CO.[18]

Smoke and Soiling Measurement

The Ringelmann Smoke Chart[19] consists of five* rectangular grids produced by black lines of definite width and spacing on a white background. When held at a distance, about 50 ft from the observer, the grids appear to give shades of gray between white and black. The grid shadings are compared with the pollution source (stack), and the grid number closest to the shade of the pollution source is recorded. About 30 observations are made in 15 min, and a weighted average is computed of the recorded Ringelmann numbers. The chart is used to determine whether smoke emissions are within the standards established by law; the applicable law is referenced to the chart. The system cannot be applied to dusts, mists, and fumes. Inspectors need training in making smoke readings. A reading of zero would correspond to all white; a reading of 5 would correspond to all black.

The Ringelmann chart is being replaced by a determination of the *percent opacity* of a particular emission as seen by a trained observer.† For example, a Ringelmann reading of No. 1 would correspond to an opacity of about 20 percent.

The Public Health Service Guide for Smoke Inspectors uses a piece of photographic film, corresponding to Ringelmann numbers, with four adjacent rectangular degrees of darkness plus a fifth black area.

Tape Sampler—Soiling. Soiling can be indicated as RUDS (Reflectance Units of Dirt Shade). One RUDS is defined as an optical reflectance of 0.01 caused by ten thousand linear feet of air passing through 0.786 square inches (1-in. diameter circle) of filter paper. A vacuum pump draws the air to be sampled through the filter tape. The particles collected soil a spot on the tape. The tape is advanced automatically after a 2-hour period; the air flow rate used is 0.455 cfm. A filter is used with the light source which admits light with a wavelength of approximately 400 millimicrons to measure the light reference, which information can be sent to a monitor. The sampling time period and air flow rate were chosen to conform with American Society of Testing Materials (ASTM) standards.

*Reduced to four grids or charts in the United States. The width or thickness of lines and their spacing in each grid or chart varies. A handy reduction of the Ringelmann Smoke Chart is the Power's Micro-Ringelmann available from Power, 33 West 42nd Street, New York, N.Y. 10036.

†USEPA Method 9, Appendix A, 40 CFR 60.

[19]Bureau of Mines Information Circular No. 7718, Publications Distribution Section, 4800 Forbes Street, Pittsburgh, Pa., August 1955.

Tape Sampler—Coefficient of Haze (COH). The tape sampler can be designed to measure light transmittance rather than reflectance. This will produce soiling measurements expressed as COH, an index of contaminant concentration, which is the USEPA preferred method. The method is similar to that outlined above except the photocell is under the tape. White light is used. It is necessary to automatically re-zero the instrument near each spot to compensate for tape thickness variation. The compensation is performed by solid state electronics.

The automated filter tape air sampler can also be used to monitor some gases. Special filter tapes are used to measure hydrogen sulfide, fluorides, and other gases. The spots produced by the gaseous pollutant are chemically treated and evaluated using the reflectance or transmission method.

Stack Sampling

The collection of stack samples, such as fly ash and dust emissions, requires special filters or thimbles of known weight and a measure of the volume of gases sampled. The sample must be collected at the same velocity at which the gases normally pass through the stack. The gain in weight divided by the volume of gases sampled corrected to 0°C (or 21°C) temperature and 760 mm mercury gives a measure of the dust and fly ash going out of the stack, usually as grains per cubic foot. When a series of samples is to be collected or measurements made, a "sampling train" is put together. It may consist of a sampling nozzle, several impingers, a freeze-out train, a weighed paper thimble, dry gas meter, thermometers, and pump.

A common piece of equipment for boiler and incinerator stack sampling is the Orsat apparatus. By passing a sample of the stack gas through each of three different solutions the percent carbon dioxide, carbon monoxide, and oxygen constituents in the flue gas is measured. The remainder of the gas in the mixture is usually assumed to be nitrogen.

Tracer materials may be placed in a stack to indicate the effect of a pollution source on the surrounding area. The tracer may be a fluorescent material, a dye, a compound that can be made radioactive, a special substance or chemical, or a characteristic-odor-producing material. The tracer technique can be used in reverse, that is, to detect the source of a particular pollutant, provided there are no interfering sources.

Measurement of Materials' Degradation

The direct effects of air pollution can be observed by exposing various materials to the air at selected monitoring stations. The degradation of materials is measured for a selected period on a scale of 1 to 10 with 1.0 representing the least degradation and 10.0 the worst, as related to the sample showing the least degradation. Materials exposed and conditions measured include steel corrosion, dyed fabric (nonspecific) for color fading, dyed fabric (NO_x-sensitive), dyed fabric (ozone-sensitive), dyed fabric (fabric soiling), dyed fabric (SO_2-sensitive), silver tarnishing, nylon deterioration, rubber cracking (crack depth), leather deterioration, copper pitting, and others. The samples are exposed for a selected period, such as rubber, 7 days at a time; silver, 30 days at a time; nylon 30 or 90 days at a time; cotton, 90 days at a time; steel, 90 days and one year at a time; zinc, one year at a time. Shrubs, trees, and other plants sensitive to certain contaminants or pollutants can also be used to monitor the effects of air pollution.

ENVIRONMENTAL FACTORS

The behavior of pollutants released to the atmosphere is subject to diverse and complex environmental factors associated with meteorology and topography, Meteorology involves the physics, chemistry, and dynamics of the atmosphere and includes many direct effects of the atmosphere on the earth's surface, ocean, and life. Topography refers to both the natural and man-made features of the earth's surface. The pollutants can be either accumulated or diluted, depending on the nature and degree of the physical processes of transport, dispersion, and removal and the chemical changes taking place. Because of the complexities of pollutant behavior in the atmosphere, it is important to distinguish between the activity of short-range primary pollutants (total suspended solids, sulfur dioxide) to which micrometeorology applies and to long-range secondary pollutants (ozone, acid rain) to which regional meteorology applies.

Within the scope of this text, it is not intended to provide a complete technical understanding of all the meteorological and topographical factors involved, but to provide only an insight into the relationships to air pollution of the more important processes.

Meteorology

The meteorological elements that have the most direct and significant effects on the distribution of the air pollutants are wind speed and direction, solar radiation, stability, and precipitation. It is therefore important to have a continuing base line of meteorological data, including these elements, to interpret and anticipate probable effects of air pollution emissions. Data on temperature, humidity, wind speed and direction, and precipitation are generally available through official government weather agencies. The National Weather Service (formerly U.S. Weather Bureau), Asheville, North Carolina, is a major source of information. Other potential sources of information are local airports, stations of the state fire weather service, military installations, public utilities and industrial complexes, colleges, and universities.

Wind

Wind is the motion of the air relative to the earth's surface. Although it is three-dimensional in its movement, generally only the horizontal components are denoted when used because the vertical component is very much smaller than the horizontal. This motion derives from the unequal heating of the earth's surface and the adjacent air, which in turn gives rise to a horizontal variation in temperature and pressure. The variation in pressure (pressure gradient) constitutes an imbalance in forces so that air motion from high toward low pressure is generated.

The uneven heating of the surface occurs over various magnitudes of space resulting in different magnitudes of organized air motions (circulations) in the atmosphere. Briefly, in descending order of importance, these are:

1. The primary or general (global) circulation associated with the large-scale hemispheric motions between the tropical and polar regions.
2. The secondary circulation associated with the relatively large-scale motions of migrating pressure systems (highs and lows) developed by the unequal distribution of large land and water masses.
3. The tertiary circulation (local) associated with small-scale variations in heating, such as valley winds and land and sea breezes.

For a particular area, the total effect of these various circulations establishes the hourly, daily, and seasonal variation in wind speed and direction. With respect to a known source or distribution of sources of pollutants, the frequency distribution of wind direction will indicate toward which areas the pollutants will be most frequently transported. It is customary to present long-term wind data at a given location graphically in the form of a "wind rose," an example of which is shown in Figure 6-1.

The concentration resulting from a continuous emission of a pollutant is inversely proportional to wind speed. The higher the wind speed the greater the separation of the particles or molecules of the pollutant as they are emitted, and vice versa. This is shown graphically in Figure 6-2*a*. The wind speed therefore is an indicator of the degree of dispersion of the pollutant and contributes to the determination of the area most adversely affected by an emission. Although an area may be located in the most frequently occurring downwind direction from a source, the wind speeds associated with this direction may be quite high so that resulting pollutant concentrations will be low as compared to another direction occurring less frequently but with lower wind speeds.

Smaller in scale than the tertiary circulation mentioned above there is a scale of air motion that is extremely significant in the dispersion of pollutants. This is referred to as the micrometeorological scale and consists of the very short-term, on the order of seconds and minutes, fluctuations in speed and direction. As opposed to the "organized" circulations discussed above, these air motions are rapid and random and constitute the wind characteristic called "turbulence." The turbulent nature of the wind is readily evident upon watching the rapid movements of a wind vane. These air motions provide the most effective mechanism for the dispersion or dilution of a cloud or plume of pollutants. The turbulent fluctuations occur in both the horizontal and vertical directions. The dispersive effect of fluctuations in horizontal wind direction is shown graphically in Figure 6-2*b*.

Turbulent motions are induced in the air flow in two ways: by thermal con-

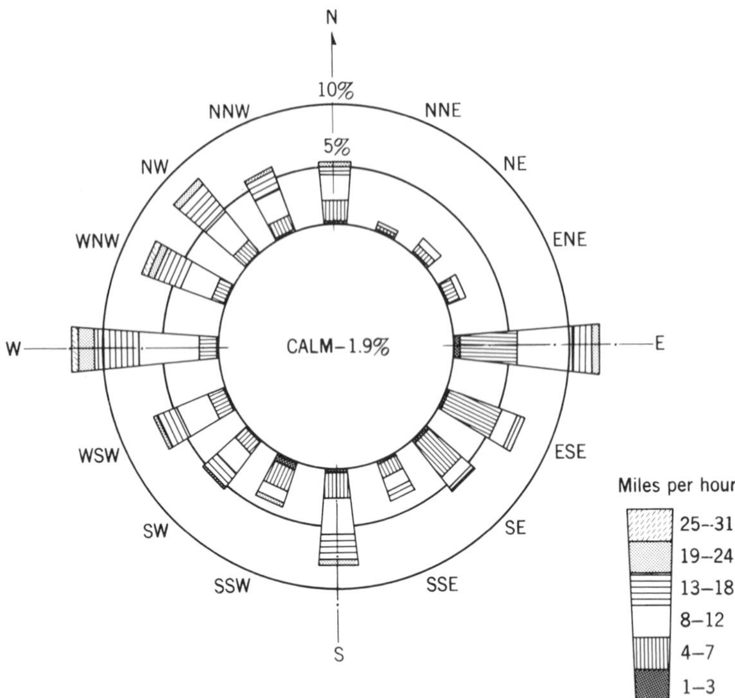

Figure 6-1 Example of wind rose, for a designated period of time, by month, season, or year. The positions of the spokes show the direction from which the wind was blowing. The total length of the spoke is the percent of the time, for the reporting period, that the wind was blowing from that direction. The length of the segments into which each spoke is divided is the percent time the wind was blowing from that direction at the indicated speed in mph. Horizontal wind speed and direction can vary with height.

vective currents resulting from heating from below (thermal turbulence) and by disturbances or eddies resulting from the passage of air over irregular, rough ground surfaces (mechanical turbulence).

It may be generally expected that turbulent motion and, in turn, the dispersive ability of the atmosphere would be greatly enhanced during a period of good solar heating and over relatively rough terrain.

Another characteristic of the wind that should be noted is that wind speed generally increases with height in the lower levels. This is due to the decrease with height of the "frictional drag" effect of the underlying ground surface features.

Stability and Instability

The stability of the atmosphere is its ability to enhance or suppress vertical air motions. Under unstable conditions the air motion is enhanced, and under stable conditions the air motion is suppressed. The conditions are determined by the vertical distribution of temperature.

In vertical motion, parcels of air are displaced. Due to the decrease of pressure with height, a parcel displaced upward will encounter decreased pressure and

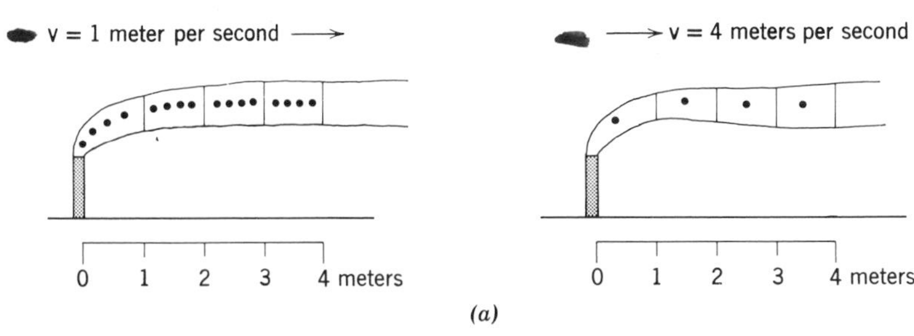

Figure 6-2a Effect of wind speed on pollutant concentration from constant source. (Continuous emission of 4 units per second.)

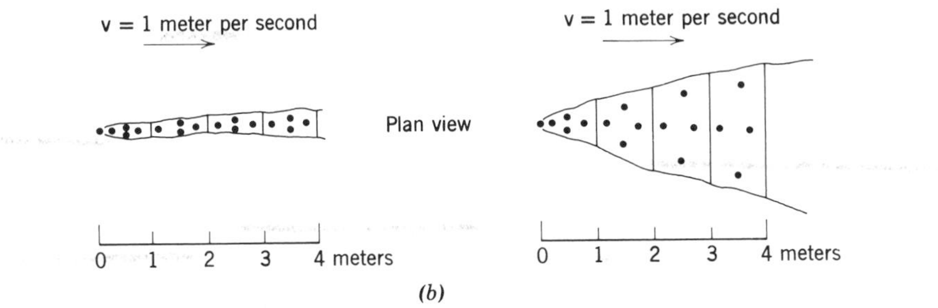

Figure 6-2b Effect of variability of wind direction on pollutant concentration from constant source. (Continuous emission of 4 units per second.)

expand. If this expansion process is relatively rapid or over a large area so that there is little or no exchange of heat with the surrounding air or by a change of state of water vapor, the process is dry adiabatic and the parcel of air will be cooled. Likewise, if the displacement is downward so that an increase in pressure and compression is experienced, the parcel of air will be heated.

The rate of cooling with height is the *dry adiabatic process lapse rate* and is approximately −1°C/100 m (−5.4°F/1000 ft).

The *prevailing* or *environmental lapse rate* is the decrease of temperature with height that may exist at any particular time and place. It can be shown that if the decrease of temperature with height is greater than −5.4°F/1000 ft, parcels displaced upward will attain temperatures higher than their surroundings. Air parcels displaced downward will attain lower temperatures than their surroundings. The displaced parcels will tend to continue in the direction of displacement. Under these conditions, the vertical motions are enhanced and the layer of air is defined as "unstable."

Furthermore, if the decrease of temperature with height is less than −5.4°F/1000 ft, it can be shown that air parcels displaced upward attain temperatures lower than their surroundings and will tend to return to their original positions. Air parcels displaced downward attain higher temperatures than their surroundings and also tend to return to their original position. Under these conditions, vertical motions are suppressed and the layer of air is defined as "stable."

Finally, if the decrease of temperature with height is equal to $-5.4°F/1000$ ft, displaced air parcels attain temperatures equal to their surroundings and tend to remain at their position of displacement. This is called "neutral stability."

Inversions

Up to this point, the prevailing temperature distribution in the vertical has been referred to as a "lapse rate," which indicates a decrease of temperature with height. However, under certain meteorological conditions, the distribution can be such that the temperature increases with height within a layer of air. This is called an "inversion" and constitutes an extremely stable condition.

There are three types of inversions that develop in the atmosphere: radiational (surface), subsidence (aloft), and frontal (aloft).

Radiational inversion is a phenomenon that develops at night under conditions of relatively clear skies and very light winds. The earth's surface cools by reradiating the heat absorbed during the day. In turn the adjacent air is also cooled from below so that within the surface layer of air there is an increase of temperature with height.

Subsidence inversion develops in high-pressure systems (generally associated with fair weather) within a layer of air aloft when the air layer sinks to replace air that has spread out at the surface. Upon descent the air heats adiabatically, attaining temperatures greater than the air below.

A condition of particular significance is the subsidence inversion that develops with a stagnating high-pressure system. Under these conditions the pressure gradient becomes progressively weaker so that the winds become very light, resulting in a great reduction in the horizontal transport and dispersion of pollutants. At the same time the subsidence inversion aloft continuously descends, acting as a barrier (lid) to the vertical dispersion of the pollutants. These conditions can persist for several days so that the resulting accumulation of pollutants can cause a serious health hazard.

Frontal inversion forms when air masses of different temperature characteristics meet and interact so that warm air overruns cold air.

There are many and varied effects of stability conditions and inversions on the transport and dispersion of pollutants in the atmosphere. In general, enhanced vertical motions under unstable conditions increase the turbulent motions, thereby enhancing the dispersion of the pollutants. Obviously the stable conditions have the opposite effect.

Stack emissions in inversions—depending on the elevation of emission with respect to the distribution of stability in the lower layers of air, the behavior of the plumes can be affected in many different ways. Pollutants emitted within the layer of a surface-based (radiational) inversion by low stacks can develop very high and hazardous concentrations at the surface level. On the other hand, when pollutants are emitted from stacks at a level aloft within the surface inversion, the stability of the air tends to maintain the pollutant at this level, preventing it from reaching the surface. However, after sunrise and continued radiation from the sun resulting in heating of the earth's surface and adjacent air, the inversion is "burned off." Once this condition is reached, the lower layer of air becomes unstable and all of the pollutant that has accumulated at the level aloft is rapidly dispersed downward to the surface. This behavior is called "fumigation" and can result in very high concentrations during the period. See Figure 6-3.

Precipitation

Precipitation constitutes an effective cleansing process of pollutants in the atmosphere in three ways: the washing out or scavenging of large particles by falling raindrops or snowflakes (washout); accumulation of small particles in the formation of raindrops or snowflakes in clouds (rainout); and removal of gaseous pollutants by dissolution and absorption.

The most effective and prevalent process is the washout of large particles, particularly in the lower layer of the atmosphere where most of the pollutants are released. The efficiencies of the various processes depend on complex relationships between properties of the pollutants and the characteristics of the precipitation.

Topography

The topographic features of a region include both the natural (hills, valleys, oceans, rivers, lakes, foliages, and so on) and man-made (cities, bridges, roads, canals, and so on) elements distributed within the region. These elements, per se, have little direct effect on pollutants in the atmosphere. The prime significance of topography is its effects on the meteorological elements. As stated previously, the variation in the distribution of land and water masses gives rise to various types of circulations. Of particular significance are the local or small-scale circulations that develop. These circulations can contribute either favorably or unfavorably to the transport and dispersion of the pollutants.

Along a coastline during periods of weak pressure gradient, intense heating of the land surface, as opposed to the lesser heating of the contiguous water surface, develops a temperature and pressure differential that generates an onshore air circulation. This circulation can extend to a considerable distance inland. At times during stagnating high-pressure systems, when the transport and dispersion of pollutants have been greatly reduced, this short-period afternoon increase in airflow may well prevent the critical accumulation of pollutants.

In valley regions, particularly in the winter, intense *surface inversions* are developed by the drainage down the slopes of air cooled by the radiationally cooled valley wall surfaces. Bottom valley areas that are significantly populated and industrialized can be subject to critical accumulation of pollutants during these periods.

The increased roughness of the surface created by the widespread distribution of buildings throughout a city can significantly enhance the turbulence of the airflow over the city, thereby improving the dispersion of the pollutants emitted. But at the same time the concrete, stone, and brick buildings and asphalt streets of the city act as a heat reservior for the radiation received from the sun during

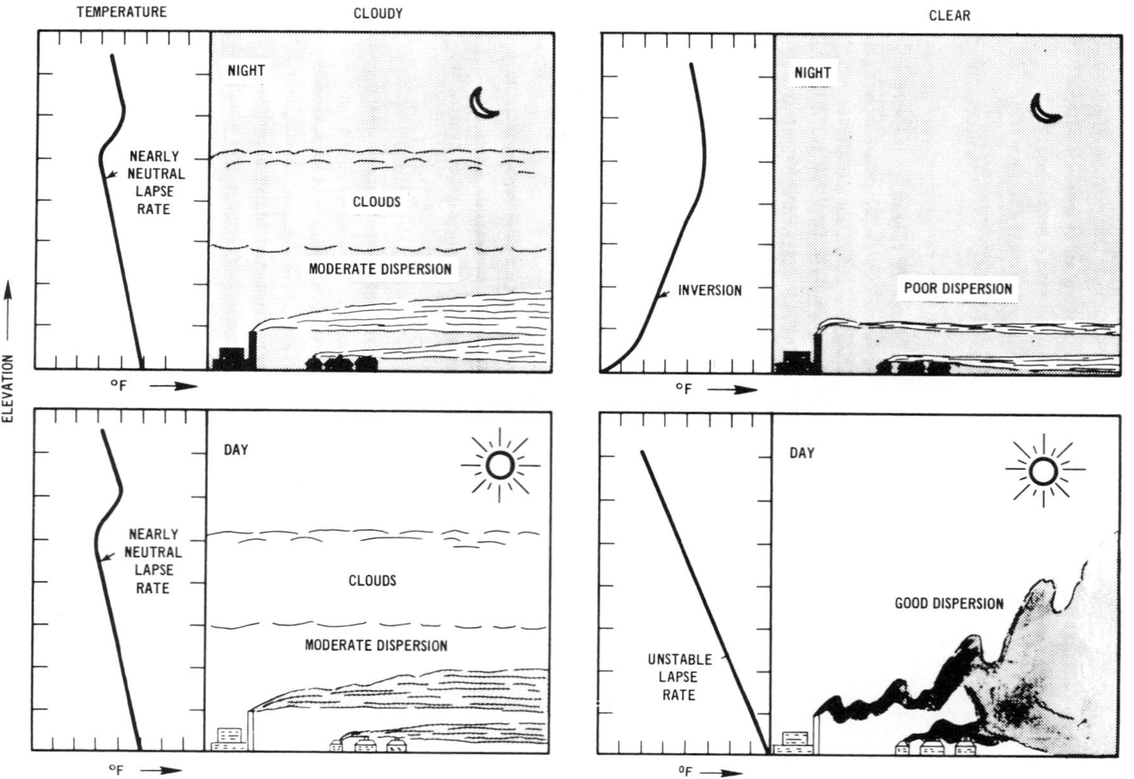

Figure 6-3 Diurnal and nocturnal variation of vertical mixing. Source: *Field Operation and Enforcement Manual for Air Pollution, Vol. 1, Organization and Basic Procedures*, USEPA, Office of Air Programs, Research Triangle Park, N.C. 27711, p. 1.24.

the day. This, plus the added heat from nighttime space heating during the cool months of the year, creates a temperature and pressure differential between the city and the surrounding rural area so that a local circulation inward to the city is developed. The circulation tends to concentrate the pollutants in the city. This phenomena is called the "urban heat island effect."

Areas on the windward side of mountain ranges can expect added precipitation due to the forced rising, expansion, and cooling of the moving air mass with resultant release of available moisture. This increased precipitation serves to increase the removal of the pollutants.

It is apparent, then, that topographical features can have many and diverse effects in the meteorological elements and the behavior of pollutants in the atmosphere.

AIR POLLUTION SURVEYS

An air pollution survey of a region having common topographical and meteorological characteristics is a necessary first step before a meaningful air resources management plan and program can be established. The survey includes an inventory of source emissions and a contaminant and meteorological sampling network, supplemented by study of basic demographic, economic, land use, and social factors.

Inventory

The inventory includes the location, height, exit velocity, and temperature of emission sources and identification of the processes involved; the air pollution control devices installed and their effectiveness; the pounds or tons of specific air pollutants emitted per day, week, month, and year, together with daily and seasonal variations in production. Inventories of area sources, i.e., home heating, small dry cleaners, etc. can be done simplistically through fuel use and solvent sales data. The emissions are then calculated from emission tables or by material balance. An estimate can then be made of the total pollution burden on the atmospheric resources of any given air basin.[20] Tables have been developed to assist in the calculation of the amounts and types of contaminants released; they can also be used to check on information received through personal visits, questionnaries, telephone calls, government reports, and the technical and scientific literature.[21] Additional sources of information are the complaint files of the health department, municipal and private agencies, published information, university studies, state and local chamber of commerce reports and files, and results of traffic surveys; also the Census of Housing, local fuel and gasoline sales.

[20]W. E. Jackson and H. C. Wohlers, "You need an air-pollution inventory," *Am. City*, (October 1967).

[21]Harry H. Hovey, Arnold Risman, and John F. Cunnan, "The Development of Air Contaminant Emission Tables for Nonprocess Emissions," *J. Air Pollut. Control Assoc.*, **16**, 362–366 (July 1966).

Air Sampling

Air and meteorological sampling equipment located in the survey area will vary, depending on a number of factors such as land area, topography, population densities, industrial complexes, and manpower and budget available. A minimum number of stations are necessary to obtain meaningful data.

Specific sampling sites for a comprehensive survey or for monitoring are selected on the basis of objective, scope, and budgetary limitations; accessibility for year-round operation, availability of reliable electrical power, amount and type of equipment available, program duration, and personnel available to operate stations; meteorology of the area, topography, adjacent obstructions, and vertical and horizontal distribution of equipment; and sampler operator problems, space requirements, protection of equipment and site, possible hazards, and public attitude toward the program.[22] The USEPA can provide monitoring and siting guidance.[23,24] Careful attention must also be given to the elimination of sampling bias and variables as related to size of sample, rate of sampling, collection and equipment limitations, and analytical limitations.

Basic Studies and Analyses

Basic studies include population densities and projections; land use analysis; mapping; economic studies and proposals including industrialization, transportation systems, community institutions, environmental health and engineering considerations, relationship to federal, state, and local planning, and related factors. Liaison with other planning agencies can be helpful in obtaining needed information that may already be available. See Chapter 2.

When all the data from the emission inventory, air sampling, and basic studies are collected, analyzed, and evaluated, a report is usually prepared. The analysis step should include calculations to show how the pollutants released to the atmosphere are dispersed and their possible effects under existing conditions and with future development.

Mathematical models could be developed, based on certain assumptions and on the data collected, to estimate the pollution levels that might result under various emission, topographical, and meteorological conditions. A data bank and a system for the collection and retrieval of information would generally be indicated. The approximate cost to achieve selected levels of air quality (for health, aesthetic, plant, and animal considerations) and the possible effect on industrial expansion, transportation modes and systems, availability and cost of fuel, and community goals and objectives should be determined. See also Planning and Zoning, this chapter.

The report would recommend air quality objectives based on EPA standards for areas in the region studied based on existing and proposed land uses. This

[22]"Introduction to Air Pollution Control," New York State Dept. of Health, Albany, March 1964.
[23]*Guidance for Air Quality Monitoring Network Design and Instrument Siting—CO Siting*, USEPA Office of Air Quality Planning and Standards Report No. 1.2-012, Supplement A, September 1975, and related reports.
[24]F. L. Ludwig, "Siting air monitoring stations," *Environ. Sci. Technol.*, July 1978, pp. 774–778.

will require consultation and coordination with state and local planning agencies.[25]

Short- and long-term objectives and priorities should be established to achieve the desired air quality. Recommendations to reduce air pollution might include control of pollutant emissions and limits in designated areas and under hazardous weather conditions and predictions; time schedules for starting control actions; control of fuel composition; requirement of plans for new or altered emission sources and approval of construction for compliance with emission standards; denial of certain plan approvals and prohibition of activities, or requirement of certain types of control devices; and performance standards to be met by existing and new structures and facilities.

The report is then formally submitted to the regulatory agency, board, or commission for further action. It would generally include recommendations for needed laws, rules, and regulations and administrative organization and staffing for the control of existing and new sources of air pollution.

AMBIENT AIR QUALITY STANDARDS*

Topographic, meteorological, and land use characteristics of areas within an air region will vary. The social and economic development of an area will result in different degrees of air pollution and demands for air quality. Because of this it is practical and reasonable to establish different levels of air purity for certain areas within a region. However, any standards adopted must assure, at a very minimum, no adverse effects on human health.†

Federal Standards

The Federal Air Quality Act of 1967 (Public Law 90-148) was amended in 1970, 1974, and in 1977. The act requires that the administrator of the USEPA develop

*"Ambient air" means that portion of the atmosphere, external to buildings, to which the general public has access.

†In the United States national primary and secondary ambient air quality standards were promulgated effective April 30, 1971. *Primary* ambient air quality standards are those that, in the judgment of the USEPA Administrator, based on the air quality criteria and allowing an adequate margin of safety, are requisite to protect the public health. *Secondary* ambient air quality standards are those that, in the judgment of the administrator, based on the air quality criteria, are requisite to protect the public welfare from any known or anticipated adverse effects associated with the presence of air pollutants in the ambient air (on soil, water, vegetation, materials, animals, weather, visibility, personal comfort and well-being).

In England (Ministry of Housing and Local Government 1966B) the standard states, in part: "No emission discharged in such amount or manner as to constitute a demonstrable health hazard in either the short or long term can be tolerated. Emissions, in terms of both concentration and mass rate of emission, must be reduced to the lowest practicable amount."

In the Soviet Union the goal is protection from any agent in the atmosphere that can be demonstrated to produce physiological effect, even if the effect cannot be shown to be harmful.

[25]Vernon G. MacKenzie, "Management of Our Air Resources," paper presented at the Growth Conference on Air, Land, and Water Resources, University of California at Riverside, October 7, 1963.

and issue to the states criteria of air quality for the protection of public health and welfare and further specifies that such criteria shall reflect the latest scientific knowledge useful in indicating the kind and extent of all identifiable effects on health and welfare that may be expected from the presence of an air contaminant, or combination of contaminants, in varying quantities.

The act requires the administrator to designate interstate or intrastate air quality control regions throughout the country as considered necessary to ensure adequate implementation of air quality standards. These regions are to be designated on the basis of meteorological, social, and political factors, which suggests that a group of communities should be treated as a unit.

The Federal Clean Air Act, as amended, requires that the administrator of the USEPA promulgate national ambient air quality standards for sulfur oxides, particulate matter, carbon monoxide, photochemical oxidants, hydrocarbons, and nitrogen oxides. These standards are included in Table 6-6.

The act requires each state to adopt

> a plan which provides for the implementation, maintenance, and enforcement of such national ambient air quality standards within each air quality control region (or portion thereof) within the State.

States are expected to attain the national primary ambient air quality standards, after approval by the administrator of the state plan, by December 1982. Both primary and secondary federal standards apply nationwide; however, state standards may be more stringent, except for motor vehicle emission standards, which are prescribed by law (California is exempt). States have been given until December 1987 to obtain the carbon monoxide and oxidant standard from cars if steps are taken to reduce pollution (by auto emission inspection and maintenance program, establishing mass transit systems, encouraging car pools, etc.).

The 1977 amendments to the Clean Air Act allow each state to classify clean air areas as Class I, where air quality has to remain virtually unchanged; Class II, where moderate industrial growth would be allowed; or Class III, where more intensive industrial activity would be permitted.

Class I areas *shall* include international parks, national wilderness areas exceeding 5000 acres, national memorial parks exceeding 5000 acres, and national parks exceeding 6000 acres. This classification and designation was made by Congress.

Pollutant Standards Index (PSI)

The Pollutant Standards Index (PSI) is a uniform method recommended* to classify and report urban air quality. Five criteria pollutants are judged for

*Recommendation of task force consisting of the Council on Environmental Quality, USEPA, Dept. of Commerce, National Oceanic and Atmospheric Administration, and the National Bureau of Standards.

Table 6-6 Federal Air Quality Standards

Pollutant	Primary[a]	Secondary[b]
Sulfur oxides (sulfur dioxide)		
Annual arithmetic mean	80 μg/m³ (0.03 ppm)	60 μg/m³ (0.02 ppm)
Maximum 24-hr concentration[c]	365 μg/m³ (0.14 ppm)	260 μg/m³ (0.1 ppm)
Maximum 3-hr concentration[c]	—	1,300 μg/m³ (0.5 ppm)
Particulate matter		
Annual geometric mean	75 μg/m³	60 μg/m³
Maximum 24-hr concentration[c]	260 μg/m³	150 μg/m³
Carbon monoxide		
Maximum 8-hr concentration[c]	10 mg/m³ (9 ppm)	10 mg/m³
Maximum 1-hr concentration[c]	40 mg/m³ (35 ppm)	40 mg/m³
Ozone		
Maximum 1-hr concentration[c]	240 μg/m³ (0.12 ppm)	240 μg/m³
Hydrocarbons		
Maximum 3-hr concen. (6-9 a.m.)[c]	160 μg/m³ (0.24 ppm)	160 μg/m³
Nitrogen dioxide		
Annual arithmetic mean	100 μg/m³ (0.05 ppm)	100 μg/m³
Maximum 24-hr concen.[c]	250 μg/m³ (0.125 ppm)	250 μg/m³
Lead		
Maximum ave. over 3-mo period	1.5 μg/m³	

Note: Sampling and analyzing shall be made in accordance with the "reference method" for the air pollutant as described in an appendix to Part 50—*National Primary and Secondary Ambient Air Quality Standards, Fed. Reg.*, **36,** No. 288 (Nov. 25, 1971) or by an equivalent method that can be demonstrated to the administrator's satisfaction to have a consistent relationship to the reference method. All measurements are corrected to a reference temperature of 25° C and a pressure of 760 millimeters of mercury.

[a] To protect the public health.

[b] To protect the public welfare.

[c] Not to be exceeded more than once per year.

amount and adverse effects on human health as shown in Table 6-7. On that basis the air quality evaluated is designated as presenting "hazardous conditions" if the PSI is greater than 300; "very unhealthful conditions" if the PSI is between 200 and 300; "unhealthful conditions" if the PSI is between 100 and 200; "moderate" if the PSI is 50 to 100; and "good" if the PSI is between 0 and 50. The PSI for one day rises above 100, that is, to the "Alert" level or higher, when any one of the five criteria pollutants reaches a level that may be judged to have adverse effects on human health.

Table 6-7 Comparison of PSI Values with Pollutant Concentrations and Health Effects

Index value	Air quality level	Pollutant levels: TSP, 24-hr ($\mu g/m^3$)	SO$_2$, 24-hr ($\mu g/m^3$)	CO, 8-hr ($\mu g/m^3$)	O$_3$, 1-hr ($\mu g/m^3$)	NO$_2$, 1-hr ($\mu g/m^3$)	Health effect descriptor	General health effects	Cautionary statements
500	Significant harm	1000	2620	57.5	1200	3750		Premature death of ill and elderly. Healthy people will experience adverse symptoms that affect their normal activity.	All persons should remain indoors, keeping windows and doors closed. All persons should minimize physical exertion and avoid traffic.
400	Emergency	875	2100	46.0	1000	3000	Hazardous	Premature onset of certain diseases in addition to significant aggravation of symptoms and decreased exercise tolerance in healthy persons.	Elderly and persons with existing diseases should stay indoors and avoid physical exertion. General populations should avoid outdoor activity.

300	Warning	825	1600	34.0	800	2260	Very unhealthful	Significant aggravation of symptoms and decreased exercise tolerance in persons with heart or lung disease, with widespread symptoms in the healthy population.	Elderly and persons with existing heart or lung disease should stay indoors and reduce physical activity.
200	Alert	375	800	17.0	400[a]	1130	Unhealthful	Mild aggravation of symptoms in susceptible persons, with irritation symptoms in the healthy population.	Persons with existing heart or respiratory ailments should reduce physical exertion and outdoor activity.
100	NAAQS	260	365	10.0	160	[b]			
50	50 percent of NAAQS	75[c]	80[c]	5.0	80	[b]	Moderate		
0		0	0	0	0	[b]	Good		

Source: *Guideline for Public Reporting of Daily Air Quality—Pollutant Standards Index* (*PSI*), USEPA, EPA-450/2—76-013, OAQPS 12.-044, Research Triangle Park, N.C., 1976, Table 3, p. 10.

[a]No index values reported at concentrations below those specified by alert level criteria.

[b]400 $\mu g/m^3$ was used instead of the O_3 alert level of 200 $\mu g/m^3$.

[c]Annual primary NAAQS. (National Ambient Air Quality Standards). See Table 6-6.

CONTROLS

Air pollution involves a source such as a power-generating plant burning heavy fuel oil; a production by-product or waste such as particulates, vapors, or gases; release of pollutants into the atmosphere, such as smoke or sulfur dioxide; transmission by airflows; and receptors who are affected, such as people, animals, plants, structures, and clothing. Controls can be applied at one or more points between the source and the receptor, starting preferably at the source. The application of control procedures and devices is more effective when supported by public information and motivation, production and process revision, installation of proper air-cleaning equipment, regulatory persuasion, and, if necessary, legal action.

Source Control

Processes that are sources of air pollution include chemical reaction, evaporation, crushing and grinding, drying and baking, and combinations of these operations.

For stationary combustion installations, such as fossil fuel-fired electric generating stations and plants generating steam for space heating or processes, the amounts and types of pollutants can be kept to a minimum using a fuel with less air pollution potential. Some examples of the types and amounts of contaminants from different types of fuels are given in Table 6-8.

Processes can also be designed and modified to reduce waste and the pollutants produced at the source. This has been a fundamental step in the reduction of industrial wastewater pollution and can certainly be applied to air pollution control.

The internal combustion engine is a major producer of air pollutants. A change from gasoline to another fuel or major improvement in the efficiency of

Table 6-8 Contaminant Emissions (LB/1,000,000 BTU Of Fuel)

Contaminant	Bituminous Coal	Anthracite Coal	Residual Fuel Oil	Distillate Fuel Oil	Natural Gas
Solids	0.630 (A)[a]	0.800	0.112	0.085	0.018
SO_2	1.407 (S)[b]	1.52 (S)	1.046 (S)	1.120 (S)	<0.001
NO_2	0.769	0.160	0.439	0.365	0.200
Organics	0.037	0.039	0.020	0.021	0.020
Organic Acids	1.150	0.595	0.714	0.765	0.003
Aldehydes	<0.001	<0.001	0.007	0.014	0.005
NH_3	0.078	0.040	0.047	0.050	0.020
CO	0.074	2.00	0.001	0.014	0.004

Source: Personal communication from E. W. Davis, Division of Air Resources, New York State Dept. of Environmental Conservation, Albany, 1972.

[a]Contaminant emission in pounds = .630 × ash content in percent.

[b]Contaminant emission in pounds = 1.407 × sulfur content in percent.

the gasoline engine would attack that problem at the source. Inspection of cars and light trucks for compliance with exhaust emissions standards can cut significantly hydrocarbon and carbon monoxide levels in the ambient air. Heavy-duty gasoline trucks also add a large percentage of carbon monoxide and hydrocarbons but their reduction will require phasing out old trucks and catalytic converter installation on new trucks (after 1983). Reducing the lead content of gasoline, capturing and recycling gasoline vapors resulting from control of lead in gasoline, and gasoline evaporation during handling, from filling stations, petroleum storage tanks, auto tanks, and carburetors are other means of control at the source. Improved mass transit, use of bus lanes, reduced travel by personal car, better traffic control for faster vehicle travel, and less stop-and-go are other means to reduce emissions.

Significant air pollution control can be achieved by process and material changes, by the reuse and recycling of waste materials, or by product recovery as by collection of combustion product particles of value. See also Chapter 5, Resource Recovery and Hazardous Wastes.

Proper design of basic equipment, provision of adequate solid waste collection service, elimination of open burning, and the upgrading or elimination of inefficient apartment house, municipal, and commercial incinerators also attack the problem at the source.

Proper operation and maintenance of production facilities and equipment will not only often reduce air pollution but will save money. For example, air-fuel ratios can determine the amount of unburned fuel going up the stack, combustion temperature can affect the strain placed on equipment when operated beyond rated capacity, and the competency of supervision can determine the quantity and type of pollutants released and the quality of the product.

Emission Control Equipment

Emission control equipment is designed to remove or reduce particulates, aerosols (solids and liquid forms), and gaseous by-products from various sources, and, in some instances, emissions result from inefficient design and operation.

The operating principles of aerosol collection equipment include:

1. Inertial entrapment by altering the direction and velocity of the effluent
2. Increasing the size of the particles through conglomeration or liquid mist entrainment so as to subject the particles to inertial and gravitational forces within the operational range of the control device
3. Impingement of particles on impact surfaces, baffles, or filters
4. Precipitation of contaminants in electrical fields or by thermal convection.[26]

The collection of gases and vapors is based on the particular physical and chemical properties of the gases to be controlled.

[26] *Air Pollution Control Field Operations Manual*, PHS, Pub. 937, DHEW, Washington, D.C., 20201, 1962.

Particulate Collectors and Separators

Some of the more common collectors and separators are identified below. These have application in mechanical operations for dust control such as in pulverizing, grinding, blending, woodworking, and in handling flour; also at power stations, incinerators, cement plants, heavy metallurgical operations, and other dusty operations. In general, collector efficiencies increase with particle size and from a low efficiency with baffled settling chambers, increasing with cyclones, electrostatic precipitators, spray towers, scrubbers, and baghouses, depending also on design and operation.

Settling chambers cause velocity reduction, usually to slower than 10 fps, and the settling of particles larger than 40 μm in diameter in trays that can be removed for cleaning. Special designs can intercept particles as small as 10 μm.

Cyclones impose a spiraling downward movement on the tangentially directed incoming dust-laden gas, causing separation of particles by centrifugal force and collection at the bottom of the cone. Particle sizes collected range from 5 to 200 μm at gas flows of 30 to 25,000 ft^3/min. Removal efficiency below 10 μm particle size is low. Cyclones can be placed in series or combined with other devices to increase removal efficiency. See Figures 6-4 and 6-5.

Sonic collectors can be used to facilitate separation of liquid or solid particles in settling chambers or cyclones. High-frequency sound-pressure waves cause

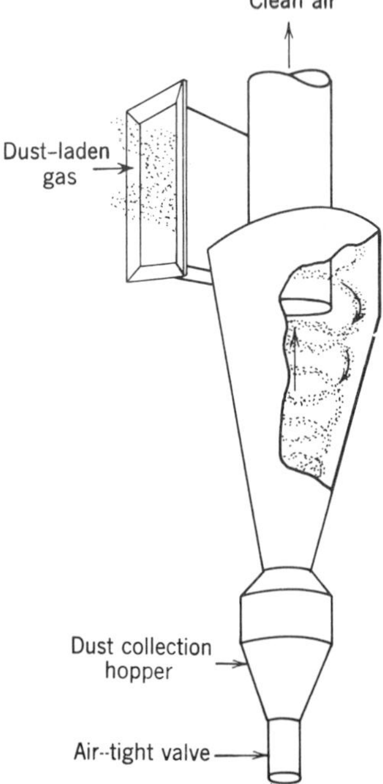

Figure 6-4 Flow of dust through cyclone. (Adapted from *Air Pollution Control Field Operations Manual*, PHS Pub. No. 937, DHEW, Washington, D.C., 1962.)

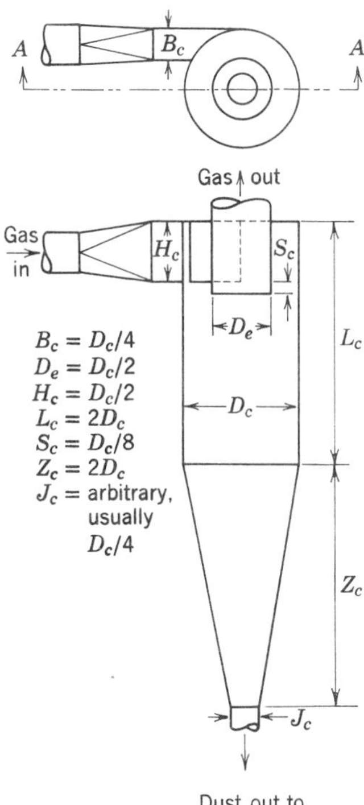

Figure 6-5 Diagram of cyclone separator. (From *Air Pollution Control Field Operations Manual*, PHS Pub. No. 937, DHEW, Washington, D.C., 1962.)

particles to vigorously vibrate, collide, and coalesce. Collectors can be designed to remove particles smaller than 10μm.

Filters are of two general types: the baghouse and cloth-screen. The filter medium governs the temperature of the gas to be filtered, particle size removed, capacity and loading, and durability of the filter. Filter operating temperatures vary from about 200°F for wool or cotton to 450 to 500°F for glass fiber.

A *baghouse filter* is shown in Figure 6-6. The tubular bags are 5 to 18 in. in diameter and from 2 to 30 ft in length. The dust-laden gas stream to be filtered passes through the bags where the particles build up on the inside and, in so doing, increase the filtering efficiency. Periodic shaking of the bags (tubes) causes the collected dust to fall off and restore the filtering capacity. The baghouse filter has particular application in cement plants, heavy metallurgical operations, and other dusty operations. Efficiencies exceeding 99 percent and particle removal below 10 μm in size are reported, depending on the major form and buildup.

Cloth-screen filters are used in the smaller grinding, tumbling, and abrasive cleaning operations. Dust-laden air passes through one or more cloth screens in series. The screens are replaced as needed. Other types of filters use packed fibers, filter beds, granules, and oil baths.

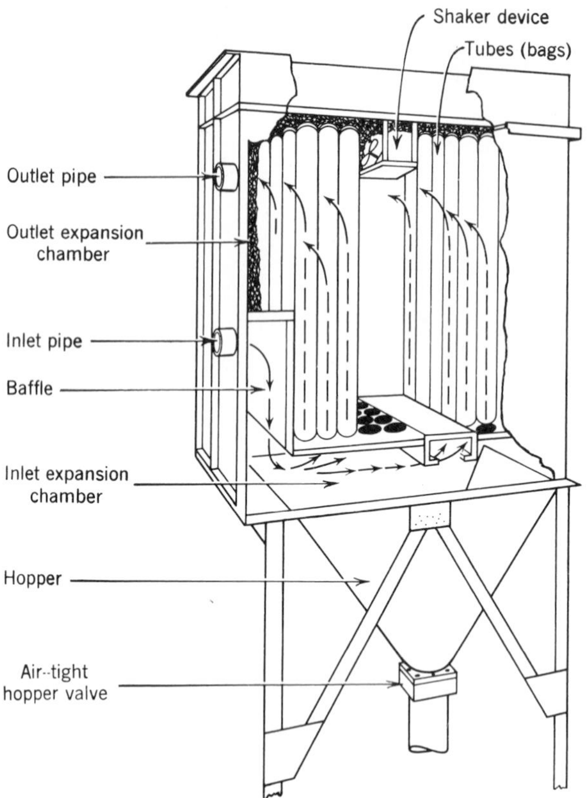

Figure 6-6 Simplified diagram of a baghouse. (From *Air Pollution Control Field Operations Manual*, PHS Pub. 937, DHEW, Washington, D.C., 1962.)

Electrostatic precipitators have application in power plants, in cement plants, and in metallurgical, refining, and heavy chemical industries for the collection of fumes, dusts, and acid mists. Particles, in passing through a high-voltage electrical field, are charged and then attracted to a plate of the opposite charge, where they collect. The accumulated material falls into a hopper when vibrated. See Figure 6-7.

The gases treated may be cold or at a temperature as high as 1100° F; but 600° F or less is more common. Precipitators are efficient for the collection of particles less than 0.5 μm in size; hence cyclones and settling chambers, which are better for the removal of larger particles, are sometimes used ahead of precipitators. Single-stage units operate at voltages of 25,000 or higher; two-stage units (used in air conditioning) operate at 12,000 V in the first or ionizing unit and at 6000 V in the second collection unit.

Electrostatic precipitators are commonly used at large power stations and incinerators to remove particulates from flue gases. They are considered one of the most effective devices for this purpose.

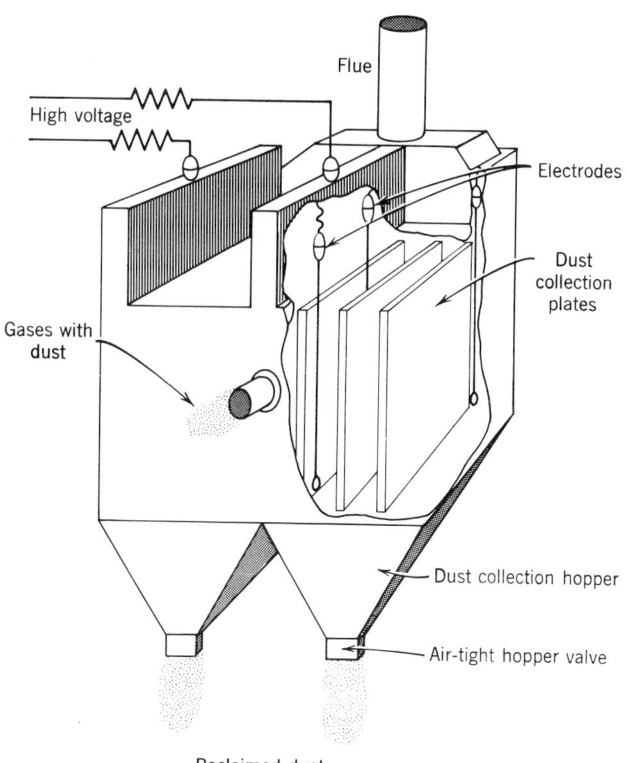

Figure 6-7 Diagram of plate-type electrostatic precipitator used to collect catalyst dust. (Adapted from *Air Pollution Control Field Operations Manual*, PHS Pub. 937, DHEW, Washington, D.C., 1962.)

Scrubbers are wet collectors generally used to remove particles that form as a dust, fog, or mist. A high-pressure liquid spray is applied to the gas passing through the washer, filter, venturi, or other device. In so doing the gas is cooled and cleaned. Although water is usually used as the spray, a caustic may be added if the gas stream is acidic. Where the spray water is recirculated, corrosion of the scrubber, fan, and pump impeller can be a serious problem. Particle size collected may range from 40 μm to as low as 1 μm with efficiency as high as 99 percent depending on the collector design. See Figures 6-8 and 6-9.

Gaseous Collectors and Treatment Devices

The release of gases and vapors to the atmosphere can be controlled by combustion, condensation, absorption, and adsorption. Combustion devices include thermal afterburners, catalytic afterburners, furnaces, and flares.

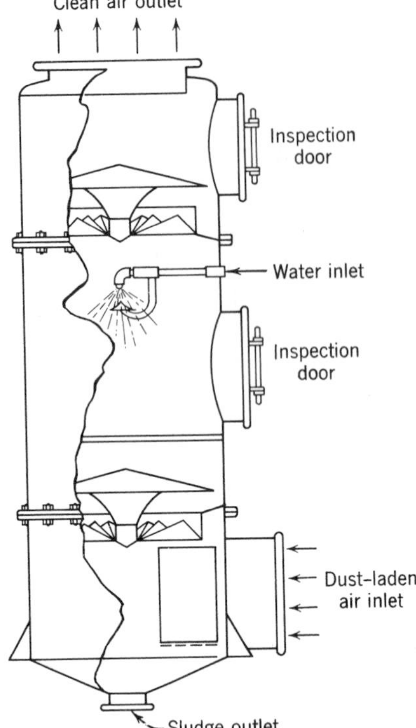

Figure 6-8 Centrifugal wash collector. (From *Air Pollution Control Field Operations Manual*, PHS Pub. 937, DHEW, Washington, D.C., 1962.)

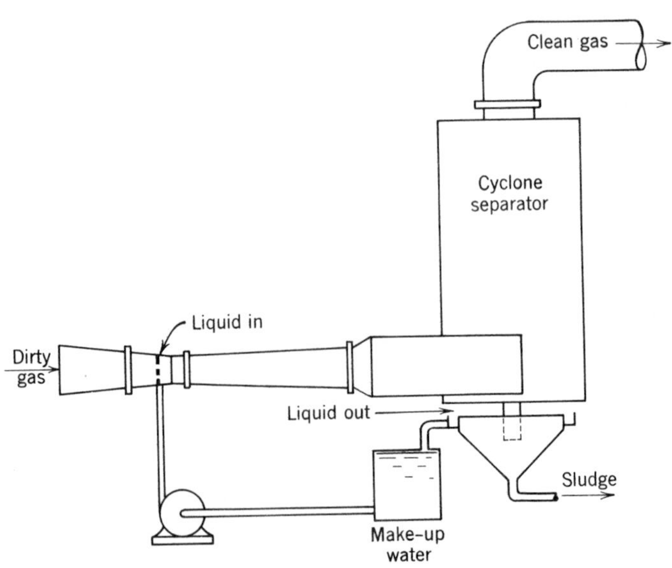

Figure 6-9 Venturi scrubber. (From *Air Pollution Control Field Operations Manual*, PHS, Pub. 937, DHEW, Washington, D.C., 1962.)

Thermal afterburners are used to complete the combustion of unburned fuel, such as smoke and particulate matter, and to burn odorous combustible gases. Apartment house and commercial incinerators and meat-packing plant smokehouses are examples of smoke and particulate emitters. Rendering, packing house, refinery, and paint and varnish operations, fish processing, and coffee roasting are examples of odor-producing operations. Afterburners usually operate at around 1200° F, but may range from 900 to 1600° F depending on the ignition temperature of the contaminant to be burned.

Catalytic afterburners may be used for the burning of lean mixtures of combustible gaseous air contaminants.

Condensers are best used to remove vapors by condensation, generally prior to passage to other air pollution control equipment and thus reduce the load on this equipment. Condensers are of the surface and contact types. In the surface condenser the vapor comes into contact with a horizontal cool surface and condenses to form liquid droplets with a pure saturated vapor or, more commonly, a film. In the contact condenser the coolant, vapors, and condensate are all in intimate direct contact.

Adsorbers are of the fixed-bed stationary or rotating type, in horizontal or vertical cylinders, usually with activated carbon beds or supported screens, through which the gas stream passes. In adsorption the molecules of a fluid such as a gas, liquid, or dissolved substance to be treated are brought into contact with the adsorbent, such as activated carbon, aluminas, silicates, char, or gels which collect the contaminant in the pores or capillaries. The material adsorbed is called the adsorbate. In some cases the adsorbent, such as activated carbon, is regenerated by superheated steam at about 650° F; the contaminant is condensed and collected for proper disposal. In other cases the adsorbent and adsorbate are separated from the fluid and discarded. Solid adsorbents have very large surface-to-volume ratios and different adsorptive abilities dependent on the particular adsorbate. The life of an activated carbon adsorption bed is reduced if particulate matter is not first removed.

In *absorption*, the gaseous emission to be treated is passed through a packed tower, spray or plate tower, and venturi absorbers, where it comes in contact with a liquid absorbing medium or spray that selectively dissolves or reacts with the air contaminants to be removed. For example, oxides of nitrogen can be absorbed by water; hydrogen fluoride, by water or an alkaline water solution. Absorption is generally also used to control emissions of sulfur dioxide, hydrogen sulfide, hydrogen chloride, chlorine, and some hydrocarbons.

Vapor conservation equipment is used to prevent vapors escaping from the storage of volatile organic compounds such as gasoline. A storage tank with a sealed floating roof cover or a vapor recovery system connected to a storage tank is used. Vapors that can be condensed are returned to the storage tank.

Dilution by Stack Height

Since wind speed increases with height in the lower layer of the atmosphere, the release of pollutants through a tall stack enhances the transport and diffusion of the material. The elevated plume is rapidly transported and diffused downwind. This generally occurs at a rate faster than that of the diffusion toward the ground. The resulting downwind distribution of pollutant concentrations at the ground level is such that concentrations are virtually zero at the base of the stack, increase to a maximum at some downwind distance, and then decrease to negligible concentrations thereafter. This distribution and the difference due to stack height are shown schematically in Figure 6-10. This applies to uncomplicated weather and level terrain. Obviously, if the plume is transported to hill areas, the surfaces will be closer to the center of the plume and hence will experience higher concentrations.

Meteorological conditions will determine the type of diffusion the pollutant plume will follow. See Figure 6-3. With heavy atmospheric turbulence associated with an unstable lapse rate, the plume will "loop" as it travels downwind. With lesser turbulence associated with a neutral lapse rate, the plume will form a series of extended, overlapping cones called "coning." With stable air conditions and little turbulence associated with an inversion, the plume will "fan" out gradually. With the discharge of a plume below an inversion, the plume will be dispersed rapidly downward to the ground surface causing "fumigation." With the discharge of a plume within the inversion layer, the plume will spread out horizontally as it moves downwind with little dispersion toward the ground. Erratic weather conditions can cause high concentrations of pollutants at ground level if the plume is transported to the ground.

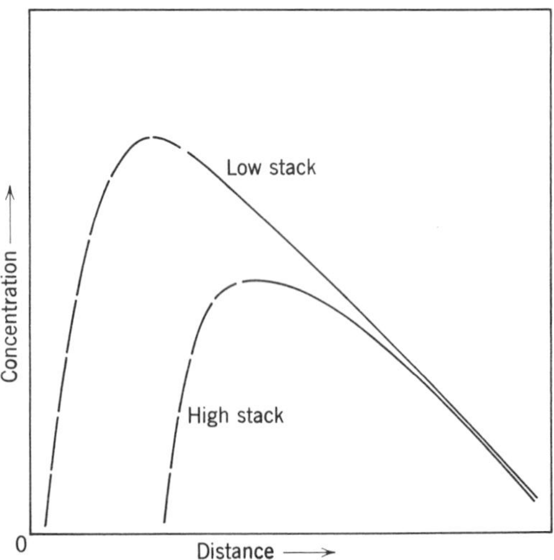

Figure 6-10 Variation of ground-level pollutant concentration with downwind distance. The distance may be hundreds of miles.

It has been general practice to use high stacks for the emission of large quantities of pollutants, such as in fossil-fueled power production, to reduce the relatively close-in ground-level effects of the pollutants. Stacks of 250 to 350 ft in height are not unusual, and some are as high as 800 to 1250 ft. It should be recognized, however, that there is a practical limit to height beyond which cost becomes excessive and the additional dilution obtained is not significant.

Although local conditions are improved where a tall stack is used, adverse environmental effects continue to be associated with the distant (long-range) transport of pollutants. For example, the pollutants contribute to acid rain, heavy metal particle deposition, and toxic metal dissolution from surrounding or downwind soils and rocks into surface and groundwaters which adversely affect the flora and fauna hundreds or more miles away as previously noted. Emphasis therefore should be placed on reduction of emission concentrations, rather than on dispersion from a tall stack, to improve ambient air quality.

Open Space

The vegetation and soils in open spaces can, through absorption, adsorption, impingement, and deposition, reduce the amount of pollutants passing through. The outer 65- to 85-ft edge of an uncrowded forest or greenbelt can, under certain conditions of temperature, humidity, wind velocity, soil conditions, and plant diversity and density, remove significant amounts of sulfur dioxide and carbon monoxide and reduce particulates 40 to 50 percent. However all pollutants are not removed; lead particulates for example are not removed. The effects of pollutant accumulation on soil fertility, groundwater quality, and food chains are not known. On the "plus" side, greenbelts also serve as noise buffers, recreation and wildlife refuge areas, and pleasing screens for visual blight.[27]

Planning and Zoning

The implementation of planning and zoning controls requires professional stimulation and the cooperation of the state and regional planning agencies and the local county, city, village, and town units of government.

Although air pollution control should start at the source, the devices available and the economic, social, and political factors involved will limit the amount of control that can be realistically achieved in many instances. Hence a combination of planning and zoning means should be considered when adequate pollution elimination at the source is not feasible. These means could include plant siting downwind from residential, work, and recreation areas, with consideration given to climate and meteorological factors, frequency of inversions, topography, air movement, stack height, and adjacent land uses. Additional factors

[27]"Nature's Scrubber," *Conserv. News*, August 15, 1977, p. 11 reported in USEPA study "Open Space As An Air Resource Management Measure."

are distance separation, open-space buffers, designation of industrial areas, traffic and transportation control, and possible regulation of plant raw materials and processes. All these controls must recognize the present and future land use and especially the air quality needed for health and comfort, regardless of the land ownership.

The maintenance of air quality that meets established criteria therefore requires regulation of the location, density, and/or type of plants and plant emissions that could cause contravention of air quality standards. This calls for local and regional land-use control and cooperation so that the construction of plants permitted would incorporate practices and control equipment that would not emit pollution that could adversely influence the air quality of the community in the airshed. See Tables 6-6 and 6-7 and Ambient Air Quality Standards, earlier in this chapter.

Monitoring of the air at carefully selected locations would continually inform and alert the regulatory agency of the need for additional source control and enforcement of emission standards. Conceivably, under certain unusually adverse weather conditions, a plant may have to take previously planned emergency actions to reduce or practically eliminate emissions for a period of time.

Air zoning establishes different air quality standards for different areas based on the most desirable and feasible use of land. As discussed earlier in this chapter, the 1977 amendments to the Clean Air Act allow each state to classify air areas as either Class I, II, or III. Class I areas would remain virtually unchanged and Class III could permit intensive industrial activity. Specific standards are established for each classification level. In all levels, however, protection of the health of people is paramount. Insofar as air zoning is concerned, an industry should be able to choose its location and types of emission controls so long as the air quality standards are not violated.

Although air zoning provides a system or basis for land use and development, sound planning can assist in greatly minimizing the effects of air pollution. A WHO Expert Committee suggests the following:[28]

1. The siting of new towns should be undertaken only after a thorough study of local topography and meteorology.
2. New industries using materials or processes likely to produce air contaminants should be so located as to minimize the effects of air pollution.
3. Satellite (dormitory) towns should restrict the use of pollution-producing fuels.
4. Provision should be made for green belts and open spaces to facilitate the dilution and dispersion of unavoidable pollution.
5. Greater use should be made of hydroelectric and atomic power and of natural gas for industrial processes and domestic purposes, thereby reducing the pollution resulting from the use of conventional fossil fuels.
6. Greater use should be made of central plants for the provision of both heat and hot water for entire (commercial or industrial) districts.
7. As motor transport is a major source of pollution, traffic planning can materially affect the level of pollution in residential areas.

[28] "Environmental Health Aspects of Metropolitan Planning and Development," *WHO Tech. Rep. Ser.* **297**, 1965.

It is apparent that more needs to be learned and applied concerning open spaces, bodies of water, and trees and other vegetation to assist in air pollution control. For example parks and greenbelts appear to be desirable locations for expressways because vegetation, in the presence of light, will utilize the carbon dioxide given off by automobiles and release oxygen. In addition highway designers must give consideration to such factors as road grades, speeds and elevations, natural and artificial barriers, interchange locations, and adjacent land uses as means of reducing the amounts and effects of automobile noise and emissions.

Air Quality Modeling

It is possible to calculate and predict, *within limits*, the approximate effects of existing and proposed air pollution sources on the ambient air quality.[29–32] A wide variety of models are used to estimate the air quality impacts of sources on receptors, to prepare or review new industrial and other source applications, and to develop air quality management plans for an area or region.

Air quality models can be categorized into four classes.

1. *Gaussian.* Most often used for estimating the ground level impact of nonreactive pollutants from stationary sources in a smooth terrain.
2. *Numerical.* Most often used for estimating the impact of reactive and nonreactive pollutants in complex terrain.
3. *Statistical.* Employed in situations where physical or chemical processes are not well understood.
4. *Physical.* Involves experimental investigation of source impact in a wind tunnel facility.

Because of the almost limitless variety of situations for which modeling may be employed, no single model can be considered "best." Instead, the user is encouraged to examine the strengths and weaknesses of the various models available and select the one best suited to the particular job at hand.

The USEPA has made a number of models available to the general public through its User Network for Applied Modeling of Air Pollution (UNAMAP). These models can be obtained from the National Technical Information Service (NTIS).

The information needed to use an air quality model includes source emissions data, meteorological data, and pollutant concentration data.

[29]Maynard Smith, *Recommended Guide for the Prediction of the Dispersion of Airborne Effluents*, American Society of Mechanical Engineers (ASME), New York, 1968.

[30]Arthur C. Stern, Henry C. Wohlers, Richard W. Boubel, and William P. Lowry, *Fundamentals of Air Pollution*, Academic Press, N.Y., 1973, pp. 274–287.

[31]R. E. Munn, "Air Pollution Meteorology," *Manual on Urban Air Quality Management*, WHO Regional Publication European Series No. 1, M. J. Suess and S. R. Craxford, Eds., Copenhagen, 1976, pp. 101–126.

[32]A. T. Rossano and T. A. Rolander, "The Preparation of an Air Pollution Source Inventory," *Manual on Urban Air Quality Management*, WHO Regional Publication European Series No. 1, M. J. Suess and S. R. Craxford, Eds., Copenhagen, 1976, pp. 127–152.

Source Emission Data

Sources of pollutants can generally be classified as point, line, or area sources. Point sources are individual stacks and are identified by location, type and rate of emission, and stack parameters (stack height, diameter, exit gas velocity, and temperature). Line sources are generally confined to roadways and can be located by the ends of roadway segments. Area sources include all the minor point and line sources which are too small to require individual consideration. These sources are usually treated as a grid network of square areas, with pollutant emissions totalled and distributed uniformly within each grid square.

Meteorological Data

The data needed to represent the meteorological characteristics of a given area consist of (as a minimum) wind direction, wind speed, atmospheric stability, and mixing height. The representativeness of the data for a given location will be dependent upon the proximity of the meteorological monitoring site to the area being studied, the period of time during which data are collected, and the complexity of terrain in the area. Local universities, industries, airports, and government agencies can all be used as sources of such data.

Pollutant Concentration Data

In order to assess the accuracy of a model for a particular application, predicted concentrations must be compared against observed values. This can be done by obtaining historical pollutant concentration data from air quality monitors located in the study area. Air quality data from monitors located in remote areas should also be obtained to determine if a background concentration must be included in the model. Data should be verified using appropriate statistical procedures.

The accuracy of the model used depends upon the following factors:

1. How closely do the assumptions upon which the model is based correspond to the actual conditions for which the model is being used? For example, a model which assumes that the area being modeled is a flat plain of infinite extent may work well in Kansas but not in Wyoming.
2. How accurate is the information being used as input for the model? Of particular importance here is verifying the accuracy of source emission data. Some points to consider are:
   a. Should the source emission data be given in terms of potential, actual or allowable emissions? "Actual" emissions should always be used for model verification.
   b. Does emission rate vary by time of day or time of year?
   c. What level of production, percent availability, etc. should be assumed for each emission source? The emission rates for industrial sources will often decline significantly during periods of economic recession. Similarly, stationary fuel combustion sources (for space heating) will vary according to the severity of the winter.
   d. Are stack parameters correct? Are there nearby structures or terrain features which could influence the dispersion patterns of individual sources?
   e. Is the source location correctly identified?
   f. How reliable is the pollution control equipment installed on each emission source?

The user will often find that the job of verifying the input data is the most difficult and time-consuming part of the modeling process.

As the cost of computer services continues to decline, it is expected that air quality modeling will become an available technology for smaller agencies such as local health and planning departments. The person who performs this modeling will have to be knowledgeable not only in traditional air pollution control engineering, but also in the fields of air pollution meteorology and computer programming.

PROGRAM AND ENFORCEMENT

General

A program for air resources management should be based on a comprehensive area-wide air pollution survey including air sampling, basic studies and analyses, and recommendations for ambient air quality standards. The study should be followed by an immediate and long-term plan to achieve the community air quality goals and objectives, coupled with a surveillance and monitoring system and regulation of emissions.

MacKenzie proposes the following conclusions and decisions for the implementation of a study:[25]

1. Select air quality standards—possibly with variations in various parts of the area.
2. Cooperate with other community planners in allocating land uses.
3. Design remedial measures calculated to bring about the air quality desired. Such measures might include several or all of the following: limitations on pollutant emissions; variable emission limits for certain weather conditions and predictions; special emission limits for certain areas; time schedules for commencing certain control actions; control of fuel composition; control of future sources by requiring plan approvals; prohibition of certain plan approvals; prohibition of certain activities or requirements for certain types of control equipment; and performance standards for new land uses.
4. Outline needs for future studies pertaining to air quality and pollutant emissions and design systems for collection, storage, and retrieval of the resultant data.
5. Establish priorities among program elements and set dates for implementation.
6. Prepare specific recommendations as to administrative organization needed to implement the program; desirable legislative changes; relationships with other agencies and programs in the area and adjoining areas and at higher governmental levels; and funds, facilities, and staff required.

As in most studies, a continual program of education and public information supplemented by periodic updating is necessary.[33] People must learn that air pollution can be a serious hazard and must be motivated to support the need for its control. In addition, surveys and studies must be kept current; otherwise the air resources management activities may be based on false or outdated premises.

International treaties, interstate compacts or agreements, and regional organizations are sometimes also needed to resolve air pollution problems that cross jurisdictional boundaries. This becomes more important as industrialization increases and as people become more concerned about the quality of their environment.

[33]Alexander Rihm, Jr., "Public Relations in Air Pollution Control," paper presented at meeting of Public Relations Society of America, Albany, N.Y., November 16, 1967.

It becomes apparent that the various levels of government each have important complementary and cooperative roles to play in air pollution control.

The federal government role includes research into the causes and effects of air pollution as well as the control of international and interstate air pollution on request of the affected parties. It should also have responsibility for a national air-sampling network, for training, for preparation of manuals and dissemination of information, and for assisting the states and local governments. In the United States this is done primarily through the USEPA. Other federal agencies making major contributions are the U.S. Weather Service; the Nuclear Regulatory Commission, in relation to the effects of radioactivity; the Department of Agriculture, in relation to the effects of air pollution on livestock and crops; the Department of Interior; the Department of Commerce, including the National Bureau of Standards; and the Civil Aeronautics Administration.

The state role is similar to the federal role. It would include, in addition, the setting of statewide standards and establishment of a sampling network; the authority to declare emergencies and possession of appropriate powers during emergencies; the delegation of powers to local agencies for control programs; and the conducting of surveys, demonstration projects, public hearings, and special investigations.

The role of local government is that delegated to it by the state and could include complete program implementation and enforcement.

Organization and Staffing

Organization and staffing will vary with the level of government, the legislated responsibility, funds provided, government commitment, extent of the air pollution, and other factors. Generally air pollution programs are organized and staffed on the state, county, and large-city, in addition to federal, levels. In some instances limited programs of smoke and nuisance abatement are carried out in small cities, towns, and villages as part of a health, building, or fire department program. Because of the complexities involved, competent direction, staff, and laboratory support is needed to carry out an effective and comprehensive program. A small community usually cannot afford and in fact might not have need for a full staff; but it could play a needed supporting role to the county and state programs. In this way uniform policy guidance and technical support could be provided and local on-the-spot assistance utilized. The local government should be assigned all those responsibilities it is capable of handling effectively.

An organization chart for an Air Resources Management Agency is shown in Figure 6-11. There are many variations.

Regulation

A combination of methods and techniques is generally used to prevent and control air pollution, after a program is developed, air quality objectives es-

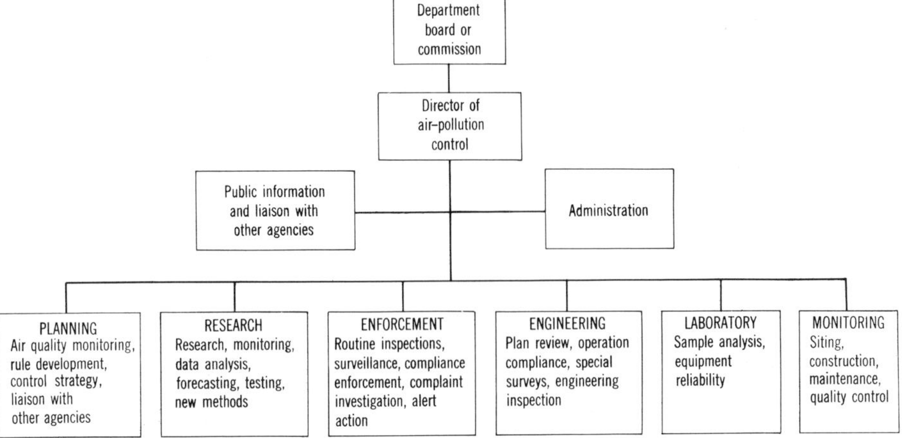

Figure 6-11 Air resources management functional organization chart.

tablished, and problem areas defined. These include:

1. Public information and education.
2. Source registration.
3. Plan review and construction–operation approval.
4. Emission standards.
5. Monitoring and surveillance.
6. Technical assistance and training.
7. Inspection and compliance follow-up.
8. Conference, persuasion, and administrative hearing.
9. Rescinding or suspension of operation permit.
10. Legal action—fine, imprisonment, misdemeanor, injunction.

Effective regulation requires the development and retention of competent staff and the assignment of responsibilities. In a small community the responsibilities would probably be limited to source location and surveillance, data collection, smoke and other visible particulate detection, complaint investigation, and abatement as an arm of a county, regional, or state enforcement unit.

Regulatory agencies usually develop their own procedures, forms, and techniques to carry out the functions listed above. Staffing, in addition to the director of air pollution control, may include one or more of the following: engineers, scientists, sanitarians, chemists, toxicologists, epidemiologists, public information specialists, technicians, inspectors, attorneys, administrative assistants, statisticians, meteorologists, electronic data processing specialists, and personnel in supporting services.

Detailed information on inspection and enforcement is given in *Air Pollution Field Operations Manual*[34] and in *Community Action Guides for Public Officials—Air Pollution Control.*[35] Additional information is also given in Chapter 12.

[34]Melvin I. Weisburd and S. Smith Griswold, PHS Pub. 937, DHEW, Washington, D.C., 1962.
[35]National Association of Counties Research Foundation, Washington, D.C.

Important in regulation is the development of working relationships and memoranda of agreements with various public and private agencies. For instance government construction, equipment, and vehicles could set examples of air pollution prevention. The building department would ensure that new incinerators and heating plants have the proper air pollution control equipment. The police would enforce vehicular air pollution control requirements. The fire department would carry out fire prevention and perhaps boiler inspections. The planning and zoning boards would rely on the director of air pollution control and the director's staff for technical support, guidance, and testimony at hearings. Equipment manufacturers would agree to sell only machinery, equipment, and devices that complied with the emission standards. The education department would incorporate air pollution prevention and control in its environmental health curriculum. Industry, realty, and chainstore management would agree to abide by the rules and to police itself. Cooperative training and education programs would be provided for personnel responsibile for operating boilers, equipment, and other facilities that may contribute to air pollution. These are but a few examples. With ingenuity many more voluntary arrangements can be devised to make regulation more acceptable and effective.

NOISE CONTROL

One of the most important tasks of architects, builders, acoustic engineers, urban planners, industrial hygiene engineers, equipment manufacturers, and public health personnel is to insure that noise and vibration are kept to an acceptable level in the general environment, in the work place, and inside dwellings. Noise is of special concern in occupational health where hearing loss has been documented.

The discussion that follows will touch upon some of the fundamentals of noise and its effects, measurement, reduction, and control. Special problems should involve experts such as acoustical consultants.

Definitions and Explanation of Selected Terms and Properties of Sound

Sound Sound, and therefore all noise, is physically a rapid alteration of air pressure above and below atmospheric pressure. Basically, all sounds travel as sound pressure waves from a vibrating body such as a human larynx, radio, TV, or record player speaker, or a vibrating machine.

A sound that contains only one frequency is a *pure tone* which is expressed in Figure 6-12 as a sine curve. Most sounds contain many frequencies. In general, the waves travel outward from the source in three dimensions. The *pitch* of a sound is determined primarily by frequency: vibrations per second. The amplitude or magnitude of sound is the *sound pressure.*

The distance that a sound wave travels in one cycle or period is the wave-

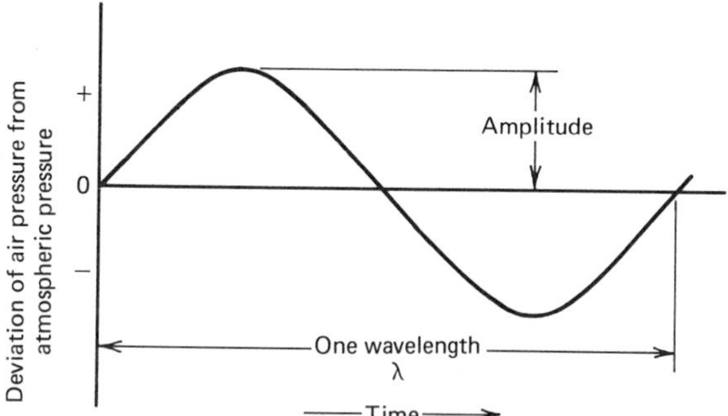

Figure 6-12 Pure tone, sine wave.

length of the sound. This is illustrated in Figure 6-12. Wavelength is given by the equation

$$\lambda = \frac{c}{f}$$

in which

λ = wavelength, in ft
f = frequency, in Hz (cycles per sec)
c = speed of sound, in ft per sec

Sound travels through gases, liquids, and solids, but not through a vacuum. The speed with which sound travels through a particular medium is dependent on the compressibility of and density of the medium. Our own voice reaches us primarily through the bony structures in our head. Most sound reaches us through the air and less frequently through solids and liquids. The speed of sound through various media is given in Table 6-9.

As sound travels through a medium, it loses energy or amplitude in two ways: molecular heating and geometric spreading. Drapes for example absorb sound, releasing the energy as heat to the surrounding air. Air itself also absorbs sound to a smaller degree because it is not perfectly elastic. Plane waves emitted from a large distant source travel in a plane or front perpendicular to their direction of travel. There is no geometric spreading or energy loss in plane waves, neglecting molecular heating. Spherical waves, resulting from a small vibrating sphere in close proximity, spread in three dimensions. They lose energy according to the inverse square law, given by

$$I_{\text{ave}} = \frac{W}{4\pi r^2}$$

where

I = the sound intensity (watts/cm^2)
r = the distance to the source (cm)
W = the total source power (watts)

Table 6-9 Speed of Sound in Various Media

Media	Speed (m/s)	Speed (fps)
Air, 21°C (69.8°F)	344	1129
0°C	331	1086
Alcohol	1213	3980
Lead	1220	4003
Hydrogen, 0°C	1269	4164
Water, fresh	1480	4856
Water, salt, 21°C, at 3.5% salinity	1520	4987
Human body	1558	5112
Plexiglas	1800	5906
Wood, soft	3350	10,991
Concrete	3400	11,155
Fir timber	3800	12,468
Mild steel	5050	16,570
Aluminum	5150	16,897
Glass	5200	17,061
Gypsum board	6800	22,310
(Copper)	(3970)	(13,026)

Source: A. J. Schneider, *Noise and Vibration Rocket Handbook*, Bruel & Kjaer, 5111 W. 164th St., Cleveland, Ohio 44124, p. 18.

For every doubling of distance, the intensity is reduced by a factor of 4 or 6 dB. The sound from an infinite line source spreads geometrically in two dimensions so that energy is halved or loses 3 dB when the source distance doubles. When reflecting objects are near, a more complex sound field results.

Noise Noise is unwanted sound. It may be unwanted for a variety of reasons: causing hearing loss, interfering with communication, causing loss of sleep, adverse effects on human physiology, or just plain annoyance.

Noise pollution Noise pollution is the condition where noise has characteristics and duration injurious to public health and welfare, or which unreasonably interfere with the comfortable enjoyment of life and property in such areas as are affected by the noise.

Ambient Noise Ambient noise is the total noise in a given situation or environment.

Noise Level Noise level is the weighted sound pressure level in dBA obtained by the use of an approved type [American National Standards Institute (ANSI)] sound level meter. See "Decibel," "Sound Pressure," and, under Measurement of Noise, "Sound Level Meter."

Frequency Frequency of sound is the number of times a complete cycle of pressure variation occurs in one second; both an elevation and a depression below atmospheric pressure. The frequency of a sound determines its *pitch*. Frequency is expressed in hertz (Hz) which is the metric unit for cycles per second (cps). For example, sounds with a frequency of 30 Hz are considered

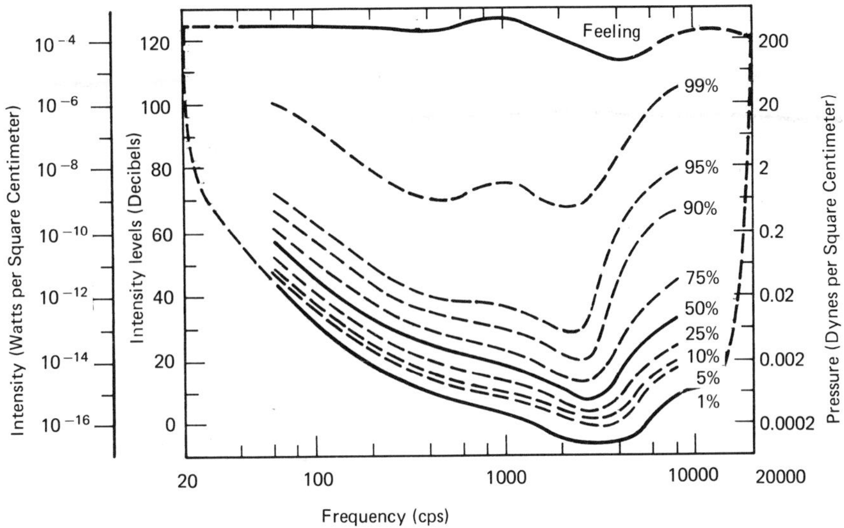

Figure 6-13 Absolute auditory threshold for a typical group of Americans. Curves are labeled by per cent of group that could hear tones below the indicated level. (Source: *Toward a Quieter City*, A Report of the Mayor's Task Force on Noise Control, New York City, 1970.)

very low pitch; sound with a frequency of 15,000 Hz are very high pitch. A young healthy ear can detect frequencies over a range of about 20 to 20,000 Hz; but the most common sensitive hearing range is between 1000 and 6000 Hz. Normal speech is in the range of 250 to 3000 Hz. However the audibility of sound is dependent on both frequency and sound pressure level. This is illustrated in Figure 6-13 for a typical group of Americans. Since most sounds are made up of several frequencies, a narrow band analyzer is used to determine the various frequencies in a sound. Most sounds are in the sonic frequency range of 20 to 20,000 Hz. Ultrasonic range is 20,000 Hz and above. Infrasonic range is 20 Hz and below. See "Sound Analyzer" and "Octave-Band Analyzer," under Measurement of Noise.

Decibel Decibel (dB) is a dimensionless unit to express physical intensity or sound pressure levels. The starting or reference point for noise level measurement is 0 dBA, the threshold of hearing for a young person with very good hearing. The threshold of pain is 120 dBA. The decibel is one-tenth of the bel, a unit using common logarithms named for Alexander Graham Bell.

Sound Pressure The sound pressure level is expressed by the relationship:

$$\text{Sound pressure level (SPL) in dB} = 20 \log_{10} \frac{P}{P_0}$$

in which

P = pressure of measured sound, in micropascals (μPa)
P_0 = sound pressure reference level of 20 μPa;* for measurements in air; this is the threshold of human hearing at 1000 Hz.

*Equals 10^{-12} W for sound power and 10^{-12} W/m^2 for intensity.

The sound pressure level is measured by a standard sound level meter. The meter has built into it electrical characteristics or weighting which simulate the way that the ear actually hears sound. The *A*-weighting is most commonly used and the sound levels are read in dBA. See "Sound level meter" under Measurement of Noise, for further discussion.

Table 6-10 shows the calculated sound pressure levels in dB for selected sound pressure values.

To add sound level values, it is first necessary to convert each decibel reading to sound intensity using the formulae:

$$\text{Sound intensity level in dB} = 10 \log_{10} \frac{I}{I_0}$$

$$= 10 \log_{10} \frac{I_1 + I_2}{I_0}$$

in which:

I = unknown sound intensity, in watt/m^2
I_0 = sound intensity reference base = 10^{-12} W/m^2
I_1 = sound intensity from source 1
I_2 = sound intensity from source 2

Table 6-10 Sound Pressures for Selected Decibel Values

Sound Pressure[a]		Sound Pressure Level (SPL)
(μbar)	(μPa)	(dB)[b]
0.0002	20	0[c]
0.00063	63	10
0.002	200	20
0.0063	630	30
0.02	2000	40
0.063	6300	50
0.2	20,000	60
0.63	63,000	70
1.0	100,000	74
2.0	200,000	80
6.3	630,000	90
20	2,000,000	100
63	6,300,000	110
200	20,000,000	120
2000	200,000,000	140

[a]0.0002 microbars (μbar) for sound pressure in air = 20 μPa = 0.00002 newtons/m^2 (20 μN/m^2) = 2.9 × 10^{-9} psi = 0.0002 dyne/cm^2.
[b]Relative to 20 μPa or 0.0002 μbar = standard reference value.
[c]0 dB = 2.9 × 10^{-9} psi = 10^{16} W/cm^2 = 10^{-12} W/m^2 for sound intensity = threshold of human hearing.

Pascal (Pa) is a unit of pressure corresponding to a force of 1 N acting uniformly upon an area of 1 m^2; 1 Pa = 1 N/m^2.

Newton (N) is the force required to accelerate 1 kg mass at 1 m/s^2. It is approximately equal to the gravitational force on a 100-g mass.

All sound intensities are added and then the sum is converted to a resultant decibel reading. A similar procedure is followed to subtract the numbers of decibels. For example, to add two sound levels dB_1 and dB_2 find the I_1 corresponding to dB_1; find I_2 corresponding to dB_2 and add to I_1 yielding I; then reconvert to dB using the above formulae. This rather complex process is much simplified by use of Table 6-11. For example, consider the summation of a 50-dB sound with a 56-dB sound. For a difference of 6 dB we find from Table 6-11 that 1 dB is added to the higher of the two sounds. The combined sound level is 57 dB. In adding several sound levels, start with the lowest.

Consider another example involving three noise sources. An industrial safety engineer wants to compute the total sound pressure level in a work area from the machinery nearby. An air compressor, a drill press, and ventilation fans contribute 85 dB, 81 dB, and 75 dB sound pressure level, respectively. Starting with the lowest, according to Table 6-11, an 81-dB level and a 75-dB level sum to 82 dB. The 82-dB level and the 85-dB level sum to 86.8 dB. Note that if the 75-dB level were missing, the total would have been 86.5 dB, almost the same. A noise contribution more than 10 dB lower than the other noise contributions can usually be neglected.

It should be noted that in using the above formula a generalization can be made: namely, any two *identical* sound levels will have the effect of increasing the overall level by 3 dB and that any three will increase the overall level by 4.8 dB.

Intensity Intensity of a sound wave is the energy transferred per unit time (sec) through a unit area normal to the direction of propagation. It is commonly measured in W/m^2 or W/cm^2. For a pure tone (single frequency) there is a one-to-one correspondence between loudness and intensity. However,

Table 6-11 Approximate Increase when Combining Two Sound Levels

Difference Between Levels (dB)	No. of dB to Be Added to Higher Level
0	3.0
1	2.6
2	2.1
3	1.8
4	1.5
5	1.2
6	1.0
7	0.8
8	0.6
10	0.4
12	0.3
14	0.2
16	0.1

Source: A. C. Hosey (Ed.), *Industrial Noise, A Guide to Its Evaluation and Control*, PHS Pub. 1572, DHEW, Washington, D.C., 1967.

almost all sound contains multiple frequencies. The relationship is not simple because of the interference effects of the sound waves.[36] For example, increasing sound pressure level by 3 dB is equivalent to increasing the intensity by a factor of two. Increasing sound pressure level by 10 dB is equivalent to increasing the intensity by a factor of 10, and increasing the sound pressure level by 20 dB is equivalent to increasing the intensity by a factor of 100. Expressed in another way, whereas 10 decibels is 10 times more intense than one decibel, 20 decibels is 100 times (10 × 10) more intense, and 30 decibels 1000 times (10 × 10 × 10) more intense.

Loudness Loudness or amplitude of sound is sound level or sound pressure level as perceived by an observer. The apparent loudness varies with the sound pressure and frequency (pitch) of the sound. This is illustrated in Figure 6-14. It is specified in sones or phons. For a pure tone, each time the sound pressure level increases by 10 dB the loudness doubles (sones increase by a factor of two). Sound levels of the same intensity may not sound the same since the ear does not respond the same to all types of sound.

A 1000-Hz pure tone, 40 dB above the listener's hearing threshold (0 dB), produces a loudness of one *sone*, which is a unit of loudness[37,38] This loudness of one sone is equal to 40 phons. Loudness levels are usually expressed in phons. For practical purposes, each doubling of the sones increases the phons by 10, that is, 1 sone = 40 phons; 2 sones = 50 phons; 4 sones = 60 phons. Also for pure tones, a 10-dB increase in sound level would be perceived as a 10-phon increase in loudness by a person with good hearing in the frequency range of 600 to 2000 Hz.

Take for example a human listener with normal hearing who hears a 100-Hz pure tone with a sound pressure level (SPL) of 90 dB. What loudness does the listener perceive?

From Figure 6-14, a SPL of 40 dB at approximately 100 Hz equals a loudness of 10 phons. Since a 50 dB increase in SPL is equivalent to a 50 phons increase in loudness, the tone's loudness is 60 phons or 4 sones.

Noys Noys is a measure of the perceived noise level (PNL) in dB in relation to the noisiness or acceptability of a sound level. Although similar to loudness, the ratings by observers when tested were different.

Procedures for the calculation of loudness and noisiness are given in standard texts.[39]

Effects of Noise—A Health Hazard

Noise pollution is an environmental and workplace problem. Excessive noise can cause permanent or temporary loss of hearing. Loud sounds affect the circu-

[36] M. I. Davis, *Air Resources Management Primer*, ASCE, New York, August 1973.

[37] *The Industrial Environment . . . Its Evaluation and Control*, "Sound and Noise," by Charles D. Jaffe, PHS Pub. 614, GPO, Washington, D.C., 1958, p. B-20, 2.

[38] *Environmental Health Criteria 12, Noise*, WHO, 1980, pp. 24–25.

[39] Arnold P. G. Peterson and Erwin E. Gross, Jr., *Handbook of Noise Measurement*, General Radio Company, Concord, Mass., 1974, pp. 23–35.

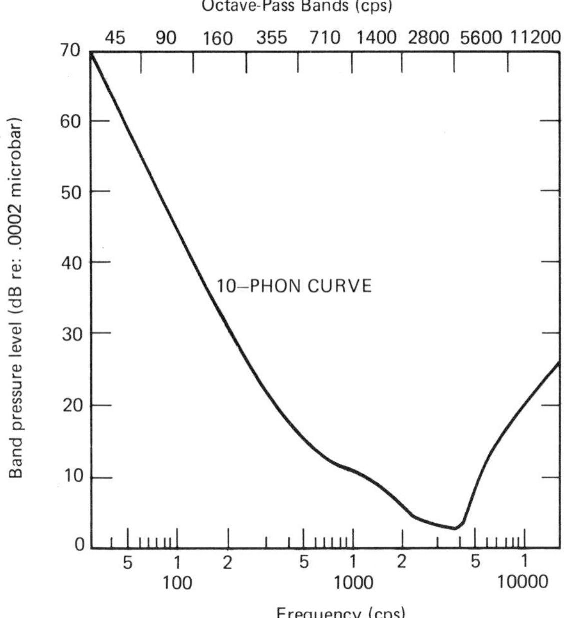

Figure 6-14 Equal loudness contour. (Source: *Toward a Quieter City*, A report of the Mayor's Task Force on Noise Control, New York City, 1970.)

latory and nervous systems, although the effects are difficult to assess. It interferes with speech, radio, and TV listening; disturbs sleep and relaxation; affects performance as reduced work precision and increased reaction time; causes annoyance, irritation, and public nuisance. Sonic boom can cause physical damage to structures. With the WHO definition of health as "a state of complete physical, mental and social well-being and not merely the absence of disease or infirmity," then excessive noise is clearly a health problem. David G. Hawkins, assistant USEPA administrator reported that

> A recent poll conducted by the U.S. Bureau of the Census showed that noise is considered to be the most undesirable neighborhood condition—more irritating than crime and deteriorating housing.[40]

Criteria for hearing protection and conservation have been established primarily for the worker. The major factors related to hearing loss are intensity (sound pressure levels in dB), frequency content, time duration of exposure, and repeated impact (a single pressure peak incident). In measuring the potential harm of high-level noise, frequency distribution as well as intensity must be considered. Continuous exposure to high-level noise is more harmful than intermittent or occasional exposure. High- and middle-frequency sounds at high levels generally are more harmful than low-frequency sounds at the same levels. Greater harm is done with increased time of exposure.

[40]"E.P.A., Battling Noise Pollution, Tells of Extent of Damage to Ears," United Press International, *New York Times*, November 11, 1979.

Individuals react differently to noise depending on age, sex, and socioeconomic background. The relation of noise to productivity or performance is contradictory and not well established.

For workers, a sound level over 85 dBA calls for study of the cause. A level above 90 dBA should be considered unsafe for daily exposure over a period of months and calls for noise reduction or personal ear protection if this is practical. A level of 120 dBA causes discomfort; levels of 120 to 140 dBA pain and possibly nausea and dizziness.

A USEPA report identified a 24-hr exposure level of 70 dBA as the level of environmental noise which will prevent any measurable hearing loss over a lifetime. Levels of 55 dBA outdoors and 45 dBA indoors are identified as preventing annoyance and not interfering with spoken conversation and other activities such as sleeping, working, and recreation.[41] Some common sound levels and human responses are noted in Figure 6-15.

Other effects of noise are reduced property values; increased compensation benefits and possible accidents, inefficiency, and absenteeism; and increased building construction costs.

Sources of Noise

Transportation, industrial, urban, and commercial activities are the major sources of noise, plus the contributions made by household appliances and equipment. The major sources of transportation noise are motor vehicles including buses and trucks, aircraft, motorcycles, and snowmobiles.

Industrial, urban, and commercial noises emanate from factories, equipment serving commercial establishments, and construction activities. Construction equipment sources are power tools, air compressors, earthmovers, dump trucks, diesel cranes, pneumatic drills, and chain saws; also garbage collection trucks. Compactor trucks manufactured after October 1, 1980 may not exceed a noise level of 79 decibels and may not exceed 76 decibels after July 1, 1982 measured on the *A*-weighted scale, seven meters from the front, side, and rear of the vehicle while empty and operating.

Residential noise is associated with dishwashers, garbage disposal units, air conditioners, power lawn mowers, and home music amplifier units.

Measurement of Noise

Noise measurement equipment selection is dependent upon the task to be performed. For an initial survey, a sound level meter is adequate for a rapid evaluation and identification of potential problem areas. To study and also determine the characteristics of a noise problem area, a sound level meter, fre-

[41] *Information on Levels of Environmental Noise Requisite to Protect Public Health and Welfare with an Adequate Margin of Safety*, USEPA Report 550/9-74-004, March 1974.

	Noise Level	Response	Conversational Relationships
	150		
Carrier Deck, Jet Operation	140		
		Painfully loud	
	130	Limit Amplified Speech	
Jet Takeoff (200 ft)	120	Maximum Vocal Effect	
Discotheque			
Auto Horn (3 ft)			
Riveting Machine	110		
Jet Takeoff (2,000 ft)			
Garbage Truck	100		Shouting in Ear
New York Subway Station		Very Annoying	
Heavy Truck (50 ft)	90	Hearing Damage (8 hr)	Shouting at 2 ft
Pneumatic Drill (50 ft)			
Alarm Clock	80	Annoying	Very Loud Conversation 2 ft
Freight Train (50 ft)			
Freeway Traffic (50 ft)	70	Telephone Use Difficult	Loud Conversation, 2 ft—Possible contribution to hearing impairment begins
Air-Conditioning Unit (20 ft)	60	Intrusive	Loud Conversation, 4 ft
Light Auto Traffic (100 ft)	50	Quiet	Normal Conversation, 12 ft
Living Room			
Bedroom	40		
Soft Whisper (15 ft)	30	Very Quiet	
Broadcasting Studio	20		
	10	Just Audible	
	0	Threshold of Hearing	

Figure 6-15 Sound levels and human response. (Source: *Sound Levels and Human Responses*, Office of Planning Management, USEPA, July 1973.)

quency analyzer, and recorder are needed to determine sound pressure distribution with frequency and time. More sophisticated equipment would be needed for research or solution of special noise problems.

Sound Level Meter A sound level meter is used to measure the sound pressure level; it is the basic instrument for noise measurement.

Meters are available to cover the range of 20 to 180 dB. The specifications usually refer to the American National Standards Institute (ANSI) and particularly to the "American National Standard Specification for Sound Level Meters," ANSI S1.4-1971. Three weighting networks, A, B, and C, are provided to give a number which best approximates the total loudness level for a particular situation, with consideration of the sound frequency, intensity, and impact levels. There are three types of meters. Type I is highest quality; Type III lowest and not suitable for public health professionals. Type II is the most common type used by public health officials. Most noise laws and regulations permit either Type I or II but not Type III meters.

The B and C networks are no longer normally used. The *A*-weighted scale is most commonly used. It discriminates against frequencies below 500 Hz and most nearly encompasses the most sensitive hearing range of sound, that is, 1000 to 6000 Hz. The symbol dBA is used to designate the *A*-weighted decibel scale which combines both frequency and pressure levels; it measures environmental noise and should be supplemented by the time or duration to determine the total quantity of sound affecting people. The sound level meter provides settings for "F" (fast time response) and "S" (slow time response).

The most important part of the equipment is a calibrator which generates a known decibel standard. Without a calibration before and after a measurement, the measurement is suspect.

Noise Dosimeter The noise dosimeter will measure the amount of potentially injurious noise to which an individual is exposed over a period of time. A dosimeter can be set to the desired level and will then total the exposure time to noise above the set level.

Sound Analyzer A frequency analyzer may be necessary to measure complex sound and sound pressure according to frequency distribution. It will supplement readings obtained with a sound level meter. Noise analyzers cover different frequency bands. The octave-band analyzer is the most common. The impact noise analyzer is used to measure the peak level and duration of impact noise. Examples of impact noises are drop hammer machines and gun fire.

Cathode-ray Oscillograph This makes possible observing the wave form of a noise and pattern. The magnetic tape recorder makes possible collection of noise information in the field and subsequent analysis of the data in the office or laboratory. Environmental noise monitors are now available which can be located in a community and will retain noise levels in a memory.

Octave-Band Analyzer This has filters which usually divide a noise into eight possible frequency categories. Each category is called an octave-band, with frequency ranges of 45 to 90, 90 to 180, 180 to 355, 355 to 710, 710 to 1400, 1400 to 2800, 2800 to 5600, and 5600 to 11,200 Hz or cps. The bands are identified by their center or mid-frequencies: 63, 125, 250, 500, 1000, 2000, 4000, and 8000 Hz. With center frequency bands at 31.5 and 16,000 Hz, the audible frequency range of 20 to 20,000 Hz is then covered with 10 octave bands.

Background Noise Background noise is noise in the absence of the sound being measured that may contribute to and obscure the sound being measured. A rough correction could be made by applying the correction factors given in Table 6-12. However such subtractions typically introduce significant error in the final result. The message to be obtained from Table 6-12 is that the background noise should be at least 10 dB lower than the noise being measured. This will introduce negligible error (less than 0.5 decibel) due to interfering background.

Table 6-12 Correction for Background Noise

Total Noise Level Less Background Level (dB)	Subtract from the Total Noise Level to Get the Noise Level Due to the Source (dB)
10	0.5
9	0.6
8	0.7
7	1.0
6	1.2
5	1.6
4	2.2
3	3.0
2	4.3
1	6.9

Source: Herbert H. Jones, "Noise Measurement," *The Industrial Environment . . . Its Evaluation and Control*, PHS Pub. 614, DHEW, Washington, D.C., 1958, p. B-21.

Noise Control

Noise can be controlled at the source, in its path of transmission (through a solid, air, or liquid) or where it is received. Sometimes, because no one method is sufficiently effective, controls are instituted at two or at all three steps in the path of noise travel from the source to the receptor. In general, it is best to reduce the noise at the source. This should include establishment of clear, reasonable, and enforceable noise design objectives for manufacturers.

Noise control generally involves adoption and effective enforcement of reasonable and workable regulations; protection of workers from hazardous occupational noise levels; building quieter machines, use of vibration isolators, new product regulation, and product labeling for consumer information; improved building construction and use of rubber sleeves, gaskets, and noise barriers; compatible land use planning and zoning; and informing the public of harmful effects of noise and methods to reduce noise to acceptable levels. Regulations may encompass ambient noise in general and industrial noise, motor vehicle noise, and aircraft noise; also building and construction codes, housing occupancy codes, sanitary codes, and nuisance codes.

A WHO Expert Committee[42] suggests the following preventive measures to control noise and vibration.

(a) General measures: locating noisy industrial plants, airports, landing fields for helicoptors, railway stations and junctions, superhighways, etc., outside city limits.

(b) Improving technical processes and industrial installations with a view to reducing noise and vibration; installing noise suppressors (mufflers) on automobiles, motorcycles, etc.

[42] "Environmental Health Aspects of Metropolitan Planning and Development," *WHO Tech. Rep. Ser.* **297,** 1965, p. 52.

(c) Improving the quality of surface highways and urban streets. (Also tire tread designs.)
(d) Creating green spaces in each neighborhood district.
(e) Perfecting procedures for acoustic insulation.
(f) Adopting administrative regulations with a view to limiting the intensity of background noise within the urban environment.

The Committee recommends close international collaboration and close co-operation between metropolitan planners and environmental health personnel to reduce to a minimum noise and vibration.

Control of Industrial Noise

Noise control should start in the planning of a new plant, or when planning to modernize an existing plant. Consideration should be given at that time to minimizing the effects of noise on the workers, on office personnel, and on nearby residents. The control of an existing noise problem first requires suitable noise standards and an identification of the location, extent, and type of noise sources. This would be followed by the application of needed noise control measures to achieve the required or desired levels.

Factors to be taken into consideration in industrial noise control are:[43]

1. Selection of building site which is isolated or an area where there is a high background noise level. Topography and prevailing winds should be considered, as well as the use of landscaping and embankments, to reduce the noise travel whereby it may cause a nuisance.
2. Building layout to separate and isolate noisy operations from quiet areas.
3. Substitution of low noise level processes for noisy operations, such as welding instead of riveting, metal pressing instead of rolling or forging, compression riveting instead of pneumatic riveting, and belt drives in place of gears.
4. Selection of new equipment with the lowest possible noise level. (Also modification of existing equipment with better mufflers.)
5. Reduction of noise at its source through maintenance of machinery, covers and safety shields, and replacement of worn parts; reduction of driving forces; reduction of response of vibrating surfaces; intake and discharge sound attenuation and flexible connections or collars; use of total or partial enclosures, with sound absorbing materials (also coatings or sound absorbing materials on metals to dampen vibration noise); isolation of vibration and its transmission. See Noise Control and Noise Reduction.
6. Use of acoustic absorption materials to prevent noise reflections.
7. Control of noise in ventilation ducts or conveyor systems.
8. Use of personnel shelters.

Sometimes the practical and economical method of noise control is through the use of personal protective devices. These may also be a supplement to the applied engineering controls. Personal ear protector types include properly fitted earplugs and earmuffs and helmets providing a good seal around the ear. They

[43]A. D. Hosey and C. H. Powell (Eds.), *Industrial Noise, A Guide to Its Evaluation and Control*, "Principles of Noise Control," H. H. Jones, PHS Pub. 1572, DHEW, Washington, D.C., 1967, pp. N-10-1 to N-10-5.

should meet established criteria for comfort, tension, sound attenuation, simplicity, durability, and so on. To be effective however, the worker must cooperate by wearing the protective device where needed. Dry cotton plugs do not provide significant sound attenuation.

Control of Transport Noise

Noise from various forms of transport and its transmission into the home may be reduced:[44]

1. At the source, i.e., by controlling the *emission* of noise.
2. By means of town and country planning and traffic engineering, i.e., by controlling the *transmission* of noise.
3. In the home, i.e., by controlling the *reception* of noise by the occupants.

Some specific measures to reduce the effect of highway noise include:

1. Enclosure of highways going through residential areas.
2. Wider rights-of-way, i.e., separation or buffer zone between the source and the receptor.
3. Walls designed to deflect or absorb noise.
4. Changes in highway alignment and grade to avoid sensitive areas; minimizing stop-and-go traffic, and shifting to low gears.
5. Setting lower speed limits for certain sections of a highway.
6. Adjacent barriers, nonresidential buildings in sound transmission path, earth embankments or berms, elevation or depression of highways. It is reported however that barriers provide little attenuation of low frequency sounds and that a thick band of deciduous trees 200 to 300 ft in width is relatively ineffective in cutting down traffic noises, reducing them only on the order of 4 or 5 dB.[45] Separation distance is most effective in reducing noise from highways.
7. Establishing alternate truck routes.
8. Building codes requiring building insulation to limit interior transmission of noise. Additional measures are masonry walls, elimination of windows, use of double windows or glazing, soundproofing of ceilings, thick carpeting, over-stuffed furniture, and heavy drapes.

Noise Reduction

Sound Absorption

The amount of sound energy a material can absorb (soak up) is a function of its absorption coefficient (α) at a specified frequency. The sound absorption coefficient is the fractional part of the energy of an incident sound wave that is absorbed by a material. A material with an absorption coefficient of 0.8 will absorb 80 percent of the incident sound energy. A material that absorbs all incident energy, such as an open window, has an absorption coefficient of one. The sound absorption of a surface is measured in sabins. A surface hav-

[44] *Health Hazards of the Human Environment*, WHO, 1972, p. 265.
[45] J. E. Heer, Jr., D. J. Hagerty, and J. L. Pavoni, "Noise in the Urban Environment," *Pub. Works*, October 1971, pp. 60–64.

ing an area of 100 ft^2 made of material having an absorption coefficient of 0.06 has an absorption of 6 sabin units. To determine the noise reduction in a room, the floor, walls, and ceiling surface areas multiplied by the absorption coefficient of each surface, at a given frequency, before and after treatment, must be added to obtain the total room surface absorption in sabin units.

The *noise reduction* (NR) in dB at a given frequency of a surface before and after treatment can be determined by

$$NR = 10 \log_{10} \frac{A_2}{A_1}$$

in which

A_2 = total room surfaces absorption in sabins after treatment
A_1 = total room surfaces absorption in sabins before treatment

Incremental noise reduction from a piece of machinery can be obtained by a rigid, sealed enclosure, plus vibration isolation of a machine from the floor using spring mounts or absorbent mounts and pads, plus acoustical absorbing material on the inside of the enclosure, plus mounting the enclosure on vibration isolators and enclosing, without contact, in another enclosure having inside acoustical absorbing material. If machinery air cooling and air circulation is needed, provide baffled air intakes. Insert a flexible connector, if a physical pipe or duct connection is needed between the machinery and other building piping or duct work, to reduce noise transmission.

Sound energy however can go around or through a particular material (go around corners) or pass through openings (cracks, windows, ducts) and thereby nullify the sound absorption as well as transmission reduction efforts. For example, one square inch of opening transmits as much sound as about 100 ft^2 of a 40-dB wall.[46] This emphasizes the importance of sealing all cracks, pipe and conduit sleeves, or openings with nonsetting caulking compound.

Sound absorptive materials include rugs, carpets with felt pads, heavy drapes, ceiling and wall acoustical materials designed to absorb sound, and stuffed furniture. These materials absorb high frequency sounds much more effectively than low. Sound absorptive materials are most effective to the occupant when used in and near the areas of high level noise. These materials can control interior noise, sound reflection, and reverberation; but noise easily passes through. Hard, smooth, impervious materials reflect sound. Some absorption coefficients at 1000 Hz are plate glass 0.03, brick wall 0.01, linoleum, asphalt or rubber tile on concrete 0.03, smooth plaster on brick or hollow tile 0.03, plywood paneling $\frac{3}{8}$ in. 0.09, felt-lined carpet on concrete 0.69, velour (14 oz/yd^2 0.75, concrete block painted 0.70.

Sound Transmission

Sound transmission loss (TL) is the ratio of the energy passing through a wall, floor, or ceiling to the energy striking it. The sound transmission varies

[46]Joseph N. Boaz, Ed., *Architectural Graphic Standards*, John Wiley & Sons, New York, 1970, p. 509, pp. 504–512.

with the frequency of the sound, the weight or mass, and the stiffness of the construction. Hence reduction of the transmission of noise from outside to inside a building is accomplished through control of the design, thickness, and weight of wall, floor, and ceiling materials. Improved design of building equipment, its installation, noise and vibration isolation, and discontinuance or gaps in structural members are interior factors to also be considered. The transmission loss increases as the frequency increases.

Mechanical equipment, household appliances, and other stationary sources of noise should be isolated from the floor, wall, or mountings by means of rubber or similar resilient pads to prevent or reduce sound transmission to the structure, as noted above under Sound Absorption.

Sound transmission class (STC) loss ratings for various types of materials are given in decibels in design handbooks,[46] texts and standards such as the National Bureau of Standards, Building Materials and Structures Report BMS 144 for "Insulation of Wall and Floor Construction." For example, 4-in. cinder block weighing 25 lb/ft^2 has an average approximate STC loss rating of 25 dB; if the block is plastered on one side its rating is 40 dB. A 4-in. brick wall weighing 40 lb/ft^2 has a rating of approximately 45 dB. A 4-in. concrete slab with a resiliently suspended ceiling has a rating of 55 dB. A $\frac{1}{4}$-in. plywood sheet nailed to studs has a STC rating of 24 dB; $\frac{1}{2}$-in. gypsum board on studs has a rating of 32 dB. The frequency of the sound affects the sound transmission loss. Theoretically, transmission loss increases at the rate of 6 dB per doubling of the weight of the construction. Some building codes recognize the need to prevent sound transmission between apartments in a multiple dwelling or in row houses. A sound pressure level reduction of about 50 decibels in the normal speaking range (250 to 3000 Hz) is suggested.

Since a room floor, wall, and ceiling is usually of different construction materials, an average transmission coefficient must be calculated taking into consideration the coefficient for each material (including doors, windows, and vents) and its area to determine the room noise insulation factor in dB. The total noise reduction level accomplished by a wall, or other divider, is a function of the wall transmission loss, the room absorption characteristics, and the absorption in the rooms separated. It is determined by measuring the difference in sound levels in the rooms. The types of windows (single or double-hung) and doors can have a major effect on the overall noise insulation factor. For example, opening a window can double the interior noise.

Numerous sample calculations for sound and vibration control situations are given in various texts including the *American Society of Heating, Refrigerating and Air Conditioning Engineers Guide and Data Book, Systems, 1970.*

Mechanical noises such as high velocity noises require proper design of ventilation systems and plumbing systems to reduce flow velocity. Hammering noise in a plumbing system is usually due to a quick-closing valve in the plumbing system requiring installation of an air chamber on the line to absorb the pressure created when the momentum of the flowing water is suddenly stopped.

Separation distance between the sound source and receptor should be emphasized and not overlooked in the planning stages as a practical noise reduction method. In general, if there are no sound reflecting surfaces in the vicinity, a sound pressure level will be reduced approximately 6 dB for each doubling of the distance. Doubling the air space between panels increases the transmission loss by about 5 dB.

Federal Regulations

Maximum acceptable or permissible noise levels are established for certain categories by federal or state regulations or by local ordinances. Some guides are given in Table 6-13.

The Department of Labor in May 1969 issued the first federal standards for occupational exposure to noise. OSHA sets and enforces regulations, under the Occupational Safety and Health Act of 1970, for the protection of workers' hearing. See Table 6-14. The federal regulatory approach is to start control at the point of manufacture.

The Federal Highway Act of 1970 led to design noise levels related to land use as a condition to federal aid participation. If design noise levels shown in Table 6-15 are exceeded, noise abatement measures are required in the highway design. Federal highway funds may also be used to abate noise on previously approved highway projects.

The Noise Control Act of 1972 (P.L. 92-574) requires regulation of noise from a broad range of sources and products. The USEPA and the Department of Transportation have been given responsibilities to implement the law. The USEPA estimates that 16 million people are presently exposed to aircraft noise levels with effects ranging from moderate to very severe.

The Federal Aviation Administration (FAA), in the department of transportation, has primary authority for aircraft noise regulations and standards. The FAA has adopted noise emission standards for new aircraft and has a plan to retrofit older aircraft.

Table 6-13 Some Guides for Maximum Acceptable Sound Levels

Space	Sound Level (dBA)
Private offices	40 to 45
Small conference room	35 to 40
Secretarial offices	55 to 60
Drafting rooms	55
School rooms	30 to 40
Hospital rooms	40
Hotel rooms	45
Libraries	40 to 45
Restaurants	50
Auditoriums	30 to 45
Movie theaters	35 to 45

Table 6-14 Permissible Noise Levels in Plants[a]

Duration per Day (hours)	Sound Level at Slow Response (dBA)	
	Existing	Proposed
(10)	(87)	82
8	90[b]	85
6	92	87
4	95	90
3	97	
2	100	95
$1\frac{1}{2}$	102	
1	105	100
$\frac{1}{2}$	110	
$\frac{1}{4}$ or less	115	

Source: *Guidelines To The Department of Labor's Occupational Noise Standards for Federal Supply Contracts*, U.S. Dept. of Labor, Washington, D.C., December 4, 1970.
Note: Workers must not be exposed to sound levels above 115 dBA. Impact noise lasting less than one second should not exceed 140 dB (unweighted).
[a]Occupational Safety and Health Act of 1970 extends Walsh-Healey Public Contracts Act of May 1969 to contractors with federal contracts of $10,000 or more.
[b]The upper limit of a daily dose which will not produce disabling loss of hearing in more than 20 percent of the exposed population.

The Department of Housing and Urban Development (HUD) has criteria for the sound insulation characteristics of walls and floors in row houses, nursing homes, and multifamily housing units. These criteria must be met by housing of this type in order to qualify for HUD mortgage insurance.

The National Bureau of Standards and the National Science Foundation are concerned with research in noise control and abatement in factories, homes, offices, and commercial work areas.

Table 6-15 Design Noise Level—Land Use Relationships

Design Noise Level (dBA)	Description of Land Use Category
60 (Exterior)	Areas such as amphitheaters, certain parks or open spaces in which local officials agree serenity and quiet are of extraordinary significance.
70 (Exterior)	Residences, motels, hotels, public meeting rooms, schools, churches, libraries, hospitals, recreation areas.
75 (Exterior)	Developed land; properties or activities not included in above two categories.
55 (Interior)	Residences, motels, hotels, public meeting rooms, schools, churches, libraries, hospitals and auditoriums.

Source: U.S. Dept. of Transportation, Policy Procedure Memorandum 90-2 Appendix B, Transmittal 279, February 8, 1973.

Table 6-16 Noise Levels for Sleeping Quarters in New Structures

Exterior	Interior
Does not exceed 45 dBA for more than 30 min per 24 hours (Acceptable)	Not greater than 55 dBA for more than an accumulation of 60 min in any 24-hour day
Does not exceed 65 dBA for more than 8 hours per 24 hours (Normally Acceptable)	Not greater than 45 dBA for more than 30 min during nighttime sleeping hours 11 p.m. to 7 a.m., and Not greater than 45 dBA for more than an accumulation of 8 hours in any 24-hour day

Source: HUD Circular 1390, amended Sept. 1, 1971.
Note: Not greater than 30 dBA preferred for bedrooms.

The USEPA has issued noise control regulations for interstate trucks, for interstate railroad carriers, for new medium and heavy-duty trucks, and for new air compressors. USEPA and DOT regulations establish a maximum noise level of 90 dBA for interstate trucks and buses over 10,000 pounds in speed zones over 35 mph and 86 dBA at 35 mph or less, measured 50 ft from the center line of the lane of travel. For new trucks less than 10,000 pounds, the USEPA has proposed 83, 80, and 75 dBA after January 1977, 1981, and 1983 respectively.

The USEPA program for certain noise-emitting and noise-reducing products requires a noise rating giving the number of decibels (dBA) a product emits and a noise reduction rating.

HUD noise levels for new sleeping quarters are given in Table 6-16.

State and Local Regulations

New York State enacted a state highway anti-noise law in 1965 and California followed in 1967. Chicago put into effect a comprehensive noise control program in July 1971. Regulations require reduced noise levels after 1979 for vehicles, construction machinery, home powered equipment, and the like manufactured equipment. St. Louis County has a noise code which limits noise in residential areas to 55 dBA and in industrial areas to 80 dBA. New Jersey enacted comprehensive noise legislation January 1972. Most states in the snow belt have established a maximum noise level for snowmobiles of 78 dBA at 50 feet. Some 12,000 states and municipalities have noise control legislation but enforcement has been weak and spotty.

Local regulations which are consistent with federal and state laws and which are enforced locally are encouraged as being more practical for enforcement.

Model noise control ordinances are available to assist local communities in the development of a local program.[47–49]

Maximum acceptable sound levels for different situations are given in Tables 6-13, 6-15, and 6-16. Maximum permissible sound levels for workers in industrial plants and factories regulated by the Occupational Safety and Health Act are given in Table 6-14.

BIBLIOGRAPHY

Air pollution

"Air Pollution," *WHO Monogr. Ser.* **46,** Columbia University Press, New York (1961).

Air Pollution and Your Health, USEPA, Washington, D.C. 20460, March 1969.

Air Pollution Control Field Operations Manual, compiled and edited by Melvin I. Weisburd and S. Smith Griswold, PHS Pub. 937, DHEW, Washington, D.C. 20201, 1962.

Air Pollution Engineering Manual, compiled and edited by John A. Danielson, Pub. AP-40, USEPA, Office of Air Quality Planning and Standards, Research Triangle Park, N.C. 27711, May 1973.

Air Pollution Primer, National Tuberculosis and Respiratory Association, New York, 1969.

Cleaning Our Environment—A Chemical Perspective, a report by the Subcommittee on Environmental Improvement, American Chemical Society, Washington, D.C., 1978.

Community Action Guide for Public Officials—Air Pollution Control, National Association of Counties Research Foundation, Washington, D.C., 1969.

Davis, Mackenzie L., *Air Resources Management Primer,* ASCE, New York, 1973.

The Effects of Air Pollution, PHS, DHEW, Washington, D.C., 20201, 1967.

Interim Guidelines on Air Quality Models, USEPA, Office of Air Quality Planning and Standards Pub. 1.2-080, Research Triangle Park, N.C. 27711, October 1977.

Meteorological Aspects of Air Pollution, U.S. Dept. of HEW, CP & EHS, NAPCA, Box 12055, Research Triangle Park, N.C. 27711.

Recommended Guide for the Prediction of the Dispersion of Airborne Effluents, edited by M. E. Smith, American Society of Mechanical Engineers, United Engineering Center, 345 East 47th Street, New York, 10017, May 1968.

"Research into Environmental Pollution," *WHO Tech. Rep. Ser.,* **406,** Geneva, 1968.

Rihm, Alexander, "Air Pollution," in *Dangerous Properties of Industrial Materials,* N. Irving Sax, Ed., Van Nostrand Reinhold Co., New York, 1969.

Stern, Arthur C., Ed., *Air Pollution,* Academic Press, New York.
  Volume I *Air Pollutants, Their Transformation, and Transport,* 1976.
  Volume II, *The Effects of Air Pollution,* 1977.
  Volume III *Measuring, Monitoring, and Surveillance of Air Pollution,* 1976.
  Volume IV, *Engineering Control of Air Pollution,* 1977.
  Volume V, *Air Quality Management,* 1977.

Stern, Arthur C., Henry C. Wohlers, Richard W. Boubel, and William P. Lowry, *Fundamentals of Air Pollution,* Academic Press, New York, 1973.

Suess, M. J., and S. R. Craxford, *Manual on Urban Air Quality Management,* WHO, Regional Office for Europe, Copenhagen, 1976.

[47]"A Model for Community Noise Control," *J. Environ. Health,* July/August 1977, pp. 24–44.

[48]*Local Noise Ordinance Handbook,* Bureau of Noise Control, Division of Air Resources, N. Y. State Dept. of Environmental Conservation, Albany, N.Y. 12233, Revised October 1977.

[49]USEPA, Washington, D.C. 20460 (Maintains up-to-date compilation of city and state noise control ordinances).

Noise Control

"A Model for Community Noise Control," National Environmental Health Association Noise Committee, *J. Environ. Health*, July/August 1977, 25–44.

ASHRAE Guide and Data Book, Systems, 1970, Chapter 33, Sound and Vibration Control, American Society of Heating, Refrigerating, and Air-Conditioning Engineers, Inc., 345 East 47th Street, New York, N.Y. 10017.

Berendt, Raymond D., Edith L. R. Corliss, and Morris S. Ojalvo, *Quieting: A Practical Guide to Noise Control*, U.S. Dept. of Commerce, National Bureau of Standards, July 1976, GPO, Washington, D.C. 20402.

Burns, W., *Noise and Man*, J. B. Lippincott Co., Philadelphia, 1973.

Crocker, M. J., and A. J. Price, *Noise and Noise Control*, Vol. 1 and 2, CRC Press, 18901 Cranwood Parkway, Cleveland, Ohio 44128.

Environmental Health Criteria 12, Noise, WHO, Geneva, 1980.

The Industrial Environment—its Evaluation & Control, DHEW, PHS, CDC, National Institute for Occupational Safety and Health, 1973, GPO, Washington, D.C. 20402.

Information on Levels of Environmental Noise Requisite to Protect Public Health and Welfare with an Adequate Margin of Safety, USEPA Report 550/9-74-004, Washington, D.C., March 1974.

Local Noise Ordinance Handbook, Bureau of Noise Control, Division of Air Resources, N.Y. State Dept. of Environmental Conservation, Albany, New York, 12233, Revised October, 1977.

Peterson, Arnold P. G., and Ervin E. Gross, *Handbook of Noise Measurement*, General Radio Company, Concord, Mass., 1974.

7

RADIATION USES AND PROTECTION

RADIATION FUNDAMENTALS

Definitions

Radiation is the emission of energy from a point of origin. Any electromagnetic or particulate radiation capable of producing ions, directly or indirectly, by interaction with matter is referred to as ionizing radiation. The medical uses generally involve X rays and gamma rays as well as high-energy electrons and beta particles. Corpuscular emission from radioactive substances of other sources, commonly alpha particles, beta particles, and neutrons, are also forms of radiation.

The more common types of radiation are classified according to wavelength. The relationship of radiation wavelength, velocity, and frequency is shown by:

$$\text{wavelength of radiation} = \frac{\text{velocity of radiation}}{\text{frequency of radiation}}$$

The velocity of radiation varies with the medium through which it travels. The frequency of radiation is dependent on its source. Frequency is the number of occurrences of a periodic quantity as waves, vibrations, or oscillations in a unit of time, usually expressed as cycles per second. Particles are not included.

The shorter the wavelength, the higher the frequency and energy. The longer the wavelength, the lower the frequency and energy. In general, the shorter the wavelength of a particular type of radiation, the greater is its power of penetration. This is apparent by inspection of Figure 7-1, which also shows radiation sources in the electromagnetic spectrum.

The energy of ionizing radiation is measured as electron volts (eV), thousands of electron volts (keV), or millions of electron volts (MeV), whichever is applicable. An electron volt is the energy an electron gains in passing through a difference of potential of one volt and is equivalent to 1.6×10^{-12} ergs. The intensity of a beam of radiation is usually expressed in terms of ergs per cm^2 per

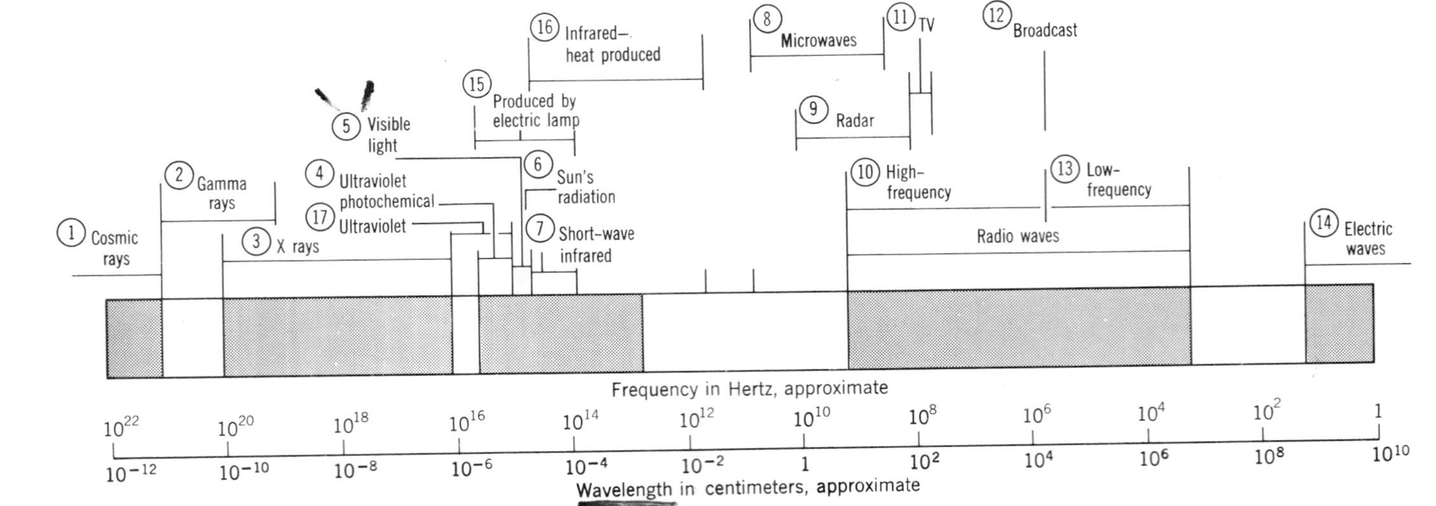

Figure 7-1 Electromagnetic radiations.
(1) Cosmic rays. Reaching earth from sky.
(2) Gamma rays.
(3) X-rays. High-frequency oscillations produced by X-ray tubes.
(4) Ultraviolet. Photochemical-photoelectric and fluorescent effects. Germicidal action and health maintenance by virtue of radiation absorbed by bacteria.
(5) Visible. Seeing—discrimination of color and detail.
(6) Sun's radiation reaching the earth.
(7) Short-wave infrared. Heat therapy—drying.
(8) Microwaves.
(9) Radar.
(10) High-frequency.
(11) Television.
(12) Broadcast.
(13) Low-frequency.
(14) Electric waves. Produced by electric generators.
(15) Produced by electric lamps
(16) Infrared. Produced by heat.
(17) Ultraviolet.

Adapted from *Lamps and the Spectrum*, General Electric, Nela Park, Cleveland, Ohio, and other sources. The velocity or speed of radiation throughout the spectrum is that of light, or about 186,000 mi/sec. The nature of all radiation is the same, and the difference lies only in the frequency and wavelength, i.e., frequency × wavelength = velocity. The range in wavelengths is from about 10^{-12} cm for cosmic rays to a length of 3100 miles for 60-cycle electric current. One micron (μm) = 10^{-4} cm = 10^{4} angstroms = one micrometer. One cm = 10^{4} microns μm = 10^{8} Angstrom units (Å).

sec. It is the quantity of energy passing through a known area per unit of time. The quantity of radiation may be stated as ergs per square centimeter.

Roentgen (R) The absorption of about 86 ergs of energy per gram of air represents one roentgen. It is equal to one electrostatic unit (esu) per 0.001293 grams of dry air. The roentgen is a measure of the ionization in air produced by exposure to X rays or gamma rays and is not applicable to particle radiations such as alphas, betas, and neutrons. Survey meters usually read in roentgens or milliroentgens per hour.

Rad One rad, or *r*adiation *a*bsorbed *d*ose, represents the absorption of 100 ergs per gram of medium from any type of ionizing radiation. For water and soft tissue, with X rays and gamma rays having energies of up to 3 MeV, an exposure to one roentgen is roughly equal to an absorbed dose of about 0.93 to 0.98 rad.[1] (The rad replaced the rep as a unit of measure of radiation absorption.) An ordinary chest X ray produces an exposure of about 0.1 rad; a very heavy diagnostic series, about 10 rads.

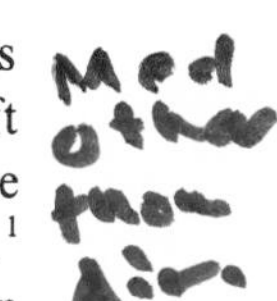

Rem The rem, short for *r*oentgen *e*quivalent *m*an, is intended to take into consideration the biological effect of different kinds of radiation from the same dose in rads. For medical radiation purposes, covering only X and gamma radiation, the number of rems may be considered equivalent to the number of rads or to the number of roentgens. A quality or damage factor is used to convert rads to rems. Some practical quality factors (QF) are[2] 1.0 for X rays, gamma rays, electrons or positrons, energy >0.03 MeV; 1.0 for electrons or positrons, energy <0.03 MeV; 3 for neutrons, energy <10 keV; 10 for neutrons, energy >10 keV; 1 to 10 for protons; 1 to 20 for alpha particles; and 20 for fission fragments recoil nuclei.

$$\text{rems} = \text{rads} \times QF.$$

Person-rem is a term used to show the exposure of large populations to low-level radiation. For example, 5000 persons exposed to a background radiation of 0.1 rem per year would represent a person-rem exposure of 500.

Curie (Ci) The rate at which atoms of radioactive sources (radionuclides) disintegrate is measured in curies. One curie is defined as 37,000,000,000 (3.7×10^{10}) disintegrations per sec. The radioactivity of one gram of radium is approximately one curie. A picocurie (pCi) yields 2.22 disintegrations per minute, equals 10^{-12} Ci.*

*A nanocurie (nCi) is a billionth (10^{-9}) of a curie; a microcurie (μCi) is a millionth (10^{-6}) of a curie; a millicurie (mCi) is a thousandth (10^{-3}) of a curie; a kilocurie (kCi) is a thousand (10^{3}) curies; and a megacurie (MCi) is a million (10^{6}) curies.

[1] C. B. Braestrup and K. J. Vikterlof, *Manual on radiation protection in hospitals and general practice*, WHO, 1974, p. 16.

[2] *Basic Radiation Protection Criteria*, Rep. No. 39, National Council on Radiation Protection and Measurement, 4201 Connecticut Avenue, N.W., Washington, D.C., January 15, 1971, p. 83.

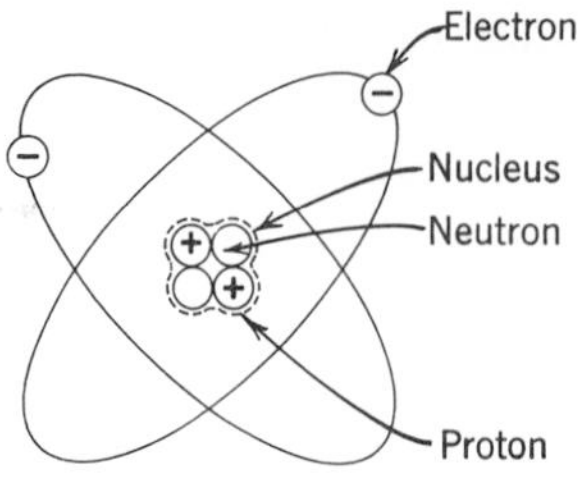

Figure 7-2 Atomic structure. The electrons move in the electron cloud, orbit, or sheath and have a negative electrical charge. The protons have a positive electrical charge; the neutrons are neutral. The number of protons (Z) = atomic number of the element. The number of protons plus the number of neutrons is equal to the mass number (A). Isotopes of the same element have as many mass numbers as there are isotopes of the element, but the same atomic (Z) number. Atoms in their natural state are electrically neutral; the number of electrons outside the nucleus are the same as the number of protons within the nucleus.

The Atom

A brief elementary review of the characteristics of the atom is useful for understanding the effects of radiation. All atoms are composed of protons, electrons, and neutrons except the simple hydrogen atom, which contains a proton and an electron. The structure and components of an atom are shown in Figure 7-2.

When atoms combine to form compounds, the resulting molecule usually has the same number of electrons as it does protons and it is also electrically neutral. Since the electron is the lightest part of the atom, and each electron is not bound as tightly to the nucleus as the protons and neutrons are bound to each other in the nucleus, the electron is more mobile and can be removed from the atom or molecule without expending much energy. (The energy binding the electron to the nucleus is small.) When an electron is removed from an atom or a molecule, the resulting component has an electrical charge because the protons are then in excess of the electrons. The atom or molecule with the electron removed has an electrical charge and is, by definition, an ion. When the nucleus of a heavy atom such as uranium is split or fissioned, part of the energy holding the protons and neutrons together is released as heat. In fusion, the nuclei of two very light atoms such as tritium and deuterium are joined; large amounts of energy are also produced in the process.

Ions

Ions may be charged positively or negatively and may exist as crystals, liquids, and gases. Positive ions are produced by removing electrons from neutral atoms and molecules; negative ions are created when electrons attach themselves to neutral atoms or molecules. Any process by which a neutral atom or molecule loses or gains electrons, thereby resulting in a net charge, is known as ionization. Figure 7-3 illustrates the process.

The ionization resulting from radiation absorption has significant public health aspects. The production of ions within tissues can injure people, animals, and plants and cause genetic as well as somatic damage. The radiation is detected

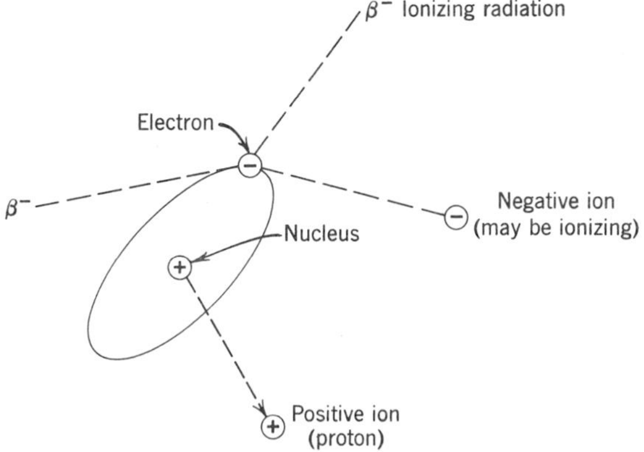

Figure 7-3 Ionization of a hydrogen atom by a beta particle β^-.

and measured through the ions created by the energy transfer. In practice, however, exposure to ionizing radiation is controlled by the use of shields of lead, concrete, and other materials.

Isotopes

There are also different members, or isotopes, of an element. The isotopes have the same atomic number (number of protons), but different mass numbers (different number of neutrons in the nucleus). For example hydrogen (^1_1H), deuterium (^2_1H), and tritium (^3_1H) all have the same atomic number, of 1, but different numbers of neutrons. But they are all members of the hydrogen family. Uranium 238 contains 92 protons and 146 neutrons, while the isotope uranium 235 contains 92 protons and 143 neutrons. Oxygen has six known isotopes; tin has at least 23. The chemical properties of all isotopes of an element are, for practical purposes, the same; the physical properties may vary somewhat.

Types of Radiation

The common types of radiation are X rays, gamma rays, alpha particles, beta particles, and neutrons. Each has its own characteristics. X rays and gamma rays are similar. The types of radiation emitted by various radionuclides is given in Table 7-1.

Alpha particles have large specific ionization values. Since they create many ions per unit of path length, they dissipate their energy rapidly and penetrate only 3 to 5 cm of air. A thin sheet of paper will stop alpha particles. Alpha particles have a positive electric charge, exactly twice that of an electron, and a mass of four, the same as a helium atom. The particles are normally a hazard to health only in the form of internal radiation through ingestion, inhalation, or open wounds. Alpha emitters are heavy elements, such as uranium, lead, plutonium, and radium, and are also toxic in themselves as heavy metals. However, alpha

Table 7-1 Radioactive Sources of Educational Interest

Radionuclide	Half-Life	Type of Radiation	Specific Gamma-Ray Constant R/mCi-h at 1 cm[a]	"Generally Licensed," or "Exempt," Quantities	
				Activity (μCi)	Exposure rate, mR/hr at 1 m
Carbon 14	5730 yr	Beta		50	
Cesium 137	30 yr	Beta and gamma	3.0	1	0.0003
Cobalt 60	5.3 yr	Beta and gamma	12.9	1	0.0013
Gold 198	2.7 days	Beta and gamma	2.27	10	0.0023
Hydrogen 3 (Tritium)	12.3 yr	Beta		250	
Iodine 131	8.1 days	Beta and gamma	2.20	10	0.0022
Iridium 192	74.4 days	Beta and gamma	4.83	10	0.0048
Phosphorus 32	14.2 days	Beta		10	
Polonium 210	138 days	Alpha		0.1	
Silver 111	7.5 days	Beta and gamma	0.13	10	0.0002
Sodium 24	0.63 days	Beta and gamma	18.2	10	0.0187
Strontium 90	28 yr	Beta		0.1	
Sulfur 35	87 days	Beta		50	
Radium (in equilibrium with Radium A, B, C)	1620 yr	Alpha, beta, gamma	8.25 with 0.5 mm Pt filter[b]	0.1	0.00008 with 0.5 mm Pt filter[b]

Source: *Radiation Protection in Educational Institutions*, Rep. No. 32, National Council on Radiation Protection and Measurements, Washington, D.C., 1966.
[a]Roentgens per hour for a point source of 1 millicurie at a distance of 1 centimeter.
[b]Platinum (Pt).

particles are not an external hazard since they cannot penetrate outer layers of skin.

Beta particles are light in weight and carry single charges. They are high-speed electrons that originate in the nucleus. Their specific ionization values are intermediate between those of alpha particles and gamma and X rays. They ionize slightly, dissipate their energies rather quickly, and are moderately penetrative but are stopped by a few millimeters of aluminum. Beta particles can be a health hazard either as internal or external radiation due to the ionization in the tissues.

X rays and gamma rays move with the speed of light. The only difference in this respect between gamma rays, X rays, and visible light is their frequency. See Figure 7-1. X rays and gamma rays ionize slightly in travel and are very penetrating compared to alpha and beta particles. They constitute the chief health hazard of external radiation, although gamma rays can be a hazard also as internal radiation. Whereas gamma rays come from the nucleus of an atom, X rays come from the electrons around the nucleus and are produced by electron bombardment. As is commonly known, when X rays pass through an object, they give a shadow picture of the denser portions on film or a special screen.

Neutrons are uncharged high-energy particles that may be given off under certain conditions and can have both physiological effects and the ability to make other substances radioactive. Neutrons can present major problems in areas around nuclear reactors and particle accelerators.

A *radionuclide* is a radioactive isotope of an element. A radionuclide can be produced by placing material in a nuclear reactor or particle accelerator (cyclotron, betatron, Van de Graff generator) and bombarding it with neutrons. Radionuclides have particular application as tracers in many areas of medicine, industry, and research. Mixing a radionuclide with a stable substance makes possible study of the path it follows and the physical and chemical changes the substance goes through. Naturally occurring radionuclides consist of unstable isotopes such as potassium 40, rubidium 87, thorium 232, uranium 235.

A *radioisotope* is an artificially created radioactive isotope of a chemical element that is normally not radioactive; they are also used in medical therapy.

Half-Life ($T_{1/2}$)

A characteristic common to all radionuclides is the decay or loss of activity as it ages. This loss is measured as the time it takes a radionuclide to lose half its activity and is referred to as its *radioactive half-life*. In general, seven half-lives will reduce the radioactivity of a radionuclide to about 1 percent of what it was when first measured. The loss of activity is usually measured in terms of disintegrations per sec or min, which can be readily converted to curies or millicuries. For example, if the radioactivity of a material with a half-life of 24 hours is a gamma emitter and emits radiation equivalent to 100 milliroentgens per hour (mR/h), it will be 50 mR/h after 24 hr, 25 mR/h after the second 24 hr, and 0.78 mR/h after the seventh 24-hr period. If the radioactivity of a material is not known, periodic measurements can be made and a curve constructed with radioactivity plotted on the vertical axis and time on the horizontal axis. The time for the material to lose half its radioactivity can then be read off the graph. The measure of half-life is useful to identify a particular radionuclide. Some common radioactive materials and characteristics are given in Table 7-1.

A distinction should be made between biological and effective half-life. The *biological half-life* is the time in which a living tissue, organ, or individual eliminates, through biologic processes, one-half of a given amount of substance that has been introduced into it. The *effective half-life* is the time in which a radioactive substance fixed in an animal body is reduced to half the original amount, resulting from the combined actions of natural radioactive decay and biological elimination. The relationship of effective half-life, radioactive half-life, and biological half-life is expressed by the following formula:

$$\text{effective half-life} = \frac{\text{biological half-life} \times \text{radioactive half-life}}{\text{biological half-life} + \text{radioactive half-life}}$$

Sources of Exposure

Sources of radiation are natural background, radioactive fallout from nuclear testing or use of nuclear devices, radiation from medical diagnosis and treatment, and radiation from industrial and other man-made sources. It is estimated that 240 million dental and medical X rays are taken annually and that 15 million tests using radioactive materials as tracers in the human body are also made.[3]

Some radiations to which individuals are or might be exposed are given in Table 7-2 to help understand and better appreciate the significance of radiation dosages.

X-Ray Machines and Equipment

One of the common types of ionizing radiation is the X ray. It is produced by fluoroscopic equipment, radiographic equipment (fixed and mobile), X-ray therapy equipment, dental X-ray machines, X-ray diffraction apparatuses, industrial X-ray machines, and other ionizing, radiation-producing equipment. Gamma-beam therapy equipment, including a sealed source capsule in a protective housing, is another source of ionizing radiation (gamma rays); gamma rays differ from X rays in the manner in which they are produced, the source (cesium 137, cobalt 60, gold 198, iridium 192, radium 226), or from where in the atom they originate.

The X-ray machine is essentially a particle accelerator in which the electrons

[3]Richard D. Lyons, "Public Fears Over Nuclear Hazards Are Increasing," *New York Times*, July 1, 1979, p. 28.

Table 7-2 Some Radiations to Which We Are or Might Be Exposed

Peacetime occupational allowable exposure	500 mR/yr with a max. of 300 mR/wk, avg in 13 wk
Cosmic rays	0.1 R/day (gamma)
Naturally occurring radioactive substances in water, air, soil	0.1 to 0.3 mR/day[a]
Natural environment—Denver	1.4 mR/day, incl. cosmic
San Francisco	0.4 mR/day, incl. cosmic
Human red blood cells	0.1 mR/day
Sun glasses—cornea glasses tinted with uranium oxide	1 to 10 mR/hr (beta and alpha)
Chinaware	5 mR/hr (beta and gamma)
Radium dial watch	100 mR/day (beta and gamma at surface)
Common X rays[b]	
Chest	10 to 250 mR/film or higher
Chest fluoroscopic	1000 mR/film
Pregnancy examination[c]	9000 mR
Lumbar spine, lateral	5700 mR
Lumbar spine, AP	1500 mR
Abdomen	1300 mR
Fluoroscopic	10,000 mR/min
Gastrointestinal examination	5000 to 50,000 mR
Dental[d]	200 to 1000 mR/film
Shoe-fitting fluoroscopic machine[e]	500 to 5000 mR/sec
Nuclear fallout (through 1962)	100 mR/30 yr, (0.02 to 0.5R)
In wooden homes	0.25 to 0.50 mR/day
In masonry homes, brick, stone, concrete blocks[f]	0.30 to 1.50 mR/day

Source: Various sources. See definitions of roentgens, rems, and rads in this chapter; can be assumed to be equivalent for this purpose. Exposures are approximate.

Note: The mean active bone marrow dose and mean gonad dose per examination are more significant measures and are considerably less than indicated here. (U.S. Environmental Protection Agency, ORP, *Radiological Quality of the Environment in the U.S. 1977*, EPA 520/1-77-009).

[a]Exposure increases slightly with altitude and varies with soils and rock natural radioactivity content. The main sources of background radiation are potassium 40, thorium, uranium, and their decay products, including radium and cosmic rays. Average 125 mrem (100 to 400) per year total body.

[b]May be exceeded with poor practice or equipment, or if medically indicated.

[c]Only if medically indicated.

[d]A whole-mouth X ray giving a dose of 50 R can be reduced to 5 R if an aluminum filter (2mm) is used in the X-ray machine.

[e]Is prohibited.

[f]Will vary with radiation content of the material.

(negative charge) are first generated by a hot filament, then accelerated through a vacuum by means of a high voltage, and abruptly stopped by a "target" that usually consists of a material of high atomic number. Tungsten is usually used because of its high melting point and high atomic number. See Figure 7-4.

As electrons enter the surface layers of the target, they are slowed down by collision with the nuclei and electrons in the target and are diverted from their

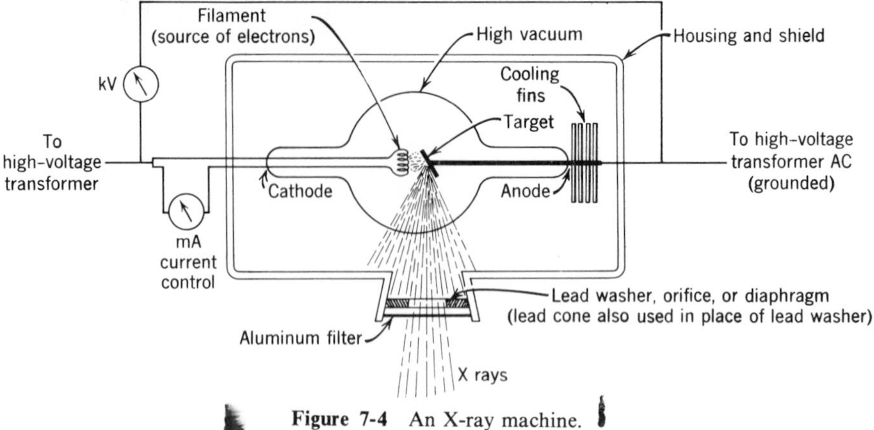

Figure 7-4 An X-ray machine.

original direction of motion. Each time the electron suffers a deceleration, energy in the form of X rays is produced. The energy of the X ray that results will depend on the deceleration that occurs. If brought to rest in a single collision, all its energy will appear in one burst of high energy and short wavelengths. If the electron encounters a less drastic collision, a longer wavelength results and lower-energy rays will be produced. Therefore energies of a wide range are produced from an X-ray tube.

The strength and quantity of X rays are usually expressed in terms of electrical current between the filament and target within the X-ray tube. The energy is measured in thousands of volts or kilovolts (kV). The quantity of X rays produced is directly proportional to the current through the tube and is measured as thousandths of an ampere or milliampers (mA). The dose or energy absorbed by an irradiated object is a function of the kV (machine setting) and mA of the machine as well as other factors.

Sources of X rays that can operate at a voltage above 10 kV produce more penetrating radiation and therefore may be more hazardous than sources less than 10 kV, *if not adequately shielded.* These include high-voltage television projection systems, color television sets, electron microscopes and power sources, high-power amplifying tubes producing intense microwave fields, radio-transmitting tubes, high-voltage rectifier tubes, and other devices producing penetrating electromagnetic radiation.

Radioactive Materials

Other sources of ionizing radiation are radionuclides used in medicine, research, and industry as previously mentioned. Radium in equilibrium with its daughter products was widely used but is replaced by other radionuclides. It is an alpha, beta, and gamma emitter and requires special shielding and care in handling. Sealed radium sources are used primarily as gamma emitters because the alpha and beta components of the radiation are stopped by the encapsulation material. Some of the other commonly used, naturally occuring radioactive materials are uranium, polonium, actinium, and thorium.

Artificially produced radionuclides generally include those made either through the fissioning process in a nuclear reactor or by bombardment of nonradioactive isotopes in high-energy accelerators or reactors. Many hundreds of isotopes have been produced.

The widespread use of radioactive materials and radionuclides in research, medical diagnosis, laboratories, industry, and educational institutions requires that precautions be taken in their storage, use, and disposal. Ingestion may occur through contaminated hands, cigarettes, or food. Radioactive vapor, dust, gas, or spray may be inhaled. Medical uses can present a hazard to technicians, patients, and others through the improper handling of radionuclides and contaminated wastes.

BIOLOGICAL EFFECTS OF RADIATION 

Ionizing Radiations

Radiation which is capable of producing ionization and which is absorbed can cause injury or other damage. As the radiation penetrates the organism or cell, it comes into contact with atoms and molecules in the living protoplasm, altering the structure or electrical charges with the formation of other molecules and substances which in turn may cause other effects. The radiation may pass through the cell causing no damage; or if damaged the cell may partially repair itself. But if a cell is damaged and not repaired it may reproduce in damaged form, or the cell may be destroyed. The unknown long-term effects of radiation, including possible cancers, birth defects, and hereditary changes that may be passed on to future generations, heighten public concern and individual emotions.

Somatic and Genetic Effects

The biological effects of radiation on all living organisms, including human beings, are termed *somatic*, meaning effects occuring during the lifetime of the exposed organism, as opposed to *genetic*, which are the effects on generations yet unborn[4] Somatic or genetic effects from radiation exposure may not be evident for months, years, or a lifetime. Even relatively large doses usually show no immediate apparent injury. But many cells, tissues, and organs of the body are interdependent, and the destruction of one may eventually result in the poor functioning or death of the other. All cells, with the exception of sperm cells and

[4] *Biological Effects of Radiation*, PHS, DHEW, Washington, D.C., 1963.

nerve cells, can apparently replace themselves or recover to some extent from radiation exposure, if the dose is not excessive and is administered in small increments over an extended period of time. The sixth report of the Royal Commission on Environmental Pollution noted that

> at levels of radiation likely to be permitted in relation to possible somatic effects, the genetic effects should be of little concern.

Hence the somatic risks should govern the required standards of radiological protection.[5]

In assessing radiation hazard the sum total of all exposures should be considered, that is, natural background, medical, and occupational exposures, and radiation ingested through the air, water, and food. Exposure of the gonads (ovaries or testes), egg, or sperm is necessary to cause genetic effects from ionizing radiations. Everyone is subjected to natural background radiation, as well as heat and trace amounts of chemicals that cause a small number of so-called spontaneous mutations in genes. The genes determine all inherited characteristics. Hence any additional radiation of the reproductive cells should be avoided to keep down further mutations. The genetic harm done by radiation is cumulative up to the end of the reproductive period and depends on the total accumulated gonadal dose. Not all mutant genes are equally harmful. Some may cause serious harm, most cause very little or no apparent damage. Nevertheless, radiation exposure of large population groups should be kept as low as practicable since any mutation is generally considered harmful. It is estimated, for example, that a person receives on the average 125 millirems/yr from natural sources[6] and 50 to 70 millirems/yr (genetically significant dose—GSD)* from medical exposures.[2] The actual amounts received in different parts of the world will vary with altitude, background, medical practices, and other factors. For example, a roundtrip flight between New York and Los Angeles will produce an exposure of 5 mR. It is estimated that 75 to 230 additional cancer deaths per rad exposure would result in a hypothetical population of one million persons exposed to a *one-time* radiation dose of 10 rads. The national average is less than one-fifth rad per year, mostly cosmic rays and medical X rays.[7] The death rate due to all types of cancer in the U.S. in 1975 was 1717 per million, or 343,400 in a population of 200 million.

In another example of the relative health hazard associated with natural radiation, fallout radiation, and similar effects from all other causes in the United

*The genetically significant dose is the dose which, if received by every member of the population, would be expected to produce the same total genetic injury to the population as do the actual doses received by the various individuals. (*Basic Radiation Protection Guide*, Rep. No. 39, National Council on Radiation Protection and Measurement, Washington, D.C., January 15, 1971, p. 13.)

[5]Brian Wade, "Energy Policy and the Public Health—The Safety of Nuclear Power," *J. R. Soc. Health*, December 1979, pp. 239–246.

[6]Report of the United Nations Scientific Committee on the Effects of Atomic Radiation, 1962.

[7]National Academy of Sciences Committee on Biological Effects of Ionizing Radiation (BEIR III), Washington, Associated Press, July 30, 1980.

States, the Federal Radiation Council gave the following summary estimates for the next 70 years:[2]

	Leukemia	Bone Cancer
Total number of cases from all causes	840,000	140,000
Estimated number caused by natural radiation	0 to 84,000	0 to 14,000
Estimated number of additional cases from all nuclear tests through 1962	0 to 2,000	0 to 700

Knowledge of the long-term effects on human beings of both acute doses (large doses received in a short period of time) and chronic doses (those received frequently or continuously over a long period of time), from either external or internal sources of radiation, is far from complete. Relatively speaking however, more is known about radiation effects than effects of trace amounts of toxic substances.

Factors to Consider

Some of the factors that determine the effect of radiation on the body are the following:

1. Total amount and kind of radiation absorbed, external and internal.
2. Rate of absorption.
3. Amount of the body exposed.
4. Individual variability.
5. Relative sensitivity of cells and tissues; parts of the body exposed.
6. Nutrition, oxygen tension, metabolic state.

For example, a dose of 600 R whole-body X radiation in one day would mean almost certain death. On the other hand, 600 R administered in weekly increments over a period of 30 years may result in little or no detectable effect. This would amount to

$$\frac{600\ \text{R}}{30\ \text{yr} \times 52\ \text{wk}} = 0.385\text{R/wk, or } 385\ \text{mR/wk.}$$

Large doses of radiation can be applied to local areas, as in therapy, with little danger. A person could expose a finger to 1000 R and experience a localized injury with subsequent healing and scar formation.

Effects of whole-body radiation are given in Table 7-3. It is known that the blood-forming organs—spleen, lymph nodes, and bone marrow—are the most radiosensitive organs when the whole body is irradiated and that protection of these organs from exposure lessens the whole-body effect.

Table 7-3 Effects of Total-Body Radiation

Dose (roentgens)[a]	Probable Effect
0 to 50	No obvious effect except possible minor blood changes.
150 to 250	Vomiting and nausea for about 1 day, followed by other symptoms of radiation sickness in about 50% of personnel. No deaths expected.
200 to 400	Vomiting and nausea for about 1 day, followed by other symptoms of radiation sickness in about 50% of personnel. 10% deaths expected.
350 to 550	Vomiting and nausea in nearly all personnel on first day, followed by other symptoms of radiation sickness. About 25% deaths expected.
550 to 750	Vomiting and nausea in all personnel on first day, followed by other symptoms of radiation sickness. About 50% deaths within 1 month.
1000	Vomiting and nausea in all personnel within 1 to 2 hr. Probably no survivors.
5000	Almost immediate incapacitation. All personnel will be fatalities within 1 wk.

Source: *Fundamentals of Nuclear Warfare*, Special Text 3-154, U.S. Army Chemical Corps Training Command, Fort McClellan, Ala., January 1961, p. 16.

Note: A lower dose may produce the indicated effect.

[a]And/or rem in the case of neutrons.

Table 7-4 Maximum Permissible Dose Equivalent Values (MPD)

	Maximum 13-week dose (rem)[a]	Maximum yearly dose (rem)[a]	Maximum accumulated dose (rem)[a]
Controlled areas[b]			
Whole body, gonads, lenses of eye, red bone marrow	3	5	5(*N*-18)[c]
Skin (other than hands and forearms)		15	
Hands	25	75	
Forearms	10	30	
Other organs, tissues, and other organ systems	5	15	
Noncontrolled areas[d]			
Occasional exposure		0.05	
Population dose limit		0.17 ave.	

Source: *Basic Radiation Protection Criteria*, Rep. No. 39, The National Council on Radiation Protection and Measurement, Washington, D.C., 1971, pp. 88–107.

[a]The numerical value of the dose equivalent in rems may be assumed to be equal to the numerical value of the exposure in roentgens for this purpose. MPD for fertile women = 0.5 rem in gestation period.

[b]Occupational exposure.

[c]N = age in years and is greater than 18. When the previous occupational history of an individual is not definitely known, it shall be assumed that he has already received the MPD permitted by the expression 5 (N-18).

[d]Whole-body, or critical organs. See Report No. 39 for conditions governing each value.

Table 7-5 Maximum Permissible Doses and Dose Limits Set by the International Commission on Radiological Protection, 1966

Organ or Tissue	Adult Radiation Staff[a] (rems per year)	Members of the Public (rems per year)
Gonads, red bone marrow[b]	5[c]	0.5
Skin, bone	30	3.0
Thyroid	30	3.0[d]
Extremities	75	7.5
Other single organs	15	1.5

Source: C. B. Braestrup and K. J. Vikterlöf, *Manual on radiation protection in hospitals and general practice, Volume 1, Basic protection requirements*, WHO, 1974, p. 32.
[a]Occupational.
[b]Whole body.
[c]For pregnant women the dose to the fetus accumulated during the remaining period of pregnancy after the diagnosis should not exceed 1 rem.
[d]1.5 rems in a year to the thyroid of children up to 16 years of age.

Avoiding Unnecessary Radiation

Ideally, exposure to any ionizing radiation should be avoided. In real life this is not possible or in some instances desirable. Background radiation cannot be eliminated, and there are times when medical uses of radiation may be indicated. Research and industrial uses are also usually indicated with proper controls.

To receive the many social and health benefits of radiation and keep the risk minimal but within acceptable limits, maximum permissible doses (MPD) have been proposed by the International Commission on Radiological Protection (ICRP), the National Council on Radiation Protection (NCRP), the Federal Radiation Council (FRC), and others. The functions of the FRC have been assumed by the USEPA under the President's Reorganization Plan Number 3 of 1970. Radiation protection guides are summarized in Table 7-4. Maximum permissible body burdens for certain body organs are given in Table 7-5. It is believed by some researchers that the hazards of low-level radiation (0.1 to 10 rads) may be worse than previously predicted, supporting the principle that "X ray should be used only when there is a good medical reason."[8] Apparently there is no "threshold below which radiation ceases to have adverse effects on humans, but those effects can be so small in a large population that they are masked by more prominent causes."[7]

TYPES OF X-RAY UNITS AND COMMON HAZARDS

Medical and dental diagnostic examinations constitute a major source of radiation exposure to the individual. The National Academy of Sciences—

[8]Erwin D. Bross, Marcella Ball, and Steven Falen, "A Dosage Response Curve for the One-Rad Range: Adult Risks from Diagnostic Radiation," *Am. J. Pub. Health*, February 1979, pp. 130–136.

National Research Council (NAS-NRC) Committee on Biological Effects of Ionizing Radiation reported in 1972 that medical diagnostic radiology, including radiopharmaceuticals, accounts for at least 90 percent of the total man-made radiation dose to which the U.S. population is exposed. Therapeutic uses involve special considerations. Some of the common hazards associated with the use of medical and dental diagnostic and therapeutic machines are outlined below. There is far too much carelessness, familiarity, and ignorance involved in the use of radiation-producing equipment that should not be condoned. Ward mobile X-ray equipment should not be used except in extreme emergencies and then only with special care and precautions. Education and motivation on the part of the medical and dental profession and their X-ray technicians can do more than anything else to reduce medical exposure.[9] Personnel exposure and workload records should also be kept.

Dental X-ray Machines

These units are usually operated at 50 to 100 kVp (kilovolt peak) at a tube current of 7 mA. Machine voltage should be properly calibrated. Fast, sensitive film and proper film-developing techniques are important. A 90-kVp unit will require 0.5 sec exposure time compared to 1.5 sec for a 50 to 75-kVp unit. A fast film will reduce exposure at skin surface by a factor of approximately 3 for a machine below 60 kVp and by approximately 9 for an 80-kVp or above machine as compared to slow film.[10] It is necessary to establish the correct exposure time for the kilovoltage, milliamperage, and source-to-skin distance.[11] Common deficiencies are the following:

1. Diameter of X-ray beam exceeds 3.0 in. at the end of the cone. Requires use of proper diaphragm (lead washer) or collimators to keep entire beam within edges of film.
2. Exposure to the direct beam, if pointed toward occupied adjacent rooms, to the ceiling, or to the floor.
3. Scatter radiation, especially that from the patient. Plastic cones contribute to scatter; open-ended devices are recommended. Check old X-ray tube housing for leakage.
4. Filters missing; use minimum 1.5 mm aluminum for 50 to 70-kVp unit, and 2.5 mm aluminum for unit above 70 kVp.
5. Multiple sources, particularly if several units are in the same room.
6. Operator in the same room as patient; no extension cord on timer button to make possible distance (at least 6 ft) protection.
7. Timer not accurate or of a dead-man type. Calibrate equipment frequently.
8. Use of outdated or slow emulsion speed film. Use of old chemicals; no time and temperature control.
9. Failure to use lead apron to shield the neck and gonads of the patient, especially children.

[9]Karl Z. Morgan, "Common Sources of Radiation Exposure," *Am. Eng.*, July 1968, pp. 38–42.
[10]K. L. Travis and J. E. Hickey, "A State Program for Reducing Radiation Exposure from Dental X-ray Machines," *Am. J. Pub. Health*, August 1970, pp. 1522–1527.
[11]*Exposure and Processing Guide For Dental Radiography*, DHEW, PHS, FDA, Rockville, MD, 20857, revised February 1978.

10. Improper target-skin distance. Use minimum of 4 in. at 50 kVp or below; 7 in. with equipment operating above 50 kVp.
11. Improper film processing. Film processing demands strict attention to recommended time-temperature procedures in order to obtain optimum film quality.
12. Overexposing film and compensating by underdeveloping. This results in poor quality film and excessive radiation to the patient.
13. Inadequate dark room conditions, (light leaks, improper safe-lighting) can affect sensitivity of films.
14. Do not sight develop!

Fluoroscopy

Fluoroscopy units usually operate at voltages up to 125 kVp, usually about 80, at 3 to 5 mA. They should be used *only* where X-ray film will not provide the information. A film gives greater detail and a record for study, with only a small fraction (5 percent) of the patient exposure when compared to fluoroscopy. Common deficiencies are the following:

1. The useful beam may extend beyond the fluoroscopic screen and its lead glass at maximum beam size and target screen distance. Shutter opening not controlled to minimum size needed for examination. If a full screen is needed, the shutter should be limited so there will be a visible dark margin of at least $\frac{1}{4}$ in. around the perimeter of the screen when the screen is about 14 in. from the table top.
2. Scatter radiation from the patient and the undersurface of the tabletop.
3. Timing device not functioning properly; maximum exposure to the patient is not kept at or below 10 R/min. at maximum kVp and mA. Calibrate frequently.
4. Absence of a cone between tube housing and tabletop.
5. Inadequate shielding of the tube enclosure against the direct beam.
6. Too short a target table distance. Source-skin-distance inadequate.
7. Absence of leaded apron and gloves for physician and attendant.
8. Filter too thin or totally lacking. Use minimum 0.5 mm aluminum below 50 kVp, 1.5 mm for 50 to 70 kVp, and 2.5 mm for above 70 kVp. A total of at least 3 mm of aluminum is recommended.
9. Increased time of use due to failure to allow for eye accommodation (10 to 20 min.). Intensifying screens should be used. An image intensifier will permit use of a much lower $\frac{1}{2}$ to $\frac{1}{10}$) dose rate.
10. Dead-man switch* not used, inoperable, unavailable.
11. Exposure not terminated by barrier removal. The useful beam must always be intercepted by a primary barrier (usually lead glass screen or image intensifier assembly) irrespective of the panel screen distance. The exposure should terminate when the barrier is removed from the useful beam.

Radiography

Usual operating conditions are 40 to 100 kVp at currents up to 1000 mA. The machine should be equipped with calibrated voltage meter and calibrated timing

*Circuit closing contact can be maintained only by continuous pressure on the switch by the operator.

device. Fast, sensitive film and proper developing techniques should be used. Lead shielding may be necessary behind the cassette holder for chest and upright X rays, or at any other primary beam areas, if the "useful" beam penetrates into a waiting room or other occupied areas. Common deficiencies are the following:

1. Failure to confine the X-ray beam to the part of the patient's body being examined. Collimation inadequate.
2. Scatter radiation around or over protective screens or into control booth where no door is provided.
3. Scattering under doors and at junction of walls with floor and ceiling if no lead baffle is provided.
4. Useful beam and scatter radiation passing through windows and into occupied regions nearby.
5. Holding the patient or film during exposure.
6. Filters missing or inadequate. Use minimum 0.5 mm aluminum below 50 kVp, 1.5 mm for 50 to 70 kVp, and 2.5 mm for above 70 kVp.
7. Exposure switch or timer inadequate.
8. Failure to use gonadal shielding if gonads lie within the primary field, or within close proximity (about 5 cm), despite proper beam limitation. Shaped contact shield, flat contact shield, and shadow shield should be available and used.
9. The exposure switch is so arranged that it can be operated with any part of the operator's body outside the shielded area.
10. The operator is not able to communicate with and/or view the patient during an exposure while remaining in a shielded position.

Therapy Units

Therapy units operate at energies up to 10 MeV and higher, depending on the type of equipment. Common deficiencies are the following:

1. Scattering around or into control booth, where no door is provided.
2. Scattering under doors and at junctions of walls with floor and ceiling if no lead baffle is provided.
3. Scattering from nearby buildings or from the floor of the room below if the treatment room floor is insufficiently shielded.
4. Leakage around doors and observation windows.
5. Filtration, exposure control, calibration, interlocks inadequate.

In addition to the hazards stated above, it is not unusual to find in X-ray installations loss of adequate shielding, removal or failure of safety devices, infrequent calibration, and electrical hazards. Lack of door interlocks or insufficient primary or secondary shielding may also present a hazard with therapy machines.

When developing a control program, many X-ray machines, although properly installed originally, will be found to be hazardous because of alterations to the building and removal of the original shielding. Filters become lost and lead diaphragms are missing. A concentration of X-ray machines in one building may also increase the scattered radiation to unsatisfactory levels. In some instances equipment may not have current and voltage meters.

Also used in medical therapy is the radioisotope (such as cobalt 60 and cesium 137) teletherapy unit and particle accelerator. These units require specially designed rooms, facilities, shielding, and operating procedures.

CAT Scanner

The computerized axial tomography (CAT) scanner is a combination computer and X-ray machine which, together with auxiliary equipment, can weigh several tons. The major advantage of a CAT scanner is its ability to produce pictures of soft tissue which is not possible by the ordinary X-ray machine unless some opaque chemical such as barium is ingested and photographed as it passes through the organ being studied. Pictures taken by the CAT scanner in close sequence (tomograms) can also show depth. Scanners are used to locate tumors, blood clots, anatomical malformations, and the like thereby avoiding exploratory surgery.[12]

X-ray and Other Radiation Sources Used in Industrial and Commercial Establishments

Industrial and commercial X-ray devices include primarily the following:

1. Radiographic and fluoroscopic units used for the determination of defects in welded joints and in casting fabricated structures and molds.
2. Fluoroscopic units used for the detection of foreign material, as in packaged foods.
3. Ionizing radiation being considered for the pasteurization of foods, sterilization of medical supplies, and other purposes.
4. Antitheft and antisabotage fluoroscopic examinations. These devices should not be recommended or used, except for inanimate objects.
5. Sealed sources for measuring the density or thickness of products and for determining liquid levels in closed tanks.

Radiation machines used in industry, training, and research include Van de Graff accelerators, Cockcroft-Walton, particle accelerators, neutron generators, and other special X-ray units. These high voltage machines and the sources mentioned have the same hazards associated with them as do the medical and dental X-ray equipment. Radiographic units having sealed sources use cobalt 60, cesium 137, and iridium 192. The industrial radiographic X-ray machine has a poor operator and surrounding worker operating safety record.

Some industries have used antitheft, full-body fluoroscopic examinations of employees and similar examinations for antisabotage purposes. Unless such examinations are absolutely necessary and records kept as to exposure of personnel, it is possible for employees to receive excessive doses of radiation. Such devices should not be used.

[12]Harry Schwartz, "The Government Puts a Damper on the Scanner Bonanza," *New York Times*, December 8, 1977.

Other

Projection television tubes, diffraction analyzers, and, for that matter, any equipment using high voltages, are capable of generating X rays. Even the home television tubes generate X rays, but have not become a hazard since the "soft" X rays generated are filtered out by the glass of the tube. However, if higher voltages are used than are allowed in order to get good images in color, or if designed with improper shielding, these tubes may also become a hazard. The radiation should not exceed 0.5 mR/hr at any accessible point 5 cm (about 2 in.) from the surface under the most adverse operating conditions.* Viewing at least 6 ft from the tube is advised in any case.

General Preventive Measures

Initiation of a routine inspection program will reveal a number of common defects. For instance, many dental installations may lack filters in the X-ray beam. Lead cones to define the area of the X-ray beam will not be found in some medical installations. In many cases X-ray beams on fluoroscopes will extend past the fluoroscopic screen and protective lead glass, thereby exposing the operator to the direct beam.

All these defects are very simple to find and need no instruments whatever. However, their importance should not be underestimated. For instance, the addition of filters to dental machines can reduce the surface dose to the patients by a factor of up to 10, as well as the resultant exposure to the operator. A filter in the tube port providing the equivalent filtration of approximately 2 mm of aluminum will absorb the soft, or less penetrating, radiations. This will reduce the radiation that is absorbed by the patient, reduce the stray radiation, and eliminate scatter haze on the film. The use of open-ended pointing devices on X-ray machines can eliminate much unnecessary irradiation of the patient by reducing the exposed area to the particular field of interest. These are very simple adjustments needing very little technical knowledge, but do a great amount of good.

Patient and individual exposure can be greatly reduced. The basic measures include proper collimation to limit the X-ray beam to the area needed (can reduce the average gonadal dose by 65 percent), also adequate beam filtration, beam alignment, and tight tube housing; the use of fast film when film is used; gonadal shielding; better-trained technicians; and continued education of those using or prescribing X ray. The more general use of machines with automatic collimation and other controls will also do much to reduce unnecessary exposure.

*Recommended by the International Commission on Radiological Protection and the National Council on Radiation Protection. Also limiting value in the United Kingdom.

Federal Guidelines To Reduce Radiation Exposure

A presidential directive for the guidance of federal agencies to eliminate clinically unproductive examinations; use optimal techniques when examinations are performed; and employ appropriate X-ray equipment, is equally applicable to nonfederal institutions. It is quoted below for reference.[13]

1. General radiographic or fluoroscopic examinations should be prescribed only by licensable Doctors of Medicine or Osteopathy or, for specified limited procedures, postgraduate physician trainees and qualified allied medical professionals under their direct supervision; specialized studies should be prescribed only by those physicians with expertise to evaluate examinations in the particular specialty. Exception for specified procedures may be made for dentists and podiatrists.

2. Prescription of X-ray studies should be for the purpose of obtaining diagnostic information, should be based on clinical evaluation of symptomatic patients, and should state the diagnostic objective and detail relevant medical history.

3. Routine or screening examinations, in which no prior clinical evaluation of the patient is made, should not be performed unless exception has been made for specified groups of people on the basis of a careful consideration of the magnitude and medical benefit of the diagnostic yield, radiation risk, and economic and social factors. Examples of examinations that should not be routinely performed unless such exception is made are:

   a. chest and lower back X-ray examinations in routine physical examinations or as a routine requirement for employment,
   b. tuberculosis screening by chest radiography,
   c. chest X rays for routine hospital admission of patients under age 20 or lateral chest X rays for patients under age 40 unless a clinical indication of chest disease exists,
   d. chest radiography in routine prenatal care, and
   e. mammography examinations of women under age 50 who neither exhibit symptoms nor have a personal or strong family history of breast cancer.

4. Prescription of X-ray examinations of pregnant or possibly pregnant patients should assure that medical consideration has been given to possible fetal exposure and appropriate protective measures are applied.

5. The number, sequence, and types of standard views for an examination should be clinically oriented and kept to a minimum. Diagnosticians should closely monitor the performance of X-ray examinations and, where practicable, direct examinations to obtain the diagnostic objectives stated by clinicians through appropriate deletion, substitution, or addition of prescribed views. Technique protocols for performing medical and dental X-ray examinations should detail the operational procedures for all standard radiographic projections, patient preparation requirements, use of technique charts, and image receptor specifications.

6. X-ray equipment used in Federal facilities should meet the Federal Diagnostic X-ray Equipment Performance Standard, or as a minimum for equipment manufactured prior to August 1, 1974, the Suggested State Regulations for Control of Radiation (40 FR 29749). General purpose fluoroscopy units should provide image-intensification; fluoroscopy units for nonradiology specialty use should have electronic image-holding features unless such use is demonstrated to be impracticable for the clinical use involved. Photofluorographic X-ray equipment should not be used for chest radiography.

7. X-ray facilities should have quality assurance programs designed to produce radiographs that satisfy diagnostic requirements with minimal patient exposure; such programs should contain

[13] *Fed. Reg.*, Vol. 43, No. 22, February 1, 1978.

material and equipment specifications, equipment calibration and preventive maintenance requirements, quality control of image processing, and operational procedures to reduce retake and duplicate examinations.

8. Operation of medical or dental X-ray equipment should be by individuals who have demonstrated proficiency to produce diagnostic-quality radiographs with the minimum of exposure required; such proficiency should be assessed through national performance-oriented evaluation procedures or by didactic training and practical experience identical to, equivalent to, or greater than training programs and examination requirements of recognized credentialing organizations.

9. Proper collimation should be used to restrict the X-ray beam as much as practicable to the clinical area of interest and within the dimensions of the image receptor; shielding should be used to further limit the exposure of the fetus and the gonads of patients with reproductive potential (21 CFR Part 1000,50) when such exclusion does not interfere with the examination being conducted.

10. Technique appropriate to the equipment and materials available should be used to maintain exposure as low as is reasonably achievable without loss of requisite diagnostic information; measures should be undertaken to evaluate and reduce, where practicable, exposures for routine nonspecialty examinations which exceed the following Entrance Skin Exposure Guides (ESEG):

Examination (Projection)	ESEG (milliroentgens)[a]
Chest (P/A)	30
Skull (Lateral)	300
Abdomen (A/P)	750
Cervical Spine (A/P)	250
Thoracic Spine (A/P)	900
Full Spine (A/P)	300
Lumbo-Sacral Spine (A/P)	1000
Retrograde Pyelogram (A/P)	900
Feet (D/P)	270
Dental (Bitewing or Periapical)	700

[a]Entrance skin exposure determined by the Nationwide Evaluation of X-Ray trends program for a patient having the following body part/thickness: head/15 cm, neck/13 cm, thorax/23 cm, abdomen/23 cm, and foot/8 cm.

11. X-ray examinations for dental purposes should be prescribed only by licensable Doctors of Dental Surgery or Dental Medicine or properly supervised postgraduate dentists on the basis of prior clinical evaluation or pertinent history; neither a full-mouth series nor bitewing radiographs should be used as a routine screening tool in the absence of clinical evaluation in preventive dental care. Exception may be made for justifiable forensic purposes.

12. Open-ended shielded position-indicating devices should be used with the paralleling technique to perform routine intra-oral radiography and should restrict the X-ray beam to as near the size of the image receptor as practicable.

RADIUM AND OTHER RADIOACTIVE MATERIALS

Radium occurs naturally. It was used extensively in medicine and in luminous compounds for markers and watch and instrument dials. Radium has a very long half-life (1622 yr), which on decay produces radon. Radon is an alpha emitter; in addition, the radon daughter decay products are alpha, beta, and gamma emitters. Radium as used in medicine is in the form of a loose powdered salt placed in a sealed capsule. There is a danger of the capsule rupturing or leaking due partly to the pressure exerted by the radon gas or by careless handling.

Because of the hazards associated with the use of radium, and loss of the capsule, an accountability procedure should be in effect at places where radium is used and stored. This should be supplemented by radiation protection equipment and techniques and periodic radiation safety surveys. If a capsule is lost or broken, the state and local health departments should be immediately notified.

Medical personnel involved in radium radiation therapy will be close to the source. They may therefore be unnecessarily exposed during the handling of the radium needle, during insertion or removal, and during the time the patient is being treated. It is important therefore that medical personnel receive special training, experience, and direct supervision if their duties include assisting in the therapeutic use of radium or of the radionuclides that have replaced radium.

The radioactive substances used in medicine may be either sealed or unsealed. Sealed sources (needles), in addition to radium 226, include cobalt 60, gold 198, radon 222, yttrium 90, and cesium 137. Unsealed sources consist of radioactive materials in liquid form, such as solutions of iodine 131, gold 198, and phosphorus 32.[14]

The general safety precautions for radium apply equally to unsealed and to other sealed, encapsulated, or otherwise contained radioactive sources.

RADIATION PROTECTION

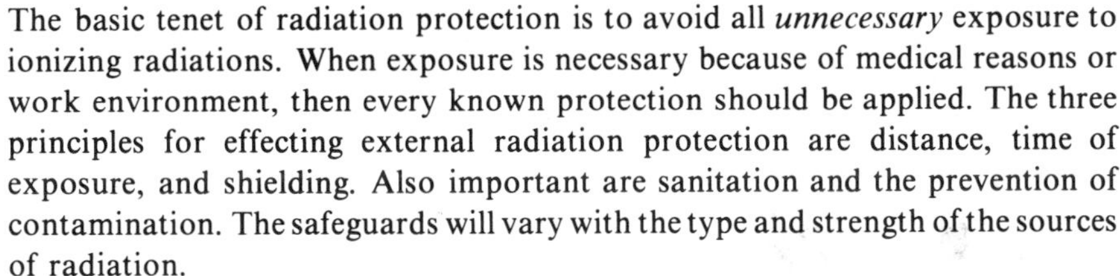

The basic tenet of radiation protection is to avoid all *unnecessary* exposure to ionizing radiations. When exposure is necessary because of medical reasons or work environment, then every known protection should be applied. The three principles for effecting external radiation protection are distance, time of exposure, and shielding. Also important are sanitation and the prevention of contamination. The safeguards will vary with the type and strength of the sources of radiation.

Distance

The further away a person is from a radiation source, the less will be the exposure he or she receives. This is particularly true for a point source of radiation, as the subject exposure decreases inversely with the square of the distance from the source. For example, if at a distance of 5 ft one is exposed to A mR, at a distance of 10 ft the exposure would be $\frac{1}{4}A$ mR. This is known as the Inverse Square Law and is expressed as $I_1/I_2 = R_2^2/R_1^2$ in which R_1 and R_2 are any two distances, and I_1 and I_2 are the values of the intensity at the distances.

In practice the above principle also applies to other than point sources of radiation in view of the small sizes of the radiation sources and relatively large distances generally involved.

[14] C. B. Braestrup and K. J. Vikterlöf, *Manual on radiation protection in hospitals and general practice*, WHO, 1974, pp. 14, 40–44.

Time of Exposure

When exposure to radiation is necessary, the time of such exposure should be kept as low as practicable to accomplish a particular task. In the case of occupational or similar situations, the total exposure shall be kept below an individual's maximum permissible dose. This might mean transfer to another assignment where she or he will receive no radiation so as to stay within the limits shown in Tables 7-4 and 7-5.

Contamination Prevention

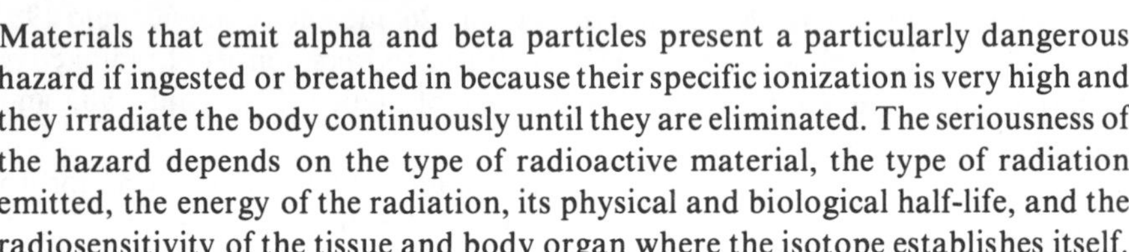

Materials that emit alpha and beta particles present a particularly dangerous hazard if ingested or breathed in because their specific ionization is very high and they irradiate the body continuously until they are eliminated. The seriousness of the hazard depends on the type of radioactive material, the type of radiation emitted, the energy of the radiation, its physical and biological half-life, and the radiosensitivity of the tissue and body organ where the isotope establishes itself. See Tables 7-1, 7-4, and 7-5.

The objective must therefore be to keep the radioactive materials out of the body. This is accomplished by the use of proper procedures and good practices, such as using laboratory hoods, air filters, and exhaust systems; eliminating dry sweeping; making dry runs and wearing protective clothing; using respirators when indicated; using proper monitoring and survey instruments; and prohibiting eating and smoking where radioactive materials are handled or used.

Shielding

Shielding is the interposition of a dense attenuating material between a source of radiation and the surroundings so as to adequately reduce or practically stop the travel of radiation. A shield may be used for different purposes, such as to absorb X rays, gamma rays, neutrons, or intense heat. For neutrons, a hydrogenous material such as paraffin is used. The energy and type of the radiation to be attenuated and occupancy or use are basic factors in the selection of the shielding material or materials, their size, and thickness. A primary protective barrier is of a material and thickness to attenuate the useful beam to the required degree; a secondary barrier attenuates stray radiation. See Table 7-6.

The computation of shielding or barrier requirements is simplified by the use of tables and graphs.[15,16] For example, if the distance from the X-ray tube target to the wall (the controlled area to be protected) to be protected is 7 ft, the kVp of the X-ray therapy machine is 250, the workload is 20,000 mA min per week, and

[15] *Structural Shielding Design and Evaluation for Medical Use of X Rays and Gamma Rays of Energies up to 10 MeV*, Rep. No. 49, NCRP, Washington, D.C., 1976.
[16] *Dental X-Ray Protection*, Rep. No. 35, NCRP, Washington, D.C., March 9, 1970, pp. 29–37.

Table 7-6 Thickness of Lead Required for a Primary Barrier Located 5 cm from the Focal Spot

Anode Current (mA)	Thickness of lead (mm) 50 kVp	70 kVp	100 kVp
20	1.5	5.6	7.7
40	1.6	5.8	7.9
80	1.6	5.9	
160	1.7		

Source: National Bureau of Standards - NBS Handbook 114, *General Safety Standard for Installations Using Nonmedical X-ray and Sealed Gamma-ray Sources, Energies up to 10 Mev*, Washington, D.C.

the fraction of the workload during which the radiation under consideration is directed at a particular barrier (use factor) is $\frac{1}{4}$, the required primary barrier thickness is 7.9 mm lead or 14.7 in. concrete for a controlled area. If the radiation is applied obliquely incident on a barrier, the required barrier thickness will be less. With a 200-kVp machine under the same conditions, 4.4 mm lead and 12.7 in. concrete would be required, and with a 100-kVp therapy machine, a workload of 4000 mA min per week and a use factor of $\frac{1}{4}$, 1.65 mm lead and 5.2 in. concrete would be required.[17] See NCRP Report No. 49[15] for computations.

The term "half-value layer" (HVL) is used to designate the thickness of a particular material that will reduce by one-half the intensity of radiation passing through the material. Lead and concrete are commonly used to shield X and gamma radiations. Glass or plastic are commonly used to eliminate beta radiation. See Figure 7-5.

Radiation, however, will scatter and bounce off the floor and ceiling and from wall to wall. It can therefore be reflected around shields, around corners, over and under doors, and through ventilating transoms and windows. Shields must be placed so as to protect anyone who might have access to spaces above or below, or on any side of the source. It is generally better to place shields close to the source of radiation because this will reduce the required area and weight of the shield.

Some of the characteristics of alpha and beta particles, X rays and gamma rays, and neutrons are given earlier in this chapter. The ranges of alpha and beta particles in air are shown in Tables 7-7 and 7-8. Neutron shielding presents special problems around nuclear reactors and particle accelerators.

Calculation of structural shielding requirements takes into consideration the work load, use factor, occupancy factor, and whether the area exposed is under control of the person responsible for radiation protection.

The work load, usually expressed in milliampere seconds per week, is the milliamperage of the current passing through the X-ray tube times the sum of the number of seconds the tube is in operation in one week.

The use factor is the part of the time a machine is in use that the useful beam

[17]NCRP Report No. 49, pp. 64, 66, 67.

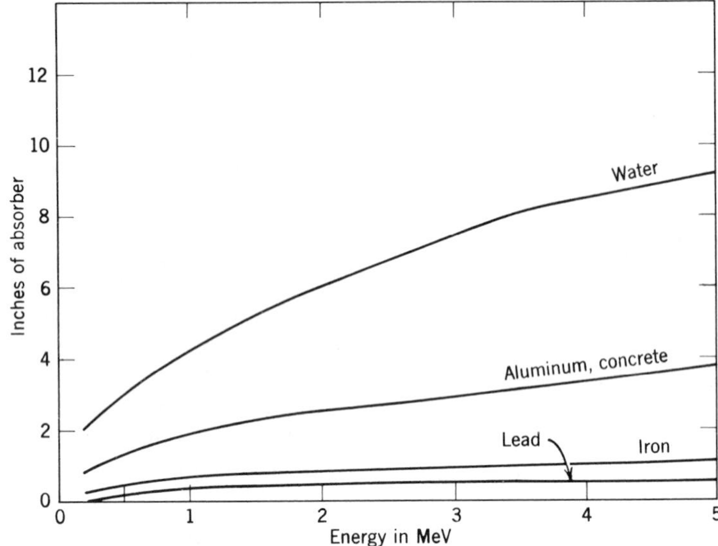

Figure 7-5 Half-value layers for different electromagnetic energies as related to thickness (in.) of iron and lead, aluminum, concrete, and water. [From *Atomic Radiation*, RCA Service Co., Government Services, Camden, N.J., 1959 (Contract Nr. AF 33(616)-3665).]

may strike the wall, ceiling, or floor to be shielded. The occupancy factor is the time a person on the other side of a shielded wall, floor, or ceiling will be exposed to radiation when a unit is in operation. This factor will vary from 1 when there are people living or working in the adjoining space to $\frac{1}{16}$ when people may be exposed in a stairway or elevator.

Information for the calculation of shielding requirements is given in *Structural Shielding Design and Evaluation for Medical Use of X-Rays and Gamma-Rays of Energies Up to 10 MeV*[15] and in *Dental X-Ray Protection.*[16]

Controlled Area[18]

Areas which may cause or permit exposure to radiation or radioactive materials, through occupancy, work, or access to the area, are required to be controlled to protect the public as well as the individual. See Table 7-4. Such area or radiation installation may be in a hospital; medical, dental, chiropractic, osteopathic, podiatric, or veterinary institution, clinic or office; educational institution; commercial, private, or research laboratory performing diagnostic procedures or handling equipment or material for medical uses; or any trucking, storage, messenger, or delivery service establishment. High radiation areas are posted "Radiation Area," "Radiation Zone," or "Restricted Area" as required by the

[18] "Ionizing Radiation," *State Sanitary Code*, New York State Dept. of Health, October 31, 1977, Section 16.2.

Table 7-7 Range of Beta Particles in Air

Energy (millions of electron volts)	Maximum Range of Beta Rays in Air(m)
0.01	0.0022
0.02	0.0072
0.03	0.015
0.04	0.024
0.05	0.037
0.06	0.050
0.07	0.064
0.08	0.080
0.09	0.095
0.10	0.11
0.15	0.21
0.2	0.36
0.3	0.65
0.4	1.0
0.6	1.8
0.8	2.8
1.0	3.7
1.5	6.1
2.0	8.4
3.0	13.0
4.0	16.0
5.0	19.0

Source: *Atomic Radiation*, RCA Service Co., Government Services, Camden, N.J., 1959 [Contract Nr. AF 33(616)-3665].

Nuclear Regulatory Commission (NRC) and state regulations. These and similar type installations and areas should be under the supervision of a person qualified by training and experience in radiological health to evaluate the radiation hazards of such an installation and to establish and administer an adequate radiation protection program.

Table 7-8 Emission Energy and Ranges of Alpha Particles in Air

Radioisotope	Alpha Emission Energy (millions of electron volts)	Alpha Emission Range (cm)
Thorium (Th^{232})	3.97	2.8
Radium (Ra^{226})	4.97	3.3
Radiothorium (Th^{228})	5.41	3.9
Radium A (Po^{218})	5.99	4.6
Thorium A (Po^{216})	6.77	5.6
Radium C′ (Po^{214})	7.68	6.9
Thorium C′ (Po^{212})	8.78	8.6

Source: *Atomic Radiation*, RCA Service Co., Government Services, Camden, N.J., 1959 [Contract Nr. AF 33(616)-3665].

Table 7-9 Ranges of Transient Rates of Intake (Picocuries per Day) for Use in Graded Scale of Action—Ranges I, II, III

Radionuclides	Range I[a]	Range II[b]	Range III[c]
Radium 226	0 to 2	2 to 20	20 to 200
Iodine 131[d]	0 to 10	10 to 100	100 to 1000
Strontium 89	0 to 200	200 to 2000	2000 to 20,000
Strontium 90	0 to 20	20 to 200	200 to 2000

Source: Adapted from *Public Health Service Drinking Water Standards 1962*, PHS Pub. 956, DHEW, Washington, D.C., 1963. The range for specific radionuclides is as recommended by the FRC. The radionuclide intake ranges recommended by the FRC are the sum of radioactivity from air, food, and water. Daily intakes were prescribed with the provision that dose rates be averaged over a period of one year.

[a]Periodic confirmatory surveillance as necessary in this range.

[b]Quantitative surveillance and routine control of useful applications of radiation and atomic energy such that expected average exposures of suitable samples of an exposed population group will not exceed the upper value of this range.

[c]Evaluation and application of additional control measures as necessary to reduce the levels to Range II or lower to provide stability at lower levels.

[d]In the case of iodine 131, the suitable sample would include only small children. For adults, the Radiation Protection Guide for the thyroid would not be exceeded by rates of intake higher by a factor of 10 than those applicable to small children.

Radiation Protection Guides for Water, Food, and Air, and Radioactivity Releases

The controlled use of nuclear energy will result in the release of small amounts of radioactivity to the environment. Nuclear weapons testing will add to this. Although there is government control of nuclear power plants and the industrial and medical uses of radioactive materials and devices, it is nevertheless essential to limit body intake and exposure from such sources to protect people who might be affected. The FRC, to meet this need, has provided guidelines for federal agencies carrying out activities that might affect individuals or groups of people. These guidelines have general application and are widely used.[19]

The FRC guides consider three ranges of daily intake of radioactivity. The ranges and actions recommended to be taken with reference to specific radionuclides are given in Table 7-9. The council emphasizes that

> there can be no single permissible or acceptable level of exposure without regard to the reason for permitting the exposure. It should be general practice to reduce exposure to radiation, and positive effort should be carried out to fulfill the sense of these recommendations. It is basic that exposure to radiation should result from a real determination of its necessity.

The graded scale of action for Range I is periodic confirmatory surveillance as necessary; for Range II, quantitative surveillance and routine control; and for Range III, evaluation and application of additional control measures as

[19]Radiation Protection Guide Reports No. 1 and 2, and Protective Action Guide Reports No. 5 and 7.

necessary. These guides are under review by the FDA and may be lowered. The FRC functions have been assigned to the USEPA.

Water

The maximum permissible radioactivity levels in drinking water are given in Table 3-5. See Chapter 3 for radioactivity removal.

Food

The amount of radioactivity in the diet should be kept at levels that will ensure protection of the general population in accordance with FRC reports, as stated earlier and in Table 7-9.

In Great Britain "working levels for assessment" are used in the same manner as "radiation protection guides" are used in the United States.[20] The working level for strontium is given as 130 pCi strontium 90/gram of calcium in the total diet of the general population. For individuals or small groups the working level may be three times that for the general population. In the case of iodine 131, the annual average exposure has been set at 130 pCi per liter of milk.

The main sources of radiation in the diet are probably from man-made fallout on agricultural land by iodine 131, strontium 89, strontium 90, and cesium 137. By comparison the amount contributed by inhalation and drinking water would be negligible except when there is a direct discharge into the water.

Milk would be the major source of iodine 131 and the isotopes of strontium and cesium. The radiation dose from the ingestion of strontium 90 is primarily determined by the ratio of strontium 90 to calcium in the total food intake. Although strontium 90 follows approximately the same route as calcium through the food chain, the deposition of strontium 90 in the bone depends on the amount of calcium in the diet. Both strontium and calcium will enter plants through the root system; but only one-seventh to one-tenth of the strontium as compared to calcium will be transferred by cattle to their milk, and that reaching the bone is about one-quarter of that in the diet.[20] Meat is also a significant source of cesium 137. According to *FRC Report No. 2*, strontium 90 should not exceed 200 pCi/1, assuming a person drinks one liter of milk per day.

Radioiodine (iodine 131) may enter the body through consumption of fluid milk and fresh vegetables. It concentrates in the thyroid and increases the risk of thyroid cancer. Milk consumed within a few days after contaminated pasture grass is eaten by dairy cows presents the greater public health hazard. However, since iodine 131 has a relatively short half-life (8.14 days, physical), milk could be directed into manufacturing processes, such as in powdered milk, butter and cheese where it could be stored until safe. An alternative would be to put cattle on stored feed, when iodine 131 levels in their milk exceeds 100 pCi/1. *FRC Report No. 5*, Table 1, indicates that when the level of iodine exceeds 4200 pCi/1 per day its use by infants and pregnant women should be discontinued. *FRC Report No. 5* also established a Protective Action Guide for iodine 131 ingestion of 30 rem to the individual and 10 rem to a suitable sample of the population per year.

[20] *Radiation Levels: Air, Water and Food*, Royal Society of Health, London, S.W. 1, 1964.

Strontium 90 is also hazardous because of its long biological half-life and because it deposits in the bone, increasing the risk of bone tumors and leukemia. Cesium 137 is not normally retained in the body for a long time, but it adds a small increment of exposure and increases the probability of genetic damage.[9,21]

Assessment of the significance of the dietary intake of radioactivity must take into consideration the total exposure including air, water, food (including milk and fish), occupational, background, and medical exposures. In the United States, the PHS and some state health departments routinely monitor the milk supply and the total dietary intake of radioactivity in controlled population groups up to age 18. The FDA also samples other foods. The USEPA monitors the environment and activates additional air monitoring stations when nuclear weapons are detonated in the atmosphere.

Air

Contamination of the air by radioactive materials can come from nuclear explosions (peaceful uses, testing, and war), nuclear reactors and fuels processing, accidental releases, and natural sources. Insofar as the general population is concerned, the maximum permissible exposure should be in the order of that due to the natural background level, approximately 100 to 150 mR/yr. Any activity in excess calls for action to reduce releases to achieve this goal. With this constraint in mind, the International Commission on Radiological Protection recommends, based on certain generally accepted assumptions, that occupational air exposure not exceed the following and that exposure of a small part of a population not exceed one-tenth of these concentrations (not including background, medical, or dental exposures) under the circumstances stated.[20]*

10^{-9} μCi/ml—no alpha-emitting radionuclides and beta-emitting radionuclides (Sr^{90}, I^{129}, (1000 pCi/m^3) Pb^{210}, Ac^{227}, Ra^{228}, Pa^{230}, Pu^{241}, Bk^{249}) present.

10^{-10} μCi/ml—no alpha-emitting radionuclides and beta-emitting radionuclides (Pb^{210}, Ac^{227}, (100 pCi/m^3) Ra^{228}, Pu^{241}) present.

4×10^{-13} μCi/ml—no information as to composition of radioactivity. (0.4 pCi/m^3)

In nuclear power plants air pollution is possible from the release of radioactive xenon and krypton produced by fission; the activity induced in gases in the air; and the release of fission products as a result of fuel-cladding ruptures. The significant releases are likely to be iodine 131, krypton 85, and tritium. Coal-fired power plants are reported to release somewhat more radiation than nuclear reactors.

Nuclear reactors. It is estimated that there will be a world total of 322 reactors by 1980.[22] There were 126 reactors in existence and 156 planned for

*μCi/ml = microcurie per milliliter; pCi/m^3 = picocurie per cubic meter.

[21] Herman E. Hilleboe and Granville W. Larimore, *Preventive Medicine*, W. B. Saunders Co., Philadephia, 1965.

[22] Anthony J. Parisi, "The Nuclear Slowdown: Concern and Elation," *New York Times International Economic Survey*, February 5, 1978, p. 11.

Europe as of 1980, including the Soviet Union.[23] The increasing use of nuclear reactors for power production requires the use of more cooling water. Although significant contamination of the water is extremely unlikely, the water used for this purpose will increase in temperature. It may first have to pass through a cooling tower before discharge to a stream or lake (if located inland) if the receiving body of water is unable to assimilate the added heat without causing ecological harm. It is also possible, however, that the heat energy going to waste can be put to productive use, such as for space heating; controlled fish propagation; irrigation to extend the agricultural growing season and increase crop production; water desalting; sewage evaporation; steam generation (boiling water reactor); and recreational purposes. Cooling water released to the environment should contain no radioactivity above that in natural background or that in Table 3-5. Over a period of four years, ten accidents have been reported involving nuclear power plants or radioactive fuels that power them. Included were overexposure of two workers, a nonnuclear explosion with no injuries or radioactive release, a nonnuclear explosion with one serious overexposure, an over-turned truck on a highway, X-ray equipment, a death and six injuries due to overexposure, and a fire.[24] The most serious accident was at Three Mile Island Nuclear Reactor (pressurized water reactor) in Pennsylvania. There was an insignificant release of radioactivity to the environment, but considerable anxiety and concern was created.

Radioactive contaminants, with their half-lives, from the detonation of nuclear weapons include iodine 131 (8 days); strontium 90 (28 years); strontium 89 (53 days); cesium 137 (30 years); and cerium 144 (275 days). Atmospheric detonations should obviously be prohibited by international agreements.

Nuclear fuels processing plants also present special problems due to the reprocessing of irradiated nuclear reactor fuel in which uranium, strontium 90, cesium 137, and plutonium 239 are separated from the other fission products. In the process it is possible for radioactive liquid, gases, and particulates to be released. Reliance is placed on process control, filtration, dilution by high-stack emission, liquid retention, and dilution of low-level liquid wastes to keep releases to a minimum and within acceptable limits.

The production of nuclear fuels involves the washing and concentration of the ore with the danger of radon and thoron contamination of dust. In the separation of the isotopes of uranium and thorium, dusts of uranium, thorium oxides or gases, and iodine 131 add to the potential hazard. Surveillance of waste discharges is required, as is environmental surveillance of the transportation and "permanent" disposal of radioactive wastes.

WASTE DISPOSAL

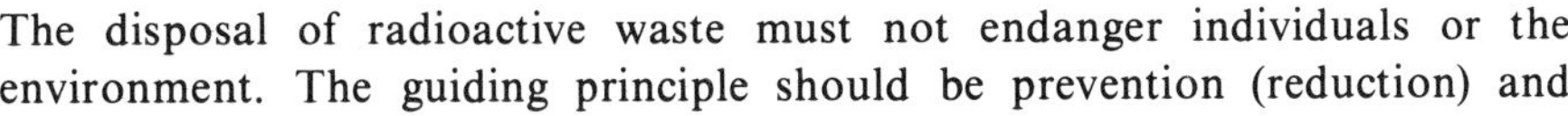

The disposal of radioactive waste must not endanger individuals or the environment. The guiding principle should be prevention (reduction) and

[23]John Vinocur, "Swedish Vote Encourages Other Pronuclear European Efforts," *New York Times*, March 30, 1980, p. E3.

[24]*New York Times*, April 1, 1979, p. 30.

control of gaseous, liquid, and solid waste at the source followed by segregation and the indicated collection, treatment, and storage or disposal.

Collection and Treatment

Liquid and solid wastes may originate in radioisotope laboratories, chemical processing plants, water-cooled reactors, change rooms, and decontamination laundries. Solid wastes are collected in paper- or plastic-lined containers, if of low activity, and then disposed of in an approved procedure such as by commercial waste disposal or incineration. High-activity solid wastes are placed in shielded containers and stored or disposed of in an approved manner.

Radioactive wastes may be treated by physical and chemical methods depending on the characteristics of the waste. Low-level liquid and gaseous wastes are usually disposed of under known conditions by dilution and dispersion to reduce activity to below maximum permissible levels. High-level long-lived liquid or solid wastes are concentrated and stored. Liquid wastes are commonly concentrated by evaporation, by filtration, by precipitation of a soluble material with an insoluble precipitate, and by ion exchange. Biological assimilation is also possible. The reprocessing of spent nuclear fuels would reduce the amount of radioactive wastes to be disposed of.

Radioactive wastes are classified as low-level, intermediate-level, and high-level depending on their potential hazard.[25] *Low-level radioactive wastes* containing less than one microcurie of radioactivity per gallon can be disposed of to the environment if diluted to less than 10 percent of the maximum permissible concentration for continuous occupational exposure.

Solid wastes may be burned provided radioactivity in the gaseous effluent does not exceed Nuclear Regulatory Commission limits. Wastes may be disposed of to areas the NRC has licensed for disposal by burial. *Intermediate-level wastes* have been defined as containing "10^4 to 10^6 times their maximum permissible concentration." They may be treated by evaporation, ion exchange, or flocculation to separate high-level and low-level components. *High-level wastes* are mostly liquid wastes from fuel reprocessing plants. Disposal of these wastes from nuclear electric power plants is the major unresolved problem because of the long-lived radioactivity present. Decay of the wastes to background level may require hundreds to possibly several thousands years if long-lived alpha emitters are present.

Storage and Disposal

After treatment and concentration high-level radioactive wastes are carefully packaged and, when indicated, fixed in an inert solid material in preparation for

[25] *Cleaning Our Environment: A Chemical Perspective*, American Chemical Society, Washington, D.C., 1978, pp. 419–422.

storage or disposal in a restricted area. The secret of radioactive waste disposal according to Eliassen is the separation and concentration of radioactive fission products to a small volume and the conversion of these materials to a glassy solid.[26] Special glass containers used to store wastes, maintained at 100 to 150°C, are believed to stand up for millions of years; but 300°C and pressures of 2000 to 4000 psi cause the glass to corrode and flake off.[27]

Concentrated high-activity liquid wastes are stored in a specially constructed tank within a tank or a tank over a secondary container of equal volume that can be monitored for leakage. These require careful monitoring and a well-developed emergency plan, as container failures have occurred. Liquid radioactive wastes may also be collected for storage and retention until sufficient decay has taken place to permit controlled discharge and disposal by dilution.

Ground disposal of low- or intermediate-level wastes may be permitted under approved soil, rock, and groundwater conditions. Other disposal methods considered include dry natural caverns, deep mines, salt cavities, deep-well disposal, and ocean disposal; but each of these requires careful evaluation before being permitted. There appears to be no consensus for a completely acceptable permanent solution to the disposal of concentrated solid high-level radioactive wastes as yet, although deep salt mines and other dry, stable geologic formations seem to offer the safest alternative. However, special precautions would have to be taken in salt mines. The U.S. Geological Survey found that storage of canisters holding radioactive wastes at 200 to 500°C in deep salt formations would permit the fluid in salt crystals, up to 10 percent, to move through the salt toward sources of heat. The fluid, which is extremely corrosive, would attack the canister and permit the escape of radioactive fluid. Under the circumstances underground storage is proposed when waste temperatures are 100°C or less.[28]

Plutonium 239, with a half-life of 24,400 years, will require storage of at least 500,000 years (special processing could reduce this time to 1000 years) and strontium 90 and cesium 137 take about 700 years to no longer be a problem. Angino has evaluated geologic disposal, seabed disposal, burial in Antarctica, and waste storage in arid zones.[29] Dallaire has summarized known waste disposal methods.[30]

Disposal of radioactive wastes is under the jurisdiction of the Nuclear Regulatory Commission, state health department, or other state agency, if the NRC has an agreement with the state. NRC regulations require solidification of high-level radioactive wastes within 5 years of their production and then transfer to a federal storage site (when available) within 5 years. Steps have been taken

[26]Rolf Eliassen, "The search for energy self-sufficiency," *Am. City*, February 1974, pp. 33, 34.

[27]Barry Scheetz, quoted by Gene Dallaire in "Nuclear waste disposal: is there a safe solution?," *Civ. Eng., ASCE,* May 1979, p. 76.

[28]J. Dale, "U.S. Geological Survey—the other water quality agency," *J. Water Pollut. Control Fed.*, April 1980, pp. 650–655.

[29]Ernest E. Angino, "High-Level and Long-Lived Radioactive Waste Disposal," *Science*, December 2, 1977, pp. 885–890.

[30]Gene Dallaire, "Nuclear waste disposal: is there a safe solution?," *Civ. Eng., ASCE*, May 1979, pp. 72–79.

(1980) to establish monitored federal interim disposal sites for spent nuclear fuels to be safe for a century.

ENVIRONMENTAL RADIATION SURVEILLANCE AND MONITORING

Nuclear Facilities Siting

The siting of nuclear facilities is subject to extensive regulation and licensing by the Nuclear Regulatory Commission. The Commission's jurisdiction extends to design, construction, and operation for compliance with its regulations including the possession, transportation, and use of radioactive materials, both natural and man-made. The National Environmental Policy Act (NEPA) requires that the environmental aspects be analyzed and that alternative sites be considered. See Chapter 2. The Clean Water Act and Clean Air Act, as amended, and state laws must also be considered in the siting and operation of nuclear facilities. The NRC and the USEPA have numerous guides, criteria, and regulations governing siting and operation. A Task Committee of the ASCE[31] has reviewed the nuclear facilities siting regulations, the fuel cycle, site evaluation factors and methodology, and the issues involved. Their report recommends that redundant and extended regulatory procedures be minimized, that solutions be found to the disposal of high activity wastes, that the problems associated with inactive mill tailings be corrected, and that objective and quantitative standards for safety be adopted.

A permit to construct or install a nuclear reactor must be obtained from the NRC, and then a separate operating license is needed before the fuel is inserted and the reactor is placed in operation. Preoperation and postoperation environmental surveillance and monitoring are usually conducted cooperatively by the NRC, USEPA, state health and environmental protection departments, other state agencies concerned, and the industry involved. The preoperational survey should be started at least one year before the proposed date of operation and should cover an area within a 20-mi radius of the facility.

Monitoring Stations and Sampling

An environmental monitoring program preceded by a preoperational survey in and around large nuclear facilities serves as a check on the various operations that might produce contamination.[32] It also permits evaluation of the exposure to the surrounding area and the need for preventive action. The program should

[31] "Nuclear Facilities Siting," Task Committee on Nuclear Effects of the Committee on Air Resources and Environmental Effects Management of the Environmental Engineering Division, *J. Environ. Eng. Div., ASCE,* June 1979, pp. 443–502.

[32] Dade W. Moeller and Abraham S. Goldin, "Environmental Protection for Nuclear Applications," *J. Sanit. Eng. Div., ASCE,* June 1969, pp. 373–385.

be a cooperative one between health and other regulatory authorities and the industry.

Industry has responsibility for internal housekeeping and for monitoring all its waste discharges in terms of types and quantities. Peripheral monitoring should be shared between the industry and the regulatory authorities; distant monitoring up to 40 mi (100 km), is done by the regulatory agency. Maximum fallout from a stack will usually occur immediately after emission within a 1-mi (1.6 km) radius; the area should be secured by the plant.

Solid, liquid, and gaseous radioactive materials can be carried considerable distances by air and water and may adversely affect plant and animal life as well as people. It is essential therefore that a surveillance and monitoring program be maintained in and around installations from which radiation release is possible. These include nuclear power plants, fuel processing plants, uranium milling industries, university reactors, and certain industries and laboratories. The surveillance monitoring should be carefully planned with the advice and assistance of radiological health specialists and physicists, biologists, meteorologists, environmental engineers, geologists, and others familiar with the problem and the local area. The sampling for resulting contamination should include air, water, milk, food, biota, soil, and people. Interpretation of the sampling requires mapping of the surrounding area showing salient geographical and topographical features. This information makes possible assessment of the risk associated with any release over a large area under various meteorological conditions, when coupled with previously prepared air diffusion models and overlays. Mapping should show surface and groundwater hydrology, types of soil and vegetation, population centers, transportation systems, sources of water supply, recreation areas, and other land uses. Past meteorological conditions would include prevailing winds and speeds, temperature, and rainfall.

At a fuel processing plant there is a potential for discharge of iodine 131, krypton 85, and tritium. Release of iodine 131 is minimized by storage of fuel elements at least 100 days before processing. Krypton release is greater than from a nuclear reactor and cannot as yet be effectively removed before release to a stack. Tritium is not yet removable and is discharged as a liquid or gaseous waste.

Careful surveillance is therefore necessary to ensure that discharges of these radionuclides are kept to a minimum and within acceptable limits. Sampling may include any combination of the following, with consideration of the facility monitored:

Air Around the installation and at some distance, as noted above.

Water At plant outfalls, receiving stream and downstream; also groundwater from nearby wells.

Soil Immediately around the installation and at some distance, including stream bottom muds.

Biological specimens Fish, deer, possibly cows, rodents, plants—especially those used for food; also, where available, shellfish, ducks, plankton, and other plant life.

Milk From dairy farms in vicinity.

Spaces Indoor spaces, containers, and conduits.

Table 7-10 gives suggested frequency of sampling and determinations.

Table 7-10 Suggested Monitoring Around a Nuclear Reactor or a Fuel Reprocessing Plant

Sample[a]	Frequency	Determination	Remarks[b]
Water intake and effluent	Daily where indicated, composite or weekly grab	Gross alpha and beta of dissolved and suspended solids and gamma scans	Quarterly composite shall not exceed the standard established by NRC. Make complete isotopic analyses if exceeded
Receiving waters	Daily	Same as above	
Fish, shellfish, mud, stream, water supplies, plankton	Quarterly[c] upstream and downstream from outfalls	Gross gamma and gross beta activity	Identification of radionuclides if standard exceeded. Sample above and below outfall
Milk, farms in vicinity	Weekly	Iodine 131, strontium 90, tritium, and cesium 137	Identification of radionuclides if standard exceeded
Air around facility	Continuous monitors, daily	Iodine 131, krypton 85, gross beta on particulates	Shall not exceed standard established by NRC. If exceeded or approached, monitor environment and eliminate source
Soil	Annually	Strontium 90 and cesium 137 or gross beta	Sample farm soil downwind
Vegetation, animals	Growing season[c]	Gross beta	Same as for milk
Fallout, downwind	Daily cumulation, biweekly	Gross beta	Relate to fallout from weapons or other testing

[a]Include meteorological data and stream flows.

[b]See Table 7-8 and Radiation Protection Guides for Water, Food, and Air, in this chapter for ranges and limits.

[c]Quarterly or annually for radionuclides of long half-life; for short-lived nuclides, no more than two or three half-life intervals.

Sampling stations are usually located on the site being monitored, in the immediate vicinity, and at some distance. Those onsite and in the immediate vicinity should preferably be fixed and be continuous monitors. Stations at a distance could be used for periodic grab sampling, selected if possible with reference to other stations maintained by the Weather Bureau, health department, a university, U.S. Geological Survey, radio station, or airport that could provide supporting data. Stream samples include water, mud, and biota. Tiles, stones, or slides are suitable for the collection of biological attachments.

Table 7-11 Some Radioactivity Concentrations in Stream Samples

	Gross Alpha (pCi/l)	Gross Beta (pCi/l)
Stream sample	0 to 21	0 to 334
(relatively clean)	0 to 3	0 to 72
	0 to 220	7 to 370
	0 to 33	7 to 67
Stream sample	4700	5500
(contaminated)	300,000	25,000
Water supply—raw	0 to 25	3 to 440
treated	0 to 11	3 to 144
	(pCi/gm, dry)	(pCi/gm, dry)
Algae		
Clean water	0 to 25	41 to 322
Clean water	2 to 83	0 to 220
Contaminated water	to 2500	to 10,000
Mud		
Clean water	13 to 30	18 to 71
Contaminated water	660	6600

Note: Rainfall usually contains more radioactivity than surface water.

A monitoring station might contain a continuous water sampler, high-volume air sampler, silver nitrate filter, film badge, adhesive paper, silica gel, rain gauge, ionization chamber, and such other equipment or devices as may be indicated by a safety analysis and the nature of the facility being monitored. Table 7-10 shows the type of monitoring indicated around a nuclear reactor or fuels processing plant.[33] Table 7-11 shows some radioactivity concentrations that might be found in clean and contaminated surface waters, algae, and bottom muds.

The air monitoring around a boiling-water reactor is described by Thomas.[34] Within a 1-mi radius distributed on a 500-ft grid will be 50 thermoluminescent dosimeters; 3 air monitoring stations to sample airborne particulates, radioiodine, heavy particulate fallout, and rainwater. Air filters will be scanned continuously by a beta-gamma-sensitive G-M tube, and readings will be reported at the control center. In addition, four air monitors about 10 mi away in an urban area will provide continuous reports to the control center. Weekly composite samples will also be collected from stations about 40 mi from the plant. The air monitoring is supplemented by the sampling of fish, mussels, algae, and other biota as well as milk, water, vegetation, and soil. State and federal agencies also have responsibilities for public health protection.

Careful plans should also be made by the operators of reactors and by state agencies for immediate emergency monitoring and action in case of an accidental

[33]Sherwood Davies, "Environmental Radiation Surveillance at a Nuclear Fuel Reprocessing Plant," *Am. J. Pub. Health*, December 1968, pp. 2251–2260.

[34]Fred W. Thomas, "TVA's Air Quality Management Program," *J. Power Div., ASCE*, March 1969, pp. 131–143.

discharge to the environment. This is necessary to protect the public while an assessment is being made. Evacuation within 10 mi of a plant and special precautions downwind within 40 to 50 mi might be indicated in extreme situations. Gross activity levels are useful for rapid measurements, as are samples from fixed air and water monitoring stations.

The FDA, the USEPA, the NRC, and some state environmental and health departments routinely sample the air, water, and food supplies. Reports on the surveillance maintained are issued periodically by these agencies. Surveillance is also maintained over fallout.

Space and Personnel Monitoring

A radiation installation should be responsible for its own monitoring program. This should include the facility work and storage areas, sources, waste disposal, emergency procedures, and personnel monitoring. A personnel monitoring system, such as film badges and thermoluminescent dosimetry, will determine individual exposure received and effectiveness of the control measures being taken. This may be supplemented in special instances by measurement of the radioactivity in the body or in the excreta, or by monitoring the work environment, and by health surveillance. The film badges should be processed and interpreted by a specialized laboratory or government agency.[35,36] A competent health physics safety unit should be established and given responsibility for the monitoring and authority to take prompt corrective action whenever indicated. In-house safety rules governing operating procedures are helpful.

Instruments for detecting and measuring radiation are necessary for an effective monitoring program, and these must be selected with consideration of their sensitivity and purpose to be served. A summary of some radiation-detecting devices giving their characteristics, advantages, and disadvantages is given in Table 7-12. Thermoluminescent dosimeters and photoluminescent glass devices are also available.

RADIATION PROTECTION PROGRAM

General

A state radiation protection program may involve control of X-ray units, radioactive materials (medical and nonmedical), waste disposal, and environmental monitoring and surveillance. Particle accelerators and other radiation

[35] C. B. Braestrup and K. J. Vikterlöf, *Manual on radiation protection in hospitals and general practice*, WHO, 1974, pp. 68–72.

[36] *Dental X-Ray Protection*, Rep. No. 35, NCRP, Washington, D.C., March 9, 1970, pp. 22–23.

Table 7-12 Radiation Detecting Devices

Detector	Types of Radiation Measured	Typical Full Scale Readings	Use	Minimum Energy Measured	Directional Dependence	Advantages	Possible Disadvantages
Scintillation counter	Beta, X, gamma, neutrons	0.02 mR/h to 20 mR/h	Survey	20 keV for X rays. Variable for betas.	Low for X or gamma	1. High sensitivity 2. Rapid response	1. Fragile 2. Relatively expensive
Geiger-Müller counter	Beta, X, gamma	0.2 to 20 mR/h or 800 to 80,000 counts/min	Survey	20 keV for X rays. 150 keV for betas.	Low for X or gamma	Rapid response	1. Strong energy dependence 2. Possible paralysis of response at high-count rates or exposure rates 3. Sensitive to micro-wave fields 4. May be affected by ultraviolet light
Ionization chamber (Cutie pie)	Beta, X, gamma	3 mR/h to 500 R/h	Survey	20 keV for X rays. Variable for betas.	Low for X or gamma	Low energy dependence	1. Relatively low sensitivity 2. May be slow to respond
Alpha counter	Alpha	100 to 10,000 alpha/min	Survey	Variable	High	Designed especially for alpha particles	1. Slow response 2. Fragile window
Film	Beta, X, gamma, neutrons	10 mR and up	Survey and monitoring	20 keV for X rays. 200 keV for betas.	Moderate	1. Inexpensive 2. Gives estimate of integrated dose 3. Provides permanent record	1. False readings produced by heat, certain vapors, and pressure 2. Great variations with film type and batch 3. Strong energy dependence for low-energy X rays
Pocket ionization chamber and dosimeter	X, gamma	200 mR to 200 R	Survey and monitoring	50 keV	Low	1. Relatively inexpensive 2. Gives estimate of integrated dose 3. Small size	1. Subject to accidental discharge
B F_3 Counter	Neutrons	0 to 100,000 counts/min	Survey	Thermal		Designed especially for neutrons	

Source: *Radiation Protection in Educational Institutions*, Rep. No. 32, NCRP, Washington, D.C., July 1, 1966. All radiation survey meters should be standardized periodically against Bureau of Standards instruments.

machines, certain radionuclides produced artificially and in small quantity, radium, and other naturally occuring radionuclides are included. In all cases the goal must be the elimination of all unnecessary exposure to ionizing radiation. Toward this end the Bureau of Radiological Health/FDA, in cooperation with participating states, conducts surveys to determine if the public is being exposed to hazardous or excessive radiation. Examples of such surveys are the National Evaluation X-Ray Trends (NEXT), Dental Exposure Normalization Techniques (DENT), and Breast Exposure Nationwide Trends (BENT). The FDA survey shows that in 1977 about one-third of the X rays made were unnecessary, but that equipment was generally quite reliable. A 45-state survey in 1979 found that one-third of the dental X-ray machines and almost half of the breast X-ray machines examined were emitting unacceptable levels of radiation because of poor operation.

The licensing and regulation of nuclear reactors and reactor-produced radionuclides is the sole responsibility of the Nuclear Regulatory Commission (NRC). However the control of reactor-produced radionuclides can be specifically delegated to a state government adopting control procedures acceptable to the NRC. The possession and use of certain reactor-produced radionuclides requires a general license for stated small quantities and a specific license for a particular use not coming under a general license.

Planning

The planning of a state or local radiation program requires that the problem be defined and means for bringing it under control be developed and implemented.

The first step, then is to conduct a survey to determine:

1. The location, number, and types of radiation sources within the area of jurisdiction. State professional licensing agencies, professional societies, telephone-book yellow pages, equipment manufacturers, and supply houses may be helpful.
2. The legislative and regulatory needs.
3. The types of problems that can be anticipated and are to be resolved.
4. The training and education needs of staff.
5. The required workload, staff, equipment, and budget.

Reference to the following may be helpful in planning a local program, keeping in mind that it is offered only as a guide.[37]

Manpower allocation
  X-ray equipment—70 percent
  Radioactive materials—30 percent

[37] "The Responsibility of Local Health Agencies in the Control of Ionizing Radiation," report of the Committee on Local Control of Ionizing Radiation of the Conference of Municipal Public Health Engineers, November 1962.

Manpower allocation to X-ray equipment
- Medical—60 percent
  - Therapy—1 percent
  - Fluoroscopic—24 percent
  - Dental and radiographic—35 percent
- Nonmedical—10 percent
  - Fluoroscopic—5 percent
  - Radiographic—5 percent

Manpower allocation to radioactive materials
- Medical—10 percent
  - NRC—5 percent
  - NonNRC—5 percent (radium)
- Nonmedical—20 percent
  - NRC—15 percent
  - NonNRC—5 percent (radium)

The time required to make the initial complete survey of a particular installation has been estimated as follows:[37]

Dental X ray	10.5 man-hours
Chiropody X ray	10.5
Radiographic (chest X ray)	11.5
Medical fluoroscope	11.5
Miscellaneous isotope	8.0

These estimates include travel, inspection, data analysis, and report writing for a complete and comprehensive survey by a two-man team. The time can be cut in half by using one experienced radiological health specialist. A comprehensive reinspection, however, would generally require in all less than two man-hours.

Control Program

A control program should include registration of all radiation sources; education of the inspectors and users, licensing or certification of X-ray technicians, and the dissemination of informational literature; inspection of all equipment and the procedures used; discussion of deficiencies observed and their correction with the operator and then confirmation of the recommendations in writing; follow-up inspections to ensure compliance and administrative enforcement when indicated; reinspection every two to three years. This would include medical and dental, veterinarian, industrial, laboratory, university, and hospital uses.

Coordination should be maintained with other agencies, such as with the building department (including plumbing and electrical), for referral of applications involving installation of X-ray equipment or construction of a laboratory planning to use radioactive materials. Liaison should also be maintained with the department of public works in connection with possible problems and need for monitoring at the sewage treatment plant, incinerator, or sanitary landfill.

In some instances when a large number of X-ray installations are to be controlled, consideration might be given to a requirement that the installation be routinely surveyed by a licensed individual. A report of satisfactory compliance with statutory regulations would be submitted annually to the regulatory agency. The agency would establish licensing requirements, make periodic evaluations on a statistical random-sampling basis, and take whatever enforcement action that would be indicated. This approach may not significantly reduce the size of the regulatory staff but would strengthen the program. As of 1980 11 states and Puerto Rico had licensing requirements for machine operators.[38] Between 40,000 and 50,000 operators of medical X-ray machines do not have certification.

Improper Use of X-ray Equipment

It is usually the opinion of health departments that the use of radiation on humans should be limited to usage by persons licensed to practice medicine, dentistry, podiatry, or osteopathy, and in some instances chiropody. This would eliminate, for example, the exposure of humans by unorthodox procedures such as fluoroscopy for the purpose of fitting shoes. This particular use of X rays, besides being considered completely unnecessary by orthopedic experts, has been used almost exclusively on children, whose rapidly developing bodies are most susceptible to radiation damage. It is prohibited. Fluoroscopic equipment has also been used as an advertising technique and on music students to study throat-muscle development. These and similar uses should not be permitted.

Coordination with Other Agencies

Federal, state, and local program activities should be carefully integrated to avoid duplication and ensure complete coverage. Each should perform, in a previously agreed-on systematic manner, what each is best equipped to do by virtue of its legal responsibility and resources, its physical location, and its staffing. Educational institutions, research centers, the Federal Radiation Council, the Bureau of Standards, and the NRC play an important role in education, surveillance, standards setting, and research. The Federal Emergency Management Agency also participates in case of a serious accident.

The state usually has a basic responsibility to promote local programs when indicated, to carry out training, to provide assistance, to provide laboratory support, to evaluate and make recommendations for program improvement, and to require that a competent program be conducted.

The local agency, if given the resources, should make routine inspections of X-ray equipment and other facilities, conduct education, carry out enforcement when necessary, and report the work done, problems, and accomplishments to

[38]A. O. Sulzberger, Jr., "House Panel Reports Excess Radiation in X-Ray Tests," *New York Times*, August 24, 1980, p. 35.

the state. If a proper program cannot be carried out locally, the state agency having the overall responsibility will have to find a way to carry out the program.

OTHER RADIATION SOURCES, NONIONIZING

The electromagnetic spectrum, Figure 7-1, shows the relationship of various types of radiation and their approximate wavelengths. Ionizing radiation has been previously discussed. This section will concentrate on certain mechanical and electromagnetic nonionizing radiations.

Mechanical radiation, or energy, travels at relatively slow velocities through some medium such as the air, water, soil, or a solid material. Major forms are infrasound emitted by mechanical vibrations at a frequency usually too low to be heard (20 Hz and below) by the human ear; sound which is in the range of human hearing of approximately 20 to 20,000 Hz (20 kHz);* and ultrasound which is in the range of 20 kHz and 10 megahertz (MHz). Ultrasound is used for medical therapy and diagnosis and cleaning of teeth and metals.[39]

Electromagnetic radiations include microwave and intense electromagnetic field radiation such as found in microwave ovens and medical diathermy units; infrared radiation—a direct beam can damage the eye and gonads; visible light for seeing detail and color; ultraviolet light used for sterilization, disinfection, and for diagnosis and therapy—can cause damage to the eyes and skin, and form ozone; lasers and masers, intense beams used in medicine, industry, military, and research—direct or reflected laser beams can cause damage to eyes, gonads, and central nervous system. Lasers can travel great distances and can be used for cutting, drilling, and welding; also in communications systems, in surveying to establish lines and grades, in computer systems, to detect pollution, and in certain surgical procedures.

Some electronic products can emit harmful radiation if not properly designed, constructed, installed, or used. The FDA, Department of Health and Human Services, has responsibility under the Radiation Control for Health and Safety Act to protect the public from unnecessary exposure to radiation from all types of electronic products including sonic, infrasonic, and ultrasonic waves, and ultraviolet light. Electronic products, in addition to those mentioned below, but excluding those producing ionizing electromagnetic radiation, include sanitizing and sterilizing devices, welding equipment, alarm systems, vacuum condensers, voltage regulators, vacuum switches, rectifiers, tanning lamps, black light sources, radar microwaves, and other sources of intense magnetic fields.

The FDA has issued performance standards or recommendations for laser products, mercury vapor lamps, microwave ovens, ultrasound therapy devices, sunlamps, and other electronic products.

*Discussed in Chapter 6 under Noise Control.

[39]Karl Z. Morgan, "Exposure To Nonionizing Radiation," in *Environmental Problems In Medicine*, William D. McKee, Ed., Charles C. Thomas, Springfield, Illinois, 1974.

Microwaves

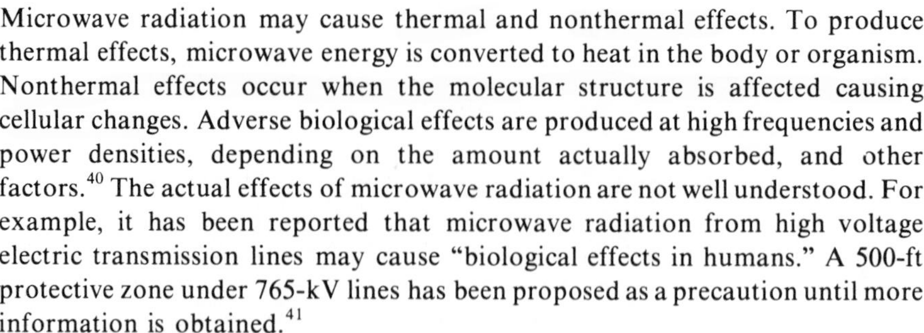

Microwave radiation may cause thermal and nonthermal effects. To produce thermal effects, microwave energy is converted to heat in the body or organism. Nonthermal effects occur when the molecular structure is affected causing cellular changes. Adverse biological effects are produced at high frequencies and power densities, depending on the amount actually absorbed, and other factors.[40] The actual effects of microwave radiation are not well understood. For example, it has been reported that microwave radiation from high voltage electric transmission lines may cause "biological effects in humans." A 500-ft protective zone under 765-kV lines has been proposed as a precaution until more information is obtained.[41]

Microwaves are used in industry to test for flaws in equipment, in the home to operate TV remote controls and garage door openers, in radar at fixed installations and on police cars, in transmitters, in power systems, and in communications using the telephone via satellites. Industry also uses microwaves to cure plywood, to cure rubber and resins, to raise bread and doughnuts, and to cook potato chips.

Microwave ovens and other electromagnetic energy sources (diathermy units, some radio frequencies, neon lights, and gasoline engines and their ignition systems) may interfere with the proper functioning of certain unshielded cardiac pacemakers. Pacemaker users should be alerted to the potential interference possibilities from outside electromagnetic sources. New pacemakers are supposed to be shielded.

Microwave ovens and similar devices have magnetron tubes which use electrical energy to generate high frequency short wave (microwave) energy. The Federal Communication Commission (FCC) has approved for ovens the use of 2450 and 915 MHz. Microwaves pass through paper, plastic, glass, and other clear materials; they are reflected by metals but are absorbed, refracted, or transmitted by nonmetallic materials such as water, food, and human tissue (eyes, internal organs, other) with the production of heat. By the time burning or pain is felt, the damage is done, but large amounts of microwave radiation are needed. Hence metal, metal dishware, or dishware containing metal parts should not be used; they may cause damage to the magnetron tube due to arcing or flashing.

Microwave ovens with poorly fitting doors, which can operate with the door open or partly open, or with a safety interlock system that can be bypassed, can cause serious harm to the user. Microwave ovens are used in the home, in mass-feeding facilities, in restaurants, and as an adjunct to quickly heat foods from automatic vending machines. Dirt, grease, or metal particles on door seals can

[40] John P. Lambert, "Biological Hazards of Microwave Radiation," *J. Food Prot.*, August 1980, pp. 625–628.

[41] Harold Faber, "P.S.C. Finds 'Biological Effects' Caused by High-Voltage Conduits," *New York Times*, September 17, 1977.

Table 7-13 Microwave Standards

Maximum Exposure [(in milliwatts per square centimeter (mW/cm^2)]	Reference
10[a]	Underwriter Laboratories, U.S.
10[a]	U.S.A. Standards Institute
10, continuous	U.S. Army and Air Force
10 to 100, limited	U.S. Army and Air Force
10, continuous	Great Britain (post office regulation)
1.0, prolonged	Sweden
0.01, continuous	U.S.S.R.
0.1, 2-hr maximum	U.S.S.R.
1.0, 15 to 20 minutes[b]	U.S.S.R.
1.0[c]	PHS/FDA
5.0[d]	PHS/FDA

[a]For 8-hour workday and 40-hour workweeks (OSHA) at 300 MHz to 300 GHz, however less than 1 mW/cm^2 is advised for prolonged exposures.
[b]Protective goggles required.
[c]For microwave ovens at time of sale. Maximum leakage rate.
[d]Not more than 1 mW/cm^2 prior to oven sale and not more than 5 mW/cm^2 throughout the useful life of a microwave oven measured 2 in. (5 cm) from the external oven surface. (Radiation Control for Health and Safety Act of 1968, P.L. 90-602.)

permit microwave leakage, hence cleanliness is essential. The door must form a tight seal.

A densiometer or power density meter, although not entirely adequate, is used to monitor radiation leakage. Two independently operated safety locks are required on oven doors to stop microwave generation when the door is opened. The FDA issues information leaflets on microwave ovens.[42]

Some microwave exposure standards are given in Table 7-13.

Medical diathermy. In medical diathermy, heat produced by microwave radiation is used for therapeutic treatment. The heat can penetrate muscles as much as 2 in. increasing the flow of blood and nutrients to the area being treated through dilation of the blood vessels. Pain is reduced and healing promoted. Microwave diathermy is being used experimentally in efforts to destroy cancerous tumors. Improper use of microwave equipment can cause cataracts in patients and severe burns of the skin or underlying tissue. Shortwave and ultrasonic diathermy devices are also used for medical therapy. The FDA has responsibility for publication of performance standards for microwave diathermy units.[43]

[42]*Consumer Memo, Microwave Oven Radiation*, DHEW Pub. (FDA) 79-8058, PHS, Rockville, Maryland, March 1979.
[43]James Greene, "Microwave Diathermy: The Invisible Healer," *FDA Consumer*, February 1979, p. 7.

Ultraviolet Light

Ultraviolet radiation (UVR) may emanate from a natural or artificial source. The sun is the principal natural source. Ordinary window glass and most transparent materials absorb the ultraviolet radiation. Quartz and fluorite allow UVR to pass through. Artificial sources include mercury vapor lamps and various products which make use of the germicidal properties of UVR. These products are used in hospitals, laboratories, schools, and certain industries. Special lamps are used for the treatment of skin diseases and for the prevention of vitamin D deficiency. Ultraviolet therapy produces a tan or sunburn. Improper therapeutic, occupational, or recreational uses, improper operation, or failure to wear protective goggles can harm the skin and eyes.[44] The FDA requires that sunlamps, which radiate ultraviolet rays, be equipped with: a timer that automatically shuts off the lamp after 10 min or less, a manual switch, a base for the sunlamp bulb that cannot be used in a regular light socket, and protective goggles for the maximum number of people who can use the lamp at one time.[45]

Mercury vapor lamps can cause severe eye irritation and skin burns at distances of 30 ft if the outer globe is broken and the inner part continues to operate, permitting ultraviolet radiation to escape. Outbreaks of conjunctivitis and skin erythema traced to broken (unshielded) mercury vapor lamps in public gymnasiums have been reported.[46] Skin cancer can result following repeated exposure.

A current cut-off is required to provide protection when the outer globe is punctured or broken. Use of the lamps in gymnasiums, schools, stores, and industrial facilities without the cut-off switch introduce a danger of possible injury from hazardous shortwave ultraviolet radiation.[47,48]

BIBLIOGRAPHY

Basic Radiation Protection Criteria, Rep. No. 39, National Council on Radiation Protection and Measurements (NCRP), Washington, D.C., January 15, 1971.

Braestrup, C. B., and K. J. Vikterlöf, *Manual on radiation protection in hospitals and general practice*, WHO, Geneva, 1974.

Dental X-Ray Protection, Rep. No. 35, NCRP, Washington, D.C., March 9, 1970.

Environmental Radiation Measurements, NCRP, Washington, D.C., 1976.

Instrumentation and Monitoring Methods for Radiation Protection, NCRP, Washington, D.C., 1978.

[44] *Environmental Health Criteria 14: Ultraviolet radiation*, WHO, 1979.

[45] "Safeguards on Sunlamps Ordered by U.S. Agency," *New York Times*, Washington, Nov. 10, 1980, Associated Press.

[46] "Conjunctivitis Caused by Unshielded Mercury-Vapor Lamps—Michigan, New Jersey," *MMWR*, CDC, Atlanta, Ga., March 23, 1979, p. 121.

[47] "F.D.A., Citing Radiation Peril, Acts For Mercury Vapor Lamp Safety," Washington, April 21, 1978, United Press International.

[48] "Mercury Lamp Standards Set," *FDA Consumer*, November 1979, p. 23.

Ionizing Radiation, American Public Health Association, Inc. 1015 Fifteenth Street, N.W., Washington, D.C. 20005, 1966.

Medical X-Ray and Gamma-Ray Protection for Energies up to 10 MeV—Equipment Design and Use, Rep. No. 33, NCRP, Washington, D.C., February 1, 1968.

Morgan, Karl S., "Exposure To Non-Ionizing Radiation," in *Environmental Problems in Medicine*, William D. McKee, Ed., Charles C. Thompson, Springfield, Ill., 1974.

Palmer, P.E.S., *Radiology and Primary Care*, Sci. Pub. No. 357, Pan American Health Organization, WHO, Washington, D.C., 1978.

Radiation Exposure from Consumer Products and Miscellaneous Sources, Rep. No. 56, NCRP, Washington, D.C., 1977.

Radiation Levels: Air, Water and Food, Royal Society of Health, London, S.W. 1, England, 1964.

Radiation Protection in Educational Institutions, Rep. No. 32, NCRP, Washington, D.C., July 1, 1966.

Radiation Protection in Veterinary Medicine, Rep. No. 36, NCRP, Washington, D.C.

Radiological Health Handbook, PHS, DHEW, GPO, Washington, D.C., January 1970.

Structural Shielding Design and Evaluation for Medical Use of X-Rays and Gamma Rays of Energies up to 10 MeV, Rep. No. 49, NCRP, Washington, D.C., 1976.

8

FOOD PROTECTION

The health and disease aspect of food and the investigation of foodborne illnesses are summarized in Figure 1-2 and discussed in Chapter 1. Program administration, including inspection, evaluation, education, and enforcement, is discussed in Chapter 12. This chapter concentrates on the technical and sanitation aspects.

In 1977 sales of food and drink in restaurants, taverns, cafeterias, hotels, motels, and vending machines amounted to $86 billion. In 1980 consumers spent $90 billion at 350,000 restaurants.[1] Food service sales are estimated to reach $18 billion in 1981.[2] The FDA estimates that Americans spend almost $145 billion annually for food and beverages consumed both in the home and outside the home.[3] On the other hand, the annual costs attributable to illnesses due to human cases of salmonellosis alone has been estimated at $1.2 billion.[4]

DESIGN OF STRUCTURES

Locating and Planning

The essential elements of a preliminary investigation to determine the suitability of a site for a given purpose were discussed in Chapter 2. Time and money spent in study, before a property is purchased, is a good investment and sound planning. For example, a food processing plant that requires millions of gallons of cooling or processing water would not be located at too great a distance from a lake or clear stream unless it was demonstrated that an unlimited supply of satisfactory well water or public water was available at a reasonable cost. An industry having as an integral part of its process a liquid, solid, or gaseous waste would not locate where adequate dilution or a disposal facility was not available, following economical waste recovery or treatment. Existing regulations, effect on the environment, and public attitudes need to be studied and cost of compliance determined.

[1] "For Your Dining Pleasure—A Model Ordinance," *FDA Consumer*, February 1980, pp. 17–19.

[2] Arnold S. Roseman, "Problems and Opportunities for the Food Service Industry," *J. Food Prot.*, November 1978, pp. 907–909.

[3] *Food Service Sanitation Manual*, 1976, DHEW, PHS/FDA, Washington, D.C., p. 1.

[4] Frank L. Bryan, "Impact of foodborne diseases and methods of evaluating control programs," *J. Environ. Health*, **40** (6) (1978), pp. 315–323.

Such factors as topography, drainage, highways, railroads, airports, watercourses, raw materials, exposure, flooding, swamps, prevailing winds, noise, rainfall, dust, odors, insect and rodent prevalence, type of rock and soil, availability and adequacy of public utilities and manpower, the need for a separate power plant, and water, sewage, or waste treatment works should all be considered and evaluated before selecting a site for a particular use. An environmental impact analysis of the proposed structures and uses on the surroundings and the prevention of deleterious effects in the planning stage will minimize future costs, nuisance conditions, and environmental hazards.

Structural and Architectural Details 

The design of structures is properly an engineering and architectural function that should be delegated to individuals or firms that have become expert in such matters. Certain design and construction details will be shown and discussed here as they apply particularly to food establishments, although the basic principles will apply equally as well to other places.

The construction material used is dependent on the type of structure and operation. Some materials can be used to greater advantage because of location, availability of raw materials, labor costs, type of skilled labor available, local building codes, climate, and other factors.

Floors

There are many types of floor construction; a few of them will be briefly discussed. Floors in food processing plants, dairy plants, kitchens, toilet rooms, and similar places should be sloped $\frac{1}{8}$ to $\frac{1}{4}$ in./ft to a drain. A trapped floor drain is needed for very 400 ft^2 of floor area, with the length of travel to the drain not more than 15 ft.

Concrete Floors. Wear causes dust and pitting. New floors require at least seven days for curing and must be kept wet during this time. Concrete can be colored and requires a good seal. Special finishes include rubber resin enamel, concrete paint or sealer, chemical treatment, wax over paint, or sealer. Warm linseed oil treatment makes concrete acid-resistant. Wire-mesh reinforcement will reduce cracking, particularly in coolers and driers where large temperature variations are expected. Floors in receiving rooms should be armored. Trucks and dollys with rubber or composition wheels reduce concrete wear and noise. Use alkaline cleaner.

Wood Strip or Block Flooring must be close-grained, seasoned, and carefully laid. A new floor requires sanding, two coats of a penetrating-type sealer or primer, buffing, and a wax, resin, or other finish. Avoid use of water.

Asphalt or Mastic Tile, Vinyl Tile & Rubber Tile. Use water sparingly. Strong soap, lye, varnish, or lacquer causes covering to shrink, curl, and crack. Use proper sealer, mild cleaner, and wax, usually water-based; mop dry. Can be used below grade with waterproof adhesive. Asphalt and rubber tile are affected by oil and grease.

Terrazzo and mosaic floors require six months for curing. Clean with warm water and treat with terrazzo seal. Solvents, oils, strong alkalis, and abrasives cause discoloration and destruction.

Vitreous tile, ceramic tile, & packing-house brick tile are hard, nonabsorbent, resist acid, and are not scratched by steel or sand. Cement joints should be full $\frac{1}{16}$ in. Ceramic or quarry tile (nonslip) properly laid is best for kitchens, dairies, breweries, bakeries, shower rooms, and similar places. Wash with hot water and neutral cleaner to keep floor in good condition. Soap and hard water cause discoloration. Do not use unglazed tile.

Linoleum floors require wax treatment and damp mopping. Water, alkali, or oil compounds cause linoleum to break down. Dampness causes linoleum to raise. Linoleum provides insect harborage and is not recommended for use in food-preparation rooms. It is no longer widely available. Newer tile and sheet linoleum are more resistant.

Cork floors are treated with a filler and water-emulsion wax. Avoid use of water. Provide furniture rests. Do not use directly on basement concrete floors. Not suitable for heavy traffic. Has good sound-deadening qualities.

Oxychloride, magnesite, & composition cement flooring has been used for marine decking, industrial building flooring, and other purposes. A penetrating-type dressing about every six months and waxing is the usual maintenance. Excessive water causes deterioration.

New or renovated floors require a conditioning or primary treatment. This usually involves a preliminary cleaning, then sealing and finishing to prevent harborage of vermin and penetration of moisture, soil, fungi, and bacteria and also to improve the wearing qualities of the floor. Following the conditioning the floor should be cleaned periodically and properly maintained to keep down dust and the microorganism population, prevent accidents due to slippage, and remove fire hazards. Manufacturers of flooring can give specific advice for the sealing and care of their particular material.

A good floor cleaner is all-purpose and neutral, requiring no rinsing; it should dissolve completely, have a low surface tension so as to pick up soil, and be economical and easy to use. A proper maintainer has a nonoil base and does not harm the floor material, furniture, walls, or equipment. It must not soften the floor finish; it should be applied either as a spray or powder, reduce slippage, be noninflammable, be easily removed from the mop or applicator, impart a pleasant aroma, and have good spreading or coverage quality. Ensure proper ventilation when using chemical cleaners. Rope off wet floors until dry.

Walls and Juncture with Floor

The line or point of juncture between the wall and floor, and with built-in equipment, should form a tight, sanitary cove and a smooth and flush connection. Some typical base details are illustrated in Figure 8-1. Walls should be smooth, washable, and kept clean. Many materials have been used for interior walls and wall finishes. These include glass blocks, plaster, cement, clay tile and cement, concrete tile and block, marble, concrete, ceramic tile, glazed brick and tile, and architectural terra cotta wall blocks. Absorbent materials such as wallboard and rough plaster require careful treatment and maintenance where moisture or water is an integral part of a process. These surfacings are to be avoided in such cases if possible. The wall finish should be a light color in work and processing rooms. Windowsills that are sloped down at an angle of about 30 deg simplify cleanliness.

Hollow walls and partitions, hung ceilings, and boxed-in pipes and equipment will provide a channel of communication, shelter, and protection for insects and rodents. They should be eliminated or, if this is not possible, special entrance openings or ducts should be provided to make possible simple and direct fumigation. The use of gypsum blocks is not recommended where water or dampness might be encountered.

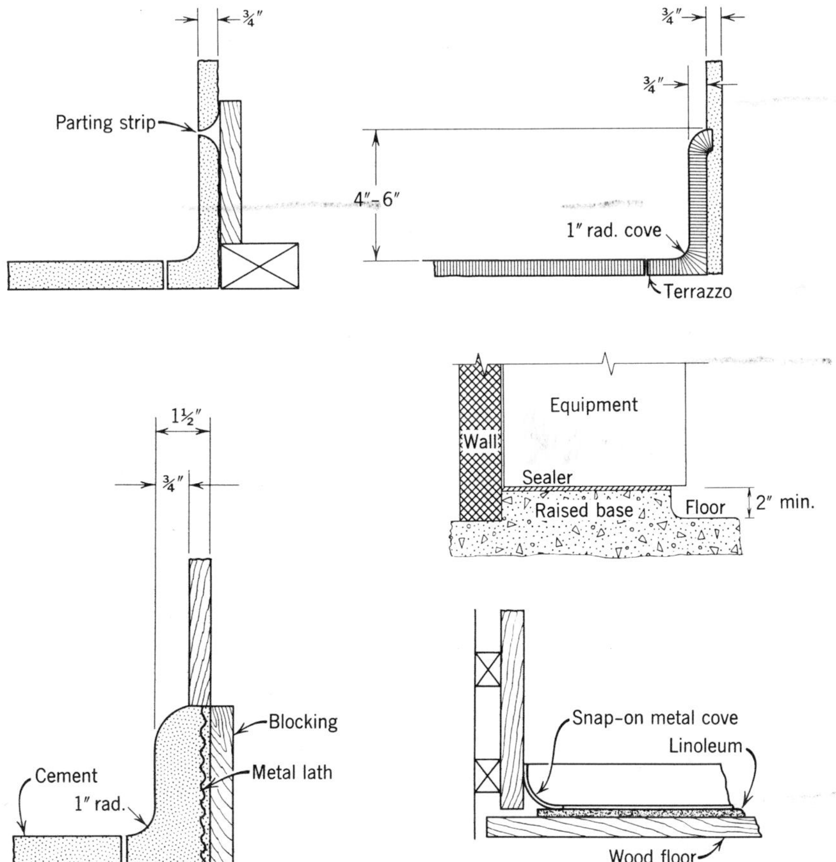

Figure 8-1 Floor and wall-base details.

Mechanical Details

Consideration should be given to high-volume and -pressure air supply for high-pressure or foam cleaning of equipment, and special parts washer sinks with integral circulation pumps and washing jets, automatically cleaned ovens, and hoods over stoves, cookers, and fryers. Use good building and plumbing codes.

Condensation

A difficult problem in food plants and restaurants is the condensation of moisture on cold pipes or surfaces, with resultant dampness, annoying dripping of water, and possible contamination of food and food-preparation tables. Condensation is caused by warm, humid air in contact with cool pipes or surfaces below the dew-point temperature. Lowering the humidity of the air, cooling the air, or increasing the temperature of the surface above the dew point will prevent condensation. Dehumidification can be accomplished by mechanical means and indirectly by good ventilation, which also would cause a lowering of the air temperature. The increase of the temperature of the surface is generally accomplished by insulation; the type and thickness of insulation is determined by the particular air temperature and humidity.

Lighting

Adequate light, with consideration to interior surface finish and color, is essential to proper operation and the maintenance of cleanliness. A minimum of 30 ft-c and up to 100 ft-c is recommended on work surfaces, dependent on the tasks to be performed. (See Table 8-7.)

Washing Facilities, Toilets, and Locker Rooms

The provision of adequate, convenient, and attractive sanitary facilities including dressing rooms or locker rooms is not only required by most health, industrial hygiene, and labor departments, but also is a good investment. Contented employees make better workers, and the habitual practice of personal hygiene in food establishments especially is essential. Wall-hung fixtures and toilet partitions in washrooms and the proper selection of floor and wall materials substantially reduce cleaning costs.

The number of plumbing fixtures to provide depends on the type of establishment and the probable usage. Local standards are usually available; but in the absence of regulations, the suggestions given in Table 11-6 for public buildings can be used as a guide. Washbasins and showers should be connected with both hot and cold running water or with tempered water. Soap or a suitable cleanser and sanitary towels are standard accessories. Extra washbasins with warm water, soap dispensers, and paper towels at convenient locations in all workrooms help promote cleanliness.

Water Supply

An improperly constructed and protected water supply is a common deficiency at food establishments and particularly at dairy farms located in areas not served by community water systems. This condition prevails probably because the water supply is related to the production of milk and food only indirectly and hence is considered unimportant in the spread of disease or product quality. It is also possible that the average inspector, being inexperienced in water supply sanitation, overlooks weaknesses in the development and protection of the water supply source.

A safe and adequate water supply under pressure is fundamental to the promotion of hygiene and sanitary practices at a restaurant, dairy farm, pasteurization plant, and other types of food-processing plants. It is good practice to oversize the piping and provide extra tees at regular intervals to permit future connections. Where a contaminated or potentially contaminated water supply under pressure is also available, it must not be connected to the plant potable water system unless special approval is obtained from the health department and required backflow preventers are properly installed. The availability of an unsafe water supply in a plant is dangerous; it should be eliminated if at all possible. An inadequate water supply will hamper and in many instances make impractical good sanitary practices and procedures. Contaminated water can become a link in the chain of infection and, in addition, lead to the production of food of poor quality. The development, protection, and location of wells and springs and the treatment of water are discussed in Chapter 3.

Liquid and Solid Waste Disposal

Where the disposal or treatment of industrial wastes is a problem, the sanitary sewer and waste drainage systems should normally be separate. Clear, unpolluted water should not as a rule be mixed with the sewage or industrial wastes and thus aggravate these problems, but can be disposed of separately without treatment. The waste treatment problem can usually be reduced by good plant housekeeping. Solid wastes are collected and stored for separate disposal; liquid wastes are kept at a minimum by careful operation and supervision; and where possible wastes are salvaged. Information concerning the treatment and disposal of sewage and industrial wastes is given in Chapter 4. Solid waste management is discussed in Chapter 5.

FOOD PROTECTION, QUALITY, AND STORAGE

Disease Control

The reader is referred to Chapter 1 and Figure 1-2 for more detailed information on the causes, prevention, and control of foodborne illnesses.

Food Handling

Temperature control and clean practices should be the rule in kitchens and food-processing plants if contamination is to be kept to a minimum. All food contact surfaces and equipment used in preparation must be kept clean and in good repair. Frozen meat, poultry, and other bulk frozen foods should be thawed in a refrigerator at 45° F (36° F to 38° F is ideal) and *not left to stand at room temperature overnight to thaw.* A small turkey under 12 lb will require about 2 days; a 16-lb turkey about 3 days; and a 24-lb turkey about 4 days. As an alternate, the frozen turkey in its unopened bag may be placed in potable running water at a temperature of 70° F or less with good velocity to the overflow. The thawed turkey must then be immediately cooked or refrigerated. A microwave oven may be used if the food is immediately cooked to completion. Frozen vegetables and chops need not be thawed but can be cooked directly. Prepared foods, especially protein-types, should be served immediately, kept temporarily at a temperature of 45° F or less, or on a warming table maintained at a temperature above 140° F until served. If not to be served, the food should be refrigerated in shallow pans to a depth of 2 to 3 in. within 30 min at a temperature of 45° F. Bulk foods should be cut into smaller pieces and refrigerated within 30 min of preparation unless immediately served. Spoilage of hot foods is prevented by prompt refrigeration. (Food short-term and long-term storage temperatures and periods are given in Table 8-10.)

Personal Hygiene and Sanitary Practices

All employees have a basic responsibility to maintain a high degree of personal cleanliness and observe hygienic and safe practices. The owner or manager must insist that employees comply or be removed. One or more washbasins located in the kitchen or workroom supplied with warm running water, a soap dispenser, and individual paper towels is necessary and conducive to greater personal cleanliness. Standard instructions to food handlers, including chefs, should include the following precautions.

1. Keep perishable foods covered and in the refrigerator until used. Cook, or reheat, foods to proper internal temperature and hold at proper temperature until served, if not served immediately, or refrigerate. Do not prepare food far in advance of intended service time.
2. Wash hands thoroughly after using the toilet, after handling raw foods such as poultry and meat, and when soiled. Use plenty of soap and warm water; rub hands together for at least 30 sec and clean thoroughly between fingers and around fingernails.
3. Do not pick up food with the fingers during food preparation unless absolutely necessary, and never when serving. Use a serving spoon, fork, spatula, or tongs.
4. Keep hands clean and fingernails short and clean. Keep fingers out of food, nose, and hair and off face.
5. Keep body and clothes clean and wear a head covering or hairnet.
6. Pick up cups, spoons, knives, and forks by the handles and keep fingers out of glasses, cups, soup bowls, and dishes.
7. Cover nose and mouth with a paper tissue when sneezing or coughing, then discard tissue and wash hands thoroughly. Do not smoke where food is prepared.

8. Report to a doctor at the first sign of a cold, sore throat, boils, vomiting, running sores, fever, or loose bowels. Stay at home! See a doctor if symptoms persist.

9. Help keep the entire premises clean. Store foods in a clean, dry place, protected from overhead drippage and animal, human, rodent, and insect contamination. Keep food preparation tables and utensils clean; avoid cross-contamination between food and unclean surfaces.

Food Handler Examination and Responsibilities*

The requirement for a food handler to have a medical examination prior to employment and periodically thereafter has been discussed and debated for some time. The medical examination, even if thoroughly done with laboratory support, does not and cannot give the assurance expected. The results are only valid for the time the examination is made. Fecal examinations may give false negative results because the laboratory does not always detect all the pathogens, or the pathogens are not being excreted at the time, or because the organisms are not uniformly dispersed in the specimen examined. It is believed that the time and expense involved in the medical examination and laboratory analyses of throat cultures, blood, urine, and feces are not commensurate with the public health protection obtained. Effort can be better devoted to improving food handler attitude, hygiene, and sanitary practices, and by daily *inspection* of the food handler by management. The food handler should be relieved from duty with pay when suffering from a pyogenic skin infection on the face, arms, or hands, a bad cold, or any other obvious infection or disease that could be transmitted directly or through food. However, the medical examination and specimen collections are an important part of a foodborne illness investigation.

The importance of individual personal hygiene and sanitation is emphasized by the Association of Public Health Inspectors.[5] it states in part,

> Every food handler must take all the precautions necessary to protect food and drink from the risk of contamination and the Food Hygiene (General) Regulations, 1970 make it an offence liable to a fine of up to 100 pounds and/or three months imprisonment to expose food to risks of contamination. This places specific responsibility upon each individual food handler and stresses the care which must be taken in the preparation, handling, storage, display, and service of food and drink.

Food Inspection

Food for human consumption is expected to be clean, wholesome, and sanitary. Decayed, insect-infested, moldy, or musty food is not acceptable. In any case, purchased food should be in full weight and measure and of the indicated grade. The best time to check this is on delivery against the order or invoice, before acceptance. Although special training is needed for expert determination, some general guides are summarized in Table 8-1.

*See also "Health Examination of Food Handlers—Europe," *MMWR*, CDC, Atlanta, Ga., June 12, 1981, p. 267.

[5] *Hygiene In Public Houses and Licensed Premises*, The Association of Public Health Inspectors, 19 Grosvenor Place, London, S.W. 1, Practice Notes No. 14, October 1973.

Table 8-1 Tests for Food Spoilage or Poor Quality

Canned food—Grade A, B or C

1. Swelled top and bottom, or one end only.[a]
2. Dents along side seam, deep rust.
3. Off-odor or molds.
4. Foam, leaks.
5. Milkiness of juice.

This applies to canned vegetables, meats, fish, and poultry. Home-canned foods should be cooked thoroughly. Do not taste!

Fresh fish

1. Off-odor.
2. Gray or greenish gills.
3. Eyes sunken, dull; pupils gray.
4. Flesh easily pulled away from bones.
5. Mark of fingernail indentation remains in flesh.
6. Not rigid, scales dry and dull.
7. Oyster and clam shells not tightly closed. Shell gives dull sound on tapping.

Raw shrimp

1. Pink color on upper fins and near tail. Soft, dull, sticky.
2. Off-odor similar to ammonia.

Some types of shrimp are naturally pink. Cooked shrimp are also pink. Both are wholesome if the odor is not abnormal.

Meat—Grade Prime, Choice, Good

1. Off-odor, sourness, taint.
2. Slimy to touch.
3. Beef soft, dark, coarse-grained; soft fat, yellow.
4. Lamb or veal flesh dark, fat yellow.

Beef usually spoils first on the surface. Pork spoils first at meeting point of bone and flesh in the inner portions. To test for spoiled beef or pork use a pointed knife to reach the interior of the meat. An off-odor on the knife means spoilage. Fresh meat should be firm and elastic when pressed; not discolored.

Leftover food

1. Discoloration.
2. Off-color.
3. Mold.

Any food that has not been refrigerated below 45° F may be considered slightly spoiled. The off-odor of spoiled food is not always apparent. Do not keep cooked food such as ground meats, hollandaise sauce, cream fillings, cream sauces, custards, or ham, chicken, egg, or fish salads.

Bacterial spoilage of food begins as soon as it becomes warm. Refrigeration will delay this spoilage, but will not destroy toxins previously formed due to contamination and improper storage.

Salads and desserts

Chicken salad, tuna, and other fish salads, nonacid potato salad, all types of custard-filled pastries, and some types of cold cuts must be kept refrigerated at all times. All have been touched with the hands during their manufacture and may be considered slightly contaminated.

Refrigeration will keep contamination from increasing. Spoilage is often impossible to detect until foods are totally spoiled. Serve salads immediately after taking from refrigerator.

Frozen foods—Grade U.S. A, B, C (USDA)

Frozen foods will spoil if kept out of the refrigerator. Spoilage is caused by growth of bacteria on the food. Thaw poultry and roasts in refrigerator for 2 to 3 days before use.

Cook frozen vegetables thoroughy before serving to destroy any contamination that may be present. Do not overcook. Heat ready-to-eat dinners thoroughly, 165° F, before use.

Do not use refrozen fish or shellfish.

Avoid soft, mushy, discolored foods.

Fruits and vegetables—Fancy, No. 1, No. 2

1. White or grayish powder around stems of fruit and at juncture of leaves and stems of cabbage, cauliflower, celery, and lettuce.
2. Poorly developed leaves, dry or yellow, soft, loose, coarse, sprouted, slime, mold, insect infested, unclean.
3. Fruits soft, blemishes, molded, decayed, worms; citrus fruits soft, light.
   The powder indicates spray residues.

Most of the chemicals used by growers are not dangerous but some may be. All fruits and

Table 8-1 (*Continued*)

vegetables must be washed before eaten or cooked. Cooking will not destroy the spray chemicals.

Dressed poultry—Grade A, B (USDA)

1. Stickiness or rancidity under wing, at the point where legs and body join, and on upper surface of the tail. Sunken eyes.
2. Darkening of wing tips, soft flabby flesh.

Dressed poultry should be washed thoroughly before cooking. Wash your hands after handling.

Cereal

1. Insects in cereal.
2. Lumps, mustiness, mildew.

Spread the cereal on brown paper. If insects are present they will be easily seen. If even one is observed destroy the entire batch of cereal.

These insects are not dangerous, but neither are they appetizing.

Smoked meats

1. Pale color, soft, moist, flabby.
2. Sausage, bologna, frankfurters slimy, moldy, discolored spots, internal greening.

Cured fish

1. Offensive odor, soft flesh.

Eggs—Grade AA, A, B (USDA)

1. Candling shows large air cell.
   Jumbo 28 oz/doz, extra-large 27 oz, large 24 oz, medium 21 oz, small 18 oz.
2. Cracked, dirty or leaky shell.
3. Frozen, liquid, or dried powder not pasteurized; should be free of salmonella.[b]

Butter—Grade AA, A, B (USDA)

1. Off-flavor, fishy, stale, unclean, rancid or cheesy flavor—scored 83 to 87½.
2. Slightly objectionable flavor—scored 89 to 91½.
3. Made from unpasteurized cream.

A score of 92 indicates a clean, sweet butter lacking in rich creamy flavor. A score of 93 to 100 indicates a clean, sweet, creamy, excellent butter. Butter made from raw milk is dangerous.

Instant nonfat dry milk—U.S. Extra Grade (USDA)

Milk should have a sweet and pleasing flavor and dissolve immediately.

Cheese

1. Evidence of contamination or infestation; mold not characteristic of the product.
2. Made from unpasteurized product or not stored at least 30 days above 35° F. Should be made from pasteurized product. Salmonella in milk in large numbers can survive extended storage in Cheddar cheese of high pH.[c]

Chemical tests[d]

1. Solution of malachite dye mixed with ground meat will turn bright red if sodium sulfite added to preserve or mask decomposition of meat. Zinc plus dilute mineral acid gives off H_2S in presence of sulfite.
2. The filtrate, when alcohol is added to meat, will turn red or pink if artificial coloring has been added.
3. Shucked oysters with a pH of 5.8 to 5.4 suspicious, pH can be 5 in oysters from certain beds. Use a drop of oyster liquid on chlorophenol red test paper and compare with standard. If washed oyster meat turns persistent red when methyl red indicator solution used, oysters are spoiled. Teaspoonful of crab meat in water plus 0.5 ml Nessler Reagent turns deep yellow or brown if meat is spoiled.
4. Cadmium-plated utensils detected by rubbing a swab moistened with 10 percent nitric acid and placing on moistened filter paper impregnated with 20 percent solution of sodium sulfite. A canary yellow stain indicates cadmium.
5. Cyanide is detected by a special test paper that turns orange and then brick red in 5 to 10 min when suspended in air space of bottle containing suspected polish if cyanide exceeds 0.5 percent.
6. Rodent stains fluoresce in ultraviolet light. Cook's test paper placed on stain turns black when moistened.
7. Arsenic is detected by a test similar to the sulfite determination.
8. The presence of fluoride is detected by a

Toxin will not be distroyed by cooking

Table 8-1 (*Continued*)

special test paper in the presence of citric acid.

9. Sulfites and bisulfites are not permitted in meats.

Custard pastries

Custard-filled pies, puffs, eclairs, and similar pastries, to be considered safe, should be:

1. Prepared with custard or cream filling that has been heated and maintained for at least 10 min at 190° F or 30 min at 150° F or
2. Rebaked for 20 min at an oven temperature of 425° F or 30 min at 375° F and
3. Cooled to 45° F within 1 hr after heating and maintained below 45° F until consumed.

Only properly pasteurized milk or cream should be used; filling equipment should be cleanable and sterilized before each use, and no cloth filling bags are to be used.

Source: Adapted from Ohio Dept. of Health and other sources.
[a] Sweller—both ends stay bulged when pressed. Springer—a sweller in which the end will give when pressed. Flipper—both ends flat, end will bulge out when end is pressed.
[b] D. H. Bergquist, "Sanitary Processing Egg Products," *J. Food Prot.*, July 1979, pp. 591–595.
[c] Charles H. White and Edward W. Custer, "Survival of Salmonella in Cheddar Cheese," *J. Milk Food Techol.*, May 1976, pp. 328–331.
[d] Walter D. Tiedeman and Nicholas A. Milone, *Laboratory Manual and Notes for E. H. 220*, Sanitary Practice Laboratory, School of Public Health, University of Michigan, Ann Arbor, 1952, and as revised 1971.

Questions are sometimes raised concerning the microbiological quality of wines. Wines typically have a high acidity, low pH, and high alcohol content compared to other foods. These characteristics discourage the growth of pathogens and most nonpathogens. Organisms that might be encountered include yeasts, molds, lactic acid organisms, and acetic acid bacteria. Although these organisms may cause spoilage, they are not of public health significance.[6]

Food Preservation

Canning. The Codex Alimentarius Commission advises,

> Canned products with an equilibrium pH above 4.5 shall have received a processing treatment sufficient to destroy all spores of clostridium botulinum unless growth of surviving spores is permanently prevented by product characteristics other than pH.

Other food preservation methods are: pasteurization, sterilization, blanching, freezing, drying, fermentation, and pickling; and chemical preservation using proper concentrations of sugar, salt, spices, sulfur dioxide, carbon dioxide, benzoic acid, fumigation, antibiotics, and irradiation. Safety of canned hams depends on the combination of pasteurizing heat treatment and the presence of curing salts.[7]

Clostridium botulinum vegetative forms, are killed in 10 to 15 min at 80° C

[6] A. C. Rice, "Wine—Its Impact as a Food Commodity," *N. Y. State Assoc. Milk Food Sanit. News.* Cornell University, N.Y., January 1980.
[7] A. D. Bostock, "Food Control," *J. R. Soc. Health*, October 1974, p. 35.

(176° F). Two percent vinegar in pickled food, or 8 to 10 percent salt will kill the spores.[8]*

Microbiological and Chemical Standards, Guidelines, or Criteria

Natural microbial variations in different foods, food composition, and the statistical aspects of sampling present considerable difficulties in the establishment of firm standards. In addition, because of normal errors inherent in laboratory techniques, it is practical to allow some leeway in the standards proposed and refer to them as guidelines. One might use the geometric mean of say, ten samples or allow one substandard sample out of four. Parameters used include total aerobic count, toxigenic molds, number of coliforms, number of *E. coli*, coagulase-positive staphylococci, salmonellas, shigella, *Clostridium perfringens, Clostridium botulinum,* and beta-hemolytic streptococci as indicated.

The FDA has established action levels for poisonous or deleterious substances in human food or animal feed. The action level is that limit at or above which FDA will act to remove the product from the market. A list of substances for which action levels have been established is shown in Table 8-2.[9]

The FDA action level for aflatoxins in foods and feeds is 20 ppb, which may be lowered to 15 ppb. However, the action level for milk shipped in interstate commerce is 0.5 ppb. The U.S. Department of Agriculture is involved in the control of such products as peanuts and grains. Other countries also have regulatory programs for the control of aflatoxins in food and feeds.[10]

A study reported by Senn[11] showed that standard plate counts of less than 25,000 for ham and meat sandwiches and 40,000 for salad sandwiches, 0 to 3 for shelled eggs in the laboratory, and 24,000 to 2,800,000 (average 1,600,000) after shelling at a good commercial egg-breaking plant prior to freezing liquid eggs can be obtained.

Celery has been found to be a major source of salad contamination. Immersion in boiling water for 30 sec, then chilling under cold tap water reduces the total bacterial count, coliform, enterococci, and staphylococcus to acceptable levels.[12]

*For commercial sterilization, the time-temperature relationships for a desired *Cl. botulinum* spore reduction are: 2.45 min at 250° F, 8.79 min at 240° F, 31.55 min at 230° F, and 114.58 min at 220° F. (John M. Last, ed., *Maxey-Rosenau Public Health and Preventive Medicine*, 11th ed., Appleton-Century-Crofts, New York, 1980, p. 894.)

[8]Herbert L. Dupont and Richard B. Hornick, "Infectious Disease from Food," *Environmental Problems in Medicine*, William C. McKee, Ed., Charles C. Thomas, 1977.

[9]"Food Contaminants," *FDA Consumer*, Dec. 1978–Jan. 1979. p. 9.

[10]L. B. Bullerman, "Significance of Mycotoxins to Food Safety and Human Health," *J. Food Prot.*, January 1979, pp. 65–86.

[11]Charles Senn, "Microbiology—An Essential in Food Inspection," *J. Environ. Health*, September/October 1964.

[12]Syed A. Shahid, et al., "Celery Implicated in High Bacteria Count Studies," *J. Environ. Health*, May-June 1970.

Table 8-2 Action Levels for Poisonous or Deleterious Substances in Human Food or Animal Feed

Substance	Commodity	Action Level
Aflatoxin	Peanuts and peanut products, grain products, animal feed	20 ppb
	Milk	0.5 ppb
Aldrin and dieldrin	Eggs, animal feed	0.03 ppm
	Blackberries, blueberries	0.05 ppm
	Butter, fish (smoked, frozen, canned), milk	0.3 ppm
Endrin	Apples, apricots, grapes, citrus fruits	0.05 ppm
	Eggs	0.03 ppm
	Milk, vegetable oils, fats	0.03 ppm
Kepone	Crabs (frozen or canned)	0.4 ppm
	Fish, oysters, clams, mussels (smoked, frozen, canned)	0.3 ppm
Lead	Evaporated milk	0.5 ppm
Mercury	Fish, oysters, clams, mussels	1.0 ppm
	Wheat	1.0 ppm
Mirex	Fish	0.1 ppm
PBBs	Milk and dairy products, meat, eggs, animal feed	0.3 ppm
PCBs	Paper food-packaging material intended for or used with human food, animal feed	10 ppm
	Milk and dairy products	1.5 ppm
	Poultry	3.0 ppm
	Eggs	0.3 ppm
	Fish	2.0 ppm

Source: *FDA Consumer*, Dec. 1978–Jan. 1979, p. 9.

Washing of celery clean and immersion in a 20 mg/l chlorine solution has also been found effective.

Numerous microbiological standards, actionable levels, guidelines, good manufacturing practices, and criteria have been proposed and are in use for industry and regulatory control purposes to indicate food quality and the need for investigative action. Some are summarized in Table 8-3. Analytical procedures and interpretation of laboratory results are given in various publications.[13–16] Abrahamson has pointed out, based on his many years of experience in New York City, that the publication of administrative microbiological standards is an excellent educational service. Successful results are obtained more through understanding than by fiat.

[13] Marvin L. Speck, Ed., *Compendium of Methods for the Microbiological Examination of Foods*, Am. Pub. Health Assoc., 1015 Fifteenth St. NW, Washington, D.C., 1976.

[14] *Official Methods of Analysis*, Assoc. of Official Analytical Chemists, Washington, D.C., 1970.

[15] *International Commission on Microbiological Specifications for Foods*, "Microorganisms in foods," "1. Their significance and methods of enumeration, 1977," and "2. Sampling for microbiological analysis: principles and specific applications, 1974," University of Toronto Press.

[16] "Microbiological aspects of food hygiene," Report of WHO Expert Committee with the participation of FAO, *Tech. Rep. Ser.* **598**, 1976.

Table 8-3 Some Microbiological Guidelines for Foods[a]

Food	Total Count per Gram Not to Be Exceeded					References
	E. coli	Plate Count	Coliform	Staphylococcus	Salmonella	
Prepared foods[b]	Negative	100,000	20	100	Negative	
Crab cakes and crabs	3.6	10,000	3.6	3.6		1
Frozen cream-type pies		50,000	50			6
Gelatin		3000	20			6
Frozen cooked seafoods and meals		100,000	20	100		2
Cheese	1000		1500	1000		3
Meat						
frozen, fresh, ground	50	5,000,000				4
cooked, smoked	10	1,000,000				4
Beef						
ground unfrozen	100	10,000,000		100	Negative	5
ground frozen	100	1,000,000		100	Negative	5
Oysters, fresh, shucked, frozen		500,000	230			7
Shrimp, raw, breaded, frozen		500,000	50			7
Dehydrated food—dependent on item	Negative	25,000 to 200,000	10 to 40		Negative	7
Uncooked poultry	20	1,000,000	100 to 1000	10	10	

[a] Results should be interpreted in the light of a complete sanitation inspection. The counts noted are maximums.

[b] Fecal streptococcus should not exceed 1000, yeast and molds 20; and *Clostridium perfringens* should be negative.

[1] Robert Angelotti, "Catering Convenience Foods—Production and Distribution Problems and Microbiological Standards," *J. Milk Food Technol.*, May 1971, p. 231. Administrative guidelines for crab cakes—cooked, frozen; crabs—deviled, cooked, frozen. Examination of a minimum of 10 subs with the indicated count not exceeded in 20 percent or more of the subs; however the plate count is the geometric average of the subs.

[2] *An Evaluation of Public Health Hazards from Microbiological Contamination of Foods*, National Academy of Sciences-National Research Council Pub. 1195, Washington, D.C., 1964, pp. 52–58. Conclusions of the Committee on Food Microbiology and Hygiene of the International Association of Microbiological Societies—Conference, Montreal, Canada, August 16–18, 1962. Counts considered attainable. Coliform count of 100 permitted by some standards.

[3] D. L. Collins-Thompson, I. E. Erdman, M. E. Milling, U. T. Purvis, A. Loit, and R. M. Coulter, "Microbiological Standards for Cheese," *J. Food Prot.*, June 1977, pp. 411–414. The counts listed are considered "unacceptable." "Acceptable" counts are: total coliform—500, fecal coliform—100, and *Staphylococcus aureus*—100.

[4] Kenneth E. Carl, "Oregon's Experience With Microbiological Standards for Meat," *J. Milk Food Technol.*, August 1975, pp. 483–486. The counts are used primarily as a tool to improve sanitation practices.

[5] "Proposed Microbial Standards for Ground Meat in Canada," *J. Milk Food Technol.*, December 1975, p. 639. Salmonella count is per 25 g.

[6] *Fed. Reg.*, August 2, 1973. The counts for pies are for banana, coconut, lemon, and chocolate cream-type. Counts are the geometric mean of 10 analytical units.

[7] Edmund M. Powers, "Microbiological Criteria for Food in Military and Federal Specifications," *J. Milk Food Technol.*, January 1976, pages 55–58. The count for oysters is MPN fecal coliforms per 100g at 44.5° C.

Ice

Water used in the manufacture of ice must be from a source meeting drinking water standards. The manufacturing process, handling, storage and distribution must assure protection of the ice from contamination. The plant should meet the basic sanitation requirements given in reference[9] on page 768. The same principles apply to ice-making machines in restaurants, hotels, and public places.

Freezing does not kill bacteria. A WHO publication [17] points out that

> when water freezes, impurities present in the water tend to be "squeezed" toward the middle. Taking advantage of this phenomenon, ice manufacturers pass bubbles of air through the water during the freezing process, keeping it in motion and preventing precipitated chemicals, solids, and bacteria from freezing into the ice crystals. The impurities are thus concentrated in the core section (i.e., the unfrozen water in the middle of the block of ice). In the final stages of manufacture the core is removed (pulled) and fresh water added in order to reduce the concentration of foreign matter before the final freezing. Inspection of a block of ice shows how well the core is pulled; discoloration in the middle of the block indicates contamination. Some minerals and chemicals that cause problems in ice manufacture are iron, manganese, calcium and magnesium carbonates, aluminum oxide, and silica.

Dry Food Storage

Practically all foods, whether canned, pickled, dried, or chemically preserved, deteriorate on storage. They change in color and texture, develop off-odors or off-flavors, and lose nutritional value. Storage temperature is the single most important factor that affects the storage life of food items. A temperature of 40° F is good for the storage of canned and dehydrated foods 3 to 5 years. A temperature up to 70° F is acceptable for short-term storage; but the useful storage life of most products at 100° F is 6 months or less. Dried fruits are best stored at 32 to 40° F and 55 percent relative humidity. It is also of practical value to know in connection with food storage that insects become active at a temperature of around 48° F. There are no such things as nonperishable or semiperishable foods; they begin to deteriorate as soon as prepared or packaged. This was a conclusion of the Quartermaster Food and Container Institute for the Armed Forces, sponsored by the Department of the Army Office of the Quartermaster in March 1955. Refrigeration however will prolong storage life of canned food.

Food storage rooms generally adjoin unloading platforms and connect as directly as possible with the food preparation areas of the kitchen. The storage room that provides a floor area of approximately 1 to $1\frac{1}{2}$ ft^2 per meal served per day is usually adequate in size. Storerooms should be clean, cool, dry, dark, ventilated, protected against the entrance of insects and rodents, and organized. Artificial lighting should provide 10 to 20 ft-c of light. The floor should be well

[17]"Surveillance of Drinking Water Quality," *WHO Monog. Ser.* **63**, (1976), p. 65.

drained but without a floor drain, cleanable, and above any possible high-water or sewage flooding. Food must not be stored in basements that might be flooded. If sewer or waste pipes must pass through storerooms, the pipes should be provided with drip pans to carry off possible leakage or condensation and food must not be stored directly beneath the pipes. Good ventilation is important. It can be obtained with screened windows opened at the top and bottom, an exhaust fan, a circulating fan, and upper and lower wall and door louvered vents that do not open into a hot kitchen.

Storerooms should be furnished with shelves, pallets, and covered containers. With shelves 10- to 20-in. deep and at least 18-in. away from the walls, so as to be accessible from all sides, and the bottom shelf 12 in. or more above the floor, stock inventory and cleanliness are simplified. Food storage platforms should be movable, at least 12 in. off the floor and away from the wall. Bulk items are stored in the original packing on pallets, and loose foods on shelves, in some predetermined order. Sugar, coffee, flour, rice, beans, and other dry stores are stored in metal or other waterproof containers with tight-fitting covers. Old stock should of course be used first; the dating of new supplies will help accomplish this objective. Provide a $3\frac{1}{2}$-ft aisle space between shelving. Do not store pesticides, cleaning compounds, disinfectants, or lubricating oil in food storage rooms. (Some of the precautions to take in the care of fruits and vegetables are given in Table 8-10.) See also Refrigeration this chapter.

Compliance with and Enforcement of Sanitary Regulations

The training of foodhandlers and managers of food service establishments is considered by many as an important part of a food protection program. This can be a formidable and unending task. Experience shows that the effectiveness of foodhandler training is limited and of short duration at best. Training of managers has been considered of greater value. However, Cook and Casey[18] found in a limited study that

> management enrollment in and completion of the . . . course did not produce significantly higher . . . sanitation scores than those of equivalent establishments without . . . trained managers.

The effort expended should be carefully evaluated to determine if the time and effort could be better devoted to more productive activities.

Management Responsibility[19]

Management has a primary and continuing responsibility for the education and supervision of employees in order to insure that basic health requirements are met. Employees who cannot be trusted to carry out hygienic and sanitary

[18] Charlotte C. Cook and Ralph Casey, "Assessment of a Foodservice Management Sanitation Course," *J. Environ. Health*, March/April 1979, pp. 281–284.

[19] J. A. Salvato, *Guide To Sanitation In Tourist Establishments*, WHO, 1976, pp. 81–82.

procedures at all times should be replaced by reliable personnel. The health authorities are responsible for insuring that all food establishments comply with regulations adopted to protect the public health and they should be empowered to enforce the regulations in a reasonable manner.

Strengthening the roles of health authorities and managers or operators in relation to the provision of safe and sanitary food services could offer mutual benefits since each party could assume greater responsibility within its particular sphere of competence. The health agency would then probably need fewer, but very competent, sanitarians. Establishments controlled by a reliable and informed owner or operator would be subject to fewer inspections and the management would have full control over and responsibility for the hygienic and sanitation procedures in the establishment. The system proposed is not entirely new and is already in operation in various degrees. The procedure is as follows.

1. A legal responsibility is created for the health agency to approve or provide for the initial qualification and certification of the operator or manager of every food establishment. The certification is valid for a stated period of time subject to suspension or revocation after a formal inquiry. Qualification of the individual requires special education and training in the basic principles of sanitation and hygiene, and so on, and periodic attendance at refresher courses. Operation of an establishment by unqualified persons would be illegal.
2. The health agency is required to provide the special education and training and refresher courses, either using the agency staff or through contractual agreements with a college or university, in addition to maintaining a food service inspection and public education program.
3. After certification the operator or manager has sole responsibility for insuring full compliance with established sanitary regulations within the establishment.
4. The health agency is required to maintain a registry of all food establishments and make periodic *comprehensive* evaluations of the quality of food and food services. Evaluations should be based on statistical sampling and should cover all places at least once every 3 years, but there should be a separate microbiological food quality sampling evaluation program for selected potentially hazardous foods.
5. An operator's certificate is suspended by the health agency on evidence of a major violation of the sanitary regulations, and the establishment must be closed pending a formal hearing to determine if the certificate should be revoked.

Food Protection Program Objectives

Measures which will reinforce traditional sanitary inspection and quality control programs to reduce the widespread occurrence of foodborne diseases are summarized in the following American Public Health Association Position Paper Objectives:[20]

1. A renewed awareness that preventable foodborne diseases are costly and occur more frequently than necessary;
2. recognition that the usual sanitary inspections and quality assurance programs are no longer sufficient to reduce the incidence of foodborne illnesses in the United States;

[20] "Protection of the Public Against Foods and Beverages That Are Unfit for Human Consumption," Policy Statements Adopted by the Governing Council of the Am. Publ. Health Assoc., November 7, 1979, *Am. J. Pub. Health*, March 1980, pp. 312–314.

3. an appreciation of the benefits to be gained by supplementing established food-sanitation activities with other efforts such as: a) certification of food-establishment managers, b) increased epidemiological investigation of suspected foodborne disease outbreaks, c) hazards analysis and critical control point evaluations of food establishments, and d) standardized evaluation of food quality and safety;
4. acceptance of microbiological criteria as useful, supplemental measures for: a) maintaining the level of cleanliness in food-handling operations that will satisfy consumer expectations, b) detecting exposure of foods to insanitary conditions that may result in the spread of foodborne diseases, as well as increased spoilage, c) discouraging the use of microbiologically inferior ingredients in commercially formulated foods, and d) establishing definitive microbiological limits for acceptance or rejection of some foods.

MILK SOURCE, TRANSPORTATION, PROCESSING, AND CONTROL TESTS

Milk Quality

The quality of milk reaching the ultimate consumer is largely determined at the farm where the milk is produced. The type of herd, feed, and health of the cows are important; but unless the raw milk is obtained, handled, and stored in a sanitary manner, from the cow to the processing plant and ultimate consumer, the final product will be mediocre or even unacceptable.

A healthy milk herd is expected to be mastitis-, tuberculosis-, and brucellosis*-free. Requirements for testing, accreditation, and disposition of reactors or segregation of unsound animals are established by the Animal and Plant Inspection Service, U.S. Dept. of Agriculture (USDA). Decisions of the federal agency are usually accepted by the state agency having jurisdiction. The milk produced should not be bloody, stringy, or otherwise abnormal, nor should the milk be used within 15 days before calving or 5 days after calving. Milk from cows treated with antibiotics should be withheld from the market until at least 3 days have elapsed.

The reservoir of bovine tuberculosis, although greatly reduced, has not been eliminated; hence the dairy farmer, department of agriculture, and health department must be constantly on the alert against the introduction of the disease in a clean herd. Other illnesses that can be transmitted to man via milk include brucellosis, Q fever, salmonellosis, shigellosis, infectious hepatitis, diphtheria, staphylococcic infections, and streptococcic infections. See Chapter 1. Inspection of meat for tubercle lesions and blood testing at the slaughter house can lead to identification of infected herds. Calves can be vaccinated against brucellosis (about 65 percent effective). Staphylococci, streptococci, and coliform microorganisms are associated with mastitis. Every person having anything to do with the processing of milk should be free of communicable diseases or running sores.

*Also known as Bang's disease, cause of contagious abortion of cattle, spreads rapidly through a herd. Blood tests before importation or addition to a herd can prevent introduction of infection.

The production of a clean milk requires that abnormal milk be discarded; that milk be promptly cooled; and that the utensils and equipment used, the udders and teats of all milking cows, as well as the flanks, bellies, tails, the milker's hands and clothing, and the milking area be clean. In addition, the milker's hands, milk utensils and equipment, and the udders and teats should be rinsed or wiped with an approved bactericidal solution. Although these elementary principles may be obvious, they require continual emphasis and reemphasis with some dairy farmers. If after repeated instruction a dairy farmer cannot be relied on to produce a clean milk, his milk should be excluded from the market.

Dairy Farm Sanitation

Routine inspections of dairy farms by representatives of industry, and occasionally by official agencies, should be made with the owner or manager so that unsatisfactory conditions and practices can be pointed out and discussed. Control programs usually include the collection of milk samples for bacterial examinations. A low-count milk can be produced at every farm. Experience shows that attention to the following procedures will result in a good-quality, low-count milk.

1. Keep milking equipment and utensils clean.
    a. Take all milking equipment apart immediately after each milking and rinse in lukewarm water. Replace equipment and utensils having open seams.
    b. Scrub all parts, including dairy utensils, in hot water with a brush and good washing compound, then disinfect. Hot water at a temperature of 180 to 190° F is very good, or use a strong (200 mg/l) chlorine solution according to the manufacturer's directions to disinfect rubber, valves, and other parts. Store all equipment on a clean rack in the milkhouse.
    c. Remove milk-stone deposits with an acid cleaner. One oz of phosphoric or gluconic acid or similar milk-stone remover per gal of disinfecting water at 145° F will help prevent milk-stone and water-hardness deposits.
    d. Replace all worn rubber parts. Two sets of inflations used every other week will last longer.
    e. Boil rubber parts for 15 min once a week in a solution of 2 tbsp of lye in a gal of water. Soak rubber gaskets and teat cups, and other rubber pieces, for $\frac{1}{2}$ hour or until next milking in a solution of hot water containing 1 oz of lye per gal, or immerse for 5 min in water at a temperature of at least 170° F. Rinse the parts and store in a cool, dry, dark place. Disinfect before reuse.
2. Cool promptly all milk in a clean milkroom or milkhouse.
    a. Maintain water level above milk level in can cooler, but do not submerge cans. Change water in cooler frequently. Maintain water temperature at 40° F or less and provide circulation of water in cooler, or
    b. Cool milk in a refrigerated farm bulk-milk tank and keep at 36 to 38° F.
3. Prevent external contamination.
    a. Brush down cobwebs in barn and control dust.
    b. Keep barn and cows clean. Clip and clean flanks, tails, and udders of milking herd. Whitewash barn walls and ceilings at least once a year.
    c. Assure proper sewage, wastewater, and manure disposal. Control flies and keep yard drained.
    d. Use a protected water supply that is safe and adequate. Contamination with psychrophilic bacteria can affect milk bacterial count and shorten raw-milk storage life. Treatment (chlorination) may be necessary if psychrophiles are normally present in the water supply.
    e. Feed cows odorous feeds after milking.

4. Use good milking procedures.
   a. Massage udders with a warm disinfecting solution just prior to milking.
   b. Prohibit wet-hand milkings. Wash hands thoroughly; disinfect with a quaternary solution or equivalent and dry before milking.
   c. Use single-service pads for straining milk.
5. Keep the milking herd healthy.
   a. Make monthly or quarterly chemical milk-screening tests on herd milk. Cows should be free from tuberculosis, brucellosis (Bang's disease), mastitis, and other diseases. Obtain veterinary diagnosis when mastitis is suspected and treat if necessary.
   b. Use a strip cup, check for abnormal milk, and use proper milking procedures.
   c. Provide ample, clean, dry bedding.

Detailed inspection report forms and explanations of regulations are available from the FDA, state and local health departments, and in some instances agriculture departments. Some industries provide supplementary information.

Barn

The milking barn should have 20 ft-c of artificial light, with 50 ft-c at the cow's udder, as well as natural lighting. Adequate air space and ventilation are necessary to prevent overcrowding and condensation, excessive odors, and dampness. Hinged windows tilting open at the top and other vertical or horizontal means of ventilation are suitable. A window area of 4 $ft^2/60\ ft^2$ of floor surface and a minimum of 500 ft^3 of air space per animal are recommended. Concrete manure gutters and walks at least 3-ft wide with a 6-in. coved curb at walls integral with the floor are easy to keep clean. A hose bib connection (protected against freezing if necessary) in the milking barn, stable, or parlor, connected with a safe water under pressure, simplifies the dairy cleaning operation. The water outlet should be equipped with a backflow preventer. In any case, the milking area must be clean and walls, ceilings, and windows constructed of cleanable material that keeps out dirt, flies, and odors. Manure must be spread on the fields or otherwise disposed of at least every four days in the fly season. Temporarily, it may be stored in a drained area not closer than 25 ft to any building, although daily removal is preferred. Storage pits, where provided, must be flytight.

Plans and assistance for animal management are usually available from the regulatory agency, USDA, extension service, and equipment manufacturers.

Pen Stabling

Pen stabling, also referred to as loafing barn, loose housing, and straw shed, is a method in which the cattle are permitted to roam at will within a bedded area and to an adjoining open space. A feeding and milking area in addition to the milkhouse are necessary adjuncts. The loafing area should be well drained and allow 70 ft^2 bedded area per animal, plus an additional 25 ft^2 per animal for a paved feeding area. Mixed manure and fresh bedding as needed to keep bedding dry may be permitted to accumulate for 3 to 6 months to a year and then removed. A more frequent cleaning interval and manure removal before the fly season are recommended. The feeding area usually requires weekly cleaning. The milking area is separated from the loafing area and should be so constructed as to be cleanable. Centrally located milking areas or centers are encouraged. It is claimed that pen stables reduce the risk of injury to cattle, are less expensive to build, save labor, and result in greater milk production since the cattle are free to browse, eat, and rest at will.

Pipeline Milker

A laborsaving device on the farm is the pipeline milker. Milk flows from the teat-cup assembly and hoses of the milking machine to a special glass or stainless steel pipeline terminating in a refrigerated bulk-milk storage tank or other container. This equipment requires special attention to assure it remains clean. Off-flavors have been attributed to pipelines. A water heater of adequate capacity and equipment for circulating detergent solution mixed with air at a velocity of at least 5 fps are needed to clean the pipeline in place. At this velocity a 10-gal can is filled in 14 sec by a 2-in. stainless steel line, in 12.5 sec by a 2-in. glass pipeline, and in 22 sec by a $1\frac{1}{2}$-in. glass pipeline. Turbulent flows give better cleaning. Other parts, including the milking machine and teat-cup assemblies, the air lines, air chambers, valves, fittings, and dead ends, are disassembled and immediately brush-cleaned by hand. Pipelines can be cleaned as follows.

1. Prerinse with lukewarm water (100 to 120° F) immediately after use until water at outlet is clear. Do not recirculate. Milk fats dissolve at about 90° F.
2. Wash for 10 min by circulating an adequate volume of a proper detergent solution (130 to 140° F or as recommended by detergent manufacturer) through the system at a velocity (5 fps) that will give turbulent flow. Use an acid detergent solution once a week if alkaline detergent is normally used. The type of washing solution and cleaning procedure should take into consideration the characteristics of the water supply. Rinse out detergent with warm water.
3. Before reuse, rinse with 170 to 180° F water measured at outlet for at least 5 min (a separate booster heater for this purpose is recommended), 75 to 90° F water containing 50 to 100 mg/l available chlorine solution, or other approved disinfectant for at least 2 min; open valves momentarily to allow contact with interior of valve. Cap the pipelines when finished.
4. If the total length of piping is less than about 50 ft, it will probably be easier and more economical to take the piping apart rather than attempt in-place cleaning. The system should be self-draining or dismantled.
5. Equipment must be thoroughly cleaned after each use and sanitized prior to reuse to assure low bacteria counts.

The Northeast Dairy Practices Council[21] recommends the following minimum water heating facilities for all pipeline installations, with water pressure at 20 to 25 psi:

[21] *Guidelines for the Installation of Milking Systems, 1978,* The Northeast Dairy Practices Council, 124 Stocking Hall, Ithaca, N.Y., 14853.

Type Heater	Storage (gal)	Recovery Rate (gal per 100° F rise)
Electric	80	18
Gas-fired	40	35
Oil-fired	30	50

Dairy Equipment  

The need for standards for milk equipment which would be acceptable to all having jurisdiction over milk quality led to the development of 3-A Sanitary Standards. The standards are developed through the collaboration of manufacturers of dairy equipment, users of the equipment, the International Association of Milk, Food, and Environmental Sanitarians, and the PHS/FDA. Anyone can participate; but no one is bound by a standard. However, voluntary compliance or acceptance of the standards by the industry, the fabricator, and the sanitarian is in the mutual interest of all. It makes possible reference to a nationally accepted standard, sanitary design and construction of equipment, reduced fabrication costs, and simplified, more efficient and effective enforcement in the interest of the consumer.[22]

Milkhouse

A milkhouse of proper size and construction is a necessary part of a dairy farm. The provision of adequate facilities for the cleansing, disinfection, and storage of utensils and milking equipment and for the refrigeration of milk to a temperature of 38° F are basic essentials. A milkhouse floor plan including details is shown in Figure 8-2. Plans and assistance are usually available from the USDA, state agricultural college, county extension service, power company, or building supply association.

Milking Parlors

Milking parlors are rooms for the milking of dairy animals in a sanitary, methodical, and efficient manner. Different milking arrangements are used, such as herringbone, individual, rotary tandem, and rotary herringbone. A milking parlor is considered when the herd size approaches 60. General features include:

1. Animals are elevated 30 to 36 in. above the operator's floor level.
2. Adequate water supply, water heating, and waste disposal facilities are required.
3. Constructed of easily cleaned, water resistant, and durable materials. Consideration given to moisture barrier and heat conductance. Vapor barrier directly under the exterior wall finish.

[22]Henry V. Atherton, "3A Sanitary Standards Their History and Development," *Food Fieldman*, (International Association of Milk, Food, and Environmental Sanitarians), June 1980, pp. 12–15.

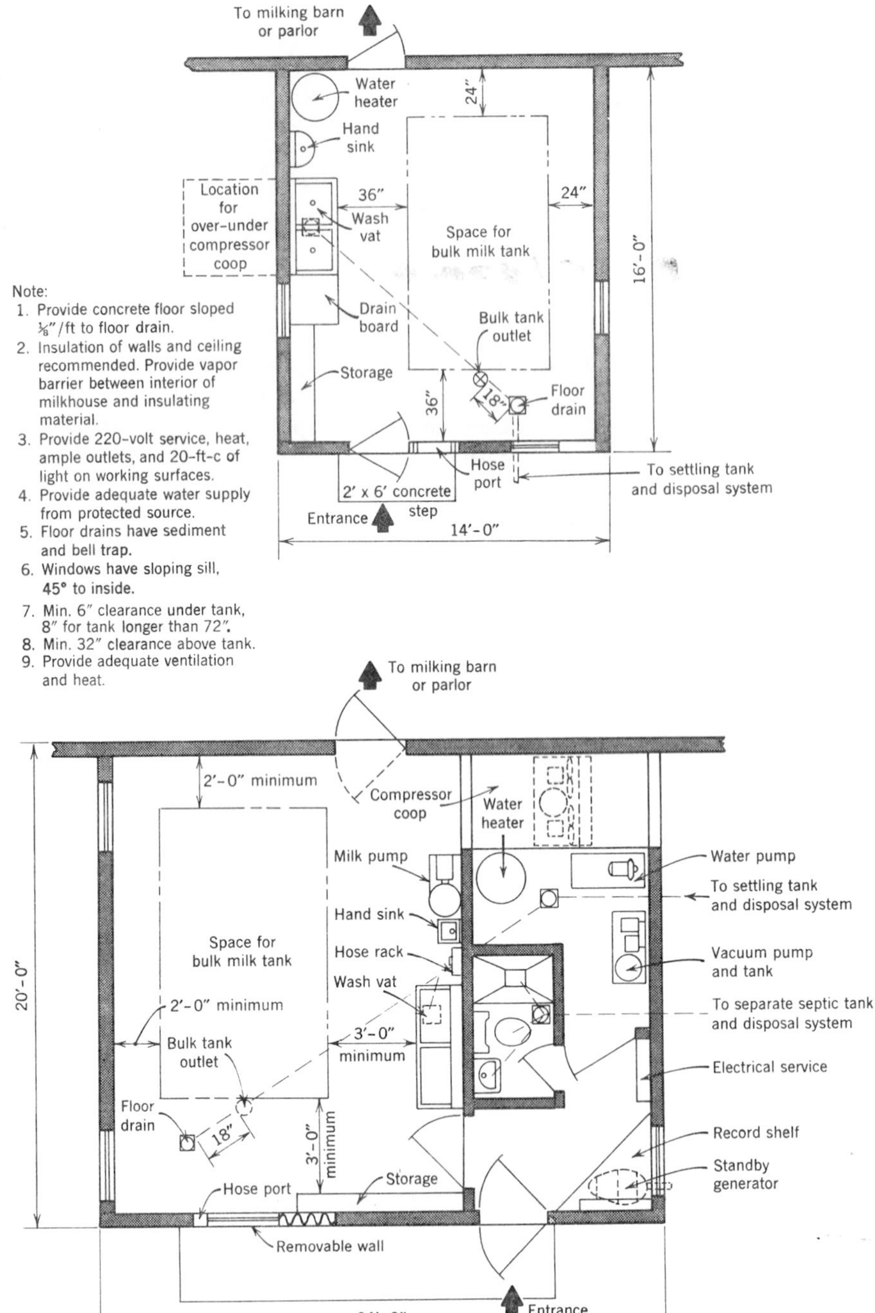

Figure 8-2 Milkhouse layouts show the arrangement of equipment on the floor plan for the small 14′ × 16′ milkhouse or large 20′ × 24′ milkroom with utility room. Size of bulk milk tank largely determines size of milkhouse. Tank sizes of 500 to 3000 gal are not uncommon. (Adapted from R. W. Guest, W. W. Irish, and R P. March, *Milkhouse for Bulk Tanks*, Agricultural Engineering Extension, Bull. 326, Cornell University, Ithaca, N.Y., p. 5.)

4. Readily accessible to dairy barn and milkhouse; expandable and sited to avoid traffic.
5. Floor sloped at least $\frac{1}{4}$ in./ft to drains at the end and in the corners of both the cow platform and the operator's pit of fixed parlors, or to drains under the cow platform from rotating platform and operator's pit.
6. Adequate wiring for all intended uses; one weatherproof 20-A grounded duplex outlet at each end of the operator's pit, with ground-fault interrupter protection; in compliance with national and local electric codes.
7. Provision of 50 ft-c at level of cow's udder plus supplemental lighting as needed.
8. Adequate ventilation for fresh air and for moisture removal; heating and cooling for operator comfort.

Various publications are available from regulatory agencies, extension services, universities, and equipment manufacturers. The Northeast Dairy Practices Council (NDPC) Guidelines are among the best.[23]

Precooling of fresh milk as it leaves the cow (96 to 100° F) can reduce the cooling load on the refrigerated bulk milk tank and help ensure rapid cooling of the fresh milk and blended milk to the desired temperature of less than 40° F. Prolonged agitation is avoided and herd size can be increased somewhat without overloading the bulk tank refrigeration system. Precooling by means of plate coolers is accomplished using well water and/or chilled water. The temperature of the milk as it leaves the cow can be reduced 10 to 40° F, depending on the plate assembly.

Bulk Cooling and Storage

Bulk cooling and storage of milk on the farm reduces handling and the possibilities for contamination, results in rapid cooling of the milk, and requires less space than cans. The cooling is accomplished in a stainless-steel-lined unit that also serves as a storage tank. It is placed 2 to 3 ft from walls or fixtures. The tank is insulated, mechanically cooled, and equipped with a motor-driven agitator, thermometer, thermostat bulb, outlet valve, and measuring stick.

A direct-expansion cooling system would require about 1 hp ($1\frac{1}{4}$ hp with air-cooled condenser) of compressor motor capacity per 50 gal of milk to be cooled at each milking. A chilled water or ice-bank system would require about $\frac{1}{3}$ hp of compressor motor capacity for each 50 gal of milk to be cooled. A can cooler would require $\frac{1}{8}$ hp per can for cooling both night and morning milk. In any case, farm-refrigerated milk-storage tanks should be designed to cool the milk to 45° F or less within 1 hr and to 36 to 38° F in 2 hr and should comply with health or agriculture department regulations.

The theoretical milk-cooling capacity of the cooling system = temperature reduction in degrees F × 8.6 × 0.93 = Btu/hr/gal. Add 15 percent to obtain recommended capacity. Milk with $3\frac{1}{2}$ percent fat and 8.6 percent nonfat solids

[23] *No. 1—Dairy Freestall Housing Systems; No. 12—Installation of Milking Systems; No. 14—Structural and Environmental Considerations For Milking Parlors; No. 15—Handling Liquid Effluents From Milking Centers; No. 26—Guidelines For Milking Parlor Types and Selection*, The Northeast Dairy Practices Headquarters, 118 Stocking Hall, Ithaca, N.Y. 14853.

has a specific gravity at 68° F of 1.033 and weighs 8.60 lb/gal. Milk has a specific heat of 0.93 Btu/lb. The temperature reduction = 90° F − 37° F = 53° F; average = 27° F. The theoretical milk cooling capacity = 27 × 8.6 × 0.93 = 216 Btu/hr/gal. Recommended capacity = 216 + 32 = 248 Btu/hr/gal in a direct-expansion bulk-milk cooling system. In an ice-bank system, 1 lb of ice takes 144 Btu to melt. To cool a gallon of milk from 90° F to 37° F will require the removal of (53 × 8.6 × 0.93) = 424 Btu from the milk, that is, the melting of 3 lb of ice.

Failure to properly cool milk in a farm bulk milk tank can lead to the production of heat-stable enterotoxin in milk from a *Staphylococcus aureus* infected herd. The resulting food poisoning in individuals drinking the milk calls attention to the value of a recording thermometer for bulk milk storage tanks for control purposes.[24] The temperature recorder can also control the cooling system and activate an alarm.

Milk is pumped out of a farm milk tank directly into a tank truck through a special hose carried by the hauler. The ends of the hose and outlet valve are sanitized by the hauler, who also rinses the farm tank thoroughly with lukewarm water after all milk is pumped out. Hot and cold water under pressure and a water hose in the milkhouse are practically essential to do a proper job. As soon as possible after rinsing, the agitator, thermometers, outlet valve plug, measuring stick, and strainers should be removed, brushed, and then washed in a warm chlorinated alkaline cleaning solution followed by a cold-water rinse. The tank's inside surfaces and covers require the same treatment. The parts removed are reassembled and the entire tank is sanitized before being placed in use again. Make a visual inspection with a strong light. Use an acid rinse if needed. Farm bulk milk haulers should be trained and certified in view of their responsibility for milk transfer, weighing, and sampling raw milk.

Transporation

Milk should be hauled in insulated trucks to the receiving station or milk-processing plant as soon as it is removed from the cooler. In this way milk that has been carefully produced will not deteriorate in quality while in transit.

The transportation of bulk milk is accomplished by insulated tank trucks designed and built to meet established standards. Equipment used to sample, fill, and empty the tank must also be of sanitary design and construction. This equipment and the transportation tank are required to be cleaned and sanitized immediately after being emptied, in a room provided for this purpose. The room should be clean, heated, drained, and provided with facilities for washing the milk tank, including an adequate supply of hot and cold water under pressure. Water supply must be of satisfactory sanitary quality and drainage disposal in accordance with local requirements and should not cause a nuisance.

[24] C. Joseph Hansberry, "Farm Tank Temperature Recording: Its Effect on Quality and Marketability of Fluid Milk," *J. Milk Food Technol.*, February 1975, pp. 105–107.

Pasteurization

The PHS/FDA[25] defines "pasteurization," "pasteurized," and similar terms to mean the process of heating every particle of milk or milk product in properly designed and operated equipment to one of the temperatures given in the following table and held continuously at or above that temperature for at least the corresponding specified time.

Temperature	Time
145° F (63° C)*	30 min
161° F (72° C)*	15 sec
191° F (89° C)	1.0 sec
194° F (90° C)	0.5 sec
201° F (94° C)	0.1 sec
204° F (96° C)	0.05 sec
212° F (100° C)	0.01 sec

The asterisks in the table indicate that if the fat content of the milk product is 10 percent or more, or if it contains added sweeteners, the specified temperature shall be increased by 5° F (3° C), *provided*, that eggnog shall be heated to at least the following temperature and time specifications.

Temperature	Time
155° F (69° C)	30 min
175° F (80° C)	25 sec
180° F (83° C)	15 sec

Other pasteurization processes may be approved by the FDA and the regulatory agency. The heat treatment should be followed by prompt cooling.

The effectiveness of pasteurization in the prevention of illnesses that may be transmitted through milkborne disease organisms has been demonstrated beyond any doubt. The continued sale and consumption of raw milk must therefore be attributed to ignorance of these facts. Pasteurization does not eliminate pesticide residues, anthrax spores, or toxins given off by certain staphylococci; but the production of toxins is nil when milk is properly refrigerated. The equipment used to pasteurize milk is described below.

Testing of pasteurization equipment and controls requires special expertise. Details are given in the *Grade A Pasteurized Milk Ordinance* (PMO).[25]

Holder Pasteurizer

Holder pasteurizers are referred to as batch, pocket, and continuous flow-type pasteurizers. They have a capacity of 100 to 500 gal or more. In the batch-type pasteurizer (Figure 8-4) the milk is heated in the pasteurizer vat or tank and held

[25] *Grade A Pasteurized Milk Ordinance*, 1978 Recommendations of the PHS/FDA, DHEW; Pub. 229, Washington, D.C., p. 21.

at 145° F for 30 min, and then the milk is run over a surface cooler and bottled. The milk may be precooled in the pasteurizer vat to about 115° F and then run over the surface cooler. The pocket, or multiple tank-type pasteurizer is no longer in general use. In the continuous flow-type pasteurizer, which consists of a series of tubes, heated milk is pumped by a calibrated pump from the lowest tube and passed out at the top from the highest tube. All tubes are jacketed.

High-Temperature Short-Time Pasteurizer

The high-temperature short-time (HTST) pasteurizer uses accurate and dependable controls, including the flow-diversion valve or device, calibrated pump, and heat-exchanger equipment to assure proper pasteurization of large quantities of milk in a short time. In this type of pasteurizer a large number of plates are carefully clamped together, with water and milk on alternate sides. It is compact and makes possible heating, holding, regenerating, and cooling all in one unit. HTST systems include Milk-to-Milk Regeneration Homogenizer Upstream From Holder, Milk-to-Milk Regeneration-Surface Cooler, Milk-to-Milk Regeneration-Booster Pump, and Milk-to-Milk Regeneration Homogenizer, and Vacuum Chambers Downstream from Flow Diversion Device. See Figures 8-3, 8-4, and 8-5.

In-the-Bottle Pasteurizer

In-the-bottle pasteurization is a method for heating milk in the final bottle with hot water or steam to accomplish pasteurization as in the holder-type, followed by cooling. Indicating and recording thermometers, circulation of the hot water, auxiliary heating of the air above the milk level, and holding of the bottle and milk inside the bottle at the proper temperature are frequently wanting, making this method of pasteurization questionable unless carefully supervised. In an emergency raw milk can be rendered safe to drink if heated in a water bath to a temperature of 165° F and then immediately cooled. This method is no longer used.

Ultra-High Temperature Pasteurizer

Ultra-high temperature (UHT) pasteurization of milk and milk products is heat treatment at a temperature of 191 to 212° F, with a holding time of 1.0 to 0.01 sec as previously noted. Timing in the holding tube and speed of response of the recorder-controller flow-diversion valve system are critical, requiring sensitive equipment and competent supervision. This process requires built-in safety factors to ensure adequate heat treatment. Because of variations of particle velocity in laminar flow, the holding time or tube length is generally made twice the calculated hold.

Ultra-Pasteurized

A milk or milk product is considered ultra-pasteurized when it has been thermally processed at or above 280° F (138° C) for at least 2 sec, either before or after packaging, so as to produce a product which has an extended shelf life

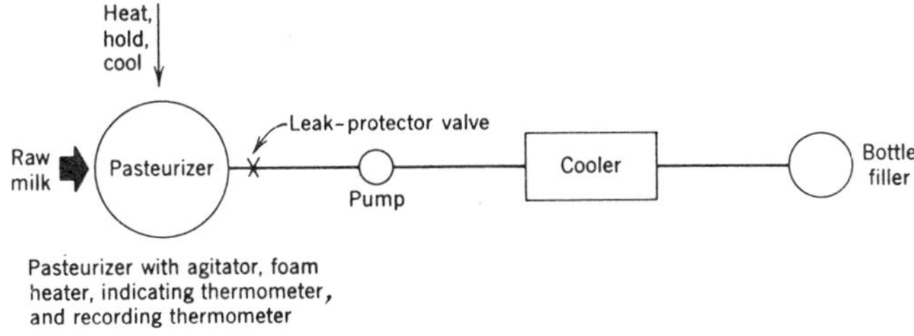

Batch-type pasteurizer

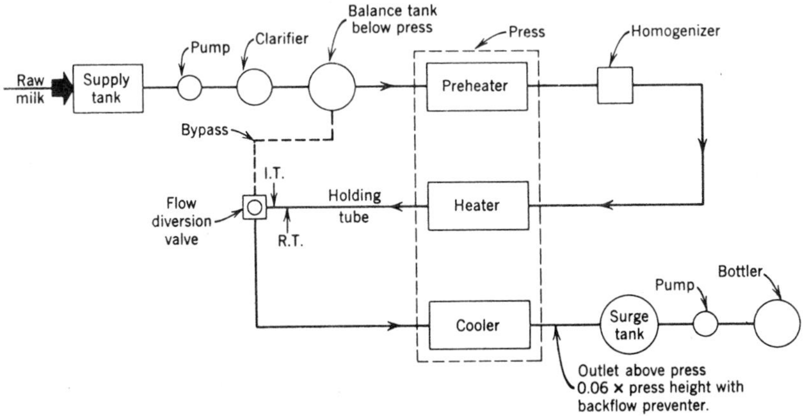

High-temperature short-time pasteurizer

Figure 8-3 Simplified pasteurizer flow diagrams. Testing of holding time involves the following steps. Complete details are given in the *Grade A Pasteurized Milk Ordinance*, 1978 Recommendations.

1. I.T. = Indicating thermometer. R.T. = Recording thermometer. Holding tube slopes up at least $\frac{1}{4}''$ per ft. Pump may be substituted for homogenizer. Flow-diversion valve is set to bypass milk when the temperature of the milk drops to $161\frac{1}{2}°$ F, and to pasteurize milk when the temperature of the milk raises to 162° F.
2. Capacity of holding tube $= \dfrac{\pi \times D^2}{4} \times L \times 7.48 =$ "A" gal.

   D = inside dia. of tube in ft; L = length of tube in ft.
3. Theoretical time to fill a 10-gal can and provide 16-sec hold $= x$; $\dfrac{16}{A} = \dfrac{x}{10}$; $x = \dfrac{160}{A}$,
4. Test holding time by means of salt conductivity or improved test for precision.
5. Holding time for milk $= \dfrac{1.032(TMw)}{Ww}$, in which:

   1.032 = specific gravity for milk with 4% fat at 68° F; 1.013 with 20% fat; and 0.995 with 40% fat.
   T = average holding time for water in sec.
   Mw = average time to deliver a measured weight of milk in sec.
   Ww = average time to deliver an equal weight of water in sec.

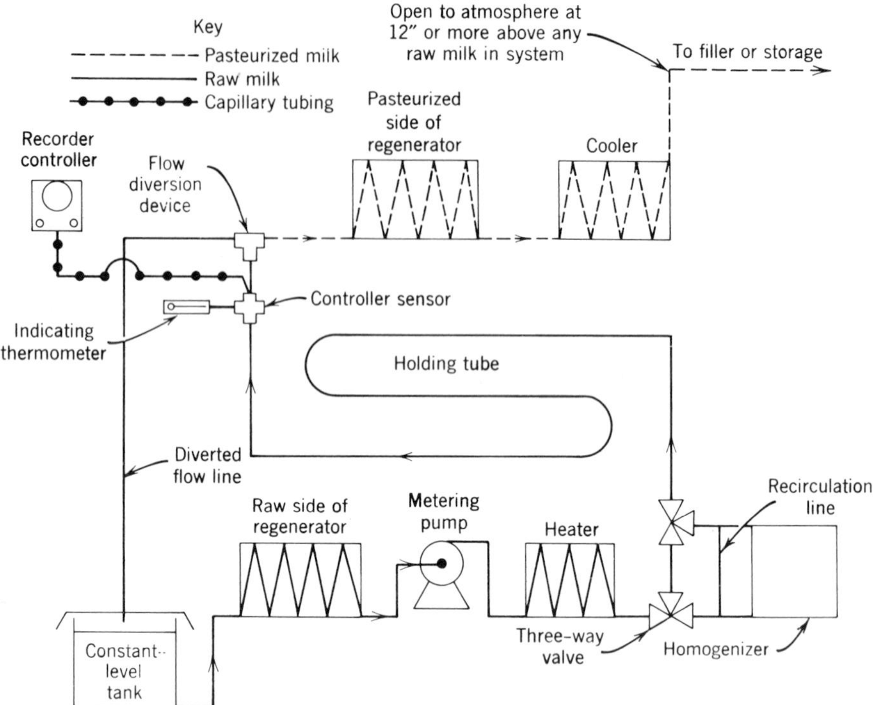

Figure 8-4 Milk-to-milk regeneration pasteurizers. Homogenizer upstream from holder. From *Grade A Pasteurized Milk Ordinance*, 1978 Recommendation of the PHS/FDA, DHEW, Pub. 229, Washington, D.C.

under refrigerated conditions.[26] The product has a limited shelf life when unrefrigerated.

Cooler

Following pasteurization the milk is promptly cooled below 45°F to keep the bacteria count from materially increasing. This is accomplished by a cooler. A common type is the surface- or external-tubular cooler. It may be a single series of tubes or a cabinet tube-type. A surface cooler is usually divided into an upper and lower section. The upper section is generally cooled by water and sometimes chilled raw milk flowing inside the tubes. The lower section usually circulates brine, ammonia, ice water, or "sweet" water that has been cooled. The warm milk is distributed by means of a perforated pipe or trough over the top of the cooler and is cooled as it flows in a thin sheet over the surface of the cooler.

Another type of cooler is the internal-tubular cooler. It may be a single series of

[26]"Part 18 Milk and Cream, Definitions and Standards of Identity," 21CFR 18.1, *Food Drug Cosmetic Law Rep.*, December 16, 1974, p. 621.

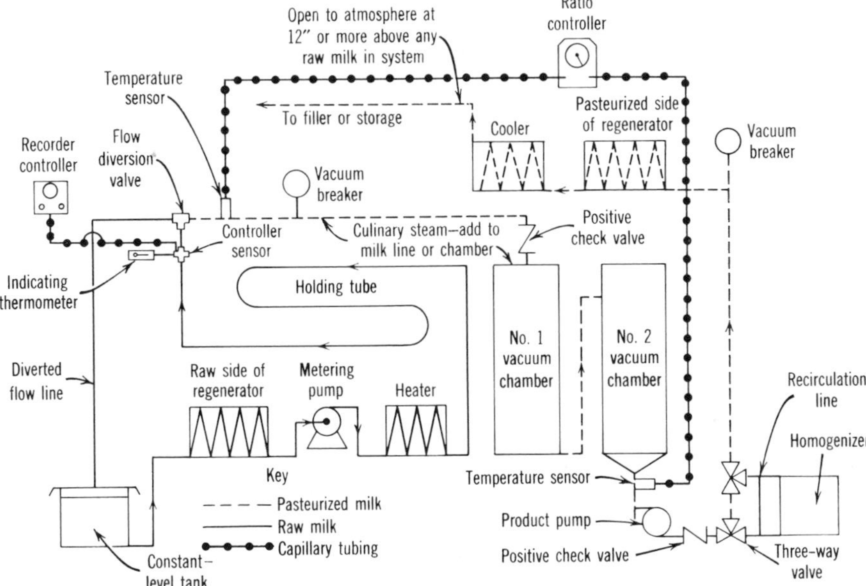

Figure 8-5 Milk-to-milk regeneration pasteurizers. Homogenizer and vacuum chambers downstream from flow diversion valve. From *Grade A Pasteurized Milk Ordinance*, 1978 Recommendation of the PHS/FDA, DHEW, Pub. 229, Washington, D.C.

tubes or a cabinet tube type. In this method there are two concentric tubes; the milk flows in the center tube and the refrigerant flows in the outside tube.

When large volumes of milk are to be cooled, a plate heat exchanger or regenerator is generally used (Figures 8-3, 8-4, and 8-5). Hot pasteurized milk on one side of the plate is cooled by cold raw milk, water, or brine on the opposite side of the plate.

The cooled milk is conducted to a filler and capper, where the milk is bottled or packaged. The milk containers are filled by means of a rotary-gravity or vacuum-type filler. Empty containers are fed automatically from a conveyor to the filler. Bottled milk and filled cartons are placed in cases that are stored in a cooler until delivered. In very small operations a hand-operated bottle filler and capper is used.

Cleaning and Sanitizing Milk Plant Equipment[27]

The maintenance of high quality milk which will continuously meet regulatory bacteriological standards requires that all equipment coming in contact with milk be cleaned and sanitized after each use.

A fundamental step in the cleaning process is prompt rinsing of all equipment

[27] *Guidelines for Cleaning and Sanitizing in Fluid Milk Processing Plants*, Bull. NDPC 29, April 1978, Northeast Dairy Practices Council (NDPC), 124 Stocking Hall, Ithaca, N.Y. 14853.

and parts after each use with warm potable water at approximately 100° F. This will remove most of the soil and minimize the formation of fat and protein organic films and inorganic films such as milk-stone and iron which harbor and permit the growth of bacteria and also reduce heat transfer where this is a factor. The wash step that follows requires the use of a suitable detergent, at the proper concentration and temperature, for a sufficient time to remove remaining soil. The detergent manufacturer's representative can provide advice for specific applications such as manual cleaning, out-of-place cleaning in a recirculation tank, in-place cleaning, spray devices, and high pressure-low volume. Although equipment may be cleaned-in-place, fittings and parts such as valves, gaskets, siphon breakers, and small lines need to be taken apart and cleaned by hand. The next step is the post-rinse to thoroughly clean and rinse out remaining soil and detergent, preferably with acidified water. The assembled equipment is then sanitized just prior to being placed in operation—usually the next day—with hot water for not less than 5 min at 170° F at the end of the system. Other methods include steam for at least 5 min at not less than 200° F at the end of a closed system, a chemical solution at a temperature of 75 to 95° F containing hypochlorite and at least 50 mg/l available chlorine, or an iodophor containing at least 2 to 5 mg/l available iodine and, at 50 to 70° F, a chlorine and bromine compound, a quaternary ammonium compound,* or an acid sanitizer. Chloramine compounds are slow-acting and their use is not recommended. Quaternaries are affected by water hardness, but are noncorrosive and less affected by organic matter. The use of hypochlorite (200 mg/l available chlorine) is usually preferred. Compounds containing phenols or heavy metal salts such as mercury, silver, lead, zinc, copper, or chromium are not acceptable for use in milk and food processing plants. Ultraviolet radiation and hydrogen peroxide have special auxiliary applications but must be used with care.

Detailed information on the cleaning and sanitizing of equipment is available from the NDPC,[27] the regulatory agency, extension service, and detergent manufacturers.

Bottle Washer

Although the bottle washer is not generally used, the trend to returnable bottles suggests that its operation is understood. Bottle washers are of various types, depending usually on the total number of bottles to be washed.

1. At small plants bottle washing is a manual operation. The bottles are soaked, washed by hand on a revolving brush, rinsed, and then sanitized by steam or a hot-water spray in the case.
2. In the larger plant a case washer is usually used. As the name implies, bottles placed in the case are individually subjected to water-pressure cleansing by a washing solution, a rinse, and then a final sanitizing rinse.

*The FDA considers food products containing residues of quaternary ammonium compounds to be adulterated.

3. At large plants a soaker-type washer is used. Bottles are placed in the machine slot or pocket and subjected to a prerinse, soaking in alkali solution for several minutes, an inside and outside alkali spray, sometimes automatic brushing, then a final rinse. The highest temperature reached is about 160° F, which is gradually reduced. The final rinse is fresh cold water, which usually contains a chemical sanitizer.

4. The bottle-washing solution used frequently determines the effectiveness of the entire washing operation. A good washing compound dissolves dried milk and foreign matter, rinses easily, and is an effective disinfectant. Experience shows that sodium hydroxide (caustic soda) is the compound of choice for machine use. It is used as a 2 to 3 percent solution. In hard-water areas the water should first be softened or a suitable detergent used. Addition of $\frac{3}{4}$ lb of 76 percent commercial caustic soda to 1 ft^3 of water, or $7\frac{1}{2}$ gal, will make a 1 percent solution of sodium hydroxide.

Cooler and Boiler Capacity

Adequate power and refrigeration capacities are necessary to the processing of milk. The advice of persons experienced in the design and operation of these units should be obtained when constructing a new plant or when making major modifications. A table of small boiler and refrigeration capacities for different size plants is given below as a general guide.

Coolers for the storage of milk are of the walk-in type. Refrigeration is obtained by coils on walls circulating brine, by direct expansion of ammonia in coils on walls, and by unit coolers equipped with coils and a fan. The unit cooler keeps the room drier and provides better air circulation. Freon, carbon dioxide, methyl chloride, sulfur dioxide, and ammonia are common refrigerants.

Plant Capacity (gal of milk)	Steam Boiler Rated Horse-power Capacity	Refrigeration Unit Rated Capacity tons*
150	15	5
300	27	11
750	42	20
1500	72	23
2500	92	32

Quality Control

Quality control involves herd health, milk handling, refrigeration, transportation, processing, and distribution. Field and laboratory testing coupled with inspection, supervision, education, surveillance, enforcement, and evaluation are the major methods used. Surveillance of bulk milk sampling from the tank to the laboratory is essential so that interpretation of laboratory results may be valid.[28]

*A ton equals 288,000 Btu in 24 hr. This is also the amount of refrigeration accomplished by the melting of 1 ton of ice.

[28] Raymond A. Belknap, "Procedures for Surveillance of Bulk Milk Sampling," *J. Milk and Food Technol.*, May 1976, pp. 362–366.

The tests used to control the quality of milk are explained in detail in *Standard Methods for the Examination of Dairy Products.*[29] The major tests are discussed here. Raw milk quality is determined by temperature; sediment; odor and flavor; appearance; reduction time; direct microscopic counts, including clumps of bacteria, leukocytes, and streptococci; standard plate counts, abnormal milk tests; freedom from antibiotics; and thermoduric determination. Tests for brucellosis and animal health are also made. Pasteurized milk quality is indicated by the standard plate count, direct microscopic count, phosphatase test, coliform test, and taste and odor tests. Other common tests are for butterfat, total solids, and specific gravity. Quality standards and sampling frequency are summarized in Tables 8-4 and 8-5.

Temperature. Milk should be promptly cooled to 45° F or less within 2 hr after milking* and maintained at that temperature until delivered to the receiving station or pasteurizing plant, unless delivered within 2 hr after milking is completed. The importance of proper cooling on the rate of bacterial growth is demonstrated by the results of studies at the Michigan Agricultural Station on freshly drawn milk. Milk having an initial count of about 4300 bacteria/ml decreased to 4100 at 40° F and increased to 14,000 at 50° F and to 1,6000,000/ml at 60° F after 24-hr storage. It is apparent therefore that milk that reaches the pasteurizing plant or receiving station 24 hr after milking, at a temperature above 50 to 60° F, will probably have a high bacteria count. Although fresh raw milk has bacteriostatic properties that help retard bacterial growth, the duration of this factor is variable.

The *sediment test* shows the amount of extraneous material in milk but will not show dissolved material. A pint of milk from the bottom of an unstirred 40-qt can is strained through a Lintine or similar disc of absorbent cotton. When milk in a farm bulk-milk tank, plant storage tank, or transport tank is to be tested, a 1-gal, well mixed sample warmed to 90 to 100° F is collected. The color of the stain produced, from yellow to brownish black, and particles retained are a simple visual indication of the amount of dirt and abnormal substances in the milk. The test and its interpretation is given in *Standard Methods*. Failure to properly wash a cow's udder just prior to milking is the common cause of high-sediment test results. A clarified or strained milk will not show the dirt, hence the test may only encourage straining of the milk at the farm rather than cleaner milking procedures. However, other tests will also reveal insanitary practices. The test is sometimes used on processed milk.

Odor, flavor, and appearance are physical tests usually made at the receiving station or at the bulk-milk tank at the same time sediment tests are made. Off-odor or off-taste is more pronounced at a temperature of 60 to 70° F or

*Provided, that the blend temperature after the first milking and subsequent milkings does not exceed 50° F.

[29] *Standard Methods for the Examination of Dairy Products*, 15th ed., Am. Pub. Health Assoc. 1015 Fifteenth Street, Washington, D.C., 1981. Also *Official Methods of Analysis of the Association of Official Analytical Chemists.*

Table 8-4 Maximum Allowable Temperature, Bacteriological, and Chemical Standards for Milk and Milk Products

Product	Temperature	Bacterial Limit	Coliform Limit	Antibiotics[a]	Phosphatase	Somatic Cell Count
Raw milk for pasteurization	45° F[b]	100,000[c] 300,000[d]	—	None	Not applicable	1,500,000
Pasteurized milk and milk products[e]	45° F	20,000[f]	10[g]	None	<1 μg[h]	—
Pasteurized cultured products	45° F	Not applicable	10	None	<1 μg[h]	—
Sterilized milk and milk products	—	<1	<1	None	Not applicable	—

Notes:

1. Bacterial, coliform, and phosphatase results are reported as per milliliter. The bacterial limit may be a standard plate, direct microscopic count (DMC), or somatic cell count.
2. Rinse of empty milk bottle, carton, or other container shall be free of coliform organisms and contain less than 1 bacteria per ml. Swab counts from 5 different 8-in.^{2} areas shall be free of coliforms and not exceed 250 colonies/40 in.^{2}.
3. Sediment. Sample from off-bottom unstirred 40-qt can of milk shall contain less than 2 mg of sediment when compared to photographs of standards. (Obtainable from Photography Division, Office of Information, U.S. Dept. of Agriculture, Washington, D.C.) Sample of stirred milk from a can, weigh tank, farm bulk-milk tank, plant storage tank, or transportation tank shall not exceed the 1-mg standard.
4. All milk samples must be representative, carefully collected using sterile technique and equipment, shipped in a proper container which will prevent contamination and will maintain the sample at proper temperature for examination in an approved laboratory. Sample must arrive at 32 to 40° F and be examined within 36 hr of collection.
5. When one or two of the last four consecutive bacteria counts, somatic cell counts, coliform results, or cooling temperatures taken on separate days exceeds the standard, send written notice. When three of last five exceed standard, suspend permit and withdraw product.
6. When phosphatase standard is exceeded, immediately determine cause. Milk involved shall not be sold. With the Scharer Rapid Phosphatase Test, a value of one microgram or more of phenol per ml of milk, milk product, reconstituted milk product, or cheese extract would indicate improper pasteurization or contamination with unpasteurized products. See *Standard Methods for the Examination of Dairy Products* for other tests.
7. When an antibiotic or pesticide residue test is positive the cause shall be determined and corrected. An additional sample shall be collected and no milk shall be offered for sale until subsequent sample is free of antibiotic or pesticide residues or below the actionable level established for such residues. The *Bacillus stearothermophilus* test is reported to be a more sensitive test to detect the presence of antibiotics in bulk milk tanks. Pesticide residues of up to 0.05 ppm in whole milk and 1.25 ppm on a milk-fat basis in manufactured dairy products are permitted under FDA tolerance levels. The maximums apply to DDT and its chemical degradation products DDD and DDE or any combination.

[a] By the *Bacillus subtilis* method or equivalent for individual producer milk; by *Sarcina lutea*[9] Cylinder Plate Method or equivalent for commingled milk.
[b] Provided that the blend temperature after the first and subsequent milkings does not exceed 50° F.
[c] Individual producer milk.
[d] Commingled milk. Counts of less than 10,000 are readily obtainable in a well run operation. Manufacturing-grade milk is usually of poorer quality, which influences the quality of the final product (Class I, 500,000 DMC per ml; Class II, 500,000 to 3 million; Class III, greater than 3 million).
[e] Except cultured products but including imitation milk.
[f] If greater than 10,000 review handling prior to pasteurization.
[g] Shall not exceed 100 per ml in bulk milk transport shipments.
[h] By Scharer Rapid Method or equivalent as micrograms of phenol per milliliter of milk.

Table 8-5 Sampling Frequency for Milk and Milk Products

Milk or Milk Product	Number of Samples
Raw milk for pasteurization from each producer[a]	At least four during any consecutive six months
Raw milk for pasteurization from each milk plant prior to pasteurization[a]	Same
Pasteurized milk from each milk plant[b]	Same
Each milk product from each milk plant[b]	Same
Milk and milk products from retail stores[b]	As required by the regulatory agency

[a] Perform bacterial counts, somatic cell counts, and cooling temperature checks. Also antibiotic tests on each producer's milk or on commingled raw milk at least 4 times during any consecutive six months; all sources tested if results positive.
[b] Perform bacterial counts, antibiotic tests, coliform determinations, phosphatase tests, and cooling temperature checks.

higher. Experienced inspectors can make very rapid and accurate determinations by sight and smell on the quality of a milk. Sour milk, dirt, and odors associated with mastitis, improper cooling, dirty utensils, horses, consumption of leeks and other weeds, or a disinfectant are easily detected. Feed flavors can be prevented by the feeding of cows after milking or at least 5 hr before milking; removal of cows from a lush pasture 3 hr before milking; elimination from pastures of wild onions, leeks, skunk cabbage, and certain mustard weeds; and storage of silage or strong-smelling feed away from the cows. Barny or musty flavors and odors can be controlled by keeping the cows, milkers, and stables clean. Good ventiliation and dry conditions without dust are important. Salty flavors are due to the use of milk from cows infected with mastitis or from cows being dried off (strippers). This milk should be discarded. Rancid milk is caused by use of milk from stripper cows, milk from cows late in lactation, slow cooling of milk, and especially excessive agitation. A malty flavor is caused by high bacteria count, dirty utensils and cows, poor cooling, and slow cooling. Very high bacteria count, dirty utensils, and slow or poor cooling cause sour milk. An oxidized or cardboard flavor is caused by milk coming in contact with copper, copper alloys, or iron equipment; exposure of milk to sunlight; copper or iron in water supply; and possibly milk from special cows. Medicinal flavors are due to certain medications for the teats, creosote-based disinfectant barn sprays, and insect sprays used just before or during milking. Clean udders and milk handling equipment, together with rapid cooling and good barn ventilation will eliminate the major causes of off-flavors and off-odors in milk.

Reduction tests. When it is not possible to collect and examine a large number of samples for bacterial examinations, an indication of the sanitary condition of the milk can be obtained by means of the methylene blue *reduction test* or the *resazurin test*.

In the methylene blue reduction test 1 ml of fresh methylene blue thiocyanate solution is added to a 10-ml sample of milk and incubated at 36° C ± 1° C (98.6° F). Observations for loss of color are made after 30 min and at hourly intervals thereafter, and results are recorded as MBRT in whole hours. A proposed USDA standard for manufacturing-grade milk sets limits of $2\frac{1}{2}$ and $4\frac{1}{2}$ hr. A reduction time of 6 to 8 hr is considered acceptable for Grade A milk.

In the resazurin reduction test, resazurin dye is used in place of methylene blue. The reduction time to 5P$\frac{7}{4}$ Munsell color end-point (triple reading test requiring 3 hr) should not be less than $2\frac{3}{4}$ hr for Grade A milk. A proposed USDA standard for manufacturing-grade raw milk establishes reduction time limits of $1\frac{1}{2}$ and $2\frac{1}{4}$ hr. Advantages of the resazurin test are that it takes less time to perform; the presence of leukocytes is indicated, although not reliably; and color readings are more distinct.

Comparisons between the reduction time and microscopic and plate counts have been attempted, but exact agreement is not expected. In general the time for a sample to change from blue to white with the methylene blue test, or purple to pink with the resazurin test, is inversely proportional to the bacterial content of the sample when incubation starts. The resazurin test must be used on fresh milk, not over 24-hr old. The reduction tests are not especially suitable on low-count, well-cooled milk, but are effective in detecting poor-quality milk.

The *direct microscopic count* (DMC) tells the number of isolated bacteria and groups of bacteria in stained films of milk dried on glass slides, as determined with the aid of a compound microscope. It is a rapid test used on small batches of raw milk to show the types of bacteria present and the possible source as dirty utensils (bacterial clumps), infected udders, or poor cooling (bacteria in pairs) provided the source, age, and temperature of the milk are known. On small volumes of pasteurized milk it is possible to detect poor-quality raw milk, the presence of leukocytes and sometimes thermoduric, psychrophilic, thermophilic, and other types of bacteria that do not normally grow on standard agar plates, and contamination after pasteurization. Generally most dead bacteria disintegrate within several hours after pasteurization; those that remain do not stain well, and hence only a few of the dead bacteria are counted when the direct microscopic count is made. *Streptococcus lactis* in pairs, three, or double pairs of oblong cocci generally indicate poor cooling or unclean equipment. Masses of bacteria indicate unclean utensils, milking machines, or tubing.

This test is of limited value when made on mixed milk from bulk-milk tanks in view of the large dilution and generally effective cooling. The DMC method should not be used on very low-count milk.

The *direct microscopic somatic cell count* (DMSCC) is a more precise test than the DMC. It gives the total leukocyte and epithelial cells in a sample of raw milk. A leukocyte cell count over 500,000/ml or a total cell count of 1,000,000/ml, together with long-chain streptococci, gives strong indication of infected udder and effects of mastitis. Enforcement action is usually based on a somatic cell count of 1,500,000/ml or more.

Screening tests. There are various screening tests for the detection of abnormal milk.[30] Tests used are the California Mastitis Test (CMT), the Wisconsin Mastitis Test (WMT), the Modified Whiteside Test (MWT), the Catalase Test, and the Electronic Somatic Cell Count. However, only the DMSCC should be used as a basis for legal action and possibly the shutting off of a farm milk supply. The CMT, WMT, and MWT can be used as field tests. Elimination of the cause of high results includes milking hygiene, milking machine functioning control, dipping of teats in 4 percent hypochlorite solution after milking, and penicillin treatment of all quarters at the time of drying off. Veterinary assistance may be needed to bring difficult problems under control, especially when the leukocyte count exceeds 1,500,000/ml.

The *standard plate count* (SPC) shows the approximate number of bacteria and clumps of bacteria that will grow in 48 hr on a standard medium held at a temperature of 32° C ± 1° C. Wide variations are common. The plate count is especially suited to products having low bacterial densities and as a measure of the bacterial quality of certified milk and pasteurized milk and milk products, except fermented milk products. It is also recommended for process sampling. High counts on freshly pasteurized milk suggest that thermoduric bacterial counts be made on producer samples to learn the source. Studies show that the SPC is superior to the DMC in detecting poor-quality raw milk and hence is the method of choice, particularly with bulk-tank milk. But excessive numbers of psychrophilic bacteria are not detected by the routine plate count. Incubation at a temperature of 7° C ± 1° C (44.6° F) for 7 to 10 days is suggested. For these reasons, an improved test is needed to measure the quality of both raw and pasteurized milk.

A preliminary incubation count of raw milk at 55° F for 18 hr can give an indication of the quality of the milk at a farm. Counts greater than 1,000,000/ml suggest dirty cows, poor udder-washing practices, slow cooling or storage temperatures above 40° F, failure to clean equipment twice each day, failure to sanitize equipment before use, or a contaminated water supply.[31] The filler may be a major source of contamination in pasteurized milk.

With the *Coliform test* practically all raw milk will show the presence of the coliform group of organisms, usually less than 100/ml in high-quality milk. Where pipeline milkers and farm bulk-milk tanks are used, the coliform count is a measure of utensil sanitization, udder cleanliness, and milking hygiene. Properly pasteurized milk will usually show the absence of coliform organisms. A positive test for coliform organisms in pasteurized milk greater than 10/ml (should be less than 3) is an indication of improper processing or excessive contamination following pasteurization by improperly cleaned and disinfected equipment, utensils, or dripping of condensate into pasteurized milk. Dust, exposure of the pasteurized milk to the air, flies, or other insects, and poor

[30] W. A. Gordon, H. A. Morris, and V. Packard, "Methods to Detect Abnormal Milk—A Review," *J. Food Prot.*, January 1980, pp. 58–64.

[31] Sidney E. Barnard, "Getting good preliminary-incubation counts," *J. Food Prot.*, October 1979, p. 836.

storage may also contribute coliforms. Some authorities claim that strains of coliform organisms exist that are heat-resistant and are not completely destroyed by pasteurization.

The *phosphatase test* shows whether milk has been properly pasteurized and whether it has been contaminated with raw milk after pasteurization. Phosphatase is an enzyme present in all raw cow's milk and is almost completely inactivated by proper pasteurization. The enzymes surviving pasteurization release phenol from disodium phosphate substrate. Upon the addition of 2, 6 dichloroquinonechloroimide (CQC), the liberated phenol reacts with CQC, producing an indophenol blue color, the intensity of which depends on the amount of enzyme present and hence the degree of pasteurization or raw milk added. In the Scharer rapid phosphatase test, a reading of 1 microgram (μg) or more of phenol per milliliter of milk, milk product, reconstituted milk product, or cheese extract is an indication of improper pasteurization or contamination with raw milk or milk product. Other methods are described in *Standard Methods*. Sale of underpasteurized products should be prohibited.

A false positive phosphatase test may be obtained on HTST or UHT pasteurized milk or milk products such as chocolate milk, cream or other high-fat products, and old cream having a high bacteria count, particularly when not continuously or adequately refrigerated. Differentiation of *reactivation*, from residual phosphatase, is explained in *Standard Methods*. Sterile milk products that have been allowed to warm up may require microbiological tests to determine the quality.

An overpasteurized milk could conceivably have added to it a very small quantity of raw milk that would not be detected by the phosphatase test; but this would undoubtedly introduce coliform organisms that would be detected by the coliform test. Pathogenic bacteria likely to be found in raw milk are destroyed more rapidly than the phosphatase enzyme.

The phosphatase test is the best single indicator of the safety of milk. When combined with the coliform test and bacteria count, a fairly complete indication is obtained of the sanitary quality of milk.

Psychrophiles or *cryophiles* are a species of bacteria that can grow within the approximate range of 35 to 50°F. Raw milk and cream are particularly susceptible if stored for 2 or 3 or more days. Psycrophiles are introduced into the milk by organisms (Pseudomonas) on dirty equipment, unclean hands, or in a contaminated water supply used to rinse the equipment. Milk stored at a temperature below 40°F will retard the growth of psychrophiles; higher temperatures will favor their growth, depending on genera and species. Pasteurization will normally destroy most psychrophiles present in raw milk. Their presence in pasteurized milk is generally an indication of postcontamination, but may also be due to an excessive number of psychrophiles in the raw milk, inadequately cleaned and sanitized contact surfaces, or use of a contaminated water supply. They also cause flavors in milk and poor keeping quality. This becomes important in the holding of raw milk and every-other-day milk delivery. Coliform microorganisms grow well but slowly with time in refrigerated pasteurized milk. Proper cleaning and sanitizing of equipment in the pasteurization plant, protection of the equipment from contamination, use of a water supply of satisfactory sanitary quality, and milk storage below 40° F will control psychrophiles. Use of a chlorinated wash-water supply containing 5 to 10 mg/l available chlorine is effective in controlling these bacteria.[32]

Thermoduric bacteria withstand pasteurization at 145° F and 161° F. They grow best at a temperature of 70 to 98° F. Milk that is not properly refrigerated at the farm would permit the growth of thermodurics. The cow's udders, improperly cleaned utensils and milking machines, feed, manure, bedding, and dust are sources of thermoduric bacteria. Preheating equipment, pasteurizers, bottles, and piping that have not been properly washed and sanitized can harbor thermoduric bacteria.

When high SPC are reported on pasteurized milk or cream, laboratory pasteurization of individual producer raw milk samples contributing to a batch will reveal the presence and producer source of thermoduric bacteria.

Thermophilic bacteria grow at a temperature of 113 to 158° F. Their optimum temperature for growth is 131° F but they can grow at 98.6° F or lower. The presence of thermophiles in large numbers may be due to repasteurization of milk or cream, prolonged holding of milk or cream in vats at pasteurization temperatures, stagnant milk in blind ends of piping at pasteurization temperatures, continuous use of preheaters, long-flow holders or vats for more than 2 to 5 hr without periodically flushing out equipment with hot water, passage of hot milk through filter cloths for more than 1 to 2 hr without replacing cloth, residual foam on milk that remains in vats when emptied at end of each 30-min holding period, and growth of thermophiles in milk residues on surfaces of pasteurizing equipment.

Mesophilic bacteria grow at a temperature of 59 to 113° F. They prefer a temperature of 98.5° F. Pathogenic bacteria are in this group.

Agglutination-type tests including the milk ring test are used to determine the presence of brucellosis. The ring test is made on samples of mixed milk from a herd, usually as delivered to the receiving station. A positive milk ring test can then be followed up by individual blood testing of the infected herd to determine the responsible animals. This makes possible more effective use of the veterinarian's time. The milk ring test will detect one infected animal in a herd of 40 when a sample of the pooled milk is examined.

Antibiotics are widely used to treat mastitis and other dairy cattle diseases, also for prophylaxis, and for growth promotion. They include penicillin, ampicillin, chloramphenicol, streptomycin, and tetracycline. Farm milk containing antibiotics and salmonellae which is not pasteurized, inadequately pasteurized, or contaminated after pasteurization presents a public health hazard when used as fluid milk or in cheese. Some individuals are sensitive to some of the drugs. In addition, antibiotics interfere with the growth of bacteria needed for the processing of cheese and cultured products, but do not affect salmonellae.

[32] J. C. Olson, Jr., R. B. Parker, and W. S. Mueller, "The Nature, Significance, and Control of Psychrophilic Bacteria in Dairy Products," *J. Milk Food Technol.*, **18**, No. 8, 200–203 (August 1955).

Manufacturer's directions should be carefully followed to assure proper antibiotic usage, dosage, and segregation of treated animals.

MILK PROGRAM ADMINISTRATION

The milk industry and the regulatory agencies have a joint responsibility in ensuring that all milk and milk products consistently meet the standards established for protection of the public health. Inspection duplication should be avoided; instead there should be a deliberate synergism of effort. With proper planning and cooperation the industry, local, state, and federal systems can actually strengthen the protection afforded the consumer. The role of industry and official agencies to accomplish the objective stated is described below.

Certified Industry Inspection

Industry quality-control inspectors are qualified by the official agency (usually health or agriculture) based on education, experience, and examination to make dairy farm inspection pursuant to the milk code or ordinance. Certificates are issued for a stated period of time of 1 to 3 years, may be revoked for cause, are renewed based on a satisfactory work record, and may require participation in an annual refresher course. Copies of all inspections, field tests, veterinary examinations, and laboratory reports are promptly forwarded to the official agency or to an agreed-on place and kept on file at least one year.

Cooperative State PHS/FDA Program for Certification of Interstate Milk Shippers (IMS)

The voluntary federal-state program is commonly referred to as the IMS Program. A state milk sanitation rating officer certified by the FDA makes a rating of a milk supply. Included are producers, receiving and transfer stations, and plants. The name of the supply and rating is published quarterly by the FDA. If the milk and milk products are produced and pasteurized under regulations that are substantially equivalent to the *Grade A Pasteurized Milk Ordinance*, and are given an acceptable milk sanitation compliance and enforcement rating of at least 90 percent, they may be shipped to another area of jurisdiction that is participating in the IMS Program without further inspection.

The procedures for rating a milk supply are carefully designed with detailed instructions to be followed by the industry and the rating officer.[33] Independent evaluations are made by FDA rating officers every 14 to 18 months to confirm ratings given, or changes since the last rating, and to ensure reproducibility of state rating results.

[33] *Methods of Making Sanitation Ratings of Milk Supplies*, 1978 ed., DHEW, PHS/FDA, Washington, D.C.

Official Local Program Supervision and Inspection

The official agency makes regular review of the industry inspection files mentioned above, takes whatever action is indicated, and at least annually makes joint inspections with the industry inspector of a randomly selected significant number of dairy farms, including receiving stations. The quality of work done is reviewed, the need for special training is determined, and recommendation concerning certificate renewal is made to the permit-issuing official. A similar review is also made of the sample collection, transportation, and the procedures, equipment, and personnel in the laboratory making the routine milk and water examinations.

In addition the local regulatory agency collects official samples as required by the state milk code, advises the industry having jurisdiction of the results and corrective action required, participates in training sessions, and serves as the state agent in securing compliance with the state milk code. The authorized local city or county agency usually has responsibility for the routine inspection of processing plants, sample collection, and overall program supervision for compliance with the code. The agency sanitarian may serve as a consultant to the industry in the resolution of the more difficult technical, operational, and laboratory problems. This whole procedure makes possible better use of the qualified industry inspector and the professionally trained sanitarian, with better direct supervision over dairy farms and pasteurization plants and more effective surveillance of milk quality. In some states the local activities are carried out in whole or in part by the state regulatory agency.

Official State Surveillance and Program Evaluation

The state department of health, and in some instances the state department of agriculture, share responsibility for milk sanitation, wholesomeness, and adulteration. The responsibility is usually given in state law, sanitary code, or milk code, and in rules and regulations promulgated pursuant to authority in the law. Most states have adopted the *Grade A Pasteurized Milk Ordinance*, or a code that is substantially the same. This makes possible a reasonable basis for uniformity in both interstate and intrastate regulations and interpretation. However, short-term and alleged economic factors frequently limit reciprocity and interstate movement of milk. Both industry and regulatory agencies need to cooperate in the elimination of milk codes as trade barriers. There usually is no objection to reciprocity where milk-quality compliance and enforcement is certified under the IMS Program.[34] However, not all states or farms participate in the program.

The IMS Program also gives the state regulatory agency a valuable tool to objectively evaluate the effectiveness of the local routine inspection and

[34] *Sanitation Compliance and Enforcement Ratings of Interstate Milk Shippers*, and *Methods of Making Sanitation Ratings of Milk Supplies*, 1978 ed., PHS/FDA, Washington, D.C.

enforcement activities. The types of additional training needed by the qualified industry inspector, the assistance the dairy farmer should have, and the supervisory training needed by the regulatory agency sanitarian become apparent. Changes in technology and practices are noted, and the need for clarification of regulations, laws, and policies are made known to the state agency.

A common function of a state regulatory agency is periodic in-depth evaluation of local milk programs. This includes the effectiveness with which the local unit is carrying out its delegated responsibilities as described above, quality of work done, staff competency and adequacy, reliability of the official laboratory work, record keeping, equipment and facilities available, number of inspections and reinspections made of dairy farms and processing plants, and their adequacy or excessive frequency. The state agency usually has responsibility for approving all equipment used in milk production and service from the cow to the consumer. The standards recommended by national organizations are generally used as a basis for the acceptance of equipment.

Federal Marketing Orders For Milk

The Agricultural Marketing Agreement Act is a pricing system which encourages dairymen to maintain large enough herds to meet drinking milk needs during the season of lowest production. Excess milk, usually during the spring and summer months, goes into ice cream, butter, cheese, cottage cheese, and milk powder. Orders under the act regulate only Grade *A* milk known as Class 1, which is the fluid milk used for drinking. Excess Grade *A* milk is priced at a lower level which is related to the market prices for the milk products mentioned. The orders are issued by the Secretary of Agriculture after a request is made to him by dairy farmers or dairy organizations. The Department may then hold a public hearing where anyone affected by the proposal is heard. A "recommended" decision is made by USDA and then a "final decision and order" is issued after all comments heard are considered. The dairy farmers vote individually or through their cooperative on the decision. The price is put into effect as an order if at least two-thirds of the dairy farmers approve. The order has the force of law and USDA's Agricultural Marketing Service is responsible for its compliance.

HOSPITAL INFANT FORMULA

Diarrhea of the newborn has been responsible for deaths in hospital nurseries. Enteropathogenic *Escherichia coli* is commonly recovered from the stool. Enteroviruses and *salmonella* are also recovered; *shigella* rarely. Sanitary engineers and sanitarians have been effective in applying their knowledge and experience in the control of water-, milk- and foodborne illness to hospital formula-room sanitation.

Where formula is prepared at a hospital the control program should initially be presented as a consultation service furnished through the cooperation of the hospital administrator and the health department. The health department would make available its combined epidemiological, nursing, medical, sanitation, and laboratory resources. After the initial surveys and compliance with accepted standards, the hospital would be expected to carry on its own inspection program; but the health department team would be utilized whenever a problem appeared. A baby formula inspection program should include the following:

1. Establish appropriate lines of communication between the hospital personnel and the health department to ensure continuing supervision and consultation service.
2. Assist in establishing procedures to minimize contamination of baby formulas, bottles, and nipples during the handling and bottling processes, and assist in the control of formula constituents.
3. Assist in establishing procedures designed to ensure that terminal sterilization is adequately carried out and that infections are not introduced into the nursery through poor hygiene and sanitation practices.
4. Assist in establishing procedures to ensure proper handling of the terminally sterilized product until time of consumption by the infants.
5. Arrange for routine weekly bacteriological testing of one bottle of each product out of every 50 prepared, the test specimen to be selected and collected by health department personnel for laboratory examination. Emphasis is to be placed on the sanitary survey and day-to-day surveillance, the bacteriological testing being only ancillary and confirmatory to the techniques followed and equipment provided.

The *Control of Communicable Diseases in Man* gives additional information.[35] A total plate count of less than 10/ml in terminally heated formula is easily obtainable, and a count of 3 or less is practical.[36] Rinse samples of 8-oz bottles that have only been properly washed should show a plate count of less than 300/ml, with no coliform microorganisms present.

Precautions to follow include:

1. Discard scratched and chipped bottles as well as old and porous nipples.
2. Thoroughly rinse all bottles and nipples with lukewarm water immediately after use to simplify and make more effective the subsequent cleansing operation.
3. Each week treat all bottles and nipples by soaking in a 1-percent acid solution to prevent the buildup of mineral deposits, by the use of a milk-stone remover found effective in the dairy industry.
4. Use a suitable detergent, followed by a clear water rinse, based on the water characteristics, in washing bottles and nipples (never use green soap or bar soap).
5. After washing and rinsing nipples, boil in water for 5 min.
6. Use a maximum-registering thermometer in each batch of formulas terminally heated.
7. Assure that bottles used in the nursery are not mixed with baby bottles from other parts of the hospital.
8. See that daily records are kept showing the maximum temperature of the maximum-registering thermometer for each batch of formulas together with the pressure and temperature during terminal heating, the number of formulas and other fluid bottles heat-treated, and the temperature of the refrigerator in which formulas are stored.
9. Assure that materials for formula preparation are stored in a separate locked room.

[35] *The Control of Communicable Diseases in Man*, Am. Pub. Health Assoc., Abram S. Benenson, Ed., 1015 Fifteenth St. NW, Washington, D.C. 20005, 1975, pp. 98–101.

[36] J. A. Salvato and Louis Lanzillo, "Infant Formula Inspection Program as an Aid in the Prevention of Diarrhea of the Newborn," *J. Milk Food Technol.*, May 1957, pp. 127–131.

Commercial preparation, sterilization, and distribution of formulas in single-service bottles simplify control. The infant formula inspection and control activity is only one part of the program for the prevention of diarrhea of the newborn. It should be integrated with special medical and nursing preventive techniques as well as the environmental protection aspects of water supply, air sanitation, cross-connection and backflow prevention control, infection control, and food sanitation.

Where baby formula and other infant foods are reconstituted, either at a central point or in the home, it is extremely important that potable water, clean sterile bottles, and adequate refrigeration be available, particularly in underdeveloped areas of the world. Their lack may set the stage for diarrheal diseases which not only debilitate the infant, but cause malnutrition and possibly death.*

REGULATION OF RESTAURANTS, SLAUGHTERHOUSES, POULTRY DRESSING PLANTS, AND OTHER FOOD ESTABLISHMENTS

A major reason for the supervision of food establishments is the prevention of foodborne illnesses as discussed in Chapter 1. Federal, state, and local agencies, in cooperation with the food processing and preparation industry, share this responsibility in addition to assuring wholesomeness of the food and maintenance of its nutritional value.

Supervision of food establishments such as restaurants, caterers, delicatessens, commissaries, pasteurizing plants, frozen-dessert plants, frozen-prepared-food plants, vending-machine centers, slaughterhouses, poultry processing plants, bakeries, shellfish shucking and packing plants, and similar places is in the public interest. This responsibility is usually vested in the state and local health and agriculture departments and also in the FDA, PHS, USDA, and U.S. Dept. of Interior. The industry affected as a rule also recognizes its fundamental responsibility.

Basic Requirements

It is to be noted that certain basic sanitation requirements are common to all places where food is processed. McGlasson has proposed a set of standards under the following headings:[37]

1. Location, construction, facilities, and maintenance
   a Grounds and premises
   b Construction and maintenance
   c Lighting

*See International Code of Marketing of Breastmilk Substitutes, World Health Assembly, WHO and UNICEF, 1981.

[37] E. D. McGlasson, "Proposed Sanitation Standards for Food-Processing Establishments," *Assoc. Food Drug Off. U.S., Q. Bull.* Vol. 31, No. 2, April 1967.

   d Ventilation
   e Dressing rooms and lockers
2. Sanitary facilities and controls
   a Water supply
   b Sewage disposal
   c Plumbing
   d Restroom facilities
   e Hand-washing facilities
   f Food wastes and rubbish disposal
   g Vermin control
3. Food-product equipment and utensils
   a Sanitary design, construction, and installation of equipment and utensils
   b Cleanliness of equipment and utensils
4. Food, food products, and ingredients
   a Source of supply
   b Protection of food, food products, and ingredients
5. Personal
   a Health and disease control
   b Cleanliness.

A similar set of basic standards was published in the *Federal Register* in April 1969.[38] The above general sanitation requirements should be supplemented by specific regulations applicable to a particular establishment or operation. Excellent codes, compliance guides, and inspection report forms are available. Some additional valuable sources of information are the following.

1. *Grade A Pasteurized Milk Ordinance*, 1978 Recommendations of the PHS/FDA, Pub. 229.
2. *Food Service Sanitation Manual*, 1976 Recommendations of the FDA, DHEW Pub. (FDA) 78-2081.
3. *Frozen Desserts Ordinance and Code*, recommended by the PHS, 1958.
4. *Poultry Ordinance*, 1955, ed., PHS Pub. 444, *Recommendations Developed by the Public Health Service in Cooperation with Interested States and Federal Agencies and the Poultry Industry.*
5. *AFDOUS Frozen Food Code*, Association of Food & Drug Officials of the United States, Vol. XXVI, No. 1, January 1962 (Adopted June 22, 1961.)
6. *Recommendations and Requirements for Slaughtering Plants*, D. E. Brady, Merle L. Esmay, and John McCutchen, Missouri College of Agriculture, Bull. No. 634, September 1954.
7. *Manual of Recommended Practice for Sanitary Control of the Shellfish Industry, U.S. Public Health Service*, Part I, 1959, Part II, 1962.
8. *The Vending of Food and Beverages*, 1978 Recommendation of the FDA, Washington, D.C.
9. *A Sanitary Standard for Manufactured Ice*, 1964 Recommendations of the PHS Relating to Manufacture, Processing, Storage and Transportation, DHEW, Washington, D.C.
10. *Sanitation Standards for Smoked Fish Processing*, Part I, Fish Smoking Establishment 1967 Recommendations, U.S. Dept. of the Interior, Fish and Wildlife and Parks, and DHEW, PHS, Washington, D.C.
11. *Voluntary Minimum Standards for Retail Food Store Refrigerators*, Health and Sanitation, (CR-S1-67), Commercial Refrigeration Manufacturers Association, 111 West Washington St., Chicago, Ill.
12. *Sanitary Standards for Food-Processing Establishments*, 1969 Recommendations for study purposes only, DHEW, PHS, Environmental Sanitation Program, NCUI, Cincinnati, Ohio.
13. *Bakery Establishments*, 1969 Recommendations, Addendum B for study purposes only, DHEW, PHS, Environmental Sanitation Program, NCUI, Cincinnati, Ohio.

[38] Title 21—Food and Drugs, Chapter 1—FDA, DHEW, Part 128—Human Foods; Current Good Manufacturing, Processing, Packing, or Holding, *Fed. Reg.*, April 25, 1969.

14. *Beverage Establishments*, 1969 Recommendations, Addendum C for study purposes only, DHEW, PHS, Environmental Sanitation Program, NCUI, Cincinnati, Ohio.

15. *Regulations Governing the Inspection of Eggs and Egg Products*, 1975, USDA, 7CFR, Part 59, June 30, 1975.

16. *Manual of Inspection Procedures of the U.S. Dept. of Agriculture*, Consumer and Marketing Service, Meat Inspection, July 1968, Washington, D.C.

17. *Methods of Making Sanitation Ratings of Milk Supplies*, PHS/FDA, 1978.

18. Association of Food and Drug Officials of the United States Retail Food Market Code-1973.

19. *Model Retail Food Store Sanitation Ordinance*, FDA, Washington, D.C., 1980.

20. *Model Food Salvage Ordinance*, Department of Health and Human Services, FDA, Rockville, MD, 20857, 1980.

Manuals and guides are not a substitute for the exercise of intelligent and mature judgment. They are, however, indispensable administrative aids that, with their continual revision, and with the supervision of field personnel, will help maintain uniform quality enforcement of a sanitary code. The reasonable and intelligent interpretation of the regulations and what represents good practice requires an understanding of the basic microbiological, engineering, and administrative principles involved. These are discussed below and throughout this text.

Inspection and Inspection Forms

Routine or frequent inspections of food-processing and food-service establishments alone will not be adequate to ensure the maintenance of proper levels of sanitation. This must be supplemented by education, motivation, persuasion, legal action, management supervision, qualified workers, laboratory sampling at critical points, and quality (including flavor) control. The primary responsibility for cleanliness and sanitation rests with management. Inspection frequency, enforcement, administration, and program evaluation are discussed in Chapter 12. See also Compliance with and Enforcement of Sanitary Regulations, this chapter.

Inspection report forms are valuable tools to assure complete investigations and uniform policy procedures. See Figures 8-6 through 8-10. They can also be made available to the industry to help in self-policing and in greater cooperation between the industry and official agencies. Preparation of an inspection form that is self-explanatory in showing what is acceptable, supplemented by a satisfactory compliance guide, can have a very desirable educational effect. A report form checked off entirely in the "Yes" column would indicate the establishment is in substantial compliance with existing regulations. Items checked "No" would indicate deficiencies that should be followed up on subsequent reinspections. Deficiencies are documented on a separate sheet. When these conditions are corrected, the third column, "CM," would be checked, indicating correction made of a previously reported deficiency. The date could be inserted in the "CM" square in place of a check mark to show when the correction was made. It is usually good practice to leave a copy of the original

Establishment Name ____________ Location ____________
Owner ____________ Address ____________ Date of Inspection ____________
Type ____________ Maximum No. Meals served daily ____________ Seating Capacity ____________
Item check 'Yes' satisfactory; 'No' unsatisfactory; 'CM' correction made. Explanation of defects numbered on back.

Item	Yes	No	CM
1. *Food-drink*			
(*a*) Storage—no contamination by overhead pipes; not on floor subject to flooding, dust, depredation by rodents			
(*b*) Protection—readily perishable foods stored at 45° F or less or 140° F or above; Water from refrig. equip. properly drained			
(*c*) Display—no open displays			
(*d*) Handling—minimum manual contact with food or drink			
(*e*) Food—wholesome from approved source; milk, etc., served in original containers; dispensed from bulk dispenser; oysters, clams, mussels, app'd. source: if shucked, kept in original container			
2. *Utensils* (multiuse, single serv.)			
(*a*) Construction—all multiuse utensils easily cleanable, kept in good repair; no corrosion, open seams, chipped, or cracked utensils; no cadmium or lead utensils			
(*b*) Cleanliness—eating and drinking utensils thoroughly cleaned after each use; other utensils cleaned each day; single service utensils used once			
(*c*) Storage and handling—stored above floor in clean place protected from splash, dust, flies, etc., inverted or covered where practical; no handling of contact surfaces; single service utensils stored and handled properly; dispensing spoons, dippers kept clean in running water			
3. *Equipment*			
(*a*) Construction—easily cleanable, no corrosion, good repair			
(*b*) Cleanliness—clean cases, counters, shelves, tables, meat blocks, slicers, refrigerators, stoves, hoods			
(*c*) Location—properly located to facilitate cleaning around and beneath the equipment			
4. *Washing facilities*			
(*a*) Manual type—3 compartment sink and long-handled wire baskets: water for disinfection maintained at 180° F or higher for 2 min; satis. chem. disinfection where hot water is not practical			
(*b*) Mechanical type—wash water maintained at 150–160° F; wash cycle not less than 40 sec.; rinse water maintained at 180° F or higher; rinse cycle not less than 10 sec; rinse water flow pressure controlled at 20 psi; wash sol. OK			
5. *Premises*			
(*a*) Floors—easily cleanable construction, smooth, good repair; clean			
(*b*) Walls and ceiling—all: clean, good repair, walls smooth and washable			
(*c*) Lighting and ventilation—all rooms in which food is prepared, stored, or in which utensils are washed shall be well lighted and ventilated			
(*d*) Clean, sanitary, orderly			
(*e*) Doors and windows—all openings into outer air shall be effectively screened and doors shall be self-closing			
(*f*) Insect and rodent control—flies, roaches, and rats under control; structure rat-proofed; poisons stored in locked cabinet			
6. *Employees* (Male__Female__) Clean outer garments; hands, arms clean; no tobacco used where food is prepared			

Figure 8-6 Eating place, inspection form. (Deficiencies are explained on back.)

	Yes	No	CM	
7. *Toilet, lavatory facilities* Comply with plumbing code and OSHA; adequate, conveniently located for employees; good repair, clean, warm and cold running water; well lighted; outside ventilation; self-closing door; soap and individual towel service				Inspection substantially satisf. Yes_____ No_____ Reinspect _____ Water heater(s): Storage in gal _____ Recovery capacity in Btu per hr _____ Therm. setting _____ °F Adequate _____
8. *Garbage, refuse* Stored in tight nonabsorbent washable receptacles; tightly fitted covers provided; receptacles covered; proper disposal				Refrig. capacity: Tons per 24 hr _____ Cubic feet walk-in _____ Cubic feet freezer _____ Cubic feet upright _____ Adequate _____
9. *Water supply* (Public_____ (Private_____) Running water hot and cold; supply adequate and of a safe sanitary quality; no submerged inlets or cross-connections				Dishwashing: Manual—No. compartments _____ —Disinfection by hot water _____ — " " Chemical _____ —Thermometer or test kit provided _____ —Type of booster heater & cap. _____
10. *Disposal facilities* (Public_____ Private_____) Liquid wastes into public sewer or satis. private system; no interconnection with washing machines or sinks, kettles, warm tables, coolers				—Name of disinfectant _____ " " detergent _____ —Adequate _____ Machine—Mfg., No. type _____ —No. tanks _____ —Pres. gauge on rinse___ reading___ psi —Temp. gauge on wash_____ rinse_____
11. *Disease control* No person at work with any communicable disease, sores, or infected wounds				—Name of detergent_____ feeder_____ —Adequate _____ Length of: soiled-dish counter_____ft drainboard_____ft;
12. *Ice* Source, storage, handling satis.				Adequate _____ Inspected by _____

Figure 8-6 (*Continued*)

inspection report form with the owner, operator, or other responsible person. When subsequent reinspections fail to show improvement, it is customary to confirm the deficiencies in a letter, with reference to previous inspections and notifications and a scheduled date for final inspection before taking further action. In any case imminent health hazards require immediate correction or closure of the operation.

Regulatory agencies usually have standard inspection report forms covering activities under their supervision. However, some forms for the sanitary inspection of eating places, slaughterhouses, poultry processing plants, and bakeries are reproduced here for convenience.

Shellfish

Shellfish include oysters, soft-shell and hard-shell (quahaug) clams, and mussels, as well as the crustacea, lobsters, crabs, and shrimp. For control purposes

Trade name ______________________ Operator ______________________

Address ______________________ Veterinarian ______________________

Personnel: M____F____Permit No.______Killed Annually: Beef____Hogs____Sheep____Other____

	Yes	No	CM
1. *Building*			
(*a*) Site drained, graded, clean			
(*b*) Structure in repair, location meets zoning restrictions			
(*c*) Rooms 12 ft high, space for storing meat and washing utensils; slaughtering separated from processing			
(*d*) Floors concrete or equal, coved at walls up 12″, graded to trapped drain, drip gutter under dressing rail, separate drain for bleeding and viscera hanging area			
(*e*) Walls and ceilings cleanable, concrete, tile or equal, wall impervious to 5 ft height, ceiling tight and cleanable			
(*f*) Pens have concrete floor, are drained, clean, 150 ft to corral			
(*g*) Insects and rodents absent, building ratproofed			
(*h*) Dressing rooms and lockers clean, separated from work			
2. *Lighting and Ventilation*			
(*a*) Adjustable windows, ducts, or fans; hoods over vats, no condensation or excessive odors, heating adequate			
(*b*) Windows 15% floor area, no glare, 10 ft candles of light min. and 25 at grading and insp. areas, light-colored walls, ceilings			
(*c*) Openings screened, flies control			
3. *Water supply*—100 gal per animal			
(*a*) Source protected, construction satisf. and adequate, rooms accessible with hose, no cross-connections, quality ok			
4. *Toilet facilities*			
(*a*) Location convenient, screened, ventilated, lighted, tissue, backflow preventers, clean			
(*b*) Lavatories with warm running water, soap, paper towels			
5. *Refuse*			
Flytight receptacles for offal manure, blood, condemned meat			
6. *Sewage and waste disposal*			
Sewage to sewer or disposal system, other liquid wastes separate, operation proper			
7. *Animals*			
Disease-free from ante-mortem veterinary inspection			
8. *Equipment*			
(*a*) Work tables, blocks, vats are cleanable and free from cracks			
(*b*) Chilling room stores carcasses 6 hr, air circulation in cooler; floors, walls, hooks, trays cleanable; thermometers			
(*c*) Washing and flushing (150°F) water under pressure, tanks, brushes, cleansing powders adequate			
(*d*) Transporting vehicles covered, cloths and waterproof paper			
9. *Handling, refrig., disinfection*			
(*a*) Veterinarian post-mortem sup, carcasses stamped, washed, protected from contamination			
(*b*) Processed meat heated to 170°F for 30 min., meat therm. used			
(*c*) Carcasses separated in refrig. supported off floor, placed immediately in chilling room at approx. 40° and stored below 34° after 24 hr, therm. used, no spoiled meat			
(*d*) Hot water, brushes and cleansing powders used, equip. clean			
(*e*) Utensils submerged 2 min. in 170° water or 200 ppm chlorine			
(*f*) Transportation vehicles clean used only for edible meats, meats wrapped and protected			
(*g*) Personnel clean, have washable outer garments, aprons, rubber boots, gloves; have clean habits and in good health			

Remarks and Explanation of Defects: (CM indicates correction made) ______________________

____

____ (over)

Inspected by ______________________ Date ______________________

Figure 8-7 Slaughterhouse inspection form.

Name ______________________ Operator ____________ Permit No. ________

Address ______________________________ Personnel: M ________ F ________

	Yes	No	CM
1. Plant rodentproof and free from rodents			
2. Structure in repair, location meets zoning restrictions			
3. Structure screened, free of flies, roaches, other vermin			
4. Light in workrooms 30–50 ft candles 30 in. above the floor			
5. Ventilation prevents condensation and odors			
6. Water supply satisfactory sanitary quality and adequate			
7. Work rooms accessible with hose supplied by hot and cold water and steam; hot water supply adequate			
8. Floors in killing and dressing area smooth concrete or equal, coved at walls up 12", graded to trapped drain			
9. Walls and ceilings cleanable concrete, tile or equal; walls impervious to 5 ft height			
10. Floors free of blood and vent and crop material; washed as often as necessary to prevent accumulations			
11. Sewage, wastes, offal, and refuse properly disposed			

	Yes	No	CM
12. Ice used for dressing birds clean and handled in clean manner			
13. Holding cages and batteries cleaned daily, washing facilities provided, all kept clean			
14. Clean scalding water provided (170–180°F) and scalders have water changed frequently to prevent accumulation, spray chambers have nozzles clear			
15. Crops and vents of birds cleaned and birds washed clean			
16. Spray and scrubber type washers and coolers provided			
17. Superchlorinated water (10–20 ppm) available and used for water flushing of birds, for cavity and bird washing after eviscerating, giblet washing			
18. All tables, buckets, cans, shackles, hooks, trays, cookers and other equipment kept clean			
19. Convenient handwashing facilities for workers			
20. Toilet, washroom, and dressing room facilities adequate, and clean			
21. Cutting, wrapping, weighing, packing, and cooler equipment clean and adequate			

Remarks and Explanation of Defects (CM denotes correction made) ____________

____

____

____

____

____

See Poultry Plant Sanitation, Supplement No. 1, Institute of American Poultry Industries, 59 East Madison, Chicago 2, Illinois.

Inspected by ______________________________ Date ____________

Figure 8-8 Poultry dressing plant inspection form.

oysters, clams, and mussels (molluscan shellfish) are given primary consideration. Since oysters are frequently eaten raw and in large quantities, they are more frequently involved in outbreaks. Mussels from the Pacific Coast, England, and Europe contain at certain times of the year a chemical poison that is not destroyed by cooking. Because of this, California has a quarantine on mussels from June to September. Shellfish can transmit infectious hepatitis, typhoid fever, cholera, dysentery, and gastroenteritis and can concentrate toxic chemicals such as mercury, radioactive materials, pesticides, and certain marine biotoxins. Lobsters, shrimp, crabs, and scallops (mussels) are less likely to cause illness as they are usually thoroughly cooked before being served. Improper handling of these shellfish before or after service can, however, introduce contamination.

The pollution of shellfish waters with sewage and industrial wastes decreases the fish and shellfish population and introduces a disease-transmission hazard.

Name ______________________ Operator ______________________

Address ______________________ Wholesale ______ Retail ______

Personnel: M_____ F_____ Products______________________ Permit No.____________

	Yes	No	CM
1. Location on or above ground level, in clean surroundings			
2. Structure in repair, rodent-proof, screened, window display enclosed, baked goods protected, lighting adequate			
3. Floor finish smooth, cleanable, and nonabsorbent, sloped to drain that is trapped, kept clean and in repair			
4. Walls and ceilings cleanable, kept clean and in repair			
5. Ventilation adequate to prevent condensation, remove smoke and odors. Hoods equipped with proper exhaust fans and ducts			
6. Rats, mice, roaches, flies, silverfish, beetles, weevils, lice, mites, moths, ants, effectively controlled			
7. Extermination and cleaning program in effect with responsibility established in a sanitary supervisor who has authority to make corrections			
8. Water supply satisfactory, of sanitary quality, easily accessible, with adequate volume and pressure			
9. Wastes disposed of to proper sewer or private sewage disposal or treatment system			
10. Toilets and washrooms convenient, have wash basins with warm water, soap, and individual towels (1 w.b. to 10 to 15); adequate waterclosets (1 wc. to 10 to 15);† handwashing sign; cleanable floors, walls, ceilings; adequate light and ventilation; tissue; kept clean			
11. Locker and dressing space, storage place for clean and soiled linen, uniform, and aprons; lighted and clean			
12. Personnel clean, head covering, clean habits, good health, supervised			
13. Design and construction of equip. cleanable and kept in repair (pans, kettles, molders and rollers, dough troughs, mixers, beaters, flour bins and chutes, conveyors, deep fat fryers, ovens, retarder and proof bins, sieves), equipment cleaned at end of each day and kept clean*			
14. Bakery and confectionery products free of vermin and other filth; flour, meal, farina, etc. stored in original pkg. at least 6″ off floor or in tight receptacles; raw ingredients from approved source; perish-			

Figure 8-9 Bakery and confectionary inspection form.

	Yes	No	CM		Yes	No	CM
able products stored at 45° F or less and frozen products at 0° F or less				kept below 45° F. Frozen prod. at 0° F.			
15. Equip. and facilities adequate for cleaning and disinfection; includes wash sinks, automatic hot water, detergents, and disinfectants, scrapers, and brushes				17. Disposition of leftovers satisfactory			
16. Refrigeration adequate for perishables; custards and creams refrigerated, perishable products so labeled and marked				18. Adequate containers for storage of refuse, clean, no accumulations			
				19. Delivery trucks clean, baked goods protected, perishables refrigerated			
				20. Display cases and racks clean, self-service products completely wrapped			

*Equipment meets standards of Bakery Industry Sanitation Committee, 3 "A" Standards Committee, National Sanitation Foundation.

†Ratio may be reduced to 1 to 40 with over 150 employees.

(Ultraviolet light, "black light," shows urine stains. Microanalysis of food and ingredients reveals presence of rodent hairs, dirt, and fragments of pellets and insects.)

Remarks: ______________________ Inspection substan. satisf. Yes____ No____

______________________ Reinspect ______________

Inspected by ______________________ Date ______________

(over)

Figure 8-9 (*Continued*)

Typhoid fever outbreaks traced to contaminated shellfish between 1900 and 1925 led to PHS Certification of dealers involved in interstate shipment and to control by state and local regulatory agencies.[39]

The FDA evaluates compliance with minimum standards by inspection of a representative number of handling and processing plants. The state regulating agencies should adopt adequate laws and regulations for the sanitary control of shellfish from the source to the consumer. The local industry is expected to comply with established procedures regarding identification information on each package of shellfish, record-keeping from the point of origin to the point of ultimate sale, and sanitary control of shellfish. Numbered certificates issued by the authorized state agencies are accepted by the FDA for approved producers and dealers engaged in interstate commerce. The FDA maintains a current list of approved shellfish shippers.

Records of all shellfish shipments must be kept; this includes reshipments. The name of the shipper and consignee, the date harvested, state and water source, the type of shellfish, and the permit certificate number of the packer and reshipper must appear on each carton or lot.

Spawning takes place when the water is warm, usually between May and August. The shellfish seed or spat is released at this time, attaches itself to some hard surface, and grows. Oyster shells dumped back into shellfish beds provide a

[39]National Shellfish Safety Program, DHEW, FDA, *Fed. Reg.*, Thursday, June 19, 1975, pp. 25916–25935.

ready growing surface. The parent shellfish is not as meaty or of best flavor during the spawning season but is edible. Problems in refrigeration during warm weather mitigate against the production and distribution of shellfish in areas lacking refrigeration equipment and facilities.

Oysters grow best at a pH of 6.2 to 6.8 and at a temperature of 41° F. The pH can be as low as 5.0. The best salinity of the water is 0.24 to 0.27 percent. A salinity greater than 0.31 percent is unsuitable.

Oyster-shell stock or shucked oysters sampled at the source (wholesale market) should not show a MPN of 230 or more fecal coliform organisms per 100 grams of shellfish or more than 500,000 total plate-count per gram. A MPN of

Name of establishment ______________ Certificate No. ______________

Address ______________ Operator ______________

Shellfish obtained from area ______________ Products ______________

	Yes	No	CM
1. *Water storage*			
(*a*) Shellfish stored in fill and draw or continuous flow tanks			
(*b*) Water meets PHS drinking water standards and is of proper salinity			
(*c*) Fill and draw tank emptied within 24 hours and water maintained of satisfactory sanitary quality by chlorination			
(*d*) Continuous flow tanks supplied with satisfactory chlorinated water at rate to replace water in not more than 24 hours			
(*e*) Tanks side walls extend 4 in. above surrounding floor and 1 ft above high water			
(*f*) Tanks maintained in a sanitary condition and employees wading in tanks wear clean rubber boots			

	Yes	No	CM
2. *Shellfish from approved source*			
(*a*) Shellfish packaged by person having permit. If in interstate traffic, person has permit from other state whose program is certified by the PHS			
(*b*) Shellfish tagged with name, address, and certificate or permit number of shipper, the contents and source of shellfish, date of shipment, name and address of consignee			
(*c*) Receiver of shellfish keeps record of quantity and source and sale			
(*d*) Boats clean, bilge water disposed away from shellfish, watertight receptacles for excreta, receptacles emptied on shore in sanitary manner			
3. *Buildings*			
(*a*) All buildings kept clean, free of debris and unused material			
(*b*) Adequate toilet facilities			
(*c*) Toilets clean, separate from workrooms; sewage disposal proper			
(*d*) Washbasins with hot and cold water, soap and sanitary towels, convenient, "Wash Hands" signs posted			
(*e*) Buildings adequately lighted, ventilated, screened			

Remarks (explain "No" items on back) (CM indicates correction made)

Inspection substantially satisf. Yes ____ No ____

Reinspect ______________

____

____

____

Inspected by ______________ Date ______________

Figure 8-10 Shellfish control inspection form.

	Yes	No	CM		Yes	No	CM
(*f*) All walls, ceilings, and floors cleanable and floors drained				(*g*) Refrigeration for shucked shellfish above freezing and below 40° F			
(*g*) Approved water supply				(*h*) Shucked shellfish stored and shipped at temperature between 32° and 40° F, not in contact with ice, in clean and new metal or equal containers, adequately sealed			
(*h*) Cleanable shell stock storage bins and rooms							
4. *Shucking*							
(*a*) Shucking and packing rooms separated, cleanable, free of insects				(*i*) Containers embossed with certificate number of shipper preceded by state			
(*b*) Shucking benches nonabsorbent, cleanable, disinfected, and cleaned				5. *General*			
(*c*) Shucking blocks free of cracks, cleaned				(*a*) Personnel clean and have clean habits			
(*d*) Containers for waste material, and waste material disposed of in a sanitary manner				(*b*) No person in plant ill or has communicable disease			
(*e*) Shucked shellfish washed with cold water of satisfactory sanitary quality				(*c*) Shellfish handled and shipped in clean containers and at temperature to keep them alive			
(*f*) All shucking equipment cleanable				(*d*) Cars and trucks in which shellfish shipped clean			

Figure 8-10 (*Continued*)

2400/100 g may be occasionally permitted, provided needed corrections are promptly made.

Shucked shellfish are shellfish that are taken from the shell and washed and packed in clean containers constructed of impervious material. The operation should be conducted under controlled sanitary conditions. The packer's certificate number and state, name and address of the distributor, and contents are permanently stamped on the container. Shucked shellfish should be stored at a temperature of 45° F or less.

Microbiological limits for seafood are shown in Table 8-6. Other information dealing with shellfish sanitation is available in government publications.[40]

Shellfish-growing-area water should have a coliform median MPN not exceeding 70/100 ml, with not more than 10 percent of the counts showing coliform MPN of greater than 230/100 ml when using five 10 ml dilution tubes, or 330/100 ml when using three 10 ml dilution tubes.[40–42] The USEPA criterion for shellfish harvesting water is

[40] *National Shellfish Sanitation Program Manual of Operations*, Part I, "Sanitation of Shellfish Growing Areas," PHS Pub. No. 33, 1965. Part II, "Sanitation of the Harvesting and Processing of Shellfish," PHS Pub. No. 33, 1965. Part III, "Public Health Service Appraisal of State Shellfish Sanitation Programs," Sup. of Documents, GPO, Washington, D.C.

[41] *Recommended Methods for the Examination of Seawater and Shellfish*, Daniel A. Hunt, Ed., American Public Health Association, 1015 Fifteenth St., N.W., Washington, D.C. 20005, expected 1982.

[42] *Compendium of Methods for the Microbiological Examination of Foods*, Marvin L. Speck, Ed., American Public Health Association, 1015 Fifteenth Street, N.W., Washington, D.C. 20005, 1976.

Table 8-6 Microbiological Limits for Seafoods (New York City Health Department)

Organism	Crabmeat	Shellfish	Prepared Fish and Crabmeat
Staphylococcus aureus	<100/g		
Coliform	<100/g		
Enterococci	<1000/g		
Standard plate count	<100,000/g	≤500,000/g	≤100,000/g
Fecal coliform		<230/100 g	None
Salmonella			None

> not to exceed a median *fecal* coliform bacterial concentration of 14 MPN per 100 ml with not more than 10 percent of samples exceeding 43 MPN per 100 ml for the taking of shellfish.[43]

The Canadian standard for shellfish growing waters is a mean fecal coliform MPN of 14/100 ml.

To be representative, water samples should be taken during different seasons of the year, and at varying tides, depths, and watershed runoffs. Examinations must be made in accordance with the procedures described in *Standard Methods for the Examination of Water and Wastewater.*[44] Sanitary surveys of shellfish growing areas and tributary watersheds, including chemical and bacterial examinations of the waters and inspections of shellfish plants, should be made periodically by the state shellfish regulatory agency.

> The sanitary survey involves the evaluation of all factors having a bearing on the sanitary quality of a shellfish growing area, including sources of pollution, the effects of wind, tides, and currents in the distribution and dilution of the polluting materials, the reliability of nearby wastewater treatment plants, and the bacteriological quality of the water.[40]

The quality of the water source of shellfish will determine the need for special treatment and a cleansing period (24 to 48 hours) for shellfish before being placed on the market. Cleansing or purification processes, for shellfish from *restricted* or *prohibited* areas, referred to as "controlled purification," involve the placing of the bivalves in tanks fed with clean seawater, under the direction of the control agency. An area is designated "restricted" when direct consumption of shellfish might be hazardous due to radionuclide or industrial wastes pollution and the median coliform MPN of the water does not exceed 700/100 ml and not more than 10 percent of samples exceed MPN of 2300/100 ml. An area is designated as

[43] *Quality Criteria for Water*, USEPA, Washington, D.C., July 26, 1976, p. 79.
[44] *Standard Methods for the Examination of Water and Wastewater*, 15th ed., American Public Health Association, Washington, D.C., 1980.

"prohibited" when consumption of shellfish might be hazardous due to radionuclides or industrial wastes pollution or the median coliform MPN exceeds 700/100 ml or more than 10 percent of the samples have coliform MPN in excess of 2300/100 ml.[40]

Sanitary handling and processing must be assured to maintain a satisfactory product from the source to the consumer. This includes thorough washing at the source, clean packing, and storage at temperatures that will keep the shellfish alive and active without loss of shell liquor. Sand trapped in the shell will irritate and cause a loss of shell liquor, with general weakening of the bivalve. Oysters free themselves of contaminating viruses and bacteria within 12 to 24 hours of exposure in purified seawater.[45]

Burlap and jute sacks used to transport shellfish must be kept clean and dry because damp or wet sacks will cause bacteria to multiply rapidly. Clean barrels, boxes, or baskets are better shipping containers.

Fresh oysters can keep for two weeks or somewhat longer on ice. Frozen oysters can be kept six months or more. Soft-shell clams and mussels should not be kept more than 2 days. Shellfish shipped to market in the shell are required to be handled in a sanitary manner, stored at a temperature of 36 to 40° F, and kept alive. Repacking of fresh oysters as breaded frozen oysters or a breakdown in sanitary precautions introduces health hazards. An effective shellfish control program requires the active participation of state and local regulatory agencies. Shellfish should be obtained only from certified dealers.

DESIGN DETAILS

Typical Kitchen Floor Plans

Some floor plans are included here to suggest the factors involved and the approach that should be used when a new establishment is constructed or major alterations are proposed. Many details, equipment, facilities, and trades are involved, all of which require specialized knowledge.[46] The preparation of scale drawings and plans by a competent person, such as a registered architect or engineer who has had the related experience, invariably saves money and anguishing frustrations. It is strongly recommended that preliminary plans be reviewed by the local building department, health department, or other agencies responsible for inspection and supervision of the particular operations regardless whether this is a requirement of any sanitary code, regulation or law. Equipment manufacturers may be of assistance. Most official agencies are happy to be

[45] *New York Times*, November 7, 1965, report on research done at the Gulf Coast Shellfish Research Center at Dauphin Island, Ala.

[46] *Reference Guide—Sanitation Aspects of Food Service Facility Plan Preparation and Review*, National Sanitation Foundation, NSF Building, P.O. Box 1468, Ann Arbor, Mich. 48106.

consulted in advance rather than require major corrections after a job is completed and business started.

General Design Guides

Restaurant and equipment design, including equipment arrangement, is a specialized field. In many instances, however, basic sanitation problems are incorporated rather than eliminated in the design. Food handling, storage, preparation, and service should receive careful consideration. Some guides in planning that result in efficient, economical, and sanitary operations are summarized below and illustrated in Figures 8-11, 8-12, and 8-13. Planning starts with study of food flow and type of service.

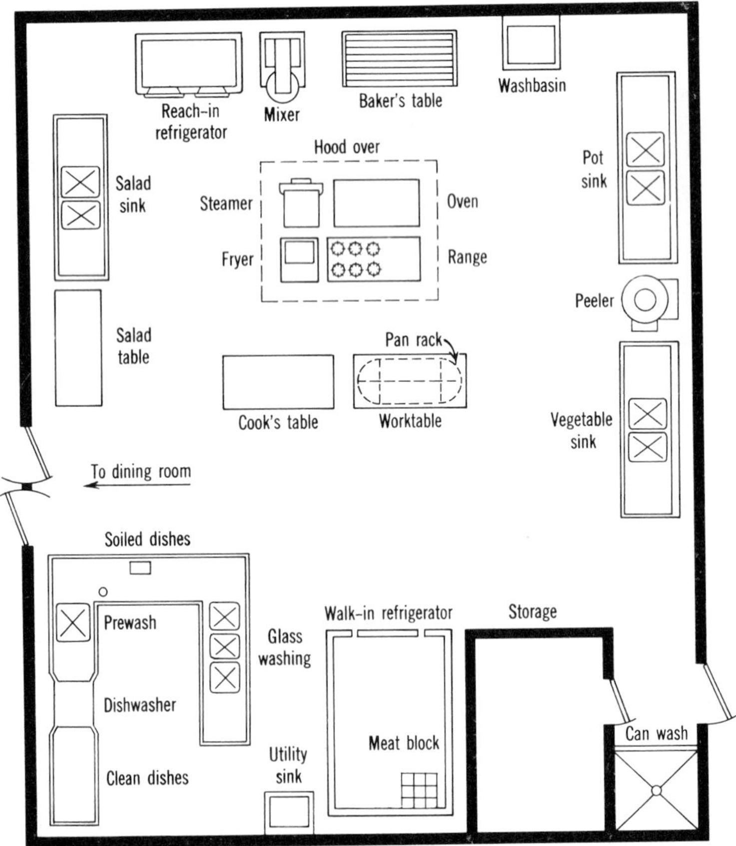

Figure 8-11 Layout of kitchen equipment. From *Environmental Health Practice in Recreational Areas*, PHS Pub. 1195, DHEW, Washington, D.C. 1965, p. 70.

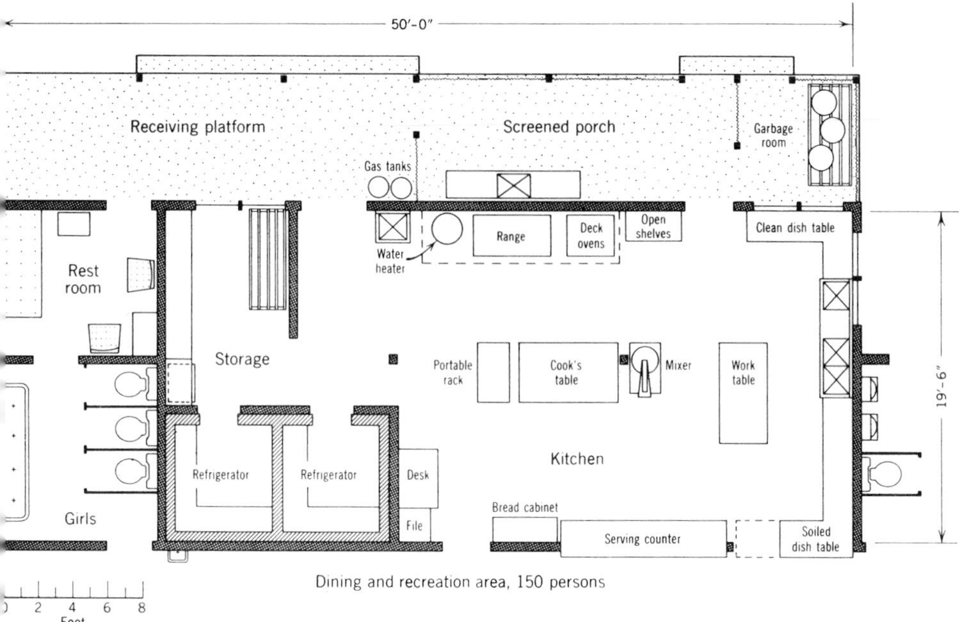

Figure 8-12 Plan of kitchen receiving platform and porch, and food storage rooms. Dishwashing center along wall at extreme right. From *Cornell Miscellaneous Bull.*, No. 14, New York State Colleges of Agriculture and Home Economics, Ithaca, March 1953.

1. Aisle spaces should be not less than 36 in.

2. Unless built-in in a sanitary manner, equipment should be spaced 18 to 30 in. from other equipment or walls and 12 to 18 in. above the floor to make possible cleaning. A smaller space that is still completely accessible for cleaning may be permitted.

3. Built-in or stationary equipment that eliminates open spaces, crevices, rat and insect harborage, and provides a coved or smooth joint where equipment abuts other equipment, the wall, ceiling, or floor will simplify sanitary maintenance.

4. The floor finish or covering selected for different spaces should recognize the type of traffic and uses to which the spaces will be put. The same would apply to wall and ceiling finishes. Linoleum or asphalt tile that is not grease-resistant, for example, is not a suitable floor covering for restaurant kitchens; quarry tile is ideal. The limitations of various foor coverings are discussed under Structural and Architectural Details. Consideration should be given to effective floor drainage.

5. Wooden shelving should be 12 to 18 in. above the floor, planed smooth, but not painted, shellacked, or varnished. Oilcloth or paper lining complicates the housekeeping job.

6. Bins, cutting and work boards, meat hooks, shelving in walk-in and reach-in refrigerators, and other equipment that comes into contact with food should be removable for cleaning. Wood work surfaces should be replaced with nonabsorbent cleanable synthetic materials.

7. Provision should be made in the kitchen for a washbasin with soap and paper towels and a utility sink, connected with hot and cold running water, for the convenience of food-handlers.

8. A garbage-can storage room is a practical necessity in the larger restaurants. This problem is discussed in greater detail in Chapter 5.

9. Garbage grinders are very desirable in the dishwashing area, in the vegetable preparation space, and in the pot-washing area.

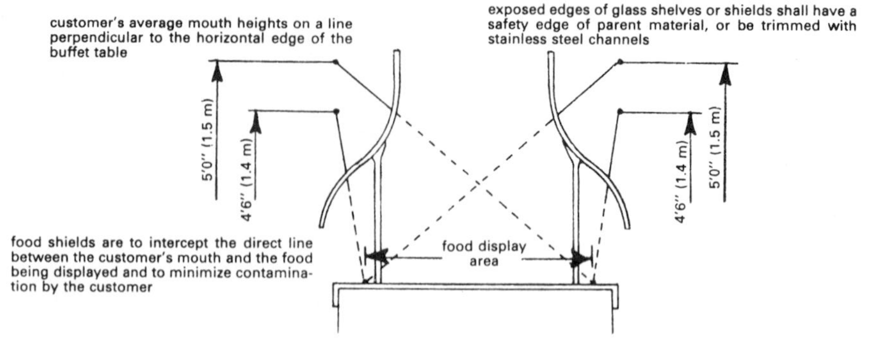

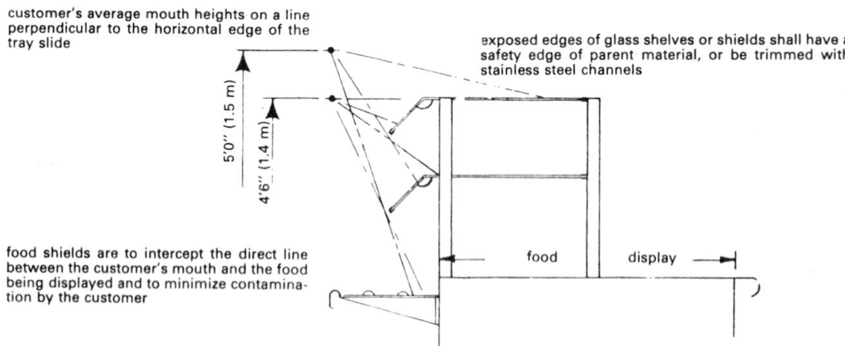

Figure 8-13 Equipment design details. Source: *Food Service Equipment Standards*, National Sanitation Foundation, NSF Building, Ann Arbor, Michigan 48105, September 1978, pp. 26, 30, 69.

10. Avoid submerged water inlets; provide back-flow prevention devices on water lines to fixtures or equipment that might permit backflow. Inlet lines should be $2\frac{1}{2}$ pipe diameters and not less than 1 in. above the overflow rim. Run wastelines from potato-peeling machine, egg boiler, dishwashing machine, steam table, refrigerator, air-conditioning equipment, and kettles to an open drain or to a special sink. See the section on plumbing for greater details.

11. Pipes passing through ceilings should have pipe sleeves with protective caps extending 2 in. above the floor surface.

Lighting

Proper lighting is essential for cleanliness and accident prevention and to reduce fatigue. Proper lighting takes into consideration the task, spacing of light sources, elimination of shadows and glare, control of outside light, and lightness of working surfaces, walls, ceilings, fixtures, trim, and floors. Some desirable lighting standards are given in Table 8-7. Minimum lighting is 20 ft-c on all food

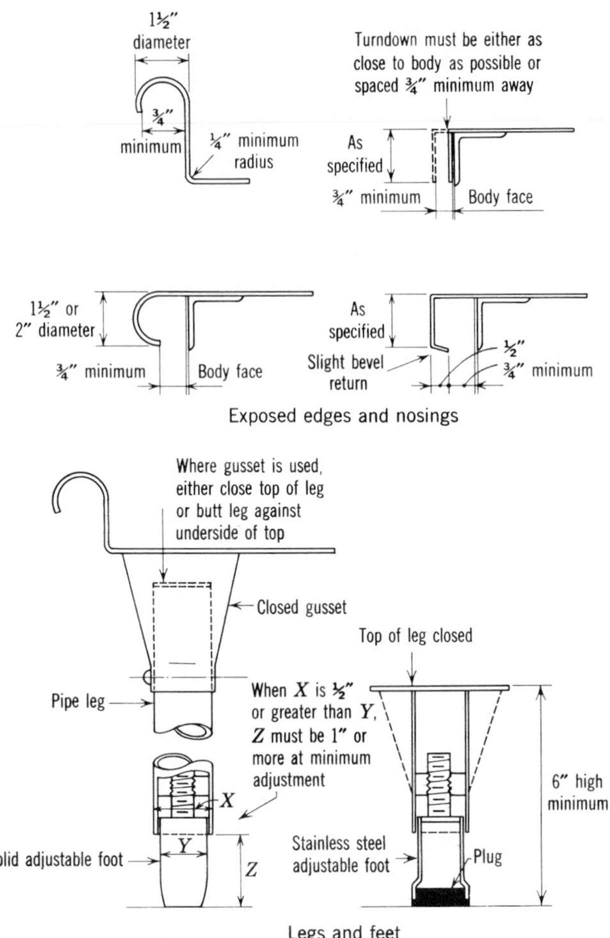

Figure 8-13 (*Continued*)

preparation surfaces and at equipment or utensil-washing work levels, in utensil and equipment storage areas, and in lavatory and toilet areas; 10 ft-c in walk-in refrigerating units and dry food storage areas.[47]

Typical color reflection factors are: white, 80 to 87%; cream, 70 to 80%; buff, 63 to 75%; aluminum, 60%; cement, 25%; brick, 13%.

Ventilation

Ventilation is considered adequate when cooking fumes and odors are readily removed and condensation is prevented. This will require sufficient windows that

[47] *Food Service Sanitation Manual*, PHS/FDA, Washington, D.C., 1976. p. 60.

Table 8-7 Lighting Guidelines

Space	Ft-c	Space	Ft-c
Bakery	30 to 50	Restaurant	
Brewery	30 to 50	Dining room	3 to 30
Candy making	50 to 100	Kitchen	70
Canning		Dishwashing	30 to 50
General	50 to 100	Quick-service	50 to 100
Inspection	100	Food quality control	50 to 70
Dairy		Cooler, walk-in	30
Pasteurization	30	Storage rooms	20
Bottle washer	100	Toilet rooms	20 to 30
Milkhouse	20 to 30	Dressing rooms	20 to 30
Inspection	100	Offices	70 to 100
Laboratory	100	Corridors, stairways	20
Gages, on face	50	Other spaces, minimum	5
Milking, at cow's udder	50	Recreation rooms	20 to 30
Meat packing			
Slaughtering	30		
Cleaning, cutting, etc.	100		

Note: The lighting should be measured on the working surface; in walk-in coolers and storerooms at the floor; and in other areas 30 in. above the floor.

can be opened both from the top and bottom or ducts that will induce ventilation. Hoods with removable filters and exhaust fans should be provided with the base of the hood normally 3 to 4 ft above the cooking surface.[48] The hood should extend 6 in. beyond the open sides of ranges, ovens, deep fryers, rendering vats, steam kettles, and over any other equipment that gives off large amounts of heat, steam, grease, or soot. Mechanical ventilation can be provided for kitchens with fans having the capacity of 15 to 25 air changes per hr, depending on the amount of cooking. It is advisable to draw fresh air in from a high point, as far away as possible from the exhaust fan, to prevent drafts and inefficient ventilation. Hood ventilation is given in Table 8-8. Hoods and filters require regular cleaning. They can become fire hazards. Grease accumulates on a filter, runs down the face, and drops into a drip tray from which it should drain into an enclosed metal container not exceeding 1 gal capacity.[48]

Uniform Design and Equipment Standards

There is a distinct need for reasonably uniform standards and criteria to guide the manufacturing industry as well as design architects and engineers. In this way food service equipment, materials, and construction would meet the requirements of most regulatory agencies, and manufacturers would not need special

[48] *Design of Grease Filter Equipped Exhaust Systems*, Research Products Corp., Madison, Wis. 53701.

Table 8-8 Equipment Ventilation

Equipment	Rate of Ventilation (cfm/ft^2)
Range hoods, four exposed sides	150
Range hoods, three exposed sides	100
Range hoods, two exposed sides	85
Hoods for steam tables	60

Note: Required exhaust fan = hood length × width × cfm in table, for canopy hoods less than 8-ft long. For canopy hoods 8-ft or longer, the required exhaust fan in cfm = 200 × perimeter of the open sides of the hood in ft. Check proposed design with local fire, building, health, and labor departments, and local gas company. For national standards and regulations see the National Fire Protection Association, The Underwriters' Laboratories, and the American Gas Association.

equipment design for different parts of the country. Such standards and criteria have been established by joint committees representing industry, business, the consumer, professional and technical organizations, and regulatory agencies and many have received wide acceptance.[49] They are highly recommended for reference and design purposes.

Space and Storage Requirements

A common problem in existing restaurants is the inadequate space in the kitchen for the number of meals served. Inefficiency and the resultant confusion compound the problem. Some suggested average space requirements are listed in

[49] 1. *Food Service Equipment Standards*, National Sanitation Foundation, Ann Arbor, Michigan, September 1978.

2. *3-A Sanitary Standards*, Committee on Sanitary Procedures, International Assoc. of Milk, Food and Environmental Sanitarians, Inc.; the Sanitary Standards Subcommittee, Dairy Industry Committee; the PHS/FDA.

3. *BISSC Sanitation Standards*, Baking Industry Sanitation Standards Committee.

4. Meat Inspection Division, USDA, Washington, D.C.

5. Poultry Inspection Division, USDA, Washington, D.C.

6. Bureau of Commercial Fisheries, U.S. Dept. of Interior, Washington, D.C.

7. *National Plumbing Code*, American Standards Association, ASA A40.8-1955, American Society of Mechanical Engineers, New York.

8. *National Electrical Code*, National Fire Protection Association, 470 Atlantic Ave., Boston, Mass., 1975.

9. *National Building Code*, American Insurance Association, 85 John Street, New York, 1967.

10. *Vending Machine Evaluation Manual*, National Merchandising Assoc., 7 South Dearborn St., Chicago, Ill. 60603, October 6, 1978.

11. *Vending of Food and Beverages*, PHS/FDA, Washington, D.C., 1978, DHEW Pub. No. (FDA) 78-2091.

12. *BOCA Basic Building Code—1975*, Building Officials & Code Administrators, Chicago, Ill.

13. *Uniform Plumbing Code*, International Assoc. of Plumbing and Mechanical Officials, Los Angeles, Calif., amended 1974.

14. See also state and local standards, regulations, and guidelines.

Table 8-9 Suggested Space Requirements for Kitchens

Type of Food Service	Kitchen[a] (ft^2 per seat)	Dining Room (ft^2 per seat)
Children's camp	7 to 9	12
School lunchroom	$5\frac{1}{2}$ to $6\frac{1}{2}$	10
Restaurant	$7\frac{1}{2}$ to $9\frac{1}{2}$	14
Lunchroom (counter, service, and tables)	8 to 12	15 to 20[b]
Hospitals	20 to 30 per bed	12 to 15 per bed
Cafeteria	12	15[b]

[a] Includes storage, receiving, and dishwashing spaces, also employees' toilet rooms and locker space; deduct 4 to 5 ft^2 per seat if not included. Larger areas are suggested for the smaller places. The greater the turnover per hour the larger the kitchen per seat. Another basis for kitchen area—food preparation, refrigeration, serving, dishwashing, and office—is $1\frac{1}{2}$ to 2 ft^2 per meal served per day.
[b] Includes serving area. Experience shows that with experienced help a cafeteria can serve 300 people per hr in the line and 600 people per hr with a limited menu. With counter service, allowing 2 ft per stool, 4 persons can be served per hr. With table service, the turnover is 3 to 4 times per 2 hr. (Ned Greene, *How to Serve Food at the Fair*, Pyramid Books, New York, 1963.)

Table 8-9; however, complete scale drawings should be made to fit individual situations. See also pp. 779 to 782. How the space is used is as important as the space available.

Refrigeration

Refrigeration slows down bacterial activity and preserves food, including many microorganisms. It is frequently stated that food stored at a temperature of 50° F is adequately protected against spoilage. This generalization can be very misleading unless such factors as storage method, frequency of opening and closing of the refrigerator door, insulation, type of food stored, length of storage, box outside-temperature, compressor capacity, temperature, weight, depth and thickness of food placed in the storage box, maintenance, and similar variables are taken into careful consideration. Actually, a temperature of 50° F will favor the growth of *Salmonella enteriditis*. Studies show that the growth of salmonellae and staphylococci in perishable foods is inhibited when the internal temperature is at or below 42° F.[50] Hence it is extremely important to distinguish between the air temperature in a refrigerator and the internal temperature of the food. A temperature of 38 to 40° F has been found more satisfactory in practice for the storage of most foods that are used within 24 or 48 hr. Frozen foods should be stored compactly below 0° F.

Low-temperature heat-treated hermetically packed foods, such as cured hams with salt, also require refrigeration. All canned food is not necessarily sterilized. In some instances the can merely serves as a container, for a meat or other food

[50] Robert Angelotti, Milton J. Foter, and Keith H. Lewis, "Time-Temperature Effects on Salmonellae and Staphylococci in Foods," *Am. J. Pub. Health*, January, 1961, pp. 76–88.

product, that has been subjected to sufficient heat to only inhibit the growth of spoilage organisms. Improper refrigeration of such foods may cause the growth of pathogenic organisms without showing swelling of the can.

In general, fresh meats should be stored at a temperature of 34 to 38° F; fresh fish at 34 to 38° F; poultry at 34 to 38° F; fresh fruits at 35 to 40° F; dairy products, including eggs and salads, at a temperature of 40 to 45° F; leafy green and yellow vegetables at 40° F; tomatoes, potatoes, onions, lemons, cucumbers, and melons at around 50° F; cured meats at 40° F in a dry, well-ventilated space; butter is best held at 40° F. A more exact breakdown of recommended storage temperatures with time limitations is given in Table 8-10.

Accurate outside thermometers and warning bells with thermometer stems located in the warmest compartments of refrigerators will show the adequacy of refrigeration. Sandwich and salad mixtures and cut or leftover food should be placed in shallow pans at a depth not greater than 3 in. to accelerate the rapid cooling of such foods. Hot foods are usually cooled down to about 140 or 150° F before being placed in the refrigerator, but in any case should be refrigerated within 30 min. In this way bacterial growth is retarded and the cooling capacity of the refrigerator is not exceeded. Foods in shallow pans placed in the freezing compartment will cool rapidly.

The storage of fish, ham, bacon, cabbage, onions, and similar odorous foods in the same cooler or refrigerator with milk, butter, eggs, salads, desserts, and fruit drinks that pick up odors and flavors should be avoided. Dairy products and meats are best stored in a separate refrigerator. Large pieces of meat should be hung so that they do not touch, and all foods including roasts, chops, and steaks should be arranged so that air can circulate freely. Wrapping should be removed and the food kept loosely covered. Smoked meats may be kept tightly wrapped. If not stored separately, wrap butter, cheese, and fish tightly.

Refrigeration in a food establishment, camp, or institution is adequate when it is possible to store and maintain all perishable foods at a proper temperature, at least over weekends, without packing and crowding. The refrigerator should be relatively dry and permit air to circulate freely.

The size refrigerator to provide for a particular purpose is dependent on many variables, most of which can be controlled. The number and type of meals served, the delivery schedules, and the purchasing methods all affect the required storage space. As a guide, a refrigerator capacity of 0.15 to 0.3 ft^3 per meal served per day, with 0.25 as a good average, has been found adequate to prevent overloading over weekends and holidays. Where food purchase is done twice a week, the refrigerator should provide, $2\frac{1}{2}$ ft^3 per person served three meals a day; 3 to $3\frac{1}{2}$ ft^3 is preferred. In practice many foods are purchased or delivered five days a week, hence the required refrigerator volume should be adjusted accordingly. Additional storage space is needed for bottled-beverage cooling and for frozen foods. Allow 1 ft^3 for 50 to 75 half-pints of milk and 0.1 to 0.3 ft^3 (30 to 35 lb/ft^3) per meal served per day for frozen food storage.

The storage racks in a refrigerator or walk-in cooler should be slotted to permit circulation of cold air in the box and should be removable for cleaning. If the box

Table 8-10 Storage Data

Product	Storage Temperature		Specific Heat	Approximate Storage Period	Common Measure
	Long Storage	Short Storage	Btu/lb/°F[a]		
Apples	31 to 33	35 to 40	0.90	2 to 8 months	1 bushel = 45 to 50 lb $2\frac{1}{2}$ bushel = barrel 5 ft^3 per barrel
Asparagus	32 to 34	40 to 45	0.95	2 to 4 months	
Bananas	56 to 60	35 to 40	0.81	7 to 10 days	50 lb per bunch
Blackberries	32 to 34	35 to 40	0.89	7 to 10 days	
Beans	32 to 34	40 to 45	0.92	3 to 4 weeks	
Beef (fresh)	32 to 34	35 to 40	0.77	3 months	750 lb average beef
Butter	40 to 45	40 to 45	0.64	5 months	50 to 60 lb per tub
Cabbage	32 to 34	35 to 45	0.93	3 to 4 months	
Candy	60	65 to 70	0.93	3 months	
Carrots	32 to 34	40 to 45	0.87	2 to 4 months	
Cauliflower	32 to 34	35 to 40	0.90	2 to 3 weeks	
Celery	31 to 33	35 to 40	0.91	2 to 4 months	
Cheese	34 to 36	40 to 45	0.67	4 months	
Corn	32 to 34	35 to 40	0.80	1 to 2 weeks	
Cranberries	34 to 36	40 to 45	0.85	1 to 3 months	
Cucumbers	50 to 60	50 to 60	0.93	6 to 8 days	54 lb = bushel
Eggs (fresh)	31 to 33	40 to 45	0.76	6 months	30 doz = case = 50 lb case 12″ × 12″ × 25″

Fish (iced fresh)	30 to 32	34 to 38	0.82		
Grapes	30 to 32	35 to 40	0.90	3 to 5 months	
Lamb	32 to 34	34 to 38	0.67		75 lb avg. weight
Lard	32 to 34	45 to 50	0.52		
Lemons	50 to 60	50 to 60	0.92	1 to 4 months	
Lettuce	32 to 34	35 to 40	0.90	2 to 7 days	
Melons	50 to 55	50 to 55	0.91	1 to 3 weeks	
Nuts (dried)	32 to 34	35 to 40	0.30	8 to 12 months	
Onions	32 to 34	50 to 55	0.90	5 to 6 months	bushel = 50 ib. bag = 70 lb
Oranges	32 to 34	40 to 45	0.90	1 to 2 months	crate 15″ × 15″ × 30″
Oysters, liquid	32 to 38	40 to 43	0.90		
Peaches	32 to 34	35 to 40	0.92	1 to 2 weeks	bushel = 56 lb
Pears	32 to 34	35 to 40	0.91	1 to 4 months	box = 40 lb
Peas (green)	32 to 34	40 to 45	0.80	1 to 3 weeks	
Pork (fresh)	32 to 34	34 to 38	0.70	1 month	Avg. wt of hog 250 lb
Potatoes	36 to 50	45 to 50	0.77	3 to 5 months	150 lb/bag, 60 lb/bushel
Poultry	28 to 34	34 to 38	0.80		
Sausage—fresh	32 to 34	35 to 40	0.89		
—smoked	33 to 35	40 to 45	0.68		
Strawberries	32 to 34	35 to 40	0.89	7 to 10 days	24-qt case = 45 lb
Tomatoes	50 to 55	50 to 55	0.95	7 to 10 days	Bushel = 50 lb
Veal	31 to 33	34 to 38	0.71		

Source: *Kramer Engineering Data*, Catalog No. R-114, Kramer-Trenton C., Trenton, N.J.

Note: A temperature of 36 to 42° F and humidity of 80 to 95 percent make an excellent environment for the growth of mold and yeast.

[a] The amount of heat required to lower the temperature of 1 lb of the product 1° F up to freezing temperature.

is provided with a drain, it must not connect directly to a sewer or waste line, but rather to an open sink or drain that is properly trapped.

If cans or boxes are stored in a cooler or refrigerator it is advisable to provide duckboards or racks on the floor. This makes possible better air circulation and helps keep the floor and supplies clean. In any case, refrigerators need daily cleaning and inspection for possible food spoilage. Washing the inside at frequent intervals with a warm mild detergent solution, followed by a warm-water rinse containing borax or baking soda and then drying will keep a box sweet and clean. Wipe the door gaskets daily. Do not use vinegar, caustic, or salt in cleaning solutions.

Many types of refrigerators on the market use different principles and refrigerants. When possible, provide entrance to a walk-in refrigerator through an air lock to prevent the cold air inside from rolling out and the outside warm air from pouring in. An efficient arrangement for refrigerators is a plan providing a group of walk-in coolers opening into a central room that can serve as a general cold storage room for certain foods that do not require a low temperature and as an access room from the unloading platform or kitchen. Another alternative would be two boxes in series with entrance to the freezer room through the cold room.

Thawing Frozen Foods

There is no need to thaw small cuts of meat or fish. Large pieces such as roasts, turkeys, and chickens should be thawed under refrigeration, or placed in a plastic bag in running water. A 20-lb turkey may require 72 hr to thaw under refrigeration. Do not thaw at room temperature.

A reduction in the microbial population occurs at low temperatures. For vegetables approximately 25 to 30 percent of the microorganisms survived at a storage temperature of −18° C (0° F) after 24 months of storage. Salmonella in chicken chow mein was reduced to less than 40 percent in 14 days and to less than 20 percent after 50 days at storage temperature of −25.5° C (−14° F).[51]

Refrigeration Design

The proper refrigeration of food is one of the most important precautions for the prevention of food poisonings and infections, as well as for the preservation of food. Refrigeration is the extraction of heat from a mass, thereby lowering its temperature. Ice has been used for many years to preserve food and is the standard that serves as a base for the measure of refrigeration. One pound of ice in melting will absorb 144 Btu of heat. A ton of refrigeration is 288,000 Btu/24 hr = 2000 lb of ice × 144 Btu.

The amount of heat required to lower one pound of a product one degree Fahrenheit is known as the specific heat. For example, the amount of heat to be extracted from 1 lb of water to lower its temperature 1 degree is 1 Btu, up to 32° F.

[51] L. P. Hall, "Food Control: The effect of the E.E.C. Directives, (d) Microbiological Aspects," *J. R. Soc. Health*, Oct. 1974, pp. 253–256.

To change the 1 lb of water to ice requires the extraction of 144 Btu from the water; this is known as its latent heat of fusion or the heat absorbed due to a change from the liquid to the solid state. But to lower the temperature of ice at 32° F requires the extraction 0.5 Btu per pound of ice per degree of drop in temperature. Similarly, each food product has its own specific heat, before freezing and after freezing, and its latent heat of fusion. For example, it is generally assumed for calculation purposes that the freezing point of most food products is about 28° F. Fresh pork has a specific heat of 0.6 Btu/lb before freezing, 0.38 Btu after freezing, and a latent heat of fusion of 66 Btu/lb. One cubic foot of air at a temperature of 90° F and 60 percent relative humidity that is cooled to 35° F will release 2.43 Btu.

Calculation of refrigeration compressor capacity takes into consideration the sources of heat through wall losses, air changes and leakage, pounds of products refrigerated, and miscellaneous losses such as electric light bulbs, men working, and heat released due to respiration of fruits and vegetables.

Heat transmission through walls, floors, and ceilings is by convection, conduction, and radiation. The amount of heat transferred depends largely on the overall coefficient of heat transmission of the wall materials in Btu per hr per ft^2 of surface per degree difference in temperature. Computation of heat transmission coefficients is explained in detail in *The American Society of Heating, Refrigeration, and Air-Conditioning Engineers Guide (ASHRAE Guide)*[52] and other related standard texts. For example, 1-in. typical corkboard has a conductivity of 0.30 Btu per hr per ft^2 per degree Fahrenheit or 7.2 Btu/24 hr. Sawdust (dry) has a conductivity of 0.41 Btu per in. thickness or 9.84 Btu per 24 hr per ft^2 per degree. One in. of cement plaster has a conductivity of 8.0 Btu or 192 Btu/24 hr.

Air changes in a refrigerator each time the door is opened. Cold air rolls out and warm air enters, thereby raising the temperature of the air in the refrigerator. A walk-in cooler having a volume of 300 ft^3 will probably have about 35 air changes per 24 hr due to door opening and infiltration. This will amount to 10,500 ft^3 of air. If the air has a temperature of 90° F and 60 percent relative humidity and is to be cooled to 35° F, it will represent 2.43 Btu/ft^3 of air, or in the instance cited (10,500 × 2.43) = 25,500 Btu of heat per 24 hr to be removed by the refrigerating unit. In restaurants, where refrigerators normally receive heavy usage, it is good practice to estimate a 50 percent increase over normal in the number of air changes.

When food is placed in a refrigerator it is usually necessary to remove heat in the food in order to cool it to the desired temperature. The heat load is dependent on the weight of the product, its specific heat, the temperature reduction, and the speed with which the product is to be cooled. If 100 lb of fresh poultry is to be cooled from 65 to 35° F in 6 hr, the heat load will be:

$$\frac{100 \text{ lb} \times 0.80 \text{ sp heat} \times 30^\circ \text{ temperature reduction} \times 24 \text{ hr}}{6 \text{ hr}} = 9600 \text{ Btu/24 hr}$$

[52] ASHRAE, 345 East 47th Street, New York, N.Y., 1970.

If the products refrigerated are fresh fruits or vegetables, adjustment must also be made for the heat released by these foods in storage. It is estimated for example that strawberries stored at 40° F give off about 3 Btu/lb per 24 hr.

Miscellaneous losses include 3.42 Btu per hour per watt of light bulb, 3000 Btu per hour for each motor horsepower, and 750 Btu per hour per man working in a box. A safety factor of at least 10 percent is added to the total load, and compressor capacity is based on a 16-hr operating day to provide for a defrosting cycle. A horsepower hour is 2546 Btu or 0.7457 kilowatt hour. Keep motor and refrigerating unit free of lint and dirt to permit air circulation, and be sure door gaskets keep cold air from escaping.

An example will serve to illustrate the refrigeration design principles and data for the calculation of refrigeration load. Given: a box 10 ft × 10 ft × 8 ft outside, 3-in. cork insulation standard construction with two layers of insulation paper and sheathing inside and outside, room temperature 95° F, and box temperature 35° F, 100 lb of fresh poultry received at 65° and 150 lb of lean beef received at 65° cooled to 35° in 24 hr and 50 w electrical load.

Solution:

1. Box outside surface area: 520 ft^2
2. Temperature reduction: 60°
3. Wall loss = 1.8 Btu per ft^2 per degree temperature reduction (see Table 8-11)
   = 1.8 × 520 = 60 = 56,200 Btu/24 hr
4. Air change = 9 × 9 × 7 ft (assume inside dimension 1 ft less than outside)
   = 567 ft^3 × 23 air changes per 24 hr (see Table 8-14)
   = 13,040 ft^3
5. Btu to cool 1 ft^3 of air from 95 to 35° = 2.79 (see Table 8-12)
   Hence to cool total air change requires

   $$13{,}040 \times 2.79 = 36{,}500 \text{ Btu/24 hr}$$

6. Product load of poultry and beef
   Specific heat of poultry = 0.80 Btu/lb (see Table 8-13)
   Specific heat of beef = 0.77 Btu/lb

   $$\text{Product load} = \left[\frac{100 \times 0.8 \times (65\text{–}35°)}{24} + \frac{150 \times 0.77(65\text{–}35°)}{24}\right]24$$
   $$= 2400 + 3460 = 5860 \text{ Btu/24 hr}$$

Table 8-11 Heat Transmission Coefficients

Insulation Material	Wall Heat Loss[a] per °F Room Minus Cooler Temperature
Corkboard, 1″	7.2
Single glass	27
Double glass	11
Corkboard paper and sheathing on both sides	3.6 for 1″ cork
	2.4 for 2″ cork
	1.8 for 3″ cork
	1.5 for 4″ cork

[a] Btu/24 hr/ft^2 of outside surface. See *ASHRAE Guide* for other coefficients.

Table 8-12 BTU per Cubic Foot of Air Removed in Cooling to Storage Conditions[a]

Storage Room Temp. (° F)	Temperature of Air Outside, Assumed at 60 percent Relative Humidity			
	85°	90°	95°	100°
65	0.85	1.17	1.54	1.95
60	1.03	1.37	1.74	2.15
55	1.34	1.66	2.01	2.44
50	1.54	1.87	2.22	2.65
45	1.73	2.06	2.42	2.85
40	1.92	2.26	2.62	3.06
35	2.09	2.43	2.79	3.24

[a] Data for the calculation of refrigeration load adapted from *Kramer Engineering Data,* Catalog No. R-114, Kramer-Trenton Co., Trenton, N.J.

Table 8-13 Specific Heat, BTU per Pound[a]

Product	Btu per lb per ° F before Freezing	Btu per lb per ° F after Freezing	Latent Heat of Fusion, Btu per lb
Asparagus	0.95	0.44	134
Berries, fresh	0.89	0.46	125
Beans, string	0.92	0.47	128
Cabbage	0.93	0.47	130
Carrots	0.87	0.45	120
Dried fruit	0.42	0.27	32
Peas, green	0.80	0.42	108
Potatoes	0.77	0.44	105
Sauerkraut	0.91	0.47	129
Tomatoes	0.95	0.49	135
Fish, fresh	0.82	0.41	105
Fish, dried	0.56	0.34	65
Oyster, shell	0.84	0.44	115
Bacon	0.55	0.31	30
Beef, lean	0.77	0.40	100
Beef, fat	0.60	0.35	79
Beef, dried	0.34	0.26	22
Liver, fresh	0.72	0.40	94
Mutton	0.81	0.39	96
Poultry	0.80	0.41	99
Pork, fresh	0.60	0.38	66
Veal	0.71	0.39	91
Eggs	0.76	0.40	98
Milk	0.90	0.46	124
Honey	0.35	0.26	26
Water	1.00	0.50	144

[a] Data for the calculation of refrigeration load adapted from *Kramer Engineering Data*, Catalog No. R-114, Kramer-Trenton Co., Trenton, N.J.

Table 8-14 Air Change for Storage Rooms (32° to 50°) Caused by Door Opening and Infiltration[a]

Cubic Feet	Air Change per 24 hr	Cubic Feet	Air Change per 24 hr
250	38.0	800	20.0
300	34.5	1,000	17.5
400	29.5	1,500	14.0
500	26.0	2,000	12.0
600	23.0	4,000	8.2

[a]Data for the calculation of refrigeration load adapted from *Kramer Engineering Data*, Catalog No. R-114, Kramer-Trenton Co., Trenton, N.J.

7. Electrical load = 3.42 Btu per watt per hour. Assume 8-hr lighting
   = 3.42 × 50 × 8 = 1368 Btu/24 hr

   Total load = 56,200 (wall) + 36,500 (air change) + 5,860 (product) +1368 (electric)
   = 99,928 Btu/24 hr plus 10 percent safety factor 0.1 (99,928)
   = 99,928 + 9993 = 109,921 Btu/24 hr

   Compressor capacity, based on 16-hr operation,

$$\frac{109{,}921}{16} = 6875 \text{ Btu/hr}$$

Approximate calculation:

Usage heat loss = 1.52 Btu per ft³ per deg F per 24 hr (see Table 8-15 on p. 803)
= 1.52 × 567 ft³ × 60° = 52,000 Btu/24 hr

Total = 52,000 (usage) + 56,200 (wall) = 108,200 Btu/24 hr

Compressor capacity, based on 16-hr operation,

$$\frac{108{,}200}{16} = 6760 \text{ Btu/hr}$$

Cleansing

The effectiveness of cleansing, including bactericidal treatment of equipment and utensils, can be determined by visual inspection and by laboratory methods.[53,54] A standard measure is the bacteriological swab test. One swab for each group of five or more similar utensils, cups, glasses, or five 8-in.^{2} surface areas is used. A standard plate count (37°C for 48 hr) of less than 100 per utensil, or not more than $12\frac{1}{2}$ colonies per in.^{2} of surface examined, and the absence of coliform organisms indicates a satisfactory procedure. The Replicate Organism Direct

[53]William G. Walter, Editor, *Standard Methods for the Examination of Dairy Products*, 13th ed., American Pub. Health Association, Washington, D.C., 1972.

[54]Marvin L. Speck, Ed., *Compendium of Methods for the Microbiological Examination of Foods*. American Public Health Association, 1015 Eighteenth St., NW, Washington, D.C. 20005, 1976, pp. 95–100. See also PHS Pub. 1631, "Procedures for the Bacteriological Examination of Food Utensils and/or Food Equipment Surfaces," FDA, Division of Retail Food Protection, 200 C St. S.W., Washington, D.C. 20204.

Agar Contact (RODAC) agar plate (with slightly convex surface) method is preferred for flat surfaces. Disposable plastic plates are available. The acetate tape and filter pad method also has wide use. In the rinse solution method a measured volume of sterile solution or nutrient broth is flushed or circulated over or through a surface. The bacterial population is found by plating or with a membrane filter. Laboratory testing of dishware and equipment is not intended for routine use, but is of value for special studies and to provide specific information. There are many field methods to determine the cleanliness of surfaces and effectiveness of cleansing materials and operations.[55-57] Some are discussed below.

Single-service eating and drinking items usually have very low bacteria counts and contribute to proper sanitation in food service. They have particular application in emergency situations and in fast food service operations.

1. The suitability of a detergent can be roughly determined by rinsing it off the hands (if not a caustic) in the available water. It should not leave a greasy or sticky feeling. A normal solution in warm water should completely dissolve without any precipitate. Dishes and utensils should show no spots or stains. A measure of rinsibility is the alcohol test. A drop of alcohol on the dry surface allowed to evaporate should leave no deposit. Another field test uses phenolphthalein. A drop of standard phenolphthalein is added to the wet surface of the utensil to be tested. If the water turns red, it is an indication of residual alkalinity and hence poor rinsing. If the pH of the tap water is above 8.3, the surface to be tested should be wetted with distilled water. The cleanliness of a glass can be observed by filling it with water and watching it drain. A clean glass would show a continuous unbroken water film; breaks in the water film would indicate unremoved soil. Metal parts, however, could have a uniform coating of soil that would not give a water break. Salt sprinkled over the utensil or equipment to be examined, while still wet, will not adhere to surfaces that have not been cleaned. A simple field test is the pouring of plain soda water in a glass. Gas bubbles will cling to unclean areas. Other aids are a good flashlight, paper tissues, filter pads, a spatula, and studied individual observation.

2. A fluorochromatic method uses a nearly colorless fluorochrome such as brilliant yellow uranine, fluorescent violet 2G, ultraviolet light (black light), and others. Dishes, for example, would be preflushed, immersed a few seconds in a fluorochrome solution, and washed in the usual manner. The dishes on the clean-dish table are then examined under ultraviolet light. Any soil remaining would be fluorescent.

3. The efficiency of cleansing materials and operations can be determined by various methods with greater accuracy in a laboratory. Methods particularly suited for research use a standard soil, radioactive tracers, and fluorescent dyes.

4. Pots, pans, and other equipment require brushing in the detergent washing operation. Rancidity, staleness, and other off-flavors and odors in prepared foods can frequently be traced to incompletely removed food or mineral films on food preparation equipment. Three-compartment sinks, each 24″ × 24″ or 24″ × 30″, with 3-ft drainboards, and a steam or electric immersion heater or gas booster under the last two sinks are practical necessities. A mechanical pot and pan washer is also used.

5. A method of detecting residual grease and protein or starch films on china, plastic, glass, or

[55] Walter D. Tiedeman and Nicholas A. Milone, *Laboratory Manual and Notes for E. H. 220*, Sanitary Practice Laboratory, School of Public Health, University of Michigan, Ann Arbor, 1952.

[56] J. D. Baldock, "Microbiological Monitoring of the Food Plant: Methods to Assess Bacterial Contamination on Surfaces," *J. Milk Food Technol.*, July 1974, pp. 361–368.

[57] Bertha Yanis Litsky, *Food Service Sanitation*, Modern Hospital Press, McGraw Hill Pub. Co., Chicago, Ill. 1973, pp. 145–153.

silverware is the use of a dry mixture of 80 percent talcum dust (not commercial powder) and 20 percent safranin by weight. The mixture is placed in a saltshaker covered with cheesecloth to limit the flow and dusted over the surface of the utensil or article to be tested. It must be dry. Then hold the utensil under an open tap for at least 30 sec, tilt to drain off water, and examine for any red color. The intensity of the color is an indication of the amount of residual soil present.[58]

Proper cleansing involves the use of a washing compound or detergent suitable for the hardness, and pH and the available water should have a temperature that will result in the removal of fats, proteins, and sugars, usually referred to as "soil" or organic matter. The detergent must be used in adequate amounts. Too little does not accomplish cleansing; too much means greater expense and more care and time for satisfactory rinsing.

In some cases chemical disinfectants[59] and combination cleanser-sanitizers are used because an adequate supply of hot water is not available in the first place. As a result, some operators of eating and drinking places resort to chemical compounds in an effort to obtain clean, disinfected dishes and utensils.

Chemical disinfection is less reliable than 170 to 180° F hot water and retards air drying. Chemical sanitizers are not effective on dishes or utensils that are not properly cleansed. When properly used, hypochlorite solutions can produce satisfactory results. A solution strength of 100 mg/l available chlorine should be used. It can be prepared by mixing ¼ oz of 5¼ percent bleach with one gal of water. Immersion for at least one minute is required. The solution should be replaced when the concentration falls to 50 mg/l or when the water cools below 75° F. The equivalent of a 3-compartment sink is essential to rinse the soil off between the wash and disinfection step, as organic matter and milk use up chlorine, thereby making it ineffective as a disinfectant. A 200 mg/l solution is recommended when the hypochlorite is used as a spray. At a pH of 7.0 a 250-mg/l solution for a 2-min contact period is necessary for effective disinfection. At a pH of 8.5 a 250-mg/l solution for a 20-min contact period is necessary for effective disinfection. It should also be pointed out that chlorine turns silver black. Hypochlorite solutions affect stainless steel. Immersion for at least 1 min in a clean solution containing at least 12.5 mg/l of available iodine and having a pH not higher than 5.0 and at a temperature of at least 75° F is also acceptable.[60]

Other chemical disinfectants, such as quaternary ammonium compounds, have definite limitations. Bicarbonates, sulphates, magnesium chloride, calcium chloride, and ferrous bicarbonate interfere with the bactericidal quality of quaternary ammonium compounds. Quaternaries should therefore only be used where their successful performance has been demonstrated in the establishment concerned. The limits within which satisfactory results can be assured should be specified by the manufacturer of the sanitizer, as water chemical quality can be expected to vary from community to community. Immersion in a 200 mg/l solution for 1 min is usually required.

The so-called detergent-sanitizer combinations have been reported to be effective under certain conditions. They are more likely to be weakened in cleansing and disinfecting power by organic matter and by incompatibility under conditions of use.

If chemical disinfection is used, equipment should be provided and tests made by the operator to assure that the sanitizer is maintained at the proper strength. This objective is rarely achieved in practice.

The type of dishwashing method or machine that may be used depends on the number of dishes or pieces to be washed per hour, the money available, hot-water

[58] E. H. Armbruster, "Method for Detecting Residual Soil on Cleaned Surfaces," *J. Am. Dietetic Assoc.*, **39**, 228–230 (1961).

[59] *Food Service Sanitation Manual*, PHS/FDA, Washington, D.C., 1976, pp. 83–87.

[60] Ibid, p. 43.

storage and heating facilities, individual preferences, and other factors. For example, if fewer than 400 pieces are to be washed per hour, a 3-compartment sink or single-tank dishwashing machine can be used. If 1200 pieces are to be washed per hour, then three 3-compartment sink arrangements or one single-tank dishwashing machine may be used. If 2400 pieces are to be washed per hour, a 2-tank dishwashing machine may be the best answer. It soon becomes apparent that the cost of labor alone becomes the factor to be weighed against the cost of a dishwashing machine. Manufacturers of approved-type dishwashing machines are in a position to make recommendations for specific installations.

Soiled silverware requires special handling to assure removal of soil. It has been found that silverware that has been allowed to soak at least 15 min at 130 to 140° F in a good detergent can be hand or machine washed with uniformly good results. Pot, pan, and utensil commercial spray-type washing machines are available.

Mold Control

The relatively high humidity in most food establishments and the incompleteness of many routine cleaning operations favor the development of molds and odors. Molds grow in refrigerators and on foods, floors, walls, ceilings, and equipment. Since mold spores are carried by air currents, packing, and dirt, they are difficult to keep out; hence their control depends on cleaning, humidity reduction, and treatment. A recommended mold-control program includes cleansing of the affected area with an alkaline detergent, spraying with a 5000-mg/l sodium hypochlorite solution with care to materials that might be harmed, drying of the hypochlorite, spraying with a 5000-mg/l quaternary ammonium compound, and then respraying of the entire section with a 400 to 500-mg/l quaternary solution every week or two.[61,62] Practically all off-odors in food establishments are due to decomposing food, grease, or other organic material. Their control rests on proper design and routine cleanliness. The use of deodorants or special odorous disinfectants temporarily masks the odor, but cannot substitute for a warm detergent solution and "elbow grease."

Hand Dishwashing, Floor Plans, and Designs

For hand washing, a 3-compartment sink and long-handled wire baskets are required. See Figure 8-14. The following steps are recommended.

1. *Scrape and flush* all dishes and utensils promptly—before the food soil has a chance to dry. A spray-type flusher delivering water under pressure at a temperature of 120° F, or a pan of hot water and brush, will remove most of the heavy soil.
2. *Wash* in the first compartment with water at a temperature of 110 to 120° F using an effective

[61] Thomas D. Laughlin, *Modern Sanitation Practices*, Institutions Div. of Klenzade Products, Inc., Beloit, Wis., 1954.

[62] Austin K. Pryor and Regina S. Brown, "Quaternary Ammonium Disinfectants," *J. Environ. Health*, January/February, 1975.

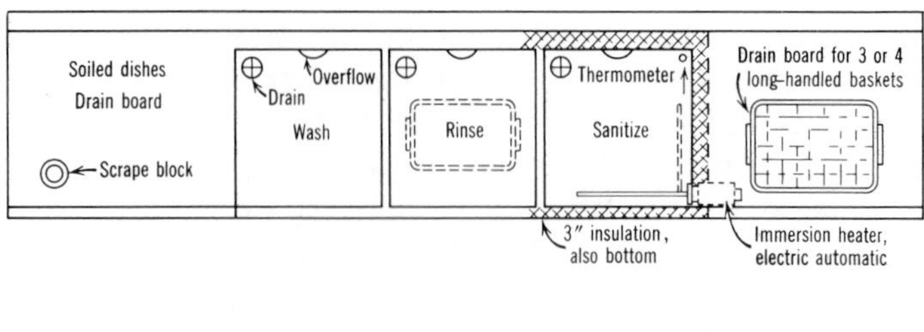

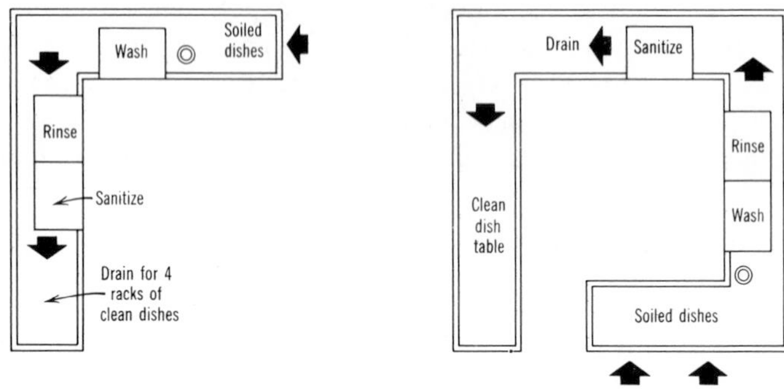

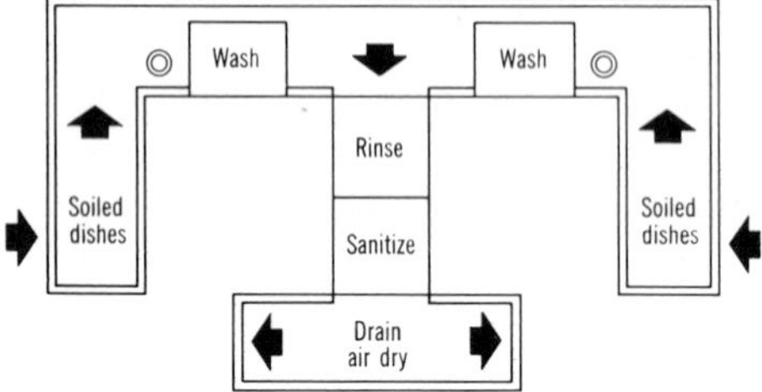

Figure 8-14 Hand dishwashing plans for small to medium operations. Dishwashing machine is recommended where possible. Dish counters 24- to 36-in. wide.

washing compound (detergent). The detergent solution strength must be maintained. Glasses will require brushing; wash, rinse, and sanitize separately from dishes.

3. *Rinse* in the second compartment by immersion in hot water. The dishes, cups, and utensils from the washing operation can be placed directly in the wire basket, previously submerged in this sink, to save an extra handling. Place cups and glasses in a venting position. A hot water spray is also suitable for rinsing if thoroughly used and the drain is kept open.

4. *Sanitize* in the third compartment by immersing the basket of dishes in water at a temperature of 170 to 180° F or higher for at least 30 sec, although 2 min is generally recommended. Steps 3 and 4 can also be accomplished by thorough spraying of carefully racked dishes with 180° F water in a properly designed and maintained single tank-stationary rack dishwashing machine or in an improvised spray-rinse cabinet.

5. *Air dry*—do not towel! Towelling is an expensive, time-consuming, and insanitary operation. Towels become moist, warm, and collectors and spreaders of germs. Store dishes, cups, and utensils in a clean dry place and handle them in a sanitary manner. Cups should face down; knives, forks, and spoons with handles up. Dishes and utensils properly sanitized with scalding hot water will dry in less than one minute. Plastic dishes will take longer unless a drying agent is used in the rinse water.

Glass washing becomes a special problem where facilities are inadequate. The glass, being transparent, is subject to critical visual inspection on being raised to one's lips, thereby creating demand for a glass that looks clean. In any case, glasses should receive cleansing and sanitization treatment equal to that given dishes and utensils. A procedure that has been found effective requires the washing of glasses separate from dishes and includes the following steps.

1. *Remove* cigarette butts, papers, ice, liquid, and straws. Then prerinse the glasses with warm water to remove milk and syrup. This will also warm up the glasses and prevent them from cracking when placed in hot water. If glasses cannot be prerinsed promptly, soak them under warm water to prevent the milk, syrup, and soil from hardening.
2. *Wash* glasses in warm (110 to 120° F) water containing an organic acid detergent using a stiff bristle brush to remove lipstick and stubborn soil. Stationary and motor-driven cluster-type brushes that are placed in the bottom of wash tanks are available.
3. *Rinse* the glasses in clean warm water or in a hot-water spray. If a basket is used care should be taken to stack the glasses in a venting position. Use a special basket that holds glasses in place.
4. *Sanitize* the glasses as described for dishes (Item 4), or in clear hot water containing $\frac{1}{4}$ oz of $5\frac{1}{4}$ percent hypochlorite solution per gal (100 mg/l) of water in the sink. Make up a fresh tank of hot water and chlorine solution when the water cools to below 75° F or the solution strength is less than 50 mg/l available chlorine. A water softener or a special solution added to the final rinse tank may prevent the formation of spots. Drying the bottom of the glass will also help prevent the formation of spots. Quaternaries and iodophors may also be reliable disinfectants under controlled conditions.
5. *Air dry* the glasses by placing them bottoms up on a wire or plastic draining and drying rack after the disinfecting rinse. Do not towel the glasses dry.

Some establishments have found it more convenient to wash and sanitize glasses in the kitchen sinks or dishwasher instead of at the fountain or bar. Although this procedure will entail keeping a somewhat larger stock of glasses, it is offset by increased efficiency, less confusion at the counter, and improved service. Where a dishwasher is used, the glasses should be washed before the dishes. If this is not possible, the dishwasher should be thoroughly cleaned, new detergent solution made up for the wash tank, and fresh rinse water added in a 2-tank machine before the glasses are washed. Special hot-water glass washers and sanitizers that greatly simplify the entire cleansing routine are also available. In either case, glasses should first be washed in a sink with stationary or rotating brushes and washing solution to assure the consistent removal of all lipstick and stubborn soil. Glasses that have become coated or clouded can be made sparkling clean by soaking in an organic acid solution similar to that used in the dairy farm or milk plant for milk-stone removal, or by the use of an organic acid detergent. Where equipment and personnel are not available to produce clean glasses, paper cups should be used. Paper service for drinking water, ice cream, milk, and other drinks will greatly reduce the dishwashing problem and frequently improve public relations.

A fundamental objective in the design of a dishwashing space is the continuous movement of dirty dishes through the cleansing operation to the point of distribution or storage without interruption or pile-up. This requires an efficient floor plan that eliminates confusion and provides adequate soiled-dish drainboard; space for air drying of long-handled baskets; an adequate number of long-handled baskets; sufficient personnel; and properly designed water-heating facilities. Some floor plans or flow diagrams are shown in Figure 8-14. In operation, an extra long-handled basket can be kept in the rinse compartment when the wash and rinse tanks are adjoining. Then as dishes are washed they can be placed directly in the basket to be rinsed, sanitized, and air-dried without further individual handling. If the wash and rinse tanks are not adjoining, the basket can be placed on the drainboard alongside the wash tank. In small establishments one man can perform all operations. During rush periods, or when the number of dishes is large, the tasks can be divided between two persons, extra wash tanks provided as illustrated, or duplicate 3-compartment sink arrangements can be installed. About 400 pieces can be washed per hour.

Hot-Water Requirements for Hand Dishwashing

Maintenance of the water in the sanitizing compartment at the proper temperature requires auxiliary heat. This is also desirable in the wash and rinse compartments, unless the water is replaced frequently. A gas burner beneath an insulated sink or an electric immersion heater in or under the sink are the common methods of keeping the water in the third compartment at 180° F or higher. Other heat sources are gas- or oil-burning units, steam and hot-water jackets, "side arm" heaters, or the equivalent. Consult with manufacturer's representative.

Thermostats that automatically control water temperatures, and tank insulation, are useful accessories that help maintain proper water temperatures and also reduce heat losses. A thermometer installed on the third compartment is necessary to show the dish washer when the water is at disinfection temperature. Thermometers on the wash and rinse tanks are also useful. When an open flame is used beneath the sink to heat the water, a copper sheet placed against the sink bottom will prevent burning out the bottom of the sink. Dial thermometers and gages are frequently inaccurate. They should be checked against a reliable standard. If an adequate number of sinks is not available, a 20-gal pot of hot water can be kept on a stove and dishes immersed in boiling water with the aid of a special basket as a temporary expedient. The 3-can method used by the U.S. Army is very satisfactory for field use. The first can contains hot water with a detergent and brush, the second can boiling water for rinsing, and the third can boiling water for sterilization. An undersink gas heater requires a vent.

An example is given below showing calculation of the amount of energy required to heat water in sinks.

Given: A sink 18″ × 24″ containing 12 in. of water at a temperature of 120° F. The sink is open at the top and uninsulated. One ft^3 of water = 62.4 lb. Room temperature = 80° F. Water surface loss is approximately 1430 Btu for 100° F temperature difference per hour per square foot. For 80° temperature difference the loss is 940 Btu, and for 120° the loss is 2320 Btu. Failure to insulate the tank results in a heat loss of 215 Btu per ft^2 per hr when the temperature of the hot water is 180° F and the air temperature is 70° F, or about 1.95 Btu per hr per deg per ft^2.

Required: The heat energy to keep the water temperature at 180° F, with air temperature at 80° F, to disinfect 100 to 400 pieces per hour. Assume the temperature of the wash water is 100° F, and a dish, saucer, and cup weigh 2 to 3 lb—say 3 lb or an average of 1 lb per piece. The specific heat, or the amount of heat required to raise 1 lb of crockery 1 deg is 0.22 Btu. The specific heat of water is 1.00 Btu per lb, at 39° F (4° C).

1. The heat loss through the sides and bottom of the sink with no insulation, in Btu/hr = ft^2 of surface area × wall loss in Btu/per hr per deg × deg temperature difference between the water and room
   = 10 × 1.95 × 100
   = 1950 Btu/hr
2. The heat loss from an open water surface, in Btu/hr = ft^2 of water surface × surface loss in Btu for 100° temperature difference between water and room
   = 3 × 1430
   = 4290 Btu/hr
3. Heat absorbed by 100 dishes, in Btu/hr = weight of dishes in lb × specific heat of crockery in Btu × temperature rise in degs
   = 100 × 0.22 = 100
   = 2200 Btu/hr
4. The heat required to raise the temperature of water in a sink 18 × 24 in. containing 12 in. water from 120 to 180° F, in Btu/hr = lb of water × specific heat of water × total temperature rise (rather than average)
   = (1.5 × 2 × 1)62.4 × 1 × 60
   = 11,250 Btu/hr

Therefore: The total heat loss is the sum of the loss from the sink sidewalls and bottom, water surface, and heat absorbed by the crockery sanitized in 1 hr.

Total heat loss for (after temperature of 180° F is reached)
—100 pieces = 1950 + 4290 + 2200 = 8440 Btu/hr
—200 pieces = 6240 + 2(2200) = 10,640 Btu/hr
—300 pieces = 6240 + 3(2200) = 12,840 Btu hr
—400 pieces = 6240 + 4(2200) = 15,040 Btu hr

The heater provided should be at least capable of conducting 11,250 Btu/hr to the water in the sink. This heat will be adequate to disinfect about 250 pieces/hr as noted above. If 400 pieces are to be disinfected/hr, the heater should be capable of conducting 15,040 Btu/hr. Hence an electric immersion heater would have a capacity $= \frac{15{,}040}{3412} = 4.5$ kW-hr.

An open-gas flame heater, being only about 50 percent efficient overall would have a capacity $= \frac{15{,}040}{0.50} = 30{,}080$ Btu/hr.

NOTE: The wall heat loss of the sink can be reduced to approximately 0.24 Btu per degree temperature difference per hour with 3-in insulation. The water surface heat loss of 4290 Btu during buildup of water temperature can be reduced to 429 Btu (1/10) by providing an insulating cover.

Electric immersion heaters for sinks require a 115-V circuit for the lower capacities of 1500 to 3000 W and a 230-V circuit for the higher capacities; thermostat control; variable or three-point heat switch; and burnout protection or low-water cutoff controls. Immersion heaters are commercially available. If sink bottom and wall insulation is omitted, radiation heat losses will be increased by about eight times and must be compensated for by increased wattage.

Where a gas or oil burner is used beneath a sink, *precautions must be taken to fire-retard the space around the burner.* Consult with the local utility company. This space should be vented to the outer air and a flame guard provided. In addition, a copper sheet placed between the flame and bottom of stainless steel or other type of sink will prevent burning out of the bottom. The American Gas Association recommends, for furnishing 180°-water in sinks supplied with 140°-inlet water, the following gas burner inputs in Btu per hour for under-sink burners.

26,000 Btu/hr for 16 × 16 to 18 × 18-in. sinks
30,000 Btu/hr for 18 × 20 to 20 × 20-in. sinks
34,000 Btu/hr for 20 × 20 to 22 × 22-in. sinks
38,000 Btu/hr for 22 × 24 to 24 × 24-in. sinks[63]

For example, a sink 24 × 24 in. with 12 in. of water will hold (2 × 2 × 1)62.4 = 249.6 lb of water. If the water enters at 120° F and is to be heated to 180° F, the heat required = 249.6 × 60 = 14,976.0 Btu/hr. The heat loss through bottom and sides of tank without insulation = surface area × wall loss × degrees temperature difference between water and room = 12 × 1.95 × 100 = 2,340 Btu/hr. Heat loss from open water surface = 1,430 Btu/ft^2 for a 100° temperature difference, which for 4 ft^2 = 5720 Btu. The total heat loss, assuming a 1-hr temperature buildup = $14{,}976 + \frac{2340}{2} + \frac{5720}{2} = 19{,}006$ Btu. With an open-gas flame heater assumed to have an overall efficiency of 50 percent, the heat input = 19,006/0.50 = 38,000 Btu/hr.

Machine Dishwashing, Floor Plans, and Designs

When more than about 400 dishes and cups must be washed per hour, either a duplicate 3-compartment sink and drainboard installation is required or a mechanical dishwasher should be provided. Of course, a dishwashing machine can be provided regardless of the number of dishes washed. Machines are available to serve the average household and up to 50, 125, 250, 400, or 3000 persons.* In any case the equipment, including the hot-water supply, must always be proper and adequate to meet peak operating requirements. Since the

*Machines are available with operating capacities as high as 450 racks (20 × 20) and 15,000 to 22,000 dishes per hour. (*Listing of Food Service Equipment*, National Sanitation Foundation, Ann Arbor, Mich., Jan. 1977.)

[63] J. Stanford Setchell, *Enough Hot Water—Hot Enough*, American Gas Association, 420 Lexington Avenue, New York, N.Y. 10017.

average person does not know which dishwashing machines will produce satisfactory results, he should be guided by the recommendations of the National Sanitation Foundation and purchase a machine that contains their seal of approval.

For proper machine dishwashing the following steps are recommended to help assure consistently satisfactory results.

1. *Scrape* dishes clean of food and soil with the hand or a rubber scraper. This operation can also be accomplished by a pressure water spray or a large recirculated stream of water at a temperature of about 110° F. A garbage grinder or food-waste disposer is a very desirable unit to incorporate in this operation. The dishwashing layout should provide a soiled-dish table equipped with a removable strainer the width of the table, just ahead of the machine, to intercept liquid garbage before it enters the dishwashing machine.

2. *Rack dishes.* Place dishes in special wooden or wire racks or flat on the conveyor belt. Do not stack or otherwise overcrowd the dishes. Spray the rack of dishes with a manual preflusher delivering water at a temperature of 140° F under pressure or preflush each dish by hand. See Figures 8-15 and 8-16.

3. *Wash and sanitize.* Feed racked dishes into the dishwashing machine at the established speed for the machine. The time for the wash and rinse cycles, the volume and temperature of water required, and other details for each type of machine are given in Tables 8-16 and 8-17. For satisfactory results, the proper use of a good washing compound or detergent that is suitable for the water hardness is necessary. The concentration of detergent should be maintained at 0.2 to 0.3 percent. This can be obtained at the beginning with about 1 lb or $1\frac{2}{3}$ cups of washing powder to 50 gal of fresh wash water. But the detergent must be supplemented during the dishwashing operation to compensate for the soil, rinse water, and make-up water entering the wash-water tank and diluting the detergent concentration. This is accomplished by automatic detergent feeders or by the hand feeding of detergent. Follow the manufacturers' directions. Studies show that the detergent concentration becomes ineffectual after 10 to 15 racks of dishes are washed. An automatic detergent feeder requires cleaning after each dishwashing period, just like a dishwashing machine. Another problem concerns the wash- and rinse-water temperatures. Provision for heating the water in the dishwashing machine wash tank or tanks is often made by the manufacturer, but this should be checked under operating conditions. The connection of an adequate volume of 180° F water to the machine final-spray rinse line is usually under a separate contract and is often neglected, unless the machine representative, designing architect or engineer, health official, or plumbing contractor calls attention to the need. This problem and its solutions are discussed separately under Hot Water for General Utility Purposes.

4. *Air dry* the dishes while in the racks. If rinsed at the proper temperature the china dishes will dry in less than 1 min. Plastic cups, dishes, and trays dry slowly. Plastic is hydrophobic, causing water to

Table 8-15 Usage Heat Loss, Interior Capacity per ° F Difference Between Room and Box Temperature, per 24 hr

Volume (ft^3)	Heat Loss (Btu per ft^3—Service) Normal	Heat Loss (Btu per ft^3—Service) Heavy	Volume (ft^3)	Heat Loss (Btu per ft^3—Service) Normal	Heat Loss (Btu per ft^3—Service) Heavy
15	2.70	3.35	400	1.62	2.38
50	2.42	3.10	600	1.52	2.28
100	2.12	2.85	800	1.48	2.22
200	1.85	2.60	1000	1.42	2.15
300	1.70	2.45	1200	1.38	2.10

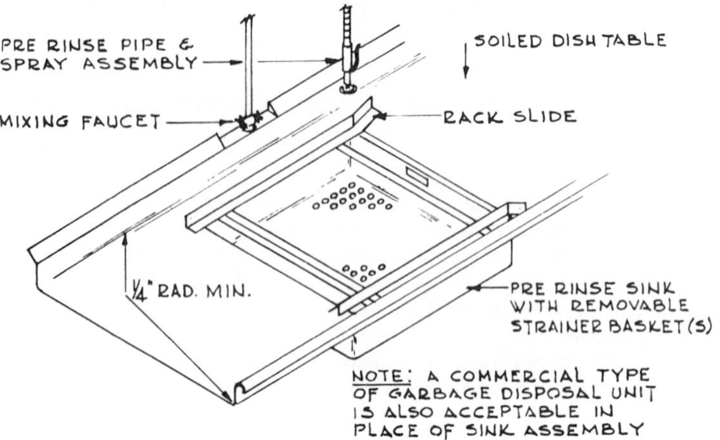

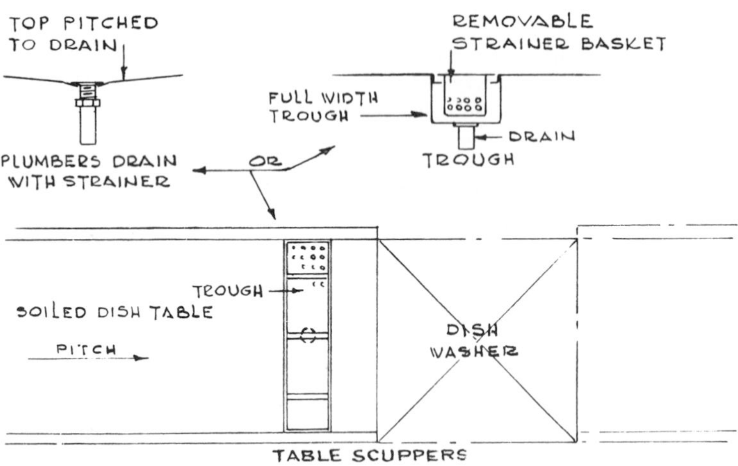

Figure 8-15 Cleaning facilities and accessories. From *Food Service Equipment Standards*, National Sanitation Foundation, NSF Building, Ann Arbor, Michigan, 48105, September 1978, pp. 45–46.

form in droplets that take longer to evaporate than a film. Drying agents are now available that can be fed into the rinse water, thereby overcoming this difficulty. Sufficient space on an airy dish table is necessary for the air drying to take place promptly. Consideration should be given to adequate ventilation to prevent condensation. The use of lower temperature final rinse water will prolong air drying and introduce potential problems of storage, bacterial growth, and service of wet dishes and utensils.

Design should provide for some wasted rinse water due to inefficiency in operation; in practice machines are not operated without interruption, nor are

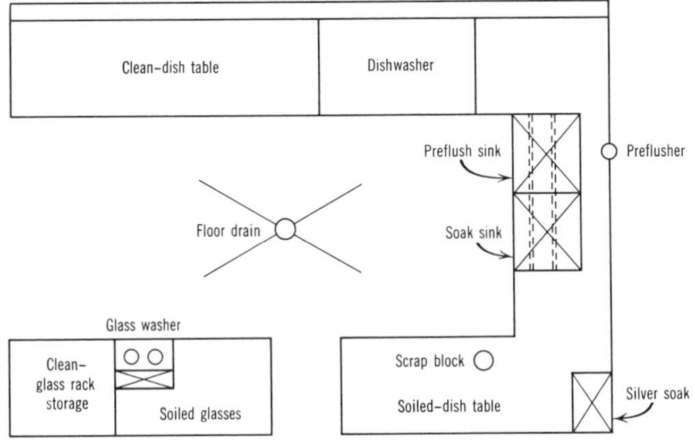

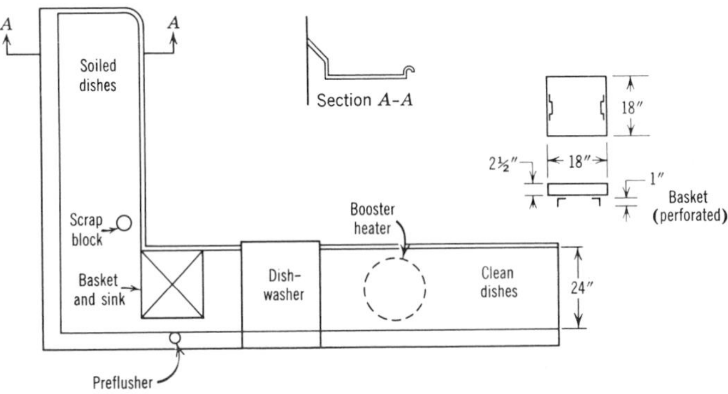

Figure 8-16 Dishwashing arrangements—machine.

racks always fully loaded. Some designers say the machine is actually in operation only about two-thirds of any hour, but then add 50 percent to the hourly flow to compensate for wasted 180° F-rinse water such as with partially filled racks or conveyors. Check actual rinse-water usage with manufacturer and actual water flow under operating conditions. Flow is based on 20 psi at machine. Do not use a push-through-type machine as the amount of rinse water depends on the "human element;" require automatic wash and rinse timers. Assume a 20 × 20-in. rack holds 24 pieces, an 18 × 18-in. rack 20 pieces, and a 16 × 16-in. rack 16 pieces. A common estimate is that an average complete restaurant meal results in 8 dishes or pieces, a luncheon in 6 pieces, a cafeteria meal in 4 pieces, and an elaborate dinner 10 pieces.

An efficient floor plan adapted to each establishment is necessary to realize the advantages associated with the use of a dishwashing plan. Some plans that show the essential elements are illustrated in Figures 8-11, and 8-16.

Table 8-16 Stationary Rack Dishwashing Machine, Wash and Rinse Water, Volume, Temperature, and Time

	150 to 160° F Wash Cycle per Rack[a]			Alternate Wash Cycle per Rack[b]		Rinse, 180° F to 195° F[c]	
Rack	Water Volume (gal)	Time; not less than (sec)	Pump Capacity (gpm)	Water Volume (gal)	Min. Pump Capacity (gpm)	Time; not less than (sec)	Water Volume (gal) 20 psi
20″ × 20″	92	40	140	92	75	10	1.73
18″ × 18″	75	40	112	75	50	10	1.44
16″ × 16″	60	40	90	60	40	10	1.15
Other	0.23 × rack area in square inches	40	0.35 × rack area in square inches	0.23 × rack area in square inches	0.15 × rack area in square inches	10	0.43 gal/100 in^2 of rack area

Source: Adapted from National Sanitation Foundation *Food Service Equipment Standards*, September 1978, pp. 87–89. See Standard No. 5 for hot-water equipment.

Note: This is illustrative. See manufacturer and local health department for specific requirements; also PHS/FDA *Food Service Sanitation Manual*, 1976.

[a] Wash water temperature thermostatically maintained at 150 to 160° F. Temperature not less than 165° F in single temperature door-type machine.

[b] Time of wash cycle increased beyond 40 sec to apply not less than the indicated volume of wash water per rack.

[c] Rinse water flow pressure on line at machine controlled between 15 and 25 lb. (A pressure control valve adjusted to keep the line pressure at 20 psi at all times is strongly recommended. If pressure in line can fall below 15 psi a hydraulic analysis and correction is required.) In single temperature door-type machine, minimum rinse time shall be 30 sec and spray volume 23 gal for each 400 $in.^2$ of rack area, wash and rinse water temperature 165° F; used rinse water is discharged to waste.

Table 8-17 Conveyor Dishwashing Machine, Wash and Rinse Water, Volume, Temperature, and Time

	Wash Cycle[a]			Recirculated Rinse, 160° F			Final Rinse, 180° to 195° F
Type Machine	Volume of Water per Linear Inch of Conveyor length	Minimum Period of Wash (sec)	Pump Capacity (gpm)	Volume of Water per Linear Inch of Conveyor length	Minimum Period of Rinse (sec)	Pump Capacity (gpm)	Minimum Flow[b]
Single-tank conveyor (dishes prewashed or water-scraped).	3 gal per 20-in. width; 2.7 gal per 18-in. width; 3.3 gal per 22-in. width; and 3.6 gal per 24-in. width (160° min.)	15	140	None	—	—	6.94 gpm across conveyor to cover space of at least 6 in. in direction of travel measured 5 in. above conveyor, with rack or conveyor width of 20 in.
Multiple-tank conveyor with dishes in inclined position or in racks (dishes prewashed or water-scraped).	1.65 gal per 20-in. width; 1.48 gal per 18-in. width; 1.82 gal per 22-in. width; 1.98 gal per 24-in. width (150° F min.)	7	125	1.65 gal per 20-in. width; 1.48 gal per 18-in. width; 1.82 gal per 22-in. width; 1.98 gal per 24-in. width	7	125	4.62 gpm across conveyor to cover space of at least 3 in. in direction of travel measured 5 in. above conveyor with conveyor width of 20-in.

Source: Adapted from National Sanitation Foundation *Food Service Equipment Standards*, September 1978, pp. 90–92. See Standard No. 5 for hot-water supply and equipment.

Note: This is illustrative. See manufacturer and local health department for specific requirements; also PHS FDA *Food Service Sanitation Manual*, 1976.

[a] Each linear inch sprayed from above and below.

[b] Rinse-water flow pressure on line at machine controlled between 15 and 25 lb. (A pressure control valve adjusted to keep the line pressure at 20 psi at all times is strongly recommended. If pressure in line can fall below 15 psi, a hydraulic analysis and correction is required.) Conveyor speed 7 fpm maximum for single-tank and 15 fpm maximum for multiple-tank conveyor.

Hot-Water Requirements for Machine Dishwashing[64]

To meet the hot-water demands of a dishwashing machine it is necessary to have a separate heater and storage tank, a two-stage heater, a separate instantaneous heater, or a booster heater. Thermostatic hot-water controls, including a flow control valve or pressure-reducing valve close to the dishwashing machine, and a hot-water circulation pump, will help assure a dependable water supply at proper temperature. A vacuum breaker is necessary at the high point before the hot water line enters the machine to prevent possible backflow of wastewater. The machine drain must not as a rule connect directly to a waste, soil, or drain pipe. See Figures 8-17 and 8-18. The required volume of hot wash and rinse water is indicated in Table 8-16, and 8-17. To this should be added the amount of water required to fill the wash and recirculating rinse tanks.

1. Fluctuations in water pressure due to large demands and hydraulic friction losses will cause variations in the rate of flow. This is apparent from the basic hydraulic formula $Q = VA$, in which Q is the rate of flow in cubic feet per second (cfs), V is the velocity of the water in feet per second (fps), and A is the area of the pipe in square feet. But $V = \sqrt{2gh}$ and $h = g/w$, in which $g = 32.2$ feet per second per second, h = head of water in feet, p = water pressure in pounds per square foot, and $w = 62.4$ pounds for water. It can be seen therefore that a change in velocity, pipe diameter, or water pressure will cause a change in flow, and similarly a change in water flow will cause a change in velocity or pressure. It should be cautioned that many waters are hard or corrosive. Most hard waters contain minerals that form a scale on the inside of pipe when heated and corrosive waters eat away the inside of pipe, both of which conditions increase the friction and hence reduce the flow of water to the dishwashing machine. The removal of hardness and control of corrosion are discussed in Chapter 3.
2. The provision of a pressure-reducing valve on the 180° F rinse line at the machine will assure that a larger volume of 180° F hot water will not be used by the machine than that for which the water heater was designed, or is needed, for proper functioning of the machine. The pressure-reducing valve should be set to maintain the water pressure at the machine preferably at 20 psi. If a larger volume of hot water is permitted to be used by the machine, due to a higher water pressure or manipulation of valves, the water heater will be overloaded beyond the design capacity, resulting in a higher water consumption and hence lower than 180° F-water being delivered to the dishwashing machine.
3. On the other hand, if the water pressure in the rinse line is less than 15 psi under conditions of maximum use, there will be an inadequate volume of 180° F rinse water delivered to the dishwashing machine to do the intended rinsing and disinfection job. This would call for a hydraulic analysis and perhaps a booster pump. A pressure in excess of 25 psi will result in atomization or a fine mist spray at the nozzles, which is not effective in rinsing. By maintaining the water pressure within the established limits, the hot-water rinse line of a known length, and the diameter usually not less than 1 in., it is possible to assure the required rate of hot rinse-water flow. Hence, failure to deliver 180° F rinse water will immediately indicate an inadequate heater or heater and storage tank, or, in an old installation, excessive friction in the water lines. There should be no shutoff valves or drawoffs between the heater and dishwashing machine. A partial closure of a valve on the line will result in a higher hot-water temperature or inadequate flow of rinse water. Drawoffs between the heater and dishwashing machine will reduce the water pressure in the line and the available volume of 180° F water to the machine.

[64] For design details see *ASHRAE Guide and Data Book Systems 1970*, American Society of Heating, Refrigerating and Air-Conditioning Engineers, Inc., 345 East 47th Street, New York, pp. 543–572. Also Standard No. 5, *Commercial Gas-Fired and Electrically-Heated Hot-Water Generating Equipment*, National Sanitation Foundation, Ann Arbor, Mich., 1980.

The required size or capacity of the water heater to serve a dishwashing machine, kitchen fixtures, hand washing sinks, and miscellaneous outlets can be estimated. Two design temperatures are used, 140 to 150° F for general utility purposes and 180 to 200° F for the dishwashing machine. The heating system must be designed to meet the estimated peak hot-water demands for its duration as well as normal hot-water needs. Several methods for computing hot-water requirements are given here.

1. One method suggested by a manufacturer of heating equipment is based on the experience that the amount of hot water required per meal including the peak hot-water demand will average 1.8 gal.[65] This is divided as follows:

28 percent for food preparation, pot and pan washing, etc., lasting 2 to $2\frac{1}{2}$ hr.
55 percent for dishwashing and related activities carried on at same time lasting 1 hr or more.
17 percent for utensil and equipment cleaning and general cleanup lasting 1 to $1\frac{1}{2}$ hr.

An average usage of 1.8 gal per person per meal with a very similar distribution is suggested in the *Cornell Miscellaneous Bulletin* 14.[66] With a 1-hr dishwashing period, it is apparent that the maximum demand to be designed for is 55 percent of 1.8 gal or 1.0 gal per hour per meal served during the heaviest 1-hr feeding period. If the dishwashing period for the same number of dishes is extended over a 2-hr period, the rinse-water requirment would be at the rate of 0.5 gal per hour per meal. At restaurants serving light meals, the peak flow is estimated at 0.7 gal per hour per meal; at school lunchrooms and cafeterias 0.7 to 0.8 gal; and 1.2 gal where elaborate meals are served.

2. Another method of estimating the hourly hot-water needs is to count the maximum number of dish racks to be washed and rinsed in 1 hr. Knowing the amount of rinse water used per rack from Table 8-16 or the manufacturer, it is a simple matter to compute the total amount of 180° F rinse water used per hour. Add to this the amount of water required to fill the dishwasher tank or tanks. For example, if 1.25 gal of 180° F rinse water is used per rack 18 × 18-in. and a rack is washed every 50 sec, which is the maximum possible in a timed single-tank machine set to wash 40 sec and rinse 10 sec, then a maximum of 90 gal will be used in 1 hr. If the wash-water tank holds 20 gal, the total amount of water used in a maximum hour would be 110 gal. In practice, it is sometimes assumed that racks can be fed at a maximum rate of 1/min in a single-tank stationary-rack dishwasher, or 60 racks/hr, which would mean a water consumption of 75 gph, with 18 × 18-in. racks, rather than 90 gal.

3. The American Gas Association suggests that 0.8 gal of 180° F water is used per meal in a single-tank dishwashing machine and 0.5 gal per meal in a two-tank dishwashing machine, in addition to 1.2 gal of 140° F-water per meal. It is assumed that eight dishes are used per meal.

4. *Design calculations.* The required heater capacity in Btu can be calculated from the formula:

$$\text{Btu} = \frac{\text{gph} \times 8.3 \times \text{temperature rise}}{\text{overall efficiency}}$$

where Btu = required input rating of the heater
gph = gallons per hour of 180° F water required, usually heated to 190° F when the heater is located some distance from the dishwasher
8.3 = weight of a gallon of water in pounds
Temperature rise = temperature of water as it leaves the heater minus the temperature of the water entering the heater

[65] *Hot Water Requirements for Food Service Establishments*, Bull. No. HDH, 987A, A. O. Smith Corporation, Kankakee, Ill.

[66] "A Central Camp Building for Administration and Food Service," *Cornell Miscellaneous Bulletin 14*, New York State Colleges of Agriculture and Home Economics, Cornell University, Ithaca, N.Y., March 1953.

Table 8-18 Heat Loss from Pipe Carrying Hot Water

Nominal Pipe Size (in)	Btu per Hour Heat Loss per Linear Foot of Horizontal Pipe in Still 70° F Air Carrying Hot Water at Temperature Stated[a]					
	Bare Iron Pipe			Insulated Pipe, 85% Magnesia, 1-in.		
	120° F	150° F	180° F	120° F	150° F	180° F
$\frac{1}{2}$	27.2	45.8	66.6	8.2	13.3	18.5
$\frac{3}{4}$	33.0	55.2	80.2	9.3	15.0	20.9
1	39.5	66.3	96.6	10.5	17.1	23.8
$1\frac{1}{4}$	48.9	81.8	119.5	12.2	19.8	27.6
$1\frac{1}{2}$	54.5	92.0	134.1	13.3	21.5	30.0

[a] Does not include Btu needed to reheat water in pipes. For example, 10 ft of 1-in. pipe holds 1 gal and will require 415 Btu to raise its temperature 50° F, or wasting of water until 180° F water is obtained at the machine or outlet.

Overall efficiency* = the overall heating-system efficiency of gas may be taken at 75 percent provided the heater and storage tank, if provided, are insulated. Water heaters hold between 2 to 5 and 10 gal in the coils, or surrounding heating tubes; this water is available to meet a momentary demand. The heat loss from long runs of pipe can be appreciable. Table 8-18 gives the heat loss from bare and insulated pipe.

5. *Example:* A restaurant serves a maximum of 450 meals. Dishes are to be washed over a period of $1\frac{1}{2}$ hr. What type of dishwasher should be used? Several types of dishwashers are shown in Figures 8-17 to 8-19. What heater and storage-tank capacities are needed if the temperature of the incoming water to the heater is 40° F?

If an average meal uses 6 pieces and a 20 × 20-in. rack is used that holds 24 pieces, each rack will hold the dishes from 4 meals. With dishes from 450 meals to be washed in $1\frac{1}{2}$ hr or 300 meals in 1 hr, a total of 300/4 = 75 racks are to be washed per hour. If 18 × 18-in. racks that hold 20 pieces are used, each rack will hold the dishes from 20/6 or 3.33 meals. With dishes from 300 meals to be washed in 1 hr, a total of 300/3.33 = 90 racks are to be washed per hour. The dishwashing machine that meets these requirements should be selected.

The machine selected may be a single-tank rack-conveyor dishwasher or a multiple-tank rack-conveyor dishwasher. Assume that a single-tank 18 × 18-in. rack-conveyor dishwasher is selected which uses 1.73 gal of 180° F rinse water per rack of dishes. With 90 racks to be washed per hour, the hot rinse-water requirement would be 90 × 1.73 = 156 gph. (At the mechanical capacity this machine would use 416 gph and process 240 racks.) To this should be added the water to fill the dishwashing machine wash tank, say 25 gal, making a total of 181 gph. If no storge is provided, the required gas

$$\text{heater} = \frac{181 \times 8.3 \times (190 - 40)}{0.75} = 300{,}500 \text{ Btu/hr}$$

If an electric heater were designed for, its capacity would be

$$= \frac{181 \times 8.3 \times (190 - 40)}{0.98} = 230{,}000 \text{ Btu/hr},$$

or since 1 kW-hr = 3,412 Btu, the electric heater would have a capacity of $\frac{230{,}000}{3{,}412} = 67.4$ kW-hr. It

*The thermal efficiency of electric heaters may be taken at 98 percent; gas heaters at 75 percent; steam boilers at 80 percent; oil heaters at 60 percent; and coal-fired heaters at 50 percent. One ft^3 of natural gas = 1050 Btu. See local utility company for other ratings. Btu ratings can be reduced 4 percent for each 1000 ft above sea level, higher than 2000 ft.

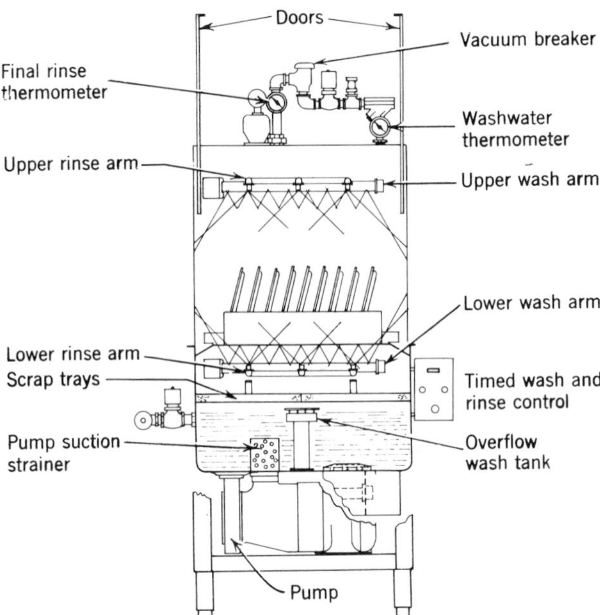

Figure 8-17 Stationary rack door-type machine. Illustrative only. From *Food Service Equipment Standards*, National Sanitation Foundation, Ann Arbor, Mich., 48105, September 1978, p. 88.

may be more practical to use an electric heater as a booster adjacent to the dishwashing machine to raise the rinse water from 140 to 180° F.

If a 120-gal storage tank is provided and the interval between meal-periods is 4 hr, a lesser capacity heater would be required, as follows: A 120-gal storage tank has an available storage of about 75 percent, or 90 gal, leaving (181–90) or 91 gph to be met instantaneously by the heater. The required gas-heater capacity (adequate for $1\frac{1}{2}$-hr dishwashing) under these circumstances would be:

$$\frac{(\frac{90}{4} + 91) \times 8.3 \times (190 - 40)}{0.75} = 188{,}000 \text{ Btu/hr}$$

If a multiple-tank rack-conveyor dishwasher is used, say, 0.5 gal of 180° F rinse water is used per rack of dishes or equivalent. With 90 racks of dishes to be washed per hr, the hot rinse-water requirement would be 90 × 0.5 = 45 gph. Adding the water needed to fill the two dishwasher tanks of 60 gal (see manufacturer's catalog), a total of 105 gal must be designed for. If no storage tank is provided the required heater would be

$$\frac{105 \times 8.3 \times (190 - 40)}{0.75} = 174{,}000 \text{ Btu/hr}$$

If a 120-gal storage tank is provded (90 gal available) and the interval between meals is 4 hr, the required heater would be

$$\frac{(\frac{90}{4} + 105 - 90) \times 8.3 \times (190 - 40)}{0.75} = 62{,}000 \text{ Btu/hr}$$

This is adequate for a dishwashing period of $1\frac{1}{3}$ hr, assuming 180° water is used to fill the two dishwasher tanks. If it is considered that the 120-gal storage tank must be heated to obtain 90 gal, the

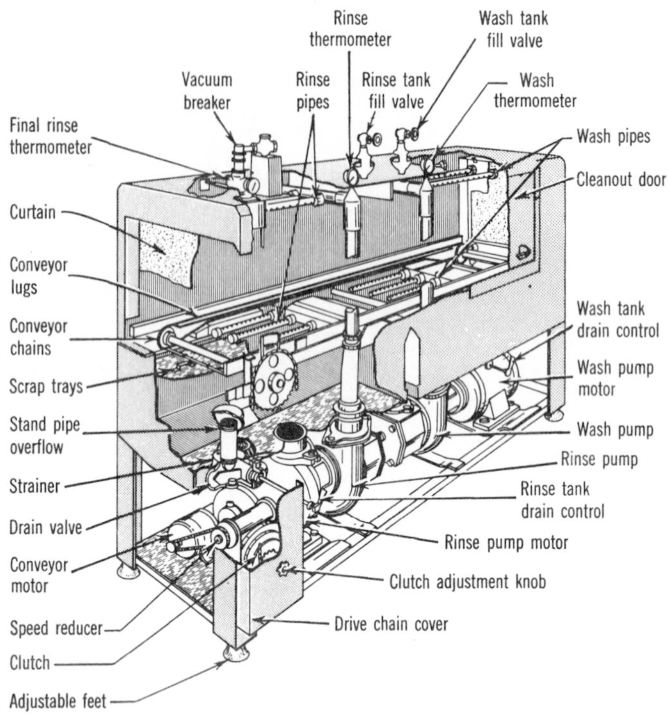

Figure 8-18 Single-tank conveyor-type machine. Illustrative only. From *National Sanitation Foundation Standards*, Standard No. 3, "Commercial Spray-type Dishwashing Machines," Ann Arbor, Mich., 48105, April 1965.

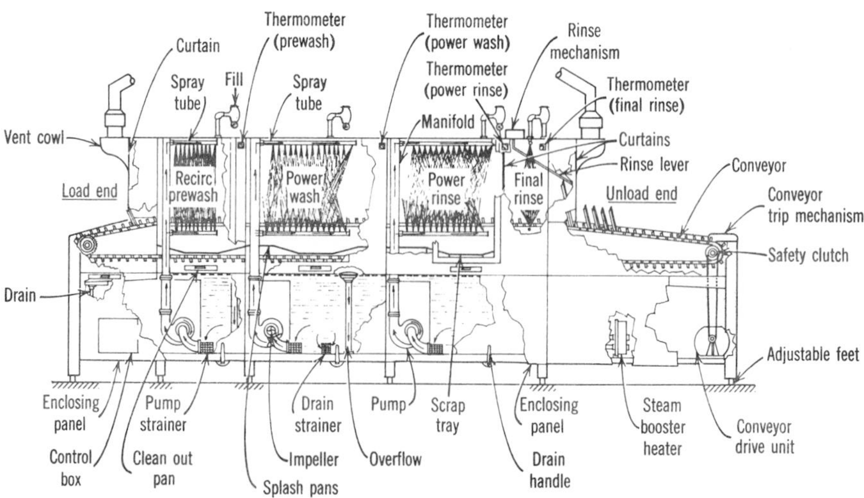

Figure 8-19 Multiple-tank rackless conveyor-type machine. Illustrative only. From *National Sanitation Foundation Standards*, Standard No. 3, "Commercial Spray-type Dishwashing Machines," Ann Arbor, Mich., 48105, April 1965.

required heater would be

$$\frac{(\frac{120}{4} + 105 - 90) \times 8.3 \times (190 - 40)}{0.75} = 75{,}000 \text{ Btu/hr}$$

This would be adequate for a 2-hr dishwashing period.

6. Some designers add a factor of safety of 50 to 100 percent to the required heater capacity or, preferably, design for the mechanical capacity of the dishwashing machine. A common practice is the provision of a storage tank for the storage of all water at a temperature of 140° F. This tank supplies the 140° F needs and feeds the heater or booster supplying 180° F water. Several water-heater flow diagrams are given in Figure 8-20.

7. For the problem given, in which dishes from 300 meals are washed, method (1) would call for 300 gal of 180° F water/hr; method (5) would require 181 gal; and method (3) would require 240 gal with a single-tank machine, and 150 gal with a two-tank machine, plus if not included, water to fill the one or two tanks. In view of the possible variations, it is important that the most probable hot-water requirements peculiar to the establishment under consideration be carefully studied and the best possible estimate made for design purposes. Such factors as type of meals, number of dishes, equipment available, possibilities for hot-water waste, type of personnel and supervision, dishes to be washed per hour, water pressure, and type of dishwashing machine all have an effect on the hot-water requirements. National Sanitation Foundation Standard 5 gives sample calculations and graphs to simplify selection of gas and electric heaters for rinse water.

Hot Water for General Utility Purposes

The hot-water requirements for general utility purposes include the needs for the hand washing of dishes, pots, pans, and utensils; for wall, floor, and equipment cleanup; for personal hygiene and toilet rooms; for food preparation; and for waste. One estimate for hand washing is 30 gal per sink per wash-up, with the dishwashing requirements broken down to prewash—0.25 to 0.5 gal/meal served; detergent washing—0.75 to 1.0 gal/meal served; and final rinsing—0.5 to 1.5 gal/meal served.[67] Total usage adds up to 1.5 to 3.0 gal/meal served. A factor of safety addition of 50 to 100 percent is suggested, depending on the type of restaurant and seasonal demands. Water temperature of 120 to 140° F is generally adequate for utility purposes.

If it is assumed that 2 gal of 140° F water are used per meal, that 50 meals are served, and that hand dishwashing is completed in 1 hr, then the total hot-water demand per hour will be 100 gal, or 150 gal if a 50 percent factor of safety is added. This can be obtained from a water heater having a recovery rate to produce 150 gal of 140° F water in 1 hr. (A separate under-sink booster is also required.) If a storage tank is provided capable of supplying the total demand, and 3 hr are available to heat the water, then the heater recovery rate would be

$$\frac{150}{0.75} \times \frac{1}{3} = 67 \text{ gph}$$

[67] R. M. McLaughlin, paper presented at the Metropolitan Gas Heating and Air Conditioning Council, New York, December 14, 1950.

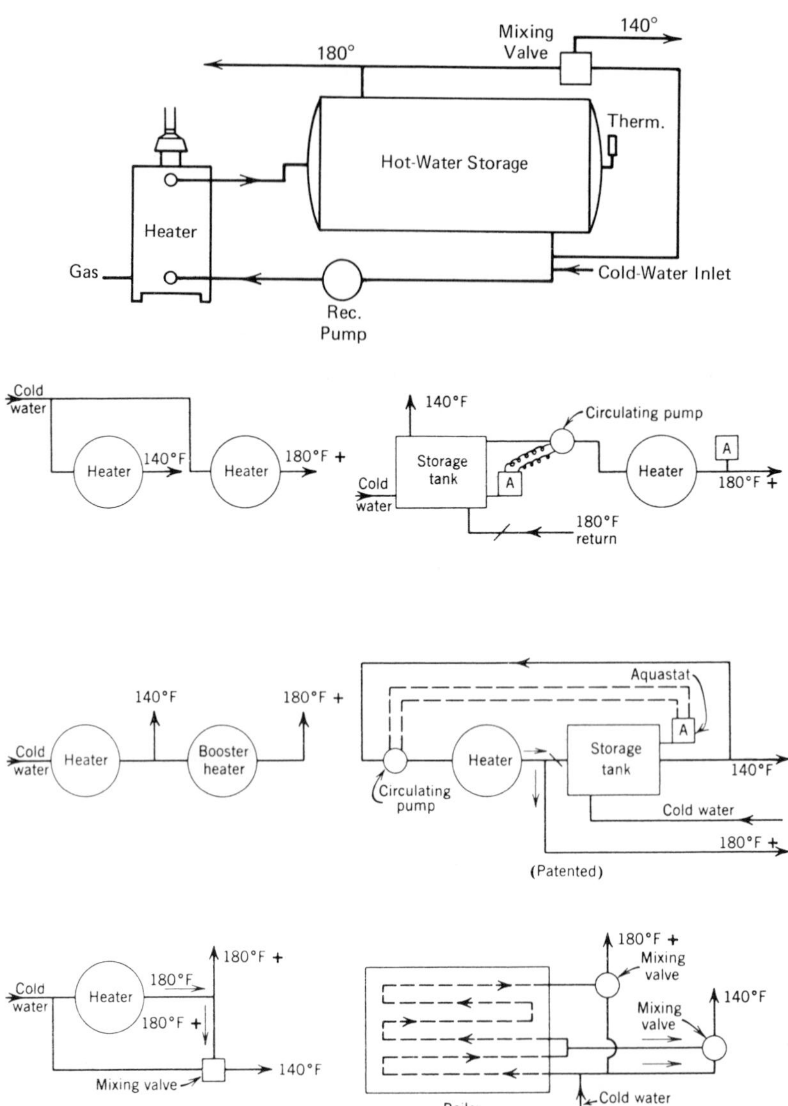

Figure 8-20 Water heater flow diagrams. The heater capacity must equal the requirements for 180° F water plus 140° F water over and above the available storage. A return circulating line is necessary on a heater located at a distance from a dishwashing machine. Thermometers, control valves, vacuum breakers, pressure-reducing flow control valves, strainer, and relief valves omitted from drawing. To supply 180°F water to a dishwashing machine requires heating the water to about 190°F. Insulate heater, storage tank, and water lines.

The factor 0.75 represents 75 percent of the storage-tank capacity that may be withdrawn before the incoming cold water cools the remaining stored hot water below the desired temperature.

Another example might be the hot-water requirements of 300 gph in a kitchen, plus that of three washbasins and two showers. The washbasins are used by 30 persons in 1 hr and the two showers by four persons in 1 hr. Say that it takes a person 3 min to wash at a basin and 10 min to take a shower. A basin faucet will flow at the rate of 3 gpm, of which 2 gpm can be taken as hot water. A shower head will flow at the rate of 5 gpm, of which $\frac{3}{4}$ or $3\frac{1}{2}$ gpm can be taken as hot water. Large variation can be expected depending on the usage, fixture, and water pressure. The required heater and storage-tank capacities can be obtained as follows:

$$\begin{aligned}
&\text{2 showers} = 4 \text{ persons} \times 10 \text{ min} \times 3\tfrac{1}{2} \text{ gpm} = 180 \text{ gal}\\
&\text{3 basins} = 30 \text{ persons} \times 3 \text{ min} \times 2 \text{ gpm} = 180 \text{ gal}\\
&\text{Kitchen usage} = 300 \text{ gal}\\
&\text{Total} = 620 \text{ gph}
\end{aligned}$$

If a storage tank is provided to supply 620 gal of hot water it would have a capacity of $\frac{620}{0.75} = 827$ gallons. With a heat-up period of 3 hr, the required heater would have a recovery rate of $\frac{827}{3} = 276$ gph.

If a storage tank has a 450-gal capacity, of which $450 \times 0.75 = 337$ gal is available, the required heater would have a recovery capacity of $620 - 337 = 283$ gph. At this recovery rate the storage tank would be reheated in $450/283 = 1.6$ hr.

In practice the nearest standard-size storage tank and heater would be selected.

BIBLIOGRAPHY

Angelotti, Robert, Milton J. Foter, and Keith H. Lewis, *Time-Temperature Effects on Salmonellae and Staphylococci in Foods*, PHS, Tech. Rep. F60-5, DHEW, Cincinnati, Ohio, 1960.

Boaz, Joseph N., Ed., *Ramsey and Sleeper Architectural Graphic Standards*, John Wiley & Sons, New York, 1970. See also 7th ed., 1981.

Bryan, Frank L., *Diseases Transmitted by Foods* (A Classification and Summary), DHEW, PHS, CDC, Atlanta, Ga. 30333, 1971

Bryan, Frank L., *Guide For Investigating Foodborne Disease Outbreaks and Analyzing Surveillance Data*, DHEW, PHS, CDC, Atlanta, Ga. 30333, 1975.

Donovan, Anne Claire, and Orville B. Ives, *Hospital Dietary Services*, PHS Pub. 930-C-11, DHEW, Washington, D.C., 1966.

Establishing and Operating a Restaurant, Industrial (Small Business) Ser. No. 39, GPO, Washington, D.C., 1946.

Food Service Sanitation Manual, 1976 Recommendations of the FDA, DHEW, PHS/FDA, Washington, D.C. 20204, June 1978.

Grade A Pasteurized Milk Ordinance, 1978 Recommendations, PHS/FDA, DHEW, Pub. 229, Washington, D.C.

"Joint FAO/WHO Expert Committee on Milk Hygiene," *WHO Tech. Rep. Ser.*, **453**, 1970.

Kotschevar, Lendal H., and Margaret E. Terrell, *Food Service Planning*, John Wiley & Sons, New York, 1963.

Litsky, Bertha Yanis, *Food Service Sanitation*, Modern Hospital Press, McGraw-Hill Pub. Co., Chicago, Ill., 1973.

Longree, Karla, *Quantity Food Sanitation*, 2nd ed., Wiley-Interscience, New York, 1972.

Longree, Karla, and Gertrude G. Blaker, *Sanitary Techniques in Food Service*, John Wiley & Sons, New York, 1971.

"Milk Hygiene," *WHO Monogr. Ser.*, **48**, (1962).

NSF Food Service Equipment Standards, National Sanitation Foundation, Ann Arbor, Mich., September 1978.

Procedures To Investigate Foodborne Illnesses, 3rd ed., International Association of Milk, Food, and Environmental Sanitarians, Inc., P.O. Box 701, Ames, Iowa 50010, 1976.

Reference Guide—Sanitation Aspects of Food Service Facility Plan Preparation and Review, National Sanitation Foundation, NSF Building, P.O. Box 1468, Ann Arbor, Mich. 48106.

Rishoi, Don C., *Food Store Sanitation*, Chain Store Age Books, New York, 1976.

Speck, Marvin L., ed., *Compendium of Methods for the Microbiological Examination of Foods*, Am. Pub. Health Assoc., Inc., Washington, D.C., 1976.

Standard Methods for the Examination of Dairy Products, 13th ed., Am. Pub. Health Assoc., Inc., Washington, D.C., 1972.

Taylor, Joan, Ed., *Bacterial Food Poisoning*, Royal Society of Health, London, England, 1970.

West, Bessie Brooks, Levelle Wood, and Virginia F. Harger, *Food Service in Institutions*, John Wiley & Sons, New York, 1967.

9

RECREATION AREAS AND TEMPORARY RESIDENCES

BEACH AND POOL STANDARDS AND REGULATIONS

The sanitation of bathing places is dictated by health and aesthetic standards. Few people would knowingly swim or water-ski in polluted water, and insanitary surroundings are not conducive to the enjoyment of "a day at the beach." People demand more and cleaner beaches and pools, and a camp, motel, hotel, club, or resort without a pool or beach is not nearly as popular as one so equipped.

Health Considerations

From our knowledge of disease transmission, it is known that certain illnesses although uncommon can be transmitted by improperly operated or located swimming pools and beaches through contact and ingestion of polluted water. Among these are typhoid fever, dysentery, infectious hepatitis, and other gastrointestinal illnesses; conjuctivitis, trachoma, leptospirosis, ringworm and other skin infections; schistosomiasis, or swimmer's itch; upper respiratory tract diseases such as sinus infection, septic sore throat, and middle-ear infection. The repeated flushing of the mucous coatings of the eyes, ears, and throat and the excessive use of alum or lack of pH control expose the unprotected surfaces to possible inflammation, irritation, and infection. Contraction of the skin on immersion in water may make possible the direct entrance of contaminated water into the nose and eyes.

Bathing Beaches

Stevenson reported that

> an appreciably higher over-all illness incidence may be expected in the swimming group over that in the nonswimming group regardless of the bathing water quality.[1]

[1] Albert H. Stevenson, "Studies of Bathing Water Quality and Health," presented before the second session of the Engineering Section of the Am. Pub. Health Assoc. at the 80th Annual Meeting, in Cleveland, Ohio, October 23, 1952.

It was further stated in his studies that

> eye, ear, nose, and throat ailments may be expected to represent more than half of the over-all illness incidence, gastrointestinal disturbances up to one-fith and skin irritations and other illnesses the balance.

Although based on limited data swimming in lake water with an average coliform content of 2300/100 ml caused "a significant increase in illness incidence . . . " and swimming in river water "having a median coliform density of 2700/100 ml appears to have caused a significant increase in such (gastrointestinal) illness." The study also showed the greatest amount of swimming was done by persons 5 to 19 years of age.

In a study of illness symptom rates among swimmers and nonswimmers at "barely acceptable" (BA) and "relatively unpolluted" (RU) beaches in New York City, Cabelli et al.[2] found a

> statistically significant swimming-associated rate of gastrointestinal symptomatology at a "barely acceptable" but not at a "relatively unpolluted" beach.

Gastrointestinal and respiratory symptoms were higher among swimmers than among nonswimmers. The geometric mean densities of total coliforms (MPN per 100 ml) were 1213 for the BA beach and 43.2 at the RU beach. The corresponding fecal coliforms were 565 and 28.4 respectively.

A significant association between diarrheal illness and swimming in river water having a mean fecal coliform of 17,500 organisms per 100 ml was demonstrated by Rosenberg et al. The same type *Shigella sonnei* was isolated from 6 swimmers and from the river water. A river-water sample collected from the swimming area one month after swimming was banned showed the presence of *S. sonnei*. Thirty-one of 45 cases of shigellosis were traced to swimming in the river at a location about 5 mi below the Dubuque, Iowa sewage treatment plant which was providing inadequate chlorination.[3]

In another study, conducted on Long Island saltwater beaches, no relationship was found between water quality and illness.[4] There are apparently no data correlating fecal coliform densities to enteric disease in bathing populations.[5] Other organisms proposed as possible indicators of bathing water quality are fecal streptococci and *Pseudomonas aeruginosa*. *P. aeruginosa* is more resistant

[2] Victor J. Cabelli, Alfred P. Dufour, Morris A. Levin, Leland J. McCabe, and Paul W. Haberman, "Relationship of Microbial Indicators to Health Effects at Marine Bathing Beaches," *Am. J. Pub. Health*, July 1979, pp. 690–696.

[3] Mark L. Rosenberg, Kenneth K. Hazlet, and John Schaefer, "Shigellosis from Swimming," presented at Am. Pub. Health Assoc. Meeting, November 19, 1975, *J. Am. Med. Assoc.*, October 18, 1976, p. 1846.

[4] R. S. Smith and T. D. Woolsey, *Bathing Water Quality and Health III—Coastal Water*, PHS, Cincinnati, Ohio, 1961.

[5] Victor J. Cabelli, Harriet Kennedy, and Morris A. Levin, "Pseudomonas aeruginosa—fecal coliform relationships in estuarine and fresh recreational waters," *J. Water Pollut. Control Fed.*, February 1976, pp. 367–376.

than coliforms and is primarily associated with human feces. *Clostridium perfringens* (*C. welchi*), shigellae, and *salmonella* are also used in making special studies of bathing beach water quality.[2]

Many other studies have been made to relate bathing water bacterial quality to disease transmission with inconclusive or negative results.[6]

British investigators have drawn the following general conclusions:[7]

(i) That bathing in sewage-polluted sea water carries only a negligible risk to health, even on beaches that are aesthetically very unsatisfactory.

(ii) That the minimal risk attending such bathing is probably associated with chance contact with intact aggregates of faecal material that happen to have come from infected persons.

(iii) That the isolation of pathogenic organisms from sewage-contaminated sea water is more important as evidence of an existing hazard in the populations from which the sewage is derived than as evidence of a further risk of infection in bathers.

(iv) That, since a serious risk of contracting disease through bathing in sewage-polluted sea water is probably not incurred unless the water is so fouled as to be aesthetically revolting, public health requirements would seem to be reasonably met by a general policy of improving grossly insanitary bathing waters and of preventing so far as possible the pollution of bathing beaches with undisintegrated faecal matter during the bathing season.

The findings of the Public Health Activities Committee of the ASCE Sanitary Engineering Division are summarized in the following abstract:[8]

Coliform standards are a major public health factor in judging the sanitary quality of recreational waters. There is little, if any, conclusive proof that disease hazards are directly associated with large numbers of coliform organisms. Comprehensive research is recommended to provide data for establishing sanitary standards for recreational waters on a more rational or sound public health basis. British investigations show that even finding typhoid organisms and other pathogens in recreational waters is not indicative of a health hazard to bathers but is only indicative of the presence of these diseases in the population producing the sewage. The Committee recommends that beaches not be closed and other decisive action not be taken because current microbiological standards are not met except when evidence of fresh sewage or epidemiological data would support such action.

A major factor which contributes to the bacteriological quality of Milwaukee bathing beaches on Lake Michigan was found to be the rainfall intensity, and the subsequent water drainage inflow pattern. Based upon long-term beach water bacteriological history, it was possible to predict the number of hours that a

[6] "Sewage Contamination of Coastal Bathing Waters in England and Wales," Committee on Bathing Beach Contamination of the Public Health Laboratory Service, *J. Hygiene*, **57**, No. 4, 435–472 (December 1959); John M. Henderson, "Enteric Disease Criteria for Recreational Waters," *J. Sanit. Eng. Div., ASCE*, SA6, Proc. Paper 6320, 1253–1276 (December 1968); "Coliform Standards for Recreational Water," Committee Report, *J. Sanit. Eng. Div., ASCE*, SA4, Proc. Paper 3617, 57–94. (August 1963).

[7] Ibid., "Sewage Contamination."

[8] "Coliform Standards," op. cit. (see reference 6).

beach should remain closed following the end of a 0.1-in. to 0.19-in. rainfall (12 hr) to 1.5 in. or more rainfall (96 hr).[9]

Diesch and McCulloch and others, summarized incidences of leptospirosis in persons swimming in waters contaminated by discharges of domestic and wild animals, including cattle, swine, foxes, raccoons, muskrats, and mice.[10–12] Pathogenic leptospires were isolated from natural waters, confirming the inadvisability of swimming in streams and farm ponds receiving drainage from cattle or swine pastures.

A case-control study showed that seven children acquired leptospirosis after swimming in a creek which ran through a pasture. However, samples of creek water four weeks after the outbreak did not yield the organism and tested serum specimens from cattle did not implicate any herd. Common-source leptospirosis outbreaks have been frequently reported in Europe.[12] Schiemann[13] summarized 10 outbreaks of leptospirosis involving swimming in natural waters.

Joyce and Weiser report that enteroviruses that are found in human or animal excreta, if introduced into a farm pond by drainage or direct flow, can constitute a serious public health hazard if used for recreation, drinking, or other domestic purposes.[14]

Primary amebic meningoencephalitis (PAM), a rare disease that is generally fatal to man, has been linked with swimming or bathing that involves immersion of the head in contaminated fresh and brackish water. Heated swimming pools, hot geothermal waters, and warm-water swimming areas associated with high ambient air temperatures have been associated with the disease. The causative agent is believed to be *Naegleria fowleri* which is a free-living ameboflagellate. It is a common species in soil, decaying vegetation, and natural fresh water. The naegleria cyst is reported to survive a temperature of 56° C when suspended in water.[15–18] Acanthamoeba, another free-living ameba, generally causes subacute or chronic infections less severe than PAM.[19]

[9]George A Kupfer, "Control of Swimming at Milwaukee's Beaches by Pollution Prediction Formula," paper presented at the American Public Health Association Annual Meeting, November 19, 1975, Chicago, Illinois.

[10]Stanley L. Diesch and William F. McCulloch, "Isolation of Pathogenic Leptospires from Waters Used for Recreation," *Pub. Health Rep.*, April 1966, pp. 299–304.

[11]"Leptospirosis," *MMWR*, CDC, Atlanta, Ga., Nov. 14, 1959, p. 45.

[12]"Leptospirosis—Tennessee," *MMWR*, CDC, Atlanta, Ga., March 19, 1976, p. 84.

[13]D. A. Schiemann, "Leptospirosis May Result From Swimming in Streams and Lakes," *J. Environ. Health.*, July/August 1973, pp. 70-73.

[14]Gayle Joyce and H. H. Weiser, "Survival of Enteroviruses and Bacteriophage, in Farm Pond Waters," *J. Am. Water Works Assoc.*, April 1967, pp. 491–501.

[15]Gunther F. Craun, "Waterborne outbreaks," *J. Water Pollut. Control Fed.*, June 1977, pp. 1268–1279.

[16]"Primary Amebic Meningoencephalitis—California, Florida, New York," *MMWR*, CDC, Atlanta, Ga., September 15, 1978, pp. 343–344.

[17]J. A. Salvato, *Guide To Sanitation In Tourist Establishments*, WHO, 1976, p. 86.

[18]"Primary Amebic Meningoenchephalitis—United States," MMWR, CDC, Atlanta, Ga., August 29, 1980, pp. 405–407.

[19]*Water-related Disease Outbreaks*, Annual Summary 1978, U.S. Department of Health and Human Resources, PHS, CDC, Atlanta, Ga. 30333, pp. 20–22.

Preventive PAM measures suggested at suspect warm-water pools include the diversion of surface water drainage and the display of prominent warnings against diving, jumping, underwater swimming, or other activities likely to cause water to be driven up into the nose. In addition the pool water should be given complete treatment including *continuous* recirculation, filtration, and free residual chlorination (at least 1.0 mg/l), whether the pool is heated or not.[17] If a satisfactory level of free available residual chlorine cannot be maintained because of the temperature, presence of dissolved chemicals, or flow-through of water, the pool should be designated for bathing purposes only and should have a depth of not more than $2\frac{1}{2}$ ft. If the pool is an artificial one it should be emptied at least daily and the bottom, sides, overflow gutters, and decks regularly scrubbed using a suitable disinfectant.[20] Raising the pool water salt concentration to 0.7 percent is reported to kill the amebae; but 10 mg/l chlorine has been ineffective.[21]

In view of the available information, emphasis should be placed on elimination of sources of pollution (sewage, storm water, land drainage), effective disinfection of treated wastewaters, and the proper interpretation of bacterial examinations of samples collected from representative locations.

Swimming Pools

In swimming pools there is a possibility of direct or indirect transmission of eye, ear, respiratory tract, and skin infections and other illnesses from one bather to another, particularly if the water does not have an active disinfectant such as free available chlorine. However, data documenting pool-acquired illnesses is sparse and often only circumstantial. Skin infections caused by *Mycobacterium marinum* and leading to a granuloma have been reported following swimming in public pools. The organism was found to be resistant to free available chlorine in the 1.5 mg/l range.[22] Other bacterial and viral infections including conjunctivitis and fever have also been associated with swimming pools.[23]

Swimming pools receive not only body discharges such as mucous from the nose, saliva from the mouth, sweat, and traces of fecal matter, urine, and dead skin, but also street and workplace soil that collects on the skin, and various body lotions, oils, and creams all of which can contribute to pollution of pool water.[24] Outdoor pools in addition receive dust, pollens, and other air pollutants. This emphasizes the importance of proper swimming pool design, including recircula-

[20] Private communication (1973) from Dr. R. R. L. Harcourt, Division of Public Health, Dept. of Health, New Zealand.

[21] *Control of Communicable Diseases in Man*, Abram S. Benenson, Ed., American Public Health Association, Washington, D.C., 1975, p. 6.

[22] Ue Kang Park and William S. Brewer, "The Recovery of *Mycobacterium Marinum* From Swimming Pool Water and its Resistance to Chlorine," *J. Environ. Health*, May/June 1976, pp. 390–392.

[23] Karen Van Dusen and Gary Fraser, *Swimming Pool Program Study*, Dept. of Social and Health Services, Olympia, Wash., October 1977, pp. 12–22.

[24] *The Purification of the Water of Swimming Pools*, Department of the Environment, Her Majesty's Stationery Office, London, 1975, pp. 2–3.

tion, filtration, and disinfection, and proper operation, as well as enforcing use of the toilet and warm water cleansing showers before entering the pool and after using the toilet, not only for health and hygienic reasons, but also for aesthetic reasons. These factors are discussed in this chapter.

Whirlpools, Spas, and Hot Tubs

Whirlpools are found at health spas, hospitals, and at recreation facilities usually associated with swimming pools. Skin infections or rashes due to *Pseudomonas aeruginosa* have been reported in several instances.[25,26] The relatively high water temperatures and agitation in whirlpools make maintenance of adequate free residual chlorine difficult, and hence the likely survival of pseudomonas and other pathogens. A study of eight whirlpool baths showed that *P. aeruginosa* can be recovered from whirlpool waters having 2 to 3 mg/l *total* residual chlorine. The high concentration of total organic carbon, total Kjeldhal and ammonia nitrogen, usually found present would hinder maintenance of free available residual chlorine.[27] However, continuous recirculation and filtration with automatic hypochlorination and maintenance of at least 1.0 to 3.0 mg/l free residual chlorine should prevent infections.[26,28] The water needs to be replaced and the whirlpool cleaned periodically and when the total dissolved solids build up to the point of producing cloudy water. The large evaporation associated with whirlpools, hot tubs, and spas results in a build-up of total dissolved solids.

Similar precautions apply to hot tubs. A minimum of 2.5 mg/l free chlorine residual, a total alkalinity of 120 to 140 mg/l, and weekly cleaning including superchlorination at 10 mg/l for 10 hr has been recommended for redwood hot tubs.[29]

The Consumer Product Safety Commission (CPSC) reported ten deaths, apparently due to drowning, associated with use of hot tubs.[30] Maintenance of water temperature at 100° F for healthy adults and 98° F for children under 5 is advised. Temperatures above 104° F for adults, and 102° F for women during the first three months of pregnancy, may be hazardous. Individuals under medical care should check with their doctor. Inspection of the electrical system for

[25] "Pseudomonas Infection Traced to Motel Whirlpool," *J. Environ. Health*, March/April 1975, pp. 455–459.

[26] "Rash Associated with Use of Whirlpools—Maine," *MMWR*, CDC, Atlanta, Ga., April 1979, pp. 132–133.

[27] Beverly J. Kush and A. W. Hoadley, "A Preliminary Survey of the Association of *Pseudomonas aeruginosa with Commerical Whirlpool Bath Waters," Am. J. Pub. Health*, March 1980, pp. 279–281.

[28] *Recommended Standards for Swimming Pools*, Great Lakes-Upper Mississippi River Board of State Sanitary Engineers, 1977 ed., Sec. 14.0, Health Education Service, P.O. Box 7283, Albany, N.Y. 12224.

[29] Storm C. Goranson, "An Environmental Field Study—A Redwood Hot Tub Spa," *J. Environ. Health*, January/February 1979, pp. 189–192.

[30] "Users of Hot Tubs Warned of Dangers," Washington, United Press International, *Los Angeles Times*, Jan. 3, 1980, Part 1, p. 21.

possible shock, thermostatic controls, and an alarm system are recommended by the CPSC.[31] Detailed design and operational standards are available.[28,32,33]

Proper operation* is indicated by very low to negative total plate counts, coliform bacteria, and *Pseudomonas aeruginosa*; a free chlorine of at least 1.0 to 1.5 mg/l; a pH of 7.5 to 7.6; a total alkalinity as $CaCO_3$ of 100 to 120 mg/l; and a water clarity of 0.5 JTUs or less. The water should be turned over at least once every 30 min.

Regulations and Standards

Health departments generally require that swimming pools and bathing beaches be operated under permit. Usually swimming pools are not to be constructed or altered until plans and specifications prepared by a licensed professional engineer or registered architect are submitted and approved by the commissioner of health. Regulations and standards cover water quality, design, construction, operation, maintenance, sanitary facilities, and related factors.

Beaches

Regulations governing the use of bathing beaches or natural partly artificial pools, and evaluation of their suitability, are based on sanitary survey of the drainage area to the beach or pool; the water quality including meteorological factors; epidemiological data linking illnesses to the bathing area; and water circulation and dilution.

The *sanitary survey* takes into consideration geographic factors and probable sources of pollution on the watershed tributary to the bathing beach. This includes sewage and industrial wastewater discharges, storm-water overflows, bird and animal populations, commercial and agricultural drainage, and their relationship to the beach. The location and volume of the pollution and its chemical, bacterial, and physical characteristics; the volume and quality of the diluting water; water depth; water surface area; tides; time of day and year; thermal and salinity stratification; confluence of tributaries; water currents; rainfall; and prevailing winds are all evaluated in interpreting water sampling data to determine the suitability of the water for bathing. Also important are safety hazards such as fast currents, strong tides, submerged objects, beach slope and sharp drop-offs, and an uneven or unstable beach bottom in the wading area, and the water depth in the diving area. The New York State Public Health Law,

*Less than 200 bacteria/ml, coliform positive in not more than 10 percent of samples, no two consecutive samples positive, and chlorine tests made hourly.

[31] "Be careful in that hot tub," *Changing Times*, April 1980, p. 48.

[32] *Minimum Standards for Public Spas*, National Swimming Pool Institute, 2000 K Street, N.W., Washington, D.C. 20006, April 1, 1978.

[33] *Suggested Health and Safety Guidelines for Public Spas and Hot Tubs*, CDC, Environmental Health Services Division, Atlanta, Ga. 30333, 1980.

for example, prohibits the maintenance of a bathing beach within 500 ft of a sewer outfall.

Bathing-beach water quality standards vary throughout the country. They range from no standard to a permissible total coliform count of 50/100 ml to 2400/100 ml or higher; the preponderant number relate results to a sanitary survey of the bathing area, epidemiological data, and judgment. Interpretation of bacteriological results should take into consideration possibly associated illnesses, the sanitary survey, and meteorological factors such as rainfall, tidal currents, and prevailing winds as mentioned above.

Bathing-beach standards usually refer to the most probable number (MPN) of the coliform group of organisms per 100 ml of sample. This group includes the *Escherichia coli*, which usually inhabit human and animal intestines, and the *Aerobacter aerogenes* and the *Aerobacter cloacae*, which are frequently found in many types of vegetation and in pipe joint material, pipelines, and valves. The intermediate-aerogenes-cloacae (IAC) subgroups are sometimes found in fecal discharges, but usually in smaller numbers than the *E. coli*. *A. aerogenes* and intermediate types of organisms are also commonly present in soil and in waters polluted sometime in the past. It is apparent therefore that all surface waters can be expected to be polluted to a greater or lesser extent and that bathing in "unpolluted" water is a practical impossibility. It is also apparent that the total coliform test does not distinguish between the more hazardous recent human and animal pollution and the soil, vegetation, and old or past human and animal pollution.

The coliform group of organisms includes the fecal coliforms (*Escherichia coli* at 44.5° C). The presence of fecal coliforms is thought to be more indicative of recent human and animal pollution and hence the presence of pathogenic organisms. The fecal coliforms may average 15 to 20 percent of the total coliforms in stream samples. See also Chapter 3, page 197. A recommended standard for water used for wading, swimming, water-skiing and surfing states:[34]

> Fecal coliforms should be used as the indicator organism for evaluating the microbiological suitability of recreation waters. As determined by the multiple-tube fermentation or membrane filter procedures and based on a minimum of not less than five samples taken over not more than a 30-day period, the fecal coliform content of primary contact recreation waters shall not exceed a log mean of 200/100 ml, nor shall more than 10 percent of total samples during any 30-day period exceed 400/100 ml.[35]
>
> . . . the pH should be within the range of 6.5–8.3 except when due to natural causes and in no case shall be less than 5.0 nor more than 9.0.
>
> . . . the clarity should be such that a Secchi disc (20 cm in diameter divided into four quadrants painted alternating black and white) is visible at a minimum depth of 4 feet. In "learn to swim" areas the clarity should be such that a Secchi disc on the bottom is visible. In diving areas the clarity shall equal the minimum required by safety standards, depending on the height of the diving platform or board.

[34] *Report of the Committee on Water Quality Criteria*, Federal Water Pollution Control Administration, U.S. Dept. of the Interior, Washington, D.C., April 1, 1968, pp. 8–14.

[35] Criterion for Bathing Waters, in *Quality Criteria for Water*, USEPA, Washington, D.C. 20460, July 26, 1976, p. 79.

Water quality requirements for bathing waters adopted by the European Economic Community Council are shown in Table 9-1.

Swimming Pool Water Quality

The sanitary quality of swimming pool water is determined by certain microbiological, chemical, and physical tests. Examinations are made in accordance with the analytical procedures described in *Standard Methods* by competent laboratory personnel.[36]

The American Public Health Association Public Health Joint Committee on Swimming Pools and Bathing Places, in *Public Swimming Pools* (see Bibliography), recommends the following:

pH	7.2 to 8.0*.
Alkalinity (methylorange, MO)	at least 50 mg/l total but not greater than 150.
Clarity	6-in. diameter black and white disc readily visible at deepest point when viewed from side of pool.†
Plate count (agar, 24 hr at 35° C)	not more than 15 percent of the samples over any 30-day period while pool is in use contain more than 200 colonies per ml. (A count of less than 100 colonies per ml can be maintained in a properly designed, constructed, and operated pool.)
Coliform organisms (dechlorinated sample)	not more than 15 percent of the samples over any 30-day period while pool is in use show positive (confirmed test) in any of the five 10-ml portions of a sample in multiple fermentation tube method or more than 1.0 coliform organisms per 50 ml in membrane filter test.
Staphylococcal group (if made)	not more than 50 organisms per 100 ml.

In terms of Jackson Turbidity Units (JTU) as a measure of water clarity, 0.5 JTU or greater is very noticeable, 0.1 JTU or less is excellent; but up to 0.5 JTU is acceptable. The near absence of turbidity assures the effectiveness of disinfection.

Cloudy pool water may be due to improper filtration (unclean sand and cracks in bed, need for alum with sand filter, or improper precoat in diatomite filter, high alkalinity). Reddish brown color may be due to iron; brownish black color

*See also Swimming Pool Operation, this Chapter.

†The National Swimming Pool Institute recommends a water clarity of 0.5 or less JTU. (*Minimum Standards for Public Swimming Pools*, April 1, 1977, p. 18.)

[36] *Standard Methods for the Examination of Water and Wastewater*, American Public Health Association, 1015 Fifteenth Street, N.W., Washington, D.C., 1980.

Table 9-1 European Economic Community Quality Requirements for Bathing Waters

Parameters		G (Guide)	I (Mandatory)	Minimum sampling frequency
Microbiological				
1 Total coliforms (/100ml)		500	10,000	Fortnightly[a]
2 Faecal coliforms (/100ml)		100	2,000	Fortnightly[a]
3 Faecal streptococci (/100ml)		100	—	*b*
4 Salmonella(/1)		—	0	*b*
5 Enteroviruses (PFU/101)		—	0	*b*
Physico-chemical				
6 pH		—	6 to 9 (0)	*b*
7 Colour		—	No abnormal change in colour (0)	Fortnightly[a]
		—	—	*b*
8 Mineral oils	(mg/l)	—	No film visible on the surface of the water and no odour	Fortnightly[a]
		≤0.3	—	*b*
9 Surface-active substances reacting with methylene blue	(mg/l) (lauryl-sulphate)	—	No lasting foam	Fortnightly[a]
		≤0.3	—	*b*
10 Phenols (phenol indices)	(mg/l) C_5H_5OH	—	No specific odour	Fortnightly[a]
		≤0.005	≤0.05	*b*
11 Transparency	(m)	2	1 (0)	Fortnightly[a]
12 Dissolved oxygen	(% satn O_2)	80 to 120	—	*b*
13 Tarry residues and floating materials such as wood, plastic articles, bottles, containers of glass, plastic, rubber or any other substance. Waste or splinters.		Absence		Fortnightly[a]

Source: "Britainia waives the rules" *World Water*, June 1978, pp. 24–27.
Adopted by the European Economic Community Council (EEC) December 8, 1975.

[a] When a sampling taken in previous years produced results which are appreciably better than those in this table and when no new factor likely to lower the quality of the water has appeared, the competent authorities may reduce the sampling frequency by a factor of 2.

[b] Concentration to be checked by the competent authorities when an inspection in the bathing area shows that the substance may be present or that the quality of the water has deteriorated.

due to manganese; and blue green color due to copper corrosion. Algae growths may cause slime, and green or brown coloring of the water. These problems and their solution are discussed in Chapter 3 and in this chapter.

Other Recreation Water

For surface waters for general recreational use, not involving significant risk of ingestion and in the absence of local epidemiological evidence to the contrary, a standard of "an average not to exceed 2000 fecal coliforms per 100 ml and a maximum of 4000 per 100 ml, except in specified mixing zones adjacent to outfalls" is suggested.[34]

For waters where the probability of ingesting appreciable quantities is minimal,

> the fecal coliform content, as determined by either the multiple tube fermentation or membrane filter technique, should not exceed a log mean of 1000/100 ml, nor equal or exceed 2000/100 ml in more than 10 percent of the samples."[34]

Sample Collection

Water samples should be collected in wide-mouth bottles by plunging the bottle downward and then forward until filled while the bathing area or swimming pool is in use. The sampling points should be in the bathing area of a beach or near the pool outlet or outlets and at such representative points as will indicate the quality of the bathing water. A sample per 300 ft of beach in about a 2-ft depth of water is suggested. Samples of chlorinated water must be collected in sodium-thiosulfate-treated bottles so as to dechlorinate the sample and thus give a true indication of the quality of the water at the time of collection. Sodium thiosulfate should not be flushed out. In any case, the sample should be returned to the laboratory for examination as soon as possible after collection.

Accident Prevention

According to the CPSC National Electronic Injury Surveillance System* an estimated 65,000 persons required emergency room treatment in 1975 for injuries associated with swimming pools, swimming pool slides, and diving boards. Most accidents were due to slippery walkways, decks, diving boards, and ladders; striking the bottom or sides of a pool because of inadequate depth for diving or sliding; and drowning when swimming alone or without adult supervision.

In a study made by Van Dusen and Fraser[37] it was reported that there were 8100 deaths due to drowning in 1975 in the United States. Of this number approximately 3100 were associated with swimming, and 5000 with other water-

*Based on data collected from admissions to 119 selected hospital emergency departments for injury and consumer product-associated injury. The hospitals represent a statistical sample from 5939 hospitals in the United States and its territories.

[37]Karen Van Dusen and Gary Fraser, *Swimming Pool Program Study*, Dept. of Social and Health Services, Olympia, Wash., October 1977, p. 5.

based recreation. Of the 3100 swimming pool drownings, approximately 600 occurred in residential pools (19 percent). The other deaths occurred at motel-hotel pools (10 percent), apartment or condominium pools (10 percent), public pools (20 percent), and in pools under sponsorship of schools, voluntary organizations, and private groups such as country clubs (41 percent). Swimming pool diving boards and water slides were a significant source of injuries.

In a special study[38] of 72 swimming pool injury cases, 52 "retrospective" cases in which injuries were sustained before January 1, 1976 and 20 "prospective" cases involving injuries after January 1, 1976, although not representative of the 65,000 annual pool related accidents, the findings were informative. The major types of injury (57) were quadriplegic.* Diving and striking the head on the bottom was the primary cause of the injury—18 diving in in-ground pools with 4 ft or less water depth and 18 diving in above-ground pools with 4 ft or less water depth. An additional 8 accidents were due to water slides; 6 head-first entry in water of $3\frac{1}{2}$ ft or less depth. The remainder were miscellaneous injuries. In 54 of the injuries there was no designated person in charge at the time of the injury. Forty residential pools; 23 hotel, motel and apartment pools; 6 public school and city pools; and 2 private club pools were involved. Significantly, 32 of the pools were vinyl-lined; hands slipped as individual struck the bottom. Also, 38 of the pools were a concrete basin with variable depth and 20 were above-ground vinyl-lined pools 5 ft or less in depth; 46 did not have depth markers on the pool coping or interior walls; 43 had no warning signs posted in the pool area; 59 had no bottom markings in the pool; 51 occurred where water depth was 4 ft or less; and 15 where water depth was 5 ft 1 in. or deeper. Shallow water, the absence of bottom markings, the lack of water depth markings, and the failure to have warning signs were the most prevalent conditions.

Other studies show that a smooth bottom in the pool diving and up-slope or run-out area will minimize possible head and neck injuries, that depth markings are often ignored, and that most injuries by far occur in the 13 or higher age group. Injuries to the cervical spine can occur at impact speeds equal to or greater than 2 to 4 fps. There appears to be no practical way to build a pool deep enough or large enough to rely on hydrodynamic forces alone to slow a diver to a safe velocity.[38,39] A diving training and education program, in addition to proper pool depth, geometry, and construction, offers the best means for reducing diving related accidents.

The elimination of tripping or slipping hazards in pools, runways, and decks, and controlled area bathing with adequate lifeguard supervision will do much to prevent accidents and drownings. Attention to accident hazards and protection of life are musts at all public beaches and pools, including those at resorts,

*Total care and wages loss estimated at $1,000,000 during the average life span.

[38] *Medical Analysis of Swimming Pool Injuries*, University of Miami School of Medicine and Nova University, CPSC, Washington, D.C., December 1977.

[39] Richard S. Stone and John D. States, "Why Do Diving Accidents Cause Quadraplegia? A Biomechanical Study and Analysis," Dr. Richard S. Stone, 15 Acorn Park, Cambridge, Massachusetts 02140, October 1980.

schools, clubs, associations, and so forth. Failure to do so may cause serious injury and place the management in an untenable position in case of lawsuits.

The American Red Cross suggests that one guard may be sufficient for each 100 bathers provided they are in a confined area.[40] On surf bathing beaches one guard in a tower for every 100 yd of beach plus a guard in a boat every 200 yd in the swimming area is recommended. Double this number may be needed on weekends and holidays. One lifeguard director for every 75 swimmers plus one trained "life saver" for every 10 persons in the water has also been recommended at recreational camps and similar places. Bathing areas and pools should be marked for swimmers and nonswimmers. Lifeboats with anchors and bamboo poles 12 to 15 ft long, surfboards, torpedo buoys, heaving lines of $\frac{3}{16}$-in. manila line made up in coils containing 75 ft of line attached to 15-in. diameter ring buoys, and grappling irons are standard equipment for bathing beaches.

Life-saving equipment per 2000 ft^2 of pool-water area should include a bamboo pole 12 to 15 ft long or shepherd's crook and a ring buoy not more than 15 in. in diameter, with 60 ft of $\frac{3}{16}$ in. rope or two tightly rolled balls of $\frac{1}{4}$-in. heaving line equal to $1\frac{1}{2}$ times the pool width. This equipment should be conspicuously distributed around the pool. An elevated lifeguard tower should be required for a pool 2000 or 2250 to 4000 ft^2, and additional towers should be required for larger pools. Where a lifeguard is not on duty, a sign in plain view should state in clear letters, at least 4-in. high, "Warning—No Lifeguard on Duty." Another sign should state, "Children Under Age Sixteen Should Not Use Pool Without An Adult in Attendance," and "Adults Should Not Swim Alone."[41] Check state and local laws; many prohibit bathing at public beaches and pools unless a lifeguard is on duty. All pools should be fenced and kept locked when not under supervision.

Where the water is not crystal clear at all times, as in natural ponds and dammed-up streams, it is especially important that bathing be carefully organized. A system of checking such as "roll call" or "tag board" to determine who is in the water and who comes out, and a "cap" system to distinguish nonswimmers, beginners, and swimmers, or a "buddy" system whereby bathers are paired according to their abilities are accepted and highly recommended safety practices.

A telephone should be readily accessible and the telephone numbers for respirators, ambulances, doctors, and hospitals conspicuously posted for use in case of emergency. Also needed are a 24-unit first-aid kit, cot and blankets, and a trained first-aid man.

Artificial pools, lakes, coastal beaches, and ponds may have holes, steep sloping bottoms, sudden drops, large rocks, stumps, heavy weed growths, tin cans, bedsprings, broken glass, and miscellaneous debris that can cause injury or

[40] *Life Saving and Water Safety*, American Red Cross, Blakiston Co., Philadelphia, 1937; also, Bull. No. 27, New York State Dept. of Health, Albany, 1976, pp. 2–5 and *Public Swimming Pools*, American Public Health Association, Washington, D.C., 1980.

[41] *Public Swimming Pools*, Recommended Regulations for Design and Construction, Operation and Maintenance, American Public Health Association, 1015 Fifteenth St., N.W., Washington, D.C., 1981.

drowning. These should of course be removed or, if not practical, carefully marked "Off Limits." The bottom should be gently sloping, clean, clear, and firm at least out to a depth of 5 ft. Silt, muck, or quicksand bottoms are not suitable. Where it is not feasible to remove the silt, muck, or mud and replace it with sand, a polyethylene sheet covered with 12 in. of sand or a paved bottom might be substituted, at least in the wading area. This will help keep the water clear and reduce the drowning hazard.

Electric Hazards

Electrical hazards at swimming pools have caused serious disability and death. Use of Underwriters Laboratories, Inc. approved equipment (UL Label) and installation in accordance with the National Electrical Code (1975) or equal should greatly minimize if not eliminate the hazards. A ground-fault circuit-breaker or interrupter which will immediately shut off the electric power at the source should be standard equipment on all electrical circuits in public and residential pool areas. This device is automatically activated when a defective wire, connection, or equipment (such as an electric motor, vacuum equipment, underwater light, electric panel box, light switch, radio wire, or ungrounded electric receptacle) permits current to flow or leak into the ground directly or indirectly. The power to an outlet or other connection is cut off when a preset amperage (5/1000 of an ampere, or 5 milliamperes) is exceeded, which amperage is much less than that required to activate the usual fuse or circuit-breaker. All pool equipment and accessories, including lights on metal poles, diving board stands and platforms and handrails, underwater and surface lights, step handrails, and ladders are required to be properly grounded. Under no circumstances should electric wires or cables be permitted to cross a swimming pool.

SWIMMING POOL TYPES AND DESIGN

Swimming pools are generally classified as "artificial pools" and "partly artificial pools." Artificial pools are usually constructed of reinforced concrete or gunite, steel, aluminum, fiberglass, and with vinyl lining. Pools are designed and operated as recirculating pools with filtration and disinfection, or rarely and *if permitted*, with disinfection only, as fill-and-draw pools, or as flow-through pools. Saline water swimming pools should meet the same requirements and standards as fresh water pools. The major concern is greater corrosion of metal piping and parts.

Recirculating Swimming Pool

In the typical recirculating-type swimming pool a pump takes water out of the pool, passes it through a filter, chlorine is added, and the water is returned to the pool. Water lost by evaporation, splashing, and backwashing of the filters is replaced by fresh water. This type of pool permits use by a maximum number of bathers; a minimum amount of water is wasted, and fuel for heating the water, where desired, is saved because the filtered pool water is reused. If the filters are omitted or operated intermittently the organic and dirt load cumulate, residual chlorine control is difficult, and frequent change of water and cleaning is necessary. Filtration is provided by means of gravity rapid-sand filters, pressure sand and anthracite filters, or diatomaceous earth pressure and vacuum filters. The pressure filters are most commonly used on swimming pools.

Fill-and-Draw Pool

The fill-and-draw pool is filled with fresh water, used for some period of time, then emptied, cleaned, and refilled. Such pools are nothing more than common bathtubs in which the pollution introduced is circulated among the bathers. Their use is generally prohibited.

Flow-Through Pool

The flow-through pool is fed by a continuous supply of fresh water, without treatment, which causes an equal volume of pool water to overflow to waste. Although the bacterial pollution is reduced, it is not completely flushed out of the pool, as shown in Table 9-2. For example, a pool 60′ × 20′, holding 55,000 gal and provided with 166,500 gal of water per day to displace the pool water in 8 hr, can, according to the formula $Q = 6.25T^2$, accommodate 17 bathers per hour.* Q = quantity of fresh water (no disinfectant) required per bather, which in this case is 400 gal, and T = the turnover period in hours, 8 hr in this example. On the other hand, this same pool would accommodate 48 bathers at any one time if the water were continuously chlorinated, filtered, and recirculated, the restriction being a desired pool area of 25 ft^2 per bather in the pool, maintenance of 0.6 mg/l free available chlorine, and water clarity. The pool capacity could be 120 if it is 5 ft or less in depth and only 10 ft^2 is required per bather. Flow-through pools require large controlled volumes of clean water and strict control of the number of bathers and microbiological water quality, which is not normally possible.

Partly Artificial Pool

The partly artificial pool is made by damming a stream, causing water to back up to form a small pond or lake. If the size of the artificial pool is small, the permissible number of bathers per hour would be governed largely by the dilution or volume of water entering the pool. But if a large lake is formed such as

*$\dfrac{166{,}500 \text{ gpd}}{24 \text{ hr}} = 6938$ gph; $\dfrac{6938 \text{ gph}}{400 \text{ gal}} = 17$ bathers per hr. Also $Q = 6.25 \times 8^2 = 400$ gal.

Table 9-2 Effectiveness of Continuously Flowing Fresh Diluting Water in Removing Pollution from a Swimming Pool in the Absence of Disinfection

1 Number of Times Each 24 hr Water is Displaced by Fresh Water	2 Displacement Period (Hr) *T*	3 Computed Quantity in Fresh Water per Bather *Q*	4 Percent of Pollution of Water Remaining in Pool	5 Number of Days for Pollution to Reach Uniform Values Shown in Column 4
1	24	3,600	58	9
2	12	900	16	4
3	8	400	5	3
4	6	225	2	2

Source: New York State Health Dept., Bull. 31, Albany, 1950.
Note: Values given in columns 4 and 5 are in accordance with the "dilution law." $Q = 6.25T^2$. If the displacement period is 4 hr only 0.3 percent pollution would remain.
Assumption: Daily increment of pollution equals that initially present.

to permit the natural laws of purification to operate, the permissible number of bathers would be governed by the results of bacteriological examination of a series of samples of water collected during the bathing period and the sanitary survey.

Summary of Pool Design*

Site Selection and Layout

1. Accessible to users; space for parking, recreation, picnicking, bathhouse, and purification equipment.
2. Adequate and satisfactory water supply.
3. Adequate sanitary sewer or separate disposal system.
4. Pool drainage and wastewater disposal proper. High water-table relief.
5. Site drained, gently sloping, 100 yd or more from roads, railroads, factories, and wooded areas.
6. Pool area enclosed with high wall or fence.

Pool Area and Depths

1. Provide for diving, swimming, and nonswimming areas. A minimum width of 20 ft and length of 60 ft is suggested for public pools.

A water depth of at least $8\frac{1}{2}$ ft is required for a deck level to 2-ft diving board, 10 ft for a 3 ft 3-in. or 1-m board, and 12 ft for a 10-ft (9 ft 9.84-in.) or 3-m board. The minimum length of the diving well plus diving board overhang is 15 ft for the $8\frac{1}{2}$-ft depth, 17 for the 10 ft-depth, and 25 ft for the 12-ft depth. In all cases, the bottom run-out from the diving well is on a slope of at least 1 vertical to 3 horizontal to the 5-ft depth, and 1 vertical to 10 horizontal in the shallow area beyond the 5-ft depth.[42]

Diving boards should provide a clear height above the board of at least 13 ft extending at least 16 ft forward of the plummet, 8 ft behind the plummet and 8 ft to both sides of the plummet. Follow

*See state and local regulations.

manufacturer's directions if greater distance is recommended. Diving boards up to 1 m in height above the water level are located at least 10 ft from an adjacent diving board, center to center, and 10 ft from the pool side wall. Boards higher than 1 m are located at least 10 ft from adjacent boards and 12 ft from pool side walls.[42]

Because of the danger of an inexperienced person making a "perfect dive" and hitting the bottom of the pool at a dangerous velocity,[43] the installation of diving boards should in general be prohibited or discouraged unless special diving instruction and supervision can be provided.

2. Mark water depth on pool perimeter and sidewall above water surface, at 2-ft changes in water depth, and every 25 ft or less.

3. Use Table 9-3 as a guide in determining pool size, in the absence of state or local regulations.

4. Allow at least 10 ft^2 of deck space per bather and include a minimum 5-ft wide strip around an indoor pool and 8 ft for outdoor pool, plus lounging space. Provide floor drain for at least each 400 ft^2. Pave deck space sloped $\frac{1}{4}$ to $\frac{3}{8}$ in./ft to drain; do not use grass or sand. Deck space at least 3 to 4 times pool-water area is preferred. Outdoor carpeting, if permitted, should contain no natural fibers, be resistant to rotting and attack by fungus, be well drained, and not hold water. The carpeting should also be resistant to fading and freezing temperature where this is applicable and meet National Sanitation Foundation Standards or equal.

Source of Fresh Water

1. Use municipal source if available.

2. Use an existing stream or lake, or saline water for saltwater pool, if clean to fill pool through the filters.

3. Develop a well or spring if necessary.

Recirculation System

1. Design to replace water in 6 to 8 hr.* For a private lightly loaded pool, 12 hr may be acceptable. Check state and local regluations. Design for at least 30 percent recirculation from main drain.

2. Provide inlets on four sides, at least 12 in. below water surface, not more than 15- to 20-ft apart and one inlet within 5 ft from each corner; use directional inlets with gate valve or similar control. Provide one inlet with adjustable orifice at least 12 in. below water line for each 600 ft^2 of water surface or 15,000 gal of pool volume, whichever is greater. Include sidewall inlets in pools over 60-ft long and multiple inlets at shallow end if more than 20 ft wide. Maximum pipe velocity 6 fps in suction and 8 to 10 fps in pressure lines, and 3 fps in gravity flow.

3. Pool drain(s) permits pool to be emptied preferably in 4 hr or less. No direct connection to sewer.

4. Outlet opening or grating area is at least 4 times drain pipe width. Use multiple outlets if pool is more than 30-ft wide. Grate velocity $1\frac{1}{2}$ fps maximum.

5. Design for head loss of 5 to 7 lb/in.^{2} in pressure sand filter, 10 to 15 lb/in.^{2} in high-rate, 30 lb/in.^{2} in diatomite filter, and 15 in. of mercury with vacuum diatomite filter. Add for piping head loss.

6. Select correct pump for type of filter static head and head loss in recirculation piping and

*A 3- to 4-hr turnover is recommended where pools are expected to be well used. (*The Purification of the Water of Swimming Pools*, Dept. of the Environment, London, p. 29, 1975.)

[42]For complete details see *Public Swimming Pools*, op. cit., and *Recommended Standards for Swimming Pools*, Great Lakes-Upper Mississippi River Board of State Sanitary Engineers, 1977 ed., Health Education Service, P.O. Box 7283, Albany, N.Y. 12224, which states "these diving area dimensions do not meet the requirements of NCAA or AAU. Dimensions for diving pools shall be in accordance with the Standards of International Amateur Swimming and Diving Federation (FINA)."

[43]*Medical Analysis of Selected Swimming Pool Injuries*, Summary Report for the CPSC, Washington, D.C., December 1977, pp. 41–43.

Table 9-3 Swimming Pool Design Bathing Load Limits

Authority	Area (ft^2) Diving	Swimming	Nonswimmers
APHA[a]			
Bathers in pool enclosure	1-m board or less, 8 ft each side of board and 16 ft forward; for higher board, 10 ft each side of board and 20 ft forward; 10 persons per diving board.	25	15
GLUMR Board[b]			
Bathers in pool enclosure	300 ft^2 around each diving board or platform.	25	10, 15 if pool depth 5 ft or less
National Swimming Pool Institute[c]	300 ft^2 per diving board	20	15
New York State			
Bathers actually in water	—	25	25
Bathers present in pool area	—	$12\frac{1}{2}$	$12\frac{1}{2}$
Illinois			
Bathers actually in water[d]	—	30	15
Iowa State College	For cities under 30,000; maximum day = 5 to 10 percent of population and maximum day at any one time = one-third maximum day.		
National Recreation Assoc.	Provide swimming and bathing space for 3 percent of population. Allow 15 ft^2 of water area per person.		
Tile Council of America	600 ft^2 per 1000 for communities of 4000 people or less to 320 ft^2 per 1000 for communities up to 90,000 population.		
Anonymous	Pool area required = 1/5 × total population served (for communities greater than 5000 population).		

Note: Adjust required pool area for class use such as in schools, clubs, and for special local conditions. The water area 5 ft or less in depth is considered nonswimmer area; greater than 5 ft in depth is swimming area except for diving area.

[a] *Public Swimming Pools, Recommended Regulations for Design and Construction, Operation and Maintenance*, Am. Pub. Health Assoc. 1015 Fifteenth St., N.W., Washington, D.C. 20005, 1981.

[b] *Recommended Standards for Swimming Pools*, Great Lakes-Upper Mississippi River Board of State Sanitary Engineers, 1977ed., Health Education Service, P.O. Box 7283, Albany, N.Y. 12224.

[c] *Minimum Standards for Public Swimming Pools*, National Swimming Pool Institute, 2000 K Street, N.W., Washington, D.C. 20006, 1977 p. 1. (With minimum 6-ft continuous unobstructed deck space width for public pool.)

[d] Capacity may be increased by 35 percent in an outdoor pool, if deck space is adequate. Post maximum pool and pool enclosure capacity.

fittings. Pump to provide design pool turnover at an operating head of approximately 50 ft for pressure sand and 80 ft for pressure diatomite filter. See (5) above. Make hydraulic analysis.

Accessories

1. Water heater with automatic thermal control for indoor and some outdoor pools.
2. Locate water heater, if provided, on water flowing from filters to pool. Use cold water for filter-washing. Temperature of water entering pool not greater than 110° F. Provide thermometer on heater outlet and near pool inlet.

3. Hair and lint catcher. Provide spare on bypass, with valves. Area of strainer openings at least 4 times area of main recirculating line.

4. Vacuum cleaner connected to portable pump or recirculating pump suction line, with proper connections in pool sides at least 8 in. but not more than 12 in. below the pool water surface.

5. Space heaters and ventilation for indoor pools.

6. Residual chlorine and pH testing kits. Chlorine range 0.3 to 3.0; pH range 6.8 to 8.2, phenol red, most common. If cyanurates used, test kit should measure to 100 mg/l with increments of 25 mg/l.

7. Spare parts for chlorinator, including ammonia bottle and self-contained respirator or an approved-type gas mask where gas machine used.

8. Hose bibs provided to hose down walks and floors.

9. Steps and ladders for pools more than 30 ft long.

Disinfection

1. Chlorination is most common. Bromine, iodine, and chloro-iso-cyanurates are also used. Disinfectants used must be registered with the USEPA.

2. Chlorinator capacity adequate to dose indoor pool water at 5 mg/l and outdoor pool at 10 mg/l. Manufacturers recommend gas chlorinator capacity of 15 lb for first 100,000 gal plus 10 lb for each additional 100,000 gal of pool volume. Another design basis is 1 lb of chlorine per 8 hr per 10,000 gal of water in the pool.

3. Gas chlorinator housed in separate room; can be mechanically ventilated with minimal danger to attendants, bathers, and adjoining areas. Provide self-contained breathing apparatus in a separate location for emergency use. (See Figure 9-6; also Chapter 3, Gas Chlorinator.) Use of hypochlorinator is recommended for safety reasons. Store sodium hypochlorite and other disinfectants in a cool, dry, well ventilated, and secured place.

4. Compare first cost, operation, *safety*, and maintenance of positive feed hypochlorinator versus solution feed gas chlorinator. If safety of gas chlorinator to bathers, and public, cannot be assured because of topography, meteorology, and other factors, do not use.

Clarification of Swimming Pool Water

1. Pools require filtration equipment. See Table 9-4.

2. Provision made for batch treatment with alum, soda ash, or other chemical by means of solution feed pump, or through the make-up water or surge tank. Assure that several minutes' coagulation-flocculation can take place before water reaches filters.

3. Alkalinity adjustment by means of soda ash (sodium carbonate), added by dry chemical feeder, solution feeder, solution pot, or soda-ash briquets. Calcite or calcium carbonate graded sand may be used in filter. Sodium bicarbonate, $1\frac{1}{2}$ lb per 10,000 gallons, will increase alkalinity 10 mg/l. Add $1\frac{1}{2}$ pints of muriatic acid to lower alkalinity 10 mg/l.

4. Filter sand 30-in. deep, clean and sharp, effective size 0.4 to 0.55 mm, uniformity coefficient not greater than 1.75. Anthracite 0.6- to 0.8-mm size. Filters provide at least 18-in. freeboard between top of sand and overflow.

5. Rate of filtration 2.0 to 3.0 gpm/ft^2 with sand or anthrafilt, and 1.0 to 2.0 gpm with diatomite. The lower rates are recommended; but up to 2.5 gpm/ft^2 may be permitted with continuous body feed. Rates up to 15 gpm may be permitted for pressure sand filters. With cartridge filters, up to 3 gpm/ft^2 may be permitted for the depth type and up to 0.375 gpm/ft^2 for the surface type. Design for recirculation from main drains and perimeter overflow or surface skimmers.

6. Wash-water rate for sand filter, 15 gpm/ft^2 desirable; 9 to 12 gpm is acceptable. Use 9 gpm with anthracite.

7. Continuous feed of diatomite said to result in less diatomite use.

8. Provide on filter system a rate-of-flow indicator for recirculation and back-wash measurement, air-release valve on each filter shell at top, pressure gauges on influent and effluent lines and on either side of hair catcher, two baskets for hair catcher, sight glass on waste-discharge line.

Table 9-4 Swimming Pool Filtration and Capacity Data

Diatomite Filter Septa		Vertical Pressure Filters			
Diameter (in.)	Cylindrical Area (ft^2 per ft of length)	Diameter (in.)	Surface Area (ft^2)	Inlet (in.)	Waste to Sewer (in.)
2	0.524	30	4.9	$1\frac{1}{2}$	2
3	0.785	36	7.1	2	$2\frac{1}{2}$
4	1.047	42	9.6	2	$2\frac{1}{2}$
5	1.309	48	12.6	$2\frac{1}{2}$	3
6	1.571	54	15.9	$2\frac{1}{2}$	3
		60	19.6	3	4
		66	23.8	3	4
		72	28.3	4	5
		78	33.2	4	5
		84	38.5	4	5
		90	44.2	4	5
		96	50.3	5	6
		102	56.8	5	6
		108	63.6	5	6
		114	70.9	5	6
		120	78.5	5	6

Filter rate —2.0 to 3.0 gpm/ft^2 with sand or anthrafilt; 15 gpm/ft^2 high-rate.
—3.0 gpm/ft^2 for depth-type cartridge; 0.375 gpm/ft^2 for surface-type.
—1.0 to 2.0 gpm/ft^2 with diatomite, continuous feed.

Wash-water rate —15 gpm/ft^2 for sand filter. Use 9 gpm/ft^2 when using anthrafilt since it has about one-half the density of sand. Water use should not exceed 2 percent of the water filtered.

Recirculating pump max. capacity (gpm) = total filter area in ft^2 × filter rate in gpm/ft^2

$$\text{Hours for one pool turnover} = \frac{\text{pool capacity in gal}}{\text{recirculating pump capacity in gpm}} \times \frac{1}{60}$$

= 6 to 8 hr for public pool; 12 hr for private pool.

Recirculating pump capacity = sum of flows from each recirculation inlet, which is made proportional to the volume of water in that part of the pool, and still provide design pool turnover.

Overflow Gutters and Skimmers

1. Overflow gutters perfectly level completely around pool, if provided; a 12- to 18-in. flat gutter sloping away from pool also used. See Figure 9-1. Overflows easily accessible for cleaning and inspection.

2. Surge capacity of 1 gal/ft^2 of pool area provided in special gutters, surge tank, or elsewhere. Gutters and skimmers have the hydraulic capacity to carry away the overflow water, without flooding, due to swimmers entering the pool. A swimmer displaces on the average 17 to 20 gal of water.

3. Outlet drains not less than $2\frac{1}{2}$ in., 15 ft on centers. Drainage to recirculating pump or balancing tank.

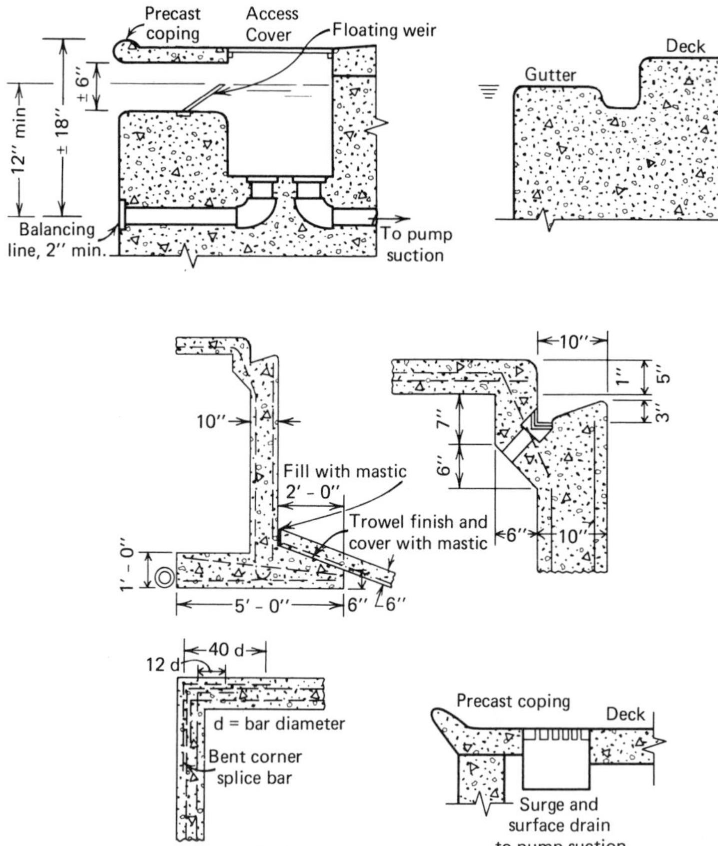

Figure 9-1 Swimming pool wall and gutter details. Prefabricated metal, precast gutters, and surface skimmers are available.

4. Pools less than 30-ft wide may be provided with skimmers built into the side and corners of the pool to take the place of gutters if acceptable to the regulatory agency. Skimmer design and construction meets National Sanitation Foundation Standards. Minimum 2-in. line 12-in. below lip of skimmer to prevent air entering pump suction.

5. At least 50 percent of recirculation from gutters or skimmers. Capacity of gutters 100 to 125 percent of pump recirculation. One skimmer for each 500 ft^2 of pool area, minimum of two, flow rate at least 30 gpm or 3.75 gpm/linear in. of weir; total capacity less than total recirculation capacity.

No Connection with Potable Water Supply

1. Design does not provide opportunity for pool water to enter drinking water system.

2. Introduce fresh water directly to the pool through an air break or into the suction side of the recirculation pump, preferably through a balancing tank, or through an approved backflow preventer. See Figure 9-2.

Bathhouse

1. Facilities include separate dressing rooms, showers, toilets, and wash basins for each sex in proportion shown in Table 9-5. Showers have 95 to 105° F water. See Figure 9-3.

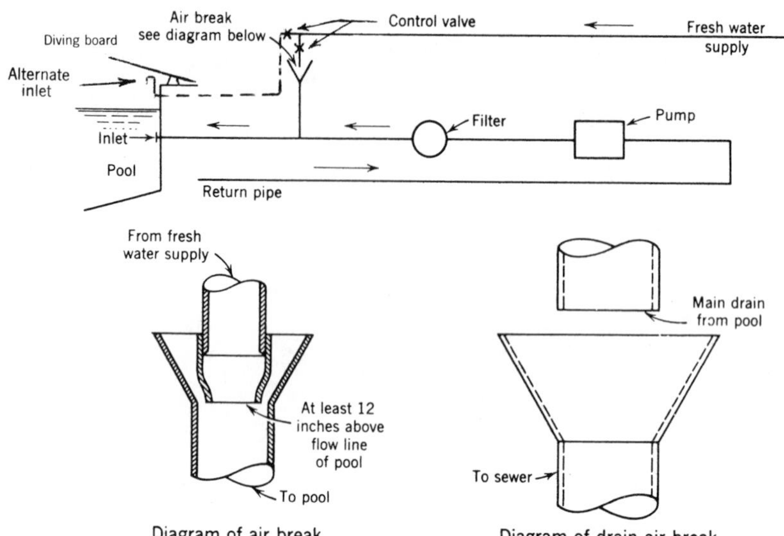

Figure 9-2 Acceptable means for adding fresh water to a pool and for draining wastewater.

2. Dressing rooms provide 7 ft^2/female and 3.5 ft^2/male patron expected at maximum periods. Well ventilated and lighted. Provide for direct entrance of sunlight where possible; ordinary window glass and most transparent materials absorb the ultraviolet radiation.
3. Floor drains not over 25-ft apart and floor slopes between $\frac{1}{8}$ and $\frac{1}{2}$ in./ft.
4. Entrance and exit at shallow end of pool.
5. Bathers from dressing room must pass toilets and go through shower room before entering pool.
6. All electrical work, indoor and outdoor, shall be installed, comply with, and be maintained in accordance with the standards of the National Electrical Code. See Accident Prevention.

Table 9-5 Plumbing Fixtures Recommended at Swimming Pools

	Number Men Served				Number Women Served			
Fixture	*a*	*b*	*c*	*d*	*a*	*b*	*c*	*d*
1 water closet	75	60	15	60	50	40	10	40
1 urinal	75	60	20	60	(if provided)			60
1 washbasin	100	60	15	60	100	60	10	60
1 shower	50	40	4	40	50	40	4	40
1 service sink	At least one				At least one			

Note: For classes, the number of showers equals one-third the number of pupils in the maximum class. Provide at least one drinking fountain in pool area. Showers to provide 3 gpm; water temperature 95° to 105° F through single valve. See local regulations.

[a] *Public Swimming Pools, Recommended Regualtions for Design and Construction, Operation and Maintenance*, Pub. Health Assoc., 1015 Fifteenth Street, N.W., Washington, D.C. 20005, 1981.

[b] New York State Health Dept., Bull. 27, Albany, 1972.

[c] Class in school, YMCA, and similar pool for 1-hr class. Multiply number of persons by $2\frac{1}{2}$ if class is of 2-hr duration.

[d] *Swimming Pools*, PHS/DHEW, Washington, D.C., June 1976.

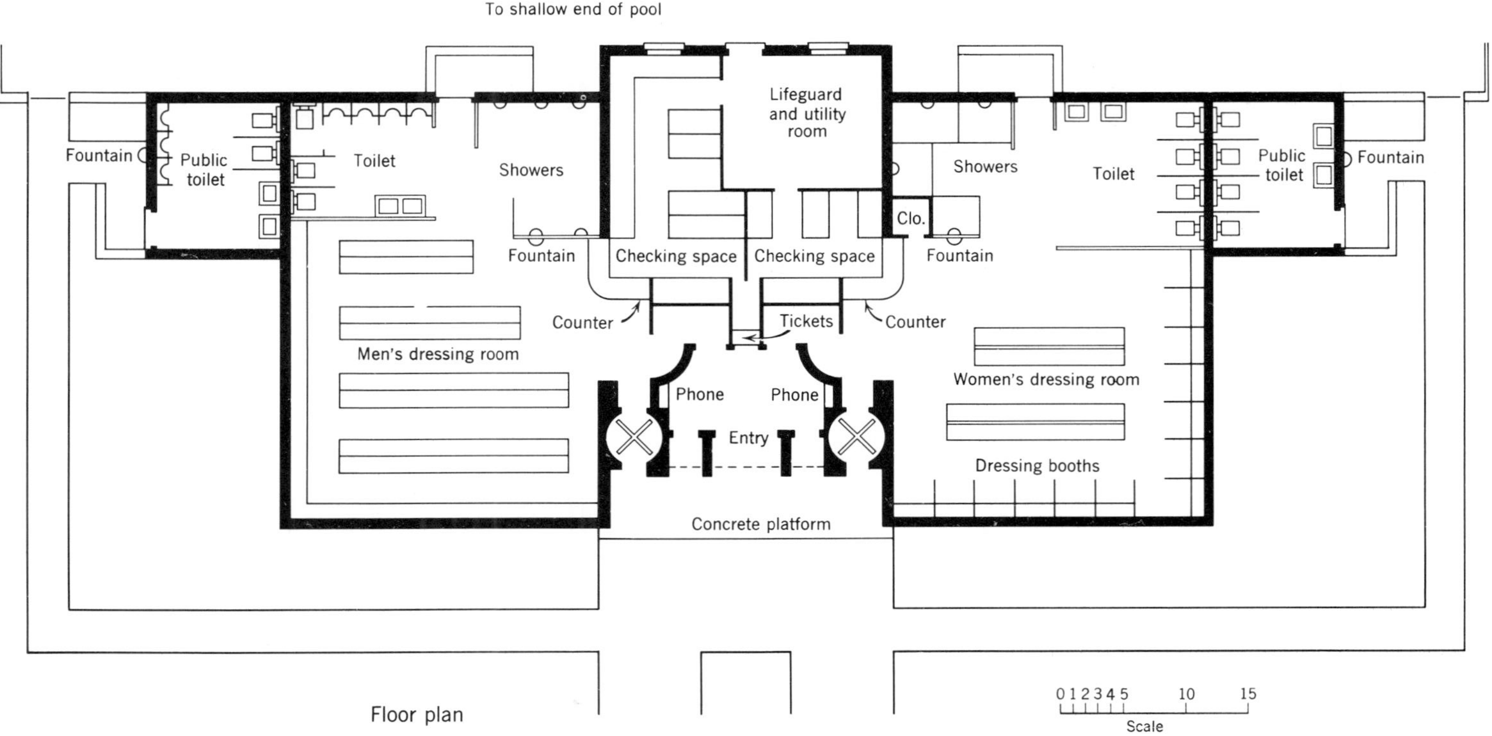

Figure 9-3 Suggested layout for a bathhouse for a 40 × 100-ft pool.

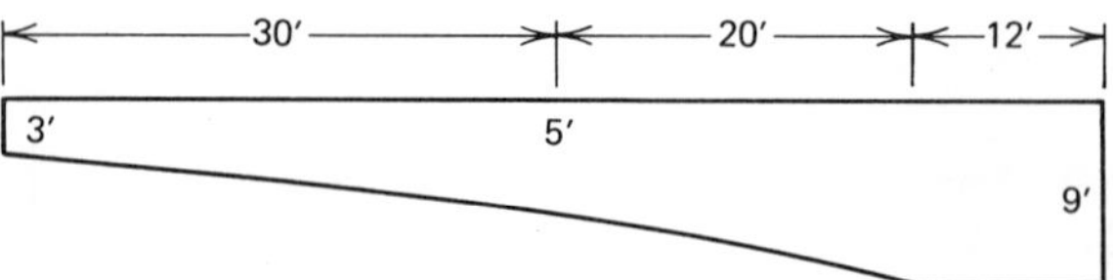

Figure 9-4 Typical summer resort swimming-pool design, without diving board.

A Small Pool Design

To illustrate some of the design principles an example is given below. Figure 9-4 shows a cross-section of the pool.

Size: $62' \times 30' = 1{,}860\ \text{ft}^2$

No. persons: $\dfrac{1860}{25} = 75$ people at any one time; 150 in pool enclosure

Capacity: $\left[\dfrac{(3+5)}{2}\,30 + \dfrac{(5+9)}{2}\,20 + 9 \times 10\right] 30 \times 7.5 = 78{,}750$ gal

Recirculation period: assume 8 hr

Required capacity of recirculating pump $= \dfrac{78{,}750}{8 \times 60} = 164$, say 175 gpm

Required electric motor (1.750 rpm, with magnetic starter and under voltage protection)

$$= \frac{\text{gpm} \times \text{total head in ft}}{3{,}960 \times \text{pump and motor efficiency}}$$

$$= \frac{175 \times (16' \text{ in filter} + 16' \text{ in recir. system})}{3960 \times 0.60 \times 0.80}$$

= 2.95, say, 3 hp with pressure sand filters.

$$= \frac{175 \times (69' \text{ in filter} + 16' \text{ in recir. system})}{3960 \times 0.60 \times 0.85}$$

= 7.4, say $7\frac{1}{2}$ hp with pressure diatomite filters (minimum)

Piping: Main lines to and from recirculating pump = 4 in.; lines to pool inlets = 2 in.; provide minimum of 4 pool inlets, more preferred. Locate inlets 18- to 24-in. below water surface. Provide 2-in. vacuum cleaner connections and flow measurement. All valves are gate valves. See Chapter 3, Design of Small Water Systems, for hydraulic analyses.

Hair catcher: 4-in.; coagulant feeder: one; soda ash feeder: one; one vacuum cleaner, and one flow indicator.

Chlorinator type: hypochlorinator, positive feed.

Filters: required area = $\frac{175}{3} = 58\ \text{ft}^2$; 3 5-ft dia. filters = $3 \times 19.6 = 58.8\ \text{ft}^2$. Provide 3 5-ft dia. pressure sand filters; *or* one diatomite filter with total septum area of 88 ft^2, with air release at top of each filter shell consisting of $\frac{1}{2}$-in. line with globe valve, and pressure gauges on inlet and outlet lines.

Pool water heating:

$$
\begin{aligned}
H &= \text{heater size in Btu/hr} \\
Q &= \text{volume of water in pool in gal} \\
T &= \text{degrees F water temperature increased} \\
8.33 &= \text{weight of 1 gal of water in lb}
\end{aligned}
$$

Then:

$$H = Q \times 8.3 \times T$$

A pool 30 × 60 ft in area and with a depth of 3 ft to 5 ft in the shallow area and up to 9 ft in the deep area holds approximately 80,000 gal of water (actually 78,750 gal). The temperature of the pool water is cooled to 58° F, and it is required that the temperature of this water be raised to 78° F. Determine required heater output.

$$
\begin{aligned}
H &= 80{,}000 \times 8.3 \times 20 \\
H &= 13{,}280{,}000 \text{ Btu/hr} \\
H &= 664{,}000 \text{ Btu/hr with a 20-hr pool heat-up time}
\end{aligned}
$$

If the heat loss from the water surface is 12 Btu per ft^2 per hour per degree temperature difference, and the average air temperature is 58° F, the water surface heat loss is:

$$30 \times 62 \times 12 \times (78 - 58) = 446{,}400 \text{ Btu/hr}$$

Since the heater required will have an output of 664,000 Btu/hr and exceeds the water surface heat loss of 446,000 Btu/hr, it will be adequate for the example given to maintain the water temperature, and allow for heat loss due to moderate wind velocity.

The heater output capacity can also be estimated by assuming a heat requirement of 7 to 10 Btu/hr/gal of water recirculated in an indoor pool. Another basis is 15 Btu/hr/ft^2 of pool surface area per F° temperature rise. The average air temperature is an important factor.

A recommended filtering and disinfecting system for an indoor pool including surge tank for make-up water, steam heating, and other appurtenances is illustrated in Figures 9-5 and 9-6.

SWIMMING POOL OPERATION

Bacterial, Chemical, and Physical Water Quality

The bacterial quality of artificial swimming pool water approaches, and the clarity may exceed, that of the Federal Safe Drinking Water Act. Maintenance of such water quality will be affected by many factors, many of which can be controlled by the pool personnel. These include the following.

1. The operation and maintenance of purification equipment provided, such as strainers, pump, filters, chemical feed apparatus, and chlorinator.
2. The bathing load, or number of persons permitted to be in the pool at any one time and in any one day, based on the capacity of the purification equipment and pool area.

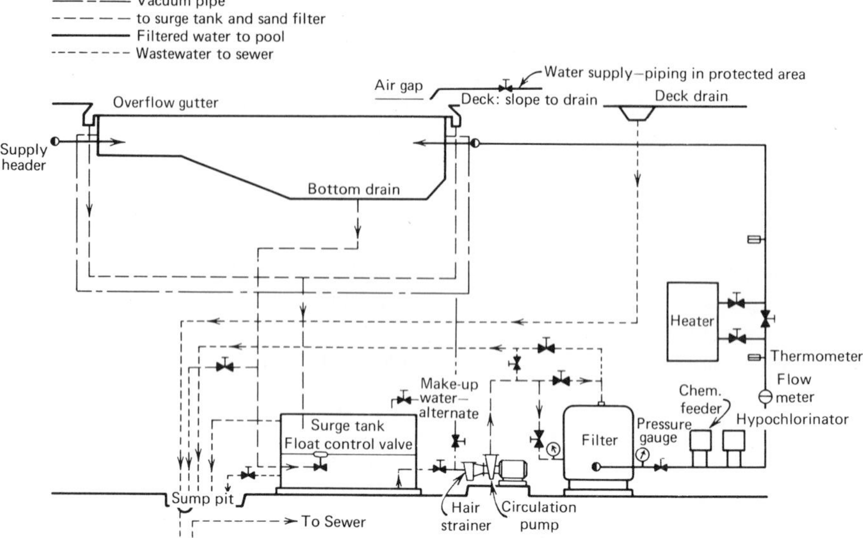

Figure 9-5 Diagram of pool with recirculating system and purification equipment. (Adapted from *Swimming Pools and Bathing Beaches*, Bull. 27, New York State Dept. of Health, Albany, N.Y., June 1972, p. 22.)

3. Pollution introduced by the bathers and enforcement of warm water and soap cleansing showers before entering or reentering the pool enclosure.
4. Supervision of pool personnel and bathers.
5. Maintenance of pool decks, general cleanliness, and separation of recreation and picnic areas from the pool. Prohibition of food and spectators from pool enclosure.
6. Tests for residual chlorine and pH at least three times a day, water clarity (and cyanuric acid if used) measured daily, and tests made periodically for alkalinity, bacterial plate and coliform counts. Keep daily operation record and include also number of bathers, peak bathing load, recirculation period, rate of flow, amounts of chemicals used, volume of make-up water, and other maintenance. Most health agencies have a standard form.

The accepted and effective method of maintaining satisfactory pool water quality in a properly designed pool is by continuous recirculation, chlorination, and filtration over a 24-hr period. Attempts to economize by intermittent operation invariably lead to ineffective or nonuniform chlorination, algal growths, and reduced water clarity. This is particularly true at outdoor pools and at pools subject to periodic heavy bathing loads. Proper water level must be maintained in a pool if the scum gutters and/or skimmers are to function as intended.

Where adequate laboratory facilities are available, samples should be collected at least once a week for bacterial analyses made at an approved laboratory near the pool, if possible, to reduce the time lapse between collection and examination of the samples. The tests made may include the total coliform and fecal coliform tests, which indicate the presence of intestinal-type organisms; the total bacteria plate (35° C) count, which indicates the concentration of bacteria in the pool

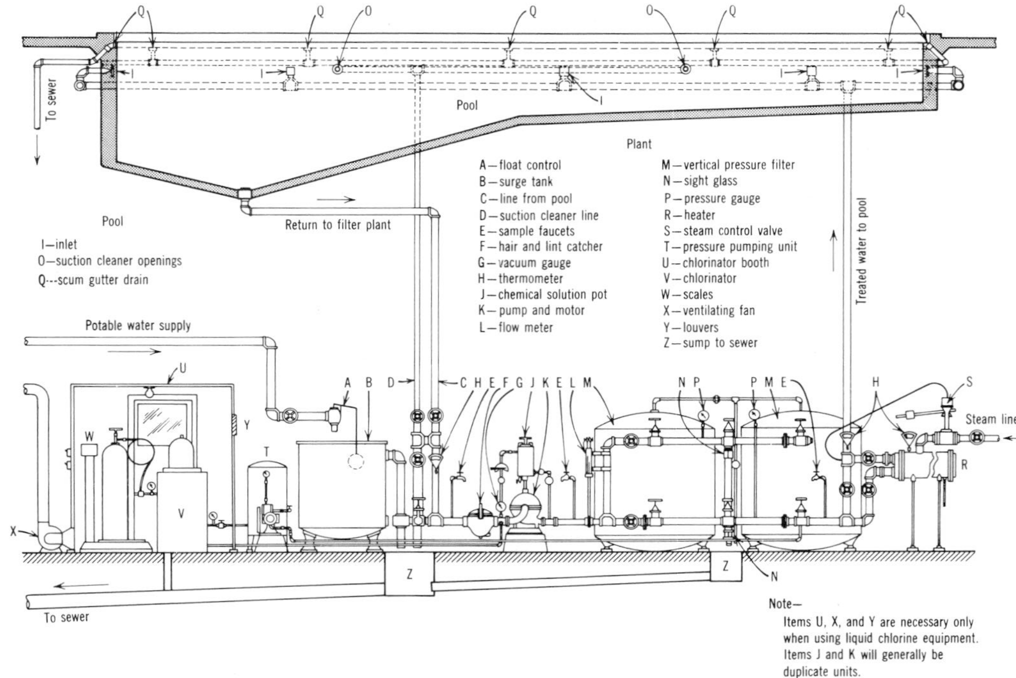

Figure 9-6 Complete pressure filter plant of recommended type, with simplified plumbing system. For purposes of illustration, only two filter units are shown instead of three or more units. (From Michigan Dept. of Health, Bureau of Engineering, Bull. No. 18, Lansing, 1963.)

water; and the test for enterococci (streptococci), which indicates the presence of organisms associated with respiratory illnesses. Other specific tests of interest are for staphylococci (*Staphylococcus aureus*), fecal streptococci (*Streptococcus faecalis*), *Pseudomonas aeruginosa*, and *Clostridium perfringens.* Of the tests mentioned the coliform tests and the plate count are most valuable for general use, although the others give additional information. Streptococcal bacteria found in pools are more resistant to chlorine than coliform bacteria. *Pseudomonas aeruginosa* is as sensitive to chlorine as *Escherichia coli*; *Staphylococcus aureus* is more resistant.

As in other activities, the results of bacterial analyses must be interpreted in the light of a sanitary survey of the facility. Representative samples are an indicator or check on the effectiveness of pool operation at times of peak use. Deficiencies in equipment or operation usually become apparent before they become too serious. Hence unsatisfactory results should be immediately investigated to eliminate their cause before any harm is done.

Disinfection

Chlorine, bromine, iodine, chlorinated cyanurates, and ultraviolet ray lamps have been used to disinfect swimming pool water. Chlorine, and to some extent bromine, are the chemicals of choice. Chlorinated cyanurates and ultraviolet ray lamps have distinct limitations. Iodine has not been used to any extent but is reported to be a satisfactory pool disinfectant and more stable than chlorine in outdoor pools. Bromine as bromine chloride reacts with water and ammonia to form bromamines which are reported to be superior to chloramines against bacteria and viruses. Chlorine however has been the chemical of choice for swimming pool water disinfection.

The maintenance of at least 0.6 mg/l free available residual chlorine (without cyanurates) and a pH of 7.2 to 7.6 in a pool water will produce consistently satisfactory bacteriological results. Free available chlorine is the sum of chlorine as hypochlorous acid (HOCl) and hypochlorite ion (OCl^-). The HOCl component is the markedly superior disinfectant, but the proportion is largely dependent on the pH of the water and the water temperature. For example, at a pH of 7.2, 62 percent of the available chlorine is in the form of HOCl; at pH 7.4, 52 percent is available as HOCl; at pH 7.6, 42 percent is available as HOCl; at pH 7.8, 32 percent is available as HOCl; and at a pH of 8.0, 22 percent is available as HOCl, all at a temperature of 68° F. See Table 3-16.

Eye irritation is said to be caused by prolonged swimming in water with a pH below 7.4. A pH of 6.5 to 8.3 can be tolerated for a limited period of time because of the high buffering capacity of the lacrimal fluid.[44] A pH of 7.5 to 7.6 is probably optimal for minimal eye irritation together with a free available chlorine residual of 0.6 mg/l for adequate disinfection. A pH above 7.6 would

[44] Eric W. Mood, "The Role of Some Physico-Chemical Properties of Water as Causative Agents for Eye Irritation of Swimmers," *Report of Committee on Water Quality Criteria*, Federal Water Pollution Control Administration, U.S. Dept. of the Interior, Washington, D.C., April 1, 1968. p. 15.

require 1.0 mg/l free available residual chlorine for equivalent disinfection. The pH of the pool water should be kept below 8.0 since the amount of active chlorine present decreases greatly with increasing pH as noted above; only about 8 percent of the chlorine is as HOCl at pH 8.5.

If use of cyanuric acid or chlorinated isocyanurates to stabilize the residual chlorine is *permitted*, the recommended minimum free available residual chlorine is 1.0 mg/l with a pH of 7.2 to 7.5, 1.25 with pH 7.6, 1.5 with pH 7.7, 1.75 with pH 7.8, 2.0 with pH 7.9, and 2.5 with pH 8.0. The cyanuric acid concentration should be maintained between 30 and 100 mg/l. Dilution or replacement water may become necessary if the concentration is exceeded, as the apparent free chlorine available is slower-acting, however it is more persistent than free chlorine in the presence of sunlight.[41]

Ammonium alum or free ammonia should not be used as it will be impractical to maintain a free available chlorine in the pool water because the ammonia would combine with the chlorine to form chloramines or combined chlorine. Combined residual chlorine is a slow-acting disinfectant and very ineffective as a bactericide in swimming pools, where it is important to immediately neutralize any contamination introduced by the bathers. With continuous recirculation and adequate chlorination, this pollution can be neutralized by adding sufficient chlorine to maintain an adequate free available chlorine in the pool water. When the combined residual chlorine reaches 0.2 mg/l it should be neutralized by heavier chlorination.

The chlorination of water containing iron in solution will cause discoloration of the water and staining of pool walls. See page 297.

When bromine is used as the disinfectant, to find the bromine residual with a chlorine comparator multiply the residual chlorine reading by 2.25 to convert it to bromine. A residual of 2.0 mg/l should be maintained. Pure bromine at room temperature is a deep red color with a strong odor. It fumes readily and is extremely corrosive. It is a very effective disinfectant.

Control of pH, Corrosion, and Scale

The chemical quality of the pool water is generally measured by tests for the pH value and chlorine residuals. Occasional laboratory determination of calcium hardness, alkalinity, free carbon dioxide, and total dissolved solids is desirable, but pH and residual chlorine tests should be made at least three times a day, before and during peak loads. The maintenance of a proper pH and alkalinity ratio is important for possible corrosion reasons and others. Water coagulation, disinfection with chlorine, and eye irritation are affected by the acidity-alkalinity balance.

pH Adjustment

A sudden jump in pH may be due to algal growths as result of inadequate or interrupted chlorination. Algal control is discussed in Chapter 3. In practice the pH of the pool water will drop rapidly, where chlorine gas is used. Alum if used also tends to lower the pH. If not corrected, the water may become corrosive and cause eye irritation. Alum may also appear in a pool as a floc if the pH of the water is not controlled. Its concentration in the pool water should be less than 0.1 mg/l.

To raise the pH it is common to add sodium carbonate, also known as soda ash, to the pool water. This is done by means of a chemical solution feed machine which introduces the chemical, usually in the recirculating pump suction line, or by the placing of soda ash briquets near the pool recirculation outlets. Sodium hydroxide, known as caustic soda or lye, will also produce the desired result, but this material can cause severe burns, hence its use is not recommended unless competent and reasonably intelligent personnel are available and proper precautions are taken in handling hazardous chemicals. Sodium bicarbonate will also raise the pH but its usual purpose is to raise total alkalinity.

If the pH gets too high, it may promote scale formation. It should be lowered by diluting with fresh water, if its pH is lower. Acid such as hydrochloric acid, also called muriatic acid, or dilute sulfuric acid could be used. Sodium bisulfate is recommended as it is safer to use. Mix the powder in water, then sprinkle around pool or add to pump suction. Handle acids with care. Never add acids to water; add water to acid. The alkalinity will also be lowered on addition of acid. Adjustment of the alkalinity first to about 80 mg/l will simplify pH adjustment. Keep the pH from falling below 7.2; 7.5 to 7.6 is optimum.

Alkalinity Adjustment

Chlorine as hypochlorous acid and hypochloric acid reacts with alkalinity in the water at a rate of 1 part chlorine to 1.2 parts alkalinity. This can be compensated for by adding 1.2 mg/l alkalinity as calcium carbonate. On the other hand, chlorination by means of sodium or calcium hypochlorite will add some alkalinity to pool water. To increase alkalinity as calcium carbonate by 10 mg/l add $1\frac{1}{2}$ lbs of sodium bicarbonate (baking soda) per 10,000 gal of water. To lower total alkalinity by 10 mg/l add 30 oz of sodium bisulfate or $1\frac{1}{2}$ pts of muriatic acid, in small amounts at any one time, to 10,000 gal of water. Also 1 lb of calcium chloride per 10,000 gal water will raise hardness as calcium carbonate about 11 mg/l.[45]

Scale Control

Scale control ordinarily requires a balance between calcium hardness as calcium carbonate, total alkalinity as calcium carbonate, and pH for the prevailing water temperature. Calcium carbonate is the principal scale former. It will form a deposit on filter media, in piping, on pool accessories, and on pool walls. Sodium and calcium hypochlorite, soda ash, and caustic soda tend to support scale formation unless otherwise controlled. Once formed scale is

[45] *Public Swimming Pools, Recommended Regulations for Design and Construction, Operation and Maintenance*, American Public Health Association, 1015 Fifteenth St., N.W., Washington, D.C., 1981.

difficult to remove. Professional help may be needed. The goal is a calcium carbonate saturation which results in a stabilized water with neither corrosion nor precipitation of calcium carbonate. Several methods for determining the approximate level of calcium carbonate stability are discussed in Chapter 3 under Corrosion Cause and Control and in standard publications dealing with water supply and treatment. The *Public Swimming Pools*[45] gives a sample method for calculating the Langlier Index or saturation index to determine if a water tends to be corrosive or scale-forming. Scale control is simplified by maintaining control over the pool water pH.

Clarity

The physical quality of the water—its appearance and clarity—is determined by sight tests previously described. The clarity of an artificial swimming pool water is maintained by continuous filtration. Where sand or anthrafilt filters are used, alum is the chemical usually used to coagulate and trap the suspended matter and color in the water. For best results, in addition to pH control previously discussed, the alum must be fed in small controlled quantities, be well mixed with the recirculated water, and then be allowed to flocculate before it reaches the filter. One way in which this is done is to add filter alum (aluminum sulfate) into a reaction or make-up tank and then introduce the alum into the suction side of the pump, where it is thoroughly mixed with the water in passing through the pump and flocculated before reaching the filters. A dosage of 3 to 4 lb of alum per 10,000 gal of pool water is normally used. At some pools, excellent results are obtained by slowly adding a small quantity of alum immediately following backwash of the filters, to replace the flocculant mat on top of the sand bed, following which the chemical feed is stopped. Diatomite filters use 0.10 to 0.15 lb/ft^2 of diatomaceous earth for precoat plus continuous feed to maintain porosity of filter.

Swimming Pool Water Temperature

At indoor pools, it is desirable to control the temperature of the pool water and air. An air temperature about 5° F warmer than the temperature of the water in the pool is recommended. The maximum water temperature suggested for general use is 80° F; 74 to 76° F is comfortable; 76 to 78° F is ideal. Outdoor pools in hot climates may require water replacement or the addition of large quantities of ice to keep the water from getting too warm. Spray aeration of recirculated water to the pool can lower the water temperature 5 to 10° F. Cold water make-up water, heat exchange equipment, a shade structure designed to withstand wind, and a cooling tower can also be used. The water temperature should not exceed 85° F (30° C). Swimming or bathing in water above this temperature for any length of time has a temporary debilitating effect.

Swimming in 40° F water causes fatigue and breathing difficulty in a short time—10 min or less.

Condensation Control

In indoor pools condensation can be a serious problem if ventilation and humidity control are inadequate. Unit heaters, which are combination fans and heaters, offer a practical relief, but serious problems require professional consultation.

Testing for Free Available Chlorine and pH

The residual chlorine test and the test for pH of swimming pool water require a special procedure and technique to give better accuracy. Combined chlorine is of little value as a disinfectant in swimming pool water and can produce a false sense of security.

If the test for residual chlorine is not carefully made, it may appear to show the presence of free available chlorine, whereas the chlorine present may in fact be in the combined form. This occurs when the water temperature is above 35° F, especially when the water temperature is above 68° F. At a temperature below 35° F, free available chlorine will be readily detected, as orthotolidine reacts immediately with free available chlorine to produce the characteristic color independent of the temperature. However, combined chlorine, if present, reacts slowly with orthotolidine at this low temperature. Therefore, to determine the free available chlorine present in a pool water using orthotolidine, it is first necessary to cool the sample to below 35° F, before the reagents are added. Because free available chlorine reacts readily with soil or organic matter, it becomes extremely important to use clean glassware and a clean technique when making the test.

It really is not practical to measure accurately poolside free available chlorine using the orthotolidine method described above. In addition, orthotolidine is listed as a carcinogen. These problems can be overcome however by using the diethyl-p-phenylenediamine (DPD) method developed by A. T. Palin. Commercial test kits for field use are available which permit simple comparison with a color standard to accurately determine free, combined, and total residual chlorine.* The DPD test compensates for interference by copper and dissolved oxygen. It is not affected by nitrite nitrogen up to 5 mg/l. Oxidized manganese if present can interfere with the color reading but a procedure to correct for this is available.[46]

*Test kits for measuring pH, residual chlorine, and cyanuric acid should comply with National Sanitation Foundation Standard No. 38 or equal.

[46] *The Purification of the Water of Swimming Pools*, Dept. of the Environment, Her Majesty's Stationery Office, London, 1975, pp. 12–13.

When high free available chlorine is maintained in a pool water, the chlorine will bleach the pH color indicator, giving an improper reading. If the free available chlorine is first removed, a proper reading can be obtained. This can be done by adding one drop of $\frac{1}{4}$ percent sodium thiosulfate solution to the pool water sample. This will destroy the free available chlorine present without appreciable effect on the pH value. The glassware used must be clean and the sample must not be contaminated by lime, alum, or other soil. Phenol red (range 6.8 to 8.4) as the indicator and a color comparator are used in the field. The optimum pH is 7.5 to 7.6.

Algae Control

Algae development in a pool causes a slimy growth on the walls and bottom, reduced water clarity, increased chlorine consumption, and a rapid rise in the pH of the pool water in one day. The wall and bottom growths penetrate into cracks and crevices, making them difficult to remove once they become attached. The suspended or floating type are more easily treated. However, the best and simplest control method is to prevent the algae from developing, and this can be accomplished by maintaining a free chlorine residual of at least 0.6 mg/l in the pool water at all times. This cannot be accomplished by intermittent recirculation and chlorination.

If algae difficulties are anticipated or experienced, several control measures can be tried. These include heavy chlorination, copper sulfate treatment, quaternary ammonium treatment, emptying the pool and scrubbing the walls and bottom with a strong chlorine, caustic soda, or copper sulfate solution, and combinations of these methods. Caustic soda can cause severe burns; its use by a lay person is not recommended.

Heavy chlorination, or superchlorination, is the preferred treatment. Dosage of the pool water so as to maintain a free chlorine residual of 0.6 to 2 mg/l, together with recirculation and filtration, should be effective in preventing and destroying the algae growths normally found in a pool. Hand application of $2\frac{1}{2}$ to 5 gal of 14 percent sodium hypochlorite to a 100,000-gal pool will speed up the treatment.* Other forms of chlorine, such as 25 or 70 percent calcium hypochlorite, can of course also be made up into 10 to 15 percent solution and the clear supernatant evenly distributed in the pool water. Bathers should not be permitted to use the pool until the residual chlorine drops to less than 4 mg/l. Iron, manganese, and also hydrogen sulfide if present will precipitate out of solution in the presence of free chlorine.

Copper sulfate has long been known for its ability to control algal growths. Most algae are destroyed by a dosage of 5 lb per million gal; but practically all forms are killed at a dosage of 2 mg/l or 16.6 lb of copper sulfate per million gal of

$$^*\frac{5 \text{ gal} \times 8.34 \text{ lb} \times .14}{100{,}000 \text{ gal} \times 8.34 \text{ lb}} = \frac{X}{1{,}000{,}000}; \; X = 7 \text{ mg/l dosage}$$

water. The copper sulfate crystals can be easily dissolved in the pool water by dragging the required amount, placed in a burlap bag, around the pool. There are dangers in the use of copper sulfate with certain waters. If the pool water has a high alkalinity, a milky precipitate will be formed that also interferes with the action of the copper ion, and waters high in sulfur or hydrogen sulfide will react and produce a black coloration upon the addition of copper sulfate. If in doubt, try dosing a batch of the water in a barrel or pail. Too high a concentration of copper sulfate will tend to discolor one's hair or bathing suit; hence pool water that has been overdosed should be diluted.

Quaternary ammonium compounds have been suggested for algae control and some satisfactory results have been reported, but free residual chlorination is made difficult. Care must be used not to overdose because foaming will be produced.

If the methods mentioned above are not effective, it will be necessary to drain the pool and scrub the bottom and walls with a 5 percent solution of copper sulfate, or spray and scrub with a 1 percent solution of calcium or sodium hypochlorite, followed by rinsing with a hose after about 15 min.

Prevention of Ringworm and Other Skin Infections

The prevention and control of ringworm infections, including floor treatment, are discussed in Chapter 1.

At one time, the provision of a footbath containing a strong solution of chlorine or other fungicide at the entrances to a pool and in gym shower rooms was required by health officials. Experience has shown, however, that footbath solutions were rarely properly maintained, with the result that they became in fact incubators and spreaders of the infection rather than inhibitors. The contact time between the feet and fungicide solution was too brief to do much good. The fungus on the feet is usually so imbedded in the skin that the solution could not penetrate to the spores.

Where footbaths are built into the floor, they can serve the very valuable purpose of preventing the tracking of dust, sand, and dirt into the pool. They must, however, be provided with a continuously running spray and an open drain.

Rented towels and bathing suits used at public pools can also be the means whereby ringworm and other infections are spread. It is therefore essential that towels and suits be carefully laundered and sanitized after each use. This can probably be done best by a public laundry. If sanitization is done at the pool, equipment and supervision must be provided to assure that towels and suits will be washed in hot water and suitable soap or detergent, rinsed in clean water, and then dried for at least 30 min by artificial heat at a temperature above 175° F, which will destroy bacteria and the fungus spores. A water temperature of 160° F for 25 min will kill nearly all microorganisms other than spores. Chemical disinfection may also be effective, although the heat treatment is believed to be more reliable. Woolens, silks, nylons, and elasticized materials will require special handling. Cold-water disinfection can be accomplished by soaking suits 5 min in a 1 : 1000 dilution of alkyl-dimethyl-benzyl-ammonium chloride or other quaternary ammonium compound. This is reported to show "no adverse effect on dyes or materials commonly used in swimming suits." Temperatures above 110° F will shrink most woolens.

Personnel

As in most other instances, the designation of one competent person who is responsible for satisfactory operation of the entire swimming pool is a basic necessity. Only in this way can safe, clean, and economic operation be assured. In large pools this person would be the manager, and he should have under his or her supervision the bathhouse attendants, lifeguards, and water purification plant operator. A competent manager should be a good administrator who is familiar with all phases of pool operation, including the water-treatment plant, collection of water samples, making of routine control tests, and keeping of operation reports. At summer resorts, camps, and similar places, the senior lifeguard may be given the overall responsibility for pool operation for reasons of economy. This does not usually prove to be satisfactory unless the lifeguard can be trained before the summer season in the fundamentals of pool operation.

Most states have laws or sanitary codes regulating the operation of swimming pools and bathing beaches. The owner has a legal responsibility to comply with these laws and rules. Failure to do so would make the owner liable to prosecution for being negligent. It is to the owner's interest and protection that such records be kept as will show proper operation of the swimming pool purification equipment, clarity of pool water, lifeguards on duty, elimination of accident hazards, and any other measures taken to protect the users of the pool. Swimming pool operation report forms are available from most health departments. To be of value these reports must be carefully and accurately filled out each day. Analysis of reports will indicate where failure is occurring and need for new equipment. The operator should be provided with a residual chlorine test kit having a range of 0.2 to 3.0 mg/l and a pH test kit having a range between 6.8 and 8.4.

Pool Regulations

Regulations should be as logical and concise as possible. The following are samples.

1. Urinating, spitting, or blowing the nose in the pool can spread disease to other bathers. Use the overflow gutter for expectoration and toilet facilities to help keep your pool clean.
2. Persons having skin disease, running sores, a cold, or other infection endanger the health of their fellow bathers; hence they cannot be permitted to use the pool until well.

3. Spectators carry dirt on their feet that would be tracked into the pool. This lowers the quality of the pool water; therefore spectators are not permitted inside the pool enclosure. Pets, except a seeing eye dog with a blind person, should be excluded.

4. A pool is a common bathtub in which the water is kept clean by continuous purification and replacement of the water. Dirt and bacteria may build up in the pool faster than the purification equipment can take it out if bathers do not remove perspiration, dirt, and dust from their bodies. Therefore, take a warm water and soap shower before entering the pool and after using the toilet. Bathers with long hair are required to wear a bathing cap.

5. The maintenance of order is necessary to prevent accidents and drownings and permit maximum enjoyment of the facilities. Obey the lifeguard and attendants promptly. There shall be no swimming in the absence of an attendant.

SWIMMING POOL MAINTENANCE

Recirculating Pump

1. Inspect and service regularly.
2. Check motor bearings for lateral and vertical wear. Brushes and commutators should wear evenly; clean commutator with fine sandpaper when placing in operation and turn down commutator if uneven.
3. Pump impeller has little end play or slippage; check suction and pressure produced by pump, lubrication, and stability of pump mounting. No entrance of air in suction. Slight leakage from packing gland is permissible.
4. Check pump capacity against operating head.
5. Compute backwashing, recirculating rate, and pool turnover against design. A new pump may be needed.
6. Determine water clarity, uniformity of chlorination, and bacteriological quality of pool water.
7. If repairs are needed call in competent pump man or electrician at beginning of season or during closed periods.
8. Drain pump casing, cover pump and motor with canvas, and remove fuses from switch boxes at end of season. If subject to flooding, remove.

Hair Catcher

1. Maintain a regular daily cleaning schedule.
2. Keep a replacement basket strainer on hand. A hair catcher in parallel with proper valves is very desirable.

Filters

1. Open air valve at top of each filter shell to release air. If problem recurs frequently find and eliminate cause. Check tightness of valves, packings, fittings, gaskets.
2. At end of each season after backwashing, drain each filter, open manhole, and inspect sand. The sand should be level, without dirt, hair, holes, or cracks. Open inlet valve to filter to slowly admit water. If sand comes up unevenly or boils in spots, sand may have to be cleaned or replaced or the underdrain system may be partially clogged.
3. Sample of sand at depth of 6 to 8 in. should be clean.

4. If backwash rate is not adequate to clean sand bed, anthracite may be substituted, which requires only about two-thirds the backwash rate for sand.

5. Deep raking of sand surface may permit washing out of mud and dirt.

6. Sodium hydroxide treatment may be used to clean sand bed. Add 1 to 2 lb per ft^2 of filter area to 12 in. of water over the sand. Lower the water level to just above the sand surface; let stand 12 hr, drain out solution, and repeat. Backwash and observe condition of sand as explained in (2). A strong solution of chlorine, sulfur dioxide (sulfuric acid), salt, or detergent may also be used to remove organic matter.

7. Excessive use of soda ash and alum with a hard water may cause cementation of sand grains and improper water filtration. Break up incrustations or replace sand. Use of a metaphosphate with a hypochlorinator may prevent this difficulty.

8. Failure to use a filter aid with a diatomite filter or pretreatment of water high in iron or manganese will cause clogging of the filter tubes. Check head loss every few weeks; remove and clean tubes whenever necessary, soak in 8 percent solution of sodium hexametaphosphate for 2 hr and scrub; replace damaged tubes; use a new gasket for filter shell head when it is opened.

9. Backwash and drain filters at end of summer season to prevent freezing. Check for needed repairs.

10. See that valves operate easily and properly. Grease stems and lubricate moving parts. Check packing and repack if necessary. Drain and inspect all piping and fittings at end of each season.

Chemical Feed Equipment

1. Dismantle at least once a year; inspect and replace worn parts. Special attention should be given to gaskets, check valves, shutoff valves and seats, tubing and piping, diaphragms, pistons, and special gears.
2. Rinse with clear water and drain before storing.
3. Coat metal parts subject to corrosion with petroleum jelly, plastic, or paint.
4. Check oil level and lubricant cups in accordance with manufacturer's instructions.

Pool Structure

1. Overflow troughs, drains, walls, floors, walks, lockers, shower rooms, toilets, and fixtures kept clean and in good repair.
2. At least one person is delegated responsibility for twice a day cleanup. Proper equipment and cleaning compounds provided.
3. Depressions in walks leveled; broken or cracked floors repaired.
4. Vacuum cleaning equipment provided and used. Pool walls and floors clean, without growths, smooth but not slippery.
5. Winter protection of a pool includes drainage of all equipment, piping, tanks, floor drains, and fixture traps. Trapped floor drains should be sealed watertight. If pool is designed to withstand ice pressure due to freezing water, the pool need not be drained. Pool walls with an outward batter of 1 to $1\frac{1}{2}$ in. per foot of height will permit ice to expand and raise.

WADING POOLS

Definitions of wading pools vary. All are meant to be shallow, up to 24 in. maximum depth, and specifically for the use of children. They are usually included with the design of a swimming pool as an independent structure.

Public health and recreational agencies have recognized for some time that wading pools can actually be the cause of more illness than swimming pools if not properly operated and maintained as noted below. The irresponsible child may relieve himself in the pool and drink the same water with innocent abandon. Dirt, sand, grass, food, and other debris are carried into the wading pool in such quantity as to make purification of the wading pool water a special problem.

Types

Wading pools have been constructed and operated as spray showers with open drain so as not to collect water, as flow-through types, as recirculation and filtration with chlorination types, and as fill-and-draw pools. The fill-and-draw wading pool cumulates pollution and is rarely kept clean; it should not be used at public places. The recirculation, with continuous chlorination, -type of pool may be suitable if the pollution load is not too great and careful supervision can be given to the operation. The recirculation should include filtration and addition of sufficient chlorine to maintain at least 0.6 mg/l free residual chlorine in the pool water. Because of the heavy pollution, the wading pool should not normally be connected with the swimming pool recirculation system. A 1-hr turnover and 10 ft^2 per child is recommended.[42] Of all the types of wading pools mentioned, the spray type with an open drain is the safest and cleanest. Children seem to enjoy the sprays immensely; the danger of drowning and infection from polluted water is eliminated. Existing wading pools can easily be converted to spray-type pools in which there is no standing water.

The design details of a wading pool are shown in Figure 9-7. A gently sloping floor, say 6 in. in 10 ft, a drain provided with a sand trap, and a drained concrete apron at least 10-ft wide completely around the pool are elements common to all wading pools. A 6-in. drain is adequate for most wading pools; it must be so graded as to make the backing up of sewage impossible. Where a sand play area is provided, and this is very desirable, constant maintenance will be needed. This would include frequent raking to remove foreign matter and turnover of the sand to prevent the development of odors, particularly where the sand tends to remain wet and is in the shade. An underdrain system in such cases is necessary. Disinfection of the sand by sprinkling with a chlorine solution, followed by a fresh-water spraying from a garden hose may be found periodically desirable. Toilet facilities should be convenient to the wading pool users to help reduce pollution of the wading pool and sand play area.

BATHING BEACHES

The location of bathing beaches, sanitary surveys, bacterial standards, health considerations, safety, and accident prevention were previously discussed. A suggested camp waterfront layout which includes numerous safety features is show in Figure 9-8.

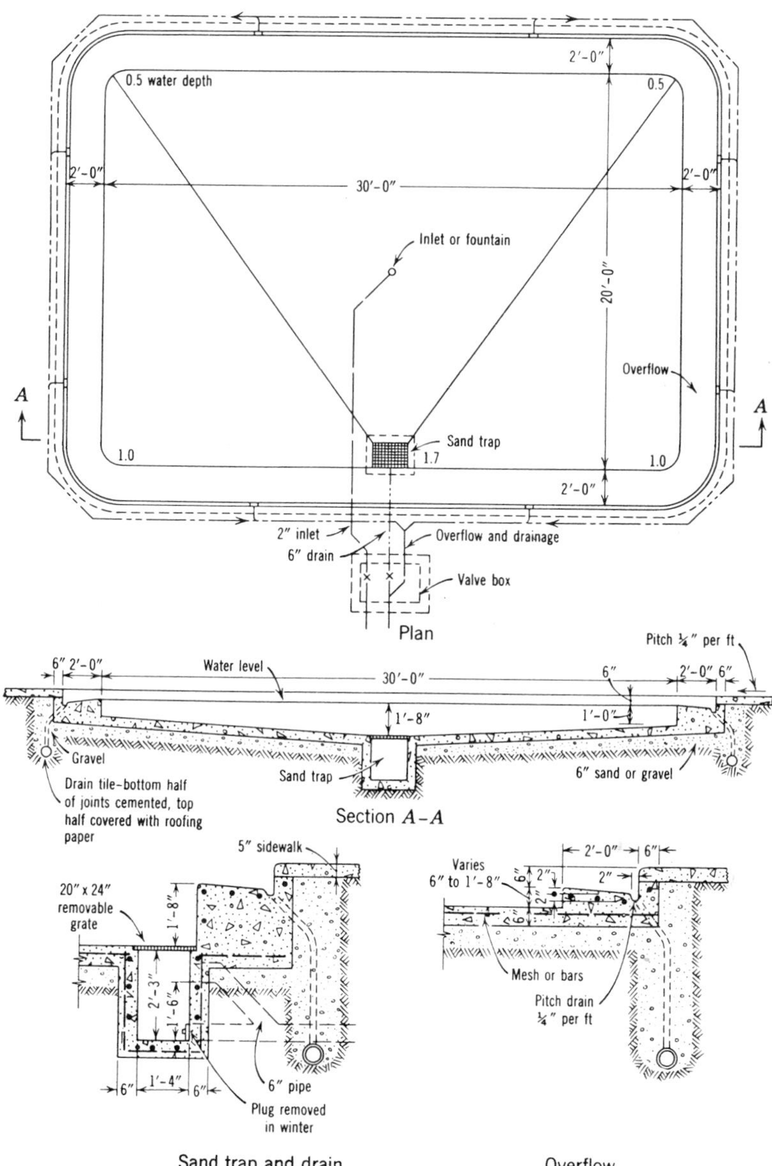

Figure 9-7 Suggested design for wading pool.

Where bottom conditions are hazardous, or unstable, or the water depth excessive, consideration might be given to construction of a "floating pool" on the shore. The pool would have a solid bottom, and sides which permit the free movement of water in and out, also an ample deck space.[47] In any case, diving areas are expected to be cleared of stumps, rocks, and other obstacles, and have

[47] "Floating swimming pool for the blind," *Civ. Eng., ASCE,* August 1979, p. 76.

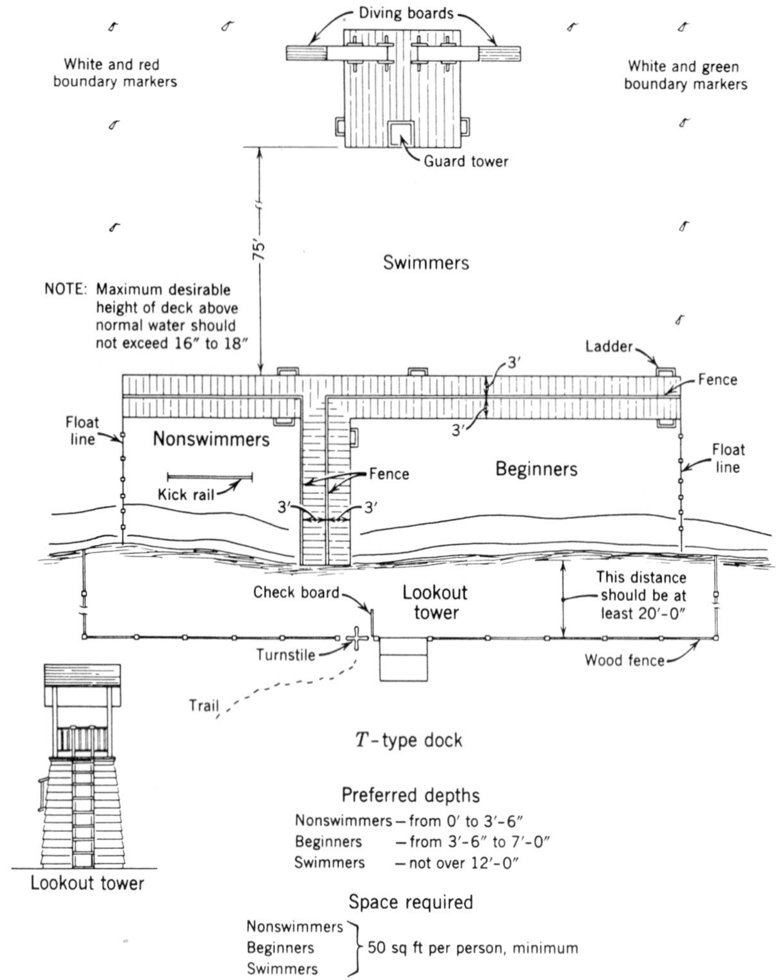

Figure 9-8 Suggested waterfront layout. Drawing not to scale. (From Boy Scouts of America, Division of Program, Engineering Service.)

minimum water depths and areas meeting those specified for swimming pool diving boards or platforms.

Public beaches usually attract large numbers of people for extended periods of time. Toilet facilities are a necessity. Table 9-6 suggests the number of plumbing facilities to be provided.

Disinfection and Water Quality

Unsatisfactory laboratory reports on a bathing beach water call for a sanitary survey and evaluation to determine their significance. If a beach is being

Table 9-6 Plumbing Fixtures Required for Natural Bathing Places

Number of Fixtures	Commodes		Urinals (Males)	Lavatory Per Sex	Showers Per Sex
	Male	Female			
1	1 to 199	1 to 99	1 to 199	1 to 199	1 to 199
2	200 to 399	100 to 199	200 to 399	200 to 399	100 to 199
3	400 to 600	200 to 399	400 to 600	400 to 750	200 to 299
4		400 to 600			
	Over 600, one fixture for each additional 300 persons		Over 600, one fixture for each 300 males	Over 750, one for each additional 500 persons	Over 299, one for each additional 100 persons

Source: *Environmental Health Practice in Recreational Areas*, PHS/DHEW, Washington, D.C., Jan. 1978, p. 49.

excessively polluted by sewage or other wastes, the obvious solution is reduction of the pollution at the source to the point at which it does not adversely affect the water quality. The importance of a sanitary survey and its intelligent interpretation in such cases is of major significance. When the remaining pollution reaching the bathing area or introduced by the bathers is small, or when the dilution provided (as determined by the dilution formula $Q = 6.25T^2$, in which Q = quantity of water per bather per day in gallons and T = the replacement period in hours) is not adequate to meet bacteriological standards, disinfection of the water may be possible.

The method for applying chlorine is determined by such factors as the type of beach, water current, size, and facilities available. If a sewer outfall is the culprit, and it cannot be removed, the simplest emergency procedure might be heavy chlorination, thorough mixing, and 15- to 30-min retention of the disinfected sewage before discharge to the receiving body of water. Chlorine can also be added to the bathing area at one time to treat a given volume of water several times a day; it can be added directly and continuously to the bathing water by means of submerged orifices, or the chlorine can be added in high concentration to water recirculated back to the bathing area. Each method has its limitations and requires careful planning and supervision. Prohibition of bathing may be indicated, but the actual health risk and social implications of such action during a hot summer of social unrest for example should also be taken into consideration. See Health Considerations earlier in this chapter.

The batch treatment or continuous treatment of a given volume of water is a special problem in each case. The chlorine solution, up to 10 or 14 percent, can be added before each bathing period from a motorboat or rowboat as a spray or by means of a belt-driven solution feeder so as to traverse the bathing area. Since a large quantity of organic matter will be present to absorb the chlorine, a sufficient quantity of chlorine solution will have to be added to satisfy, at least partially, the initial chlorine demand. One might start with a dosage of 5 mg/l and be guided by the results of residual chlorine and bacteriological tests. It will be exceedingly difficult to maintain a residual chlorine for any extended period of time.

The continuous addition of chlorine, using a gas solution feed machine, through an underwater distribution system with orifices has been tried with partial success. Rubber or plastic piping should be used. If there is a current, the perforated chlorine distribution piping system would of course be located upstream. Where a small stream is dammed up and there is a measurable flow through the bathing area, chlorine can be added by means of one or more improvised drip chlorinators strategically located.

Good results can be obtained by continuous chlorination and recirculation of the water in the bathing area. The capacity of the pump or pumps, number and location of inlets and outlets, and capacity of the chlorinator must be determined to fit individual cases.

The effectiveness of chlorination will be readily apparent by observing the clarity of the water and presence or absence of algae. The algae may appear as a green coloring in the water resembling pea soup, as a green scum, or as a dark flaky deposit. Experience has repeatedly shown that when a free residual chlorine is maintained in the water at all times, the bathing water is relatively clear of algal growths and attractive. In all cases technical control over the treatment, including the maintenance of adequate daily operation reports, is necessary.

Control of Algae

The control of algae is discussed in some detail in Chapter 3.

Under average conditions in recreation areas, a copper sulfate dosage of 5 lb per million gal evenly applied at intervals of 2 to 4 weeks will prevent the development of most microorganisms, but may result in some fish kill. In the eastern part of the United States, the treatment should start in April or, if microscopic examinations are made of the water, when more than 300 organisms, or areal standard units, per ml of sample are reported.

Chlorination treatment can also be effective in a partly artificial pool, such as that formed by damming up a small stream. However, to be effective, it must be under good technical control and at least 0.6 mg/l free residual chlorine maintained in the water at all times. This will not only prevent the growth of most algae, but will also help keep the water cleaner looking and relatively clear.

Control of Aquatic Weeds

Aquatic weeds are objectionable in bathing areas. They become entwined around the legs and arms of bathers, interfere with swimming, and provide harborage for snails.

One of the simplest ways of controlling the growth of aquatic weeds, where possible, is lowering the water level sufficiently to cause the plants to dry out. If this is followed by burning and removal of the remaining debris, reasonably good control will result. The physical cutting and removal of the weeds is also temporarily effective. Where aquatic weeds grow above the water surface chemical control by the use of weed killers is possible.

Sodium arsenite as As_2O_3 at a rate of 4 to 7.5 mg/l has been very effective to control submerged rooted or anchored aquatic weeds, but its use is no longer advised. The chemical is very toxic to man and dangerous to the applicator. Other chemicals such as Endothall, Acrolein, Diquat, and 2,4-D granules or pellets may be used *if approved* by the control agency. Diquat, 2,4-D low-volatile esters, or Amitrol-T may be permitted for floating weeds. The manufacturer's directions should be carefully followed, and when the water is located on a water supply watershed permission must first be obtained from the water supply officials. Sodium arsenite is not usually allowed. Endothall and Diquat are permitted by USEPA if applied according to directions.

The health, agriculture, and conservation departments should also be consulted because some of the chemicals available are toxic to humans and fish. In many instances knowledge regarding persistence and effect on the ecology may be limited. A permit to apply chemicals to public waters is usually required. Only approved chemicals should be used and only when necessary. See also Chapter 10.

Control of Swimmer's Itch

Swimmer's itch is a nonhuman form of schistosomiasis known as schistosome dermatitis. The larvae of certain schistosomes are found in lakes and salt water beaches in many parts of the world including North America. Migratory fowl, small mammals, and other birds carry and distribute the nonhuman schistosomes. Droppings containing the eggs fall in water. The eggs hatch and the larvae (mircadia) must enter an appropriate species of snail* within 36 hr or die. The organism develops in the snail to the larval form which is released in the water to produce the fork-tailed schistosome cercariae. The cercariae may infect the natural hosts, or bore into the bather's skin where they die causing a rash and severe itching, if not rubbed off with a rough towel *before* the water film dries on the skin surface. An immediate fresh-water shower after leaving the water is also effective. These schistosomes do not mature in man.

Elimination of the snail, which is necessary to the life cycle of schistosomes, will break the life chain and prevent schistosome dermatitis. The control of debris and aquatic growths in shallow water, to which snails become attached, will reduce snail harborage. But actual destruction of the snail offers the best protection. The application of copper sulfate, sodium pentachlorophenate, copper pentachlorophenate, or copper carbonate so as to provide a dosage of 10 mg/l to the water will kill snails in the water, if the water can be impounded for 48 hr, as well as the free-swimming cercariae. Mackenthun reports that the application of a mixture of 2 lb of granular copper sulfate and 1 lb of copper carbonate for each 1000 ft^2 of bottom to be treated will control the snails for a season or two in lake waters with a total methyl orange alkalinity of 50 mg/l or greater. With a lesser alkalinity 2 lb of copper carbonate is successful.[48] A solution may be applied to the bottom being treated by means of a hose dragged behind a boat. The optimum time in north central U.S. lake states is between mid-June and July 4. It must be remembered that the dosages mentioned here will kill fish life and perhaps bleach swimming suits; hence special precautions should be taken. Under certain circumstances acidity control of the water might be possible. In such cases maintenance of a pH of 7.0 or less will discourage snail growth. Regular drainage and drying of areas with exposure to sunlight will kill the snails and eggs. Chlorination to provide a residual of 0.5 mg/l for 20 min in a

***Stagnicola emarginata* and *Physa parkeri* beach snails, and *Lymnaea stagnalis* and *Stagnicola palustris* swamp snails; also graulus, physa, lymnaea.

[48] Kenneth M. Mackenthun, *Toward a Cleaner Aquatic Environment*, USEPA, Washington, D.C., 1973.

lake bathing area will kill the cercariae. If snail control is not possible, swimming or wading in endemic areas should be prohibited.

Cost of chemical treatment to control *schistosomiasis-causing snails*, found in many parts of the world,* is high and requires repeated applications based on field observation and testing. It is better to first prevent and minimize the conditions which are favorable to the growth of snails and aquatic growths which promote schistosomiasis. Proposed impoundments such as dams (Aswan) should first evaluate possible deleterious effects and their prevention. See also Chapter 1 for other control measures.

The molluscicides of choice are niclosamide and N-tritylmorpholine, and yurimin in Japan. Sodium pentachlorophenate and copper sulfate are still in limited use as molluscicides.[49] Granular bayluscide has also been found effective to kill snails.

Evaluation of Bathing Beach Safety and the Impact of Closing a Beach

The obvious action to protect the public from a "polluted" beach is to prohibit bathing. Enforcement of such a determination is not always simple and may pose serious problems. Many factors determine the suitability of natural waters for bathing purposes, and their importance varies. Strict standards for water quality and safety must therefore be interpreted in the light of:

1. A sanitary survey of the area to identify pollution sources, patterns of water circulation and sources of dilution, effects of various rainfall intensities, the potential for selfpurification, and related factors.
2. The microbiological, physical, and chemical quality of the water.
3. Epidemiological data indicating a significant incidence of related illnesses.
4. Economic and social, including psychological, impact.

Items (1), (2), and (3) have been discussed earlier in this chapter under Health Considerations and Regulations and Standards. The social and economic impact of closing a beach can be far-reaching and should also be carefully weighed in interpreting the scientific data available in view of the inconclusive disease transmission evidence available. Nevertheless, it is considered unwise to use waters known to be grossly polluted for bathing or swimming, as a relationship and hence potential hazard exists.

Closing of a beach could have serious economic consequences. Businesses such as restaurants, hotels, motels, gas stations, and suppliers and employment dependent on the visitors would suffer immediate and possibly long-term substantial loss. Vacationers and tourists would patronize other areas and

*Africa, the Arabian peninsula, northeastern and eastern South America, the Caribbean area, Middle East, part of India, Orient. The disease causing schistosomes are not indigenous to North America.

[49]Schistosomiasis Control," *WHO Tech. Rep. Ser.*, **515**, 1973; also Frederick Eugene McJunkin, *Engineering Measures for Control of Schistosomiasis*, Agency for International Development, Washington, D.C., 1970.

survival of the resort area could be jeopardized. The closing of a beach, especially during a hot summer of social unrest, could trigger rioting, looting, and property damage, not to mention possible injuries and deaths. Public safety and fire services would also have to be placed on alert with consequent increased municipal costs. It is apparent that the closing of a beach cannot be taken lightly. A careful evaluation must be made of the public health, safety, social, and economic risks and benefits, before a decision is made.

TEMPORARY RESIDENCES

Operation of Temporary Residences

Camp, motel, hotel, and resort management is a very specialized field, requiring more than just the ability to provide an entertaining program for children and adults. There are more than 11,000 children's camps and an unknown number of summer family camp colonies, dude ranches, hostels, boarding houses, motels, hotels, and campgrounds scattered throughout the United States. In many cases these places are communities unto themselves, with their own utilities and services. The operation problems approach those of small villages; some have a capacity of more than 1000 persons.

Camp leadership training courses usually include study of programming, nature lore, arts and crafts, counselor leadership, camping and woodcraft, fishing, aquatics, music, and related activities. The provision of basic sanitary and safety services and facilities, and an environment conducive to good health is frequently overlooked, underemphasized or taken for granted. Every camp director should have the basic knowledge to understand the significance of the multiple health and sanitation problems that may arise. Camp directors should know the resources available and be able to decide promptly the proper steps that may have to be taken to satisfactorily solve the problems. The health departments are anxious to help. They too are desirous of seeing that the health of the campers and guests is protected. Consult with the local and state health departments and seek their advice in health and sanitation matters.

The fundamentals and problems associated with disease transmission; location and planning; water supply; sewage and solid waste disposal; swimming pools and bathing beaches; food, including milk; insects; rodents; noxious weeds; and housing are described elsewhere in this text and should be referred to for more detailed information. Although the discussion that immediately follows concentrates on children's camps, the same principles apply to other types of temporary residences.

Camps

It has been estimated that more than 8,000,000 children in the United States have the opportunity of going to 11,000 summer camps each year. Probably an equal number of adults visit resort hotels and camps. The importance of maintaining minimum standards of sanitation, health, and safety at these establishments is recognized by federal guidelines,[50] and state and local health department laws. Camps include overnight, summer day, and traveling day camps.

The development of children and the extension of their education through planned and supervised activities are major objectives of many camps. The camping philosophy followed should be determined and understood before interpreting compliance and the satisfactoriness of camp operation, maintenance, and supplied facilities.

A typical inspection report form is shown in Figure 9-9. Many variations and more detailed versions are used to help determine if an establishment is in compliance with the intent of existing regulations. Actually, trained judgment is necessary to determine what is adequate or satisfactory under the particular conditions of operation. A more detailed explanation of the items can be found in the text under the appropriate headings. Each permitting authority has specific regulations to be complied with. State and local laws, ordinances, and regulations dealing with fire, electrical, and recreation safety, in addition to sanitation and hygiene must be met. The more common items numbered as in Figure 9-9 are briefly explained here to guide and help obtain better understanding of the intent of the laws and regulations.

Compliance Guide[50-57]

1. *Permit.* A current permit signed by the health officer and displayed in a prominent place would indicate compliance with the sanitary code at the time of issuance. The person to whom the permit is issued is required to comply with the provisions of the permit.

2. *Competent sanitation supervision.* At least one competent person who is very familiar with all sanitary facilities at the camp and who is charged with the responsibility of keeping the grounds and

[50] *Youth Camp Safety & Health*, Suggested State Statute & Regulations, USDHEW, PHS, CDC, Atlanta, Ga. 30333, revised August 1977.

[51] National L.P. Gas Association, 79 West Monroe Street, Chicago, IL 60603, pamphlet on L.P. gas safety.

[52] National Fire Protection Association, 470 Atlantic Avenue, Boston, MA 02210: 1973 Life Safety Code, 101; 1973 Flammable Liquids Code, No. 30; 1972 Homes Forest Areas, No. 224; 1973 Fire Doors and Windows, No. 80; 1969 Gas Appliances, Piping, No. 54.

[53] U.S. CPSC, Washington, D.C.: Fact Sheet No. 4, Tent Flammability; No. 6, Stairs, Ramps, Handrails and Landings; No. 19, Glass Doors and Windows; No. 16, Extension Cords and Wall Outlets; No. 22, Playground Equipment. Gen. Publication No. CPSC-75-620-9.

[54] American Camping Association, Bradford Woods, Martinsville, IN 46151: Camp Staff Application Forms; Camp Health Record Forms; Camp Health Exam Forms; Camping & Woodcraft—Pub. No. CC07.

[55] Underwriters Laboratories, Inc., Publication Stock, 333 Pefingsten Road, Northbrook, IL 60062; Building Materials Directory; Fire Resistance Index.

[56] Fire Equipment Manufacturer's Association, Inc., 1803 S. Busse Road, Mount Prospect, IL 60056: Owner's Instruction Manual.

[57] DHEW, PHS, CDC, Environmental Health Services Division, Atlanta, Ga. 30333—separate pamphlets available on: Watercrafts, Archery, Horseback Riding, Scuba and Skin Diving, Caving, Firearms, On the Trail.

NAME .... TYPE ....

LOCATION .... T.V.C. ....

Person interviewed .... Title ....

Capacity .... Occupants: Adults .... Children .... Illness ....

No.	Item	Yes	No	CM	No.	Item	Yes	No	CM
	General					*Water supply*			
1.	Permit and code posted				23.	Apparently safe			
2.	Competent sanitation supervision				24.	Adequate quantity, hot and cold			
					25.	Operation reports satisfactory			
3.	Medical and nursing care adequate				26.	Sources—protected			
					27.	No cross- or inter-connections			
4.	Adult for child care available				28.	Storage protected			
5.	Soil drainage satisfactory				29.	Common cup prohibited			
6.	General conditions sanitary; insects, rodents, weeds controlled				30.	Drinking fountain satisfactory			
						Toilets and wastewater disposal			
	Housing				31.	Separate toilet facilities			
					32.	Toilets convenient and adequate			
7.	Structurally safe				33.	Privy location, construction, maintenance, satisfactory			
8.	Adequate size								
9.	Easy to keep clean				34.	Disposal systems satisfactory			
10.	Watertight roof and sides				35.	Location satisfactory			
11.	Lean-tos exclude rain					*Other sanitation*			
12.	Cleanable floors provided								
13.	Stoves fire protected and vented				36.	Pasteurized milk used			
14.	Buildings, grounds, etc., clean				37.	Refrigeration adequate			
15.	Fire escapes and exits provided				38.	Food storage satisfactory			
16.	Sleeping quarters adequate				39.	Food handling satisfactory			
17.	Buildings adequately ventilated				40.	Dishes washed, cleansed, sanitized			
	Kitchen and dining rooms				41.	Garbage storage satisfactory			
18.	Separate from dormitory and toilet				42.	Garbage disposal satisfactory			
					43.	Shower facilities provided			
19.	Screening adequate				44.	Pools, etc., conform to sanitary code			
20.	Equip. adequate for food prep.								
21.	Floors, walls, ceilings clean				45.	Communicable diseases reported			
22.	Ventilation adequate								

NOTE: CM indicates correction made.

EXPLANATION OF OBSERVED DEFECTS BY NUMBER

....

....

....

....

....

....

....

....

....

To the operator of the above described property: You are notified that the items checked in the "No" column are in violation of the requirements of the State Sanitary Code. Immediate elimination of any or all violations is required.

Received by .... Inspected by: ....

Title ....

Date .... Title ....

Figure 9-9 Reinspection form for camps and resorts. (See text for interpretation of each numbered item.)

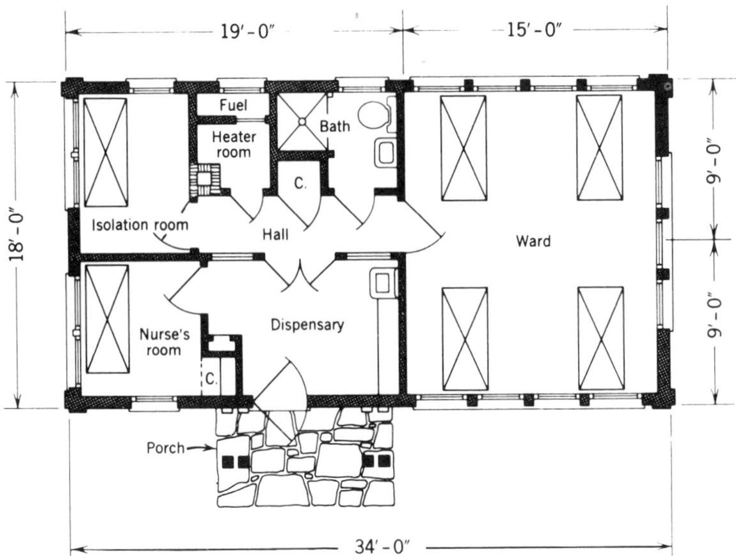

Figure 9-10 A typical summer camp, health lodge. (From *Organized Camp Facilities*, National Park Service, U.S. Dept. of the Interior, reprint from *Park and Recreation Structures*.)

buildings clean, the water supply, swimming pool, and sewerage systems working properly, and the hot-water systems, refrigerators, and dishwashing facilities functioning should be available at all times the property is open for use. At hotels and motels a record should also be kept of guests' names, addresses, and car license numbers.

3. *Medical and nursing supervision.* There should be at least a registered professional resident nurse in attendance at all camps where children are not physically normal. At all camps there should be a licensed physician on call, and either a physician's assistant, a licensed or practical nurse, or a person at the camp trained in first aid acceptable to the permit-issuing official, an equipped first-aid cabinet, and telephone. An infirmary including hot and cold running water, examining room, isolation and convalescent space, and bathroom with flush toilets are practically essential at the larger camps. In the smaller camps a tent with a fly and mosquito netting and floor are the minimum facilities. See Figure 9-10.

Every person attending camp should have on file in the dispensary an immunization record including a dated history of live measles (Rubeola) vaccination at or after 12 months of age or a physician-documented history of the disease,[58] also diphtheria, mumps, polio, Rubella, and tetanus.

An indication of the need for nursing supervision and the application of preventive principles is shown in a study of illnesses and injuries in a summer camp.[59] Of 8873 visits to the infirmary by 320 campers and 92 staff members during an eight-week season, 40 percent were for the administration of drugs. The majority of the visits involved the upper respiratory tract, the skin, the skeletal-muscular system, the central nervous system, the ears, and the lower respiratory tract.[58]

In 1974 the U.S. Public Health Service Center for Disease Control had record of 25 deaths, 1448 injuries, and 1223 serious illnesses while children were at camp. But these figures are acknowledged to be low being based on news clippings and questionnaires.[60]

[58] *MMWR*, CDC, Atlanta, Ga., July 20, 1979, p. 335.
[59] Charles B. Rotman and Elizabeth Schmalz, "Illnesses and Injuries in a Summer Camp," *Am. J. Nurs.*, may 1977, pp. 821–822.
[60] "Camp Safety," *Parade*, August 4, 1974.

4. *Adult supervision.* A competent camp director, or director's assistant, should be available at all times at camps accommodating children under 16 years of age, in addition to other adult supervision as may be needed. It is common to have one counselor for every 6 to 10 campers dependent on camper age, plus a service and activities staff. At least one counselor should be present during sleeping hours on every level of a building used for sleeping. The camping philosophy will govern. In any case, the director or deputy should assume the personal responsibility to make daily inspections of all environmental factors, including the cleanliness of food handlers, food servers, refrigeration, water supply treatment, sewage and refuse collection and disposal, sleeping quarters, swimming and bathing areas, fire-fighting facilities, and accident hazards. The director's competence should include practical and educational experience; usually at least 25 years of age, college graduate and at least 3 years experience in camping including administration. The counselors should be at least 18 years old.

The supervisor of specialized activities such as aquatics, mountain climbing, horseback riding, and scuba diving must be qualified by specialized training and experience. Where a rifle or archery range is established, be sure that its use is permitted only under competent supervision and where established precautions can be taken.

Campers must not be permitted in open trucks. All camp vehicles and vehicles used for transportation of campers must be maintained in safe operating condition. Drivers must possess the required license and have at least one year's experience.

5. *Surface drainage.* If surface water is not carried away or readily absorbed by the soil, or if groundwater, rock, or clay is close to the surface, making the subsurface disposal of sewage and other liquid wastes impractical, surface drainage would be considered unsatisfactory unless special provisions are made. The grounds should be kept reasonably dry. The Soil Conservation Service may be of assistance on drainage problems.

6. *General conditions sanitary and safe.* Where papers, tin cans, bottles, garbage, or other refuse is strewn about in buildings or on the ground, where numerous flies are present, where building interiors or exteriors are not clean looking, or where there is sewage or other liquid-waste overflow, conditions are not considered to be sanitary.

This is also interpreted to mean that poison ivy, oak, or sumac, and ragweed, mosquitoes, ticks, chiggers, mice, rats, or other insects and rodents that may transmit disease or cause injury should be effectively controlled or eliminated. Pesticides must be stored in a locked cabinet and used only by qualified persons. The control of insects, rodents, and noxious weeds is discussed in Chapter 10. Power equipment protection, watercraft, and vehicles must meet federal and state regulations. Open wells or pits must be closed and steep cliffs barricaded. Gasoline, fuel oil, and bottled gas is stored away from habitation and in a protected area. Electrical wiring including swimming pool and area installations must comply with the latest edition of the National Electrical Code. Ask for a certificate from the Board of Fire Underwriters or licensed electrician. See also Accident Prevention, Swimming Pools Type and Design, and Bathing Beaches, this chapter.

7. *Structurally safe.* A building is considered structurally safe when it is capable of supporting $2\frac{1}{2}$ to 4 times the loads and stresses to which it is or may be subjected and comply with applicable state and local building codes. The piers, foundations, girders, beams, flooring, and stairways must be sound and sturdy. Stairways and balconies should include handrails and balusters. The floors should not give significantly when stamped on. A licensed professional engineer or registered architect is qualified to determine the structural safety of a building.

8. *Adequate size.* Adequacy depends on the use to which the building is put. For example, sleeping quarters should provide at least 40 ft^2 of floor area per bed, including about 10 ft^2 for a footlocker and clothes closet as shown in Figure 9-11, with a ceiling height of 7 ft over at least 80 percent of the floor area. Kitchens should provide sufficient space for storage and to avoid overcrowding and confusion. Dining rooms should allow $1\frac{1}{2}$ to 3 ft around tables for serving and free movement of chairs and benches. The provision of 20 to 25 ft^2 per person in recreation and assembly rooms, 12 to 15 ft^2 per person in mess halls, and 7 to 9 ft^2 per person in kitchen areas, including storage and dishwashing, has been found adequate, with the smaller areas being more suitable for camps accommodating more than 100 and the larger areas for camps of 100 or less. See Figures 9-12 and 9-13.

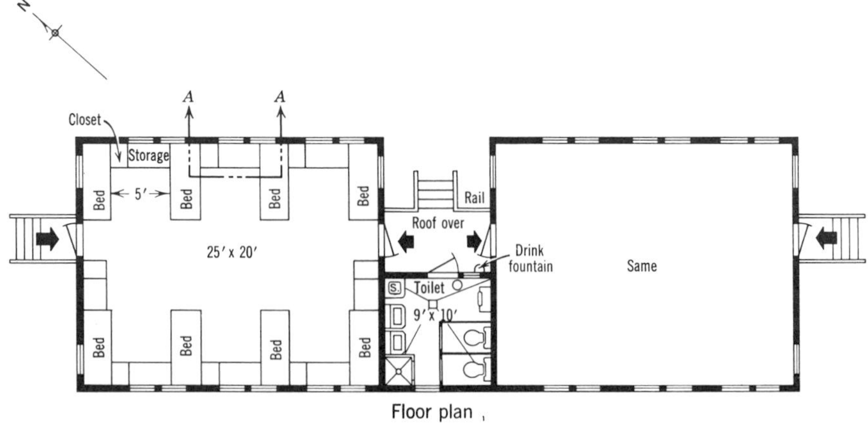

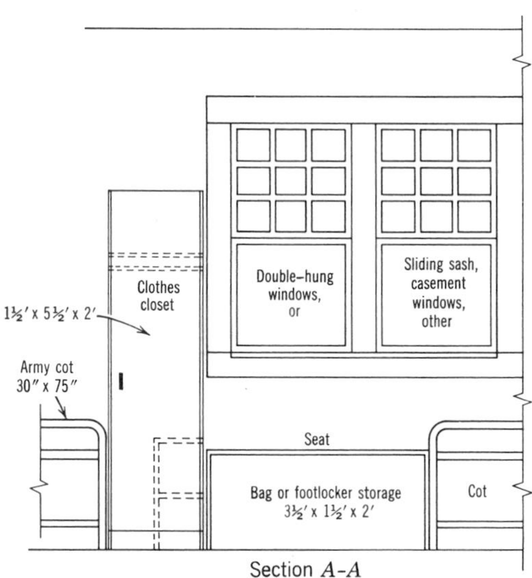

Figure 9-11 A double-cabin housing plan with connecting toilet. Note: (1) All dimensions are approximate; (2) Closets and cabinets are built in place; (3) Roof eaves overhang about 24 in; (4) Cabins are built 2 to 3 ft above ground; (5) Cabin floors are tight, first-grade wood or cement, composition, or equal material; (6) All doors are self-closing; (7) All windows, louvres, and doors are screened.—Windows provide a glass area equal to not less than 15 percent of the floor area; (8) Toilet floors, including shower stalls, are cement, tile, or terrazzo with sanitary cove base integral with floor; (9) Toilet stalls are open at bottom; (10) Louvres are provided in roof; (11) Ceilings have a height of 8 ft not including peak; (12) Bathroom includes utility sink, waste paper basket, 2 flush toilets, 2 wash basins, 1 urinal, 1 shower, paper towels, and soap or soap dispensers.

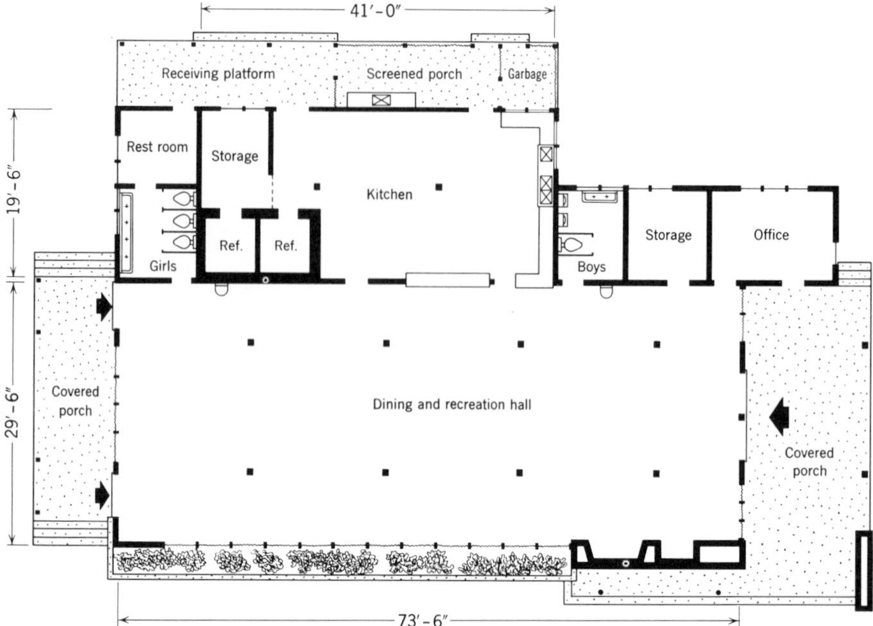

Figure 9-12 Plan of Cornell central camp building. [From *Cornell Miscellaneous Bull.* 14, New York State Colleges of Agriculture and Home Economics, Cornell University, Ithaca (March 1953).]

9. *Easy to keep clean.* Smooth, washable finishes that are light colored, readily drained, and well lighted are easier to keep clean. More detailed information on floor and wall construction and lighting is given in Chapter 8 and in the following paragraphs.

10. *Watertight roof and sides.* Dormitories, cabins, tents, dining halls, and similar structures should have watertight roofs and sides protected by overhanging eaves, storm canopies, or windows to keep out the rain. Floors should be so constructed as to remain dry and free of dampness. To accomplish this objective the floors, including joists, are raised above the ground with the space beneath ventilated. Masonry floors built on the ground require damp-proofing.

11. *Lean-tos exclude rain.* Lean-tos are short-term shelters; nevertheless where provided they should be constructed so as to keep regularly used spaces relatively dry.

12. *Cleanable floors provided.* Where floors are frequently wet, hard acid brick or quarry tile laid with full $\frac{1}{16}$-in. Portland cement or equal hard joints, sloped to a drain, make good floors. Where floors are wet infrequently, top-grade maple wood floors, well laid, are preferred. Other types of floor materials and coverings that do not entrain dust, grease, or other soil and that can be kept clean are satisfactory. All corners should be rounded and a sanitary flush cove base provided around the floor at the walls. See Chapter 8.

13. *Stoves fire-protected and vented.* Heating stoves and baking ovens should rest 4 in. above a fire-resistant floor. In existing buildings the equipment can be placed on a built-up fire-resistant floor consisting of 22 gauge sheet metal, $\frac{1}{4}$-in. layer of asbestos, and 4-in. of masonry, all extending about 24 in. in all directions beyond the stove. Clearance from combustible walls will vary, from 12 in. for a circulating-type wood stove which has a double wall with air vents top and bottom, to 36 in. for a radiant-type. Stove pipes are at least 18 in. from the ceiling. Lesser spacing is permitted with suitable fire protection. (See National Fire Protection Association No. 89M, 1971) Portable stoves should be prohibited. All cook stoves, water heaters, furnaces, and space heaters should be properly vented to the outer air and provision made for adequate dilution air and back-draft protection. Fuel-burning water heaters should not be allowed in sleeping rooms and the use of kerosene cooking and heating stoves should be prohibited. Follow National Fire Protection Association and building department standards and specifications. Get expert advice.

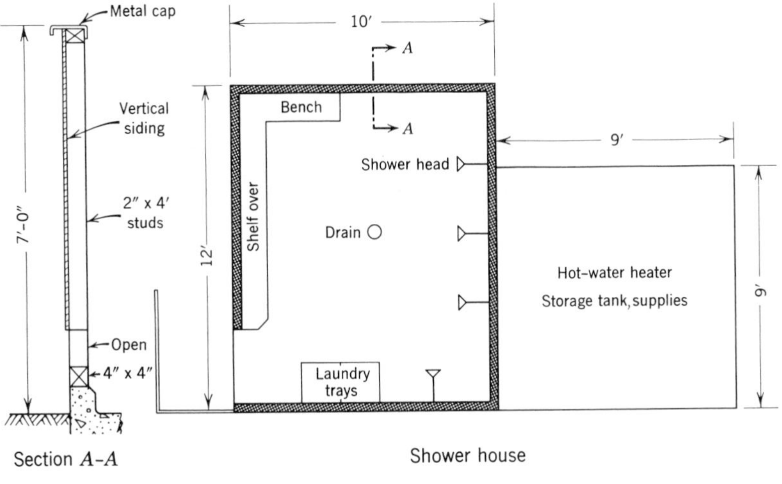

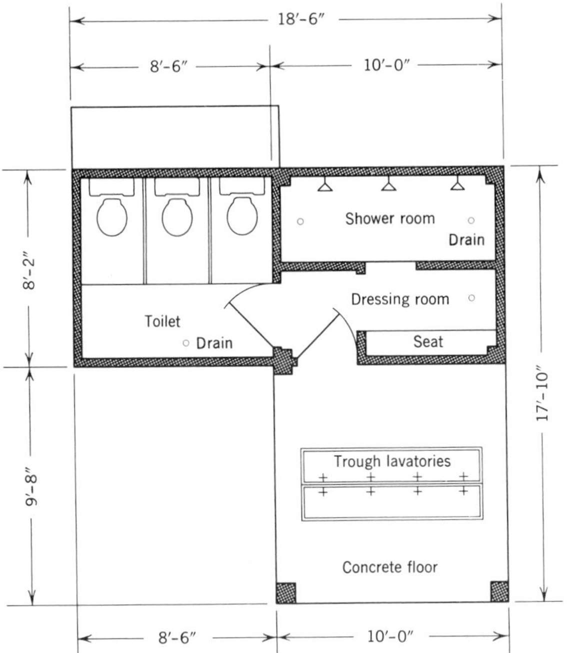

Figure 9-13 Latrine and shower-house floor plans (Unit latrine from *Organized Camp Facilities*, U.S. Dept. of Interior, Washington, D.C., 1938.)

Hoods over kitchen ranges should provide at least a 9-in. space between the hood and combustible material. Vents should extend to a properly constructed lined masonry chimney or an exterior stack extending at least 2 ft out from the building and 2 ft above the roof in such a manner that no danger can occur if grease in the hood and vent should catch fire. Horizontal runs should be not longer than 20 ft and slope up at least $\frac{1}{4}$ in. per foot be provided with a back-draft diverter, and the connection made to the chimney should terminate flush with the interior of the chimney lining. Asbestos cement pipe is not suitable as a flue unless completely enclosed in masonry. Terra cotta clay pipe is standard.

Hoods, exhaust ducts, or vents, fans, and filters require regular cleaning. Stoves burning wood, coal, gas, or oil must be vented to the outside air. A spark screen should be installed on chimney caps where a solid fuel is used. (See also Figure 11-15.)

14. *Fire prevention.* Suitable fire extinguishers should be readily available at all times. A foam- or carbon dioxide-type of extinguisher is recommended for gasoline, paint, oil, or grease fires. The loaded stream- or dry chemical-type may also be used. A carbon dioxide extinguisher is recommended for electrical fires. The dry chemical-type is also suitable. A water-type extinguisher is suitable for wood, paper, textile, rubber, and rubbish fires; do not use on grease, oil, or electrical fires. Water extinguishers can be protected from freezing by adding calcium chloride containing a corrosion inhibitor. Five lb to 9 qt of water will protect to 10° F. Eight lb six oz of calcium chloride to 2 gal of water will protect to −20° F, and 10 lb will protect to −40° F.

One $2\frac{1}{2}$ or two $1\frac{1}{2}$-gal fire extinguishers per 1000 ft^2 of floor space accessible within 100 ft is recommended. Provide one additional $2\frac{1}{2}$-gal or two $1\frac{1}{2}$-gal extinguishers for each additional 2000 ft^2. For each transformer or power generator provide one 4-lb carbon dioxide extinguisher; and in each kitchen or other area where flammable grease or liquids are used, stored, or dispensed, provide one 12-lb carbon dioxide or dry chemical extinguisher. In any case, *be guided by local fire and building department regulations* where they exist, and the National Fire Protection Association suggestions.

At small camps, it is also good practice to have available 50 to 100 ft of 1-in. garden hose with nozzle and convenient hose-bib water outlets within reach of all structures. Adequate water volume and pressure are of course essential. At large camps, a community-type water system as described in Chapter 3 should be provided. All extinguishers require annual inspection. The soda and acid-type and foam-type should be discharged, refilled, and tagged. The water-tank type should be operated to assure it is not clogged and refilled. Other-type extinguishers should be weighed, refilled if necessary, and tagged. Do not overlook the usefulness of such simple devices as a covered water bucket in tent and cabin areas, fire shovel, ground-fire beater, stirrup pump near streams, and back pump.

All tents or canopies should be fiber-impregnated flame-retardant and no plastic tents of any type allowed.

15. *Buildings and grounds clean.* Trash, weeds, brush, poison ivy, or manure should not be permitted to accumulate. Barns, stables, picket lines, and so on should be kept clean. Existing structures should be kept in repair or torn down.

16. *Fire escapes and exits provided.* In general, there should be more than one exit from each floor of a building with doors hung to swing in the direction of egress. When rooms are provided above the ground floor, at least two separated unobstructed exits should be indicated that are accessible from all rooms through approaches 3-ft or more in width. Stair halls, stairs, or passages connecting the entrance or first story with the second story should be separated by means of a self-closing fire-resistant door. Exits should be plainly marked. Buildings with more than two floors should not be used unless special approved fire protection and escape facilities are provided, including outside iron stairways or enclosed fire-resistant stairways with self-closing doors. Also provide automatic fire or smoke detection devices in all buildings used for sleeping. In addition, fire alarm systems and horn or siren should be provided in buildings sleeping more than 50 persons and in all multistory buildings. Have a fire plan for emergency (including drill), and determine and comply with state and local regulations. Dormitories should be spaced at least 40 ft from other buildings, 25 ft back from the access road curb, and with one side at least 60 ft from any other building. In any case, *conform with state and local fire protection regulations.*

17. *Sleeping quarters adequate.* Beds that are spaced 5 ft apart, with an allowance of at least 50 ft^2 of floor space per bed, prevent overcrowding and provide for clothing storage.[61] Where this is not possible, at least 40 ft^2 per person and head-to-foot sleeping should be used, with provision for aisle space, ready exit in case of fire, and clothing storage. If a minimum spacing between beds of 3 ft cannot be obtained, temporary housing, such as under fiber-impregnated flame-retardant canvas, should be furnished to alleviate overcrowding. Sleeping quarters should provide at least 400 to 600 ft^3 of air space per bed and a window area equal to not less than 10 to 15 percent of the floor area. In special cases a lesser volume than 400 ft^3 may be permitted. Double-decker beds are considered two

[61] *Military Sanitation*, FM21-10, AFM160-46, Dept. of the Army and the Air Force, May 1957, states in reference to barracks, "There should be a space allowance of from 60 to 72 square feet per person."

beds; they are not recommended since an accident hazard is introduced and the spread of upper-respiratory diseases is facilitated.

18. *Buildings adequately ventilated.* This is accomplished when the accumulation of body odors is prevented, usually when an air change of 10 ft^3/min is provided per person. The necessary air change is frequently secured in the summer by opening windows and by air leakage or seepage through the walls, ceilings, floor, windows, and doors. An area of window glass equal to 10 to 15 percent of the floor area preferably divided between opposite walls, with one-half of the area openable, is desirable. Toilet rooms should have openable screened windows (which are used) or separate forced-air or gravity ventilation providing 4 to 6 air changes per hour.

19. *Kitchens and dining rooms separate from dormitory and toilet.* This is meant to prohibit sleeping in the kitchen and dining room and to provide a separation between the kitchen and toilet room. A hand-washing vestibule ahead of the toilet compartment accomplishes this objective.

20. *Screening adequate.* Rooms in which food is prepared or served should have all usable windows and other openings equipped with full-length 16-mesh wire screening during the fly season. Sleeping rooms, dining rooms, recreation halls, and theaters are frequently screened for protection against mosquitoes and other insect pests. In endemic insectborne disease areas screening is necessary. The number of flies and mosquitoes actually present, the control measures used, and the effectiveness of residual insecticides applied also determine the need for screening.

21. *Equipment adequate for food preparation.* Utensils, tables, cutlery, dispensers, sinks, stoves, and so forth used in the preparation of food should be sufficient in number and smooth and seamless or with flush seams. Stainless steel is highly recommended. Tables should have nonabsorbent tops; spatulas, scoops, butter forks, tongs, relish jars, mustard jars, and ketchup bottles should be adequate and easily cleanable. See Chapter 8 and National Sanitation Foundation Standards. Stoves and ovens should be large enough to prepare the food for a usual meal.

22. *Floors, walls, ceilings clean.* The walls, floors, and ceilings of kitchens, dining rooms, and workrooms should be so constructed as to permit them to be readily cleaned, and they should be kept clean and in good repair. Light colors, tightness, and smoothness facilitate cleanliness. A concrete, tile, or composition floor that is badly cracked; a wooden floor that is slivered, warped, or poorly laid so as to leave spaces between strips where dirt and water could collect; or a floor covering that is cracked and with holes worn through would not be satisfactory. The floor should be well drained, even, smooth (but not slippery), tight, and water-repellent without cracks, holes, or slivers. Walls should be washable to splash height.

Equipment placed 18 in. away from the walls or other stationary equipment and 8 in. above the floor will simplify cleaning.

Proper lighting simplifies the working and cleaning operations and reduces accident hazards. Lighting should meet the recommendations tabulated in Chapter 8. A more efficient design results when considerations are given to the spacing of light sources, elimination of glare, shielding, control of outside light, lightness of working surfaces, walls, ceilings, fixtures, trim, and floors.

23. *Ventilation adequate.* The ventilation is adequate when cooking fumes and odors are readily removed and when excessive heat condensation is prevented. This will require windows that can be opened both from the top and bottom and, in most cases, ducts that will induce ventilation. Hoods with exhaust fans should be provided over ranges, ovens, deep-fat fryers, rendering vats, steam kettles, and any other equipment that will give off large amounts of steam or soot.

24. *Water supply safe.* A water supply is considered safe when a complete sanitary survey of the water system shows that the water is obtained from a source adequately purified by natural agencies or adequately protected by artificial treatment. The water supply system should also be free from sanitary defects and health hazards. In addition, water samples collected from representative points on the distribution system and examined by an approved water laboratory should consistently show the absence of coliform microorganisms. The physical and chemical characteristics of the water should not be objectionable.

The explanations and sketches given in Chapter 3 describe in greater detail what is satisfactory and what is good construction.

If ice is used for beverages, only that which is obtained from an approved source and which is distributed in a clean manner should be used. Tongs, scoops, or an automatic ice dispenser, and not the hands, should be used for the serving of ice.

Table 9-7 Hot-Water Storage Tank and Heater Sizes

Facility	Capacity (persons)	Approximate Tank Capacity[a] (gal)	Heater Size (Btu per hr)
Central kitchen	125 to 150	225	150,000[b]
	200 to 300	500	190,000[c]
Central showers	125 to 150	250	60,000[d]
	200 to 300	400	90,000[d]
Health lodge	—	30	Home size

Source: From *Camp Sites and Facilities*, Boy Scouts of America, New Brunswick, N.J., 1950.
[a] Use nearest commercial size; specify minimum working pressure of 85 psi for galvanized steel tank, with standard tappings. Insulate tank with 2-in. asbestos cement on 1-in. mesh chicken wire or expanded metal wrapped around tank. Also insulate hot-water delivery lines.
[b] Rated to raise 140 gal/hr 90° C for hand-type dishwashing. Add booster heater to deliver 180° water to disinfecting rinse sink, and an immersion or under-sink heater to keep water at about 200° F.
[c] Rated to raise 200 gal/hr 90° for general purposes, and as a booster to furnish 180 to 200° F water to spray-type dishwashing machine with circulator.
[d] Provide mixing valve, inaccessible to campers, to limit water temperature at shower heads to 95 to 105° F maximum. A hot-water temperature of 135° F can cause third-degree burns in 10 to 15 sec.

Water available at any public establishment at any tap, faucet, or hose is assumed to be, and should be, safe to drink unless it is made inaccessible.

The provision made to secure water on hikes or camping trips should be investigated. Iodine or chlorine water-purification tablets should be available and their use understood.

25. *Adequate quantity of hot and cold water.* The quantity of water should be ample for drinking, handwashing, cooking, and dishwashing, for flush toilets if provided, and for bathing and laundry purposes. This total will probably vary from 30 to 75 or more gal per person per day, depending on the number of plumbing fixtures, water pressure, type of establishment and guests, location, water usage, and other variables. A water outlet should be available within 100 ft of any structure or recreational area.

Computation of the probable total water demand, storage, pipe sizes, and pump sizes is discussed in Chapter 3. The design of the hot-water supply and demand for kitchen and general utility purposes is discussed in Chapter 8. It should be remembered that peak water demands may be 5 to 10 times average demands and that pneumatic water-storage tanks can make available only about 20 percent of the tank volume.

Some suggested hot-water storage tank and heater sizes suitable for camps are shown in Table 9-7 for comparison and reference purposes. Various rule-of-thumb guides are given in providing hot water. At camps a minimum of 3 to 5 gal per person per day is suggested for showers alone. Another value used is 5 to 7 gal per person for all purposes with 40 to 70 percent in storage and about 30 percent provided by the heater per hour. (See also Table 11-9.) Coin-operated, *electrically activated* shower controls require a ground-fault interrupter between the power source and the shower head, if unit operates on a voltage in excess of 15 volts, approved by a recognized testing agency such as the Fire Underwriters.[62]

26. *Operation reports satisfactory.* Where a water supply is treated, accurate and complete daily reports on the operation of the treatment plant must be kept and submitted monthly to the health department. A special form that includes the desired specific information is usually provided by the health department. The report will show whether the resort owner or manager is doing his or her job properly and also serve as evidence that reasonable care has been taken to treat the water. To be of value, however, all entries must be accurate. The same principles apply to sewage treatment and swimming pool operation reports.

27. *Sources of water protected.* See Chapter 3.

[62] New York State Dept. of Health News Release, August 18, 1977, Albany, N.Y.

28. *No cross-connections or backflow permitted.* This subject is illustrated and discussed in greater detail in Chapter 3 and in Chapter 11 under Plumbing.

29. *Water storage protected.* Proper construction is illustrated in Chapter 3.

30. *Common drinking cup prohibited.* A glass or cup should not be left near a water faucet whereby anyone may use it. Provide paper cups in sanitary dispensers or sanitary drinking fountains in the ratio of one to 50 to 75 persons.

31. *Drinking fountains satisfactory.* A satisfactory drinking fountain is one having a protected nozzle extending above the rim of the bowl in which the nozzle throws an inclined, uniform jet of water. Wastewater should be drained away to a disposal system or seepage pit to prevent the collection of pools of water. Use American Standards Assoc. standard Z4.2 for sanitary drinking fountain.

32. *Separate toilet facilities provided.* Separate toilet facilities for campers and visitors are required for each sex. These may be flush toilets or sanitary privies.

33. *Toilets convenient and adequate.* Toilets provided should be near the sleeping quarters and centers of activities and not further away than 150 ft. Flush toilets for food handlers and kitchen help should adjoin the kitchen. Washbasins with warm running water, soap, and paper or other type of sanitary towels are essential for food handlers, preferably in or adjoining the kitchen.

Table 9-8 shows the minimum recommended number of plumbing fixtures to provide at residential camps, with hot and cold running water at showers, washbasins, service sinks, and laundry tubs.

The height of water closets and washbasins should be adjusted to the age of the children served. Suggested heights are given below.

	Height to Rim of Fixture	
Age	Washbasin	Water Closet
Up to 6	1′ 9″	10″
7 to 9	2′ 1″	10″
9 to 12	2′ 3″	1′ 0″
Over 12	2′ 6″	1′ 2″

The walks and toilet rooms or latrines should be clear, well marked, and lighted at night. See Figure 9-14.

34. *Privy location, construction, and maintenance satisfactory.* Privies should be located not less than 100 ft from any well, stream, or lake and not less than 50 ft from any sleeping area; also at least 200 ft from places where food is prepared or served, and convenient to the users—within 200 ft. The pit should be at least 2 ft above seasonal high groundwater. Other types of privies and toilets, and details concerning location, construction, and maintenance are given in Chapter 4. Privies that are properly constructed, located, and kept clean are relatively inoffensive and satisfactory sanitary devices.

35. *Disposal systems satisfactory.* Sewage disposal systems are considered to be satisfactory when the overflow or exposure of inadequately treated sewage, including bath, sink, and laundry wastes, on the surface of the ground is prevented and when surface and groundwater pollution is prevented. When there is an overflow or when an overflow is imminent, immediate steps should be taken to correct this dangerous condition.

To protect the owner and comply with the requirements of most sanitary codes, a licensed professional engineer or architect competent in matters of sewage disposal and treatment should be engaged to prepare plans for an adequate system. The plans should be comprehensive, with construction details of existing and proposed work, including structures served, water system, sewerage system, bathing area, roads, topography, soil profiles, and any other features necessary for proper design and review of the plans. Following examination and approval of the plans by the health department sanitary engineer, the job can be let out for competitive bids. Construction of the sewage disposal or treatment system should be under the supervision of the designing engineer or architect to assure compliance with the approved plans.

Most camp sewerage systems include one or more septic tanks, which should be preceded by a

Table 9-8 Plumbing Fixtures Recommended for Residential Childrens Camps

	Toilet Facilities[a]		Urinals	Handwashing Facilities	Shower Heads
Individuals of Each Sex to be Served	Male	Female	Male	Male or Female	Male or Female
1 to 10	1	1		1	1
11 to 18	1	2	1	2	2
19 to 33	2	2	1	3	2
34 to 48	2	3	2	3	3
49 to 63	3	4	2	4	4
64 to 79	3	5	3	4	5
80 to 95	4	6	3	5	6

Source: *Youth Camp Safety & Health*, PHS, DHEW, Atlanta, Ga., May 1975, pp. 14–15.
[a] Flush urinals may be substituted for not more than one-half the required number of toilet facilities for girls.

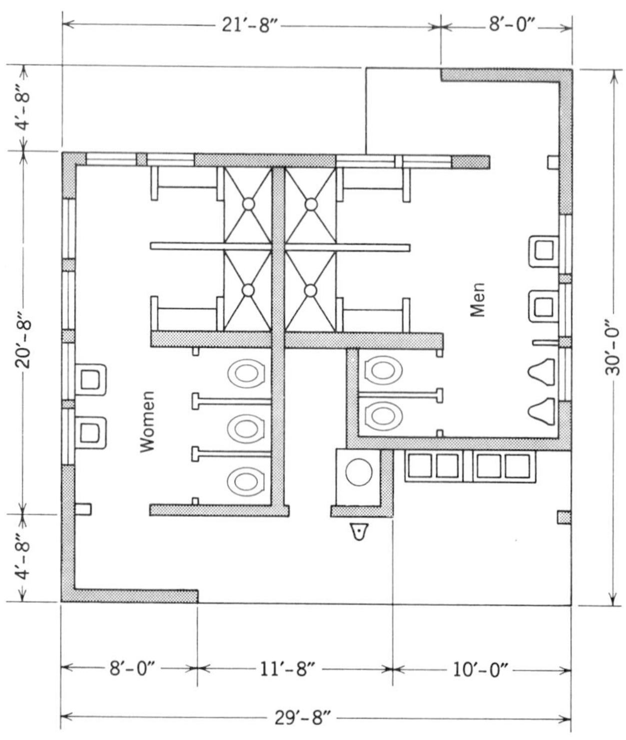

Figure 9-14 Layout of a permanent-type of comfort station. (From *Environmental Health Practice in Recreational Areas,* PHS Pub. 1195, DHEW, Washington, D.C., revised 1977.)

grease trap if designed to receive kitchen wastes. Septic tanks should be inspected each year before the camping season and cleaned when the sludge or scum approaches the depth given in Table 4-8. Grease traps require frequent cleaning, depending on the size and type. Under-sink grease traps are not advised. Outside septic-tank-type grease traps should be cleaned at least once a month. More frequent inspections should be made to determine the cleaning schedule for each establishment if the carry-over of grease and clogging of sewers is to be prevented. Where other more elaborate treatment systems are provided, each unit should be maintained in proper working order so as not to lower the overall efficiency of the treatment process and so as to prolong the useful life of the system.

36. *Location of sewage disposal or treatment system satisfactory.* Sewage disposal systems are satisfactorily located when they are at a lower elevation, at least 100 and preferably 200 ft, from sources of water supply and 50 to 100 ft from lakes, streams, or swamps. Many other factors, as explained in Chapters 3 and 4, must be taken into consideration.

37. *Pasteurized milk used.* Only pasteurized milk, cream, cottage cheese, or other dairy products obtained from an inspected and approved source should be used.

38. *Refrigeration adequate.* Refrigeration is adequate when it is possible to store and maintain all perishable foods at the proper temperature over weekends, without packing or crowding. The box should be relatively dry and should permit air to circulate freely. There is never too much refrigeration space provided at camps.

Refrigerator sizes, temperatures, food storage, and maintenance are discussed in Chapter 8. Camps should provide at least $2\frac{1}{2}$ ft^3 of refrigeration space per person accommodated; 3 ft^3 is better.

39. *Food storage satisfactory.* If perishable fresh or prepared foods are left standing outside of refrigerators, food storage is unsatisfactory. If food is packed or disorganized in refrigerators, stored on the floor, or uncovered, the storage is not satisfactory. (Foods are packed in freezers.) Storerooms and storage spaces should be provided that are rodent-, insect-, and damp-proof, or food and dry stores should be kept in containers providing this protection. Food and drink should be wholesome. Details are given in Chapter 8.

40. *Food handling satisfactory.* Tongs, ladles, spatulas, or forks should be used in place of the hands to handle food whenever possible, and clean practices must be the rule if food poisonings and infections are to be prevented. Frozen meat, poultry, and other frozen foods, except vegetables and chops, should be thawed slowly in the refrigerator and not thawed by letting them stand overnight in a warm kitchen. Prepared foods should be either served immediately, kept on a warming table maintained at a temperature above 140° F, or refrigerated in pans to a depth of 2 to 3 in. at or below 45° F. See Chapter 8.

Food handlers are expected to have clean hygienic habits, to wash their hands thoroughly with soap and warm water after soiling them and after visiting the latrine, to wear clean clothing, to bathe daily, and to be free from communicable diseases.

Milk is one food that is particularly susceptible to contamination. Satisfactory handling and protection of milk can be obtained by serving it in quart, pint, or half-pint prepackaged containers. Bulk-milk dispensers in which a special can with sanitary outlet is cleansed, disinfected, filled, and sealed at the pasteurizing plant and eliminates the use of the dipper and bulk milk are available. A prior understanding is necessary before the pasteurizing plant will agree to handle this equipment. The dispenser must be an approved type. Leftover food including milk returned from the dining room must not be reused. It should be thrown away.

Other precautions to prevent food poisoning and infection are given in Chapters 1 and 8. Pesticides should be carefully stored and used. See Chapter 10.

41. *Dishes washed, cleansed, and disinfected.* The dishwashing operation and equipment are discussed in considerable detail in Chapter 8. Dishes and utensils should look and feel clean.

42. *Garbage storage satisfactory.* There should be available an adequate number of covered metal receptacles to store all the garbage and other soiled refuse from one meal. See Chapter 5.

All the liquid wastes should be discharged to a properly designed disposal system through a grease trap.

43. *Garbage disposal satisfactory.* Probably the most satisfactory method for the disposal of garbage is to make arrangements for the use of a municipal incinerator or sanitary fill. When this is not practical, all refuse, including garbage, kitchen wastes, tin cans, bottles, ashes, and rubbish

should be disposed of in a camp or resort-operated sanitary fill. A description of the sanitary fill operation is given in Chapter 5.

An open dump invariably encourages fires and the breeding of rats and flies. It is always unsatisfactory. Selling garbage for hog feeding encourages the spread of trichinosis. It should be discontinued in the interest of public health unless provision is made by the collector to first boil all garbage 30 min.

44. *Shower facilities provided.* Bathing facilities consisting of showers supplied with hot and cold running water, preferably with tempered water, are satisfactory. Floors should be nonskid, graded to a trapped drain, and the walls constructed of impervious material to splash height. Shower facilities convenient to kitchen personnel are a major consideration to help assure cleanliness. The availability of a swimming pool or bathing beach is not a substitute for warm water and soap cleansing showers, but is further reason for providing warm water showers to prevent greater pollution of the swimming pool. In tick-infested areas daily showers should be encouraged. Common wash pans should be discarded as they are just as much disease spreaders as common drinking cups or common bath and dish towels. See Figure 9-13. Adjust shower water temperature to 95 to 105° F.

45. *Bathing areas conform with sanitary code.* Pools and bathing beaches are operated and maintained in conformity with sanitary-code regulations. No bathing should be permitted at any time unless under the supervision of a competent person trained in life-saving procedures.* Construction of pools and bathing beaches, the life-saving procedures established, records kept, operation, and maintenance are covered earlier in this chapter. The importance of assuring that effective safety and life-saving plans are actually being practiced cannot be sufficiently emphasized. Pools should be fenced in. See this chapter for details.

46. *Reports of communicable diseases.* Persons in charge of a camp must report cases of diseases that are presumably communicable. It is a duty of the person in charge of a camp and the doctor who may be in attendance to report immediately to the health officer the name and address of any individual in the camp known to have or suspected of having a communicable disease. Strict isolation shall be maintained until official action has been taken. The method of isolation should be approved by the health officer. The person in charge should not allow such individual to leave or be removed without permission of the health officer.

Whenever an outbreak of suspected food poisoning or an unusual prevalence of any illness in which fever, diarrhea, sore throat, vomiting, or jaundice is a prominent symptom, it is the obligation of the person in charge of a camp to report immediately by telephone, telegram, or in person the existence of such illness to the health officer having jurisdiction. Health officers are required to investigate such outbreaks or the unusual prevalence of disease to determine the cause and prevent its repetition or spread. See Chapter 1, Investigation of Water- and Foodborne Disease Outbreaks.

47. *Duty to enforce regulations.* It is the responsibility of the person operating a camp or the person to whom a permit is issued to see that all regulations are faithfully observed at all times.

Travel-Trailer Parks

A travel trailer is

> a vehicular, portable structure built on a chassis, designed as a temporary dwelling for travel, recreation and vacation, having body width not exceeding 8 feet and its body length does not exceed 32 feet.[63]

*Should be over 21, have American Red Cross water-safety instructor's certificate or equivalent, and be assisted by a certified senior lifesaver for each 25 swimmers.

[63] *Environmental Health Guide for Travel Trailer Parking Areas,* prepared by PHS, DHEW, Washington, D.C.; published by Mobile Home Manufacturers Association, 6650 North NW Highway, Chicago, Ill., January 1966. *Environmental Health Guide for Mobile Home Parks*, (same source as above), *Mobile Home Court Development Guide*, U.S. Dept. of Housing and Urban Development, Washington, D.C., January 1970. Also *Environmental Health Practice in Recreational Areas*, PHS, DHEW, Washington, D.C., revised 1977, and *Standard for Recreational Vehicle Parks*, National Fire Protection Association, NFPA No. 501D, Boston, Mass., 1975.

There has been a great increase in vacation travel by trailer and motorhome in the United States and abroad. The more than 2000 privately owned parks and 1200 national and state parks and campgrounds are frequently overtaxed during the vacationing season. Proper planning, designing, operation, and maintenance of trailer parks and campgrounds are essential. Most states recognize the potential hazards involved where insanitary conditions are permitted to exist and have adopted regulations to protect the users. Accepted standards to promote good sanitary practices at travel-trailer parks are summarized below for easy reference.[62]

Site. Adequate size, well drained, no water accumulations or breeding places for insects or rodents, relatively free of dust, smoke, soot, noise, and odors. If possible select site accessible to public water and sewerage systems, main highways, and service areas. Comply with zoning, building, health department, and other regulations. See Chapter 2.

Roads and parking areas. Roads should be at least 18-ft wide (12 ft-wide if serving less than 25 trailers), continuous, and dust-controlled. Make roads 24-ft wide if two-way and 34-ft wide if cars parked on both sides. Provide space for off-street parking and maneuvering. See Figures 9-15 and 9-16.

Trailer space. Not more than 25 trailer spaces/acre; at least 10 ft from other trailer or structure. Add recreational area of not less than 8 percent of the gross area of the site, not less than 2500 ft_Q.

Plans. Submit plans to health and zoning or planning agencies for approval. Show to scale the location and area of the trailer park, topography, the location of coach spaces, roads, service buildings, and other structures, water lines, fire hydrants, sanitary sewers, storm-water drains, catch basins, refuse storage racks, private wells, and sewage disposal or treatment systems. Include service building floor plan and fixtures, details of the water system and sewage disposal or treatment works, trailer water, electric and sewer connections, cold- and hot-water storage, and heater capacity.

Service building. Provide at least one. Access walks surfaced and lighted with at least 5 ft-c of illumination. Building within 300 ft of spaces served. Construction cleanable; floors drained to sewerage system; maintained clean, screened, ventilated, windows as high as practicable, providing at least 10 percent of floor area; laundry room and mirrored areas have 40 ft-c of illumination, heated to 70° F in cold weather; separate men's and women's toilet rooms; hot and cold water for every lavatory, sink, tub, shower, and clothes washer. A floor plan is shown in Figure 9-17.

Minimum plumbing facilities are discussed earlier in this chapter. The number may be reduced if trailer-park use is limited to only self-contained trailers.

Water supply. At each trailer space and service building, connect a public water supply if available or an approved private supply that provides an adequate volume of water. The source should provide at least 100 gal per trailer space per day for individual connections and 50 gal per space per day without connections. The distribution system, including storage and pumping equipment, should be designed for a maximum momentary water demand based on Table 9-9.

A special water supply outlet in the ratio of one to each 100 spaces is needed for filling trailer water-storage tanks. A flexible hose, shutoff valve, and backflow preventer are provided.

A trailer water-supply connection extending at least 4 in. above the ground with a $\frac{3}{4}$-in. valved outlet is needed at each space capable of supplying water at a minimum pressure of 20 lb/in.^{2}. See Figure 9-18. Sanitary-type drinking fountains should be provided at service building and recreation areas. Surface water or questionable groundwater supplies should not be used unless use and treatment have been approved by health authorities. Daily records are required.

See Chapter 3 for water supply design, storage, construction, and protection details.

Sewerage system. Plumbing complies with state and local regulations. Special attention is given to possibility of back-siphonage; no connection permitted where possibility exists. Provide a 4-in. sewer connection extending 4 in. above ground surface and trapped below frost for each trailer space as illustrated in Figure 9-19. No wastewater shall discharge to the ground surface.

Sewer lines are 10 ft from water lines and are designed for 2-fps velocity when flowing full. See

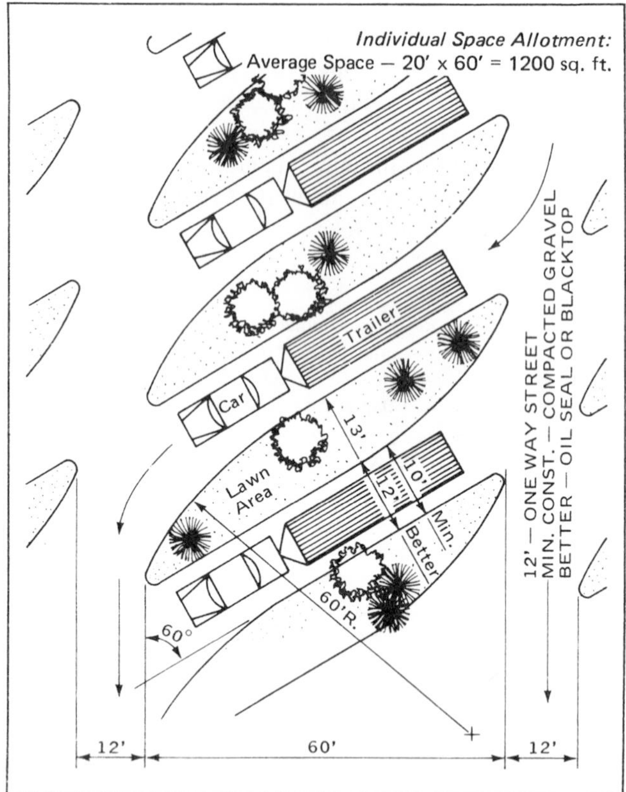

Figure 9-15 Typical transient (overnight) recreational vehicle park. From *Environmental Health Practice in Recreational Areas,* PHS Pub. 1195, DHEW, Washington, D.C., revised 1977, p. 53. Recommended Facilities for Overnight Parks:
(1) Absolute Minimum—Central recreational vehicle sanitary & water stations, and toilets. (2) Fair—Individual electrical outlets, central recreational vehicle sanitary & water stations, and toilets. (3) Good—Individual electrical outlets, central recreational vehicle sanitary & water stations, toilets, and showers. (4) Better—Individual electrical & water outlets, several individual sewer connections, one or more central recreational vehicle sanitary stations, toilets, showers, and coin-operated laundry. (5) Best—Individual electrical, water, & sewer connections, toilets & showers, coin-operated laundry, and picnic tables.

Chapter 4. Joints are watertight, infiltration minimum, and surface water excluded. Manholes not more than 400-ft apart, at juncture of 2 or more sewers and at change in direction. Capped cleanouts with plugs to grade are permitted on 4- and 6-in. lines in place of manholes if at least 100-ft apart. Vents are needed on airtight system. Minimum 6-in. sewers are recommended.

Connect to public sewer. If this is not possible, design the treatment plant for a minimum flow of 100 gal per trailer space per day and so as not to cause a health hazard or nuisance. Design of plant must meet state and local requirements. See Chapter 4 for design and construction details. Separate provision should be made for storm and surface water drainage.

Refuse handling. Tightly covered containers providing 30-gal capacity per trailer space should be adequate. Bulk containers can also be used. Can storage rack is cleanable and located within 150 ft of any trailer. See Chapter 5 for acceptable disposal methods if municipal or contract service is not available.

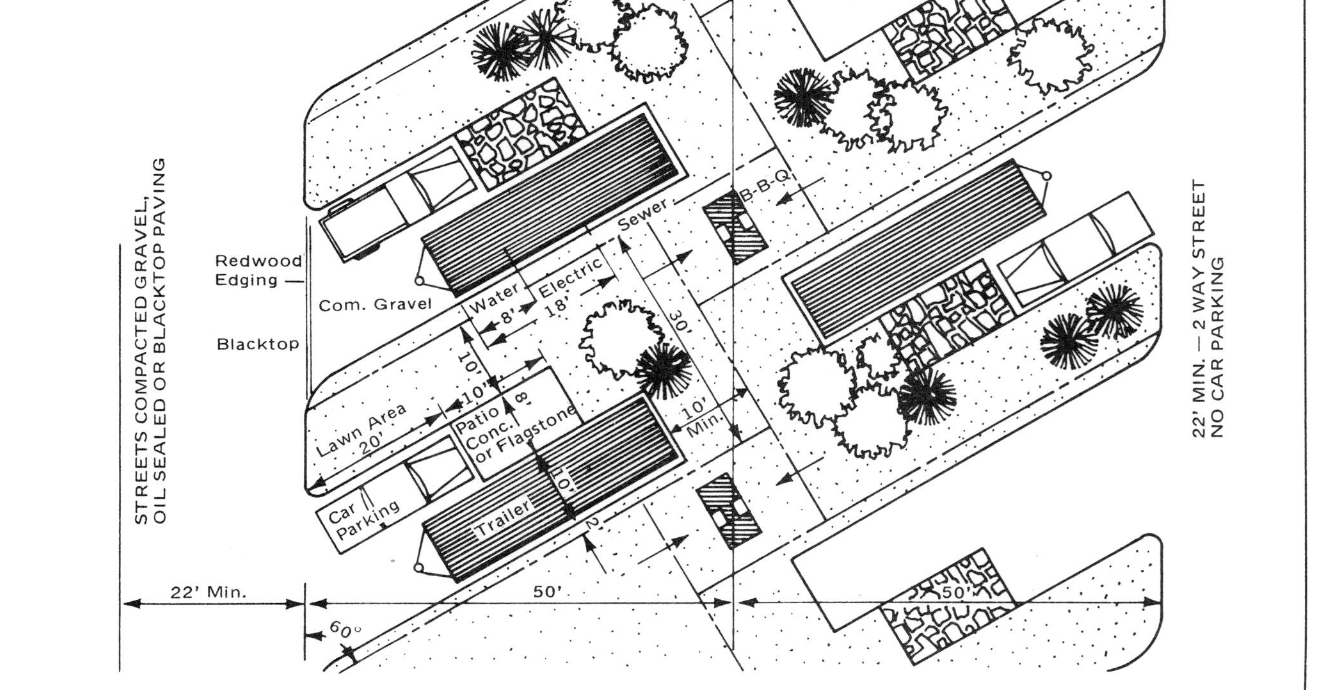

Figure 9-16 Typical resort (destination) recreational vehicle parking area. From *Environmental Health Practice in Recreational Areas*, PHS Pub. 1195, DHEW, Washington, D.C., revised 1971, p. 53.

Recommended Facilities for Destination Parks:

(1) Absolute Minimum—Back-in parking, individual electrical outlets, central recreational vehicle sanitary & water stations, and toilets & showers. (2) Fair—Back-in parking, individual electrical & water connections, central recreational vehicle sanitary stations, toilets & showers. (3) Good—Drive-through parking, individual electrical & water connections, central recreational vehicle sanitary stations, toilets, showers, coin-operated laundry, and picnic tables. (4) Better—Drive-through parking, individual electrical, water & sewer connections, toilets, showers, coin-operated laundry, picnic tables, grocery. Also, barbecue, bottled gas, recreational vehicle parts for sale, plus bait & other fishing and sport accessories. Recreation building and swimming pool may be on a "pay as you go" basis.

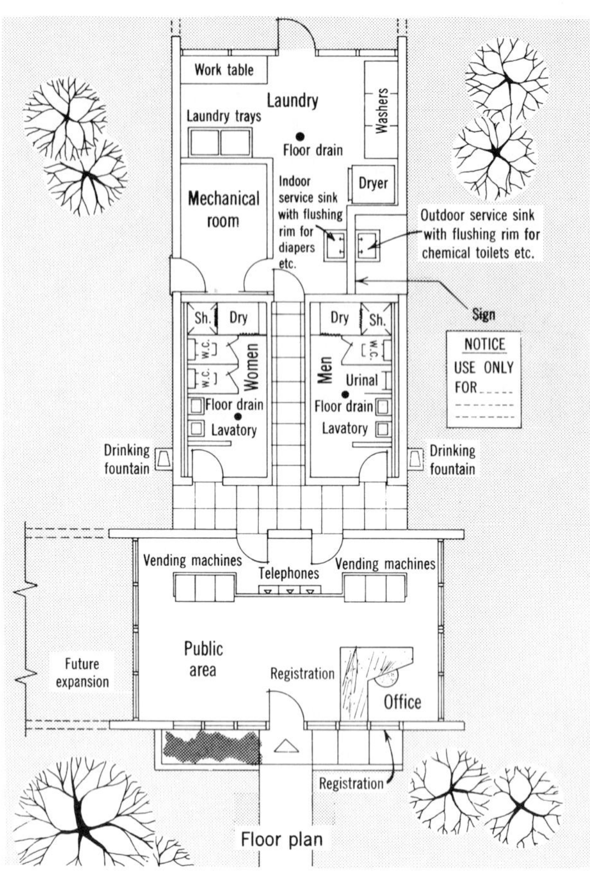

Figure 9-17 Service building. (From *Environmental Health Guide for Travel Trailer Parking Areas*, PHS, DHEW, Washington, D.C.; published by Mobile Homes Manufacturers Assoc., 6650 North Northwest Highway, Chicago, Ill., 1966.

Table 9-9 Estimated Maximum Water Demand for Travel-Trailer and Mobile-Home Parks

Number of Spaces	Demand Load (gpm) Travel[a]	Home[b]
25	42	65
50	65	115
75	85	155
100	115	180
150	150	235
200	180	285
250	215	325
300	230	370

[a] Travel-trailer park at 4 fixture units per space.
[b] Mobile-home park at 8 fixture units per provided space.

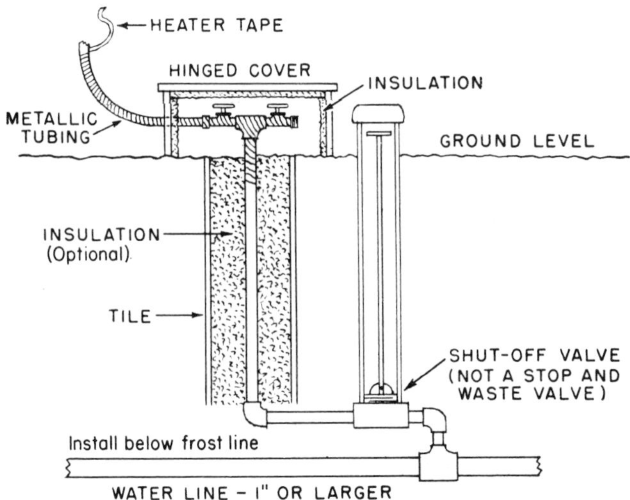

Figure 9-18 Typical water supply connection showing a method of winter protection. Terminate riser 12 to 18 in. above ground surface and equip with anti-siphon backflow preventer. (From *Environmental Health Practice in Recreational Areas*, PHS Pub. 1195, DHEW, Washington, D.C., revised 1977, p. 54.)

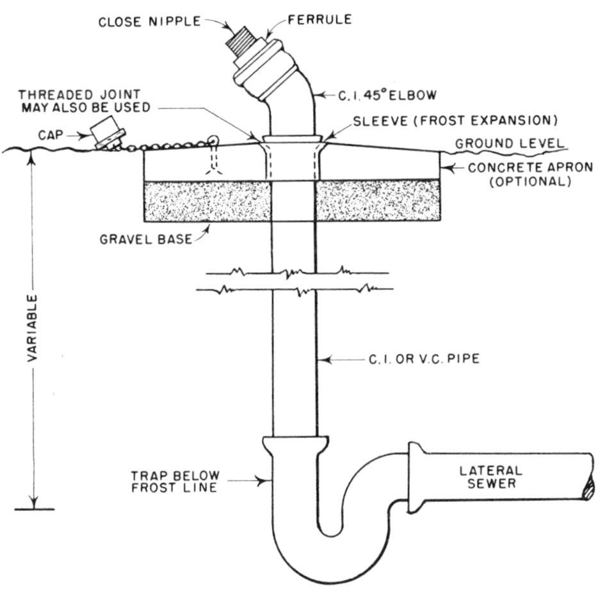

Figure 9-19 Typical sewer connection. Use 4-in. riser, and minimum 3-in. sewer connection, extending at least 4 in. above ground surface. Concrete apron or curb 3-in. thick and 12 in. from riser. (From *Environmental Health Practice in Recreational Areas*, PHS, Pub. 1195, DHEW, Washington, D.C., revised 1977, p. 54.)

Insect and rodent control. Mosquitoes, flies, roaches, rats, mice, and so forth are eliminated or kept under control. See Chapter 10.

Miscellaneous. Electrical wiring and apparatus complies with local codes and National Electrical Code.[64] Fuel oil and bottled gas tanks are protected and securely connected to stoves or heaters by means of metallic tubing.[65] Take fire prevention and protection precautions. Include smoke detectors and fire extinguishers.[66] Make alterations, repairs, and additions in conformance with good practices. Keep pets under control and conditions sanitary. Maintain a register of occupants and vehicles. Eating and drinking establishments are constructed and maintained in compliance with PHS, state, and local ordinances. See Chapter 8. Tiedowns will minimize overturning during windstorms.

Campgrounds[67]

Privately owned, national, and state campgrounds serve hikers and all types of recreation vehicles (RV) involving millions of campers. Recreation vehicles include travel trailers, campers, motor homes and vans, tent trailers, and truck-mounted shells. The amenities provided with the vehicle vary widely from self-sufficient and luxurious accommodations to merely sleeping and storage spaces providing shelter. The number of recreation vehicles in use in the United States in 1973 was estimated at 5 million.

Site Selection

Environmental factors to be considered in site selection and planning are discussed in Chapter 2.

Location

The location of the campground should relate to the highway system. A location within 2 mi of a main highway interchange and on a road leading to it is suggested.

A traffic count of campers and tourists traveling the nearest main highway, along with interviews, will give an indication of the number and type of users that might be expected. A location near a population center is more popular.

The site must have a relatively high degree of privacy. Areas adjacent to commercial and industrial sites may encounter problems of noise, air pollution, odors, and other nuisances that will seriously detract from the attractiveness of the site and the comfort of its residents. Vacationers using camps and other recreation facilities are interested in benefiting from outdoor living, and the resources of the area should be such as to contribute to this experience.

[64] *National Electrical Code*, C1-1962, American Standards Association, 10 East 40th Street, New York, 1962. Also National Fire Protection Association (NFPA), Boston, 1971.

[65] *Standard for Fire Protection in Trailer Courts*, NFPA No. 501A, National Fire Protection Association, 470 Atlantic Avenue, Boston, Mass., 1964. See also *National Fire Protection Standards for Recreation Vehicle Park*, NFPA No. 501D, Boston, Mass., 1975.

[66] *Standard for Fire Protection in Mobile Homes and Travel Trailers*, NFPA No. 501B, NFPA, Boston, Mass. Also *Standards for Mobile Home Parks*, NFPA No. 501A, 1971.

[67] Adapted from Joseph A. Salvato, Jr., *Guide to Sanitation In Tourist Establishments*, WHO, 1976, pp. 22–24 and 54–57.

Area Requirements

The land area required will depend upon the type of establishment, construction, topography, usable area, population to be served and many other factors. An investment analysis and traffic count as noted above would give an indication of the potential users and the area needed. This is a good procedure to follow, but it is not always done.

A minimum number of tent sites, recreation vehicle spaces, and perhaps motel units or cabin units is needed to produce a reasonable return on an investment, and in estimating this number the anticipated extent of usage must be taken into account. For example, about 100 tent sites may justify running a camping ground as a full-time summer activity. This number, the type of construction, and the facilities to be provided will give an indication of the land area required. Land area requirements can obviously vary widely, depending on the type of camping site. However, some guidelines can be given based on what experience has shown to be good practice.

A committee of experts on public health considered it advisable to limit the size of a camping site to 25 acres, with a capacity of not more than 2000 persons. In addition, the density should not exceed 60 tent or recreation vehicle (RV) sites or 80 persons per acre. This would correspond to a gross area of 540 ft^2 per person or 1800 ft^2 per site with an average of 3.3 persons per site.

Another guideline[68] for planning vacation-type camping grounds suggests an area of about 2500 ft^2 per site with 5000 ft^2 as being more desirable. For overnight sites, a minimum of 1500 ft^2 is suggested. These areas would include space for car and recreation vehicle parking. For level areas, 12 sites per acre is considered optimum and would include roads.

The area requirements for other types of tourist accommodation and recreation areas could be quite variable, depending on the type of facility and construction. A summary of some area guidelines is given in Table 9-10.

Plumbing Fixtures

Adequate numbers of plumbing fixtures should be provided. Lack of facilities will result in misuse of the area, the practice of undesirable health habits, difficulties in maintaining the area in a sanitary condition, and adverse criticism by campers. Without the cooperation of campers and supervision by the management the maintenance of a healthful and pleasant environment is practically impossible.

Commercially available fixtures are generally made with smooth, impervious, easily cleaned surfaces and no concealed crevices. Plastic is being introduced to a limited degree, especially for bathtubs and showers, and the use of precast concrete for laundry tub units and shower bases is not uncommon. Bathtubs are not suitable for camping and recreation areas. Rough usage may be expected and an effort should be made to select the strongest types of fixture. The use of

[68] *Guidelines for Planning Vacation-Type Campgrounds*, USDA, Soil Conservation Service, Syracuse, N.Y., 1970 (Recreation Technical Note NY-5).

Table 9-10 Area Guidelines for Camp and Recreation Places

Camping site density including roads	12 to 24 per acre
Population density including roads	50 to 80 persons per acre
Camping site area, overnight	1200 to 1500 ft^2
Camping site area, family vacation, or travel-trailer	1800 to 5000 ft^2

Source: Adapted from J. A. Salvato, *Guide to Sanitation in Tourist Establishments*, WHO, 1976, p. 24.

Note: Where local standards are more stringent, they shall apply. Recommended areas are intended to be usable areas.

heavy-duty institutional-type fixtures securely anchored will minimize problems of vandalism.

Fixtures should be installed in such a way that cleaning presents no problem. Piping should be buried or kept as far above the floor as possible and run to the nearest wall to minimize the amount of exposed piping. Wall-hung closet bowls, for example, simplify floor cleaning procedures. Water closet seats should be of smooth nonabsorbent material of the open-front type. The elongated type of water closet bowl with a greater surface area of water is preferred since it is less likely to become soiled. Urinals should be of the individual wall-hung type with open trapways (no strainers) and visible water seals. Trough and wall urinals are difficult to maintain in a sanitary condition unless regularly flushed, cleaned, and washed down together with the adjoining walls and floor.

Hot water should be supplied to all laundry tubs, kitchen sinks, hand washbasins or troughs, and showers through mixing-type faucets that permit the temperature of the water to be adjusted to suit the user. A controlled temperature of 95° to 105° F is practical and safe for showers and hand basins. Water heaters and hot-water storage tanks should have an adequate capacity to meet the anticipated demand.

Comfort Stations

Central facilities (comfort stations or service buildings) are essential in camping and similar recreation areas. The facilities should be of permanent construction provided with an interior finish of moisture-resistant materials that will withstand heavy use, frequent washing and cleaning. With floor drains and concrete floors the entire area can be hosed down rapidly and easily. The use of ceramic tile floors and wainscot reduces maintenance costs and is believed to discourage vandalism. Partitions should terminate about 1 ft above the floor. Adequate lighting, both natural and artificial, and ventilation are essential. Windows should be located above eye-level for privacy and should be screened with 1.5 mm (16-mesh) screen during the insect season. All doors should open outwards and be self-closing. Separate facilities should be provided for men and women. Savings in construction costs can be made by grouping the facilities for both sexes under one roof. If this is done, clearly marked separate entrances, preferably at opposite ends of the building, should be provided (see Figures 9-14 and 9-17). The interior wall separating the facilities must be of durable soundproof construction. There is some advantage in having a double dividing wall so that the interior space can be used as a pipe chase where all piping is readily available for maintenance.

The accessibility of toilet facilities is most important; if they are too far from tents, trailers, or camping areas, improper practices will occur, resulting in nuisances and health hazards. It is recommended that no camping, tent, or trailer area be located more than 300 ft from a comfort station. This distance may be rather excessive during the night for young children and older people. In such cases individual "portable toilet units" might be used in emergencies and a separate facility consisting of a flush-rim service sink should be provided in an enclosure attached to the comfort station for disposing of wastes and cleaning the portable unit.

Mirrors, shelves, and power outlets for electric shavers are also required. Shaver outlets should be provided at the rate of one outlet for every 20 camp sites, but to avoid congestion in washrooms they should not be located above hand washbasins. Toilet paper and/or paper holders, soap dispensers, and paper towels should be available at all times. Special marked containers are needed for the disposal of paper towels, sanitary napkins, and paper diapers to help in preventing blockages of water closets. This waste should be disposed of along with other refuse. Cotton or linen roller towels with an automatic rewind mechanism for the soiled portion, or hot air blowers, are sometimes provided for hand drying but paper towels are preferred.

Recommended ratios of users to fixtures are shown in Table 9-11 as a guide for design purposes. A comfort station with well balanced facilities should contain at least 2 water closets, 2 hand wash basins, and 1 urinal for males; 3 water closets and 2 hand washbasins for females; 1 flush-rim service sink; and 1 laundry sink. Piped hot and cold water should be provided to all fixtures. The use of coin-operated laundry machines and dryers in comfort stations is gaining acceptance; the machines are installed in a room separate from the toilet facilities. Where washing machines are not provided, laundry trays and wringers in the recommended ratio (see Table 9-11, footnote *c*) and a protected clothes drying area should be available for the use of visitors.

Other

Water supply, sewage and other wastewater disposal, solid waste storage and disposal, food protection, and vector control factors applicable to campground design and operation are also discussed in Chapters 3, 4, 5, 8, and 10.

Migrant-Labor Camps

The problems associated with recruitment, labor relations, minimum wages, housing, unemployment compensation, health and welfare, child labor, and

Table 9-11 Minimum number of plumbing fixtures for various types of tourist accommodations

Type of occupancy	No. of sites	No. of water closets M	No. of water closets F	No. of urinals M[a]	No. of wash-hand basins M	No. of wash-hand basins F	No. of showers M	No. of showers F
Picnic areas[b]	1 to 40	1	2	1	1	1	—	—
	41 to 80	2	4	2	2	2	—	—
	81 to 120	3	6	3	3	3	—	—
	over 120 add 1 fixture for the additional number of sites shown in parentheses	1 (60)	1 (60)	1 (100)	1 (100)	1 (100)	—	—
Camping areas[b,c]	1 to 20	1	1	1	1	1	1	1
	21 to 30	1	2	1	2	2	1	1
	31 to 40	2	2	1	3	3	1	1
	41 to 50	2	3	2	3	3	2	2
	51 to 75	3	4	2	4	4	2	2
	76 to 100	3	4	2	4	4	3	3
	over 100 add 1 fixture for the additional number of sites shown in parentheses	1 (40)	1 (40)	1 (100)	1 (40)	1 (40)	1 (50)	1 (50)
Caravan or trailer camps[b,c] (1 service sink with a flushing rim should be provided for the disposal of liquid wastes. Where camps permit only self-contained caravans it is only necessary to provide 1 flush toilet and 1 wash-hand basin for each sex per 100 spaces or every fraction of 100.)	1 to 15	1	1	1	1	1	1	1
	16 to 30	1	2	1	2	2	1	1
	31 to 45	2	2	1	3	3	1	1
	46 to 60	2	3	2	3	3	2	2
	61 to 80	3	4	2	4	4	2	2
	81 to 100	3	4	2	4	4	3	3
	over 100 add 1 fixture for the additional number of sites shown in parentheses	1 (30)	1 (30)	1 (100)	1 (30)	1 (30)	1 (40)	1 (40)
Boarding and lodging houses, tourist homes, hotels, dormitories, without private bath[b,c]	1 fixture for the number of persons shown in parentheses	1 (10)	1 (8)	1 (25) for over 150 men 1 (50)	1 (8)	1 (8)	1 (8) for over 150 persons 1 (20)	1 (8)

Source: Joseph A. Salvato, Jr., *Guide to Sanitation in Tourist Establishments*, WHO, 1976, p. 56.

[a] The use of special urinals for women is being recommended for recreation and similar areas. Where they are used the same number should be provided as for men, and the number of water closets reduced proportionately.

[b] Provide one sanitary drinking fountain outside each toilet room or for every 50 persons; for beaches provide one for every 30 m of beach.

[c] Provide one laundry tray or clothes washing machine and a kitchen sink or dish washing trough with piped hot and cold water for every 30 sites or 60 persons. A foot-shower is also useful in camping areas.

education at migrant-labor camps have received nationwide and international[69] study. Many of the problems are related to economic exploitation, poor housing and facilities, exposure to pesticides, tuberculosis, parasitic diseases, malaria (outside the United States), mental health, nutrition, drug abuse, institutional arrangements, and environmental health. Reports and publicity have been generally unfavorable. The problems are difficult but not impossible to ameliorate. This area of work will tax the ingenuity and patience of the most dedicated people in public health, as well as those in other official and voluntary agencies having an interest in this challenging field.

An indication of the attitude of some enforcement officials and operators of farm-labor camps is obtained from the following comments relating to environmental sanitation.

If bathing facilities are provided at farm-labor camps the people will not use them; the tubs will be used to store and collect trash.

Do not provide showers—the soil is too tight. Sewage will overflow onto the ground surface and create a more dangerous public health hazard.

The farm laborer does not expect bathing facilities. He never had any where he came from anyway.

A galvanized iron tub is provided which is perfectly adequate for taking a sponge bath. That's what we used when we were kids.

A washbasin and pitcher of water is all you need to keep clean.

The water supply is inadequate now—how can we provide laundry tubs and showers with running water?

The health department should first prove that our well-water supply is adequate before asking us to put in showers.

Showers are too expensive; we cannot afford to put them in.

The growers will not go along with the health department.

Political influence will be used to prevent the health department from carrying on a progressive farm-labor program.

The local health department will be voted out of existence.

You will not get the support of your own staff.

There is no health hazard involved in the failure to provide showers.

The opposition expressed by these statements is really no different from that which one would expect when embarking on any new environmental sanitation program. The half-truths must be attacked with education, painstaking engineering investigation, conferences, sound advice, and patience. When this is coordinated with a long-range plan to improve the environmental sanitation conditions and the work of other agencies, a successful program will result. The ultimate solution of the problems rests on elimination of the migrant-labor camp system, as the very system itself exposes the migrant to increased health risks and exploitation.

Specific items relating to health and sanitation at farm-labor camps are similar to those discussed previously under Temporary Residences and under the various headings in the remainder of the text. Good information, including inspection

[69] Boris Velimirovic, "Forgotten People—Health of the Migrants," *Bull. Pan Am. Health Organ.*, 13(1), 1979, pp. 66–85.

Table 9-12 Some Construction Guides and Facilities at Farm-Labor Camps

Use	Area, Number, or Volume Recommended
Kitchen area	8 ft^2/person including service counter, storage shelves, refrigerator, sink with hot and cold water, cook stove.
Dining area	10 to 12 ft^2/person.
Sleeping area	50 ft^2/person plus storage space; beds spaced 36 in.; double-deck bunks spaced 48 in.
Showers and washbasins, with 95 to 105°F water	3 to 5 gal/person; 30% min. provided per hour by heater and 2 gal/person in storage.
Kitchen—hot water	3/4 to 1 gal/person per hour.
Water closet or privy seat	1/15 men, and 1/15 women.
Urinal	1/20 men.
Shower	1/10 persons.
Washbasin	1/6 persons with hot and cold water.
Laundry tub or machine	1/30 persons; min. of one tub.
Sleeping and cooking rooms	100 ft^2/person.
Common stove	1/10 persons or 2 families per stove
Water supply	35 gal/person per day for cooking, bathing, and laundry at peak rate of $2\frac{1}{2}$ times ave. hr demand.
Light and ventilation	windows not less than 1/10 fl area, $\frac{1}{2}$ openable, screened; 30 ft-c in kitchen and living areas, 20 elsewhere.
Building heating	equipment capable to maintain 70°F in cold weather.

reports and compliance guides, is also available from most state health departments, extension services, the PHS, and the Department of Labor. Some construction and facilities guides are given in Table 9-12. Labor camps are under the jurisdiction of the U.S. Department of Labor and state agencies. (A five-year enforcement program is graphically illustrated in Figure 12-2.)

Mass Gatherings

The assemblage of large numbers of people in a limited area requires that certain minimum facilities be provided for the protection of the health, safety, and welfare of the people. The gatherings can vary from several thousand to several hundred thousand with and without overnight accommodations and facilities. The types of events may include fairs, jamborees, auto races, music festivals, carnivals, and similar "happenings" held in open areas. The magnitude of such events can completely overwhelm all but the largest communities unless the highways, space, communication, sanitation, and related facilities are designed, operated, and maintained to handle the mass of people likely to descend upon the site. Such events should require a permit from the health department or other designated authority to indicate that adequate preparations have been made.

Some guidelines and information needed to assist in preparation for mass gatherings follow.[70]

1. Estimated maximum attendance and area reached by advertising.
2. Expected opening and closing date.
3. Name, owner, and location of property.
4. Name of operating "person."
5. Statement from state department of transportation giving capacity of connecting highways.
6. Statement from state and local police that traffic control plan is adequate to serve 20,000 persons per hour.
7. Statement from local civil defense director approving plans for the assemblage and its evacuation.
8. Statement from local fire authority approving the fire-protection plan.
9. Emergency medical plan, including medical personnel, first-aid, hospital, and ambulance arrangements, and emergency evacuation facilities such as a helicopter.
10. Plan showing area of site and location in relation to adjoining towns within 20 mi.
11. Statement certifying all required construction, facilities, services, utilities, and operational plans are functional and approved 48 hr before assemblage time.
12. Security plan acceptable to state police ensuring crowd control, security enforcement, narcotics and drug control.
13. Plan for site showing boundaries, roads, toilet facilities, water supply, assemblage areas, first-aid stations, food service areas, refuse storage and disposal, sewage and wastewater disposal, water storage tanks, lakes, ponds, streams, wells, electric service, telephones, radio communications, emergency access and egress roads, parking areas, lighting, and other services and facilities necessary for operation.
14. Space—50 ft^2 per person.
15. Water supply—1 pt/person/hr; source, distribution, quality, and quantity comply with state health department standards.
16. Toilet facilities, for each sex—one seat/100 persons and one lavatory/100 persons; 40 percent of required seats for men should be urinals. Where portable toilets are used, obtain adequate cleanup and servicing schedule, including permit for waste disposal.
17. Refuse storage and disposal—provide in areas of assemblage 25 ft^3 of storage receptacle per 100 persons per day (24 hr), including policing and approval of disposal site.
18. Food service—provide detailed plan for food service including food sources, menu, mandatory use of single-service dishes and utensils, *refrigeration facilities*, food handling.
19. Vector control—where mosquito and biting fly population are in excess of 15 specimens per trap per night or other potential disease vectors are found, vector populations shall be reduced to a satisfactory level 48 hr before people assemble and shall be maintained for duration of the event.
20. Noxious weeds such as poison ivy or poison oak shall be removed from accessible areas.
21. Signs—show to the extent needed location of all facilities including roads, toilet facilities, first-aid stations, fire-fighting equipment, parking areas, camping sites, eating places, exits.
22. Drinking fountains—one sanitary-type drinking fountain including waste disposal facility per 500 persons.
23. Sewage and other wastewater disposal—complies with health department standards and permit obtained for collection system and disposal.

[70] Adapted from "Guidelines for Preparation of Engineering Report to Accompany Application for Permit to Conduct a Mass Gathering in New York State," New York State Dept. of Health, Albany, June 30, 1970; and "A Resolution to License and Regulate Persons Engaged Within the Boundaries of the County, but Outside the Limits of the City, Villages and Incorporated Towns, in the Business of Providing Entertainment or Recreation, or Providing for the Lodging of Transients; adopted September 13, 1949, as amended," County Board of Supervisors of Jackson County, Ill., February 1970.

24. Bathing areas—show location and safety measures; permit from health department.
25. Lighting—illuminate public areas of site at all times with no reflection beyond boundary of site.
26. Noise control—sound level not to exceed 70 dBA at perimeter of site.
27. Parking—provide usable space at rate of 100 passenger cars per acre and 30 buses per acre located off public roadways.
28. Public liability and property damage insurance as needed and reasonable in relation to the potential risks and hazards involved.
29. Performance bond—provide security sufficient to execute plans submitted and cover reasonable contingencies.
30. Communication center—at least one for operator and one for permit-issuing official.
31. Operation and maintenance—adequate staff, equipment, facilities, and spare parts, as well as communications and security.

There is a special need for the permit-issuing official to ensure coordination and cooperation between state and local agencies. The agencies having a responsibility and role in any mass gathering include the local government, state and local police, health, fire, transportation, and civil defense. Each should clearly understand in advance its particular responsibilities and plan accordingly. One individual should be in charge and serve as liaison officer, with the permit-issuing official having overall responsibility.

Highway Rest Areas

Rest areas on superhighways may be simple or elaborate. They may include parking areas, rest rooms, water, and sheltered picnic areas with provisions for outdoor cooking, restaurant, fast-food service, tourist information, dog trails, scenic views and trails, and gasoline station. Toilet facilities may vary from privies to flush toilets with washbasins, drinking fountains, and heat. It is apparent that the water supply must be safe and adequate; the food service safe; the sewage and excreta disposal sanitary; the refuse storage and disposal proper; flies, mosquitoes, ticks, and noxious weeds controlled; and cooking spaces, tables and benches adequate.

Marinas and Marine Sanitation[71]

The Problem

Expansion of interest in tourism and recreational activities generally has involved increased use of pleasure boats and land-based boat servicing facilities—public and private marinas, yacht clubs, sleeping and eating facilities, boat launching sites, and so on. As is usual when large numbers of people congregate (whether on land or water), increased public health hazards are associated with the growing popularity of boating, sailing, and cruising. Agencies concerned

[71] Joseph A. Salvato, Jr., *Guide to Sanitation in Tourist Establishments*, WHO, 1976, pp. 113–117.

with health, water pollution control, conservation, and recreation, along with boat owners, resort managements, and those providing onshore services, therefore have the responsibility of ensuring that problems connected with boating are minimized or prevented.

Boats and ships of all kinds (recreational, commercial, industrial, and government-owned), contribute to the water pollution problem. Pollutants may be discharged not only into the water but also into the air or on land.

1. Water pollutants include gasoline, oil, and lead compounds; human wastes, galley wastes, wastewater from toilets, sinks, and showers (all considered as sewage); garbage, cans and bottles, and other solid wastes; bilge and ballast water, grease, chemicals, and accidental cargo spillages. Galley, sink, and shower wastes are also referred to as gray water.

2. Air pollutants include fuel hydrocarbons, oxides of carbon and other gases; soot and lead compounds discharged below water level; gaseous, particulate, and thermal emissions from marine-sewage incinerating devices; and unpleasant odors and noise.

3. Land pollutants include oil residues, grease, garbage, and miscellaneous debris stranded along shorelines, which not only cause economic loss but also make the area unattractive to both residents and visitors. Also included in this category are wastewater and solid wastes deposited on land where servicing facilities for watercraft are not provided.

In addition, depending whether the boating activities take place on the sea, an estuary, a river or a lake, drinking water supplies, fishing grounds, and recreation areas may be polluted and fish, including shellfish, contaminated. Serious health hazards are associated with the ingestion of contaminated drinking water and shellfish. Contact with polluted water during bathing, skin diving, and water-skiing activities should be avoided.

The pollution problem associated with water craft can be especially acute at marinas, popular fishing sites, and sheltered inlets providing attractive mooring places for weekends and holidays; at regattas and similar events; and at bathing beaches where large numbers of watercraft congregate. Since privately owned recreational watercraft in the United States are estimated to carry an average of 3.1 to 3.6 persons, the amount of sewage discharged at these places may be equivalent to that produced by a sizeable community. The magnitude of the problem is indicated by the report of the USEPA for 1973 which stated that enforcement of its standards would affect some 600,000 U.S. vessels, including about 550,000 recreational craft, as well as a large number of foreign ships using U.S. national waters.[72]

Control of Pollution from Watercraft

The control of pollution at marinas and from watercraft should be based on the general principles given in this text but regulations should also cover the following points.

1. Installation and operation of marine toilets, preferably with no discharge to the water. Secondary treatment, including the treatment of gray water should be required as a minimum if

[72] *Action for Environmental Quality*, USEPA, Washington, D.C., 1973.

permitted. Devices available for watercraft include portable toilets, recirculating toilets (manual and electrical) with an integral or separate holding tank, and incinerating toilets. Wastewater treatment systems with discharge to the water include maceration/disinfection units and physicochemical treatment systems. Approval of such devices, if given, should be for certain large vessels only and should be based on certification from a recognized national testing laboratory that there is no cross-connection with a potable water supply and that the construction, safety, and operating procedures are in accordance with the requirements of the regulating agency. Enforcement of discharge standards is difficult, if not impracticable, except when vessels are berthed at marinas or moored in confined bodies of water. Flow-through systems should be considered inadequate in the vicinity of water supply intakes, shellfish beds, fishing grounds, and bathing waters. Notices and buoys are required to mark restricted discharge areas.

2. Onshore pump-out stations for emptying boats' holding tanks and toilets and providing for the proper disposal of wastewater to a municipal or private collecting and treatment system or onshore storage. The provision of pump-out facilities should be encouraged; septic-tank and cesspool pump-out vehicles might be adapted for this purpose. Prior approval of the arrangement should be obtained from the regulating agency. Pump-out units may be stationary or portable and mechanically, manually, or electrically (coin-in-the-slot) operated. A positive displacement pump is generally used. Motors, electrical connections, and switches should be explosion-proof. Special pump-out fittings, flush deck flanges and caps, and hose connections are needed to pump out boats' holding tanks and toilets.

3. Watercraft engines that emit only minimum permissible amounts of air and water pollutants. Two-stroke outboard motors that discharge crankcase drainage, raw fuel, or other pollutants to the receiving waters should be phased out.

4. Proper disposal of bilge, grease, ballast waters, and cleaning chemicals.

5. The use only of new or thoroughly cleaned drums for the flotation of docks, floats, houseboats, and similar structures in order to prevent possible leakage of polluting chemicals.

Present-day knowledge and experience shows that the on-board holding tank that can be pumped out to a municipal or other shore-based collection and treatment system is the best method of controlling sewage and other wastewater disposal from watercraft. This concept is particularly applicable to inland fresh waters and coastal waters.

Shore-based Support Facilities

Suggested services and facilities requiring supervision may include the following (see Figure 9-20).

1. Marinas (for boat docking, mooring, storage, leasing, servicing, toilet recharge chemicals, provisioning), boat launching ramps, and pump-out facilities. Hoses for pumping out boats' holding tanks and toilets should be distinctly labelled and colored; brown is a suitable color. Special hoses are also needed for rinsing out holding tanks; they should also be clearly marked and distinctively colored, and equipped with approved backflow prevention devices. These hoses should not be used for any other purpose. Marinas may be required to obtain an annual license or permit for purposes of regulation and to ensure compliance with established standards.

2. Public toilets with a service sink for emptying portable toilets and a proper wastewater disposal system. Public toilets should include water closets and urinals, washbasins, and showers supplied with hot and cold water, and clothes washers in separate enclosures; the number of these facilities being based on the number of moorings or boat slips, docking and boat servicing facilities, restaurants, picnic areas, and estimated use.

3. Potable water supply outlets equipped with approved backflow prevention devices. Hoses should be distinctly labelled and colored; blue is a suitable color. Hose nozzles should be protected from contamination and stored in a special cabinet when not in use.

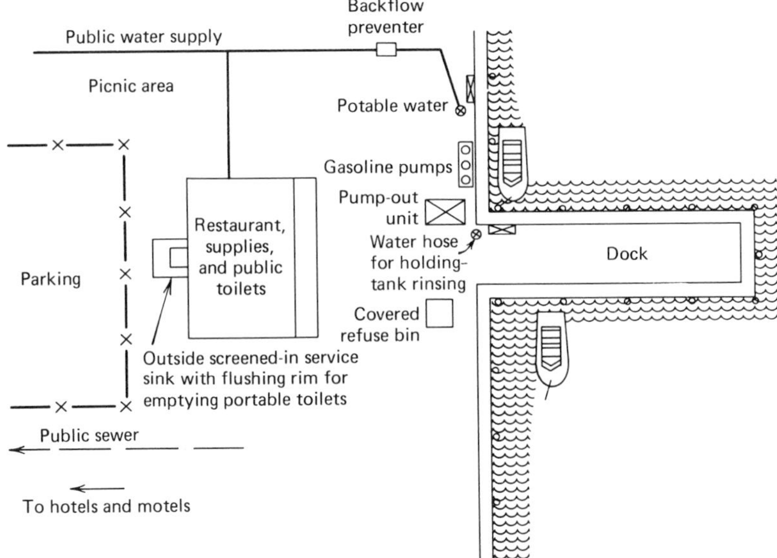

Figure 9-20 Marina with pump-out unit and other facilities. (From Joseph A. Salvato, Jr., *Guide to Sanitation in Tourist Establishments*, WHO, 1976, p. 116.)

4. Refuse collection receptacles or bins and a refuse disposal service.
5. Food service, catering, and provisioning facilities.
6. Parking areas.
7. Picnic areas and fish cleaning facilities.
8. Hotels and motels.
9. Bathing beaches.
10. Club houses and other recreational facilities.

BIBLIOGRAPHY

American Camping Association Standards, Bradford Woods, Martinsville, Ind. 46151.

Black, A. P., M. A. Keirn, J. J. Smith, G. M. Dykes, and W. E. Harlan, "The Disinfection of Swimming Pool Waters, Part II," *Am. J. Pub. Health*, April 1970, pp. 740–750.

Black, A. P., R. N. Kinman, M. A. Keirn, J. J. Smith, and W. E. Harland, "The Disinfection of Swimming Pool Waters, Part I," *Am. J. Pub. Health*, March 1970, pp. 535–544.

Environmental Health Guide for Travel Trailer Parking Areas, prepared by PHS, DHEW, Washington, D.C.; published by the Mobile Home Manufacturers Association, 6650 North Northwest Highway, Chicago, Ill., 1966.

Environmental Health Practice in Recreational Areas, PHS, DHEW, Washington, D.C., revised 1977.

Mallman, W. L., "Public Health Aspects of Beaches and Pools," *Critical Reviews in Environmental Control*, Chemical Rubber Co., Cleveland, Ohio, June 1970, pp. 221–255.

Minimum Property Standards for Swimming Pools, FHA 4516.1, U.S. Dept. of Housing and Urban Development, Washington, D.C., June 1970.

Public Swimming Pools, Recommended Regulations for Design and Construction, Operation and Maintenance, American Public Health Association, 1015 Fifteenth St., N.W., Washington, D.C. 20005, 1981.

Recommended Standards for Bathing Beaches, Great Lakes-Upper Mississippi River Board of State Sanitary Engineers, 1975 ed., Health Education Service, P.O. Box 7283, Albany, N.Y. 12224.

Recommended Standards for Swimming Pools, Great Lakes-Upper Mississippi River Board of State Sanitary Engineers, 1977 ed., Health Education Service, P.O. Box 7283, Albany, N.Y. 12224.

Salomon, Julian Harris, *Camp Site Development*, Girl Scouts of America, 830 Third Avenue, New York, 1959.

Salvato, J. A., *Guide to Sanitation In Tourist Establishments*, WHO, Geneva, 1976.

Swimming Pool Operation, Eng. Bull. No. 18, Michigan Dept of Health, Lansing, 1963.

Swimming Pools, DHEW, Atlanta, Ga., June 1976.

Swimming Pools and Bathing Beaches, New York State Dept. of Health, Albany, N.Y., June 1972.

Van Dusen, Karen and Gary Fraser, *Swimming Pool Program Study*, Department of Social and Health Services, Olympia, Washington, October 1977.

Youth Camp Safety & Health, Suggested State Statute & Regulations, PHS, DHEW, CDC, Atlanta, Georgia 30333, Revised August 1977.

10

VECTOR AND WEED CONTROL AND PESTICIDE USE

As a general principle, the control of arthropods, rodents, and weeds where needed should take into consideration the life cycle and the conditions favorable to growth. For example, in the development of a multipurpose reservoir or recreation area, the results of entomological, biological, and engineering studies should identify potential mosquito, aquatic weed, schistosome, or other related problems and the best means for their prevention or control. *The full potential of naturalistic and source-reduction measures should be applied before considering chemical means.*

In the control of houseflies, roaches, and rats, the primary emphasis should be on basic community environmental sanitation to eliminate the conditions that make possible their survival and reproduction. After this has been conscientiously done, and if the problem has not been brought under adequate control, then the judicious use of the proper pesticide may be considered.

There are situations where chemo-sterilants, synthetic attractants, synthetic repellants, parasites and predators (staphylinid beetle to attack the pineapple souring beetle and the control of the banana skipper by an insect from Thailand)[1], release of sterile male flies (screwworm fly from pupae treated with gamma radiation), and quarantine have application. The sterile male technique, which eradicated screwworm flies, a major livestock pest, in the southeastern United States, is used to control reinfestation in the southwestern United States and Mexico and is also being adapted to control the tsetse fly of Africa.[2]

PESTICIDES

Pesticide Use 

There has been much public pressure to eliminate the use of persistent pesticides because of their accumulation in the food chain in concentrations sufficient

[1]Press Release, Honolulu (Associated Press) by Cile Sinnex, April 29, 1976.
[2]Deborah Takiff Smith, "Challenging the Tsetse Fly," *War on Hunger*, Washington, D.C., April 1977.

to reduce reproduction and to affect the offspring. Lake trout, robins, bald eagles, and osprey are examples of wildlife affected. This has led to controls in the United States, Great Britain, Sweden, Canada, the European Economic Community, and other countries to restrict the use of certain pesticides and completely ban others. An exception may be made to permit DDT and other chemicals needed for emergencies and for the prevention or control of human, plant, and animal diseases and other essential uses for which no alternative is available. The WHO considers DDT irreplaceable in public health at the present time for the control of some of the most important vectorborne diseases of man. In addition, DDT is considered to be essential in the eradication of malaria and in the control of trypanosomiasis, onchocerciasis, louseborne typhus, and bubonic plague, although resistence to the insecticide has been reported in 62 of the 107 nations having malaria.[3] When a complete ban is imposed, there is a danger of promoting the use of a pesticide that has a higher acute toxicity, with a greater direct risk to man, wildlife, and livestock. In such instances a persistent pesticide that does not leave undesirable residues on a crop or elsewhere in the environment or cause concentration to harmful levels through the food chains may be preferable. Table 10-1 indicates the relative oral and dermal toxicity of selected pesticides. The persistence of pesticides in soils is discussed in Chapter 5 under Hazardous Wastes, Treatment and Disposal.

The decision to use any pesticide (or any control technique), whether persistent or not, should weigh the expected good against the undesirable effects. This requires an environmental assessment and comprehensive knowledge. Some of the factors to consider are:[4]

1. The development of strains of pests that are resistant to pesticides.
2. The need for repeat treatments.
3. The possible flourishing of other pests because their natural enemies have been destroyed.
4. The undesirable side effects on fish, birds, and other wildlife; on parasites and predators; on honey bees, and other beneficial insects; on man and domestic animals; and on food crops.
5. The hazard of residual pesticides on food crops.
6. The direct hazard to the workers involved in pesticide application.
7. The effect of reducing and simplifying the biota; the advantages of preserving a diverse stable and natural environment as compared to controlled manipulation of the environment.

The objective in pest control should be integrated pest management involving the use of a combination of educational, cultural, biological, physical, chemical, and legal measures, as appropriate. For example, well established cultural control measures which seem to be forgotten include crop rotation to control nematodes, certain insects, and diseases in the soil; planting late or harvesting early to avoid maturing insects; and maintenance of vegetated strips to minimize

[3]Jim Toedtman, "World Agencies Battle Rise in Malaria," *Los Angeles Times*, Oct. 15, 1978, p. 2, Part 1-A.

[4]Ray F. Smith, "Pesticides: Their Use and Limitations in Pest Management," talk presented at a research and training conference on "Concepts of Pest Management," March 27–29, 1970, in Raleigh, N.C. The conference was sponsored by the North Carolina State University and the Entomological Society of America.

Table 10-1 Acute Oral and Dermal LD^{50} Toxicity Values of Selected Pesticides to Female Rats (mg/kg.)[a]

Chemical Class	Pesticide	Oral	Dermal
Organochlorine	chlordane	430	690
	DDT	118	2,510
	dieldrin	46	60
	lindane	91	900
	methoxychlor	6,000[b]	>6,000
Organophosphorus	Abate	13,000	>4,000
	Diazinon	285	455
	dichlorvos (DDVP, Vapona)	56	75
	dimethoate (cygon)	245	610
	chlorpyrifos (Dursban)	82	202[b]
	fenthion (Baytex, Entex)	245	330
	rabon (Gardona)	1,125	>4,000
	malathion	1,000	>4,444
	naled (Dibrom)	250[c]	800[c]
	parathion (ethyl)	3.6	6.8
	parathion (methyl)	24	67
Carbamate	propoxur (Baygon)	86	>2,400
	carbaryl (Sevin)	500	>4,000
Arsenical	Paris green	100	2,400
Botanical	pyrethrum	263[d]	—
Miscellaneous	diphacinone	1.9[d]	—
	pindone	280[d]	—
	warfarin	3[d]	—

Source: "Public Health Pesticides," National Communicable Disease Center, PHS, DHEW, Savannah, Ga., 1970. See also "Insecticides for the Control of Insects of Public Health Importance," Harry D. Pratt and Kent S. Littig, PHS, DHEW, CDC, Atlanta, Ga. 30333, 1974, p. 51.

[a]All figures taken from tests performed by the Atlanta Toxicology Branch, Division of Pesticides, FDA.
[b]Sex not specified.
[c]Date for male specimens.
[d]Unpublished data.

migration of insects to crop fields. The objective is to achieve the best results with the least amount of a selective chemical poison and a minimum hazard to man and wildlife. Included is the use of pest parasites, pathogens, sex attractants (pheromones), predators, and resistant crops. One should carefully consider the possible spread of chemical contamination via the air, water, food, and soil, the household, transportation and handling, accidents, and occupational exposure.

It is extremely important to store pesticides in a secure place, to use them properly in accordance with the directions on the containers, and to dispose of the containers or leftover pesticides safely. See Chapter 5.

If control cannot be achieved without the use of chemicals, a chemical should be selected that is specific and has the least effect on nontarget species. Only the area affected should be treated, and care should be taken to avoid streams and lakes. The treatment should be timed to coincide with the presence of the pest, and only the amount of chemical needed and type needed to achieve the result in the least time should be used. Select the formulation that will limit the range of the chemical, that is, granules for concentrated toxic action, emulsions within bodies of water, and solutions or suspensions in oil on surfaces. Consider the use of repellants or noisemakers to move nontarget species from the area during the effective period of the chemical.[5]

Legal Controls

The distribution and use of pesticides are controlled by federal laws. In many instances, state and local laws and regulations parallel the federal laws.

The Toxic Substance Control Act (TSCA) gives the USEPA authority to obtain data and regulate existing and new potentially harmful chemicals. The USEPA can require that industry provide information on the production, distribution, and use of a new chemical and its health and environmental effects. Industry must notify the USEPA and give support information 90 days before a new chemical is manufactured. In addition, records must be kept of significant adverse health or environmental effects. The TSCA therefore is intended to be a preventive measure; but the task involved is tremendous when about 1000 new chemicals are produced each year. For example, the USEPA published an initial inventory of about 43,500 which are exempt as of June 1979 from the premanufacturing notification.

The Federal Insecticide, Fungicide, and Rodenticide Act, as amended, requires a USEPA registration number on every pesticide product. The number is the public's assurance that the product has been reviewed by the USEPA and may be safe and effective when used as directed on the label. The act also provides for registration of pesticides manufacturing and formulating, and for state certification of applicators who are qualified to use certain restricted pesticides. A "Restricted Classification" pesticide is highly toxic or requires special knowledge or equipment for application and, therefore, will generally not be available to the homeowner. A "General Classification" pesticide is for use by the general public. However, the certification examinations for applicators vary from state to state calling for competence evaluation and national minimum standards plus continual education of applicators.

For a pesticide to be permitted for application on a raw agricultural food or feed, the residue cannot exceed the tolerance established for the product by the DHEW (now Health and Human Services) under the conditions of use. A product exceeding the tolerance may be considered adulterated. In addition, a pesticide added intentionally or incidentally to a processed food is considered an additive under the Food, Drug, and Cosmetic Act, Food Additive Regulations, which require, among other conditions, that no additive be used that can cause disease, such as cancer, under the conditions of use. These regulations, including those relating to adulterated and misbranded food, are enforced by the

[5] J. Henry Wills, "Pesticides—A Balanced View," *Health News*, Vol. 44, New York State Dept. of Health, Albany, April 1967.

FDA, except that the USDA enforces the law and regulations for meat and poultry products.

The USEPA has cancelled for most uses or cancelled registration for the following pesticides:

Aldrin	Endrin
Bandane	Heptachlor
BHC	Mercury compounds
Chlordane	Selenites and selenates
DDD, TDE	Strobane
DDT	Thallium
Dieldrin	Toxaphene
Silvex*	2, 4, 5-T*

Certain pesticides are permitted only after review of the specific use proposed; only for specific nursery, structural, or agricultural purposes and concentrations; or for use by certified applicators. However, any restricted pesticide may be used to cope with a public health emergency or to enforce a federal or state quarantine.

In New York State, for example, the purchase, distribution, sale, use, and possession of many pesticides is restricted.[6] Certain pesticides may be used, only upon issuance of a commercial or purchase permit, as listed on the approved label as registered with the New York State Department of Environmental Conservation.

Pesticide Groupings

A pesticide may be defined as any substance or mixture of substances intended for eliminating or reducing local populations of, or for preventing or decreasing nuisance from, any insect, rodent, fungus, weed, or other form of plant or animal life or viruses, except viruses, microorganisms, or other parasites on or in living man or animals; and any substance or mixture intended to use as a regulator of plant growth and development, a defoliant or a desiccant, but not materials intended primarily for use as plant nutrients, trace elements, nutritional chemicals, plant inoculants and soil amendments.

A grouping of pesticides is given below for convenient reference.[7] There are thousands of formulations. Repellants, attractants, impregnants, and auxiliaries are also mentioned.†

Inorganic contact poisons

Sulfur group—sulfur, lime-sulfur

Mercury group—calomel, corrosive sublimate

*Registration suspended February 28, 1979.

†State or local regulations may impose certain restrictions on the use of these compounds; therefore the individual should consult local or state authorities on the accepted-use practice.

[6]*Circular 364, Part 326—Rules and Regulations Relating to Restricted Pesticides*, New York State Dept. of Environmental Conservation, Albany, N.Y., revised 1973.

[7]Harry D. Pratt and Kent S. Littig, *Insecticides for the Control of Insects of Public Health Importance*, DHEW, PHS, CDC, Atlanta, Ga. 30333, Sept. 1977, DHEW Pub. (CDC) 77-8229.

Organic contact poisons
  Synthetic
    Nonsubstituted hydrocarbons—formaldehyde, ferbam, naphthalene, benzly-benzoate, xanthone
    Chlorinated hydrocarbons—pentachlorophenol, BHC, lindane, methoxychlor, DDT, DDD, heptachlor, chlordane, aldrin, dieldrin, toxaphene, endrin, dilan, dimite
    Organic phosphorus group—malathion, diazinon, parathion, TEPP, EPN, HETP, DDVP (dichlorvos),* systox, potasan, schradan, ronnel, dicapthon, dimethoate
    Organic nitrogen group—dinitrophenol, DNOC, diphenylamine, prolan, dinitronaphthol
    Organic sulfur group—aramite, sulfenone, pyrolan, sevin, isolan, ovex
    Organic thiocyanate group—lethane, thanite, loro
  Natural
    Petroleum—fuel oil, kerosene
    Alkaloids—nicotine, anabasine, hellebore, ryania
    Esters—pyrethrum, pyrethrins, allethrin, scarbrin, rhododendron, mammein
    Rotenoids—rotenone, elliptone, sumatrol, malaccol, degulin, milletia
    Resins—turpentine, sage oil, basil oil, croton oil, amorphin, phellodendrin
Stomach poisons
  Arsenic group—arsenic trioxide, lead arsenate, copper arsenate, Paris green
  Fluorine group—sodium fluoride, 1080, cryolite, sodium fluosilicate
  Other groups—phosphorus paste, borax, tartar emetic, thalium sulfate, SALP
Fumigants
  Naphthalene, dichlorobenzene
  Hydrogen cyanide, methyl bromide, carbon tetrachloride, ethide, sulfure dioxide
Repellents
  Diethyltoluamide, indalone, Rutgers 612, dimethylphthalate, dibutylphthalate
Attractant
  Geranoil, eugenol, ammonia, saccharine, anethol
Impregnants
  Geranoil, eugenol, ammonia, saccharine, anethol
Impregnants
  Benzil, benzlbenzoate, eulan, M-1960, benzocaine, mitin, eulan, boconizer
Auxiliaries
  Synergists, inhibitors, solvents, emulsifiers, wetting agents, spreaders, perfumes, dilutents.

THE HOUSEFLY

The housefly is the most abundant and familiar nonbiting insect about human habitations during the warm months of the year. The female lays its eggs, as many as 2700 in 30 days, in almost any moist, rotting, or fermenting material within 4 to 12 days after it is full grown. Horse, hog, chicken, and cow manure, human excrement and sewage, animal runs, pet droppings, garbage, poorly digested sewage sludge and scum, rotting grass clippings, fruits and vegetables, decaying grains, soiled rags, and papers serve as good places for eggs to mature. In warm weather, the larvae, or maggots, hatch out from the eggs in 10 to 24 hr, become pupae in about 4 to 10 days, and adult insects in 3 to 6 more. Cold or lack of food and adverse moisture conditions will prolong this cycle from an average of about 12 to 30 days. See Figure 10-1.

*Vapor dispensers of these pesticides should not be used where food is prepared or in the home, institutions, or other continuously occupied spaces.

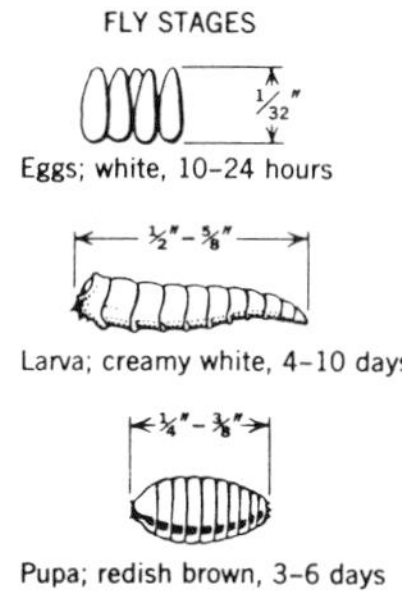

Figure 10-1 Fly stages.

Although warmth, moisture, and suitable food are required for growing maggots, relatively cool and dry materials are benefical during the pupal stage. The optimum temperature for breeding is 80 to 90° F. Temperatures of 115° F or above will kill eggs and larvae; below 44° F the fly is inactive. Larvae and emerging adults crawl through loose debris toward light. The full-grown flies are attracted by odors. Although "tagged" houseflies have been found 7 to 20 mi from breeding places, a significant number travel only 1 to 2 mi. Blowflies have been found 15 to 28 mi from breeding places, but are not numerous beyond 3 to 4 mi away.[8] Flies are most numerous in the late summer and early fall. In warm climates and in heated buildings, breeding can take place throughout the year.

Spread of Diseases

The fly has a hairy body and sticky padded feet. A single fly may carry as many as 6,500,000 microorganisms. Bacteria may be carried in the digestive system of the fly for as long as 4 weeks; the bacteria can be transmitted to succeeding generations. The fly's instinct for survival and attraction to odors leads it to open privies, garbage dumps, manure, overflowing sewage-disposal units, or garbage cans, and to grossly sewage-soiled banks of streams in search of the partially digested or decaying food that is to its liking. In feeding on the filthy material, the fly covers its legs, body, and wings with germs. After the main course, rich in broken-down proteins, the fly is ready for its dessert. For this course the fly may alight on a milk carton, cup, pudding, fruit, cake, bread, baby's bottle, face, or mouth. Since the fly is on a liquid diet, it is necessary for it to transform the cake frosting or pudding, for example, into a liquid. It is well-equipped to do this by regurgitating some of the liquids already swallowed, such as sewage and microbe-laden saliva, until the relatively solid material is softened sufficiently to be swallowed. In so doing, only part of the regurgitated liquid and saliva can be taken in again. Hence the fly leaves behind part of its vomit, germs from its legs, and incidentally, its excretions. If undisturbed, the vomit and feces dry to form the familiar black specks frequently seen on some restaurant kitchen walls.

The housefly can be a significant agent or mechanical carrier in the spread of such diseases as typhoid fever, paratyphoid and other salmonella infections,

[8]Herbert F. Schoof, "How Far Do Flies Fly?," *Pest Control* (April 1959).

bacillary and amebic dysentery, cholera, anthrax, diarrhea or gastroenteritis, conjunctivitis, trachoma, possibly pulmonary tuberculosis, and poliomyelitis. The role played by houseflies in the transmission of many disease microorganisms and parasitic worms has been demonstrated in the laboratory and by epidemiological circumstantial evidence. A community-wide fly-control program in Hidalgo County, Texas, resulted in a significant reduction in the incidence of shigellosis and salmonellosis in the population.

Fly Control

Fly control requires the elimination of fly breeding places, destruction of larvae and adults, and making food inaccessible. The application of good sanitary practices in an area throughout the year will be most effective in the prevention of fly breeding and in making a control program successful. Fly population surveys are sometimes made to give an indication of the problem and as a means to evaluate the effectiveness of control measures. This can be accomplished by means of fly-trap surveys using baited traps, flypaper strips and cone traps, or fly grills. Fly grills give quick and valid results. See Figure 10-2.

The elimination of fly breeding depends primarily on environmental sanitation and secondarily on the proper use of insecticides, poisoned baits, electric grids, and traps. This involves cleanliness, good housekeeping, and the application of recommended sanitary control measures. Garbage should be wrapped in

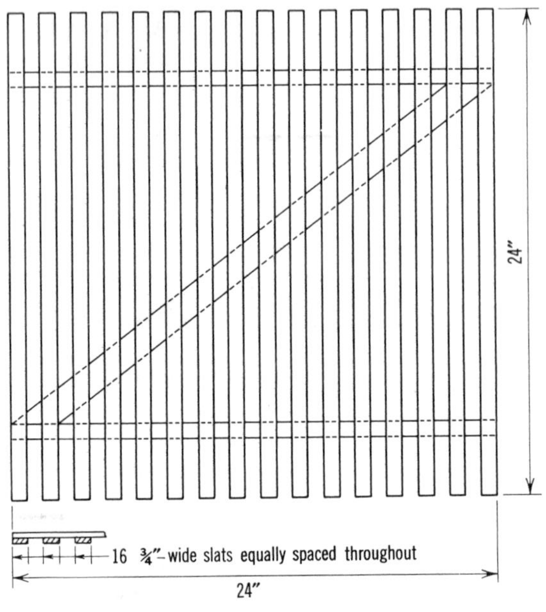

Figure 10-2 Fly-counting grill.

several layers of newspaper and placed in covered, washable, and cleaned containers. The receptacles should be stored in a cool or airy place and be of sufficient number and size to prevent spillage. In a warm climate, or if freezing weather is expected, a separate room kept at a temperature of 40 to 45° F should be provided at commercial establishments. A room with concrete or equal sidewalls and floor, equipped with a floor drain and hot water, promotes cleanliness. Soiled garbage cans and bulk garbage storage containers should be cleansed with hot water and a cleaning compound such as lye to keep down odors and prevent fly breeding. Soiling of the garbage can and bulk container will be reduced by wrapping drained garbage in paper or by using a special water-resistant lining. Spilled garbage and indiscriminately discarded food or trash should of course be promptly collected and the area cleaned.

Garbage and trash may be completely burned in a properly designed incinerator or disposed of in a sanitary fill as explained in Chapter 5. Garbage grinders in kitchen sinks and as part of the dish-scraping tables in restaurants are highly recommended where local conditions permit. However a rich source of food for rats, roaches, or other vermin is provided if they can gain entrance to sewers.

Privies, to be sanitary, must be constructed so that flies and rodents cannot gain access to the pit. Sewage and wastewater should be discharged into a sewerage system so that exposure of inadequately treated sewage on the surface of the ground or into streams is prevented. Incompletely digested sludge and septic-tank or cesspool cleanings are usually required to be buried at least 200 to 500 ft from any watercourse, swamp, or lake in trenches, at least 2 ft above seasonal high groundwater, with at least 2 ft of earth cover. Thorough spraying with kerosene, fuel oil, or other larvicide or dusting with borax before and while covering will make the breeding of flies less likely.

Although the elimination of fly breeding and destruction of fly larvae will reduce greatly the number of adult flies, it is not usually possible to eliminate flies completely. Those flies that do develop or come from adjoining places can be killed on alighting by spraying an approved residual insecticide on screens, walls, and resting places just before the fly season. In addition to this residual spray, all buildings, particularly kitchens, nurseries, infirmaries, and those places where food is prepared or served, should be completely screened. Do not spray food, utensils, pets, stoves, or open fires. Fly-repellant-type fans and air curtains may be used in place of screening. Screen doors should open outward so as not to drive flies into a room when entering.

Insecticides suitable for use in fly control of food-processing plants, dairies, poultry houses, and for outdoor space spraying are summarized in Table 10-2. Pyrethrin, 0.25 percent plus 1 to 2 percent of synergist, is a good space spray. Insecticide bait may be used in such places as unscreened farm buildings, except milk rooms, and in poultry houses. Paint-on bait gives good control in animal pens.

Filth encourages the presence of flies. Complete dependence for killing flies cannot be placed on insecticides alone, as strains of houseflies will develop that are resistant to insecticides. Careful attention to the basic sanitation principles outlined here and good housekeeping will ensure success in controlling these disease-carrying insect pests. Store insecticides in their original containers out of the kitchen and where they cannot be mistaken for flour, baking powder, syrup, or other foods. Use with care.

Houseflies that develop in landfills can penetrate and emerge through five ft of uncompacted cover or through nearly six in. of compacted cover.[9] This emphasizes the importance of prompt collection, disposal, and thorough compaction of solid waste as it is being deposited, followed by a daily earth cover which is compacted.

MOSQUITO CONTROL

The most important diseases transmitted by mosquitoes are dengue or breakbone fever, encephalitis, filariasis, malaria, Rift Vally fever, and yellow fever. Only malaria and the encephalitis diseases are found in the United States, although malaria is under control. A summary of these diseases including etiologic agent, reservoir, transmission, incubation period, and control is given in Table 1-10.

In many parts of the world mosquitoborne diseases are a major cause of debility and death. In others the mosquito is a pest, cause for considerable annoyance and discomfort.

Malaria

The WHO reported that new confirmed malaria cases increased from 3,252,000 in 1972 to 7,517,000 in 1976. China, Vietnam, and Africa are excluded in these totals. About one million people, mostly children, die in Africa

[9] *Comprehensive Studies of Solid Waste Management*, PHS, DHEW, Bureau of Solid Waste Management Pub. 2039, First and Second Annual Reports, 1970, p. 89, GPO, Washington, D.C. 20402.

Table 10-2 Organophosphorus Insecticides Used as Baits, Space Sprays, and Larvicides in Fly Control.[a]

Type Application	Toxicant	Formulation	Remarks
Bait	diazinon	1# 25% WP plus 24# sugar; 2 fl oz 25% EC plus 3# sugar in 3 gal of water.	Normal application is 3 to 4 oz (dry) or 1 to 3 gal (wet) per 1000 ft^2 in areas of high fly concentration. Repeat 1 to 6 times per week as required. Avoid application of bait to dirt or litter.
	dichlorvos (DDVP, Vapona)	3 to 6 fl oz 10% EC plus 3# sugar in 3 gal water.	The use of permanent bait stations will prolong the efficacy of each treatment.
	malathion	2# 25% WP plus 23# sugar.	Available as commercial baits labeled for use in

Table 10-2 *(Continued)*

Type Application	Toxicant	Formulation	Remarks
	naled (Dibrom)	1.0 fl oz 50% EC plus 25# sugar in 2.5 gal water.	dairies and in food processing plants[b] except at sites where food is exposed. None of these baits should be employed inside homes nor should diazinon and trichlorfon be used in poultry houses. Diazinon also is not to be used in dairy barns. *Do not contaminate feed or watering troughs.*
	ronnel (Korlan)	2 pt 25% EC plus 3# sugar in 3 gal water.	
	trichlorfon (Dipterex)	1# 50% SP plus 4# sugar in 4 gal water.	
Outdoor Space Spray[c]	diazinon[d] fenthion	11 gal 25% EC in 34 gal water.	Application rate is 15 gal per mi.
	dichlorvos	6 gal 50% EC in 44 gal water.	Application rate is 15 gal per mi.
	dimethoate[d]	3 or 6 gal 50% EC in 50 gal water.	Application rate is 20 or 10 gal per mi.
	malathion	5 gal 55% EC in 41 gal water.	Application rate is 20 gal per mi.
	naled	1.5 gal 65% EC in 50 gal water.	Application rate is 15 or 20 gal per mi.
Larvicide	diazinon	1 fl oz 25% EC to 1 gal of water.	Application rates are 7 to 14 gal per 1000 ft^2 as a coarse spray. Repeat as necessary, usually every 10 days or less. For chicken droppings, use only where birds are caged. Diazinon is not labeled for use in poultry houses. *Avoid contamination of feed or water and drift of spray on animals.*
	dichlorvos	2 fl oz 10% EC to 1 gal of water.	
	dimethoate	0.5 pt 43% EC to 2.5 gal of water.	
	malathion	5 fl oz 55% EC to 3 gal of water.	
	ronnel	1 pt 25% EC to 3 gal of water.	

Source: "Public Health Pesticides," CDC, PHS, DHEW, Savannah, Ga., 31402, 1973. For guidance only. Consult with state and local authorities on restrictions in effect.

Notes: State regulations may impose certain restrictions on the use of these toxicants in dairies or at other specified sites; therefore, the individual should be certain that usage conforms with local restrictions. EC = emulsifiable concentrate. WP = wettable powder. SP = soluble powder.

[a]For information on chemicals to be used against livestock and crop pests and for their residue tolerances on crops, consult your state agricultural experiment station or extension service.

[b]Includes dairies, milk rooms, restaurants, canneries, food stores, and warehouses, and similar establishments.

[c]Based on swath width of 200 ft.

[d]Not specifically labeled for outdoor space applications.

alone each year. Antimalaria programs have been losing their effectiveness in many countries because of inadequate funding, adaptation of mosquitoes to new environments, resistance to certain insecticides, resistance of *Plasmodium falciparum* parasites to drugs used to prevent and treat malarial infections, inadequate research, and many other factors.[10,11] If an intensive program of source elimination, chemical treatment, disease detection and control, and education is interrupted *before eradication is accomplished*, or if surveillance and control of treated areas is abandoned, the mosquito vector and the disease will soon (within 5 years) become reestablished. Sterilization of the male anopheles mosquito with a 99 percent population reduction was not enough to prevent reestablishment of a normal population[12] within 3 to 4 years. On the other hand, the Agency for International Development (AID) reports that between 1960 and 1980 malaria has been eliminated in 12 countries and territories of the Western Hemisphere. In all, malaria was eradicated in 37 countries, out of 143 countries or areas where malaria was originally endemic, removing the risk in 500 million people.[13]

Areas where malaria is known to exist include parts of Mexico, Haiti, Central America, South America, Africa, the Middle East, the Indian subcontinent, Southeast Asia, Korea, Indonesia, and Oceania.[14]

The disease is spread when certain species of the female anopheles mosquito bite a person or animal infected with malaria parasites. The parasites go through changes in the body of the female mosquito until its cycle is completed. Then, when the mosquito bites a susceptible person in good health, the parasites are injected to develop the disease in the individual, who in turn becomes an additional reservoir of the disease until cured. It therefore becomes important in areas where the disease is prevalent to prevent the mosquito from gaining access to man.

The risk of acquiring malaria when traveling in malarious areas can be reduced by spraying living quarters with a suitable insecticide and by remaining in well screened areas between dusk and dawn—the feeding time of the anopheles mosquitoes. Wearing clothing that covers the arms and legs and the application of mosquito repellant to exposed areas of the skin provides further protection. Travelers to malarious areas should obtain specific recommendations for malaria chemoprophylaxis from their state department of health or the Centers for Disease Control, U.S. Dept. of Health and Human Services, PHS, Atlanta, Ga.

Encephalitis

Encephalitis has replaced malaria as the major mosquitoborne disease in North America. Wild mammals and/or birds are reservoirs; humans, horses, or

[10]Joel E. Cohen, "Malaria—A Moving Target," *New York Times*, July 1, 1979.
[11]"Constraint to Malaria Control," *WHO Tech. Rep. Ser.*, **640,** 1979.
[12]"Malaria Prevention. Experiment in El Salvador, Ca.," *New York Times*, January 4, 1976.
[13]Susan Super, "The fight against malaria isn't over," *Agenda*, AID, Washington, D.C., January/February, 1980, pp. 10–13. Also "Constraint to Malaria Control," WHO Expert Committee on Malaria: seventeenth report, *WHO Tech. Rep. Ser.*, **640,** 1979, pp. 10–11.
[14]*MMWR*, CDC, Atlanta, Ga., March 10, 1978, pp. 81–90.

pheasants are susceptible hosts.[15] The most important vectors are probably *Culiseta melanura* between birds, and certain aedes species between birds or animals and man for eastern equine encephalitis; *Culex tarsalis* for western equine; *Culex pipiens, Culex tarsalis* and others for St. Louis equine; and *Aedes canadiens* and *A. triseriatus* for California equine encephalitis. The California virus is transmitted from mammal to man. Equine encephalitis is also found from Japan to the Philippines, in eastern Asia, parts of Australia and New Guinea, Central and South America, and the Caribbean Islands.[16] The disease is not communicable directly from man to man, but only by the bite of infected mosquitoes. Horses, as is apparent from the name, develop the disease.

Preventive and control measures are similar to those for malaria. Immediate fogging and spraying from aircraft, with an acceptable and effective insecticide, has been found successful to promptly bring mosquitoes under control when surveillance network findings show high adult and larval abundance. Control should start with the elimination of breeding sites.

Life Cycle and Characteristics

Mosquitoes breed in water. Weeds, tall grass, and shrubbery offer harborage to the adults. Eliminate standing water and the mosquito will be prevented from reproducing. Actually this is very difficult to accomplish because the mosquito will breed on the edges of ponds and streams, and in salt marshes, depressions that collect water, improperly designed and maintained irrigation and drainage ditches, holes in stumps or trees, open cesspools, open septic tanks, overflowing sewage, water barrels, street catch basins, clogged eaves troughs, empty tin cans and jars, and similar places.

Mosquito eggs hatch into the larva or wiggler stage in a matter of hours to 2 days. After 6 to 8 days of growth they enter the pupal or tumbler stage, from which the adult mosquito emerges in about 2 days.

The female mosquito feeds only on the blood of man and animals. The male mosquito feeds on the nectar from flowers. If mosquitos are very numerous the chances are they are breeding close by, although they may travel several thousand yards in search of a meal. Of course airplanes, trains, boats, and automobiles can and do carry mosquitoes long distances.

Mosquitoes have different biological characteristics and breeding habits. The *Aedes aegpti*, for example, have a flight range of less than ½ mi, whereas *Aedes sollicitans* and *taeniorhynchus* may range for many miles. Eggs of some aedes laid in damp or even dry soil are known to survive for months without water and then hatch out when flooded. Ponding of water after floods or where drainage is poor provides good breeding places for aedes and culex mosquitoes.

[15]E. M. Bosler, "Arbovirus and Pest Mosquitoes," *Proceedings of the Seminar on Environmental Pests and Disease Vector Control*, Karl Westphal, Ed., New York State Dept. of Health, Albany, N.Y., January 20–24, 1975, pp. 105–115.

[16]*Control of Communicable Diseases in Man*, Abram S. Benenson, Ed., American Public Health Association, Washington, D.C., 1975, pp. 22–25.

Polluted waters favor the breeding of *Culex pipens*. Because of variations such as these, control measures should be based on the characteristics and life cycle of the particular mosquito species identified in the area. See Figure 10-3.

Control Program

A mosquito-control program should emphasize sanitation, education, *environmental and engineering control* (physical and biological), and should start with a sanitary survey. This would locate and map existing and potential breeding places, permit the collection and identification of larvae and adult mosquitoes, and make possible appraisal of the problem and planning of a control program. Light traps at key locations make possible collection of adults and an estimation of mosquito density. Supplemental surveys and surveillance are required throughout the year to determine sites suitable for early spring mosquito breeding, likely sites favorable for later breeding, and discovery of mosquito species not previously identified, that might require special control. Available reports and studies by local universities and state and county agencies may also identify and locate species of mosquitoes present, concentration indices, and disease reported in the past. Continual surveillance of the various vector species and evaluation of results should provide the basis for program direction and control. Water management is an extremely important aspect of mosquito control.

The field surveys will identify the problem areas that require primary attention. A plan can then be developed for the timing and application of the most appropriate control measure, that is, source reduction, biological, naturalistic, or chemical (last) methods. A combination of methods may be the best in some instances.

Malaria control would include systematic distribution of drugs with special attention to high-risk groups, while trying to obtain its eradication.

Physical source reduction is simplified by the use of topographical maps. Survey data spotted on the map will show the relative location of problem areas and their relationships to habitation, recreation areas, and proposed new realty subdivisions or other developments. The need for and feasibility of tide gates, draining, diking, and pumping; filling, surface grading, and diversion of water; the relocation of streams and ditches; subsoil drainage; and brush clearing becomes readily apparent by studying the map and making field investigations.

Drainage involves the making of a survey to determine the breeding places, preparation of a detail map, determination of volume of water runoff, and type, cross-sectional area, and slope of main ditches. Main ditches should be staked out to approximate line and grade. Laterals may be constructed in the field by eye or line level as needed. Some typical ditch and covered drain cross sections are illustrated in Figure 10-4. It should be remembered that drainage plans, excavations, and construction must include regular maintenance to remain effective and keep the cost of mosquito control within reasonable limits.

When a pond, lake, or reservoir is involved, the fluctuations in water level and resultant shallow areas should be located. Places conducive to mosquito breeding should be cleared, deepened where possible, and exposed to wave action. Areas subject to seasonal flooding or water backup should be graded to stone-lined channels draining to the pond, lake, or reservoir. Reservoir water management including a schedule for spring surcharge, rapid drawdown to strand floating material, and gradual withdrawl throughout the summer, and a minimum clean 2-ft shore water depth will minimize mosquito breeding.[17]

Where the filling of breeding places is possible, and the groundwater and wildlife are not endangered, a sanitary landfill can combine source elimination and economical solid waste disposal with land reclamation. Hydraulic fill can also be used to deepen channels and fill low areas.

Biological measures including the stocking of ponds with *Gambusia affinis*, the guppy, (poecillia), goldfish, or other "top feeders" that feed on larvae. The removal or encouragement of shade, depending on the specie of mosquito to be controlled, is also useful. For tidal marshes the stocking of ditches with the killifish, a saltwater minnow, and maintenance of water circulation are effective against larvae.

Naturalistic measures include rendering the water unsuitable for mosquito breeding by changing the chemical or physical characteristics of the water. For example, acid waters, a changing salinity of water, muddying water, changing water level, and agitation of water surface do not favor the growth of mosquito larvae.

Chemical control should be reserved for situations in which the other measures are not feasible. One should keep in mind the need for repeat treatment, the cost, and the limitations and hazards of pesticide use. Pesticides used in mosquito control are shown in Table 10-3.

Preflood or prehatch larvicide treatment of areas identified in surveys known to produce an early generation of mosquitoes should be treated late in winter or early spring if the site cannot be eliminated. The control of mosquitoes in the larval stage should receive a high priority.

For larvicide spraying of small water surfaces, a compression garden-type sprayer is adequate. A power-spray unit of the orchard type is preferable for larger water surfaces. Space spraying is accomplished by hand or power sprayers, fogging apparatus, smoke or aerosol generators, and airplanes. In marshes airboats, helicopters, and special tractors and vehicles are used.

Other larvicides, in addition to kerosene, No. 2 fuel oil, or Paris green, include methoxychlor, borax, and soap emulsions of pyrethrum extract in kerosene. Paris green (copper-aceto-arsenite) is especially suited for control of malaria-carrying species. Borax at the rate of 2 oz/gal of water in fire buckets will prevent

[17]Richard O. Hayes, et. al., "Lewis and Clark Lake Mosquito Control Recommendations," *J. Environ. Eng. Div., ASCE,* August 1978, pp. 701–716.

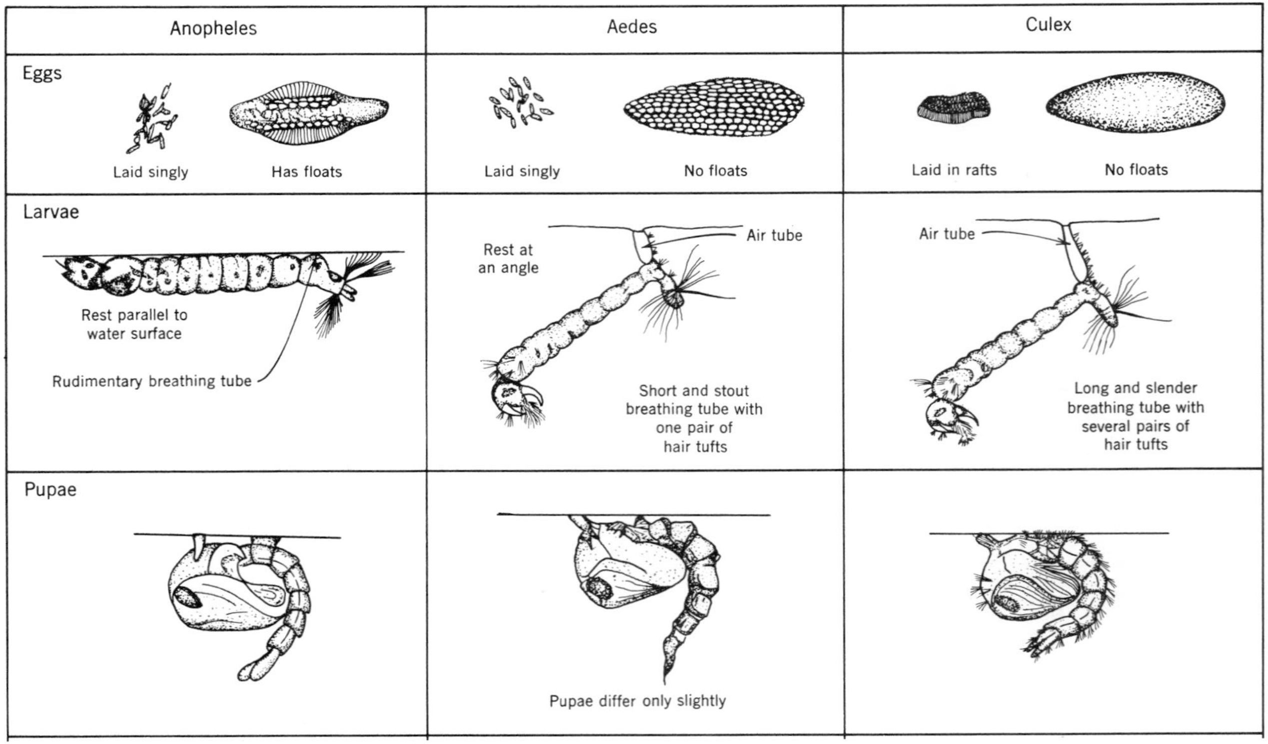

Figure 10-3 Principal characters for identifying the three mosquito genera. (From *Military Entomology Operational Handbook*, TM 5-632, Dept. of the Army, Washington, D.C., June 1965.)

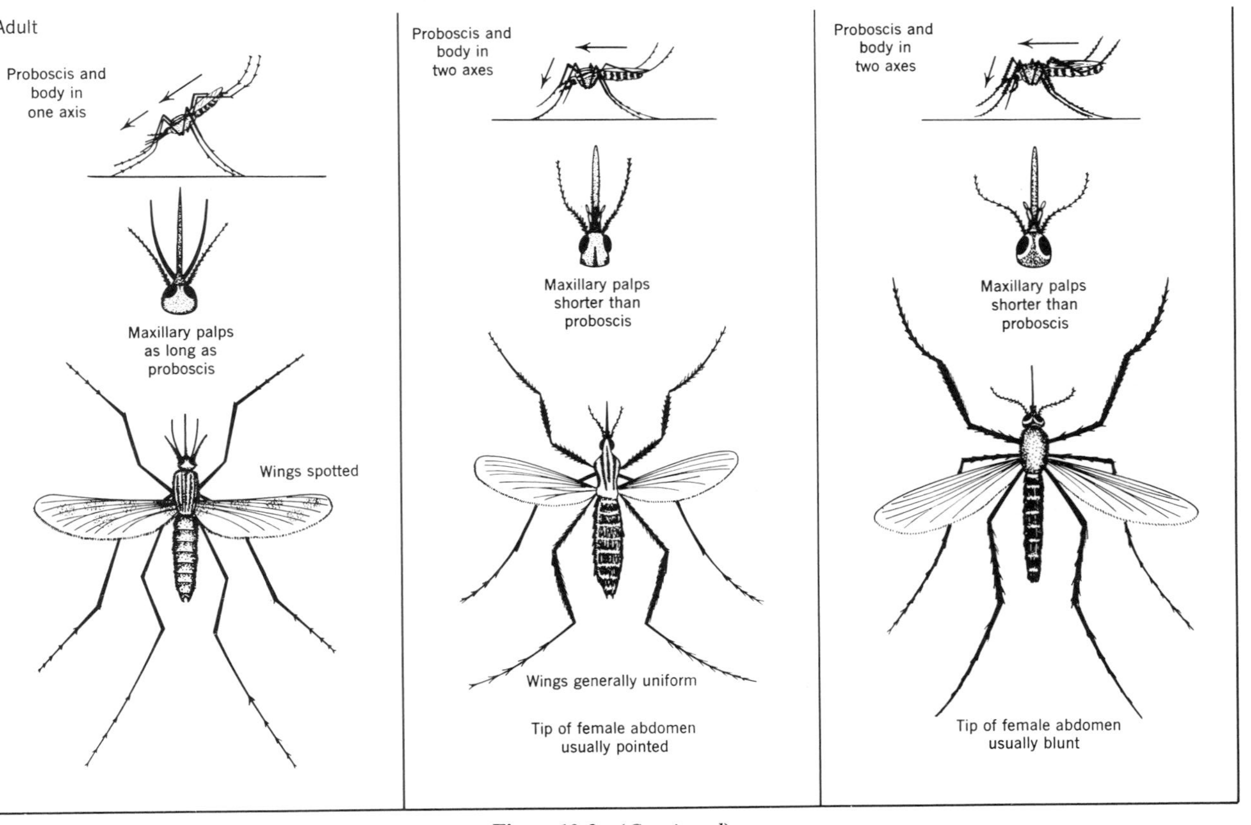

Figure 10-3 (*Continued*)

Table 10-3 Pesticides Currently Employed in Mosquito Control[a]

Type Application	Toxicant[b]	Dosage	Remarks
Residual Spray		**(Mg/ft^2)**	For use in United States as an interior house treatment.
	malathion	100 or 200	Particularly persistent on wood surfaces and remains effective for 3 to 5 months. *For use in overseas zones as a standard application for*
	BHC[c]	25 or 50	*treating the interior of homes in malarious areas.* A suspension formulation is most effective. Dosage and
	DDT[c]	100 or 200	cycle of retreatment depend on the vector, geographic area, and transmission period. DDT and dieldrin
	dieldrin[c]	25 or 50	are effective for 6 to 12 months, BHC for 3 months. When the vectors are resistant to these organochlorine
	propoxur (Baygon)	100 or 200	compounds, malathion or propoxur should be used. Their efficacy is from 2.5 to 6 months.
Continuous Vapor Treatment	dichlorvos (DDVP, Vapona)	1 dispenser per 1000 ft^3	Formulated in resin in United States. Dispensers are suspended from ceiling or roof supports. Provides 2½ to 3½ months of satisfactory kills of adult mosquitoes. Do not use where infants, ill, or aged persons are confined or in areas where food is prepared or served.
		1 dispenser per catch basin.	Dispensers are suspended 12″ below catch basin cover.
Outdoor Ground-Applied Space Spray		**(Lb/acre)**	Dosage based on estimated swath width of 300 ft. Mists
	carbaryl (Sevin)	0.2 to 1.0	or fogs are applied during the dusk to dawn period. Mists are usually dispersed at rates of 7 to 25 gal per mi at a vehicle speed of 5 mph.
	chlorpyrifos[d] (Dursban)	0.0125	Fogs are applied at a rate of 40 gal/hr dispersed from a vehicle moving at the same speed; occasionally at much higher rates and greater
	fenthion[e] (Baytex)	0.001 to 0.1	speeds. Finished formulations contain from 0.5 to 8 oz/gal actual insecticide in oil, or in the case of the nonthermal fog generator, in a
	malathion	0.075 to 0.2	water emulsion. Dusts also

Table 10-3 (*Continued*)

Type Application	Toxicant[b]	Dosage	Remarks
	naled (Dibrom)	0.02 to 0.1	can be used. For ground ULV application[f] technical grade malathion is dispersed at a rate of 1 to 1.5 fl oz/min and a vehicle speed of 5 mph or at a rate of 2 to 3 fl oz/min and 10 mph. For synergized pyrethrins the discharge rate
	pyrethrins (synergized)	0.002 to 0.0025	is 2 to 2.25 fl oz/min at a vehicle speed of 5 mph.
Larvicide		**(Lb/acre)**	Apply by ground equipment
	Abate	0.05 to 0.1	or airplane at rates up to 10 qt of formulation per acre depending upon con-
	chlorpyrifos	0.0125 to 0.05	centration employed. Use oil or water emulsion formula-
	EPN[e]	0.075 to 0.1	tion in areas with minimum
	fenthion[g]	0.05 to 0.1	vegetative cover. Where vegetative cover is heavy, use granular formulations.
	malathion	0.2 to 0.5	*Do not apply parathion in urban areas.* For prehatch
	methoxychlor	0.05 to 0.2	treatment on an area basis, use methoxychlor (1 to 5 lb/acre) or chlorpyrifos (0.1 lb/acre). Chlorpyrifos and
	parathion[c], ethyl or methyl	0.1	fenthion provide prolonged effectiveness in contaminated water at dosages 5 to 10 times those listed. Apply paris green pellets
	paris green	0.75	(5%) at rate of 15 lb/acre with ground equipment or airplane. Apply to cover water surface in catch basins or at a rate of 15 to 20 gal/acre in open
	fuel or petroleum oil	2 to 20 gal/acre	courses. With a spreading agent at a rate of 0.5%, the volume can be reduced to 2 to 3 gal/acre.

Source: "Public Health Pesticides," PHS, DHEW, CDC, Savannah, Ga. 31402, 1973. For guidance only. Consult state and local authorities on restrictions in effect.

Note: State or local regulations may impose certain restrictions on the use of these compounds; therefore, the individual should consult local or state authorities on the accepted use practices.

[a] When insecticides are to be applied to crop lands, pasture, range land, or uncultivated lands, consult agricultural authorities as to acceptable compounds and application procedures.

[b] Other compounds, such as Thanite, Lethan 384, propoxur and ronnel, may have uses in certain of the categories mentioned. If so, follow label directions.

[c] USEPA has cancelled most uses of this toxicant.

[d] Not to be applied to waters containing valuable fish, crabs, or shrimp.

[e] For use by trained mosquito control personnel only.

[f] Adhere strictly to label specifications and directions for use.

[g] Label requires a 3-week interval between applications, except for fog treatments.

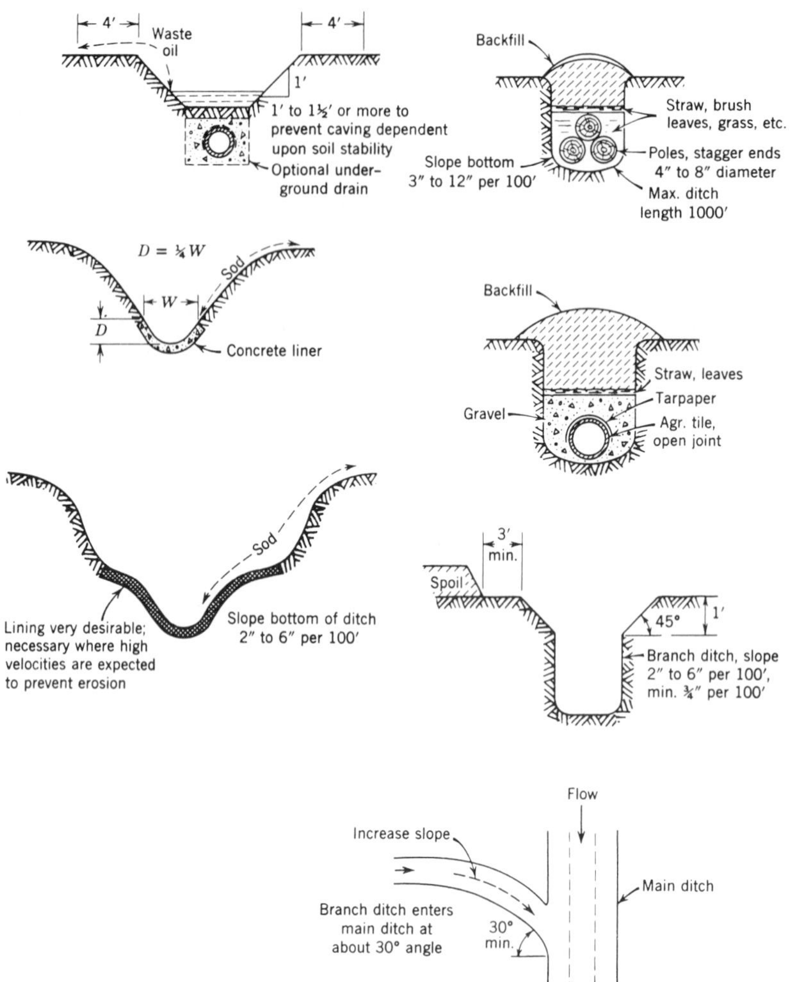

Figure 10-4 Drainage measures. A velocity greater than 2 fps is necessary to prevent desposition and less than 4 fps to prevent erosion. Pumping station may also be needed to supplement gravity flow, to drain low areas. Sizes of ditches based on hydraulic analysis. Design to drain an area in 2 to 3 days to prevent larvae and pupae developing into adult mosquitoes.

breeding. Drip cans are satisfactory in slow-moving streams. A pyrethrum emulsion can be prepared using 8 oz of 40 percent liquid soap to 1 gal of water and 2 gal of kerosene pyrethrum extract. For effectiveness against larvae of the culex and aedes species, which are bottom feeders in contrast to the anopheles larvae, which are surface feeders, add Paris green to moist sand and scatter the sand in breeding pools if Paris green pellets are not available. See Table 10-3.

Domestic Mosquito Control

In many cases mosquitoes are home grown and hence can be controlled in a community through combined individual action based on the life cycle of the particular species. Fundamental is the elimination of standing water in depressions, mud flats, flood plains, barrels, tin cans, drain-clogged or sagged roof gutters, and flat roofs. Inspect septic tanks, cesspools, and sewage drain fields for overflow and for openings or leaks, and repair if defective. Stack containers that can accumulate water upside down and cut up or have collected old tires and cans. Treat sewer catch basins and areas that cannot be drained or filled by weekly spraying with fuel oil, kerosene, or other larvicide.

The spraying or fogging of shrubbery and tall grasses around habitation in the early morning or evening will give temporary relief from adult mosquitoes for a week or two. The application of pesticide should be based on knowledge of the species involved, the source, and seasonal prevalence. A pyrethrin or dichlorvos spray beneath and around lawn furniture will help keep down mosquitoes for several hours; Dichlorvos plastic strips and pyrethrin space sprays are useful for indoor control.

For detailed information on pesticides for control of adult mosquitoes and larvae see Table 10-3. Use care and follow directions when applying pesticides.

Temporary relief from mosquito biting can be obtained by the use of repellents applied to the exposed skin and to the clothing. Common repellants are diethyltoluamide, indalone, dimenthylphthalate, and Repellent 612.

CONTROL OF MISCELLANEOUS ARTHROPODS

Bedbugs

Bedbugs (Figure 10-5) disturb the rest and comfort of individuals, and their continued presence in sleeping quarters is an indication of poor housekeeping. Bedbugs are nocturnal in their habits. They are transported and distributed by clothing, laundry, baggage, bedding, and old furniture. The following extermination procedure is recommended:

Determine the Source. Find where the bedbugs are hiding. Habitual hiding places are usually made evident by black or brown irregular spots, which indicate that bedbugs are or have been present in that place. These spots are usually close to the victim's sleeping quarters and are easily detected. Bedbugs live and lay their eggs in the crevices and cracks of floors, in the seams and folds of infested mattresses and bedclothes, and in the walls of buildings not far from the individual who supplies the blood meal needed for their reproduction. Favorite hiding places also are cracks in posts and bed frames, the interior of springs, and crevices of bedsteads.

Destroy Eggs, Larvae, and Adults. Many liquid insecticides and dusts are effective in killing bedbugs. Malathion, pyrethrum, ronnel, and lindane

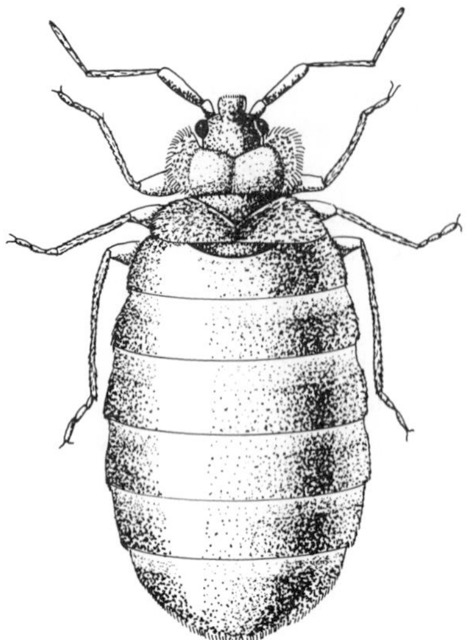

Figure 10-5 Bedbug (dorsal view). (From *Military Entomology Operational Handbook*, TM 5-632, Dept. of the Army, Washington, D.C., June 1965.)

sprays are some of the insecticides commonly available. A residual oil solution or emulsion is preferred until the bedbugs develop resistance, in which case lindane is effective. Walls, floors, bed frames, and bedding should be thoroughly sprayed. Pyrethrin sprays help drive bedbugs out from cracks and crevices. It may be necessary to repeat the application two or three times to kill eggs hatching out after initial spraying. Care should be used in selecting an insecticide that does not stain. Bedding and mattresses that have been sprayed should be aired 4 to 8 hr before use. Do not treat mattresses or other bedding with spray containing lindane or malathion.

Infested clothing and bedding should be exposed daily, if possible, to fresh air and sunlight. Hand-picking, brushing with a stiff bristled brush, and the shaking of blankets is also recommended, particularly if insecticides are not readily available. Steam and scalding hot water are effective when they can be applied with safety.

Ticks

Infected wood ticks and dog ticks (Figure 10-6) are carriers of Rocky Mountain spotted fever, Q fever, relapsing fever, and tularemia. The adult wood tick appears during the spring and early summer in the north western states, and the dog tick appears throughout the summer in the eastern and southern states. The period of greatest prevalence is April to September. Rabbits, squirrels, woodchucks, dogs, horses, sheep, and cattle are frequently infested with ticks. The disease-causing organism is conveyed to man through the bite of the tick or by contact with crushed tick blood or feces through a scratch or wound, but not all ticks carry the disease organism.

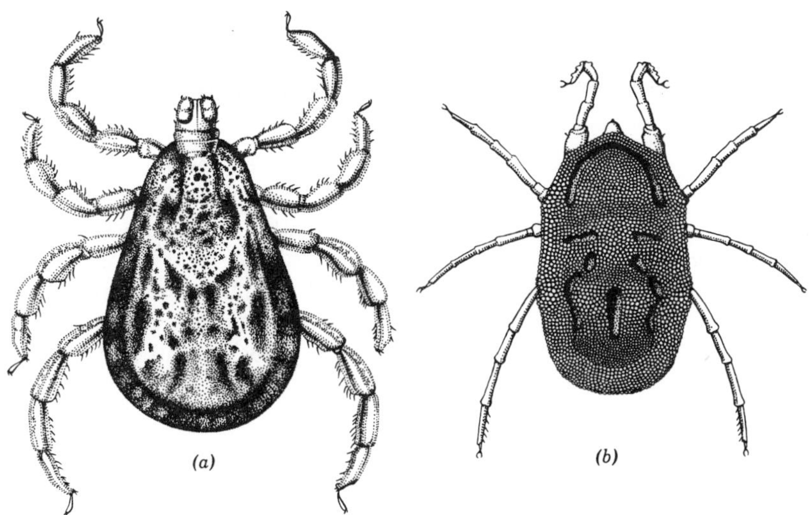

Figure 10-6 (*a*) Common wood tick; (*b*) relapsing fever tick.

Identification. Mature ticks are reddish brown in color and may have white markings on the back. They are usually one-quarter of an inch long, are oblong or seed-shaped, and have eight prominent legs. Collection of several ticks in a pill bottle containing a damp pad can make possible identification and incrimination of the tick.

Tick-Bite Prevention. One of the best ways of preventing infection is to stay away from tick-infested areas during the tick season. If that is not possible, wear clothing that completely protects the legs, arms, and hands. Inspect the body and clothing when in the field during rest periods and immediately remove any ticks found, being careful not to crush them. On returning to camp from an infested area undress, thoroughly inspect the body, clothing, and hiking pack, and immediately remove and destroy any ticks. Inspection of the neck, back, head, and so on by another person is helpful because no tick should be overlooked. It is most important that ticks be removed within a few hours, before infection can occur. One does not usually feel it when the tick is in the act of biting, and those attached to the body can only be found by a complete and careful inspection.

Removal of Ticks. Proper method of removal and destroying of ticks is important. Use fine-pointed tweezers or a penknife, if available, for removal by insertion under the tick. *Do not crush* the tick on your body or between the fingers. Apply gentle but firm traction on the tick, being careful not to leave the mouth parts in the skin. Do not use force; a slow steady pull is required. The tick will soon let go. Or apply a few drops of oil, turpentine, kerosene, or gasoline between the skin and the undersurface of the tick. This application will kill the tick in a few minutes. A tick may also loosen its hold if a lighted cigarette is applied. The use of heat or oil is not recommended by some authorities because the irritation is said to cause the tick to regurgitate into the wound. The hands should be thoroughly washed with soap and warm water after handling ticks, and an antiseptic applied to the bites.

Control. Area control can be obtained with naled, propoxur, and pyrethrum. Rotation of chemicals reduces possible development of tick resistance to a pesticide. If found indoors, apply diazinon, malathion, or lindane into cracks around windows, baseboards, and floors. Infested dogs and kennels may be dusted with lindane or malathion. Do not use on cats; use 1 percent rotenone. A dog dip can be made by adding a tablespoon of 50 percent emulsifiable malathion concentrate to each gallon of water, being careful not to dip the dog's head; sponge the ears if necessary. For a large dog, soak thoroughly the hair to the skin.[18]

Chiggers

The chigger is a small, bright-red mite, almost invisible. The chigger mite larva has 6 legs; the nymph and adult, 8. The chigger is the larval stage of the mite and is known as the "red bug" or "red spider." American mites do not cause disease, although the severe itching from chigger bites causes great annoyance and results in loss of sleep and fatigue. The oriental mites found in the South Pacific, however, transmit a serious disease called scrub typhus.

The adult chigger lays its eggs in the spring of the year when warm weather comes. The adult then dies; but the eggs hatch and chiggers emerge. Since the larvae have but one ambition, to find a host, they soon attach themselves to some passing or resting animal. The larva sinks its mouth parts into the skin, injects an irritating fluid, and sucks up the broken-down skin with some blood. After feeding for 3 or more days, if not scratched off, the larva drops off and enters the nymph stage, from which the adult develops. The nymph and adult feed on organic matter; the larva only feeds on humans and other animals, including reptiles and birds.

When possible, mite-infested areas should be marked off and avoided or cleared. Since this is sometimes impractical, certain precautions can be taken when hiking or camping in infested areas. Individual protection can be obtained by wearing clothing that fits snugly around the wrists, ankles, and neck. Clothing impregnated with 2 g of diphenyl carbonate or benzil per ft^2 of cloth remains effective in repelling chiggers even after several washings. Lindane, 5 percent powders of benzil, dimethylphthalate, sulfur, pyrethrum and derris powder applied to clothing also kill or repel chiggers. A shower with soap and warm water will lessen the effect of chigger bites, and hanging infested clothing in the sun will cause chiggers to fall off or die. Ammonia, rubbing alcohol,

[18] *Controlling Household Pests*, Home and Garden Bull. 96, USDA, Washington, D.C., June 1977, p. 27.

vaseline, iodine, or a weak lysol solution will relieve the irritation from chigger bites.

Many chemicals are available to control chiggers on the ground, lawns, and brush. Wettable sulfur applied as 50 percent suspension in water at the rate of $2\frac{1}{2}$ lb/1000 ft^2, or as a powder at the rate of 1 lb/1000 ft^2, controls chiggers for about 2 weeks. Lindane at the rate of 0.5 lb/acre thoroughly applied to foliage, is another recommended insecticide spray. Additional control can be obtained by cutting grass and brush to the ground and either hauling away or burning the brush when dry. Burning over a camp area with a power oil burner is especially desirable when possible.

Lice

The presence of lice on the body or in the clothing is called pediculosis or lousiness. Such infestation may be due to the head louse, body louse, or pubic louse, also known as the crab louse (Figure 10-7). Lice require human blood to live; they obtain it by puncturing the skin, thereby causing itching and irritation. The resulting scratching may also cause secondary infections such as impetigo.

The body louse and head louse develop in about 17 days and the crab louse in 22 to 25 days. Body lice are found in clothing, along seams, and occasionally on body hairs. Head lice prefer hairy parts of the body, such as the hair behind the ears. Crab lice infest hair around the groin, armpits, and eyebrows. The adult female body louse begins to lay eggs 4 days after maturity, at the rate of from 5 to 10 eggs a day. This will continue for 30 days under favorable conditions.

Lousiness is spread by contact with infested persons, blankets, clothing, common brushes, combs, and hats and possibly by toilet seats. Crowding, insufficient bathing and laundry facilities, and low standards of personal hygiene, chiefly infrequent bathing, favor the spread of lice. Schools, camps, and institutions are the types of places more often involved.

The control of pediculosis begins with detection of the sources of infestation and treatment of those affected. Routine medical inspection of children should include scrutiny of the head under a strong light. If lice are found, a careful search should be made of the clothing and body of all children and adults living, training, working, or playing together. The U.S. Army recommends that all persons in a unit be deloused if inspection reveals 5 percent or more are infested.

Delousing may be accomplished by the proper application of powder or liquid insecticides prepared for the purpose to hairy parts of the body, clothing, and bedding. Repeat the application if necessary after 1 week and after 2 weeks, or until no further lice are found. Most of the insecticide louse powders contain 10 percent DDT or 1 percent lindane or malathion as the active ingredient. Pyrethrum (0.2 percent) and allethrin (0.3 percent) dust, powder, or liquid are effective. Abate dust is also reported to be very effective. When in-

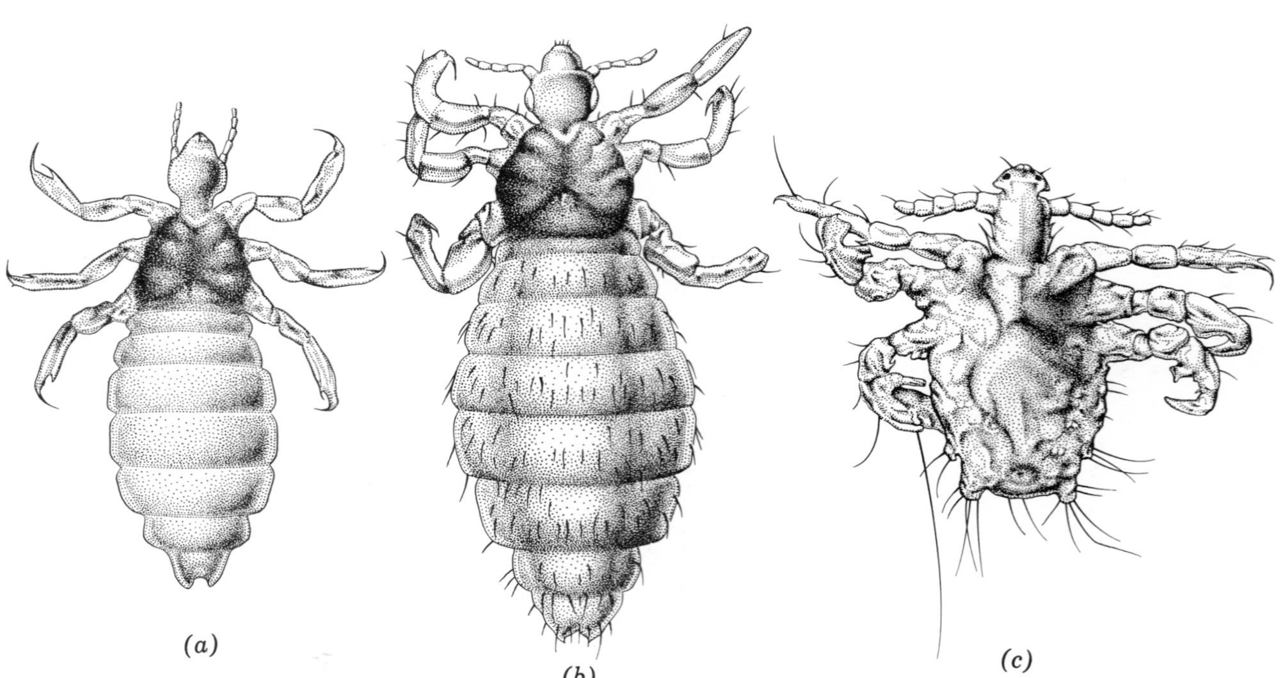

Figure 10-7 (*a*) Head louse, adult female; (*b*) body louse, adult female; (*c*) crab louse, adult male. (From *Insect and Rodent Control*, TM 5-632, Dept. of the Army, Washington, D.C., October, 1947).

secticides are not available, shaving or close clipping of the infested areas, followed by repeated scrubbing with soap and hot water will be effective. When the head is involved, clipping the hair short, the application of vinegar and kerosene to the scalp if free of cuts or sores, and combing with a fine-toothed comb to remove the nits has been found effective. However, as previously stated, use of an insecticide is preferred. Clothing and bedding may be fumigated with methyl bromide if competent supervision is available, and also with dry heat or steam. Storage for at least 30 days will cause lice to die of starvation. Pressing of clothing, especially the seams, with a hot iron is also effective.

In any case, attention should be directed to personal hygiene and home cleanliness as the best preventive measures. Frequent bathing with hot water and soap, clean clothing, and clean surroundings will discourage the infestation and survival of lice.

Mites (Scabies)[19–21]

Scabies is an infectious disease of the skin caused by burrowing of the female mite *Sarcoptes scabiei* into skin where it deposits its eggs. Identification of the mite requires microscopic examination of skin scrapings. The resulting itching may be severe and scratching often leads to secondary infection.

Scabies more often affects preschool and elementary school children (finger webs, wrists, elbows, forearm, and upper arm near the armpit), adolescents and young adults (also through sexual contact), and elderly debilitated patients (chest, abdomen, back) in nursing homes and institutions. Transmission is primarily by transfer of the mites through close bodily and hand contact with an infested individual, also possibly through freshly contaminated bedding or undergarments.

Control involves education in personal hygiene, use of clean undergarments and bedding, treatment with a scabicide, thorough cleansing baths, and isolation of infested individuals, their families, and close contacts until free of mites. Treatment of the secondary infection may also be necessary. The disease is not uncommon in poor countries and during war where frequent bathing is impractical, and where there is a breakdown in personal hygiene.

Fleas

Fleas are associated with house pets, such as cats and dogs, and rats. Fleas can transmit disease, as explained in Chapter 1, although this is unlikely in

[19] *Control of Communicable Diseases in Man*, Abram S. Benenson, Ed., 12th ed., 1975, Am. Pub. Health Assoc., 1015 Fifteenth Street NW, Washington, D.C., pp. 280-81.

[20] Louis J. Lanzillo, Rensselaer County Health Dep., New York, January 3, 1980, personal communication.

[21] *Maxcy-Rosenau Preventive Medicine and Public Health*, Philip E. Sartwell, Ed., Appleton-Century Crofts, New York, 1965, pp. 467–470.

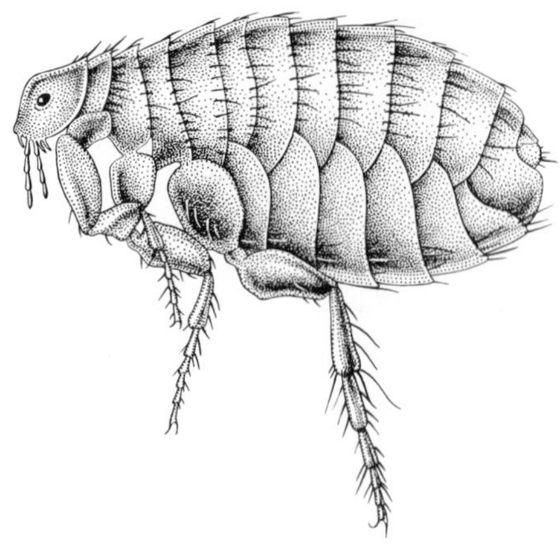

Figure 10-8 The human flea (much enlarged).

the United States. The adult fleas attach themselves to the animals, and sometimes humans, and feed on their blood. They breed in the fur or hair of pets, and the larvae hatch and live in dust, in cracks, under carpets, and in cat or dog bedding. Figure 10-8 shows the human flea.

Where rats are present, a dust applied to rat runs and harborage areas will be picked up by the rat on its fur and be carried to the flea. Methoxychlor, pyrethrum, or malathion spray is effective against soil and house infestation by cat and dog fleas. Dusts of malathion and carbaryl can also be used outdoors.

Oil solutions must not be applied to domestic animals as the poisons will be absorbed by the skin. Infested dogs and cats can be dusted with 1 percent pyrethrum, 1 percent rotenone, or carbaryl (Sevin).

Roaches

Roaches have been found to be mechanical carriers of many disease organisms including *staphylococcus*, *Escherichia coli*, *salmonella*, and *streptococcus*.[22] Pathogens have also been identified in the gut and feces of cockroaches.[23] Although roaches are associated with filth, epidemiological evidence showing actual disease transmission is not well documented.[24]

The continued presence of cockroaches in any establishment is due to poor housekeeping in the premises or in adjoining premises. These pests can be controlled by the maintenance of scrupulous cleanliness in kitchens, dining rooms, and food-storage places, and by elimination of breeding places, the removal of sources of food and water for the insects, proper storage of garbage and other refuse, and the periodic application of dusts and sprays as needed. Cockroaches are illustrated in Figure 10-9. The following are sanitary measures to apply.

[22]Austin M. Frishman and I. Edward Alcamo, "Domestic Cockroaches and Human Bacterial Disease," *Pest Control*, June 1977.

[23]James Busvine, "Urban Pests of Public Health Importance," *J. R. Soc. Health*, June 1977, pp. 130–134.

[24]Gary W. Bennett, "The Domestic Cockroach and Human Bacterial Disease," *Pest Control*, June 1977.

Clean Premises and Close Openings. Remove temporarily all food and equipment from the place to be treated. Scrub the entire room clean, starting from the rafters or ceiling and ending with the floor. Include shelves, window sills, tops and bottoms of tables, corners, and floors under refrigerators, vegetable bins, refrigerators, sinks, around motors, compressors, and other equipment. Close all holes and cracks in floors, walls, ceilings, and around pipes and plumbing.

Treatment. Insecticides used in roach control are given in Table 10-4. Treat walls, starting from the ceiling and working down to the floor. The cracks around doors, windows, pipes; the space around kitchen cabinets, refrigerators, stoves, sinks, wood framing, baseboards, floor covering; and the bracing under tables and benches should be thoroughly covered. A pyrethrum spray applied

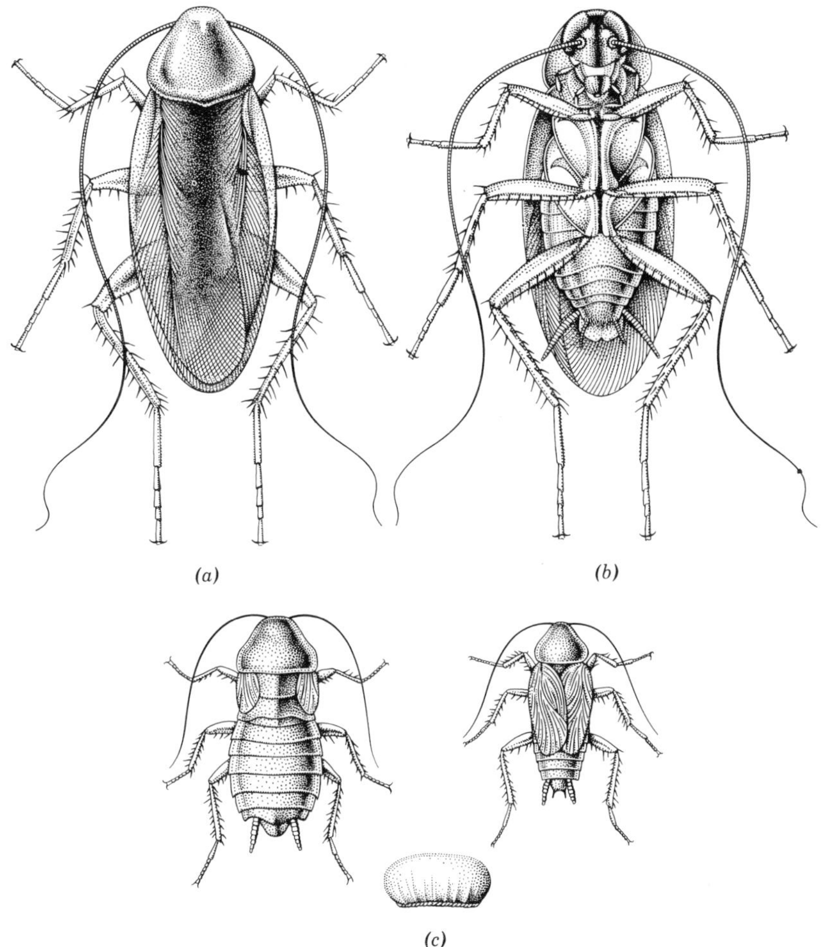

Figure 10-9 American cockroach: (*a*) view from above; (*b*) view from beneath; (*c*) Oriental cockroach, female and male, about natural size. (From *Insect and Rodent Control*, TM 5-632, Dept. of the Army, Washington, D.C., October 1947.)

Table 10-4 Insecticides Used in Cockroach Control[a]

Insecticide	Formulation	Percent Concentration[b]
propoxur (Baygon)	Spray	1.0
	Bait	2.0
Diazinon	Spray	0.5[c]
	Dust	1.0[c]
boric acid	Dust	100.0
Dursban[d] (chlorpyrifos)	Spray	0.5
malathion	Spray or dust	5.0

Adapted from "Public Health Pesticides," DHEW, CDC, Savannah, Ga. 31402, 1973.

[a] Consult with state and local authorities on restrictions in effect.

[b] Maximum allowable.

[c] 1% spray and 2 to 5% dust for pest control operators only.

[d] Pest control operators only.

before a residual treatment will drive roaches out of cracks and crevices and give a quick kill where heavy infestations are encountered. Boric acid powder applied to infested areas and between walls and floors with a bulb duster or powder dispenser provides good control and will remain effective as long as it is dry. The powder sticks to the feet and body, is ingested when the roaches groom themselves, and also is carried to nesting places. The powder is toxic; it should not be accessible to children or pets.

Ants

Ants are a nuisance and spoil food they infest. Some establish nests in rotting woodwork, inside a partition, under trash, and sometimes outdoors. Control should start with observation of the ants' habits, location and removal of the nest if possible, and the sealing of openings permitting entrance from the outside. Baits are useful to help locate nests. This can be followed by the application of a suitable insecticide to the areas involved. In buildings, insecticide can be applied with a sprayer or paintbrush to places where ants are seen crawling and at points of entry. Malathion, lindane, propoxur, or diazinon may be used to control ants in lawns and ant mounds. Dusts and Paris green bait, 5 percent in brown sugar, and a mixture of boric acid powder and sugar are also effective.

Punkies and Sand Flies

The culicoides species apparently breeds in moist soil and is found along coastal areas, stream and pond shores, and in tree holes. Members of this species are called punkies, no-see-ums, or biting midges.

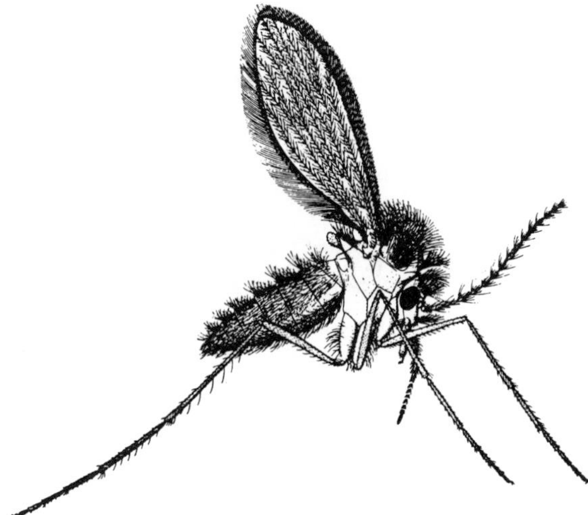

Figure 10-10 Sand fly (phlebotomus). (From *Military Entomology Operational Handbook*, TM 5-632, Dept. of the Army, Washington, D.C., June 1965.)

Salt-marsh sand flies (phlebotomus), shown in Figure 10-10, can be controlled by use of fly spray. Parks, camping areas, and living quarters can be treated with a residual spray. Ordinary screens do not keep out these sand flies, but painting the screens with a 5 percent methoxychlor or malathion emulsion will give temporary protection.

The phlebotomus species is a disease problem in southern Asia, the Mediterranean region, and in the high altitudes of South America. Pappataci fever, kala azar, verruga peruana, and oriental sore can be spread. Some species are found in the United States, but they pose no threat. Residual spraying of buildings, stone walls, and other resting places gives long-term control.

Blackflies

Blackflies (Figure 10-11) are blood-sucking insects belonging to the simuliidae species, and are commonly called buffalo gnats. Only a few of the species are annoying and unbearable biters. They are a disease vector for onchocerciasis in parts of Mexico, Central and South America, and Africa; they are pests in Alaska, northern United States, and Canada and attack both man and animals.

Although some temporary relief can be obtained from mosquito-type spraying, better control is possible by killing the larvae. Individual control is impractical; area-wide control is necessary. The larvae of blackflies are found attached to rocks and vegetation in flowing streams. Application of methoxychlor or dibrom to a stream has been found effective as a larvicide. Abate is also considered a potentially useful blackfly larvicide. Excellent control has been obtained by emulsion application. Weekly larvicide treatments for the 10 to 12

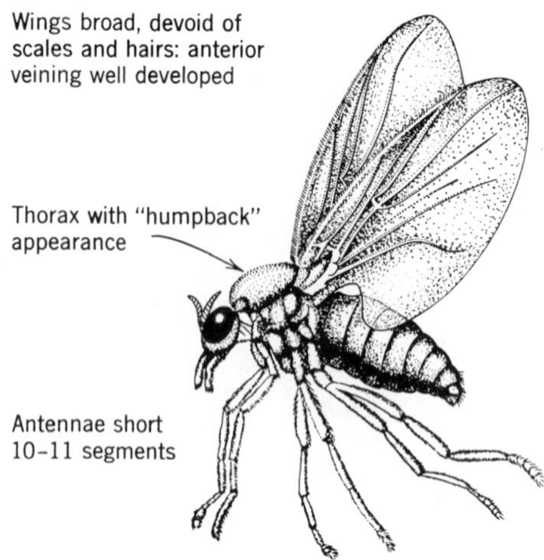

Figure 10-11 Adult blackfly (Simulium). (From *Military Entomology Operational Handbook*, TM 5-632, Dept. of the Army, Washington, D.C., June 1965.)

weeks breeding season are needed for onchocerciasis control where this disease is endemic.[25]

Airplane treatment with methoxychlor oil solution plus 0.5 to 0.75 percent emulsifier such as Triton X 100 is effective when applied over large areas. A spray mixture of 1 gal of 15 percent methoxychlor per flight-mile applied in approximately 100-ft swaths spaced $\frac{1}{4}$ mi on centers, to cross streams at right angles, is used. Two to three treatments annually during the breeding season should reduce the annoyance and biting of blackflies.

Termites

There are two major groups of termites in the United States: the subterranean and the above-ground. Subterranean termites travel from underground colonies to wood above ground by means of the connecting tunnels that they build. The above-ground termites may be the powder-post, dry-wood, damp-wood, or rotten-wood varieties. Termites, to survive, need food such as wood or other cellulose materials, moisture such as moist soil, and particularly protection from ants, which can enter termite tunnels and galleries.

Control requires elimination of buried wood from ground and prevention of contact of wood structural members, skirting, or siding with the ground (6-in. clearance). Good construction practices should be followed, including space ventilation and drainage; solid masonry construction, such as reinforced concrete; solid or filled hollow-block construction; and use of treated wood posts, piers, and steps. Screening (18 × 18 mesh) of windows, louvers, eaves, venti-

[25]Hugo Jamnback, "Recent Developments in Control of Blackflies," *Annual Review of Entomology*, Vol. 18, 1973.

lators, and doorways and caulking of cracks, joints, and other openings will help keep out the above-ground termites.

The soil around buildings can also be treated by injection of a suitable pesticide where needed to provide a barrier next to foundations, piers, and walls and under concrete slabs. Children and pets should be kept away from the treated soil, and the chemicals should not be applied in the vicinity of well-water supplies, streams, or lakes.

Wasps and Honey Bees

"Wasp" is a general term which includes yellow jackets, cicada killers, and hornets. Wasps nest in the ground and in elevated nests, often around dwellings. Yellow jackets generally build their nests in the ground; hornets nest in gray oblong-shaped hives suspended in trees or in shrubs. Wasps are attracted by garbage, outdoor barbecues, and picnic foods, especially those containing sugar, and to fruit orchards. Bright-colored, smooth-textured clothing and hair oils attract wasps and should be avoided in situations where they are likely to be present. For most people a wasp sting causes some pain and swelling lasting for several hours or perhaps longer. An ice pack will reduce the swelling and accompanying itching. People who have a history of asthma, hay fever, or other allergy may develop a serious reaction to a sting and should immediately notify or see their doctor. A wasp does not leave its stinger. If stung by a honey bee, remove the stinger as rapidly as possible by carefully scraping off with a knife or fingernail in a manner that does not force the sac venom into the wound and save if possible for identification.

Honey bees are kept by commercial beekeepers. They may be found in hives, also in hollow trees or between walls in a house. Nests near human habitation or activity can be a hazard. To eliminate the nest, spray or dust the opening using a pesticide registered for the control of wasps. Carbaryl (Sevin) is toxic to honey bees. This should be done *after dark* to minimize the danger of being stung. Wear protective clothing. If the nest is on the ground, cover the nest opening with a shovelful of earth after treatment. Call in professional assistance when in doubt. Destruction of bees should be balanced against their usefulness.

CONTROL OF RATS AND MICE

Rat Control Program 

Rats are carriers of disease germs, fleas, lice, mites, and intestinal parasites. They destroy food and crops before harvest and cause an estimated damage of $900,000,000 per year in the United States, with higher losses in less developed countries. They breed prolifically; the gestation period averages about 22 days. A female Norway rat becomes sexually mature in 2 to 3 months, can potentially produce 4 to 7 litters per year, with 8 to 12 young per litter; but normally only about 20 survive. A rat has a life expectancy of 1 to 3 years, with an average of about 1 year in the wild. A rat drops 25 to 150 pellets of feces, 10 to 20 ml of urine, and several hundred hairs per day. Rats migrate when there is a lack of food, water, and shelter; when there are floods or crop failures; when dumps are abandoned; when buildings are vacated; and when warm weather and then cold weather come. The Norway rat has a normal range of 100 to 150 ft. Rats see poorly but can detect contrast of motion well. They have a keen sense of smell, hearing, taste, and touch. Peaks in breeding occur in the spring and fall in temperate zones.

Rattus norvegicus is also referred to as the Norway rat, sewer rat, house rat, barn rat, wharf rat, and brown rat; *Rattus rattus* is also called the roof rat and black rat. The common terms may not accurately identify the actual species.

Rats can be controlled by community participation and coordinated premises sanitation through removal of food, water, and harborage. The number of rats is largely determined by the condition or capacity of an existing environment to support a prevailing population. Rat control is an integral part of a housing and community sanitation program. Public cooperation and individual hygienic practices on a continual basis are essential. Start with a community plan and program for action to include the following.

1. *Community Survey.* Identify areas infested and degree of infestation. Determine condition of premises' sanitation, refuse storage and collection frequency, residential and business structure dilapidation and deterioration (housing and census data). Record data on inspection forms, maps, coded cards; analyze data and make charts, summaries, and graphs to show trends. Use these data as a basis for program planning, education and operation, followed by evaluation and program redirection. Establish a priority plan, administrative organization, and trained staff. Attack the rat infestation on a contiguous block-by-block, neighborhood-by-neighborhood basis. The survey and effectiveness evaluation should always be made by the regulatory agency; the control work may be done by the agency or by contract. Premises and block survey forms are very useful in identifying and analyzing the problem.[26]

2. *Public Participation.* Solicit assistance of local neighborhood organizations, residents, and property owners in pointing out infested areas and in all planning and implementation phases of the program. Determine planning priorities and the extent of the participation of local people in eliminating the causes of the rats. An effective tool is the photonovel (*photonovela* in Spanish). Photographs of local scenes and people are used instead of drawings to show community problems and corrective action. The design, writing, and production of the booklet is done with community participation. This facilitates implementation and sustained community effort.[27]

3. *Education and Information.* Inform local officials, the mass media, and the community. Employment of local residents on a part-time or full-time basis to explain to their neighbors what brings rats and how to eliminate them can be one of the most effective measures that can be taken, particularly in run-down neighborhoods. These people have been designated as health guides, health educator aides, and sanitation aides, or by similar titles. They are given special training in rodent control and general environmental sanitation and work under the direction of a professionally trained supervisor. This should be supplemented by community education through the

[26] Kent S. Littig, B. F. Bjornson, H. D. Pratt, and C. F. Fehn, *Urban Rat Surveys*, Environmental Control Program, PHS, DHEW, Washington, D.C., February 1969; Joseph A. Salvato, Jr., *Control of Rats and Mice*, New York State Dept. of Health, Albany, October 1969.

[27] Stephen C. Frantz, Project Director, *A Working Neighborhood—What does it take?*, New York State Dept. of Health, Bureau of Community Sanitation, Albany, N.Y., May 1978.

schools, news media, demonstrations, and other means to help the people learn how to help themselves. A public information example is shown in Figure 10-12.

4. *Building, Housing, and Sanitary Code Enforcement.* Make sure that modern codes have been adopted and that they include sound, rodent-proof construction and maintenance. Competent and adequate staffs to ensure proper construction in the first instance and then systematic inspection to ensure that use and maintenance of structures and premises are in conformance with housing and sanitary codes are essential. This includes sheds, garages, alleys, backyards, empty lots, and vacant buildings. Abandoned structures should be rehabilitated if sound or demolished and removed.

5. *Rat Eradication and Harborage Removal.* The killing of rats should be planned on a block-by-block and neighborhood basis just before permanent control measures are applied, but may follow cleanup where rat populations are not excessive. The need for concurrent ectoparasite con-

- Rats live on garbage. Wherever they can get it. But no rat ever chewed through galvanized steel.
- That's why your best defense against rats is a garbage can. With the lid on. Tight.

The directions are simple. Keep the lid on your garbage can, tight. With your garbage inside. Always.

Rats are smart. If a hungry rat finds there's no loose garbage on your premises, he won't hang around for long. Neither will his brother rats. They'll get the message fast.

Each rat consumes some 17 pounds of garbage a year. To get at it, he'll climb brick walls, swim half a mile underwater, gnaw through cement and swing from exposed ceiling beams. In his spare time he creates more rats, with a single pair producing 880 descendants a year.

To fight the rat, attack the loose garbage problem on two fronts.

One is block-by-block. The other is a campaign urging every household to stow garbage in the can, with the lid on, at all times.

Figure 10-12 Educational item, "Starve a rat today." (From *Control of Rats and Mice*, New York State Dept. of Health, Albany, 1969.)

trol must be determined. Rat eradication is only a temporary expedient if not accompanied by the permanent elimination of water, food, and harborage. Vacated buildings must be treated promptly to prevent the migration of rats in search of food to adjacent structures. This may also be necessary prior to demolition if reinfestation occurs. Local pest-control operators can play an important role in this activity.

6. *Adequate Municipal Refuse-Collection Service.* All communities should have an adequate municipal or contract collection service. This should include public, business, and residential areas. Twice-a-week collection during the fly-breeding season and more often in problem neighborhoods is sometimes necessary. Proper storage, collection, and disposal are essential parts of a satisfactory refuse-collection service and a rat control program.

7. *Community Sanitation and Good Housekeeping.* It should be recognized that without the maintenance of community sanitation and good housekeeping, a rat poisoning and killing program will produce only a temporary improvement and may produce resistance. When the poisoning and killing are suspended, the rat population will again multiply as high as the available food supply and environment will support. Adequate refuse storage, collection, and disposal; rat-stoppage; housing code enforcement; and good housekeeping practices, including proper food and refuse storage and environmental hygiene, must be emphasized. Education of local residents in rat control measures is necessary so that they do not contribute to and become part of the problem.

8. *Control of Rats in Sewers, Waterfront, Dumps.* Arrangements should be made for a sewer inspection and maintenance program, waterfront treatment and cleanup, and conversion of dumps to sanitary landfills preceded by poisoning. These aspects will be discussed further.

9. *Administration and Interagency Coordination.* All existing local agencies should cooperate and coordinate their activities to eradicate rats. One agency, such as the health department, could serve as the coordinator to ensure a planned comprehensive approach and complementary effort. Agencies involved might include, in addition to the health department, the housing code enforcement, urban renewal, sanitation, public works, buildings, fire, education, social services, and extension service agencies, as well as federal and state agencies having related programs and responsibilities.

10. *Training and Employment of Residents of Substandard Housing Areas.* It has been found that residents of problem neighborhoods can communicate more effectively with the local people and obtain their cooperation in the application of rat control techniques. Such individuals can be given special instructions in basic environmental sanitation, lead poisoning control, and in the availability of community services to give them competency to serve as health guides or sanitation aides in the neighborhood. They can also assist in obtaining utilization of health services, correcting housing code deficiencies, and giving advice on better and more nutritious diets. The experience gained by the health guide or sanitation aide in rodent control and in community education can lead to other employment and advancement.

11. *Evaluation.* Evaluation involves the setting of objectives, ways of measuring achievement of the objectives, periodic measurement of the success obtained, and redirection of effort as indicated by the results. Evaluation of rodent control programs is essential if expenditure of money and effort is not to go on endlessly.[28] This activity should be carried out by the health department or other designated control agency. The evaluation should include the counting of confirmed rat bites and the measurement of recurring rat signs through resurvey, complaint investigation, bait stations, trapping, sanitation maintenance, and the possible development of bait resistance. Rat signs would include evidence of droppings, urine, smudge marks, runs, tracks, burrows, gnawings, seeing rats, nests and food caches, rat odors, and pet excitement.

Rat Control Techniques

A brief outline of some specific rat control measures is given below.

[28]Clyde Fehn and Bayard F. Bjornson, "Developmental Evaluation Techniques for Community Rat Control Programs," Environmental Control Administration, Consumer Protection and Environmental Health Service, PHS, Washington, D.C., November 1, 1968.

Starve Rats

Starve rats by eliminating their supply of food and water. An adult rat eats about 1 oz of food and drinks 1 oz of water per day.

1. Store garbage in tightly covered metal containers.
2. Store foods in tightly covered containers of metal or glass. Sweep floors and stairways free of bits of food.
3. Store fruits and vegetables in refrigerator or in rat-proof room.
4. Keep laundry soap, candles, wax paper, and so on where rats cannot get at them.
5. Keep garbage cans, storage areas, and picnic areas neat and free of spilled food and litter.
6. Ensure adequate refuse storage, collection, and disposal and cleanliness of streets, alleys, empty lots, and premises.
7. Eliminate standing water. Fill in depressions and drain water to sewer, street, or soakage pit. Fix leaks.

Build Rats Out

Rat-proof buildings and remove rat harborages. See Figure 10-13.

1. Find the holes rats use to enter buildings. Seal all holes in basement walls and floors with good cement or concrete. See that all doors and windows fit tightly, and screen all windows that are kept open. Make all doors self-closing and outward-opening and see that they are kept closed.
2. Remove infested wooden basement floors and replace with concrete.
3. Pile boards or lumber at least 18 in. above the ground. Remove old sheds and storage bins.
4. Keep premises, including yards and alleys, clean and free of accumulations of junk and debris.

Kill Rats by Using Poisons

Poisoning is not a substitute for basic environmental sanitation, only a supplement. Poisoning alone will only provide a temporary reduction in the rat population. Some common methods follow.

1. Mix fortified red squill or an anticoagulant rodenticide such as Warfarin, Pival, Diphacinone, Fumarin, or zinc phosphide poison with fish, cereal, grain, or meat. Other poisons are chloropicrin liquid and calcium cyanide solid disks (releases hydrocyanic acid on exposure to moist air). These poisons should be used only by certified applicators and should not be used in populated areas. Vary the type of fresh bait. Red squill, zinc phosphide, and the anticoagulants are relatively harmless in the concentrations used. Zinc phosphide is quick-acting (48 hrs) against the Norway rat: its use more than once a year is not advised. Place anticoagulants and zinc phosphide in bait boxes so as not to be accessible to cats and dogs. Red squill is not effective against roof rats or house mice.

   Anticoagulants are the preferred rat poisons for most purposes, but continuous feedings are required for 3 to 10 days. They cause capillaries to break down, resulting in internal bleeding, and they also prevent normal coagulation of the blood. Bait should be available for at least 2 weeks. Rodenticides and bait concentrations to poison rats—*Rattus rattus* and *Rattus norvegicus*—are given in Tables 10-5 and 10-6.

   Sodium fluoroacetate,* also known as "1080," is the most effective fast-acting rodenticide; but its use is prohibited in residences. It is allowed in sewers. Fluoracetamide* (1081) is in the same group. It is very toxic and should be used only by professional exterminators (certified applicators) under carefully controlled conditions. Zinc phosphide should also be used with care. To protect pets add 3 parts tartar emetic to 8 parts zinc phosphide by weight even though bait acceptability is reduced.
2. Place poisoned baits every 10 to 20 ft along rat runways; 3 to 5 ft for mice. Use closed bait boxes

*The use of 1080 or 1081 is not permitted in some states.

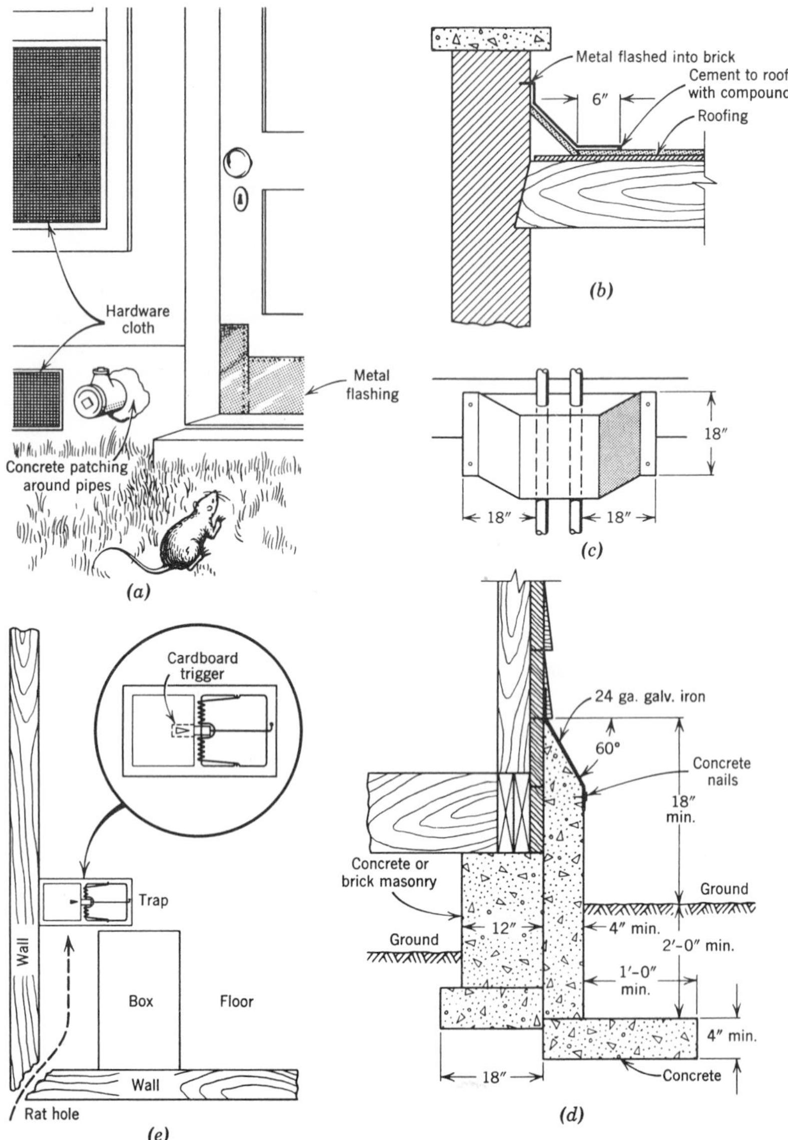

Figure 10-13 Rat-proofing and trapping: (*a*) rat-proofing; (*b*) roof protection; (*c*) cable guard; (*d*) curtain wall; (*e*) trapping. Door metal flashing should form a channel around door bottom and edge. (Parts *a* and *e* from "Feed People Not Rats," U.S. Dept. of Interior, Fish and Wildlife Service.)

at floor level indoors in restaurants and homes. Use bait boxes or burrow-baiting outdoors but, if exposed to rain, anticoagulants should be mixed in paraffin. CAUTION: Handle With Care. Protect children and pets. Place bait in pipes, underneath buildings, in basements, and in other relatively inaccessible places.

3. Pick up and destroy poisoned baits, except anticoagulants, after 24 to 72 hr of exposure.
4. Consult a reliable pest-control operator in difficult situations. State and local health departments,

Table 10-5 Multiple-dose Rodenticides Employed Against Mice and Commensal Rats

Rodenticide	Percent Concentration
chlorophacinone[a]	0.005
diphacinone[b]	0.005
Fumarin	0.025
pival	0.025
warfarin[c]	0.025
brodifacoum[d]	0.005

Source: Stephen C. Frantz, "Integrated Pest Management," National Association of Housing and Rehabilitation Officials (NAHRO) Maintenance Clinic, 3–4 September 1980, Albany, N.Y., New York State Dept. of Health, Albany, N.Y.

Note: For guidance only. Consult with state and local authorities on restrictions in effect.

There are no restrictions on the use of brodifacoum bait indoors (nonagricultural building); chlorophacinone bait indoors and outdoors; diphacinone bait indoors and outdoors; Fumarin bait indoors and outdoors; or warfarin bait indoors and outdoors.

[a]Also used as a tracking powder for mice and rats as a 0.2 concentration. May kill after a single dose. Restricted for use as stated only by certified applicators for tracking powder in indoor use.

[b]Restricted for use as stated only by certified applicators for tracking powder in indoor and outdoor use.

[c]Restricted for use as stated only by certified applicators for tracking powder in indoor use.

[d]Restricted for use as stated only by certified applicators for bait in outdoor use.

the state extension service, and U.S. Fish and Wildlife Service, the USEPA, and the USDA can also provide guidance.

5. Use only USEPA-registered poison products; they note on the package the killing ingredients, the antidote, the word "POISON," the skull and crossbones (not on anticoagulant baits), and the name and address of the manufacturer. Confirm actual quality of bait used by laboratory tests.

6. Tracking powders may be blown in burrows and within wall spaces; placed in runways, or on the floor of bait boxes.

7. Some recommended bait formulations for large-scale control are given below.

Trap Rats

1. Use common wooden-base or steel snap traps and plenty of them. Provide at least 10 at a time. Tie down traps.

2. Set the traps at a right angle to and against the wall. Use fresh baits and change them daily, except for anticoagulants. See Figure 10-13. Try bacon strips, peanut butter, doughnuts, apple, fish, or bacon-scented oatmeal.

3. Enlarge the trigger of the common snap trap by fastening a 2-in. square of cardboard over the trigger using scotch tape. No bait is needed.

Table 10-6 Single-Dose Rodenticide Used Against Rats and Mice

	Percent Concentration in Baits		
Rodenticide	*Ratus norvegicus*	*Rattus rattus*	*Mus musculus*
ANTU[a]	2.0 to 3.75[f]	[e]	[e]
fluoroacetamide[b] (1081)	2.0 to 3.0[f,g]	2.0 to 3.0[f,g]	[e]
red squill (fortified)	3.5 to 10.0[f,h]	[e]	[e]
strychnine[c]	[e]	[e]	0.3 to 0.5[f]
sodium fluoroacetate[c] (1080)	0.2 to 0.39[g]	0.2 to 0.39[g]	0.2 to 0.39[g]
zinc phosphide[d]	1.0 to 2.0[f]	1.0 to 2.0[f]	1.0 to 2.0[f]

Sources: *Public Health Pesticides*, National Center for Disease Control, Dept. of HEW, 1973, and Stephen C. Frantz, "Integrated Pest Management," NAHRO Maintenance Clinic, September 3–4, 1980, New York State Department of Health, Albany, N.Y.

Note: For guidance only. Consult with state and local authorities on restrictions in effect.

There are no restrictions on the use of red squill bait indoor and outdoor, zinc phosphide 1 to 2 percent bait indoor and outdoor. Certain rodenticides may not be used in or around residences by anyone. These include arsenic trioxide bait 1.5 percent or less, phosphorous bait, sodium fluoroacetate (1080) bait.

Fumigants restricted for use as stated only by a certified applicator include calcium cyanide powder for indoor and burrow use, chloropicrin liquid for indoor and burrow use, hydrocyanic acid solid disk for indoor use, methyl bromide liquid for indoor use, sodium chlorate and sodium cyanide solid for indoor use.

There are no restrictions on the use of carbon tetrachloride, ethylene dichloride, and paradichlorobenzene for burrow use, and sodium nitrate, sulfur, charcoal (gas cartridge) solid for burrow use.

[a]Bait or tracking powder in indoor, outdoor, and burrow use.

[b]Concentrate for dilution in outdoor use.

[c]Bait for indoor and outdoor use.

[d]Bait not greater than 2 percent concentrate in indoor and outdoor use and 10 percent tracking powder in indoor use.

[e]Not used against these species.

[f]Solid bait.

[g]Liquid bait.

[h]The percent concentration used depends on the toxicity of the red squill, *i.e.*, a formulation with an oral LD_{50} to rats of 500 mg/kg would be prepared at the 10 percent level.

Maintain Constant Vigilance

1. Rat control measures must be continuous to be effective.

2. Make periodic surveys of buildings and neighborhoods, keeping in mind sources of rat food and water, probable harborage and nests, and what can be done to eliminate rats and prevent reinfestation. Maintain and check bait stations.

3. Public and individual cooperation in the maintenance of a clean community and home are necessary. If there is a relaxation in sanitation, renew educational campaign.

4. Ensure that a modern housing code, including rodent control, is adopted and enforced. See Chapter 11.

Rat Control at Open Dumps

Rats will travel to nearby farms and residential areas if the food supply at an open dump is reduced or cut off. This would occur when a site is abandoned or converted to a sanitary landfill. A poison-bait application program should be started ten days before a site is abandoned or converted and before earth-moving equipment is brought to the site. Post the entrance and warn nearby residents to keep children and pets at home. Baiting should be continued for 3 to 10 days after the open dump is completely converted to a sanitary landfill or closed and covered with at least 2 ft of compacted earth. This will prevent the migration and breeding of rats. A properly operated sanitary landfill will not support rats.

How to Proceed

Make a sketch of the open dump. Place bait under boxes along rat runs on the top surface, face, and sides of the dump, and in all burrow openings using a long-handled spoon. Mark bait locations on the sketch and check every two or three days; replace bait eaten as needed. Apply bait at the rate of $\frac{1}{4}$ lb/yd^2 of open dump area including dump face. Treat the dump when there is the least activity; late Friday afternoon followed by a quiet weekend should give good results. Do not apply bait during or before rain or snow is forecast.[29]

Recommended Bait Formulations

Ideally, a poisoning sequence of fumigation with calcium cyanide or chloropicrin to quickly reduce the rat population, followed by an acute poison such as zinc phosphide, and then an anticoagulant will do a thorough job. Red squill and ANTU are also satisfactory acute poisons. In any case, Compound 1080 or 1081, or thallium sulfate should not normally be used because of their extreme toxicity and associated hazards.

The following formulation is recommended.

1. Lean ground horse meat, ground fresh fish, or canned cat food; poultry mash during freezing weather—70 lb.
2. Rolled oats (acts as an extender, but may be omitted; if so, increase meat or fish by 20 lb)—20 lb.
3. Flour if using ANTU or zinc phosphide—5 lb.
4. Water—add enough to make bait quite moist. A large amount is needed if rolled oats are used; corn or peanut oil (about 2 gal) should be used as a binder during freezing weather with poultry mash instead of the meat or fish.
5. Add *one* of the following poisons to the above bait.
    a. Fortified red squill (minimum toxicity 500 mg/kg, LD 50)—10 lb.
    b. ANTU, use once a year, (for brown rats only)—$1\frac{1}{2}$ lb.
    c. Zinc phosphide (63%) rat poison (with antimonypotassium tartrate)—$1\frac{1}{2}$ lb.
    d. Anticoagulant (0.5 percent concentrate), Warfarin, Pival, Fumarin,—5 lb.
    e. Diaphacinone, chlorophacinone—1 lb.

[29]G. C. Oderkirk, E. M. Mills, and M. Caroline, "Rat Control on Public Dumps," *Pest Control*, August 1955; "Controlling Rats on Dumps," Fish and Wildlife Service, U.S. Dept. of the Interior, May 1960.

The anticoagulants are more expensive and need to have an edible oil added to prevent drying out. They must be eaten over a period of 3 to 10 days (requiring a bait exposure of at least 2 weeks) to cause death. Commercially prepared baits and wax-treated bait blocks may be purchased though the latter are generally not as palatable to rats. Poison bait (5.a, b, c above) should be exposed 7 to 14 days.

Canned dog or cat food in the ratio of 90 lb of meat to 10 lb of corn meal is a simple and rapid bait base. Prebaiting will determine the location of heavy infestation, suggest the amount of poisoned bait to use, and increase percentage kill. See also Tables 10-5 and 10-6.

Rats may be imported with refuse dumped at a refuse disposal site after it is converted to a sanitary landfill. However, proper daily operation, including prompt spreading and compaction of the refuse as dumped, daily 6-in. earth cover, and a final cover of 2 ft of compacted earth will eliminate the rats brought in.

Control of Rats in Sewers

Sanitary and combined sewers provide food, water, and harborage, permitting sewer-rat populations to reproduce until the environment can no longer support the growth. Population pressures then force the rats to move out through burrows, broken drains and sewers, manholes, catch basins, and house laterals to streets, yards, plumbing, and buildings. Rats can also return to sewers by the same routes when the cold weather sets in. Hence, sewer rats are a constant reservoir for community reinfestation unless controlled. Poison baits are normally placed on manhole ledges or suspended from manholes or catch basins. Zinc phosphide and anticoagulant baits are used, but spoil in a few days unless coated or mixed with paraffin. At least three applications per year are needed, followed by a monitoring system.[30] Lower baits into sewer manholes or basins from the surface and attach wire fastening to steps at top. Entrance of sewer manholes or basins is *dangerous*. See Chapter 4, Safety. Baits should be checked at least twice during the winter months. The USEPA has registered 1081 for sewer application. Every effort should be made to repair and maintain sewers and manholes to prevent entrance of rats in the first place.

Control of Rats in Open Areas

Waterfronts, streams in urban and suburban areas, railroad yards, stockyards, granaries, and farms may also provide food, water, and harborage to support a rat population. These places also require surveillance and control, as described above, as well as mowing of weeds and grass.

[30]W. C. Hickling and J. W. Peterson, "Rat Control as a Public Works Problem," *Public Works*, August 1968.

Control of Mice

House mice are found in most parts of the world. They consume 3 to 5 grams of food and 1 to 2 ml of water daily. A female can have 6 to 10 litters with 5 to 6 mice in a litter; the gestation period is 19 to 21 days. Mouse droppings and hairs are a common contaminant of grains and other food crops.[31]

In general, the control of mice *Mus musculus Linnaeus* is based on the same techniques as those used in the control of rats. House-mice populations may increase as rat populations are reduced. Trapping, unless the infestation is heavy, will often be sufficient. This will provide safe and quick control and no odors from dead mice. Trap baits include peanut butter, fried bacon, cereal, gumdrops, and rolled oats. Mice can pass through a $\frac{1}{2}$-in. diameter hole. Zinc phosphide bait containing antimonypotassium tartrate emetic, anticoagulants, and strychnine alkaloid impregnated grain give satisfactory results. Strychnine is highly toxic and its use around habitation should be avoided. See Tables 10-5 and 10-6 for rodenticides and bait concentrations to poison mice. Primary control measures include: cleanliness, storage of food in metal or glass containers or in the refrigerator, placing of garbage in covered metal cans, sealing of cracks and openings around pipes, floor and walls, screening of openings with $\frac{1}{8}$-in. wire mesh, and keeping outside doors closed. The use of snap traps, multiple-catch traps, glue boards, and tracking powders will generally give better control than poisons. Paradichlorobenzene and naphthalene repel mice.

Resistance

A rat control program should continually evaluate the possible development of rat and mouse resistance to anticoagulant bait and the need for alternate baiting. Resistance has been reported in Scotland, Denmark, England, Wales,[32] France, the Netherlands, and the United States. Significant levels of resistance have been found in Norway rats in numerous cities along the eastern seaboard and extending to Chicago and Houston. Resistance levels vary from zero to as high as 76 percent with 19 out of 40 city areas sampled having rat resistant rates of 8.5 percent or greater.[33]

Where resistance to anticoagulant rodenticides is developing, it is advisable to switch to a single-dose poison such as fortified red squill, zinc phosphide, or ANTU every 6 to 12 months to kill as many as possible of the resistant rats. Calcium cyanide can also be used for burrow gassing in situations where it is safe to use the chemical.[34]

[31]"Mouse Control," *Pest Control*, August 1976, pp. 27–35.

[32]E. W. Bentley, "Developments in Methods for the Control of Rodents," *J. R. Soc. Health*, May–June 1970, p. 129. Resistance found in North Carolina 1971 and New York State 1972.

[33]S. C. Frantz and C. M. Padula, "Recent developments in anticoagulant rodenticide resistance studies: Surveillance and application in the United States," *Proc. Ninth Vert. Pest Conf.*, J. P. Clark, Ed., University of California Press, Davis, 1980, pp. 80–88.

[34]Harry D. Pratt, Bayard F. Bjornson, and Kent S. Littig, *Control of Domestic Rats & Mice*, PHS, DHEW, CDC, Atlanta, Ga. 30333, October 1979, pp. 26–27.

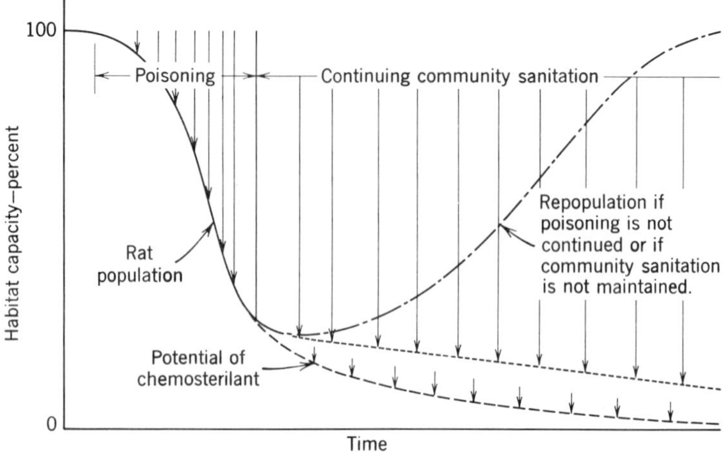

Figure 10-14 Rat population reduction and eradication using different techniques.

The increasing resistance to anticoagulant baits again emphasizes the importance of community sanitation and cooperation as basic and fundamental to the lasting control of rats.

Rat Control by Use of Chemosterilants

Various antifertility agents have been proposed to prevent the reproduction of rats. This technique has promise and can be applied as an *adjunct* to basic environmental sanitation and rat poisoning, particularly in those places where it is not practical to remove food, water, and harborage. Infested combined sewers, sanitary sewers, streams, wharfs, and farms are examples. But first an effective bait formulation that rats will accept must be developed. Studies with the use of a male chemosterilant show that even with 90-percent sterility, the remaining fertile males are quite adequate to impregnate the females with no resultant reduction in the rat population. Female chemosterilants seem to offer the most promise.[35]

In practice, it is rare to find community sanitation that is 100 percent effective, and complete dependence on area-wide poisoning for rat control is expensive and in many cases impractical. Figure 10-14 illustrates how a combination of methods, including chemosterilants, might be applied to eradicate a rat population.

If a rat control program is suspended and community sanitation is relaxed, the rat population will become reestablished within 12 months.* Community sanitation is the key to permanent reduction of rat population.

*New York State Health Department experience in Buffalo, N.Y., 1976–1977.

[35]Joe E. Brooks and Alan Bowerman, staff reports, New York State Dept. of Health, Albany, 1968–1971.

PIGEON CONTROL

The Health Hazard

Although the public health significance of pigeons has yet to be fully determined,[36] they do pose economic and aesthetic problems and are a potential disease threat to man.[37] The hazard is greatest to workers engaged in the removal of pigeon manure and nests; suitable dust respirators and protective clothing should be used. The diseases that may be spread to man by pigeon droppings and dust are ornithosis (a mild form of psittacosis), histoplasmosis, cryptococcosis, and others; but the probability is remote except where there are large pigeon populations in close proximity to man.

Control Measures

Pigeon control should start with the removal of food, water, and harborage. Emphasis should be placed on sanitation and pigeon stoppage. The following measures have been used where permitted by state and local laws and where there is no serious public opposition.

1. Eliminate sources of food. Prohibit feeding. Remove animal wastes and keep animal shelters clean. Clean up spilled grain and other food sources. Properly store and collect refuse. Convert open dumps to sanitary landfills. Clean empty lots of weeds and debris.
2. Trap and humanely dispose of pigeons. This is a slow and expensive operation.
3. Locate and enclose nesting places. Close openings in buildings, sheds, attics, eaves, steeples. Screen with $1\frac{1}{2}$-in. durable mesh hardware cloth. To keep out sparrows use $\frac{3}{4}$-in. mesh; use $\frac{1}{2}$-in. mesh to keep out rats and $\frac{1}{4}$-in. mesh to keep out mice and bats. Slope ledges and windowsills on angle of 45 deg or more.
4. Place repellents on roosting places such as ledges, signs, ridges, roof gutters, and cornices. Special chemicals and naphthalene, calcium chloride, and lye have been used but they need frequent replacing.
5. Use professional sharpshooters.
6. Use poisoned bait (strychnine-treated corn) when "beneficial" birds have migrated, where permitted.
7. Use treated bait to temporarily anesthetize (alpha chloralose) the pigeon. Then pick up and remove; release songbirds and humanely dispose of pigeons. Ornitrol, a chemosterilant has a USEPA approval label for use where it is difficult to eliminate all pigeons. Ornitrol mixed with corn is said to be effective for about 6 months.
8. Cats discourage pigeon roosting. Owls, hawks, mice, and rats prey on pigeons.
9. Mist sprays (5 percent ammonia in water with a detergent) penetrate the feathers and wet the body, causing freezing.
10. Grounded electric wires or screens can be used. Noisemakers and explosives are of temporary effectiveness.

[36]E. S. McDonough, Ann L. Lewis, and L. A. Penn, "Relationship of Cryptococcus neoformans to Pigeons in Milwaukee, Wisconsin," *Pub. Health Rep.*, **81,** 12 (December 1966).
[37]Harold George Scott, "Pigeon-Borne Disease Control through Sanitation and Pigeon Stoppage," *Pest Control*, **32,** 9 (September 1964).

A control program to eliminate nesting places should be preceded by ectoparasite control. It is necessary to destroy the mites, ticks, lice, fleas, and other bugs in the nests so they do not migrate when the pigeons fail to return. Infested areas can be treated with malathion emulsion or dust. These principles can also be applied to the control of nuisance sparrows and starlings.

Bat Control[38]

Although bats serve a useful purpose in keeping down the number of insects, some bats become infected with rabies—probably less than one percent—and pose a threat to both humans and animals. Bats are also objectionable because of the odor from their droppings and urine and the noises they make in roosting areas.

Anyone bitten by a bat should immediately wash the wound thoroughly with soap and water and go to a hospital emergency room or see a doctor without delay. Try to capture the bat without destroying the head and, *using gloves* place the body in a plastic or heavy paper bag, take it to the nearest health department. The saliva of dead or live rabid animals may contain the rabies virus, and can enter the body through scratches or open cuts. Aerosol of urine can enter through the nose or mouth and may be infective through inhalation but not ingestion. Anyone bitten by a bat which has not been captured should receive rabies shots without delay. If the bat can be captured and *tested immediately* for rabies, antirabies treatment may be delayed until the laboratory results are determined.

The best way to keep bats out of buildings is to close openings that allow them to enter their roosts, usually attics. Observing bats entering or leaving a building just before dawn or at dusk will help locate openings which can be closed with screening, wood, or sheet metal. Elimination of dark spaces such as behind shutters by adding wooden blocks between the building and shutters to provide a 2-in. space will discourage roosting. Keeping low-wattage light bulbs or fluorescent bulbs on 24 hours a day for several weeks during the early spring and summer months when bats are returning to summer roosts may be effective when all openings in an attic cannot be closed. Naphthalene flakes (100% granules) or moth balls spread over the floor of the roosting space at the rate of 5 lb per 2000 ft^3 helps to repel bats in areas with relatively poor ventilation.[39] Bat-proofing should be done in the late fall through winter, when bats are hibernating in caves. Night, when they are away from their roost, is also a good time. Bat-proofing should not be conducted in late spring or early summer because baby bats can be trapped inside.

[38]"Control of Bats," New York State Dept. of Health, Albany, N.Y., 1979.

[39]William G. Smith, "The Little Brown Bat," Cooperative Extension, Cornell University, Ithaca, N.Y., March 1976.

CONTROL OF POISON IVY, POISON OAK, AND POISON SUMAC

The number of persons poisoned by these plants in the United States each year has been estimated at at least 500,000. In the many suburban communities where poison ivy or poison oak grow unchecked, it is a continual source of parental concern. The best preventive is recognition of the plant and its avoidance. See Figure 10-15.

The poison of poison ivy, oak, and sumac is an oleoresin (urushoil) that is found in the leaves, bark, flowers, roots, and fruit, but not in the wood. It is effective even after months of drying. The oily poison is nonvolatile; it is not found in pollen. Birds, livestock, pets, tennis balls, golf clubs and balls, shoes, gloves, auto tires, and tools that have come into contact with the plant serve to spread the poison.

Control measures for poison ivy are generally applicable to poison oak and poison sumac. Clothing that has come in contact with the plant should be handled with care and dry cleaned; the cleaner should be advised.

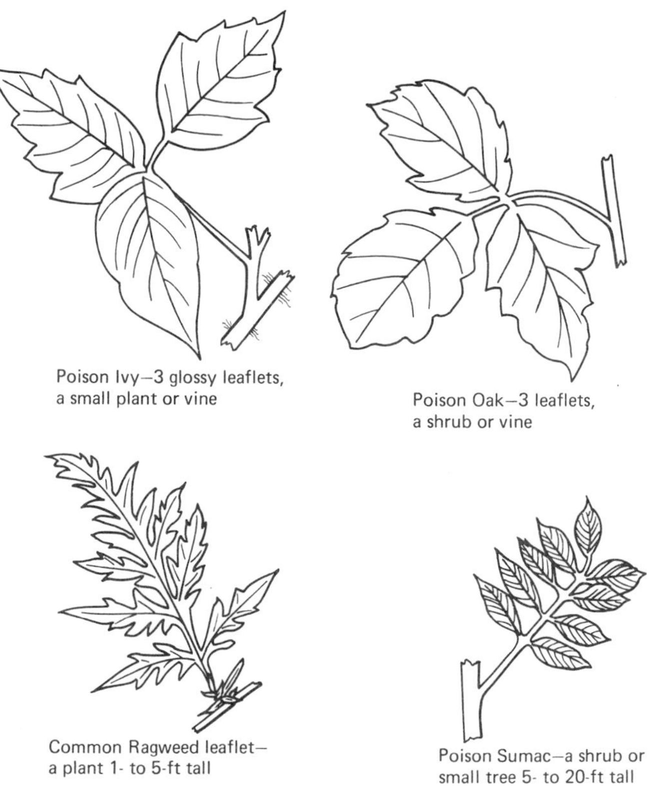

Figure 10-15 Noxious weeds.

Eradiction

Grubbing the plant out by the roots is possible although there is danger of infection. Part of the root system and debris are often left behind. It is an expensive method and is impractical to remove completely.

Burning off the plant is not advisable because smoke will carry particles long distances and spread the infection.

Chemical treatment is the recommended method of controlling these noxious weeds, but the chemical must not be applied where it can drain into a reservoir, stream, lake, or other body of water. Check with health and conservation agencies. Carefully follow the manufacturers' recommendations. Some of the better known weed killers are listed.

1. Ammonium sulfamate, known commercially as Ammate, does a good job; it will kill all vegetation but is somewhat corrosive to spray equipment. Application of a solution of 1 lb of Ammate crystals to 1 gal of water gives excellent control in 48 hr.
2. Borax is also effective but will damage surrounding plants if not carefully applied. The recommended dosage is 4 lb/100 ft^2.
3. The weed killer, 2,4-D, dichlorophenoxyacetic acid, takes 2 to 3 weeks. Here again care must be used not to damage other broad-leaf plants. It is not too effective in shaded areas.
4. A refinement of 2,4-D is 2,4,5-T (trichlorophenoxyacetic acid). It is a hormone-type selective herbicide that can be applied to stems or leaves during the growing season. The spray must do a thorough wetting job to be effective. A 2,4,5-T–diesel-oil solution sprayed thoroughly around the base of a plant in late winter or early spring or summer does a good job. Equal amounts of 2,4-D and 2,4,5-T make a good overall herbicide, but 2,4,5-T may contain minute quantities of dioxin, a deadly poison and unintentional contaminant. *The USEPA suspended use of 2,4,5-T* as of March 1, 1979.
5. A solution consisting of 3 lb of salt in 1 gal of water applied at the rate of 1 gal to 800 ft^2 does a good sterilizing job.
6. Calcium chlorate made up by 1 lb in 1 to 2 gal water is also effective. However, it is highly inflammable and will cause burns.
7. Repeated ploughing and cultivation will cause vines to gradually die out.
8. Sodium arsenite formulated with 1 lb of chemical in 5 gal of water is very effective but must be used with care as it is very poisonous to man and fish.
9. Amino triazole (Amitrole) is a very selective weed killer for poison ivy. It is available as a liquid concentrate, a water-soluble powder, or in a pressurized aerosol can. The chemical kills the leaves and roots in 10 to 14 days.

Remedies

Various remedies have been suggested in the past; the best is recognition and avoidance of the plant. Better treatment methods can be developed. Some of the remedies used are listed below.

1. An old remedy that has been found effective if done *immediately* after exposure is repeated thorough scrubbing with soap and water. Prior cleansing with alcohol is also advised. A good detergent should be equally effective in removing the oily irritant. The irritant penetrates the skin within 30 min.
2. Aluminum acetate and lead acetate—a solution of a zirconium oxide combined with the antihistamine drug, phenyltoloxamine dihydrogen citrate, applied to exposed areas of the skin within 8

hr is reported to alleviate or prevent the symptoms. This should be applied under medical supervision. It may cause granulomas (small hard lumps) in sensitive individuals.

3. A 10 percent solution of tannic acid dissolved in alcohol repeated at 6-hr intervals has been recommended. Do not use on face or genitals.
4. A 5 percent solution of ferrous sulfate or copperas is a better preventive treatment.
5. Calamine lotion with about 2 percent phenol, Burrow's solution, compresses, and zinc oxide lotion relieve itching.
6. Jewelweed plant juice swabbed on blisters is found effective by some persons.
7. A medication injected as a liquid that is said to provide immunity against poison ivy and poison oak for up to 12 months is available.

The disease tends to disappear in 10 days to 3 weeks, and in uncomplicated cases in 5 to 7 days. The symptoms may appear after a few hours to 7 days. Only mild cases of dermatitis should be treated by lay persons; refer problem cases to a physician.

CONTROL OF RAGWEED AND OTHER NOXIOUS WEEDS

In general, there are three kinds of weeds. *Annuals* live only one year, but are propagated by their seeds if not destroyed. Ragweed and crabgrass are examples of annuals. *Biennials* grow slowly the first year but develop a taproot and growth of leaves close to the ground. The second year rising stems and seeds are produced. Formation of seeds should be prevented to control this group. Examples are burdock and wild carrot. *Perennials* live three years or longer, each year producing runners, underground stems, roots, or bulbs. Poison ivy, Canada thistle, milkweed, bindweed, wild onions, and dandelions are examples of perennials. To control perennials requires complete destruction of all roots.

The control of weeds is accomplished by preventing spread of weeds into new areas, destruction of top weeds and underground parts of weeds, and destruction of weed seeds in the soil. Frequent mowing or cutting will prevent seed formation. Chemicals offer one of the most effective means of controlling weeds where physical control is impractical. Some are selective in killing weeds but not crops or grasses. There are three general types of weedicides: soil sterilants which are absorbed by the plant root systems; contact weedicides which are toxic to the plant living cells it covers; and hormone or growth regulators, which are absorbed by the plant surfaces above ground and carried through the plant to the root system, sometimes also referred to as systemic weedicides. The best time to apply a weedicide is in the late spring or early summer when plant growth and development is taking place.

Weeds take over when land is neglected. Some of the conditions that encourage the growth of weeds are soil abuse; overgrazing; erosion, uncontrolled drainage, and flooding; overcultivation; deforestation; same kind of trees; and abandoned farms. It may take 10 to 15 years before plant growth progresses from the weed stage, by plant succession, to a stabilized shrub and tree phase; however, a tremendous amount of damage may be done in the interim. These conditions and practices conducive to weed growth should therefore be eliminated. Time of planting and mulching can minimize weed growth.

Reasons for Control

It has been reported that 15 million Americans suffer from hay fever. Approximately 5 percent of the persons living in the Northeastern United States have pollen hay fever and 80 to 90 percent are sensitive to ragweed pollen. Ragweed pollen and other plant materials such as mold spores (aeroallergens) either singly or in combination with man-made air pollutants cause an estimated 18 million Americans significant discomfort and disability.[40]

Although mild cases do not cause great inconvenience, it can cause complications with asthma. Asthma occurs in about two-thirds of the persons who have suffered from hay fever for 25 or more seasons.[41] Hay fever can also lead to more serious illnesses. Thousands of persons are incapacitated annually during the pollen season.

Weeds disfigure the landscape, introduce driving hazards by reducing visibility, detract from recreational areas, clog ditches, and encourage mosquito breeding. The loss due to weeds and costs to farmers to control weeds runs into the billions of dollars.

Hay Fever Symptoms and Pollen Sampling

Hay fever, more correctly, pollenosis, causes sneezing, itchy runny eyes, swelling of the nasal passages, and an accompanying water discharge. Inhalation of air laden with pollen causes the symptoms to a greater or lesser degree depending on type, susceptibility, and, apparently, heredity. Relief may be obtained by acquired immunity, moving to a pollen-free area, air conditioning,* air purification, and weed control. Sources of pollen are trees in the spring, grasses in the summer, and weeds in the fall. Mold and fungus spores such as found around dead leaves, hay, and straw contribute to the problem. Ragweed is a major culprit and is given major discussion here.

A mature ragweed plant can produce up to one billion pollen grains in one season. Concentrations of less than 25 pollen grains/yd^3 of air in 24 hr usually do not produce allergic reactions. One method of making pollen counts is to read a light oil-, silicone grease-, or clear petroleum jelly-coated slide (glass 1″ × 3″ microscope slide with one end frosted), kept protected from the rain, after 24-hr exposure. The pollen grains on a 1.8 sq cm area are counted. A pollen count of 25

*Requires a filter that will remove particles greater than 10 μm.

[40] *Basic Concepts of Environmental Health*, DHEW Pub. (NIH)77-1254, National Institute of Environmental Health Sciences, Research Triangle Park, N.C. 27709.

[41] Alfred H. Fletcher, "Procedures of Promoting and Operating Ragweed Control Program," *Pub. Health News*, N.J. Dept. of Health, May 1955, pp. 170–175.

or more equals 1 point; the slide having the maximum count for a day is read, and 1 point is allowed for each 100 count; the total seasonal count is divided by 200 to give additional points. The sum of all these points is the pollen index for the area. An index of less than 5 is considered practically free of ragweed pollen contamination. An index of 5 to 15 is moderately free, and an index greater than 15 is indicative of a heavy concentration of pollen. Another generally accepted standard uses a 2 cm^2 area (2cm × 1cm) for counting. In this method a ragweed pollen grain count of $7/cm^2$ is sufficient to show the symptoms of pollinosis or hay fever. Readings of 25 or more signify heavy contamination. The Durham sampler is generally used. There is need for improved methods, equipment, and techniques to give more precise results. In any case, counts should separate pollens into the various genera. A count that does not do this is of little value.

Problems

1. Some pollens are carried long distances but are diluted in the atmosphere. Most ragweed pollen grains settle to the ground within about 200 ft of the source.
2. Responsibility for control may be vague.
3. Legal support may be lacking.
4. Many weeds, grasses, and trees contribute pollen that causes hay fever.
5. Measuring devices to measure the amount of pollen in air are not accurate.
6. Concentration and type of pollen causing specific allergic reactions is not well established.

A Control Program

1. The most practical way to control hay fever is to treat the environment rather than the patient. Therefore plan a broad program and attack on ragweed, the principal culprit.
2. Make a survey to locate ragweed. Use police, Boy Scouts and Girl Scouts, older boys, civic clubs, schools. Locate ragweed areas on a map, scale 1 in. = 400 to 600 ft. Fifty square miles can be mapped by two men in a car in a few days.
3. Have an ordinance adopted prohibiting ragweed. Solicit the support of hay-fever sufferers in key position.
4. Educate the public in possible control methods; pull out before flowering, spray, and so forth. Give cost estimates.
5. Study results of the survey, publicize program, and obtain budget allowance for a long-range program.
6. Enlist the cooperation of adjoining communities and conservation and highway departments.
7. Use existing equipment such as mosquito spray equipment and purchase additional equipment within budget limitations.
8. Start the spray program in early summer. Proceed with demonstrations, publicity—use newspapers, posters, circulars, schools, civic organizations, official support, and private agency support. Ragweed appears in early spring. Give periodic reports to the press. See Figure 10-16.
9. Seed with grass, Japanese honeysuckle, and so on, to replace ragweed. Issue leaflets dealing with identification and control of plants detrimental to health.
10. Spraying in early or middle summer will prevent pollenation. A spraying program using 2,4-D or other herbicide centrally directed and administered has proven effective in urban areas. Conduct training program for field and supervisory personnel.

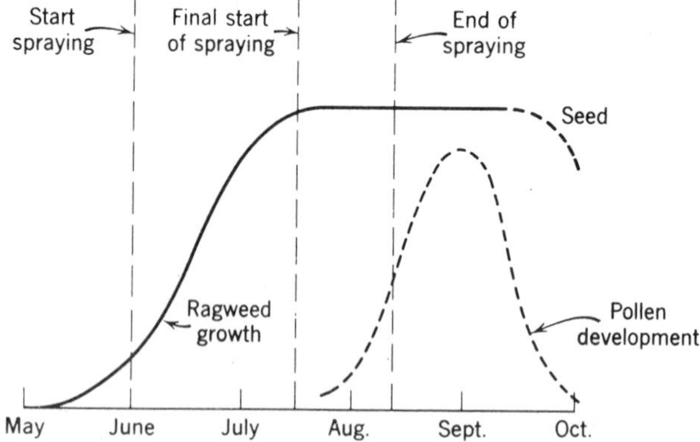

Figure 10-16 Typical ragweed and pollen development.

11. NOTE! Weeds will not grow where the ground is occupied with grass or other acceptable vegetation. Grass that is kept cut or grazed will not cause appreciable hay fever, provided it is not permitted to grow and get out of control. Spraying with 2,4-D kills ragweed, prevents seeds from maturing the following year, and encourages grasses and some plants to grow, thus discouraging weed growths. Integrate the program with tree spraying, caterpillar control, and work of other departments. Start cutting weeds and grass on streets and highways.

12. Start the program in area where most people will benefit.

13. Control ragweed along highways, particularly where shoulders are not seeded to grass. Ragweed is the principal pollen-producing offender. Periodic mowing of weeds is effective.

14. Program should be under supervision of the health department, for it is conducted for the health of the citizens. A coordinating committee should contain a health officer, integrated pest management and agricultural specialist, highway superintendent, conservationist, public health educator, and public health engineer or sanitarian.

15. Intelligent field supervision is necessary to prevent damage to valuable plants, human hazard due to spray drift, and contamination of reservoirs, surface and ground waters. Consider wind, droplet size, and spray height above ground to avoid drift.

16. Evaluate control measures by determining ragweed area reduction as result of spraying. Areas not sprayed will determine work areas in the next year. Pollen reduction, as measured by pollen sampling stations over wide areas and at different atmospheric levels, correlated with time of year, day, weather, and wind direction, will also indicate the progress being made if sufficient readings to be representative are taken.

Eradication

Pull out by the roots before flowering.

Prohibit by ordinance the growing of noxious weeds on private property and provide for removal or spraying of any private property where the owner has permitted the noxious weeds to grow.

Use a special burner or flame thrower for nonselective control such as on reservoir shore and on ditch or stream banks where chemicals cannot be used and mowing is not possible.

Place greater emphasis on physical removal of weeds and on tillage, mowing, and mulching as appropriate thus eliminating or minimizing herbicide usage.

Chemical treatment, centrally directed and administered, is effective. A good weed killer should be selective, readily absorbed by plants, nontoxic to humans, and effective in minute concentrations; 2,4-D is the chemical of choice. Some weed killers are listed below.

1. 2,4-D was discovered in 1944.* It is available as a powder, liquid, or tablet. It is a selective weed killer. 2,4-D will control annual, biennial, and perennial weeds. Several general types are available. The type used depends on cost, weed, and equipment. Broadleaf weeds and bent grasses, including most decorative plants, are killed, thereby encouraging the growth of grasses and other plants that in turn compete with and discourage the growth of weeds. The chemical works on water plants too. It is more effective in the light than in the shade.
2. Sodium chlorate is a good patch killer; but the chemical presents a fire hazard.† Use 200 to 300 lb of dry sodium chlorate per acre, or a solution of 1½ oz in 1 gal water for each 100 ft^2.
3. Sodium arsenite is very toxic to humans and animals.
4. Kerosene and other petroleum products.†
5. Borax applied at the rate of about 2000 lb/acre will persist for 2 years.† Also other boron compounds.
6. Ammonium sulfamate (known commercially as AMS or Ammate).
7. MCPA, 2,4-DB, 2,4-DEP, TCA, IPC. (All synthetic hormones.)
8. Ametryne, Atrazine, Benefin, Bensulfide,† Paraquat, DCPA, Dichlobenil, Diuron, DSMA, Fenuron, Monuron,† Sesone, and many others.

Education and promotion are essential to assure continuation of appropriations and public support. Exhibits, news releases, radio programs, demonstrations, and other health education measures should be used on a planned basis.

TCDD (dioxin) is likely to be a contaminant in 2,4,5-T, 2,4,5-TP (Silvex), Erbon, Ronnel, and 2,4,5-trichlorophenol, and an ingredient in the production of the antiseptic hexachlorophene. The USEPA suspended registration of certain uses of 2,4,5-T and Silvex on February 28, 1979.[42,43]

Equipment for Chemical Treatment

Use a mounted 100- to 1000-gal tank, a 15- to 30-gpm pump, capable of producing 50 to 100 psi; two 100-ft lengths of ½- or ¾-in. pressure hose. A nozzle diameter of about 0.08 in. is satisfactory. A man to handle each hose, one man to help, and one driver will be needed. One of the men should be responsible for keeping written reports of the work done each day. Converted street flushers have been used.

A knapsack sprayer of 3- or 5-gal capacity, sprinkling cans, garden sprayer, and hand carts equipped with perforated pipe distributor are suitable for patch treatment.

The use of a fogging machine or aerial sprayer is not recommended except for

*Synthetic hormone.

†Soil sterilant.

[42] *Dioxin: Sources, Transport, Exposure and Control*, USEPA Office of Toxic Substances, Washington, D.C., April 1979.

[43] *Environmental Quality—1979*, The Tenth Annual Report of the Council on Environmental Quality, December 1979, pp. 210–213, GPO, Washington, D.C. 20402.

large isolated areas since it cannot be controlled for spot treatment. A jeep with tank trailer and pressure sprayer is very useful.

Chemical Dosages

In the early season, when the weeds are 4- to 6-in. high, use for example, a 0.1 percent 2,4-D solution applied at the rate of 150 gal/acre or 1 gal to about 300 ft^2. Use 300 gal/acre when the weeds are in full leaf. One pound of 2,4-D to 100 gal of water makes a 0.12-percent solution. In all cases the weeds should be thoroughly wetted to produce satisfactory results.

One four-man crew can spray 2 to 6 acres per day depending on size and proximity of plots. As much as 30 acres can be covered in open country.

Control of Aquatic Weeds

The control of algae and other aquatic organisms and aquatic weeds is discussed in Chapter 3. Aquatic weed control as it relates to bathing beaches is discussed in Chapter 9.

BIBLIOGRAPHY

Agricultural Chemicals, Manufacturing Chemists' Association, Inc., 1825 Connecticut Avenue, N.W., Washington, D.C., 1963.

Apply Pesticides Correctly, USEPA, Washington, D.C. 20460.

Bottrell, Dale R., *Integrated Pest Management*, GPO, Washington, D.C. 20402, December 1979.

Cleaning our Environment—A Chemical Perspective, A Report by the Committee on Environmental Improvement, American Chemical Society, Washington, D.C., 1978, pp. 320–377.

Controlling Household Pests, Home and Garden Bull. 96, USDA, Washington, D.C., June 1977.

DeVaney, Thomas E., *Chemical Vegetation Control Manual for Fish and Wildlife Management Programs*, U.S. Dept. of the Interior, Washington, D.C., January 1968.

Hayes, Wayland J., Jr., *Clinical Handbook on Economic Poisons*, PHS Pub. 476, DHEW, Washington, D.C., 1963.

Headquarters, Dept. of the Army, *Military Entomology Operational Handbook*, TM 5-632, Washington, D.C., 1965.

Pratt, Harry D., Bayard F. Bjornson, and Kent S. Littig, *Control of Domestic Rats and Mice*, PHS, DHEW, CDC, Atlanta, Ga. 30333, October 1979.

Pratt, Harry D. and Kent S. Littig, *Insecticides for the Control of Insects of Public Health Importance*, PHS, DHEW, CDC, Atlanta, Ga. 30333.

Pratt, Harry D., Kent S. Littig, and George Scott, *Household and Stored-Food Insects of Public Health Importance and Their Control*, PHS, DHEW, CDC, Atlanta, Ga. 30333, July 1977.

Proceedings of the Seminar on Environmental Pests and Disease Vector Control, Karl Westphal, Ed., New York State Dept. of Health, Albany, N.Y., January 20–24, 1975.

Proceedings of the 1977 Seminar on Cockroach Control, Karl Westphal, Ed., New York State Dept. of Health, Albany, N.Y., March 28–31, 1977.

Specifications for pesticides used in public health, 5th ed., WHO, Geneva, 1979.

11

THE RESIDENTIAL AND INSTITUTIONAL ENVIRONMENT

The WHO Expert Committee on the Public Health Aspects of Housing defined housing (residential environment) as

> the physical structure that man uses for shelter and the environs of that structure including all necessary services, facilities, equipment and devices needed or desired for the physical and mental health and social well-being of the family and individual.[1]

Every family and individual has a basic right to a *decent home* and a *suitable living environment*. However, large segments of population in urban and rural areas throughout the world do not enjoy one or both of these fundamental needs. Housing therefore must be considered within the context of and relative to the total environment in which it is situated, together with the structure, supplied facilities and services, and conditions of occupancy.

The realization of a decent home in a suitable living environment requires clean air, pure water and food, adequate shelter, and unpolluted land. Also required are freedom from excessive noise and odors, adequate recreation and neighborhood facilities, and convenient community services in an environment that provides safety, comfort, and privacy. These objectives are not achieved by accident but require careful planning of new communities and conservation, maintenance, and redevelopment of existing communities to ensure that the public does not inherit conditions that are impossible or very costly to correct. In so doing, that which is good or sound, be it a structure or a natural condition, should be retained, restored, and reused.

SUBSTANDARD HOUSING AND ITS EFFECTS

Growth of the Problem

Practically all urban and rural areas contain substandard, slum, and blighted areas.* The causes are numerous; they are not easily detected in the early stages or for that matter easily controlled.

[1] *WHO Tech. Rep. Ser.*, **225**, 1961, p. 6.

*Substandard housing is said to exist when there are 1.51 or more persons per room in a dwelling unit, when the dwelling unit has no private bath or is dilapidated, or when the dwelling unit has no

There has been in the United States a rapid growth in population and in the size of urban areas, with most of the growth taking place in the suburbs. Many of the older cities that have not enlarged their boundaries are no longer experiencing a population increase.† Many have shown a decline. The city of Buffalo and Erie County, New York, population trend graphs (Figure 11-1) are typical of population shifts in old cities and metropolitan areas.

Between 1800 and 1910 a very rapid growth took place in the cities of the United States due to the mass movement of people from rural to urban areas and as a result of heavy immigration. Many of the newer immigrants sought out their relatives and compatriots, who usually lived in cities, thereby straining the available housing resources.

With the movement of large numbers of people to cities, urban areas became congested; desirable housing became unobtainable. Inadequate facilities for transporting people rapidly and cheaply to and from work made it necessary for many people to accept less desirable housing in the cities, close to their work. The inability of the ordinary wage earner to economically afford satisfactory housing left little choice but to accept what housing was available. Some landlords and speculators took advantage of the situation by breaking up large apartments into smaller dwelling units, by minimizing maintenance, and by constructing "cheap" housing.

Unfortunately, assistance or leadership from local government units to control potential problems is slow; there is usually a lag between the creation of housing evils and the enactment of suitable corrective legislation, and enforcement. For example, Tenement House Law applicable to New York City was passed by the New York State legislature in 1867. However, subsequent amendments and discretionary powers vested in the board of health resulted in nullifying the law to a large extent. A new Tenement House Law of 1901 was made mandatory for New York City and Buffalo. But many amendments were made to the law within the next 10 years, as a result of powerful pressure from corporate property owners, which practically defeated the original intent of the law. The experience in New York City and Buffalo shows the wisdom of having an informed public opinion to support legislation for the public good and to combat the pressures of vested interests. It also shows the practical dangers of discretionary powers.

---

running water. Other bases used are described under Appraisal of Quality of Living, this chapter. An extensive definition of substandard buildings is given in the *Uniform Housing Code* (International Conference of Building Officials, 5360 South Workman Mill Road, Whittier, California 90601, 1976 Edition, pp. 22–24).

A slum is "a highly congested, usually urban, residential area characterized by deteriorated unsanitary buildings, poverty, and social disorganization" (*Webster's Third International Dictionary*, 1966). A slum is a neighborhood in which dwellings lack private inside toilet and bath facilities, hot and cold running water, adequate light, heat, ventilation, quiet, clean air, and space for the number of persons housed. It is also a heavily populated area in which housing and other living conditions are extremely poor.

To blight is to "prevent the growth and fertility of; hence to ruin; frustrate" (Webster). A blighted area is an area of no growth in which buildings are permitted to deteriorate.

†The 1980 U.S. Census of Housing shows a general population increase in the "Sun Belt" states in the South and the Southwest, but a decline or stabilizing in many of the other states.

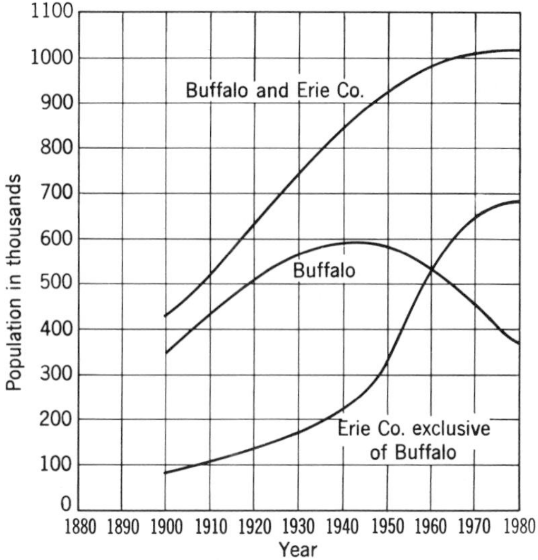

Figure 11-1 Population trends in city of Buffalo and in Erie County, N.Y.

	Population		
Year	Buffalo	Erie Co.	Total
1900	352,387	81,399	433,686
1910	423,715		
1920	506,775	127,913	634,688
1930	573,076	189,332	762,408
1940	575,901	222,476	798,377
1950	580,132	319,106	899,238
1960	532,759	530,931	1,064,688
1970	462,768	650,723	1,113,491
1980	357,381	657,545	1,014,926

In 1912 the legislature strengthened the law. The Tenement House Law applied only to dwellings with three or more families who did their own cooking on the premises. A Multiple Dwelling Law with wider coverage became effective in New York City in 1929, replacing the Tenement House Law. It became effective in Buffalo in 1949. It is to be noted that between 1867 and 1912 the existent laws were of doubtful effectiveness, yet in the 32-year period between 1880 and 1912 the New York City population increased from about 2 million to 5 million and that of Buffalo from 155,000 to 425,000.

Obsolescence is another factor that causes growth of slums. Land and property used for purposes for which they may have been well suited in the first place may no longer be suitable for that purpose. An example is the slum frequently found on the rim of a central business area, originally a good residential district convenient to business. This may start with the expansion or spill-over of business into the contiguous residential areas, thereby making the housing less desirable. People next door or in the same building, desiring quiet and privacy, move. Owners are hesitant to continue maintenance work, causing buildings to deteriorate. The landlord, to maintain income, must either convert the entire building to commerical use or lower rents to attract lower-income tenants. If the landlord converts, more people leave. If the landlord lowers the rents, maintenance of the building is reduced still further and overcrowding of apartments frequently follows. The progressive degradation from blight to slum is almost inevitable. As blight spreads so does crime, delinquency, disease, fires, housing decay, and welfare payments.

There are areas that are slums from the start. The absence of or failure to enforce suitable zoning, building, sanitary, and health regulations leads to the development of "shanty towns" or poor-housing areas. Add to this small lots, cheap, new, and converted dwellings and tenements that are poorly located, designed, and constructed, just barely meeting what minimum requirements may exist, and a basis for future slums is assured.

An indifferent or uninformed public can permit the slums to develop. The absence of immediate and long-range planning and zoning, lack of parks and playgrounds, poor street layout, weak laws, inadequate trained personnel to enforce laws, pressure groups, lack of leadership from public officials and local key citizens, and poor support from the courts and press make the development of slums and other social problems only a matter of time.

The population growth has been taking place outside the major older cities, but the rate of housing construction and rehabilitation has not kept pace with the needs of population growth. Federal grants and loans encouraging rehabilitation of sound housing; the high cost of new housing, facilities, and services; and the high cost of gasoline have slowed down the migration of people from the cities to the suburbs. The loss of housing due to obsolescence, abandonment, decay, and demolition further compound the problem. The National Association of Home Builders estimates that more than 2 million new housing units are needed each year in the 1980s, but construction is not keeping pace.

According to a 1974 housing survey, lack of plumbing, leaking roofs, inadequate heating, and generally bad housing repair, common problems in the late 40's, have almost been eliminated. Generally, however, low-income families still occupy homes with defects, and housing and neighborhoods in sections of U.S. cities show obvious signs of deterioration.[2]

It is apparent that unless the rate of new-home construction is accelerated, the rehabilitation of sound substandard dwelling units strongly encouraged, and the conservation and maintenance of existing housing required, a decent home for every American family will never be realized. Added to this is the continued need to provide public housing and financial assistance to the low-income family.

Critical Period

There comes a time in the ownership of income property, particularly multiple dwellings, when the return begins to drop off. This may be due to obsolescence

[2] *Environmental Quality*, the eighth annual report of the Council on Environmental Quality, December 1977, Sup. of Documents, GPO, Washington, D.C. 20402, pp. 308–310.

and reduced rents or increase in the cost of operation and maintenance. At this point, the property may be sold (unloaded), repairs may be made to prevent further deterioration of the property, the property may be sold and demolished for a more appropriate use, or a minimum of repairs made consistent with a maximum return, and tax payments delayed. This is a critical time and will determine the subsequent character of a neighborhood. In situations where repairs are not made or where a property is sold and repairs are not made, and taxes are not paid, the annual rental from substandard property may equal or exceed the assessed valuation of the property. A complete return on one's investment in five to seven years is not considered unusual in view of the so-called risks involved. Because of this, housing ordinances should be diligently enforced and require owners to reinvest a reasonable part of the income from a property in its conservation and rehabilitation, *at the first signs of deterioration*. This would tend to prevent the "milking" of a property and nonpayment of taxes. The burden on the community to acquire and demolish a worthless structure for nonpayment of taxes, or maintain an eyesore and fire and accident hazard, would be lessened. Shortening the time, from the usual 5 years to 2 or 3 years, required for the initiation of "in rem" proceedings to foreclose for real estate tax delinquencies, while the property still has value, would reduce abandonment if coupled with firm but reasonable code enforcement. Cause for further property devaluation and extension of the blighting influence would also be reduced. See Figures 11-2 and 11-3.

Figure 11-2 Run-down, filthy, vermin-infested backyards present many real health hazards. (Buffalo Municipal Housing Authority.)

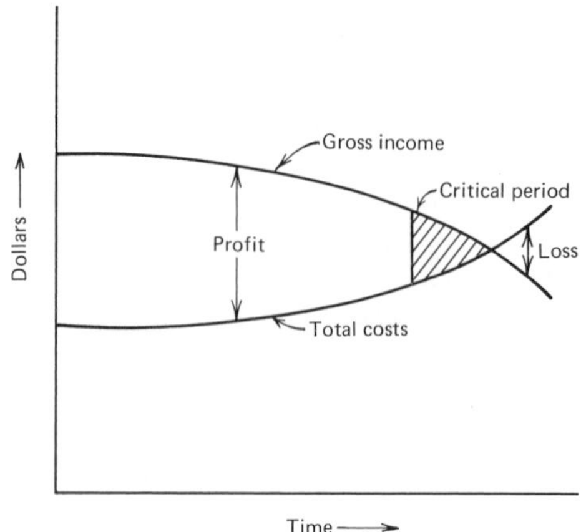

Figure 11-3 Rental housing gross income vs. total costs with time, showing critical period. Gross income tends to go down and total costs up as property ages.

It is an unfortunate practical fact, because of the complexity of the problem, that effective code enforcement for housing conservation and rehabilitation is in many places not being accomplished. Efforts suffer from frustration and lack of support, aided and abetted by government apathy or sympathy and financial assistance through welfare payments. A greater return can be realized by giving greater assistance to those communities and property owners demonstrating a sincere desire to conserve and renew basically sound areas. Evidence of actual maintenance and improvement of the existing housing supply, code enforcement, encouragement of private building, low-interest mortgages, and provision of low- and middle-income housing are some of the facts that should guide the extent and amount of assistance a community receives.

Rent controls, however, can place the owners of rental properties in poor neighborhoods in a financially untenable position. An inadequate return on an investment usually leads to reduced services and maintenance and to property deterioration. Since a significant number of renters in poor neighboorhoods may be on welfare, it would appear sound to provide higher welfare rent allowances and rent subsidies tied to property maintenance and housing code compliance. But an adequate return or profit may be distorted by the sale and resale of property by speculators to dummy corporations at increasing cost. Hence caution is necessary.

Another factor may add to the housing problem. When there is a shortage of dwelling units and rentals are high, there is a pressure to purchase rather than rent. There is also an incentive to purchase since an income tax deduction can be taken as a home owner, which is not available to a renter. However, there is likely to be a concurrent rise in property values when this is taking place, sometimes doubling in a short time, because of the shortage of rental units at a moderate price. These factors favor the conversion of existing multifamily units to condominiums and construction of new condominiums thereby excluding from the housing market many who are poor, or not sufficiently affluent, and those on fixed incomes who cannot afford the higher cost of a condominium. Those displaced may be forced into less desirable housing and possible exploitation, thus contributing to the spreading of slums.

Health, Economic, and Social Effects

The interrelationship of housing and health is complex and is not subject to exact statistical analysis. For example, poverty, malnutrition, and lack of education and medical care also have important effects on health. These may mean long hours of work with resultant fatigue, improper food, and lack of knowledge relating to disease prevention, sanitation, and personal hygiene. The problem is compounded by poor job and income opportunities and by the slum itself through the feelings of inferiority and resentment of the residents against others who are in a better position. In addition, slums are characterized as having high delinquency, prostitution, broken homes, and other social problems. Who can say if people are sick because they are poor, or poor because they are sick. Although a real association is perceived to exist between poor health and substandard housing, it has not been possible to definitely incriminate housing as *the* cause of a specific illness. Many factors contribute to the physical and mental health and social well-being of the family and individual, of which housing and the housing environment is one.[3] Studies show that as a matter of practical fact many factors associated with substandard housing are profoundly detrimental to the life, health, and welfare of a community. The results of a few studies and reports are summarized in Table 11-1.

The higher morbidity and mortality rates and the lower life expectancy associated with bad housing are also believed to be the cumulative effect or result of continual pressures on the human body. Dubos points out in a related discussion,

> Many of man's medical problems have their origin in the biological and mental adaptive responses that allowed him earlier in life to cope with environmental threats.[4]

He adds,

> The delayed results of tolerance to air pollutants symbolize the indirect dangers inherent in many forms of adaptation, encompassing adaptation to toxic substances, microbial pathogens, the various forms of malnutrition, noise, or other excessive stimuli, crowding or isolation, the tensions of competitive life, the disturbances of physiological cycles, and all other uncontrolled deleterious agencies typical of urbanized and technicized societies. Under normal circumstances, the modern environment rarely destroys human life, but frequently it spoils its later years.

[3] *Proceedings of The First Invitational Conference on Health Research in Housing and its Environment*, Airlie House, Warrenton, Virginia, March 17–19, 1970, PHS, DHEW, Environmental Health Service, Washington, D.C.
[4] Rene Dubos, paper delivered at the Smithsonian Institution Annual Symposium, February 16–18, 1967; *The Fitness of Man's Environment*, Smithsonian Institution Press, Washington, D.C., 1968.

Table 11-1 Effects Associated with Substandard Housing

Diseases	Tuberculosis	Health and Other	Police	City Costs
CD rate 65% higher, VD rate 13 times higher, CD death rate as high as 50 years ago.[a] Intestinal disease rate 100% higher in homes lacking priv. flush toilet.[b] Meningococcis rate $5\frac{1}{2}$ times higher.[c] Infective and parasitic disease death rate 6.6 times higher.[d] Pneumonia and influenza death rate 2 times higher.[c] Lead poisoning in children higher. (Rat and roach infestation higher.) Carbon monoxide poisoning.	Half of cases from $\frac{1}{4}$ of population.[a] TB rate 8 times higher.[b] Secondary attack rate 200%.[e] Death rate 8.6 times higher.[c] 20% of area in city accounts for 60% of cases.[f]	Source of 40% of mentally ill in state institutions.[a] Infant death rate 5 times higher.[a] Infant mortality as high as 50 years ago.[a] Life expectancy 6.7 years less.[d] 64% of out-of-wedlock cases.[g] Interest rate for mortgages higher in blighted areas. Fire insurance rates higher. Accidental death rate 2.3 times higher.[d] Injuries, burns, and accidental poisoning 5–8 times the national average.[h]	Juvenile delinquency twice as high.[a] 20% of area in city accounts for 50% of arrests, 45% of major crimes, 50% of juvenile delinquency.[f] 2.6 times more arrests, 1.9 times more police calls, 2.9 times more criminal cases, 3.7 times more juvenile delinquents[i] 50% of murders, 60% of manslaughters, 49% of robberies.[g]	20% of area in city brings in 6% of real estate tax.[f] Contributes $5\frac{1}{2}$% of real-estate tax but takes 53% of city services.[f] Slums cost $88 more per person than they yield; good areas yield $108.[j] 20% of area in city accounts for 45% of city service costs, 35% of fires.[f] 1.5 times more fires, 15.7 times more families on welfare, 4 times more nursing visits.[i] Account for most of the welfare benefits. Slums yielded $\frac{1}{2}$ cost of serices required.[k]

[a] Release by Dr. Leonard Scheele, Surgeon General, PHS Pub. 27, 1949, regarding 6 cities having slums.
[b] National Health Survey by U.S. Health Service, 1935–1936.
[c] Dr. Bernard Blum, *J. Am. Pub. Health Assoc.*, **39**, 1571–1577 (December 1949).
[d] Erie County Health Dept., N.Y., 1953 Annual Report.
[e] Paper on housing and progress in public health by W. P. Dearing, M.D., presented at the University of North Carolina, April 16, 1950.
[f] Miscellaneous city studies.
[g] The Twenty-seventh Annual Report of the Buffalo Urban League, Inc., N.Y.
[h] Albert H. Stevenson, Airlie House Conference, DHEW, Washington, D.C., March 17–19, 1970.
[i] Report from city of Louisville, Ky.
[j] Raymond M. Foley, "To Eradicate our Vast Slum Blight," *New York Times*, Magazine section, November 27, 1949.
[k] John P. Callahan, "Local Units Fight Problem of Slums," *New York Times*, July 22, 1956.

Emphasis must be on preventive sanitation, medicine, architecture, and engineering to avoid some of the contributory causes of early and late disability and premature death. This avoidance includes the insidious, cumulative, long-term insults to the human body and spirit, as well as maintenance and improvement of those factors in the environment that enhance the well-being and aspirations of people.

Although this discussion concentrates on the environmental health aspects of housing, it is extremely important that concurrent emphasis be placed on the elimination of poverty and on improved education for those living in poor housing and neighborhoods. It is essential that the causes of poverty and low income be attacked at the source, that usable skills be taught, and that educational levels be raised. In this way more individuals can become more productive and self-sufficient and develop greater pride in themselves and in their communities.

APPRAISAL OF QUALITY OF LIVING

APHA Appraisal Method

The American Public Health Association (APHA) appraisal method for measuring the quality of housing was developed by the Committee on the Hygiene of Housing between 1944 and 1950. This method attempts to eliminate or minimize individual opinion so as to arrive at a numerical value of the quality of housing that may be compared with results in other cities and may be reproduced in the same city by different evaluators using the same system. It is also of value to measure the quality of housing in a selected area, say at 5-year intervals, and to evaluate the effects of an enforcement program or lack of an enforcement program. The appraisal method measures the quality of the dwellings and dwelling units as well as the environment in which they are located

The items included in the APHA dwelling appraisal, Tables 11-2 and 11-3, are grouped under "Facilities," "Maintenance," and "Occupancy." Additional information obtained includes rent, income of family, number of lodgers, race, type of structure, number of dwelling units, and commercial or business use. The environmental survey, Table 11-4, reflects the proximity and effects of industry, heavy traffic, recreational facilities, schools, churches, business and shopping centers, smoke, noise, dust, and other factors that determine the suitability of an area for residential use.

The rating of housing quality is based on a penalty scoring system, shown in Table 11-2. A theoretical maximum penalty score is 600. The practical maximum is 300; the median is around 75. A score of zero would indicate all standards are met. An interpretation of the dwelling and environmental scores is shown in Table 11-5. It is apparent therefore that according to this scoring system, either the sum of dwelling and environmental scores or a dwelling or environmental score of 200 or greater would classify the housing as unfit.

Table 11-2 Appraisal Items and Maximum Standard Penalty Scores (APHA)

Item	Maximum Score	Item	Maximum Score
A. Facilities		17. Rooms lacking window[a]	30
		18. Rooms lacking closet	8
Structure:		19. Rooms of substandard area	10
1. Main access[a]	6	20. Combined room facilities[c]	
2. Water supply[a] (source)	25		
3. Sewer connection[a]	25	*B. Maintenance*	
4. Daylight obstruction	20	21. Toilet condition index	12
5. Stairs and fire escapes	30	22. Deterioration index[a] (structure, unit)[b]	50
6. Public hall lighting	18	23. Infestation index (structure, unit)[d]	15
Unit:			
7. Location in structure	8	24. Sanitary index (structure, unit)[d]	30
8. Kitchen facilities	24	25. Basement condition index	13
9. Toilet[a] (location, type, sharing)[b]	45		
10. Bath[a] (location, type, sharing)[b]	20	*C. Occupancy*	
11. Water supply[a] (location and type)	15	26. Room crowding: persons per room[a]	30
12. Washing facilities	8	27. Room crowding: persons per sleeping room[a]	25
13. Dual egress[a]	30	28. Area crowding: sleeping area per person[a]	30
14. Electric lighting[a]	15		
15. Central heating	3	29. Area crowding: nonsleeping area per person	25
16. Rooms lacking installed heat[a]	20	30. Doubling of basic families	10

Source: *An Appraisal Method for Measuring the Quality of Housing*, Part II, "Appraisal of Dwelling Conditions," Am. Pub. Health Assoc., Washington, D.C., 1946.

Note: 1. Maximum theoretical total dwelling score is 600, broken down as:
Facilities 360 Maintenance 120 Occupancy 120

2. Housing total = dwelling total + environmental total.

[a]Condition constituting a basic deficiency.

[b]Item score is total subscores for location, type, and sharing of toilet or bath facilities.

[c]Item score is total of scores for items 16–19 inclusive. This duplicate score is not included in the total for a dwelling but is recorded for analysis.

[d]Item score is total of subscores for structure and unit.

Application of the APHA appraisal method requires trained personnel and experienced supervison.* The survey staff should be divorced from other routine work so as to concentrate on the job at hand and produce information that can be put to use before it becomes out-of-date. In practice, it is found desirable to select a limited area or areas for pilot study. The information thus obtained can be used as a basis for determining the need for extension of the survey, need for new or revised minimum housing standards, extent of the housing problem, develop-

*A trained sanitarian can inspect about ten dwelling units per day. For every four inspectors there should be one trained field supervisor, three office clerks, and one office supervisor.

Table 11-3 Basic Deficiencies of Dwellings (APHA)

Item[a]	Condition Constituting a Basic Deficiency[b]
A. Facilities	
2.	Source of water supply specifically disapproved by local health department.
3.	Means of sewage disposal specifically disapproved by local health department.
9.	Toilet shared with other dwelling unit, outside structure, or of disapproved type (flush hopper or nonstandard privy).
10.	Installed bath lacking, shared with other dwelling unit, or outside structure.
11.	Water supply outside dwelling unit.
13.	Dual egress from unit lacking.
14.	No electric lighting installed in unit.
16.	Three-fourths or more of rooms in unit lacking installed heater.[c]
17.	Outside window lacking in any room unit.[c]
B. Maintenance	
22.	Deterioration of ciass 2 or 3 (penalty score, by composite index, of 15 points or over).
C. Occupancy	
26.	Room crowding: over 1.5 persons/room.
27.	Room crowding: number of occupants equals or exceeds 2 times the number of sleeping rooms plus 2.
28.	Area crowding: less than 40 ft^2 of sleeping area/person.

Note: Some authorities include as a basic deficiency unvented gas space heater, unvented gas hot-water heater, open gas burner for heating, lack of hot and cold running water.

[a] Numbers refer to items in Table 11-2.

[b] Of the 13 defects that can be designated basic deficiencies, 11 are so classified when the item penalty score equals or exceeds 10 points. Bath (item 10) becomes a basic deficiency at 8 points for reasons involving comparability to the U.S. Housing Census; deterioration (item 22) at 15 points for reasons internal to that item.

[c] The criterion of basic deficiency for this item is adjusted for number of rooms in the unit.

ment of coordination between existing official and voluntary agencies, the part private enterprise and public works can play, public information needs, and so forth.

Census Data

Much valuable information is collected and summarized by the U.S. Bureau of the Census. The 1980 Census includes number of one family and multifamily dwelling units, trailers and mobile homes, and condominiums; the population per owner- and renter-occupied unit;* the number of dwelling units with private

*The 1980 Census Population and Housing Unit Cost shows an average of 2.6 persons per housing unit in contrast to 2.96 in 1970 thereby reflecting changes in life style.

Table 11-4 Environmental Survey—Standard Penalty Scores (APHA)

	Item	Maximum Penalty Score[a]
A.	Land crowding	
	1. Coverage by structures—70% or more of block area covered.	24
	2. Residential building density—ratio of residential floor area to total = 4 or more.	20
	3. Population density—gross residential floor area per person 150 ft^2 or less.	10
	4. Residential yard areas—less than 20-ft wide and 625 ft^2 in 70% of residences.	16
B.	Nonresidential land areas	
	5. Areal incidence of nonresidential land use—50% or more non-residential.	13
	6. Linear incidence of nonresidential land use—50% or more non-residential.	13
	7. Specific nonresidential nuisances and hazards—noise and vibration, objectionable odors, fire or explosion, vermin, rodents, insects, smoke or dust, night glare, dilapidated structure, insanitary lot.	30
	8. Hazards to morals and the public peace—poolrooms, gambling places, bars, prostitution, liquor stores, nightclubs.	10
	9. Smoke incidence—industries, docks, railroad yards, soft-coal use.	6
C.	Hazards and nuisances from transportation system	
	10. Street traffic—type of traffic, dwelling setback, width of streets.	20
	11. Railroads or switchyards—amount of noise, vibration, smoke, trains.	24
	12. Airports or airlines—location of dwelling with respect to runways and approaches.	20
D.	Hazards and nuisances from natural causes	
	13. Surface flooding—rivers, streams, tide, groundwater, drainage, annual or more.	20
	14. Swamps or marshes—within 1000 yd, malarial mosquitoes.	24
	15. Topography—pits, rock outcrops, steep slopes, slides.	16
E.	Inadequate utilities and sanitation	
	16. Sanitary sewerage system—available (within 300 ft), adequate.	24
	17. Public water supply—available, adequate pressure and quantity.	20
	18. Streets and walks—grade, pavement, curbs, grass, sidewalks.	10
F.	Inadequate basic community facilities	
	19. Elementary public schools—beyond $\frac{2}{3}$ mi, 3 or more dangerous crossings.	10
	20. Public playgrounds—less than 0.75 acres/1000 persons.	8
	21. Public playfields—less than 1.25 acres/1000 persons.	4
	22. Other public parks—less than 1.00 acres/1000 persons.	8
	23. Public transportation—beyond $\frac{2}{3}$ mi, less than 2 buses/hr.	12
	24. Food stores—dairy, vegetable, meat, grocery, bread, more than $\frac{1}{3}$ mi.	6

Source: *An Appraisal Method for Measuring the Quality of Housing*, Part III, "Appraisal of Neighborhood Environment," Am. Pub. Health Assoc., Washington, D.C., 1950.
[a] Maximum environment total = 368.

Table 11-5 Housing Quality Scores (APHA)

Factor	*A*—Good	*B*—Acceptable	*C*—Border line	*D*—Sub standard	*E*—Unfit
Dwelling score	0 to 29	30 to 59	60 to 89	90 to 119	120 or greater
Environmental score	0 to 19	20 to 39	40 to 59	60 to 79	80 or greater
Sum of dwelling and environmental scores	0 to 49	50 to 99	100 to 149	150 to 199	200 or greater

bath, including hot and cold piped water as well as flush toilet and bathtub or shower, and the number lacking some or all these facilities; the number of dwelling units occupied by persons of Spanish/Hispanic origin or descent; the number of rooms and bedrooms per dwelling unit; the number on public water supply, drilled well, dug well, spring or other; the number on public sewerage, septic tank or cesspool, or other system; the type of heating and fuel used; size of plot; information on occupation and income; the monthly rental; and the value or sale price of owner-occupied one-family homes. Other statistics on selected population characteristics for areas with 2500 or more inhabitants and for counties are available. This information, if not too old, can be used as additional criteria to supplement reasons for specific program planning. Plotting the data on maps or overlays will show concentrations sometimes not discernible by other means.

The accuracy of census data for measuring housing quality has been questioned and hence should be checked, particularly if it is to be used for appraisal or redevelopment purposes. It is nevertheless a good tool in the absence of a better one.

Health, Economic, and Social Factors

It is frequently possible to obtain morbidity and mortality data for specific diseases or causes by census tracts or selected areas. Also available may be the location of cases of juvenile delinquency, public and private assistance, and probation cases. Sources of fires, nuisances, and rodent infestation, and areas of social unrest give additional information. Health, fire, police, and welfare departments and social agencies should have this information readily available. See Table 11-1. Tabulation and plotting of these data will be useful in establishing priorities for action programs.

Planning

The location of existing and proposed recreational areas, business districts, shopping centers, churches, schools, parkways and thruways, housing projects, residential areas, zoning restrictions, redevelopment areas, airports, railroads, industries, lakes, rivers, and other natural boundaries help to determine the best usage of property. Where planning agencies are established and are active, maps giving this as well as additional information are usually well developed. Analysis and comprehensive planning on a continuing basis are essential to the proper development of cities, villages, towns, counties, and metropolitan areas. The availability of state and federal aid for community-wide planning should be investigated. Plans for urban development, housing-code enforcement, rehabilitation, and conservation should be carefully integrated with other community and state plans before decisions are made. These subjects are discussed in greater detail in Chapter 2.

Environmental Sanitation and Hygiene Indices

Most modern city and county health departments can carry on a housing-inspection program based on minimum housing standards, provided competent personnel is assigned. Where housing inspections are made on a routine basis and records are kept on a data processing punch-card system, visible card file, or by combination methods, problem streets and areas can be detected with little difficulty. A survey and follow-up form based on a modern housing ordinance is shown in Figure 11-4. It should list recommendations for correction based on what is practical from field observation so as to serve as the basis for an accurate confirmatory letter to the owner. The type of deficiencies found, such as the lack of a private bath and toilet, dwelling in disrepair, or lack of hot and cold running water, can give a wealth of information to guide program planning. To this can usually be added the origin of complaints, which if plotted on a map will give a visual picture of the heavy work-load areas. Reliable cost estimates for complete rehabilitation of selected buildings and a simple foot survey should be made to confirm administrative judgment and decisions before any major action is taken. Many apparently well-thought-out plans have fallen down under this simple test. For example, deterioration may have proceeded to the point where rehabilitation is no longer economically or structurally feasible under the existing market conditions.

Other Survey Methods

These include aerial surveys, external ground-level surveys, and the Public Health Service Neighborhood Environmental Evaluation and Decision System (NEEDS).[5] The selection of a small number of significant environmental variables can also provide a basis for a rapid survey if checked against a more comprehensive survey system.

NEEDS

This is a five-staged systems technique designed to provide a rapid and reliable measure of neighborhood environmental quality. The data collected are adapted

[5]NEEDS, Neighborhood Environmental Evaluation and Decision System, Bureau of Community Environmental Management, PHS, DHEW, Washington, D.C., 1970.

Dates inspected ____________________

Address ____________________ Type of Structure and Occupancy ____________________

(frame, stucco, brick veneer, solid brick; residential, factory, store)

Owner and Address ____________________

No.	Item	Yes	No	CM
1.	Water supply in each apt. satisfact. quality and quantity (no X-conn).			
2.	Private in each apt.			
	(*a*) water closet			
	(*b*) washbasin			
	(*c*) shower/tub			
	(*d*) kitchen sink			
	(*e*) cabinets and counter			
	(*f*) refrig. and stove			
3.	Piped hot water for			
	(*a*) washbasin			
	(*b*) shower/tub			
	(*c*) kitchen			
4.	Plumbing, heating, electricity, and fixtures properly installed and maintained.			
5.	Water-repellent floor and base in toilet room and bathroom.			
6.	Window $\frac{1}{10}$ floor area in every room; openable, adequate light and air or induced ventilation for bath.			
7.	Dwelling unit provides 150 ft^2 for one and 100 ft^2 area for each additional occupant.			
8.	Dwelling can be heated to 68°F.			
9.	Sleeping rooms provide 70 ft^2 for one person and 50 ft^2 for each additional person.			
10.	Every habitable room has 2 electric outlets; bathroom, w.c. stall, laundry, and hall have a min. 10ft-c on floor.			
11.	Occupant keeps dwelling unit and fixtures clean and sanitary.			
12.	Space and water heaters adequate, properly connected, and vented to outer air; back-draft guard.			
13.	Premises free of rodent and vermin infestation; rodent-proof.			
14.	Refuse, garbage, and ash storage proper and adequate.			
15.	One or more apartments above 2nd floor have 2 means of egress.			
16.	Public halls and stairs lighted, daylight and artificial in MD.			
17.	Property and dwelling properly drained and sewered.			
18.	Owner keeps public areas of building and premises clean.			
19.	Living in cellar prohibited.			
20.	Dwelling in good repair, safe, sanitary, and weatherproof (handrails, stairs, walls, wiring, floors, siding, doors, frames, plaster, porch, eaves, roof, foundation beams firm and sound).			
21.	Lodging house has one washbasin, shower or tub, and water closet per 6 persons.			
22.	Lodging house supplies clean linen and towels prior to letting and weekly.			
23.	Cooking in lodging house done in approved and lawful kitchen or kitchenette only.			

NOTE: Explain each "No" item on back by item number and follow with recommendation for correction. "CM" is checked or dated when correction is made. "MD" denotes three or more dwelling units.

Floor	Apt.	Total Hab. Area	Total Hab. Rooms	Bedrooms		Persons	Shelter monthly rental
				No.	Area		
			{				
			{				
			{				
			{				
			{				

Remarks: (tenant names, agent, change in ownership)

____

____

____

____

Inspected by: ____________________

EH-49,5

Figure 11-4 Dwelling survey.

for electronic data processing to reduce the time lapse between data collection, analysis, program planning, and implementation.

In Stage I an exterior sidewalk survey is made of 10 to 20 blocks in each problem neighborhood of a city to determine which are in greatest need of upgrading. The conditions analyzed include structural overcrowding, population crowding, premises conditions, structural condition of housing and other buildings, environmental stresses, condition of streets and utilities, natural deficiencies, public transportation, natural hazards and deficiencies, shopping facilities, parks and playgrounds, and airport noise. (Time: one man-hour per block.)

In Stage II the neighborhood(s) selected are surveyed in some depth to determine the physical and social environmental problems facing the residents. About 300 dwelling units and families are sampled in the study area. This phase includes interviews to determine demographic characteristics, health problems, health services, interior housing conditions, and resident attitudes.

In Stage III the data collected in Stage II are computer-processed and analyzed with local government and community leaders. Problems are identified and priorities for action are established.

In Stage IV programs are developed to carry out the decisions made in Stage III. The community participates and is kept aware of the study results and action proposed.

In Stage V the decisions made and programs developed are implemented. Federal fiscal assistance is also solicited, and the information and support made possible by this technique are used to strengthen applications for grants.[5]

APHA Criteria

The American Public Health Association Committee on the Hygiene of Housing has listed the criteria to be met for the promotion of physical, mental, and social health on the farm as well as in the city dwelling. Thirty basic principles, with specific requirements and suggested methods of attainment for each are reported in *Basic Principles of Healthful Housing*, originally published by the American Public Health Association in 1938.

The "basic principles" have been expanded to reflect progress made and present-day aspirations of people to help achieve total health goals such as defined by the WHO. The American Public Health Association Program Area Committee on Housing and Health, prepared what is probably the best comprehensive statement of principles available to guide public policy and goal formulation.[6] These can also serve as a basis for performance standards to replace specification standards for building construction, living conditions, and community development. The major headings, or objectives, of the committee re-

[6] *Housing: Basic Health Principles & Recommended Ordinance*, American Public Health Association, 1015 Fifteenth Street, N.W., Washington, D.C. 20005, 1971.

port are quoted below. The reader is referred to the committee report for further explanation of the items listed.

Basic Health Principles of Housing and Its Environment

I. *Living Unit and Structure*

"Housing" includes the living unit for man and family, the immediate surroundings, and the related community services and facilities; the total is referred to as the "residential environment." The following are basic health principles for the residential environment, together with summaries of factors that relate to the importance and applicability of each principle.

A. Human Factors

1. Shelter against the elements.
2. Maintenance of a thermal environment that will avoid undue but permit adequate heat loss from the human body.
3. Indoor air of acceptable quality.
4. Daylight, sunlight, and artificial illumination.
5. In family units, facilities for sanitary storage, refrigeration, preparation, and service of nutritional and satisfactory foods and meals.
6. Adequate space, privacy, and facilities for the individual and arrangement and separation for normal family living.
7. Opportunities and facilities for home recreation and social life.
8. Protection from noise from without, other units, and certain other rooms and control of reverberation noises within housing structures.
9. Design, materials, and equipment that facilitate performance of household tasks and functions without undue physical and mental fatigue.
10. Design, facilities, surroundings, and maintenance to produce a sense of mental well-being.
11. Control of health aspects of materials.

B. Sanitation and Maintenance

1. Design, materials, and equipment to facilitate clean, orderly, and sanitary maintenance of the dwelling and personal hygiene of the occupants.
2. Water piping of approved, safe materials with installed and supplied fixtures that avoid introducing contamination.
3. Adequate private sanitary toilet facilities within family units.
4. Plumbing and drainage system designed, installed, and maintained so as to protect against leakage, stoppage, or overflow and escape of odors.
5. Facilities for sanitary disposal of food waste, storage of refuse, and sanitary maintenance of premises to reduce the hazard of vermin and nuisances.
6. Design and arrangement to properly drain roofs, yards, and premises and conduct such drainage from the buildings and premises.
7. Design and maintenance to exclude and facilitate control of rodents and insects.
8. Facilities for the suitable storage of belongings.
9. Program to assure maintenance of the structure, facilities, and premises in good repair and in a safe and sanitary condition.

C. Safety and Injury Prevention
  1. Construction, design, and materials of a quality necessary to withstand all anticipated forces that affect structural stability.
  2. Construction, installation materials, arrangement, facilities, and maintenance to minimize danger of explosions and fires or their spread.
  3. Design, arrangement, and maintenance to facilitate ready escape in case of fire or other emergency.
  4. Protection against all electrical hazards, including shocks and burns.
  5. Design, installation, and maintenance of fuel-burning and heating equipment to minimize exposure to hazardous or undesirable products of combustion, fires, or explosions and to protect persons against being burned.
  6. Design, maintenance, and arrangement of facilities, including lighting, to minimize hazards of falls, slipping, and tripping.
  7. Facilities for safe and proper storage of drugs, insecticides, poisons, detergents, and deleterious substances.
  8. Facilities and arrangements to promote security of the person and belongings.

II. *Residential Environment*

The community facilities and services and the environment in which the living unit is located are essential elements in healthful housing and are part of the total residential environment.

A. Community or Individual Facilities
  1. An approved community water supply or, where not possible, an approved individual water supply system.
  2. An approved sanitary sewerage system or, where not possible, an approved individual sewage disposal system.
  3. An approved community refuse collection and disposal system or, where not possible, arrangements for its sanitary storage and disposal.
  4. Avoidance of building on land subject to periodic flooding and adequate provision for surface drainage to protect against flooding and prevent mosquito breeding.
  5. Provision for vehicular and pedestrian circulation for freedom of movement and contact with community residents while adequately separating pedestrian from vehicular traffic.
  6. Street and through-highway location and traffic arrangements to minimize accidents, noise, and air pollution.
  7. Provision of such other services and facilities as may be applicable to the particular area, including public transportation, schools, police, fire protection, and electric power, health, community and emergency services.
  8. Community housekeeping and maintenance services, like street cleaning, tree and parkway maintenance, weed and rubbish control, and other services requisite to a clean and aesthetically satisfactory environment.

B. Quality of the Environment
  1. Development controls and incentives to protect and enhance the residential environment.
  2. Arrangement, orientation, and spacing of buildings to provide for adequate light, ventilation, and admission of sunlight.
  3. Provision of conveniently located space and facilities for off-street storage of vehicles.

4. Provision of useful, well designed, properly located space for play, relaxation, and community activities for daytime and evening use in all seasons.
5. Provision for grass and trees.
6. Improved streets, gutters, walks, and access ways.
7. Suitable lighting facilities for streets, walks, and public areas.

C. Environmental Control Programs

To promote maintenance of a healthful environment necessitates an educational and enforcement program to accomplish the following.

1. Control sources of air and water pollution and local sources of ionizing radiation.
2. Control rodent and insect propagation, pests, domestic animals, and livestock.
3. Inspect, educate, and enforce so that premises and structures are maintained in such condition and appearance as not to be a blighting influence on the neighborhood.
4. Community noise control and abatement.
5. Building and development regulations.

Minimum Standards Housing Ordinance

Building divisions of local governments have traditional responsibility over the construction of new buildings and their structural, fire, and other safety provisions as specified in a building code. Fire departments have responsibility for fire safety. The health department and such other agencies as have an interest and responsibility are concerned with the supplied utilities and facilities, their maintenance, and the occupancy of dwellings and dwelling units for more healthful living. This latter responsibility is best carried out by the adoption and enforcement of a housing ordinance. See Chapter 2 for code definitions, also this chapter.

In an enforcement program, major problems of structural safety or alterations involving structural changes for which plans are required would be referred to the building division. Serious problems of fire safety would be referred to the fire department. Interdepartmental agreements and understanding can be mutually beneficial and make possible the best use of the available expertise. This requires competent staffing and day-to-day cooperation.

The *APHA-PHS Recommended Housing Maintenance and Occupancy Ordinance* has been prepared for local adoption.[7] It updates and replaces the APHA's *A Proposed Housing Ordinance* of 1952 but retains most of the fundamental principles and adds certain important administrative and enforcement features. The ordinance should apply to *all* existing, altered, and new housing. Other model ordinances are also available.[8]

[7] *APHA-PHS Recommended Housing Maintenance and Occupancy Ordinance*, 1975 revision, DHEW, PHS, CDC, Atlanta, Ga. 30333, DHEW Pub. (CDC) 75-8299, GPO, Washington, D.C. 20402.

[8] *Uniform Housing Code*, 1976 ed., International Conference of Building Officials, 5360 South Workman Mill Road, Whittier, California 90601.

The essential features of the housing ordinance to be complied with before occupancy are summarized here for easy reference. These are minimum standards and in many instances should be exceeded.

Summary of *APHA-PHS Recommended Housing Maintenance and Occupancy Ordinance*[9]

A. *Responsibilities of Owners and Occupants*

1. Dwelling unit and premises not occupied or let unless it is clean, sanitary, fit for human occupancy, and complies with state and local requirements.
2. Owner of dwelling containing two or more units keeps shared or public areas of dwelling and premises clean and sanitary.
3. Occupant keeps parts he or she occupies and controls clean and sanitary.
4. No rat harborages in or about any dwelling unit.
5. Occupant stores and disposes of all his or her rubbish in a clean, sanitary, safe manner.
6. Occupant stores and disposes of all her or his garbage and other refuse in a clean, sanitary, safe manner. Cans and containers are rat-proof, insect-proof, watertight, strong, covered, maintained in clean sanitary condition.
7. Cans and containers stored on concrete slab or platform at least 18 in. above ground, or on movable platform.
8. Bulk storage containers placed on concrete platforms equipped with drains connected to approved sewer. Water faucet at site for cleaning container unless other provisions made. Containers have self-closing lids, are stable, and cannot be tipped by a hanging force of 191 pounds or a horizontal force of 70 pounds.
9. Garbage and other refuse containers provide capacity to meet needs of occupants.
10. Owner provides containers or facilities for garbage and other refuse storage in dwelling containing three or more units. Occupant provides container when there are two or fewer units.
11. Owner provides and hangs screens and storm doors and windows when required.
12. Occupant of single-dwelling unit keeps unit free of vermin, and occupant of dwelling unit, which is one of several units, if infested keeps own unit free; owner responsible when two or more units infested or when maintenance is needed.
13. Occupant keeps all plumbing fixtures clean, sanitary, and operable.
14. Supplied heat maintains dwelling unit at at least 68°F.
15. Dwelling or dwelling unit free from hazards to health due to presence of toxic substances, e.g., lead-based paint; owner responsibility.
16. Owner or occupant does not apply lead-based paint to any surface.

B. *Minimum Standards for Basic Equipment and Facilities*

1. Kitchen sink properly connected and operating.
2. Flush toilet and washbasin properly operating and connected.
3. Tub or shower properly operating and connected.
4. Kitchen cabinets, shelving, and counters adequate for permissible occupancy.

[9]See *APHA-PHS Recommended Housing Maintenance and Occupancy Ordinance* for definition of terms and complete explanation of all regulations.

5. Proper cook stove and refrigerator in operation or adequate space for them if provided by occupant.

6. Suitable facilities for safe storage of drugs and poisonous household substances.

7. Exterior doors of dwelling or dwelling unit have safe locks.

8. Kitchen sink, washbasin, and tub or shower connected with adequate hot- and cold-water lines.

9. Water heater properly installed and supplies adequate water at 120 to 130°F.*

10. Sound handrails for steps with four or more risers and porches 3-ft or higher.

11. Every dwelling unit has dual safe means of egress leading to ground level or as required by local law. Access or egress without passing through another unit.

C. *Minimum Standards for Light and Ventilation*

1. Every habitable room, bathroom, and water closet compartment has window to the outdoors 10 percent of floor area, or

   a. Skylight on top floor, 10 percent of floor area, as substitute for window.

   b. Mechanical ventilation in kitchen, bathroom, or water closet compartment as substitute for window.

2. Rooms have openable windows equal to at least 45 percent of minimum window area or other approved ventilation.

3. Where electric service is within 300 ft, each dwelling unit supplied with separate 15 ampere circuit; habitable rooms provide at least two separate duplex electric outlets or one duplex outlet and one fixture; water closet compartment, bathroom, laundry room, furnace room, public hall, and kitchen have at least one light fixture, all properly connected and maintained. Switches conveniently located. No temporary wiring or extension cords.

4. Public hall and stairway adequately lighted and provide at least 10 ft-c of light on tread, in three-family or more dwelling at all times; if less than three families, light switches acceptable.

D. *Minimum Thermal Standards*

1. Dwelling can be heated to temperature of 68°F measured 36 in. above floor; equipment properly installed and maintained.

2. No space or hot-water heater using a flame unless properly vented and supplied with fresh air.

E. *General Requirements Relating to the Safe and Sanitary Maintenance of Parts of Dwelling and Dwelling Units*

1. Structure is sound, weather-tight, damp-free, watertight, rodent-proof, affords privacy, and is in good repair. Premises drained, clean, safe, and sanitary.

2. Every window, interior floor, exterior door, and entranceway near ground level is rodent-proof, and in good repair.

3. Stairways, porch, etc., safe, sound, and in good repair.

4. Plumbing fixtures, water and waste pipes, chimney, flue and smoke pipe, facility, or equipment properly installed and in good condition. Fixtures supplied with safe water.

5. Water closet, bathroom, and kitchen floor reasonably impervious and cleanable.

6. Accessory structures sound, maintained in good repair, free of insects and rats.

*A lower temperature (110°F) is advised to prevent scalding.

7. Service, facility, equipment, or utility not discontinued except for repair.

8. Exterior walls, foundations, basements, roofs, pipes, porches, wires, skirting, lattice, doors, grilles, windows below or within 48″ of grade accessible to rats, and other openings or accessory structures or exterior stored materials are rat-proofed.

9. All construction and materials, ways and means of egress, installation and use of equipment, conform with state and local fire protection regulations.

F. *Maximum Density, Minimum Space, Use and Location Requirements*

1. Dwelling unit provides at least 150 ft^2 floor space for first occupant and 100 ft^2 per additional occupant. Maximum occupancy is less than 2 times number of habitable rooms.

2. No more than one family plus two boarders per dwelling unit unless under permit or license.

3. Dwelling unit has at least 4 ft^2 of floor-to-ceiling closet space or equivalent additional space for each occupant.

4. Dwelling unit of two or more rooms, every room for sleeping provides at least 70 ft^2 for first person and 50 ft^2 for each additional.

5. Access to two or more sleeping rooms, a bathroom, or water closet compartment, without going through a bedroom or bathroom or water closet compartment.

6. At least one-half of floor area 7 ft high; floor area with less than 5-ft ceiling not part of total floor area.

7. Space more than 4 ft below grade not used as habitable room or dwelling unit.

G. *Rooming House, Dormitory Rooms, and Rooming Units**

1. Annual license issued; may be suspended on inspection.

2. Sanitary facilities provided in ratio of:

   a. One flush water closet to six roomers. (Flush urinal may equal one-half of water closets in rooming house for males.)

   b. One washbasin to six roomers.

   c. One bath or shower to six roomers.

   d. Fixtures properly connected and operating.

   e. Facilities are in dwelling and reasonably accessible from common passageway.

3. Washbasin, tub, or shower supplied with hot and cold water.

4. No flush water closet, washbasin, bath, or shower below grade.

5. Bed linen and towels supplied, changed once a week and prior to letting to new occupant.

6. Bedding clean and sanitary.

7. Rooming unit provides 110 ft^2 for one person and 90 ft^2 for each additional; also 4 ft^2 of closet space at least 5 ft high for each occupant.

8. No cooking in rooms; communal cooking only if approved and complies with applicable building, fire, and food service regulations.

9. Access doors to rooming unit have proper locks.

10. Safe dual means of egress to ground level as required by local law, without passing through any other room.

11. Access to or egress from each rooming unit not through another unit.

*Requirements are in addition to other applicable items.

Additional sections in the ordinance deal with the adoption of plans of inspection; inspections—powers and duties; licensing of the operation of multiple dwellings and rooming houses; authority to adopt rules and regulations; notice of violation; penalties; repairs, designation of unfit units and/or structures and other corrective action—demolition; collection and dissemination of information; appeals; emergencies; conflict of ordinances—effect of partial invalidity; effective date; and legal notes.

Smoke detectors have been found to be effective safety devices which belong in all dwellings and dwelling units. These devices afford major protection to life and property, particularly when coupled with alarms and sprinklers in multiple dwellings.

Energy conservation concerns have emphasized the importance of insulation materials to prevent heat losses from buildings. It is necessary to assure that such materials are not flammable, that they do not release toxic fumes or materials when installed or when subjected to high temperatures.

The administrative section of a housing code should permit the phased improvement of sound dwellings and dwelling units first in selected areas having identifiable grossly substandard living conditions. The expected level of compliance would be consistent with the intent of the housing code regulations and the achievable environmental quality of the neighborhood. Low-interest loans, fiscal and technical guidance, limited grants, craftsmen and labor assistance, tax relief, and other inducements are essential to encourage rehabilitation, stem the tide of deterioration and blight, and make possible compliance with the housing code. Concurrently, community services and facilities would need to be improved and maintained at an adequate level.

HOUSING PROGRAM

Approach

Housing can be a complex human, social, and economic problem that awakens the emotions and interests of a multitude of agencies and people within a community. Government agencies must decide in what way they can most effectively produce action; that is, whether in addition to their own efforts it is necessary to give leadership, encouragement, and support to other agencies that also have a job to do in housing.

Resolution of the housing problem involves just about all official and nonofficial groups or agencies. Organization begins with the housing coordinator, as representative of the mayor or community executive officer. The groups involved include an interagency coordinating committee; citizens advisory committee; health, building, public works, law, and fire departments; urban development agency; housing authority; financing clinic; planning board; air pollution control agency; office of community relations; public library; rent-control office; universities and technical institutes; welfare department; council of social agencies; neighborhood and community improvement associations; banking institutions; real-estate groups; consulting engineers and architects; general contractors; builders and subcontractors; press; and service organizations.

Components for a Good Housing Program

Good housing does not just happen but is the result of far-sighted thinking by individuals in many walks of life. In some cases a few houses are built here and there and areas grow with no apparent thought being given to the future pattern being established. This is typical of small subdivisions of land and developments where there is no planning. In other instances, typical of large-scale developments, construction proceeds in accordance with a well defined plan. The growth of existing communities, the quality of services and facilities provided, and the condition in which they are maintained usually depend on the leadership and coordination provided by the chief executive officer and legislative body and the controls or guides followed. The more common support activities are described below.

Planning Board

A planning board can define the area under control and, by means of maps, locate existing facilities and utilities. Included are highways, railroads, streams, recreational areas, schools, churches, shopping centers, residential areas, commercial areas, industrial areas, water lines, sewer lines, and so forth. In addition, plans are made for future revisions or expansion for maximum benefit to the community. A zoning plan is needed to delineate and enforce the use to which land shall be put, such as for residential, farm, industrial, and commercial purposes. This also provides for protection of land values. Subdivision regulations defining the minimum size of lots, width and grading of roads, drainage, and utilities to be provided may also be adopted. New roads or developments would not be accepted unless in compliance with the subdivision standards. The planning board, if supported, can guide community changes and, if established early, can direct the growth of a community, all in the best interest of the people.

Department of Public Works

A public works department usually has jurisdiction over building, water, sewers, streets, refuse, and, if not a separate department, fire prevention and protection.

The building division's traditional responsibility is regulation of all types of building construction through the enforcement of a building code. A building code includes regulations pertaining to structural and architectural features, plumbing, heating, ventilation and air conditioning, electricity, elevators, sprinklers, and related items. Fire structural regulations are usually incorporated in a building code, although a separate or supplementary fire-prevention code may also be prepared. Certain health and sanitary regulations such as minimum room sizes, plumbing fixtures, and hot- and cold-water connections should of course be an integral part of a building code so that new construction and alterations will comply with the health department's housing-code minimum standards regulating occupancy, maintenance, and supplied facilities.

The water and sewer division would have the fundamental responsibility of making available and maintaining public water supply and sewerage services where these sanitary facilities are accessible. In new subdivisions of land the provision of these facilities, particularly when located outside the corporate limits of a city or village, is usually the responsibility of the developer.

The division of streets would maintain streets, including snowplowing, cleaning, rebuilding old roads, and assuring the proper drainage of surface water. In unincorporated communities, the highway department assumes these functions. New roads would not be maintained unless dedicated to the city, village, or township and are of acceptable design and construction.

Refuse collection would include garbage, rubbish, and ashes. This function may be handled by the municipality, by contract, or by the individual.

Public Housing and Urban Development

In existing urban communities the housing authority generally has responsibility for the construction and operation of dwellings for low-income families and for the rehousing of families displaced by slum clearance. The urban development agency assembles and clears land for reuse in the best interest of the community and rehouses persons displaced as a result of its actions. In basically good neighborhoods, unsalvable housing is demolished, good housing is protected, and sound substandard housing is rehabilitated. Large-scale demolition involving the destruction of viable neighborhoods having structurally sound housing should in general be avoided.

Health Department

State and local health departments have the fundamental responsibility of protecting the life, health, and welfare of the people. Although most cities have health departments, many areas outside of cities do not have the services of a completely staffed county or city-county health department. However, where provided, the health department responsibilities are given in a public health law and sanitary code. In addition to communicable disease control, maternal and child health care, clinics, nursing services, and environmental sanitation, the health department should have supervision over housing occupancy, maintenance and facilities, food sanitation, water supply, sewage and solid waste disposal, pollution abatement, air pollution control, recreation, sanitation, control over radiation hazards, and the sanitary engineering phases of land subdivision. The environmental sanitation activities, being related to the planning, public works, housing, and redevelopment activities, should be integrated with the other municipal functions. For maximum effectiveness and in the interest of the people, it is also equally proper that the services and talents available in the modern health department be consulted and utilized by the other municipal and private agencies.

Health departments have the ideal opportunity to redirect and guide nuisance and complaint investigations; lead-poisoning elimination; carbon monoxide poisoning prevention; and insect, rodent, and refuse control activities into a planned and systematic community sanitation and housing hygiene program. By coordination with nursing, medical care, and epidemiological activities, as well as with those of other agencies, the department is in a position to constructively participate in elimination of the causes that contribute to the conditions associated with slums and substandard housing. The health department should also assist in the training of building, sanitation, welfare, and fire inspectors and others who have a responsibility or interest in the maintenance and improvement of living conditions.

Private Construction

New construction and rehabilitation of housing in accordance with a good building and housing code are essential to meet the normal needs of the people. Private construction and rehabilitation of sound structures should be encouraged and controlled to meet community goals and objectives in accordance with an adopted plan. In complete rehabilitation of a sound structure, the supporting walls, facade, and floor beams are preserved and new kitchens, bathrooms, bedrooms, living rooms, public areas, and utilities are installed. The availability of streets, water, sewers, and other utilities minimizes the total cost of providing "new" rental units, particularly if supported by some federal grant and low-interest loans.

Loan Insurance

Mortgage loan insurance by the government to stimulate homeownership has been fundamental to the construction and preservation of good housing. However, private financial institutions also have a major function and obligation. The reluctance or refusal of some lending institutions to make loans in certain urban areas for housing rehabilitation or purchase, known as "redlining," and the similar practice by insurance companies for fire and property insurance,[10] confound efforts to upgrade neighborhoods. Such practices should be reviewed.

[10] Alfred E. Clark, "H.U.D. says Insurers Redlining in Urban Areas," *New York Times*, June 4, 1978, based on H.U.D. report "Insurance Crisis in Urban America."

Outlining of a Housing Program

Several approaches have been suggested and used in the development of a housing program. They all have several things in common and generally include most of the following steps.[11]

1. Establishment of a committee or committees with representation from official agencies, voluntary groups, the business community, and outstanding individuals as previously described.
2. Identification of the problem—the physical, social, and economic aspects and development of a plan to attack each.
3. Informing the community of survey results, housing needs, and the recommended action.
4. Designation of a board or commission to coordinate and delegate specific functions to be carried out by the appropriate agency; for example, urban renewal, redevelopment, public housing, code enforcement, rehousing, rehabilitation, refinancing.
5. Appropriation of adequate funds to support staffing and training of personnel.
6. Appraisal of housing and neighborhoods and designation of urban, suburban, and rural areas for (a) clearance and redevelopment, (b) rehabilitation, (c) conservation and maintenance. This involves identification of structures and sites selected for preservation, interim code enforcement, rehabilitation of basically sound structures not up to code standards, spot clearance, provision or upgrading of public facilities and services, and land-use control.
7. Preparation and adoption of an enforceable housing code and other regulations that will upgrade the living conditions and provide a decent home in a suitable living environment.
8. Institution of a systematic and planned code-enforcement program, including education of tenants and landlords, *prompt* encouragement and *requirement* of housing maintenance, improvement, and rehabilitation where indicated.
9. Concurrent provision and upgrading of public facilities and services where needed.
10. Provision of new and rehabilitated housing units through public housing, private enterprise, nonprofit organizations, individual owners, and other means.[12]
11. Aid in securing financial and technical assistance for homeowners.
12. Liaison with federal, state, and local housing agencies, associations, and organizations.
13. Requirement that payments to welfare recipients not be used to subsidize housing that does not meet minimum healthful standards.
14. Evaluation of progress made and continual adjustment of methods and techniques as may be indicated to achieve the housing program goals and objectives.

These efforts need to be supplemented by control of new building construction, land subdivision, and mobile home parks. Also to be included are migrant-labor camps, camp and resort housing, commercial properties, and housing for the aged, chronically ill, handicapped, and those on public assistance.

[11]Joseph G. Molner and Morton S. Hilbert, "Responsibilities of Public Health Administration in the Field of Housing," and Charles L. Senn, "Planning of Housing Programmes," *Housing Programmes: The Role of Public Health Agencies*, Public Health Papers 25, WHO, 1964; "Expert Committee on the Public Health Aspects of Housing," *WHO Tech. Rep. Ser.*, **225**, 1961; "Appraisal of the Hygienic Quality of Housing and its Environment," *WHO Tech. Rep. Ser.*, **353**, 1967; Subcommittee of the Program Area Committee on Housing and Health, *Guide for Health Administrators in Housing Hygiene*, Am. Pub. Health Assoc., New York, 1967.

[12]Kenneth G. MacIntosh, *Lower-Income Home Ownership Through Urban Rehabilitation—a new venture for private business*, National Gypsum Co., Dept. of Urban Rehabilitation, Buffalo, N.Y.

Solutions to the Problem

The more obvious solutions to the housing problem are production of new housing, redevelopment, slum clearance, and public housing for low-income families. Increasing emphasis is being placed on cooperative ownership and the conservation and rehabilitation of existing sound housing to prevent or slow down blight.* It is probably the most economical way of providing additional, more healthful dwelling units and at the same time protect the existing surrounding supply of good housing. Privately financed housing, redevelopment, and public housing cannot reach their full usefulness unless the neighborhood in which they are carried out is also brought up to a satisfactory minimum standard and is protected against degradation.

Federal, state, and local governments have an essential role in providing leadership and support. Federal programs make assistance available to finance housing for low- and moderate-income families, for urban development programs including rehabilitation and conservation, and for redevelopment of urban communities. Included are federal funds to help coordinate and finance the rebuilding, both physically and socially, of sections of cities. A local legally constituted agency is necessary to represent the municipality.

For example, eligibility for urban redevelopment funds may require formulation of a program to effectively deal with the problem of urban slums and blight within the community and to establish and preserve a well planned community with well organized residential neighborhoods of decent homes and suitable living environment for family life. The program can also make a municipality eligible for assistance for concentrated code enforcement, special help to blighted areas, demolition grants, rehabilitation grants and loans, and neighborhood facilities improvement and development. A workable program would require the following.

1. Adoption of adequate minimum standards of health, sanitation, and safety through a comprehensive system of codes and ordinances effectively enforced.
2. Formulation of a "comprehensive community plan" or a "general plan"—implying long-range concepts—and including land use, thoroughfare, and community facilities plans; a public improvement program; zoning and subdivision regulations. See Chapter 2.
3. Identification of blighted neighborhoods and analysis for extent and intensity of blight and causes of deterioration to aid in delineation for clearance or other remedial action.
4. Setting up an adequate administrative organization, including legal authority, to carry on the urban renewal program.
5. Development of means for meeting the financial obligations and requirements for carrying out the program.
6. Provision of decent, sanitary housing for all families displaced by urban renewal or other government activities.
7. Development of active citizen support and understanding of the urban renewal program.

*Careful analysis is advised as the hidden costs and problems associated with major building rehabilitation cannot as a rule be fully anticipated. (Michael Federman, "Building rehabilitation: the last resort," *Civ. Eng. ASCE*, July 1, 1981, pp. 72–73.)

The restoration of a substandard housing area in an otherwise sound and healthy neighborhood must include removal of the blighting influences to the extent feasible and where indicated. These include heavy traffic, air pollution, poor streets, lack of parks and trees, poor lighting, dirty streets and spaces, inadequate refuse storage and collection, inadequate water supply and sewerage, unpainted buildings, noisy businesses and industries, and tax increases for housing maintenance and improvement.

Further explanations and specifics to carry out some of the above program functions are given in the sections that follow.

Selection of Work Areas

A housing program must keep in proper perspective community short- and long-term plans, special surveys and reports, area studies, pilot block enforcement, and the routine inspection work to enforce a minimum-standards housing ordinance. Continual evaluation of the program is necessary to assure that the control of blight and deterioration of buildings and neighborhoods, as well as the rehabilitation of substandard dwellings, is being carried out where encountered. When indicated and possible, such action should be carried out on a planned block or area basis. This makes accomplishments more apparent and also awakens community pride, which can become "infectious." A blighted block does not always have a neon sign at the head of the street. The very insidiousness of blight makes it difficult to recognize and will have to be deliberately sought out and attacked. It may be only one or two houses in a block, which are recognized by a dilapidated outward appearance, by the inspector's intimate knowledge of the neighborhood and people, by office records showing a history of violations, or by routine survey reports. Signs of deterioration and dilapidation should be attacked immediately before decay and blight take over and incentive to make and support repairs becomes an almost impossible task.

The selection of conservation, rehabilitation, clearance, and redevelopment areas should take into consideration and adapt the following criteria.

1. The grading of socioeconomic areas using a weighted composite consisting of overcrowding, lack of or dilapidated private bath, lack of running water, and other measurable factors reported by the Bureau of the Census, including income and education.
2. The grading of areas using health indices and mortality and morbidity data.
3. The grading of areas on the basis of social problems, juvenile delinquency, welfare and private agency case load, adult probation, early venereal diseases, social unrest, and so forth.
4. The areas having the high, average, and low assessed valuation, dilapidation, and overcrowding.
5. Grading of neighborhoods using the APHA or equal appraisal system, health and building department surveys, and plotted data. A sampling of 20 percent of the dwellings can give fairly good information.
6. Results of surveys and plans made by planning boards, housing authorities, and redevelopment boards.

7. Selection of work areas with reference to these criteria as well as existing and planned housing projects, parks, redevelopment areas, parkways or thruways, railroads, and industrial or commercial areas; also existing barriers such as streams or lakes, good housing areas, swamps, and mountains.
8. Health department environmental sanitation and nursing division office records and personal knowledge of staff.
9. Visual foot surveys and combinations of criteria 1–8.

The superimposition of map overlays showing the above characteristics will bring out areas having the most problems.

Enforcement Program

Enforcement of a housing conservation and rehabilitation ordinance involves use of the same procedures and techniques that have been effective in carrying out other environmental sanitation programs. The development of a proper attitude and philosophy of the intent of the law and its fair enforcement should be the fundamental theme in a continuing in-service training program. A housing enforcement program can proceed along the following lines.

1. Inspection of a pilot area to develop and perfect inspection techniques and learn of problems and practical solutions to obtain rehabilitation of substandard areas.
2. Routine inspection of all hotels and rooming and boarding houses for compliance with minimum-standards housing ordinance. A satisfactory report may be made a condition to the issuance of a permit and a license to operate.
3. Routine inspection of multiple dwellings. An initial inspection on a two- or three-year plan will reveal places requiring reinspection.
4. Inspections throughout a city, village, or town on an individual structure, block, and area basis to obtain rehabilitation and conservation of sound housing or demolition of unsalvable structures.
5. Concentration of inspection in and around salvable areas to spread the border of improved housing so that it will merge with a satisfactory area.
6. Redirection of complaint inspections to complete housing surveys when practical. This can make available a large reservoir of personnel for more productive work.
7. Continuing in-service training with emphasis on law enforcement through education and persuasion, alteration and reconstruction, letter reporting, and financing to obtain substantial rehabilitation—not patchwork.
8. Close liaison with all city departments, especially city planning and building divisions, urban renewal agency, welfare, and the courts.
9. Continuing public education, including involvement of community and neighborhood organizations, to support the housing program and help or guide owners to conserve and rehabilitate their homes.

The enforcement program should take into consideration the problems, attitudes, and behavior of the people living in the dwelling units and the changes that need to be effected with the help of other agencies. The enforcement program must also recognize that, to be effective, subsidies and other forms of assistance to low-income families will usually be necessary to support needed alterations and rehabilitation.

Staffing Patterns

Various staffing patterns have been suggested to administer a housing-code enforcement program. One basis is a program director supported by one assistant to supervise up to eight inspectors, each making an estimated 1000 inspections per year. Slavet and Levin suggest the following ratios.[13]

1. One inspector per 10,000 population or one inspector per 1000 substandard dwelling units. This assumes inspection and reinspection to secure compliance at an average of 200 substandard dwelling units per year over a 5-year period.
2. One inspector per 3000 standard dwelling units, assuming inspection of 600 units per year over a 5-year period, in addition to the staff needed to handle complaints.
3. One financial specialist for every three or four housing inspectors.
4. One community relations specialist for every three or four housing inspectors.
5. One rehabilitation specialist for every two or three inspectors.
6. One clerk for every three or four inspectors.
7. One supervisor for every six to eight inspectors.

Some Contradictions

An enforcement program will reveal many contradictory facts relating to human nature. These only serve to emphasize that the housing problem is a complex one. The social scientist and anthropologist could study the problems and assist in their solution provided he works in the field with the sanitarians and engineers.

1. For example, it has been found that a person or family living in a substandard housing area may be very reluctant to move away from friends and relatives to a good housing area where the customs, religion, race, and very environment are different.
2. A housing survey may show that a structure is not worth repairing, and the expert construction engineer or architect may be able to easily prove that conclusion. But the owner of a one-, two- or three-family building will rarely agree. He will go on to make certain minimum repairs or even extensive repairs to approach the standards established in the housing ordinance, leaving you wondering if there really is any such thing in fact as a nonsalvable dwelling.
3. There is the not infrequent situation where two old structures are located on the same lot, one facing the street and the other on the rear-lot line. If one structure is demolished, another cannot be built on the same lot because zoning ordinances usually prohibit such intensive lot usage by new structures. The owner therefore, rather than lose the vested right due to prior existence of the structure, with respect to the zoning ordinance, may choose to practically rebuild the dwelling at great cost rather than tear it down.
4. Another common occurrence is the tendency for some landlords to seize upon health or building department letters recommending improvements to bring the building into compliance with the housing ordinance as the opportunity to evict tenants and charge higher rentals. This is not to say that the owner is not entitled to a fair return on an investment; but the intent is that no one should profit from human misery. (A fair return on one's investment has been given as 10 percent of the assessed valuation or purchase price of the property.) Since the department's objective is to improve the living conditions of the people, it should not become a party to such actions. As a matter of fact, it is the unusual situation where needed housing improvements cannot be made with the tenant living in the dwelling unit, even though some temporary inconvenience may result.

[13]Joseph S. Salvet and Melvin R. Levin, *New Approaches to Housing Code Administration*, National Commission on Urban Problems, Research Rep. 17, Washington, D.C., 1969.

Hence in such situations it is proper for the department to state that it is its opinion that the needed repairs can be made without evicting tenants, thereby leaving the final determination in the exceptional cases with the owner and the courts.

5. A rather unexpected development may be the situation where an owner agrees to make repairs and improvements, such as new kitchen sink and provision of a three-piece bath with hot as well as cold water under pressure, only to be refused admittance by the tenant. For the tenant knows that the rent may be increased and suddenly decides that a new bathroom is a luxury he can get along without. In such case the courts have granted eviction orders and have sanctioned the reasonable increase in rent.

Fiscal Aspects

The financing of improvements is sometimes an insurmountable obstacle to the lay person. As an aid to the people affected by a housing enforcement program, a "financing clinic" can be formed to advise homeowners in difficult cases. The clinic may be a committee consisting of a banker, an architect, a general contractor, a plumbing contractor, and a representative of the building division, redevelopment board, federal agencies, and community. The department representative would help the owner present the problem. If the homeowner is instructed to submit estimates of cost from two or three reliable contractors, a sounder basis for assistance is established.

An enforcement program must recognize that many investors use the criterion that a dwelling is not worth purchasing unless it yields a gross annual income $\frac{1}{4}$ to $\frac{1}{5}$ of the selling price. Another rule of thumb, for rehabilitation of a dwelling, is that the cost of alteration plus selling or purchase price shall not be more than five times the gross annual income.

Getting Started

The report that appeared in *The New York Times*, October 6, 1957, is an excellent example of a sound approach to initiate a housing conservation and rehabilitation. It is quoted below.

> Home owners whose properties are in Milwaukee's first "conservation" area are not likely to complain to the Mayor when their houses are inspected in the door-to-door canvass being made by the city. Mayor Frank Zeidler's home is in the area, too.
>
> The inspections are planned to find out whether the district's middle-aged homes meet housing code standards. Although the courts can enforce compliance with the code, the drive is aimed primarily at informing owners of dangerous deterioration, with the expectation that they will correct the conditions within a reasonable time.
>
> Mayor Zeidler told his neighbors that "this is in the nature of getting free advice." According to the National Association of Housing and Redevelopment Officials, the drive can save a neighborhood from becoming a slum which, experience shows, costs taxpayers more than it repays in tax revenues.

Enforcement Procedures

The fair and reasonable enforcement of the *intent* of a minimum-standards housing ordinance requires the exercise of a great deal of trained judgment by the administrators, supervisors, and field inspectors. Continuing in-service training of the housing staff is essential to carry out the purpose of the housing ordinance in an effective manner.[14] A trained educator can offer valuable assistance and guidance in planning the training sessions and in interpreting the program to the public and legislators. Health department sanitarians are admirably suited to carry out the housing program because of their broad knowledge in the basic sciences and their ability to deal effectively with the public.

The enforcement measures and procedures can be summarized as follows.

1. Preparation of a housing operating manual giving a background of the housing program, guides for conduct, inspection report form based on the housing ordinance, interpretation of sections of the ordinance needing clarification for field application, suggested form paragraphs for routine letters, including typical violations and recommendations for their correction, follow-up form letters, building construction details, and related information. A few form paragraphs, and architectural details are included in the text to suggest the type of material that can be included in a housing manual to obtain a certain amount of uniformity and organization in a housing enforcement program.
2. Complete inspection of all types of premises and dwelling units occupied as living units. A comprehensive checklist or survey form based on the housing ordinance similar to Figure 11-4 could be used.
3. Preparation of a complete rough-draft letter by the sanitarian based on field inspection, listing what was found wrong, together with recommendations for correction.
4. Review of the inspection report and suggested letter by a supervising sanitarian for completeness and accuracy. The letter is signed by the supervisor and a tickler date confirmed for a follow-up inspection. See Figure 11-5.
5. Maintenance of a visible card-index file on each multiple dwelling, boarding house, rooming house, and hotel, in addition to a file on each dwelling, including one- and two-family structures.
6. A reinspection letter based on the findings reported in the first letter and remaining uncorrected. This may take the form of an abridged original letter listing what remains to be done, a letter calling for an informal hearing, a letter giving 30 days in which to show substantial progress, or a special letter. See Figure 11-6.
7. In some instances substantial improvements are requested that justify a reasonable rent adjustment. This should be brought out at the time of informal hearings when financial hardship is pleaded. Informal administrative hearings are very useful.
8. Office consultation with owners to investigate long-range improvement programs within the limits of the law and the federal, state and private assistance available.
9. Maintenance of a list of competent contractors who perform alteration work.
10. Review of cost estimates with the owner for completeness.
11. Arranging for expert consultation to finance improvements, including a hearing before a "financing and construction clinic," which should have representation from the lending institutions as well as from the contractors, architects, and engineers.
12. In recalcitrant cases, a letter from the chief city judge or justice providing for a pretrial hearing to show cause why a summons should not be issued, followed by the hearing. A similar procedure could be followed by the corporation counsel or county attorney. The courts could also establish a special session to hear housing cases.
13. Issuance of a summons by the corporation counsel at the request of the enforcement agency when other measures have failed, if authorized by local law. A record of appearances and results should be kept.

[14] *Basic Housing Inspection*, PHS, DHEW Pub. (CDC) 76-8315, GPO, Washington, D.C. 20402, 1975 revision.

ERIE COUNTY DEPARTMENT OF HEALTH

Division of Environmental Sanitation
601 City Hall, Buffalo 2, N.Y.

MOhawk 2800
Extension 28

Re:

C. Buffalo, New York

Dear

The City of Buffalo recently adopted a minimum standards housing ordinance to assist in the conservation and rehabilitation of existing housing. In this connection Mr. of this Department made a survey of dwellings in your neighborhood on Listed below are the conditions found at your building which were in need of improvement or correction.

We shall be glad to discuss any questions or difficulties you may have in complying with the minimum housing standards. Would you please advise this office in writing within 10 days of the action you propose to take?*

Your cooperation will help prevent the deterioration of property values in your neighborhood and help make Buffalo a more healthful place in which to live.

Very truly yours,

J. A. Salvato, Jr., P.E.,

Chief, Bureau of General Sanitation

By..

* This letter is not to be construed as reason for removal of tenants, unless specifically stated therein.

Alterations or additions must be made in accordance with all applicable laws. Consult with the Division of Buildings, Room 325, City Hall for a permit to perform building, plumbing or electrical work required here. A building unlawfully occupied or with an increased number of living units must be brought into compliance, or if this is not permitted by law, it must revert to its original use.

JAS/er (1) 9/54:1000

Figure 11-5 Form letter to confirm findings at time of first inspection.

COUNTY OF ERIE
HEALTH DEPARTMENT
601 CITY HALL
BUFFALO 2, N.Y.

Re:

C. Buffalo, New York

Dear

This department notified you by letter on of the housing deficiencies existing on your premises and the need for making corrections. However, a reinspection showed that no substantial progress had been made, and to date we have received no reply from you.

Inasmuch as the conditions enumerated in our letters to you are violations of the Buffalo Minimum Housing Standards Ordinance, it is imperative that the needed improvements be made. Your failure to reply must be interpreted as an indication of lack of good faith or misunderstanding of the intent of our letters.

Under the circumstances, you are requested to appear in Room 605, City Hall, at to show cause why legal action should not be started. It is hoped that you will come prepared with an itemized report, indicating when the needed improvements will be completed, so as to make unnecessary further action by this department.

We wish to emphasize that it is our intention and obligation to follow up on violations of the Minimum Housing Standards Ordinance until such time as all work is done. It is only in this way that existing housing in the City of Buffalo can be conserved and rehabilitated.

Very truly yours,

J. A. Salvato, Jr., P. E.,

Chief, Bureau of General Sanitation

By..

JAS/er(3)
9/54:500

Figure 11-6 Informal hearing letter.

14. Issuance of a summons by an authorized member of the enforcement agency in emergencies, if authorized by local law.

15. Cooperation with the press with a view toward obtaining occasional special reports, feature articles, and community support.

16. Cooperation and liaison with all private and public agencies having an interest in housing.

17. Notification of welfare department, when public assistance is being given for rental, because payment for housing should be made only when the facilities and conditions of maintenance and occupancy meet minimum standards.

18. Education of the legal department so as to obtain an aggressive, trained, and competent attorney who can become experienced in housing problems and be able to guide the legal enforcement phase of housing work.

19. Stimulate resurgence of neighborhood pride in the rehabilitation and conservation of all buildings on a block or area basis. Information releases and bulletins explaining what community groups and individuals can do to improve their housing is an excellent approach.

20. Discourage profit at the expense of human misery by making it possible through state law to file a lien in the county clerk's office against a property that is in violation of health ordinances, for the information of all interested and affected persons.

21. Notification of the mortgage holders and insurance carriers of existing conditions.

22. Legislation requiring those who purchase city-owned property at public auctions to first list all the property they have owned in the past five years, properties they own with delinquent city taxes, the name and address of all major stockholders if a corporation, whether the purchaser owns vacant or abandoned buildings, and whether the city had ever taken title to the purchaser's property for back taxes.

23. Legislation requiring in the case of corporate ownership that the name, address and the number or percentage of shares owned in the corporation be filed in the county clerk's office.

24. Registration and periodic reregistration of current ownership to facilitate the service of notices and enforcement. Require an in-state agent for receipt of legal notices.

25. Requirement of a certificate of occupancy each time before an apartment is rented to a new tenant.

26. Legislation authorizing payment of rent into an escrow account to correct code violations, after which the account is turned over to the owner.

27. A municipal emergency repair program funded through attachment of rents, or a property lien.

28. Low interest loans and grants to make repairs.

29. Legislation permitting tax foreclosure proceedings within two years or less to permit acquisition of property while it still has value, before it has deteriorated beyond hope of economic rehabilitation, with safeguards for the sincere resident owner who is temporarily in financial difficulty. This requires a systematic inspection system, early identification of deficiencies, prompt and vigorous enforcement, or foreclosure, where indicated.

The enforcement procedure generally includes news releases and information bulletins, inspection, notification, reinspection, second notification, third notification, informal hearings, formal hearings, pretrial hearings, and summonses. In brief, therefore, the enforcement procedure consists of education and persuasion and legal action only as a last resort. See Legal Action, Chapter 12.

Not to be forgotten is stimulation of the possible role of nongovernment agencies having a direct or indirect interest and concern in the maintenance of good housing and in housing rehabilitation. The role could consist of seminars and individual assistance for home improvement given by banks; material supply companies; builder and contractor organizations; community colleges; urban development and redevelopment agencies; and fire insurance companies. The assistance could include financing methods for improvements; available federal (and state) grants and loans; filing of applications for assistance; owner-contractor relations; value of architectural services; information on materials, supplies, and equipment for home repairs and improvements, their advantages, disadvantages, and costs.

HOUSING FORM PARAGRAPHS FOR LETTERS

It is necessary to have some reasonable uniformity and accuracy when writing letters or reports to housing-code violators. This becomes particularly important when a housing-code enforcement unit has a large staff and when court action is taken. By using a dwelling-inspection form such as Figure 11-4, and the form paragraphs listed below, it should be possible for the inspector to readily prepare a draft of a suitable letter for review by that inspector's supervisor. The form paragraphs can and should be modified as needed to reflect more precisely the unsatisfactory conditions actually observed and practical suggestions for their correction. Reference to the pages that follow will be helpful in describing violations.

Structural Safety

To be considered structurally safe a building must be able to support 2½ to 4 times the loads and stresses to which it is or may be subjected.

Certain conditions that may be deemed dangerous or unsafe need explanation. For example, a 12-in. beam that has sagged or slanted more than one-quarter out of the horizontal plane of the depth of floor structural members in any 10-ft distance would be more than 3 in. out of level in 10 ft and hence unsafe. An interior wall consisting of 2-in. × 4-in. studs or 4-in. terra-cotta tile blocks more than one-half out of the vertical plane of the thickness of those members between any two floors would be more than 2 in. out of plumb and hence unsafe. See Figure 11-7.

A stair, stairway, or approach is safe to use when it is free of holes, grooves, and cracks that are large enough to constitute a possible accident hazard. Rails and balustrades are expected to be firmly fastened and maintained in good condition. Stairs or approaches should not have rotting or deteriorating supports, and stairs that have settled more than 1 in. or pulled away from the supporting or adjacent structure may be dangerous. Stair treads must be of uniform height, sound, and securely fastened in position. Every approach should have a sound floor and every tread should be strong enough to bear a concentrated load of at least 400 lb without danger of breaking through. See Figure 11-8.

Incomplete Bathroom

The (first floor rear apartment) did not have a tub or shower. A complete bathroom including a water closet, tub or shower, and washbasin connected with hot and cold running water is required to serve each family. See Figures 11-9 and 11-10. The bathroom shall have a window or skylight not less than 10 per-

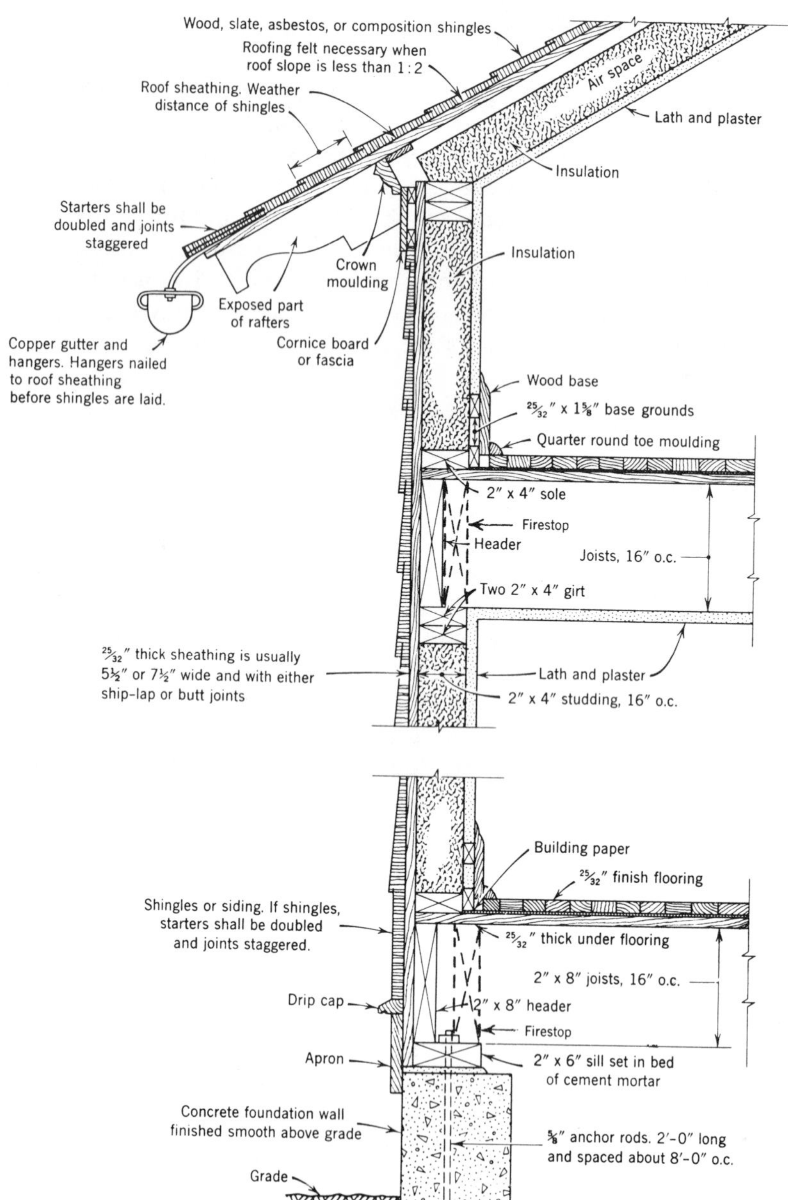

Figure 11-7 Wall construction.

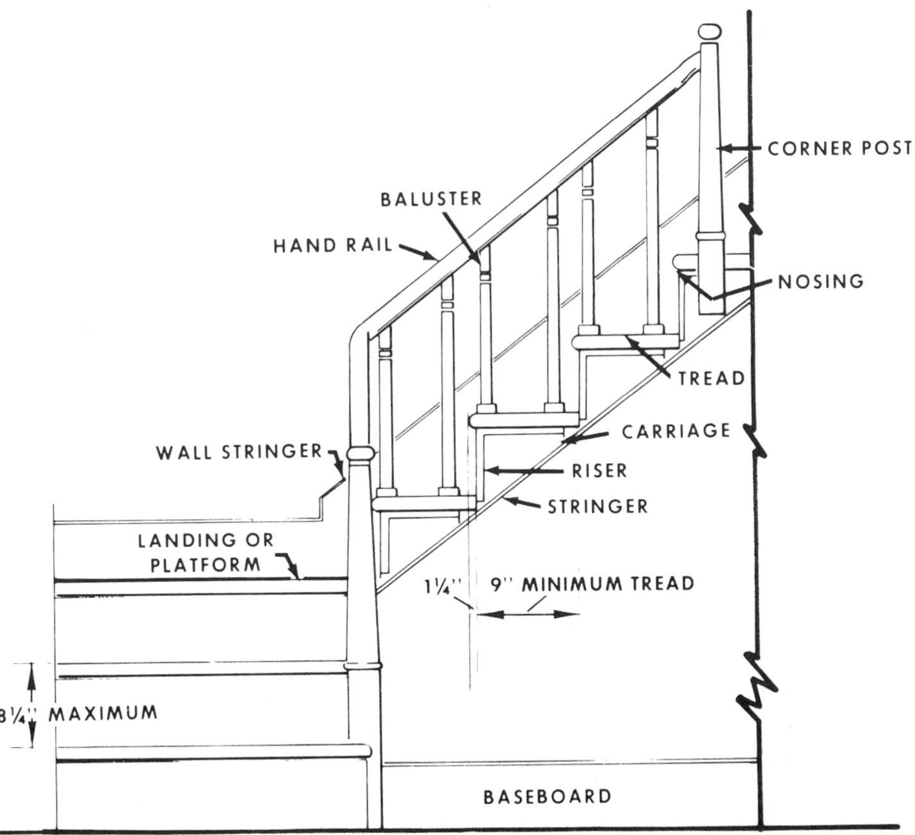

Figure 11-8 Stairway details. [Source: *Basic Housing Inspection*, DHEW Pub. (CDC) 76-8315, p. 38, GPO, Washington, D.C. 20402.]

cent of the floor area, with at least 45 percent openable, providing adequate light and ventilation. A water-repellent floor with a sanitary cove base or equivalent is necessary. A ventilation system may be approved in lieu of a window or skylight.

No Hot Water

There was no piped hot water in the (kitchen of the first floor front apartment). This apartment shall be provided with hot water or water-heating facilities of adequate capacity, properly installed and vented. The heater shall be capable of heating water so as to permit water to be drawn at every required kitchen sink, lavatory basin, bathtub, or shower at a temperature of approximately 110°F.

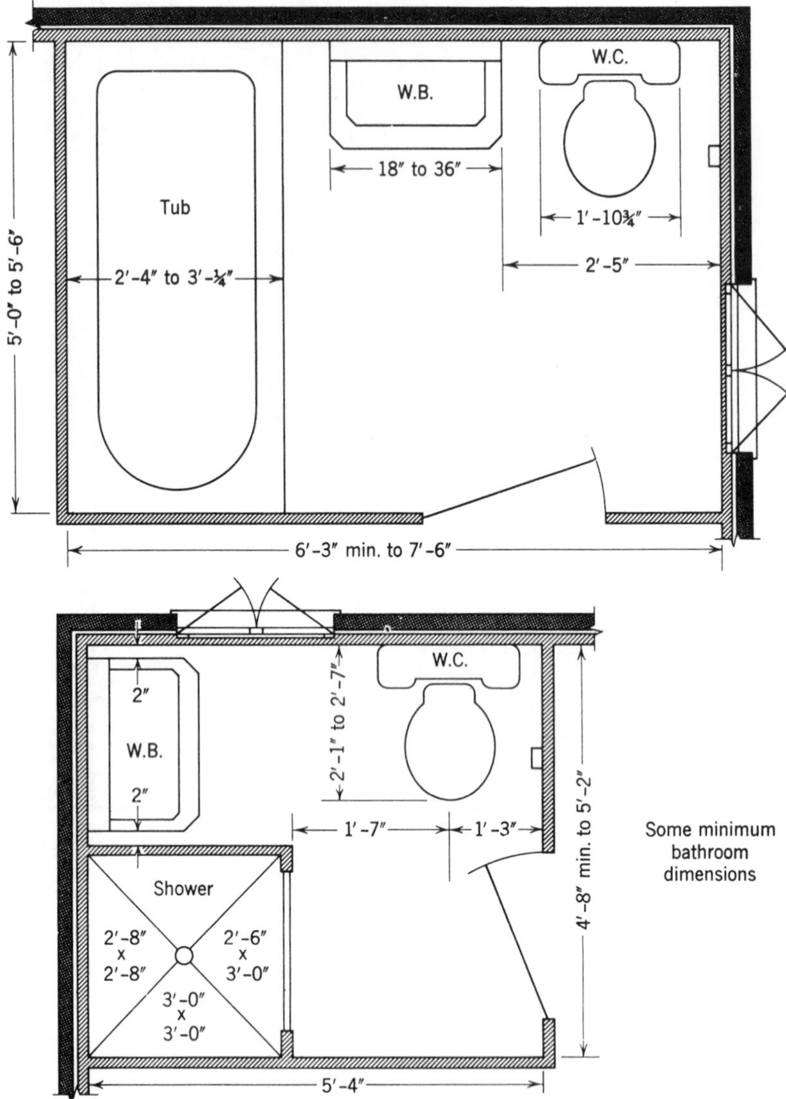

Figure 11-9 Three-piece bathroom showing minimum dimensions.

Leaking Water Closet

The water-closet bowl in the (describe location) apartment was (loose) (leaking) at the floor. (When the toilet is flushed the water drains onto the floor and seeps through the ceiling to the lower apartment.) The water closet should be securely fastened to the floor, floor flange, and soil pipe so that it will be

Figure 11-10 Alteration for shower and washbasin addition.

firm and not leak when flushed. (It will also be necessary to repair the loose ceiling plaster in the lower floor and repaint or repaper as needed.)

Floors Not Water-Repellent

The bathroom (water-closet compartment) (floor covering) was (worn through) (broken) (bare wood with open joints). The floor should be (repaired) (made reasonably watertight), and the floor covering should extend about 6 in. up the wall to provide a sanitary cove base. Satisfactory material for the floor covering is inlaid linoleum, rubber or composition tile, smooth cement concrete, tile, terrazzo, and dense wood with tightly fitted joints covered with varnish, lacquer, or other similar coating providing a surface that is reasonably impervious to water and easily cleaned.

Exterior Paint Needed

The exterior paint has (peeled, worn off) exposing the bare wood, rusting the nails, causing splitting and warping of the siding. This will lead to the entrance of rainwater, rotting of the siding, sheathing, and studs, as well as inside dampness and falling of the plaster! You are urged to immediately investigate this condition and make the necessary repairs, including painting or other weather- and decay-resistant treatment of the house, to prevent major repairs and expenses in the future.

Rotted and Missing Siding

The (shingles, siding, apron, cornice, exposed rafters) (on the north side of the house at the second-floor windows and foundation) was (were) (rotted and missing). Decayed material should of course all be removed, the sheathing repaired wherever necessary, and the (shingles, siding, etc.) replaced. Following the carpentry work, all unpainted or unprotected material exposed to the elements should be treated to prolong its life.

Sagging Wall

The (door frames and window frames) in the (location) were out of level, making complete closure of the doors and windows impossible. Outside light could be easily seen through the openings around the (window rails and door jambs). The supporting (beams, girders, posts, and studs) should be carefully inspected as there was evidence that some of these members were rotted, causing the outside wall to sag. The building should be shored and made level wherever necessary. The unsound material should be replaced, and the improperly fitting (doors, windows, and framing) repaired so as to fit and open properly. See Figures 11-7 and 11-11.

Loose Plaster

The plaster was (loose) (and buckled) (and had fallen) from the living room ceiling and walls in the (name apartment or other location) over an area of approximately (10 ft^2). All loose plaster should be removed and the wall replastered; following curing it should be painted or papered to produce a cleanable, smooth, and tight surface.

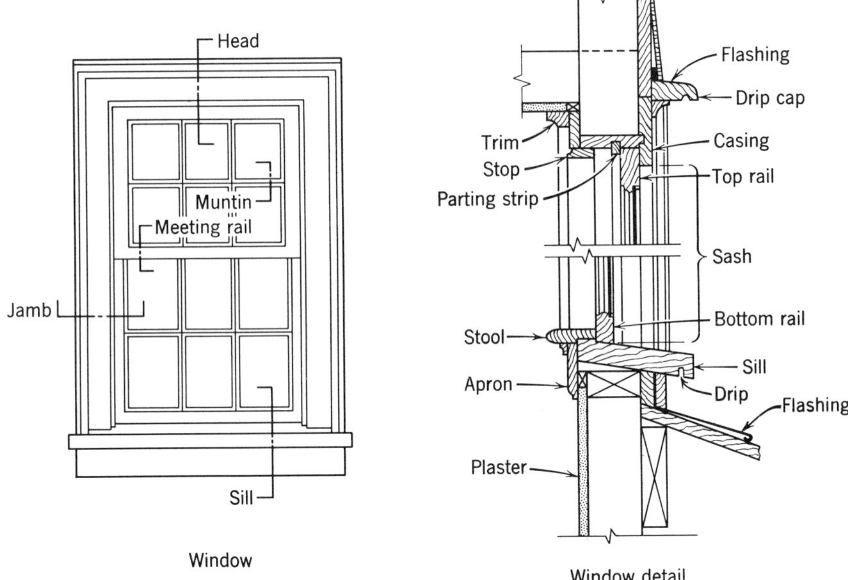

Figure 11-11 Window details.

Leaking Roof

There was evidence of the roof leaking over or near the (kitchen, living room, etc., in the tenant apartment). The (paint, paper, was stained and peeling). It is essential that the leak be found and repaired, not only to prevent the entrance of water and moisture in the apartment but also to prevent loosening of the plaster, rotting of the timbers, and extension of the damage to your property.

No Gutter or Rain Leaders

There were no gutters or rain leaders on this building. Gutters and rain leaders should be placed around the entire building and connected (to the sewer if permitted) to ensure proper drainage of rainwater. This will also make less likely rotting and seepage of water through siding and window frames and entrance of water into the (cellar) (basement).

No Handrails

There are no handrails in the stairway between the (first and second floor at the rear). This is a common cause of preventable serious accidents. Handrails should be provided and securely fastened at a height of 30 to 32 in., measured above the stair tread.

Refuse in Attic

There were (rags, refuse, paper, and trash) in attic. These materials are a fire hazard and provide harborage for mice and other vermin. All rags, paper, and trash must be removed from the attic and the attic maintained in a clean and sanitary condition at all times.

Water-Closet Flush Tank Not Operating Correctly

The (water ran continuously) in the water-closet flush tank in the (John Jones's apartment); OR the water closet in the (John Jones's apartment) could not be flushed. The (broken, worn, missing) (ball-cock valve, ball-cock float, flush-valve ball, flush lever, flush handle) should be repaired or replaced to permit proper flushing of the water closet. See Figure 11-12.

Garbage Stored in Paper Box or Bag

Garbage is being stored in (open, uncovered baskets) (paper bags) (paper boxes) (in the rear yard). This encourages rodent, fly, and vermin breeding. All garbage should be drained, wrapped, and properly stored in tightly covered containers. It will be necessary for you to procure needed receptacles for the proper storage of all garbage until collected.

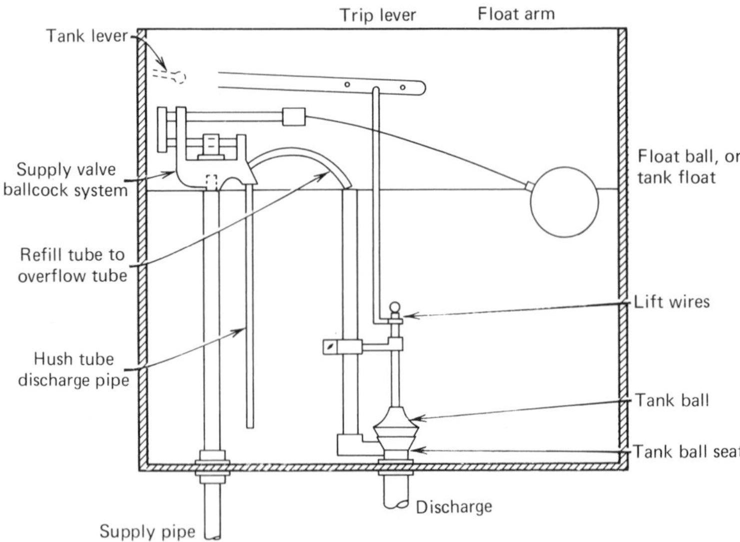

Figure 11-12 Water-closet tank. (Flapper valve can replace tank ball. Unvented supply valve requires backflow preventer.)

Dilapidated Garbage Shed

The garbage shed in the (specify location) was in a dilapidated, rotted, and insanitary condition. Garbage sheds tend to accumulate garbage and encourage rodent, fly, and vermin breeding. This dilapidated garbage shed should be removed and the premises cleaned. Store the garbage cans on an elevated rack or concrete platform. (Enclose pamphlet showing some suggested storage racks). (See pages 541 and 542.)

Debris in Yard or Vacant Lot

The vacant lot located at (specify location) was found littered with (old lumber, tin cans, and rubbish). This is unsightly and may serve as a rat harborage and as an invitation to dump on the property. It is requested that you make a personal investigation of the conditions reported and arrange to have the lot cleared and cleaned. It is also recommended that you post a "No Dumping" sign to discourage future littering of the property.

Dirty Apartment

The apartment on the (second floor) is in a very insanitary condition. (Describe.) Every occupant is expected to keep his apartment and the premises he controls in a clean and sanitary condition at all times. (Copy of letter to tenant.)

Overcrowded Sleeping Room

The bedroom(s) in the (specify location) were overcrowded. There were (three) persons in a room having an area of (80) ft^2 and (four) persons in a room (85) ft^2. Every room occupied for sleeping purposes shall contain at least 70 ft^2 of floor space for one person and 50 ft^2 for each additional person. (Suggest correction.) This apartment should not be rerented for occupancy by more persons than can be accommodated in accordance with this standard.

No Window in Habitable Room

There was no window to the outside air provided in the (living room, kitchen, bedroom). Every room used or intended to be used as a (living room, kitchen, bedroom) is required to have a total unobstructed window area of at least 10 percent of the floor area. Consideration should be given to the possibility of (cutting in a new window) (providing a skylight) if the room is to be continued in use as a (bedroom, living room, or kitchen).

Unlawful Third-Floor Occupancy

The third floor of this building had been converted and was occupied for living purposes. The conversion or alteration of a third floor or attic in a frame building for living or sleeping purposes is prohibited by Chapter X of the city ordinances. This is a major hazard in case of fire. Discontinue the use of the third floor immediately. (Refer copy of letter to division of buildings.)

Unlawful Cellar Occupancy

The cellar was being used for (sleeping, living, purposes). A cellar may not be used for living purposes, hence this space must be permanently vacated. (The housing ordinance defines a cellar as "a room or groups of rooms totally below ground level and usually under a building."

Clogged Sewer

The (soil stack, building drain, or sewer) was apparently clogged, for sewage from the upper apartment(s) backed into the (kitchen sink, water closet) in the (first-floor front apartment). The clogged sewer must be cleared and, if necessary, repaired to eliminate cause for future complaint.

Unvented Heater

The gas water heater(s) (burning carbonaceous fuel) in the (name room or space and locate) was (were) not vented. Unvented heaters in bathrooms and sleeping rooms have been the cause of asphyxiation, carbon monoxide poisoning, and death. These heaters must be properly vented to the chimney or outside air, supplied with sufficient air to continuously support combustion of the fuel, and be protected to prevent fires and minimize accidental burns. See Venting of Heating Units, this chapter.

Furnace Flue Defective

The furnace flue had rusted through in several places (and the connection to the chimney was loose), causing waste gases to escape into the basement. Since such gases rise and seep into the upper apartment(s) and have been known to cause asphyxiation, it is imperative that the flue be repaired and the collar sealed to prevent leakage of any waste gases. This should also improve the efficiency of the furnace.

Rubber-Hose Gas Connection

The gas heater(s) in the (tenant apartment) has (have) (plastic pipe, rubber hose) connection(s). Such materials eventually leak and may cause death in the household. It will be necessary to replace all plastic and rubber-hose connections with rigid, metal pipe.

Rat Infestation

There was evidence of a very bad rat condition existing in this building as indicated by (explain condition). All holes in the foundation (floors) should be sealed with cement mortar and openings around wood framing closed with metal flashing or with cement mortar where possible. Traps and repeated use of a rodenticide such as warfarin are suggested to kill rats inside the building. All sources of food and harborages must be eliminated. Such control measures should be continuous for at least two or three weeks to be effective. (Enclose pamphlet giving additional details dealing with accepted control measures.)

Roach Infestation

The apartment was apparently infested with roaches as indicated by the roachy odor, roaches observed hiding under the sink, baseboard, moldings (stains in the kitchen cabinet, pellets of excrement in the dish cabinet). Roaches are sometimes brought in with boxes of food, baskets, or bags; dirt and filth encourage their reproduction in large numbers. Thorough cleaning, filling of cracks around frames with plaster or plastic wood, followed by the proper application of an insecticide in selected places and in accordance with the manufacturer's directions should bring the problem under control. (The enclosed pamphlet gives additional detailed information.)

Overflowing Sewage Disposal System

The sewage disposal system serving your dwelling was (seeping out onto the surface) (discharging into the ditch in front of your home). This is a health hazard not only to you but also to your neighbors, children, and pets. Immediate steps should be taken to determine the cause and make corrections. (See Chapter 4 for more details.)

Improperly Protected Well-Water Supply

A sanitary survey of your well-water supply shows it to be subject to contamination. The well (is uncapped) (has a hole around the casing where surface water can drain down and into the well) (does not have a tight seal at the point where the pump line(s) pass into the casing as noted by drippage observable from looking into the well). The necessary repairs should be made to prevent contamination of your water supply, and then the well should be disinfected as explained in the enclosed instructions. (See Chapter 3 for more details.)

Major Repairs

In view of the major repairs and improvements needed, only some of which having been reported above, plans prepared by a registered architect should be submitted showing the existing conditions and all proposed alterations, for approval by this Department and the Division of Buildings, before any work is done. This procedure makes possible the receipt of comparable bids from several contractors and usually results in more orderly prosecution of the work at a minimum cost.

Minor Repairs

In view of the repairs and improvements needed, a sketch drawn to scale should be prepared showing existing and proposed work, to assure that the work can be done as intended. The sketch should be submitted to and be approved by the Division of Buildings and this Department before any work is done. This procedure makes possible the receipt of comparable bids from several contractors and usually results in more orderly prosecution of the work at a minimum cost.

Obtain at Least Three Estimates

We urge you to obtain at least three estimates from reputable contractors before having any work done. Written bids should be requested and assurance obtained from the contractor that the estimate is all-inclusive.

PLUMBING

Plumbing Code

Sanitary plumbing principles that are based on the latest scientific studies should be fundamentally similar, but will be varied in application depending on the local conditions. Some plumbing designs and standards currently in existence are based on an unsound old rule of thumb or prejudice. They could be reviewed with profit in the light of present-day knowledge.

The *National Plumbing Code*, ASA A40.8-1955[15] and PHS Pub. 1038,[16] and *Uniform Plumbing Code*, amended 1974[17] are comprehensive standard codes of minimum requirements for use throughout the country. The interested person would do well to have copies in a reference file. The sizing of water supply, drainage, vent, and storm-drain piping is concisely covered.

One term used frequently is *plumbing fixture*. This term means "installed receptacles, devices, or appliances either supplied with water or receiving on discharge liquids or liquidborne wastes or both." The bathtub, sink, water closet, dishwasher, and drinking fountain are examples of plumbing fixtures. A *fixture unit* is the load-producing flow effect from different plumbing fixtures, usually taken as 7½ gpm per fixture unit.

Health departments can accomplish more in the interest of public health by seeing that proper standards of plumbing exist and are enforced than by actually doing the plumbing inspection.

For example, the health department can see to it that plumbing and building codes prohibit dangerous cross-connections and interconnections and require a private three-piece bathroom and kitchen sink served by hot and cold water in every new dwelling unit. This is fundamental to the prevention of disease, the promotion of personal hygiene, and sanitation. Plumbing codes should prohibit use of lead piping for water distribution and the use of tin-lead (50:50 and 60:40) solder for joining copper piping, especially in areas where the drinking water is known to be soft.

Plumbing codes should specify an adequate number of fixtures for private, public, and industrial use, all properly supplied, trapped, vented, and sewered as noted in Tables 11-6 to 11-9. Water connections with unsafe or questionable water supplies would be prevented, and connections or conditions whereby used or unsafe water could flow back into the potable water system would be prohibited. A safe water supply and proper sewage disposal should of course be assured.

Housing codes would be expected to make reference to a modern plumbing code. Housing codes would be of little value unless they were applicable to all new, altered, and existing one-, two-, or more family dwellings, hotels, boarding houses, and rooming houses.

The health department should serve as a consultant to the building, plumbing, water, and sewer divisions. Any new or revised codes or regulations should first be reviewed and approved by the health department before being considered for adoption to assure that fundamental principles of public health and sanitary engineering are not violated.

Some plumbing details, with particular emphasis on backflow prevention, recommended minimum number of plumbing fixtures, the application of in-

[15]*National Plumbing Code*, ASA A40.8-1955, ASME, 29 West 39th Street, New York, N.Y. (1955).

[16]*Report of Public Health Service Technical Committee on Plumbing Standards*, PHS Pub. 1038, DHEW, Washington, D.C., September 1962).

[17]*Uniform Plumbing Code*, International Association of Plumbing and Mechanical Officials, 5032 Alhambra Ave., Los Angeles, Calif, 90032, amended 1974.

Table 11-6 Minimum Number of Plumbing Fixtures

Type of Building Occupancy	Type of Fixture: Water Closets		Urinals		Lavatories		Bathtubs or Showers	Drinking Fountains	Other Fixtures
	Number of persons	*Number of fixtures*	*Number of persons*	*Number of fixtures*	*Number of persons*	*Number of fixtures*			
Assembly—places of worship	150 women 300 men	1 1	300 men[a]	1	1		—	1	
Assembly—other than places of worship (auditoriums, theaters, convention halls)	1-100 101-200 201-400 Over 400, add 1 fixture for each additional 500 men and 1 for each 300 women	1 2 3	1-200 201-400 401-600 Over 600, add 1 fixture for each 300 men[a]	1 2 3	1-200 201-400 401-750 Over 750, add 1 fixture for each 500 persons	1 2 3	—	1 for each 300 persons	1 slop sink
Dormitories—school or labor, also institutional	Men: 1 for each 10 persons Women: 1 for each 8 persons		1 for each 25 men; over 150, add 1 fixture for each 50 men[a]		1 for each 12 persons. (Separate dental lavatories should be provided in community toilet rooms. A ratio of 1 dental lavatory to each 50 persons is recommended.)		1 for each 8 persons. For women's dormitories, additional bathtubs should be installed at the ratio of 1 for each 30 women. Over 150 persons add 1 fixture for each 20 persons.	1 for each 75 persons	Laundry trays, 1 for each 50 persons Slop sinks, 1 for each 100 persons
Dwellings—one and two-family	1 for each dwelling unit		—		1 for each dwelling unit		1 for each dwelling unit	—	Kitchen sink, 1 for each

Dwellings—multiple or apartment	1 for each dwelling unit or apartment	—	1 for each dwelling unit or apartment	1 for each dwelling unit or apartment	—	Kitchen sink, 1 for each dwelling unit or apartment[b]
Industrial—factories, warehouses, foundries, and similar establishments	*Number of each sex* — *Number of fixtures* 1-10 — 1 11-25 — 2 26-50 — 3 51-75 — 4 76-100 — 5 1 fixture for each additional 30 employees	Where more than 10 men are employed: *Number of men* — *Number of urinals* 11-30 — 1 31-80 — 2 81-160 — 3 161-240 — 4	*Number of persons* — *Number of fixtures* 1-100 persons — 1 to 10 Over 100 persons — 1 to 15	1 shower for each 15 persons exposed to excessive heat or to occupational hazard from poisonous, infectious, or irritating material	1 for each 75 persons	
Institutional—other than hospitals or penal institutions (on each occupied story.)	1 for each 25 men 1 for each 20 women	1 for each 50 men[a]	1 for each 10 persons	1 for each 10 persons	1 for each 50 persons	
Hospitals						
Individual room	1	—	1	1	—	1 slop sink/floor
Wards	1 for each 8 patients	—	1 for each 10 patients	1 for each 20 patients	1 for each 100 patients	
Waiting rooms	1	—	1	—	—	
Employees	Same as public	Same as public	Same as public	—	Same as public	
Penal institutions	1 in each cell	1 in each exercise room	1 in each cell	1 on each cell block floor	1 on each cell block floor	1 slop sink/floor
Prisoners	1 in each exercise room	—	1 in each exercise area	—	1 in each exercise area	

Table 11-6 (*Continued*)

Type of Building Occupancy	Type of Fixture: Water Closets		Urinals	Lavatories		Bathtubs or Showers	Drinking Fountains	Other Fixtures
Penal institutions (*Continued*)								
Employees	Same as public		Same as public	Same as public		—	Same as public	
Public buildings, offices; business, mercantile, storage, and institutional employees	*Number of each sex*	*Number of fixtures*	Urinals may be provided in men's[a] toilet rooms in lieu of water closets but for not more than $\frac{1}{3}$ of the required number of water closets	*Number of employees*	*Number of fixtures*	—	1 for each 75 persons	1 slop sink/floor
	1-15	1		1-15	1			
	16-35	2		16-35	2			
	36-55	3		31-60	3			
	56-80	4		61-90	4			
	81-110	5		91-125	5			
	111-150	6		1 fixture for each additional 45 persons				
	1 fixture for each additional 40 employees							
Schools	*Boys*	*Girls*						
Elementary	1/40	1/35	1/30 boys	1/50 pupils		In gym or pool shower rooms, $\frac{1}{5}$ pupils *of a class*	1/100 pupils but at least 1 per floor	Slop sinks, 1 on each floor
Secondary	1/40	1/45	1/30 boys	1/50 pupils				
Working men, temporary facilities	1/30 working men		1/30 working men	1/30 working men		—	1 fixture or equivalent for each 100 working men	

Source: *Report of Public Health Service Technical Committee on Plumbing Standards*, September 1962, PHS Pub. 1038, DHEW, Washington, D.C., 1963.

[a] Where urinals are provided for the women, the same number shall be provided as for men.

[b] For apartments or multiple dwelling units in excess of 10 apartments or units, 1 double laundry tray for each 10 units or 1 automatic laundry washing machine for each 20 units.

Table 11-7 Size of Fixture Supply and Drain

Type of Fixture or Device	Pipe Size (in.)		Type of Fixture or Device	Pipe Size (in.)	
	Supply	Drain		Supply	Drain
Bathtubs	$\frac{1}{2}$	$1\frac{1}{2}$	Sink, service, P trap	$\frac{1}{2}$	2
Combination sink and			Sink, service, floor trap	$\frac{3}{4}$	3
tray	$\frac{1}{2}$	$1\frac{1}{2}$	Urinal, flush tank, wall	$\frac{1}{2}$	$1\frac{1}{2}$
Drinking fountain	$\frac{3}{8}$	1	Urinal, direct,		
Dishwasher, domestic	$\frac{1}{2}$	$1\frac{1}{2}$	flush valve	$\frac{3}{4}$	2
Kitchen sink,			pedestal, flush valve	1	3
residential	$\frac{1}{2}$	$1\frac{1}{2}$	Water closet,		
commercial	$\frac{3}{4}$	$1\frac{1}{2}$	tank-type	$\frac{3}{8}$	3
Lavatory	$\frac{3}{8}$	$1\frac{1}{4}$	Water closet,		
Laundry tray, 1 or 2			flush valve-type	1	3
compartments	$\frac{1}{2}$	$1\frac{1}{2}$	Hose bibbs	$\frac{1}{2}$	—
Shower, single head	$\frac{1}{2}$	$1\frac{1}{2}$	Wall hydrant	$\frac{1}{2}$	—

Source: Adapted from PHS Pub. 1038, DHEW, Washington, D.C. See also Chapter 4 for trap sizes and fixture unit ratings.
Note: The minimum water pressure at the outlet, at times of maximum demand, shall not be less than 8 psi, except for direct flush valves, where 15 psi is required, and where special equipment requires other pressure.

Table 11-8 Distance of Fixture Trap from Vent

Size of Fixture Drain (in.)	Distance, (ft) Maximum
$1\frac{1}{4}$	$2\frac{1}{2}$
$1\frac{1}{2}$	$3\frac{1}{2}$
2	5
3	6
4	10

direct waste piping, and other details are given in the Tables 11-6 to 11-9 and are discussed below.

Backflow Prevention[18,19]

The backflow of polluted or contaminated water or other fluid or substance into a water-distribution piping system through backpressure or backsiphonage is a very real possibility. The best way to eliminate the danger is to prohibit any connections between the water system and any other system, fixture, vat, or tank containing polluted or questionable water. This can be accomplished by

[18] *Cross-Connection Control Manual*, USEPA, Office of Water Programs, Water Supply Division, 1973.
[19] Gustave J. Angele Sr., *Cross-Connections and Backflow Prevention*, Am. Water Works Assoc., Denver, Colorado, 1970.

Table 11-9 Hot-Water Demands and Use for Various Types of Buildings

Type of Building	Maximum Hour	Maximum Day	Average Day
Men's dormitories	3.8 gal/student	22.0 gal/student	13.1 gal/student
Women's dormitories	5.0 gal/student	26.5 gal/student	12.3 gal/student
Motels: number of units[a]			
20 or less	6.0 gal/unit	35.0 gal/unit	20.0 gal/unit
60	5.0 gal/unit	25.0 gal/unit	14.0 gal/unit
100 or more	4.0 gal/unit	15.0 gal/unit	10.0 gal/unit
Nursing homes	4.5 gal/bed	30.0 gal/bed	18.4 gal/bed
Office buildings	0.4 gal/person	2.0 gal/person	1.0 gal/person
Food service establishments:			
Type A—full-meal restaurants and cafeterias	1.5 gal/max. meals/hr	11.0 gal/max. meals/hr	2.4 gal/avg. meals/day[b]
Type B—drive-ins, grilles, luncheonettes, sandwich and snack shops	0.7 gal/max. meals/hr	6.0 gal/max. meals/hr	0.7 gal/avg. meals/day[b]
Apartment houses: number of apartments			
20 or less	12.0 gal/apt.	80.0 gal/apt.	42.0 gal/apt.
50	10.0 gal/apt.	73.0 gal/apt.	40.0 gal/apt.
75	8.5 gal/apt.	66.0 gal/apt.	38.0 gal/apt.
100	7.0 gal/apt.	60.0 gal/apt.	37.0 gal/apt.
130 or more	5.0 gal/apt.	50.0 gal/apt.	35.0 gal/apt.
Elementary schools	0.6 gal/student	1.5 gal/student	0.6 gal/student[b]
Junior and senior high schools	1.0 gal/student	3.6 gal/student	1.8 gal/student[b]

Source: Copyright by the American Society of Heating, Refrigerating and Air Conditioning Engineers, Inc. Reprinted by permission from *ASHRAE Guide and Data Book*, New York, 1970. Heaters should be preset to deliver water at 130°F.

[a]Interpolate for intermediate values.

[b]Per day of operation.

terminating the water supply inlet or faucet a safe distance above the flood-level rim of the fixture. The distance, referred to as the air gap, is 1 in. for a $\frac{1}{2}$-in. or smaller diameter faucet or inlet pipe, $1\frac{1}{2}$ in. for a $\frac{3}{4}$-in. diameter faucet, 2 in. for a 1-in. diameter faucet, and twice the effective opening (cross-sectional area at point of water supply discharge) when its diameter is greater than 1 in. When the inside edge of the faucet or pipe is close to a wall, that is, within 3 or 4 times the diameter of the effective opening, the air gap should be increased by 50 percent.

Sometimes, as with water closets and urinals equipped with flushometer valves, it is not possible or practical to provide an air gap. Under such circumstances, where the water connection is not subject to backpressure, an approved-type of backflow preventer, such as that shown in Figure 11-13, may be used to prevent backsiphonage. The backflow preventer should be installed on the outlet side of the control valve, at a distance not less than 4 times the nominal diameter of the inlet, measured from the control valve to the flood-level rim of the figure, and in no cases less than 4 in.

In some instances an air gap cannot be installed and it is found necessary to connect a potable water supply to a line, fixture, tank, vat pump or other equipment which may permit backflow due to backpressure. Under such circumstances, an approved reduced-pressure-principle backflow preventer *may* be permitted by the regulatory authority.

Cross-connection control and use of backflow preventers is also discussed in Chapter 3.

Indirect Waste Piping

Waste pipes from fixtures or units in which food or drink is stored, prepared, served, or processed must not connect directly to a sewer or drain. Stoppage in the receiving sewer or drain would permit polluted water to back up into the fixture or unit. Waste piping from refrigerators, iceboxes, food rinse sinks, cooling or refrigerating coils, laundry washers, extractors, steam tables, egg boilers, steam kettles, coffee urns, dishwashing machines, sterilizers, stills, and similar units should discharge to an open water-supplied sink or receptacle so that the end of the waste pipe terminates at least 2 in. above the rim of the sink or receptacle which is directly connected to the drainage system.

A commercial dishwashing machine waste pipe may be connected to the sewer side of a floor drain trap when the floor drain is located next to the dishwashing machine, if permitted by the regulatory agency.

An alternate to the installation of a water-supplied sink waste receptor is the provision of an air gap in the fixture waste line, at least twice the effective diameter of the drain served, located between the fixture and the trap.

The water-supplied sink or air-gap waste receptor should be in an accessible and ventilated space and not in a toilet room.

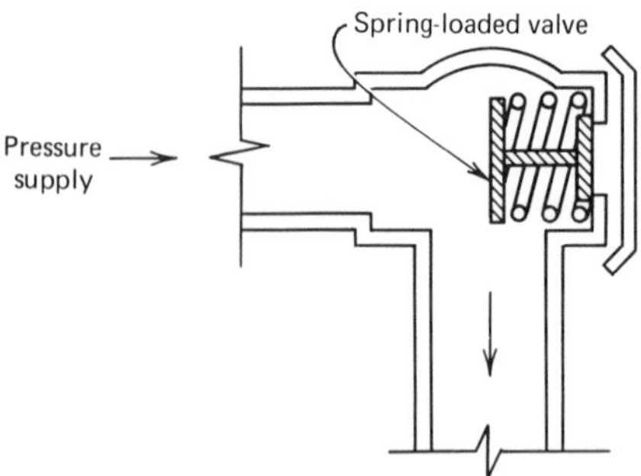

Figure 11-13*a* Pressure-type vacuum breaker, installed *before* the fixture, under water pressure.

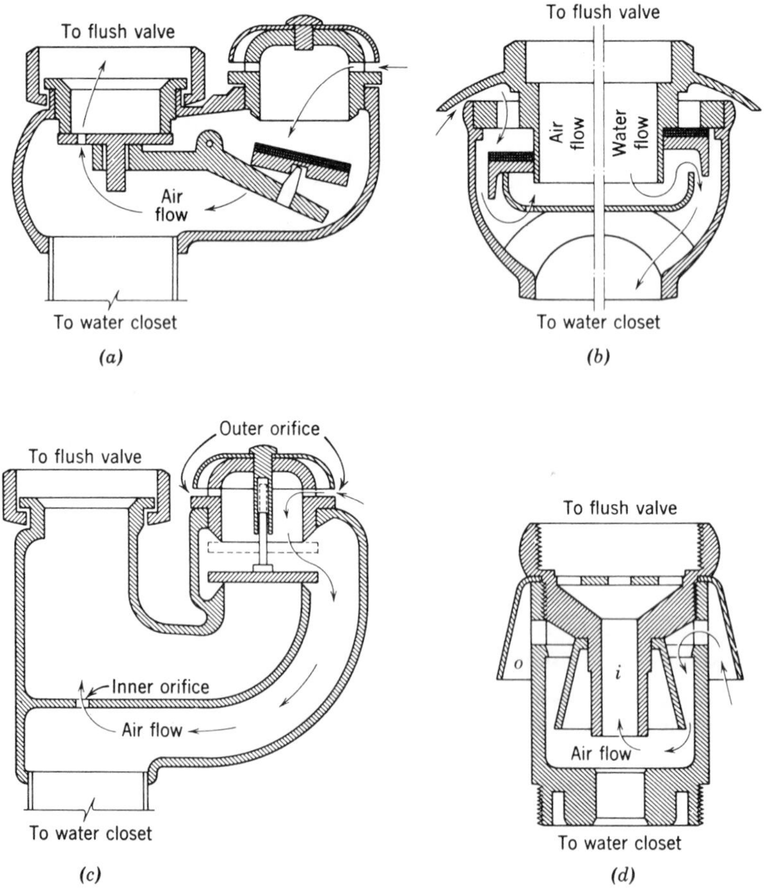

Figure 11-13*b* Vacuum, nonpressure-type siphon-breakers: (*a*), (*b*), (*c*) moving parts; (*d*) nonmoving part. Installed *after* fixture valve. [From Roy B. Hunter, Gene E. Golden, and Herbert N. Eaton. "Cross-connections in Plumbing Systems," Research Paper RP 1086, *J. Res. Natl. Bur. of Standard.*, **20** (April 1938).]

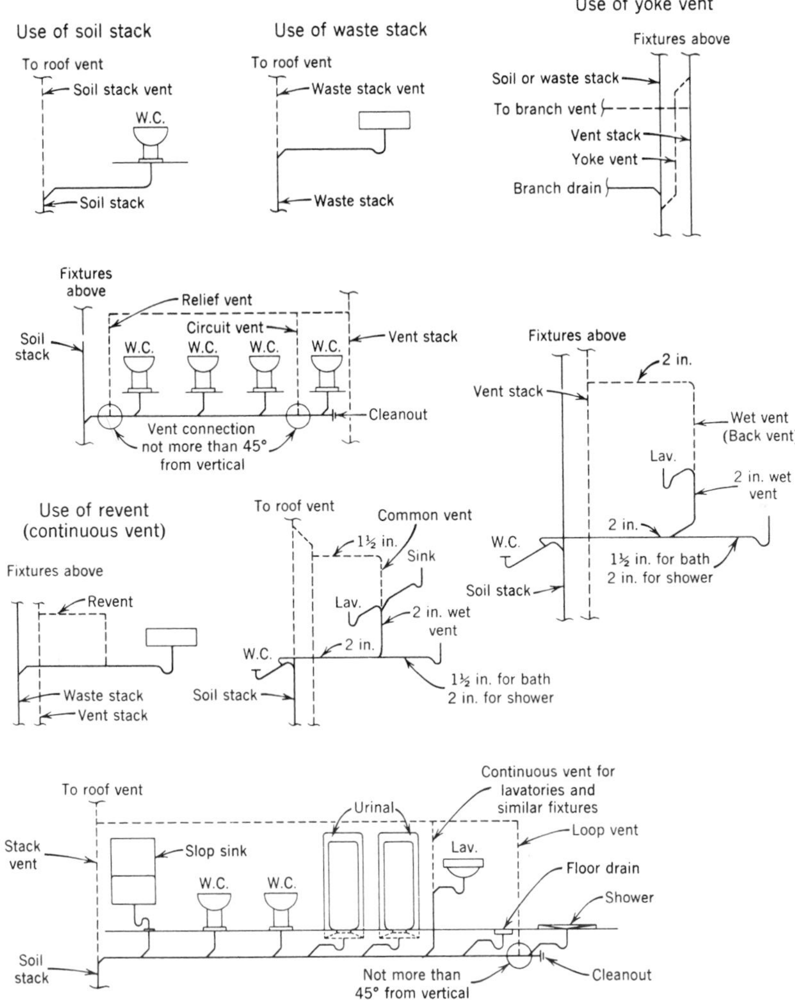

Figure 11-14 Some fixture plumbing details. (From *Code Manual for State Building Construction Code*, Division of Housing and Community Renewal, N.Y. August 1, 1977, pp. 5-14, 5-17 to 5-20.)

Plumbing Details

A few typical details and principles are illustrated for convenient reference and as guides to good practice. Many variations are to be found, dependent on local conditions and regulations. See Figures 11-14.

Other

See Tables 4-3, 4-4, and 11-7 for fixture unit ratings for sewage flow computations, trap sizes, drain sizes, special fixture values, and fixture unit load to building drains and sewers. Figure 3-25 gives curves for estimating probable water demand of a sum of fixture units. Table 3-23 lists weights of fixtures, and Table 3-24 gives the size of fixture supply pipe, rate of flow, and required pressure during flow for different fixtures.

VENTILATION

Spread of Respiratory Diseases

The danger of spreading respiratory diseases can be reduced by preventing overcrowding, by applying dust-control measures, and by preventing accumulation of the microorganism concentration through recirculation of the same air. This, of course, is in addition to good personal hygiene, reduction of contact, disinfection of eating and drinking utensils, sanitary disposal of mouth and nose discharges, and maintenance of natural resistance. Airborne infections are discussed in Chapter 1.

Formaldehyde released by some building material and insulation under certain conditions can cause eye and upper respiratory irritation. Asbestos fibers from disintegrating building materials and coatings are also a problem, being associated with possible cancer.

Thermal and Moisture Requirements

Good ventilation requires that air contain a suitable amount of moisture and that it be in gentle motion, cool, and free from offensive body odors, poisonous and offensive fumes, and large amounts of dust. Comfort zones for certain conditions of temperature, humidity, and air movement are given by the American Society of Heating, Refrigerating and Air Conditioning Engineers, Inc.[20] Relative humidities below 70 percent are reported to have no influence on human comfort. Air movement, radiant heat, the individual, and the tasks being performed must also be taken into consideration.

There is no one temperature and humidity at which everyone is comfortable. People's sensations, health, sex, activity, and age all enter into the comfort standard. McNall recommends a temperature range between 73 and 77°F and humidity between 20 and 60 percent for lightly clothed adults engaging in sedentary activities in residences.[21] Lubart gave as the comfort level a range of 68°F at inside relative humidity of 50 percent to 76°F with 10 percent relative humidity.[22] Indoor relative humidity of 60 percent or higher would cause excessive condensation and greater mildew, corrosion, and decay. Ordinarily, how-

[20] *Handbook of Fundamentals*, ASHRAE, 345 East 47th Street, New York, 1967, p. 122.

[21] Preston E. McNall, Jr., report in *Proceedings of the First Invitational Conference on Health Research in Housing and its Environment*, Airline House, Warrenton, Va., March 17–19, 1970, PHS, DHEW, Washington, D.C., p. 27.

[22] Joseph Lubart, "Winter Heating an Enemy Within," *Med. Sci.*, March 1967, pp. 33–37.

ever, only a temperature and ventilation control is used in the home, with no attempt being made to measure or control the relative humidity.

Space Ventilation and Indoor Air Pollution Control

In cold weather, when forced ventilation is combined with heating by means of ducts and registers or openings, uncomfortable drafts and large temperature variation should be avoided in the design of the ventilation system. There is also danger of spreading airborne infections by recirculating dust and pathogenic bacteria or viruses in dry, used air. Recirculation will cause an increase in the concentration of these organisms due to the cumulative effect of recirculating the same used, contaminated air. This will occur unless sufficient clean fresh air is admitted and part of the used air is exhausted to waste. During cold months of the year greater fuel consumption can be expected. The usual method of air purification by washing and filtration is relatively inefficient in removing bacteria or viruses from used air, although it can be effective in removing dust. Heating appliances producing or using dangerous gases must be vented adequately to the outside air through tight flue connections.

A minimum of one to two air changes per hour can often be secured by normal traffic and leakage through walls, ceilings, floors, and through or around doors and windows. Under ordinary circumstances, proper ventilation can be obtained by natural means with properly designed windows, both in the winter and summer. Openable windows, louvers, or doors are needed to ventilate and keep relatively dry attics, basements, pipe spaces, and cellars. Tops of windows should extend as close to the ceiling as possible, with consideration to roof overhang, to permit a greater portion of the room being exposed to controlled sunlight. The net glass space should at least equal 10 to 15 percent of the floor area and the openable window area at least 45 percent.

In schools, separate venting of each classroom to the outside is preferred. Good standards specify that the ventilating system provide a minimum air change of 10 to 15 ft^3 per min per pupil to remove carbon monoxide and odors, without drafts. The air movement should not exceed 25 ft per min, and the vertical temperature gradient should not vary more than 5°F in the space within 5 ft of the floor. Air conditioning or electric fans may be required to meet special requirements. Temperature control should be automatic.

In recreation halls, theaters, churches, meeting rooms, and other places of temporary assembly, the requirement of at least 10 ft^3 of fresh, clean air per min per person cannot ordinarily be met without some system of mechanical or induced ventilation. Any system of ventilation used should prevent short-circuiting and uncomfortable drafts. National, state, and local building codes specify the ventilation required for various space uses.

The concern for energy conservation and improved building insulation may result in the accumulation of indoor air pollutants. Pollutants of concern are carbon monoxide, nitrous oxides, ozone, sulfur oxides, and particles. Included are microorganisms, gases produced by stoves and heaters, and radon gas released by certain mineral building materials, pointing to the need for controlled ventilation.

Toilet Ventilation

Bathroom and toilet-room ventilation is usually accomplished by means of windows or ventilating ducts. The common specification is that the window area shall be at least 10 percent of the floor area and not less than 3 ft^2, of which 45 percent shall be openable. For gravity ventilation provide vents or ducts at least 72 $in.^2$ area and a minimum of 2 ft^3 of fresh air per ft^2 of room area. A system of mechanical exhaust ventilation maintained and operated to provide at least five changes per hour of the air volume of the bathroom or toilet room during the hours of probable use is usually specified for ventilation where windows, vents, or ducts are not relied on or are not available for ventilation. Fans that are activated by the opening and closing of doors would not provide satisfactory ventilation.

Venting of Heating Units

Proper venting is the removal of all the products of combustion through a designated channel or flue to the outside air with maximum efficiency and safety. Gravity-type venting relies largely on having the vent gases inside the vent hotter (thus lighter) than the surrounding air. The hotter the vent gases, the lighter they are and the greater their movement up through the vent. Thus, in order to keep the vent gases hot so that they may work at maximum efficiency, proper installation and insulation are necessary.

Factors that prevent proper venting are abrupt turns; downhill runs; common vents to small, uninsulated vent pipes; conditions that cause back drafts; obstructions in the flue or chimney to which a furnace, heater, or stove is connected such as birds nests, soot and debris, broken mortar and chimney lining, and old rags; and unlined masonry chimneys. Stained and loose paper or falling plaster around a chimney is due to poor construction. A masonry chimney will absorb a great deal of the heat given off by the vent gases, thus causing the temperature in the chimney to fall below the dew point. The high moisture in vent gases condenses inside the chimney, forming sulfuric acid. This acid attacks the lime in the mortar, leaching it out and creating leaks and eventual destruction of the chimney. It is therefore necessary to line a masonry chimney with an insulating pipe, preferably terra-cotta flue lining.

Figure 11-15 shows chimney conditions apt to result in back drafts. The flue or vent should extend high enough above the building or other neighboring obstructions so that the wind from any direction will not strike the flue or vent from an angle above horizontal. Unless the obstruction is within 30 ft or unusually large, a flue or vent extended at least 3 ft above flat roofs or 2 ft above the highest part of wall parapets and peaked roof ridges will be reasonably free from downdrafts.

To ensure proper venting as well as proper combustion, sufficient amounts of fresh air are required, as shown in Figure 11-15. An opening of 100 to 200 in.^{2} will usually provide sufficient fresh air under ordinary household conditions; this opening is needed to float the flue gases upward and ensure proper combustion in the fire box. Proper venting and an adequate supply of fresh air

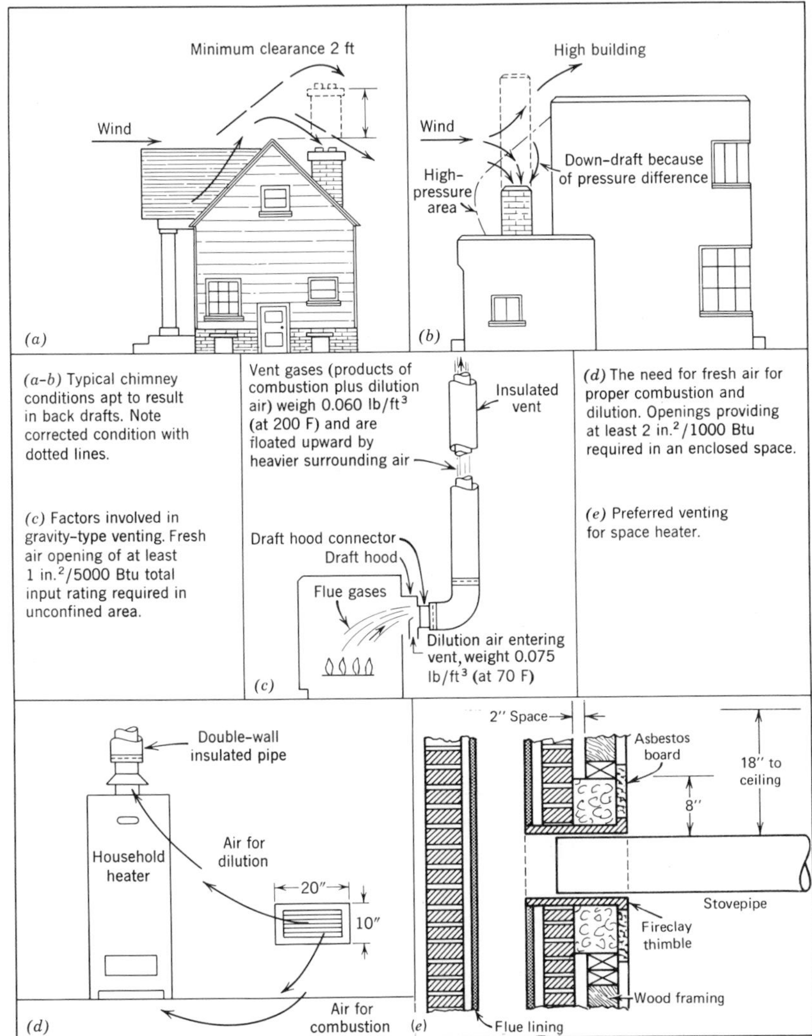

Figure 11-15 Some venting details. (Drawings are typical and not necessarily in full accordance with any code.) See state and local building and mechanical codes. (Figure (*e*) from J. P. Lassoie and L. D. Baker, "Heating with Wood," Cooperative Extension, Cornell University, Ithaca, N.Y., Oct. 1979.)

are also necessary for the prevention of carbon monoxide poisoning or asphyxiation.

The connection (breeching) between the furnace or stove and chimney should be tight-fitting and slope up to the chimney at least $\frac{1}{4}$ in./ft. Chimneys are usually constructed of masonry with a clay tile flue liner, or of prefabricated metal with concentric walls with air space or insulation in between and be Underwriter's Laboratories-approved. All furnaces and stoves should be equipped with a draft hood, either in the breeching or built into the furnace or stove, as required, for proper draft. See Figure 11-15.

Before making any vent installations or installing any gas- or oil-fired appliances, consult the building code and the local gas or utility company. Standards for chimneys, fireplaces, and venting systems, including heating appliances and incinerators, are given in the National Fire Protection Association's Pub. No. 211,[23] building codes and other publications.

MOBILE HOME PARKS

Mobile homes are defined as

> transportable, single-family dwelling units suitable for year-round occupancy and containing the same water supply, waste disposal, and electrical conveniences as immobile housing.[24]

A mobile home is also defined as a "manufactured relocatable living unit."[25] Mobile homes produced since 1954 are 10-, 12-, and 14-ft wide and up to 60- or 70-ft long. Wider units are assembled at the home site by combining sections to form double- and triple-wide units.[26,27]

The 1975 housing survey identified 3,342,000 mobile home units in the United States as year-round housing occupied by approximately 8.5 million residents.[28] Annual sales dropped to 212,000 units in 1975 and increased to 300,000 units in 1977.[27] The peak sales of 625,000 units occurred in 1973.[26] The typical mobile home is 14-ft wide and 65-ft long. In 1974 about one-third of the buyers were married couples under 35 years of age and one-third were retired and over 65.[26] Modern mobile home parks have all utilities, swimming

[23] *Chimneys, Fireplaces, Venting Systems*, NFPA, 60 Batterymarch Street, Boston, Mass. 02110, 1970.

[24] *Environmental Health Guide for Mobile Home Parks*, prepared by PHS, DHEW; published by Mobile Home Manufacturers Association, 6650 North Northwest Highway, Chicago, January 1966, p. 3.

[25] *Environmental Health Guide for Mobile Home Communities*, prepared by PHS, DHEW; published by Mobile Home Manufacturers Association, 6650 North Northwest Highway, Chicago, Ill., revised January 1971.

[26] "Mobile homes move fast to fill low-cost housing gap," *Eng. News Rec.*, January 10, 1974, pp. 16–17.

[27] "Mobile Homes Growing Up," *New York Times*, Sunday, May 22, 1977, p. F9.

[28] *Guidelines For Improving The Mobile Home Living Environment*, HUD, GPO, Washington, DC, August 1977, p. 5.

pool and other recreation facilities, laundry, community buildings, paved streets with curbs or gutters, trees and landscaping, and patio slab, and look like an established housing development. Lots are typically 4000 to 4500 ft^2. Lots larger than 5000 ft^2 permit more flexibility in exposure and siting the mobile home. It must be kept in mind that, unless the plot is owned by the mobile home owner, continued occupancy is dependent on the desire and future plans of the park owner. Hence a mobile home park cannot be considered a realty subdivision. Because of the risk of property sale, consideration has been given to cooperative ownership and contractual arrangements.

The parks established have all of the potential environmental sanitation problems of a small community. Because of this, standards have been prepared to guide mobile home manufacturers, operators, owners, and regulatory agencies to help promote good, uniform sanitary practices. Compliance with these standards is facilitated by reference to the pertinent chapters in this text; to the guides cited in the footnotes, which include recommended ordinances; and to guides published by the U.S. Department of Housing and Urban Development.[29,30] (See Chapter 9 for design and related details.) Other precautions include tiedowns at platforms to minimize overturning during windstorms, smoke detectors and fire extinguishers in homes, with two exits from each unit, protection of water connections against freezing, and sufficient electric power for all electrical equipment including approved-type inside wiring.

INSTITUTION SANITATION

Definition

An institution is a complete property and its building, facilities, and services, having a social, education, or religious purpose. This includes schools; colleges or universities; hospitals; nursing homes; homes for the aged; jails and prisons; reformatories; and the various types of federal, state, city, and county welfare, mental, and detention homes or facilities.

Institutions as Small Communities

Most institutions are communities unto themselves. They have certain basic characteristics in common that require careful planning, design, construction, operation, and maintenance. These include site selection, planning and development for the proposed use, including subsoil investigation, accessibil-

[29] *Mobile Home Court Development Guide*, HUD, January 1970.

[30] *Guidelines For Improving The Mobile Home Living Environment*, HUD, August 1977. The American National Standards Institute, National Fire Protection Association, and Building Officials & Code Administrators International, Inc. also have suggested standards for adoption by local governments.

ity, and proximity to sources of noise and air pollution; a safe, adequate, and suitable water supply for fire protection as well as for institutional use; sewers and a wastewater disposal system; roads and a storm-water drainage system; facilities for the storage, collection, and disposal of all solid wastes generated in the institution; boilers and incinerators with equipment and devices to control air pollution; food preparation and service facilities; fire-resistant housing and facilities for the resident population; laundry facilities, and insect, rodent, and noxious weed control. In addition, depending on the particular institution, they might have recreation facilities, such as a swimming pool or bathing beach. A hospital or educational institution might have its own laundry. A state training school or institution might have a dairy farm or produce farm, pasteurization plant, and food-processing plant. The environmental engineering and sanitation concerns at all these places are in many instances quite extensive and complex. The reader is referred to the appropriate subject matter throughout this text.

The material that follows will highlight environmental engineering and sanitation factors at various types of institutions. The institutions have many environmental factors in common.

Hospitals and Nursing Homes

Hospital-acquired, or nosocomial, infections result in additional morbidity, mortality, and costs pointing to the need for greater infection surveillance and control.[31] The majority of nosocomial infections are endemic. They may affect not only the patient who develops the infection, but other patients, the hospital staff, and the community as well. Data accumulated over past years indicate that under certain conditions as much as half of all nosocomial infections may be preventable.[32]

The hospital is expected to provide an environment that will expedite the recovery and speedy release of the patient. Carelessness can introduce contaminants and infections that delay recovery and may overburden the weakened patient, thereby endangering the patient's survival. It has been estimated that 1.5 million patients, out of some 300 million, incur infections in hospitals annually.[33] Numerous reasons have been offered for hospital acquired infections such as increase in the number of older patients with chronic diseases; increase in high-risk patients and surgical procedures such as open heart surgery and organ transplants; innovations in diagnostic and therapeutic procedures including widespread use of antibiotics, indwelling catheters, and arti-

[31]"Infection Surveillance and Control Programs in U.S. Hospitals: An Assessment, 1976, *MMWR*, CDC, Atlanta, Ga., April 28, 1978, p. 139.

[32]John V. Bennett and Philip S. Brachman, *Hospital Infections*, Little, Brown, and Co., Boston, 1979.

[33]Robert L. Elston, "The Role of Hands in the Spread of Diseases," *Am. J. Pub. Health*, November 1970, p. 2211.

ficial kidneys; inadequate disinfection or sterilization of respiratory-therapy and other equipment; prevalence of *Staphylococcus aureus* and Group-A streptococci, and the increasing identification of gram-negative organisms such as *Pseudomonas aeruginosa*, proteus, *Escherichia coli*, klebsiella, and *A. aerogenes*,[34] also the gram-positive, toxin producing *Clostidium dificile.*[35]

Basic to the prevention of complications are hygienic medical, nursing, and staff practices; equipment sterilization; and food, water, plumbing, air, laundering and linen handling, and housekeeping sanitation. A major control mechanism is the establishment of a representative Infection Control Committee and appointment of a full-time Environmental Control Officer with comprehensive responsibility and authority to coordinate and ensure that medical, nursing, housekeeping, maintenance, and ancillary staffs are following good practices and procedures including nosocomial infection surveillance and control, safety, and occupational health protection.

The duties of the Environmental Control Officer would include assurance of "satisfactory" responses to all of the items listed in Figure 11-16. Since the survey form is merely suggestive of the broad scope of each subject, it is apparent that the Environmental Control Officer must be broadly trained by education and experience to recognize and appreciate the full impact of conditions observed and their possible risk to patients, staff and visitors, and the promptness with which unsatisfactory conditions must be corrected. Suggested preparation would include a graduate degree from a recognized institution in environmental health science or a related degree with an internship or training in institutional health management, administrative techniques, and environmental control.

Some of the medical infection control activities would include the following.

1. Case finding and investigation of patients with potential nosocomial infections. Attention must also be given to patients with infection on admission and on discharge, to infected members of staff or hospital personnel, and to environmental factors such as listed in Figure 11-16. Infections found must of course be reported and treated.
2. Review of microbiologic cultures, and fever charts on wards.
3. Patient, personnel, supplies, equipment, and environmental microbiologic sampling as part of an outbreak investigation, or to evaluate contamination known to be associated with risk of disease; or possibly for staff education; but not for just routine sampling such as air and surfaces. Minimal environmental sampling however is indicated to check steam, gas, and dry heat sterilizers with live spores; weekly checks of hospital-prepared infant formulas; and checks of items used in direct patient care that should preferably be sterilized but are not.[36]
4. In-service education of staff on asepsis and infection control.
5. Reporting and isolation of infectious patients.
6. Discouraging indiscriminate use of antibiotics.
7. Changing of respiratory tubing, urinary catheters, and intravenous cannulae.
8. Continued emphasis on thorough prolonged hand washing, disinfection of anesthesia equip-

[34]Personal communication, Mindaugas Jatulis, Chief, Environmental Control Section, New York State Dept. of Health, Albany, N.Y.

[35]Don G. Brown, "Environmental Health and Safety Concerns in the Health Care Setting," *J. Environ. Health*, July/August 1980, pp. 11–13.

[36]Proposed Policy Statements, 7519(PP): "Environmental Microbiologic Sampling in the Hospital," *Nation's Health*, Am. Pub. Health Assoc., Washington, D.C., September 1979.

Name ______________________ Address ______________________ T.V.C. ________
Operator ______________________ Persons interviewed ______________________
Inst. No. ________ No. Cert. Beds ______ Inspected by ______________ Date ________

Item	S	U
Structure and grounds		
1. Location		
2. Buildings and grounds		
3. Accessible by emergency vehicles		
4. Service entrances		
5. Elevators		
Water supply		
6. Supply and pressure, hospital and fire		
7. Treatment, physical, biological, chemicals		
8. Quality—hospital, protect, and surveillance; incl. water carafes		
9. Quality—special, medical, and lab.		
10. Hot water		
Liquid and solid wastes		
11. Sewage piping and disposal		
12. Biological wastes collection, storage, disposal		
13. Solid wastes collection, storage, disposal		
Plumbing		
14. Toilets, lavatories, tubs, showers, sinks		
15. Cross-connection and backflow control		
16. Drinking fountains		
Emergency power and light		
17. Power to vital services		
18. Lighting to vital areas		
Ventilation, heat, and air conditioning		
19. Air flows, rates, pressure, differential		
20. Air filtration		
21. Air temperature		
22. Air humidity		
23. Intakes and exhausts		
Laundry		
24. Soiled linen handling and transportation		
25. Laundering		
26. Clean linen handling and transportation		
Dietary services		
27. Food sources		
28. Refrigerated food storage		
29. Dry food storage		
30. Food preparation		
31. Food serving, including water		
32. Food utensils		
33. Food equipment		
34. Ice making and handling		
35. Vending machines		
Insect and rodent control		
36. Physical controls		
37. Chemical controls		
Infant formula		
38. Equipment, supplies, technicians, records		
39. Approved source and handling		
Space provisions		
40. Patient rooms		
41. Isolation rooms		
42. Bath and toilet rooms		
43. Nursing service areas		
44. Other services—rooms, spaces		
45. Central and general storage		
Fire safety		
46. Fire-resistive construction		
47. Interior finishes		
48. Smoke and fire doors, open, ducts		
49. Fire-resistant enclosures—chutes, shafts, stairs, kitchen, boiler, and incinerator rooms		
50. Exit doors, access, stair, hall, signs		
51. Flame-retardant fabrics, drapes		
52. Flammable liquids		
53. Flammable anesthetics		
54. Oxygen and nitrous oxides		
55. Fire hydrants, hoses, standpipes		
56. Fire extinguishers, portable		
57. Automatic sprinklers in chutes, soiled linen, trash, and storage rooms		
58. Automatic extinguish. in kitchen hoods, shops, rooms, and halls		
59. Fire-detection systems in boiler, kitchen, labs, laundry, pantry, garage		
60. Fire-alarm system, internal		
61. Alarms connected to fire department or station		
62. Electrical hazards controlled		
63. Anesthesia areas—electrical safeguards		
64. No smoking—signs, supervision		
65. Fire plans—posting, drills, training		
Accident safety		
66. Handrails and grab bars—corridors, stairs, ramps, toilets, bath		
67. No obstacles—corridors, ramps, stairs		
68. Floors—nonslip and nontrip		

Figure 11-16 Hospital environmental health survey form.

Item	S	U	Item	S	U
69. Burn protection—heaters, hot water			*Supporting services*		
70. Patient furnishings and equipment			81. Housekeeping		
71. Electrical safety hazards			82. Plant maintenance		
72. Accident reporting, records			*Refrigeration (nondietary)*		
73. Lighting levels			83. Pharmaceutical and blood storage		
Disinfection and sterilization			84. Morgue		
74. Infection control committee			*Pools (therapy or swimming)*		
75. Sterilization facilities			85. Recirculation and filtration		
76. Sterilization—procedure, surveillance, records			86. Disinfection		
			Radiations and hazardous wastes		
77. Disinfection of anesthesia apparatus			87. Diagnostic X-ray units		
78. Disinfection of patient-used articles			88. Therapeutic X-ray units		
79. Disinfection of operating, delivery, isolation, rms., nursery			89. Teletherapy units		
			90. Radioact. materials—storage, use		
80. Disinfection of inhalation therapy apparatus			91. Microwave ovens		
			92. Proper mgt. of haz. wastes		

NOTE:
S means satisfactory equipment, construction, operation, maintenance.
U means unsatisfactory.
Contract Services ______ Housekeeping ______ Laundry ______ Dietary ______ Other ______. Give name.

Figure 11-16 (*Continued*)

ment and other medical apparatus, and proper operation of pressure steam and hot-air sterilizers; ethylene oxide gas and sub-atmospheric steam sterilizers require bacteriological tests for each load.[37]

Water for kidney dialysis must be very soft, low in minerals and dissolved solids, and of good physical and microbiological quality. Deionization or reverse-osmosis treatment is usually required, as ordinary potable water may be toxic for the dialysis patient. The physician involved should determine the water quality standards for water used in kidney dialysis.

Wash water temperatures in the hospital laundry have been studied by numerous investigators. A minimum temperature of 160 to 167°F is generally specified. It appears that the temperature can be reduced to 140°F for lightly soiled hospital linens from nonisolation areas; but more investigation is needed to assure that this temperature is adequate for the laundering of *all* linens, including isolation linens. In any case, proper handling of laundered linen to prevent cross-contamination and contamination in handling is essential.[38]

Extensive infection surveillance and control program guidelines have been published by the Joint Commission on Accreditation of Hospitals and others.[39–41]

[37]M. T. Parker, "The Hospital Environment as a Source of Septic Infection," *J. R. Soc. Health*, October 1978, pp. 203–209.

[38]Donald R. Battles and Donald Vesely, "Wash Water Temperatures and Sanitation in the Hospital Laundry," *J. Environ. Health*, March/April 1981, pp. 244–250.

[39]*Joint Commission on Accreditation of Hospitals*: *Accreditation Manual for Hospitals*, 3rd ed., Chicago, Illinois, 1976.

[40]*Infection Control in the Hospital, American Hospital*, Am. Hospital Assoc., Chicago, Illinois.

[41]*Isolation Techniques for Use in Hospitals*, PHS, CDC, Atlanta, Ga.

An extension of the hospital is the extended-care facility, where a patient can recuperate under medical and nursing supervision. This is supplemented by the skilled nursing home, where full-time nursing care and needed medications are provided under medical direction and supervision of a registered professional nurse. A third type of facility is the intermediate-care home, which provides physical assistance and care for people who cannot take complete care of themselves.

A great deal of special emphasis has been placed on hospital and nursing home construction, equipment, and inspection or survey of operations and services. Federal and state standards are required to be met with respect to fitness and adequacy of the premises, equipment, personnel, rules and bylaws, standards of medical care, and hospital services. Plans for new structures and for additions or modifications of existing facilities are also reviewed for compliance with federal and state requirements to help ensure the best possible facilities for medical care.

The survey or inspection of hospitals and nursing homes takes into consideration the administration, fire prevention, medical, nursing, environmental sanitation, nutrition, accident prevention and safety, and related matters. A proper initial survey of these diverse matters calls for a professionally trained team consisting of a physician, sanitarian, engineer, nurse, nutritionist, and hospital administrator. Figures 11-16 and 11-17 show suggested hospital and nursing home environmental survey items. They can of course be greatly amplified. The proper interpretation of the environmental health survey form items requires a well-rounded educational background and specialized training as previously noted, with support from consultants when indicated. It is good procedure to coordinate the inspection program with the work of other agencies and develop continuing liaison with the county medical society, accreditation groups, local nursing home association, local hospitals, fire department, building department, social welfare services, and others who may be involved. This can strengthen compliance and avoid embarrassment resulting from conflicting recommendations.

Name ____________ Address ____________ T.V.C. ______
Operator ____________ Person interviewed ____________
Capacity ____ Bed patients ____ Amb. patients ____ Personnel: RN ____ Pract. nurses ____ Other ____

Housing	Yes	No	CM		Yes	No	CM
1. Building, floors, stairs structurally sound				6. Beds 4-ft apart and 1 ft from walls			
2. Rooms, kitchen, cellar, attic, yard clean				7. Windows, doors, openings screened			
3. Rooms, including kitchen, well lighted				8. Free of rat and vermin infestation			
4. Rooms, including kitchen, well ventilated				9. Bedding clean, storage satisfactory			
5. Rooms, basement, closets free of debris				10. Heating suitable and adequate			

Figure 11-17 Nursing home inspection form.

	Yes	No	CM
11. Fire-resistive construction			
12. Smoke and fire doors provided			
13. Boiler rm. and incin. fire-resistant			
14. Exit door signs posted			
15. Flame-retard. drapes, fabrics			
16. Fire alarm system connec. to fire dept.			
17. Fire plans posted; drills			
18. Flammables secured			
19. Elec. hazards controlled			
Water supply			
20. Water supply approved; fire hydrant within 500 ft			
21. Adequate supply hot and cold			
22. Sanitary drinking fountain or paper cups			
23. Cross-connection and interconnection prohibited			
24. Backflow preventers on WC, flushers—bedpan			
Sewage and toilet facilities			
25. Sewage disposal satisfactory			
26. Waterclosets—1 for 8 beds			
27. Washbasins—1 for 8 beds			
28. Shower or tub—1 for 12 beds			
29. Slop sink—each floor, and 1 for 24 beds			
30. Bedpan sterilizer—1 for 24 bed patients			
31. Toilet room on each floor			
32. Separate toilet and shower for help convenient, washbasin with sanitary towels			
Food and nutrition			
33. Equipment and facilities sanitary and adequate for food storage, preparation, and serving			
34. Dishes, trays, utensils clean and disinfected			
35. Foodhandlers clean, have lockers			
36. Refrigeration adequate			

	Yes	No	CM
37. Satisfactory food storeroom.			
38. Pasteurized milk served in indiv. carton			
39. Refuse storage and disposal satisfactory			
40. No poisons stored in kitchen			
41. Menu planned week in advance			
42. Meals nutritionally balanced			
43. Food service attractive			
44. Leftovers used promptly			
Medical and nursing service			
45. Mental and TB cases prohibited			
46. Patient accidents and CD reported			
47. RN always in attendance			
48. Adequate competent help available			
49. Employees free of CD			
50. Helpless patients get special attention			
51. Records kept giving:			
(*a*) Name of MD for patient			
(*b*) Name and address of patient, sex, color, age, marital status, occupation, place of birth, admission, diagnosis, discharge, relative			
(*c*) Clinical history and treatment, medications, signed MD's orders			
(*d*) Qualification of nurses and attendants			
(*e*) Narcotics on hand, dispensed			
52. Narcotics stored in locked cabinet			
53. Drugs properly labelled in cabinet			
54. Adequate dressings and supplies			
55. Recreation facilities for patients			
56. Occupational therapy provided			

Explain "No" items, use back

NOTE: CM indicates correction made.

REMARKS ____

____

Date inspected ______________________ By ______________________

Figure 11-17 (*Continued*)

Name ____________ Location ____________ T.V.C. ________

Principal ____________ Supt. of Schools ____________ Date ________

Grades ____ No. classrooms ____ No. boys ____ No. girls ____ Public ________

Private ____ Parochial ____ Boarding ____

Item	Yes	No	CM
1. Water system approved by HD			
2. One sanitary drinking fountain for every 100 children, or			
3. One sanitary paper-cup dispenser for every 100 children, where needed			
4. One washbasin with warm and cold water for every 50 students			
5. Soap, paper towels, and mirrors provided			
6. One toilet including tissue and partition for every 35–45 girls, and one toilet including tissue and one urinal for every 60 boys; separate.			
7. Toilet and lavatory rooms convenient, clean, free of odors, ventilated; floors impervious and drained			
8. Shower and locker room clean, drained; adeq. warm water			
9. Sewage and excreta disposal satisfactory			
10. Buildings of fire-resistive construction			
11. Corridors, stairs, exits, doors marked and provide safe and ready escape from building in case of fire			
12. Flammables stored in metal cans			
13. Fire extinguishers, sprinkler heads, fire hydrants, fire hose, fire alarm, fire escapes, panic bolts operable and tested every 6 months.			
14. Fire drills conducted, each floor emptied in 2 min.			
15. Poisons, etc., labeled and secured			
16. Seats in classrooms face away from window or light sources; no glare			
17. Natural and artificial light provided			
(*a*) 20-ft-c in classrooms, libraries, offices, shops, laboratories, gymnasium, pool			
(*b*) 30-ft-c in sewing, drafting, and arts and crafts rooms			
(*c*) 40-ft-c in sightsavings classrooms			
(*d*) 10-ft-c in auditorium, assembly rooms, cafeterias			
(*e*) 5-ft-c in locker rooms, corridors, stairs, toilet rooms			
18. Thermometer at seat level, provided, which reads in winter at			
(*a*) 68–72°F in classrooms, auditoriums, offices, cafeterias			
(*b*) 66–70°F in corridors, stairways, shops, laboratories, kitchen			
(*c*) 60–70°F in gymnasium			
(*d*) 76–80°F in locker and shower room			
(*e*) 80–86°F in swimming pool			
19. In nonheating season, when outdoor temperature reads 80°, 90°, 95°F, inside temperature reads 75°, 78°, 80°F, respectively.			
20. Ventilation and heating satisfactory (10–30 ft^3 per min per person); drafts and excessive heat prevented			
21. Place provided to store clothes, lunch boxes, rubbers			
22. A separate adjustable and movable seat and desk available for each child			
23. Class room provides 20–25 ft^2; per pupil			

Figure 11-18 School sanitation inspection form.

	Yes	No	CM		Yes	No	CM
24. Buildings, windows, rooms, chalk boards, lights, fixtures, corridors, walls, ceilings, etc., clean; grounds free of litter, insects, rodents, weeds, pools of water				28. Solid wastes properly stored and disposal satisfactory			
25. Swimming pool operated and maintained in conformance with sanitary code requirements*				29. Air pollution prevented			
26. Dining room or cafeteria operated and maintained in conformance with sanitary code requirements*				30. A competent person assigned responsibility to see that all environmental hygiene precautions are observed by teachers and students. (Incorporate in curricula.)			
27. Safety precautions taken in shops, laboratories, play area				31. Floors, walls, ceilings in good repair			
				NOTE: CM indicates correction made.			

*Use separate inspection form. Explain "No" items.

REMARKS ______________________________

____

Date inspected ______________ By ______________

Figure 11-18 (*Continued*)

Special attention should be given to non-fire-resistive hospitals and nursing homes. Until such places can be replaced with fire-resistive structures, they should be protected against possible fires. This would include automatic sprinklers and alarms; horizontal and vertical fire-stopping of partitions; enclosure and protection of the boiler room; outside fire escapes; fire doors in passageways, vertical openings and stairways; fire detectors, smoke detectors, and fire extinguishers; a fire-evacuation plan and drills; and the housing of nonambulatory patients on the ground floor. These comments are also generally applicable to other health care facilities.

Schools, Colleges, and Universities

Schools, colleges, and universities may incorporate a full spectrum of facilities and services not unlike a community. Involved, in addition to basic facilities such as water supply, sewage and other wastewater disposal, plumbing, solid waste management, and air quality, are control of food preparation and service, onsite and offsite housing, hospital or dispensary, swimming pool, radiation installations and radioisotopes, insect and rodent infestations, and safety and occupational health in structures, laboratories, and work areas including fire safety, electrical hazards, noise, and hazardous materials. In view of their complexity and their affect on life and health, all institutions should have a professionally trained environmental health and safety officer and staff responsible for the enforcement of standards, such as in a sanitary code, encompassing the areas of concern noted above. Such personnel can work closely with federal, state, and local health and safety regulatory officials and thus provide maximum protection for the student population, and teaching, research, and custodial staffs.[42–44] Figures 11-18, and 11-19 suggest the broad areas to be considered when making an inspection. Guidance as to what is considered satisfactory compliance can be found in this text under the appropriate headings and also in federal, state, and local publications.

Correctional Institutions

Correctional institutions include short-term jails, long-term prisons, and various types of detention facilities. The health care services may include primary health care services; secondary care services; health care services for women offenders; mental health care; dental care; environmental concerns; nutrition and food services; pharmacy services; health records; evaluation services; and staffing. Environmental health concerns are discussed below.

Incarceration may result in, or intensify, the need for health care services. The provision of a safe and healthful environment, services, and facilities would minimize the need. Food poisoning, poor and insufficient food, vermin infestations, inadequate work and recreation programs, and overcrowding are known causes of prison unrest and illnesses. Overcrowding, poor food quality, and food service are major problems at many jails and prisons. Walker[45] summarizes the problem very clearly in pointing out that

> overcrowded conditions often overtax the ventilation system and sanitary facilities, minimize privacy and personal space. Without privacy and personal space, the basic psychological and physiological needs of the residents are not met; tension and hostility grow; security requirements increase; and a negative cycle is put into play.

The elimination of overcrowding and improvement in the wholesomeness, quantity, and sanitation of food service can eliminate major causes of grievances.

The environmental aspects of correctional institutions are in many respects similar to those of other institutions and are concerned with many of the same basic environmental engineering and sanitation facets of a community.

Designs for new construction and major alterations should be reviewed and approved by the regulatory agencies having jurisdiction and in any case comply with nationally recognized standards. Regulatory agencies should make annual

[42]Richard Bond, et al., "Environmental Health Needs in Colleges and Universities," *Am. J. Pub. Health*, April 1961, pp. 523–530.

[43]Roger L. DeRoos, "Environmental Health and Safety in the Academic Setting," *Am. J. Pub. Health*, September 1977, pp. 851–854.

[44]Robert E. Hunt, "A campus is a total community," *J. Environ. Health*, November/December, 1979, pp. 108–118.

[45]Bailus Walker and Theodore J. Gordon, "Administrative Aspects of Environmental Health in Correctional Institutions," *J. Environ. Health.*, November/December 1976, pp. 192–195.

Name ______ Address ______ T.V.C. ______
Operator ______ Persons interviewed ______
Capacity ____ Men ____ Women ____ Inspected by ______ Date ______

Item	S	U	Item	S	U
Water supply			*Dietary*		
1. Quality meets drinking water standards			23. Food sources approved		
2. Quality—yield, storage, pressure adequate—hot and cold			24. Refrigerator storage temperature, space, clean		
3. Operation, maintenance, and reports satisfactory, no backflow			25. Dry storage clean, dry, space		
4. Qualified operator			26. Food preparation, handling, cooling proper		
5. On routine sampling schedule			27. Food service temperature and protection satisfactory		
			28. Utensils and equipment type, condition, satisfactory		
Sewage and toilet facilities			29. Dishwashing—dishes, utensils		
6. Flush toilets adequate			30. Handwashing facil. adequate, convenient		
7. Wash basins adequate					
8. Showers adequate			*Structure and grounds*		
9. Service sinks adequate			31. Location suitable		
10. Treatment meets stream standards			32. Buildings and grounds well drained		
11. Operation, maintenance, and reports satisfactory			33. Accessible by emergency vehicles		
12. Qualified operator			34. Service entrances convenient		
			35. Elevators serve all floors		
Air pollution control			*Radiation*		
13. Incinerator emissions meet standards			36. Diagnostic X-ray units satisfactory		
14. Boiler emissions meet standards			37. Therapeutic X-ray units satisfactory		
15. Process emissions meet standards			38. Teletherapy units satisfactory		
16. Fuel composition and use acceptable			39. Radioactive materials properly stored, used		
			40. Microwave units satisfactory		
Solid wastes					
17. Garbage storage and collection satisfactory			*Housing and safety*		
18. Refuse storage and collection satisfactory			41. Rooms clean, lighted, ventilated		
19. Disposal satisfactory			42. Fire escape from rooms		
			43. Adequate space for occupancy		
Swimming pool and bathing beach			44. Insect and rodent control effective		
20. Life-saving equipment and lifeguards adequate			45. Clean bedding		
			46. Heating safe and adequate for intended use		
21. Adequate clarity			47. Fire protection adequate		
22. Adequate treatment and reports			48. Fire-resistive construction		
			49. OSHA standards met		

S means substantially satisfactory equipment, construction, operation, maintenance.
U means unsatisfactory. Use available codes, rules, and regulations for compliance. Mark items NA if not applicable.
Supplied services ____ Water Supply ____ Sewerage ____ Refuse Collection ____ Dietary ____ Other ____. Give name of contractor or supplier.

Figure 11-19 Abbreviated institution environmental health inspection form. (For comprehensive interpretations see pertinent sections of this text.)

inspections and reports of the facilities and services in the same manner as is done for other state, municipal, and public facilities and establishments.

Environment inspection and report outline

A comprehensive inspection and report would involve investigation of the following items.

1. Grounds and Structures
   a. Location, accessibility, service entrances, cleanliness, noise.
   b. Protection from flooding; drainage.
   c. Construction materials and maintenance; dampness, drafts, leaks; sound and in good repair.
   d. Fire protection, municipal and onsite; adequacy; water supply; alarms.
   e. Safety, accident prevention, road signs, lighting.
2. Utilities
   a. Water supply: source, treatment, storage and distribution, quality, quantity, pressure; quality surveillance and compliance with federal and state standards; operation control.
   b. Wastewater collection and disposal: sewage and all other liquid wastes collection, treatment, and disposal; compliance with federal and state standards; operation control.
   c. Solid wastes: storage, collection and disposal; storage areas or rooms, cans, bins; on-site processing and disposal; hazardous wastes handling, storage, and disposal; compliance with federal and state standards.
   d. Heating, electricity and air conditioning; adequacy; safety.
   e. Air quality: power plant, incinerator, institution operations; compliance with federal and state standards.
   f. Emergency power and disaster planning; power to vital services and lighting to vital areas.
3. Shelter
   a. Temperature control: heat, ventilation, humidity control, cooling.
   b. Lighting: walkways, assembly areas, cells, kitchens, work areas, special uses and facilities.
   c. Space requirements: cells, assembly areas, recreation areas, dining rooms, visiting areas.
   d. Fire safety: fire-resistive construction; compartmentation; interior finishes; enclosures, doors, stairs; extinguishers and extinguisher systems, sprinklers, detection and alarm systems; fire water supply; fire plans and drills.
   e. Accident prevention: physical design, working conditions, fire and electrical hazards; occupational exposures and recreation facilities; also, occupational health standards (OSHA) as applied to jails and prisons. Drugs, pesticides, flammables, and other hazardous materials stored in a secure place.
   f. Housekeeping: general cleanliness and maintenance; facility interior surfaces (walls, floors, ceilings, facilities, equipment); equipment and facilities maintenance; grounds and spaces; roster and cleaning schedule.
   g. Noise: interior, exterior; mechanical equipment, work areas comply with OSHA standards.
4. Services and Facilities
   a. Food and protection: wholesomeness, refrigeration, storage, preparation, transportation, service; processing; equipment; food handler inspection; ice; vending machines.
   b. Radiation protection: diagnostic, therapeutic, teletherapy, X-ray units; radioactive materials storage and use; microwave ovens; industrial uses.
   c. Vermin control: rodents, insects and other arthropods; physical and chemical controls; pesticide storage secure and used as directed on label.
   d. Laundry facilities: soiled linen and clean linen separate storage, handling, and transportation; laundering process.
   e. Plumbing: water, soil, and waste lines, drains, toilets, wash basins; adequacy of hot and cold water for all purposes; service sinks; cross-connection and backflow control.
   f. Recreational facilities: bathing beach, swimming pool; other; life-saving equipment and life guards; water clarity and quality; accident control; maintenance, operation and sanitation facilities; safety.

   g. Institutional operations: canning, slaughtering, dairy, pasteurization; other farm operations; manufacturing; vocational training; hospital; laundry; bakery.
   h. Facilities available for public and staff: toilets; dressing rooms; visiting areas.
   i Medical care facility area: storage of drugs; disinfection and sterilization; refrigeration of blood and drugs; morgue.
5. Personal Hygiene
   a. Personal hygiene: infestations and disinfestation; showers, towels, clothes, toiletries, etc.
   b. Bedding: mattresses, pillows, sheets, blankets, beds.
   c. Toilet and bathing facilities: number and type of water closets, squatting plates, wash basins, showers; removable pail privies.
   d. Barber and beauty shops: room designated, equipped, staffed.
6. Personnel and Supervision
   a. In-service training: staff and inmates having environmental sanitation responsibilities.
   b. Self-inspection: qualified person designated, responsible to administrator.
   c. Regulatory agencies: inspection and approval of facilities and services annually.

Because of the diverse facilities and services involved, it is essential that the regulatory person assigned to make inspections be broadly trained and have the experience and maturity to know when to call upon a specialist to investigate in greater detail and to resolve complex problems. Many resources including specially trained consultants, laboratory facilities, regulations, and inspection services are available from various departments and agencies of government (including federal), as well as national organizations. These should be utilized to identify and help resolve potential and actual deficiencies.[46]

The basic principle involved, the public health rationale, and the basis for satisfactory compliance for each item listed above is given in *Standards For Health Services in Correctional Institutions.*[46]

BIBLIOGRAPHY

Andrzejewski, A., K. G. Berjusov, P. Ganewatte, M. S. Hilbert, W. A. Karumaratne, J. G. Molner, A. S. Perockaja, and C. L. Senn, *Housing Programmes: The Role of Public Health Agencies*, Pub. Health Papers 25, WHO, Geneva, 1964.

APHA-CDC Recommended Housing and Maintenance Ordinance, 1975 Revision, PHS, DHEW, CDC, Atlanta, Ga. 30333.

APHA Program Area Committee on Housing and Health, "Basic Health Principles of Housing and Its Environment," *Am. J. Pub. Health*, May 1969, pp. 841–852.

An Appraisal Method for Measuring the Quality of Housing: Part I, "Nature and Uses of the Method," 1945; Part II, "Appraisal of Dwelling Conditions," Vols. A, B, C, 1946; Part III, "Appraisal of Neighborhood Environment," 1950; American Public Health Association, 1015 Fifteenth Street, N.W., Washington, D.C.

"Appraisal of the Hygienic Quality of Housing and its Environment," *WHO Tech. Rep. Ser.*, **353,** 1967.

ASHRAE Guide and Data Book: 1970 Systems Vol., 1969 Equipment Vol., 1968 Applications Vol., 1967 Handbook of Fundamentals; American Society of Heating, Refrigerating and Air-Conditioning Engineers, Inc., 345 East 47th Street, New York.

Basic Housing Inspection, PHS, DHEW Pub. (CDC)76-8315, 1976, GPO, Washington, D.C. 20402.

[46]*Standards For Health Services In Correctional Institutions*, prepared by Jails and Prisons Task Force, Am. Pub. Health Assoc., 1015 Fifteenth Street, N.W., Washington, D.C. 20005, p. 51 and 51–88.

Basic Principles of Healthful Housing, Am. Pub. Health Assoc., Committee on the Hygiene of Housing, Washington, D.C., 1946.

The BOCA Building Code/1975, Building Officials and Code Administrators, Inc., 1313 East 60th Street, Chicago, Ill. 60637.

Bond, Richard G., George S. Michaelson, and Roger L. DeRoos, Eds., *Environmental Health and Safety in Health-Care Facilities*, Macmillan Publishing Co., Inc., New York, 1973.

Environmental Engineering for the School, PHS Bull. 856 and Office of Education Pub. OE 21014, DHEW, Washington, D.C., 1961.

Environmental Health Guide for Mobile Home Communities, prepared by PHS, DHEW; published by Mobile Homes Manufacturers Association, 6650 North Northwest Highway, Chicago, Ill., revised January 1971.

"Expert Committee on the Public Health Aspects of Housing," *WHO Tech. Rep. Ser.*, **225**, Geneva, 1961.

General Standards of Construction and Equipment for Hospitals and Medical Facilities, PHS Pub. 930-A-7, DHEW, Washington, D.C., December 1967.

Goromosov, M. S., *The Physiological Basis of Health Standards for Dwellings*, Public Health Papers No. 33, WHO, Geneva, 1968.

Housing an Aging Population, Committee on the Hygiene of Housing, American Public Health Association, Washington, D.C., 1953.

Housing Policy Report, Supplement to *House & Home*, January 1954.

Infection Control in the Hospital, Am. Hospital Assoc., Chicago, IL, revised edition, 1970.

Knittel, Robert E., *Organization of Community Groups in Support of the Planning Process and Code Enforcement Administration*, PHS, DHEW, Washington, D.C., 1970.

Koren, H., *Handbook of Environmental Health and Safety*, Vol. 1 and 2, Pergamon Press, Inc., New York, 1980.

Krasnowiecki, Jan, *Model State Housing Societies Law*, PHS Pub. 2025, DHEW, Washington, D.C., 1970.

Litsky, B. Y., *Hospital Sanitation—An Administrative Problem*, Clessols Publishing Company, 1966.

Mobile Home Court Development Guide, HUD, Washington, D.C., January 1970.

National Electrical Code, Am. Standards Assoc., (ASA), Cl-1962, 10 East 40th Street, New York, 1962.

National Plumbing Code, ASA A40.8-1955, ASME, 29 West 39th Street, New York, 1955.

"Neighborhood Environment," *Ohio's Health*, November-December 1969.

New Approaches to Housing Code Administration, National Commission on Urban Problems, Research Rep. 17, Washington, D.C., 1969.

Nursing Homes: Environmental Health Factors—A Syllabus, PHS, DHEW, Washington, D.C., March 1963.

Parratt, Spencer, *Housing Code Administration and Enforcement*, PHS, Pub. 1999, DHEW, Washington, D.C., 1970.

Principles for Healthful Rural Housing, Am. Pub. Health Assoc., Committee on the Hygiene of Housing, Washington, D.C., 1957.

Proceedings of The First Invitational Conference on Health Research and Its Environment, Airlie House, Warrenton, Va., March 17–19, 1970; sponsored by the American Public Health Association, and PHS, DHEW, Washington, D.C.

Ramsey, C. G., and H. R. Sleeper, *Architectural Graphic Standards*, John Wiley & Sons, New York, 1970. 7th Ed. 1981.

Rehabilitation Guide for Residential Properties, HUD, Washington, D.C., January 1968.

Standards For Health Services In Correctional Institutions, American Public Health Association, 1015 Fifteenth Street, N.W., Washington, D.C. 20005, 1976.

Subcommittee of the Program Area Committee on Housing and Health, *Guide for Health Administrators in Housing Hygiene*, American Public Health Association, Washington, D.C., 1967.

Tenants' Rights: Legal Tools for Better Housing, HUD, Washington, D.C. 1967.

Uniform Building Code, International Conference of Building Officials, Pasadena, Calif., 1961.

Uniform Housing Code, 1976 ed., International Conference of Building Officials, 5360 South Workman Mill Road, Whittier, Calif. 90601.

U.S. Senate, Subcommittee on Housing and Urban Affairs, Committee on Banking and Currency, *Congress and American Housing*, Washington, D.C., 1968.

Walton, Graham, *Institutional Sanitation*, U.S. Bureau of Prisons, Washington, D.C., June 1965.

Williams, R. E. O., and R. A. Shooter, *Infection in Hospitals*, F. A. Davis Company, 1963.

Woodward, H. C. A., "The 1969 Housing Act, the Qualification Certificate," pp. 198–201, and W. Combey, "The Rehabilitation of Houses," pp. 202–207, *J. R. Soc. Health*, **90,** 4, (July/August 1970).

12

ADMINISTRATION

ORGANIZATION

Introduction

The administrative structure on the federal, state, and local levels of government for the provision of environmental control services to the people has been changing.* This has been brought about by public recognition of the environmental pollution problems that have not been resolved and the desire for a higher quality of life.

The development and implementation of an effective environmental protection program will require consideration of the scientific, social, political, and economic constraints involved. A balanced appraisal and adjudication of competing objectives are then necessary. The planning process briefly described below and in Chapter 2 will be found useful to gather and analyze the facts, make plans, establish priorities, and make decisions for action programs. A report of a WHO Expert Committee will also be found informative.[1] It reviews trends in environmental health, planning, organization and administration, international collaboration, and technical needs and problems requiring future action. The PHS has also prepared a document to guide environmental health administrators and planners in the development of appropriate and effective programs.[2]

No one organizational structure is necessarily the best for a country, state, or local unit of government because the problems existing, purposes to be served, type of government, and methods considered acceptable will vary. However, the national structure will usually influence the state and local structure, and the state structure will influence local structure.

A distinction should be made between the actual provision and operation of a service, such as refuse collection or water supply, and the control function to ensure that the service is satisfactory and meets established laws, rules, regula-

*"We tend to meet any new situation by reorganizing, and a wonderful method it can be for creating the illusion of progress while producing confusion, inefficiency and demoralization." Petronius Arbiter, 54 A.D., *Newsletter*, Sanit. Eng. Div., ASCE, January 1971. Actually the major changes relate to governmental, fiscal, and enforcement policy, priority, and support.

[1]"National Environmental Health Programmes: Their Planning, Organization, and Administration," *WHO Tech. Rep. Ser.*, **439**, 1970.

[2]*Environmental Health Planning*, PHS Pub. 2120, DHEW, Washington, D.C., 1971.

tions, and standards for the protection of the general health, safety, and welfare. The environmental control functions may be carried out on a national, state, or local basis or on some combined or shared basis, depending on the source of a potential problem and on how widespread its effects are. For example, controls for food and drugs, surface waters, and air that move across state boundaries, would be federal functions; but if such movement is confined within a state, the controls could be state functions. A sharing of responsibilities coupled with uniform standards would eliminate duplication and make for more efficient use of manpower.

The federal, state, and local control agencies are not in competition with one another, but rather are complementary. Proper utilization of the available resources at each government level can make possible the provision of a more complete service to the people, usually at less cost. Failure to realize administrative understanding among official agencies or between federal and state or state and local agencies is probably due to lack of cooperation, understanding, confidence, and two-way communication.

State and Local Programs

The activities conducted by a state or local government control agency are determined by the responsibilities delegated by the state and by the local governing body. This is usually done through the public health law, environmental protection law, sanitary code, and local ordinances. The possible scope of activities is shown on pages 1033 and 1034.

A state, city, or county department or division of environmental control can provide basic environmental protection services through an organizational structure of bureaus, sections, and units, depending on the population served. The bureau director would serve as program administrator, with assistance from section chiefs as consultants or activity specialists. The director would plan, organize, direct, and supervise the activities for which the bureau is responsible, see that the required job is done, and that quality is maintained. In a small department or division, the director and assistant director might share responsibility for a combination of programs and activities.

Where district or regional offices are maintained by a state or a large local organization, the director should be a broadly trained person. This person would be expected to have administrative and technical ability and to be qualified to oversee the review and approval of architectural and engineering plans. Where qualified specialists are not available locally, the scientific, technical, planning, engineering, and architectural problems would be referred to a central office staff.

The organizational plan should generally promote decentralization of responsibility to the most competent level capable of delivering generalized services directly to the people as the need and acceptance develop. Variations and modifications are necessary, depending on the local situation. It is known, however, that a good generalized program of inspection and supervision requires competent personnel, including a number of specialists, depending on the program complexities. The retention of competent personnel requires adequate salaries, a dynamic and challenging program, recognition of the dignity of the individual, and pleasant, stimulating working conditions.

Manpower

The work in environmental control is carried out by a diverse group in and outside of government. Those who have been carrying the major burden are the sanitary or environmental engineers, sanitarians, sanitary chemists, microbiologists, veterinarians, entomologists, public health inspectors, environmental scientists, environmental health technicians, and various other program specialists. Program specialists in environmental engineering and sanitation include, among others, personnel engaged in industrial hygiene, radiation protection, air pollution control, solid wastes management, water pollution control, water supply, hospital engineering and sanitation, vector control, milk and food technology, research, and related activities. These groups are being supported by biologists, attorneys, planners, economists, administrators, and many others having a professional or lay interest in public health and conservation and in improvement of the quality of the environment.

The minimum educational requirement for environmental engineers, sanitarians, scientists, and other environmental specialists is the baccalaureate degree. However, the trend is toward requirement of graduate education in one of the basic disciplines or in an area of categorical program specialization. In several basic disciplines the qualifying professional degree is the doctorate. The sanitarian's baccalaureate degree is usually with a major in environmental health or in the physical or biological sciences.[3] The engineer's baccalaureate or graduate degree is usually with a major in sanitary engineering, environmental engineering, or environmental health engineering. Undergraduate education in civil, chemical, mechanical, electrical, or nuclear engineering could, with appropriate supplementary courses, provide a satisfactory entry base.

Descriptions of tasks that may be carried out by personnel engaged in environmental engineering and sanitation activities may be useful and are quoted below:

Environmental or Sanitary Engineer

Environmental engineering is that branch of engineering which is concerned with:

1. the protection of human population from the effects of adverse environmental factors
2. the protection of environments—both local and global—from the potentially deleterious effects of human activities
3. the improvement of environmental quality for human health and well-being

[3] *Health Resources Statistics—Health Manpower and Facilities*, 1969, National Center for Health Statistics, DHEW, Rockville, Md. 20852, May 1970.

4. application of engineering principles in
   a. air pollution control
   b. collection, treatment and distribution of water supplies
   c. collection, treatment and disposal of domestic and industrial wastes
   d. control of land and water pollutants
   e. collection, treatment, recycling, and disposal of solid wastes
   f. radiation protection
   g. industrial hygiene
   h. milk and food sanitation, housing and institutional sanitation, vector control, and rural and camp sanitation
   i. other related categories
5. professional teaching, research, or development in the fields listed under item (4) above.[4]

Sanitarian

The sanitarian applies knowledge of the principles of physical, biological, and social sciences to the improvement, control, and management of man's environment.[3]

The U.S. Department of Labor goes further in this description by adding,

plans, develops, and executes environmental health programs. Organizes and conducts training program in environmental health practices for schools and other groups. Determines and sets health and sanitation standards and enforces regulations concerned with food processing and serving, collection and disposal of solid wastes, (water supply), sewage treatment and disposal, plumbing, vector control, recreational areas, hospitals, and other institutions, noise, ventilation, air pollution, radiation and other areas. Confers with government, community, industrial, civil defense, and private organizations to interpret and promote environmental health programs. Collaborates with other health personnel in epidemiological investigations and control. Advises civic and other officials in development of environmental health laws and regulations.[5]

Environmental Health Technician and Aide

Environmental health technicians and aides provide technical support and assistance to the sanitarian or other health specialists (e.g., sanitary engineer, health physicist, health officer). They conduct surveys and implement measures to control the spread of diseases and other health hazards or conditions (e.g., food contamination, air and water pollutants, insect and rodent harborages). They take samples of such materials as water, food, and air, and perform or assist sanitarians in performing tests to determine contamination. They explain how to repair, install, or construct sanitation facilities (e.g., water systems, sewage disposal systems, plumbing) as well as how to maintain and utilize individual facilities. They investigate public and private establishments (e.g., food markets, restaurants, dairy plants, water supplies, medical care facilities) to determine compliance with or violation of public sanitation laws and regulations. However, when the primary purpose of the position is to perform the latter duty, it should be allocated to the appropriate series in the investigation group.

At the higher levels, many of the assignments made to technicians require the same depth of analysis as sanitarian positions. They differ from sanitarian assignments in

[4] *The Diplomate*, American Academy of Environmental Engineers, Box 1278, Rockville, Md. 20850, January 1980; Minutes of Board of Trustees, October 18–19, 1979.

[5] *Dictionary of Occupational Titles*, 4th ed., 1977, U.S. Dept. of Labor, Employment and Training Administration, Sup. of Documents, GPO, Washington, D.C. 20402.

that the technician is not required to resolve problems that require the application of new methods and techniques or those that require action beyond the specific work assignment. On the other hand, the knowledges and abilities required for sanitarian work may be different in kind and breadth from those required for technician work, but not necessarily different in grade level. For example, technician work may require a high level of technical or administrative qualifications applicable to specific work assignments based on a comprehensive background of practical experience, training, and skill in applying knowledge of precedents, guides, and techniques.

While all positions require a practical knowledge of basic environmental health concepts, principles, methods, and techniques, the experienced technician must have a detailed knowledge of the laws and regulations governing environmental health practices as well as what constitutes a good environmental health program in one or more of the following or other comparable environmental health areas:

Milk and food.
Water supply.
Waste.
Insect and rodent.
Shellfish.
Recreation, housing, care facilities or other institutions.

The title for trainee or developmental jobs is *Environmental Health Aid.* Environmental health aids collect and record adequate data on existing environmental sanitation conditions and initiate corrective action on the health hazards that are fully covered by written guidelines.[6]

The minimum appropriate educational requirement for the environmental technician is an associate degree in environmental health, radiologic technology, water and wastewater, or related areas of specialization. A number of junior colleges and technical institutions offer technical training in environmental health or similar areas. The environmental aide normally is a high school graduate with varying amounts of appropriate short-course training in specialized subjects.

Salaries and Salary Surveys

Competent direction is essential to a divisional program's effectiveness. As has been well established in private business, it is false economy to pay inadequate or marginal salaries. The money spent for an entire program is in jeopardy of being totally misdirected and hence wasted when placed in the hands of a mediocre person. If such a person must be "led by the hand" or continually checked, other more valuable time is being misdirected. It is fundamental that high salaries over a period of time attract the most competent people.

[6]United States Civil Service Commission Position—Classification Standards, October 1969; From "Tech Curriculum Developed at NEHA Sponsored Meet," *J. Environ. Health*, September/October 1970, p. 158.

Surveys of professional positions by "experts" for the purpose of establishing equitable salaries are fraught with danger. In some cases it has been found politically expedient to employ a consultant to make a salary survey and thus shift the blame for questionable decisions to an outsider. Since the greatest number of people are usually employed in the lower-salaried nonprofessional positions, which also have counterparts in private industry, and which also represent a larger number of votes, the work done is more clearly appreciated and more likely to be acknowledged by upward salary adjustment. But the professional administrative positions are relatively few; the value of the work done is not understood unless the expert is well acquainted with environmental engineering and sanitation practices. It is more probable therefore that professional people will suffer rather than gain from a salary survey by experts. Every effort should therefore be made to separate salary surveys into professional and nonprofessional groups. Make certain the professional salaries are surveyed by people who understand environmental control programs.

ENVIRONMENTAL CONTROL PROGRAM PLANNING

In order to plan properly for an environmental control program it is important to know what the people want, what they need, what they can afford, and what is possible. This requires involvement and participation of the people affected in all stages of the planning process. The answers to these questions can evolve out of the planning process as explained in Chapter 2, Comprehensive Environmental Engineering and Health Planning. An example is outlined below.

Statement of Goals and Objectives*

This statement should recognize the environmental quality aspirations and needs of the community. These might include decent housing and adequate and safe water supply, clean air, proper sewage collection and disposal, proper solid waste collection and disposal, adequate and safe parks and recreation facilities, clean food service establishments, elimination of mosquitoes and rats, and other environmental quality matters of pressing concern to the people. Having agreed on broad goals, the next step is the establishment of specific measurable objectives to be achieved within a specified time frame. The terms "goals" and "objectives" are sometimes used interchangeably. Goals should be considered the final purpose or aim, the ends to which a design tends, the ideal which is rarely realized. Objectives are the realistically attainable ends.

Data Collection, Studies, and Analyses

This phase involves the systematic collection of basic demographic land-use, and economic data and their projection; research, study, and analysis of the

*See Chapter 2.

data; and the community institutions and government. Included would be sociological factors, characteristics and needs of the people, their expectations, and attitudes. The political structure, laws, codes and ordinances, tax structure, departmental budgets, and special background reports that may be pertinent should also be considered.

Basic data collection should identify and generally define both favorable and unfavorable natural and man-made environmental conditions and factors. These would provide background for the more in-depth environmental health and engineering study to follow and would include such matters as weather and climate; topography, hydrology, flooding, tidal effects, seismology, soils characteristics, and drainage; natural and man-made air, land, and water pollution; flora and fauna of the area; noise, vibrations, and unsightly conditions; and condition of housing, community facilities, and utilities.

This part of the study should probe in some depth those environmental factors that have or may have a direct or indirect effect on man's physical, mental, and social well-being both as an individual and as a member of a community. Epidemiological surveys including morbidity and mortality, and investigation of the adequacy of water supply, sewerage, refuse collection and disposal, recreation facilities, housing, and the many other environmental factors listed in Chapter 2 would be included.

The data collection and analyses should also take into consideration factors that may influence the plan being prepared and problem identification and resolution. This step would include evaluation of the effects of scientific and technological development; the probable increase in liquid, solid, and gaseous wastes; federal and state grants; government and private planning and construction programs; and availability of manpower.

Upon completion of the studies and analyses, the goals and objectives previously established should be reevaluated and revised. Feasible objectives should be more clearly defined within the broad goals established. It is also important to recognize and coordinate with other federal, state, local and agency program plans and make indicated adjustments.

Environmental Control Program Plan Areas and Activities

The information obtained in the step described above will usually reveal many weaknesses in need of attention through the implementation of an environmental control plan. Such a plan would identify those factors subject to environmental control and those factors requiring new institutional arrangements or construction or rehabilitation of a facility, such as a water system, in order to meet a desired objective. This calls for the presentation and study of alternatives and consideration of the advantages and disadvantages of each. The establishment of priorities and commitment of resources to achieve the agreed-upon ends require public and political confirmation and support. At the same time essential services and facilities are maintained, recommendations are made, and support is given to the agencies responsible for the design, construction, and operation of needed environmental control facilities and services.

The factors or functions subject to control include all or most of those proposed by a WHO Expert Committee quoted below.[7]

1. Water supplies, with special reference to the provisions of adequate quantities of safe water that are readily accessible to the user and to the planning, design, management, and sanitary surveillance of community water supplies, giving due consideration to other essential uses of water resources.
2. Wastewater treatment and water pollution control, including the collection, treatment, and disposal of domestic sewage and other waterborne wastes, and the control of the quality of surface water (including the sea) and groundwater.
3. Solid waste management, including sanitary handling and disposal.
4. Vector control, including the control of arthropods, mollusks, rodents, and other alternative hosts of disease.
5. Prevention or control of soil pollution by human excreta and by substances detrimental to human, animal, or plant life.
6. Food hygiene, including milk hygiene.
7. Control of air pollution.
8. Radiation control.
9. Occupational health, in particular the control of physical, chemical, and biological hazards.
10. Noise control.
11. Housing and its immediate environment, in particular the public health aspects of residential, public, and institutional buildings.
12. Urban and regional planning.
13. Environmental health aspects of air, sea, or land transport.
14. Accident prevention.
15. Public recreation and tourism, in particular the environmental health aspects of public beaches, swimming pools, camping sites, and so forth.
16. Sanitation measures associated with epidemics, emergencies, disasters, and migrations of populations.
17. Preventive measures required to ensure that the general environment is free from risk to health.

The American Public Health Association adopted a resolution and policy which incorporated the following program areas in environmental health planning.[8,9]

1. Air Quality (control of air pollution).*
2. Food Protection.
   a. Food service operations.
   b. Food and beverage vending machines.
   c. Food processing establishments.
   d. Milk sanitation.
   e. Nutrition (including food wholesomeness and additives).
   f. Labeling.
3. Water Quality.
   a. Drinking water supplies (community, noncommunity and individual water).

*Parenthetical comments are added except for (6.a), (6.g), and (7.a).

[7]"National Environmental Health Programmes: Their Planning, Organization, and Administration," *WHO Tech. Rep. Ser.*, **439,** 1970, pp. 10–11.

[8]Resolutions and Policy Statements, "Environmental Health Planning," *Am. J. Pub. Health*, January 1977, pp. 88–93.

[9]See also "A Proposed Recommendation for a National Health Services System," National Environmental Health Association, *J. Environ. Health*, November 1976, pp. 214–218.

   b. Reuse/multiple use (including control of surface and groundwater quality sources).
   c. Aquaculture.
4. Liquid Waste Management.
   a. Community and individual sewage (including noncommunity).
   b. Industrial liquid waste.
   c. Agriculture.
   d. Runoff.
5. Solid Waste Management.
   a. Residential.
   b. Commercial.
   c. Industrial.
6. Shelter (including occupancy, maintenance, structure, facilities, and services).
   a. Housing (single-family and multiple dwellings) (also conservation and rehabilitation).
   b. Hotels, motels (and tourist homes).
   c. Mobile home parks.
   d. Migrant labor camps (and other labor camps).
   e. Health care institutions and domicilary care.
   f. Schools (public and private elementary, secondary, colleges).
   g. Institutions (including jails and prisons).
   h. Emergency and temporary shelter (in connection with disasters, emergencies, and epidemics).
7. Recreation Safety and Health.
   a. Camps (public and private).
   b. Overnight recreation vechicle parks.
   c. Highway rest areas.
   d. Swimming pools and other bathing places.
   e. Public assembly areas (including tourist attractions, fairs, carnivals, and highway rest areas).
   f. Parks and playgrounds.
8. Rural and Urban Planning (including environmental health considerations).
   a. Land use.
   b. Transportation.
   c. Environmental impact and conservation.
   d. Demographic analysis.
   e. Growth planning.
9. Vector Control (including arthropods, molluscs, rodents).
10. Occupational Health (physical, chemical, thermal, and biological hazards).
11. Injury Control.
12. Radiological Health (including ionizing and nonionizing sources, and emergency response planning).
13. Hazardous Substances Control.
14. Noise Control.
15. Animal Control.

Each of the program activities pursued should have a built-in evaluative system including specific objectives and their measurements, as well as a public information component showing the need and benefits to the individual. Numerous methods and techniques are used to achieve program objectives. These include inspection, education, and enforcement; planning, evaluation and systems analysis; legislation, standards, and guidelines; incentives, conferences, and training; and demonstration, construction, and research.

The program activities listed above are traditional environmental health engineering and sanitation activities. They are usually carried out, unfortunately, in a fragmented and limited manner by a combination of health, environmental protection, agriculture, labor, water resources, building, and other agencies. In such instances, there is a real possibility for gaps in coverage, overlapping, controversy over jurisdictional control, and complaints by those regulated. It is apparent that agency coordination and cooperation must be achieved on a day-to-day basis to avoid legislative intervention and to better serve the public.

To survive, regulatory agencies must adjust their emphases and provide leadership in the development of new program activities in areas where the need exists. Determination of need requires continual evaluation of the goals and objectives of an agency. A tool that can be used in the environmental health engineering and sanitation area is the Comprehensive Environmental Engineering and Health Planning study outlined in Chapter 2. This will help to identify community service and facility deficiencies which need attention and strengthening. Suggested new program activities are discussed below.

Elements of a Program Plan

A program plan should be developed for each of the program activities. The plan should list for each program activity the following:

1. *Need for the program*—as determined by analyses of the data collected, trends in technology, and social development.
2. *Goals and objectives*—long-term goals (5 to 20 years), and short-term objectives (1 to 5 years), with the specific tasks or change to be accomplished each year for the first five years.
3. *Law*—applicable public health law, sanitary code, rules and regulations. Program authority and responsibility.
4. *Work load and methods to meet program objectives*—program size and activities to be performed by surveys, inspections, technical consultation, sampling, engineering and architectural plan review, certification, permit, education, legal action, surveillance, conference, evaluation. Develop size indicators and their measurement. Indicate role of industry and its participation.
5. *Resource requirements*—manpower and data needs, coding and electronic data processing support, standards development, registry categorizing types of place, consultant services, support from other agencies, laboratory support, research, and development.
6. *Evaluation*—effectiveness measures (must meet test of accuracy and validity and be measurable), that is, percent of places, facilities, and samples meeting program objectives, regulations, or standards; reduction in violations, unsatisfactory conditions, complaints; reduction in morbidity and mortality; other tangible and intangible accomplishments; all within a specific time frame. See Evaluation, this chapter.
7. *Financing*—estimated funding requirements, sources of funds (federal, state, local, grants, other). Capital construction needs (regular, first instance).

Administration

The implementation of a control program requires a management organization to administer a total environmental control program and to serve as liaison with other agencies and with the executive and legislative branches of government. This would include professional direction and management support, such as fiscal, legal, personnel, planning, education and public information, research, overall monitoring, surveillance, and evaluation. Administration should also have the competence and authority to develop staff support, to make changes, and redirect program efforts as needed.

Standards Setting

Standards setting is a very complex process. Many standards have been arbitrarily established, or accepted because of historical precedent; they seem to satisfy the current needs. Changing public aspirations and awareness, and new knowledge can and should cause revision of standards. Since standards, when adopted by a government agency, become legal requirements (which may involve the expenditure of thousands or millions of dollars) it is natural and proper that they be debated and questioned to determine the public health, safety, and welfare basis for, for example, a particular number in a standard.

A standard is perceived as providing safety, but in fact a guarantee of safety or zero-risk does not exist. This is so because generally all the factors involved to make possible absolute safety are not known, are variable, are not precise, are not controllable, or are not directly measurable. This requires public understanding and participation and the making of certain informed assumptions to arrive at an acceptable morbidity, mortality, or environmental risk factor; but its accurate quantification is very very difficult, as is the quantification of benefits.

Standards must be scientifically supportable, understandable, attainable, and enforceable and consider the technical and economic feasibility of compliance. The development of standards requires consideration and evaluation of the total body burden from *all* sources, that is, air, water, food, and drugs (also psychological stresses and existing diseases) to determine the full impact of a particular pollutant on man and the environment, together with the possible additive, synergistic, and neutralizing effects when combined with other pollutants. The process must recognize what is known, what is unknown, the extrapolation limitations of animal studies to evaluate human exposure and risk and to after-the-fact epidemiological studies, the limitations of laboratory analyses and analytical methods, the health benefits or reduced risks if a certain standard is adopted, and the cost or risk if not adopted. Obviously, all this information is difficult to obtain. There are those who believe that standards inhibit thinking and should be replaced by performance objectives. Performance objectives and criteria can however be the basis of standards.

Inherent in the implementation of a regulatory control program is the enforcement of standards. From a government administrative and legal standpoint, standards make program enforcement more manageable. But to be practical, deviation or variance from a standard should be permitted to avoid hardship or unreasonableness based on substantiating evidence, a hearing, and regulatory approval, provided the intent is still achieved.

The standards development process should apply to scientific and technical as well as government organizations involved in setting standards or in recommending standards. They must exercise great restraint and care to assure competent and balanced input by those affected, and resort to the consensus process when necessary.[10] The goal should be the protection of present and future generations and of the environment to the extent feasible, with cost being a secondary consideration. This requires wisdom, that is, making the proper decision with inadequate information, but based on knowledge available, experience, and understanding.

Future Preventive Environmental Program Activities

Future improvement in health is dependent on maintaining the proven preventive procedures, such as immunization, sanitation, and nutrition, and on strengthening programs for control of environmental pollutants and prevention, minimizing self-imposed risks, and application of new knowledge dealing with human biology.

Self-imposed risks or life-style factors subject to control include lung cancer from cigarette smoking; esophageal and liver cancer from heavy drinking; skin cancer from excessive exposure to sunlight; improper diet, which contributes to malnutrition, obesity, and probably other types of cancer and cardiovascular disease; drug abuse, which contributes to accidents, malnutrition, cirrhosis, suicide, homicide, and mental disorders; failure to use seat belts, which exacerbates the traumatic effects of accidents; and failure to maintain physical fitness making the individual more prone to illness and disability.

It must also be remembered that many aesthetic factors, although not directly related to any particular illness, may have a very salutory influence on health and well-being. Improvement in the quality of life* has probably done as much or more to prevent disease, illness, injury, and disability as organized community efforts to improve the public health.

Preventive action with emphasis on the control of environmental pollutants should be based upon existing knowledge and the extent to which the cause or factor to be attacked can be identified and hence eliminated or controlled. However, where a positive cause and effect relation does not exist but a strong association is proven, prudence dictates that precautionary measures be taken.

Traditional activities. These activities should be maintained at the level necessary in a particular area to control and prevent recrudescence of disease and to improve the quality of life. The traditional environmental health, engineering, and sanitation program activities suggested by the WHO and the American Public Health Association have been listed earlier in this chapter.

New and emphasized program activities. These activities would include those related to the control of communicable diseases which need attention or strengthening, the control of the noncommunicable or noninfectious diseases, and the

*See quality of life (QOL) under Statement of Goals and Objectives, Chapter 2.

[10]"ASCE Rules for Standard Committees," *Civ. Eng. Div., ASCE,* June 1980, pp. 96–99.

improvement of the quality of life. Many environmental pollutants are believed to contribute to the noninfectious diseases. In some instances the health effects are classified as "definite"; in others the effects are classified as "possible."[11,12] See Tables 1-12 and 1-13 and Figure 1-4. Some examples of program activities that might be undertaken to control the environmental factors, agents, or sources involved follow.

1. Eliminate or minimize environmental pollutants in the air, food and drinking water associated with chronic* and degenerative† diseases. For example, arsenic and mycotoxins (aflatoxins) in food; sulfur oxides, nitrogen oxides, arsenic, lead, hydrocarbons, particulate matter, ozone, carbon monoxide, and mercury in air; and nitrates, lead, mercury and other toxic inorganic and organic substances in drinking water and food. This calls for sound standards and more effective pollution control and enforcement. See Chapters 1–8 and 10.

2. Eliminate protozoa and viruses, in addition to other pathogens, from drinking water. Provide, in addition to final disinfection, coagulation, flocculation, sedimentation, and filtration of all surface water sources of drinking water to remove *Giardia lamblia*, enteropathogenic *Escherichia coli*, and the virus causing infectious hepatitis, as well as other pathogens. Also remove, where indicated, chemical organic and inorganic pollutants, with emphasis on elimination of pollution at the source and on watershed control. See Chapter 3.

3. Conservation and rehabilitation of housing and the residential and institutional environment. Involved are comprehensive community land-use planning and development and a systematic program of housing inspection to maintain standards for occupancy, maintenance, and supplied facilities, including control of carbon monoxide, asbestos, lead poisoning, rats, roaches, and other vermin, and refuse storage, collection, and transportation. Redirect nuisance and complaint inspections to complete housing and premises inspection so as to also eliminate observed insanitary conditions, future causes of complaints, and duplicate inspections, and thus also improve inspection efficiency and effectiveness. See Chapters 2, 5, 10, and 11.

4. Concentrate more on elimination of accident hazards in the home, community, work, and recreation environment during the course of routine environmental sanitation inspections. Update existing sanitary codes and other regulations to incorporate safety standards. See all chapters.

5. Adopt and enforce regulations to control excessive indoor and outdoor noise, indoor heat and cold, indoor air pollutants released by materials and products, and inadequate ventilation. Give the same attention to the inside air as to the outside air. See Chapters 6 and 11.

6. Continued emphasis on reduction of ambient air pollutants, with standards based upon sound data, taking into account the variations and vulnerability in the aged, infirm, and infants, and their reactions to different levels of pollution, in addition to economic and social factors. See Chapter 6.

7. Participate in community planning activities to obtain recognition and consideration of environmental health engineering and sanitation preventive and improvement factors by other agencies, to prevent causes for future complaints, communicable and noninfectious diseases, and thus improve the quality of life. See Chapter 2.

8. Maintain, using information from all available sources, a profile of environmental quality indicators showing the changing status and effectiveness of environmental protection programs and their possible relationship to improved quality of life, morbidity, and mortality on an area-wide basis.[13] See Tables 2-2, 2-3, 3-3, 3-4, Figure 2-4, Tables 6-6, 6-7, 11-4, and 11-5 for environmental quality indicators.

*A *chronic condition* is a condition described by the individual affected as having been first noticed more than 3 months before the week of the interview (as distinguished from *acute*).

†A *degenerative condition* is the result of deterioration or breaking down of a tissue or part of the body (aging).

[11] *Statistics Needed for Determining the Effects of the Environment on Health*, DHEW Pub. (HRA) 77-1457 National Center for Health Statistics, Hyattsville, Md., July 1977, pp. 27–29, 32.

[12] *Health Hazards of the Human Environment*, WHO, 1972.

[13] *Health Hazards of the Human Environment*, WHO, 1972, pp. 269–338.

9. More effective nutritional improvement with greater attention to food additives having possible deleterious effects. Include the principles of food sanitation in nutrition programs. Incorporate elements of food sanitation and nutrition in elementary and secondary school education. See Chapters 1 and 8.

10. Continued promotion of fluoridation of community drinking water. See Chapters 1 and 3.

11. Greater control over heavy metals entering the food chain, such as cadmium in soil used for food crops or in water used as a source of drinking water. Concentrate more on pollution source control and elimination, and on watershed protection. See Chapters 1, 3–5, and 10.

12. Greater participation in occupational health and safety; elimination or minimization of exposure to inorganic and organic pollutants causing disease; supplement federal and state inspections of factories and other work places on an advisory and/or regulatory basis. See Chapter 1.

13. Concern for possible health effects and elimination or reduction of unnecessary exposure to ionizing radiation, ultraviolet radiation, infrared radiation, microwaves, extremely low frequency waves, noise, vibration, ultrasonics, heat, and cold. See Chapter 7.

14. Routine inspection of correctional institutions including jails, prisons, detention homes; also day care centers, homes for the aged, nursing homes, and children's homes, with the same attention as given to other community and public establishments. See Chapter 11.

15. Greater protection of the groundwater resources as sources of drinking water with greater attention to control of landfills, ponds, pits, lagoons, unauthorized sites, and injection wells for the disposal of wastes, and to improvement of onsite well construction and small community and onsite water systems. See Chapters 3 and 4.

16. Promotion of onsite sewage disposal where soil conditions are suitable and where public sewerage is not feasible, economical, or desired. Give greater consideration to alternate methods of sewage disposal where appropriate. See Chapter 4.

17. Greater public participation and environmental health education, especially in the medical, nursing, and dental schools, elementary school, and secondary school curricula, as well as for the general public.

18. Assurance of competent, professionally trained personnel; cooperation with academic institutions; continual in-service training; and broadening the disciplines, expertises, and backgrounds of the manpower resources. See Manpower, this chapter.

19. Consolidation and coordination of environmental health engineering and sanitation inspection-regulatory activities so as to provide a comprehensive review of *all* environmental health factors involved in a particular operation, including land-use planning and community development. See Chapter 2 and this chapter.

20. Study and follow-up of selected hospital admissions. The medical history on admission should include an environmental health and sanitation profile on the patient for possible correlation of the illness, disability, or injury cause with the home, recreation site, place of work, or activity where the illness, disability, or injury occurred.

21. Continual evaluation and program redirection as indicated.

PROGRAM SUPERVISION

Evaluation

Evaluation is the administrative process of measuring and analyzing results in relation to specific goals, objectives, aims, or targets. It measures (to the extent feasible) benefits against costs, or outputs against inputs. It can also determine the relevance of programs to health problems. The terms assessment and evaluation are often used interchangeably; they express both results and goals in quantitative, qualitative, and temporal terms.[14]

[14] *Who Tech. Rep. Ser.*, **439,** 1970, pp. 37 and 43.

Evaluation involves the following steps.

1. Statement of the program objectives. An objective to be meaningful must:
   a. define the problem
   b. be measurable
   c. identify the target population
   d. give the existing status of the program
   e. specify the desired change
   f. establish a timetable to accomplish the change
2. Selection of indicators to measure achievement of the program objectives.
3. Comparison of program achievement with the stated objectives.
4. Measurement of program cost.
5. Measurement of productivity, with consideration of the expenditure and effectiveness or degree to which a program has achieved the stated objectives.
6. Analysis of the above findings, their overall program impact, and adjustment of the program and objectives as needed.

In order to perform an evaluation it is necessary that the data or information collected be *reliable* and *valid*. To be reliable the information must be consistent, reproducible, and precise. To be valid the information must be accurate, or correct, and measure what it is supposed to according to established procedure. For example, the number of inspections reported may be correct, but if the inspections were cursory, the information is not reliable. A sample for BOD determination may be representative (reliable), but if the incubator temperature is 5° off, the result is not valid.

Objective evaluation of the environmental engineering and sanitation program conducted by state and local agencies serves many purposes. It can provide the basis for integrating, adjusting, and balancing the program; it can be used to demonstrate the need for obtaining and retaining competent personnel; it aids the administrator of the program, and the supervisor of one or a group of activities, to determine whether available personnel are being utilized to do the work considered most important; and if the established objectives are being accomplished. In addition, it can provide cost data and facts for supporting program recommendations and policy determinations. It has been found effective in showing which program activities need more inspection time, which are receiving too much inspection time, and which are not producing results.[15]

To be of value evaluation studies should consider the work load, work done, quality of work, and its effectiveness. The data assembled must be interpreted in the light of the thoroughness and competence of the inspections. Evaluation should start in the planning stage, before a program gets started. Goals are set, plans are made to reach feasible objectives, and analyses or studies are made to see how close one has come to the goals and objectives and what changes may be indicated. Figures 12-1 and 12-2 illustrate the type of data needed for program evaluation. Productivity, efficiency, and performance are discussed later.

[15] J. A. Salvato, Jr., "Evaluation of Sanitation Programs in a City-County Health Department," *Pub. Health Rep.*, **68**, 6, 595–599 (June 1953).

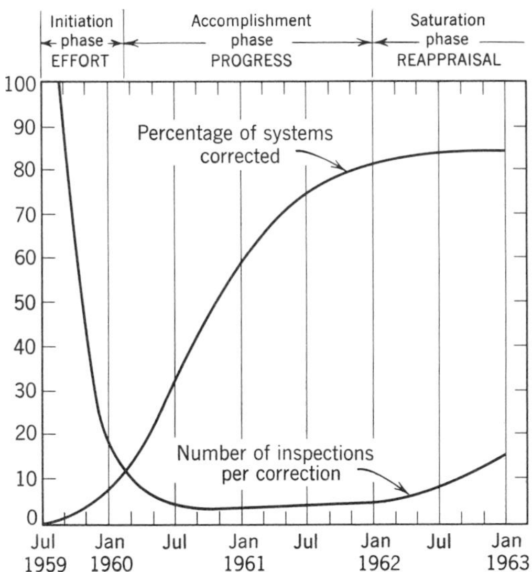

Figure 12-1 Percentage of individual systems corrected and number of inspections per correction in a community sewerage survey.

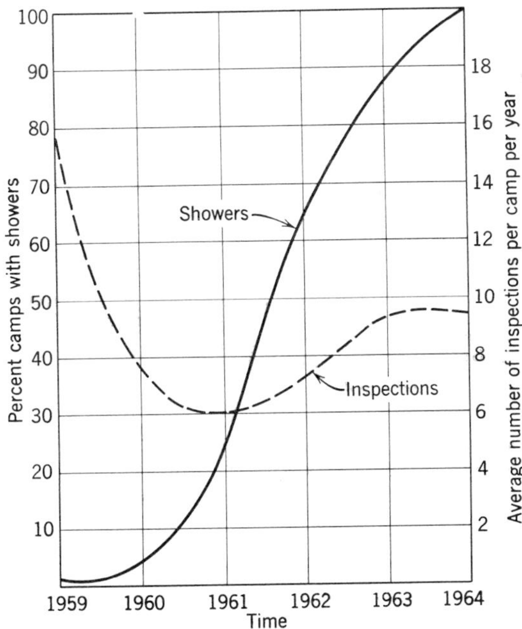

Figure 12-2 Farm-labor camps provided with showers, a planned program.

Reporting

The reporting of work done is a problem that must be realistically faced by every organization. Field personnel usually consider this an unnecessary chore, probably because the reason for keeping the records is not understood. It is important therefore that field personnel fully appreciate that their status, salary, and very existence are intimately tied up with accurate reporting of field visits and other work accomplished. The organization's monthly, quarterly, or annual report can be no more accurate than the reporting used. It is essential, therefore, that the system in effect be as simple as possible to give the desired information. Inspection and similar reports amenable to optical scanning simplify record keeping and data processing.

Daily Activity Report

A daily activity report can serve practically all the reporting needs of an organization for statistical purposes. It can be greatly simplified by using a coding system, thereby adapting it for a mechanical record-keeping system. A form the activity report can take is shown in Figure 12-3.

Coding

Personnel should all be given a code number such as 01, 02, 03, and so on. In setting up a code system it is suggested that numbers be allotted for future personnel. The date, hours worked, time in office, time in field, and time on leave are self-explanatory. Under "Field" the employee writes the name of each premise visited. The time spent on each assignment, including travel, is also noted.

Location is coded by giving a code number to each township, village, city, or district in the department jurisdiction. With this information in hand the field man simply puts down the code 01, 12, or whatever the case may be.

The reason for a field visit is of interest in determining how much of the work done is as a result of planning to carry out the sanitary code responsibility. This is a routine inspection. How much time is spent as the result of requests by the public for service, such as requests for assistance in the design of a private sewage disposal system; for examination of a sample of water from a school, private home, or camp; or for help at a pasteurizing plant or restaurant would be service requests. Then there is the fieldwork that is necessary because of a garbage complaint, overflowing cesspool, dirty lot, standing water, barking dog, pigeons, and so on. These inspections would be made by reason of a reported nuisance or complaint. The "Reason" for a field visit and its coding would be as follows.

Reason	Code
Routine inspection	01
Service request	02
Complaint or nuisance	03

DAILY ACTIVITY REPORT Page ____ of ____

Employee Code ________________ (1–6) Date ____________ (7–12) Name ____________________ (13–34)

Hours Worked: Office ____________ (36–38) Field ____________ (40–42) Leave ____________ (44–46)

FIELD	CODING								OFFICE	
NAME OF PLACE	Location (F-O) 1	Reason (F) 5	Activity Program (F-O) 7	Kind of Field Activity (F) 9	Finding (F) 11	Time Spent* (F-O) 14	Plans Reviewed or Approved (O) 17	Office Function 19	Name of place, person involved in correspondence, office conference, telephone call, plan review or approval or enforcement action. 79	
										02
										03
										04
										05
										06
										07
										08
										09
										10
										11
										12
										13
										14
										15
										16

REMARKS	

F = Field O = Office

*Time recorded to the nearest 15 minutes as 0.25 and hours as 1.00 for example.

Signed ____________________

Figure 12-3 Division of environmental control daily activity report.

The next item shown on the daily activity form is "Program Activity." The program activities carried on depend on the extent of the division program. A rather complete listing is given in the earlier discussion of Environmental Control Program Plan Areas and Activities, and in Table 12-1, together with a suggested coding. A more extensive breakdown may be desired.

Following "Program Activity" on the daily activity report is "Kind of Field Activity." This is meant to show the kind of visit made, which can be tabulated as follows.

Kind of visit	Code
Initial inspection of year	01
Reinspection	02
Inspection during construction	03
Conference in field	04
Person not at home	05
Collection of sample only	06
Training	07

"Persons not at home" incidents can be reduced by making telephone appointments.

The listing under "Finding" indicates the status of the place, which can be reported as shown below. However, if the visit was actually a field conference or if no one was at home, there would be no finding.

Finding	Code
Substantially satisfactory	01
Unsatisfactory—Imminent health hazard	02
Continuing unsatisfactory	03
System installed (private sewerage)	04

Every effort should be made, while personnel are in the field, to routinely have an inspection and report made, rather than to just have a field conference. This is not to minimize the importance and value of a field conference under certain circumstances, but it should normally be part of an inspection rather than just a conference. The term "substantially satisfactory" as used above is applied to a place where the violations observed are few in number; are not deemed to be serious, repeated, or persistent violations; and, in the judgment of the regulatory authority or authorized representative, do not constitute an imminent health hazard requiring immediate correction or prompt follow-up within a scheduled period of time. The term "unsatisfactory—imminent health hazard" means a condition which has associated with it an impending threat or risk of consequential disease, injury, or death thereby requiring immediate corrective action. "Continuing unsatisfactory" would apply to a place which has the same violations on repeat inspections or is found to have numerous (three or more) violations on reinspections. This would call for appropriate enforcement action in view of the repeated or persistent violations, if significant.

The coding of "Office Function" is shown below. Space is also provided in the right-hand column for recording the names of persons spoken to, telephone numbers, and comments to assist in follow-up.

Table 12-1 A Division of Environmental Control Activity Coding—Field

Activity Type	Code
Air sanitation, public	01
Industrial	02
Private	03
Bathing—beaches	04
Swimming pools	05
Carnivals and fairgrounds	06
Emergency sanitation, civil defense	07
Food—eating places	08
Processing establishments[a]	09
Distributing and vending[b]	10
Abattoirs—meat	11
Abattoirs—poultry	12
Other, specify	13
Hospitals	14
Nursing homes	15
Convalescent homes	16
Other medical care facilities	17
Housing—multiple family	18
Hotels, rooming houses	19
One- and two-family	20
Mobile homes	21
Lead poisoning	22
Industrial and business places	23
Insect, rodent, and other vermin control	24
Institutions—state	25
County	26
City	27
Private	28
Ionizing radiation—medical[c]	29
Dental	30
Chiropractic, veterinarian	31
Research, radioactive materials	32
Milk—dairy farms, can	33
Dairy farms, vat	34
Dairy farms, bulk, machine	35
Dairy farms, bulk, pipeline	36
Transportation tank truck	37
Pasteurizing plants, HTST	38
Pasteurizing plants, vat	39
Milk processing, other	40
Quality control supervision	41
Nuisances—water supply-pressure, taste, odor, appearance	42
Sewage—plumbing, leaks, clogging	43
Air pollution, noise, odors	44
Animals—dogs, cats, pigeons	45
Vermin—insects and rodents	46
Refuse—garbage, rubbish	47
Vacant lot—weeds, drainage	48
Insanitary housing—premises	49
Other	50
Parks and recreation—state	51
Municipal	52
Private	53
Places of public assembly	54
Rabies control	55
Realty subdivisions—private water supply and sewerage	56
Public water	57
Public sewer	58
Public water and sewerage	59
Refuse—incinerators	60
Sanitary landfills	61
Open dumps	62
Scavengers, swill, vehicles	63
Other	64
Schools—elementary and secondary	65
Colleges and universities	66
Sewage and wastes—municipal	67
Industrial	68
Private dwelling	69
Other	70
Stream pollution	71
Temporary residences—day camps	72
Children's camps, overnight	73
Motels, hotels	74
Campsites	75
Lodging, boarding homes	76
Mobile-home parks	77
Migrant-labor camps	78
Training, given	79
Received	80
Water supply—community	81
Noncommunity	82
Cross-connection	83
Watering points—land, sea, air	84
Factory, industrial plant	85
Private dwelling	86
Bottled water	87
Public springs or wells	88
Weed control	89
Algae control	90
Special survey, specify	91
Nonionizing radiation	92
Ice manufacture	93
Emergencies	94
	95

[a]Includes bakery, delicatessen, caterer, frozen meal preparation.

[b]Includes vending machine, supermarket.

[c]Includes radiographic, fluoroscopic, therapeutic, and teletherapeutic machines.

Office Function	Code
Office Conference and Meetings	01
Enforcement Action	02
Correspondence, records	03
(Rough-draft letter attached to time report)	

The number of plans reviewed or approved are coded by type, as shown below. The listings can be reduced or added to depending on local conditions and the desired information. The number and type of samples collected can be added to the "Remarks" space on the time sheet and coded as noted below.

Sample Type	Code
Water supply*—community	01
private dwellings	02
camps, restaurants, hotels, schools, bottled milk plants, etc.	03
Wastewater*	04
Industrial wastes*	05
Stream pollution*	06
Swabs, restaurants	07
Milk—pasteurizing plants	08
street, route	09
other, specify	10
farms, bulk	11
temperature	12
chemical screening test	13
sediment	14
baby formulae	15
Food*	16
Rinses—containers, equipment	17
Swimming pools or beaches	18
Air—high volume	19
fallout	20
pollen	21
stack	22
materials' degradation	23
Radiologic—water	24
air	25
food	26
Other, specify	27

Plan Type	Code
Subdivision	01
Water supply and treatment	02
Sewerage and treatment	03
Swimming pool	04
Housing	05
Restaurant	06
Other, specify	07
Environmental impact analysis	08

*Could break down into specific physical, chemical, microscopic, and microbiological tests.

Record Keeping and Data Processing

The daily activity report should be completed and submitted daily to the supervisor or director. Attached should be an inspection report form, rough-draft letter, or memorandum covering each field visit. This material is reviewed each day by a supervisor for completeness and indicated quality of the visits. The activity report is then referred to the clerical section for posting and statistical recording. Other records kept on a daily basis would include the results of water, food, and milk examinations and such other information as may be indicated or desired for special study purposes.

Records for statistical recording can be kept by a manual or mechanical system (electronic data processing), depending on the volume and resources available. Since each field visit requires the processing of one card or form, unless a hand tally is maintained, the number by the end of each reporting period can be quite large and will determine the need for a mechanical or manual sorting system. Local representatives of large data processing corporations are available who can explain adaptation of the coding system to their particular operation. A marginal punch card for manual sorting is shown in Figure 12-4. Special forms can be developed to fit the system used, and the number of entries can be adapted to satisfy particular needs.

It will soon become apparent to the person using a manual or mechanical system that a tremendous amount of information is available. The combinations possible, from the basic items punched or recorded, can give more data than can possibly be used. IBM and Remington Rand are examples of mechanical systems; McBee is manual. The forms can be adapted as inspection reports if desired to expedite transfer of data directly to a computer. The McBee system is suitable when the maximum number of cards per sorting does not exceed about 500. When large numbers of inspections are involved, electronic data processing should be used for data recording and evaluation. Data collection forms and types should be designed with the help of specialists.

0 0 0 0	0 0 0 0	0 0	0 0 0 0	0 0 0	0 0 0 0	0 0 0 0	0 0 0
units	units	tens	units	units	tens	units	
1 2 4 7	1 2 4 7	1 2	1 2 4 7	1 2 4	1 2 4 7	1 2 4 7	1 2 4
Personnel	Month	Location		Reason	Type		Visit

		Samples				Time in		
Finding	Action	Number		Type		Office	Field	Office activity
1 2 4	1 2 4	1 2 4	1 2 4 7	1 2	1 2 4 7	1 2 4 7	1 2 4 7	1 2 4 7
units	units	tens	units	tens	units	units	units	units
0 0 0	0 0 0	0 0 0	0 0 0 0	0 0	0 0 0 0	0 0 0 0	0 0 0 0	0 0 0 0

Figure 12-4 Punch card for manual statistical recording. (Reduced.)

A visible card file, or its equivalent, is an indispensable administrative aid. The system can be as detailed or as simple as desired. If a file is kept on each establishment or facility under supervision, the visible card file can be very simple. If information is being assembled for a special study, a detailed breakdown may be justified. Visual display equipment could substitute for visible card files if resources are available to enter data in the first place. *It is essential to determine the use to which the information is to be put* before asking the record-processing unit and clerical staff to spend the many tedious hours required to keep records month after month or year after year. A few accurate figures are more useful than a mass of figures of questionable accuracy.

A simple visible card file, which should be kept current or on an annual basis, three-year basis, or such interval as is practical depending on the inspection frequency established, is illustrated in Figure 12-5. The city, village, or town is placed in the upper right-hand corner, and the name of the place on the left. The card does not leave the office.

A colored tab is placed over "Permit" to show that a permit has been issued, if required. Different colors can be used to indicate a temporary or conditional permit. Another colored tab is placed over the month an inspection or reinspection is made; a standard color such as green can be used to designate a place found in substantial compliance and a color such as red can be used for one found deficient. Deficiencies found at the time of inspection are listed by item number as numbered on the inspection report. If the inspection found conditions substantially satisfactory the "S" column is checked; if unsatisfactory the "U" column is checked. If several hundred or thousand places are being supervised under an activity, it is possible, by pulling out a visible

NAME OF PROPERTY	LOCATION	TYPE	CAP.	GAZETTEER AND FACILITY CODES	CO.	C.T.V.	FACIL. NO.

NAME OF OPERATOR	TELEPHONE NUMBERS SEASONAL	PERMANENT
ADDRESS: PERMANENT ZIP	SEASONAL	ZIP

FACILITIES:

Water Supply: Municipal ☐ On-Site Supply Chlorinated Yes ☐ No ☐

Solid Waste Disposal: Municipal ☐ On–Site ☐

Sewage Disposal: Municipal ☐ On-Site System Chlorinated Yes ☐ No ☐

Food Served ☐ Swimming Pool ☐ Bathing Beach ☐ OTHER (Specify)

INSPECTIONS

DATE	S	U	UNSAT. ITEM NOS.	DATE	S	U	UNSAT. ITEM NOS.	DATE	S	U	UNSAT. ITEM NOS.

(Details of facilities and summaries of laboratory results on back or in premises file.)

NAME OF PROPERTY	MO./YR. EXPIR.	TYPE OF PERMIT	MONTH INSPECTED											
		Ann. ☐ Cond. ☐	J	F	M	A	M	J	J	A	S	O	N	D

Figure 12-5 Visible file card.

card-file drawer (the cards are overlapping with just the name of property and bottom line showing), to see at a glance places that have not yet been inspected, places operating without any permit or with a conditional permit, the month a place was last inspected, and the places that have substantial deficiencies. In this way it is possible in a very short time to analyze all or several program activities from an administrative or supervisory point of view and determine where time and effort should be directed.

Where a data processing system is used, a program can be prepared to print out or display on a viewer if equipment is available, the desired information on, say, a quarterly basis, provided the inspection report form has been coded and the information recorded.

Statistical Report For Program Evaluation

The workload of a division or department is the total program and miscellaneous activities for which it is responsible. The program must take into consideration the responsibilities legally delegated, the demands of public officials and the people, and the public health needs and promotional work to be done. These activities should be documented and reported to elected officials and the public to help obtain understanding of the services performed and budgetary support when needed.

The preparation of a monthly, quarterly, or annual statistical report to summarize the work of a division or department is a common requirement. Since it is not possible to reduce adequately all the work done to figures, and since such information usually needs interpretation, it is customary to accompany statistical data with a narrative report. A statistical report form that has been used is shown in Table 12-2.

Data Collection, Accuracy and Reliability

The daily activity report, Figure 12-3, is an essential tool for data collection. However, the data recorded must be accurate and reliable as previously noted so that the information provided for any reports, such as Table 12-2, can be analyzed with confidence. Information of this type is used to determine program activity effectiveness, production, efficiency, manpower needs, problem areas, training needs, program activity costs, program redirection needs, and time allocation to office and field activities, and to provide other information for policy decisions. If electronic data processing facilities are available other "programs" can be written and printouts obtained for analysis based on the information provided in the daily activity report. But the program director must be careful not to ask for any more information on a routine basis than he actually needs and can intelligently analyze and evaluate to better understand and improve the operation. There is a temptation and danger to accumu-

Table 12-2 Division of Environmental Control Statistical Report for Program Evaluation

Name of Unit ______________ Report Period ______________ 1980 Population ______________

Program Activities[a]	Total Man-Days Expended				Premises				Field Activity				Plans		Complaints or Service Requests Investigated	Office Conferences & Meetings	Training Man-Days	Cost of Program ($1000)
	Engineer	Sanitarian	Technician	Other	On Record	Inspected	Unsatisfactory End of Report Period	Enforcement Actions	Inspections	Inspections Unsatisfactory	Conferences & Meetings	Sampling Visits	Reviewed	Approved				
Radiological Health																		
1. X-Ray Installations																		
2. Radioactive Materials																		
Institutions																		
3. State																		
4. Licensed by State																		
5. Local																		
6. Day Care Center																		

7. Schools & Colleges																		
Food																		
8. Service Food Estab.																		
9. Food for the Aged																		
Camps & Recreation																		
10. Children's Camps																		
11. Hotels/Motels																		
12. Campsites (Proprietary)																		
13. Public Gathering Sites																		
14. Parks & Campsites (Gov't.)																		
15. Swimming Pools																		
16. Bathing Beaches																		
Community Sanitation																		
17. Indiv. Sewage Disposal																		
18. Indiv. Water Supply																		
19. Realty Subdivisions																		
20. Mobile Home Parks																		
Housing Hygiene																		
21. Residential Housing																		

Table 12-2 (*Continued*)

Program Activities[a]	Total Man-Days Expended				Premises				Field Activity				Plans		Complaints or Service Requests Investigated	Office Conferences & Meetings	Training Man-Days	Cost of Program ($1000)
	Engineer	Sanitarian	Technician	Other	On Record	Inspected	Unsatisfactory End of Report Period	Enforcement Actions	Inspections	Inspections Unsatisfactory	Conferences & Meetings	Sampling Visits	Reviewed	Approved				
Housing Hygiene (*Continued*)																		
22. Lead Poisoning																		
23. Labor Camps																		
24. Nuisances																		
Vermin Control																		
25. Rats & Mice																		
26. Other Vermin																		
27. Animal Bites (Rabies)																		

Water Supply																		
28. Community Systems																		
29. Noncommunity Systems																		
30. Bottled & Bulk Water																		
31. Ice Manuf. & Distribution																		
32. Cross-Connection Control																		
33. Operator Qualifications																		
34. Emergencies																		
Miscellaneous																		
35. Planning Proj. Reviews																		
36. Administration																		
37. Other																		
38.																		
39.																		
40.																		
TOTALS																		

Table 12-2 (*Continued*)

Expenditures for Environmental Health

Cost Items	Amount
Professional & Technical Personnel	
Clerical & Stenographic Personnel	
Laboratory Services	
Other Costs—Overhead & fringe benefits	
Total Expenditures	

Remarks

Submitted by ____________________ Date __________

[a]See Table 12-1 for a more complete listing of possible program activities; see also Environmental Control Program Plan Areas and Activities discussion.

late many many pounds of printouts giving information that is nice to have but which is never really used. This is a waste of technical, professional, and administrative manpower and of machine time, which of course should be avoided. It is far better to have a program written to be used when and if certain specific information, not otherwise available, is needed.

An explanation of each of the horizontal headings in Table 12-2 follows; the totals recorded should be cumulative so that a second quarter report, for example, is the sum of the first and second quarter.

The "Total Man-Days Expended" will show the type and amount of manpower allocated to each of the program activities. Imbalances or improper assignments become apparent and adjustments can be easily made by changes in workload assignments.

The "Premises On Record" include the total number of premises or initial services in each program activity for which the local health unit has a continuing responsibility. A premise under routine supervision must be listed in the record-keeping system; a premise folder and card file for each place is the minimum system. New premises added for supervision and premises discontinued will have to be reported individually in order to keep current the number of premises in each activity category and the total on record.

A premise not under routine or continuing supervision but subject to special inspection because of animal bite, lead poisoning, nuisance complaint, and so on, would be recorded in the "On Record" column on initial inspection. This total figure is cumulative as the premises are inspected during each reporting period, and is started new each year. Such premises should be identified in the record-keeping system.

The sum of the "Premises On Record" for each program activity is an indication of the work load in connection with places for which the department has a routine responsibility. The total work load is determined by adding to this total the number of special program services provided not otherwise listed, such as talks given, special requests for service, special sampling programs, plans reviewed and approved, conferences, educational efforts, planning, evaluation, environmental impact analyses, and so forth.

"Inspected" refers to the number of premises where an annual, routine, or special inspection was made for the first time during the year. Follow-up inspections and visits are recorded under "Field Activity."

The "Inspected" places show what part of the work ("Premises on Record") has been done and hence what part remains to be done if each place is to receive an annual inspection. If an activity includes a large number of premises, such as multiple dwellings, restaurants, or dairy farms, and personnel are not available or not desired to make annual inspections, this activity may be set up as a 2- or 3-year inspection program. This information is very valuable in planning future work.

"Unsatisfactory at End of Report Period"—see "Inspections Unsatisfactory" below. Enter those premises remaining unsatisfactory at the end of the report period, using a separate summary for each program activity category.

"Enforcement Actions" refers to the number of formal office conferences prior to legal action at which a written report is prepared summarizing conclusions reached. It includes administrative hearings and court cases.

"Inspections" include the sum total of the annual or initial inspections of the year, reinspections, routine and special inspections, investigations, and sanitary surveys where a comprehensive report is prepared. A standard report form should be used.

When an inspection is made of a premise containing multiple facilities (water, sewage, food, pool, beach, or any combination), one inspection would be reported for the *major program activity* containing these facilities, that is, the inspection of food services, sewage disposal, water supply, swimming pool, and other facilities associated with a children's camp would be reported under the Children's Camp program on line 10. Similarly, such facilities or combination of facilities associated with a state institution would be reported on line 3 pertaining to state institutions. Nuisance investigations or inspections *not* associated with a specific program would be listed on line 24, "Nuisances."

The number in "Inspections" tells where inspection time is being spent and can be used to determine the average number of inspections each place has received when compared to the number of "Premises Inspected." This information may indicate the need for redirecting inspection time or for greater field supervision. It must of course be considered in relation to the problems, the needs, and the results produced. For example, a premise may have been inspected during the year about 5½ times. This figure might be considered high, but the low percentage of places remaining "Unsatisfactory" may tend to confirm that the time was well spent. This could be investigated further to determine specifically what type of corrections were made. In addition, by looking at the visible card file, if one is maintained, one can direct subsequent inspections the following year first to those places having deficiencies and thus reduce the total number of inspections.

"Inspections Unsatisfactory" refers to those places where *imminent public health hazards are found requiring immediate correction.* Those places where, in the judgment of the inspector, a follow-up inspection is indicated within a protracted scheduled time period to protect the public health and safety would be considered substantially satisfactory.

The number of "Inspections Unsatisfactory" compared to the number of "Inspections" for a program activity show the progress being made and the condition of the premises under supervision. If a place removes all deficiencies and on subsequent inspection is found to have slipped back, it would again be listed under "Inspections Unsatisfactory." When progress is not being made, additional inspection, supervision, or review of the program may be indicated. The key may be lack of direct supervision, poor quality of supervision, departmental policy, lack of staff promotional opportunities, need for in-service training or additional public health education, or poor morale. Interpretation or definition of unsatisfactory deficiencies, of course, may vary with the individual. The variations can be minimized with adequate in-service training, by the employment and retention of competent personnel, and by the development and use of compliance guides.

When "Inspections Unsatisfactory" is subtracted from "Inspections," the difference is the number of premise inspections that were satisfactory. The number of premise "Inspections Unsatisfactory" divided by the number of "Inspections" will give a ratio indicative of the program activity having problems and needing greater attention. When the "Inspections Unsatisfactory" is compared to the number of premises "Unsatisfactory End of Report Period" the progress made in getting corrections made and the need for other enforcement action is indicated.

"Conferences and Meetings" include professional meetings, conferences, presentations for program promotion, and evening meetings; they are reported in this column.

"Sampling Visits" for the primary purpose of collecting a sample are recorded in this column. This visit must *not* be reported as an "Inspection."

"Plans Reviewed and Approved"—plans and sketches are usually reviewed several times before approval. Each complete review and approval relating to a major program should be reported separately for that program. Plans of a school, children's camp, mobile home park, institution or other type of premise which include separate or attached plans for water supply, sewage disposal, drainage, and other facilities would be reported on the appropriate line for the school, children's camp, mobile home park, institution or other type of premise as a plan review or approval.

Review of a home well-water supply or a home subsurface sewage disposal plan would be reported as an individual water supply or sewage disposal.

The actual number of plans reviewed and the actual number approved should be shown in the appropriate vertical column.

"Complaints or Service Requests Investigated"—record the number investigated for the appropriate program activity. Inspections, investigations, visits, and so on are to be recorded under "Field Activity." See the previous explanation regarding recording of "On Record" premises.

Several complaints regarding the *same* problem should be recorded as one complaint.

"Office Conferences and Meetings"—this column identifies the total number of office and telephone conferences including meetings in the office relating to each program activity. Conferences with an alleged violator requested in writing to appear at the office and discuss what action will be taken to correct a violation(s) are to be recorded.

What is an Inspection?

There appears to be a need to clarify what is meant by an "inspection" to ensure that inspections are meaningful, effective, and adequate. In general it can be said that there are three types of inspections: initial or complete inspections, routine inspections, and reinspections. Each is discussed separately below. It must be clearly understood that an inspection is a means to an end; it is not an end in itself. The objective is continued compliance to protect the public health regardless of inspection.

Initial or Complete Inspection

This is a detailed type of inspection, which should be made when a property or establishment is first brought under department control to determine compliance with the law, code, rules, and regulations. The inspection is very comprehensive and includes the collection of basic data and information concerning ownership, operation, physical conditions, and equipment and facilities at the establishment. Sketch plot plans, floor plans, flow diagrams, equipment, and structure layout are included. See Figures 3-21, 4-4, 4-33 to 4-37, 5-13, 8-3, 8-11, 9-5, and 9-8. Inasmuch as conditions do not remain static, it is advisable to have an initial or complete comprehensive inspection made every three to five years and when major changes are made.

Routine Inspection

Routine inspections are usually scheduled inspections or annual inspections of establishments, or other type inspections based on a field inspection form to determine compliance with the law, state or local sanitary code, or rules and regulations. All items on the inspection form should be completed and specific deficiencies explained at the bottom of the form or on the back under "Remarks." Experience shows that if the written explanation of the violation is followed by a recommendation suggesting how the violation can be corrected, while the inspector is at the site, more practical recommendations are made. This can also serve as the basis for informal discussions with the operator of the establishment, thereby setting the basis for immediate correction of the violation. In any case, significant violations noted should be confirmed by letter with a timetable for correction preferably as previously agreed upon with the operator at the time of the inspection. The report and letter involving important violations and recommendations should be reviewed by the program director or supervisor before the letter is sent out. Figures 8-6 to 8-10, 9-9, 11-4, and 11-16 to 11-19 are examples of routine inspection forms.

Reinspection

A reinspection would be made as a follow-up on a routine or initial inspection, particularly when significant violations are noted, to determine whether the needed corrections have been made. When a reinspection is made, the inspection report form should be completed in its entirety. In other words, all items should be inspected and not only those that were found unsatisfactory at the time of a previous inspection. A partial reinspection might be indicated when only one or two specific items are to be rechecked and other conditions are not likely to change since the last reinspection. It is emphasized, however, that in general all items on the inspection report form should be reviewed and not only those that were found unsatisfactory. This will take relatively little additional time and will ensure that deficiencies overlooked at the time of the routine inspection are observed. It will also leave the impression that a thorough inspection has been made.

There may be exceptions to the above procedures based on special circumstances. However, a uniform policy should be followed by the field offices to help ensure that thorough, competent inspections are being made and that the requirements of the law, sanitary code, and rules and regulations are being interpreted in a reasonably uniform manner.

Right of Inspection

Decisions of the U.S. Supreme Court have reaffirmed the right under Article IV of the Constitution of the United States, that

> The right of the people to be secure in their persons, houses, papers, and effects, against unreasonable searches and seizures shall not be violated, and no warrants shall issue, but upon probable cause, supported by oath or affirmation, and particularly describing the place to be searched, and the persons or things to be seized.[16]

The conduct of sanitation, housing, health, and safety inspections of nonpublic areas can be made by consent specifically granted by the owner or occupant for the intended purpose. If not granted, an inspection can only be forced under an issued search or inspection warrant or equivalent court order. Any conditions observed without specific authority cannot be used as a basis for the swearing out of a search warrant. States lacking the legal authority for the issuance of a search warrant, other than in connection with criminal proceedings, need to amend state law to make possible inspections required by law and still safeguard the fundamental right of the people. The courts have suggested that the regulatory agency obtain an "administrative warrant" which can be issued on the basis of a *general administrative plan* for particular industries across a given geographical area.[17]

Frequency of Inspection

In the interest of professional program direction, no arbitrary frequency of inspection should be followed unless required by law or special agreement. The emphasis should be on getting proper operation rather than on the number of inspections per se. Kaplan and El-Ahraf,[18] in a limited study, concluded that

[16] *Camara* v. *Municipal Court of the City and County of San Francisco*, 387 U.S. 541 (1967), and *See* v. *City of Seattle*, 387 U.S. 541 (1967). For further discussion see Frank Grad, *Public Health Law*, Am. Public Health Assoc., 1015 Fifteenth St., N.W. Washington, D.C. 20005, 1976, pp. 90–107.

[17] William J. Curran, "Administrative Warrants for Health and Safety Inspections," *Am. J. Public Health*, October 1978, pp. 1029–1030.

[18] O. Benjamin Kaplan and Amer El-Ahraf, "Relative Risk Ratios of Foodborne Illness in Foodservice Establishments: An Aid in Deployment of Environmental Health Manpower," *J. Food Prot.*, May 1979, pp. 446–447.

there was no logical basis for the inspection of all types of establishments a given number of times. Zaki, et al.,[19] in a study of 450 retail food establishments found that the frequency of inspection of food establishments could best be determined by the compliance history of the establishment. It was also found that numerical scoring had little value in determining necessity for reinspection or the frequency of inspection. Bader, et al.,[20] found that reducing restaurant inspection frequency from quarterly to once a year, while maintaining complaint inspections, resulted in some decline of the standard of sanitation, but the level may still be considered acceptable in simple menu establishments.

The effectiveness of a program is largely a function of the attitude and technical competence of the assigned personnel, the quality of work done, and the direction it receives, rather than of the number of inspections made per establishment. The success of a program will depend largely on the extent to which the owner, operator, manager, and employees accept their responsibilities under the law. Effort should be concentrated on places of greatest risk of illness or safety hazard and should, among other factors, consider the volume of services and number of persons exposed. The frequency of inspection should vary according to circumstances. In some instances an establishment may be inspected frequently until either compliance is achieved or legal evidence is accumulated for legal action.

To specify an arbitrary frequency of inspection for a program after an initial inventory has been made, fails to recognize that one place may not need an inspection, whereas other places in the same program may be sorely in need of help and inspection. For some programs, an *average* of three inspections per year is needed to maintain a satisfactory level of sanitation. Other programs may require on the average only one or two inspections. An average can also mean that some places receive one inspection in two years and that some places may receive six inspections a year. A program, for various reasons, may be deliberately planned to be covered on a 2- or 3-year basis. Where annual permits are issued, a minimum of one inspection per year is necessary. Where there is a high risk of illness, injury, or death a more frequent inspection schedule is indicated.

It should be remembered that there are other, perhaps more effective ways of obtaining compliance with good sanitation and safety practices besides arbitrary inspections. These include continual evaluation and comparison of effort versus accomplishment, seminars and conferences, industry and public participation and education, manager and supervisor training, results of surveillance surveys and sampling, and enforcement when indicated. To not utilize other techniques such as these is to proceed blindly and to fail to recognize that inspection is only one means to help achieve the program goals and objec-

[19]Mahfouz H. Zaki, George S. Miller, Mary C. McLaughlin, and Sidney B. Weinberg, "A Progressive Approach to the Problem Of Foodborne Infections," *Am. J. Pub. Health*, January 1977, pp. 44–49.

[20]Max Bader, Eugene Blonder, James Henriksen, and Walter Strong, "A Study of Food Service Establishment Sanitation Inspection Frequency," *Am. J. Pub. Health*, April 1978, pp. 408–410.

tives. See Education, Persuasion and Motivation, and Legal Action later in this chapter.

Inspection Supervision

Review of inspection reports by trained, experienced, and competent sanitary engineers, sanitarians, or scientists, who are directly responsible for the supervision of specific activities, will aid in evaluating the fieldwork from day to day. Incomplete reports, a high percentage of "No Violations," sketchy explanations of deficiencies observed, and no recommendations or vague and nonspecific ones readily become apparent. An unusually large or small number of inspections made in a day may indicate whether an inspector is trying to do a good job. This is discussed further under Efficiency.

The ratio of supervising engineers, scientists, sanitarians, or technicians to field personnel will depend on such factors as the difficulty of the work, its newness, and the degree of progress already made in obtaining satisfactory compliance. In a going housing or restaurant program, the ratio may be 1 supervisor to 6 or 8 inspectors; in a migrant-labor camp or recreation camp inspection program it may be 1:4; a special housing appraisal or stream pollution survey may require a ratio of 1:2 or 1:3. Since environmental engineering and sanitation activities are of a wide range of difficulty, broad generalizations need to be adjusted in individual situations.

Inspection supervision can be extended to encourage establishments under supervision to appoint a sanitation supervisor. Responsibility would include self-inspection and assurance of compliance with legal requirements as well as company policies. It is in the interest of food-processing, food market, hotel, motel, restaurant, and fast-food chain companies to employ a sanitation supervisor or environmental control officer to oversee at all times all aspects of health, sanitation, and safety under their jurisdiction.

The program inspection activities need not all be carried out with equal intensity at the same time. This is rarely possible or for that matter, necessary. Gibson in fact pointed out the necessity of sanitation program planning so as to assure that greater effort is deliberately directed to those activities that are actual or potential problems.[21] For example, use of raw milk, lack of community water supply or sewerage, presence of malaria or rabies, radiation hazards, and a large number of public requests for a particular service are all areas of work that should receive priority attention. Traditional areas of work such as restaurant supervision and complaint investigation may receive inspections out of proportion to the relative need, simply because they are usually well established and easier to perform. However, it is not enough to determine the key problems for emphasis. Those who are directly concerned, such as the enforcing agencies, the local elected officials, and the operators of establishments, must recognize that a

[21]William C. Gibson, "Sanitation Planning," paper presented before the Engineering Section of the Am. Pub. Health Assoc., 81st Annual Meeting, New York, November 12, 1953.

problem exists and accept responsibility for taking the action that may be indicated. It is therefore necessary to enlist the participation of those individuals or occupation groups directly involved with the carrying out of good sanitary practices so as to help make a sanitation program effective. These groups would include food and milk handlers; water, swimming pool, and sewage treatment plant operators; local plumbing and building inspectors; well drillers and plumbers; resort operators, school administrators, pest control operators, equipment manufacturers, architects, engineers, land planners, and surveyors; and others. Not to be forgotten is the importance of keeping the public and community leaders informed so as to obtain individual and organized community support when needed.

Production

Information useful in the planning and management of an operating program can be obtained from a minimum of data usually available in all departments. The number of inspections made is a record commonly kept. The number of man-days used in a program can be determined from payroll and attendance records. The number of productive hours per work day may be obtained by subtracting the hours spent in "nonproductive" work from the total working hours. A typical day might show the following nonproductive work: 1 hr in writing reports, discussing special problems, making appointments, receiving in-service training, and so forth; ½ hr in getting to the first assignment; and ½ hr in maintaining good public relations. Therefore, of a 7-hr workday (exclusive of the lunch hour), 5 hr were spent in "productive" work. From the data given it is possible to determine:

1. Average number of inspections per man-day $= \dfrac{\text{Number of inspections}}{\text{Number of man-days}}$

2. Average number of hours per inspection $= \dfrac{\text{Number of productive hours per days}}{\text{Number of inspections per man-day}}$

An example will illustrate the procedures for the determinations made above. During a one-year period 25,000 inspections were made in a city environmental sanitation program and 10,000 in a county suburban and rural program. There were 5067 man-days on duty in the city and 2777 in the suburban and rural areas. The net average annual work year corrected for holidays, vacations, and sick leave was 220 days. Thus:

1. Average number of city inspections per man-day $= \dfrac{25{,}000}{5067} = 4.9$

2. Average number of county inspections per man-day $= \dfrac{10{,}000}{2777} = 3.6$

3. Average number of hours per inspection in city $= \dfrac{5}{4.9} = 1.02$

4. Average number of hours per inspection in county $= \dfrac{5}{3.6} = 1.39$

These figures, which may be determined for a total program or for one activity, have many uses. Annual comparison of this information on an overall activity basis will indicate trends and may show where special attention should be directed. The figures can also be used to determine the approximate number of personnel needed to carry out an existing or new inspection responsibility. An adjustment should be made if an activity is of more or less than "average difficulty."

For example, an analysis of a city environmental sanitation program showed that it had an annual workload of 17,685 places, and that 11,271 of them were inspected during the year a total of 35,145 times. Each place was therefore inspected an average of 3.1 times. The places not inspected amounted to 17,685 − 11,271, or 6414. If an average of 3.1 inspections per place was required for reasonable control, 6414 × 3.1, or 19,883 additional inspections would appear to be needed. Since this work was in a city where experience showed that 4.9 inspections could be made per man-day, 19,883 ÷ 4.9, or 4058 additional man-days or 4058 ÷ 220 = 19 men would be needed in addition to supplementary supervisory staff. This assumes that one man-year equals 220 man-days.

If, of the 6414 places not inspected, 2164, or one-third, represent an activity that is of minor public health importance, it may be possible to deemphasize inspections in this category to release men for other work. It is also probable that some licensing agency may be making inspections for the activity, thereby providing good reason for avoiding duplication and making only special inspections. Hence the number of men needed might be reduced by one-third.

If another activity has been brought under satisfactory control, the reduced number of inspections necessary may release two or three men. It may also be decided that the inspection of some activities could be conducted on a 2- or 3-year planned inspection interval rather than on an annual basis, thereby reducing the needed personnel still further. Under the circumstances, therefore, instead of asking for 18 additional men it may be possible to do an adequate job with four or six additional men properly used. This is not to say that one should always try to carry out a comprehensive environmental program with a fraction of the required staff. But it is meant to combat a tendency to keep asking for additional people without first thoroughly reevaluating the uses being made of available personnel. It may be found that what is needed is not only additional personnel, but policy changes in the light of environmental engineering and sanitation progress made and the work not done.

Efficiency

The inspection data may also be used in evaluating the efficiency of the sanitation program. It is first necessary to have determined by an impartial expert the average amount of time that should be required to make each type of inspection. One need not make an elaborate time study; an informed estimate with field checks is adequate. What is more important is retention of the figure without change. Multiplying the number of inspections actually made by the average time required for each type of inspection will give the total time that should have been spent to do the work reported. The ratio of the time required to the time actually spent will give the percentage efficiency. It is not inconceivable that an efficiency of 150 or 200 percent may be found when quantity rather than quality has been emphasized. Annual comparison of efficiencies can tell a very interesting story over a period of years. The numerical figure obtained is not nearly as important as the relative change from year to year.

Another measure of effectiveness is graphically illustrated in Figures 12-1 and 12-2. By plotting on an annual basis inspections versus corrections or improvements, trends are readily apparent. The point of diminishing returns can be seen and the need for revising inspectional procedures or administrative philosophy is shown.

Performance

Another way of evaluating a program or activity is by determining its cost from year to year and the services received. The amount of money required to perform a service is something everybody can understand. If services are related to cost, it can be said that an environmental program cost, say, A dollars per year, the amount provided and spent in the budget. If a division performs a total of B weighted services* per year:

$$\text{Cost per service} = \frac{\text{Budget in dollars } (A)}{\text{Weighted services performed } (B)}$$

Hence, if a budget is cut X dollars, and the cost per service is known approximately, one can say that the number of weighted services that can be performed will be reduced by an amount that can be computed.

For example, an environmental control budget is $200,000 and the weighted services performed = 12,500. The cost per weighted service = $200,000/12,500 = $16.00. If the budget is reduced to $180,000, the weighted services anticipated will be reduced to $180,000/$16.00 = 11,250, or by 1250. This can be carried further by saying that in the previous year, for example, the restaurant inspection activity required 1000 weighted services, the private water supply survey and sampling activity required 1500 weighted services, the dairy farm inspection activity required 2500 weighted services, the milk sampling program required 500 weighted services, and so forth. If the budget is cut $20,000, the weighted services that can be performed will be reduced by 1250 as previously computed. Hence, either the restaurant activity can be eliminated, or the private water supply survey and sampling activity practically eliminated, or the dairy farm inspection activity cut in half, or the milk sampling control and restaurant inspection activity drastically reduced. If the budget director, finance committee, or others involved are confronted with the necessary reduction in services accompanying a budget cut, a better understanding of the services is obtained. To be effective, however, the officials must understand fully the principles involved; otherwise this will only be an administrative tool.

*See Table 12-2 and the explanation that follows.

Using the same reasoning, one can also say that an increase in the budget by $20,000 will result in a corresponding increase in the services performed, if needed personnel can be employed. For example, with a budget of $220,000 and a weighted service cost of $16.00, the number of weighted services rendered will be $220,000/$16.00 = 13,750, or an increase of 1250. This might mean a new service, such as ionizing radiation control, or strengthening of existing services.

A performance-type of evaluation based on cost requires the reduction of all tangible services to comparable units. This would be a weighted unit of service. For example, a bathing beach inspection could be given a weight of one. The collection of a milk sample could be given a weight of 0.2. Other activities can be similarly weighted, as shown in Table 12-3. This summation can take the place of a statistical report. These are average weights and recognize the fact that the time to make inspections within each activity will vary but also tend to balance one another when a large number of inspections are made or samples collected. This list has been abbreviated; it can be greatly expanded to include a breakdown of weighted original inspections, reinspections, and partial inspections, as well as weights for different types of food establishments, temporary residences, nuisances, and so forth. This exactness may be desirable but is not always required. There is so much inherent variation within the particular inspection activity that a greater degree of accuracy in the final figure would be of little value. In other words the final accuracy cannot and need not be any greater than the components.

The weighted forecast shown in Table 12-3 would represent the goal established at the beginning of a reporting period, such as a year. By completing this summary each quarter, semiannually, and annually, it is possible to determine whether the goal is being achieved and, at the end of the year, if it has been achieved. The administrator, however, should be prepared to change plans at the end of a quarter, if conditions warrant, so that the overall program reflects the needs as they may develop and as the administrator interprets them.

It is apparent that an artificial or "paper" increase in units of work reported that is not based on a complete and competent service or inspection can easily produce a misleading lowered cost per weighted unit of work. As previously stated, quality inspection must be emphasized and assured through the employment of competent personnel, by training, and by supervision. The most important element, therefore, in obtaining the greatest value for the taxpayers' dollar is high-quality administration. In a health department this means the best health commissioner and program directors money can buy. And money alone cannot buy top administrators for long unless it is accompanied by pleasant and satisfying working conditions. This staff nucleus, plus essential auxiliary personnel, equipment, and the plant or building is therefore a basic or "fixed" cost in the establishment of a health department. It largely determines the quality of the health service. Other costs are considered "variable" costs.

A state health department parceling out of state aid to local health departments might do well to help guarantee adequate salaries for the "fixed" cost item. For nothing will better assure best use of state-allotted funds than the people in responsible charge. Of course other things must be considered, but this question of adequate salaries for dedicated professional career program administrators is frequently overlooked. The establishment by law or sanitary code of *minimum* qualifications to be met for the employment of health department personnel is a step in the proper direction. The lowering of qualifications is a disservice to the taxpayer.

As previously pointed out, the cost per weighted service at the end of a year can be easily determined. The amount of money spent can be readily obtained. The number of weighted services rendered is known. It is therefore a simple matter to determine the actual cost of a weighted service. Comparison of the cost of a weighted service from year to year is an additional evaluation criteria. One should be cautioned against trying to compare programs of different departments on a cost basis. There are many, many variables that make such comparisons misleading and incorrect. Such variables as community needs, type of program, number of years a program is in effect, progress made, adequacy and competence of staff, prevailing salaries, quality of administrative direction, morale, extent of political infiltration, quality of work and supervision, and many other variables would first have to be adjusted before valid comparisons can be made. On the other hand, comparison of a program within the same department from year to year is reasonable.

The principles outlined under performance budgeting have direct application to industry. It is also useful in determining the cost of carrying out an existing activity, a new activity, or a new environmental program.

ENFORCEMENT

Enforcement Philosophy

The administrator is constantly looking for ways to make more effective the limited time and personnel available to carry out the department or division program. Frequently doubts are raised as to whether the time and effort spent to obtain improvements in accordance with law or good practice are worth the

Table 12-3 Performance Report Based on Weighted Work Units

Type of Service	Unit Weight	Actual Number Units	Actual Weighted Units	Forecast Weighted Units
Bathing beach, inspection	1.0	37	37	36
Swimming pool, inspection	1.7	38	65	38
Food establishment, inspection	1.2	900	1,080	1,320
Fairground or carnival, inspection	4.0	6	24	24
Hospital—maternity, inspection	2.0	196	392	150
Housing, inspection	1.2	296	355	450
Industrial or business place, inspection	1.0	24	24	20
Institution, inspection	2.0	2	4	4
Ionizing radiation, inspection	1.0	160	160	50
Milk—dairy farm, inspection	1.0	885	885	515
Milk—processing plant, inspection	2.0	234	468	244

Table 12-3 (*Continued*)

Type of Service	Unit Weight	Number Units	Actual Weighted Units	Forecast Weighted Units
Nuisance, investigation	1.0	247	247	500
Picnic ground, inspection	1.2	32	38	24
Rabies, investigation	1.0	137	137	186
Realty subdivision—rural, investigation	4.0	50	200	160
Realty subdivision—urban, investigation	1.0	12	12	8
Refuse disposal facility, inspection	1.0	52	52	40
Rodent or insect control, investigation	3.0	10	30	18
School—central, inspection	2.3	122	281	138
School—rural, inspection	1.0	53	53	80
Sewage—municipal, inspection	2.0	36	72	48
Sewage, wastes—industrial, inspection	2.0	6	12	10
Sewage—private, inspection	1.0	988	988	1,000
State parks, inspection	2.0	2	4	4
Stream pollution, investigation	3.0	19	57	36
Temporary residence—camp, inspection	3.0	76	228	270
Temporary residence—other, inspection	1.0	246	246	180
Water supply—municipal, inspection	2.0	224	448	400
Water supply—quasi-public, inspection	1.0	63	63	60
Water supply—carriers, inspection	1.0	6	6	6
Water supply—private, inspection	1.0	901	901	1,000
Special survey, investigation	3.0	6	18	15
Field conference, visit	1.0	2,330	2,330	2,000
Not at home, visit	0.3	476	143	120
Unclassified	1.0	59	59	100
Collection of sample—milk, water, other	0.2	10,198	2,040	1,532
Plan review—subdivision, pool, sewerage, water	2.0	258	516	375
Total	—	—	12,675	11,161

Recapitulation

Estimated weighted units of work to be performed (at beginning of year)	11,161
Estimated budget expenditures—total costs (variable costs = \$182,272.00, fixed costs = \$23,726.00)	\$205,998
Estimated total cost per weighted work unit	\$18.46
Total actual weighted units of work performed	12,675
Total man-days at work, corrected for vacation, sickness, training, etc.	2,954
Average time in field	65%
Average time in office	35%
Average weighted units of work per man-day	4.3
Actual budget expenditures—total costs (variable cost = \$171,928.00, fixed costs = \$23,726.00)	\$195,654
Actual total cost per weighted work unit	\$15.44
Cost of each activity = weighted unit × \$15.44. For example, the milk control activity cost is farm and plant inspections plus sampling = [885 + 468 + 480 (wt. samples) × \$15.44	= \$28,300

Note: Fixed costs are those costs that are directly related to the initial operation, i.e., rent, light, heat, commissioner, program directors and secretaries, cleaning, etc. Variable costs are primarily for staff and their maintenance. Additions should be made for retirement benefits, hospitalization, workman's compensation, and social security.

results. In this connection education, persuasion, and motivation are fundamental to accomplish the bulk of the objectives. Legal action is resorted to when all other means have failed in the time period considered reasonable. It is sometimes said that to resort to legal action is to admit failure. Perhaps it would be fairer to say that a program based *solely* on legal action is doomed to failure.

Hollis stated that

> the most primitive approach by the sanitarian was to carry a big stick. A more sophisticated approach was to speak softly and carry the big stick in a velvet glove. With the advent of epidemiology, it proved effective to speak cogently and carry a slide rule (pocket calculator). . . . The pressure to comply with approved sanitation practice now rises less from a fear of epidemics or of legal sanctions and more from a desire for good living and common realization of mutual interest.[22]

Stated another way, the primitive approach to sanitary regulation is police enforcement; the civilized approach is environmental health education; the sophisticated approach to law enforcement is self-inspection under regulatory surveillance and control, with effectiveness evaluation by industry management and the regulatory agency. In any case, compliance with regulations will be more effective when the regulated know that prompt legal action will follow when necessary. One might ask, "Is the enforcement agency considered a friend or foe?" "How can the most lasting public health protection be obtained?" It is apparent that a delicate balance must be maintained.

Shattuck, in his classic report, stated that

> local Boards of Health endeavor to carry into effect all their orders and regulations in a conciliatory manner; and they resort to compulsory process only when the public good requires.[23]

Sir Edwin Chadwick, a pioneer sanitarian (1854), was removed from office because he tried to force his ideas of cleanliness on the people of London. These are not isolated examples.

Public officials, including the health officer, environmental control officer, and supervisor, have a duty and obligation to enforce the law. However, the action taken, or not taken, should consider possible consequences such as malfeasance, misfeasance, or nonfeasance. Grad[24] defines *malfeasance* as "doing of an act that is wrongful and that is known to be wholly unauthorized by the official." *Misfeasance* "is the doing of an authorized act in an unauthorized manner," as closing down an establishment without first giving a required statutory notice. *Nonfeasance* "is the failure to perform an official duty without sufficient excuse," as when an establishment, known to have imminent health hazards, is not closed. On the other hand, the official may be subject to civil or

[22]Mark Hollis, "Public, Professional, Industrial Allies in Sanitation," *Pub. Health Rep.*, August 1953, pp. 805–810.

[23]Lemuel Shattuck, *Report of the Sanitary Commission of Massachusetts, 1850*, Harvard University Press, Cambridge, Mass., 1948.

[24]Frank P. Grad, *Public Health Law Manual*, 5th printing, 1976, American Public Health Association, Washington, D.C., p. 177.

criminal liability for actions taken, but the law also affords far-reaching protection for officials in carrying out responsibilities.

Enforcement techniques are described below and under Enforcement Program in Chapter 11.

Performance Objectives and Specification Standards

There has been considerable discussion among administrators regarding the advantages and disadvantages of performance and specification standards to carry out environmental control responsibilities. Some of this discussion has been useful in clarifying the objectives of a program and how inspection and enforcement is to be carried out.

From a legal standpoint, the specification-type of standard is easier to enforce than the performance standard. However, the objective of a program is not strictly legal enforcement, but is directed toward obtaining compliance (performance) with the *intent* of the law by elevating the level of operation through attitudinal changes, education, and persuasion. The interpretation of a standard should in most instances vary with the type of facility, equipment available, people employed, geographic location, construction, and many other variables. In most cases it is difficult to establish a definite standard that can be applied uniformly without causing some unnecessary expense and undue hardship. It must also be kept in mind that when a standard is incorporated into law, administrative discretion is lost if provision is not made for the granting of a waiver, particularly when the deficiency is not an imminent danger to the public. On the other hand, where it is possible to establish an enforceable standard that has broad general application, this should be done.

For example, a standard for dishwashing could state that the dishes and utensils shall be so cleansed and bactericidally treated as to have a total bacteria count of not more than 100 per utensil surface examined, as determined by a test made in a laboratory approved for the purpose by the state commissioner of health. This would actually be a performance objective. On the other hand if an attempt were to be made to explain how dishes shall be washed, rinsed, and bactericidally treated, using a two-compartment sink, a three-compartment sink, a single-tank door-type dishwashing machine, a single-tank conveyor-type dishwashing machine, a two-tank conveyor-type dishwashing machine, for example, very detailed specifications would have to be written and continually updated. The same analogy can be used in determining whether a water supply is satisfactory. We can use the statement that a water supply shall be of safe sanitary quality, and of adequate quantity and pressure. What is adequate and satisfactory requires detailed explanations, depending on the type of establishment, source of water, type of treatment, and many many other variables, all of which require professional judgment. A very common method of handling this problem is to include in the law the performance objectives that the person responsible is expected to achieve and then make reference to certain publications that have official or semi-official status, such as *Standard Methods*, or *Recommended*

*Standards for Water Works.** Stream pollution control standards may require that wastewater treatment plants meet certain effluent standards and freshwater or saltwater classifications.

It is extremely important to avoid the establishment of rigid specification standards unless some flexibility can be built in. A rigid specification is sometimes inadequate, and under certain circumstances, quite ridiculous. An example might be a requirement of 35 gal or even 100 gal of water per person per day at a camp or institution without regard to the type of operation, number and type of plumbing fixtures, peak demands, storage, and capacity of the water source.

For many years efforts have been made to get away from the old-fashioned "inspector" who simply completes an inspection checklist, finds what is wrong, and issues a summons without making any attempt to interpret the regulations and their intent in light of the operation. An investigation followed by explanation of the findings with the responsible person with a view toward getting at the source of difficulty so that the cause can be eliminated and the establishment operated in compliance with the objectives of the law would appear to be a more effective way to operate on a routine basis and thus hopefully obtain lasting benefits.

Specification-type standards are extremely valuable when they have universal application. In such cases they should be written into law. However, when a specification standard cannot be established because of the nature of the operation or lack of a reliable measurement tool, it is better to rely on the performance objective. It would then be up to the individual responsible to determine how to comply with the objective established. This individual might be referred to certain publications or educational materials for guidance or perhaps in some instances to a consultant. This does not prevent the enforcing agency from recommending or even requiring, when this is indicated, that certain definite procedures be followed when dealing with some recalcitrant and willful violators, when the public health or safety is endangered.

Correspondence

Letter and report writing can be one of the most effective means of obtaining improvements or correction of deficiencies. Continuity of effort can be maintained; differences of opinion or misunderstandings can be reduced; and other interested people can be kept abreast by copies of letters and thus lend support to sound objectives. It is also a means of acquainting local officials with the diversity of problems they are actually being relieved of, with resultant greater appreciation of the services being rendered.

In many cases the only contact people have with the control agency is through a letter. The impression left may therefore be good or bad, depending on the tone of the letter, its appearance, the choice of words, completeness, and other factors. The letter can also indirectly serve as an educational piece and as a base for

*See Chapter 3.

obtaining cooperation. If improperly worded or if inaccurate, the letter can be a source of embarrassment.

Form paragraphs for different activities have been found to be of great assistance in letter writing to confirm the results of large numbers of inspections. If letters are individually written to fit specific situations, and the form paragraphs are varied to apply accurately, a good letter can result. It is then possible to have the field inspector prepare a draft of the letter he or she feels should be sent based on the conditions actually observed. The letter would then reflect department policy and leave to supervisory personnel the task of reviewing the proposed letter and then signing the letter after it is typed. Form paragraphs should be periodically reviewed and improved. Suggested form paragraphs and letters were given in Chapter 11 (as applied to housing).

There will of course be instances when a letter is unnecessary. Many departments follow the practice of having the owner or operator sign the inspection report checklist to acknowledge receipt of a copy of the report. This also serves to officially advise the responsible person of any deficiencies and the need for making prompt corrections. A confirmatory letter could follow only when serious deficiencies were in evidence.

Compliance Guides

Satisfactory compliance guides explaining in some detail and with examples what is "adequate," "satisfactory," "proper," "good," "safe," and so on, can be very helpful to field personnel. Many parts of this text can serve this need. The sections on milk, food, camps, recreation, and others can be used as compliance guides. The PHS/FDA *Grade A Pasteurized Milk Ordinance* and *Food Service Sanitation Manual* (see Bibliography, chapter 8) incorporate the public health reasons for the regulations and what would constitute satisfactory compliance. The American Public Health Association *Standards for Health Services in Correctional Institutions* is a similar document (see Bibliography, chapter 11).

It is possible under such circumstances to obtain some of the best thinking and reasonably uniform interpretation of regulations. This is essential to prevent chaotic enforcement. On the other hand it is also essential that such compliance guides be reviewed periodically and revised as the need exists.

Compliance guides can be developed with the assistance of personnel responsible for a new activity. By so doing, a better understanding of the governing regulations is obtained by the personnel actually involved in the enforcement. It then also becomes an opportunity for in-service training applied to a specific problem. Such experience is usually much more effective and of lasting benefit. Representatives of the industry involved should also be requested to participate to help obtain acceptance and application of the principles involved.

In-Service Training

Every division having environmental control responsibilities should maintain a continuing program of in-service staff training. In a large division it could be the full-time duty of one person who is qualified to carry out this very important function. In a small division it would have to be an additional duty of the director or a member of the director's staff. Regardless of division size or work load, to neglect in-service training is to grind to a standstill. Diligence on the part of administrative personnel cannot produce effective results unless the entire staff, and particularly field personnel, are kept abreast of new developments and new or changing policies. Opportunities for discussion, attendance at short courses, conferences, conventions, and participation in committee activities, should be encouraged for their stimulating effects and for improved individual performance.

In-service or on-the-job training of new personnel is a very necessary extension of a student's formal education. Unless provided as an integral planned part of a college curriculum, the young graduate is at a considerable disadvantage when on the job. It will probably take 2 or 3 years before this worker can function effectively with little direct supervision. Opportunities for field application and learning should be provided to the student by the college or university during summer vacations and long holidays, by term papers, and by project-oriented theses. Cooperative college-industry programs have also shown the advantages of this type of training. State and full-time local environmental control agencies should assist in this approach by providing opportunities for on-the-job training of engineers, scientists, biologists, veterinarians, sanitarians, technicians, and sanitary inspectors, even though this means extra work for those in charge. This could also include assignment of an employee to work in a restaurant, in a water treatment plant, in a wastewater treatment plant, in industry, at a swimming pool, with a pest control operator, with a consulting engineer, at a solid waste facility, with a well driller, or with a septic-tank installer for, say, a 3 to 4 week period to obtain first-hand experience and understanding of the particular operation. Colleges and universities should, in fairness to their students, seek out opportunities to give them the field experience they need, as in coop programs.

Not to be forgotten is the value of technical or occupational group training of those persons directly responsible for carrying out good sanitary practices. Supervisors of food handlers; resort managers; operators of swimming pools, water and sewage treatment plants, and milk plants should also be given special in-service training. Attorneys, realtors, architects, consulting engineers, contractors, and others can also benefit from and be helpful in implementing special programs through the conducting of workshops or special conferences. The cooperation of a local educational institution can be of great assistance. See also Inspection Supervision, this chapter.

Education

The educator and public information specialist are important members of a control agency, voluntary agency, and other related organization. Their services as consultants can be extremely valuable in making more effective a department or division program for training, public participation, community organization, interpretation to the community and legislators, release of educational material, and special reports. Other services can be rendered depending on the competency of the control agency staff and organization. In the final analysis, the division director and staff will have to provide most of the basic technical information for the guidance and assistance of the educator and information specialist. This is as it should be, for the professional educator or information specialist is not expected to be an expert in restaurant sanitation, septic-tank systems, stream pollution abatement, or bonding, for example; but educational techniques can be applied to make the specialized knowledge of the division staff more effective in obtaining public understanding and approval of a new regulation, health practice, or environmental facility improvement.

Education should start with planned intensive in-service training of all field and office personnel. The extent of the training would depend on the competency of the personnel; but it should nevertheless be a continuing activity.

Field engineers, sanitarians, or inspectors are, in addition to their primary duties, educators. Every time they talk to a person they are acting in the role of an educator. They explain what is wrong or what can be improved and discuss how corrections and improvements can be made. The occasion may frequently arise when other department programs and services can be mentioned and explained. Every inspection or field conference is an opportunity to point out the services rendered and enhance the prestige of the department, thereby making it easier to put across the total program for the benefit of the people.

The preparation of suitable literature, including leaflets, pamphlets, and media releases adapted to the local conditions, can do much to obtain acceptance of good environmental sanitation practices. It also helps to avoid confusion in the interpretation of recommendations. For example, a leaflet giving precautions to be taken before buying a home in a rural area can emphasize the superiority of public sewers and water supply or of having to carefully look into the well-water supply quality and quantity and the construction of an approved septic-tank absorption system. A bulletin on rural water supply can illustrate and explain good well and spring location, construction, and protection. A leaflet on food poisoning gives opportunity to bring out the necessity of refrigeration and hygienic food-handling practices. A release on septic tanks can point out the importance of inspecting and cleaning them. Typical construction details of septic tanks, distribution boxes, absorption fields, and leaching pits will help the contractor installing the system to do the job in accordance with the best practices so that approval can be readily given. The many illustrations in this book are adaptable for educational use. Most state and local agencies have pamphlets and leaflets available to meet the needs that exist.

Another important aspect of education and public relations involves the neutralizing of unfounded or emotional opposition and inaccurate information. In this way, for example, support can be developed for a bond issue to construct sewers and a treatment plant or to obtain acceptance of a sanitary landfill site or incinerator. Public information activity should start in the planning phase and be continuous, culminating at a critical time, such as before voting on approval of a bond issue.

Other education opportunities are presented to the educator when asked to give talks to students, visitors, and civic and professional groups and when participating in meetings and conferences with representatives of other organizations. Attendance at short courses and scientific conferences should be encouraged, not only for the new information gained, but also for the stimulating effect on the individual and enthusiasm transmitted to others. Every situation should be used to point out the services available and how individuals can help themselves, and how, as members of the community, they can help improve the environment in which they are living.

Persuasion and Motivation

Although many people are ready to grasp new ideas that will improve their standard of living, there are perhaps just as many who resist change. This requires the application of a little more effort to convince them and get them to do what is indicated. Here the personal contact can do much. The benefits to be derived, the reason for a particular regulation, and the economic, aesthetic, and disease dangers involved can be explained in some detail. Examples of how others have solved the problem and a visit to demonstrate the solution can be powerful motivating influences.

A written report to the responsible person formalizing a violation is always a good persuasive step. Compliance is encouraged if the letter is accompanied with sound recommendations and, if possible, illustrations. It is usually helpful if copies of letters are sent to interested persons such as the local health officer, town official, official agencies, the mortgage holder, and so forth as appropriate. Letters should be clear and direct but polite and persuasive.

In many cases a permit for operation, that incidentally promotes compliance and inspection supervision, is required. If an annual permit is needed its issuance can and should be withheld, with notification to the applicant, including reasons, until all legal requirements are complied with. If a continuing permit is issued it can be suspended and revoked after a hearing. It is well known that a regulation against operation without a permit is readily enforceable. Most operators will want to avoid the threat of prosecution. As a rule, operators of public places are reasonable and will cooperate when confronted with letters and accumulated evidence by the enforcing agency. Sometimes the cost of legal services or effect on public relations is the determinant.

Another technique, when a formal permit or license is not required by law, is an informal certification system. In this process the health officer or other official certifies that a place meets certain established standards or perhaps higher than minimum standards. The certification is withdrawn when the standards are not met. The periodic publication of "certified places" can provide a competitive incentive and voluntarily raise industry and commerical facility and operation standards. This is made particularly effective when many of the operators subscribe to the system and when government agencies, corporations, and individuals require the certification as a prerequisite to doing business. The

Voluntary Interstate Milk Shippers Program and certain restaurant, camp, hotel, and motel association ratings are examples of the system.

Reinspections and additional field conferences are sometimes indicated to educate and persuade the offending party. The approach, attitude, patience, and ability of the inspector is frequently the deciding factor. In some cases a new inspector might be able to accomplish the desired objectives. A joint inspection with a supervisor might also have the desired effect. Certainly these approaches should be given a trial before deciding on other measures.

It cannot be denied that there will be some situations where solution of the problem appears impossible. A more forceful approach may be indicated in such cases. A registered letter signed by another person in higher authority sometimes brings the desired result. If this fails, an office conference may be called to give the violator an opportunity to explain the problems. Perhaps a step-by-step program of improvements can be agreed upon, with immediate attention being given to the dangerous conditions. Another approach would be the calling of an administrative or formal hearing, and finally legal action.

Legal Action

Legal proceedings to enforce a law, depending on the statutory authority, may include issuance of an order to eliminate a violation; suspension or revocation of a permit or license, which should discourage continued operation in violation of the law; civil sanctions such as a hearing and fine or lien on a property, or possibly a jail sentence for failure to pay the fine; penal sanctions involving misdemeanors or offenses, as designated in many state and local laws and codes, leading to a jail sentence or more often to a fine, since the violator really had no criminal intent; summary abatement, such as when a nuisance presents a serious and immediate hazard to the public health, which may require obtaining legal access to a property and removal of the nuisance contrary to the owner's wishes; and injunctions in which a court issues an order, on the basis of sufficient information, directing that a certain action be taken or that a certain action not be taken, which action is necessary to protect public health, such as an order prohibiting the opening of a childrens' camp because the only water supply available is well known to be polluted with sewage.

An additional sanction sometimes used is the *embargo:* an action authorized by law to restrict or prevent the movement of goods in commerce, or a prohibition of trade in a particular commodity, for the protection of the public health, safety, and welfare. A regulatory agency may, upon written notice to the owner or person in charge, specifying with particularity the reasons thereof, place a "hold order" on any food which it believes is in violation, such as food which is spoiled, filthy, or otherwise contaminated rendering it unsafe for human consumption. Such food is tagged or labelled and set aside and stored until either brought into compliance or ordered destroyed. A hearing may be held, or the food destroyed if storage is not possible and the public health is at risk.

In general, the advice and assistance of the legal department should be enlisted before considering the appropriate remedies and legal action; but the decision for the proper course of action should rest with the director having the program responsibility, and understanding of the probable effects of the action to be taken. If any general compliance sluggishness or resistance is detected, an educational program would be indicated.

It should not be expected that the corporation counsel, county attorney, or other public prosecutor would have the same understanding of the problems and their effect on the public as the department head, program director, professional engineer, or sanitarian. If this is recognized, it would be well to keep the legal department advised of any new legal and policy decisions in the field and to seek informal assistance and interpretations. Joint field visits provide an excellent opportunity to get out of the office and to get acquainted firsthand with the broad environmental control program—its objectives and problems. Both the attorney and the inspector should of course carefully review all pertinent information prior to going into court. The inspector should be prepared and familiar with his function as a witness. See Preparation for Legal Action (Legal Conduct of Sanitarians and Engineers), this chapter.

If a division is well organized, with sound administrative procedures in effect, the file on a particular establishment should be replete with factual information. The educational and persuasive steps previously taken, including inspection reports and letters, will show the reasonableness of the division's actions. If this can be supplemented with office conferences or informal hearings that are confirmed in writing, the accumulated evidence alone is usually sufficient to convince the violator or the violator's attorney to come to a prompt agreement. This information is also invaluable to the department counsel and makes possible a strong, clear-cut case that is relatively easy to prosecute. In all instances, cases should be carefully prepared; all information must be factual, supportable, and documented; and prehearing conferences held between the department attorney and the inspector.

The administrator of a division or department must use a fine sense of judgment, statesmanship, and diplomacy in bringing up cases for legal action. Such actions should be kept at a minimum if one is not to be overwhelmed with the preparation of cases and appearances for legal action to the detriment of the total program.

An approach found effective, when issuance of an order or suspension or revocation of a permit or license have failed, is the holding of a formal hearing before the board of health, commissioner, or preferably before an authorized hearing officer or administrative tribunal. This would be a civil action leading possibly to a fine and would thus overcome the objections to a criminal proceeding, when a violation is called a misdemeanor. The department attorney should be present at that time to guide and present the department's case. The outcome of this and previous persuasive attempts would determine the need for other court action. Once a case is turned over to the legal department the administrator should refer all further communications and requests for conferences, meetings, compromises, and so on, to the department attorney. The

DEPARTMENT OF HEALTH

Enforcement Warning Letter

Date

Mr. John Smith
600 Broadway
Troler, New York

RE: John Smith d/b/a
The Villa Restaurant

Dear Mr. Smith:

Recent inspections of your (establishment), copies of which are enclosed (or copies of which were previously left with or sent to you), have shown continued and serious violations of Part ________________ of the ________________ State Sanitary Code.

In view of the above, you are advised that such violations subject you to:

1. A maximum penalty of ($250–$1000) per day for each violation for each day it continues after the third day.

2. Suspension and/or revocation of any permit previously issued.

3. Injunction against continued operation.

4. Criminal proceedings under Public Health Law Sec. __________ which can result in imprisonment for not more than 15 days or fines of ($250) per violation or both.

In order to avoid formal legal action, you are invited to attend an informal office conference to be held in this office on ____________________ at which time we will discuss your continued violation of the State Sanitary Code and your liability as set forth above.

Very truly yours,

Figure 12-6 Enforcement warning letter. (Adapted from New York State Dept. of Health and Sanitation Manual.)

administrator should not, however, forget about the action. Close liaison should be maintained with the attorney assigned until the case is settled.

The procedures mentioned above are suggested for use in connection with routine legal enforcement. The administrator may find from experience that some steps can be omitted. Most departments have or can obtain hearing, affidavit of service, permit withheld, permit denied, order of board of health, and similar forms or letters to expedite compliance with sanitary code and law requirements. Figures 12-6 to 12-9 show some of the forms.

A system of issuing a departmental summons in accordance with a procedure previously agreed upon by the court may be possible in some areas. This privilege, however, should not be abused as it will only result in a greater number of adjournments and suspended sentences and may discourage educational and persuasive efforts by staff. The net result will be loss of prestige, a great deal of time lost in court, poor legal representation, and only a passing success at best.

Another standard approach in rural and suburban areas is the filing of information before the local justice of the peace. In some cases this works well, *particularly when the aggrieved local people can be present.* However, punishment of a violation as a *misdemeanor* is sometimes not effective for, as previously pointed out, there is a reluctance on the part of the courts to brand one found guilty of a misdemeanor as a criminal. Civil action, including an administrative hearing and a fine, would be preferable in such situations. Other types of penalties and sanctions should be explored with the agency counsel.

Injunction proceedings to close or cease an operation, or a mandatory injunction to require a certain action or improvement, are extremely effective techniques if permitted under existing law and allowed by the court.

The press and the courts can be valuable allies to a department that is sincerely trying to do a good job. It takes time to develop this prestige and acceptance. The department must demonstrate beyond any doubt that it is carrying on an intelligent, planned enforcement program, and not a day-to-day panicky program based solely on court actions.

There will always be situations where the slow educational and persuasive approach is not practical. Such cases may arise when a truly dangerous condition exists and when cooperation is nonexistent. When confronted with such situations, the administrator must act summarily, in concert with the legal advisor, to eliminate with dispatch the dangerous condition, as provided for in the existing laws and administrative procedures.

Preparation for Legal Action[25]

1. Keep a record of every visit or inspection of the premises involved.
   a. Date, time, place, location.
   b. With whom did the witness talk at the premises; name, title.

[25]"Guide and Checklist for Public Health Personnel Witnesses," prepared by Emanuel Bund as training notes to guide New York State Dept. of Health engineers and sanitarians and related personnel, Albany, N.Y., 1966.

DEPARTMENT OF HEALTH

NOTICE OF SUSPENSION OF PERMIT
TO CONDUCT A.......................................
AND OF TIME OF HEARING THEREON.

To..
....

NOTICE IS HEREBY GIVEN that the permit heretofore issued to you by the Health Department of the ....to operate a................................ .... at .... Street, in the.................................., is hereby suspended in accordance with the provisions of Section .... of the .... Sanitary Code.

Such suspension is made upon the grounds and for the reason that said place is operated and conducted in an extremely unsanitary manner and condition, thereby endangering the health of the citizens of the ....

YOU ARE FURTHER NOTIFIED that on the .... day of ...., 19........, at the hour of ....M., there will be conducted before the Board of Health in the board room of said Board, on the .... floor at .... Street in the city of ...., a hearing upon the question of revocation or further suspension of the above named permit on the grounds and for the reason above set out, at which time and place you may appear and show cause, if any you have, why such proposed action should not be taken by said Board.

YOU ARE FURTHER NOTIFIED that you may not operate said ....hereafter until the time of said hearing unless, because of the abatement of the conditions causing the suspension of your permit, the foregoing order of suspension is cancelled and proceeding before the Board is dismissed.

Commissioner of Health

.... M.D.

Effective ....M., ...19....

By..

Served this .... day of
...................................., 19....

by leaving with...
Title ....

Figure 12-7 Notice of suspension of permit. (Adapted from City of Los Angeles Health Dept. Form SH-55.)

c. Measurement, heights, weights, depths, temperatures.
d. Who owns the premises: private, partnership, corporation; exact names.
e. Who was with the witness at the time of the inspection; name, title, department.

2. Take samples if the circumstances of the matter warrant it.
   a. Samples must be taken in a proper container; be properly identified, by labelling if at all possible.
   b. Indicate date, place taken from and initial.
   c. Keep a record of how a sample was handled and the temperature control: the disposition of the sample; to whom it was delivered.

DEPARTMENT OF HEALTH

Notice to Discontinue Violations

To Date

Pursuant to the Public Health Law, the State Sanitary Code, or the Administrative Rules and Regulations adopted by the ________________ State Department of Health,

YOU ARE HEREBY ORDERED:

To discontinue the violations marked in the attached Inspection Report within __________ days.

Failure to comply with this Order may result in immediate suspension of any permit previously issued.

An opportunity for reconsideration of this Order or the inspector's findings will be provided upon written request filed by you within the time established for correction of violations.

Please Take Further Notice that violations of the Public Health Law, the State Sanitary Code, or the Administrative Rules and Regulations of the State Department of Health may subject you to a civil penalty or fine for each violation in addition to any other remedies or civil or criminal penalties allowed by the Public Health Law or local or other State laws.

Received by ______________________ ____________________

(Inspector)

Figure 12-8 Notice to discontinue violation. (Adapted from New York State Dept. of Health Engineering and Sanitation Manual.)

d. A receipt may be given for the sample if requested.
e. Do not take anything of substantial value as a sample, unless accompanied by an associate, or by payment.
f. Samples are taken either for evidentiary purposes or for laboratory tests, the reports of which may be used as evidence.
g. Human or animal wastes or other offensive materials need not be taken for evidentiary purposes—but may be taken for laboratory tests if the circumstances require.

AFFIDAVIT OF SERVICE OF NOTICE OR CITATION

State of.. }
County of..................................... } ss:
.... of .... }

...., being duly sworn, says that he is over twenty-one years of age; that he served the annexed notice (or order) upon.......................... ....and that he knew the person so served to be the same person mentioned and described in said notice (or order).

....

Sworn to before me this
.... day of .... 19........
....
....

Figure 12-9 Affidavit of service of notice or citation.

3. Chemical and laboratory tests should be made in every instance where pollution, contamination or deterioration is or may be in issue. The reports must be available at the hearing. The reports should properly identify the sample.
4. Prepare a map or diagram for use at the hearing in every case involving land, waters, buildings, and physical premises. A diagram need not be to scale if the person who made it is present and can identify it and testify that it is a fair and true representation of the area depicted. Maps are particularly valuable. They are inexpensive and should be marked to indicate the areas being discussed.
5. Photographs are invaluable and should be available whenever possible.
6. Each witness should read the entire file before the hearing and be familiar with every part of it, including the reports.
7. The entire file must be in the hearing room for availability.
8. Once a hearing notice has been issued by Central Office, the matter should not be discussed with persons not affiliated with the agency. All inquiries should be referred to Central Office or the Office of Counsel or to both.
9. A final inspection of the premises should be made as near to the date of the hearing as possible and additional final samples and laboratory tests made.
10. All personnel, including personnel of local or other state agencies having any familiarity with the facts and circumstances or with the premises, should be alerted to be available to serve as witnesses—even if they may not be called by Counsel. They, too, should have their files available.
11. Other state or local agencies should be asked if they have any relevant information—that is,

relevant to the issues. For instance, it is quite likely that the conservation, water resources, and environmental protection authorities may have facts and laboratory results on bodies of water which could be useful in a health department proceeding.

12. Central Office and Counsel should be apprised of the names, titles, and departments of persons who may be called as witnesses—in advance of the hearing, i.e., game wardens, state troopers, technicians, inspectors, investigators, engineers, sanitarians.
13. Consideration should be given to the calling of expert witnesses—persons who do not necessarily have personal knowledge of the facts and circumstances but whose training, experience, and background can be used to support a particular position or contention.
14. Scientists, Engineers, Sanitarians, and if the circumstances warrant it, District Office Directors, and Regional Directors (if not serving as hearing officers) should be available to assist Council with technical questions and propositions and to serve as "last minute" expert witnesses if the need arises during the hearing.
15. The witness (engineer or sanitarian) should be prepared to testify as to the nature and character of the surrounding area, i.e., changing neighborhoods, best usage—resort, residential, industrial, commercial, fishing, etc.
16. Once a hearing notice has been served, or other legal action has been initiated, do not fraternize with respondents or other interested nondepartmental persons. It is understandable that in many instances friendly relations exist and perhaps should be maintained, but department personnel who may be called as witnesses can very well politely and inoffensively indicate "the matter is out of may hands," or "I do not know the answer."

Legal Conduct of Sanitarians, Inspectors, Technicians, and Engineers[26]

1. Sanitarians, inspectors, technicians, and engineers must be aware at all times that they are public servants, that they represent the commissioner and the department; that their misconduct will result in loss of prestige for the department.

2. As witnesses for the people in court, and in all of their official activities, they should show a fair, impartial, courteous, and objective attitude.

3. To indicate the efficiency of their department in court, they must be thoroughly prepared to testify. Such preparation depends, basically, on the thoroughness of the inspection and the records made of such inspection. Every inspection should be made with the thought that it may form the basis for a court trial (memorandum to refresh recollection).

4. They must be punctual in court and present in courtroom when case is called. Papers should be reviewed in morning before court session begins to make certain that they are in proper form and signed and to refresh recollection about case.

5. Avoid talking to defendant or defendant's attorney in court unless instructed to do so by the judge. If they have any questions refer them to the corporation counsel or prosecutor for the people. Do not discuss the case outside the office until it is terminated.

6. Above all, avoid showing any personal interest or vindictiveness or officious attitude. The court is inclined to take your word for any situation you describe in an expert, official manner. But as soon as you show that you have a personal feeling, the court feels that you have a grudge against the defendant.

7. You have the burden of proving your case beyond a reasonable doubt.

8. It is a common court tactic, as in sports, to get your opponents riled so that they will lose their heads. The defendant's attorney may try that on you.

9. Do not pretend to be an expert on everything. Use ordinary, simple language—remember that the judge is not a sanitarian. If you do not understand a question, ask that it be explained. Answer questions precisely and simply. Do not give opinions or conclusions, unless asked; just state the facts based on personal observations. Do not volunteer answers or information beyond the exact scope of the question.

10. In preparing a complaint, carefully read the section or regulation applicable and as succinctly

[26]Adapted from New York State Health Dept. Field Training Center Notes, prepared by Irving Witlin.

as possible indicate how that section was violated. Remember that every word in a law has been carefully thought of to give a certain common and usual meaning. Do not try to give it your own personal interpretation.

11. If you do not recall the facts and wish to refer to personal notes or the file, ask to do so; but the attorney on either side may ask to see the notes or file.

EMERGENCY SANITATION

Emergencies and natural disasters may be associated with war, displaced persons, epidemics, floods, hurricanes, earthquakes, volcanic eruptions, fires, major accidents, extended interruption of utilities and services, or other crises. On such occasions large masses of people may be at the mercy of nature. Their survival in an uncontrolled environment is dependent on the extent to which individuals can help themselves and upon how soon relief can be mustered and a favorable or controlled environment restored. The speed with which action can be taken is largely dependent on the prior plans made to cope with emergencies. This must include the delegation of specific authority and responsibilities, assignment of selected manpower resources, stockpiling, and inventory of supplies and equipment. In the interim, the people affected are under tremendous psychological and physical stress.

The Federal Emergency Management Agency and local governmental organizations (including the Army, Navy, and Air Force), and the Red Cross are usually involved in dealing with major disasters. International agencies that have provided assistance are the League of Red Cross Societies and the United Nations, including the World Health Organization (WHO), United Nations Children's Fund (UNICEF), Food and Agriculture Organization of the United Nations (FAO), and the World Food Programme, International Rescue Committee, and others.

A comprehensive publication has been prepared on emergency sanitation by the WHO. It is recommended as a reference. Numerous other technical manuals, bulletins, and guides are available through the national or local civil defense and emergency planning organizations and health departments.[27]

Emergency Sanitation Bulletin

An emergency sanitation bulletin should include the following.[28]

[27]M. Assar, *Guide to Sanitation in Natural Disasters*, WHO, 1971. Also *Red Cross Disaster Relief Handbook*, League of Red Cross Societies, Geneva, 1970; *An Outline Guide Covering Sanitation Aspects of Mass Evacuation*, PHS Pub. 498, DHEW, Washington, D.C., 1956; *Military Sanitation*, FM 21-10, AFM 160-46, Depts. Army and Air Force, Washington, D.C., May 1957; *Emergency Water Services and Environmental Sanitation*, Dept. of National Health and Welfare, Ottawa, 1965. S Rajagopalan and M. A. Shiffman, *Guide to Simple Sanitary Measures for the Control of Enteric Diseases*, WHO, Geneva, 1974.

[28]*Emergency Sanitation*, New York State Dept. of Health, Albany, N.Y., 1978.

Drinking Water—Use Only Safe Water

Assume all sources are unsafe for use until approved by your local health department. Meanwhile, use health department-approved bottled water, or water distributed by health department-approved tank truck.

Or, disinfect water using a household liquid bleach disinfectant containing at least 5 percent available chlorine by weight. To one gal of clear water, mix 6 to 8 drops of bleach (one teaspoonful to 10 gal); let the solution stand for 30 min before drinking. If the water is cloudy and contains particles, allow the particles to settle. Pour off the clear water into a separate container and add to it double the amount of bleach mentioned above; or heat water at a rapid boil for 2 min (10 minutes in areas where amebiasis or giardiasis are endemic); or disinfect water using purification tablets. Carbonated beverages are satisfactory.

Follow water utility and health department directions to conserve water.

To Disinfect Your Well

First, pump well until water is clear; then:

1. Mix one qt of liquid bleach in 5 gal water. (For driven wells, use 6 tablespoons to 5 gal.)
2. Pour solution into well through top of casing or use vent pipe where available.
3. Pump until chlorine odor appears at all taps, then close taps.
4. If possible, reopen one tap and recirculate water back into well by means of a garden hose for at least 1 hour; then close tap.
5. Repeat steps 1 through 3 to assure that the chlorinated water has circulated through the entire system. Let well stand idle for at least 8 hours, except that water may be used for occasional toilet flushing.
6. Pump heavily chlorinated water to waste, away from grass and shrubbery.

Contact your health department regarding analysis of your well water.

Milk and Food

Use only pasteurized milk. When the safety of milk is in doubt, or when only raw milk is available, bring the milk to a boil and cool in a clean container. If boiling is not possible, canned or powdered milk mixed with safe water can be used.

Cook all foods thoroughly. Frozen foods that have been thawed should be discarded if not consumed immediately or kept refrigerated at 45° F or lower.

Avoid creamed dishes, hash, puddings, meat, poultry, and pre-prepared salads unless refrigeration (45° F or lower) can be assured.

Utensils can be used if clean and not exposed to flood waters.

Raw foods exposed to flood waters should be avoided unless cooked because of possible contamination. If raw foods must be used, clean thoroughly and soak in chlorinated water ($\frac{1}{4}$ oz or 2 teaspoonsful of 5 percent liquid bleach solution per gal of water) at least 5 min.

Destroy the contents of crown-capped bottles and foods in glass jars.

Destroy canned foods in tins where swelling, leaking, rusting, or serious denting is visible.

Wash off and immerse all hermetically sealed cans in good condition in a bleach disinfecting solution ($\frac{1}{4}$ oz or 2 teaspoonsful of 5 percent liquid bleach solution per gal of water) for at least 5 min prior to use.

Clean and then disinfect all utensils and equipment prior to use in the same manner as above.

Silverware should be immersed in boiling water for at least 2 min. (CAUTION: Disinfection with a bleach solution will cause tarnishing).

Use single-service paper or plastic dishes, cups, and utensils where possible.

Sewage and Excreta Disposal

When flush toilets cannot be used, use a small covered watertight bucket with a removable plastic liner to receive body wastes. Use a small amount of chlorine bleach, chlorinated lime, or any commercially available preparation which disinfects and deodorizes; follow manufacturer's instructions. Remove the plastic liner as necessary; tie and place it into a 10-gal or larger watertight, rodent- and insect-proof garbage can with a tight cover for temporary storage. The wastes may be disposed with the garbage when normal collection is restored, or by burial in a deep hole with at least 18 in of dirt cover located at a distance of not less than 100 ft from any well or surface waters, or, when and where feasible, by empyting the contents into the flush toilet and carefully disposing of the plastic bag with the garbage.

Use the temporary public toilets established in the disaster area; they may be located over sewer manholes.

Insect and Rodent Control

To prevent breeding, store and dispose of garbage, sewage, and dead animals according to instructions. (See Refuse Collection, Storage, and Disposal.)

Drain standing water where possible, including containers of stagnant water.

Treat standing water with kerosene, fuel oil, or a commercially available larvicide during mosquito breeding season; apply 3 oz (6 tablespoons) per 100 ft^2 of water surface or in accordance with the manufacturer's instructions.

To kill rats, when trapping proves unsatisfactory, use commercially available "quick kill" poisons—such as zinc phosphide or red squill. Follow directions on the label. USE WITH CARE! Keep these poisons out of the reach of children and away from animals. Store in a locked cabinet.

Refuse Collection, Storage, and Disposal

Drain and wrap garbage in several layers of newspaper to absorb moisture.

Store garbage in watertight, rodent- and insect-proof containers having a tight-fitting cover. Use plastic liner if available.

Bury garbage when necessary in a hole deep enough to provide for at least 18 in of dirt cover and located at least 100 ft from any well or surface water.

Store all other refuse in a convenient place until normal collection services are provided, or take to the nearest designated place established by local authorities.

Bury all dead animals as quickly as possible at a distance of at least 100 ft from any well or surface water and cover to a depth of at least 2 ft.

Household Cleanup Procedure

Cellars—If possible, wait for the groundwater level to drop below floor level. Otherwise, drain or pump water from flooded cellars, being careful to avoid the collapse of walls caused by the pressure of the water-saturated ground around the basement. Postpone pumping if groundwater level is high. Wash down the walls, floors, and other areas exposed to floodwaters; keep windows and doors open for ventilation.

Disinfect the washed area by applying a solution of bleach with a broom or stiff brush. Allow the solution to remain in contact for approximately 10 min. The solution is prepared by adding 2 oz of bleach to 5 gal of water. Rinse with cold water as soon as possible to minimize staining. Any commercially available disinfectant can be used.

Floor Coverings—Flush rugs and carpets with hose and squeegee, then wash with lukewarm water containing a detergent. Rinse and dry in sun. CAUTION: Wool fibers will shrink more than synthetic materials.

Furniture—Clean and then wash metal, plastic, and leather surfaces with mild soap and water and *wipe dry immediately*. Some upholstery may be washed on the surface with soap and water and wiped dry. Expose to open air and sunshine.

Safety Precautions

Entering Damaged Buildings—Proceed with caution. If you have any doubts about the structural safety of a building, seek professional advice or assistance before going in. Check for visible signs of buckled walls, loose bricks, cracks, or shifting of foundation.

Gas or Electrical Services—Follow instructions of utility company or appliance service dealer concerning restoration of service. Obtain their assistance. Be sure electrical appliances are dry and in good condition before using. Do not turn on or try to fix gas-fired units. Avoid contact with electrical fixtures when standing in water or on a damp floor.

Keep poisons, other chemicals, and flammable substances in a place not accessible to children.

Rubber gloves should be worn while scrubbing damaged interiors with bleach solutions.

Do not neglect supposedly minor cuts, scratches, or other injuries or sickness experienced during the emergency.

Planning Ahead

The following supplies and equipment should be stored in a safe place for use during an emergency.

Bottled drinking water—2 gal

Canned foods—milk, juice, soup, vegetables, fish and meat

Packaged cereals and dried foods including chocolate bars*
Crackers or biscuits*
Paper or plastic plates and cups, including paper towels
Plastic utensils and can opener
Covered sauce pan—2 qt size
First-aid kit
Flashlight with extra bulb and dry cells; candles
Transistor radio with extra batteries
Canned-heat burner and extra fuel, or bottled gas camp stove
Matches in waterproof container
Shovel, hammer, nails, knife, cord or rope
Disinfecting deodorant spray
Toilet tissue, plastic bags, 5-gal pail.

Folding portable toilets are also available.

Typhoid Vaccination

There is frequently public concern for typhoid vaccination following floods and other natural disasters. The low level of typhoid in the United States, and the little evidence in the world's literature showing a greater risk of typhoid fever, make vaccination unnecessary. Besides, vaccination would not provide protection at the time of greatest risk as the procedure would require immunization with 2 injections several weeks apart. It is far better to direct the limited resources usually available to water purification and to emergency health and sanitation measures including disease surveillance.[29] The same principles would apply to other diseases as a basic, primary preventive effort.

TRAVEL AND CARRIER SANITATION AND HYGIENE

The mobility of people and the ease of travel has encouraged the use of various means of transportation. Involved is not only day travel, but also overnight and extended travel requiring food service, water supply, sewage and refuse disposal, toilet facilities, vector control, housing in some instances, and general cleanliness. Illnesses[30–33] associated with such travel are not uncommon, emphasizing the need for strict adherence to the basic principles of hygiene and sanitation which have been demonstrated to be effective in controlling communi-

*Store in plastic bag; inspect and replace occasionally, also batteries.

[29] *MMWR*, CDC, Atlanta, Ga., April 15, 1977. Also *MMWR*, December 15, 1978, pp. 500–501.

[30] Diane Roberts, Betty C. Hobbs, "Feeding the Traveler," paper presented at meeting of the Royal Society of Health, October 19, 1973, *J. R. Soc. Health*, **3**, 1974, pp. 114–117.

[31] "Current Trends Survey of the Incidence of Gastrointestinal Illness in Cruise Ship Passengers," *MMWR*, CDC, Atlanta, Ga., February 6, 1974, p. 65–66.

[32] S. Benson Werner, et. al., "Gastroenteritis on a Cruise Ship—A Recurring Problem," *Pub. Health Rep.*, October 1976, pp. 433–436.

[33] Frank L. Bryan, "Time-Temperature Observations of Food and Equipment in Airline Catering Operations," *J. Food Prot.*, February 1978, pp. 80–92.

cable diseases. Regulatory surveillance programs should include inspection of aircraft, buses, trains, and ships, and their terminals and the caterers serving them. Attention must also be given to sources of water and food, as well as to rest areas, ports of call, and the areas traversed, and to the prevalent epidemic and endemic diseases and the associated vectors and vehicles.

Applicable principles of hygiene and sanitation are covered in this text. Specific standards and guidelines can be found in the following publications.

1. James Bailey, *Guide to Hygiene and Sanitation in Aviation*, WHO, 1977.
2. Joseph A. Salvato, *Guide to Sanitation in Tourist Establishments*, WHO, 1976.
3. *Railroad Passenger Car Construction*, PHS Pub. 95, GPO, Washington, D.C., 1951.
4. *Dining Cars in Operation*, PHS Pub. 83, GPO, Washington, D.C., 1952.
5. *Railroad Servicing Areas*, PHS Pub. 66, GPO, Washington, D.C., 1951.
6. *Airlines*, PHS Pub. 308, GPO, Washington, D.C., 1953.
7. *Vessels in Operation*, PHS Pub. 68, GPO, Washington, D.C., revised 1963.
8. *Public Health Service Interstate Carrier Sanitation Program Manual or Operations*, DHEW, PHS, New York, N.Y., May 1964.
9. V. B. Lamoureux, *Guide to ship sanitation*, WHO, 1967.

Tourist information dealing with personal hygiene, sanitation, immunization, and self-protection in general is available from national, state, and local health agencies, from travel bureaus, and others.[34-37] Individual precautionary measures can provide the best protection.

BIBLIOGRAPHY

Changing Environmental Hazards, Am. Pub. Health Assoc., Washington, D.C., 1967.

"Environmental Health in Community Growth—An Administrative Guide to Metropolitan Area Sanitation Practices," *Am. J. Pub. Health*, May 1963, pp. 802–822.

Environmental Health Planning, PHS Pub. 2120, DHEW, Washington, D.C., 1971.

Franklin, Jack L., and Jean H. Thrasher, *An Introduction to Program Evaluation*, John Wiley & Sons, New York, 1976.

Grad, Frank P., *Public Health Law Manual*, Am. Pub. Health Assoc., Washington, D.C., 1976.

Hanlon, John J., *Public Health Administration and Practice*, C. V. Mosby Co., Saint Louis, Mo., 1979.

Herman, Harold, and Mary Elisabeth McKay, *Community Health Services*, Institute for Training in Municipal Administration, International City Managers' Association, 1140 Connecticut Avenue, N.W., Washington, D.C., 1968.

Manual for Public Health Inspectors, Canadian Pub. Health Assoc., 1255 Yonge Street, Toronto, 1961.

Maxcy-Rosenau Public Health and Preventive Medicine, John M. Last, Ed., Appleton-Century-Crofts, New York, 1980.

[34]Stanley S. K. Seah, *Health Guide for Travelers to Warm Climates*, Canadian Pub. Health Assoc., 1335 Carling, Suite 210, Ottawa, Ontario, Canada K1Z 8N8, 1979.

[35]Kevin M. Cahill, *Medical Advice for the Traveler*, Holt, Rinehart and Winston, New York, 1970.

[36]Anthony C. Turner, "The Health of the Traveler and the Spread of Disease," *J. R. Soc. Health*, **5**, 1977, pp. 210–213.

[37]*Health Information for International Travel 1980*, HHS Pub. (CDC) 80-8280, DHHS, Supt. of Documents, GPO, Washington, D.C. 20402.

"National Environmental Health Programmes: Their Planning, Organization, and Administration," Report of a WHO Expert Committee, *WHO Tech. Rep. Ser.*, **439,** 1970.

Oviatt, Vinson R., *Proposed Instruction Manual for Sanitarians Using the McBee Marginal Punch Card System of Recording*, Michigan Dept. of Health, Lansing, November 1955.

Public Information Handbook, No. M0017, Water Pollut. Control Fed., 2626 Pennsylvania Ave., N.W., Washington, D.C. 20037.

Salvato, J. A., Jr., "Evaluation of Sanitation Programs in a City-County Health Department," *Pub. Health Rep.*, June 1953, pp. 595–599.

APPENDIX I

MISCELLANEOUS DATA*

Degrees C° = $\frac{5}{9}$(F − 32°); Degrees F° = $\frac{9}{5}$C° + 32

Circumference of circle	$= 2\pi r = \pi D$;
	$\pi = 3.1416$; r = radius; D = diameter
Area of circle	$= \dfrac{\pi D^2}{4} = 0.7854D^2$
Area of sphere	$= \pi D^2$
Volume of sphere	$= \dfrac{\pi D^3}{6} = 0.5236D^3$
Volume of cone	= area of base × $\frac{1}{3}$ altitude
Volume of pyramid	= area of base × $\frac{1}{3}$ altitude
Area of triangle	= $\frac{1}{2}$ altitude × base
Area of trapezoid	= $\frac{1}{2}$ (sum of parallel sides) × altitude
1 acre	= 43,560 ft^2
1 square mile (mi^2)	= 640 acres
1 inch (in.)	= 2.54 centimeters (cm)
1 ppm (in water)	= 1 mg per liter = 8.34 lb per million gal
1 cubic foot (ft^3)	= 7.48 gal = 62.4 lb
1 U.S. gallon (gal)	= 8.345 lb = 3.785 liters = 231 $in.^3$
	= 0.833 British Imperial gallons
1 pound (lb)	= 7000 grains = 0.4536 kg = 16 oz
1 grain per gallon	= 17.12 parts per million
1 pound per square inch (psi)	= 2.31 ft vertical head of water
	= 2.04 in. of mercury
	= 6.9 kilonewtons per square meter
Atmospheric pressure	= 33.9 ft of water at sea level
	= 31.6 ft of water at 2000 ft
	= 28.3 ft of water at 5000 ft
	= 23.4 ft of water at 10,000 ft
1 gallon per minute (gpm)	= 1440 gal per day
	= 0.133 cubic feet per minute
	= 0.0038 cubic meters per minute

*See Appendix II for English–metric conversion factors.

1 cubic foot per second (cfs)	= 0.646 million gallons per day
1 million gallons per day (mgd)	= 1.547 cfs
1 horsepower hour (hp-hr)	= 0.746 kW-hr = 2546 Btu
	= (33,000 × 60) ft-lb
	= 0.065 gal of diesel oil (approx.)
	= 0.110 gal of gasoline (approx.)
	= 10–20 ft^3 of gas
1 British thermal unit (Btu)	= quantity of heat required to raise the temperature of one pound of water one degree F
1 Btu per minute	= 17.57 watts
1 Btu	= 778 ft-lb
	= 0.000393 hp-hr
1 ton of refrigeration	= 228,000 Btu per 24 hr
	= 2000 lb of ice × 144 Btu; 1 lb of ice absorbs 144.3 Btu of heat in melting
1 pound of water evaporated from 212° F	= 0.284 kW-hr = 970.3 Btu
1 kilowatt-hour (kW-hr)	= 3413 Btu per hour
1 boiler horsepower	= 33,479 Btu per hour

Pump Efficiencies

Deep-well displacement pumps have an efficiency of 35–40 percent. Small centrifugal pumps (10–40 gpm) have an efficiency of 20–50 percent. Larger centrifugal pumps (50–500 gpm) have an efficiency of 50–80 percent.

Small duplex, triplex, and reciprocating pumps in general have efficiences of 30–60 percent; but large pumps have efficiencies of 60–80 percent.

Horsepower to Pump Water

Horsepower to pump water:

$$= \frac{\text{gal of water pumped per min} \times \text{total head pumped against in ft} \times \text{weight of one gal of water in lb}}{\text{33,000 ft-lb per minute} \times \text{pump efficiency (as a decimal)}}$$

$$\text{Horsepower of the motor drive} = \frac{\text{gal per min} \times \text{total head in ft}}{3{,}960 \times \text{pump efficiency} \times \text{motor efficiency}}$$

Power input to a motor in watthours

$$= \frac{3{,}600 \times \text{number meter disk revolutions} \times \text{disk constant}}{\text{number seconds}}$$

Power input to a motor in watthours

$$= \text{voltage} \times \text{amperage} \times \text{power factor} \times 3{,}600 \times \begin{pmatrix} 1 \text{ for single-phase} \\ 1.41 \text{ for two-phase} \\ 1.73 \text{ for three-phase} \end{pmatrix}.$$

NOTE: One rotation of meter disk = watts marked on disk.

Waterfall

Power of waterfall = 0.114 × cfs × head in ft × eff. (eff. = ± 88%).

Channel Flow

$$V = \frac{1.486 R^{2/3} S^{1/2}}{n} \quad \text{(Manning formula)}$$

$$\text{where } R = \frac{\text{cross-sectional area of water (ft}^2\text{)}}{\text{perimeter wet (ft)}} \quad \text{(hydraulic radius)}$$

S = slope of water surface (ft per ft)
n = 0.015± (See hydraulic texts)

$Q = VA$; in which, Q = quantity of water, in cubic feet per second (cfs)
V = average velocity of flow, in feet per second (fps)
A = area of pipe or conduit, in square feet.

For a stream having a trapezoidal section:

$$A = \tfrac{1}{2}(a + b)h$$

$V_{(\text{average})}$ = 0.85 × surface velocity. Surface velocity is time for a floating object to travel a known distance in a fairly straight section of stream.

Weir Discharge

Discharge over 90° weir = $Q = 2.53H^{5/2}$

Discharge over rectangular weir = $Q = 3.33(b - 0.2H)H^{3/2}$
(with end contractions)

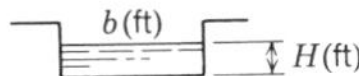

Table A Motor Wiring for Various Horsepowers, Motor Types, and Currents

	3-Phase Squirrel-Cage Induction Motors						Single-Phase Induction Motors						Direct Current Motors						
	220 Volts			440 Volts			115 Volts			230 Volts			115 Volts			230 Volts			
Horsepower	Approximate Full-Load Amperes	Wire Size, min AWG	Branch Circuit Fuse, Amperes	Approximate Full-Load Amperes	Wire Size, min AWG	Branch Circuit Fuse, Amperes	Approximate Full-Load Amperes	Wire Size, min AWG	Branch Circuit Fuse, Amperes	Approximate Full-Load Amperes	Wire Size, min AWG	Branch Circuit Fuse, Amperes	Approximate Full-Load Amperes	Wire Size, min AWG	Branch Circuit Fuse, Amperes	Approximate Full-Load Amperes	Wire Size, min AWG	Branch Circuit Fuse, Amperes	Horsepower
$\frac{1}{2}$							7.4	14	25	3.7	14	15							$\frac{1}{2}$
$\frac{3}{4}$							10.2	14	35	5.1	14	15							$\frac{3}{4}$
1	3.5	14	15	1.8	14	15	13	12	40	6.5	14	20	8.6	14	15	4.3	14	15	1
$1\frac{1}{2}$	5	14	15	2.5	14	15	18.4	10	60	9.2	14	30	12.6	12	20	6.3	14	15	$1\frac{1}{2}$
2	6.5	14	20	3.3	14	15	24	10	80	12	14	40	16.4	10	25	8.2	14	15	2
3	9	14	30	4.5	14	15	34	6[a]	110	17	10	60	24	10	40	12	14	20	3
5	15	12	45	7.5	14	25				28	8[a]	90	40	6	60	20	10	30	5
$7\frac{1}{2}$	22	10	70	11	14	35							58	3[a]	90	29	8	45	$7\frac{1}{2}$
10	27	8[a]	80	14	12	45							76	2[a]	125	38	6	60	10
15	40	6	125	20	10	60							112	00[a]	175	56	4	90	15
20	52	4[a]	175	26	8[a]	80							148	4/0[a]	225	74	2[a]	125	20
25	64	3[a]	200	32	8	100													25
30	78	1[a]	250	39	6	125													30
40	104	00[a]	350	52	4[a]	175													40
50	125	000[a]	400	63	3[a]	200													50
60	150	4/0[a]	450	75	2[a]	250													60

Source: Adapted from General Electric Co. 1955 Diary. See *National Electrical Code* for additional details.

Note: Values given are for not more than three conductors in a cable. Single-phase and direct current motors use two conductors.

[a] Next smaller size may be used with RH-type insulation.

With no end contractions $Q = 3.33bH^{3/2}$; b = at least $3H$

NOTE: Place hook gage at least 3 ft back of weir. No or low velocity of approach.

Water Pressure

$$h = \frac{p}{w}$$

where p = pounds per ft^2
w = pounds per ft^3 = 62.4 for water

$$h = \text{head of water (ft)} = \frac{p \times 144}{62.4} = 2.3p, \text{ where } p \text{ is in psi}$$

$$h = \frac{V^2}{2g}$$

where g = 32.2 ft per sec per sec and V = velocity in fps.

Flow of Water in Pipe

$$h = \frac{flV^2}{d2g} \qquad \text{(Darcy–Weisbach formula)}$$

where h = ft head loss
l = ft length
d = ft diameter
f = friction factor = 0.02 for cast iron (see hydraulic texts and Moody diagram).

Bernoulli's equation:

$$\frac{p^1}{w} + Z_1 + \frac{V_1^2}{2g} = \frac{p^2}{w} + Z_2 + \frac{V_2^2}{2g} + \text{loss head (ft)}$$

where Z = ft available potential energy; elevation of center line of pipe.

Flow from full-flowing horizontal pipe, in gpm $= \dfrac{2.83D^2X}{\sqrt{Y}}$;

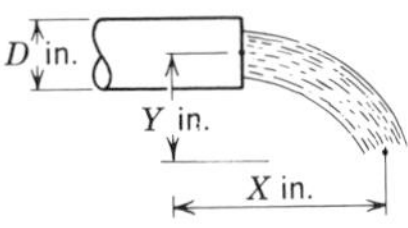

To determine flow in pressure conduit or conduit flowing full, i.e., sewers and small channels.

$$V = 1.318\ CR^{0.63}S^{0.54} \qquad \text{(Hazen–Williams formula)}$$

where V = average velocity of flow in fps

R = hydraulic radius; $\dfrac{\text{cross-sectional area (ft}^2\text{)}}{\text{wetted perimeter (ft)}}$

S = slope of hydraulic grade line; loss of head in ft divided by its corresponding horizontal length

C = Hazen–Williams friction coefficient
= 140 for cement asbestos; 40–120 for old unlined cast iron; 130–150 for cement lined cast iron, brass, copper; 140–150 for bitumastic enamel lined, plastic, and new unlined steel; 135 for centrifugally spun concrete and rubber-lined fire hose; 145–150 for coal-tar enamel lined; 100–140 for vitrified pipe. (See hydraulic texts)

OR, substituting in $Q = VA$ and converting Q into gpm, and pipe diameter into inches:

$$Q = 0.285Cd^{2.63}S^{0.54} \quad \text{(Use Hazen–Williams nomogram)}$$

FINANCE OR COST COMPARISONS

Compound Interest

$$A = P(1 + i)^n;$$

where A = amount or sum of principal and interest
P = principal or amount borrowed
i = rate of annual interest, or value of money
n = number of years money is drawing interest.

Present Worth

$$P_w = \frac{A}{(1 + i)^n};$$

where P_w = present worth of a sum of money A due in n years at compound interest rate i.

Sinking Fund (annuity)

$$S = \left[\frac{(1 + i)^n - 1}{i}\right]d;$$

where S = sinking fund or an amount that will accumulate in n years by making equal annual installments d, at interest rate i

or

$$d = \frac{Si}{(1+i)^n - 1}$$

Bond Issue

$$d = B\left[\frac{i}{(1+i)^n - 1} + i\right];$$

where B = bond issue or present sum of money, to be paid off in equal annual installments d in n years at interest rate i. The bond issue includes principal, legal and engineering fees, and interest.

Capitalized Cost—Economic Comparison

$$S_C = C + \frac{O}{i} + \frac{C - \text{salvage}}{(1+i)^n - 1} + \frac{R}{(1+i)^x - 1};$$

where S_C = capitalized cost of a project having a useful life of n years. It includes renewal or first cost C, annual cost of maintenance and operation $\frac{O}{i}$, and depreciation $\frac{C - \text{salvage}}{(1+i)^n - 1}$ at interest rate i, and major repairs $\frac{R}{(1+i)^x - 1}$ every x years. Where the salvage value of project or major repairs are nil, omit.

Total Annual Cost—Economic Comparison

$$S_Ci = Ci + O + \frac{i(C - \text{salvage})}{(1+i)^n - 1} + \frac{Ri}{(1+i)^x - 1};$$

where S_Ci = total annual cost and $\frac{Ci}{(1+i)^n - 1}$ represents the annual rate of depreciation or money set aside each year at compound interest i to equal the first cost C at end of n years.

Capital Recovery

$$R = P\left[\frac{i(1+i)^n}{(1+i)^n - 1}\right]$$

where R is the annual rate of capital recovery; P is principal or first cost of installation; n is capital recovery period or useful life; i is interest rate.

CONVERSION FACTORS

Units of Weight

1 kilogram (kg) = 1,000 grams (gm)
1 milligram (mg) = 10^{-3} gm
1 microgram (μg) = 10^{-6} gm
1 nanogram (ng) = 10^{-9} gm
1 picogram (pg) = 10^{-12} gm

Units of Concentration—for Solids

1 part per million (ppm) = 1 mg/kg = 1 μg/gm
1 part per billion (ppb) = 1 μg/kg = 1 ng/gm
1 part per trillion (ppt) = 1 ng/kg = 1 pg/gm

Units of Concentration—for Water (on a weight-to-weight basis)

1 milligram per liter = 1 mg/l = approx. 1 ppm, or = 1 ppm × specific gravity
1 microgram per liter = 1 μg/l = approx. 1 ppb
1 nanogram per liter = 1 ng/l = approx. 0.001 ppb

Units of Concentration—for Air (on a weight-to-volume basis corrected for temperature and pressure)

1 microgram per cubic meter = 1 μg per cu m = 1 μg/m^3
1 nanogram per cubic meter = 1 ng per cu m = 1 ng/m^3

Power of Ten

One quintillion	= 1,000,000,000,000,000,000	= 10^{18}, exa (E)*
One quadrillion	= 1,000,000,000,000,000	= 10^{15}, peta (P)
One trillion	= 1,000,000,000,000	= 10^{12}, tera (T)
One billion	= 1,000,000,000	= 10^{9}, giga (G)
One million	= 1,000,000	= 10^{6}, mega (M)
One thousand	= 1,000	= 10^{3}, kilo (k)
One hundred	= 100	= 10^{2}, hecto (h)
Ten	= 10	= 10^{1}, deka (da)
Ten to zero power	= 1	= 10^{0}

*Prefixes and symbols

One tenth	= 0.1	= 10^{-1}, deci (d)
One hundredth	= 0.01	= 10^{-2}, centi (c)
One thousandth	= 0.001	= 10^{-3}, milli (m)
One millionth	= 0.000,001	= 10^{-6}, micro (μ)
One billionth	= 0.000,000,001	= 10^{-9}, nano (n)
One trillionth	= 0.000,000,000,001	= 10^{-12}, pico (p)
One quadrillionth	= 0.000,000,000,000,001	= 10^{-15}, femto (f)
One quintillionth	= 0.000,000,000,000,000,001	= 10^{-18}, atto (a)

APPENDIX II

ENGLISH-METRIC CONVERSION FACTORS FOR COMMON UNITS

Customary Unit	Multiplier	International System of Units (SI) Unit
acre	× 4 046.9 → ← 2.471 1 × 10^{-4} ×	m^2
acre	× 0.404 69 → ← 2.471 1 ×	ha*
acre-ft	× 1 233.5 → ← 8.107 ×	m^3
Btu	× 1.055 → ←0.947 8 ×	kJ
Btu/hr/sq ft	×3.154 → ← 0.317 0 ×	$J/m^2 \cdot s$
Btu/lb	× 2.326 → ← 0.430 0 ×	kJ/kg
bu (bushel)	× 0.036 37 → ← 27.496 ×	m^3
cfm	× 4.719 × 10^{-4}→ ← 2 119 ×	m^3/s
cfs	× 0.028 32 → ← 35.315 ×	m^3/s
cfs/acre	× 0.069 97 → ← 14.29 ×	$m^3/ha \cdot s$*
cfs/sq miles	× 1.093 × 10^{-4} → ← 9 147 ×	$m^3/ha \cdot s$*
cu ft	× 0.028 32 → ← 35.315 ×	m^3
cu ft	× 28.32 → ← 0.035 315 ×	l*
cu in.	× 16.39 → ← 0.061 02 ×	cm^3
cu yd	× 0.764 6 → ← 1.308 ×	m^3
°F	× 0.555(°F − 32) → ← 1.8 (°C) +32 ×	°C
°C	+ 273 → ← 273 —	K

Customary Unit	Multiplier	International System of Units (SI) Unit
fathom	× 1.829 → ← 0.546 8 ×	m
ft	× 0.304 8 → ← 3.281 ×	m
ft-c	× 10.764 → ← 0.929 ×	lm/m^2
ft-lb	× 1.356 → ← 0.737 6 ×	J
gal	× 3.785 → ← 0.264 2 ×	l*
gal	× 0.003 785 → ← 264.2 ×	m^3
gpd/acre	× 0.009 353 → ← 106.9 ×	$m^3/ha \cdot d$*
gpd/ft	× 0.012 42 → ← 80.52 ×	$m^3/m \cdot d$*
gpd/sq ft	× 0.040 74 → ← 24.55 ×	$m^3/m^2 \cdot d$*
gpm	$\times 6.308 \times 10^{-5}$ → ← 1.585×10^3 ×	m^3/s
gpm	× 0.063 08 → ← 15.85 ×	l/s*
gpm/sq ft	× 0.679 02 → ← 1.473 ×	$l/m^2 \cdot s$*
hp	× 0.745 7 → ← 1.341 ×	kW
hp-hr	× 2.685 → ← 0.372 5 ×	MJ
in.	× 25.400 → ← 0.039 37 ×	mm
lb (force)	× 4.448 → ← 0.224 8 ×	N
lb (mass)	× 0.453 6 → ← 2.205 ×	kg
lb/day/acre-ft	× 0.367 7 → ← 2.719 6 ×	$g/m^3 \cdot d$*
lb/1 000 cu ft	× 16.02 → ← 0.062 43	g/m^3
lb/day/acre	× 0.1121 → ← 8.922 ×	$g/m^2 \cdot d$*
lb/day/cu ft	× 16.02 → ← 0.062 43 ×	$kg/m^3 \cdot d$*
lb/day/sq ft	× 4.883 → ← 0.204 8 ×	$kg/m^2 \cdot d$*
lb/ft	× 1.488 → ← 0.672 0 ×	kg/m
lb/mil gal	× 0.119 8 → ← 8.344 ×	g/m^3
mil gal	× 3 785 → ← 2.642×10^{-4} ×	m^3

Customary Unit	Multiplier	International System of Units (SI) Unit
mgd	× 3 785 → ← 2.642 × 10^{-4} ×	m^3/d*
mgd	× 0.043 8 → ← 22.83 ×	m^3/s
mgd/acre	× 1.082 × 10^{-5} ← 9.238 × 10^4	$m^3/m^2 \cdot s$
mile	× 1.609 → ← 0.621 4 ×	km
ppb (by weight)	× 10^{-3} → ← 1 000 ×	mg/l*
ppm (by weight)	essentially → ← essentially	mg/l*
pcf (pound per ft^3)	× 16.02 → ← 0.062 43 ×	kg/m^3
psf (pound per ft^2)	× 0.047 88 → ← 20.89 ×	kN/m^2
psf	× 4.882 → ← 0.204 8 ×	kg/m^2
psi	× 6.895 → ← 0.145 0 ×	kN/m^2
psi	× 0.070 3 → ← 14.22 ×	kgf/cm^2*
sq ft	× 0.092 9 → ← 10.76 ×	m^2
sq in.	× 645.2 → ← 0.001 55 ×	mm^2
sq miles	× 2.590 0 → ← 0.386 1 ×	km^2
tons (short)	× 907.2 → ← 1,102 × 10^{-3} ×	kg
yard	× 0.914 4 → ← 1.094 ×	m

Source: *J. Water Pollut. Control Fed.*, March 1980, Part Two, pp. 101–103.

Note: SI Unit abbreviations used: cm = centimeter, °C = degree Celsius, d = day, g = gram, ha = hectare, J = joule, K = Kelvin, k = kilo, kg = kilogram, kgf = kilogram force, km = kilometer, kW = kilowatt, l = liter, lm = lumen, M = mega, m = meter, mg = milligram, mm = millimeter, N = newton, s = second.

*Not strictly an SI unit.

Index